Ethnobiology

Ethnobiology is the study of the dynamic relationship between plants, animals, people, and the environment. Academic and applied interests include ethnobotany, ethnozoology, linguistics, paleoethnobotany, zooarchaeology, ethnoecology, and many others. The field lies at a dynamic intersection between the social and biological sciences. The major contribution from the biological sciences has come from economic botany, which has a rich historical and scientific tradition. Indeed, the objectives of the colonial enterprise were as much about the quest for "green gold" –herbal medicines, spices, novel cultivars, and others—as it was for precious metals and sources of labor. The view that ethnobiology concerns mostly the discovery of new and useful biota extended into the 20th century. The social sciences have contributed to the field in both descriptive studies but also within quantitative approaches in cognitive anthropology that have led to general principles within ethnobiological classification. Ethnobiological research in recent years has focused increasingly on problem solving and hypothesis testing by means of qualitative and especially quantitative methods. It seeks to understand how culturally relevant biotas are cognitively categorized, ranked, named, and assigned meaning. It investigates the complex strategies employed by traditional societies to manage plant and animal taxa, communities, and landscapes. It explores the degree to which local ecological knowledge promotes or undermines resource conservation, and contributes to the solution of global challenges, such as community health, nutrition, and cultural heritage. It investigates the economic value and environmental sustainability to local communities of non-timber forest products, as well as the strategies through which individual ecological knowledge and practices encourage resilience to change—modernization, climate change, and many others. Most importantly, contemporary ethnobiological research is grounded in respect for all cultures, embracing the principles of prior informed consent, benefit sharing, and general mindfulness.

Nigel Smith

Amazon Fruits: An Ethnobotanical Journey

Springer

Nigel Smith
Geography
University of Florida
Gainesville, FL, USA

ISSN 2365-7553 ISSN 2365-7561 (electronic)
Ethnobiology
ISBN 978-3-031-12802-8 ISBN 978-3-031-12803-5 (eBook)
https://doi.org/10.1007/978-3-031-12803-5

This Springer imprint is published by the registered company Springer Nature Switzerland AG
The registered company address is: Gewerbestrasse 11, 6330 Cham, Switzerland

To the memory of
Paulo Bezerra Cavalcante
(1922–2006)
Brazilian botanist who worked for many years at the Museu Paraense Emilio Goeldi in Belém and whose monumental work entitled Frutas Comestíveis na Amazônia went through several editions and inspired me to undertake this book.

Preface

Most of my fieldwork, starting in 1970, has been with rural folk of mixed ethnic ancestry, sometimes referred to as *caboclos*, *ribeirinhos*, or *colonos* in Brazil. I have visited a few indigenous communities such as the Shipibo at Nueva Cajamarca along the Ucayali in Peru, Juruna and Arara villages along the middle Xingu, and Tukano and Tuyuka settlements along the Uaupés and its affluent the Tiquié. Most of the information on the uses of fruits by indigenous people, and their local names, was obtained from literature and in some cases herbarium specimens. All photographs were taken by me with consent.

Sometimes numerous and quite different names exist for a fruit tree by the same indigenous group. This may be in part because of difficulties in comprehending indigenous languages, especially by people not familiar with that language. Multiple names may also be due to linguistic diversity even among the same tribe when they live in different areas, such as the Yanomami who live in scattered villages in northern Roraima, Brazil, and contiguous areas in southern Venezuela. Some names may vary also depending on the age of the tree or may refer to different parts of the tree or bush.

Information on the distribution of fruits was obtained from my field notes, literature, and virtual (online) herbaria of the Missouri Botanical Garden (Tropicos), the New York Botanical Garden, and Kew. Herbarium specimens also provided information on habitats, local names, and how high up the species may reach in mountains.

In some parts of the Peruvian and Brazilian Amazon, colleagues accompanied me on excursions and were authorized to collect botanical specimens of fruit plants. In the case of Peru, Rodolfo Vásquez and one of his assistants (Isau Huamantupa) gathered some specimens, identified them, and deposited them in the national herbarium in Lima and herbaria of the Missouri Botanical Garden in St Louis and in Oxapampa, Peru (Appendix 1). While working at the Instituto Nacional de Pesquisas da Amazonia (INPA) in Manaus during the 1970s, I collected a few botanical specimens which are deposited in the institute's herbarium. In 2015, Juliana Lins, a doctoral student at INPA, and Luiz Carlos Batista, a parataxonomist at the Goeldi Museum in Belém, gathered botanical specimens with me along the middle and lower Xingu; these specimens are deposited at the Goeldi Museum. Luiz Carlos Batista made further collections of edible fruits during joint visits to communities in the Municipality of Gurupá, Pará, in 2016.

I would like to acknowledge the help in identifying, or at least confirming, some species via photographs I took. The following individuals graciously offered their opinion as to the scientific names of some of the plants covered here: Walter Judd, retired, botany department at the University of Florida; Luiz Carlos Batista, botanist at the Museu Goeldi in Belém; Rodolfo Vasquez, Director of the herbarium at Oxapampa, Peru, and collaborator with the Missouri Botanical Garden; Charles Clement, INPA, Manaus; Terry Pennington, Royal Botanic Gardens, Kew; Scott Mori, New York Botanical Garden; Suzanne Renner, Ludwig-Maximilians University, Munich; Helen Kennedy, University of British Columbia; Fred Ganders, University of British Columbia; Frank Almeda, California Academy of Sciences; Bruce Holst, Selby Botanical Garden, Sarasota; Peter Jorgensen, Missouri Botanical Garden; Sir Ghillean Prance, retired from the Royal Botanic Gardens, Kew; Valdely Kinupp, biology professor and herbarium curator at the Instituto Federal de Educação, Ciência e Tecnologia do Amazonas, Manaus. Any errors in plant identification are entirely my responsibility.

Gainesville, FL, USA Nigel Smith

Fruits, Culture, and Anthropogenic Landscapes

Amazonia is one of the most biodiverse regions of the world, and this is reflected in the large number of fruits consumed by people in the region. This book surveys 219 species of native trees, bushes, and herbs that produce edible fruits or nuts spanning 50 botanical families. Humans are emulating other primates in this regard. For example, woolly monkeys *Lagothrix lagotricha* along the lower Apaporis River in the Colombian Amazon consume the fruits of 185 species in 41 botanical families (Defler and Defler 1996). In temperate regions of the world, you can count on one or two hands the number of wild fruits that people gather. Emphasis is on fruits that are eaten or used for making juice. Several plants produce both pulp and edible nuts, while some others are gathered just for their nuts. Fruits used for fish bait as well as fruits that fall to the ground and attract game are also mentioned. Other uses for fruits, such as body paint and for folk remedies, are covered as well.

Fruit Diversity

Tropical forests are among the most biodiverse habitats in the world, and the Amazon Basin is a good example of that. The number of edible fruits native to the Amazon is orders of magnitude greater than in temperate regions. The great diversity of fruits in the region is not just attributed to forests, however. Many fruits consumed by locals in the Amazon come from successionary vegetation in disturbed habitats, grasslands, and scrub savannas. Forests are also far from monolithic; species composition varies according to soil type, rainfall patterns, the history of human occupation in the area, and whether they are on floodplains or uplands.

Several botanical families examined here have only one species with edible fruits, such as Achariaceae, Asteraceae, and Aptandraceae. But others contain numerous species. The family with the largest number of edible fruits examined here is Fabaceae (legume family) with 27 species. In temperate regions, the legume family is better known for herbaceous plants that produce peas and beans of various types, but in the Amazon this family is dominated by trees, some enormous such as

Hymenaea courbaril. One genus in the Fabaceae, *Inga*, has an astonishing number of species with edible fruits: 22. And that figure is by no means complete. More ethnobotanical research in the vast region is sure to uncover more species of *Inga* with fruits gathered by locals. Another family with a large number of edible fruits is Melastomataceae, a family hardly known in most temperate regions. In the Amazon, at least 20 species in this family, mainly in the genera *Clidemia*, *Miconia*, and *Mouriri*, are eaten by rural people. The Melastomataceae family embraces over 4000 species in 155 genera, mostly confined to the tropics. The American tropics has the largest assemblage of melastomes with some 3000 species. So it is likely that my list of 20 edible species is only the tip of an iceberg.

Only four species in the Anacardiaceae are covered here, but it is nevertheless an important family because of two widespread species: cashew *Anacardium occidentale* and yellow mombin *Spondias mombin*. The most famous fruit from this family is mango *Mangifera indica*, native to tropical Asia and widely planted in Africa, Latin America, the Caribbean, and Polynesia. But cashew, which is native to Brazil and best known for its nuts, also produces juicy fruits that are rich in vitamin C. Wild populations of cashew are found on sand dunes in coastal areas of northeastern Brazil and on savannas in central Brazil. In the Amazon, wild or feral cashew also grows in pockets of savannas in upland forest as well as along the sandy banks of clear and black water rivers. Yellow mombin, which is native to floodplains, is sometimes found in home gardens on account of its fruit which is used to make juice.

The Solanaceae family is best known for the potato *Solanum tuberosum*, which is grown throughout the world for its tubers which are boiled or fried. It is native to the Americas and was brought to Europe during the colonial period initially for its ornamental flowers. Two genera in this family, *Datura* and *Brugmansia*, are known for their hallucinogenic properties and are used by healers in various parts of Tropical America. In western Amazonia *Brugmansia*, known as angel trumpet in English, is used by shamans in certain areas for spiritual purposes, but the fruits of at least six species in this family are gathered in the Amazon for snacks and some appear in urban markets. The fruits of two species of *Physalis* are eaten and three species of *Solanum*.

I include fruits that also occur wild in other parts of the Neotropics, such as guava *Psidium guajava*. Some fruits, such as cacao *Theobroma cacao* and cupuaçu *Theobroma grandiflorum*, clearly arose in Amazonia, but others may have evolved in other parts of lowland South or Central America and then spread to the Amazon by natural means.

Fruit Names

Some of the fruits featured here have numerous local names. This is due to the vastness of the region and its cultural diversity. Even in the same watershed, rural inhabitants may ascribe different names to the same plant. The diversity of names for a fruit tree or bush within some indigenous cultures can be ascribed to two main reasons. First, some indigenous groups inhabit villages scattered over vast areas, such as the Kichwa, so local dialects may have different names for the same plant. And indigenous groups have different names for parts of the same plant, such as the fruit, seed, or flowers.

Anthropogenic Vegetation

Terms such as primary, pristine, primeval, or virgin forests are still used by some travelers and scientists in describing Amazonia. In 1865, the American ichthyologist Louis Agassiz used a different term when describing forests around farmhouses along the Amazon between Santarém and Óbidos: "the primitive forest about the houses is usually cleared" (Agassiz and Agassiz 1896:170). In the early twentieth century, a French naturalist who spent many years in the Brazilian Amazon also used the term primitive forest (Le Cointe 1922b:116). In Portuguese, the term is *mata virgem* or *bosque primario*, and in Spanish *monte virgen*. Throughout this book, I will instead use the term mature forest. A mature forest may have never been cleared in the past, or was cut down long ago to plant crops or construct dwellings and has since returned to tall woods rich in species. Mature forest can return to homestead sites with planted orchards when they are abandoned, thus forming a cultural forest (Smith 2001). Or forests might have been altered in more subtle ways, such as by discarded fruit seeds around campsites of people hunting or gathering forest products which then sprouted to form small groves (Zent and Zent 2002a, b).

My exploration of Amazon fruits is tied to an ongoing debate about the degree to which forests and other environments in Amazonia have been modified by human activity. For a long time, the prevailing view was that the Amazon Basin had been sparsely settled; indigenous cultures were concentrated in the Andes and highlands in Central America. Some archaeologists, such as the late Betty Meggers, suggested in the 1960s and 1970s that the Amazon rainforest was not propitious for the development of higher cultures and dense settlements. Subsequently, a growing number of archaeologists, anthropologists, botanists, and geographers have challenged this view based on field work in widely scattered locations in Amazonia (Arroyo-Kalin 2017; Clement et al. 2015; Heckenberger 2014, 2016; Heckenberger and Neves 2018; Lima et al. 2016; Macedo et al. 2019; Roberts et al. 2017; Schmidt et al. 2014; Smole 1976:202; Stahl 2015). Archaeological research has revealed numerous large archaeological sites, many with anthropogenic dark earth (anthrosol, Amazon dark earth, *terra preta de índio*) containing potsherds, plant remains, and sometimes lithic material such as stone ax heads both on uplands and floodplains. Even forested areas in the Amazon can be domesticated landscapes, such that nature in many parts of Amazonia more closely resembles a garden rather than a pristine, natural wilderness (Erickson 2008; Levis et al. 2017, 2018). The same holds for many other parts of the Americas (Denevan 1992).

Another prevailing idea for many years was that indigenous settlements were concentrated along sediment-rich rivers because they have fertile floodplains for growing crops and have more abundant fish populations. While ancient settlements may have been more common along "muddy" rivers such as the Amazon-Solimões, large archaeological sites with deep black earth have been found along black water rivers such as the Rio Negro and clear water rivers such as the Xingu.

Another old idea, now increasingly debunked, was that interfluvial forests were not settled in precontact times. But sizeable archaeological sites have been found along small tributary streams far removed from the main rivers, such as in southern Amazonia (Souza et al. 2018). Numerous archaeological sites have been found on uplands, such as the Belterra plateau south of Santarém and inland from the Tapajos River (Stenborg et al. 2018). And in Acre, over 450 geoglyphs have been identified in upland, interfluvial sites, occupying some 13,000 sq. km, and were built between 3500 and 900 years ago (Watling et al. 2017, 2018). These geoglyphs are mostly circular or square-shaped and were used for ceremonial purposes.

Far from being largely "empty" and minimally disturbed forests, a picture is emerging of relatively advanced and populous civilizations in various parts of the Amazon Basin. One estimate, for example, suggests that the indigenous population in Amazonia reached at least 8.4 million in precontact times (Dull et al. 2010), while Denevan (2014) suggests the population may have reached 10 million.

Another dimension to consider is how long people have been in Amazonia, thereby modifying various environments. For a long time, the prevailing view was that when hunters and gatherers entered South America they stuck to the coasts and mountains, such as the Andes. However, they also penetrated the Amazon Basin, mostly likely along rivers, at least 14,000 years ago (Neves 2016a, b, c). At the end of the last ice age 10,000 years ago, sea levels rose by about 100 along the coasts of South America, thus backing up the Amazon and the lower stretches of its tributaries. Many early campsites of hunters and gatherers thus became flooded and buried in sediment in Amazonia (Roosevelt et al. 2012) as well as coastal areas of South America (Goldberg et al. 2016). It is not inconceivable the hunters and gathers may have penetrated some parts of the Amazon 20,000 years ago. The oldest pottery in South America, dated at 8000 years old, was unearthed at the Taperinha archaeological site overlooking the Amazon floodplain a little downstream from Santarém (Roosevelt 1992; Roosevelt et al. 1991). The idea that plant domestication in Amazonia only began when indigenous people started making pottery is now being challenged (Neves 2016a). Dating of charcoal from sites occupied by hunters and gatherers near the Tocantins River in eastern Amazonia indicates that they lived there for some 6000 years starting 11,000 years BP (Caldarelli et al. 2005). For sure, these indigenous people altered the vegetation around their settlements even though they were not clearing the forest to plant crops.

The distribution of linguistic families in the Amazon Basin also suggests a long history of occupation and migration. Over 300 indigenous languages have been recorded in the Amazon Basin representing six major linguistic families: Aruak, Tupi, Carib, Panoan, Tucanoan, and Gê. The distribution of linguistic families in Amazonia suggests a long history of migration (Aikhenvald 2013). It is likely that some of these groups brought seeds, seedlings, and cuttings of several useful plants, including fruit trees.

Far from being a cultural backwater, then, the Amazon Basin with its diverse natural resources has served as a home to humans for millennia, and it has impacted the vegetation in a variety of ways. My ethnobotanical journey with Amazon fruits has opened my eyes to the diverse ways that people have altered the vegetation over long periods of time (Smith 2014).

As more archaeologists engage in field research in Amazonia, it is becoming increasingly clear that what was once regarded as mostly pristine forest in the region has in fact been substantially altered in many areas by indigenous peoples (Balée 2013:12; Erickson 2006). And river dwellers (*ribeirinhos* in Brazil) and small-scale farmers living away from large rivers are continuing that process.

At the time of contact with Europeans in the early sixteenth century, dense indigenous settlements were found along many of the major rivers, such as the Amazon-Solimões. But within a hundred years or so, many of those populations disappeared or were greatly reduced due to warfare, slaving raids, and introduced diseases (Hemming 2008:73, 104). Those ancient populations, now mostly gone, had profound effects on the vegetation, not triggering massive destruction rather anthropogenic vegetation that was often enriched with useful plants.

The idea that interfluvial forests were largely untouched in the past is turning out to be a myth, at least in some areas. Although often smaller, extensive archaeological sites have been found along small rivers and streams, such as in the Caxiuanã National Forest along the left bank of the Anapu River in Pará. And the idea that black water rivers were "rivers of hunger" (*rios de fome*) is another myth. In the last century, many social scientists and archaeologists assumed that black water rivers could not support large human populations because they did not have fertile floodplains. While it is true that the sandy floodplains of acidic rivers were planted sparsely if at all, they nevertheless had productive fisheries. And we now know that large indigenous settlements once thrived along black water rivers such as the Rio Negro and Urubu River in Amazonas, Brazil. These populations planted most of their crops on uplands near their villages, thereby altering the vegetation.

Second growth, known as *capoeira* or *capoeirão* in Brazil and *purma* in Peru, is another commonly used term when describing vegetation in Amazonia. The problem with that term is that many areas called "second" growth have actually undergone repeated clearing followed by regrowth. So I will use terms such as successionary or regrowth vegetation instead. The Portuguese and Spanish terms for regrowth vegetation are neutral with regard to how many times that area has experienced successionary growth.

Some forests and successionary vegetation have been enriched with useful plants, including fruit and nut trees, by people, such as when forests envelop abandoned homesites. During the collapse of the rubber boom in the early twentieth century, many rubber tappers abandoned their huts and home gardens which then became enveloped by the surrounding forest (Maia 1966:24). Also some manioc fields are deliberately planted with fruit trees so that people can harvest the fruits long after the manioc has been harvested. Eventually these fields, known as *roças* in

Brazil and *chacras* in Peru, return to forest, but they are different from the woods that were cleared because of the planted trees. Indigenous peoples have been modifying landscapes and vegetation in many parts of the Amazon for thousands of years including some forest islands in savanna and raised fields and mounds in seasonally flooded areas such as parts of the Llanos de Mojos in Bolivia and parts of Marajó Island at the mouth of the Amazon (Balée 1988).

Some fruit and nut trees have been moved around going back a long time. Extensive trade routes have been documented in pre-contact and colonial times not only from the Andes down into western Amazon, but from western Amazonia to the central Amazon (Eriksen 2011:40). For example, green frog-shaped ornaments (*muiraquitãs*) have been found in ancient burial mounds on Marajó Island suggesting trade networks in the lower Amazon, Orinoco River basin, and the Guianas (Barreto 2016). Also some Amazonian crops were taken to the Pacific Coast of Peru and Ecuador in precontact times, such as cacao *Theobroma cacao*. Trade in marine conches (shells) *Strombus gigas* started at least 7000 years ago with Jívaro and other indigenous groups in western Amazonia who used them for trumpets (Eriksen 2011:51). Trade routes also linked the Upper Rio Negro, Casiquiare, Orinoco, and Essequibo basins (Eriksen 2011:150, 205).

Extensive trade networks also developed within Amazonia, especially after people took up agriculture and settlement sizes grew. These internal trade routes are still poorly understood, but evidence of such exchange networks are manifested by certain stone tools, green jadeite pendants in the shape of frogs known as *muiraquitã* which were made along the Upper Nhamundá and Trombetas rivers (Palmatary 1960), and the distribution of some fruit trees. Both the density and distribution of some of the Amazon fruits have thus been affected by human activities extending back thousands of years.

The fruiting plants used by people in the Amazon, as in other regions, evolved for seed dispersal by various means including wind, water, birds, mammals, reptiles, or fish. In some cases, the fruits are dispersed by more than one means, such as water currents and fish. When humans arrived, they also began to disperse seeds inadvertently such as by swallowing them and defecating, tossing seeds on the ground along trails around campsites, and then later deliberately planting them in their backyards and sometimes their fields.

Some fruit-bearing plants are confined to uplands, whereas others are found only on floodplains. But some fruit bushes and trees that bear edible fruit are found on both uplands and floodplains. In the latter case, this may be a natural phenomenon. But human activity cannot be ruled out, at least for some species that straddle both environments. Given the long history of human occupation in Amazonia, it is possible that some fruit trees and bushes were brought to upland environments.

For each fruit profiled here, three possibilities concerning its status are indicated: wild, spontaneous in cultural settings, and planted. Wild can include mature forest, successionary vegetation, and grasslands/savannas. Some fruit trees or bushes occur in a variety of habitats from forest to grasslands (*campo* in Brazil) or scrub woodland (*cerrado*, *cerradão*). Spontaneous in cultural settings embraces a variety of settings from cultivated fields, trails, and home gardens. Planted means the seed or seedling was deliberately planted.

By and large, I avoid the term domesticated. That is because various definitions of plant domestication have been offered and I find them a bit confining. The spectrum from "wild" to "domesticated" is very broad. Some people classify plants as domesticated when varieties have been developed or have arisen spontaneously and are then maintained. There are many shades of gray between fruits found in forest or successionary vegetation and those occurring in home gardens and fields.

Children's Fruits

Children are a wonderful source of information about the location of fruits in the wild and the seasons when the fruits ripen. When I first started doing fieldwork in the Amazon over 40 years ago, I was single and in my twenties, and children did not seem important in my investigations. I might have noted how many children were in a family and their sexes and ages when I conducted formal interviews so that I could construct a population pyramid for the area I was working in. But I did not see them as helpful at all to my research; on the contrary, I was often a bit irritated when they tagged along with their father or mother when visiting home gardens or cultivated fields. They seemed a distraction rather than interesting people with plenty of insights.

After starting my own family and raising three children with my wife, I came to appreciate more children's personalities and refreshing perspectives. We have come a long way in the last few decades in appreciating the role of women in ethnographic and ethnobotanical studies. Now it is time to recognize more fully the contribution of children to household economies, culture, and local knowledge about natural resources. In the case of villages in the Amazon, close to half the people are under 15 years old, and they are a fountain of information about fruits, both wild and cultivated. In 2004 I recall walking along a trail in successionary vegetation near the village of Veinte de Enero along the lower Yanayacu in the Peruvian Amazon and encountering a fruit on the ground. I first asked my guide the name of the fruit, then if people eat it. "No," he replied "only monkeys eat them." But then I glanced at the kids tagging along and I saw them merrily eating them. Children's fruits tend to be either tiny and thus deemed not worth picking by adults, or larger fruits that contain precious little pulp or juice that kids nevertheless relish.

The special connection between kids and wild fruits has been noted by others in Amazonia, such as among the Ka'apor in the Brazilian Amazon (Prance et al. 1987) and among indigenous groups and mestizos in northwestern Bolivia (Blacutt-Rivero et al. 2016), and those familiar with rural lifeways in other parts of the world, often long ago. In the mid-nineteenth century, for example, the American naturalist Henry David Thoreau noted that children were particularly astute at finding ripe blueberries in the countryside of Massachusetts:

> They ripen first on the tops of hills, before they who walk in the valleys suspect it. When old folks find only one turned (ripe) here and there, children, who are best acquainted with the localities of berries, bring pailsfull to sell at their doors. (Thoreau 2001:22)

Likewise in the Amazon, children bring several wild fruits, such as guava, to sell in the streets of towns and villages. Children are not only experts at finding early ripening fruits, but also snack on a whole range of fruits largely ignored by their elders. Indeed, kids are so adept at finding wild fruits of all sizes and shapes along paths that they often pick bushes clean before I have a chance to set up my camera and flash. Like swarming ants, I sometimes have had to ask them to slow down so that I can at least capture the moment when they pluck the fruits.

Even today in urban areas in parts of North America and Europe, children still take delight in gathering wild blackberries at the edge of town or in abandoned lots. What child does not relish those wild, tart fruits, they taste all the better for being free? Some kids cannot resist the temptation of bringing at least some of the fruits home, at the risk of scalding for staining their T-Shirt or trousers. The bond between wild fruits and people, especially youngsters, is thus universal. Here we explore a few of the wild fruits gathered by children in forests, second growth, swamps, and floating meadows in the Peruvian Amazon.

Cultural Changes That Affect the Use of Native Fruits

Sadly, though, fewer and fewer children are gathering wild fruits in the Amazon and elsewhere. With rising incomes in many rural areas, especially in the Brazilian Amazon where people with modest incomes have access to the *Bolsa Familia* program, purchased snacks are increasingly consumed. Some vendors even set up their carts at school exits, such as in the vicinity of Tefé, to entice children to buy popcorn, biscuits, crackers, and candy. Village paths are increasingly littered with snack packaging instead of fruits seeds and skins. The higher standard of living in many rural areas of the Amazon is a positive development, but it has some downsides. In the Brazilian Amazon especially, child and adult obesity is on the increase in the countryside as well as urban areas, one of the downsides of "progress." When I visited the Amazon for the first time in 1970, I don't recall seeing any overweight children and very few adults. Now there appears to be an *explosão de gordura* (fat explosion) in both rural and urban areas of the Amazon, especially in Brazil. Rising obesity is a worldwide trend and brings serious health consequences.

Open-air fruit markets have essentially disappeared in the Brazilian Amazon, such as in Portel and Gurupá. In some urban areas, fruit vendors must pay a fee to sell fruit from covered stalls, such as the Ver-o-Peso market in Belém. Fruits stalls I visited in Bragança and Breves in Pará in 2015 had a few vendors with limited offerings. Fruit is increasingly sold in supermarkets and fruit stores, but most of the fruit is from other regions, such as apples, oranges, and grapes from southern Brazil, Argentina, and Chile. And the diversity and quantity of fruit available in stores are much more restricted than traditional street markets, especially in the Brazilian Amazon.

To my surprise, school snacks (*merenda escolar*) sometimes feature imported fruits such as apples, as at Paquiçamba, a Juruna community along the middle Xingu near Belo Monte. Apples are more durable than many of the Amazonian fruits, but it is ironic that schoolchildren get to snack on a temperate zone fruit when they live in one of the richest regions on earth with regard to tropical fruits.

In my earlier years of fieldwork in the Amazon, it was common to encounter boys selling popsicles of frozen fruit juices from Styrofoam containers slung across their shoulders. The youngsters typically plied ferry crossings, open-air markets, and commercial districts of towns and cities. Boys would also sell some regional fruits, such as muruci *Byrsonima crassifolia*, at speed bumps along roads and locations similar to those selling frozen popsicles. Such sights are rare today. In Portel in 2015, I purchased "avocado" ice cream from a man with a pushcart; he assured me it was made locally. It was green, but obviously from an artificial dye. It also did not taste anything like avocado. Rather it was mostly artificial ingredients so I ended up tossing it.

In major cities, such as Belém, Manaus, and Iquitos, some ice cream companies do make delicious ice creams flavored with regional fruits. But those ice creams are rarely available in smaller towns. In the Brazilian Amazon, ice cream in supermarkets, stores, and restaurants is often made in southern Brazil. Back in the 1970s and 1980s, several ice cream parlors in Belém offered a wide variety of ice creams produced locally and flavored with Amazonian fruits. Now only one, Cairu, remains. Gelar, which once produced delicious popsicles flavored with Amazonian fruits that were sold by ambulant vendors, has also gone out of business. Hopefully, more firms, both large and small, will start making more ice creams flavored with regional fruits. In the USA, Haagen Dazs once sold an ice cream flavored with fruits of the açaí *Euterpe oleracea* palm, but it has long since been withdrawn. Also, several companies once sold açaí fruit juice in US supermarkets but now only Sambazon still offers the drink. Eventually, more Amazon fruits might penetrate markets in North America, Europe, and Asia.

Environmental Destruction

Amazonian forests are under threat from a wide variety of development pressures. While cattle were introduced to the Amazon during the colonial period, they were mostly raised on naturally occurring grasslands, such as on Marajó Island. But in the last fifty years or so, extensive areas of forest have been cleared and planted to African pasture grasses to raise cattle, especially in the vicinity of major highways that began crisscrossing the region beginning in the 1960s. Even though some of these highways, such as the Transamazon, were designed to provide lots for small-scale farmers, many of the latter have subsequently switched to cattle grazing which entails maintaining permanent pastures rather than allowing forest to regenerate in the case of slash-and-burn farming. Large-scale soybean farms began penetrating southern Amazonia starting in the 1980s, again leading to landscapes where the forest does not come back. In Brazil, a moratorium on forest clearing for soybean production was declared in 2006 but that has not stopped deforestation in the Amazon. Forest clearing for soybean production is accelerating in the Bolivian Amazon.

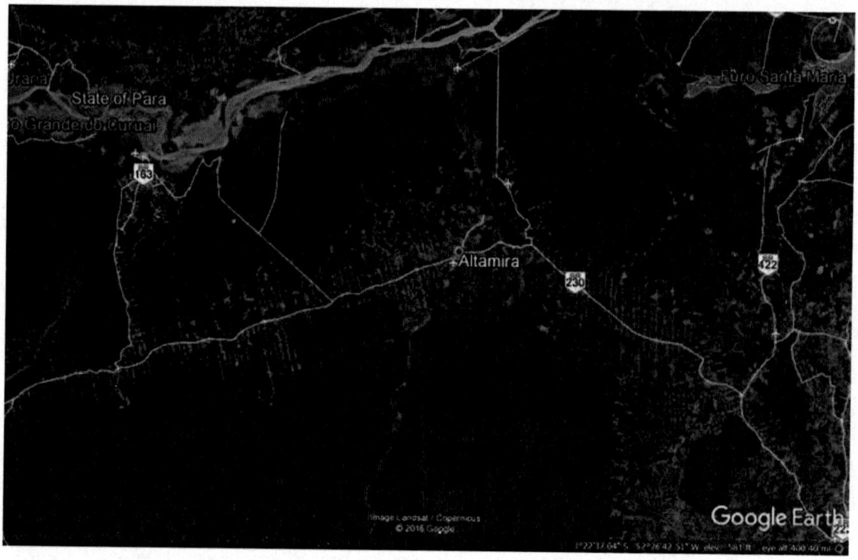

Deforestation along the Transamazon highway and around the Tucuruí reservoir along the Tocantins River (on right). Forest is better protected in indigenous reserves such as those belonging to the Arara who live along the Iriri, an affluent of the Xingu (green block along the Transamazon west from Altamira) and the Parakanã (green area surrounded by deforestation southeast from Altamira)

According to information from INPE (Instituto Nacional de Pesquisas Espaciais), deforestation in the Brazilian Amazon increased by 29% in 2016 compared to the year before. Brazil's economic problems have shrunk budgets for environmental monitoring. Apart from shrinking genepools and the loss of species, deforestation has other environmental consequences; the felling and burning of forests for a variety of economic activities are responsible for 10% of greenhouse gas emissions. The shrinking of forests in Amazonia not only represents a loss of fruit production and resources for plant breeding but is contributing to global warming.

Forest fragmentation associated with colonization by small-scale farmers along the Cuiaba-Porto Velho highway. LANDSAT image of Ariquemes, Rondônia, Brazil 6-25-96

Increased pressure from loggers has also led to the demise of many forest areas. Loggers create trails to extract timber, many of which are subsequently invaded by small-scale farmers. Maps that show large areas of the Amazon that have been set aside for conservation appear impressive, but the reality on the ground can be quite different. As they say in Brazil: "so para o Ingles ver (just for the Englishman to see)." Lack of enforcing protected areas in many cases has led to penetration by loggers, small-scale farmers, and even cattle ranchers.

Another major threat to forests in the Amazon comes from dam building to generate hydroelectric power (HEP). The Tucuruí Dam on the Tocantins River closed in the early 1980s flooding over 2000 sq. km of forest. The Belo Monte dam on the Xingu, slated to generate some 11,000 MW of electricity, closed in 2015. And some 43 HEP dams are planned for the Tapajós River further to the west. Some 140 dams have been built or are under construction in the Amazon basin (Latrubesse et al. 2017). While HEP is considered a "clean" energy source, the dams cause considerable environmental damage and social disruption. Many dams have been built in low-lying areas which means that large areas become permanently flooded. Also, riverine habitats below the dams have and will be impacted by the change in sediment load and water chemistry, which will likely adversely affect the habitat for riverine fruits.

The environmental pressures outlined above have led to the shrinking of genepools for many fruit and nut species in the Amazon, which may endanger their long-term prospects for survival. Shrinking genepools limit options for developing commercial varieties. In some cases, fruit species with limited distributions could be eliminated entirely.

References

Agassiz, L. and E. C. Agassiz. 1896. A Journey in Brazil. Houghton, Mifflin and Company, Boston

Aikhenvald, AY 2013. Amazonia: linguistic history. In: The Encyclopedia of Global Human Migration, Ness, Immanuel (ed), Blackwell, Hoboken, https://doi.org/10.1002/9781444351071.wbeghm850

Arroyo-Kalin, Manuel. 2017. Amazonian dark earths. In: *Archaeological Soil and Sediment Micromorphology*, C. Nicosia and G. Stoops (eds), Wiley and Sons, New York

Balée, W. 1988. Historical ecology: premises and postulates. In: Advances in Historical Ecology, William Balée (ed), Columbia University Press, New York, pp 13–29

Balée, W. 2013. Cultural Forests of the Amazon: A Historical Ecology of People and Their Landscapes. University of Alabama Press, Tuscaloosa

Barreto, C. 2016. O que a cerâmica Marajoara nos ensina sobre fluxo estilístico na Amazônia? In: Cerâmicas Arqueológicas da Amazônia, C. Barreto, HP Lima, and CJ Betancourt (eds), IPHAN/Museu Paraense Emilio Goeldi, Belém, pp 115–124

Blacutt-Rivero, E., M. A. Morales, M. Gruca, and H. Balslev. 2016. Local knowledge about palms (Arecaceae) among children in Bolivia. Botanical Journal of the Linnaean Society 182: 505–516

Caldarelli, Solange Bezerra; Costa, Fernanda de Araújo; Kern, Dirse Clara. 2005. Assentamentos a céu aberto de caçadores-coletores datados da transição Pleistoceno final/Holoceno inicial no Sudeste do Pará. Revista de Arqueologia 18:95–108

Clement, Charles R.; Denevan, William H.; Heckenberger, Michael J.; Junqueira, André Braga; Neves, Eduardo G.; Teixeira, Wenceslau G.; Woods, William I. 2015. The domestication of Amazonia before European conquest. Proceedings of the Royal Society of London, Ser. B. 282 https://doi.org/10.1098/rspb.2015.0813

Defler, Thomas R.; Defler, Sara B. 1996. Diet of a group of *Lagothrix lagothricha lagothrica* in southeastern Colombia. International Journal of Primatology 17(2):161–190

Denevan, W. M. 1992. The pristine myth: the landscape of the Americas in 1492. Annals of the Association of American Geographers 82(3):369–385

Denevan, W. M. 2014. Estimating Amazonian indian numbers in 1492. Journal of Latin American Geography 13(2):207–221

Dull Robert A.; Nevle, Richard J.; Woods, William I.; Bird, Dennis K.; Avnery, Shiri; Denevan, William H. 2010. The Columbian encounter and the little Ice Age: abrupt land use change, fire, and greenhouse forcing. Annals of the Association of American Geographers 100(4):1–17

Erickson, C.L. 2006. The domesticated landscapes of the Bolivian Amazon. In: Time and Complexity in Historical Ecology: Studies in the Neotropical Lowlands, Balée, William; Erickson, Clark L. (eds), Columbia University Press, New York, pp. 235–278

Erickson, C.L. 2008. Amazonia: the historical ecology of a domesticated landscape. In: Handbook of South American Archaeology, Siverman, Helaine and William H. Isbell (eds), Springer, New York, pp 157–183

Eriksen, L. 2011. *Nature and Culture in Prehistoric Amazonia.* Lund University, Lund

Goldberg, Amy; Mychajliw, Alexis N.; Hardly, Elizabeth A. 2016. Post-invasion demography of prehistoric humans in South America, https://doi.org/10.1038/nature17176

Heckenberger, M. 2014. Bio-historical diversity, sustainability and collaboration in the Xingu. Anuário Antropológico 39(2):69–97

Heckenberger, M. 2016. Xingu garden cities: domesticated forests of the southern Amazon. In: Tropical Forest Conservation: Long-Term Processes of Human Evolution, Cultural Adaptations and Consumption Patterns, UNESCO, Paris, pp 66–86

Heckenberger, M.J. and E. Neves. 2018. Archaeology of the Amazon. Cambridge University Press, Cambridge

Hemming, J. 2008. Tree of Rivers: The Story of the Amazon. Thames and Hudson, London

Latrubesse, Edgardo M.; Arima, Eugenio Y.; Dunne, Thomas; Park, Edward; Baker, Victor R.; d'Horta, Fernando M.; Wight, Charles; Wittmann, Florian; Zuanon, Jansen; Baker, Paul A.; Ribas, Camila C.; Norgaard, Richard B.; Filizola, Naziano; Ansar, Atif; et al. 2017. Damming the rivers of the Amazon basin. Nature 546: 363–369

Le Cointe, P. 1922b. *L'Amazonie Brésilienne: Le Pays-ses Habitants, ses Resources, Notes et Statistiques jusqu'en 1920.* Augustin Challamel, Paris, vol 2

Levis, C. et al. 2017. Persistent effects of pre-Columbian plant domestication on Amazonian forest composition. Science 355:925–931

Levis, Carolina; Flores, Bernardo M.; Moreira, Priscilla A.; Luize, Bruno G.; Alves, Rubana P.; Franco-Moraes, Juliano; Lins, Juliana; Konings, Evelien; Peña-Claros, Marielos; Bongers, Frans; Costa, Flavia R. C.; Clement, Charles R. 2018. How people domesticated Amazonian forests. Frontiers in Ecology and Evolution 5(171):1–21

Lima, H. P.; Barreto, C.; Betancourt, C. J. 2016. Novos olhares sobre as cerâmicas arqueológicas da Amazônia. In: Cerâmicas Arqueológicas da Amazônia, C. Barreto, HP Lima, and CJ Betancourt (eds), IPHAN/Museu Paraense Emilio Goeldi, Belém, pp 19–31

Macedo, Rodrigo Santana; Teixeira, Wenceslau Geraldes; Lima, Hedinaldo Narciso; Souza, Adriana Costa Gil de; Silva, Francisco Weliton Rocha; Encinas, Omar Cubas; Neves, Eduardo Góes. 2019. Amazonian dark earths in the fertile floodplains of the Amazon River, Brazil: an example of non-intentional formation of anthropic soils in the Central Amazon region. Boletim do Museu Paraense Emílio Goeldi, Ciencias Humanas 14(1):207–227

Maia, Álvaro. 1966. Defumadores e Porongas. Edições do Govêrno do Estado do Amazonas, Manaus

Neves, E. G. 2016a. Não existe Neolítico ao sul do Equador: as primeiras cerâmicas Amazônicas e sua falta de relação com a agricultura. In: Cerâmicas Arqueológicas da Amazônia, C. Barreto, HP Lima, and CJ Betancourt (eds), IPHAN/Museu Paraense Emilio Goeldi, Belém, pp 32–39

Neves, E. G. 2016b. L'Amazonie: une mer partagée. Dossiers d'Archeologie 373:32–37

Neves, Eduardo G. 2016c. A tale of three species or the ancient soul of tropical forests. In: Tropical Forest Conservation: Long-Term Processes of Human Evolution, Cultural Adaptations and Consumption Patterns, N. Sanz (ed.), UNESCO, Mexico, pp. 228–244

Palmatary, H.C. 1960. The Archaeology of the Lower Tapajós Valley, Brazil. Transactions of the American Philosophical Society, Philadelphia, New Series, Vol. 50, Part 3

Prance, G. T.; Balée, W.; Boom, B. M. 1987. Quantitative ethnobotany and the case for conservation in Amazonia. Conservation Biology 1(4):296–310

Roberts, Patrick; Hunt, Chris; Arroyo-Kalin, Manuel; Evans, Damian; Boivin, Nicole. 2017. The deep human prehistory of global tropical forests and its relevance for modern conservation. Nature Plants 3(17093)

Roosevelt, A. C. 1992. Secrets of the forest: an archaeologist reappraises the past-and future-of Amazonia. The Sciences 32(6):22–28

Roosevelt, Anna Curtenius; Housley, R. A.; Imazio da Silveira, M.; Maranca, S.; Johnson, R. 1991. Eighth millennium pottery from a prehistoric shell midden in the Brazilian Amazon. Science 254(5038):1621–1624

Roosevelt, Anna C.; Bevan, Douglas B.; Silveira, Maura Imazio da; Brown, Linda L. 2012. Early mounds and monumental art in ancient Amazonia. In: Early New World Monumentality, Burger, Richard L.; Rosenswig, Robert M. (eds), University Press of Florida, Gainesville, pp. 255–288

Schmidt, Morgan J.; Py-Daniel, Anne Rapp; Moraes, Claide de Paula; Valle, Raoni B. M.; Caromano, Caroline F.; Teixeira, Wenceslau G.; Barbosa, Carlos A.; Fonseca, João A.; Magalhães, Marcos P.; Santos, Daniel Silva do Carmo; Silva, Renan da Silva e; Guapindaia, Vera L.; Moraes, Bruno; Lima, Helena P.; Neves, Eduardo G.; Heckenberger, Michael J. 2014. Dark earths and the human built landscape in Amazonia: a widespread pattern of anthrosol formation. Journal of Archaeological Science 42:152–165

Smith, N. J. H. 2001. Land use dynamics in the Amazon estuary and implications for resource management. *Amazoniana* 16 (3/4): 517–537

Smith, N. 2014. A rainforest cornucopia: the cultural importance of native fruits in Amazonia. In: Amazonía: *Memorias de las Conferencias Magistrales del 3er Encuentro Internacional de Arqueolgía Amazônica*, Rostain, S. (editor), EIAA, Quito, pp. 31–50

Smole, W. J. 1976. The Yanoama Indians: A Cultural Geography. University of Texas Press, Austin

Souza, Jonas Gregorio de; Schaan, Denise Pahl; Robinson, Mark; Barbosa, Antonia Damasceno; Aragão, Luiz E. O. C.; Marimon Jr., Ben Hur; Marimon, Beatriz Schwantes; Silva, Izaias Brasil da; Khan, Salman Saeed; Nakahara, Francisco Ruji; Iriate, José. 2018. Pre-Columbian earth-builders settled along the entire southern rim of the Amazon. Nature Communications 9:1–10

Stahl, Peter W. 2015. Interpreting interfluvial landscape transformations in the pre-Columbian Amazon. The Holocene 25(10):1598–1603

Stenborg, Per; Schaan, Denise P.; Figueiredo, Camila G. 2018. Contours of the past: LiDAR data expands the limits of Pre-Columbian human settlement in the Santarém region, lower Amazon. Journal of Field Archaeology https://doi.org/10.1080/00934690.2017.1417198

Thoreau, H. D. 2001. Wild Fruit. W. W. Norton, New York

Watling, J.; Iriarte, José; Mayler, Francis E.; Schaan, Denise; Pessenda, Luiz C. R.; Loader, Neil J.; Street-Perrott, F. Alayne; Dickau, Ruth E.; Damesceno, Anontia; Ranzi, Alceu. 2017. Impact of pre-Columbian "geoglyph" builders on Amazonian forests. Proceedings of the National Academy of Sciences 114:1–6

Watling, Jennifer; Mayle, Francis E.; Schaan, Denise. 2018. Historical ecology, human niche construction and landscape in pre-Columbian Amazonia: a case study of the geoglyph builders of Acre, Brazil. Journal of Anthropological Archaeology 50:128–139

Zent, Egleé; Zent, Stanford. 2002a. Impactos ambientales generadores de biodiversidade: conductas ecológicas de los Hotï de la Sierra Maigualida, Amazonas Venezolano. Interciencia 27(1):9–20

Zent, Egleé; Zent, Stanford. 2002b. Los Jodï: sabios botánicos del Amazonas venezolano. Antropologica 97–98:29–70

Contents

Chapter 1
Achariaceae

1.1 *Carpotroche longifolia*

Brazil Canudo do pito

Colombia Choco latillo

Ecuador Ubiza muyo, achiote de monte; machanumi (Achuar), ma-sju-kè-mo-chi (Cayapa), uttetsupandiricho (Cofán), huira muyo, ñaccha muyu, punchu muyo, iramuyu caspi (Kichwa), cuacuiyo jucu, cuacuiyo soquë (Siona-Secoya), numy-nere (Shuar), meñica, nonguihue, miguinca, nonokowe, wika, ikay, migui kague, namentague (Waorani)

Peru Huira huayo, champa huayo, barba huayo, casha huayo, aceite huayo, sajino huayo, cacahuito, sacha cacahuillo, sacha chope, ojo de venado, oreja de burro, sapote de monte; dusenín, shipún, suwa dusení, saka, bukash, agkach, ankash, shawin, numi kunshigpip, súasúa (Aguaruna), shinupi (Chayahuita), wiramuyu (Kichwa)

N. Smith, *Amazon Fruits: An Ethnobotanical Journey*, Ethnobiology,
https://doi.org/10.1007/978-3-031-12803-5_1

Status Wild

Huira huayo ***Carpotroche longifolia*** **in upland forest. Grover, 13 years, is about to pick a fruit to eat the arils. Near Miraflores, lower Rio Tigre, Loreto, Peru, 6-12-04**

Carpotroche longifolia is a slender and short understory tree, usually under 8 m high, that grows in upland forests. It is found mostly in central and western Amazonia including Bolivia, the Brazilian states of Acre and Amazonas, Peru, Ecuador, and Colombia. In the western portion of its range, the tree extends up to 1000 m in the Andes (Brack 2003:113). The tree's range also extends into southern Panama. A few botanists have recorded the tree on floodplains, although it appears to be more common in areas that are not inundated periodically. *C. longifolia* has been noted as growing as high as 20 m in the Colombian and Ecuadorian Amazon (MOBOT collection numbers 15541, 1100), but this cauliflorous tree generally does not exceed 8 m in height (Vásquez 1997:354; Vásquez et al. 2010:165).

Opened fruit of huira huayo *Carpotroche longifolia* showing orange arils about to be eaten by Jairo, 12 years. Near Miraflores, lower Rio Tigre, Loreto, Peru, 6-12-04

Known as huira huayo in the Tigre watershed of the Peruvian Amazon, the tennis ball-sized fruits are born on the trunk. The pale green to almost white fruits with soft spines have a thick skin covering 20 or so seeds. Children especially eat the bright orange arils surrounding the pale gray seeds, the size of small raisins. Adults in the village of Miraflores at the mouth of the Tigre generally cannot be bothered with eating the arils, in spite of their pleasant, savory flavor. Since children have limited access to snack foods in most rural areas of the Peruvian Amazon, they are willing to take the trouble to savor the parsimonious arils and spit out the seeds. The fruits resemble in some ways those of annatto (*Bixa orellana*), hence one of the common names for the plant in the Ecuadorian Amazon: achiote del monte (annatto of the forest). Huira huayo fruits in the rainy season.

Drawing by Grover, 13 years, showing various regional fruits he likes to eat. Huira huayo *Carpotroche longifolia* **is featured second from the right. Miraflores, Boca del Tigre, Loreto, Peru, 2004**

In the Napo watershed in the Ecuadorian Amazon, some locals use the seeds to make ornaments according to the botanists John Brandbyge and Eduardo Asanza (MOBOT collection number 33217). The Maku who inhabit the Colombian Amazon and contiguous forest in northwest Amazonas in Brazil use the poison in a related species, *Carpotroche amazonica*, to hunt armadillos. The Nukak scrape the bark from the tree and heap the scrapings at the mouth of armadillo burrows. The armadillos then eat the bark and die nearby (Prance 1972).

References

Brack, A. 2003. *Frutas del Perú*. Universidad San Martin de Porres, Lima

Prance, G. T. 1972. An ethnobotanical comparison of four tribes of Amazonian Indians. *Acta Amazonica* 2(2): 7–27

Vásquez, R. 1997. *Flórula de las Reservas Biológicas de Iquitos, Perú*. Monographs in Systematic Botany Vol. 63, Missouri Botanical Garden, St. Louis.

Vásquez, R., R. Rojas, H. van der Werff. 2010. *Flora del río Cenepa, Amazonas, Perú*. Missouri Botanical Garden, St. Louis, vol. 1

Chapter 2
Anacardiaceae

2.1 *Anacardium giganteum*

Bolivia Cayuazú, cayú de monte

Brazil Cajuí, cajuí folha grande, caju açu, cajuaçú do mato, caju grande, caju da mata, caju do mato, caju bravo, caju silvestre, cajueiro da mata, moranha; ngoné (Desana), akayu (Guajá), akayu-'ɨ, akajupina'y (Ka'apor), daunlíkisu (Nambicuara), akajuhúa (Piripkura), orixiá (Suruí), akazu-ɨw-ate (Tembé), olio (Tiriyó), sololá (Tukano), akay-u (Wajãpi) oruxi hi, oru xihi, wito hi, witowito hi (Yanomami)

Colombia Marañón gigante, marañón grande, marañón de monte; uuregue (Miraña), añaaje (Muinane), anana (Witoto)

Peru Sacha cashu, sacha casho, casho moena, casho, cashillo, marañón gigante; kashu (Aguaruna), wësnid dëbiate (Matses)

Venezuela Merey, merey montero, merey del monte, merey montañero, odoi; dawíri (Baniwa), akaiyupánari (Curripaco), uli eloli jyaɨ (Hotï), witowito, weteoma nafi (Yanomami), majako shodö, odoi (Ye'kuana), akaiyuassú (Yeral)

Status Wild, planted

N. Smith, *Amazon Fruits: An Ethnobotanical Journey*, Ethnobiology,
https://doi.org/10.1007/978-3-031-12803-5_2

Vendor holding red fruits of *Anacardium giganteum*. The yellow fruits in plastic bags are yellow mombin *Spondias mombin*, another member of the cashew family. Ver-o-Pêso market, Belém, Pará, Brazil, 11-13-98

As noted by the late Alwyn Gentry when collecting botanical specimens, this formidable tree can soar to 50 m (MOBOT collection number 54662), and occurs in upland forests throughout Amazonia, the Orinoco basin in Venezuela, the Guianas, and along the Pacific coast of Colombia (Almeida et al. 1993; Araujo-Murakami et al. 2015; Batista et al. 2015; Black et al. 1950; Cavalcante 1973:10, 2010:85; Claudino et al. 2015; Condé 2011:74; Díaz Pérez et al. 2012; Lima et al. 2011, 2012; Mackensen et al. 2000; Mitchell and Mori 1987; Pereira et al. 2011; Pires and Salomão 2000; Salomão and Lisboa 1988; Silva et al. 1985, 1992). It also occurs occasionally on floodplains (Amaral et al. 2009; Batista and Gonçalves 2013; Batista et al. 2011; Black et al. 1950; Cain et al. 1956; Carim 2016:86; Díaz Pérez et al. 2012; Gama et al. 2003; Godoy et al. 1999; Kuchla et al. 2015; Miranda 2000; Silva et al. 1977:48). This relative of the cultivated cashew occurs in mature forest as well as successionary vegetation (Ebert et al. 2014; Rodrigues et al. 2004; Scoles 2009). This giant tree is also found in stunted vegetation on some sandy soils known as *varillal* in the Peruvian Amazon (Villacorta et al. 2003) and campinarana in the Brazilian Amazon (Vale 2011:39), and in seasonal forests in the watersheds of the Upper Tocantins and Araguaia Rivers in the southeastern fringe of the Amazon Basin (Haidar et al. 2013). This tree is more common in state of Pará (Cavalcante 2010:85).

The "fruits" are actually swollen pedicels with a single nut hanging from the bottom. The nut is encased with a toxic skin to discourage predation. The swollen pedicels, similar in design to the cultivated cashew *Anacardium occidentale*, have evolved to attract dispersal agents. Tapir *Tapirus terrestris*, peccaries (*Tayassu pecari* and *Pecari tajacu*), brocket deer *Mazama* spp., capuchin monkeys *Cebus apella* and *C. albifrons* and spider monkeys *Ateles* spp. are known to eat this fruit and likely disperse some of the nuts (Cummings and Read 2016; Montalván 2015:38). A species of spider monkey, *Ateles belzebuth*, has been recorded eating this fruit on Maracá Island, Roraima and likely disperses some of the seeds (Mourthé 2012:34). Tapirs, peccaries, brocket deer, and spider monkeys are hunted for their meat in Amazonia, the Orinoco Basin, and the Guianas so this fruit also contributes indirectly to the rural diet of people via game. The fruits are esteemed enough to find their way to some regional markets, such as Ver-o-Peso in Belém. This tree fruits in the dry and rainy seasons, such as in the Trombetas watershed (Knowles and Parrotta 1995).

Cajuí *Anacardium giganteum* in forest. Sitio São Benedito, side-road leading west from km 43 Santa Isabel-Vigia road, about 5 km south of Vigia, Pará, Brazil, 11-15-98

The common names for this wild relative of the cultivated cashew in the Amazon estuary at first seem contradictory. Cajuí means "small cashew" and refers to the diminutive fruit that is about a third of the size of cashew *Anacardium occidentale* fruit. Most swollen pedicels of *Anacardium giganteum* are under 7 cm long (Silva 2011b:112). Caju açu, on the other hand, means the "big cashew," and reflects the tree's enormous size. The species name, *giganteum*, also indicates that the tree reaches an impressive stature. The tree can become a forest canopy emergent so is much too tall for anyone to climb and pluck the kiwi-sized fruits from the branches. Rather, young and old alike gather the waxy-skinned fruits from the forest floor toward the end of the dry season and the beginning of the rainy period. Among the Ka'apor in eastern Amazonia, for example, *Anacardium giganteum* is one of the most important wild fruits gathered in the forest (Balée 1993b:69). The Yanomami in northern Amazonas and Roraima states in the Brazilian Amazon also gather these fruits in the forest (Albert and Le Tourneau 2007).

Those fruits not taken by canopy dispersal agents eventually fall off. Most survive the long drop to the forest floor because they are compact and light. When ripe, the delicate fruits emit a pervasive, sweet odor, evidently to attract seed dispersal agents. The Yorkshire botanist Richard Spruce described the fruit's smell as a "rather pleasant odor" when collecting a specimen along the Rio Negro in 1851 (Kew specimen number K000537398). The nut is an appendage to the fruit (technically a swollen peduncle), and birds, such as toucans, and monkeys are careful to discard the single nut covered with a protective skin impregnated with a burning chemical. The fruits of this tree are eaten by tapir *Tapirus terrestris*, peccaries (*Pecari tajacu* and *Tayassu pecari*), agouti (*Dasyprocta* spp.), paca *Agouti paca*, brocket deer (*Mazama americana* and *M. gouazoubira*), and tortoises (*Chelonoidis carbonaria* and G. *denticulata*) (Ayres and Ayres 1979; Barboza 2008:56; Fragoso and Huffman 2000; Moskovits and Bjorndal 1990; Redford et al. 1992; Salas and Fuller 1996). All of these animals are hunted for their meat throughout the Amazon, so the fruits also contribute indirectly to the diet of rural people.

Woolly monkeys *Lagothrix lagotricha*, spider monkeys *Ateles belzebuth* and *A. paniscus*, capuchin monkeys *Cebus apella*, the bald uakaris *Cacajao calvus*, coppery titis *Callicebus cupreus*, monk sakis *Pithecia monachus*, and squirrel monkeys *Saimiri sciureus* also consume the fruits (Defler and Defler 1996; Gómez-Posada 2012; Nunes 1998; Van Roosmalen 1985; Van Roosmalen and Klein 1988; Voss and Fleck 2011). All these animals are hunted in the Amazon and Guianas so cajuí also contributes indirectly to the rural diet when people eat those game animals. The Santarém marmoset *Callithrix humeralifera* also eats these fruits and although it is too small to be hunted regularly for its meat it is likely involved in dispersing the seeds within its distribution from the Tapajós River in the east to the Madeira River in the west (Stevenson and Rylands 1988).

The fiery red fruits of caju açu are a welcome treat during the latter part of the dry season when fewer trees tend to be in fruit. In addition to relishing the fresh fruits, they are also squeezed to make juice. River dwellers along the Aruá near Belém regard the juice as a good treatment for stomach and intestinal problems, such as ulcers and diarrhea (Lisboa and Silva 2009). In the Amazon estuary, the yellow pulp of the fruits is also mashed to make juice for mixing with cachaça, potent clear alcohol distilled from fermented sugarcane juice. The resulting beverage is called *batida de caju-açu*. The fruits also serve as an accompaniment to drinking cachaça straight. A bite of succulent caju-açu is a refreshing chaser after a shot of harsh sugarcane spirits. A few vendors in Belém's Ver-o-Pêso market carry the fruits in season, usually in plastic bags of ten, which weigh about 1 kg and fetch around US$ 0.85/bag. These vendors used to be strategically located close to alcoholic beverage stalls, tucked by the crumbling walls of the old fort, that sell cachaça and other hard liquors by the shot. Recently, however, vendors have shifted their produce to covered stands and one no longer finds piles of fruits on the ground for sale at Ver-o-Pêso.

Fruits of *Anacardium giganteum* would make an intriguing accompaniment to a fruit bowl at a buffet in any restaurant. Not only is the fragrance pleasant but also the fruits resemble large, smooth strawberries. But cajá-açu fruits are unlikely to adorn display tables in fine restaurants outside of Amazonia anytime soon. The tree is too large and takes too long to reach maturity to be of interest to most commercial growers. Nevertheless, the giant tree has been recruited to provide fruits around some indigenous villages, such as the Tumbira along the Curicuriarí River, an affluent of the Upper Negro (Ducke 1946). In the mid-nineteenth century, Richard Spruce noticed a group of caju açu *Anacardium spruceanum* trees, a similar species to *A. giganteum*, in upland forest at Lages downstream from Manaus at the confluence of the Negro and Solimões: "we traced out several trees, and found them so nearly of the same size that they might all have been planted by indigenous people" (Spruce 1908a, b). Spruce was probably right because Lages is an archaeological site with extensive tracts of Amazon Dark Earth, known in Brazil as *terra preta do índio*. In the Del Rey community near Paragominas in Pará state, a farmer has planted some seeds of *A. giganteum* that he obtained in the nearby forest, a testament to the ongoing recruitment of wild plants into the repertoire of crops maintained by small farmers throughout the basin. The Witoto along the Caquetá River in the Colombian Amazon plant *A. giganteum* in diverse agroforestry plots, undoubtedly an ancient practice (Veléz and Veléz 1992) as do some rural farmers in northeastern Pará (Santos et al. 2019). In the Venezuelan Amazon, the trunk is also used to make dugout canoes (MOBOT collection number 103).

Cajuí *Anacardium giganteum* spared when the forest was cleared to plant manioc. The field has now been converted to a cattle pasture. Some yellow-rumped caciques *Cacicus cela* are nesting in the branches. Outskirts of Santo Antonio de Ipapucu, Lago de Tefé, Amazonas, Brazil, 8-27-12

Farmers and small-scale ranchers sometimes spare the tree when clearing the forest so that they can gather the fruits from the ground. According to Adolpho Ducke (1946), a naturalist and ethnographer who spent his entire career in the Amazon, caju açu was also cultivated, or spared, in some of the suburbs of Belém in the late nineteenth and early twentieth centuries. But the city has since grown so much that few if any have likely survived the urban sprawl of Pará's capital. In home gardens, pastures, or manioc fields where *Anacardium giganteum* has been spared or planted yellow-rumped caciques *Cacicus cela* sometimes build their hanging nests at the end of branches. These noisy, social birds, known in Brazil as *japim*, also build their nests in other trees in home gardens such as peach palms *Bactris gasipaes*. Some indigenous groups along the middle Caquetá River in the Colombian Amazon plant this tree in their manioc fields so that they can return later to gather the fruits (Veléz 1991). Rural folk in parts of Pará plant or spare this tree in their home gardens, such as in the Comunidade Rural do Bacuri, 25 km southeast of Abaetetuba (Gonçalves and Lucas 2017), in Vila Cuera, a rural community in the vicinity of Bragança, in the municipality of Cametá in northeastern Pará (Santos et al. 2016b; Silva et al. 2017) and in the vicinity of Marabá (Miranda et al. 2012). This tree is also planted in some urban home gardens, such as in a suburb of Abaetetuba along the lower Tocantins in Pará (Lobato et al. 2017).

A young caju açu *Anacardium giganteum* tree in a home garden (medium-sized tree with light green leaves just to the right of center). Sitio Cafezal, Curuá River, FLONA Caxiuanã, Município de Melgaço, Pará, Brazil 7-2-16

The bark is also used in a variety of folk remedies. In Roraima, tea is prepared from the leaves to reduce inflammation (Luz 2001). The inner bark is soaked in water to treat inflammation of the uterus after giving birth, a condition sometimes called *mãe-do-corpo* which is attributed to supernatural causes in the Brazilian Amazon (Coelho-Ferreira 2009). Wounds are also washed with the solution to promote healing. A "tea" obtained by soaking bark segments is taken to treat stomachache, diarrhea, gastritis, and diabetes (Albert and Milliken 2009:129; Milliken and Albert 1997; Palheta et al. 2017; Rocha et al. 2013). On Colares, an island some 100 km northeast of Belém in the Amazon estuary, inhabitants prepare an infusion with the bark of this tree which is applied topically to dogs to treat mange and a tea is prepared from the bark to alleviate dogs with diarrhea (Ritter et al. 2012).

Caju açu resists anthracnose, a disease problem in some cashew plantations (Clement et al. 1982). It may be possible to transfer the gene or genes responsible for this resistance to cashew, thus boosting yields. The value of *Anacardium giganteum* as a source of disease resistance and much-appreciated wild fruit is threatened in some areas because the straight bole and hardwood make it a sought-after timber tree (Loureiro et al. 1979:92; Silva and Bentes-Gama 2008; Veríssimo et al. 1992). For example, the Kanamari, who inhabit the middle Jutaí River in the Brazilian state of Amazonas, encounter fewer of the trees now that logging operations have reached their lands (Carvalho and Reesink 1993).

2.2 *Anacardium occidentale*

Bolivia Cayú, cay, acaya, acajú, marañón; cayo (Chácobo), cayu ina (Tacana)

Brazil Cajú, cajú amarelo, cajú branco, cajú vermelho, caju roxo, cajueiro, cajueiro roxo, cajuí, acaiu; oroté (Arara), akayú (Araweté), akaiú, akayo (Baniwa), ngo'ré (Desana), akayu-rána (Guajá), awãkayu (Juruna), akayu (Ka'apor), akaju'i (Kaiabi), caju (Kayapó), ahkryt (Krahô), fĩgafegï (Kuikuru), erekiçu, eehrú (Nambikwara), zuitiá (Paresí), kaiu (Sateré-Maué), akazu (Tembé), co-a, coü (Ticuna), sõ'rá, sorã (Tukano), hiri (Waimiri-Atroari), akayu (Wayãpi)

Colombia Marañón, marañón amarillo, marañón pequeño, marañón rojo, meréy, caujil, amarillo, anacardo, anacardio, arbol de jobo, pauji; konka (Andoke), churá (Bari), sorá (Desana), sõna (Makuna), añaje-e (Miraña), añaje (Muinane), kóú (Ticuna), añay-j-ride, añana (Witoto)

Ecuador Marañón

Peru Cashu, casho, marañón, merey, cajú; kasú (Aguaruna), anáájihe (Bora), kaasho (Campa), pawa (Chayahuita), caashu, kasho (Matsigenka), casho (Shipibo), coa (Ticuna)

Venezuela Merey, merey chaparro; wito (Yanomami)

Status Wild, planted

Cashew ***Anacardium occidentale*** **tree in fruit in a home garden. San Martin Tipishca, Rio Samiria, Pacaya-Samiria, Loreto, Peru 6-29-06**

Ambrósio Fernandes Brandão, a Portuguese merchant writing from Brazil in the mid-sixteenth century, captures admirably the unique qualities of cashew:

> The bountiful cashew tree, proud of being different from the other trees, bears its fruit the opposite of the way all others do, for the nuts that other fruits hide within themselves these cashews flaunt on the outside. (Brandão 1987:217)

A cylindrical-shaped cashew *Anacardium occidentale* fruit in a home garden. Seringalzinho, Rio Jaú, Amazonas, Brazil, 10-3-12

Cashew bears juicy, vitamin C-rich fruits, technically swollen pedicels, as well as creamy-tasting nuts. The fruits vary in color from bright red to pale orange or yellow. As noted in the eighteenth century by João Daniel, a Jesuit priest, some cashew fruits are part red and part yellow (Daniel 1976:329). The yellow form has inspired the name for a variety of mango (*Mangifera indica*) that produces small, yellow fruits (manga caju) in parts of Marajó Island. Yellow varieties of cashew tend to have slightly higher levels of vitamin C, whereas red-skinned cashew fruits are higher in carotenoids (Assunção and Mercadante 2003). Fruit shape is also variable, ranging from compact and rounded to elongated and cylindrical. It is not uncommon to see trees bearing fruits of different colors and shapes in the same home garden, suggesting considerable genetic variation among cultivated forms. Cashew fruits are more commonly encountered in the dry season and are thus particularly welcome since most trees fruit during the rainy months.

Yellow cashew *Anacardium occidentale* **fruit tethered to avoid touching the toxic skin that covers the nut. Near Vigia, Pará, Brazil 11-15-98**

The nuts, which are attached to the bottom of the fruit, are encased in a green skin impregnated with highly toxic chemicals. Animals have learned to leave the nuts alone when eating the fruit. The fruits have evolved for bird and mammalian dispersal: the animals eat the fleshy part of the fruit and then drop the nut unharmed. Rural folk scorch off the skin on open fires before handling the nuts. Not all fruit-eating birds disperse the seeds, however. One farmer who lives along a side road at km 115 of the Belém-Brasília highway remarked that silver-beaked tanagers (*Ramphocelus carbo*) and blue-gray tanagers (*Thraupis episcopus*) peck at the fruits of his cashew trees, thereby ruining them. The free-loading tanagers are too small to fly off with the fruits.

Red cashew *Anacardium occidentale* **fruit gathered in a home garden. Porto Valente, lower Tefé River a few km downstream from Tefé, Amazonas, Brazil, 8-22-12**

Orange cashew *Anacardium occidentale* fruits gathered in a village home garden. Jocojó, Gurupá municipality, Pará, Brazil 9-24-14

Cashew is native to the scrub savanna (cerrado) region of central Brazil as well as to sand dunes along the Atlantic coast of the northeastern part of the country (Almeida Jr. et al. 2011; Lenza et al. 2015; Mori 1989) including the sandy coastline of northeastern Pará (Amaral et al. 2008; Bastos et al. 1995, 2003:12; Costa Neto et al. 1996; Lemos 2013:35; Quaresma and Jardim 2017; Santos and Rosário 1988; Santos 2010:41; Silva et al. 2010) and neighboring Maranhão (Freire and Monteiro 1993; Machado and Almedia 2019). It is found in scrub savanna in the southern fringes of the Amazon Basin in the states of Mato Grosso and Tocantins (Furley et al. 1978; Gomes et al. 2011; Lima et al. 2009; Medeiros and Walter 2012; Pinto et al. 2014; Rondon Neto et al. 2010; Tavares and Candeiro 2012), including rocky outcrops with a thin layer of soil (Maracahipes et al. 2011). It also occurs on laterite (*canga*) plateaus in the Serra dos Carajás in southern Pará (Mota et al. 2018) and in pockets of scrub savanna on sandy soils in other parts of the Amazon such as in the vicinity of Monte Alegre along the Amazon and Alter do Chão along the lower Tapajós in Pará as well as in Amapá and on Marajó Island at the mouth of the

Amazon (Amaral et al. 2019; Brown and Lidstone 1878:40; Magnusson et al. 2008; Miranda 1993, 1995; Sanaiotti 1996:42; Sanaiotti et al. 1997; Sanaiotti and Magnusson 1995; Silva 2008; Wallace 1889:100). It is also found in some campina vegetation growing on pockets of sandy soil, such as in the state of Pará (Ferreira 2009:84; Ferreira et al. 2013). At least one species of stingless bee, *Tetragonisca angustula*, has been recorded carrying pollen from this tree in the vicinity of Belterra and Santarém in Pará and likely pollinates this tree (Novais et al. 2014) as well as *Melipona seminigra merrillae* in Manaus (Ferreira and Absy 2017).

The exact boundaries of wild populations are unclear because of the tree's long association with humans (Mitchell and Mori 1987). Domesticated forms of cashew probably trace their origins to northeastern Brazil. In the Amazon, cashew is cultivated and encountered as wild stands on sandy soils along clear and black-water streams and rivers, such as along the Tapajós, and in well-drained parts of some savannas. Extensive stands of wild or feral cashew, some bearing red fruits others white, have been reported on the sandy plain along the coast of French Guiana (Aublet 1775:392). Whether these are truly wild or were brought there in precontact times by indigenous groups is unclear. The nut and swollen peduncle ("fruit") of cashew can float (Johnson 1973), so ocean currents may have taken cashew to the Guianas.

Cashew was taken to Central America around 900 to 400 BC (Lentz 2000) and was shipped to other continents during the colonial period (Sauer 1993:15). Cashew is an important cash crop in parts of India (Das 1985) and reached Tahiti around 1875 (Hermann et al. 1989:20). Cashew sometimes becomes feral outside its native range. In Mozambique, for example, native fruit bats dispersed the nuts after the Portuguese brought cashew there. Escaped cashew also adorns savannas and sandy shores of parts of Africa, tropical Asia, and the East Indies (Dalziel 1948:336, Kennedy 1936:188; Sauer 1988:25). Cashew is also feral in parts of the Caribbean, including Martinique (Kervégant 1937) and the sandy plains of Pinar del Río province in western Cuba (Esquivel and Hammer 1992).

How cashew reached the Amazon is a mystery. It is possible that some wild populations arose during dry climatic cycles in the Pleistocene when tongues of cerrado penetrated deep into the Amazon. Some pockets of cerrado still survive in the Amazon today, but they are being slowly engulfed by the forest. Wild stands of cashew on the Rupununi savannas bordering the Guianan shield may owe their existence to such a dry climatic period. Wild cashew stands on those savannas are sufficiently dense to make it worthwhile to undertake harvesting trips to gather the nuts (Baichoo 1981). In the savannas of Surinam, cashew is one of the most common trees (Essed 1908).

Wild cashew also occurs in patches of scrub savanna in the central Amazon. Richard Spruce, a Yorkshire botanist, remarked on the abundance of wild cashew in the vicinity of Santarém in the middle of the nineteenth century:

> The vegetation of the upland campos reminded me of an English pleasure-ground. It consisted of scattered low trees, rarely exceeding 30 feet in height, and here and there beds of gaily-flowering shrubs, with intervening grassy patches and lawns. Among the trees then in flower, the Cajú or Cashew-nut (*Anacardium occidentale*, L.) was exceedingly abundant;

> and an old Cajú, with its rough bark, its branches touching the ground on every side, its young leaves of a delicate red-brown, and its numerous pear-like yellow or red fruits. (Spruce 1908a:66)

Wild cashew also abounds on sandy upland plains in the vicinity of nearby Monte Alegre along the middle Amazon (Lima 1958). In the second half of the nineteenth century, Charles Frederick Hartt observed that:

> Cajú trees grow all over the serra, and the visitor will always gratefully remember their thirst-assuaging, acid fruit. The cajueiros of the serra are all very small, and the fruit is dwarfed and rather sour. On the sandy campos the tree is everywhere met with, and the fruit is sometimes very large and delicious. (Hartt 1874)

Hartt, a professor of Geology at Cornell University, also noted that old Indian residents (which he dubbed "Tupís") of the nearby town of Santarém were fond of a wine made from the fruits of feral cashew which was called *akayú kauín*. The wild cashew trees on the savannas in the vicinity of Santarém and Monte Alegre may be due to drier climatic cycles in the Pleistocene or introductions by people, or both.

Cashew is also found in patches of scrub savanna (cerrado) in southeastern Marajó Island at the mouth of the Amazon. How cashew got there is not known; it could have spread naturally during a dry climatic period, or was introduced by people in pre-contact times and subsequently became feral. Large swathes of the Switzerland-sized island were occupied by an advanced chiefdom, the Marajoara, for approximately a thousand years, so it is entirely possible that cashew was carried to Marajó long before the arrival of Europeans. Today, people gather cashew fruits from scrub savannas on Marajó some of which eventually reach markets, such as along the road leading into Salvaterra.

People likely had a hand in bringing cashew to the Amazon, thereby setting the stage for the establishment of feral populations. Over a century ago, Jacques Huber, a Swiss-Brazilian botanist suggested that indigenous groups took cashew inland from stands in northeastern Brazil (Huber 1904). Cashew remains have been found at archaeological sites near Monte Alegre in Pará dating to the late prehistoric period (Roosevelt et al. 1996). During the early colonial period, some 10,000–15,000 Tupinambá migrated to the Amazon from the coast of Brazil to escape slave raiding parties organized by the Portuguese; they settled along various sections of the Amazon, even reaching Peru (Sweet 1974:18). However, it is likely that cashew reached the Amazon as a crop before the Tupinambá arrived because indigenous cultures in the Amazon had extensive trade networks and social exchanges with far-flung groups long before the arrival of Europeans (Gupindaia 2008; Neves 2006; Schaan 2008). Fruits and nuts were surely traded between indigenous groups bordering Amazonia; in this manner, cashew gradually made its way inland. The brackish coast of Pará state east of Belém (known as the Bragantina Zone) was a promising avenue for introducing cultivated cashew in the Amazon. Cashew was grown by coastal tribes of Ceará and Maranhão, and groups living along the mangrove-dominated coast of the Bragantina Zone probably acquired cashew from neighboring tribes to the east. Spontaneous cashew is found on sand dunes on Algodoal, an island opposite the coastal village of Maracanã (Bastos et al. 1995); it was likely introduced to the island as a crop in pre-contact times. Cashew trees are also found wild in woodlands occupying some sand dunes along other parts of the northeast coast of Pará (Amaral et al. 2015).

Ancient cashew trees on relic sand dunes in seasonally flooded savannas in the northeastern part of Marajó Island probably arose in a similar manner. Cashew trees are common on strips of elevated sands, formed when sea levels were higher (Miranda 1909). Indeed, the largest ranch on Marajó, Fazenda Cajueiro, is named after the fruit tree. However, wild cashew fruits on Marajó are considered too astringent to eat; they "cut the mouth" (*corta a boca*) as explained by a maid in the main ranch house at Fazenda Jilva, so the fruits are employed to make sweetened juice. The wild cashew trees on Marajó fruit in September and October, during the height of the dry season.

An ancient cashew *Anacardium occidentale* tree on a relic sand dune. Fazenda Cajueiro, Chaves Municipality, Marajó Island, Pará, Brazil 8-18-00

Indigenous people are also most likely responsible for a grove of cashew trees on a patch of sandy soil at km 123 of the Santarém-Rurópolis Highway (BR 163) in Pará. This site was once an indigenous village as evidenced by the 1-meter-deep dark earth (*terra preta do índio*) which covers 0.25 hectares and is replete with potsherds and several stone axes. The former occupants of this settlement likely brought cashew to the area. The site is surrounded by an upland forest which does not contain any cashew trees. After introduced diseases devastated many indigenous groups along the Amazon and its tributaries in the sixteenth and seventeenth centuries, some of the abandoned cashew plantings became feral.

Fruits of feral cashew are still gathered on beaches that line the lower Tapajós to eat as a fresh snack, to prepare a fermented beverage, and to make preserves. The wild cashew trees, which often sprawl with sinuous branches, bear fruit in the dry season so villages and river dwellers can easily access the groves by walking along the beaches. The fruits, usually red when ripe, are smaller than cultivated forms. Feral cashew trees also adorn the beaches of the black water Arapiuns, an affluent of the lower Tapajós. A farmer on the outskirts of Vila Franca, a village at the confluence of the Arapiuns and Tapajós, explained that the wild cashew trees on beaches and savannas in the area are too sour (*azedo*) to eat fresh, so his family gathers them to make jam (*doce de caju*). Some wild cashew fruits are sweet, however, as noted by a river dweller in the Pindobal community along the lower Tapajós; Maria, a 65-year-old farmer who is fond of smoking a home-made pipe, ferments the juice of wild cashew fruits by pouring the liquid into glass bottles which she then places in the sun in her backyard for a week. The practice of fermenting the juice of cashew fruits in Brazil likely arose in pre-Columbian times. Cashew wine was esteemed by the Portuguese when they arrived in the early sixteenth century; indeed, that was one of the main motivations for taking cashew early to Mozambique and the Malabar coast of India (Johnson 1973).

River dweller gathering wild cashew ***Anacardium occidentale*** **fruits along the lower Tapajós River. Igarapé Iruçanga near Porto Novo, Pará, Brazil, 11-21-02**

Tuyuka girl snacking on a cashew *Anacardium occidentale* fruit in her mother's manioc field which is interplanted with several perennials. Near São Pedro, Rio Tiquié, Amazonas, Brazil, 11-8-12

Cashew is widely cultivated throughout Amazonia in rural and urban home gardens, around huts used for processing manioc flour (*casa de farinha, casa de forno*), and in and around manioc fields (Borges et al. 2004; Gonçalves and Pasa 2015; Lobato et al. 2017; Siviero et al. 2011; Souza and Scudeller 2009; Vieira et al. 2007; Wezel and Ohl 2006; WinklerPrins 2002). It is planted in home gardens on uplands and floodplains (Castro et al. 2009; Kawa et al. 2015; Klüppel 2006:24; Lima and Saragoussi 2000; Matos 2015:39; Silva 2011a:37, 2017:29; Souza et al. 2017), including the savanna region of Roraima (Pinho 2008:37). Cashew is most commonly interplanted with other perennials or in some cases annual food crops. One farmer in the Caxiuanã Reserve along the lower Anapú River in Pará likes to plant cashew in his manioc fields (*roças*) as the last tubers are harvested. In this manner, he and his family can return to the fields as they are being taken over by successionary vegetation and gather cashew fruits. Other farmers, such as in the municipality of Gurupá in the lower Amazon, often plant several cashew trees at the edges of their manioc field in order to provide shade as well as fruit and nuts.

Cashew *Anacardium occidentale*, **bacaba** *Oenocarpus bacaba*, **and tucumã** *Astrocaryum aculeatum* **trees around the site of a former manioc processing hut that ceased operations a decade ago. The manioc field has been converted to pasture. Bom Socorro, Paraná Tambaqui, near Lago Amanã, Amazonas, Brazil, 9-22-12**

Cashew *Anacardium occidentale* trees planted at the edge of a manioc field on an archaeological site with *terra preta do índio*. Anjo da Guarda, Rio Pracupi, affluent of the Anapú River, FLONA Caxiuanã, Pará, Brazil 5-15-15

Cashew is at home on uplands as well as on the higher parts of floodplains. Cashew is thus a highly versatile crop that thrives in a wide variety of soils, including highly leached oxisols and ultisols, as well as sandy soils that contain few nutrients and quickly dry out after rains.

Boy using a catapult to knock down cashew *Anacardium occidentale* fruits in his home garden. Bom Socorro, Paraná Tambaqui, near Lago Amanã, Amazonas, Brazil, 9-22-12

Fruits rather than the nuts are the main use of cashew in the Amazon. The fruits are easily gathered from the lower branches because the tree is short and the branches tend to spread laterally. Cashew fruits are delicate and tend to split if they fall to the ground, so youngsters often climb up the trunk to the higher branches to pick the fruits. Alternatively, the fruits are gathered by attaching a can to a long pole, as in Alter do Chão along the lower Tapajós, or by kids knocking down fruits using a catapult. The juicy fruits are eaten with the skin or squeezed to make a pungent, but pleasant tasting juice with a granular texture. In some areas, the juice is fermented to make cashew wine.

Children gathering cashew *Anacardium occidentale* **fruits in a tree in a village home garden. Pontão, Lake Canaçari, near Silves, Amazonas, Brazil, 9-21-10**

Farm boy eating a cashew *Anacardium occidentale* fruit he gathered in his home garden. Tauarí, Tefé River, Amazonas, Brazil, 9-11-12

Cashew fruits are used to garnish fish dishes in some parts of the Amazon. In the small town of Chaves on the northern coast of Marajó Island, for example, locals are fond of boiling fresh fish with cut-up cashew fruits. A highly esteemed local delicacy entails boiling salted jeju (*Erythrinus erythrinus*), a predatory fish abundant in tidal streams, with cashew fruits that are not quite ripe ("*cozinha com caju não bem amarelo, meio verde*").

Cashew *Anacardium occidentale* nuts soon to be roasted to burn off the toxic skin. The nuts were gathered from trees in a ribeirinho home garden. Sitio Timbó, 1eft bank of Rio Xingu about 45 km upstream from Altamira, Pará, Brazil, 9-17-15

Although fresh fruit and juice are the main uses of cashew in the Amazon, some rural people take the trouble to roast the nuts on fires in their backyards, mostly for local consumption, such as in the village of Guajará on Colares Island near Vigia. Along the margins of the Curuá-Una Reservoir near Santarém, a farm family likes to toast the nuts and then pound them in a wooden mortar along with manioc flour and sugar to make a dish called *paçoca*. A resident in Alter do Chão along the lower Tapajós sometimes adds a little salt when making *paçoca*; she likes to eat this regional dish in the morning with coffee or as a snack between meals. An inhabitant of Vigia explained that while some people still prepare *paçoca*, this traditional food appears to be on the decline. Home-roasted cashew nuts are partially scorched, thereby reducing their commercial value. Nevertheless, some farmers prepare plastic bags of roasted cashew nuts for sale along roads or in street markets, such as in northern Mato Grosso and in Santarém at the confluence of the Tapajós and the Amazon.

Roasting cashew *Anacardium occidentale* nuts to burn off the toxic outer skin. Rayssa, 11 years, is using a long stick to avoid inhaling the acrid smoke as she stirs the nuts in her backyard. Carrazedo, mouth of the Xingu River, Município de Gurupá, Pará, Brazil 11-1-16

Cashew ***Anacardium occidentale*** **nuts roasted in a fire to destroy the toxic skin in the backyard of a small farm. Some of the nuts will be bagged for sale. Near Juscimeira, Mato Grosso, Brazil, 10-2-93**

Cashew nuts still with their skin are also used by children in parts of the Brazilian Amazon to play a game called *biliboli*. Some half a dozen of the nuts with toxic skin are placed in a cupped hand then the nuts are tossed into the air. The child then tries to intercept as many nuts as possible on the other side of the hand before they fall to the ground. The skin can be touched without a problem. However, biting the skin around the nuts or inhaling smoke when the skin is burned off can lead to serious medical issues.

Hanna, 6 years, playing biliboli using cashew *Anacardium occidentale* nuts still with their skin cover. Carrazedo, mouth of the Xingu River, Município de Gurupá, Pará, Brazil 10-30-16

Carts selling guaraná (*Paullinia cupana*) smoothies have become a fixture in some towns in the Brazilian Amazon. A variety of flavors are typically offered, all advertised as a means to boost one's energy, and by implication sexual prowess. One vendor in Abaetetuba, a riverside town along the lower Tocantins, sticks to one recipe that features cashew nuts along with guaraná powder, guaraná syrup, catauba (*Anemopaegma mirandum*) syrup, quail eggs, peanuts, Brazil nuts *Bertholletia excelsa*, and milk powder. The concoction is placed in a blender along with ice and served in disposable cups.

In addition to fruit and nuts, cashew serves other purposes. On Marajó Island, for example, the front arch (*cabeçote*) of cowboy saddles is often fashioned from cashew wood. The wooden arch, upon which leather is attached, is treated with diesel oil to retard rotting. In Acre, this tree is sometimes planted as a living fence around cattle pastures (Franke 1999).

Cashew has long been used for medicinal purposes (Sousa 1971). For example, in the nineteenth century, a juice was prepared from the fleshy fruit along the middle Amazon to treat syphilis (Smith 1879:267). In Roraima, in the municipality of Cuiabá in Mato Grosso, and in the vicinity of Manaus, Amazonas, rural folk use the fruit to make juice for treating diarrhea, diabetes, and to heal wounds (Gonçalves and Pasa 2015; Luz 2001; Scudeller et al. 2009). In Rio dos Couros, a rural community in Mato Grosso, locals use the bark of this tree to make an infusion to treat dysentery and stomachache (Costa et al. 2017). In the Colombian Amazon, some rural people use the fruit to treat diarrhea and dermatitis (Lizcano et al. 2010) while on Marajó island in the Amazon estuary, people make a tea from the bark and leaves to treat their animals suffering from diarrhea (Monteiro et al. 2011). Rural communities along the lower Jauaperi which form the boundary between Roraima and Amazonas use the bark to make an infusion for treating diabetes and dysentery (Pedrollo et al. 2016). Families in the vicinity of the Caxiuanã National Forest in Pará steep the bark in water to wash wounds so that they heal quicker (Rocha et al. 2013) as do rural and urban folk in Roraima (Luz 2001; Pinto and Maduro 2003) and Rondônia (Santos et al. 2008). Also in Rondônia, rural folk prepare a decoction with the leaves of this tree to treat infections and inflammation (Santos et al. 2014). In the vicinity of Tefé, Amazonas, locals told me that the bark of the cashew tree is steeped in water and swallowed to control diarrhea as is the case in the village of Jocojó in the Municipality of Gurupá in the state of Pará. River dwellers in the watershed of the Jaú River, an affluent of the middle Rio Negro in Amazonas, also use the bark to make an infusion for treating gastrointestinal disturbances as do rural folk in the vicinity of Ajuruteua on the Bragantina coast of eastern Pará (Carneiro et al. 2010; Rodrigues 2006). In the village of Marudá along the coast of Pará, the inner bark of the cashew tree is also used to treat urinary infections, ovarian cysts, and to prepare a postpartum bathing solution (Coelho-Ferreira 2009), while in the same area the fruit is used to treat inflammation (Jardim and Medeiros 2006). Some residents of Imperatriz, a town in Maranhão that borders eastern Pará, use the leaves of this tree to prepare an infusion for treating inflammation (Penido et al. 2016). Also in Maranhão, rural folk of the community of Quilombo Sangrador in the municipality of Presidente Juscelino use the bark of this tree to make a solution for treating bone and muscle disorders (Monteles and Pinheiro 2007) and a powder is obtained from the leaves and spread on wounds caused by leishmaniasis (Moreira et al. 2002). In the nearby municipality of Marapanim, Pará, the flowers are used to make a tea to treat pain from diarrhea (Furtado et al. 1978). In Rio Branco, Acre, the bark of this tree is used to make a tea to treat respiratory problems (Siveiro et al. 2012). On Colares island some 100 km from Belém in the Amazon estuary locals prepare an infusion with the bark of this tree to treat wounds on cats and dogs (Ritter et al. 2012). Also on Colares, the fruit is burned to create smoke to dispel flies from chickens.

This practice of cashew use to treat ailments undoubtedly arose among indigenous groups, such as the Ticuna who live along the Upper Solimões in the Brazilian state of Amazonas (Schultes and Raffauf 1990:55). The Ticuna also make juice from the fruits to treat influenza. The Mura in the municipality of Autazes in Amazonas, Brazil, boil pieces of cashew bark in water, and once it has cooled use the liquid as a mouthwash to stem bleeding when teeth are pulled (Silva et al. 2008:209). The Mura also prepare tea from the bark to treat colic. In the early part of the twentieth century, a decoction of cashew bark was used for treating diabetes (Le Cointe 1922b:515). Yanomami shamans summon the tree spirit (*utupë*), which dwells in *Anacardium occidentale* and some other forest trees, to expel malevolent spirits that have invaded people causing them to be ill (Albert and Milliken 2009:137).

In the Bolivian Amazon, the Tacana use the leaves and fruit of cashew to prepare a decoction to stem diarrhea (Bourdy et al. 2000) and the Guarasug'we use cashew leaves to make a tea to treat the same condition (Hajdu and Hohmann 2012). Rural folk in the Bolivian Amazon eat the fresh fruit to stem diarrhea and assuage coughs (Murakami et al. 2016:47).

Cashew is also used for curing purposes in areas where it has been introduced in the Peruvian Amazon. The Chayahuita make a decoction from the bark to treat diarrhea which is taken three times a day (Odonne et al. 2013). To reduce fever, the Chayahuita steep the leaves and bark in water and drink the liquid twice a day. The Shipibo in the vicinity of Pucallpa along the Ucayali in the Peruvian Amazon use the bark to prepare a decoction for treating diarrhea, dysentery, bleeding, diabetes, and abdominal pain (Polesna et al. 2011). Rural people in the Peruvian Amazon boil the bark then take it off the fire and when the infusion becomes warm use it to disinfect vaginas at night. The nuts are crushed and applied to skin infections (Mejia and Rengifo 2000:60). In Chazuta and its vicinity along the Upper Huallaga River, people eat the fruit fresh to treat upset stomach and diarrhea (Sanz-Biset et al. 2009). In various parts of the Peruvian Amazon, rural people use this plant is used to treat a variety of other health issues. For example, to treat diabetes, the outer skin (epicarp) is removed from swollen pedicels and mashed in 1 l of water which is then left to soak for 24 hours; a cup of this solution is then drunk every 3 hours for 20–30 days (Lossio et al. 1995:85). To treat diarrhea, a handful of tender leaves (about 20 g) is soaked in 1 l of water or juice is squeezed from two ripe swollen pedicels (pseudo fruits) and a cup or glass of this liquid is then drunk with meals. To stem bleeding around teeth 20 g of the bark of this tree is boiled in 1 l of water and when the liquid cools it is used as a mouth wash. To treat vaginal inflammation, 20 g of the bark is boiled in 1 l of water and when the solution cools it is used to wash the vagina (Lossio et al. 1995:86).

Various parts of the cashew tree are used in folk medicine in several countries outside Amazonia. In neighboring Suriname, for example, Saramaccan maroons who are descendants of African slaves who were brought to the Guianas to work on sugarcane plantations along the coast use this plant to treat a variety of health issues in babies (Ruysschaert et al. 2009). Also in Suriname, the leaves and bark of this tree are sold in street markets to treat diarrhea (Van Andel et al. 2007). In Haiti, the leaves are used in a concoction to turn people into zombies (Davis 1988:121) and in Africa the stem bark is used to treat a variety of conditions from arthritis, fever, aches, and pains, to inflammation (Olajide et al. 2013).

Given the cultural importance of cashew in Amazonia and other regions it is not surprising that the tree surfaces in some lore. The Patamuna who inhabit a forest and savanna landscape in the Guiana highlands bordering Brazil recount a legend in which one of their ancestors was killed by the Makushi and subsequently transformed into a cashew tree. This supernatural tree still grows on the outskirts of their village. The Patamuna always carry cashew nuts with them on their walks which they plant in new places (Whitehead 2003). Among the Tenetehara who live along the Pindaré River in the eastern fringes of the Amazon forest in Maranhão, the tapir spirit called *tapiira azan* can enter the healer (*pajé*) when he sings at night during dancing ceremonies. When that happens, the *pajé* then heads for the river, transforms into a tapir, and swims across the river to reach a landscape that has been burned to eat cashew fruit. Upon returning to the village, the *pajé* transforms back into human form and vomits the cashew fruit (Galvão 1996:131). Another indigenous group in the eastern fringes of the Amazon forest in Maranhão, the Ka'apor, believe that the cashew was domesticated when their cultural hero Ma'ir planted branches of yaŝiamɨr *Lecythis idatimon*, an unrelated tree in the Brazil nut family, which they had gathered in the surrounding forest of their village (Balée 2013:67).

2.3 *Poupartia amazonica*

Brazil Cedro, cedro branco, uvaia, jacaiacá, taperebá açu, taperebá cedro, cedrorana, fruta de cedro

Status Wild

Yellow cedro *Poupartia amazonica* **fruits with orange fruits of tucumã** *Astrocaryum vulgare* **palm in a street market. Vigia, Pará, Brazil 5-8-99**

Like cashew fruits, *Poupartia amazonica* is employed as a chaser when drinking pure sugarcane alcohol, known as *cachaça* in Brazil (Andrade 1984:24). The tangerine-sized fruits are also used to make juice for mixing with *cachaça*, known as *batida de cedro*. Similar to a rum punch, *batida* flavored with cedro juice needs plenty of sugar to counteract the tartness. For this reason, the yellow fruits with distinctive tan mottling are almost never eaten fresh except as a *tira gosto* (to take away the taste of the harsh raw alcohol). Consumption of *cachaça* is so entwined in Brazilian culture that cedro fruits occasionally turn up in street markets, such as in Vigia, Pará. At one time, cedro fruits were employed to make ice cream in Belém (Cavalcante 2010:155).

Cedro *Poupartia amazonica* fruits gathered in pasture. Near ruins of sugarcane mill (*engenho*) of the Barão de Guajará, Ilha do Colares, near Vigia, Pará, Brazil 5-7-99

Fruits of cedro, a name that also applies to tropical cedar *Cedrela odorata*, an important timber tree, are gathered toward the end of the rainy season from the upland forest and second growth. The fruits are usually gathered from the ground because the tree can reach 30 m (FAO 1986). Although *Poupartia amazonica* is widespread in the Brazilian Amazon, from the estuary west to Acre and extending into Peru (Pennington et al. 2004:430), it is nowhere common.

Customer buying *Poupartia amazonica* fruits in a street market. Vigia, Pará, Brazil 5-10-99

2.4 *Spondias mombin*

Bolivia Casharoña, cedrillo, orocorocillo, ocorocillo, azucaré, azucaró del monte, ciruelo, sucá, turino; aquiacha (Sirionó), diji (Tacana), moco' (Tsimane)

Brazil Taperebá, taperebá de anta, taperebá de veado, taperebazeiro, cajá, cajá manga, cajá mirim, cajazeira, cajazeiro, cajázinha, cajazinho, cajuaçú da mata, peterebá, seriguela, umbu, umbuzeiro; mobé (Arara), akaya-'i (Araweté), kayuwa-'ɨwa (Asurini), tawa-wa-'ɨ (Guajá), taper-iwa-'ɨ (Ka'apor), bàrere-krã-kryre, bojrêrêkrãkrye (Kayapó), akajáb (Piripkura), oktsoto (Rikbaktsa), akai (Sateré-Maué), tawe-wa-'ɨw (Tembé), mope (Tiriyó), canaxaron (Waiká), akaya (Wayãpi), pirimë ahɨtoto (Yanomami)

Colombia Hobo, hobo arisco, jobo, jobo colorado, ciruela, ciruela calentana, ciruelo hobo, canabal, ubos, ubo, uvo, abal, pepa de morrocoy; keleté (Piapoco), djaumêru (Ticuna)

Ecuador Jobo, hobo, hobo de monte, ovo, ovo de monte, chirigua, jocote de pava, ciruela amarilla, ciruela del monte, mango ciruelo; kanú, mamantunim, kawau numi (Achuar), tsuyacho (Cofán), auru muyo, auru muyu, azua muyo, asua muyu yura, pusu muyo, pusco muyo, pepa agia (Kichwa), mamántunim (Shuar), rohí, daji, doji (Siona-Secoya), miyontahue, mientuhue, mientuhueno, miguitowe (Waorani)

Peru Ubos, ubos colorado, uvos, ciruela agria, motelo huayo, jobo, orocorocillo, mombin amarillo, mombin jaune, ciruela amarilla, shungu, tobo de la montaña, ushum, casharana, nispero; tsãpúnum (Achuar Mayna), jijuagtan, mamántunim (Aguaruna), metsoki (Asháninka), atsiyoriki, methoki, tsiyoroki (Campa), diji (Ese Eja), yawati wayu (Kichwa), shupud (Matses), tsigoroki, tsigoriaki (Matsigenka), oshon, sheshon, xexon (Shipibo), yomeru (Ticuna)

Venezuela Jobo; mojwe jyëï (Jodï); oponi (Yanomami), meje (Ye'kwana)

Status Wild, spontaneous in cultural settings, planted

Yellow mombin *Spondias mombin* tree in fruit on the Amazon floodplain. Near Itacoatiara, Amazonas, Brazil 12-8-06

Yellow mombin can reach 40 m in height in upland and floodplain forests and bears oval to oblong fruits that are much appreciated throughout the Amazon as well other lowland tropical forests of the Americas and the Caribbean (Aparício 2011:59, 92; Aquino 2005; Balslev et al. 1987; Batista and Gonçalves 2013; Peters and Hammond 1990; Vásquez 1997:83). It is also found in some mature tidal forests in the Amazon estuary, such as on Combu and Sororoca islands along the northeast coast of Pará and along the coast of Amapá (Almeida and Jardim 2012; Freitas et al. 2019; Queiroz et al. 2006; Lau and Jardim 2013; Santos 2010:41). This fruit tree is also found in successionary vegetation in the Amazon Basin, such as in the Department of Santa Cruz in Bolivia (Toledo and Salick 2006), in 20 year-old-successionary woodlands growing on uplands on abandoned agricultural plots near Santa Mercedes along Rio Putumayo in the northern Peruvian Amazon bordering Colombia (Gomez 2011:85; Muñoz 2010:34), in the Madre de Dios region of the southeastern Peruvian Amazon (Robles 2017:46), on uplands in southwestern Amapá (Guimarães et al. 2016), the tidal zone in the Amazon estuary and in the Tocantins watershed in Pará (Almeida and Jardim 2011; Batista et al. 2013; Bentes-Gama et al. 2002; Carim et al. 2008; Coelho et al. 2012; Fortini et al. 2006; Queiroz and Machado 2008; Taurino 2017:65) and in Maranhão bordering the Amazon Basin (Balée 1993c). It is also found in successionary vegetation in French Guiana (Hoff and Cremers 2005). It also grows in open forest in the eastern fringes of the Amazon Basin in the state of Maranhão (Gama et al. 2007; Muniz et al. 1994) as well as in seasonal forest and scrub savanna (*cerrado, cerradão*) in southern Amazonia such as along the mid Araguaia and Tocantins river basins (Amaral et al. 2019; Haidar et al. 2013; Medeiros et al. 2012; Oliveira et al. 2018; Ratter 1987) and in the Rio Blanco watershed in Bolivia (Paz-Rivera and Putz 2009).

The fruits are around 3.5 cm long and 2.6 cm wide but can reach 6 cm in length (Bora et al. 1991). About half the fruit is mesocarp which surrounds a single seed. Known as taperebá in the Brazilian Amazon and cajá in other parts of Brazil, the fruits are eaten fresh, especially by children, and have long been gathered in the forest by indigenous groups such as the Yanomami in northern Amazonas and Roraima states in Brazil and contiguous areas of southern Venezuela (Albert and Le Tourneau 2007). A mature tree can produce as many as 10,000 fruits (Janzen 1985). The pulp of yellow mombin is highly nutritious with 100 g providing at least one-third of the recommended daily allowance for vitamin A as well as significant amounts of potassium, magnesium, phosphorus, and copper and high antioxidant qualities (Bailão et al. 2015; Tiburski et al. 2011).

A mature *Spondias mombin* can produce as many as 10,000 fruits in a single season (Adler and Kielpinski 2000). The fruits are a bit tart, so the fruits are often mixed with sugar and water to make a refreshing drink. In the Peruvian Amazon, the fruit pulp is sometimes mixed with the juice squeezed from grated sweet manioc to make a non-alcoholic drink called cahuana (Flores 1997:274). Recently, Amazon Beer which is based in Belém has started selling *Witbier Taperebá*, a beer flavored with yellow mombin juice which has won a beer award in Brazil (Carvalho and Uchoa 2015).

The practice of using fruits to make juice can be traced to indigenous cultures, such as the Apinajé in eastern Pará (Nimuendajú 1956). The Zo'e who live in scattered communities in the municipalities of Óbidos and Alenquer in Pará ferment the juice for their festivities (Assis 2013). In addition to making juice, urban residents often put the fruits into a blender to make a milkshake or ice cream. Frozen pulp of yellow mombin is also found occasionally in supermarkets in the Brazilian Amazon. The bright yellow fruits, sometimes tinged orange, are gathered off the ground along the banks of rivers and streams or are plucked out of the water from canoes since they float. Most of the fruits on branches are out of reach from the ground because of the tree's height.

Tuyuka children eating yellow mombin *Spondias mombin* fruits in their village. It is near the end of the dry season and the fruits are not quite ripe. São Pedro, Rio Tiquié, affluent of the Uaupés, Amazonas, Brazil, 11-2-12

The precise origins of *Spondias mombin* are not known because humans have long been involved in moving the tree around. My guess is that its "natural" habitat is floodplains in Amazonia, an idea shared by Pedro Lisboa, a Brazilian botanist who worked at the Museu Goeldi in Belém before retiring (Lisboa 2012:196). Taperebá occurs in floodplain forests in the tidal zone of the Amazon estuary and in the Bragantina zone in northeastern Pará (Anderson et al. 1985; Batista et al. 2011; Cattanio et al. 2002; Queiroz et al. 2005, 2007; Maués et al. 2011), along the Araguaia River in the southeastern part of the Amazon Basin (Kurzatkowski et al. 2015), as well as in the Madre de Dios watershed in the Bolivian Amazon (Navarro 2002). It also occurs in riparian woods along streams in Manaus, Amazonas (Santos et al. 2016a). It is also found in floodplain forests along the Solimões/Amazon River as well as in the Bolivian Amazon (Luize et al. 2015; Thomas 2009). It is relatively common in floodplain forests on high river banks of the Ucayali in the vicinity of Jenaro Herrera in the Peruvian Amazon (Nebel et al. 2000). At the turn of the twentieth century, Jacques Huber, a Swiss-Brazilian botanist and former Director of the Museu Goeldi in Belém suggested that yellow mombin is native to the Amazon estuary and was subsequently spread to other areas by indigenous groups (Huber 1904). The fact that yellow mombin fruits float is a further hint that the tree is native to floodplains.

Renato, 7 years, eating yellow mombin *Spondias mombin* fruit gathered under the tree in his home garden. Sitio do Raimundo, Rio Curuá, Floresta Nacional Caxiuanã, Município de Melgaço, Pará, Brazil 5-5-15

Yellow mombin occurs in wild or feral form in other parts of northern South America as well as Central America, such as Panama (Standley 1927), Costa Rica (Holdridge et al. 1971:72; Janzen 1985), and Mexico (Herbarium records of the Royal Botanical Gardens, Kew). Fruits of yellow mombin were also dispersed by sea currents or carried by indigenous peoples to parts of the Caribbean, such as the Dominican Republic, Dominica, and Puerto Rico. Ocean currents, rather than humans, most likely dispersed yellow mombin fruits to Africa where it has become naturalized (Duvall 2006).

Yellow mombin grows in mature forests, such as in the Ecuadorian Amazon (Vickers and Plowman 1984), but is more common in disturbed sites on both uplands and floodplains (Balée 1993a, 2013:64). The tree is sometimes planted in home gardens as well as agroforestry fields, as I have observed along the Maracá River in Amapá and along the Transamazon Highway in Pará. This fruit tree is also found in some upland home gardens such as along Ramal do Pau Rosa, an offshoot of the Manaus-Boa Vista highway in the watershed of the Tarumã-mirim north of Manaus (Machado 2016:27), in the municipality of Presidente Figueiredo some 100 km NW of Manaus (Silva 2011a:37), in the São Raimundo district of Manaus (Barbosa

2018:39), in the municipalities of Rio Preto and Parintins in Amazons (Lourenço et al. 2009; Silva 2017:29), in Boa Vista, Roraima (Batista and Barbosa 2014), in the vicinity of Alenquer in western Pará (Rayol et al. 2013), in the Comunidade Rural do Bacuri, 25 km southeast of Abaetetuba, Pará (Gonçalves and Lucas 2017), in the municipality of Bonito in northeastern Pará (Vieira et al. 2007), along the Rio Guamá in the Amazon estuary (Scoles 2009), in the vicinity of Macapá, Amapá (Pereira et al. 2017) as well as near Marabá, Pará (Miranda et al. 2012). It is also found in home gardens in the municipalities of Cáceres and Cuiabá in Mato Grosso (Gonçalves and Pasa 2015; Pereira and Figueiredo Neto 2015), in the town of Palmas in Tocantins (Pinheiro et al. 2018), and in Acre in the western part of the Brazilian Amazon (Martins et al. 2012; Siviero et al. 2011). Rural folk in the community of Boa Esperança at km 43 of the Santarém to Curuá-Una highway (PA 370) in Pará also plant this tree in some of their home gardens (Garcia et al. 2017). This tree is also found in some home gardens along the Amazon River near Iquitos in the Peruvian Amazon (Balbin and Samaniego 1996).

Some farmers in the vicinity of Tomé-Açu, Pará, also plant taperebá in their agroforestry plots (Bolfe and Batistella 2011) as do some cacao famers in the vicinity of km 107 of the Altamira-Itaituba stretch of the Transamazon Highway (Leão et al. 2017) and in the municipality of São Félix do Xingu in southern Pará (Braga et al. 2019). The Ticuna, an indigenous group living in scattered villages along Solimões River in Amazonas, Brazil, and the contiguous Amazon River in Peru also plant this tree (Rengifo-Salgado et al. 2017). This tree is found in some home gardens of the Yagua who occupy several villages along the Amazon River downstream from Iquitos in Peru (Lamont et al. 1999). In some home gardens, both planted and spontaneous yellow mombin trees can be found, as at the confluence of the Aruanã and Caju-una Rivers some 5 km from Afuá on Marajó Island. At Sitio Santo Antonio at km 56 of the Santarém-Rurópolis highway in Pará, the yellow mombin tree in the home garden arose from discarded seeds. In my survey of 150 home gardens in widely scattered locations in the Brazilian Amazon 22 of them had yellow mombin trees (Lima et al. in press). *Spondias mombin* is also found occasionally in home gardens in the Bolivian Amazon (Murakami et al. 2016:43) as well as in villages in the interior of French Guiana (Gély 1986). When yellow mombin arises spontaneously in upland fields, possibly dispersed by birds or mammals, farmers typically spare the seedlings so that they can gather the fruits after annual crops have been harvested, a practice also noticed among some indigenous groups such as the Jodï in Venezuela (Zent and Zent 2002a, b). Several species of bats are known to disperse the seeds of *Spondias mombin* (Foster et al. 1986; Janzen 1985; Romo 1996).

A large variety of game animals eat the pulp of *Spondias mombin* fruits and sometimes disperse the seeds. The lowland tapir *Tapirus terrestris* has been documented dispersing the seeds of yellow mombin (Alviz and Pérez-Albarracín 2015:40; Fragoso and Huffman 2000; Herrera et al. 1999; Montenegro 2004:67; O'Farrill et al. 2013) and the Ka'apor who inhabit the eastern fringes of the Amazon forest in Maranhão suggest that brocket deer (*Mazama* spp.) also disperse the seeds (Balée 2013:68). In French Guiana and Guyana, the red brocket deer *Mazama americana* has been documented eating the pericarp and seeds of this fruit

(Cummings and Read 2016; Gayot et al. 2004). Collared peccaries *Pecari tajacu* and white-lipped peccaries *Tayassu pecari* also eat the fruits and disperse the seeds (Beck 2005a, b). Both brocket deer and peccaries are popular game animals throughout Amazonia and the Guianas. Agoutis *Dasyprocta punctata* and pacas *Agouti paca*, widely hunted in Central and South America for their meat, also eat the pulp of *Spondias mombin* but they drop the nuts on the spot rather than disperse them (Janzen 1985). Coatimundis *Nasua narica*, which occur widely in lowland tropical forests of Central and South America, are not hunted for their meat but they swallow the fruits and defecate the seeds some distance from the tree (Janzen 1985). Tapirs, brocket deer, agoutis, and paca are hunted for their meat so this fruit also contributes indirectly to the diet of rural people via game.

Several species of monkeys are also involved in dispersing the seeds, such as woolly monkeys *Lagothrix lagotricha* which are widespread in Amazonia (Aquino and Bodmer 2004; Defler and Defler 1996; Dew 2008; Stevenson 2000, 2005, 2007; Stevenson et al. 1994; Voss and Fleck 2011). On Barro Colorado Island in Panama and in the Santa Rosa National Park in Costa Rica, white-faced capuchin monkeys *Cebus capucinus* are known to feed on the fruits and disperse the seeds (Janzen 1985; Oppenheimer 1982). White-faced capuchin monkeys are confined to Central America and the extreme northwestern part of South America, but a related species, *Cebus apella*, is widespread in Amazonia and other parts of lowland South America and also consumes the fruits of *Spondias mombin* such as in the Peruvian Amazon and in the Guianas (Aquino and Bodmer 2004; Gregory 2006:55; Terborgh 1983:63). The white-capped capuchin *Cebus olivaceus*, which ranges from Venezuela through the Guianas and into Amapá and northern Pará in the Brazilian Amazon also eats this fruit and likely disperses some of the seeds (Robinson 1986). Squirrel monkeys *Saimiri sciureus*, *S. boliviensis*, and *S. vanzolinii*, which occur in various parts of Amazonia, also feed on this fruit and probably disperse some of the seeds (Aquino and Bodmer 2004; Paim et al. 2017; Pinheiro et al. 2013; Santos et al. 2018).

The mantled howler monkey *Alouatta palliata* has been observed swallowing the seeds of yellow mombin in wet and dry forests in Mexico, Costa Rica, and Panama (Estrada and Coates-Estrada 1984; Glander 1979, 1981; Milton 1980:54; Neville et al. 1988; Wehncke et al. 2004) and a related species, the red howler monkey *A. seniculus*, has been recorded eating these fruits in French Guiana and in Colombia (Guillotin et al. 1994; Rojas 2012). The red howler monkey occurs widely in western and the northern part of the Amazon Basin as well as Venezuela, Trinidad, and the Guianas. Several species of howler monkey inhabit the Amazon basin where they undoubtedly swallow and disperse the seeds.

Spider monkeys *Ateles belzebuth*, *A. chamek*, and *A. paniscus* also consume these fruits and disperse the seeds (Ahumada et al. 1998; Cruz 2018:53; Dew 2008; Felton et al. 2008, 2009, 2010; Link and Di Fiore 2006; Nunes 1995, 1998; Pozo 2004; Stevenson and Link 2010; Pontes 1997; Suarez 2006; Van Roosmalen 1985; Van Roosmalen and Klein 1988; Wallace 2005). A rare and endangered spider monkey, *Ateles hybridus*, which lives in the Magdalena river basin in northern Colombia, also feeds on this fruit (Luna et al. 2017).

In eastern Venezuela, bearded sakis *Chiropotes santanas* have also been observed feeding on the pulp of this fruit (Norconk et al. 1997); this species also occurs in parts of eastern Amazonia where it is highly endangered. Many of the above monkeys are hunted for their meat so this fruit also contributes indirectly to the rural diet of people via game.

Saddle-back tamarins *Saguinus fuscicollis illigeri* have been observed feeding on these fruits in the Peruvian Amazon (Soini 1987) and northern bearded sakis *Chiropotes satanas chiropotes* have been recorded doing the same in Surinam (Van Roosmalen et al. 1981, 1988) and both monkeys probably disperse the seeds. Although these monkeys are rather small and are rarely hunted for their meat, they are sometimes kept as pets. The seeds are dispersed in monkey feces which helps the seeds get a good start because of the fertilizer. Pygmy marmosets *Cebuella pygmaea*, which occur in western Amazonia, bite the bark of *Spondias mombin* to feed on the exudate (Soini 1988); these monkeys are too small, typically weighing between 85 and 140 g, to be worth hunting for their meat.

In Manaus, five stingless bee species, *Frieseomelitta varia*, *Melipona rufiventris paraensis*, *M. seminigra merrillae*, *Scaptotrigona fulvicutis*, and *Trigona williana*, have been observed gathering pollen of this fruit tree (Marques-Souza et al. 1995, 1996, 2002, 2007; Oliveira et al. 2009). Several species of stingless bees in the genus *Partamona* also gather pollen from this tree in the Rio Negro watershed, Amazonas (Rech and Absy 2011). The honey of several stingless bees is gathered in the wild, so this fruit tree also contributes to some local diets through wild honey.

Taperebá *Spondias mombin* intercropped with cacao *Theobroma cacao*. Farm of Elido Trevisan, Gleba 23, lote 2, km 75 Altamira-Itaituba, Transamazon highway, Municipality of Medicilandia, Pará, Brazil 9-7-15

Yellow mombin fruits have been gathered in the Amazon for thousands of years, such as on Marajó Island where the seeds are abundant in middens of the extinct Marajoara culture (Roosevelt 1991:375). On Marajó Island, yellow mombin is also found in anthropogenic forest islands on some mounds in seasonally flooded savannas, often built by indigenous cultures so that their houses would not be flooded during the rainy period (Schaan 2012:180). Yellow mombin is a good indicator that people have been in the area, as noted over a century ago (Huber 1904). Yellow mombin is found in regrowth forests on uplands near Borba along the Madeira River, possibly a vestige of ancient plantings (Kawa 2016:99). Indeed, the Tiriyó of the upper Paru de Oeste in northern Pará, Brazil, consider the groves of *Spondias mombin* on their lands as being planted by their ancestors (Frikel 1978). In the savannas of northern Colombia, the long-extinct Sinu used to plant yellow mombin on the burial mounds of their chiefs (Oyuela-Caycedo 2008).

Farm boy eating yellow mombin *Spondias mombin* fruits in the early part of the rainy season. Amazon floodplain a few kilometers downstream from Itacoatiara, Amazonas, Brazil 12-8-06

Yellow mombin has become so abundant in some areas due to human activities that places are named after the fruit, such as Taperebatuba (taperebá = yellow mombin, tuba = place) along the Urubu River in Amazonas, Brazil. The name the Ka'apor ascribe to yellow mombin, taperiwa'y, means "fruit tree of the anthropogenic forest" (Balée 2010). Yellow mombin is also considered an indicator of cultural forests in the Siriono territory in the Bolivian Amazon (Erickson and Balée 2006). Yellow mombin is common on anthropogenic dark earth in various parts of the Amazon (Clement et al. 2003; Fraser et al. 2011; Junqueira et al. 2010; McCann 1999). Accordingly, yellow mombin was considered an indicator of fertile soils for planting cacao in the early part of the twentieth century (Le Cointe 1922a:116).

Vendor in a street market placing yellow mombin *Spondias mombin* fruits in plastic bags. Macapá, Amapá, Brazil 12-13-94

Seeds of yellow mombin sprout readily when discarded in sunlit areas. Because the fruits are so widely appreciated, yellow mombin is usually spared from the ax or machete when clearing vegetation for a house site or to plant crops. Yellow mombin is also planted in home gardens using rooted cuttings, but most of the fruits consumed in Amazonia come from spontaneous trees. Yellow mombin is occasionally used to form living fences to restrain livestock because cuttings quickly take root, as in the community of Murumuru near Santarém and on Ilha Grande near Óbidos, both along the Amazon River in Pará. This practice has also been observed in Costa Rica (Sauer 1979).

The phenology of yellow mombin varies even in the same area. Near San Carlos along the Puinahua, a branch of the Ucayali River, and Pucallpa further upstream, ubos, as the fruit is known in Peru, fruits in the early part of the rainy season, whereas further west at the village of Nueva Esperanza on the Marañón River some trees produce fruit in the dry season, while others fruit during the wet period. In the vicinity of Tucuruí on the lower Tocantins, taperebá fruits from February to March, typically the wettest months of the year. Taperebá also fruits during the rainy season around Altamira on the Xingu River. In the Amazon estuary, most of the trees fruit during the dry season, with production peaking in November and December, shortly before the beginning of the rains. Occasionally, however, *Spondias mombin* trees fail to produce fruit in a given year, then resume fruiting the next year as on Barro Colorado Island in Panama (Milton 1985).

Yellow mombin *Spondias mombin* fruits in plastic bags on the ground in a market. Ver-o-Peso, Belém, Pará, Brazil 12-15-99

Fruit production is so copious in some areas that the fruits are taken to urban markets, such as Macapá and Belém at the mouth of the Amazon and Santarém at the confluence of the Amazon and Tapajos rivers (Rego et al. 2011). In Belém's Ver-o-Peso market, yellow mombin fruits were piled in the hundreds on the ground near the old fort until the municipal government had vendors display their fruit in covered stalls. Vendors also hawk yellow mombin fruits at ferry crossings and at speed bumps along major roads, although this practice has become much less common in recent years. The powerful and intriguing perfume of yellow mombin fruits wafts for several meters and often wins over customers if they have their car windows open. The delicately skinned fruits are sold in lots weighing about 2 kg, either in tall plastic bags or slender, make-shift baskets (*paneiros*) made with strips torn from palm petioles, such as buriti *Mauritia flexuosa*. The fruits are relatively inexpensive; in November 1998, for example, yellow mombin was selling for US$0.40/kg in Belém's bustling Ver-o-Peso market.

Roadside vendors with *Spondias mombin* fruits. Arapari, Pará, Brazil 12-12-99

Teddy Roosevelt, the former US president, recognized the virtues of yellow mombin nearly a century ago. During his adventures along the River of Doubt in southern Amazonia in 1914, he noted:

> The vast still forest was almost empty of life. We found old Indian signs. There were very few birds, and these in the tops of the tall trees. We saw a recent tapir-track; and under a

> cajazeira-tree by the bank there were tracks of capybaras which had been eating the fallen fruit. This fruit is delicious and would make a valuable addition to our orchards. The tree although tropical is hardy, thrives when domesticated, and propagates rapidly from shoots. The Department of Agriculture should try whether it would not grow in southern California and Florida. (Roosevelt 1926:290–291)

In addition to humans and capybara *Hydrochoerus hydrochaeris*, the latter hunted for its meat, other animals also disperse the seeds of *Spondias mombin*, including bats (Kubitzki 1985). A fisherman at Altamira on the Xingu River claims that grupité, a pacu-like fish, feeds on the fruits. In the early twentieth century, the Wapishana of the Upper Essequibo in southern Guyana and Roraima in Brazil used to toss yellow mombin fruits in the water to attract pacu (*Myleus* sp.) which they then shot using a bow and arrow (Farabee 1918:58). In the vicinity of Vigia in the southern part of the Amazon estuary, fisherman use yellow mombin fruits to catch bacu pedra (*Lithodras dorsalis*), a highly prized catfish. Bacu pedra swallow the fruits whole and are undoubtedly involved in dispersing the seeds. Turtles are also likely involved in dispersing the seeds of *Spondias mombin*; in Costa Rica, the black river turtle *Rhinoclemmys funera* is known to feed on these fruits and defecate the seeds (Moll and Jansen 1995). *Rhinoclemmys funera* is confined to Central America from Costa Rica and Honduras south to Panama but a similar species, perema *Rhinoclemmys punctularia*, occurs in the Amazon Basin and likely also disperses the seeds of yellow mombin. *Rhinoclemmys punctularia* is caught by rural people in the Amazon to be cooked and eaten.

A woodsman (*mateiro*) based in Tucuruí on the lower Tocantins asserts that tapir (*Tapirus terrestris*) and paca (*Agouti paca*) consume the fruits when they fall to the ground. Bulldog-sized pacas have also been recorded eating yellow mombin fruits in Colombia (Govoni et al. 2004). In the Peruvian Amazon, tapirs have been documented passing fruits of *Spondias* unscathed (Bodmer 1991). Youngsters in the village of Veinte de Enero along the Yanayacu River in the Peruvian Amazon claim that tortoises and deer (*Mazama americana*) also feed on the fallen fruits, an observation confirmed by ecologists in the case of the yellow-footed tortoise *Chelonoidis denticulata* (Guzmán and Stevenson 2008; Jerozolimski et al. 2009). Woolly monkeys (*Lagothrix lagotricha*) and spider monkeys (*Ateles* spp.) pick the fruits in the trees and are known to swallow the seeds (Kubitzki 1985; Stevenson et al. 2005).

Esteemed game animals such as tapir, paca, tortoises, deer, woolly, and spider monkeys, as well as several species of fish and turtle consume yellow mombin fruits so the tree contributes indirectly to the diet of rural folk. In French Guiana which borders Amapá in northeastern Amazonia yellow mombin fruits are particularly common in the diet of the lowland tapir *Tapirus terrestris* (Henry et al. 2000); tapirs have also been recorded eating the fruit in southern Venezuela (Salas and Fuller 1996). One of the local names for *Spondias mombin* in the Peruvian Amazon is motelo huayo which means tortoise fruit. The red-footed tortoise *Geochelone carbonaria*, a forest dweller, has been recorded eating the fruits of *S. mombin* (Vogt 2008:88). Two species of tortoise in the genus *Geochelone* are gathered in the wild and sometimes kept in pens until the household is ready to eat them. A popular forest game bird, various species of guan (*Penelope* spp.), are also fond of the fruits and are probably involved in dispersing the fruits such as in Mexico (Leopold 1972:208).

Livestock also feed on the fruits while foraging near homes, such as pigs on Ilha das Onças in the southern part of the Amazon estuary (Anderson et al. 1985). On Mexiana Island in the northern part of the Amazon estuary, river dwellers chop open fallen trunks of yellow mombin in the tidal zone to extract shipworms (*Neoteredo* sp.) which they consume raw.

Yellow mombin *Spondias mombin* bark with medicinal plant vendor. Mercado Belén, Iquitos, Loreto, Peru 6-17-04

Ubos *Spondias mombin* bark in a stand of medicinal plants at a street market. Mercado Belén, Iquitos, Peru 1-7-06

Yellow mombin is also used in a large variety of folk remedies. In the Municipality of Manacapuru east of Manaus, Amazonas, rural people make a tea from the bark to treat gastritis and inflammation (Vásquez et al. 2014) as do some urban residents of Boa Vista in Roraima (Pinto and Maduro 2003). In the municipalities of Xapuri in Acre and Pauini in southern Amazonas state, the bark is used to make a decoction to treat vomiting (Ferreira et al. 2015). Along Igarapé Jari, near Santarém, Pará, rural folk boil the bark in water and then use the infusion after it cools to wash wounds. Residents of Tiningú, a quilombo community 47 km southeast of Santarém, macerate the bark into water then drink the liquid to expel intestinal worms (Carvalho 2015:57). In the village of Marudá along the coast of Pará, people make a tea from the bark to treat ovarian cysts and prepare a bathing solution with the bark to alleviate vaginal itching and for post-partum cleansing (Coelho-Ferreira 2009). Folk healers take care not to ring the tree so that it can survive after removing a section of bark. A decoction of the flowers is used to treat eye irritation and laryngitis (Le Cointe 1934:427). Along the Jari River which borders Pará and Amapá, cowboys treat cattle suffering from diarrhea with liquid squeezed from the bark of taperebá trees (Lins 1991:97). Families also prepare a tea from the bark in the vicinity of the Caxiuanã National Forest in Pará to treat people who are vomiting (Rocha et al. 2013). On Colares Island, some 100 km from Belém in the Amazon estuary, locals apply powder obtained from the bark of this tree to the wounds of their dogs to promote healing (Ritter et al. 2012). In the community of Quilombo Sangrador in the municipality of Presidente Juscelino in Maranhão rural folk use the leaves of this tree to make tea to treat bone and muscular disorders (Monteles and Pinheiro 2007). In the municipality of Cuiabá in Mato Grosso some rural people macerate the inner bark and roots of this tree to treat diabetes (Gonçalves and Pasa 2015). Rural people in the north Araguaia region of Mato Grosso use the fruits as a diuretic (Ribeiro et al. 2017).

Rural people in the Bolivian Amazon prepare an infusion from the bark to treat diabetes and stomach problems (Murakami et al. 2016:43). Some indigenous people in the Bolivian Amazon mash the fruits of this tree to obtain juice which is then placed in bottles with a stopper and the liquid is taken to treat pneumonia (Ponz et al. 2005:62). In the extensive street market in the Belén district of Iquitos, ubos bark can be found for sale to treat hernia and postpartum issues. In parts of the Peruvian Amazon, a tea is made from the leaves and bark to treat diarrhea, stomach and vaginal infections, ovarian cancer, and dermatitis (Jovel et al. 1996; Peters and Hammond 1990; Tananta 2014:57). Also in the Peruvian Amazon, the bark of this tree us used to make a bathing solution to treat fungal skin infections (Wen et al. 2011). The resin obtained from the bark is applied directly to wounds to promote healing (Mejia and Rengifo 2000:202). In Bolivar state, Venezuela, which borders Roraima and Pará in the Brazilian Amazon, various indigenous groups prepare a decoction from the leaves for treating malaria (Caraballo et al. 2004). In Suriname, the bark and leaves are used to make a genital bathing solution and to treat wounds and kidney problems (Van Andel et al. 2007) while in Guyana the ashes from roasted seeds are used to treat wounds left by vampire bats, the bark and leaves are used to make a bathing solution to treat sores, and a tea is prepared from the bark to alleviate colds (Cummings and Read 2016).

The use of *Spondias mombin* in folk remedies is an ancient practice. The Tiriyó who live in the border zone between northern Pará and Surinam squeeze liquid from the roots of this tree which is then diluted in a little water and drunk by children to encourage their growth; women also apply this solution to their breasts to increase their size (Cavalcante 1973:68). The Ticuna who live along the Upper Solimões boil the bark in water and then use the infusion to alleviate pain and prevent excessive bleeding during menstruation (Schultes and Raffauf 1990:56). The Ticuna also apply the bark directly on infected skin (Rengifo-Salgado et al. 2017). The Yanomami who live in various villages in southern Venezuela and bordering areas of Brazil pour an infusion prepared with the leaves of this tree and pour it over the head and body of those suffering from fever (Milliken and Albert 1996). The Guarasug'we in the Bolivian Amazon make a tea from the bark to treat diarrhea

(Hajdu and Hohmann 2012) as do rural folk in the vicinity of Tefé in Amazonas, Brazil. The Macuxi who inhabit some savanna areas in Roraima in the northern Brazilian Amazon use the bark to make a decoction to treat malaria (Milliken 1997).

In the nineteenth century, farmers along the middle Amazon sometimes placed freshly cut branches of taperebá in manioc fields to distract the attention of leaf cutter ants (*Atta* spp.). Apparently, the leaf cutter ants prefer to take cut up leaves of taperebá to their underground nests, and thus damage to the manioc crop is avoided or at least reduced (Smith 1879:273).

At Carrazedo at the mouth of the Xingu, village boys carve toy boats from pieces of yellow mombin wood. The wood is also used for a variety of construction purposes throughout the Amazon, including small boats and poles used in house construction (Cerón and Ayala 1998:25; Guèze et al. 2014; Lorenzi 2008:28; Reyes-García 2001:237; Silva et al. 2007). The Cofán in the Ecuadorian Amazon sell the wood of this species (Cerón 1995:68). Near Abaetetuba along the lower Tocantins, some brick kilns use *Spondias mombin* wood as fuel and rural folk along the Amazon River in Peru also use the wood as fuel for cooking food (Balbin and Samaniego 1996).

Toy boat carved from yellow mombin *Spondias mombin* wood. Carrazedo, lower Rio Xingu, Municipality of Gurupá, Pará, Brazil 8-21-14

References

Adler, Gregory H.; Kielpinski, Kathleen A. 2000. Phenology of a tropical canopy tree, Spondias mombin. Biotropica 32(4):686–692

Ahumada, Jorge A.; Stevenson, Pablo R.; Quiñones, Marcella J. 1998. Ecological response of spider monkeys to temporal variation in fruit abundance: the importance of flooded forest as a keystone habitat. Primate Conservation 18:10–14

Albert, Bruce; Le Tourneau, François-Michel. 2007. Ethnogeography and resource use among the Yanomami: toward a model of "reticular space". Current Anthropology 48(4):584–592

Albert, Bruce; Milliken, William. 2009. Urihi A: a terra-floresta Yanomami. Instituto Socioambiental (ISA)/Institut de Recherche pour le Devéloppement, São Paulo
Almeida, Adrielson Furtado; Jardim, Mário Augusto Gonçalves. 2011. Florística e estrutura da comunidade arbórea de uma floresta de várzea na Ilha de Sororoca, Ananindeua, Pará, Brasil. Scientia Florestalis 39(90):191–198
Almeida, Adrielson Furtado; Jardim, Mário Augusto Gonçalves. 2012. A utilização das espécies arbóreas da floresta de várzea da Ilha de Sororoca, Ananindeua, Pará, Brasil por moradores locais. Revista Brasileira de Ciências Ambientais 23:48–54
Almeida Jr., Eduardo Bezerra de; Santos-Filho, Francisco Soares; Araújo, Elcida de Lima; Zickel, Carmen Sílvia. 2011. Structural characterization of the woody plants in restinga of Brazil. Journal of Ecology and the Natural Environment 3(3):95–103
Almeida, Samuel S.; Lisboa, Pedro Luiz B.; Silva, Antônio Sérgio. 1993. Diversidade florística de uma comunidade arbórea na estação científica "Ferreira Penna", em Caxiuanã (Pará). Boletim do Museu Paraense Emilio Goeldi, Série Botanica 9(1):93–128
Alviz, Ángela; Pérez-Albarracín, Karen. 2015. Plan para la conservación de la danta de tierras bajas (Tapirus terrestris) en los departamentos de Casanare, Arauca y Vichada. Corporinoquia-Fundación Orinoquia Biodiversa, Yopal, Colombia
Amaral, Dário Dantas do; Prost, Maria Thereza; Bastos, Maria de Nazaré do Carmo; Costa Neto, Salustiano Vilar; Santos, João Ubiratan Moreira dos. 2008. Restingas do litoral amazônico, estados do Pará e Amapá. Boletim do Museu Paraense Emílio Goeldi, Ciencias Naturais 3(1):35–67
Amaral, Dário Dantas do; Vieira, Ima Célia Guimarães; Almeida, Samuel Soares de; Salomão, Rafael Paiva; Silva, Antonio Sérgio Lima da; Jardim, Mário Augusto Gonçalves. 2009. Checklist da flora arbórea de remanescentes florestais da região metropolitana de Belém e valor histórico dos fragmentos, Pará, Brasil. Boletim do Museu Paraense Emílio Goeldi, Ciencias Naturais 4(3):231–289
Amaral, Dário Dantas do; Jardim, Mário Augusto Gonçalves; Costa-Neto, Salustriano Vilar; Bastos, Maria de Nazaré do Carmo. 2015. Síndromes de dispersão de propágulos e a influência da floresta amazônica na composição de espécies lenhosas de uma restinga no litoral norte brasileiro. Biota Amazônia 5(3):28–37
Amaral, Dário Dantas do; Rocha, Antônio Elielson; Pereira, Jorge Luis Gavina; Costa Neto, Salustiano Vilar. 2019. Identificação dos subtipos de savanas na Amazônia oriental (Pará e Amapá, Brasil) com uma chave dicotômica de individualização. Boletim do Museu Paraense Emílio Goeldi, Ciencias Humanas Naturais 14(2):183–195
Anderson, Anthony B.; Gely, A.; Strudwick, J.; Sobel, G. L.; Pinto, M. G. C. 1985. Uma sistema agroflorestal na várzea do estuário amazônico (Ilha das Onças, Município de Barcarena, Estado do Pará). Acta Amazonica, Supl. 15(1–2):195–224
Andrade, J. 1984. Folclore na Região do Salgado, Pará: Teredos na Alimentação, Profissões Ribeirinhas. Escola de Folclore, São Paulo
Aparício, Wegliane Campelo da Silva. 2011. Estrutura da vegetação em diferentes ambientes na RESEX do Rio Cajari: interações solo-floresta e relações com a produção de castanha. Doctoral Dissertation, Universidade Federal Rural de Pernambuco, Recife
Aquino, Rolando. 2005. Alimentación de mamíferos de caza en los aguajales de la Reserva Nacional Pacaya-Samiria (Iquitos, Perú). Revista Peruana de Biología 12(3):417–425
Aquino, Rolando; Bodmer, Richard E. 2004. Plantas útiles en la alimentación de primates en la cuenca del río Samiria, Amazonia Peruana. Neotropical Primates 12(1):1–6
Araujo-Murakami, Alejandro; Villarroel, Daniel; Pardo, Guido; Vos, Vincent A.; Parada, G. Alexander; Arroyo, Luzmila; Killeen, Timothy. 2015. Diversidad arbórea de los bosques de tierra firme de la Amazonía Boliviano. Kempffiana 11(1):1–28
Assis, R. 2013. *Zo'e*. Terceiro Nome, São Paulo
Assunção, Raquel B.; Mercadante, Adriana Z. 2003. Carotenoids and ascorbic acid from cashew apple (Anacardium occidentale L.): variety and geographic effects. Food Chemistry 81:495–502
Aublet, F. 1775. *Histoire des Plantes de la Guiane Française*. Pierre-François Didot, Paris

Ayres, J. M.; Ayres, C. 1979. Aspectos da caça no alto Rio Aripuanã. Acta Amazonica 9(2):287–298

Baichoo, C. S. 1981. Tropical fruit tree crop production in Guyana. In: *Proceedings of the Caribbean Workshop on Traditional and Potential Fruit Tree Crop Development, 9–14 November 1980, St. Georges, Grenada*. Inter-American Institute for Cooperation on Agriculture (IICA), San José, Costa Rica, pp. 217–237.

Bailão, Elisa Flávia Luiz Cardoso; Devilla, Ivano Alessandro; Conceição, Edemilson Cardoso da; Borges, Leonardo Luiz. 2015. Bioactive compounds found in Brazilian cerrado fruits. International Journal of Molecular Sciences 16:23760–23783

Balbin, Isabel Oré; Samaniego, Dedy Llapapasca. 1996. Huertas domesticas como sistema tradicional de cultivo en Moena Caño, Rio Amazonas, Iquitos, Peru. Folia Amazonica 8(1):91–110

Balée, W. 1993a. Biodiversidade e os índios Amazônicos. In: *Amazônia: Etnologia e História Indígena*, E. Viveiros de Castro and M. C. da Cunha (Editors), Núcleo de História Indígena e do Indigenismo da USP/FAPESP, São Paulo, pp. 385–393

Balée, W. 1993b. *Footprints of the Forest: Ka'apor Ethnobotany, the Historical Ecology of Plant Utilization by an Amazonian People*. Columbia University Press, New York

Balée, W. 1993c. Indigenous transformation of Amazonian forests: an example from Maranhão, Brazil. L'Homme 33(126–128):231–254

Balée, W. 2010. Contingent diversity on anthropic landscapes. *Diversity* 2:163–181

Balée, W. 2013. Cultural Forests of the Amazon: A Historical Ecology of People and Their Landscapes. University of Alabama Press, Tuscaloosa

Balslev, Henrik; Luteyn, James; Ollgaard, Benjamin; Holm-Nielsen, Lauritz B. 1987. Composition and structure of adjacent unflooded and floodplain forest in Amazonian Ecuador. Opera Botanica 92: 37–57

Barbosa, Cristiano de Souza. 2018. Diversidade e uso de plantas úteis nas quintais do bairro de São Raimundo, zona oeste de Manaus, AM. Master's thesis, INPA, Manaus

Barboza, M. S. L. 2008. O preço que a natureza pagou e os efeitos colaterais que sofremos para hoje se ter energia: uso dos recursos animais e percepção dos impactors entre os ribeirinhos do Lago de Tucuruí (PA). Master's thesis, Universidade Federal do Pará, Nucleo de Altos Estudos Amazônicas, Programa de Pós-Graduação em Desenvolvimento Sustentável do Trópico Úmido, Belém.

Bastos, M. N. C.; C. S. Rosário; L. C. B. Lobato. 1995. Caracterização fitofisionômica da restinga de Algodoal, Maracanã, Pará, Brasil. Boletim do Museu Paraense Emílio Goeldi, Série Botanica 11(2): 173–197

Bastos, Maria de Nazaré do Carmo; Costa, Denise Cristina Torres; Santos, João Ubiratan Moreira dos. 2003. Vegetação de restinga: aspectos botânicos e uso medicinal. Museu Paraense Emilío Goeldi Projeto Renas, Belém

Batista, Debora Lima; Barbosa, Reinaldo Imbrozio. 2014. Agrobiodiversidade urbana: composição florística, riqueza e diversidade de plantas nos quintais de Boa Vista, Roraima. Revista Brasileira de Agroecologia 9(2):130–150

Batista, Fábio de Jesus; Gonçalves, Mário Augusto. 2013. Florística e estrutura da regeneração natural arbórea de uma floresta de várzea na Reserva Extrativista Chocoaré-Mato Grosso, Pará, Brasil. Biota Amazônia 3(3):139–145

Batista, Fábio de Jesus; Jardim, Mário Augusto Gonçalves; Medeiros, Tonny David Santiago; Lopes, Isaac Luiz Magalhães. 2011. Comparação florística e estrutural de duas florestas de várzea no estuário Amazônico, Pará, Brasil. Revista Árvore 35(2):289–298

Batista, Anderson Pedro Bernardina; Santos, Vanessa Silva dos; Aparício, Wegliane Campelo da Silva; Aparício, Perseu da Silva; Silva, Diego Armando Silva da. 2013. Similaridade e gradientes de riqueza florística em um floresta de várzea na cidade de Macapá. Revista Verde de Agroecologia e Desenvolvimento Sustentável 8(4):152–158

Batista, Anderson Pedro Bernardina; Aparício, Wegliane Campelo da Silva; Santos, Vanessa Silva dos; Lima, Robson Borges de; Mello, José Marcio de. 2015. Caracterização estrutural em uma floresta de terra firme no estado do Amapá, Brasil. Pesquisa Florestal Brasileira 35(81):21–33

Beck, H. 2005a. Seed predation and dispersal by peccaries throughout the Neotropics and its consequences: a review and synthesis. In: Seed Fate: Predation, Dispersal and Seedling Establishment, Forget, Pierre-Michel; Lambert, Joanna E.; Hulme, Philip E.; Wall, Stephen B. Vander (eds), CABI, Wallingford, pp 77–115

Beck, H. T. 2005b. Caffeine, alcohol, and sweeteners. In: *The Cultural History of Plants*. Prance, G.T. and Nesbitt, M (eds), Routledge, New York, 173–190

Bentes-Gama, Michelliny de Matos; Scolforo, José Roberto Soares; Gama, João Ricardo Vasconcellos. 2002. Potencial produtivo de madeira e palmito de uma floresta secundária de várzea baixa no estuário Amazônico. Revista Árvore 26(3):311–319

Black, G. A.; Dobzhansky, T. H.; Pavan, C. 1950. Some attempts to estimate species diversity and population density of trees in Amazonian forests. Botanical Gazette 111(4):413–425

Bodmer, R. E. 1991. Strategies of seed dispersal and seed predation in Amazonian ungulates. *Biotropica* 23 (3): 255–261.

Bolfe, Édson Luis; Batistella, Mateus. 2011. Análise florística e estrutural de sistemas silviagrícolas em Tomé-Açu, Pará. Pesquisa Agropecuária Brasileira 46(10):1139–1147

Bora, Pushkar S.; Narain, Narendra; Holschuh, Heinz J.; Vasconcelos, Margarida da S. 1991. Changes in physical and chemical composition during maturation of yellow mombin (*Spondias mombin*) fruits. Food Chemistry 41:341–348

Borges, Sérgio Henrique; Filoni, Fernando; Siqueira, Iranildo Cursino. 2004. Prácticas agriculturais dos moradores do Parque Nacional do Jaú. In: Janelas para a Biodiversidade no Parque Nacional do Jaú: Uma Estratégia para o Estudo da Biodiversidade na Amazônia, Borges, Sérgio Henrique; Iwanaga, Simone; Durigan, Carlos César; Pinheiro, Marcos Roberto (eds), Fundação Vitória Amazônica, Manaus, pp 245–256

Bourdy, G.; DeWalt, S. J.; Chávez de Michel, L. R.; Roca, A.; Deharo, E.; Muñoz, V.; Balderrama, L.; Quenevo, C.; Gimenez, A. 2000. Medicinal plant uses of the Tacana, an Amazonian Bolivian ethnic group. Journal of Ethnopharmacology 70:87–109

Braga, Daniel P. P.; Domene, Frederico; Gandara, Flávio B. 2019. Shade trees composition and diversity in cacao agroforestry systems of southern Pará, Brazilian Amazon. Agroforestry Systems 93:1409–1421

Brandão, A. F. 1987. *Dialogues of the Great Things of Brazil* (*Diálogos das Grandezas do Brasil*). University of New Mexico Press, Albuquerque

Brown, C.B. and W. Lidstone. 1878. *Fifteen Thousand Miles on the Amazon and its Tributaries*. Edward Stanford, London

Cabral Freire, M.C.C. and R. Monteiro. 1993. Florística das praias da Ilha de São Luís, Estado do Maranhão (Brasil): diversidade de espécies e suas ocorrências no litoral brasileiro. Acta Amazonica, 23, pp.125–140.

Cain, Stanley A.; Castro, Gustavus M. de Oliveira; Pires, J. Murça; Silva, Nilo Tomâs. 1956. Application of some phytosociological techniques to Brazilian rain forest. American Journal of Botany 43(10):911–941

Caraballo, Alejandro; Caraballo, Brigida; Rodríguez-Acosta, Alexis. 2004. Preliminary assessment of medicinal plants used as antimalarials in the southeastern Venezuelan Amazon. Revista da Sociedade Brasileira de Medicina Tropical 37(2):185–188

Carim, Marcelo de Jesus Veiga. 2016. Estrutura, composição e diversidade em florestas alagáveis de várzea de maré e de igapó e suas relações com variáveis edáficas e o período de inundação no Amapá, Amazônia oriental, Brasil. Doctoral dissertation, Programa de Pós-Graduação em Botânica, INPA, Manaus

Carim, Marcelo de Jesus Veiga; Jardim, Mário Augusto Gonçalves; Medeiros, Tonny David Santiago. 2008. Composição florística e estrutura de floresta de várzea no Município de Mazagão, Estado do Amapá, Brasil. Scientia Florestalis 36(79):191–201

Carneiro, Diogo Borges; Barboza, Myrian Sá Leitão; Menezes, Moirah Paula. 2010. Plantas nativas úteis na Vila dos Pescadores da Reserva Extrativista Marinha Caeté-Taperaçu, Pará, Brasil. Acta Botanica Brasilica 24(4):1027–1033

Carvalho, Telma Lélia Gonçalves Schultz de. 2015. Etnofarmacologia e fisiologia de plantas medicinais do quilombo Tiningú, Santarém, Pará, Brasil. Master's thesis, Universidade Federal do Oeste do Pará, Santarém.

Carvalho, M. R. and E. B. Reesink. 1993. Ecologia e sociedade: uma breve introdução aos Kanamari. In: *Sociedades Indígenas e Transformações Ambientais*, Núcleo de Meio Ambiente, Universidade Federal do Pará, Belém, pp. 113–153

Carvalho, K. and R. Uchoa, 2015. Um brinde ao açaí: Amazon Beer leva o primeiro lugar no Festival Brasileiro da Cerveja com bebida a partir da fruta símbolo do Pará. *Way* 23: 4–7

Castro, Albejamere Pereira de; Fraxe, Therezinha de Jesus Pinto; Santiago, Jozane Lima; Matos, Raimundo Barbosa; Pinto, Ilzon Castro. 2009. Os sistemas agroflorestais como alternativa de sustentabilidade em ecosistemas de várzea no Amazonas. Acta Amazonica 39(2):279–288

Cattanio, José; Anderson, Anthony B.; Carvalho, Manoel S. 2002. Floristic composition and topographic variation in a tidal floodplain forest in the Amazon estuary. Revista Brasileira de Botanica 25(4):419–430

Cavalcante, P. B. 1973. A Farmacopéia Tiriyó: Estudo Étno-Botânico. Museu Paraense Emílio Goeldi, Publicações Avulsas No. 24

Cavalcante, P. B. 2010. *Frutas Comestíveis na Amazônia*. Museu Paraense Emílio Goeldi, Belém

Cerón, Carlos Eduardo. 1995. Etnobiologia de los Cofanes de Dureno, Provincia de Sucumbíos, Ecuador. Museo Ecuatoriano de Ciencias Naturales, Herbario Nacional, Serie Monografia 10(3), Quito

Cerón, Carlos Eduardo; Ayala, Consuelo Montalvo. 1998. Etnobotanica de los Huaorani de Quehueiri-ono Napo, Ecuador. Abya-Yala, Quito

Claudino, Wesley Vicente; Garcia Marcos Leandro; Lopes, Célia Regina Araújo Soares, Rodrigues, Lucirene; Cabral, Fabiana Ferreira; Piva, José Hypolito; Silva, Antônio Carlos Silveiro da. 2015. Análise fitossociológica da vegetação na região do arco do desmatamento. Revista de Ciências Agroambientais, Alta Floresta, MT 13(1): 43–52

Clement, C. R., C. H. Muller, and W. B. Chavez. 1982. Recursos genéticos de espécies frutiferas nativas da Amazônia Brasileira. *Acta Amazonica* 12 (4): 677–695.

Clement, C. R., J. M. McCann, and N. J. H. Smith. 2003. Agrobiodiversity in Amazonia and its relationship with dark earths. In: *Amazonian Dark Earths: Origin, Properties, Management*, J. Lehmann, D. C. Kern, B. Glaser, W. I. Woods (Editors), Kluwer, Dordrecht, pp. 150–178

Coelho, Roberta de Fátima Rodrigues; Miranda, Izildinha Souza; Mitja, Danielle. 2012. Caracterização do processo sucessional no Projeto de Assentamento Benfica, sudeste do estado do Pará, Amazônia oriental. Boletim do Museu Paraense Emílio Goeldi, Ciencias Naturais 7(3):251–282

Coelho-Ferreira, Márlia. 2009. Medicinal knowledge and plant utilization in an Amazonian coastal community of Marudá, Pará state (Brazil). Journal of Ethnopharmacology 126:159–175

Condé, Tiago Monteiro. 2011. Avaliação dos impactos da vegetação após a exploração madeireira em floresta ombrófila de terra firme no Município de Caracaraí, RR. Master's thesis, Programa de Pós-Graduação em Recursos Naturais, Universidade Federal de Roraima, Boa Vista

Costa, Ineilian B. C. da; Bonfim, Filipe P. G.; Pasa, Maria C.; Montero, Daniel A. V. 2017. Ethnobotanical survey of medicinal flora in the rural community Rio dos Couros, state of Mato Grosso, Brazil. Boletin Latinoamericano y del Caribe de Plantas Medicinales y Aromáticas 16(1): 53–67

Costa Neto, Salustiano Vilar da; Bastos, Maria de Nazaré do Carmo; Lobato, Luiz Carlos Batista. 1996. Composição florística e fitofisionomia da restinga do Crispim, município do Marapanim, Pará. Boletim do Museu Paraense Emílio Goeldi, Série Botânica 12(2): 237–249

Cruz, Raúl Federico Bello Santa. 2018. Comportamiento de monos arañas (*Ateles chamek*) reintroducidos en el sureste de la Amazonía Peruana. Master's thesis, Universidad Nacional Agraria La Molina, Lima

Cummings, Anthony R.; Read, Jane M. 2016. Drawing on traditional knowledge to identify and describe ecosystem services associated with northern Amazon's multiple-use

plants. International Journal of Biodiversity Science, Ecosystem Services & Management 12(1–2):39–56
Dalziel, J. M. 1948. *The Useful Plants of West Tropical Africa*. The Crown Agents for the Colonies, London.
Daniel, J. 1976. *Tesouro Descoberto no Rio Amazonas*. Biblioteca Nacional, Rio de Janeiro, vol. 1
Das, P. F. 1985. Status of production and trade of cashew in India. *Agricultural Situation in India* 40(6):765–770
Davis, W. 1988. *Passage of Darkness: Ethnobiology of the Haitian Zombie*. University of North Carolina Press, Chapel Hill
Defler, Thomas R.; Defler, Sara B. 1996. Diet of a group of *Lagothrix lagothricha lagothrica* in southeastern Colombia. International Journal of Primatology 17(2):161–190
Dew, J. Lawrence 2008. Spider monkeys as seed dispersers. In: Spider Monkeys: Behavior, Ecology and Evolution of the Genus *Ateles*, Campbell, Christina J. (ed), Cambridge University Press, Cambridge, pp 155–182
Díaz Pérez, Wilmer Antonio; Daza, Félix; Sarmiento, William. 2012. Composición florística, estructura y diversidad del bosque ribereños del Río Kakada, cuenca del Río Caura, estado Bolívar, Venezuela. Revista Científica UDO Agrícola 12(2):275–289
Ducke, A. 1946. *Plantas de Cultura Precolombiana na Amazônia brasileira. Notas Sôbre as Espécies ou Formas Espontâneas que Supostamente lhes teriam dado Origem*. Instituto Agronômico do Norte, Belém, Boletim Técnico 8, pp. 1–24
Duvall, C. S. 2006. In the origin of the tree *Spondias mombin* in Africa. Journal of Historical Geography 32(2):249–266
Ebert, Alexandre; Teixeira, Leandro Ribeiro; Silva, Adriana Zanirato Contini da; Costa, Reginaldo Brito da. 2014. Natural regeneration in tropical secondary forest in southern Amazonia, Brazil. Open Journal of Forestry 4(2):151–160
Erickson, C. L. and W. Balée. 2006. The historical ecology of a complex landscape in Bolivia. In: *Time and Complexity in Historical Ecology*, W. Balée and C. L. Erickson (Editors)
Esquivel, M. and K. Hammer. 1992. Native food plants and the American influence in Cuban agriculture. In: *'...y tienen faxones y fabas muy diversos de los nuestros...': Origin, Evolution and Diversity of Cuban Plant Genetic Resources*, Hammer, K., Esquivel, M., Knüpffer, H. (Editors), Institut für Pflanzengenetik und Kulturpflanzenforschung, vol. 1, pp. 4–74
Essed, E. 1908. The savannahs of Guyana. *Transactions of the Botanical Society of Edinburgh* 23:159–164
Estrada, Alejandro; Coates-Estrada, Rosamond. 1984. Fruit eating and seed dispersal by howling monkeys (*Alouatta palliata*) in the tropical rain forest of Los Tuxtlas, Mexico. American Journal of Primatology 6:77–91
FAO, 1986. *Food and Fruit-Bearing Forest Species 3: Examples from Latin America*. Food and Agriculture Organization, Rome, Forestry Paper 44 (3).
Farabee, W. C. 1918. *The Central Arawaks*. The University Museum, University of Pennsylvania, Philadelphia, vol. 10
Felton, Annika M.; Felton, Adam; Wood, Jeff T.; Lindenmayer, David B. 2008. Diet and feeding ecology of *Ateles chamek* in a Bolivian semihumid forest: the importance of Ficus as a staple food resource. International Journal of Primatology 29:379–403
Felton, Annika M.; Felton, Adam; Wood, Jeff T.; Foley, William J.; Raubenheimer, David; Wallis, Ian R.; Lindenmayer, David B. 2009. Nutritional ecology of Ateles chamek in lowland Bolivia: how macronutrient balancing influences food choices. International Journal of Primatology 30:675–696
Felton, A. M.; Felton, A.; Foley, W. J.; Lindenmayer, D. B. 2010.The role of timber tree species in the nutritional ecology of spider monkeys in a certified logging concession, Bolivia. Forest Ecology and Management 259:1642–1649
Ferreira, Carlos Alberto Cid. 2009. Análise comparativa de vegetação lenhosa do ecosistema campina na Amazonia Brasileira. Doctoral dissertation, INPA/UFAM, Manaus

Ferreira, Marcos Gonçalves; Absy, Maria Lucia. 2017. Pollen analysis of honeys of Melipona (Michmelia) seminigra merrillae and Melipona (Melikerria) interrupta (Hymenoptera: Apidae) bred in Central Amazon, Brazil. Grana 56(6):436–449

Ferreira, Leandro V.; Chaves, Priscilla P.; Cunha, Denise de A.; Rosário, Alessandro S. do; Parolin, Pia. 2013. A extração ilegal de areia como causa do desaparecimento de campinas e campinaranas no estado do Pará, Brasil. Pesquisas Botânica 64:157–173

Ferreira, Almecina Balbino; Ming, Lin Chau; Haverroth, Moacir; Daly, Douglas Charles; Caballero, Javier; Ballesté, Andrea M. 2015. Plants used to treat malaria in the regions of Rio Branco-Acre state and southern Amazonas state, Brazil. International Journal of Phytocosmetics and Natural Ingredients 2(9):1–5

Flores, Salvador. 1997. *Cultivo de Frutales Nativos Amazonicos: Manual para el Extensionista.* Tratado de Cooperación Amazónica, Lima

Fortini, L. B.; Rabelo, F. G.; Zarin, D. J. 2006. Mixed potential for sustainable forest use in the tidal floodplain of the Amazon River. Forest Ecology and Management 231:78–85

Foster, Robin B.; Arce, Javier; Wachter, Tatzyana S. 1986. Dispersal and the sequential plant communities in Amazonian Peru floodplain. In: Frugivores and Seed Dispersal, Estrada, Alejandro; Fleming, Theodore H. (eds), Dr. W. Junk Publishers, Dordrecht, pp 357–370

Fragoso, José M. V.; Huffman, Jean M. 2000. Seed-dispersal and seedling recruitment patterns by the last Neotropical megafaunal element in Amazonia, the tapir. Journal of Tropical Ecology 16:369–385

Franke, Idésio Luis. 1999. Principais usos e serviços de árvores e arbustos promissores que ocorrem em pastagens no Estado do Acre. EMBRAPA Acre Comunicado Técnico 106:1–6

Fraser, J. A., A. B. Junqueira, and C. R. Clement. 2011. Homegardens on Amazonian Darth Earths, non-anthropogenic upland, and floodplain soils along the Brazilian middle Madeira River exhibit diverging agrobiodiversity. *Economic Botany* 65(1):1–12

Freitas, João da Luz; Santos, Erick Silva dos; Cruz Junior, Francisco de Oliveira; Borges, César Henrique Alves; Santos, Adriano Castelo dos. 2019. Structural analysis and distribution patterns in lowland tropical forest, eastern Amazon. Journal of Agricultural Studies 7(4): 240–255

Frikel, P. 1978. Áreas de arboricultura pré-agrícola na Amazônia: Notas preliminares. *Revista de Antropologia* (São Paulo) 21 (1): 45–52

Furley, P. A.; Ratter, J. A.; Gifford, D. R. 1978. Observations on the vegetation of eastern Mato Grosso, Brazil. III. The woody vegetation and soils of the Morro de Fumaça, Torixoreu. Proceedings of the Royal Society of London, B 203:191–208

Furtado, Lourdes Gonçalves; Souza, Ruth Cortez de; Van den Berg, Maria Elizabeth. 1978. Notas sobre uso terapêutico de plantas pela população cabocla de Marapanim, Pará. Boletim do Museu Paraense Emílio Goeldi, Nova Série Antropologia 70:1–31

Galvão, Eduardo. 1996. Diários de Campo de Eduardo Galvão: Tenetehara, Kaioá, e Índios do Xingu. Editora UFRJ/Museo do Índio-FUNAI, Rio de Janeiro

Gama, J. R. V., S. A. Botelho, M. Bentes-Gama, and J. R. S. Scolforo 2003. Estrutura e potencial futuro de utilização da regeneração natural de floresta de várzea alta no Município de Afuá, estado do Pará. *Ciencia Florestal, Santa Maria* 13(2):71–82

Gama, João Ricardo Vasconcellos; Souza, Agostino Lopesde Souza; Calegário, Natalino; Lana, Guilherme Campos. 2007. Fitossociologia de duas fitocenoses de floresta ombrófila aberta no município de Codó, estado do Maranhão. Revista Árvore 31(3):465–477

Garcia, Bruna Naiara Rocha; Vieira, Thiago Almeida; Oliveira, Francisco de Assis. 2017. Tree and shrub diversity in agroforestry homegardens in rural community in eastern Amazon. Floresta 47(4):543–552

Gayot, Marc; Henry, Olivier; Dubost, Gérard; Sabatier, Daniel. 2004. Comparative diet of the two forest cervids of the genus *Mazama* in French Guiana. Journal of Tropical Ecology 20(1):31–43

Gély, Anne. 1986. Agriculture créole; une étude de cas: Saül (Guyane Française). Journal d'Agriculture Traditionnelle et de Botanique Appliquée 33:143–192

Glander, K. E. 1979. Howling monkey feeding behavior and plant secondary compounds: a study of strategies. In: The Ecology of Arboreal Foliovores, Montgomery, G. G. (ed), Smithsonian Institution Press, Washington, DC, pp 561–574

Glander, K. E. 1981. Feeding patterns in mantled howling monkeys. In: Foraging Behavior: Ecological, Ethological, and Psychological Approaches, Kamil, A. C.; Sargent, T. D. (eds), Garland Press, New York, pp 231–257

Godoy, Judith Rosales; Petts, Geoffrey; Salo, Jukka. 1999. Riparian flooded forests of the Orinoco and Amazon basins: a comparative review. Biodiversity and Conservation 8:551–586

Gomes, Letícia; Lenza, Eddie; Maracahipes, Leandro; Marimon, Beatriz Schwantes; Oliveira, Edmar Almeida de. 2011. Comparações florísticas e estruturais entre duas comunidades lenhosas de cerrado típico e cerrado rupestre, Mato Grosso, Brasil. Acta Botanica Brasilica 25(4):865–875

Gomez, Tedi Pacheco. 2011. Valoración económica de purmas de diferentes edades en la comunidad nativa de Santa Mercedes, río Putumayo, Perú. Master's thesis, UNAP, Iquitos

Gómez-Posada, C. 2012. Dieta y comportamiento alimentario de un grupo de mico maicero *Cebus apella* de acuerdo a la variación en la oferta de frutos y artrópodos, en la Amazonía colombiana. Acta Amazonica 42(3):363–372

Gonçalves, Janaína Pinheiro; Lucas, Flávia Cristina Araújo. 2017. Agrobiodiversidade e etnoconhecimento em quintais de Abaetetuba, Pará, Brasil. Revista Brasileira de Biociências 15(3):119–134

Gonçalves, Karina Gondolo; Pasa, Maria Corette. 2015. A etnobotânica e as plantas medicinais na Comunidade Sucuri, Cuiabá, MT, Brasil. Interações 16(2):245–256

Govoni, G., D. Fielding, and M. G. Paoletti. 2004. Rodent farming in the Amazon: experience with Amerindians in Venezuela. In: *Ecological Implications of Minilivestock: Potential of Insects, Rodents, Frogs and Snails*, M. G. Paoletti (Editor), Science Publisher, Enfield, New Hampshire, pp. 47–71

Gregory, Tremaine. 2006. Comparative socioecology of sympatric, free-ranging white-faced and bearded sakis in Brownsberg Nature Park, Suriname. Doctoral dissertation, Kent State University, Kent, Ohio

Guèze, Maximilien; Luz, Ana Catarina; Paneque-Gálvez, Jaime; Macía, Manuel J.; Orta-Martínez, Martí; Pino, Joan; Reyes-García, Victoria. 2014. Are ecologically important tree species the most useful? a case study from indigenous people in the Bolivian Amazon. Economic Botany 68(1):1–15

Guillotin, M.; Dubost, G.; Sabatier, D. 1994. Food choice and food competition among the three major primate species of French Guiana. Journal of Zoology 233:551–579

Guimarães, Felipe Silva; Bueno, Guilherme Taitson; Mendes, Débora de Sena Oliveira; Souza, Jorge Batista; Diniz, Alisson Duarte. 2016. Mapeamento das unidades de paisagem natural na bacia do Rio Demini, afluente do Rio Negro – AM. Acta Geográfica 10(24):18–33

Gupindaia, V. 2008b. Prehistoric funeral practices in the Brazilian Amazon: The Maracá urns. *Handbook of South American Archaeology*, Silverman, H. and W. H. Isbell (Editors), Springer, New York, pp. 1003–1024

Guzmán, Adriana; Stevenson, Pablo R. 2008. Seed dispersal, habitat selection and movement patterns in the Amazonian tortoise, *Geochelone denticulata*. Amphibia-Reptilia 29:463–472

Haidar, Ricardo Flores; Fagg, Jeanine Maria Felfili; Pinto, José Roberto Rodrigues; Dias, Ricardo Ribeiro; Damasco, Gabriel; Silva, Lucasde Carvalho Ramos; Fagg, Christopher William. 2013. Florestas estacionais e áreas de ecótono no estado do Tocantins, Brasil: parâmetros estruturais, classificação das fitofisionomias florestais e subsídios para conservação. Acta Amazonica 43(3):261–290

Hajdu, Z. and J. Hohmann. 2012. An ethnopharmacological survey of the traditional medicine utilized in the community of Porvenir, Bajo Paraguá Indian Reservation, Bolivia. Journal of Ethnopharmacology 139:838–857

Hartt, C. F. 1874. Contributions to the geology and physical geography of the Lower Amazonas: the Ereré-Monte Alegre district and the table-topped hills. *Bulletin of the Buffalo Society of Natural History* 26: 210–235.

Henry, Olivier; Feer, Francois; Sabatier, Daniel. 2000. Diet of the lowland tapir (*Tapirus terrestris* L.) in French Guiana. Biotropica 32(2):3364–368

Hermann, B., J-C Celhay, M. Guerin, J. M. Maclet, and J Rentier. 1989. *Fleurs et Plantes de Tahiti*. Les Editions du Pacifique, Singapore

Herrera, José Carlos; Taber, Andrew B.; Wallace, Robert B.; Painter R. Lilian E. 1999. Lowland tapir (Tapirus terrestris) behavioral ecology in a southern Amazonian tropical forest. Vida Silvestre Neotropical 8(1–2):31–37

Hoff, Michel; Cremers, Georges. 2005. Le jardin Guyanais: inventaire des plantes cultivées et des adventices des jardins de Guyane française. Journal de Botanique de la Société de France 29:3–40

Holdridge, L. R., W. C. Grenke, W. H. Hatheway, T. Liang, and J. A. Tosi. 1971. *Forest Environments in Tropical Life Zones: A Pilot Study*. Pergamon Press, Oxford

Huber, J. 1904. Notas sobre a patria e distribuição geographica das arvores fructiferas do Pará. *Boletim do Museu Goeldi (Museu Paraense) de Historia Natural e Ethnographia* 4 (2–3):375–406

Janzen, D. H. 1985. *Spondias mombin* is culturally deprived in megafauna-free forest. *Journal of Tropical Ecology* 1(2):131–155

Jardim, M.A.G. and Medeiros, T.D.S., 2006. Plantas oleaginosas do Estado do Pará: composição florística e usos medicinais. *Revista Brasileira de Farmácia* 87(4):124–127.

Jerozolimski, A., M. B. Ribeiro, and M. Martins. 2009. Are tortoises important seed dispersers in Amazonian forests? *Oecologia* 161: 517–528

Johnson, D. 1973. The botany, origin, and spread of the cashew, *Anacardium occidentale* L. *Journal of Plantation Crops* 1(1–2): 1–7.

Jovel, E. M.; Cabanillas, J.; Towers, G. H. N. 1996. An ethnobotanical study of the traditional medicine of the mestizo people of Suni Miraño, Loreto, Peru. Journal of Ethnopharmacology 53: 149–156

Junqueira, A. B., G. H. Shepard, and C. R. Clement. 2010. Secondary forests on anthropogenic soils in Brazilian Amazonia conserve agrobiodiversity. *Biodiversity Conservation* 19:1933–1961

Kawa, N. C. 2016. Amazonia in the Anthropocene: People, Soils, Plants, Forests. University of Texas Press, Austin

Kawa, Nicholas C.; Michelangeli, José A. Clavijo; Clement, Charles R. 2015. Household agrobiodiversity management on Amazonian dark earths, oxisols, and floodplain soils on the lower Madeira River, Brazil. Human Ecology 43:339–353

Kennedy, J. D. 1936. *Forest Flora of Southern Nigeria*. Government Printer, Lagos

Kervégant, D. 1937. Les plantes utiles et ornementales de la Martinique. *Bulletin Agricole de La Martinique* 6(1):1–125

Klüppel, Marina Pinheiro. 2006. Sistemas agrícolas e plantas medicinais em terra preta de índio da Amazônia Central. Master's thesis, INPA/UFAM, Manaus

Knowles, Oliver H.; Parrotta, John A. 1995. Amazonian forest restoration: an innovative system for native species selection based on phenological data and field performance indices. Commonwealth Forestry Review 74(3):230–243

Kubitzki, K. 1985. Dispersal of forest plants. In: Key Environments: Amazonia, G. T. Prance and T. E. Lovejoy (eds.), Pergamon Press, Oxford, pp. 192–206

Kuchla, Wilian José; Stepka, Thiago Floriani; Sawczuk, Alex Roberto; Lisboa, Gerson dos Santos. 2015. Florística, estrutura horizontal e distribuição diamétrica em área de transição de cerrado e floresta aluvial no município de Campos de Júlio, MT. Ambiência 11(1):13–30

Kurzatkowski, Dariusz; Leuschner, Christoph; Homeier, Jürgen. 2015. Effects of flooding on trees in the semi-deciduous transition forests of the Araguaia floodplain, Brazil. Acta Oecologica 69:21–30

Lamont, Susan R.; Eshbaugh, W. Hardy; Greenberg, Adolph M. 1999. Species composition, diversity, and use of homegardens among three Amazonian villages. Economic Botany 53(3):312–326

Lau, Arthur Viana; Jardim, Mário Augusto Gonçalves. 2013. Florística e estrutura da comunidade arbórea em uma floresta de várzea na área proteção ambiental, Ilha do Combu, Belém, Pará. Biota Amazônia 3(2):88–93

Le Cointe, P. 1922a. *L'Amazonie Brésilienne: Le Pays-ses Habitants, ses Resources, Notes et Statistiques jusqu'en 1920*. Augustin Challamel, Paris, vol 1

Le Cointe, P. 1922b. *L'Amazonie Brésilienne: Le Pays-ses Habitants, ses Resources, Notes et Statistiques jusqu'en 1920*. Augustin Challamel, Paris, vol 2

Le Cointe, P. 1934. *Amazônia Brasileira III: Árvores e Plantas Úteis*. Livraria Classica, Belém

Leão, Fábio Miranda; Dionisio, Luiz Fernandes Silva; Silva, Nágilla Gabriella Euzébio da; Oliveira, Marcelo Henrique Silva de; D'arace, Larissa Martins Barbosa; Neves, Raphael Lobato Prado. 2017. Fitossociologia em sistemas agroflorestais com diferentes idades de implantação no município de Medicilândia, PA. Revista Agroambiente 11(1):71–81

Lemos, Helena Lara. 2013. Vegetação arbustivo-arbórea em áreas de cerrado típico e cerrado rupestre no estado de Tocantins. Master's thesis, Departamento de Engenharia Florestal, Universidade de Brasília, Brasília

Lentz, D. L. 2000. Anthropocentric food webs in the Precolumbian Americas. In: *An Imperfect Balance: Landscape Transformations in the Precolumbian Americas*, D. L. Lentz (Editor), pp. 89–119, Columbia University Press, New York.

Lenza, Eddie; Santos, Josias O.; Maracahipes-Santos, Leonardo. 2015. Species composition, diversity, and vegetation structure in a gallery forest-cerrado sensu stricto transition zone in eastern Mato Grosso, Brazil. Acta Botanica Brasilica 29(3):327–338

Leopold, A. Starker. 1972. Wildlife of Mexico: The Game Birds and Mammals. University of California Press, Berkeley

Lima, D. A. 1958. Viagem aos campos de Monte Alegre, Pará. Instituto Agronômico do Norte, *Boletim Técnico* 36: 99–150

Lima, Roberval M. B. de; Saragoussi, Muriel. 2000. Floodplain home gardens on the central Amazon in Brazil. In: The Central Amazon Floodplain: Actual Use and Options for a Sustainable Management, Junk, W. J.; Ohly, J. J.; Piedade, M. T. F.; Soares, M. G. M. (eds), Backhuys Publishers, Leiden pp 243–268

Lima, Edson de Souza; Lima, Herson Souza; Ratter, James Alexander. 2009. Mudanças pós-fogo na estrutura e composição da vegetação lenhosa, em um cerrado mesotrófico, no período de cinco anos (1997–2002) em Nova Xavantina – MT. Cerne 15(4):468–480

Lima, Rosival Barrosde Andra de; Silva, José Antônio Aleixo da; Marangon, Luiz Carlos; Ferreira, Rinaldo Luiz Caraciolo; Silva, Roseane Karla Soares da. 2011. Sucessão ecológica de um trecho de floresta ombrófila densa de terras baixas, Carauari, Amazonas. Pesquisa Florestal Brasileira 31(67):161–172

Lima, R. B. A.; Silva, J. A. A.; Marangon, L. C.; Ferreira, R. L. C.; Silva, R. K. S. 2012. Fitossociologia de um trecho de floresta ombrófila densa na Reserva de Desenvolvimento Sustentável Uacari, Carauari, Amazonas. Scientia Plena 8(1):1–12

Lima, H., N. Smith, J. Lins, M. Schmidt, and B. Moraes, in press. Ethnobotany, archaeology and anthropogenic landscapes in the Amazon rainforest. In: *Examining Amazonia's past, present and future from a biocultural perspective: perspectives on health, environment and nutrition.* Louis C. Forline and Hilton P. da Silva (Editors), Springer

Link, Andres; Fiore, Anthony Di. 2006. Seed dispersal by spider monkeys and its importance in the maintenance of neotropical rain-forest diversity. Journal of Tropical Ecology 22:235–246

Lins, Cristóvão. 1991. Jari: 70 Anos de História. Dataforma, Rio de Janeiro

Lisboa, P. L. B. 2012. *A Terra dos Aruã: Uma História Ecológica do Arquipélago do Marajó*. Museu Paraense Emílio Goeldi, Belém

Lisboa, P. L. B. and M. L. Silva. 2009. O manejo dos recursos biológicos. In: *Aruá: Comunidades e Florestas*, P. L. B. Lisboa (Editor), Museu Paraense Emílio Goeldi, Belém, pp. 93–173

Lizcano, Leandro J.; Bakkali, Fadil; Ruiz-Larrea, M. Begoña; Ruiz-Sanz, José Ignacio. 2010. Antioxidant activity and polyphenol content of aqueous extracts from Colombian Amazonian plants with medicinal use. Food Chemistry 119:1566–1570

Lobato, G. de J. M.; Lucas, F. C. A.; Tavares-Martins, A. C. C.; Jardim, M. A. G.; Martorano, L. G. 2017. Diversidade de uso e aspectos socioambientais de quintais urbanos em Abaetetuba, Pará, Brasil. Revista Brasileira de Agroecologia 12(2):95–105

Lorenzi, H. 2008. Árvores Brasileiras: Manual de Identificação e Cultivo de Plantas Arbóreas Nativas do Brasil. Instituto Plantarum, Nova Odessa, São Paulo, vol. 1

Lossio, L.C., Delgado, H.S., Ruiz, A.G., Donayre, R.A., Ruiz, J.G., Panduro, M.P., Sifuentes, T.C. 1995. Plantas Medicinales de la Amazonía Peruana. Instituto Peruano de Seguridad Social/Instituto de Medicina Tradicional, Iquitos

Loureiro, Arthur A.; Silva, Marlene F. da; Alencar, Jurandyr da Cruz. 1979. Essências Madeireiras da Amazônia. Instituto Nacional de Pesquisas da Amazônia, Manaus, vol. 1

Lourenço, José Nestor de; Sousa, Silas Garcia de; Wandelli, Elisa Vieira; Lourenço, Francisneide de Sousa; Guimarães, Rosângela dos Reis; Campos, Lian da Silva; Silva, Roberto Lima da; Martins, Vivian Fernanda Carneiro. 2009. Agrobiodiversidade nos quintais agroflorestais em três assentamentos na Amazônia Central. Revista Brasileira de Agroecologia 4(2):965–969

Luize, Bruno Garcia; Venticinque, Eduardo Martins; Silva, Thiago Sanna Freire; Novo, Evlyn Marcia Leão de Moraes. 2015. A floristic survey of angiosperm species occurring at three landscapes of the central Amazon várzea, Brazil. Checklist: The Journal of Biodiversity Data 11(6):1–15

Luna, Ana Gabriela de; Link, Andrés; Montes, Andrés; Alfonso, Felipe; Mendieta, Leonardo; Di Fiore, Anthony. 2017. Increased folivory in brown spider monkeys Ateles hybridus living in a fragmented forest in Colombia. Endangered Species Research 32:123–134

Luz, Francisco Joaci F. 2001. Plantas medicinais de uso popular em Boa Vista, Roraima, Brasil. Horticultura Brasileira 19(1):88–96

Machado, Danilo de Oliveira. 2016. A agrobiodiversidade de quintais agroflorestais em propriedades agrícolas familiares na BR 174, Ramal do Pau-Rosa, Manaus, AM. Master's Thesis, INPA, Manaus

Machado, Monielle Alencar; Almedia Jr., Eduardo Bezerra de. 2019. Spatial structure, diversity, and edaphic factors of an area of Amazonian coast vegetation in Brazil. Journal of the Torrey Botanical Society 146(1):58–68

Mackensen, Jens; Tillery-Stevens, Madeleine; Klinge, Rudolf; Fölster, Horst. 2000. Site parameters, species composition, phytomass structure and element stores of a terra-firme forest in East-Amazonia, Brazil. Plant Ecology 151:101–109

Magnusson, William E.; Lima, Albertina P.; Albernaz, Ana L. K. M.; Sanaiotti, Tânia M.; Guillaumet, Jean-Louis. 2008. Composição florística e cobertura vegetal das savanas na região de Alter do Chão, Santarém-PA. Revista Brasileira de Botanica 31(1):165–177

Maracahipes, Leandro; Lenza, Eddie; Marimon, Beatriz Schwantes; Oliveira, Edmar Almeida de; Pinto, José Roberto Rodrigues; Marimon Junior, Ben Hur. 2011. Estrutura e composição florística da vegetação lenhosa em cerrado rupestre na transição cerrado-floresta Amazônica, Mato Grosso, Brasil. Biota Neotropica 11(1):133–141

Marques-Souza, Antonio Carlos; Absy, Maria Lúcia; Kerr, Warwick Estevam; Peralta, Francisco Javier Aguilera. 1995. Pólen coletado por duas espécies de Meliponíneos (Hymenoptera: Apidae) da Amazônia. Revista Brasileira de Biologia 55(4):855–864

Marques-Souza, Antonio Carlos; Moura, Cleonice de Oliveira; Nelson, Bruce Walker. 1996. Pollen collected by *Trigona williana* (Hymenoptera: Apidae) in central Amazonia. Revista de Biologia Tropical 44(2):567–573

Marques-Souza, Antonio Carlos; Miranda, Ires Paula de Andrade; Moura, Cleonice de Oliveira; Rabelo, Afonso; Barbosa, Edelcílio Marques. 2002. Características morfológicas e bioquímicas do pólen coletado por cinco espécies de Meliponíneos da Amazônia central. Acta Amazonica 32(2):217–229

Marques-Souza, Antonio Carlos; Absy, Maria Lucía; Kerr, Warwick Estevam. 2007. Pollen harvest features of the central Amazonian bee Scaptotrigona fulvicutis Moure 1964 (Apidade: Meliponinae), in Brazil. Acta Botanica Brasilica 21(1):11–20

Martins, Williane Maria de Oliveira; Martins, Lilliane Maria de Oliveira; Paiva, Fabiano Silveira; Martins, Wilton José de Oliveira; Lima Júnior, Sebastião Ferreira. 2012. Agrobiodiversidade nos quintais e roçados ribeirinhos na comunidade Boca do Môa – Acre. Biotemas 25(3):111–120

Matos, Caio Barros. 2015. Diversidade e usos de quintais agroflorestais da comunidade São Domingos, Floresta Nacional do Tapajós, Belterra, PA. Master's thesis, Universidade Federal do Oeste do Pará, Santarém.

Maués, Bernardo Antônio Rodrigues; Jardim, Mário Augusto Gonçalves; Batista, Fábio de Jesus; Medeiros, Tonny David Santiago; Quaresma, Adriano da Costa. 2011. Composição florística e estrutura do estrato inferior da floresta de várzea na área de proteção ambiental Ilha do Combu, município de Belém, estado do Pará. Revista Árvore 35(3):669–677

McCann, J. M. 1999. Before 1492, the making of the Pre-Columbian landscape, Part II: the vegetation, and implications for restoration for 2000 and beyond. *Ecological Restoration* 17(3):107–119

Medeiros, Marcelo Brilhante de; Walter, Bruno Machado Teles. 2012. Composição e estrutura de comunidades arbóreas de cerrado stricto sensu no norte do Tocantins e sul do Maranhão. Revista Árvore 36(4): 673–683

Medeiros, Marcelo Brilhante de; Walter, Bruno Machado Teles; Silva, Glocimar Pereira da; Gomes, Beatriz Machado; Lima, Isabel Lustz Portela; Silva, Suelma Ribeiro; Moser, Pamela; Washington, Luís Oliveira; Cavalcanti, Taciana Barbosa. 2012. Vascular flora of the Tocantins River middle basin, Brazil. Check List 8(5):852–885

Mejia, Kember; Rengifo, Elsa. 2000. Plantas Medicinales de Uso Popular en la Amazonía Peruana. Agencia Española de Cooperación Internacional, Lima

Milliken, William. 1997. Traditional anti-malarial medicine in Roraima, Brazil. Economic Botany 51(3):212–237

Milliken, W. and B. Albert. 1996. The use of medicinal plants by the Yanomami indians of Brazil. Economic Botany 50(1):10–25

Milliken, William; Albert, Bruce. 1997. The use of medicinal plants by the Yanomami indians of Brazil, Part II. Economic Botany 51(3):264–278

Milton, K. 1980. The Foraging Strategy of Howler Monkeys: A Study in Primate Economics. Columbia University Press, New York

Milton, K. 1985. Dietary quality and demographic regulation in a howler monkey population. In: The Ecology of a Tropical Forest. Seasonal Rhythms and Long-Term Changes, Leigh, Egbert G.; Rand, A. Stanley; Windsor, Donald M. (Editors), Smithsonian Institution Press, Washington, DC, pp 273–289

Miranda, V. C. 1909. Os campos de Marajó e a sua flora considerados sob o ponto de vista pastoril. *Boletim do Museu Goeldi (Museu Paraense) de Historia Natural e Ethnographia* 5(1): 96–151

Miranda I. S. 1993. Estrutura do estrato arbóreo do cerrado amazônico em Alter-do-Chão, Pará, Brasil. Revista Brasileira de Botânica 16(2):143–150

Miranda, I. S. 1995. Fenologia do estrato arbóreo de uma comunidade de cerrado em Alter-do-Chão, PA. Revista Brasileira de Botânica 18(2):235–240

Miranda, Izildinha S. 2000. Análise florística e estrutural da vegetação lenhosa do Rio Comemoração, Pimenta Bueno, Rondônia, Brasil. Acta Amazonica 30(3):393–422

Miranda, Regiane da Silva; Nunes, Jayane Santos; Oliveira, Ilmaione Keiza de Souza; Mororó, Dayanne Lustosa; Maneschy, Rosana Quaresma. 2012. Quintais agroflorestais como estratégia alimentar familiar no assentamento 26 de Março, Marabá, Pará. Agroecossistemas 4(1): 68–80

Mitchell, J. D. and S. A. Mori, 1987. *The Cashew and its Relatives (Anacardium: Anacardiaceae).* Memoirs of the New York Botanical Garden, vol. 42

Moll, Don; Jansen, Kevin P. 1995. Evidence for a role in seed dispersal by two tropical herbivorous turtles. Biotropica 27(1):121–127

Montalván, Nadiuska Giannina Achong. 2015. Plantas alimenticias usadas por mamíferos mayores en la cuenca alta del Río Itaya, Loreto, Perú. Bachelor thesis, UNAP, Facultad de Ciencias Biológicas, Iquitos

Monteiro, Maria Vivina Barros; Bevilaqua, Claudia Maria Leal; Palha, Maria das Dores Correia; Braga, Roberta Rocha; Schwanke, Katiane; Rodrigues, Silvane Tavares; Lameira, Osmar Alves. 2011. Ethnoveterinary knowledge of the inhabitants of Marajó Island, eastern Amazonia, Brazil. Acta Amazonica 41(2):233–242

Monteles, Ricardo; Pinheiro, Claudio Urbano B. 2007. Plantas medicinais em um quilombo maranhense: uma perspectiva etnobotânica. Revista de Biologia e Ciências da Terra 7(2):1–12

Montenegro, Olga Lucia. 2004. Natural licks as keystone resources for wildlife and people in Amazonia. Master's thesis, University of Florida, Gainesville

Moreira, Rosilene da Conceição R.; Rebêlo, José Manuel Macário; Gama, Mônica Elinor Alves; Costa, Jackson Maurício L. 2002. Nível de conhecimentos sobre Leishmaniose Tegumentar Americana (LTA) e uso de terapias alternativas por populações de uma área endêmica da Amazônia do Maranhão, Brasil. Cadernos de Saúde Pública 18(1):187

Mori, S. A. 1989. Eastern, extra-Amazonian Brazil. In: *Floristic Inventory of Tropical Countries* D. G. Campbell and H. D. Hammond (Editors), New York Botanical Garden, pp. 427–454

Moskovits, Debra K.; Bjorndal, Karen A. 1990. Diet and food preferences of the tortoises *Geochelone carbonaria* and *G. denticulata* in northwestern Brazil. Herpetologica 46(2):207–218

Mota, Nara Furtado de Oliveira; Watanabe, Mauricio Takashi Coutinho; Zappi, Daniela Cristina; Hiura, Alice Lima; Pallos, Julieta; Viveros, Raquel Stauffer; Giulietti, Ana Maria; Viana, Pedro Lage. 2018. Cangas da Amazônia: a vegetação única de Carajás evidenciada pela lista de fanerógramas. Rodriguésia 69(3):1435–1488

Mourthé, Ítalo Martins da Costa. 2012. Influência das características físico-químicas e disponibilidade dos frutos na ecologia dos primatas em uma floresta no norte da Amazônia. Doctoral dissertation, INPA, Manaus

Muniz, Francisca Helena; Cesar, Oswaldo; Monteiro, Reinaldo. 1994. Fitossociologia da vegetação arbórea da reserva florestal do Sacavém, São Luís, Maranhão (Brasil). Acta Amazonica 24 (3/4):219–236

Muñoz, Marianella Guevara. 2010. Evaluación de purmas con fines de manejo en la comunidad nativa de Santa Mercedes, Rio Putumayo, Perú. Master's thesis, Universidad Nacional de la Amazonia Peruana, Iquitos

Murakami, Alejandro Araujo; Reyes, Juan Fermandon; Milliken, William. 2016. Frutales Silvestres y Promisorios de Pando. Herencia/Museo de Historia Natural Noel Kempff Mercado, Cobija

Navarro, G. 2002. Provincia biogeográfica del Acre-Madre de Dios (Amazónica Suroccidental). In: Geografía Ecológica de Bolivia: Vegetación y Ambientes Acuáticos, Navarro, Gonzalo; Maldonado, Mabel (eds), Centro de Ecología Simón I. Patiño, Cochabamba, pp 51–98

Nebel, Gustav; Kvist, Lars Peter; Vanclay, Jerome K.; Christensen, Henning; Freitas, Luis; Ruiz, Juan. 2000. Estructura y composición florística del bosque de la llanura aluvial en la Amazonía Peruana: I. El bosque alto. Folia Amazónica 10(1–2):91–149

Neves, E. G. 2006. Tradição oral e arqueologia na historia indígena no Alto Rio Negro. *Amazônia, Além dos 500 Anos*, L. Forline, R. Murrieta, and I. Vieira (Editors), Museu Paraense Emilio Goeldi, Belém, pp. 1–37.

Neville, Melvin K.; Glander, Kenneth E.; Braza, Francisco; Rylands, Anthony B. 1988. The howling monkeys, genus *Alouatta*. In: Ecology and Behavior of Neotropical Primates, Mittermeier, Russell A.; Rylands, Anthony B.; Coimbra-Filho, Adelmar F.; Fonseca, Gustavo A.B. da (eds), World Wildlife Fund, Washington, DC, vol. 2, pp 349–453

Nimuendajú, C. 1956. Os Apinayé. *Boletim do Museu Paraense Emílio Goeldi* 12:1–146

Norconk, Marilyn A.; Wertis, Catherine; Kinzey, Warren G. 1997. Seed predation by monkeys and macaws in eastern Venezuela: preliminary findings. Primates 38(3):177–184

Novais, Jaílson S. de; Garcêz, Ana Cristina A.; Absy, Maria Lúcia; Santos, Francisco de Assis R. dos. 2014. Comparative pollen spectra of *Tetragonisca angustula* (Apidae, Meliponni) from the lower Amazon (N Brazil) and caatingas (NE Brazil). Apidologie 46:417–431

Nunes, Andréa. 1995. Um teste de germinação em sementes dispersas por macacos-Aranha em Maracá, Roraima; Brasil. Studies On Neotropical Fauna and Environment 30(1):31–36

Nunes, Andréa. 1998. Diet and feeding ecology of *Ateles belzebuth belzebuth* at Maracá Ecological Station, Roraima, Brazil. Folia Primatologica 69:61–76

Odonne, Guillaume; Valadeau, Celine; Alban-Castillo, Joaquina; Stien, Didier; Sauvain, Michel; Bourdy, Genevieve. 2013. Medical ethnobotany of the Chayahuita of the Paranapura basin (Peruvian Amazon). Journal of Ethnopharmacology 146:127–153

O'Farrill, Georgina; Galetti, Mauro; Campos-Arceiz, Ahimsa. 2013. Frugivory and seed dispersal by tapirs: an insight on their ecological role. Integrative Zoology 8:4–17

Olajide, Olumayokun A.; Aderogba, Mutalib A.; Fiebich, Bernd L. 2013. Mechanisms of anti-inflammatory property of Anacardium occidentale stem bark: inhibition of NF-KB and MAPK signaling in the microglia. Journal of Ethnopharmacology 145:42–49

Oliveira, Francisco Plácido Magalhães; Absy, Maria Lúcia; Miranda, Izildinha Souza. 2009. Recurso polínico coletado por abelhas sem ferrão (Apidae, Meliponinae) em um fragmento de floresta na região de Manaus- Amazonas. Acta Amazonica 39(3):505–518.

Oliveira, Lucicléia Mendes de; Sousa, Rhonan Martins de; Correa, Norma Elena Reynosa; Santos, André Ferreira dos; Giongo, Marcos. 2018. Florística e síndromes de dispersão de um fragmento de cerrado ao sul do estado do Tocantins. Scientia Agraria Paranaensis 17(1): 104–111

Oppenheimer, John R. 1982. *Cebus capucinus*: home range, population dynamics, and interspecific relationships. In: The Ecology of a Tropical Forest: Seasonal Rhythms and Long-term Changes, Egbert G. Leigh, A. Stanley Rand, and Donald M. Windsor (eds), Smithsonian Institution Press, Washington DC, pp. 253–272

Oyuela-Caycedo, A. 2008. Late Pre-historic chiefdoms of northern Colombia and the formation of anthropogenic landscapes. *Handbook of South American Archaeology*, H. Silverman and W. H. Isbell (Editors), Springer, New York, pp. 405–428

Paim, Fernanda P.; Chapman, Colin A.; Queiroz, Helder L. de; Paglia, Adriano P. 2017. Does resource availability affect the diet and behavior of the vulnerable squirrel monkey, *Saimiri vanzolinii*? International Journal of Primatology 38(3):572–587

Palheta, Ivanete Cardoso; Tavares-Martins, Ana Cláudia Caldeira; Lucas, Flávia Cristina Araújo; Jardim, Mário Augusto Gonçalves. 2017. Ethnobotanical study of medicinal plants in urban home gardens in the city of Abaetetuba, Pará state, Brazil. Boletín Latinoamericano y del Caribe de Plantas Medicinales y Aromáticas 16(3):206–262

Paz-Rivera, Clea; Putz, Francis E. 2009. Anthropogenic soils and tree distributions in a lowland forest in Bolivia. Biotropica 41(6):665–675

Pedrollo, Camilo Tomazini; Kinupp, Valdely Ferreira; Shepard Jr., Glenn; Heinrich, Michael. 2016. Medicinal plants at Rio Jauaperi, Brazilian Amazon: ethnobotanical survey and environmental conservation. Journal of Ethnopharmacology 186:111–124

Penido, Alexandre Batista; Morais, Selene Maia de; Ribeiro, Alan Bezerra; Silva, Ana Zélia. 2016. Ethnobotanical study of medicinal plants in Imperatriz, state of Maranhão, northeastern Brazil. Acta Amazonica 46(4):345–354

Pennington, T. D., C. Reynel, and A. Daza. 2004. *Illustrated Guide to the Trees of Peru.* David Hunt, Sherborne, England

Pereira, Paulo Vinícius Miranda; Figueiredo Neto, Leonardo Francisco. 2015. Conservação de espécies florestais: um estudo em quintais agroflorestais no município de Cáceres – MT. Revista Eletrônica em Gestão, Educação e Tecnologia Ambiental 19(3):783–793

Pereira, Luciano Araújo; Sobrinho, Felipe de Araújo; Costa Neto, Salustiano Vilar da. 2011. Florística e estrutura de uma mata de terra firme na Reserva de Desenvolvimento Sustentável no Rio Iratapuru, Amapá oriental, Brasil. Floresta 41(1):113–122

Pereira, Luciano Araujo; Lima-Barbosa, Jackson Rodrigo de; Almeida, Maria Zélia de; Guimarães, Elsie Franklin. 2017. Diversidade de plantas em quintais quilombolas, conhecimento local sobre uso e cultivo de pimentas na Amazônia oriental, Brasil. Revista de Biologia Neotropical 14(1):56–72

Peters, Charles M.; Hammond, Elysa J. 1990. Fruits from the flooded forests of Peruvian Amazon: yield estimates for natural populations of three promising species. Advances In Economic Botany 8: 159–176

Pinheiro, Tatyana; Ferrari, Stephen F.; Lopes, Maria Aparecida. 2013. Activity budget, diet, and use of space by two groups of squirrel monkeys (*Saimiri sciureus*) in eastern Amazonia. Primates 54:301–308

Pinheiro, Renato Torres; Marcelino, Dianes Gomes; Moura, Dieyson Rodrigues. 2018. Espécies arbóreas de uso múltiplo e sua importância na conservação da biodiversidade nas áreas verdes urbanas de Palmas, Tocantins. Desenvolvimento e Meio Ambiente 49: 264–282

Pinho, Rachel Camargo de. 2008. Quintais agroflorestais indígenas em áreas de savana (lavrado) na Terra Indígena Araçá, Roraima. Master's thesis, INPA/UFAM, Manaus

Pinto, Angélica Auxiliadora da Costa; Maduro, Cice Batalha. 2003. Produtos e subprodutos da medicina popular comercializados na cidade de Boa Vista, Roraima. Acta Amazonica 33(2):281–290

Pinto, José Roberto Rodrigues; Mews, Henrique Augusto; Jancoski, Halina Soares; Marimon, Beatriz Schwantes; Bomfim, Bárbara de Oliveira. 2014. Woody vegetation in a floodplain campo de murundus in central Brazil. Acta Botanica Brasilica 28(4): 519–526

Pires, João Murça; Salomão, Rafael de Paiva. 2000. Dinâmica da diversidade arbórea de um fragmento de floresta tropical primária na Amazônia oriental-1. Período: 1956 a 1992. Boletim do Museu Paraense Emílio Goeldi, Série Botânica 16(1):63–110

Polesna, Lucie; Polesny, Zbynek; Clavo, Mirella Z.; Hansson, Anders; Kokoska, Ladislav. 2011. Ethnopharmacological inventory of plants used in Coronel Portillo Province of Ucayali Department, Peru. Pharmaceutical Biology 49(2):125–136

Pontes, Antonio Rossano Mendes. 1997. Habitat partitioning among primates in Maracá Island, Roraima, northern Brazilian Amazonia. International Journal of Primatology 18(2):131–157

Ponz, Elizabeth; Carpio, Griceldo; Meo, Severo. 2005. La Medicina Tradicional de los Tacana y Machineri: Conocimientos Prácticos de las Plantas Medicinales. Investigaciones Regionales Pando, La Paz

Pozo, Wilmer E. 2004. Agrupación y dieta de *Ateles belzebuth belzebuth* en Parque Nacional Yasuní, Ecuador. Anuario de la Investigación Científica 2(1):77–102

Quaresma, Adriano Costa; Jardim, Mário Augusto G. 2017. Formations of coastal forests in the Amazon and ecological relations with vascular epiphytes. Revista Árvore 41(2):1–8

Queiroz, José Antonio Leite; Machado, Sebastião do Amaral. 2008. Fitossociologia em floresta de várzea do estuário Amazônico no estado do Amapá. Pesquisa Florestal Brasileira 57:5–20

Queiroz, José Antonio Leite de; Mochiutti, Silas; Machado, Sebastião do Amaral; Galvão, Franklin. 2005. Composição florística e estrutura de floresta em várzea alta estuarina Amazônica. Floresta 35(1):41–55

Queiroz, José Antonio Leite de; Mochiutti, Silas; Machado, Sebastião do Amaral. 2006. Distribuição diamétrica em floresta de várzea do braço norte do Rio Amazonas. Revista de Ciência Agrária 46:151–170

Queiroz, José Antonio Leite de; Machado, Sebastiã do Amaral; Hosokawa, Roberto Tuyoshi; Silva, Ivan Crespo da. 2007. Estrutura e dinâmica de floresta de várzea no estuário Amazônico no estado do Amapá. Floresta 37(3):339–352

Ratter, J. A. 1987. Notes on the vegetation of the Parque Nacional do Araguaia (Brazil). Notes from the Royal Botanic Garden, Edinburgh 44(2):311–342

Rayol, Breno Pinto; Alvino-Rayol, Fabrízia de Oliveira; Silva, Maria da Conceição Pereira da. 2013. Caracterização de quintais agroflorestais desenvolvidos por agricultores familiares do município de Alenquer, Pará. Cadernos de Agroecologia 8(2):1–5

Rech, André Rodrigo; Absy, Maria Lúcia. 2011. Pollen sources used by species of Meliponini (Hymenoptera: Apidae) along thc Rio Negro channel in Amazonas, Brazil. Grana 50(2):150–161

Redford, Kent H.; Klein, Bert; Murcia, Carolina. 1992. Incorporation of game animals into small-scale agroforestry systems in the Neotropics. In: Conservation of Neotropical Forests:

Working From Traditional Resource Use, Redford, Kent H.; Padoch, Christine (eds), Columbia University Press, New York, pp 333–348
Rego, Lyvia Julienne Sousa; Elly, Márcio Piero; Pereira, Jéssica Lira; Garcia, Bruna Naiara; Barbosa, Humberto Figueira; Vieira, Thiago Almeida. 2011. Produtos florestais não-madeireiros comercializados em feiras de Santarém, Pará. Cadernos de Agroecologia 6(2):1–6
Rengifo-Salgado, Elsa; Rios-Torres, Sandra; Malaverri, Lizardo Fachín; Vargas-Arana, Gabriel. 2017. Saberes ancestrales sobre el uso de flora y fauna en la comunidad indígena Tikuna de Cushillo Cocha, zona fronteriza Perú-Colombia-Brasil. Revista Peruana de Biología 24(1):67–78
Reyes-García, Victoria. 2001. Indigenous people, ethnobotanical knowledge, and market economy: a case study of the Tsimane' Amerindians in lowland Bolivia. Doctoral dissertation, University of Florida, Gainesville.
Ribeiro, Reginaldo Vicente; Bieski, Isanete Geraldini Costa; Balogun, Sikiru Olaitan; Martins, Domingos Tabajarade Oliveira. 2017. Ethnobotanical study of medicinal plants used by ribeirinhos in the north Araguaia microregion, Mato Grosso, Brazil. Journal of Ethnopharmacology 205: 69–102
Ritter, Rhuan Amorim; Monteiro, Maria Vivina Barros; Monteiro, Frederico Ozanan Barros; Rodrigues, Silvane Tavares; Soares, Marina Lira; Silva, Jean Carlos Ramos; Palha, Maria das Dores Correia; Biondi, Germano Francisco; Rahal, Sheila Canevese; Tourinho, Manuel Malheiros. 2012. Ethnoveterinary knowledge and practices at Colares island, Pará state, eastern Amazon, Brazil. Journal of Ethnopharmacology 144:346–352
Robinson, J. G. 1986. Seasonal variation in use of time and space by the wedge-capped capuchin monkey, *Cebus olivaceus*: implications for foraging theory. Smithsonian Contributions to Zoology Number 431, Washington, DC.
Robles, Rembrandt Ramiro Canahuire. 2017. Composición florística y estructura de la recuperación natural en un área degradado por minería en la comunidad nativa de Tres Islas, Tambopata, Madre de Dios. Master's thesis, Escuela Profesional de Ingeniería Forestal y Medio Ambiente, Puerto Maldonado, Perú
Rocha, S. F. R., M. R. C. Ferreira, and M. G. Sablayrolles. 2013. Farmacopeia vegetal da comunidade Caxiuanã, na Floresta Nacional de Caxiuanã. In: Caxiuanã: Paraíso ainda Preservado, P. L. B. Lisboa (Editor), Museu Paraense Emílio Goeldi, Belém, pp. 257–285
Rodrigues, Eliana. 2006. Plants and animals utilized as medicines in the Jaú National Park (JNP), Brazilian Amazon. Phytotherapy Research 20:378–391
Rodrigues, Ricardo Ribeiro; Martins, Sebastião Venâncio; Barros, Luiz Carlos de. 2004. Tropical rain forest regeneration in an area degraded by mining in Mato Grosso state, Brazil. Forest Ecology and Management 190: 323–333
Rojas, Natalia Alejandra Alvis. 2012. Patron de actividad, dieta, área de actividad y dispersion de semillas de *Alouatta seniculus* en un fragmento de Bosque en San Juan del Carare (Santander). Thesis, Universidad del Tolima, Facultad de Ciencias, Bogotá
Romo, Mónica C. 1996. Seasonal variation in fruit consumption and seed dispersal by canopy bats (Artibeus spp.) in a lowland forest in Peru. Vida Silvestre Neotropical 5(2):119–119
Rondon Neto, Rubens Marques; Santos, Jeneffer Soares dos; Silva, Manoel Advíncula da; Koppe, Valdinei Cristi. 2010. Potencialidades de uso de espécies arbustivas e arbóreas em diferentes fisionomias de cerrado, em Lucas do Rio Verde/MT. Revista de Biologia e Ciências da Terra 10(2):113–126
Roosevelt, T. 1926. *Through the Brazilian Wilderness*. Charles Scribner's Sons, New York.
Roosevelt, A. C. 1991. *Moundbuilders of the Amazon: Geophysical Archaeology on Marajo Island, Brazil*. Academic Press, San Diego
Roosevelt, A. C.; Costa, M. Lima da; Machado, C. Lopes; Michab, M.; Mercier, N.; Valladas, H.; Feathers, J.; Barnett, W.; Silveira, M. Imazio da; Henderson, A.; Silva, J.; Chernoff, B.; Reese, D. S.; Holman, J. A.; Toth, N.; Schick, K. 1996. Paleoindian cave dwellers in the Amazon: the peopling of the Americas. *Science* 272:373–384

Ruysschaert, Sofie; Van Andel, Tinde; Van de Putte, Kobeke; Van Damme, Patrick. 2009. Bathe the baby to make it strong and healthy: Plant use and child care among Saramaccan Maroons in Suriname. Journal of Ethnopharmacology 121:148–170

Salas, Leonardo A.; Fuller, Todd K. 1996. Diet of the lowland tapir (Tapirus terrestris L.) in the Tabaro River valley, southern Venezuela. Canadian Journal of Zoology 74: 1444–1451

Salomão, Rafael de Paiva; Lisboa, Pedro L. B. 1988. Análise ecológica da vegetação de uma floresta pluvial tropical de terra firme, Rondônia. Boletim do Museu Paraense Emílio Goeldi, Série Botânica 4(2):195–233.

Sanaiotti, Tânia Margarete. 1996. The woody flora and soils of seven Brazilian Amazonian dry savanna areas. Doctoral dissertation, University of Stirling, Scotland

Sanaiotti, Tânia M.; Magnusson, William E. 1995. Effects of annual fires on the production of fleshy fruits eaten by birds in a Brazilian Amazonian savanna. Journal of Tropical Ecology 11(1):53–65

Sanaiotti, Tânia M.; Bridgewater, Samuel; Ratter, James A. 1997. A floristic study of the savanna vegetation of the state of Amapá, Brazil, and suggestions for its conservation. Boletim do Museu Paraense Emílio Goeldi, Série Botânica 13(1):3–24

Santos, Luciana Oliveira dos. 2010. Comparação da vegetação lenhosa entre borda e interior de uma mancha de mata de restinga na RESEX Mar Caeté-Taperaçu, Bragança, Pará. Master's thesis, Universidade Federal Rural da Amazônia/Museu Paraense Emílio Goeldi, Belém

Santos, João Ubiratan M. dos; Rosário, Carlos da S. 1988. Levantamento da vegetação fixadora das dunas de Algodoal-PA. Boletim do Museu Paraense Emílio Goeldi, Série Botânica 4(1): 133–154

Santos, Maurício Reginaldo A. dos; Lima, Maria Railda de; Ferreira, Maria das Graças R. 2008. Uso de plantas medicinais pela população de Ariquemes, Rondônia. Horticultura Brasileira 26:244–250

Santos, M. R. A.; Lima, M. R.; Oliveira, C. L. L. G. 2014. Medicinal plants used in Rondônia, western Amazon, Brazil. Revista Brasileira de Plantas Medicinais 16(3):707–720

Santos, Otilene dos Anjos; Couceiro, Sheyla Regina Marques; Rezende, Alinne Costa Cavalcante; Silva, Márcia Daniela de Sousa. 2016a. Composition and richness of woody species in riparian forests in urban areas of Manaus, Amazonas, Brazil. Landscape and Urban Planning 150:70–78

Santos, Silvio Roberto Miranda dos; Kato, Osvaldo Ryohei; Tourinho, Manoel Malheiros. 2016b. Análise florística e finalidade de espécies cultivadas em sistemas agroflorestais no município de Cametá, Pará, Amazônia Oriental. Amazônia: Ciência e Desenvolvimento 13(23):9–22

Santos, Deborah Cruz dos; Corrêa Neto, José de Jesus; Alvarenga, Adriano Braga Brasileiro de. 2018. Frutos consumidos por macaco-de-cheiro *Saimiri sciureus collinsi* (Osgood, 1916) em fragmentos de florestas em uma unidade de conservação no arquipélago do Marajó, Amazônia Oriental, Brasil. Revista Brasileira de Zoociências 19(1):82–90

Santos, Silvio Roberto Miranda dos; Kato, Osvaldo Ryohei; Tourinho, Manoel Malheiros. 2019. Diversidade florística e estoque de carbono de sistemas agroflorestais em dois municípios do nordeste paraense, Brasil. Boletim do Museu Paraense Emílio Goeldi, Ciencias Naturais 14(1):31–42

Sanz-Biset, Jaume; Campos-de-la-Cruz, José; Epiquién, Mirbel A.; Cañigueral, Salvador. 2009. A first survey on the medicinal plants of the Chazuta valley (Peruvian Amazon). Journal of Ethnopharmacology 122:333–362

Sauer, J. D. 1979. Living fences in Costa Rican agriculture. *Turrialba* 29 (4): 255–261.

Sauer, J. D. 1988. *Plant Migration: The Dynamics of Geographic Patterning in Seed Plant Species*. University of California Press, Berkeley.

Sauer, J. D. 1993. Historical Geography of Crop Plants: A Selected Roster. CRC Press, Boca Raton

Schaan, D. 2008. The nonagricultural chiefdoms of Marajó Island. *Handbook of South American Archaeology*, Silverman, H. and W. H. Isbell (Editors), Springer, New York, pp. 339–357.

Schaan, D. 2012. Sacred Geographies of Ancient Amazonia: Historical Ecology of Social Complexity. Left Coast Press, Walnut Creek, California

Schultes, R. E. and R. F. Raffauf. 1990. The Healing Forest: Medicinal and Toxic Plants of the Northwest Amazonia. Dioscorides Press, Portland, Oregon

Scoles, Ricardo. 2009. El quintal y las frutas: recursos económicos y alimentares en la comunidad negra de Itacoã, Acará, Pará, Brasil. Acta Amazonica 39(1):1–12

Scudeller, Veridiana Vizoni; Veiga, Josephina Barata da; Araújo-Jorge, Lúcia Helena de. 2009. Etnoconhecimento de plantas de uso medicinal nas comunidades São João do Tupé e Central (Reserva de Desenvolvimento Sustentável do Tupé). In: Biotupé: Meio Físico, Diversidade Biológica e Sociocultural do Baixo Rio Negro, Amazônia Central, Santos-Silva, Edinaldo Nelson; Scudeller, Veridiana Vizoni (eds), UEA Edições, Manaus, vol. 2, pp 185–199

Silva, Regina Oliveira da. 2008. Biodiversidade e políticas de conservação: o caso do parque Estadual Monte Alegre, Pará. Doctoral dissertation, Universidade de Brasília

Silva, Rafael Carletti Marcolino da. 2011a. Agrobiodiversidade de propriedades agrícolas familiares no município de Presidente Figueiredo, AM. Master's thesis, INPA, Manaus

Silva, Silvestre. 2011b. Fruits of the Brazilian Amazon. Metalivros, São Paulo

Silva, Edinei Santos da. 2017. Plantas alimentícias em comunidades agrícolas no município de Rio Preto da Eva-AM. Master's thesis, INPA, Manaus

Silva, Ana Paula Ferreira Frota da; Bentes-Gama, Michelliny de Matos. 2008. Fitossociologia de uma floresta ombrófila aberta em área de assentamento rural no distrito de Jaci Paraná, Porto Velho, Rondônia. Ambiência 4(3):435–452

Silva, M. F., P. L. B. Lisbôa, R. C. L. Lisbôa 1977. Nomes Vulgares de Plantas Amazônicas. INPA, Manaus

Silva, José Natalino Macedo; Carvalho, João Olegário Pereira de; Lopes, José do Carmo Alves. 1985. Inventário florestal de uma área experimental na Floresta Nacional do Tapajós. Boletim de Pesquisa Florestal 10/11:38–110

Silva, Antônio Sérgio Lima da; Lisboa, Pedro L. B.; Maciel, Ubirajara N. 1992. Diversidade florística e estrutura em floresta densa da bacia do rio Juruá-AM. Boletim do Museu Paraense Emílio Goeldi, Série Botânica 8(2):203–258

Silva, Suzy Cristina Pedroza da; Pereira, Cloves Farias; Fraxe, Therezinha de Jesus Pinto; Witkoski, Antônio Carlos; Silva, Michele Andreza Pedroza da. 2007. In: Comunidades Ribeirinhas Amazônicas: Modos de Vida e Uso dos Recursos Naturais, Fraxe, Therezinha de Jesus Pinto; Pereira, Henrique dos Santos; Witkoski, Antônio Carlos (eds), EDUA, Manaus, pp 141–154

Silva, Aglair Gomes da; et al. 2008. Aldeias indígenas Mura. Editora da Universidade Federal do Amazonas, Manaus

Silva, Rachel Macedo da; Mehlig, Ulf; Santos, João Ubiratan Moreira dos; Menezes, Moirah Paula Machado de. 2010. The coastal restinga vegetation of Pará, Brazilian Amazonia: a synthesis. Revista Brasileira de Botanica 33(4):563–573

Silva, Hanna Kassia Machado da; Gama, João Ricardo Vasconcellos; Sousa, Randerson José de Araujo; Lameira, Mahyanny Karoline da Silva; Costa, Daniele e Lima da; Oliveira, Douglas Valente de; Rocha, Jobert Silva da; Oliveira, Thiago Gomes de Sousa. 2017. Composição florística de quintais agroflorestais na vila Cuera, Bragança, Pará. Agroecossistemas 9(2):330–338

Siveiro, A.; Delunardo, T. A.; Haverroth, M.; Oliveira, L. C.; Mendonça, A. M. S. 2012. Plantas medicinais em quintais urbanos de Rio Branco, Acre. Revista Brasileira de Plantas Medicinais 14(4):598–610

Siviero, Amauri; Delunardo, Thiago Andrés; Haverroth, Moacir; Oliveira, Luis Cláudio de; Mendonça, Ângela Maria Silva. 2011. Cultivo de espécies alimentares em quintais urbanos de Rio Branco, Acre, Brasil. Acta Botanica Brasilica 25(3):549–556

Smith, Herbert H. 1879. Brazil, the Amazons and the Coast. Charles Scribner's Sons, New York

Soini, P. 1987. Ecology of the saddle-back tamarin *Saguinus fuscicollis illigeri* on the río Pacaya, northeastern Peru. Folia Primatologica 49:11–32

Soini, P. 1988. The pygmy marmoset, genus *Cebuella*. In: Ecology and Behavior of Neotropical Primates, Mittermeier, Russell A.; Rylands, Anthony B.; Coimbra-Filho, Adelmar F.; Fonseca, Gustavo A.B. da (eds), Ecology and Behavior of Neotropical Primates, World Wildlife Fund, Washington, DC, vol. 2, pp 79–129

Sousa, G. S. 1971. *Tratado Descritivo do Brasil em 1587*. Companhia Editora Nacional/Editôra da Universidade de São Paulo, São Paulo

Souza, Célia Cristina Valero; Scudeller, Veridiana Vizoni. 2009. Plantas úteis nos quintais das comunidades ribeirinhas Julião e Agrovila, Reserva de Desenvolvimento Sustentável do Tupé, Amazônia Central. Revista Brasileira de Agroecologia 4(2): 2487–2491

Souza, Antonio Maricélio Borges de; Alves, Keila de Nazaré; Lobato, Whesley Thiago dos Santos; Leal, Ayla Julliane da Silva; Almeida, Gabriela Mourão de; Souza, Antonio Alan Santos de; Mota, Adriano Vitti. 2017. Aspectos da segurança alimentar com base em quintais agroflorestais na comunidade rural de Santa Luzia do Induá no município de Capitão Poço, PA. Agroecosistemas 9(2): 275–287

Spruce, R. 1908a. Notes of a Botanist on the Amazon and Andes. Macmillan, London, vol. 1

Spruce, R. 1908b. Notes of a Botanist on the Amazon and Andes. Macmillan, London, vol. 2

Standley, P. C. 1927. *The Flora of Barro Colorado Island, Panama*. Smithsonian Miscellaneous Collections, Smithsonian Institution, vol. 78, no. 8.

Stevenson, Pablo R. 2000. Seed dispersal by woolly monkeys (*Lagothrix lagothricha*) at Tinigua National Park, Colombia: dispersal distance, germination rates, and dispersal quantity. American Journal of Primatology 50: 275–289

Stevenson, Pablo R. 2005. Potential keystone plant species for the frugivore community at Tinigua Park, Colombia. In: Tropical Fruits and Frugivores: The Search for Strong Interactors, Dew, J. Lawrence; Boubli, Jean Philippe (eds), Springer, Dordrecht, pp 37–57

Stevenson, Pablo R. 2007. Estimates of the number of seeds dispersed by a population of primates in a lowland forest in western Amazonia. In: Seed Dispersal: Theory and its Application in a Changing World, Dennis, Andrew J.; Schupp, Eugene W.; Green, Ronda J.; Westcott, David A. (eds), CABI, Wallingford, pp 340–362

Stevenson, Pablo R.; Link, Andrés. 2010. Fruit preferences of *Ateles belzebuth* in Tinigua Park, northwestern Amazonia. International Journal of Primatology 31:393–407

Stevenson, Miranda F.; Rylands, Anthony B. 1988. The marmosets, genus *Callithrix*. In: Ecology and Behavior of Neotropical Primates, Mittermeier, Russell A.; Rylands, Anthony B.; Coimbra-Filho, Adelmar F.; Fonseca, Gustavo A.B. da (eds), Ecology and Behavior of Neotropical Primates, World Wildlife Fund, Washington, DC, vol. 2, pp 131–222

Stevenson, Pablo R.; Quiñones, Marcela J.; Ahumada, Jorge A. 1994. Ecological strategies of woolly monkeys (*Lagothrix lagotricha*) at Tinigua National Park, Colombia. American Journal of Primatology 32:123–140

Stevenson, P.R.; Pineda, M.; Samper, T. 2005. Influence of seed size on dispersal patterns of woolly monkeys (*Lagothrix lagotricha*) at Tiningua Park, Colombia. *Oikos* 110: 435–440

Suarez, Scott A. 2006. Diet and travel costs for spider monkeys in a nonseasonal, hyperdiverse environment. International Journal of Primatology 27(2):411–436

Sweet, D. G. 1974. A Rich Realm of Nature Destroyed: The Middle Amazon Valley, 1640–1750. Doctoral dissertation, University of Wisconsin, Madison

Tananta, Leslie Virginia. 2014. Análisis del conocimiento tradicional del uso de especies vegetales en tres comunidades de la cuenca baja del río Ucayali, Loreto, Peru. Master's thesis, UNAP, Iquitos

Taurino, Tássia Cristina da Conceição Barros. 2017. Recuperação natural da diversidade de espécies de plantas em florestas secundárias na Amazônia Oriental. Master's Thesis, Universidade Federal do Pará, Belém

Tavares, Lanuze Fabielly S.; Candeiro, Carlos Roberto A. 2012. Região de fronteiras: fauna e flora em Itaguatins, norte do estado do Tocantins. Brazilian Geographical Journal: Geosciences and Humanities research medium 3(2):513–519

Terborgh, J. 1983. Five New World Primates: A Study in Comparative Ecology. Princeton University Press, Princeton

Thomas, Evert. 2009. New light on the floristic composition and diversity of indigenous territory and national park Isiboro-Sécure, Bolivia. Biodiversity Conservation 18:1847–1878

Tiburski, Júlia Hauck; Rosenthal, Amauri; Deliza, Rosires. 2011. Nutritional properties of yellow mombin (*Spondias mombin* L.) pulp. Food Research International 44: 2326–2331

Toledo, Marisol; Salick, Jan. 2006. Secondary succession and indigenous management in semideciduous fallows of the Amazon basin. Biotropica 38(2):161–170

Vale, Gabriel Damasco do. 2011. O efeito do solo, inundação e topografia sobre as campinaranas inundáveis na Amazônia. Master's Thesis, INPA, Manaus

Van Andel, Tinde; Behari-Ramdas, Joelaika; Havinga, Reinout; Groenendijk, Sara. 2007. The medicinal plant trade in Suriname. Ethnobotany Research & Applications 5:351–372

Van Roosmalen, M. G. M. 1985. Habitat preferences, diet, feeding strategy and social organization of the black spider monkey (*Ateles paniscus paniscus* Linnaeus 1758) in Surinam. Acta Amazonica 15(3/4 suplemento) 1–238

Van Roosmalen, Marc G. M.; Klein, Lewis L. 1988. The spider monkeys, genus *Ateles*. In: Ecology and Behavior of Neotropical Primates, Mittermeier, Russell A.; Rylands, Anthony B.; Coimbra-Filho, Adelmar F.; Fonseca, Gustavo A.B. da (eds), Ecology and Behavior of Neotropical Primates, World Wildlife Fund, Washington, DC, vol. 2, pp 455–537

Van Roosmalen, Marc G. M.; Mittermeier, Russel A.; Milton, Katherine. 1981. The bearded sakis, genus *Chiropotes*. In: Ecology and Behavior of Neotropical Primates, Coimbra-Filho, Adelmar F.; Mittermeier, Russell A. (eds), Academia Brasileira de Ciências, Rio de Janeiro, vol. 1, pp 419–441

Van Roosmalen, Marc G. M.; Mittermeier, Russel A.; Fleagle, John G. 1988. Diet of the northern bearded saki (*Chiropotes satanas chiropotes*): a Neotropical seed predator. American Journal of Primatology 14:11–35

Vásquez, R. 1997. *Flórula de las Reservas Biológicas de Iquitos, Perú*. Monographs in Systematic Botany Vol. 63, Missouri Botanical Garden, St. Louis.

Vásquez, Silvia Patricia Flores; Mendonça, Maria Silvia de; Noda, Sandra do Nascimento. 2014. Etnobotânica de plantas medicinais em comunidades ribeirinhas do Município de Manacapuru, Amazons, Brasil. Acta Amazonica 44(4):457–472

Veléz, Germán Alonso. 1991. Los frutales amazónicos cultivados por las comunidades indígenas de la región del Medio Caquetá. Colombia Amazonica 5(2): 135–183

Veléz, Germán Alonso; Veléz, Antonio José. 1992. Sistema agroflorestal de "chagras" utilizado por las comunidades indígenas del medio Caquetá (Amazonia Colombiana). Colombia Amazonica 6(1): 101–134

Veríssimo, Adalberto; Barreto, Paulo; Mattos, Marli; Tarifa, Ricardo; Uhl, Christopher. 1992. Logging impacts and prospects for sustainable forest management in an old Amazonian frontier: the case of Paragominas. Forest Ecology and Management 55: 169–199

Vickers, W. T. and T. Plowman. 1984. *Useful Plants of the Siona and Secoya Indians*. Field Museum of Natural History, Fieldiana Botany New Series, no. 15

Vieira, Thiago Almeida; Rosa, Leonilde dos Santos; Vasconcelos, Paulo César Silva; Santos, Mônica Mota dos; Modesto, Rozi da Silva. 2007. Sistemas agroflorestais em áreas de agricultores familiares em Igarapé-Açu, Pará: caracterização florística, implantação e manejo. Acta Amazonica 37(4):549–558

Villacorta, Roosevelt García; Reátegui, Manuel Ahuite; Zumaeta, Mauro Olórtegui. 2003. Clasificación de bosques sobre arena blanca de la zona reservada Allpahuayo-Mishana. Folia Amazónica 14(1):17–33

Vogt, R. C. 2008. Amazon Turtles. Wust Ediciones, Lima

Voss, Robert S.; Fleck, David W. 2011. Mammalian diversity and Matses ethnomammalogy in Amazonian Peru Part 1: Primates. Bulletin of the American Museum of Natural History 351:1–81

Wallace, Alfred Russel. 1889. A Narrative of Travels on the Amazon and Rio Negro, with an Account of the Native Tribes, and Observations on the Climate, Geology, and Natural History of the Amazon Valley. Ward, Lock & Co., London

Wallace, R. B. 2005. Seasonal variations in diet and foraging behavior of Ateles chamek in a southern Amazonian tropical forest. International Journal of Primatology 26(5): 1053–1075

Wehncke, Elisabet V.; Valdez, Catherine Numa; Domínguez, César A. 2004. Seed dispersal and defecation patterns of *Cebus capucinus* and *Alouatta palliata*: consequences for seed dispersal effectiveness. Journal of Tropical Ecology 20:535–543

Wen, Louise; Haddad, Mohamed; Fernández, Irma; Espinoza, Giovana; Ruiz, Candy; Neyra, Edgar; Bustamante, Beatriz; Rojas, Rosario. 2011. Actividad antifúngica de cuatro plantas usadas en la medicina tradicional Peruana: Aislamiento de 3'-formil-2',4',6'-trihidroxidihidrochalcona, principio ativo de *Psidium acutangulum*. Revista de la Sociedad Química del Perú 77(3):199–204

Wezel, A.; Ohl, J. 2006. Homegarden plant diversity in relation to remoteness from urban centers: a case study from the Peruvian Amazon region. In: Tropical Homegardens: a time-tested example of sustainable agroforestry, Kumar, B. M.; Nair, P. K. R. (eds), pp 143–158, Advances in Agroforestry vol. 3, Springer, Dordrecht

Whitehead, N. L. 2003. Three Patamuna trees: landscape and history in the Guyana highlands. In: *Histories and Historicities in Amazonia*, N. L. Whitehead (editor), University of Nebraska Press, Lincoln, pp. 59–77

WinklerPrins, Antoinette M. G. A. 2002. House-lot gardens in Santarém, Pará, Brazil: Linking rural with urban. Urban Ecosystems 6: 43–65

Zent, Egleé; Zent, Stanford. 2002a. Impactos ambientales generadores de biodiversidade: conductas ecológicas de los Hotï de la Sierra Maigualida, Amazonas Venezolano. Interciencia 27(1):9–20

Zent, Egleé; Zent, Stanford. 2002b. Los Jodï: sabios botánicos del Amazonas venezolano. Antropologica 97–98:29–70

Chapter 3
Annonaceae: Alligator Pears and Their Like

The Annonaceae has been generous to inhabitants of the American tropics. Arguably the cream of the crop when it comes to tasty and fulfilling fruits, indigenous cultures in both highland and lowland South America have domesticated close to a dozen species in this family of small trees. Little known to most inhabitants of temperate lands because their skins are delicate and easily damaged during transportation, two representatives of this family, cherimoya (*Annona cherimola*) from the Andes and soursop (*Annona muricata*), native to Amazonia, are conquering new markets in North America on account of their exquisite flavors. Cherimoya was domesticated in the highlands of Peru or Ecuador and spread in precontact times to highland Central America. Fresh fruits are airfreighted from Central and South America to cities in the United States with large Hispanic populations, such as Los Angeles. Frozen pulp of soursop, known as guanábana in Spanish-speaking countries and graviola in Brazil, is dispatched to states in the United States with sizeable Latino populations, such as Florida, to make juice and ice cream. A few members of this family that are cultivated or found wild in Amazonia are sampled here. All of them produce succulent fruits that could one day charm the palates of consumers on other continents if the forests in which they thrive survive.

3.1 *Annona glabra*

Brazil Condesa, conde, araticum, araticum branco, araticum da praia, araticum da lagoa, araticum do mangue, araticum do brejo, araticu bravo, araticu cortiça, araticu da praia, araticupaná, jaca do lago, ata, jacama

Colombia Guanábana cimarrona, guanábana de pozo, baga, condessa, anón

Status Wild, spontaneous in cultural settings

N. Smith, *Amazon Fruits: An Ethnobotanical Journey*, Ethnobiology,
https://doi.org/10.1007/978-3-031-12803-5_3

Condesa *Annona glabra* tree in fruit on a ranch in the Amazon estuary. Fazenda Santa Paz, Caviana Island, Pará, Brazil, 5-21-01

Known as pond apple or custard apple in English, *Annona glabra* occurs wild along the coast of Brazil from Amapá to Santa Catarina (Amaral et al. 2008, 2015; Lorenzi et al. 2006a, b:47; Santos et al. 2003) and is widely distributed along the coasts of Central America, the Caribbean, and in the everglades of southern Florida. It has spread to other tropical regions via ocean currents, such as West Africa, parts of India, and northern Australia where it is considered an aggressive weed in mangroves. It not only prefers swampy areas near lagoons and mangroves but also occurs occasionally on sand dunes such as along the coastal zone of northeastern Pará in the Amazon estuary (Bastos et al. 1995). It is planted in parts of eastern and central Amazonia and is also cultivated in southern Taiwan (Chang et al. 2000; Major et al. 2005; Pinho 2008:37). This tree is also found in home gardens in the suburbs of Palmas in Tocantins in the southern fringe of the Amazon Basin (Pinheiro et al. 2018). In the Colombian Amazon, capuchin monkeys *Cebus apella*, which are widespread in Amazonia and other tropical forests in South America, have been observed feeding on these fruits (Gómez-Posada 2012).

Floridians apparently do not bother to gather the fruits today, although indigenous groups undoubtedly did in the distant past. The shores and islands of Lake Okeechobee in south-central Florida were once lined with extensive stands of pond apple, and some of these almost pure stands might have arisen from plantings or forest management by indigenous folk. Before diking, damming, and marina and housing developments in the twentieth century drastically altered the natural flooding cycle of Lake Okeechobee, pond apple covered some 12,600 hectares; now only dying remnants of the once extensive groves of pond apple remain (Nabhan 1989:143). For the time being at least, the stands of pond apple at the mouth of the Amazon have been spared such drastic habitat destruction.

Maria Aparecida, 7 years, picking fruit from a conde *Annona glabra* bush planted in a 2-year-old agroforestry field with numerous perennial and annual crops. Jocojó, Rio Jocojó, Município de Gurupá, Pará, Brazil 4-23-15

In the Brazilian Amazon, condesa is gathered in the wild in wetlands at the tail end of the rainy season and into the early part of the summer. Pond apple is more common in the Amazon estuary, although it has been recorded as far west as the state of Acre in Brazil (NYBG ID 40895). The fruits are easy to reach because the pond apple is usually shrubby, though it can reach 18 m. The apple-sized fruits have a shiny yellow skin that is peeled to reveal a sweet, peach-colored pulp that breaks into segments. The taste is similar to that of soursop *Annona muricata*, though a bit blander. Numerous rectangular seeds, the size of the nail on one's small finger, are embedded in the sticky pulp. The dark brown seeds are typically discarded in backyards, which explains why condesa arises spontaneously around cowboy homes on the islands of Caviana and Mexiana at the mouth of the Amazon. The seeds are sometimes used as a purgative, such as on Ilha de Algodoal on the coast of Pará (Jardim and Medeiros 2006). Also along the coast of Pará, a bathing solution is prepared from the leaves to reduce leg swelling (Bastos et al. 2003:15) and in Guyana leaves and twigs of this plant are boiled to make a tea, and this is sometimes mixed with the whole plants of *Passiflora foetida*, to treat nematodes and flatworms (Austin and Bourne 1992).

I also encountered *Annona glabra* planted in an agroforestry field on uplands on the outskirts of Jocojó village in the lower Amazon. Jocojó is an archaeological site along a black water river; when *Annona glabra* was introduced there is not clear.

Condesa is another example of a floodplain tree with edible fruit that has been recruited by people for cultivation on uplands. In the municipality of Marapanim on the northeastern coast of Pará, locals employ the leaves of this tree to make a bathing solution to treat headaches (Furtado et al. 1978).

In the Bailique archipelago in the northern part of the Amazon estuary, condesa is known as jaca do lago, which means the lake jackfruit. Jackfruit *Artocarpus heterophyllus* is a fruit tree native to the Western Ghats of India and was introduced to Brazil. Although condesa is not related to the jackfruit, the common name jaca do lago captures two distinguishing characteristics of this fruit: the segmented pulp, similar to that of jackfruit, and its propensity for water. I have never seen condesa fruits in markets.

Jaca do lago *Annona glabra* fruits gathered from a bush in tidal grassland (*campo*) in the Amazon estuary. Marinheiro, Bailique, Amapá, Brazil 5-18-02

Condesa trees are relatively short, typically less than 3 m tall, and grow in stands in areas prone to flooding during the wet season or daily when incoming tides back up brackish water. On the northeastern coast of Marajó Island at the mouth of the Amazon, *Annona glabra* sometimes occurs in thickets, as has been noted in Honduras (Standley 1930). Confined to back swamp vegetation behind mangroves (Chatrou et al. 2004), *A. glabra* is mostly confined to the estuarine area of the Amazon.

The fruits of *Annona glabra* are eaten by the big-headed Amazon river turtle *Peltocephalus dumerilianus* (Pérez-Emán and Paolillo 1997). Known as cabeçudo (big head) in the Brazilian Amazon, this turtle with its formidable bite is reminiscent of Florida's alligator snapping turtle *Macrochelys temminckii.* Cabeçudo reaches about a meter in length and is captured by rural folk to eat. Pond apple thus contributes indirectly to the human diet via turtles and likely certain fish as well.

Cabeçudo *Peltocephalus dumerilianus* feeds on fruits of *Annona glabra* and is being kept in a backyard to be eventually eaten. Barcelos, Rio Negro, Amazonas, Brazil 9-6-72

3.2 *Annona hypoglauca*

Bolivia Ahuaracatoco, chirimoya; rononosëbi, ahuaracatoco (Chácobo)

Brazil Biribá, biribarana, biribirana, graviola da várzea, graviola do igapó, araticum, envira taia

Colombia Anonilla, anoncillo, anón de pescado, guanabana de la várzea

Ecuador Chirimoya; mi'ca (Siona-Secoya), obahue, yaqueberebe (Waorani)

Peru Anonilla, anonilla huasca, anonilla del monte, anona, guanábana sacha, tortuga blanca, paujil ruru; yugkuanim (Aguaruna), anonilla (Kichwa), kéach (Mayna)

Status Wild, spontaneous in cultural settings, planted

Anonilla *Annona hypoglauca* with unripe fruit during the annual flood. Near confluence of the Yanayacu and Marañón Rivers, 4 38.92S 73 49.33W, Loreto, Peru 4-21-06

The growth habit of *Annona hypoglauca* varies from a shrub to a liana-like small tree in forest and successionary vegetation on floodplains of black, clear, and sediment-rich rivers (Assis et al. 2017; Boubli 2002; Casula 2012:24; Duivenvoorden 1994; Kalliola et al. 1991; Luize et al. 2015; Parolin 2000a, b, c; Reis 2010:30). It is also occasionally found in upland forest, such as along the middle Xingu River in Pará (Lemos et al. 2015) and in 10-year-old successionary vegetation on uplands in the vicinity of Santa Mercedes along the Rio Putumayo in the northern Peruvian Amazon bordering Colombia (Gomez 2011:91; Muñoz 2010:26). It is also cultivated in some dooryard gardens of the Chácobo in the Bolivian Amazon (Boom 1987:13) as well as in the vicinity of Leticia in the Colombian Amazon (Reitberg 2014:27). This relative of the better-known soursop (*Annona muricata*) is confined to central and western Amazonia, including the Bolivian lowlands, Ecuador, Colombia, the Upper Rio Negro in Brazil and Venezuela, the Guianas, and the Magdalena in Colombia which flows into the Caribbean (Murillo 2001; Roosmalen 1985:6).

The fruits of *Annona hypoglauca* appear during the rainy season and resemble those of the domesticated relative *Annona mucosa* because they are yellow when ripe, unlike the fruits of most species of *Annona* which are green when mature, and because of the nobly skin with its slightly pointed protrusions (muricated). Indeed, along the Madeira River, *Annona hypoglauca* is known as biribá, the same name for *A. mucosa* in the Brazilian Amazon. The latter is known as anona in the Peruvian Amazon, and anonilla, the name applied to *A. hypoglauca*, means the "little anona."

Percy, 11 years, eating anonilla *Annona hypoglauca* fruit in riparian forest. Near Yarina, Rio Yanayacu, Pacaya-Samiria National Reserve, Loreto, Peru, 3-7-04

Anonilla *Annona hypoglauca* fruits are normally 8–9 cm long by 4.5–5 cm wide but can attain 12 cm × 7 cm (Wittmann et al. 2010:63). The fruits of *A. hypoglauca* are gathered from the ground as floods recede or from canoes at high water. Each fruit has about 100 black seeds, the size of apple seeds only flatter. After savoring the creamy, white pulp, the seeds, measuring 1.5 cm long by 0.5 cm, are spat out on the ground or into the water. *A. hypoglauca* has been recorded in home gardens of the Chácobo who live along the Ivón River, a tributary of the Beni in the Bolivian Amazon (Boom 1989). Some of those plants may have arisen from discarded seeds or possibly even planted.

Percy, 11 years, eating fruit of *Annona hypoglauca* in riparian forest. He will swallow the white pulp after spitting out the black seeds. Near Yarina, S 4 44.704 W 73 59.014, Rio Yanayacu, Pacaya-Samiria National Reserve, Loreto, Peru, 3-7-04

The fruits float in water and are dispersed by currents or by several fish important in subsistence and commerce, such as tambaqui *Colossoma macropomum*, pirapitinga *Colossoma bidens*, and several species of *Brycon* and *Triportheus* (Gottsberger 1978; Goulding 1980:151). At least two species of esteemed catfish, bacu *Lithodoras dorsalis* and pirara *Phractocephalus hemiliopterus* also disperse the seeds (Goulding 1980:195; Parolin et al. 2010). The highly prized giant river turtle *Podocnemis expansa* also feeds on this fruit (Figueroa et al. 2012).

As is typical with other fruits in the Annonaceae, *Annona hypoglauca* is used in various folk remedies. In Peru, for example, the bark is used to prepare concoctions for treating diarrhea and expelling hookworms (Brack 1999:38). In the Ecuadorian Amazon, the Waorani not only eat the fruit but use the wood for making canoes and their homes, as well as for firewood (Cerón and Ayala 1998:26–27).

3.3 *Annona montana*

Bolivia Biribá de monte, chirimoya; tëto (Chácobo)

Brazil Araticum, araticum açu, araticum apé, araticum ponhe, araticum cagão, araticú do mato, araticú grande, abiurana do mato, conde, jaca de igapó; tayaha-‘a-‘i (Arawaté), araciko’a-‘ɨ (Guajá), arašiku-‘ɨ, arašiku-ran (Ka’apor), ongre (Kayapó); kumũ (Sateré-Maué), aračiku-‘ɨw (Tembé), apai (Tiriyó), araxki, araxky (Waimiri-Atroari)

Colombia Guanábana, chirimiyo, anón de caiman, anón de monte

Peru Sacha guanábana, guanábana, guanábana cimarrona, guanábana de loma, guanábana de perro, anona, anona amarilla, anonilla, turagua, carahuasca

Venezuela Catuche

Status Wild, planted

Farmer with *Annona montana* fruit harvested in her home garden. Km 12 Tarapoto-Moyobamba highway, San Martin, Peru 8-25-04

Known as mountain soursop or wild custard apple in English, *Annona montana* is a relative of the better-known soursop *Annona muricata* but is not nearly as common as the latter, probably because the fruit's flavor is less appealing. The fruits of mountain soursop, which remain green when ripe, have smaller prickles than soursop. Also, the seeds are brown instead of black. The round to oblong fruits, which can reach 26 cm long (Lorenzi et al. 2006a, b:47), are used to make juice. The white pulp surrounding the seeds is removed by hand and mixed with water and drunk. Neither the fruits nor the juice find their way to markets. The medium-sized tree, which can reach up 18 m in height, fruits in the rainy season and is occasionally planted in home gardens.

Farmer picking araticum *Annona montana* fruit from a tree in her home garden. Km 130 Manaus-Itacoatiara highway, Amazonas, Brazil 5-17-96

In the Brazilian Amazon, araticum is found mostly in cultivation in home gardens and in successionary vegetation associated usually with human settlements or activities, such as mining (Baar et al. 2004; Brienza Júnior 1999:26; Coelho et al. 2003; Ducke 1946; Gualberto et al. 2014; Machado 2016:27; Oliveira and Jardim 1998; Rodrigues et al. 2004; Santana et al. 2004; Souza 2010:23). This tree is also planted by the Yuracaré and Trinitario in the Bolivian Amazon (Thomas and Van Damme 2010). However, it is also found in upland forest, such as in the Tapajós National Forest in western Pará (Gonçalves and Santos 2008; Sandel and Carvalho 2000), along the middle Xingu (Lemos et al. 2015), in the southern fringe of the Amazon Basin in Mato Grosso (Malheiros et al. 2009), the eastern fringe of the basin in Maranhão (Muniz et al. 1994) and in western Amazonia including the Colombian Amazon near Leticia (Rudas and Prieto 1998) and the Andean foothills in Peru (Cartagena 2014:69).

This fruit tree also occurs in scrub savanna and gallery forest in the southern part of the Amazon Basin such as along the middle Tocantins River (Haidar et al. 2013; Medeiros et al. 2012), and in tidal forest on Combu Island at the mouth of the Guamá River near Belém (Amaral et al. 2009; Cattanio et al. 2002). It also grows in floodplain forest along the clear water Jari River in eastern Amazonia (Carim 2016:86). In the Colombian Amazon, this fruit tree is found in both upland and floodplain forests (Rudas and Prieto 1998). Tapirs *Tapirus terrestris* eat this fruit in the Upper Orinoco Basin in eastern Colombia and probably disperse some of the seeds (Alviz and Pérez-Albarracín 2015:41).

The precise origins of *Annona montana* are obscure, but it is probably native to the lowlands of tropical America. In the Guianas, mountain soursop is common on sandy soils and along river banks (Roosmalen 1985:8). The West Indies has been offered as another possible origin of this crop (Popenoe 1920:192).

Pedro, 10 years, with an *Annona montana* fruit gathered in regrowth forest, site of the ancient village of Arapijó. Upland archaeological site with black earth and potsherds. Inland from Carrazedo, confluence of the Amazon and Xingu Rivers, Municipio de Gurupá, Pará, Brazil 4-12-15

The seeds of *A. montana* are surrounded by a layer of air containing tissue to help them float (Waldhoff et al. 1996) so the tree's original habitat was likely in wetlands. Mountain soursop is also found in the Caribbean where it may also be indigenous (Popenoe 1920:192), or populations there could have arisen from seeds carried from Central or northern South America. Where in this vast range mountain soursop was domesticated is still a mystery but it may have been brought into cultivation in various locations in the distant past.

Preparing to make juice with the pulp of *Annona montana*. The fruit was harvested in a home garden. Boa Vista, near Santa Cruz do Arari, Marajó Island, Pará, Brazil, 5-15-99

Annona montana also has medicinal value. In rural communities near Santarém in western Pará, for example, a juice is extracted from the leaves and drunk to alleviate the venom injected by *Bothrops jararaca*, a widespread snake in Brazil (Moura et al. 2015). In the vicinity of Chazuta along the Huallaga River in the Peruvian Amazon, the seeds are crushed and applied to skin affected by scabies which is triggered by itch mites (Sanz-Biset et al. 2009). Although mountain soursop fruits have not penetrated US or European markets, the fruits are used to make dietary supplements containing acetogenins, which are allegedly helpful in preventing and treating certain cancers. Pills containing various chemicals obtained from mountain soursop can be purchased on the internet.

Davison, 7 years, drinking conde *Annona montana* juice in his rural home. Boa Vista, near Santa Cruz do Arari, Marajó Island, Pará, Brazil, 5-15-99

3.4 *Annona mucosa*

Bolivia Chirimoya

Brazil Biribá, biribazeiro, birimbazeiro, berimbá, beribá, nata, ata, conde, fruta da condesa, pinha da mata, araticum, jaca de pobre; wiriwa (Baniwa), mi'ikãn (Desana), sìp (Hupd'äh), biri (Kayapó), wiriwa (Sateré-Maué), wiriwa (Ticuna), pixká, piká, pi'kaa (Tukano), briba (Waimiri-Atroari), yviry (Wajãpi)

Colombia Anón, anón amazónico, anón de montana, anón enormes, anón silvestre, anóna, anonilla, cherimoya, cohayú, chirimoya, carahuasca, moco de padre, waituto; mikã (Makuna), dtog rai (Witoto)

Ecuador Ananas panga, anona, anúna, cabeza de negro, chirimoya arisco, chirimoya silvestre, sacha chirimoya, chirimoyuela, zambo, guanabana Amazónica; cc'aña (Cofán), yunkua (Shuar), mica (Siona-Secoya)

Peru Anona, nona, anón cimarrón, anonilla, anona cimarrona, anón amazónico, chirimoya, chirimoya del monte, masa samba; kéach, pawák (Achuar), yugkua, anúna (Aguaruna), chimi (Mayna)

Venezuela Riñón, riñón de montaña, riñón montañero, chirimoya, anón de monte, cherimoya, chirimoya, llama

Status Wild, cultivated

Anona *Annona mucosa* fruit in a street market. Mercado Belén, Iquitos, Loreto, Peru 5-9-03

Annona mucosa (formerly classified as *Rollinia mucosa*), known as biribá in Brazil and anona in Peru, occurs wild in upland and wetland forests up to 1500 m in widely scattered locations in the Neotropics from Mexico to South America (Cárdenas et al. 2009; Lorenzi et al. 2006a, b:53, Maas and Westra 1972:129). In Peru, anona is confined to the Amazon Basin, where it is commonly planted in home gardens as high as 820 m in the Andean foothills, such as in headwater tributaries of the Huallaga near Tarapoto in Peru. However, the fruits are taken to urban markets along the coast, such as Mercado Breña in Lima. The fruits also appear in urban markets in other parts of the Amazon, such as in Santarém, Pará (Rego et al. 2011).

Fruit vendor approaching a passenger and cargo boat. Biribá *Annona mucosa* fruits are by the paddle. Furo Tajapuru, western side of Marajó Island, Pará, Brazil 9-7-16

Where *Annona mucosa* was domesticated remains a mystery. The late Paulo Cavalcante, a distinguished Brazilian botanist based at Belém's Goeldi Museum, suggests that the medium-sized tree was first brought into cultivation in the border area between Brazil and Peru, or possibly in Rondônia in the southwestern portion of the Brazilian Amazon (Cavalcante 2010:67). *Annona mucosa* is encountered in some upland forests of the Bolivian Amazon (Araujo-Murakami et al. 2015), along the middle Xingu in Pará (Lemos et al. 2015), and in northern Amapá (Carim et al. 2013). Whether the trees in the Brazilian Amazon are truly wild or descendants of cultivated forms remains a mystery. Clement (1999) suggests northwestern Amazonia as the area where *A. mucosa* was domesticated. Larger fruit forms are more common in the western part of the Amazon, suggesting that the fruit has been cultivated there for a long time. At Tabatinga along the Amazon River at the border between Brazil and Colombia, for example, the fruit can attain 4 kg (Lima et al. 1984). Fruits are generally smaller in the eastern part of the basin and have characteristic pointed protuberances above each carpel segment, whereas fruits of the western landrace have a knobby rather than spiky skin (Clement 1989).

Anderson, 6 years, with a biribá *Annona mucosa* fruit gathered in a home garden. Taperebatuba, Rio Urubu, Amazonas, Brazil 4-3-15

Annona mucosa is grown from seeds, rather than grafted, so a good deal of variation is found in fruit size and shape. Fruit size ranges from 7 to 13 cm long and 7 to 13 cm wide. Some fruits are pear shaped or round, whereas others are relatively slender. Although variable in size and shape, anona fruits are easily distinguished from the other fruits in the Annonaceae by two characteristics: they are yellow when ripe, whereas most cultivated fruits in the family are green when mature, and the fruits sport thick, stud-like appendages, which in some eastern forms resemble small fingers.

The main fruiting season for *Annona mucosa* is during the rainy months, roughly from January through May in in the southern hemisphere part of Amazonia. However, some trees produce fruit well into the dry season, such in some villages along the Marañón River in Peru. Along Igarapé do Santarém, a stream that flows into the lower Maués in Amazonas, Brazil, a farmer reports that his biribá tree produces fruit year-round. A mature tree can produce some 150 fruit annually (Souza et al. 1996:56). A fully grown tree may reach 20 m, so youngsters are typically dispatched up the trees to gather fruits. Alternatively, young and old alike sometimes toss a stick to bring down fruits high up in the tree, such as in the village of Veinte de Enero along the lower Yanayacu in Loreto, Peru, although the fruits may be damaged in this way. The fruits break apart easily when ripe to reveal a sweet, smooth pulp. Children and adults relish the slippery pulp which contains appreciable amounts of vitamin C (Flores 1997). The numerous dark brown seeds, the size of currants, are spat out. The mottled-face tamarin *Saguinus inustus* has been observed feeding on this fruit along the Caquetá River in the Colombian Amazon and is likely involved in dispersing the seeds (Palacios et al. 2004) as is the case with *Phyllostomus elongatus* bats in the vicinity of Manaus in the Brazilian Amazon (Guillaumet and Reis 1983). Anona fruits have relatively thick skins, which facilitates transportation to markets. In Iquitos, the fruits sell for about US$0.30 apiece and are always eaten fresh.

Biribá *Annona mucosa* fruit gathered in a home garden. Lucas, 14 years, son of a fisherman. Igarapé Jacarécuara, Ilha de Trambioca, near Barcarena in the Amazon estuary, Pará, Brazil 3-19-15

Annona mucosa is cultivated mostly in rural and urban home gardens on both uplands and floodplains from the mouth of the Amazon to the Andean foothills. On floodplains, the trees usually grow on levees (*restingas*), where homes are generally built, and can tolerate flooding for several weeks. In the Amazon estuary, biribá survives daily flooding at high tide during the rainy season, such as near Afuá on Marajó Island. Biribá is also planted occasionally around huts used to process manioc flour (*casa de forno*, also known as *casa da farinha*) near fields, such as in the vicinity of Carrazedo at the confluence of the Xingu and Amazon Rivers.

Annona mucosa is also interplanted occasionally in fields along with other perennial crops. In 1992, near Jarú in Rondônia, for example, I saw biribá planted alongside rubber *Hevea brasiliensis* and robusta coffee *Coffea canephora*. On another small farm near Cacaulandia in Rondônia, biribá was interplanted with cacao *Theobroma cacao*. Near Machadinho Rondônia, a famer established an agroforestry plot in a former pasture containing biribá mixed with 11 other perennial crops. In 1996, at km 165 of the Altamira-Uruará stretch of the Transamazon, a farmer had interplanted two species of timber trees along with Robusta coffee and biribá. In the vicinity of Belterra on the planalto south of Santarém, another small farmer interplanted biribá with eight other perennials, both exotic and native fruits and nuts. In 2003, near the Shipibo community at Nueva Cajamarca along the Ucayali River, a 6-year-old field (*chacra*) contained anona interplanted with plantain, peach palm *Bactris gasipaes*, mahogany *Swietenia macrophylla*, sapote *Matisia cordata*, orange, and ashipa *Pachyrrhizus tuberosus*. In a *chacra* near Victoria, further up the Ucayali, a farmer created an agroforestry field with 20 perennials, including anona. The practice of intercropping *Annona mucosa* in cultivated fields probably goes back a long time. For example, the Wajãpi in the state of Amapá tend to *Annona mucosa* trees, which they call yviry, after they harvest the manioc so that they will bear fruit for many years in successionary growth (Gallois 2009:21). A similar practice occurs among the Witoto along the Caquetá River in the Colombian Amazon (Veléz and Veléz 1992).

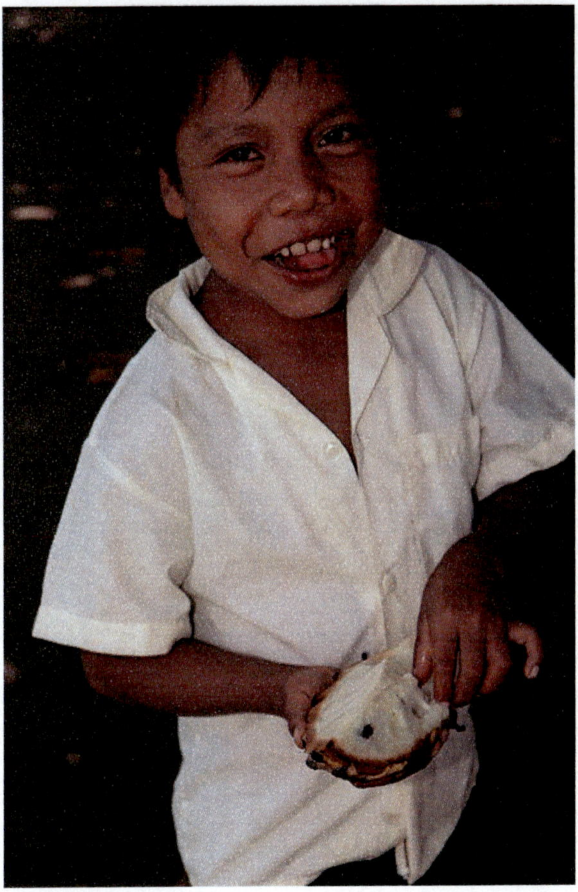

Anthony, 7 years, eating an anona *Annona mucosa* fruit in his village. Manco Capac, Rio Puinahua (a branch of the Ucayali), Reserva Nacional Pacaya-Samiria, Loreto, Peru 4-14-06

In the colonial era, the Portuguese soon learned to incorporate biribá into their home gardens. In describing Belém in the 1660s, a judge, Mauricio de Heriarte, noted that the recently established town was happy and full of fruit trees: "*alegre e cheia de arvores fructiferas, com laranjas, limões, limas, beribases*" (Heriarte 1964). Biribá was considered a "cold" fruit, well suited therefore for Europeans not accustomed to the sultry tropics. In describing life in the Amazon in the mid-eighteenth century, João Daniel, a Jesuit priest, considered the fruits of biribá as "cold" and thus most appropriate for consumption in the hot and humid Amazon: "*são muito frias, e por isso muito aptas para refrigerar os calores do Amazonas*" (Daniel 1976:328). Even today, some rural folk classify foods as hot or cold (ideas traced to the Old World), irrespective of the temperature at which they are served.

Annona mucosa also has medicinal properties. In the Municipality of São Gabriel da Cachoeira along the Upper Rio Negro in Amazonas, Brazil, rural folk use the roots to prepare a drink that is taken three times a day for 7 days to treat malaria

(Kffuri et al. 2016). Also in the Upper Rio Negro region, the Desana prescribe an infusion of biribá and abiu *Pouteria caimito*, another fruit with slippery pulp, for women who are having a difficult time giving birth. The juice purportedly helps lubricate the birth canal (Buchillet 1988). The inner bark of biribá is also used to make an infusion for the same purpose among families in the Caxiuanã National Forest in Pará (Rocha et al. 2013). Inhabitants of the rural community Julião in the Reserva de Desenvolvimento Sustentável (RDS) do Tupé, 25 km west of Manaus, eat the fruit to treat anemia (Veiga and Scudeller 2015). River dwellers along the Solimões make a bathing solution with the leaves of biribá to treat fleas and their bites as well as head lice (Silva et al. 2007b; Vásquez 2014:14; Vásquez et al. 2014) as do rural folk in various parts of Rondônia (Santos et al. 2004). In the municipalities of Xapuri in Acre and Pauini in southern Amazonas, rural folk make an infusion from the leaves of this plant to treat liver issues (Ferreira et al. 2015). In Ecuador, the leaves are used to treat rheumatism (Maas et al. 1992:131). In Abaetetuba, a town along the lower Tocantins River in Pará, residents prepare a decoction from the fruit skin or bark to treat sore throat (Palheta et al. 2017).

This widely cultivated fruit tree has other uses as well. The wood is used by some carpenters, such as to make canoes and planks (Falcão et al. 1981). The seeds are used to make necklaces (Lustosa 1976:395; Maas et al. 1992:131; Prance and da Silva 1975:52).

3.5 *Annona muricata*

Bolivia Sinini, sinini del monte, sinini de la pampa, nopocorrur, guanábana, guanábano

Brazil Graviola, gravioleira, jacama, jaca, araticum, araticu, araticu manso, fruta do conde; jaaka, wiriwa hipolelalipe (Baniwa), araticum (Ka'apor), yaca (Ticuna), iáka (Tukano), araxki (Waimiri-Atroari)

Colombia Guanábana, guanábana amazónica, guanábano, guanabarjo, zapote agrio guanabo, catuche, anón; djaca (Ticuna), amena (Witoto)

Ecuador Guanábana, chirimoya, guaba; oa-na-sa-tapé (Cayapa), keách (Shuar), na-ña-peo-mi-cá, mi-cá (Siona-Secoya)

Peru Guanábana, guanábano, chirimoya, huanábano, masasamba, corosol, cachimán, sopa agria, anona; kistian yugkua, chirimoya (Aguaruna), huana huana, huanana (Shipibo)

Venezuela Guanábana, catoche, catuche

Status Planted

Graviola ***Annona muricata*** **tree in fruit in a home garden. Thaynna, 10 years, with genipap** ***Genipa americana*** **body paint. Juruna village, Paquiçamba, Rio Xingu near Belo Monte, Pará, Brazil 9-8-15**

Annona muricata **tree in fruit in a village home garden. Nueve de Octubre, Marañón River, Loreto, Peru 7-12-10**

The melon-sized fruits of soursop (*Annona muricata*), known as guanábana in Spanish-speaking countries and graviola in Brazil, are used for making sweetened juice rather than eating fresh because the slippery mesocarp is tart, hence one of the common names for this fruit in the Peruvian Amazon: *sopa agria*, which means bitter soup (Ugent and Ochoa 2006:50). The prickly green fruits, which range in size from 15 cm × 10 cm to 35 cm × 25 cm, are produced year-round (Nascimento et al. 2002). The fruits are peeled, and the white pulp is squeezed from the numerous, shiny black seeds; water and sugar are then added to make a tantalizing and refreshing drink. In major urban centers, the pulp is also frozen to make ice cream, milkshakes, yogurt, and puddings. Guanábana ice cream is even available in some cities with sizable Latino populations in developed countries, such as Miami, Florida. Soursop fruits are nutritious since they contain appreciable amounts of vitamins B1, B2, and C (Samson 1986:275). About 85% of soursop fruit is the pulp, but they must be used within 6 days of harvest otherwise the fruit starts to deteriorate (Pareek et al. 2011). That is why this fruit is rarely encountered in stores.

Annona muricata is the best-known fruit in the Annonaceae, second only to cherimoya *Annona cherimola* which is native to the Andes. The origins of soursop are obscure. Some consider *A. muricata* as native to northern South America (Pennington et al. 2004:165), but lowland Central America or even the West Indies are also candidates. In Panama, for example, soursop grows in tropical moist forest in the Canal Zone and San Blas (Croat 1978:398). Apparently, wild specimens of soursop have been collected in various parts of Amazonia for herbaria, but some of these may be escapees from cultivation. Whatever its origins, soursop has been cultivated in the Amazon for a long time.

In the Amazon, soursop is grown mostly in rural home gardens and polycultural fields on fertile floodplains of silt-laden rivers, such as the Amazon, as well as on uplands. It is also occasionally planted in some urban home gardens, such as in Santarém along the Amazon in western Pará state (WinklerPrins 2002), Rio Branco in Acre (Siviero et al. 2011), Cuiabá in Mato Grosso (Mamede et al. 2015), Abaetetuba along the lower Tocantins in eastern Pará (Lobato et al. 2017), and in São Luiz, Maranhão (Akinnifesi et al. 2010). Soursop is occasionally interplanted with other crops in agroforestry fields in the Amazon, but overall it is not a commercially important crop. One reason that soursop is not grown on a large scale in the Amazon is that large-scale plantings are particularly susceptible to diseases and pests. Most of the commercial production of soursop in Brazil is concentrated on irrigated farms in the dry northeast part of Brazil (Clement et al. 1997). In Costa Rica, where the fruit is also grown commercially in monocrop orchards, the fruits are covered with plastic bags to reduce insect damage.

Soursop *Annona muricata* harvested in a home garden. Santana do Ituqui, Amazon River, Santarém Municipality, Pará, Brazil 9-13-97

Soursop fruits are delicate and do not travel well unless carefully boxed. To avoid damage to the skin, growers interested in selling some of their fruits often harvest them using a pole with a small sack tied to the end. Most of the soursop appearing in urban markets, such as Iquitos in Peru and Belém in Brazil, come from nearby home gardens. However, most of the fruits found in street markets generally arrive piled in the holds of boats, on the beds of trucks or pickup trucks, or stashed in the baggage compartments of local buses. In the Amazon, soursop produces fruit over several months, spanning the latter part of the rainy season through the drier months. Guanábana fruits have been available in Lima along the desert coast of Peru since the eighteenth century (Jaramillo-Arango 1952:40); they likely come from trees planted in irrigated valleys nearby.

Farmer harvesting graviola *Annona muricata* on the Amazon floodplain using a pole with an attached sack so that the fruits are not damaged. Near Itacoatiara, Amazonas, Brazil 12-8-06

Some people use parts of the tree in folk remedies in the Amazon, the Guianas, Central America, and the Caribbean. The Guarasug'we in the Bolivian Amazon drink the juice made from the fruit to treat kidney disorders and hypertension (Hajdu and Hohmann 2012). Also in the Bolivian Amazon, rural and urban folk use the leaves and ground bark to treat diabetes and the leaves to prepare an infusion to alleviate liver problems (Murakami et al. 2016:55). In the Peruvian Amazon, some rural folk mix 20 g of the bark, roots, and leaves of this tree in 1 l of water a cup of which is then drunk along with meals to treat diabetes (Lossio et al. 1995:125). To alleviate dysentery, six leaves are boiled and one cup of the solution is drunk when it cools with meals (Lossio et al. 1995:126). Also in the Peruvian Amazon, the fresh leaves are squeezed and a teaspoon of the juice is swallowed while fasting to treat intestinal parasites and the leaves, roots, and/or bark are cooked in water and a cup of the hot liquid is swallowed to treat colic disorders (Mejia and Rengifo 2000:92). In the vicinity of Chazuta along the Huallaga River in the Peruvian Amazon, people use the leaves to make an infusion which is swallowed to increase urine flow (Sanz-Biset et al. 2009). In southern Ecuador, a bark decoction is drunk to treat rheumatism (Van den Eynden et al. 1999:24). On Ilha do Algodoal along the coast of Pará, rural folk use the seeds as a purgative (Jardim and Medeiros 2006).

Ribeirinhos along the Solimões prepare a tea from the leaves to treat uterine infections, lower cholesterol, and help lose weight (Silva et al. 2007a) as is the case with rural people in Roraima (Luz 2001). The leaves of this tree are sold in the street markets of Boa Vista, the capital of Roraima, and Cuiabá, the capital of Mato Grosso, to make a tea to treat diabetes (Pinto and Maduro 2003; Pinto et al. 2013). In the Chico Mendes Extractive Reserve in Acre, rural folk prepare an infusion with the leaves of this plant to treat high blood pressure, prostate inflammation, and to serve as a calmant (Gonçalve et al. 2018). Rural folk in the vicinity of Manacapuru along the Solimões near Manaus use the leaves and bark to treat inflammation and gastritis (Vásquez 2014:14; Vásquez et al. 2014). Residents of Julião, a rural community in the Reserva de Desenvolvimento Sustentável (RDS) do Tupé some 25 km west of Manaus, use the leaves to make an infusion for treating anemia (Veiga and Scudeller 2015). Residents of Tiningú, a quilombo community 47 km southeast of Santarém, make a tea from the leaves to treat nausea and stomachache (Carvalho 2015:57). The inhabitants of the village of Marudá along the coast of Pará and in Abaetetuba along the lower Tocantins River use the leaves and flowers to make a tea to treat diabetes (Coelho-Ferreira 2009; Palheta et al. 2017). In the municipality of Marapanim along the northeastern coast of Pará, locals make an infusion from the leaves to alleviate heart pain, difficulty in breathing, and nervousness (Furtado et al. 1978). On Colares Island, some 100 km from Belém in the Amazon estuary locals macerate the leaves of this tree to treat snake bites on dogs, cats, horses, and cattle (Ritter et al. 2012). The skin of the roots and the leaves of this tree are also used to prepare a tea for treating diabetes and to calm people who are stressed (Van den Berg 2010:42).

In Surinam which borders Roraima in the Brazilian Amazon, Saramaccan maroons who are descendants of African slaves brought to the coast of the Guianas to work on sugarcane plantations use parts of *Annona muricata* in a variety of folk remedies to treat babies with health issues (Ruysschaert et al. 2009). In Guyana, boiled leaf extracts are used to make a solution to treat heart palpitations (Austin and Bourne 1992). In the West Indies, where the tree was introduced possibly in precontact times the leaves are soaked in water and the solution drunk to promote a good night's sleep, and the leaves are strewn on the bed for the same purpose (Morton 1966). The medicinal virtues of soursop, both real and imagined, are featured on a number of websites touting the fruit. Although some of the claims may be exaggerated, scientists are exploring annonaceous acetogenins, which are found in the leaves, seeds, and stems of soursop and other related fruits, because they can arrest the growth of cancerous cells without being toxic to healthy cells. It is this quality of annonaceous fruits that has spawned a number of sites selling graviola pills.

Farmer preparing to make guanábana *Annona muricata* **juice in her kitchen. Dos de Mayo, Rio Huallaga, Loreto, Peru 8-29-04**

Graviola juice and ice cream has not gained much traction in US markets, however. The Knudsen Company once featured soursop as one of its exotic tropical fruits, but the product was soon withdrawn (Slater 2002). Another better-known Amazonian fruit, açaí *Euterpe oleracea*, has also seen a decline in juice sales in the United States in recent years. Once three or four companies were disputing shelf space in supermarkets in Florida to sell the palm fruit juice, now only one remains in the major supermarket chains in the state.

3.6 *Annona nitida*

Peru Anonilla

Venezuela Guanábana rebalsera

Status Wild

Anonilla *Annona nitida* in fruit in floodplain forest during the high water season. Near Yarina, Yanayacu River, Pacaya-Samiria, Loreto, Peru 1-16-04

The fruits of *Annona nitida* closely resemble those of its domesticated cousin, soursop *Annona muricata*. Known as anonilla in Peru, the fruits of this wild tree are relatively large, about the size of a cantaloupe, with shiny green skin covered sporadically with small, prickle-like protrusions. Anonilla grows in floodplain forests along black water rivers, such as the Yanayacu in the Pacaya-Samiria area of Peru, and the fruits are typically gathered from canoes at high water. Although the tree can reach 15 m high, it is usually shorter and during the annual flood, many of the fruits can be reached by standing up in a canoe. This small tree has also been recorded in upland stunted vegetation (*campina*) growing on sandy soils, such as near Manaus,

Amazonas, where fruits mature in August and September during the dry season (Alencar 1990). *Annona nitida* produces fruits in the rainy season when growing on floodplains, but the opposite occurs on upland sandy soils suggesting a diverse gene pool for this species.

River dweller gathering fruits of anonilla in floodplain forest. Near Yarina, Yanayacu River, Pacaya-Samiria, Loreto, Peru 3-7-04

The relatively thin skin is peeled to reveal the white pulp; only rural folk get to appreciate the fruit's unique and intriguing flavor because the fruits are not taken to markets. *Annona nitida* is an ideal candidate for domestication since it already has large fruits that are not as tart as soursop so they can be eaten fresh as well as made into an appealing drink.

Girl eating anonilla fruit. Yarina, Yanayacu River, Reserva Nacional Pacaya-Samiria, Loreto, Peru 3-7-04

The fruits are probably dispersed by frugivores, such as certain monkeys, fish, and turtles. The giant South American river turtle *Podocnemis expansa* is known to eat the fruits of this species (Vogt 2008:10). This floodplain tree thus contributes indirectly to the diet of river dwellers. Some of the ingested seeds likely pass through the gut indigested and if they are passed in water can float because they are covered in a sponge-like substance known technically as parenchyma. *Annona nitida* occurs from the central Amazon around Manaus west to the Peruvian Amazon. It appears to be confined within roughly 8 degrees north and south of the equator.

3.7 *Annona squamosa*

Bolivia Chirimoya, anón, anona

Brazil Ata, ateira, pinha, fruta de conde, condesa, condeceira, cabeça de negro, araticum

Colombia Anón, anón morado, anona azucarada, anona blanca, saramuyo, viñón

Ecuador Cherimoya, chirimoya, anona; keach (Achuar)

Peru Anona, anona escamosa, anonilla

Venezuela Anón, anona, riñon, saramoyo, saramuyo

Status Cultivated

Ata *Annona squamosa* tree in fruit in a home garden on an upland archaeological site. Sitio Senhor Heladio, left bank of the Rio Xingu about 50 km upstream from Altamira, Pará, Brazil 9-17-15

Sweetsop *Annona squamosa* fruits are the size of baseballs and are easily recognized by their pale yellow-green, convoluted skins. Known as ata in the Brazilian Amazon, pinha in other tropical parts of Brazil, and anon in Spanish-speaking countries, sweetsop's relatively thick peel is pulled off to reveal a tantalizing sweet pulp, widely appreciated in both rural and urban areas. The sugar content of the fruits is high with equal values for glucose, fructose, and saccharose (Andrade et al. 2001). In the Brazilian Amazon, sweetsop fruits during the height of the rainy season, and because the rind is relatively thick for a member of the Annonaceae, the fruits travel well and often find their way to street markets where they are piled like miniature cannonballs on the ground or are stacked in baskets, such as in Alenquer along the middle Amazon.

In the Amazon, this tree is typically planted in home gardens, such as among Japanese-Brazilian farmers in the vicinity of Tomé-Açu, Pará (Yamada and Osaqui 2006), in the municipality of Bonito in northeastern Pará (Rosa et al. 2009; Vieira et al. 2012), in the vicinity of Macapá the capital of Amapá (Pereira et al. 2017), in Abaetetuba along the lower Tocantins (Lobato et al. 2017), and in Santarém at the junction of the Amazon and Tapajós rivers (WinklerPrins and Souza 2005). It is also found in home gardens in Rio Branco, Acre (Siviero et al. 2011), in the savanna region of Roraima and in Boa Vista, the capital of that state (Batista and Barbosa 2014; Pinho 2008:37), along the lower Madeira River in Amazonas (Kawa et al. 2015), and in rural home gardens near Manaus, Amazonas (Major et al. 2005; Souza 2010:23). It also grows in some home gardens in the southern fringes of the Amazon Basin, such as in the community of São Miguel in the municipality of Várzea Grande in Mato Grosso (Mamede and Pasa 2018), in Cuiabá, the capital of Mato Grosso (Mamede et al. 2015) as well as the municipality of Cuiabá (Gonçalves and Pasa 2015) and in the town of Palmas in the state of Tocantins (Pinheiro et al. 2018).

This cultivated tree also has some other uses. In Roraima, Brazil, rural people prepare a tea from the leaves to treat diarrhea and to alleviate problems with their liver and kidneys (Luz 2001; Van den Berg and Silva 1988). Residents of Tiningú, a quilombo community 47 km southeast of Santarém, make a tea from the leaves to treat high blood pressure (Carvalho 2015:57). An infusion of the leaves is used to treat headache, ulcers, fever, and intestinal worms by some residents of Imperatriz, Maranhão, a town along the Belém-Brasilia highway close to the border of eastern Pará (Penido et al. 2016). In the Caruarú community on Mosqueiro Island some 70 km northeast of Belém residents boil the roots of this plant in water and when it cools drink the fluid to alleviate vesicular rashes (Mesquita and Tavares-Martins 2018). At Vila do Algodoal in the municipality of Maracanã, along the coast of northeastern Pará, the wood is used as fuel for cooking (Coelho-Ferreira and Jardim 2005).

Sweetsop *Annona squamosa* tree with unripe fruit in a rural home garden. Paracari, near Santarém, Pará, Brazil 9-3-93

Sweetsop fruits are plucked from small trees, usually under 6 m tall, in dooryard gardens on farms and in towns and villages, such as Altamira on the banks of the Xingu and Santana do Ituqui near Santarém. It is usually planted, although it was apparently gathered in the wild in the vicinity of Santarém during the nineteenth century (Bates 1863:14). Sweetsop is usually found in uplands, but is also cultivated around homes on higher parts of the Amazon floodplain, such as along the banks of the Surubim-Açu near Santarém. It is also found occasionally in home gardens on the floodplain of the lower Solimões near its confluence with the Rio Negro (Lima and Saragoussi 2000). Sweetsop is also occasionally planted in agroforestry plots away from homes in Pará, such as in the vicinity of Murumuru and Belterra in the Municipality of Santarém, Lastancia in the Municipality of Itupiranga, and Aninduba along the lower Tocantins.

The fruits are always eaten fresh, rather than in ice creams, puddings, or cakes. A few farmers maintain small orchards of sweetsop to cater to markets, such as in

Aninduba on an upland bluff overlooking the Amazon a little upstream from Santarém and near Figueirópolis in Mato Grosso. Commercial orchards of sweetsop in the Brazilian Amazon are rare, in part because of limited demand for the fruits but also because of pest pressure. A farmer near Figueirópolis in Mato Grosso, for example, has had problems with Lepidopteran larvae ("*broca*") attacking immature fruits in his grove of 500 sweetsop trees. In home gardens, fruits appear to be less affected by diseases and pests because the trees are typically interplanted with a wide variety of other perennial as well as annual crops.

Also known in English as custard apple or sugar apple, the origins of sweetsop are obscure, but it is native to lowland tropical America or the Caribbean. It has long been widely cultivated in Latin America from Mexico south through Central America and the Caribbean to lowland tropical parts of South America (Popenoe 1920:179). It was taken during the colonial period to other continents, such as tropical Africa, Asia, and the Pacific. It is a highly popular fruit in India where vendors sell the fruits from carts in many cities. Although it might not be native to the Amazon, sweetsop has been cultivated in backyards of the region for over 200 years (Daniel 1976:327).

Sweetsop *Annona squamosa* fruits in a street market. Alenquer, Pará, Brazil 3-20-93

3.8 *Duguetia spixiana*

Bolivia Chirimoyillo, chirimorillo, sinini, pancho de chancho, chirimoya del monte, chocolatillo; dupurai, huabu midha (Tacana), pi'serej, veya (Tsimane')

Brazil Envira de jaboti, envira condurú, envira condurú amarelo, biribarana; kapôt kuben me (Kayapó)

Colombia Mano de tigre, cajaowisio; chocolatillo; toñefokai (Andoke), oimeba (Miraña), jikonogi (Witoto)

Ecuador Coniogocho tetacho (Cofán), sacha anonas (Kichwa), amero (Siona-Secoya), nangenca, nanemo, uñetahue (Waorani)

Peru Tortuga huayo, tortuga caspi, tortuga caspi blanco, tortuga de restinga, motelo caspi, espintana, espintana blanca, pintana negra, pampa tortuga, anona, carahuasca, chirimolita, chirimoya del monte, tangarana, biniánisa, tamara blanca, iakui, anona, quinilla; yais (Aguaruna), ixkin denpa bimi (Cashinahua)

Status Wild

Duguetia spixiana **in fruit in floodplain forest. Quebrada Huirurui, Rio Samiria, near San Martin Tipishca, Reserva Nacional Pacaya-Samiria, Loreto, Peru 6-30-06**

At first glance, the fruits of *Duguetia spixiana* do not look particularly appetizing since they are woody with rigid spines on the outside and a tough red skin covering the sides of the fruit segments (appressed carpels). Nevertheless, the brown fruits, measuring 6–9 cm by 4–7 cm, are prized for the sweet pulp underneath the red skin of the individual segments. The fruit segments contain a single black seed, 2 cm long. The seeds are discarded after the mesocarp is eaten. *D. spixiana* fruits during the rainy season. In the Pacaya-Samiria region of the Peruvian Amazon, the fruits are gathered from the forest floor on higher parts of the floodplain, from canoes or by climbing the tree. This tree is planted in some gardens on the floodplain of the lower Madeira River in Amazonas, Brazil (Kawa et al. 2015).

Tortuga huayo *Duguetia spixiana* fruit gathered in floodplain forest. Near Yarina, Rio Yanayacu, Reserva Nacional Pacaya-Samiria, Loreto, Peru 1-16-06

The common name for the grapefruit-sized fruit in the Pacaya-Samiria, tortuga huayo, suggests that tortoises feed on the fruits once they have fallen to the ground. Another common name for the tree near Iquitos, tortuga caspi, also means tortoise fruit (Vásquez 1997:94). One of the names for the tree in Amazonas, Brazil, is envira de jaboti, which means vine of the tortoise. Two species of tortoise are found in the forests of Amazonia: red-footed tortoise *Chelonoidis carbonaria* and *C. denticulata*, and both are gathered for human consumption. Woolly monkeys *Lagothrix lagotricha* also consume the fruits (Defler and Defler 1996) and are widely hunted for their meat. The fruits of *Duguetia spixiana* thus contribute indirectly to the diet of rural people. The equatorial saki *Pithecia aequatorialis*, which lives in the Amazonian forests of northeastern Peru and Ecuador, also eats this fruit but because of its small size is not a popular game animal. However, it may be involved in dispersing some of the seeds.

The tree has a few other uses as well. The wood is used for house construction and for firewood, such as among the Tsimane in the Bolivian Amazon, the Waorani in the Ecuadorian Amazon, and some rural folk along the Ucayali River in the Peruvian Amazon (Cerón and Ayala 1998:27; Guèze et al. 2014; Wittmann et al. 2010:64). The Tsimane also obtain fiber from the inner bark for use in house construction (Reyes-García 2001:110), while the Waorani use the inner bark to make straps for their baskets (Cerón and Ayala 1998:27). The Ticuna of the Rio Loretoyacu, an affluent of the Upper Solimões, prepare a tea from the bark and leaves for bathing serious leg ulcers (Schultes and Raffauf 1990:58). The Andoke in the Colombian

Amazon use the mashed fruits to treat tooth cavities and children are encouraged to eat these fruits before they gain their new teeth to help prevent cavities (Palacios 1987; La Rotta 1983:30). The Andoke also grate the bark of this tree in water and apply the solution to parts of the body with muscular pain. This same solution is used to bathe dogs to make them better hunters (La Rotta 1983:30). Also in the Colombian Amazon, the Miraña use this same solution to treat rheumatism. In Bolivia, a tea is made from the bark to treat scabies and fungal infections (Fournet et al. 1994). On the floodplain of the Solimões in the vicinity of its conjunction with another sediment-rich river, the Japurá, rural folk use the wood of this tree to make fences (Wittmann et al. 2009).

Duguetia spixiana is an understory tree, generally less than 15 m tall but reaching as high as 20 m on occasion in upland forests and on floodplains. Confined to lowlands below 600 m, *D. spixiana* is found in the central and western part of the Amazon Basin from Bolivia northward through Peru and Ecuador to Colombia (Arellano 2013:84; Rudas and Prieto 1998). It has been recorded in floodplain forests along the lower Juruá, Tefé and Purus rivers, southern affluents of the Solimões (Assis et al. 2017; Luize et al. 2015). This small tree is also found in some successionary vegetation, such as in the watershed of the Putumayo River which borders Peru and Colombia (Gomez 2011:85). Its range extends east to the Rio Negro watershed.

Several other wild and semi-domesticated species of *Duguetia* have edible fruits, but they rarely if ever turn up in markets and are poorly known. Paulo Cavalcante records locals eating the fruits of two species of *Duguetia* from the Brazilian portion of the Amazon: *D. marcgraviana* from upland forest and second growth, and *D. stenantha*, which is cultivated along the middle and upper Solimões (Cavalcante 2010:150, 208). The Nambikwara, who rely heavily on hunting and gathering for their sustenance, collect fruits of *D. marcgraviana* in the forests along the border between Mato Grosso and Rondônia states. (Venturieri et al. 1986). Interestingly, the Brazilian name for *D. stenantha* is jaboti, which means tortoise; evidently rural folk along the Solimões have made the same natural history observations as their counterparts in the Peruvian Amazon. The sweet yellow fruits of *D. stenantha*, measuring 12 cm by 7 cm, are gathered in upland forests and eaten in rural areas in the vicinity of Iquitos.

References

Akinnifesi, F. K.; Sileshi, G.; Costa, Joselene da; Moura, E. G. da; Silva, Rosele F. da; Ajayi, O. C.; Linhares, J. F. P.; Akinnifesi, A. I.; Araujo, Maria de; Rodrigues, M. A. I. 2010. Floristic composition and canopy structure of home-gardens in São Luís city, Maranhão state, Brazil. Journal of Horticulture and Forestry 2(4):72–86

Alencar, Jurandyr da Cruz. 1990. Interpretação fenológica de espécies lenhosas de campina na Reserva Biológica de campina do INPA ao norte de Manaus. Acta Amazonica 20:145–183

Alviz, Ángela; Pérez-Albarracín, Karen. 2015. Plan para la conservación de la danta de tierras bajas (Tapirus terrestris) en los departamentos de Casanare, Arauca y Vichada. Corporinoquia-Fundación Orinoquia Biodiversa, Yopal, Colombia

Amaral, Dário Dantas do; Prost, Maria Thereza; Bastos, Maria de Nazaré do Carmo; Costa Neto, Salustiano Vilar; Santos, João Ubiratan Moreira dos. 2008. Restingas do litoral amazônico, estados do Pará e Amapá. Boletim do Museu Paraense Emílio Goeldi, Ciencias Naturais 3(1):35–67

Amaral, Dário Dantas do; Vieira, Ima Célia Guimarães; Almeida, Samuel Soares de; Salomão, Rafael Paiva; Silva, Antonio Sérgio Lima da; Jardim, Mário Augusto Gonçalves. 2009. Checklist da flora arbórea de remanescentes florestais da região metropolitana de Belém e valor histórico dos fragmentos, Pará, Brasil. Boletim do Museu Paraense Emílio Goeldi, Ciencias Naturais 4(3):231–289

Amaral, Dário Dantas do; Jardim, Mário Augusto Gonçalves; Costa-Neto, Salustriano Vilar; Bastos, Maria de Nazaré do Carmo. 2015. Síndromes de dispersão de propágulos e a influência da floresta amazônica na composição de espécies lenhosas de uma restinga no litoral norte brasileiro. Biota Amazônia 5(3):28–37

Andrade, Eloisa Helena A.; Zoghbi, Maria das Graças B.; Maia, José Guilherme S.; Fabricius, Heinz; Marx, Friedhelm. 2001. Chemical characterization of the fruit of Annona squamosa L. occurring in the Amazon. Journal of Food Composition and Analysis 14: 227–232

Araujo-Murakami, Alejandro; Villarroel, Daniel; Pardo, Guido; Vos, Vincent A.; Parada, G. Alexander; Arroyo, Luzmila; Killeen, Timothy. 2015. Diversidad arbórea de los bosques de tierra firme de la Amazonía Boliviano. Kempffiana 11(1):1–28

Arellano, Gabriel. 2013. Patrones de diversidad, distribución y dominancia de plantas leñosas en los bosques montanos del Parque Nacional Madidi, Bolivia. Doctoral dissertation, Departamento de Biologia, Facultad de Ciencias, Universidad Autónoma de Madrid

Assis, Rafael Leandro de; Wittmann, Florian; Luize, Bruno Garcia; Haugaasen, Torbjon. 2017. Patterns of floristic diversity and composition in floodplain forests across four southern Amazon river tributaries, Brazil. Flora 229:124–140

Austin, Daniel F.; Bourne, Godfrey R. 1992. Notes on Guyana's medical ethnobotany. Economic Botany 46(3):293–298

Baar, R.; M. Cordeiro, M. Denrich; and H. Fölster. 2004. Floristic inventory of secondary vegetation in agricultural systems of East-Amazonia. *Biodiversity and Conservation* 13:501–528

Bastos, M. N. C.; C. S. Rosário; L. C. B. Lobato. 1995. Caracterização fitofisionômica da restinga de Algodoal, Maracanã, Pará, Brasil. Boletim do Museu Paraense Emílio Goeldi, Série Botanica 11(2): 173–197

Bastos, Maria de Nazaré do Carmo; Costa, Denise Cristina Torres; Santos, João Ubiratan Moreira dos. 2003. Vegetação de restinga: aspectos botânicos e uso medicinal. Museu Paraense Emilío Goeldi Projeto Renas, Belém

Bates, H. W. 1863. *The Naturalist on the River Amazons*. John Murray, London, vol. 2

Batista, Debora Lima; Barbosa, Reinaldo Imbrozio. 2014. Agrobiodiversidade urbana: composição florística, riqueza e diversidade de plantas nos quintais de Boa Vista, Roraima. Revista Brasileira de Agroecologia 9(2):130–150

Boom, B. M. 1987. Ethnobotany of the Chácobo Indians, Beni, Bolivia. The New York Botanical garden, Advances in Economic Botany, vol. 4

Boom, B. M. 1989. Use of plant resources by the Chácobo. In: *Resource Management in Amazonia: Indigenous and Folk Strategies*, D. A. Posey and W. Balée (Editors), Advances in Economic Botany, vol. 7, New York Botanical Garden, Bronx, pp. 78–96

Boubli, J. P. 2002. Lowland floristic assessment of Pico da Neblina National Park, Brazil. Plant Ecology 160:149–167

Brack, A. 1999. *Diccionario Enciclopédico de Plantas Útiles del Perú*. Centro de Estudios Regionales Andinos Bartolomé de Las Casas (CBC), Cusco

Brienza Júnior, Silvio. 1999. Biomass dynamics of fallow vegetation enriched with leguminous trees in the eastern Amazon of Brazil. Doctoral dissertation, Georg August University, Göttingen, Germany

Buchillet, D. 1988. Interpretação da doença e simbolismo ecológico entre os índios Desana. Boletim to Museu Paraense Emílio Goeldi, Série Antropologia 4 (1): 27–42

Cárdenas, Dairon; Castaño, Nicolás; Sua, Sonia. 2009. Flora de la estrella fluvial de Inírida (Guianía, Colombia). Biota Colombiana 10(1–2):1–30

Carim, Marcelo de Jesus Veiga. 2016. Estrutura, composição e diversidade em florestas alagáveis de várzea de maré e de igapó e suas relações com variáveis edáficas e o período de inundação no Amapá, Amazônia oriental, Brasil. Doctoral dissertation, Programa de Pós-Graduação em Botânica, INPA, Manaus

Carim, Marcelo de Jesus Veiga; Guillaumet, Jean-Louis Bernard; Guimarães, José Renan da Silva; Tostes, Luciedi Cássia Leôncio. 2013. Composição e estrutura de floresta ombrófila densa do extremo norte do Estado do Amapá, Brasil. Biota Amazônia 3(2):1–10

Cartagena, Susana Esther Cotito. 2014. Diversidad y composición florística de un área de bosque ribereño premontano del valle de Chanchamayo. Master's thesis, Universidad Nacional Agraria La Molina, Lima

Carvalho, Telma Lélia Gonçalves Schultz de. 2015. Etnofarmacologia e fisiologia de plantas medicinais do quilombo Tiningú, Santarém, Pará, Brasil. Master's thesis, Universidade Federal do Oeste do Pará, Santarém.

Casula, Kátia Regina. 2012. Análise da similaridade florística e estrutural das formações florestais inundáveis em um trecho do alto Rio Madeira e de seus afluentes, estado de Rondônia, Brasil. Master's Thesis, Núcleo de Ciências Exatas da Terra, Universidade Federal de Rondônia, Porto Velho

Cattanio, José; Anderson, Anthony B.; Carvalho, Manoel S. 2002. Floristic composition and topographic variation in a tidal floodplain forest in the Amazon estuary. Revista Brasileira de Botanica 25(4):419–430

Cavalcante, P. B. 2010. *Frutas Comestíveis na Amazônia*. Museu Paraense Emílio Goeldi, Belém

Cerón, Carlos Eduardo; Ayala, Consuelo Montalvo. 1998. Etnobotanica de los Huaorani de Quehueiri-ono Napo, Ecuador. Abya-Yala, Quito

Chang, F.R.; Chen, C.Y.; Cho, C.P.; Wu, J.C. 2000. Chemical constituents from Annona glabra. Journal of the Chinese Chemical Society 47:913–920

Chatrou, L. W., H. Rainer, P. J. M. Maas. 2004. Annonaceae (Soursop Family). In: *Flowering Plants of the Neotropics*, N. Smith, S. A. Mori, A. Henderson, D. W. Stevenson, and S. V. Heald (Editors), Princeton University Press, Princeton, pp. 18–20

Clement, C.R. 1989. A center of crop genetic diversity in western Amazonia: A new hypothesis of indigenous fruit-crop distribution. *Biotropica* 39(9): 624–631

Clement, C. R. 1999. 1492 and the loss of Amazonian crop genetic resources. II. Crop biogeography at contact. *Economic Botany* 53(2):203–216

Clement, C. R., S. S. Alfaia, J. H. Iriarte-Martel, K. Yuyama, J. B. M. Gomes, J. Leeuwen, L. A. G. Souza, and W. B. C. Flores. 1997. Fruteiras nativas e exóticas. In: *Duas Décadas de Contribuições do INPA à Pesquisa Agronômica no Tróprico Úmido*, Noda, Hiroshi; Souza, Luiz Augusto Gomes de (Editors), pp. 111–129, INPA, Manaus

Coelho, Roberta de Fátima Rodrigues; Zarin, Daniel Jacob; Miranda, Izildinha Souza; Tucker, Joanna Marie. 2003. Ingresso e mortalidade em uma floresta em diferentes estágios sucessionais no município de Castanhal, Pará. Acta Amazonica 33(4):619–630

Coelho-Ferreira, Márlia. 2009. Medicinal knowledge and plant utilization in an Amazonian coastal community of Marudá, Pará state (Brazil). Journal of Ethnopharmacology 126:159–175

Coelho-Ferreira, Márlia; Jardim, Mário Augusto G. 2005. Algumas espécies vegetais usadas pelos moradores da Ilha de Algodoal, Maiandeua, município de Maracanã, Pará. Boletim do Museu Paraense Emílio Goeldi, Série Ciências Naturais 1(2): 45–51

Croat, T. B. 1978. *Flora of Barro Colorado Island*. Stanford University Press, Stanford

Daniel, J. 1976. *Tesouro Descoberto no Rio Amazonas*. Biblioteca Nacional, Rio de Janeiro, vol. 1

Defler, Thomas R.; Defler, Sara B. 1996. Diet of a group of *Lagothrix lagothricha lagothrica* in southeastern Colombia. International Journal of Primatology 17(2):161–190

Ducke, A. 1946. *Plantas de Cultura Precolombiana na Amazônia brasileira. Notas Sôbre as Espécies ou Formas Espontâneas que Supostamente lhes teriam dado Origem*. Instituto Agronômico do Norte, Belém, Boletim Técnico 8, pp. 1–24

Duivenvoorden, J.F. 1994. Vascular plant species counts in the rain forests of the middle Caquetá area, Colombian Amazonia. Biodiversity and Conservation 3:685–715

Falcão, Martha de Aguiar; Lleras, Eduardo; Kerr, Warwick Estevam; Carreira, Léa Maria Medeiros. 1981. Aspectos fenológicos, ecológicos e de produtividade do biribá (Rollinia mucosa (Jacq.) Baill.). Acta Amazonica 11(2): 297–306

Ferreira, Almecina Balbino; Ming, Lin Chau; Haverroth, Moacir; Daly, Douglas Charles; Caballero, Javier; Ballesté, Andrea M. 2015. Plants used to treat malaria in the regions of Rio Branco-Acre state and southern Amazonas state, Brazil. International Journal of Phytocosmetics and Natural Ingredients 2(9):1–5

Figueroa, Ilba Carolina; Fachín-Terán; Duque, Santiago R. 2012. Componentes alimenticios de *Podocnemis unifilis* y *P. expansa* (Testudines: Podocnemididae) en el resguardo Curare-Los Ingleses, Amazonas, Colombia. Revista Colombia de Ciencia Animal 4(2):441–453

Flores, Salvador. 1997. *Cultivo de Frutales Nativos Amazonicos: Manual para el Extensionista.* Tratado de Cooperación Amazónica, Lima

Fournet, Alain; Barrios, Alcira Angelo; Muñoz, Victoria. 1994. Leishmanicidal and trypanocidal activities of Bolivian medicinal plants. Journal of Ethnopharmacology 41:19–37

Furtado, Lourdes Gonçalves; Souza, Ruth Cortez de; Van den Berg, Maria Elizabeth. 1978. Notas sobre uso terapêutico de plantas pela população cabocla de Marapanim, Pará. Boletim do Museu Paraense Emílio Goeldi, Nova Série Antropologia 70:1–31

Gallois, C. 2009. *Wajãpi rena: roças, pátios e casas.* Museu do Indio-Funai, Rio de Janeiro

Gomez, Tedi Pacheco. 2011. Valoración económica de purmas de diferentes edades en la comunidad nativa de Santa Mercedes, río Putumayo, Perú. Master's thesis, UNAP, Iquitos

Gómez-Posada, C. 2012. Dieta y comportamiento alimentario de un grupo de mico maicero *Cebus apella* de acuerdo a la variación en la oferta de frutos y artrópodos, en la Amazonía colombiana. Acta Amazonica 42(3):363–372

Gonçalve, Maiara Cristina; Paula Filho, Galdino Xavier de; Oliveira, Daniel Henrique de; Li, Meiling; Ruppenthal, Felipe R.; Barbosa, Kennedy de Araújo; Ming, Lin Chau. 2018. Medicina da Floresta: saberes e práticas populares dos moradores do núcleo de base Semitumba. In: *Experiências Etnobotânicas na Reserva Extrativista Chico Mendes*, Ming, Lin Chau; Amorozo, Maria Christina de Mello; Ferreira, Almecina Balbino (eds), Fundação de Estudos e Pesquisas Agrícolas e Florestais, Rio Branco, pp. 99–108

Gonçalves, Karina Gondolo; Pasa, Maria Corette. 2015. A etnobotânica e as plantas medicinais na Comunidade Sucuri, Cuiabá, MT, Brasil. Interações 16(2):245–256

Gonçalves, Fábio Guimarães; Santos, João Roberto dos. 2008. Composição florística e estrutura de uma unidade de manejo florestal sustentável na Floresta Nacional do Tapajós, Pará. Acta Amazonica 38(2):229–244

Gottsberger, Gerhard. 1978. Seed dispersal by fish in the inundated regions of Humaitá, Amazonia. *Biotropica* 10 (3): 170–183

Goulding, Michael. 1980. *The Fishes and the Forest: Explorations in Amazonian Natural History.* University of California Press, Berkeley

Gualberto, Midiã Larine Colares; Ribeiro, Renato Bezerra da Silva; Gama, João Ricardo Vaconcellos; Vieira, Diego dos Santos. 2014. Fitossociologia e potencial de espécies arbóreas em ecossistema sucessional na Floresta Nacional do Tapajós, Pará. Agroecossistemas 6(1):42–57

Guèze, Maximilien; Luz, Ana Catarina; Paneque-Gálvez, Jaime; Macía, Manuel J.; Orta-Martínez, Martí; Pino, Joan; Reyes-García, Victoria. 2014. Are ecologically important tree species the most useful? a case study from indigenous people in the Bolivian Amazon. Economic Botany 68(1):1–15

Guillaumet, Jean-Louis; Reis, Nélio Roberto dos. 1983. Les chauves-souris frugivores de la région de Manaus et leur rôle dans la dissémination des espèces végétales. Revue d'Écologie: La Terre et la Vie 38:147–169

Haidar, Ricardo Flores; Fagg, Jeanine Maria Felfili; Pinto, José Roberto Rodrigues; Dias, Ricardo Ribeiro; Damasco, Gabriel; Silva, Lucasde Carvalho Ramos; Fagg, Christopher William. 2013. Florestas estacionais e áreas de ecótono no estado do Tocantins, Brasil: parâmetros estruturais, classificação das fitofisionomias florestais e subsídios para conservação. Acta Amazonica 43(3):261–290

Hajdu, Z. and J. Hohmann. 2012. An ethnopharmacological survey of the traditional medicine utilized in the community of Porvenir, Bajo Paraguá Indian Reservation, Bolivia. Journal of Ethnopharmacology 139:838–857

Heriarte, Mauricio de, 1964. *Descriçam do Estado do Maranham, Para, Corupa, Rio das Amazonas (1662).* Akademische Druck und Verlagsanstalt, Graz, Austria

Jaramillo-Arango, J. 1952. *Relación Historica del Viage, que Hizo a los Reynos del Peru y Chile el Botanico d. Hipolito Ruiz en el Año de 1777 hasta el de 1788, en cuya Epoca Regreso a Madrid.* Real Academia de Ciencias Exactas, Fisicas y Naturales, Madrid, vol. 1

Jardim, M.A.G. and Medeiros, T.D.S. 2006. Plantas oleaginosas do Estado do Pará: composição florística e usos medicinais. *Revista Brasileira de Farmácia* 87(4):124–127

Kalliola, Riso; Salo, Jukka; Puhakka, Maarit; Rajasilta, Marjut. 1991. New site formation and colonizing vegetation in primary succession on the western Amazon floodplains. Journal of Ecology 79(4):877–901

Kawa, Nicholas C.; Michelangeli, José A. Clavijo; Clement, Charles R. 2015. Household agrobiodiversity management on Amazonian dark earths, oxisols, and floodplain soils on the lower Madeira River, Brazil. Human Ecology 43:339–353

Kffuri, Carolina Weber; Lopes, Moisés Ahkutó; Ming, Lin Chau; Odonne, Guillaume; Kinupp, Valderly Ferreira. 2016. Antimalarial plants used by indigenous people of the Upper Rio Negro in Amazonas, Brazil. Journal of Ethnopharmacology 178:188–198

La Rotta, C. 1983. Observaciones Etnobotanicas sobre algunas Especies Utilizadas por la Comunidad Indigena Andoque (Amazonas, Colombia). Universidad Nacional de Colombia, Bogotá

Lemos, D. A. N.; Ferreira, B. G. A.; Siqueira, J. D. P.; Oliveira, M. M.; Ferreira, A. M. 2015. Floristic and phytosociology in dense "terra firme" rainforest in the Belo Monte hydroelectric plant influence area, Pará, Brazil. Brazilian Journal of Biology 75(3):257–276

Lima, Roberval M. B. de; Saragoussi, Muriel. 2000. Floodplain home gardens on the central Amazon in Brazil. In: The Central Amazon Floodplain: Actual Use and Options for a Sustainable Management, Junk, W. J.; Ohly, J. J.; Piedade, M. T. F.; Soares, M. G. M. (eds), Backhuys Publishers, Leiden pp 243–268

Lima, R. R., S. A. Alencar, J. M. Frade, and G. R. Brandão. 1984. Coleta e avaliação de plantas amazônicas de cultura ou de exploração pré-colombiana: Recursos genéticos da região do Solimões. Anais do 1o Simpósio do Trópico Úmido, 12 a 17 de novembro de 1984, Centro de Pesquisa Agropecuária do Trópico Úmido, EMBRAPA, Belém, Pará, vol. 4, pp. 39–48

Lobato, G. de J. M.; Lucas, F. C. A.; Tavares-Martins, A. C. C.; Jardim, M. A. G.; Martorano, L. G. 2017. Diversidade de uso e aspectos socioambientais de quintais urbanos em Abaetetuba, Pará, Brasil. Revista Brasileira de Agroecologia 12(2):95–105

Lorenzi, Harri; Bacher, Luis; Lacerda, Marco; Sartori, Sergio. 2006a. Frutas Brasileiras e Exóticas Cultivadas. Instituto Plantarum de Estudos da Flora Ltda, Odessa, Sao Paulo.

Lorenzi, H., L. Bacher, M. Lacerda, S. Sartori 2006b. Brazilian Fruits and Cultivated Exotics (for Consuming in Natura). Instituto Plantarum de Estudos da Flora Ltda, Odessa, São Paulo

Lossio, L.C., Delgado, H.S., Ruiz, A.G., Donayre, R.A., Ruiz, J.G., Panduro, M.P., Sifuentes, T.C. 1995. Plantas Medicinales de la Amazonía Peruana. Instituto Peruano de Seguridad Social/Instituto de Medicina Tradicional, Iquitos

Luize, Bruno Garcia; Venticinque, Eduardo Martins; Silva, Thiago Sanna Freire; Novo, Evlyn Marcia Leão de Moraes. 2015. A floristic survey of angiosperm species occurring at three landscapes of the central Amazon várzea, Brazil. Checklist: The Journal of Biodiversity Data 11(6):1–15

Lustosa, Dom Antônio de Almeida. 1976. No Estuário Amazônico "À Margem da Visita Pastoral". Conselho Estadual de Cultura do Pará, Belém.

Luz, Francisco Joaci F. 2001. Plantas medicinais de uso popular em Boa Vista, Roraima, Brasil. Horticultura Brasileira 19(1):88–96

Maas, P.J.M.; Westra, L.Y.T. 1972. *Rollinia.* New York Botanical Garden, Flora Neotropica Monograph 57

Maas, P.J., Lubbert, Y., Westra, T., Brown Jr, K.S., Maas, P.J.M., Ter Welle, B.J.H., Webber, A.C., Le Thomas, A., Waha, M., van der Heijden, E. and Bouman, F., 1992. Rollinia. Flora Neotropica, pp.1–188.

Machado, Danilo de Oliveira. 2016. A agrobiodiversidade de quintais agroflorestais em propriedades agrícolas familiares na BR 174, Ramal do Pau-Rosa, Manaus, AM. Master's Thesis, INPA, Manaus

Major, Julie; Clement, Charles R.; DiTommaso, Antonio. 2005. Influence of market orientation on food plant diversity of farms located on Amazonian Dark Earth in the region of Manaus, Amazonas, Brazil. Economic Botany 59(1):77–86

Malheiros, Antonio Francisco; Higuchi, Niro; Santos, Joaquim dos. 2009. Análise estrutural da floresta tropical úmida do município de Alta Floresta, Mato Grosso, Brasil. Acta Amazonica 38(3):539–548

Mamede, Jeneffer Soares dos Santos; Pasa, Maria Corette. 2018. Os quintais agroflorestais na comunidade rural em Várzea Grande, Mato Grosso, Brazil. Biodiversidade 17(2):54–70

Mamede, Jeneffer Soares dos Santos; David, Margô De; Tsukamoto Filho, Antônio de Arruda; Pasa, Maria Corette. 2015. Os quintais e as manifestações culturais da comunidade de São Gonçalo Beira Rio, Cuiabá – MT. Biodiversidade 14(1):168–182

Medeiros, Marcelo Brilhante de; Walter, Bruno Machado Teles; Silva, Glocimar Pereira da; Gomes, Beatriz Machado; Lima, Isabel Lustz Portela; Silva, Suelma Ribeiro; Moser, Pamela; Washington, Luís Oliveira; Cavalcanti, Taciana Barbosa. 2012. Vascular flora of the Tocantins River middle basin, Brazil. Check List 8(5):852–885

Mejia, Kember; Rengifo, Elsa. 2000. Plantas Medicinales de Uso Popular en la Amazonía Peruana. Agencia Española de Cooperación Internacional, Lima

Mesquita, Ulliane de O.; Tavares-Martins, Ana Cláudia. 2018. Etnobotánica de plantas medicinales en la comunidad de Caruarú, Isla de Mosqueiro, Belém-PA, Brasil. Boletín Latinoamericano y del Caribe de Plantas Medicinales y Aromáticas 17(2):130–159

Morton, J. F. 1966. The soursop, or guanabana (Annona muricata Linn.) Proceedings of the Florida State Horticultural Society 79:355–366

Moura, Valéria Mourão de; Sousa, Luciana A. Freitasde; Dos-Santos, Maria Cristina; Raposo, Juliana Divina Almeida; Lima, Aline Evangelista; Oliveira, Ricardo Bezerra de; Silva, Milton Nascimento da; Mourão, Rosa Helena Veras. 2015. Plants used to treat snakebites in Santarém, western Pará, Brazil: an assessment of their effectiveness in inhibiting hemorrhagic activity induced by *Bothrops jararaca* venom. Journal of Ethnopharmacology 161:224–232

Muniz, Francisca Helena; Cesar, Oswaldo; Monteiro, Reinaldo. 1994. Fitossociologia da vegetação arbórea da reserva florestal do Sacavém, São Luís, Maranhão (Brasil). Acta Amazonica 24 (3/4):219–236

Muñoz, Marianella Guevara. 2010. Evaluación de purmas con fines de manejo en la comunidad nativa de Santa Mercedes, Rio Putumayo, Perú. Master's thesis, Universidad Nacional de la Amazonia Peruana, Iquitos

Murakami, Alejandro Araujo; Reyes, Juan Fermandon; Milliken, William. 2016. Frutales Silvestres y Promisorios de Pando. Herencia/Museo de Historia Natural Noel Kempff Mercado, Cobija

Murillo, J. 2001. Las Annonaceae de Colombia. *Biota Colombiana* 2(1): 49–58.

Nabhan, G.P. 1989. *Enduring Seeds: Native American Agriculture and Wild Plant Conservation.* University of Arizona Press, Tucson, p. 143

Nascimento, Tânia Brito do; Gazel Filho, Alderaldo B.; Santos, Jackson de A. dos. 2002. Fenologia da gravioleira (Annona muricata) em área de cerrado do Amapá, Brasil. Acta Amazonica 32(3):367–376

Oliveira, Francisco Plácido Magalhães; Jardim, Mário Augusto Gonçalves. 1998. Composição florística de uma floresta secundária no município de Igarapé-Açu, estado do Pará, Brasil. Boletim do Museu Paraense Emilio Goeldi, Série Botânica 14(2):127–145

Palacios, P. A. 1987. Analisis de los usos y formas de manejo de algunas especies vegetales empleadas por las comunidades Andoque, Huitoto y Miraña asentadas en la ribera del rio Caquetá. Colombia Amazonica 2(2): 51–56

Palacios, Erwin; Rodríguez, Adriana; Castillo, Claudia. 2004. Preliminary observations on the mottled-face tamarin (*Saguinus inustus*) on the lower Rio Caquetá, Colombian Amazon. Neotropical Primates 12 (3): 123–126

Palheta, Ivanete Cardoso; Tavares-Martins, Ana Cláudia Caldeira; Lucas, Flávia Cristina Araújo; Jardim, Mário Augusto Gonçalves. 2017. Ethnobotanical study of medicinal plants in urban home gardens in the city of Abaetetuba, Pará state, Brazil. Boletín Latinoamericano y del Caribe de Plantas Medicinales y Aromáticas 16(3):206–262

Pareek, Sunil; Yahia, Elhadi M.; Pareek, O. P.; Kaushik, R. A. 2011. Postharvest physiology and technology of *Annona* fruits. Food Research International 44: 1741–1751

Parolin, P. 2000a. Seed mass in Amazonian floodplain forests with contrasting nutrient supplies. Journal of Tropical Ecology 16(3):417–428

Parolin, P. 2000b. O uso de árvores nas florestas inundadas por água branca na Amazônia Central. Amazoniana 16(1/2):241–248

Parolin, P. 2000c. Phenology and C02-assimilation of trees in Central Amazonian floodplains. Journal of Tropical Ecology 16:465–473

Parolin, P., D. Waldhoff, and M. T. Piedade. 2010. Fruit and seed chemistry, biomass and dispersal. In: *Amazonian Floodplain Forests: Ecophysiology, Biodiversity and Sustainable Management*. W. J. Junk, M. T. Piedade, F. Wittman, and J. Schöngart (eds), Springer, Dordrecht, pp. 243–258

Penido, Alexandre Batista; Morais, Selene Maia de; Ribeiro, Alan Bezerra; Silva, Ana Zélia. 2016. Ethnobotanical study of medicinal plants in Imperatriz, state of Maranhão, northeastern Brazil. Acta Amazonica 46(4):345–354

Pennington, T. D., C. Reynel, and A. Daza. 2004. *Illustrated Guide to the Trees of Peru*. David Hunt, Sherborne, England

Pereira, Luciano Araujo; Lima-Barbosa, Jackson Rodrigo de; Almeida, Maria Zélia de; Guimarães, Elsie Franklin. 2017. Diversidade de plantas em quintais quilombolas, conhecimento local sobre uso e cultivo de pimentas na Amazônia oriental, Brasil. Revista de Biologia Neotropical 14(1):56–72

Pérez-Emán, J. L. and A. Paolillo. 1997. Diet of the pelomedusid turtle *Peltocephalus dumerilianus* in the Venezuelan Amazon. *Journal of Herpetology* 31(2):173–179

Pinheiro, Renato Torres; Marcelino, Dianes Gomes; Moura, Dieyson Rodrigues. 2018. Espécies arbóreas de uso múltiplo e sua importância na conservação da biodiversidade nas áreas verdes urbanas de Palmas, Tocantins. Desenvolvimento e Meio Ambiente 49: 264–282

Pinho, Rachel Camargo de. 2008. Quintais agroflorestais indígenas em áreas de savana (lavrado) na Terra Indígena Araçá, Roraima. Master's thesis, INPA/UFAM, Manaus

Pinto, Angélica Auxiliadora da Costa; Maduro, Cice Batalha. 2003. Produtos e subprodutos da medicina popular comercializados na cidade de Boa Vista, Roraima. Acta Amazonica 33(2):281–290

Pinto, Andréia Zelenski de Lara; Assis, Ana Flávia Silva de; Pereira, Arildo Gonçalo; Pasa, Maria Corette. 2013. Levantamento etnobotânico de plantas medicinais comercializados no mercado do porto em Cuiabá, Mato Grosso, Brasil. Flovet 5:51–70

Popenoe, W. 1920. *Manual of Tropical and Subtropical Fruits*. Macmillan, New York

Prance, G.T. and M.F. da Silva. 1975. Árvores de Manaus. INPA, Manaus

Rego, Lyvia Julienne Sousa; Elly, Márcio Piero; Pereira, Jéssica Lira; Garcia, Bruna Naiara; Barbosa, Humberto Figueira; Vieira, Thiago Almeida. 2011. Produtos florestais não-madeireiros comercializados em feiras de Santarém, Pará. Cadernos de Agroecologia 6(2):1–6

Reis, Neidiane Farias Costa. 2010. Estrutura fitossociológica e seletividade de espécies em áreas inundáveis na região do baixo Rio Jaci-Paraná, Rondônia. Master's Thesis, Universidade Federal de Rondônia, Porto Velho

Reitberg, Ellen. 2014. Várzea floodplain agriculture in the Colombian Amazon: recommendations for agroforestry in the Yahuarcaca floodplain. Bachelor thesis, Van Hall Larenstein University of Applied Sciences, Netherlands

Reyes-García, Victoria. 2001. Indigenous people, ethnobotanical knowledge, and market economy: a case study of the Tsimane' Amerindians in lowland Bolivia. Doctoral dissertation, University of Florida, Gainesville.

Ritter, Rhuan Amorim; Monteiro, Maria Vivina Barros; Monteiro, Frederico Ozanan Barros; Rodrigues, Silvane Tavares; Soares, Marina Lira; Silva, Jean Carlos Ramos; Palha, Maria das Dores Correia; Biondi, Germano Francisco; Rahal, Sheila Canevese; Tourinho, Manuel Malheiros. 2012. Ethnoveterinary knowledge and practices at Colares island, Pará state, eastern Amazon, Brazil. Journal of Ethnopharmacology 144:346–352

Rocha, S. F. R., M. R. C. Ferreira, and M. G. Sablayrolles. 2013. Farmacopeia vegetal da comunidade Caxiuanã, na Floresta Nacional de Caxiuanã. In: Caxiuanã: Paraíso ainda Preservado, P. L. B. Lisboa (Editor), Museu Paraense Emílio Goeldi, Belém, pp. 257–285

Rodrigues, Ricardo Ribeiro; Martins, Sebastião Venâncio; Barros, Luiz Carlos de. 2004. Tropical rain forest regeneration in an area degraded by mining in Mato Grosso state, Brazil. Forest Ecology and Management 190: 323–333

Roosmalen, M. G. M. van. 1985. *Fruits of the Guianan Flora.* Institute of Systematic Botany, Utrecht University, Utrecht/Silvicultural Department of Wageningen Agricultural University, Wageningen

Rosa, Leonilde dos Santos; Vieira, Thiago Almeida; Pires, Helaine Cristine Gonçalves. 2009. Quintais agroflorestais em comunidades rurais de Bonito, Pará. Revista Brasileira de Agroecologia 4(2):1310–1313

Rudas, Augustín; Prieto, Adriana. 1998. Análisis florístico del parque nacional natural Amacayacu e Isla Mocagua, Amazonas (Colombia). Caldasia 20(2):142–172

Ruysschaert, Sofie; Van Andel, Tinde; Van de Putte, Kobeke; Van Damme, Patrick. 2009. Bathe the baby to make it strong and healthy: Plant use and child care among Saramaccan Maroons in Suriname. Journal of Ethnopharmacology 121:148–170

Samson, J. A. 1986. *Tropical Fruits.* Longman, New York.

Sandel, Maureen Peggy; Carvalho, João Olegário Pereira de. 2000. Composição florística e estrutura de uma área de cinco hectares de mata alta sem babaçu na floresta nacional do Tapajós. Embrapa Amazônia Oriental, Belém, Documentos 63, pp. 1–19

Santana, José Augusto da Silva; Almeida, Wanderléa da Costa; Sousa, Luciana Karla V. dos Santos. 2004. Florística e fitossociologia em área de vegetação secundária na Amazônia oriental. Revista de Ciências Agrárias 41:105–120

Santos, João Ubiratan M. dos; Amaral, Dário Dantas do; Gorayeb, Inocêncio de Sousa; Bastos, Maria de Nazaré do C.; Secco, Ricardo de S.; Neto, Salustiano Vilar Costa; Costa, Denise Cristina Torres. 2003. Vegetação da área de proteção ambiental Jabotitiua-Jatium, Município de Viseu, Pará, Brasil. Acta Amazonica 33(3):431–444

Santos, Silvio Roberto Miranda dos; Miranda, Izildinha de Souza; Tourinho, Manoel Malheiros. 2004. Análise florística e estrutural de sistemas agroflorestais das várzeas do rio Juba, Cametá, Pará. Acta Amazonica 34(2):251–263

Sanz-Biset, Jaume; Campos-de-la-Cruz, José; Epiquién, Mirbel A.; Cañigueral, Salvador. 2009. A first survey on the medicinal plants of the Chazuta valley (Peruvian Amazon). Journal of Ethnopharmacology 122:333–362

Schultes, R. E. and R. F. Raffauf. 1990. The Healing Forest: Medicinal and Toxic Plants of the Northwest Amazonia. Dioscorides Press, Portland, Oregon

Silva, E. M.; Souza, J. N. S.; Rogez, H.; Rees, J. F.; Larondelle, Y. 2007a. Antioxidant activities and polyphenolic contents of fifteen selected plant species from Amazonian region. Food Chemistry 101: 1012–1018

Silva, Sheila Maria Garcia da; Nascimento, Keila Gardênia Silva; Fraxe, Therezinha de Jesus Pinto; Braga, Pedro Ivo Soares. 2007b. A "Saúde" nas comunidades focais do Projeto PIATAM: o etnoconhecimento e as plantas medicinais. In: Comunidades Ribeirinhas Amazônicas: Modos de Vida e Uso dos Recursos Naturais, Fraxe, Therezinha de Jesus Pinto; Pereira, Henrique dos Santos; Witkoski, Antônio Carlos (eds), EDUA, Manaus, pp 109–138

Siviero, Amauri; Delunardo, Thiago Andrés; Haverroth, Moacir; Oliveira, Luis Cláudio de; Mendonça, Ângela Maria Silva. 2011. Cultivo de espécies alimentares em quintais urbanos de Rio Branco, Acre, Brasil. Acta Botanica Brasilica 25(3):549–556

Slater, C. 2002. Myths of the rain forest/the rain forest as myth. In: *Myth: A New Symposium*, G. Schrempp and W. Hansen (Editors), Indiana University Press, Bloomington

Souza, Célia Cristina Valero. 2010. Etnobotânica de quintais em três comunidades ribeirinhas na Amazônia central, Manaus, AM. Master's thesis, INPA, Manaus
Souza, A. G. C., N. R. Sousa, S. E. L. Silva, C. D. M. Nunes, A. C. C. Canto, L. A. A. Cruz, 1996. *Fruteiras da Amazônia*. EMBRAPA (Empresa Brasileira de Agropecuária), Brasília
Standley, P. C. 1930. A second list of the trees of Honduras. *Tropical Woods* 21: 9–41
Thomas, Evert; Van Damme, Patrick. 2010. Plant use and management in homegardens and swiddens: evidence from the Bolivian Amazon. Agroforestry Systems 80:131–152
Ugent, Donald; Ochoa, Carlos M. 2006. La Etnobotánica del Perú: Desde la Prehistoria al Presente. CONCYEC (Consejo Nacional de Ciencia, Tecnología e Innovación Tecnológica), Lima
Van den Berg, M. E. 2010. Plantas Medicinais na Amazônia. Museu Paraense Emílio Goeldi, Belém
Van den Berg, Maria Elisabeth; Silva, Milton Hélio Lima da. 1988. Contribuição ao conhecimento da flora medicinal de Roraima. Acta Amazonica 18(1–2) supl.:23–35
Van den Eynden, Veerle; Cueva, Eduardo; Cabrera, Omar. 1999. Plantas Silvestres Comestibles del Sur del Ecuador. Ediciones Abya-Yala, Quito
Vásquez, R. 1997. *Flórula de las Reservas Biológicas de Iquitos, Perú*. Monographs in Systematic Botany Vol. 63, Missouri Botanical Garden, St. Louis.
Vásquez, Silvia Patricia Flores. 2014. Conhecimento, uso e conservação da diversidade vegetal em quatro comunidades ribeirinhas no Município Manacapuru, Amazonas. Doctoral dissertation, Programa de Pós-Graduação em Botânica, INPA, Manaus
Vásquez, Silvia Patricia Flores; Mendonça, Maria Silvia de; Noda, Sandra do Nascimento. 2014. Etnobotânica de plantas medicinais em comunidades ribeirinhas do Município de Manacapuru, Amazons, Brasil. Acta Amazonica 44(4):457–472
Veiga, J. B.; Scudeller, V. V. 2015. Etnobotânica e medicina popular no tratamento de malária e males asociados na comunidade ribeirinha Julião- baixo Rio Negro (Amazônia Central). Revista Brasileira de Plantas Medicinais 17(4):737–747
Veléz, Germán Alonso; Veléz, Antonio José. 1992. Sistema agroflorestal de "chagras" utilizado por las comunidades indígenas del medio Caquetá (Amazonia Colombiana). Colombia Amazonica 6(1): 101–134
Venturieri, G. A., W. A. Rodrigues, and J. M. T. Menezes. 1986. Sa Li Ki Su e Kwa Lha Ka Ta Su: duas fruteiras dos índios Nambiquara com potencial para domesticação. *Acta Amazonica*, Supl., vol. 16/17, pp. 19–25
Vieira, Thiago Almeida; Rosa, Leonilde dos Santos; Santos, Maria Marly de Lourdes Silva. 2012. Agrobiodiversidade de quintais agroflorestais no município de Bonito, Estado do Pará. Revista de Ciências Agrárias 55(3):159–166
Vogt, R. C. 2008. Amazon Turtles. Wust Ediciones, Lima
Waldhoff, D., U. Saint-Paich, and F. Bodo. 1996. Value of fruits and seeds from the floodplain forests of central Amazonia as food resource for fish. *Ecotropica* 2:143–156
WinklerPrins, Antoinette M. G. A. 2002. House-lot gardens in Santarém, Pará, Brazil: Linking rural with urban. Urban Ecosystems 6: 43–65
WinklerPrins, Antoinette M.G.A.; Souza, Perpetuo S. de. 2005. Surviving the city: urban home gardens and the economy of affection in the Brazilian Amazon. Journal of Latin American Geography 4(1):107–126
Wittmann, Florian; Schöngart, Jochen; Queiroz, Helder L.; Wittmann, Astrid de Oliveira; Conserva, Auristela dos Santos; Piedade, Maria T. F.; Kesselmeier, Jürgen; Junk, Wolfgang J. 2009. The Amazon floodplain demonstration site: sustainable timber production and management of central Amazonian white-water floodplains. Ecohydrology & Hydrobiology 9(1):41–54
Wittmann, F. W.; Schongart, Jochen; Brito, Joneide M. de; Wittman, Astride de Oliveira; Piedade, Maria Teresa 2010. Manual of Trees from Central Amazon Várzea Floodplains: Taxonomy, Ecology, and Use. INPA, Manaus
Yamada, M.; Osaqui, H. M. L. 2006. The role of homegardens in agroforestry development: lessons from Tomé-Açu, a Japanese-Brazilian settlement in the Amazon. In: Tropical Homegardens: a time-tested example of sustainable agroforestry, Kumar, B. M.; Nair, P. K. R. (eds), pp. 299–316, Advances in Agroforestry vol. 3, Springer, Dordrecht

Chapter 4
Apocynaceae

Dogbane or Milkweed Family

This is a large family, found in both the Old World and New. In Tropical America, some 100 genera are recognized encompassing 1500 species (Smith et al. 2004:23). A predominately tropical family, the Apocynaceae is probably better known for some of the species that enter the ornamental trade, such as yellow-flowered Allamanda that thrives in subtropical or even warm temperate areas, such as Florida. Members of this family exude a characteristic white latex when the fruits or bark is scratched, hence one of the common names for the family: milkweed. Nearly all taxa are toxic (Judd et al. 1999:369), but several species produce edible fruits, and at least six of them are consumed in the Amazon.

4.1 *Ambelania acida*

Brazil Pepino do mato, pepino da mata, pepino verde da mata, pepino doce, pepino braba, pepino bravo, molongó, bacuri de espinho; kamiča'ɨ (Guajá), kuyer'ɨpukuk, kuyeri-'ɨ (Ka'apor), momi hi (Yanomami)

Colombia Pubo (Ticuna)

Peru Cuchara caspi

Status Wild, spontaneous in cultural settings

N. Smith, *Amazon Fruits: An Ethnobotanical Journey*, Ethnobiology,
https://doi.org/10.1007/978-3-031-12803-5_4

Ambelania acida **in fruit. The tree arose spontaneously in a home garden from discarded seeds. Sitio do Baiano, Igarapé do Santarém, affluent of Rio Limão which in turns flows into the lower Maués, Amazonas, Brazil, 7-9-02**

Pepino do mato ("cucumber of the forest") fruits are gathered in mature and successionary forests on uplands across Amazonia and north into the Guianas (Alarcón and Peixoto 2007; Batista et al. 2015; Cavalcante 2010:207; Carim et al. 2007; Condé 2011:74; Mackensen et al. 2000; Medeiros 2016:91; Oliveira and Jardim 1998; Pires and Salomão 2000; Salomão et al. 2007; Santana et al. 2004; Prata et al. 2010; Scoles 2009). This small tree thrives in successionary growth as young as 12 years (Coelho et al. 2003b) and also thrives in savannas among rocky hills in the Guianas (Roosmalen 1985:22) and occasionally on floodplains, such as along the black water Caxiuanã and Anapu rivers in eastern Pará (Ferreira et al. 2013) and in sediment-rich waters in the Amazon estuary near Belém (Amaral et al. 2009; Silva 2008). It is also found occasionally in tidal forests in the Amazon estuary (Lau and Jardim 2013). Although pepino do mato is not strictly speaking a crop, it does arise spontaneously in home gardens from discarded seeds. The center of origin of this fruit tree is considered to be the Amazon estuary (Silva 2011:218).

Pepino do mato *Ambelania acida* (small tree in center) in a home garden. Sitio da Dona Rita, a few km upstream from Seringalzinho, Rio Jaú, Amazonas, Brazil 10-3-12

The yellow, cylindrical fruits, some 10 cm long by 4 cm wide, are ripe during the dry season and are relatively easy to pick because the understory tree does not usually exceed 18 m in height. The skin is removed before eating the pulp to avoid the sticky latex. The mesocarp is tart and crunchy with dark brown seeds the size of those passion fruit *Passiflora edulis*. The seeds are arranged in two parallel elongated chambers. In the mid-nineteenth century, the British botanist Richard Spruce tried the fruit along the Rio Negro and described it as "very milky, smelling like a ripe apple when cut across, sweet-tasted" (Spruce 1853a, b, c).

Girl eating pepino do mato *Ambelania acida* gathered in her home garden. Sitio do Baiano, Igarapé do Santarém, affluent of Rio Limão which in turns flows into the lower Maués, Amazonas, Brazil, 7-9-02

The trees are widely scattered and nowhere common, so fruits are usually consumed locally. Occasionally, a few fruits make their way to street markets, such as in Rio Preto da Eva along the Manaus-Itacoatiara Highway (AM-10) in Amazonas, Brazil. At the turn of the twentieth century, the fruits were sold from carts in Belém, the capital of Pará. The fruits were struck with a stick to soften the pulp (Huber 1904).

This species also contributes indirectly to the human diet because tapirs *Tapirus terrestris*, the largest game animal in Amazonia, also consume the fruit (Henry et al. 2000). In Amazonia and the Guianas, spider monkeys *Ateles paniscus*, capuchin monkeys *Cebus apella* and red brocket deer *Mazama americana* also eat the fruits of *Ambelania acida* and likely disperse some of the seeds (Gayot et al. 2004; Guillotin et al. 1994; Regan et al. 2001; Simmen and Sabatier 1996; Van Roosmalen and Klein 1988). Spider monkeys and red brocket deer are widely hunted for their meat.

Ambelania acida also has other uses. The Wayapí who live in several villages in southern French Guiana and contiguous Amapá, Brazil, use the wood to make spoons (Grenand 1992). Pepino do mato is used in some folk remedies. In the early twentieth century, for example, a French agronomist reported that the fruits were used for treating coughs (Le Cointe 1922:229). The Ka'apor in the eastern fringe of the Amazon forest in Maranhão drink the sweet white sap to treat sore throat (Balée 1993:108). The latex is swallowed in some areas to treat diarrhea.

4.2 *Ambelania grandiflora*

Brazil Olho de tucunaré, angelica do igapó, molongó, culhão de bode

Status Wild

Olho de tucunaré ***Ambelania grandiflora*** **fruits gathered in regrowth forest on an archaeological site with black earth and potsherds. Sitio Ponta Alegre, Baia de Caxiuanã, Rio Anapú, Pará, Brazil 5-14-15**

Known as the eye of the tucunaré *Cichla* spp. fish, this small tree found in mature and regrowth forests produces lime-sized fruits that are typically gathered by rural folk on the ground. It appears to be a relatively rare fruit because I have only encountered it once, in a floodplain forest along the black water Anapú River in Pará, in over 40 decades of fieldwork in Amazonia. Another common name for this tree in the Brazilian Amazon, angelica do igapó, confirms its preference for periodically flooded environments (Ducke 1922; Le Cointe 1945:192; Sampaio 1934:7), especially along black water rivers (Myster 2018). The Missouri Botanical Garden has only three specimens in its herbarium all from the Brazilian states of Amazonas, Maranhão, and Pará. The New York Botanical Garden has one specimen, while Kew has none. The fruits are consumed locally, usually as a snack during forest walks, and they never appear in markets.

4.3 *Couma macrocarpa*

Brazil Sorva, sorva da mata, sorva grande, sorva grande da mata, sorva açu, sorva preta, leite de sorva, sorvão, burra-leiteira, cumã açú; wídzurite (Baniwa), güëraimi (Desana), ngechi (Ticuna), ëtaimi (Tukano), kamnjeni (Waimiri-Atroari), kãnimari ãhi, operema axihi, omanama axihi (Yanomami)

Colombia Palo de leche, leche caspi, arbol leche, popa, perillo, perillo negro, pero, pendaré, juansoco, Juan zoco, Juan solo, cunuri, abichure, avichure, sorba, surba, surva, lirio, chicle, av purga, ichure; itá (Andoke), puóhoko (Cubeo), sejuco (Karijona), hwe (Maku) wãsoko (Makuna), múútsù'é, mësee (Miraña), mësëo (Muinane), tzuzi (Piapoco), wãwã (Tatuyo), uc cai, icucaú, ikikai, ikiki, do ko, ikna, iquina, pendare (Witoto), yuuchi (Yukuna)

Ecuador Wíshich (Achuar), fanscoa (Cofán), wíshich (Kichwa), wãnsoka, huasoca, sönquio märo (Siona-Secoya), opoca (Waorani)

Peru Leche huayo, leche caspi, lecheron, cashintoqui, perillo, chicle huayo, chicle caspi, moena, árbol de la leche, sorva, palo de vaca; wíshich (Achuar), dáum, úchi dáum (Aguaruna), musúhe (Bora), bimíon (Cashibo), këku (Matses), inumentsis, awa jonra, chicri jihui (Shipibo), ngexchi (Ticuna) icucaú (Witoto)

Venezuela Arbol de la vaca, chicle, caoutchouc, cajimán, pendare, umaiyek; jkalimëne jyaï, kalibele yaɨ (Hotï), kanamari, kanimari (Yanomami), kadimani, kanimari (Ye'kuana)

Status Wild, planted

Vendor holding leche huayo *Couma macrocarpa* fruits with sticky white sap characteristic of the Apocynaceae. Mercado Belén, Iquitos, Loreto, Peru 5-20-03

The common name for *Couma macrocarpa* in the Peruvian Amazon is leche huayo which means the "milk fruit" on account of the white latex that exudes from the fruit's skin. Similarly, one of the names for this sap-laden tree in Colombia is palo de leche (the "milk tree") and in Venezuela, it is known as arbol de la vaca ("cow tree"). The tangerine-sized fruits are yellow-green when ripe and are gathered by people in mature and regrowth forests on uplands. The Tikuna recount a legend of a woman, Djaê, who headed into the forest to gather the fruits of this tree on the ground (Pereira 1967:473). The Yanomami who live in scattered villages in the forests of Roraima, Brazil, and southern Venezuela also eat the fruits and appreciate the tree because it also provides food for the tapir *Tapirus terrestris* (Albert and Le Tourneau 2007; Albert and Milliken 2009:65), the largest terrestrial game animal in Amazonia. Tapirs are also known to eat this fruit in the nearby territory of the Yekwana in southern Venezuela (Salas and Fuller 1996) and in the Peruvian Amazon (Montenegro 2004:67). The Tatuyo, a Tukanoan group along the Upper Papuri in the Uaupés watershed in the Colombian Amazon, gather the fruits in the forest (Dufour 1988). Another indigenous group in the Colombian Amazon, the Makuna, also eat the seeds (Cayón 2002:82). A related species, *Couma utilis*, also bears edible fruits (Silverwood-Cope 1990:50).

The tree, which can attain 40 m, is found in forests throughout the Amazon Basin and occurs up to 900 m in the Andean foothills. It is also found in the Guianas, Venezuela, and the Pacific coast of Colombia north to Guatemala (Williams 1962). This tree occurs in upland forests as well as some riverine woodlands, such as along streams in seasonally flooded grasslands in the municipality of Faro in western Pará near the border with Amazonas state (Ducke 1910) along the lower Tefé River, an affluent of the Solimões in Amazonas (Assis et al. 2017), and river banks in the Colombian Amazon (Duarte 1979:31). *Couma macrocarpa* has also been encountered in 10-year-old successionary vegetation on uplands in the vicinity of Santa

Mercedes along the Rio Putumayo in northern Peruvian Amazon bordering Colombia (Gomez 2011:81; Muñoz 2010:22). This tree is encountered in a wide variety of soils, including sandy soils with caatinga vegetation, a stunted forest (Boubli 2002; Zárate et al. 2015). The Piaroa of the Upper Orinoco make special trips into the forest to gather the fruits during the mid-to-late rainy season (Zent 1998). The fruits are typically picked up from the forest floor on account of the tree's height. Known as sorva in the Brazilian Amazon, the tree fruits toward the end of the rainy season and through much of the dry period. This tree is occasionally planted in home gardens and the fruits sometimes turn up in street markets, especially in the Peruvian Amazon, and it is also found in some swidden fields in the municipality of Rio Preto da Eva along the Manaus-Itacoatiara highway, Amazonas (Silva 2017:29).

Some species of monkeys and parrots feed on the fruits and are involved in seed dispersal. Along the Apaporis and Caquetá rivers in the Colombian Amazon, for example, woolly monkeys *Lagothrix lagotricha* and mottled-face tamarins *Saguinus inustus* eat the fruits of *Couma macrocarpa* (Defler and Defler 1996; Izawa 1975; Palacios et al. 2004). The saddle-back tamarin *Saguinus fuscicollis* and the moustached tamarin *Saguinus mystax* are known to disperse the seeds of this tree in the Peruvian Amazon (Culot et al. 2009; Noriega 2015:19; Sánchez and Levy 2013:27, 54). Woolly monkeys have also been documented eating these fruits in forests near Tefé in Amazonas, Brazil (Peres 1994) and the Peruvian Amazon (Montalván 2015:38; Voss and Fleck 2011). In the Colombian and Peruvian Amazon, spider monkeys *Ateles belzebuth*, capuchin monkeys *Cebus apella*, bald uakaris *Cacajao calvus*, coppery titis *Callicebus cupreus*, monk sakis *Pithecia monachus*, equatorial sakis *Pithecia aequatorialis*, and squirrel monkeys *Saimiri sciureus* have been observed feeding on these fruits as well (Bowler and Bodmer 2011; Gómez-Posada 2012; Montalván 2015:38; Voss and Fleck 2011). Spider monkeys, woolly monkeys, capuchin monkeys, coppery titis, bald uakaris, and sakis are hunted for their meat in various parts of the Amazon Basin, but squirrel monkeys are generally spared because they are so small. The bald uakari *Cacajao calvus ucayalii* also eats the fruits of *C. macrocarpa* and this red-faced monkey with brown fur is hunted for its meat in the Peruvian Amazon (Aquino 1995; Aquino and Encarnación 1999). In the Upper Rio Negro in Amazonas, Brazil, near the border with Venezuela black-headed uakaris, *Cacajao melanocephalus melanocephalus* also eat the fruits (Boubli 1999). The Santarém marmoset *Callithrix humeralifera*, the saddle-back tamarin *Saguinus fuscicollis,* and the moustached tamarin *Saguinus mystax* also eat the fruits of *C. macrocarpa* (Peres 1993; Stevenson and Rylands 1988); these monkeys are rather small and not usually hunted for their meat but they are sometimes kept as pets and are likely involved in dispersing the seeds.

The latex of *Couma macrocarpa* is as important to people as the fruits. In the nineteenth century and the first half of the twentieth century, rural people would tap sorva trees to sell the latex which was used to caulk canoes and boats. Some of the liquid latex was sent to some towns and cities, such as Manaus. In the 1940s, the latex was tapped in the Colombian Amazon to make chewing gum (Schultes 1966), but nowadays most chewing gum is made using synthetic material. Sorva tappers in the vicinity of Lago Amanã along the lower Japurá used to trek through the forest in the northeastern part of that enormous black water lake to reach the headwaters of the Unini. The latex was then sent down the Unini in canoes to the Rio Negro and thence by boat to Manaus. Seventy-three-year-old Otilho, a resident in the village of Ubim on the shores of Lago Amanã, recalled using the leaves of an epiphytic member of the Cyclanthaceae, known locally as ubim parasita, to collect the dripping latex of sorva trees. Ubim parasita is so named because it resembles *Geonoma* (several species of palm used for thatch and known as ubim in Brazil) and because it grows like a "parasite" on trunks on branches.

Ubim parasita (Cyclanthaceae) on the trunk of a fallen tree in successionary forest. The river dweller, Otilho 73 years, is explaining that the leaves were once used to collect the dripping latex from sorva *Couma macrocarpa* trees. Igarapé Ubim, Lago Amanã, Amazonas, Brazil 9-17-12

In 2012, I interviewed Lucas, an 81-year-old former sorva tapper at Vila Valente near Tefé. He recalled collecting sorva latex in the watershed of the Japurá River, a north bank affluent of the Solimões. He often cut down sorva trees with an ax to bleed out the latex. He employed the leaf bases (*curumatá*) of acai *Euterpe precatoria* palms to collect the dripping latex because *curumatá* is like a miniature dugout canoe. A sorva tree can allegedly produce up to three cans (*latas* with an 18-l capacity) of latex. He also used to spare some trees and obtain the latex by cutting (*anelar*) the bark. One reason that the market for sorva dried up was because too many gatherers were cheating by diluting the sorva latex with white sap from other trees, such as Amapá or garote. Sorva gatherers worked year-round, even in the rainy season, because the higher water levels allowed them to penetrate deeper into the forest along streams. Another river dweller at Vila Valente noted that a latex gatherer was in a sorva tree along the Juruá River in the 1930s when he was attacked by a harpy eagle *Harpia harpyja*. Sorva trees can attain immense size, soaring up to 60 m, and the upper branches are evidently used by harpy eagles to build their nests.

Old tapping scars on the trunk of a sorva *Couma macrocarpa* tree. The latex was once collected to caulk boats. Vila Valente, lower Rio Tefé, Amazonas, Brazil 8-15-12

The latex has a sweet taste and in the past was stirred into water and drunk, such as in the Municipality of Faro in Pará (Ducke 1910). This mixture was also boiled to make coffee or to mix with manioc flour or bananas (Le Cointe 1934:411). The latex also has medicinal value. River dwellers in the watershed of the Jaú, an affluent of the middle Rio Negro in Amazonas, Brazil, prepare an infusion of the sticky latex to treat diarrhea (Rodrigues 2006) as do rural people in the vicinity of Iquitos in the Peruvian Amazon (Ramírez and Vásquez 1998). The Andoke in the Colombian Amazon give one or two spoonfuls of this latex to children suffering from diarrhea (La Rotta 1983:31). I have recorded the Sateré-Maué of the Maués Valley in Amazonas doing the same. Because the Sateré-Maué consume the fruits and use the latex of sorva, they spare the tree when clearing forest for their fields. Woodsmen (*mateiros*) in the Brazilian Amazon also drink the latex to stop their intestines when suffering from undisciplined bowel movements as do the Aguaruna in the Peruvian Amazon (Jernigan 2006:194). Along the Unini River, an affluent of the Rio Negro in Amazonas, Brazil, rural folk use latex to prepare a remedy for lung problems (Lago et al. 2016). In the Cauca Valley of Colombia, the latex is used to treat youngsters suffering from diarrhea. In the Peruvian Amazon, the latex-rich bark is used to make a tea to treat diarrhea (Jovel et al. 1996); the bark is tapped to obtain the latex and a tablespoon of this sticky white liquid is also taken with each meal to treat the same condition (Lossio et al. 1995:168). In various parts of the Peruvian Amazon, the leaves and branches are used to treat protozoan infestations (Rojas et al. 2009). Because of its medicinal value, this tree is sometimes planted in agroforestry fields, such as on uplands near Iquitos (Coomes and Burt 1997).

Sorva *Couma macrocarpa* tree that has been gashed to collect the latex to treat diarrhea. Eight-year-old Sateré-Maué girl, Igarapé Aruã, Urupadi River an affluent of the Maués, Amazonas, Brazil, 7-10-02

The latex of *Couma macrocarpa* is sweet and evidently less toxic than other species in the Apocynaceae. For example, large quantities of latex from *C. macrocarpa* were gathered in the Peruvian Amazon in the 1930s and 1940s to make chewing gum, but this mini extractive boom came to an end with the advent of synthetic gum made from petroleum (Chibnik 1994:61). In the nineteenth century, the Yorkshire botanist Richard Spruce recorded the use of latex of several species of *Couma* to coat manioc graters along the Içana River, a tributary of the Upper Rio Negro in Amazonas, Brazil (Field 1996). The latex is also used to prepare paint among indigenous groups along the Vaupés River in the Colombian Amazon. The Yanomami of northern Roraima and contiguous Venezuela once mixed latex of *C. macrocarpa* with ground-up charcoal to paint their bodies black before setting off to raid enemies; they still use this latex to attach large bird feathers, such as from the king vulture *Sarcoramphus papa*, as well as feather down to their hair during certain festivals (Albert and Milliken 2009:111; Lizot 1978). The Waimiri-Atroari north of Manaus, Amazonas, boil the latex of this tree and mix it with the yellow sap of *Symphonia globulifera* to form solid cylindrical tablets. Fibers are then passed over these tablets to coat them with wax before using them to attach flight feathers to arrows (Milliken et al. 1992:29).

Tukano house with a bench made from sorva *Couma macrocarpa* wood. Serra do Mucura, Rio Tiquié, affluent of the Uaupés, Amazonas, Brazil 10-28-12

Couma macrocarpa is the preferred wood for making low benches among the Tukano along the Tiquié, an affluent of the Uaupés in the northwestern part of the Brazilian Amazon. In most Tukano villages, furniture is sparse and traditionally this ethnic group, which lives in scattered villages, has used the benches for sitting down. The wood is also used in home construction in other parts of the Brazilian Amazon, but it does not appear to be a high-value timber tree (Loureiro et al. 1979:125).

Tukano girl sitting on a bench made with *Couma macrocarpa* wood. Matapí, Rio Tiquié, affluent of the Uaupés, Amazonas, Brazil 10-27-12

4.4 *Hancornia speciosa*

Bolivia Mangaba; xabá motoha (Chácobo)

Brazil Mangaba, mangabeira, mangabeira mansa, mangava, mangava mansa, fruta de doente; peny (Apinajé), piawa (Juruna), catúa (Kalapalo), pi-ô-tire (Kayapó), Pêni (Kisêdjê), katuga (Kuikuru), katêkiçú, shikiúligosi (Nambikwara), pey (Panará), kathyola (Paresi), magaua (Tapirapé), ritó (Xavante)

Status Wild, planted

Girl selling mangaba *Hancornia speciosa* fruits that she has picked in a nearby scrub savanna. Joanes, Marajó Island, Pará, Brazil 9-7-97

The thin-skinned fruits of mangaba are gathered from contorted, and often fire-blackened, trees on upland savanna in the Amazon estuary and patches of scrub savanna in other parts of the Amazon Basin such as in Amapá (Sanaiotti et al. 1997), in the upper watersheds of the Tocantins, Araguaia, and Xingu Rivers (Coimbra et al. 2002; Oliveira et al. 2018), and the watershed of the Madeira River (Braun and Ramos 1959; Ducke and Black 1954; Ferreira 2009:53). Mangaba is also found in the headwaters of the Madeira River in Bolivia and extends as far south as northern Paraguay and west to the Madre Dios river basin in Peru. This small tree occurs in both cerrado of central Brazil, including areas bordering the Amazon forest, and pockets of upland savanna in the Amazon (Amaral et al. 2019; Braga 1979; Costa et al. 2014; Felfili et al. 2002; Guarim Neto and Pasa 2009; Guarim Neto 1984; Guarim Neto et al. 1994; Kubitzki 1979; Kuhlmann 1954; Lemos 2013:35; Lenza et al. 2015; Marimon and Lima 2001; Miranda et al. 2006; Pereira and Santos 2015; Pinheiro et al. 2018; Pires and Prance 1985; Ratter 1987; Ratter et al. 1996). It is most common, however, in Brazil, especially on river banks along the coast from Amapá and Pará south to Espírito Santo (Lorenzi et al. 2006a, b:60; Martins et al. 2012a, b; Pereira et al. 2006). It also occurs in the understory of some gallery forests in the cerrado region of Mato Grosso (Marimon et al. 2002; Stefanello et al. 2009).

The soft fruits, about the size of golf balls, are easily gathered from the low branches. The fruits are relatively easy to harvest because the tree is generally small, even shrubby, and rarely attains 10 m in height, although a few trees reach 15 m tall. The fruits are nutritious because they are rich in iron, manganese, zinc, and vitamin C (Cardoso et al. 2014; Pereira et al. 2006). However, the fruit skins are fragile so they are not commonly encountered in urban markets in the Amazon.

The fine, feathery leaves of mangaba remain a bright lime green, even at the height of the dry season. When the relentless trade winds of summer and the searing equatorial sun have exacted their toll on the grasslands of Marajó, the light, yet shimmering green leaves of mangaba provide welcome visual relief to a landscape dominated by yellow and tan hues. Mangaba trees are undaunted by the periodic fires set by ranchers to improve forage or hunters after muçuã (*Kinosternon scorpioides*) turtles in the summer; their charred trunks and shiny leaves contrast vividly with the deep blue sky, devoid of any quenching clouds.

Mangaba is used in various folk remedies. The leaves of mangaba are used in various parts of Brazil to treat diabetes (Pereira et al. 2015a, b) and extracts from the leaves have been proven to help heal wounds (Geller et al. 2015). Rural folk in Quilombo Sangrador in the municipality of Presidente Juscelino, Maranhão, use the leaves of this tree to make tea for treating genital and urinary issues (Monteles and Pinheiro 2007). In the town of Nova Xavantina in Mato Grosso, the inner bark is soaked in water and then the liquid is drunk to treat anorexia and hypolipidemic conditions, including high levels of cholesterol in the blood (Silva et al. 2010). In the community of Bom Jardim some 60 km south of Cuiabá in Mato Grosso locals also drink this liquid to treat ulcers (Pasa 2011). In the rural community of Rio dos Couros, 60 km from Cuiabá, the sap is mixed with water and sugar and drunk to treat hepatitis (Costa et al. 2017). Some residents of Imperatriz in Maranhão close to the border of eastern Pará prepare a decoction with the bark of this tree to treat inflammation and stomach ulcers (Penido et al. 2016). Laboratory tests with mice and rats in Brazil have revealed that the latex of this tree has anti-inflammatory properties (Marinho et al. 2011).

Young cowboy snacking on a mangaba *Hancornia speciosa* fruit. Fazenda Anjos, Marajó Island, Pará, Brazil 12-17-01

The pale yellow skin is peeled to snack on the sweet, cream-colored pulp. About half a dozen green seeds, the size of orange seeds, only flatter, are embedded in the sticky mesocarp. However, some of the larger fruits can contain up to 42 seeds and in Mato Grosso rural people leave some of the fruits so that animals can disperse the seeds and maintain these wild orchards on cerrado landscapes (Gonçalves et al. 2013). In the southern pantanal of Mato Grosso do Sul the red-footed tortoise *Chelonoidis carbonaria* eats this fruit and disperses some of the seeds (Wang et al. 2011; this tortoise also occurs in the Amazon Basin where it also most likely disperses some of the seeds of this tree. At least three-quarters of the fruits are edible pulp (Moura et al. 2002) which increases its potential for eventually becoming an orchard fruit. The vitamin C content of this fruit is similar to that of oranges (Moura et al. 2002). The fruits are generally eaten fresh but are also made into ice cream with a pear-like flavor, such as in Salinas, Pará. The fruits are also used to make a fermented drink on Marajó and nearby islands (Lisboa 2012:194). In the sixteenth century, Ambrósio Fernandes Brandão, a Portuguese sugarcane plantation owner and merchant who worked along the coast of Brazil, described mangaba as "one of the best fruits in the world" (Brandão 1987:217). Yet nearly four centuries later, few people outside of Brazil have ever tasted the fruit.

João Betendorf, a Jesuit missionary who traveled extensively in eastern Amazonia during the latter half of the seventeenth century, noted that mangaba trees were dense enough on certain stretches of savannas on Marajó to be referred to as groves (*mangabeiras*). Father Betendorf, who was born in Luxemburg and lived to almost a hundred years old, noted that the mangaba fruits from Marajó were especially fine (Betendorf 1910:25). In the mid-eighteenth century, João Daniel, another Jesuit and astute observer of the customs and natural history of the region, noted that one of the missions on Marajó, Aldea das Mangabeiras, was named after the fruit "because of the abundance of the fruit" (Daniel 1976:69). Aldea das Mangabeiras has since been renamed Ponta das Pedras and has emerged as a thriving entrepôt for açaí (*Euterpe oleracea*) palm fruit. In the 1880s, Joanes, a fishing village on Marajó's southern coast and former garrison for Portuguese troops during the colonial period when it was known as Monforte, dispatched mangaba fruits across Marajó Bay to Belém.

In the Amazon estuary, mangaba has two fruiting seasons, one during the wetter months from February to July (called *safra de inverno*), and the other during the drier months from September to December. The fruits produced during the latter period (*safra de verão*) are especially welcome because few other wild fruits suitable for snacking are available at that time.

Mangaba fruits are sufficiently popular that some farmers have planted the tree in their home gardens in non-flooded savanna areas of Marajó, such as near Vila Camará, in the vicinity of Macapá, the capital of Amapá (Pereira et al. 2017), and in the southern fringe of Amazonia in parts of Mato Grosso (Mamede and Pasa 2018). These farmers are carrying on an ancient practice started by indigenous peoples long ago. Natives have planted numerous mangaba trees on sandy grasslands (campos) that intersperse forest in the interfluve between the Tapajós and Xingu, and east of the Tocantins River toward the fringe of the Amazon forest in Maranhão, thus forming island orchards for use when on hunting and gathering trips (Frikel 1978). The Camayurá of the Upper Xingu plant the seeds of *Hancornia speciosa* in the vicinity of their villages because the short tree produces edible fruit and a useful latex (Oberg 1953:23). The Nambicuara of northern Mato Grosso, who rely extensively on hunting and gathering for their livelihood, also depend on semi-wild mangaba orchards (Agostinho 1996). Several indigenous groups undertake expeditions on the savannas of central Brazil for the sole purpose of picking mangaba fruits (Lévi-Strauss 1952). Populations of mangaba in the cerrado region of Brazil where the tree is more common are shrinking due to human-induced habitat change (Ganga et al. 2010) and if the gene pool of this species continues to shrink options for developing this plant into a widespread crop will decline.

Mangaba *Hancornia speciosa* trees in the home garden of a rural house. Road leading to Vila Camará from the Foz do Rio Camará-Salvaterra road, Marajó Island, Pará, Brazil 5-12-99

Some locals gather the fruits for urban markets on Marajó Island as well as Belém. The fruits are generally sold in lots of about 20, contained in baskets woven from the petiole strips of caraná (*Mauritiella armata*) palm fronds and are lined with the banana-like leaves of *Heliconia* or arumã (*Ichnosiphon obliquus*) cut in successionary vegetation and forest. Trade in mangaba fruits has been going on in the Amazon estuary for over a century. In 1885, for example, a government official noted that Joanes along the southern coast of Marajó Island shipped out mangaba fruits, evidently destined for Belém (Baena 1885).

Mangaba *Hancornia speciosa* fruits packed in cylindrical baskets for sale on a roadside cart. Boca do Camará, Marajó Island, Pará, Brazil 12-16-01

Mangaba also contributes indirectly to the human diet because certain game animals consume the fruits. Both the white-lipped peccary *Tayassu pecari* and the white-collared peccary *Pecari tajacu* eat mangaba fruits when they fall to the ground (Desbiez et al. 2009; Prado 2013). The white-lipped peccary roams widely in bands of up to 150 or so, whereas the white-collared peccary occupies smaller territories and typically tends to occur in much smaller bands. Peccaries are among the most important game animals in the Amazon (Smith 1976). Mangaba is an obligate outcrosser and several species of native bees pollinate the flowers (Darrault and Schlindwein 2005) so this plant also contributes indirectly to the diet of rural people when they gather honey from the nests of wild bees

Mangaba is relatively widespread in savannas of Brazil and it has served a variety of uses, including a type of rubber. The Apinajé, who inhabit the forests and scrub savanna in the interfluvial uplands near the confluence of the Tocantins and Araguaia Rivers in Pará, used to tap the trees for the latex by gashing the bark with a stone chisel. The latex was collected in bowls fashioned from the calabash *Crescentia cujete* gourd and then used to make balls (Nimuendajú 1956). In the middle of the twentieth century, the Pareci in Mato Grosso in the southern fringes of the Amazon Basin used to make rubber balls by coating a concave piece of wood with mangaba latex. The Pareci used their fingers to press the edges of the membrane together. Air was then blown into the ball through a small hole which was patched with a thin membrane of the latex. Several additional coatings of the latex give strength to the ball. The Nambicuara also make rubber balls from the latex of this plant (Lévi-Strauss 1952). Also, mangaba rubber bands that the women wore around their legs were made on a cylindrical piece of wood (Métraux 1963).

In the early 1900s, the sticky latex of mangaba was used to trap small birds in Pará for the pet trade (Lustosa 1976:290). During the rubber boom in the latter part of the nineteenth century and early twentieth century, wild mangaba on the savannas of Brazil was tapped for its latex and was known on world markets as Pernambuco rubber (Hoehne 1923:26; Wisniewski and Melo 1982). European and North American firms purchased mangaba rubber again during the Second World War when Japanese forces cut off supplies of rubber (*Hevea brasiliensis*) from plantations in Southeast Asia (Polhaumus 1962:57). Today, the tree is no longer tapped on a commercial scale because Southeast Asia has once again flooded world markets with relatively cheap, high-grade natural rubber.

Mangaba surfaces occasionally in indigenous lore. The Kamayurá of the Upper Xingu watershed, for example, recount a myth about the origins of the Sun and Moon, in which mangaba seeds are used to whiten the teeth of the twins (Agostinho 1974:162; Villas Boas and Villas Boas 1975:58).

4.5 *Lacmellea arborescens*

Brazil Tucujá, mongoló, visgueira, sorva, sorvinha, sorva miuda, grão de galo, pau de leite, pau de colher, caramurizinho da capoeira, pipini rana, pepino do mato, cumaí, caramuri, guajaraí, fruta bombom, apuí amarelo; nua (Sateré-Maué)

Colombia Chicle, pepa de chicle, pega pega; sodanie (Miraña), cuurëo (Muinane), gemacacu (Yukuna)

Peru Chicle, chicle huayo, chicle caspi, bellaquilla caspi, palo chicle, chirique

Status Wild

Jefferson, 8 years, gathering fruits of mongoló *Lacmellea arborescens* in regrowth forest. Anjo da Guarda, Rio Pracupi, affluent of the Anapú River, FLONA Caxiuanã, Pará, Brazil 5-15-15

Tucujá *Lacmellea arborescens* grows in a variety of habitats including mature upland forest such as north of Manaus (Carneiro 2004; Oliveira and Mori 1999; Oliveira et al. 2008), along the middle and lower Xingu (Campbell et al. 1986; Lemos et al. 2015), and in the vicinity of Alter do Chão and Belterra along the lower Tapajós (Amaral et al. 2017; Sandel and Carvalho 2000). It is particularly common in upland successionary vegetation, especially along the lower Xingu near its confluence with the Amazon, and also occurs occasionally on floodplains as noted on several accession sheets in the herbarium of the Missouri Botanical Garden. It also occurs occasionally in stunted woods (*campinarana*) growing on patches of upland sandy soils, such as in the watershed of the Uatumã northeast of Manaus, Amazonas (Targhetta et al. 2015) as well as *campinarana* subject to seasonal flooding such as in Parque Nacional do Viruá in Roraima (Vale 2011:40). It is also found in stunted woodlands growing on iron-rich sedimentary rock known as *canga* at the Serra dos Carajás in Pará (Mota et al. 2018). This small tree is also found in some cacao agroforestry plots in the municipality of São Félix do Xingu in southern Pará (Braga et al. 2019).

The shrub to a small tree, which can reach 10 m in height, is used for two purposes: the fruit and the stem. The small yellow fruits, 1.5–2 cm by 0.8–1.5 cm, have a very thin translucent pulp surrounding a single brown seed that measures 1.3 cm × 0.8 cm. In spite of the slim pickings in terms of pulp, youngsters enjoy snacking on the slippery mesocarp. In the Caquetá watershed in the Colombian Amazon mottled-face tamarins *Saguinus inustus* have been observed feeding on the fruits (Palacios et al. 2004). Mature trees can reach 30 m high, so sometimes youngsters have to shimmy up the trunk or climb neighboring trees to obtain the fruits.

Arlinho, 10 years, picking tucujá *Lacmellea arborescens* fruit in regrowth vegetation arising after a manioc field was abandoned. Inland from Carrazedo, lower Xingu River, Municipio de Gurupá, Pará, Brazil 4-21-15

Lacmellea arborescens ranges widely in South America from Argentina through Bolivia, Peru, Colombia, and Brazil. In western Amazonia, it reaches up to about 1000 m in the Andean foothills. The abundance of common names for the tree in the Brazilian states of Amazonas and Pará, many obtained from herbarium specimens at the Missouri Botanical Garden (viewable via Tropicos), suggest that *L. arborescens* is used widely, especially for its fruit.

Geraldo, 7 years, eating mongoló *Lacmellea arborescens* fruit gathered in regrowth forest. Anjo da Guarda, Rio Pracupi, affluent of the Anapú River, FLONA Caxiuanã, Pará, Brazil 5-15-15

In addition to the fruit, *Lacmellea arborescens* has two other uses. Boys cut the young stems to fashion shafts for small gigs (*zagaianas*) that they use for spearing shrimp. A small metal trident is attached to one end of the shaft to jab shrimp from elevated walkways in the village of Carrazedo and from canoes. Shrimp are caught this way during the day and at night boys employ flashlights which make the shrimp easy to spot because their eyes glow in the beam of light. Adults gather the sticky white sap of *L. arborescens* and place it on branches or leaf stems of trees, especially açaí *Euterpe oleracea* to catch parakeets (mainly *Brotogeris* spp.) and curió (*Sporophila* spp.) which they then adopt or sell as pets.

Jamieson, 10 years, gathering stems of tucujá *Lacmellea arborescens* in successionary vegetation adjacent to a manioc field. He will use the stems to make gigs for catching shrimp. Near Carrazedo, lower Xingu River, Pará, Brazil 4-13-15

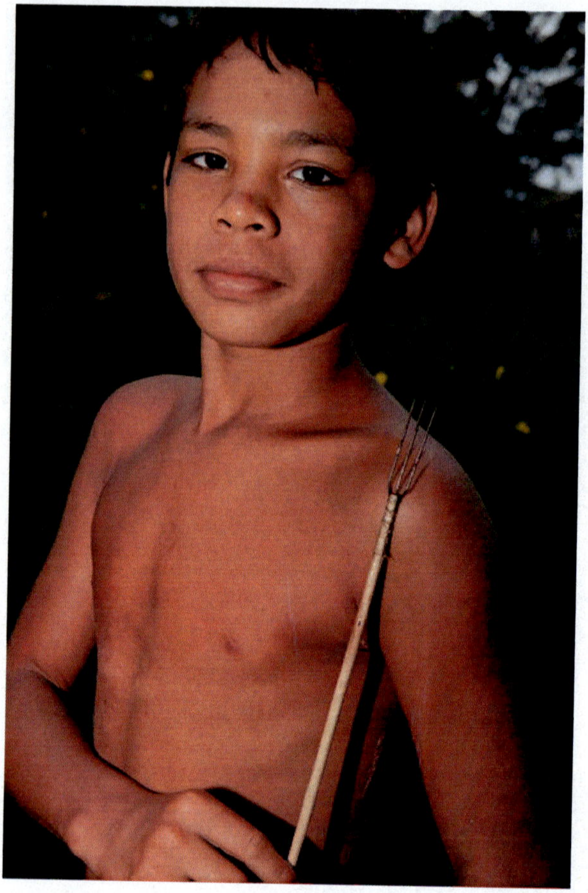

Rodrigo, 10 years, with small gig (*zagaiana*) he fashioned from a tucujá *Lacmellea arborescens* stem. He will use it to jab shrimp. Carrazedo, lower Xingu River, Pará, Brazil 4-13-15

4.6 *Lacmellea lactescens*

Brazil Tamanqueira, chicle de espinho, espinho, chico preto, sorvinha

Colombia Palo espino, chicle, chicle espinoso, chicle con espinas, juansoco con espinas, juansoquillo, bacurí; ho-ho-haru-ki (Cubeo), sodanie (Miraña), cuurëo, kuyio (Muinane)

Ecuador Chicle, ciruela; chiripi sijica, tssitssincho (Cofán), casha bimbichu, bimbicho, viti avio (Kichwa), tauch (Shuar), wëpëmöka (Waorani)

Peru Chicle, chicle huayo, chicle huayo con espinas, chicle caspi, leche guayo; jhankitausch (Aguaruna), ankí táuch (Shuar)

Status Wild

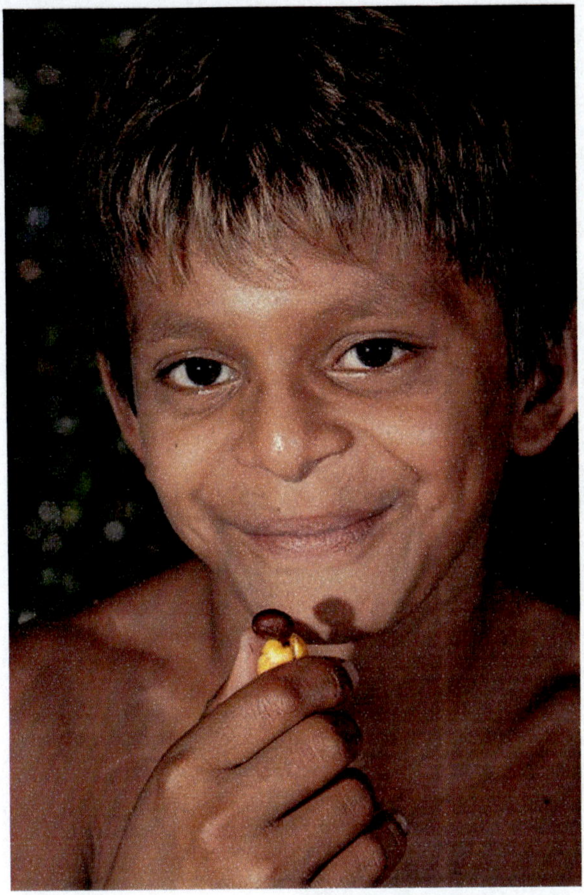

Tamanqueira *Lacmellea lactescens* **fruit gathered from a forest floor. Lago Miwá near Codajás, Rio Solimões, Amazonas, Brazil 6-24-07**

Lacmellea lactescens is a denizen of forests, including successionary woodlands, on uplands and the higher parts of floodplains in central and western Amazonia up to 600 m in the Andean foothills (Mendoza and Guerrero 2010; Santana et al. 2004). The tree can attain 35 m in height, so the yellow-skinned fruits are usually gathered from the forest floor toward the end of the rainy season and early part of the dry period. However, the Cofán in the Ecuadorian Amazon often cut the tree down to obtain the fruits (Cerón 1995:71).

The round to oblong fruits, 2–2.5 cm long by 1.5–1.75 cm in diameter, contain a single, brown seed measuring 1.5 cm by 1 cm. Although the seed is covered by a stingy layer of brown mesocarp, the fruits are abundant and typically picked up on the forest floor to snack on the spot. The fruits have so little pulp that they are not taken to market, but hundreds of fruits may litter the forest floor by the tree's studded (aculeated) trunk, similar in appearance to that of young *Hura crepitans* trees. So the fruits make up in quantity for what they lack in generosity of pulp. Evidently, tamanqueira has evolved to provide just the right amount of reward to entice dispersal agents, such as monkeys and birds, without "investing" too much physiological effort in producing large amounts of pulp per fruit. The pulp is mealy but sweet. A youngster on his way to gather açaí (*Euterpe precatoria*) palm fruits with his father and brother in the forest near his home on the shores of Miwá near Codajás compared the flavor of tamanqueira fruits to honey ("*melzinho*").

A 10-year-old snacking on tamanqueira *Lacmellea lactescens* fruits that have fallen on the floor of upland forest. Lago Miwá, near Codajás, Rio Solimões, Amazonas, Brazil 6-24-07

The fallen fruits are gathered by several indigenous groups, including the Huaorani in the Ecuadorian Amazon (Ima 2012:94). The Huaorani also use the milky latex gathered by making a cut on the trunk and take it to treat stomachache. According to rural inhabitants around Lago Amanã on the lower Japurá, some game animals also feed on the fallen fruits, including brocket deer (*Mazama* spp.) and white-lipped peccary (*Tayassu pecari*).

A related species to tamanqueira, *Lacmellea arborescens*, known as tucajá in the Brazilian Amazon, also has edible fruits that although parsimonious with regards to pulp nevertheless are produced in sufficient quantities to provide a "good meal" (Cavalcante 2010). Other species in the genus *Lacmellea* are consumed by rural people, including native peoples. For example. *L. aculeata* is an important wild fruit for the Wajãpi in Amapá, Brazil, and contiguous areas of French Guiana (Grenand 1992), and fruits of *L. peruviana* are gathered by the Achuar in the Ecuadorian Amazon (Descola 1994:73).

4.7 *Pacouria boliviensis*

Brazil Taquari da mata

Peru Paáwian (Achuar), daek pau (Aguaruna)

Status Wild

Taquari da mata ***Pacouria boliviensis*** **fruit gathered in an old secondary forest. The fruit fell from the thick liana behind Marcela, 10 years. Outskirts of Santo Antonio de Ipapucu, Lago de Tefé, Amazonas, Brazil 8-27-12**

Taquari da mata (*Pacouria boliviensis*) is a thick liana that grows in secondary and mature forests throughout Amazonia and reaches the northeastern region of Brazil (Viana et al. 2017). It produces green fruits the size of large grapefruits that fall to the ground when ripe. The fruits are nearly round, 10.5 cm × 10 cm, and have eight seeds, each of which measures 3.5 cm × 2.5 cm. The yellow pulp has a pleasant flavor, though is a bit gummy due to the latex (*leite*). In spite of the mesocarp's sticky quality of the pulp, people snack on the pulp when they encounter the fallen fruits on their way to their fields or the forest.

Taquari da mata *Pacouria boliviensis* **fruit. Outskirts of Santo Antonio de Ipapucu, Lago de Tefé, Amazonas, Brazil 8-27-12**

4.8 *Parahancornia peruviana*

Colombia Juansoquillo

Peru Naranja podrido, naranjo podrido, porotillo, caucho mashan, amapa; sieca ñi (Maijuna), diden këku (Matses)

Status Wild

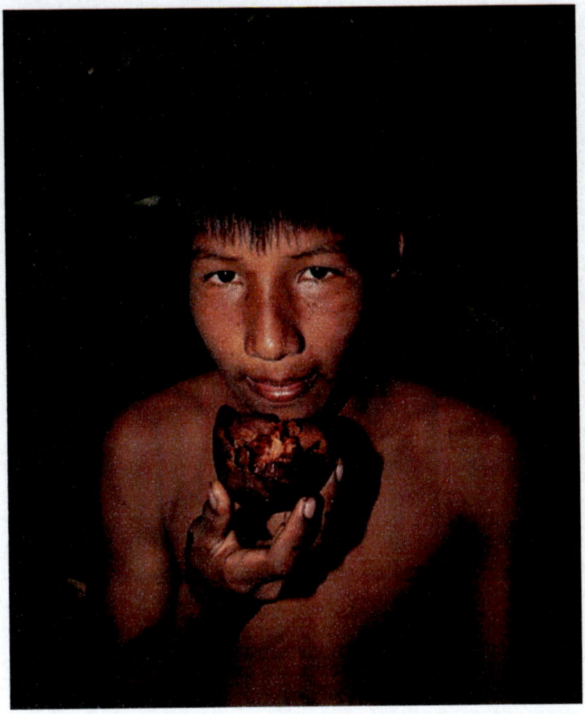

Boy snacking on naranja podrido *Parahancornia peruviana* fruit in upland forest. Near Miraflores, Lower Rio Tigre, Loreto, Peru 6-12-04

At first glance, the fruits of naranja podrido (rotten orange) are not particularly appealing since they are murky brown with a thick, woody skin. The tan pulp surrounding the 100 or so watermelon-sized seeds is not plentiful, but rural folk in the vicinity of Iquitos in the Peruvian Amazon are fond of the fruits because the mesocarp has a fermented flavor, like a rich liqueur. This understory tree fruits throughout the rainy season and is found in upland forests of western Amazonia in Amazonas and Acre states in Brazil as well as Peru and Colombia. It is also found in periodically flooded woodlands, such as near Iquitos in the Peruvian Amazon (Roucoux et al. 2013), as well as successionary forests on uplands, such as along the Putumayo River which borders Peru and Colombia (Gomez 2011:79). Woolly monkeys *Lagothrix lagotricha* and *L. poeppigii*, spider monkeys *Ateles belzebuth*, bald uakaris *Cacajao calvus ucayalii*, monk sakis *Pithecia monachus*, equatorial sakis *Pithecia aequatorialis*, and coppery titis *Callicebus cupreus* are known to eat the fruits of this tree and these monkeys are hunted for their meat in various parts of the Amazon Basin (Aquino 1995; Montalván 2015:38; Voss and Fleck 2011). The smaller saddle-back tamarin *Saguinus fuscicollis* also eats this fruit (Montalván 2015:39) but this monkey is not considered a game animal because of its small size. These monkeys also likely disperse some of the seeds.

Chopping down a naranja podrido *Parahancornia peruviana* tree in the upland forest to obtain the fruits. Near Miraflores, Lower Rio Tigre, Loreto, Peru 6-12-04

Naranja podrido is typically chopped down to feast on the fruits because the tree can grow as high as 40 m. The cutting down of trees to obtain fruits is not an uncommon practice in Amazonia. Anecdotal information from river dwellers along the Puinahua River, a branch of the Ucayali, and at the mouth of the Tigre River in Loreto, Peru, suggests that this customary way of harvesting the fruit has led to a growing scarcity of the tree. Indeed, the Peruvian botanist Rodolfo Vasquez and the late Alwyn Gentry note in their valuable analysis of wild fruit consumption in the Iquitos area that fruits of naranjo podrido no longer appear in that large urban market (Vásquez and Gentry 1989). Nevertheless, such destructive harvesting pales in comparison to the devastation wrought by the large-scale clearing of forest for cattle pasture and soybean farms, particularly common in the Brazilian Amazon.

4.9 *Tabernaemontana angulata*

Brazil Visgo, jasmin da mata, jasmin do mato, jasmin bravo, jasmin de cachorro, pimenta de lontra, picoró, pocoró, caratapê; ameno-wai-'I (Araweté), kagwaruhu-mɨra, kujeri-'iran-'y (Ka'apor), yapukuliwa (Wayãpi), akiama hi, akiã hi (Yanomami)

Status Wild

Luana Brazão de Araújo, 12 years, collecting visgo *Tabernaemontana angulata* flowers in successionary vegetation for use as earrings. Vila Santa Cruz, lower Caxiuanã River, Município de Melgaço, Pará, Brazil 7-9-16

Tabernaemontana angulata is a denizen of regrowth vegetation on uplands in the Amazon Basin such as in the watershed of the Tocantins River and in the Zona Bragantina region of Pará (Araújo et al. 2005; Brienza Júnior 1999:27; Coelho et al. 2003a, b, 2012; Jardim et al. 2013; Rayol et al. 2011; Rodrigues et al. 2007a, b). It can survive for at least 40 years in successionary forests (Rayol et al. 2006). It also occurs as a weed in some cattle pastures cleared in forests, such as in eastern Pará (Modesto Júnior and Mascarenhas 2001). This species grows in a wide range of habitats including grasslands, stunted vegetation on sandy soils (*campinarana*) and on sand dunes along the northeastern coast of Pará (Bastos et al. 1995). It also occurs occasionally in floodplain forests such as near Belém, Pará (Cain et al. 1956), along the Jaci-Paraná River in Rondônia (Reis 2010:30), and in southeastern Colombia (Cárdenas et al. 2009). Visgo can grow into a small tree in tall forests but likely succumbs to the deep shade at some point. It is endemic to the Amazon, where it is more common in the eastern part of the basin, and also occurs sporadically in the Guianas.

The fruits of *Tabernaemontana angulata* are eaten as a snack while walking through successionary growth, especially near fields, but the main use, at least by children in the Caxiuanã National Forest in Pará, are the flowers to make toy earrings. Known locally as visgo, the elongated white flowers are attached to earlobes using the plant's sticky white latex. The flowers are easy to reach because visgo is a shrub, usually under a meter in height. Inhabitants of Marudá, a village along the coast of Pará, employ the latex to make a paste which is applied to the body to alleviate back pain, pulled muscles, and abscesses. The fruits are also used to make a tea for treating coughs and catarrh, while a tea is prepared from the flowers to alleviate asthma (Coelho-Ferreira 2009). The Yanomami who inhabit several villages in southern Venezuela and contiguous areas in the northern states of Amazonas and Roraima in Brazil drink an infusion prepared from the bark of this plant to treat intestinal roundworms (Milliken and Albert 1996).

Geovana Araújo de Oliveira, 12 years, with visgo earring. Vila Santa Cruz, lower Caxiuanã River, Município de Melgaço, Pará, Brazil 7-9-16

4.10 *Tabernaemontana sananho*

Brazil Sanango, abiu bravo, molongo; guarecolo (Uaicá-Mucajaí), toru hwãtemo hi (Yanomami)

Colombia Sanango de lobo, sanango de altura

Ecuador Bolsa de perro, bolsa de mono, manzana, cunapip, pepa de leche; tssát-tecco, tsatu'cco (Cofán), tzicta, tsiczta, tsicta yura, stzicta muyo, cunapip, sikta yura, tsicta yura, tzikta (Kichwa), kunápik, kúnapi, tsektár, sticta, guajek (Shuar), pa-su'u, paisu'uhui'to; bai-su-u (Siona-Secoya), bepemunga, beyegohue, paigonka, pegunca, penoncahue, penunga, tenunca (Waorani)

Peru Sanango, sanango de bajo, ucho sanango, chiri sanango, sanango macho, lobo sanango, tahuampa sanango, jaen sanango, yacu sanango, patohuare; kúnakip (Aguaruna), patohuare' (Amuesha), tóómecóóriu (Bora), shi'nianpi (Chayahuita), mantách, tsiikat (Mayna)

Venezuela Árbol de la vaca, guaceja; ibarkádek (Pemón), waseja (Ye'kuana)

Status Wild

Tabernaemontana sananho **in fruit in a home garden. Yarina, Rio Yanayacu, Reserva Nacional Pacaya-Samiria, Loreto, Peru 6-23-06**

Tabernaemontana sananho is an understory tree in mature and successionary forests on both floodplains and uplands throughout Amazonia. Tahuampa sanango (swamp sanango), one of the common names for *T. sananho* in the department of Loreto in Peru, suggests its propensity for wetland habitats. The small tree is normally found in lowlands, but it occurs as high as 2000 m in the Andes, such as in northern Ecuador. Widespread in tropical parts of northern South America, including the Orinoco Basin and central and western parts of Amazonia, *T. sananho* also extends into Panama.

Floodplain habitat of sanango *Tabernaemontana sananho* captured in a drawing by Wilson Tello, 16 years, Veinte de Enero, Yanayacu River, Reserva Nacional Pacaya-Samiria, Loreto, Peru April 2006

The yellow fruits, measuring 3.5–5 cm long by 3–4 cm wide, ripen toward the end of the rainy season. The fruits are borne in clusters of two to nine, although technically they are carpels of a single fruit. Rural folk have a keen sense of humor as evidenced by some common names for the fruit. In Ecuador, for example, sanango is referred to as the testicles of monkeys or dogs (bolsa de mono, bolsa de perro). One of the vernacular names for the fruit in the Peruvian Amazon is sanango macho. The thick skin has three angular ridges and covers a layer of some 20 seeds. The dark reddish-brown seeds are grooved and are typically spat out after consuming the pulp.

Sanango *Tabernaemontana sananho* **fruits. Latex flows when the skin is broken. Yarina, Rio Yanayacu, Pacaya-Samiria, Loreto, Peru 4-23-06**

The exceptionally sweet pulp is located underneath the layer of pointed seeds, rather than distributed among the seeds. The evolutionary significance of this arrangement is unclear, but the seeds are likely dispersed by monkeys and birds. One of the common names for the tree in the Peruvian Amazon, lobo sanango (sanango de lobo in the Colombian Amazon), suggests that wild foxes or bush dogs (*Speothos*) also disperse the seeds. In southern Venezuela, tapirs *Tapirus terrestris*, the largest terrestrial game animal in the Amazon and Orinoco Basins, also eat the fruits and disperse the seeds (Salas and Fuller 1996). Forest tortoises (*Chelonoidis* spp.) are known to feed on the fruits and likely also disperse the seeds (Cerón and Ayala 1998:34). Throughout the Amazon tortoises are gathered for their meat and are sometimes kept in pens in the yards of rural homes.

This 10-year-old river dweller has latex on his lips from eating sanango *Tabernaemontana sananho* fruits that he has gathered in a forest. Near Veinte de Enero, Rio Yanayacu, Pacaya-Samiria, Loreto, Peru 6-28-06

The fruits are gathered mostly by pulling down the lower branches during the rainy months and into the early part of the dry season. Sanango can reach 20 m, but normally does not exceed 10 m, so branches can usually be reached easily from the ground. Furthermore, the tree starts producing fruit when only 1.5 m high, so the fruits can be easily picked when the trees are young. The high sugar content of the pulp would make this tree a promising candidate for domestication, especially since it thrives on disturbed sites in wetlands. The fruits of two closely related species, pampa sanango (*Tabernaemontana siphilitica*) and sanango de bajo (*T. markgrafiana*) are also consumed in the Pacaya-Samiria area of the Peruvian Amazon. The latter species, a shrub in successionary forest on floodplains and occasionally on uplands, has a much wider distribution, stretching from Amapá and Pará states in Brazil westward to Bolivia, Peru, Ecuador, and northward to Colombia and the Guianas.

Boys gathering sanango *Tabernaemontana sananho* fruits in floodplain forest on the outskirts of their village. Lower Rio Pucate, affluent of the Yanayacu River, Pacaya-Samiria, Loreto, Peru 6-4-04

Characteristic of the Apocynaceae, the fruit stalks and skin of *T. sananho* exude a sticky white latex. In the early nineteenth century, the German explorer Eduard Poeppig noted that hunters along the Huallaga River in the Peruvian Amazon rubbed the sap of a related species, *Tabernaemontana montana*, on the noses of hunting dogs to sharpen their sense of smell (Poeppig 2003:281). Hunters along the Upper Pastaza in the Ecuadorian Amazon still maintain this custom, using *Tabernaemontana sananho*. The Waorani in the Ecuadorian Amazon rub the latex of this tree into the noses of their dogs so that they will put on weight (Cerón and Ayala 1998:34). The Waorani also use the wood for home construction and for fuel. The Yanomamo in Roraima and bordering forest in Venezuela pound the leaves of *T. sananho* with those plucked from tingui *Clibadium sylvestre*, a cultivated shrub, for use as a piscicide in streams (Albert and Milliken 2009:70).

Tabernaemontana sananho is also employed in a variety of folk remedies, especially in the Peruvian Amazon. The Chayahuita, for example, make a powder from the dried bark and apply it as a poultice on inflamed skin and wounds (Odonne et al. 2013). They also apply the sticky latex to cuts and irritated skin. The Amuesha in the Andean foothills in the central Peruvian Amazon grate the roots and then boil them; once cooled a small glass of this solution is drunk daily to treat anemia, flu, tonsillitis, and rheumatism. This indigenous group also rubs the solution over the legs to reduce swelling (Valadeau et al. 2010). The Yanesha who live in the central Andean foothills of the western Amazon Basin in Peru boil the bark and drink the solution to treat anemia and swollen legs (Valadeau et al. 2009). In other parts of the Peruvian Amazon, the plant is used as a febrifuge, emetic, diuretic, and calmative (Van Beek et al. 1984). A tepid decoction of the leaves and bark is used to wash wounds and then the powdered bark is put over the wound; this is repeated three times a day for 10 days. A water extract of the root is drunk three times a day for 15 days to treat rheumatism and wounds. In the Chazuta district in the Huallaga valley in the foothills of the Andes in the Peruvian Amazon, the root and bark are macerated in freshwater which is trunk to treat rheumatism and intestinal parasites (Sanz-Biset and Cañigueral 2011). In the same part of the Peruvian Amazon, the leaves are mixed with the stems and leaves of *Fittonia albivenis* Acanthaceae and crushed to make a poultice which is applied to snake bites (Sanz-Biset et al. 2009). Fresh leaves of this tree are heated over a fire and then placed directly on areas of the body affected by rheumatism (Mejia and Rengifo 2000:180).

In the Ecuadorian Amazon, the Kichwa use the bark to make a topical application for treating leishmaniasis (Gachet et al. 2010) and to make an infusion to alleviate fever and diarrhea (Cerón 1996). In the Brazilian Amazon, the Ticuna, who live along the Solimões, put scrapings from the tree's bark into sugarcane alcohol (*aguardiente* in Spanish-speaking countries, *cachaça* in Brazil) and drink it to treat rheumatism; alternatively, the leaves are used to make a lukewarm bath to treat the condition. The Ticuna also dilute the latex of this small tree in water to treat eye wounds (Schultes and Raffauf 1990:76–77). In the central part of the Peruvian Amazon, the bark is used to make a decoction for drinking as part of the initiation process for folk healers (Jauregui et al. 2011).

References

Agostinho, P. 1974. Kwarìp: Mito e Ritual no Alto Xingu. Editora da Universidade de São Paulo, São Paulo

Agostinho, P. 1996. Os Nambikwara em 1975: transferências e situação. Magalhães, S. B., R. C. Britto, and E.R. Castro (Editors), *Energia na Amazônia*, Museu Paraense Emílio Goeldi/ Universidade Federal do Pará/Associação de Universidades Amazônicas, Belém, vol. 2, pp. 631–669

Alarcón, Juan Gabriel Soler; Peixoto, Ariane Luna. 2007. Florística e fitossociologia de um trecho de um hectare de floresta de terra firme, em Caracaraí, Roraima, Basil. Boletim do Museu Paraense Emílio Goeldi, Ciencias Naturais 2(2):33–60

Albert, Bruce; Le Tourneau, François-Michel. 2007. Ethnogeography and resource use among the Yanomami: toward a model of "reticular space". Current Anthropology 48(4):584–592

Albert, Bruce; Milliken, William. 2009. Urihi A: a terra-floresta Yanomami. Instituto Socioambiental (ISA)/Institut de Recherche pour le Devéloppement, São Paulo

Amaral, Dário Dantas do; Vieira, Ima Célia Guimarães; Almeida, Samuel Soares de; Salomão, Rafael Paiva; Silva, Antonio Sérgio Lima da; Jardim, Mário Augusto Gonçalves. 2009. Checklist da flora arbórea de remanescentes florestais da região metropolitana de Belém e valor histórico dos fragmentos, Pará, Brasil. Boletim do Museu Paraense Emílio Goeldi, Ciencias Naturais 4(3):231–289

Amaral, Iêda Leão do; Magnusson, William Ernest; Matos, Francisca Dionízia de Almeida; Albernaz, Ana Luisa K.; Feitosa, Yuri Oliveira; Guillaumet, Jean-Louis. 2017. Disentangling structural patterns of natural forest fragments in a savanna matrix in the eastern Brazilian Amazon. Acta Amazonica 47(2):111–122

Amaral, Dário Dantas do; Rocha, Antônio Elielson; Pereira, Jorge Luis Gavina; Costa Neto, Salustiano Vilar. 2019. Identificação dos subtipos de savanas na Amazônia oriental (Pará e Amapá, Brasil) com uma chave dicotômica de individualização. Boletim do Museu Paraense Emílio Goeldi, Ciencias Humanas Naturais 14(2):183–195

Aquino, Rolando. 1995. Conservacion de *Cacajao calvus ucayalii* en la Amazonia Peruana. Neotropical Primates 3(2):40–42

Aquino, Rolando; Encarnación, Filomeno. 1999. Observaciones preliminares sobre la dieta de *Cacajao calvus ucayalii* en el nor-oriente Peruana. Neotropical Primates 7(1):1–5

Araújo, Maristela Machado; Tucker, Joanna Marie; Vasconcellos, Joanna Marie; Zarin, Daniel Jacob; Oliveira, Wilson; Sampaio, Patrícia Delamônica; Rangel-Vasconcelos, Lívia Gabrig; Oliveira, Francisco de Assis; Coelho, Roberta de Fatima Rodrigues; Aragão, Débora Veiga; Miranda, Izildinha. 2005. Padrão e processo sucessionais em florestas secundárias de diferentes idades na Amazônia oriental. Ciencia Florestal, Santa Maria 15(4):343–357

Assis, Rafael Leandro de; Wittmann, Florian; Luize, Bruno Garcia; Haugaasen, Torbjon. 2017. Patterns of floristic diversity and composition in floodplain forests across four southern Amazon river tributaries, Brazil. Flora 229:124–140

Baena, M. 1885. *Informações sobre as Comarcas da Provincia do Pará organizadas em virtude do Aviso Circular do Ministerio da Justiça de 20 de setembro de 1883, por Manoel Baena Director da 2a Secção da Secretaria da Presidencia da mesma Provincia.* Typ. de Francisco da Costa Junior, Belém.

Balée, W. 1993. *Footprints of the Forest: Ka'apor Ethnobotany, the Historical Ecology of Plant Utilization by an Amazonian People.* Columbia University Press, New York

Bastos, M. N. C.; C. S. Rosário; L. C. B. Lobato. 1995. Caracterização fitofisionômica da restinga de Algodoal, Maracanã, Pará, Brasil. Boletim do Museu Paraense Emílio Goeldi, Série Botanica 11(2): 173–197

Batista, Anderson Pedro Bernardina; Aparício, Wegliane Campelo da Silva; Santos, Vanessa Silva dos; Lima, Robson Borges de; Mello, José Marcio de. 2015. Caracterização estrutural em uma floresta de terra firme no estado do Amapá, Brasil. Pesquisa Florestal Brasileira 35(81):21–33

Betendorf, J.F. 1910. Chronica da missão dos padres da Companhia de Jesus no Estado do Maranhão. *Revista do Instituto Historico e Geographico Brazileiro* 72(1): 1–682

Boubli, J. P. 1999. Feeding ecology of black-headed uacaris (*Cacajao melanocephalus melanocephalus*) in Pico da Neblina National Park, Brazil. International Journal of Primatology 20(5):719–749

Boubli, J. P. 2002. Lowland floristic assessment of Pico da Neblina National Park, Brazil. Plant Ecology 160:149–167

Bowler, Mark; Bodmer, Richard E. 2011. Diet and food choice in Peruvian red uakaris (*Cacajao calvus ucayalii*): selective or opportunistic seed predation? International Journal of Primatology 32:1109–1122

Braga, Pedro Ivo Soares. 1979. Subdivisão fitogeográfica, tipos de vegetação, conservação e inventário florístico da floresta amazónica. Acta Amazonica 9(4) supl.:53–80

Braga, Daniel P. P.; Domene, Frederico; Gandara, Flávio B. 2019. Shade trees composition and diversity in cacao agroforestry systems of southern Pará, Brazilian Amazon. Agroforestry Systems 93:1409–1421

Brandão, A. F. 1987. *Dialogues of the Great Things of Brazil* (*Diálogos das Grandezas do Brasil*). University of New Mexico Press, Albuquerque

Braun, E. H.; Ramos, J. R. 1959. Estudo agrogeológico dos campos Puciari-Humaitá, Estado do Amazonas e território Federal de Rondônia. Revista Brasileira de Geografia 21(4):443–497

Brienza Júnior, Silvio. 1999. Biomass dynamics of fallow vegetation enriched with leguminous trees in the eastern Amazon of Brazil. Doctoral dissertation, Georg August University, Göttingen, Germany

Cain, Stanley A.; Castro, Gustavus M. de Oliveira; Pires, J. Murça; Silva, Nilo Tomâs. 1956. Application of some phytosociological techniques to Brazilian rain forest. American Journal of Botany 43(10):911–941

Campbell, David G.; Daly, Douglas C.; Prance, Ghillean T.; Maciel, Ubirajara N. 1986. Quantitative ecological inventory of terra firme and várzea tropical forest on the Rio Xingu, Brazilian Amazon. Brittonia 38(4):369–393

Cárdenas, Dairon; Castaño, Nicolás; Sua, Sonia. 2009. Flora de la estrella fluvial de Inírida (Guianía, Colombia). Biota Colombiana 10(1–2):1–30

Cardoso, Leandro de Morais; Reis, Bárbara de Lazzari; Oliveira, Daniela da Silva; Pinheiro-Sant'Ana, Helena Maria. 2014. Mangaba (Hancornia speciosa Gomes) from the Brazilian cerrado: nutritional value, carotenoids and antioxidant vitamins. Fruits 69:89–99

Carim, Samyrams; Schwartz, Gustavo; Silva, Manoela Fernandes Ferreira da. 2007. Riqueza de espécies, estrutura e composição florística de uma floresta secundária de 40 anos no leste da Amazônia. Acta Botanica Brasilica 21(2):293–308

Carneiro, Vilany Matilla Colares. 2004. Composição florística e análise estrutural da floresta primária de terra firme na bacia do Rio Cuieiras, Manaus-AM. Master's Thesis, Universidade Federal do Amazonas (UFAM)/Instituto Nacional de Pesquisas da Amazônia (INPA), Manaus

Cavalcante, P. B. 2010. *Frutas Comestíveis na Amazônia*. Museu Paraense Emílio Goeldi, Belém

Cayón, Luis 2002. En las Aguas de Yuruparí: Cosmología y Chamanismo Makuna. Ediciones Uniandes, Estudios Antropológicos No. 5, Bogotá

Cerón, Carlos Eduardo. 1995. Etnobiologia de los Cofanes de Dureno, Provincia de Sucumbíos, Ecuador. Museo Ecuatoriano de Ciencias Naturales, Herbario Nacional, Serie Monografia 10(3), Quito

Cerón, Carlos Eduardo. 1996. Etnobotanica Quichua en la vía Hollín-Loreto, Provincia del Napo. Hombre y Ambiente 25:131–170

Cerón, Carlos Eduardo; Ayala, Consuelo Montalvo. 1998. Etnobotanica de los Huaorani de Quehueiri-ono Napo, Ecuador. Abya-Yala, Quito

Chibnik, M. 1994. *Risky Rivers: The Economics and Politics of Floodplain Farming in Amazonia*. University of Arizona Press, Tucson

Coelho, Roberta de Fátima Rodrigues; Zarin, Daniel Jacob; Miranda, Izildinha Souza; Tucker, Joanna Marie. 2003a. Análise florística e estrutural de uma floresta em diferentes estágios sucessionais no município de Castanhal, Pará. Acta Amazonica 33(4):563–582

Coelho, Roberta de Fátima Rodrigues; Zarin, Daniel Jacob; Miranda, Izildinha Souza; Tucker, Joanna Marie. 2003b. Ingresso e mortalidade em uma floresta em diferentes estágios sucessionais no município de Castanhal, Pará. Acta Amazonica 33(4):619–630

Coelho, Roberta de Fátima Rodrigues; Miranda, Izildinha Souza; Mitja, Danielle. 2012. Caracterização do processo sucessional no Projeto de Assentamento Benfica, sudeste do estado do Pará, Amazônia oriental. Boletim do Museu Paraense Emílio Goeldi, Ciencias Naturais 7(3):251–282

Coelho-Ferreira, Márlia. 2009. Medicinal knowledge and plant utilization in an Amazonian coastal community of Marudá, Pará state (Brazil). Journal of Ethnopharmacology 126:159–175

Coimbra, C. E. A.; Flowers, Nancy N.; Salzano, Francisco M.; Santos, Ricardo V. 2002. The Xavánte in Transition: Health, Ecology, and Bioanthropology in Central Brazil. University of Michigan Press, Ann Arbor.

Condé, Tiago Monteiro. 2011. Avaliação dos impactos da vegetação após a exploração madeireira em floresta ombrófila de terra firme no Município de Caracaraí, RR. Master's thesis, Programa de Pós-Graduação em Recursos Naturais, Universidade Federal de Roraima, Boa Vista

Coomes, O. T.; Burt, G. J. 1997. Indigenous market-oriented agroforestry: dissecting local diversity in western Amazonia. Agroforestry Systems 37:27–44

Costa, Reginaldo Brito da; Silveira, Ediléia Patrícia da; Souza, Thaianny Rodrigues de. 2014. Estrutura da vegetação de cerrado sensu stricto em terra indígena no noroeste do estado de Mato Grosso. Multitemas 46: 95–110

Costa, Ineilian B. C. da; Bonfim, Filipe P. G.; Pasa, Maria C.; Montero, Daniel A. V. 2017. Ethnobotanical survey of medicinal flora in the rural community Rio dos Couros, state of Mato Grosso, Brazil. Boletin Latinoamericano y del Caribe de Plantas Medicinales y Aromáticas 16(1): 53–67

Culot, Laurence; Huynen, Marie-Claude; Gérard, Paul; Heymann, Eckhard W. 2009. Short-term post-dispersal fate of seeds defecated by two small primate species (Saguinus mystax and Saguinus fuscicollis) in the Amazonian forest of Peru. Journal of Tropical Ecology 25(3):229–238

Daniel, J. 1976. *Tesouro Descoberto no Rio Amazonas*. Biblioteca Nacional, Rio de Janeiro, vol. 1

Darrault, Reisla O.; Schlindwein, Clemens. 2005. Limited fruit production in Hancornia speciosa (Apocynaceae) and pollination by noctural and diurnal insects. Biotropica 37(3):381–388

Defler, Thomas R.; Defler, Sara B. 1996. Diet of a group of *Lagothrix lagothricha lagothrica* in southeastern Colombia. International Journal of Primatology 17(2):161–190

Desbiez, A. L. J., S. A. Santos, A. Keuroghlian, and R. E. Bodmer. 2009. Niche partitioning among white-lipped peccaries (*Tayassu pecari*), collared peccaries (*Pecari tajacu*), and feral pigs (*Sus scrofa*). *Journal of Mammalogy* 90(1):119–128

Descola, P. 1994. *In the Society of Nature: A Native Ecology in Amazonia*. Cambridge University Press, Cambridge

Duarte, Luis Enrique Acero. 1979. Principales Plantas Utiles de la Amazonia Colombiana. Instituto Geografico "Augustin Codazzi", Bogotá

Ducke, A. 1910. Explorações scientificas no Estado do Pará. Boletim do Museu Goeldi (Museu Paraense) de Historia Natural e Ethnographia 7:100–200

Ducke, A. 1922. Plantes nouvelles ou peu connues de la région amazonienne (II partie). Archivos do Jardim Botanico do Rio de Janeiro 3:3–269.

Ducke, A. and G. Black. 1954. *Notas sôbre a Fitogeografia da Amazônia Brasileira*. Instituto Agronômico do Norte, Belém, Boletim Técnico 29

Dufour, D. L. 1988. The composition of some foods used in northwest Amazonia. Interciencia 13(2):83–86

Felfili, Jeanine Maria; Nogueira, Paulo Ernane; Silva Júnior, Manoel Cláudio da; Marimon, Beatriz Schwantes; Delitti, Welington Braz Carvalho. 2002. Composição florística e Fitossociologia do cerrado sentido restrito no município de Água Boa, MT. Acta Botanica Brasilica 16(1):103–112

Ferreira, Carlos Alberto Cid. 2009. Análise comparativa de vegetação lenhosa do ecosistema campina na Amazonia Brasileira. Doctoral dissertation, INPA/UFAM, Manaus

Ferreira, Leandro V.; Chaves, Priscilla P.; Cunha, Denise de A.; Matos, Darley C. L.; Parolin, Pia. 2013. Variação da riqueza e composição de espécies da comunidade de plantas entre as florestas de igapós e várzeas na Estação Científica Ferreira Penna-Caxiuanã na Amazônia oriental. Pesquisas Botânica 64:175–195

Field, D. V. 1996. Richard Spruce's economic botany collections at Kew. In: *Richard Spruce (1817–1893): Botanist and Explorer*, M. R. D. Seaward and S. M.D. Fitzgerald (editors), Royal Botanic Gardens, Kew, pp. 246–264

Frikel, P. 1978. Áreas de arboricultura pré-agrícola na Amazônia: Notas preliminares. *Revista de Antropologia* (São Paulo) 21 (1): 45–52

Gachet, María Salomé; Lecaro, Javier Salazar; Kaiser, Marcel; Brun, Reto; Navarrete, Hugo; Muñoz, Ricardo A.; Bauer, Rudolf; Schühly, Wolfgang. 2010. Assessment of anti-protozoal activity of plants traditionally used in Ecuador in the treatment of leishmaniasis. Journal of Ethnopharmacology 128:183–197

Ganga, Rita Maria Devós; Ferreira, Gislene Auxiliadora; Chaves, Lázaro José; Naves, Ronaldo Veloso; Nascimento, Jorge Luiz do. 2010. Caracterização de frutos e árvores de populações naturais de Hancornia speciosa Gomes do cerrado. Revista Brasileira de Fruticultura 32(1):101–113

Gayot, Marc; Henry, Olivier; Dubost, Gérard; Sabatier, Daniel. 2004. Comparative diet of the two forest cervids of the genus *Mazama* in French Guiana. Journal of Tropical Ecology 20(1):31–43

Geller, Fabiana Cristina; Teixeira, Marina Rodrigues; Pereira, Ana Bárbara Dias; Dourado, Luana Pereira Antunes; Souza, Danielle G.; Braga, Fernão Castro; Simões, Cláudia Maria Oliveira. 2015. Evaluation of the wound healing properties of Hancornia speciosa leaves. Phytotherapy Research 29:1887–1893

Gomez, Tedi Pacheco. 2011. Valoración económica de purmas de diferentes edades en la comunidad nativa de Santa Mercedes, río Putumayo, Perú. Master's thesis, UNAP, Iquitos

Gómez-Posada, C. 2012. Dieta y comportamiento alimentario de un grupo de mico maicero *Cebus apella* de acuerdo a la variación en la oferta de frutos y artrópodos, en la Amazonía colombiana. Acta Amazonica 42(3):363–372

Gonçalves, Laissa Gabrielle Vieira; Andrade, Fabrício Ribeiro; Marimon Junior, Ben Hur; Schossler, Thiago Rodrigo; Lenza, Eddie; Marimon, Beatriz Schwantes. 2013. Biometria de frutos e sementes de mangaba (Hancornia speciosa Gomes) em vegetação natural na região leste de Mato Grosso, Brasil. Revista de Ciências Agrárias 36(1):31–40

Grenand, P. 1992. The use and cultural significance of the secondary forest among the Wayapi Indians. In: *Sustainable Harvest and Marketing of Rain Forest Products*, M. Plotkin and L. Famolare (Editors), Island Press, Washington, DC, pp. 27–40

Guarim Neto, Germano. 1984. Plantas do Brasil: Angiospermas do Estado de Mato Grosso- 1. Rodriguésia 36(59):105–121

Guarim Neto, Germano; Pasa, Maria Corette. 2009. Estudo etnobotânico em uma área de cerrado no município de Acorizal, Mato Grosso. Flovet 1:5–32

Guarim Neto, Germano; Guarim, Vera Lucia M. S.; Prance, Ghillean T. 1994. Structure and floristic composition of the trees of an area of cerrado near Cuiabá, Mato Grosso, Brazil. Kew Bulletin 49(3):499–509

Guillotin, M.; Dubost, G.; Sabatier, D. 1994. Food choice and food competition among the three major primate species of French Guiana. Journal of Zoology 233:551–579

Henry, Olivier; Feer, Francois; Sabatier, Daniel. 2000. Diet of the lowland tapir (*Tapirus terrestris* L.) in French Guiana. Biotropica 32(2):3364–368

Hoehne, F. C. 1923. *Phytophysionomia do Estado de Matto-Grosso e Ligeiras Notas a Respeito da Composição da sua Flora*. Companhia Melhoramentos, São Paulo

Huber, J. 1904. Notas sobre a patria e distribuição geographica das arvores fructiferas do Pará. *Boletim do Museu Goeldi (Museu Paraense) de Historia Natural e Ethnographia* 4 (2–3):375–406

Ima, M. O. 2012. *Saberes Waorani y Parque Nacional Yasuní: Plantas, Salud y Bienestar en la Amazonía del Ecuador*. Iniciativa Yasuní ITT, Ministerio Coordinador de Patrimonio, Ministerio del Ambiente, Programa de las Naciones Unidas para el Desarrollo (PNUD) y Fondo para el Medio Ambiente Mundial (FMAM), Quito

Izawa, Kosei. 1975. Foods and feeding behavior of monkeys in the upper Amazon basin. Primates 16(3):295–316

Jardim, Daniel Gonçalves; Jardim, Mário Augusto G.; Quaresma, Adriano Costa; Costa, Salustiano Villar. 2013. Regeneração natural em formações florestais de uma unidade de conservação, Maracanã, Pará, Brasil. Biota Amazônia 3(2):79–87

Jauregui, X.; Clavo, Z. M.; Jovel, E. M.; Pardo-de-Santayana, M. 2011. "Plantas con madre": plants that teach and guide in shamanic initiation process in the east-central Peruvian Amazon. Journal of Ethnopharmacology 134:739–752

Jernigan, Kevin Arthur. 2006. An ethnobiological exploration of sensory and ecological aspects of tree identification among the Aguaruna Jívaro. Doctoral dissertation, University of Georgia, Athens

Jovel, E. M.; Cabanillas, J.; Towers, G. H. N. 1996. An ethnobotanical study of the traditional medicine of the mestizo people of Suni Miraño, Loreto, Peru. Journal of Ethnopharmacology 53: 149–156

Judd, W. S., Campbell, C. S., Kellogg, E. A., Stevens, P. F. and Donoghue, M. J. 1999. Plant systematics: a phylogenetic approach. Ecología mediterránea, 25(2), p.215.

Kubitzki, K. 1979. Ocorrência de *Kielmeyera* nos "campos de Humaitá" e a natureza dos "campos"—flora da Amazônia. Acta Amazonica 9(2):401–404

Kuhlmann, Edgar. 1954. A vegetação de Mato Grosso—seus reflexos na economia do estado. Revista Brasileira de Geografia 16(1): 77–122

La Rotta, C. 1983. Observaciones Etnobotanicas sobre algunas Especies Utilizadas por la Comunidad Indigena Andoque (Amazonas, Colombia). Universidad Nacional de Colombia, Bogotá

Lago, João Henrique G.; Tezoto, Jaqueline; Yazbek, Priscila B.; Cassas, Fernando; Santos, Juliana de F. L.; Rodrigues, Eliana. 2016. Exudates used as medicine by the "caboclos river-dwellers" of the Unini River, AM, Brazil--classification based in their chemical composition. Revista Brasileira de Farmacognosia 26:379–384

Lau, Arthur Viana; Jardim, Mário Augusto Gonçalves. 2013. Florística e estrutura da comunidade arbórea em uma floresta de várzea na área proteção ambiental, Ilha do Combu, Belém, Pará. Biota Amazônia 3(2):88–93

Le Cointe, P. 1922. *L'Amazonie Brésilienne: Le Pays-ses Habitants, ses Resources, Notes et Statistiques jusqu'en 1920*. Augustin Challamel, Paris, vol 2

Le Cointe, P. 1934. *Amazônia Brasileira III*: *Árvores e Plantas Úteis*. Livraria Classica, Belém

Le Cointe, P. 1945. O Estado do Pará: a Terra, a Água e o Ar, a Fauna e a Flora Mineraes. Companhia Editora Nacional, São Paulo

Lemos, Helena Lara. 2013. Vegetação arbustivo-arbórea em áreas de cerrado típico e cerrado rupestre no estado de Tocantins. Master's thesis, Departamento de Engenharia Florestal, Universidade de Brasília, Brasília

Lemos, D. A. N.; Ferreira, B. G. A.; Siqueira, J. D. P.; Oliveira, M. M.; Ferreira, A. M. 2015. Floristic and phytosociology in dense "terra firme" rainforest in the Belo Monte hydroelectric plant influence area, Pará, Brazil. Brazilian Journal of Biology 75(3):257–276

Lenza, Eddie; Santos, Josias O.; Maracahipes-Santos, Leonardo. 2015. Species composition, diversity, and vegetation structure in a gallery forest-cerrado sensu stricto transition zone in eastern Mato Grosso, Brazil. Acta Botanica Brasilica 29(3):327–338

Lévi-Strauss, C. 1952. The use of wild plants in tropical South America. *Economic Botany* 6(3):252–270

Lisboa, P. L. B. 2012. *A Terra dos Aruã: Uma História Ecológica do Arquipélago do Marajó*. Museu Paraense Emílio Goeldi, Belém

Lizot, Jacques 1978. Connaissance et usage des plantes sauvages chez les Yanomami. In: Unidad y Variedad: Ensayos Antropológicos en Homenaje a José M. Cruxent, Wagner, Erika; Zucchi, Alberta (eds), IVIC, Caracas, pp 129–171

Lorenzi, Harri; Bacher, Luis; Lacerda, Marco; Sartori, Sergio. 2006a. Frutas Brasileiras e Exóticas Cultivadas. Instituto Plantarum de Estudos da Flora Ltda, Odessa, Sao Paulo.

Lorenzi, H., L. Bacher, M. Lacerda, S. Sartori 2006b. Brazilian Fruits and Cultivated Exotics (for Consuming in Natura). Instituto Plantarum de Estudos da Flora Ltda, Odessa, São Paulo

Lossio, L.C., Delgado, H.S., Ruiz, A.G., Donayre, R.A., Ruiz, J.G., Panduro, M.P., Sifuentes, T.C. 1995. Plantas Medicinales de la Amazonía Peruana. Instituto Peruano de Seguridad Social/Instituto de Medicina Tradicional, Iquitos

Loureiro, Arthur A.; Silva, Marlene F. da; Alencar, Jurandyr da Cruz. 1979. Essências Madeireiras da Amazônia. Instituto Nacional de Pesquisas da Amazônia, Manaus, vol. 2
Lustosa, Dom Antônio de Almeida. 1976. No Estuário Amazônico "À Margem da Visita Pastoral". Conselho Estadual de Cultura do Pará, Belém.
Mackensen, Jens; Tillery-Stevens, Madeleine; Klinge, Rudolf; Fölster, Horst. 2000. Site parameters, species composition, phytomass structure and element stores of a terra-firme forest in East-Amazonia, Brazil. Plant Ecology 151:101–109
Mamede, Jeneffer Soares dos Santos; Pasa, Maria Corette. 2018. Os quintais agroflorestais na comunidade rural em Várzea Grande, Mato Grosso, Brazil. Biodiversidade 17(2):54–70
Marimon, Beatriz Schwantes; Lima, Edson de Souza. 2001. Caracterização fitofisionômica e levantamento florístico preliminar no Pantanal dos Rio Morte-Araguaia, Cocalinho, Mato GrossoActa Botanica Brasilica 15(2):213–229
Marimon, B. S.; Felfili, J. M.; Lima, E. S. 2002. Floristics and phytosociology of the gallery forest of the Bacaba stream, Nova Xavantina, Mato Grosso, Brazil. Edinburgh Journal of Botany 59(2):303–318
Marinho, Diogo Guimarães; Alviano, Daniela Sales; Matheus, Maria Eline; Alviano, Celuta Sales; Fernandes, Patricia Dias. 2011. The latex obtained from Hancornia speciosa Gomes possesses anti-inflammatory activity. Journal of Ethnopharmacology 135:530–537
Martins, Georgia Vilela; Martins, Luiza Suely Semen; Veasey, Elizabeth Ann; Lederman, Ildo Eliezer; Silva, Edson Ferreira da. 2012a. Diversity and genetic structure in natural populations of Hancornia speciosa var. speciosa in northeastern Brazil. Revista Brasileira de Fruticultura 34(4):1143–1153
Martins, Williane Maria de Oliveira; Martins, Lilliane Maria de Oliveira; Paiva, Fabiano Silveira; Martins, Wilton José de Oliveira; Lima Júnior, Sebastião Ferreira. 2012b. Agrobiodiversidade nos quintais e roçados ribeirinhos na comunidade Boca do Môa – Acre. Biotemas 25(3):111–120
Medeiros, Priscilla Sanjuan de. 2016. Indicadores ambientais para funcionalidade ecológica em florestas secundárias de diferentes idades na Amazônia Oriental. Doctoral dissertation, Universidade Federal do Pará, Belém
Mejia, Kember; Rengifo, Elsa. 2000. Plantas Medicinales de Uso Popular en la Amazonía Peruana. Agencia Española de Cooperación Internacional, Lima
Mendoza, Abel L. Monteagudo; Guerrero, Marleni Huamán. 2010. Catálogo de los arboles y afines de la selva central del Perú. Arnaldoa 17(2):202–242
Métraux, A. 1963. The tribes of Mato Grosso and eastern Bolivia. In: *Handbook of South American Indians*, J. H. Steward (Editor), Cooper Square Publishers, New York, vol. 3, pp 349–360
Milliken, W. and B. Albert. 1996. The use of medicinal plants by the Yanomami indians of Brazil. Economic Botany 50(1):10–25
Milliken, W., R. P. Miller, S. R. Pollard, and E. V. Wandelli. 1992. *The Ethnobotany of the Waimiri Atroari Indians of Brazil.* Royal Botanic Gardens, Kew
Miranda, Izildinha Souza; Almeida, Samuel Soares; Dantas, Paulo Jorge. 2006. Florística e estrutura de comunidades arbóreas em cerrados de Rondônia, Brasil. Acta Amazonica 36(4):419–430
Modesto Júnior, M. S.; Mascarenhas, R. E. B. 2001. Levantamento da infestação de plantas daninhas asociada a uma pastagem cultivada de baixa produtividade no nordeste Paraense. Planta Daninha 19(1):11–21
Montalván, Nadiuska Giannina Achong. 2015. Plantas alimenticias usadas por mamíferos mayores en la cuenca alta del Río Itaya, Loreto, Perú. Bachelor thesis, UNAP, Facultad de Ciencias Biológicas, Iquitos
Monteles, Ricardo; Pinheiro, Claudio Urbano B. 2007. Plantas medicinais em um quilombo maranhense: uma perspectiva etnobotânica. Revista de Biologia e Ciências da Terra 7(2):1–12
Montenegro, Olga Lucia. 2004. Natural licks as keystone resources for wildlife and people in Amazonia. Master's thesis, University of Florida, Gainesville
Mota, Nara Furtado de Oliveira; Watanabe, Mauricio Takashi Coutinho; Zappi, Daniela Cristina; Hiura, Alice Lima; Pallos, Julieta; Viveros, Raquel Stauffer; Giulietti, Ana Maria; Viana, Pedro

Lage. 2018. Cangas da Amazônia: a vegetação única de Carajás evidenciada pela lista de fanerógramas. Rodriguésia 69(3):1435–1488

Moura, C. F. H.; Alves, R. E.; Filgueiras, H. A. C.; Araújo, N. C. C.; Almeida, A. S. 2002. Quality of fruits native to Latin America for processing: mangaba (Hancornia speciosa Gomes). Acta Horticulturae 575: 549–554

Muñoz, Marianella Guevara. 2010. Evaluación de purmas con fines de manejo en la comunidad nativa de Santa Mercedes, Rio Putumayo, Perú. Master's thesis, Universidad Nacional de la Amazonia Peruana, Iquitos

Myster, Randall W. 2018. What we know about the composition and structure of igapó forests in the Amazon Basin. Botanical Review 84:394–410

Nimuendajú, C. 1956. Os Apinayé. *Boletim do Museu Paraense Emílio Goeldi* 12:1–146

Noriega, T. E. 2015. Algunos aspectos sobre la efectividad de Saguinus nigrifrons en la dispersión endozoocórica de semillas en la Estación Biológica Quebrada Blanco, Loreto, Perú. Thesis, Facultad de Ciencias Biológicas, Universidad Nacional de la Amazonía Peruana, Iquitos

Oberg, K. 1953. Indian Tribes of Northern Mato Grosso, Brazil. Smithsonian Institution, Institute of Social Anthropology, Publication 15, Washington, DC

Odonne, Guillaume; Valadeau, Celine; Alban-Castillo, Joaquina; Stien, Didier; Sauvain, Michel; Bourdy, Genevieve. 2013. Medical ethnobotany of the Chayahuita of the Paranapura basin (Peruvian Amazon). Journal of Ethnopharmacology 146:127–153

Oliveira, Francisco Plácido Magalhães; Jardim, Mário Augusto Gonçalves. 1998. Composição florística de uma floresta secundária no município de Igarapé-Açu, estado do Pará, Brasil. Boletim do Museu Paraense Emilio Goeldi, Série Botânica 14(2):127–145

Oliveira, Alexandre A. de; Mori, Scott A. 1999. A central Amazonian terra firme forest. I. High tree species richness on poor soils. Biodiversity and Conservation 8:1219–1244

Oliveira, Arlem Nascimento de; Amaral, Ieda Leão do; Ramos, Michele Braule Pinto; Nobre, Antônio Donato; Couto, Luciana Bovino; Sahdo, Rosana Martiniano. 2008. Composição e diversidade florístico-estrutural de um hectare de floresta densa de terra firme na Amazônia central, Amazonas, Brasil. Acta Amazonica 38(4):627–642

Oliveira, Lucicléia Mendes de; Sousa, Rhonan Martins de; Correa, Norma Elena Reynosa; Santos, André Ferreira dos; Giongo, Marcos. 2018. Florística e síndromes de dispersão de um fragmento de cerrado ao sul do estado do Tocantins. Scientia Agraria Paranaensis 17(1): 104–111

Palacios, Erwin; Rodríguez, Adriana; Castillo, Claudia. 2004. Preliminary observations on the mottled-face tamarin (*Saguinus inustus*) on the lower Rio Caquetá, Colombian Amazon. Neotropical Primates 12 (3): 123–126

Pasa, Maria Corette. 2011. Saber local e medicina popular: a etnobotânica em Cuiabá, Mato Grosso, Brasil. Boletim do Museu Paraense Emílio Goeldi, Ciencias Humanas 6(1):179–196

Penido, Alexandre Batista; Morais, Selene Maia de; Ribeiro, Alan Bezerra; Silva, Ana Zélia. 2016. Ethnobotanical study of medicinal plants in Imperatriz, state of Maranhão, northeastern Brazil. Acta Amazonica 46(4):345–354

Pereira, N. 1967. *MoronGuêta: Um Decameron Indígena*. Civilização Brasileira, Rio de Janeiro, vol. 2

Pereira, Antonio Carlos; Santos, Eduardo Ribeiro dos. 2015. Frutas nativas do Tocantins com potencial de aproveitamento econômico. Agri-Environmental Sciences 1(1):22–37

Pereira, Ailton Vitor; Pereira, Elainy Botelho Carvalho; Silva Júnior, Josué Francisco da; Silva, Dijalma Barbosa da. 2006. Mangaba. In: Frutas Nativas da Região Centro-Oeste do Brasil, Vieira, Roberto Fontes; Costa, Tânia da Silveira Agostini; Silva, Dijalma Barbosa da; Ferreira, Francisco Ricardo; Sano, Sueli Matiko (eds), EMBAPA Recursos Genéticos e Biotecnologia, Brasília, pp 188–213

Pereira, Aline C.; Pereira, Ana Bárbara D.; Moreira, Carolina C. L.; Botion, Leida M.; Lemos, Virgínia S.; Braga, Fernão C.; Cortes, Steyner F. 2015a. Hancornia speciosa Gomes (Apocynaceae) as a potential anti-diabetic drug. Journal of Ethnopharmacology 161:30–35

Pereira, Firmino Cardos; Lenza, Eddie; Maracahipes-Santos, Leonardo; Mews, Henrique Augusto; Gomes, Letícia; Lima, Stefany Lorranyny; Santos, Kelyane Silva Melo. 2015b. Comparação

dos métodos de parcelas e pontos-quadrantes para descrever uma comunidade lenhosa de cerrado típico. Biotemas 28(2): 61–72
Pereira, Luciano Araujo; Lima-Barbosa, Jackson Rodrigo de; Almeida, Maria Zélia de; Guimarães, Elsie Franklin. 2017. Diversidade de plantas em quintais quilombolas, conhecimento local sobre uso e cultivo de pimentas na Amazônia oriental, Brasil. Revista de Biologia Neotropical 14(1):56–72
Peres, C.A. 1993. Diet and feeding ecology of saddle-back (*Saguinus fuscicollis*) and moustached (S. mystax) tamarins in an Amazonian terra firme forest. Journal of Zoology 230:567–592
Peres, C.A. 1994. Diet and feeding ecology of gray woolly monkeys (*Lagothrix lagotricha cana*) in central Amazonia: comparisons with other Atelines. International Journal of Primatology 15(3):333–372
Pinheiro, Renato Torres; Marcelino, Dianes Gomes; Moura, Dieyson Rodrigues. 2018. Espécies arbóreas de uso múltiplo e sua importância na conservação da biodiversidade nas áreas verdes urbanas de Palmas, Tocantins. Desenvolvimento e Meio Ambiente 49: 264–282
Pires, J. M. and G. T. Prance. 1985. The vegetation types of the Brazilian Amazon. In: Key Environments: Amazonia, Prance, Ghillean T.; Lovejoy, Thomas E. (eds), Pergamon Press, Oxford, pp 109–145
Pires, João Murça; Salomão, Rafael de Paiva. 2000. Dinâmica da diversidade arbórea de um fragmento de floresta tropical primária na Amazônia oriental-1. Período: 1956 a 1992. Boletim do Museu Paraense Emílio Goeldi, Série Botânica 16(1):63–110
Poeppig, E.2003.*Viaje al Perú y el Río Amazonas, 1827–1832*. CETA (Centro de Estudios Teológicos de la Amazonía, Iquitos.
Polhaumus, L. G. 1962. *Rubber: Botany, Production, and Utilization*. Leonard Hill, London.
Prado, Helbert Medeiros. 2013. Feeding ecology of five Neotropical ungulates: a critical review. Oecologia Australis 17(4): 459–473
Prata, Shirley Soares; Miranda, Izildinha de Souza; Alves, Sérgio Augusto Oliveira; Farias, Flavia Cardoso; Jardim, Fernando Cristovam da Silva. 2010. Gradiente florístico das florestas secundárias do Nordeste Paraense. Acta Amazonica 40(3):523–534
Ramírez, Mauro Vásquez; Vásquez, Juan Balaurte. 1998. La extracción de productos forestales diferentes de la madera en el ambito de Iquitos, Peru. Folia Amazonica 9(1–2):69–92
Ratter, J. A. 1987. Notes on the vegetation of the Parque Nacional do Araguaia (Brazil). Notes from the Royal Botanic Garden, Edinburgh 44(2):311–342
Ratter, J. A.; Bridgewater, S.; Atkinson, R.; Ribeiro, J. F. 1996. Analysis of the floristic composition of the Brazilian cerrado vegetation II: comparison of the woody vegetation of 98 areas. Edinburgh Journal of Botany 53(2):151–180
Rayol, Breno Pinto; Silva, Manoela Ferreira Fernandes da; Alvino, Fabrízia de Oliveira. 2006. Dinâmica da diversidade florística da regeneração natural de florestas secundárias, no município de Bragança, Pará, Brasil. Boletim do Museu Paraense Emílio Goeldi, Ciencias Naturais 1(2):9–27
Rayol, Breno Pinto; Alvino-Rayol, Fabrízia de Oliveira; Silva, Manoela Ferreira Fernandes da. 2011. Similaridade florística entre o estrato arbóreo e a regeneração natural de uma floresta secundária, no município de Bragança, nordeste do estado do Pará, Brasil. Revista Brasileira de Agroecologia 6(3):107–114
Regan, B. C.; Julliot, C.; Simmen, B.; Viénot, F.; Charles-Dominique, P.; Mollon, J. D. 2001. Fruits, foliage and the evolution of primate colour vision. Philosophical Transactions of the Royal Society London B 356: 229–283
Reis, Neidiane Farias Costa. 2010. Estrutura fitossociológica e seletividade de espécies em áreas inundáveis na região do baixo Rio Jaci-Paraná, Rondônia. Master's Thesis, Universidade Federal de Rondônia, Porto Velho
Rodrigues, Eliana. 2006. Plants and animals utilized as medicines in the Jaú National Park (JNP), Brazilian Amazon. Phytotherapy Research 20:378–391

Rodrigues, Mauro Antônio Cavaleiro de Macedo; Miranda, Izildinha Souza; Kato, Maria do Socorro Andrade. 2007a. Estrutura de florestas secundárias após dois diferentes sistemas agrícolas no nordeste do estado do Pará, Amazônia Oriental. Acta Amazonica 37(4):591–598

Rodrigues, Mauro Antônio Cavaleiro Macedo; Miranda, Izildinha Souza; Kato, Maria do Socorro Andrade. 2007b. Flora e estrutura da vegetação secundária após o uso de diferentes trituradores florestais. Pesquia Agropecúaria Brasileira 42(4):459–465

Rojas, Joahan; Satalaya, Johann R.; Grandez, Maritza; Vilchez, Luis; Rengifo, Elsa; Salamanca, Efrain; Flores, Ninoska; Giménez, Alberto; Ávila, Juan Antonio; Ruiz, Grace. 2009. Actividad leismanicida de plantas medicinales de la Amazonia Peruana. Revista Boliviana de Química 26(2):43–48

Roosmalen, M. G. M. van. 1985. *Fruits of the Guianan Flora*. Institute of Systematic Botany, Utrecht University, Utrecht/Silvicultural Department of Wageningen Agricultural University, Wageningen

Roucoux, K. H.; Lawson, I. T.; Jones, T. D.; Baker, T. R.; Coronado, E. N. Honorio; Gosling, W. D.; Lähteenoja, O. 2013. Vegetation development in an Amazonian peatland. Paleogeography, Palaeoclimatology, Paleoecology 374: 242–255

Salas, Leonardo A.; Fuller, Todd K. 1996. Diet of the lowland tapir (Tapirus terrestris L.) in the Tabaro River valley, southern Venezuela. Canadian Journal of Zoology 74: 1444–1451

Salomão, Rafael de Paiva; Vieira, Ima Célia Guimarães; Suemitsu, Chieno; Rosa, Nélson de Araújo; Almeira, Samuel Soares de; Amaral, Dário Dantas do; Menezes, Moirah Paulo Machado de. 2007. As florestas de Belo Monte na grande curva do rio Xingu, Amazônia Oriental. Boletim do Museu Paraense Emílio Goeldi, Ciencias Naturais 2(3):57–153

Sampaio, A.J. 1934. Nomes vulgares de plantas na Amazonia. Boletim do Museu Nacional, Rio de Janeiro, vol. 10

Sanaiotti, Tânia M.; Bridgewater, Samuel; Ratter, James A. 1997. A floristic study of the savanna vegetation of the state of Amapá, Brazil, and suggestions for its conservation. Boletim do Museu Paraense Emílio Goeldi, Série Botânica 13(1):3–24

Sánchez, Judith Jacira Achong; Levy, Allison Licett Nuñez. 2013. Características de frutos consumidos por *Saguinus nigrifrons* (Primates, Callitrichidae), en la Estación Biológica Quebrado Blanco, Loreto, Perú. Master's thesis, Universidad Nacional de la Amazonía Peruana, Iquitos

Sandel, Maureen Peggy; Carvalho, João Olegário Pereira de. 2000. Composição florística e estrutura de uma área de cinco hectares de mata alta sem babaçu na floresta nacional do Tapajós. Embrapa Amazônia Oriental, Belém, Documentos 63, pp. 1–19

Santana, José Augusto da Silva; Almeida, Wanderléa da Costa; Sousa, Luciana Karla V. dos Santos. 2004. Florística e fitossociologia em área de vegetação secundária na Amazônia oriental. Revista de Ciências Agrárias 41:105–120

Sanz-Biset, Jaume; Cañigueral, Salvador. 2011. Plant use in the medicinal practices known as "strict diets" in Chazuta valley (Peruvian Amazon). Journal of Ethnopharmacology 137:271–288

Sanz-Biset, Jaume; Campos-de-la-Cruz, José; Epiquién, Mirbel A.; Cañigueral, Salvador. 2009. A first survey on the medicinal plants of the Chazuta valley (Peruvian Amazon). Journal of Ethnopharmacology 122:333–362

Schultes, R.E. 1966. Lactiferous plants of the Karaparaná-Igaraparaná region of Colombia. Acta Botanica Neerlandica 15: 178–189 1969,

Schultes, R. E. and R. F. Raffauf. 1990. The Healing Forest: Medicinal and Toxic Plants of the Northwest Amazonia. Dioscorides Press, Portland, Oregon

Scoles, Ricardo. 2009. El quintal y las frutas: recursos económicos y alimentares en la comunidad negra de Itacoã, Acará, Pará, Brasil. Acta Amazonica 39(1):1–12

Silva, Maria Betânia Araujo da. 2008. Composição florística e estrutura fitossociológica de uma floresta ripária situada às margens do Rio Moju, Pará. Master's thesis, Universidade Federal Rural da Amazônia, Belém

Silva, Silvestre. 2011. Fruits of the Brazilian Amazon. Metalivros, São Paulo

Silva, Edinei Santos da. 2017. Plantas alimentícias em comunidades agrícolas no município de Rio Preto da Eva-AM. Master's thesis, INPA, Manaus

Silva, Marcondes Alves B. da; Melo, Laura Valdiane L.; Ribeiro, Reginaldo V.; Souza, João Paulo M. de; Lima, Joaquim Corsino S.; Martins, Domingos Tabajara de O.; Silva, Regilane M. da. 2010. Levantamento etnobotânico de plantas utilizadas como anti-hiperlipidêmicas e anorexígenas pela população de Nova Xavantina-MT, Brasil. Revista Brasileira de Farmacognosia 20(4):549–562

Silverwood-Cope, P. L. 1990. *Os Makú: Povo Caçador do Nororeste da Amazônia*. Editora UNB, Brasília.

Simmen, Bruno; Sabatier, Daniel. 1996. Diets of some French Guianan primates: food composition and food choices. International Journal of Primatology 17(5):661–693

Smith, N. J. H. 1976. Utilization of game along Brazil's Transamazon Highway. *Acta Amazonica* 6(4):455–466

Smith, N., Mori, S.A., Henderson, A., Stevenson, D.W. and Heald, S.V., 2004. Flowering plants of the Neotropics. Princeton University Press.

Spruce, R. 1853a. Edible fruits of the Rio Negro. In: *Hooker's Journal of Botany and Kew Garden Miscellany*, W. J. Hooker (Editor), Reeve, Benham, and Reeve, London, vol. 5, pp. 183–187.

Spruce, R. 1853b. Journal of a voyage up the Amazon and Rio Negro. *Hooker's Journal of Botany and Kew Garden Miscella*ny 5:207–215

Spruce, R. 1853c. Botanical objects communicated to the Kew Museum, from the Amazon River, in 1851. In: *Hooker's Journal of Botany and Kew Garden Miscellany*, W. J. Hooker (Editor), Lovell Reeve, London, vol. 5, pp. 169–177.

Stefanello, Daniel; Fernandes-Bulhão, Clarissa; Martins, Sebastião Venâncio. 2009. Síndromes de dispersão de sementes em três trechos de vegetação ciliar (nascente, meio e foz) ao longo do Rio Pindaíba, MT. Revista Árvore 33(6):1051–1061

Stevenson, Miranda F.; Rylands, Anthony B. 1988. The marmosets, genus *Callithrix*. In: Ecology and Behavior of Neotropical Primates, Mittermeier, Russell A.; Rylands, Anthony B.; Coimbra-Filho, Adelmar F.; Fonseca, Gustavo A.B. da (eds), Ecology and Behavior of Neotropical Primates, World Wildlife Fund, Washington, DC, vol. 2, pp 131–222

Targhetta, Natália; Kesselmeier, Jürgen; Wittman, Florian. 2015. Effects of the hydroedaphic gradient on tree species composition and aboveground wood biomass of oligotrophic forest ecosystems in the central Amazon basin. Folia Geobotanica 50:185–205

Valadeau, Céline; Pabon, Adriana; Deharo, Eric; Albán-Castillo, Joaquina; Estevez, Yannick; Lores, Fransis Augusto; Rojas, Rosario; Gamboa, Dionicia; Sauvain, Michel; Castillo, Denis; Bourdy, Geneviève. 2009. Medicinal plants from the Yanesha (Peru): Evaluation of the leishmanicidal and antimalarial activity of selected extracts. Journal of Ethnopharmacology 123:413–422

Valadeau, Céline; Castillo, Joaquina Alban; Sauvain, Michel; Lores, Augusto Francis; Bourdy, Geneviève. 2010. The rainbow hurts my skin: medicinal concepts and plants uses among the Yanesha (Amuesha), an Amazonian Peruvian ethnic group. Journal of Ethnopharmacology 127:175–192

Vale, Gabriel Damasco do. 2011. O efeito do solo, inundação e topografia sobre as campinaranas inundáveis na Amazônia. Master's Thesis, INPA, Manaus

Van Beek, T. A.; Verpoorte, R.; Svendsen, A. Baerheim; Leeuwenberg, A. J. M.; Bisset, N.G. 1984. Tabernaemontana L. (Apocynaceae): a review of its taxonomy, phytochemistry, ethnobotany and pharmacology. Journal of Ethnopharmacology 10:1–156

Van Roosmalen, Marc G. M.; Klein, Lewis L. 1988. The spider monkeys, genus *Ateles*. In: Ecology and Behavior of Neotropical Primates, Mittermeier, Russell A.; Rylands, Anthony B.; Coimbra-Filho, Adelmar F.; Fonseca, Gustavo A.B. da (eds), Ecology and Behavior of Neotropical Primates, World Wildlife Fund, Washington, DC, vol. 2, pp 455–537

Vásquez, R. and A. H. Gentry. 1989. Use and misuse of forest-harvested fruits in the Iquitos area. *Conservation Biology* 3(4): 350–361.

Viana, Simone Silva; Santos, João Ubiratan Moreira dos; Simões, André Olmos. 2017. Diversidade taxonômica de Apocynaceae na ilha do Marajó, PA, Brasil. Rodriguesia 68(2): 623–652

Villas Boas, Orlando; Villas Boas, Claudio. 1975. Xingu: The Indians, Their Myths. Souvenir Press, London

Voss, Robert S.; Fleck, David W. 2011. Mammalian diversity and Matses ethnomammalogy in Amazonian Peru Part 1: Primates. Bulletin of the American Museum of Natural History 351:1–81

Wang, Ellen; Donatti, Camila I.; Ferreira, Vanda L.; Raizer, Josué; Himmelstein, Jeffrey. 2011. Food habits and notes on the biology of *Chelonoidis carbonaria* (Spix 1824) (Testudinidae, Chelonia) in the southern Pantanal, Brazil. South American Journal of Herpetology 6(1):11–19

Williams, Llewelyn. 1962. Lactiferous plants of economic importance III. Couma species. Economic Botany 16(4):251–263

Wisniewski, A. and C. F. M. Melo. 1982. *Borrachas Naturais Brasileiras, III. Borracha de Mangabeira.* EMBRAPA, Centro de Pesquisa Agropecuária do Trópico Úmido, Belém.

Zárate, Ricardo; Mori, Tony Jonatan; Ramirez, Fredy Francisco; Davila, Hilda Paulette; Gallardo, George Pepe; Cohello, Geancarlo. 2015. Lista actualizada y clave para la identificación de 219 especies arbóreas de los bosques sobre arena blanca de la Reserva Nacional Allpahuayo Mishana, Loreto, Peru. Acta Amazonica 45(2):133–156

Zent, S. 1998. Independent yet interdependent "isode": the historical ecology of traditional Piaroa settlement pattern. In: *Advances in Historical Ecology*, W. Balée (Editor), Columbia University Press, New York, pp. 251–286.

Chapter 5
Aptandraceae

5.1 *Cathedra acuminata*

Bolivia Café rama

Brazil Cajazinho, bacuri de anta, marfim

Colombia Michimango

Peru Cashillo, ushum sacha

Venezuela Cascaroncillo, yaguazil, asta blanca; sedi udu (Ye'kuana)

Status Wild

N. Smith, *Amazon Fruits: An Ethnobotanical Journey*, Ethnobiology,
https://doi.org/10.1007/978-3-031-12803-5_5

Segundo, 12 years, gathering fruits of cashillo ***Cathedra acuminata*** **from a forest floor. Near Yarina, Rio Yanayacu, Reserva Nacional Pacaya-Samiria, Loreto, Peru 1-15-06**

The bright orange fruits of *Cathedra acuminata* are typically gathered from the forest floor because the tree is tall, reaching as high as 35 m in some cases. The ovoid fruits, measuring some 2–2.5 cm long by 1.3–2 cm wide, contain a single tan-colored seed. The skin is thin and eaten along with the orange-colored pulp. The mesocarp is savory rather than sweet, with a creamy and somewhat farinaceous texture reminiscent of umarí (*Poraqueiba sericea*), an upland fruit popular throughout the Amazon. People of all ages, from children to grandparents, look forward to the season when the fruits fall, such as at the village of Yarina in the Reserva Nacional Pacaya-Samiria in Loreto, Peru, because the fruits have a unique flavor and are satisfying.

Known as cashillo along the Yanayacu, a black water affluent of the lower Marañón, *C. acuminata* fruits throughout the rainy season. The tree is widely distributed in the Amazon and Orinoco Basins, including the Guianas, especially in wetlands, but it is more common in central and western Amazonia. It is also sometimes found in stunted vegetation (*campina*) growing on sandy soils, such as in Roraima in the north-central part of the Amazon Basin (Rodrigues et al. 2017). The type specimen was collected in the middle of the nineteenth century from the Upper Negro in Venezuela, Amazonia's largest black water river. The tree is evidently adaptable to various soils for it is found on seasonally flooded sandy areas with stunted vegetation in the watershed of the Upper Rio Negro and in gallery forests that traverse savannas, such as the llanos of Venezuela.

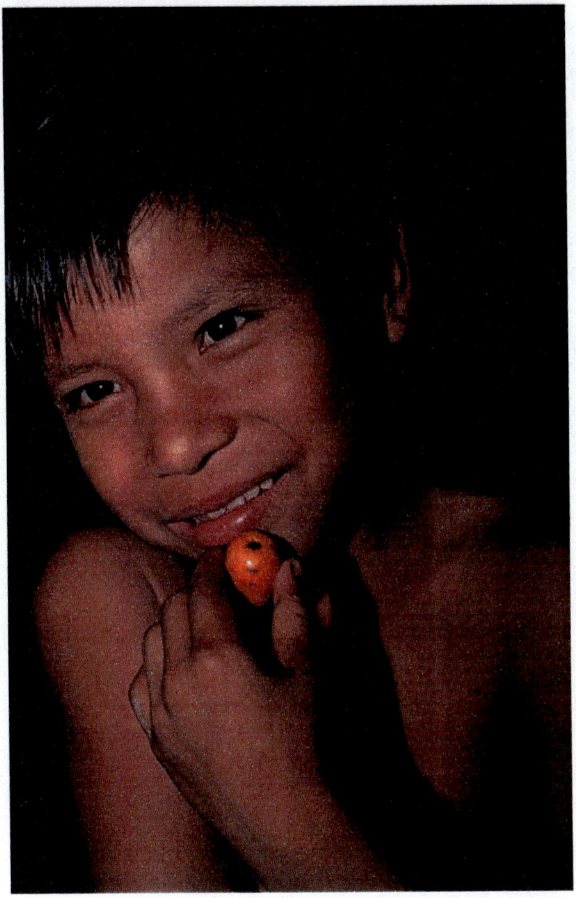

Christian, 8 years, with cashillo fruit he has gathered from the forest floor. Near Yarina, Rio Yanayacu, Reserva Nacional Pacaya-Samiria, Loreto, Peru 6-7-04

Reference

Rodrigues, Rodrigo Schütz; Perdiz, Ricardo de Oliveira; Flores, Andréia Silva. 2017. Novas ocorrências de angiospermas para o estado de Roraima, Brasil. Rodriguésia 68(2):783–790

Chapter 6
Araceae: The Aroid Family

The Araceae, known best for the large fleshy leaves often referred to as elephant ears, has provided humankind with many useful products throughout the lowland tropics of the Old and New World. Several root crops, such as taro (*Colocasia esculenta*) and cocoyam or malanga (*Xanthosoma sagittifolium*), provide sustenance to millions of rural and urban folk in Asia, Polynesia, Latin America, and the Caribbean. Several genera have worked their way into the tropical ornamental plant trade, particularly *Dieffenbachia*, and they grace the home and offices around the world. In the American tropics, several species of aroids are epiphytes on rainforest trees and their long tap roots that drape down from the canopy are used for cordage and to make baskets. The last item that one might imagine that people use from the aroid family is the fruits. Yet in the Amazon, river dwellers employ the fruits of a giant aquatic aroid, *Montrichardia arborescens*, when hunting turtles.

6.1 *Montrichardia arborescens*

Brazil Aninga, aninga mole; kahpó (Tukano)

Peru Raya balsa

Venezuela Bejuco de veinte cuatro

Status Wild

N. Smith, *Amazon Fruits: An Ethnobotanical Journey*, Ethnobiology,
https://doi.org/10.1007/978-3-031-12803-5_6

A stand of aninga *Montrichardia arborescens* **by a floating mat of mureru** *Pontederia rotundifolia.* **Amazon River near Gurupá Mirim, Municipality of Gurupá, Pará, Brazil 9-19-14**

Arum *Montrichardia arborescens* is a giant aquatic aroid that often forms sizeable strands along rivers, streams, and lakes throughout the Amazon basin, especially in sediment-rice waters. It also occurs in the Orinoco Basin, the Guianas, Central America, and parts of the Caribbean. Arum takes root in shallow water and stands of the fleshy-stemmed plant are virtually impenetrable unless one is equipped with a machete. Technically, an herb because arum is not woody, the plant can nevertheless reach 4 m high, and often grow closely together (Gibernau et al. 2003). Known as *aningais*, in Brazil, arum stands are favorite sites for hoatzins (*Opisthocomus hoazin*) to feed while they raise their broods in nearby trees. Whenever one hears the distinctive call of hoatzin, arum is sure to be nearby.

An aningal *Montrichardia arborescens* in a floodplain lake along the middle Amazon. Paraná do Cambixe, Careiro Island, near Manaus, Amazonas, Brazil 6-27-07

Islands of arum in floodplain lakes also serve as heronries and are rumored to be inhabited by many snakes, including the anaconda. Arum stands also serve as lairs of electric eels, widely respected in the region. Not surprisingly, then, arum clumps figure in some of the regional folklore surrounding menacing supernatural entities. For example, tapirê-iauara, a legendary tapir-like creature that attacks people and overcomes them with its vile stench, likes to hang out in arum clumps (Smith 1996:76).

Catraia with longline. One end is weighted with a piece of laterite rock, while the other end floats with the aid of a piece of aninga *Montrichardia arborescens* **stem. Carrazedo, lower Xingu River, Pará, Brazil 8-17-14**

In spite of their reputation as the haunting ground for some real and imagined dangers, fisher folk approach arum stands for several useful parts of the plant: the trunks, used as floats for gillnets and containers (*viveiros*) of freshwater shrimp, the fruits which are employed to attract turtles and for fish bait, and the leaves used to cover or wrap food. In the vicinity of Itacoatiara along the middle Amazon River in Brazil, for example, both subsistence and commercial fishermen cut trunks of arum into roughly 30-cm sections to buoy gillnets that are draped in lakes and floodplain forests at high water. Unfortunately, fishermen are increasingly using pieces of Styrofoam for such purposes because they last longer; however, they also contribute to the increased trash load found in rivers. Large aninga stems are sometimes used to float logs to sawmills, such as in the vicinity of Mazagao, Amapá, in the Amazon estuary (Queiroz and Machado 2007).

Sections of aninga *Montrichardia arborescens* stem used to suspend a gillnet deployed by the girl's father, a cowboy. Fazenda Cajueiro, Municipality of Chaves, Marajó Island, Pará, Brazil 8-15-00

***Viveiro* to keep shrimp alive after they have been caught in traps (*matapis*). The *viveiro* is kept afloat by sections of aninga *Montrichardia arborescens* stems strapped to the side. 15 km east of Afuá, Marajó Island, Pará 8-19-98**

Arum fruits are segmented and resemble a cross between pineapple and dragon fruits, but are the size of the West Indian form of avocado. The fruits are yellow when ripe and eventually fall into the water where they are consumed by certain fishes and turtles, especially the much appreciated *Podocnemis expansa*, known as tartaruga in Brazil and charapa in Peru, *Podocnemis unifilis*, referred to as tracajá in Brazil and taricaya in Peru, and the big-headed turtle *Peltocephalus dumerilianus* (Pérez-Emán and Paolillo 1997). Although less important in the human diet, matamatá *Chelus fimbriata* can also be found in and near stands of aninga according to a cowboy on Fazenda Aruãs in northern Marajó Island. This long-necked turtle apparently ambushes prey in the aninga stands. In the Amazon estuary, hunters after tracajá turtles gather ripe arum fruits from their canoes and break them up and toss pieces onto the surface of lakes and small rivers. The pieces of fruit float, and if there are any tracajá turtles in the vicinity they may come to investigate. As soon as they come near to the surface to snap off the fruit, the hunter hurls a specialized harpoon (*tapuã*). Turtle harpoons have an unbarbed, metal tip that penetrates the shell (carapace) of the turtle just enough to be secure. The hunter then hauls in the harpoon which he has secured by a long cord.

River dweller with specialized harpoon (*tapuã*) used for catching tracajá (*Podocnemis unifilis*) turtles. The aninga *Montrichardia arborescens* fruits in the bottom of his canoe are being used as chum to attract the turtles in this blackwater stream. Igarapé do Lago, affluent of the Maracá River, Amapá, Brazil 5-3-96

In the days when *Podocnemis expansa* turtles were more plentiful, hunters would lie in wait by the margins of arum stands and wait for tartaruga, which can attain 50 kg, to come and feed on the falling fruits. In the middle of the seventeenth century, Antonio Vieira, a Jesuit priest, noted that residents of Pará state in Brazil used to feed arum leaves to tartarugas kept in corrals along rivers. In the eighteenth century, another observant Jesuit, João Daniel, reported that arum was the principal food of tartaruga (Daniel 1976:95). Excessive egg collecting during the colonial period to obtain oil for cooking and lighting caused a population crash of this valuable resource; today, *P. expansa* is a rare delicacy (Smith 1979). The Amazonian manatee *Trichechus inunguis* is also fond of the leaves and fruit of *Montrichardia arborescens* (Domning 1980) but this large aquatic mammal is increasingly rare due to overhunting.

A stand of aninga *Montrichardia arborescens* by a river dweller's home. Limoeiro River, affluent of the lower Tocantins, Pará, Brazil 7-4-07

Because of its usefulness, aninga is often spared when clearing a home site. However, cattle ranchers on floodplains sometimes torch stands of aninga during the dry season, such as near Alenquer in Pará, to expand pasture for their cattle.

Arum fruits are also used as bait to capture fish, such as bacu pedra (*Lithodoras dorsalis*), an armored catfish. In the vicinity of Vigia in the southern part of the Amazon estuary, fishermen sometimes fry or roast arum fruits to bring out their flavor and thus render them more attractive to frugivores. Fishermen living along the Puinahua River, a branch of the Ucayali in the Peruvian Amazon, report that gamitana (*Colossoma macropomum*) are also fond of arum fruits, a feeding habit also noted in the early twentieth century along the lower Amazon (Lange 1914:10). Tambaqui have also been noted feeding on aninga fruits in the vicinity of Santarém, Pará (Lucas 2008). *Colossoma macropomum*, which can attain 20 kg, is known as tambaqui in Brazil and is one of Amazonia's premier eating fish. Arum fruits have been found in the guts of *C. macropomum* and other commercially important fish, such as pirapitinga (*Piaractus brachypomus*), on the Amazon floodplain in the vicinity of Santarém in Pará, Brazil where *aningais* are important fishing grounds (Lucas 2008). Although people do not eat the fleshy part of the fruit, the seeds are consumed in some parts of the Peruvian Amazon after roasting (Brack 2003:38).

Bacu pedra *Lithodoras dorsalis* caught on a handline along the Amazon River. Fishermen sometimes use aninga *Montrichardia arborescens* fruits to capture this catfish which has highly esteemed flesh. Itacoatiara, Amazonas, Brazil 9-22-10

The leaves of aninga are used on occasion to cover food, such as fruits and shrimp in the street market of Abaetetuba along the lower Tocantins. A taxi driver in Belém recalled that when he was a 10-year-old boy he would help his Dad cut aninga leaves on the outskirts of Belém to wrap meat sold in the São Braz market. The taxi driver, who was born in Cametá, also wrapped meat for customers in the now disbanded market. Meat sold in Belem is now mostly placed in plastic bags, thus contributing to the trash accumulation in various parts of the rapidly growing city. Children play with aninga plants on occasion. Young stems, for example, are sometimes cut into sections to create toy spears.

This large aroid is also used in several folk remedies and by some shamans. In the Amazon estuary, the sap of aninga is dripped onto stingray wounds to deaden the pain. Along the Upper Rio Negro in Amazonas, Brazil, rural people employ the roots to make a beverage for treating malaria (Kffuri et al. 2016). In the early nineteenth century, a German naturalist, Carl Friedrich Phillip von Martius, noted that rural folk in the Japurá watershed in Amazonas, Brazil, obtained fresh liquid from the leaves and roots of this plant and applied it to wounds and ulcers. A decoction of fresh roots or dried roots was also used to treat rheumatic pain, testicular tumors, and hydrothorax (Breitbach et al. 2013). In Guyana, the leaves are boiled and used in baths to treat skin problems (Austin and Bourne 1992). In the Peruvian Amazon, some shamans drink the liquid obtained by crushing shoots of aninga, known there as raya balsa, so that they have the power to travel under water and visit people who live at the bottom of some lakes and rivers (Luna 1984a, b). The idea that there are supernatural villages under water is also shared in some parts of the Brazilian Amazon (Smith 1996).

Wadrio, 8 years, taking a break from fishing for sardinha *Triportheus elongatus* to make a toy spear with the stem of a young aninga *Montrichardia arborescens*. The canoe is parked by a stand of aninga along a stream behind Carrazedo, lower Xingu River, Pará, Brazil 8-24-14

References

Austin, Daniel F.; Bourne, Godfrey R. 1992. Notes on Guyana's medical ethnobotany. Economic Botany 46(3):293–298

Brack, A. 2003. *Frutas del Perú*. Universidad San Martin de Porres, Lima

Breitbach, Ulrike B.; Niehues, Michael; Lopes, Norberto P.; Faria, Jair E. Q.; Brandão, Maria G. L. 2013. Amazonian Brazilian medicinal plants described by C.F.P. von Martius in the 19th century. Journal of Ethnopharmacology 147:180–189

Daniel, J. 1976. *Tesouro Descoberto no Rio Amazonas*. Biblioteca Nacional, Rio de Janeiro, vol. 1

Domning, D. P. 1980. Distribution and status of manatees *Trichechus* spp. near the mouth of the Amazon River, Brazil. *Biological Conservation* 19: 85–97

Gibernau, M., D. Barabé, D. Labat, P. Cerdan, and A. Dejean. 2003. Reproductive biology of *Montrichardia arborescens* (Araceae) in French Guiana. *Journal of Tropical Ecology* 19: 103–107.

Kffuri, Carolina Weber; Lopes, Moisés Ahkutó; Ming, Lin Chau; Odonne, Guillaume; Kinupp, Valderly Ferreira. 2016. Antimalarial plants used by indigenous people of the Upper Rio Negro in Amazonas, Brazil. Journal of Ethnopharmacology 178:188–198

Lange, A. 1914. *The Lower Amazon*. MacMillan, New York

Lucas, C. M. 2008. Within flood season variation in fruit consumption and seed dispersal by two characin fishes of the Amazon. *Biotropica* 40(5): 581–589.

Luna, L. E. 1984a. The healing practices of a Peruvian shaman. Journal of Ethnopharmacology 11:123–133

Luna, L. E. 1984b. The concept of plants as teachers among four mestizo shamans of Iquitos, northeastern Peru. Journal of Ethnopharmacology 11:135–156

Pérez-Emán, J. L. and A. Paolillo. 1997. Diet of the pelomedusid turtle *Peltocephalus dumerilianus* in the Venezuelan Amazon. *Journal of Herpetology* 31(2):173–179

Queiroz, José Antonio Leite de; Machado, Sebastião do Amaral. 2007. Potencial de utilização madeireiras de espécies florestais de várzea no município de Mazagão no estado do Amapá. Floresta 37(2):293–302

Smith, N. J. H. 1979. Aquatic turtles of Amazonia: an endangered resource. *Biological Conservation* 16(3): 165–176

Smith, N. J. H. 1996. *The Enchanted Amazon Rain Forest: Stories from a Vanishing World*. University Press of Florida, Gainesville.

Chapter 7
Asteraceae

7.1 *Tilesia baccata*

Brazil olho de camarão, cambará

Ecuador Mish-timu-tapé, mana-tchapé (Cayapa), chichin huasca (Kichwa), chínyachiniap (Shuar), emotokawe (Waorani)

Peru Masurnumi daek, murasa kumpari, sesaip (Aguaruna), pamau masuuri (Huambisa)

Status Wild

Cambará *Tilesia baccata* along a path through successionary vegetation. Sitio do Raimundo, Rio Curuá, Floresta Nacional Caxiuanã, Município de Melgaço, Pará, Brazil 5-5-15

N. Smith, *Amazon Fruits: An Ethnobotanical Journey*, Ethnobiology,
https://doi.org/10.1007/978-3-031-12803-5_7

Tilesia baccata is the only member of the sunflower family that has fleshy fruits. Sunflower *Helianthus annuus* is the best-known member of the Asteraceae; its large, flat showy flowers produce edible seeds that are sold in many countries for human and pet food, especially for birds and hamsters. The sunflower is one of the few plants domesticated in North America. Numerous other members of this family are also cultivated, but for their flowers, especially daisies.

Children eating fruits of olho de camarão *Tilesia baccata* at the edge of a manioc field. Comunidade Terra Preta, Rio Cariá, a left bank affluent of the lower Rio Pracupi, in turn an affluent of the Anapú River, FLONA Caxiuanã, Pará, Brazil 5-17-15

Known as cambará or olho de camarão (shrimp eye) in the Brazilian Amazon, *T. baccata* occurs throughout Brazil in forests, savannas, and coastal sand dunes. It also occurs in scrub savanna (*cerrado*) in stunted woodlands (*cerradão*), such as in the state of Maranhão (Rodrigues and Conceição 2014), as well as on ferruginous laterite (*canga*) in the Serra dos Carajás, Pará, where some of the hills are mined for the iron ore (Cruz et al. 2016; Mota et al. 2018). Its range extends as far south as Paraguay and Argentina, and north into the Guianas, Venezuela, parts of the Caribbean, and Panama. In the Andes, this sprawling bush can be found as high as 1500 m.

Almost vine-like on occasion (scandent), *T. baccata* produces multiple tiny fruits with a green, translucent pulp covering the seeds. Each fruit contains a single gray seed 0.5 cm × 0.3 cm. The fruits appear to have evolved for bird dispersal. In the Caxiuanã National Forest in eastern Pará, river dwellers swallow the sweet fruits whole and are likely involved in dispersing the seeds since they do not always resort to outhouses by their homes when they need to go to the toilet. Children especially are fond of gathering the fruits which are produced in the rainy season. In the Ecuadorian Amazon, the Huaorani consume the fruits (MOBOT Tropicos collection no. 679) and the Kichwa also eat them. The fruits are also fed to domesticated animals in Esmeraldas, a Pacific coastal province of Ecuador.

David, 4 years, eating fruits of olho de camarão *Tilesia baccata* at the edge of a manioc field. Comunidade Terra Preta, Rio Cariá, a left bank affluent of the lower Rio Pracupi, in turn an affluent of the Anapú River, FLONA Caxiuanã, Pará, Brazil 5-17-15

This plant is also used occasionally in folk remedies. In Esmeraldas, Ecuador, *T. baccata* is mixed with other plants to prevent infections in deep wounds (MOBOT Tropicos collection nos. 48050). In Suriname bordering Roraima in the Brazilian Amazon, a bathing solution is prepared from the entire plant among Saramaccan maroons, descendants of African slaves, to help babies start walking and to ward off evil (Ruysschaert et al. 2009). Also in Suriname, the leaves are sold in street markets featuring medicinal plants to treat diabetes (Van Andel et al. 2007).

References

Cruz, Ana Paula Oliveira; Viana, Pedro Lage; Santos, João Ubiratan. 2016. Flora das cangas da Serra dos Carajás, Pará, Brasil: Asteraceae. Rodriguésia 67(5):1211–1242

Mota, Nara Furtado de Oliveira; Watanabe, Mauricio Takashi Coutinho; Zappi, Daniela Cristina; Hiura, Alice Lima; Pallos, Julieta; Viveros, Raquel Stauffer; Giulietti, Ana Maria; Viana, Pedro Lage. 2018. Cangas da Amazônia: a vegetação única de Carajás evidenciada pela lista de fanerógramas. Rodriguésia 69(3):1435–1488

Rodrigues, Maira dos Santos; Conceição, Gonçalo Mendes da. 2014. Diversidade florística das diferentes fisionomias de cerrado do Parque Estadual do Mirador, Maranhão, Brasil. Brazilian Geographical Journal: Geosciences and Humanities research medium, Ituiutaba 5(1):139–156

Ruysschaert, Sofie; Van Andel, Tinde; Van de Putte, Kobeke; Van Damme, Patrick. 2009. Bathe the baby to make it strong and healthy: Plant use and child care among Saramaccan Maroons in Suriname. Journal of Ethnopharmacology 121:148–170

Van Andel, Tinde; Behari-Ramdas, Joelaika; Havinga, Reinout; Groenendijk, Sara. 2007. The medicinal plant trade in Suriname. Ethnobotany Research & Applications 5:351–372

Chapter 8
Bixaceae

8.1 *Bixa orellana*

Bolivia Urucú, achiote, manduro; maxë (Chácobo), urúku (Sirionó), mat're (Tacana), faj (Tsimane)

Brazil Urucú, urucum, urucum manso, urucum casca verde, urucuzeiro, arucu, coloral, colorau, açafrão; onôtó (Aparai), iriko'i (Araweté), urukú (Asurini), araku-ate (Guajá), häw (Hupd'äh), uãcáhá (Juruna), uruku (Ka'apor), uruku (Kaiabi), py kumrenx, py poi (Kayapó), onto (Kuikuru), uduhúkaidí, dukiçú (Nambikwara), ihap (Palikur), rukua (Parakanã), ahitê (Paresí), urukú (Parintintin), urukú (Piripkura), waakap (Sateré-Maué), uruku (Tembé), dee tane, ü'ta (Ticuna), moseg-neu, moy sá, mohsã (Tukano), kiswe (Waimiri-Atroari), uruku (Wajãpi), onot (Wayana), uluku (Wayãpi), nara xihi (Yanomami)

Colombia Achiote, achote, achote de monte, achuete, abujo, acote, anoto, onote, onoto, bicha, bija, bijo, biza, carayuru, paipai, pirrimapa, rocon, uñane, uñaño, urucu, urucurana; acósi (Andoke), musa (Barasana), monhá, muhário (Cubeo), ejuselma (Karijona), musã (Makuna), dee-tane (Ticuna), mosegneu (Tukano), monóke, nonok, nonoki, nonorai, monora (Witoto)

Ecuador Achiote, achote; yetse'p (Amuesha), pachu'a cuña, cu'a cuña, ki"a kina (Cofán), manduru, puca manduro, puka manturu (Kichwa), ipiák, ipiaku (Shuar), bayo basa, muhu bosa, su'nyo bosa, posa (Siona-Secoya), kakahue, caca (Waorani)

Peru Achiote, achote, acote, shambo; ipák, yampia, shampu, pisu (Aguaruna), yetse'p (Amuesha), acosi (Andoke), potsóti (Campa), kachapo, yobsaani (Candoshi), maxe (Cashibo), nanpiruwa (Chayahuita), rucu (Cocama), apo'e (Ese Eja), manturu (Kichwa), pite (Matsés), potsóti (Matsigenka), atase, mashe, oshin mashe (Shipibo), üxta, dee tane (Ticuna)

N. Smith, *Amazon Fruits: An Ethnobotanical Journey*, Ethnobiology,
https://doi.org/10.1007/978-3-031-12803-5_8

Venezuela Onoto, onoto montañero, onotillo, onotillo rebalsero achiote, achote, bija, bijo, caituco, caituro, esacacuya, piatu; jkulilu (Jodï), nana, yamarõ si fi, yapi kë mafe koku (Yanomami)

Status Planted, spontaneous in cultural settings

Annatto *Bixa orellana* picked from a bush planted by a manioc processing hut. Wadrio, 8 years, near Carrazedo, lower Rio Xingu, Pará, Brazil 9-6-14

Known as annatto in English, *Bixa orellana* is a bush or small tree typically planted in home gardens and around manioc processing huts. It is planted in home gardens on uplands and floodplains. Seeds for planting are typically dried in the sun and stored. Annatto is not propagated by cuttings. It is also cultivated occasionally in fields, intermixed with other crops, such as in the vicinity of Igarapé-Açu, Zona Bragantina, in the state of Pará (Vieira et al. 2007). This is an ancient practice started by indigenous peoples. For example, the Kuikuru of the Upper Xingu sometimes plants a row of annatto between manioc fields to separate fields with different owners (Carneiro 1983). That explains why it is sometimes found in successionary vegetation that is taking over old manioc fields. Indeed, the Kayapó in the Upper Xingu watershed return to abandoned manioc fields to gather annatto fruits for up to 20 years (Posey 2002:74).

Annatto was most likely domesticated in Amazonia for its bright red arils that surround the seeds, used primarily for body paint, such as among the Nukak a hunting and gathering tribe in the Colombian Amazon (Politis 1996:324). The Tapirapé who inhabit the Tapirapé watershed, an affluent of the Araguaia in eastern Amazonia, make body paint by mixing annatto with oil obtained from the cooked nuts of tucumã *Astrocaryum vulgare* palm (Wagley 1977:65). The Assurini along the middle Xingu mix ground up annatto seeds with oil obtained from the nuts of a different palm, babaçu *Attalea speciosa*, in order to paint themselves. The Assurini means the "red people," a term coined by a neighboring tribe the Juruna (Lukesch 1976b:34). The Kayapó in southern Pará paint the entire body of sick people with annatto paste to ward off evil spirits (Banner 1961). The Wayana who inhabit the Upper Rio Paru de Leste in northern Pará bordering Surinam mix oil from andiroba *Carapa guianensis* nuts to prepare annatto body paint (Velthem and Linke 2010).

During the eighteenth century, the now extinct Iquitos indigenous people who lived in the Peruvian Amazon painted their shaved heads with achiote (Maroni 1988:390). The capital of Loreto province is named after them. In the nineteenth century, at least one indigenous group along the Napo River in Ecuador painted their faces with annatto to ward off flying insects (Orton 1870:213).

Annatto is thought to have supernatural properties among some cultures. The Wayapí, for example, smear urucu paste on their bodies to make themselves invisible to the Master of Water and dangerous forest spirits (Campbell 1989:84, 128; IPHAN 2006:17). Baniwa shamans streak their arms, chests, and faces with annatto paint when they take paricá, an hallucinogenic snuff, so that they transform into jaguars (Wright 2013:65).

Annatto was, and still is, used to paint objects by indigenous cultures. Annatto was used to paint figures on some indigenous pottery (Ribeiro and Velthem 1992). The Kayapó paint leaf stems (petioles) of buriti *Mauritia flexuosa* palms to celebrate the birth of a firstborn (Lukesch 1976a:276). The Kayapó color parts of headbands with annatto dye which are used in a maize festival (Verswijver 1995:163) and the Xavante who live in a landscape dominated by scrub savanna and gallery forests in the upper Xingu and Araguaia rivers paint ceremonial clubs with annatto (Maybury-Lewis 1974:107). The Mekranoti of the Upper Xingu uses annatto to dye string for stringing beads (Werner 1984:226). The Assurini who live along the Ipixuna, an affluent of the middle Xingu, dye their hair, and women dye skirts with annatto (Balée 2013:26). The Sirionó in the Bolivian Amazon dye string used for attaching arrowheads to shafts with annatto (Holmberg 1969:33). The Cashinahua who live along the Curanja, an affluent of the Purus River in the Peruvian Amazon, paint bamboo arrowheads with geometric patterns using achiote mixed with beeswax (Dawson 1975). The Cashinahua also use achiote to paint the tubes of trumpets fashioned from bamboo with a hollowed-out armadillo tail fitted at the end. Men use these trumpets for ceremonial occasions. The red dye is also used to paint gourd masks, baskets, and wooden dolls (Rabineau 1975).

The Mayoruna in the Peruvian Amazon paint figures representing game animals using annatto dye (Romanoff et al. 2004:79). The Wapishana who occupy scattered villages in the Upper Essequibo in Guyana and contiguous Roraima state in Brazil mix annatto with tree resin to paint their pots (Farabee 1918:25). The Wapishana also dye cotton with annatto and paint their bodies with the dye that has been mixed with gum from the haiwa (*Protium heptaphyllum*) tree. The Yanomami who live in Roraima, Brazil, and contiguous forests in southern Venezuela use annatto to dye cotton twine used for tying feathers of curassows (*Crax* spp.) to shafts of arrow cane *Gynerium sagittatum* for hunting purposes (Albert and Milliken 2009:59). The Tenetehara who live in the eastern fringes of the Amazon forest in the state of Maranhão use cloth dyed with annatto to wrap around the arms or legs of youngsters during certain ceremonies (Wagley and Galvão 1961:91). Cubeo shamans along the Upper Uaupés River in the Colombian Amazon used medicine rattles made with round gourds painted with annatto during their curing rituals (Allen 1947). In the mid-nineteenth century, indigenous women at São Gabriel da Cachoeira along the Upper Rio Negro dyed hammocks with annatto that were later sold in Manaus and Belém (Schomburgk 1840). The Ticuna along the Loreto-Yacú River in the Colombian Amazon dye twine obtained from the chambira *Astrocaryum chambira* palm which is used for making hammocks, storage bags, and handles for bags (Glenboski 1983:24).

Five-year-old Sateré-Maué girl with annatto *Bixa orellana* face paint. Igarapé Aruã, Rio Urupadi, Maués River, Amazonas, Brazil 11-16-02

The oval capsules contain 5–100 seeds and are split open by hand to extract the pellet-sized seeds. Most annatto pods are red, but a variety with yellow-green pods is occasionally found in home gardens in both the Brazilian and Peruvian Amazon hence one of the common names for this plant in the Brazilian Amazon: urucum casa verde. After drying the seeds or pods, usually on the ground, the seeds are ground into a powder. Market stalls with brilliant orange-red piles of annatto powder are common in many parts of Brazil, such as Bahia, where annatto is stirred into stews and rice dishes. Bixin is found in various parts of the plants, but it is concentrated in the fleshy mesocarp surrounding the seeds. For some reason, annatto is sometimes labeled as E160b in commercial food products (Pearce 1990). In the Brazilian Amazon, women sometimes mix annatto powder with water and keep it in bottles to be used in food preparation.

A variety of annatto *Bixa orellana* with yellow pods. Igarapé Jacarécuara, Ilha de Trambioca, near Barcarena, Amazon estuary, Pará, Brazil 3-19-15

When annatto was brought into cultivation is unclear, but it may have been one of the first plants domesticated in Amazonia. The Kayapó who inhabit the watershed divide between the Tocantins and Xingu Rivers cultivate six varieties of annatto (Posey 1983) suggesting that indigenous people have been growing this plant for a long time. The wild progenitor of *Bixa orellana* is likely urucurana *B. urucurana* (Moreira et al. 2015). Urucurana is a Brazilian word that translates as "like urucú" (similar to annatto). Urucurana is a large but to a small, slender tree that grows along water courses (Baer 1976). In 2014, I saw *B. urucurana* at the edge of a flooded forest (*igapó*) behind the village of Carrazedo at the confluence of the Xingu and Amazon rivers in Pará. *Bixa orellana* is also planted in some home gardens on floodplains, such as along the lower Solimões near its confluence with the Rio

Negro (Lima and Saragoussi 2000), so *B. urucurana* is the likely progenitor of *B. orellana*. I also saw a small *B. urucurana* growing on Amazon Dark Earth (*terra preta do índio*) with potsherds on an upland site about 1 km inland from Carrazedo. Perhaps urucurana was brought there in precontact times and has become a volunteer plant in a new habitat. The seed pods of urucurana are gathered by some native groups in the wild in Ecuadorian to obtain body paint (Brücher 1989:164).

Annatto *Bixa orellana* with yellow pods. Tarumã district, Manaus, Amazonas, Brazil 4-1-15

A large forest tree, *Bixa excelsa*, is another candidate for the progenitor of annatto (Schultes 1984). If that is the case, annatto may have been domesticated in Acre in southwestern Amazonia. Reports of wild *Bixa orellana* in swamp and marsh forests along rivers in the Guianas (Roosmalen 1985:54) and along some rivers in Roraima (Milliken 1997) could in some cases be abandoned plantings or their seedlings. However, *Bixa excelsa* is a large tree, so *B. urucurana* is the more likely ancestor of urucú.

Mother and son removing seeds from red- and yellow-colored annatto *Bixa orellana* pods. Carrazedo, lower Xingu River, Municipio de Gurupá, Pará, Brazil 4-21-15

Annatto was taken long ago to other parts of lowland South and Central America where it is extensively grown for the seeds that produce bixin, a bright orange-red dye. Known as achiote in Spanish and urucú in Portuguese, annatto was probably domesticated so that people would have a ready source of body paint. Such body paint was used for decorative purposes, to ward off evil, and to go to war (Stearman 1989). The Parintintin of the Madeira Basin rub urucu on human cadavers before burning them along with the deceased's belongings in the ground within their dwellings (Freitas 1926). Many indigenous groups in Amazonia, such as the Siona-Secoya of eastern Ecuador (Vickers and Plowman 1984) and the Waika of the Upper Rio Negro (Mee 1988:111) still paint their bodies with carotenoid dye, especially during ceremonies. When babies begin to crawl, the Tucuna along the Solimões River hold a feast in which the babies are painted with urucu and while it is still moist attach pieces of parrot feathers (Nimuendajú 1963). The use of annatto for dyeing cloth and coloring food came later. Rope is sometimes made from the fibrous bark.

Annatto *Bixa orellana* powder mixed with water and kept in bottles that once contained sugarcane alcohol (*cachaça*). Cowboy home, Rio Anabiju, Marajó Island, Pará, Brazil 4-22-00

Annatto is also employed in a wide range of folk remedies (Baer 1976), a tradition that can be traced to indigenous cultures. The Tacana in the Bolivian Amazon prepare a bathing solution using the leaves of annatto to treat chicken pox (Bourdy et al. 2000) and the leaves are used to make a solution which is drunk to treat hepatitis (Ponz et al. 2005:122). Another indigenous group in the Bolivian Amazon, the Chácobo, alleviate body pains by pressing the leaves against the skin (Boom 1987:18). In the Ecuadorian Amazon, the Waorani heat the leaves of this bush in a little water then place them on their closed eyes to alleviate pain (Mondragón and Smith 1997:86). Along the Napo River in the Ecuadorian Amazon, Kichwa women bathe themselves with water that contains leaves of *Bixa orellana* after they have given birth; they also rub the leaves on their bodies to fortify and purify themselves (Iglesias 1991). In Bolivar state in Venezuela which borders Roraima and Pará in the Brazilian Amazon, various indigenous groups make a decoction from the leaves and roots to treat malaria (Caraballo et al. 2004).

In the Peruvian Amazon, the Chayahuita mince the leaves in water and pour the mucilaginous liquid into inflamed eyes (Odonne et al. 2013). Also in the Peruvian Amazon, the Amuesha soak leaf stalks of this plant in water and then place one drop in each eye to reduce pain (Valadeau et al. 2010). The Amuesha also use leaf buds to make a tea which is drunk twice daily to treat heart palpitations, inflamed prostrate, and anorexia. The Matsigenka prepare a warm bathing solution of the leaves for people assisting funeral rituals on the third day (Shepard 2002). In the central part of the Peruvian Amazon, a tea is prepared from the leaves and drunk as part of the initiation rites of traditional healers (Jauregui et al. 2011). Some rural folk in the Peruvian Amazon soak from eight to 12 leaves of this plant in water overnight and then use the liquid to treat infections and inflammation of the skin and vagina (Lossio et al. 1995:35; Mejia and Rengifo 2000:16); one cup of this same solution is drunk with meals to treat diarrhea (Lossio et al. 1995:34) as well as hepatitis (Roumy et al. 2020). In the vicinity of Chazuta along the Upper Huallaga River, the leaves are soaked in water for half an hour then a gel is squeezed from the softened leaves and placed in eyes to treat infections (Sanz-Biset et al. 2009). To treat malaria, a handful of scraped roots is boiled in a liter of water then once cooled one cup of the solution is drunk in the morning for 9 days and to heal wounds 10 leaves are mashed in water and left overnight; the solution and soaked leaves are then applied to the affected area (Lossio et al. 1995:35). The seeds are also mashed and the paste applied to alleviate burns (Lossio et al. 1995:36).

Annatto is also used in folk medicine among indigenous groups and rural people of mixed ancestry in the Brazilian Amazon. Along the Upper Rio Negro in Amazonas, Tukanos suffering from malaria lie in a hammock slung over a steam bath prepared with the roots of this plant; they also drink a decoction made with the roots of this plant three times a day to treat the same illness (Kffuri et al. 2016). The Maiongong in Roraima in the northern Brazilian Amazon drink a decoction prepared from the roots of urucu to treat malaria (Milliken 1997). Rural folk in Tiningú, a quilombo community 47 km southeast of Santarém, make a tea from the roots to treat anemia and malaria (Carvalho 2015:59). River dwellers living in the watershed of the Jaú, an affluent of the middle Rio Negro, use the bark to make an infusion to treat respiratory problems (Rodrigues 2006), while *ribeirinhos* along the lower Jauaperi, a left bank affluent of the middle Rio Negro, use the seeds to prepare a drink for treating diabetes (Pedrollo et al. 2016). The seeds of *Bixa orellana* are sold in a riverside port in Cuiabá, Mato Grosso, to treat diabetes and lower blood cholesterol (Pinto et al. 2013). The seeds of this plant are also used to make a tea to treat breathing disorders in Quilombo Sangrador, a rural community in the municipality of Presidente Juscelino in Maranhão (Monteles and Pinheiro 2007). The Kayapó who live in scattered villages in the Xingu watershed rub the leaves of annatto on the stomachs of women in labor to facilitate giving birth (Elisabetsky and Posey 1989; Posey 2002:157). Rural folk in parts of southern Amazonia use the seeds to prepare a remedy to treat fever, especially when caused by malaria (Brandão et al. 1992). In Nova Xavantina in Mato Grosso, the seeds are used to make a juice for treating anorexia and hypolipidemic conditions, such as caused by high levels of cholesterol in the blood (Silva et al. 2010a, b). In various parts of Rondônia, rural folk prepare an infusion of the leaves of this plant to alleviate throat inflammation (Santos et al. 2014a, b).

In the Caxiuanã National Forest and its vicinity in eastern Amazonia, river dwellers make a tea from urucu roots to treat anemia and coughs (Lisboa et al. 2002; Rocha et al. 2013) while in Roraima that tea is used to alleviate malaria (Luz 2001). In Amapá, a decoction of the seeds is mixed with syrup and taken to treat bronchitis (Tenório et al. 1991). Villagers inhabiting Marudá along the coast of Pará macerate the seeds and apply the past to contusions, while the roots are used to make an infusion to treat vaginal discharge (Coelho-Ferreira 2009). In the vicinity of Manacapuru along the lower Solimões near its confluence with the Rio Negro villagers use the seeds to make a tea for treating venomous snake bites (Vásquez 2014:15; Vásquez et al. 2014). In Rio Branco, Acre some home gardens contain this plant, and the fruits are macerated and made into an infusion to treat asthma and bronchitis (Siveiro et al. 2012). In eastern Marajó island in the Amazon estuary, people make an infusion from the seeds for treating wounds on their animals (Monteiro et al. 2011). The Wayãpi indians who live in various villages along the Oyapock River that borders French Guiana and Amapá in the northeastern Amazon Basin mix the seed paste of this plant with oil obtained from the seeds of *Carapa guianensis* which is rubbed on the skin of people to repel insects (Odonne et al. 2011). In neighboring French Guiana, 300 g of the roots of this plant are boiled in 500 ml of water for 20 minutes and then left to cool down before drinking to treat malaria (Bertani et al. 2005).

Laboratory tests have shown that methanol extracts of *Bixa orellana* can help control yeast infections caused by *Candida neoformans* (Braga et al. 2007a, b) and an essential oil obtained from the seeds has been shown to be effective against cutaneous leishmaniasis caused by the protozoan *Leishmania amazonensis* in lab tests using mice (Monzote et al. 2014). Conventional drug treatments for treating people infected with this pathogen which is transmitted by sandflies are expensive and require prolonged treatments.

In the past, the Yanomami who inhabit northern Roraima, Brazil, and contiguous forest in southern Venezuela, used to make sticks from annatto wood which they would rub between their hands into a wooden base to generate hot shavings to start fires while traveling away from their communal houses (Albert and Milliken 2009:90; Lizot 1978). This bush provides other benefits to rural people. Over a dozen species of stingless bees, including *Aparatrigona impunctata*, *Dolichotrigona tavaresi*, *Frieseomelitta varia*, *Melipona fasciculata*, *M. fuliginosa*, *M. fulva*, *M. grandis*, *M. melanoventer*, *M. rufiventris*, *M. seminigra*, *Partamona testacea*, *Ptilotrigona lurida*, *Scaptotrigona fulvicutis*, *Tetragona clavipes*, *T. goettei*, *Trigona dallatorreana*, *T. fulviventris*, *T. truculenta*, *T. williana*, *Trigonisca ceophloei* and *T. graeffei*, visit annatto flowers in the Amazon and people sometimes visit their nests in woodlands to remove some of the honey (Absy and Kerr 1977; Absy et al. 1980, 1984; Ferreira and Absy 2017; Heard 1999; Marques-Souza et al. 1995, 2007; Martins et al. 2011; Oliveira et al. 2013:72, 83, 99, 112, 124, 143, 171, 176, 183, 206, 216, 219; Oliveira et al. 2009a, b; Rech and Absy 2011). In the eastern fringes of Amazonia, *Melipona compressipes fasciculata* and *M. flavolineata* bees also visit the flowers of this plant and probably assist in pollination (Gostinski et al. 2018; Kerr et al. 1986). Nectaries are located outside of the flowers which are visited by ants that offer some protection to the plant (Bentley 1977).

Father serving boiled pirarucu *Arapaima gigas* fish colored with annatto *Bixa orellana* to his son. The fish stew is being served on a bed of manioc flour (*farinha d'agua*). São João do Mulato, Lago de Tefé, Amazonas, Brazil 8-29-12

Annatto was a viable cash crop in Amazonia and other parts of tropical America until the advent of synthetic food colorants in the mid-twentieth century. During the nineteenth century, large quantities of annatto powder were exported from the Amazon to Europe to color butter, cheese, and varnish (Orton 1870:168; Santa-Anna Nery 1885:100). In the 1920s, large quantities of the vermilion powder were exported from South America to Europe and the United States (Standley 1924:835). After the Second World War, Red Dye no. 3 largely replaced annatto as a colorant for food and cosmetics. The artificial dye has been implicated as a carcinogen since 1970, and in early 1990, the US Food and Drug Administration finally decided to ban its use. Red Dye no. 3 has long been defended by the food and pharmaceutical industry because of its low cost and its steadfastness in fluids. But the price to be paid for red cherries that do not bleed in fruit cocktails and pink antacids with stable color was ultimately thought to be too high. The ban on Red Dye No. 3 sparked renewed interest in annatto among growers, food processors, and cosmetic companies because the natural red dye is safe for consumption and for skin applications, including lipstick hence one of its names in English: lipstick plant (Cannon and Cannon 1994:24). Industrialized countries use bixin to color butter, cheese, ointments, and chocolate. Bixin was first identified in 1825 but has not been synthesized (Baer 1976:45). Demand for annatto is thus increasing.

As with many crops, the main producing areas are outside of the native range of the plant. Some Costa Rican farmers are dedicating sizeable fields to annatto. Annatto plantations have also been established in other tropical areas including Guadeloupe, India, Zanzibar, Southeast Asia, and parts of the Pacific. It had become a commercial crop in India by 1787 (Cannon and Cannon 1994:24). In Ghana, a decoction of annatto leaves is used medicinally and the fruit pulp is ingested to treat fevers, dysentery, and kidney ailments (Irvine 1961:72). Annatto sometimes becomes feral in introduced areas. For example, annatto occurs along the shores of Barro Colorado Island in Panama and is an escapee in El Salvador, especially in the Lempa Valley, and in the Isthmus of Tehuantepec in Mexico (Coe 1980:88; Croat 1978:618; Standley and Camerón 1927:152). Annatto persists around old homesites and in old regrowth forests in the Pacific lowlands of Costa Rica (Allen 1956:138, Bentley 1983). Spontaneous annatto plants thrive on Puerto Rico as well as Trinidad and Tobago (Britton and Wilson 1924:588; Marshall 1934).

Rodrigo, 9 years, harvesting annatto *Bixa orellana* in a garden (*sítio*) planted around a manioc processing hut (*casa de forno*). Near Carrazedo, lower Rio Xingu, Pará, Brazil 9-6-14

In the Bragantina zone of Pará, annatto fares well when planted in poor soils that have undergone many slash-and-burn cycles for food production. Some growers take advantage of accumulated fertilizer in the soil in black pepper (*Piper nigrum*) that are winding down due to Fusarium attack by interplanting annatto, or other perennial cash and food crops. For continued high yields, however, annatto plants require periodic fertilization. In backyards, organic mulches and kitchen waste provide needed nutrients, but on plantations, inorganic fertilizer is occasionally needed to shore up yields.

Annatto *Bixa orellana* bush in fruit on right in a village home garden. The house has thatch and walls of babaçu *Attalea speciosa* palm fronds. Terrawangã, an Arara village, Rio Xingu, 1 km upstream from mouth of Bacajá River, Pará, Brazil 9-10-15

8.2 *Bixa urucurana*

Bolivia Urucusillo de monte, urucúcillo, urucú; i'fo (Tsimane)

Brazil Urucurana, urucú bravo

Colombia Achiote cimarron, achiote de monte, onotillo, urucú

Ecuador Achiote de monte, Achiote blanco, achiote; shipare cuña (Cofán), boyokäwe, buyo kakahue (Waorani)

Peru Achiote, achiote caspi, sacha achiote, achiote blanco, achotillo

Venezuela Pirrimapa, tirimapa, onoto montanero

Status Wild, spontaneous in cultural settings

Urucú bravo *Bixa urucurana* adjacent to a low-lying area that floods during the high-water season. Successionary vegetation adjacent to a home garden. Floriano, 67 years, farmer and former rubber tapper. Sitio Timbó, left bank of Rio Xingu about 45 km upstream from Altamira, Pará, Brazil, 9-17-15

Urucú *Bixa orellana* was likely domesticated from urucurana *Bixa urucurana*, also known in the Brazilian Amazon as urucu bravo (Moreira et al. 2015). Urucurana means "like urucú." *Bixa urucurana* occurs on higher parts of river floodplains and along forest streams (Assis et al. 2017; Baer 1976:1; Cárdenas et al. 2009; Casula 2012:54; Ducke 1910). It is also a pioneer species in disturbed landscapes, such as in eastern Acre, Brazil (Numata et al. 2017). This large bush to small tree is widespread in lowland tropical America from Mexico south to the Orinoco and Amazon basins and occurs in the Andean foothills up to 730 m. The easiest way to distinguish *B. urucurana* from *B. orellana* is by the fruit: the fruits of the former are smaller and rusty brown color, whereas those of the latter are red or yellow. The seeds themselves, however, are similar.

Urucurana occurs spontaneously in some fields on uplands, such as on a large archaeological site with terra preta inland from Carrazedo along the lower Xingu. Urucurana is also found on a large archaeological site with anthropogenic black earth at Palmeiras along the middle Xingu. Squirrel monkeys *Saimiri sciureus* have been observed feeding on this fruit in the vicinity of the Tucuruí reservoir along the Tocantins River in eastern Pará and may disperse some of the seeds (Pinheiro et al. 2013).

Urucurana was likely brought to those upland sites by people, both indigenous and of mixed ancestry because the seeds can also be used for coloring food and for body paint. Some indigenous groups in the Ecuadorian Amazon, for example, use *B. urucurana* seeds to make body paint (Brücher 1989:164). In the Bolivian Amazon, the Tsimane also use the seeds to make a dye and the wood is used for fuel (Reyes-García 2001:239). The bark is also used to make a yellow dye. Indeed, rural folk does not usually destroy urucurana when they encounter it in or near their home gardens and fields.

Urucú bravo *Bixa urucurana* adjacent to a low-lying area that floods during the high water season. Successionary vegetation adjacent to a home garden. Sitio Timbó, 1eft bank of Rio Xingu about 45 km upstream from Altamira, Pará, Brazil, 9-17-15

References

Absy, Maria Lúcia; Kerr, Warwick Estevam. 1977. Algumas plantas visitadas para obtenção de pólen por operárias de Melipona seminigra merrillae em Manaus. Acta Amazonica 7(3):309–315

Absy, Maria Lúcia; Bezerra, Eronildo Braga; Kerr, Warwick E. 1980. Plantas nectaríferas utilizadas por duas espécies de Melipona da Amazônia. Acta Amazonica 10(2):271–281

Absy, Maria Lúcia; Camargo, João M. F.; Kerr, Warwick E.; Miranda, Ires Paula de Andrade. 1984. Espécies de plantas visitadas por Meliponinae (Hymenoptera: Apoidea) para coleta de pólen na região do Médio Amazonas. Revista Brasileira de Biologia 44(2):227–237

Albert, Bruce; Milliken, William. 2009. Urihi A: a terra-floresta Yanomami. Instituto Socioambiental (ISA)/Institut de Recherche pour le Devéloppement, São Paulo

Allen, P. H. 1947. Indians of southeastern Colombia. Geographical Review 37(4):567–582

Allen, P. H. 1956. *The Rain Forests of Golfo Dulce*. University of Florida Press, Gainesville

Assis, Rafael Leandro de; Wittmann, Florian; Luize, Bruno Garcia; Haugaasen, Torbjon. 2017. Patterns of floristic diversity and composition in floodplain forests across four southern Amazon river tributaries, Brazil. Flora 229:124–140

Baer, D. F. 1976. Systematics of the genus *Bixa* and geography of the cultivated annatto tree. Doctoral dissertation, UCLA, Los Angeles

Balée, W. 2013. Cultural Forests of the Amazon: A Historical Ecology of People and Their Landscapes. University of Alabama Press, Tuscaloosa

Banner, Horace. 1961. O índio Kayapó em seu acampamento. Boletim do Museu Paraense Emílio Goeldi, Nova Série, Antropologia, vol. 13: 1–51

Bentley, B. L. 1977. The protective function of ants vising the extrafloral nectaries of Bixa orellana (Bixaceae). Journal of Ecology 65(1):27–38
Bentley, B. L. 1983. *Bixa orellana* (achiote, annatto). In: *Costa Rican Natural History*, D. H. Janzen (Editor), University of Chicago Press, Chicago, pp. 193–194
Bertani, S.; Bourdy, G.; Landau, I.; Robinson, J. C.; Esterre, Ph.; Deharo, E. 2005. Evaluation of French Guiana traditional antimalarial remedies. Journal of Ethnopharmacology 98:45–54
Boom, B. M. 1987. Ethnobotany of the Chácobo Indians, Beni, Bolivia. The New York Botanical garden, Advances in Economic Botany, vol. 4
Bourdy, G.; DeWalt, S. J.; Chávez de Michel, L. R.; Roca, A.; Deharo, E.; Muñoz, V.; Balderrama, L.; Quenevo, C.; Gimenez, A. 2000. Medicinal plant uses of the Tacana, an Amazonian Bolivian ethnic group. Journal of Ethnopharmacology 70:87–109
Braga, Fernanda G.; Bouzada, Maria Lúcia M.; Fabri, Rodrigo L.; Matos, Magnum de O.; Moreira, Francis O.; Scio, Elita; Coimbra, Elaine S. 2007a. Antileishmanial and antifungal activity of plants used in traditional medicine in Brazil. Journal of Ethnopharmacology 111:396–402
Braga, Lúcia Filgueiras; Sousa, Marcílio Pereira; Gilberti, Sirlene; Carvalho, Marco Antônio Camillo de. 2007b. Caracterização morfométrica de sementes de castanha de sapucaia (*Lecythis pisonis* Cambess, Lecythidaceae). Revista de Ciências Agro-Ambientais, Alta Floresta 5(1):111–116
Brandão, M. G. L.; Grandi, T. S. M.; Rocha, E. M. M.; Sawyer, D. R.; Krettli, A. U. 1992. Survey of medicinal plants used as antimalarials in the Amazon. Journal of Ethnopharmacology 36: 175–182
Britton, N. L. and P. Wilson. 1924. *Botany of Porto Rico and the Virgin Islands*. New York Academy of Sciences, Vol. 5, No. 4
Brücher, H. 1989. *Useful Plants of Neotropical Origin and their Wild Relatives*. Springer, Berlin
Campbell, A. T. 1989. *To Square with Genesis: Causal Statements and Shamanic Ideas in Wayapí*. Edinburgh University Press, Edinburgh
Cannon, John; Cannon, Margaret. 1994. Dye Plants and Dyeing. Timber Press, Portland
Caraballo, Alejandro; Caraballo, Brigida; Rodríguez-Acosta, Alexis. 2004. Preliminary assessment of medicinal plants used as antimalarials in the southeastern Venezuelan Amazon. Revista da Sociedade Brasileira de Medicina Tropical 37(2):185–188
Cárdenas, Dairon; Castaño, Nicolás; Sua, Sonia. 2009. Flora de la estrella fluvial de Inírida (Guianía, Colombia). Biota Colombiana 10(1–2):1–30
Carneiro, R. L. 1983. The cultivation of manioc among the Kuikuru of the Upper Xingu. In: Adaptive Responses of Native Amazonians, R. B. Hames and W. T. Vickers (eds), Academic Press, New York, pp. 65–111
Carvalho, Telma Lélia Gonçalves Schultz de. 2015. Etnofarmacologia e fisiologia de plantas medicinais do quilombo Tiningú, Santarém, Pará, Brasil. Master's thesis, Universidade Federal do Oeste do Pará, Santarém.
Casula, Kátia Regina. 2012. Análise da similaridade florística e estrutural das formações florestais inundáveis em um trecho do alto Rio Madeira e de seus afluentes, estado de Rondônia, Brasil. Master's Thesis, Núcleo de Ciências Exatas da Terra, Universidade Federal de Rondônia, Porto Velho
Coe, M. D. 1980. *In the Land of the Olmec*. University of Texas Press, Austin, vol. 2
Coelho-Ferreira, Márlia. 2009. Medicinal knowledge and plant utilization in an Amazonian coastal community of Marudá, Pará state (Brazil). Journal of Ethnopharmacology 126:159–175
Croat, T. B. 1978. *Flora of Barro Colorado Island*. Stanford University Press, Stanford
Dawson, A. 1975. Graphic arts and design of the Cashinahua. In: The Cashinahua of Eastern Peru, Kensinger, Kennetth M.; Rabineau, Phyllis; Tanner, Helen; Ferguson, Susan; Dawson, Alice (eds), pp. 130–149, Haffenreffer Museum of Anthropology, Brown University, Providence, R.I.
Ducke, A. 1910. Explorações scientificas no Estado do Pará. Boletim do Museu Goeldi (Museu Paraense) de Historia Natural e Ethnographia 7:100–200
Elisabetsky, Elaine; Posey, Darrell Addison. 1989. Use of contraceptive and related plants by the Kayapo Indians (Brazil). Journal of Ethnopharmacology 26: 299–316

Farabee, W. C. 1918. *The Central Arawaks*. The University Museum, University of Pennsylvania, Philadelphia, vol. 10

Ferreira, Marcos Gonçalves; Absy, Maria Lucia. 2017. Pollen analysis of honeys of Melipona (Michmelia) seminigra merrillae and Melipona (Melikerria) interrupta (Hymenoptera: Apidae) bred in Central Amazon, Brazil. Grana 56(6):436–449

Freitas, J. G. 1926. Os índios Parintintin. *Journal de la Société des Américanistes* (Paris, N.S.) 18:67–73

Glenboski, L. L. 1983. *The Ethnobotany of the Tukuna Indians, Amazonas, Colombia*. Universidad Nacional de Colombia, Museo de Historia Natural, Biblioteca Jose Jeronimo Triana, Bogotá

Gostinski, Luana Fontoura; Oliveira, Fabiana França; Contrera, Felipe Andrés Léon; Albuquerque, Patrícia Maia Correia de. 2018. Nicho trófico e partição de fontes polínicas por duas espécies de Melipona (Hymenoptera, Apidae) na Amazônia oriental. Oecologia Australis 22(4):449–462

Heard, Tim A. 1999. The role of stingless bees in crop pollination. Annual Review of Entomology 44:183–206

Holmberg, Allan R. 1969. Nomads of the Long Bow: The Siriono of Eastern Bolivia. Natural History Press, New York

Iglesias, G. 1991. Medicina herbolaria de los Quichuas del Napo: la cultura fitoterapeútica de las mujeres. In: Las Plantas y El Hombre: Memorias del Primer Simposio Ecuatoriano de Etnobotánica Ecónomica, Ríos, M.; Pedersen, H. Borgtoft (eds), Ediciones Abya-Yala, Quito, pp 119–127

IPHAN 2006. Expressão Gráfica e Oralidade entre os Wajãpi do Amapá. IPHAN, Rio de Janeiro

Irvine, F. R. 1961. *Wood Plants of Ghana with Special Reference to their Uses*. Oxford University Press, Oxford

Jauregui, X.; Clavo, Z. M.; Jovel, E. M.; Pardo-de-Santayana, M. 2011. "Plantas con madre": plants that teach and guide in shamanic initiation process in the east-central Peruvian Amazon. Journal of Ethnopharmacology 134:739–752

Kerr, Warwick Estevam; Absy, Maria Lúcia; Souza, Antonio Carlos Marques. 1986. Espécies nectaríferas e poliniferas utilizadas pela abelha *Melipona compressipes fasciculata* (Meliponinae, Apidae) no Maranhão. Acta Amazonica 16:145–156

Kffuri, Carolina Weber; Lopes, Moisés Ahkutó; Ming, Lin Chau; Odonne, Guillaume; Kinupp, Valderly Ferreira. 2016. Antimalarial plants used by indigenous people of the Upper Rio Negro in Amazonas, Brazil. Journal of Ethnopharmacology 178:188–198

Lima, Roberval M. B. de; Saragoussi, Muriel. 2000. Floodplain home gardens on the central Amazon in Brazil. In: The Central Amazon Floodplain: Actual Use and Options for a Sustainable Management, Junk, W. J.; Ohly, J. J.; Piedade, M. T. F.; Soares, M. G. M. (eds), Backhuys Publishers, Leiden pp 243–268

Lisboa, P. L. B., I. A. Gomes, R. C. Lisboa, and C. V. Urbinati. 2002. O estilo Amazônico de sobreviver: manejo dos recursos naturais. In: *Natureza, Homem e Manejo de Recursos Naturais na Região de Caxiuanã, Melgaço, Pará*, P. L. B. Lisboa (ed.), Museu Paraense Emílio Goeldi, Belém, pp. 41–170

Lizot, Jacques 1978. Connaissance et usage des plantes sauvages chez les Yanomami. In: Unidad y Variedad: Ensayos Antropológicos en Homenaje a José M. Cruxent, Wagner, Erika; Zucchi, Alberta (eds), IVIC, Caracas, pp 129–171

Lossio, L.C., Delgado, H.S., Ruiz, A.G., Donayre, R.A., Ruiz, J.G., Panduro, M.P., Sifuentes, T.C. 1995. Plantas Medicinales de la Amazonía Peruana. Instituto Peruano de Seguridad Social/Instituto de Medicina Tradicional, Iquitos

Lukesch, A. 1976a. *Mito e Vida dos Índios Caiapós*. Editora da Universidade de São Paulo, São Paulo

Lukesch, A. 1976b. Bearded Indians of the Tropical Forest: The Asuriní of the Ipiaçaba, Notes and Observations on the First Contact and Living Together. Akademische Druck und Verlagsanstalt, Graz, Austria

Luz, Francisco Joaci F. 2001. Plantas medicinais de uso popular em Boa Vista, Roraima, Brasil. Horticultura Brasileira 19(1):88–96

Maroni, P. 1988. *Noticias Autenticas del Famoso rio Marañón y Misión Apostólica de la Compañia de Jesús de la Provincia de Quito en los Dilatados Bosques de dicho Rio, Escribialas por los Años de 1738*. IIAP, Iquitos.

Marques-Souza, Antonio Carlos; Absy, Maria Lúcia; Kerr, Warwick Estevam; Peralta, Francisco Javier Aguilera. 1995. Pólen coletado por duas espécies de Meliponíneos (Hymenoptera: Apidae) da Amazônia. Revista Brasileira de Biologia 55(4):855–864

Marques-Souza, Antonio Carlos; Absy, Maria Lucía; Kerr, Warwick Estevam. 2007. Pollen harvest features of the central Amazonian bee Scaptotrigona fulvicutis Moure 1964 (Apidade: Meliponinae), in Brazil. Acta Botanica Brasilica 21(1):11–20

Marshall, R. C. 1934. *Trees of Trinidad and Tobago*. Government Printing Office, Port of Spain

Martins, Adriana Cecília Lima; Rêgo, Márcia Maria Corrêa; Carreira, Léa Maria Medeiros; Albuquerque, Patricia Maia Correia de. 2011. Espectro polínico de mel de tiúba (*Melipona fasciculata* Smith, 1854, Hymenoptera, Apidae). Acta Amazonica 41(2):183–190

Maybury-Lewis, D. 1974. Akwê-Shavante Society. Oxford University Press, New York

Mee, M. 1988. *In Search of Flowers of the Amazon Forests*. Nonesuch Expeditions, Woodbridge, Suffolk, UK

Mejia, Kember; Rengifo, Elsa. 2000. Plantas Medicinales de Uso Popular en la Amazonía Peruana. Agencia Española de Cooperación Internacional, Lima

Milliken, William. 1997. Traditional anti-malarial medicine in Roraima, Brazil. Economic Botany 51(3):212–237

Mondragón, Martha L.; Smith, Randall. 1997. Bete Quiwiguimamo Salvando el bosque para vivir sano: algunas plantas y árboles utilizadas por la Nacionalidad Huaorani de la Amazonía Ecuatoriana. Abya Yala, Quito

Monteiro, Maria Vivina Barros; Bevilaqua, Claudia Maria Leal; Palha, Maria das Dores Correia; Braga, Roberta Rocha; Schwanke, Katiane; Rodrigues, Silvane Tavares; Lameira, Osmar Alves. 2011. Ethnoveterinary knowledge of the inhabitants of Marajó Island, eastern Amazonia, Brazil. Acta Amazonica 41(2):233–242

Monteles, Ricardo; Pinheiro, Claudio Urbano B. 2007. Plantas medicinais em um quilombo maranhense: uma perspectiva etnobotânica. Revista de Biologia e Ciências da Terra 7(2):1–12

Monzote, Lianet; García, Marley; Scull, Ramón; Cuellar, Armando; Setzer, William N. 2014. Antileishmanial activity of the essential oil from Bixa orellana. Phytotherapy Research 28:753–758

Moreira, Priscila Ambrósio; Lins, Juliana; Dequigiovanni, Gabriel; Veasey; Clement, Charles R. 2015. The domestication of annatto (Bixa orellana) from Bixa urucurana in Amazonia. Economic Botany 69:127–135

Nimuendajú, C. 1963. The Tucuna. In: *Handbook of South American Indians*, J. H. Steward (Editor), Cooper Square Publishers, New York, vol. 3, pp 713–725

Numata, Izaya; Silva, Sonaira S.; Cochrane, Mark A.; d'Oliveira, Marcus V. 2017. Fire and edge effects in a fragmented tropical forest landscape in the southwestern Amazon. Forest Ecology and Management 401:135–146

Odonne, Guillaume; Berger, Franck; Stien, Didier; Grenand, Pierre; Bourdy, Geneviève. 2011. Treatment of leishmaniasis in the Oyapock basin (French Guiana): a K.A.P. survey and analysis of the evolution of phytotherapy knowledge amongst Wayãpi Indians. Journal of Ethnopharmacology 137:1228–1239

Odonne, Guillaume; Valadeau, Celine; Alban-Castillo, Joaquina; Stien, Didier; Sauvain, Michel; Bourdy, Genevieve. 2013. Medical ethnobotany of the Chayahuita of the Paranapura basin (Peruvian Amazon). Journal of Ethnopharmacology 146:127–153

Oliveira, Francisco Plácido Magalhães; Absy, Maria Lúcia; Miranda, Izildinha Souza. 2009a. Recurso polínico coletado por abelhas sem ferrão (Apidae, Meliponinae) em um fragmento de floresta na região de Manaus- Amazonas. Acta Amazonica 39(3):505–518.

Oliveira, Viviana Maria Araújo; Carneiro, Ana Lucia Basilio; Cauper, Glaucia Socorro de Barros; Pohlit, Adrian Martin. 2009b. In vitro screening of Amazonian plants for hemolytic activity and inhibition of platelet aggregation in human blood. Acta Amazonica 39(4):973–980

Oliveira, Favízia Freitas de; Richers, Bárbara Tadzia Trautman; Silva, Jacson Rodrigues da; Farias, Rinéias Cunha; Matos, Tércio Alves de Lima. 2013. Guia Ilustrado das Abelhas "Sem-Ferrão" das Reservas Amanã e Mamirauá, Amazonas, Brasil (Hymenoptera, Apidae, Meliponini). IDSM, Tefé

Orton, J. 1870. The Andes and the Amazon: or, Across the Continent of South America. Harper & Brothers, New York

Pearce, F. 1990. Brazil, where the ice cream comes from. *New Scientist* 127: 45–48

Pedrollo, Camilo Tomazini; Kinupp, Valdely Ferreira; Shepard Jr., Glenn; Heinrich, Michael. 2016. Medicinal plants at Rio Jauaperi, Brazilian Amazon: ethnobotanical survey and environmental conservation. Journal of Ethnopharmacology 186:111–124

Pinheiro, Tatyana; Ferrari, Stephen F.; Lopes, Maria Aparecida. 2013. Activity budget, diet, and use of space by two groups of squirrel monkeys (*Saimiri sciureus*) in eastern Amazonia. Primates 54:301–308

Pinto, Andréia Zelenski de Lara; Assis, Ana Flávia Silva de; Pereira, Arildo Gonçalo; Pasa, Maria Corette. 2013. Levantamento etnobotânico de plantas medicinais comercializados no mercado do porto em Cuiabá, Mato Grosso, Brasil. Flovet 5:51–70

Politis, G. G. 1996. *Nukak*. Instituto Amazónico de Investigaciones Científicas, Bogotá

Ponz, Elizabeth; Carpio, Griceldo; Meo, Severo. 2005. La Medicina Tradicional de los Tacana y Machineri: Conocimientos Prácticos de las Plantas Medicinales. Investigaciones Regionales Pando, La Paz

Posey, D. A. 1983. Indigenous ecological knowledge and development of the Amazon. In: The Dilemma of Amazonian Development, Moran E.F. (ed), Westview Press, Boulder, pp 225–257

Posey, D. A. 2002. Kayapó Ethnoecology and Culture. Routledge, London

Rabincaru. 1975. Catalogue of the Cashinahua collection. In: The Cashinahua of Eastern Peru, Kensinger, Kenneth M.; Rabineau, Phyllis; Tanner, Helen; Ferguson, Susan; Dawson, Alice (eds), pp. 111–231, Haffenreffer Museum of Anthropology.

Rech, André Rodrigo; Absy, Maria Lúcia. 2011. Pollen sources used by species of Meliponini (Hymenoptera: Apidae) along the Rio Negro channel in Amazonas, Brazil. Grana 50(2):150–161

Reyes-García, Victoria. 2001. Indigenous people, ethnobotanical knowledge, and market economy: a case study of the Tsimane' Amerindians in lowland Bolivia. Doctoral dissertation, University of Florida, Gainesville.

Ribeiro, Berta G.; Velthem, Lucia H. van. 1992. Coleções etnográficas: documentos materiais para a história indígena e a etnologia. In: História dos Índios no Brasil, Cunha, Manuela Carneiro da (ed.), pp. 103–112, Companhia das Letras/Secretaria Municipal de Cultura/FAPESP, São Paulo

Rocha, S. F. R., M. R. C. Ferreira, and M. G. Sablayrolles. 2013. Farmacopeia vegetal da comunidade Caxiuanã, na Floresta Nacional de Caxiuanã. In: Caxiuanã: Paraíso ainda Preservado, P. L. B. Lisboa (Editor), Museu Paraense Emílio Goeldi, Belém, pp. 257–285

Rodrigues, Eliana. 2006. Plants and animals utilized as medicines in the Jaú National Park (JNP), Brazilian Amazon. Phytotherapy Research 20:378–391

Romanoff, S., D. M. Jiménez, F. S. Uaquí, and D. W. Fleck. 2004. *Matsesën Nampid Chuibanaid: La Vida Tradicional de los Matsés*. Centro Amazónico de Antropología y Aplicación Práctica (CAAAP), Lima

Roosmalen, M. G. M. van. 1985. *Fruits of the Guianan Flora*. Institute of Systematic Botany, Utrecht University, Utrecht/Silvicultural Department of Wageningen Agricultural University, Wageningen

Roumy, Vincent; Ruiz, Lastenia; Macedo, Juan Celidonio Ruiz; Gutierrez-Choquevilca, Andrea-Luz; Samaillie, Jennifer; Encinas, Leonor Arévalo; Mesia, Wilfredo Ruiz; Cotrina, Hivelli Ericka Ricopa; Rivière, Céline; Sahpaz, Sevser; Bordage, Simon; Garçon, Guillaume; Dubuisson, Jean; Anthérieu, Sebastien; Seron, Karin; Hennebelle, Thierry. 2020. Viral hepatitis in the Peruvian Amazon: Ethnomedical context and phytomedical resource. Journal of Ethnopharmacology 255 (112735):1–11

Santa-Anna Nery, F. J. 1885. *Le Pays des Amazones, l'El-Dorado, les Terres a Caoutchouc*. Frinzine, Paris

Santos, Gideão Costa; Tourinho, Manoel Malheiros; Mendes, Fernanda da Silva; Guimarães, Célia Maria Costa. 2014a. Fitossociologia e práticas de manejo tradicional em uma floresta de várzea em Santa Bárbara do Pará, Brasil. Revista de Ciências Agrárias 57(2):138–145

Santos, L. O., L. O. Santos, M. P. M. Menezes, C. R. Beasley, and U. Mehlig. 2014b. Composition and structure of a diverse tree community at the edges of a Brazilian Amazon rainforest island surrounded by marshes and mangroves. Plant Ecology 215: 1469–1481

Sanz-Biset, Jaume; Campos-de-la-Cruz, José; Epiquién, Mirbel A.; Cañigueral, Salvador. 2009. A first survey on the medicinal plants of the Chazuta valley (Peruvian Amazon). Journal of Ethnopharmacology 122:333–362

Schomburgk, R. H. 1840. Journey from Esmeralda, on the Orinoco, to San Carlos and Moura on the Rio Negro, and Thence by Fort San Joaquim to Demerara, in the Spring of 1839. *Journal of the Royal Geographical Society of London* 10:248–267

Schultes, R. E. 1984. Amazonian cultigens and their northward and westward migration in pre-Columbian times. In: *Pre-Columbian Plant Migration*, D. Stone (Editor), Papers of the Peabody Museum of Archeology and Ethnology No. 76, pp 19–37, Harvard University, Cambridge

Shepard, G. 2002. Three days of weeping: dreams, emotion, and death in the Peruvian Amazon. Medical Anthropology Quarterly 16(2):200–229

Silva, Marcondes Alves B. da; Melo, Laura Valdiane L.; Ribeiro, Reginaldo V.; Souza, João Paulo M. de; Lima, Joaquim Corsino S.; Martins, Domingos Tabajara de O.; Silva, Regilane M. da. 2010a. Levantamento etnobotânico de plantas utilizadas como anti-hiperlipidêmicas e anorexígenas pela população de Nova Xavantina-MT, Brasil. Revista Brasileira de Farmacognosia 20(4):549–562

Silva, Rachel Macedo da; Mehlig, Ulf; Santos, João Ubiratan Moreira dos; Menezes, Moirah Paula Machado de. 2010b. The coastal restinga vegetation of Pará, Brazilian Amazonia: a synthesis. Revista Brasileira de Botanica 33(4):563–573

Siveiro, A.; Delunardo, T. A.; Haverroth, M.; Oliveira, L. C.; Mendonça, A. M. S. 2012. Plantas medicinais em quintais urbanos de Rio Branco, Acre. Revista Brasileira de Plantas Medicinais 14(4):598–610

Standley, Paul C. 1924. Trees and Shrubs of Mexico. Contributions from the United States National Herbarium, vol. 23, Smithsonian Institutions, Washington, D. C.

Standley, P. C. and S. Camerón. 1927. *Lista Preliminar de las Plantas de El Salvador*. Tipofrafía la Unión, San Salvador

Stearman, A. 1989. *Yuquí: Forest Nomads in a Changing World*. Holt, Rinehart and Winston, New York

Tenório, Maria Alice Ramalho de Oliveira; Van den Berg, Maria Elisabeth; Menezes, Ordoenha Furtado de; Salles, Pedro. 1991. Fitoterapia: uma estratégica terapêuticas natural do Amapá. In: Medicinas Tradicionais e Medicina Occidental na Amazônia, Buchillet, Dominique (ed), Museu Paraense Emílio Goeldi, Belém

Valadeau, Céline; Castillo, Joaquina Alban; Sauvain, Michel; Lores, Augusto Francis; Bourdy, Geneviève. 2010. The rainbow hurts my skin: medicinal concepts and plants uses among the Yanesha (Amuesha), an Amazonian Peruvian ethnic group. Journal of Ethnopharmacology 127:175–192

Vásquez, Silvia Patricia Flores. 2014. Conhecimento, uso e conservação da diversidade vegetal em quatro comunidades ribeirinhas no Município Manacapuru, Amazonas. Doctoral dissertation, Programa de Pós-Graduação em Botânica, INPA, Manaus

Vásquez, Silvia Patricia Flores; Mendonça, Maria Silvia de; Noda, Sandra do Nascimento. 2014. Etnobotânica de plantas medicinais em comunidades ribeirinhas do Município de Manacapuru, Amazons, Brasil. Acta Amazonica 44(4):457–472

Velthem, L. H. and I. L. Linke 2010. *Livro da Arte Gráfica Wayana e Aparai*. Museu do Índio, FUNAI, Rio de Janeiro

Verswijver, G. 1995. *Kaiapo: Material Culture-Spiritual World.* Museum für Völkerkunde, Frankfurt

Vickers, W. T. and T. Plowman. 1984. *Useful Plants of the Siona and Secoya Indians.* Field Museum of Natural History, Fieldiana Botany New Series, no. 15

Vieira, Thiago Almeida; Rosa, Leonilde dos Santos; Vasconcelos, Paulo César Silva; Santos, Mônica Mota dos; Modesto, Rozi da Silva. 2007. Sistemas agroflorestais em áreas de agricultores familiares em Igarapé-Açu, Pará: caracterização florística, implantação e manejo. Acta Amazonica 37(4):549–558

Wagley, Charles. 1977. Welcome of Tears: The Tapirapé Indians of Central Brazil. Oxford University Press, New York

Wagley, C. and E. Galvão. 1961. *Os Índios Tenetehara: Uma Cultura em Transição.* Ministério da Educação e Cultura, Serviço de Documentação, Rio de Janeiro

Werner, D. 1984. Amazon Journey: An Anthropologist's Year among Brazil's Mekranoti Indians. Simon and Schuster, New York

Wright, R. M. 2013. *Mysteries of the Jaguar Shamans of the Northwest Amazon.* University of Nebraska Press, Lincoln

Chapter 9
Boraginaceae

9.1 *Cordia nodosa*

Bolivia Nununa, pabojóni, palo blanco, picana, testiculo del gato, huevo de mono, pabojoni, mechero; bohuëtiomohuati, pabojoni (Chácobo), tayei, yeyejna (Tsimane)

Brazil Remela de cachorro, grão de galo, grão de mucura, ovo de galo, ovo de mucura, saco de mucura, gogó de guariba, pau de formiga, buxuxi de formiga, uruáranca, uruázinho, uruá da mata, baba de boi, invireira, caramurizinho, aruá felpudo, catininga, freijó branco, freijó de formiga, freijó peludo, chapéu de sol, louro do brejo, louro; ayã-ɨ (Guajá), porehetupë, xiho xihi (Yanomami)

Colombia Laurel, bolas de mono, tumba toro peludo; kaag (Miraña), ibra choreri (Witoto)

Ecuador Chinkias, murushi numi, machakuy kaspi, araña kaspi, arbol de araña; ararats, napisaí (Achuar), ccayaju'cho (Cofán), araña caspi, abispa caspi, aguas muyo, machacui caspi, machacui misonsal, sacha curarina (Kichwa), chinkias, kawachimi (Shuar), tania-ka-to, dañagato (Siona-Secoya), nomonewe, wihinhue, huaemohue ahue, dadawe, amokapumo, nomonewe (Waorani)

Peru Sacha uva, uva blanca, uvilla blanca, huevo de gato, huevo de intuto, huevo de coto, cotoruntu, sepinda, añallo caspi, cerda caspi, hormiga caspi, pucacuru caspi, San Juan caspi, pucacuru sacha blanca, pucacuro caspi, supai chacra, cumala colorada, almendrillo, añallu caspi, añayo caspi, pinchicoróto; kujáncham sukí (Achuar), ajatsatsa, murushinim, shitapach murus, supínim (Aguaruna), paryaconchpar orrno (Amuesha), mujúshimin (Awajún), cááu (Bora), chullacaspi (Kichwa), matyagiroki, matiageriki, matiogoriki (Matsigenka), ihachoreri (Witoto)

Venezuela Alatrique peludo, pelusa; dibai, diguai (Ye'kuana)

N. Smith, *Amazon Fruits: An Ethnobotanical Journey*, Ethnobiology,
https://doi.org/10.1007/978-3-031-12803-5_9

Status Wild

Grão de galo *Cordia nodosa* tree in fruit on an archaeological site with Amazonian Dark Earth *(terra preta do índio)*. IBAMA base, mouth of the Rio Jaú, confluence with the Rio Negro, Amazonas, Brazil, 10-2-12

Cordia nodosa ranges in size from an understory bush to a small tree and is found in mature and regrowth forests throughout the Amazon Basin up to 1200 m in the Andean foothills. It also occurs in the forested portions of northeast Brazil as well as the Guianas and in Venezuela. It can be found in successionary vegetation as young as 5 years old and endures in regrowth woodlands for at least 20 years, such as in the vicinity of Santa Mercedes along the Rio Putumayo in the northern Peruvian Amazon bordering Colombia (Gomez 2011:88; Muñoz 2010:19, 37), on Fazenda Aruanã in the municipality of Itacoatiara, Amazonas (Santos and Absy 2010), and in the Tocantins watershed in eastern Pará (Taurino 2017:88). It is also found as a weed in some cattle pastures in the vicinity of Paragominas in the state of Pará

(Dantas and Rodrigues 1980). It also thrives in bamboo vegetation dominated by *Guadua sarcocarpa* and *G. weberbaueri* in the southeast Peruvian Amazon and in neighboring Acre, Brazil (Griscom et al. 2007; Maranho et al. 2016). *Cordia nodosa* occurs mostly in uplands, such as between the Purus and Madeira rivers in Amazonas (Souza 2012:47), but it is also occasionally found in low-lying areas that are subject to flooding for brief periods, such as in the Bolivian, Ecuadorian, and Peruvian Amazon (Macía 2011), in the vicinity of Humaitá in the state of Amazonas, along the Upper Madeira and some of its affluents in Rondônia, Brazil, along the lower Xingu (Salomão et al. 2007) and lower Rio Negro in Amazonas (Almeida 2012:45), and in the Orinoco River Basin in Venezuela (Araujo-Murakami et al. 2009; Assis et al. 2017; Campbell et al. 1992; Casula 2012:54; Dávila et al. 2015:42; Godoy et al. 1999; Gottsberger and Morawetz 1986; Luize et al. 2015; Navarro and Maldonado 2002:74; Nebel et al. 2000a, b, 2001; Parolin et al. 2003; Reis 2010:31; Thomas 2009; Wittmann et al. 2006, 2010, 2013). This bush to small tree is found in some stunted woodlands (*campinarana*) subject to seasonal floodings, such as in the Viruá National Park in Roraima (Vale 2011:41). It also thrives in some floodplain forests of clear and black water rivers and streams that remain flooded for much of the year, such as in the Madre de Dios watershed in the Bolivian Amazon (Navarro 2002) and along a stream 60 km north of Manaus, Amazonas, Brazil (Porto et al. 1976).

It is also found in the dry forest (*cerradão*) in the Araguaia watershed in the southeastern portion of the Amazon Basin (Ratter 1987) as well as scrub savanna (*cerrado*) in Mato Grosso (Rondon Neto et al. 2010) and natural grasslands in Roraima (Dantas and Rodrigues 1982). It grows on a wide variety of soils, including sandy soils such as near Iquitos in the Peruvian Amazon (Zarate et al. 2015) and on ferruginous laterite (*canga*) in the Serra dos Carajás, Pará (Watanabe et al. 2017). In the vicinity of Manaus, the flowers are pollinated by several species of butterfly as well as *Plebeia minima*, a stingless bee (Santos 2005:40). In other areas of the Rio Negro watershed in Amazonas, Brazil, several species of other stingless bees in the genus *Partamona* gather pollen from the flowers of this tree (Rech and Absy 2011a, b).

The pale-yellow fruits are found in clumps along the trunk and branches (cauliflorous) and are eaten by adults and children alike in various parts of the Amazon, including among the Shuar in southern Ecuador (Van den Eynden 2004). This small tree produces fruit year-round but peak production occurs in the rainy season (Santos 2005:17). The marble-sized fruits measuring 1.6 cm × 1.4 cm contain a single light tan seed 1.5 cm × 0.8 cm. The fruits likely evolved for dispersal by birds (Foster et al. 1986) and some arboreal mammals. For example, the saddle-back tamarin *Saguinus fuscicollis* and the moustached tamarin *Saguinus mystax* have been observed feeding on these fruits in the Peruvian and Bolivian Amazon (Culot et al. 2009; Sánchez and Levy 2013:28, 58; Soini 1987; Yoneda 1984; MOBOT Tropicos collection #33) as is the case with the Colombian squirrel monkey *Saimiri cassiquiarensis albigena* (Carretero-Pinzón et al. 2016). These monkeys swallow the seeds along with the pulp, thus defecating the seeds later. Dung beetles bury some of the feces dropped by these monkeys that contain the seeds of this plant

(Culot et al. 2009). Although small black ants inhabit nodes close to the pubescent fruits, they are not aggressive, at least to people. One of the common names for *C. nodosa* in Acre, Brazil, is *pau de formiga* which means the ant tree (Araujo 2015; MOBOT Tropicos collection #717). However, the ants inhabiting domatia, a shelter structure produced by the plant to provide a home to protective ants, likely keep some herbivores from eating parts of this plant (Solano et al. 2003). In the Peruvian Amazon, one of the names for this tree is *hormiga caspi* which means ant fruit (Vásquez 1998).

Cordia nodosa has evidently co-evolved with ants to provide mutual benefits. The ants that inhabit the protective nodes (domatia) in turn clear some of the plants underneath and around the small tree, thereby helping it to compete. The Machiguenga in the Peruvian Amazon believe that such clearings are the abode of spirits that provide them with game animals (Shepard 1999).

People consume the thin, translucent pulp before spitting out the seed. As with many fruits in the Amazon, humans have likely influenced the distribution and density of *C. nodosa*. The Waorani in the Napo watershed in the Ecuadorian Amazon snack on the fruits while on hunting trips (Ima 2012:82). A grão de galo tree I encountered at the mouth of the Jaú River was growing on an archaeological site that has been partially cleared for a building belonging to the Brazilian institution for environmental protection (IBAMA). The Brazilian botanist Pedro Lisboa noted the presence of this tree on mounds built by the extinct Marajoara culture (Lisboa 2012:413). The Cofán in the northern part of the Ecuadorian Amazon plant this tree in some of their agricultural fields. The Kurupukari in central Guyana also consume these fruits (Johnston and Colquhoun 1996); indigenous people have likely had an impact on the distribution and density of this species in some areas.

The tree also has various medicinal uses. In the Napo region of Ecuador, for example, locals chew the bark or soak it and then apply it to bites of spiders and ants, such as among the Kichwa (Cerón 1996). In the Sucumbios region of the Ecuadorian Amazon adjacent to Colombia, the Cofán boil the bark in water and then after it cools drink the solution to treat coughs (Cerón 1995:81). Further south in the Pastaza province of the Ecuadorian Amazon, the Kichwa employ the leaves and bark to prepare remedies for treating infections, snake bites, and diarrhea (Saltos et al. 2016). In both the Ecuadorian and Peruvian Amazon, the young leaves are chewed, or a solution is made from the tree bark and drunk or rubbed on snake bites. The bark is also boiled in water and the broth is taken to treat coughs. The Amuesha in the Andean foothills in the central Amazonian region of Peru prepare a bathing solution from the leaves to treat psychic disorder (Valadeau et al. 2010). Along the Upper Uaupés River in the Colombian Amazon, indigenous groups make a paste from the leaves to treat larvae of bot flies embedded under their skin (Schultes and Raffauf 1990:114). Villagers living in Marudá along the coast of Pará at the mouth of the Amazon squeeze the leaves and apply the juice to the head to treat eczema (Coelho-Ferreira 2009). The Tsimane in the Bolivian Amazon and the Shuar in the Ecuadorian Amazon use wood for fuel (Bennett et al. 2002:133; Guèze et al. 2014).

9.2 *Cordia scabrifolia*

Brazil Cama de anta, uruazeiro, aruá, chapéu de sol, inviza do campo, freijó, freijó branco, freijorana, grão de galo, maria preta; táiwi-'i (Arawaté), apitere-wɨwa (Asurini), kicu-kowa-'y, wiriri-'ɨ (Guajá), ape'y-tuwyr (Ka'apor)

Peru Chúkchu, kuchúkuch (Mayna)

Status Wild

Jamieson, 10 years, picking *Cordia scabrifolia* fruit in regrowth vegetation in a harvested manioc field on uplands. Inland from Carrazedo, lower Xingu River, Gurupá municipality, Pará, Brazil 4-13-15

Cordia scabrifolia occurs in successionary vegetation and forest on uplands as well as wetlands throughout the Amazon Basin and the Guianas (Abreu et al. 2006; Amaral et al. 2009; Araujo et al. 2001; Batista et al. 2011; Carim 2016:87; Coelho et al. 2013; Dantas et al. 1980; Lemos et al. 2015; Medeiros 2016:92; Santana et al. 2004; Vieira 1996:108). When in early regrowth vegetation, *cama de anta* (tapir's bed) is usually a bush and the fruits are therefore easy to pick; children eagerly seek them out when they are going to and from their family manioc fields. When mature, however, it can reach 25 m in forests, and the fruits are then usually ignored because they do not drop off when ripe. The yellow fruits, 1.4 cm × 1.8 cm, contain a gummy sweet pulp surrounding a single brown seed 1 cm × 0.6 cm. This species, like many others, fruits mainly in the rainy season but in Maranhão in the eastern fringe of the Amazon Basin, this tree fruits in the dry season (Muniz 2008). The highly endangered Ka'apor capuchin monkey *Cebus kaapori* has been recorded eating this fruit in eastern Pará (Oliveira et al. 2014) and likely disperses some of the seeds. Squirrel monkeys *Saimiri sciureus* have also been observed feeding on this fruit in the vicinity of the Tucuruí reservoir along the Tocantins River in eastern Pará and may disperse some of the seeds (Pinheiro et al. 2013).

Boys snacking on fruits of *Cordia scabrifolia* in successionary vegetation adjacent to a manioc field. Carrazedo, confluence of the Xingu and Amazonas rivers, Municipio de Gurupa, Para, Brazil 4-13-15

9.3 *Cordia tetrandra*

Bolivia Picana blanca, picana negra, mechero, juparedycasmu; jurapety casmo (Tsimane)

Brazil Uruá, uveira, parápará, chapéu de sol, chapeleira, freijó, grão de galo, bomboqueiras, araba

Colombia Candilero, candelero

Ecuador Tutumbe, laurel prieto

Peru Farina seco, paujilruro negro, tahuampa caspi, macambillo

Venezuela Caujaro de agua, paragua, alatrique

Status Wild

Uveira *Cordia tetrandra* tree in fruit in a village built on a floodplain. Vila Macacoari, Rio Macacoari, Amapá, Brazil 5-15-02

Cordia tetrandra, known by a variety of names even within a relatively small area such as the Amazon estuary, is used primarily for fish bait. A wide variety of fish are fond of this fruit including tambaqui *Colossoma macropomum*, pirapitinga *Colossoma bidens*, and mandí catfish (Lucas 2008; Smith, 1999a, b). The tree is usually found on floodplains, especially along the lower Amazon and its tributaries (Batista et al. 2011; Cain et al. 1956; Carim 2016:87; Carim et al. 2008; Cattanio et al. 2002; Prance 1989; Ducke 1910), and people generally spare it when they clear woodlands for their homesites. One of the local names for this tree in the Peruvian Amazon is tahuampa caspi which means swamp fruit.

The tree is widespread in the Amazon and Orinoco basins and in the Guianas. The tree, which can attain some 18 m, also grows occasionally in mature upland forests, such as in the watershed of the Upper Juruá in Acre, Brazil (NYBG collection #230034) and along the Jari River in Amapá (Pereira et al. 2011). In forests, the tree tends to grow taller, whereas, in second growth and other more exposed locations, mature *Cordia tetrandra* are generally smaller, under 10 m.

During the flood season, the broad-canopied tree sports hundreds of fruit bunches each containing several dozen blueberry-sized fruits. Locals along the Rio Macacoari in Amapá, an affluent of the main northern channel of the Amazon at its mouth, refer to *Cordia tetrandra* as uveira, the "grape tree," on account of the fruit bunches that resemble those of grape vines, only with smaller fruits. The golden-green fruits are sweet and gummy and children in particular snack on them.

In Guyana, boys take advantage of the sticky nature of the fruit pulp to coat their paper kites (MOBOT Tropicos specimen # 2312); Brazilian boys use discarded plastic to make their *pipas*. Uruá sprouts spontaneously in some home gardens as youngsters discard the single seed in each fruit. *C. tetrandra* is one of several fruit trees found on mounds built in wetlands by the extinct Marajoara culture; perhaps they had a role in increasing the density of this species as well.

Renata, 4 years, with uveira *Cordia tetrandra* fruits she has just picked. Vila Macacoari, Rio Macacoari, Amapá, Brazil 5-15-02

Livestock and game animals also feed on the fruits of *Cordia tetrandra*. In the vicinity of Cachoeira do Arari and Santa Cruz do Arari on Marajó Island, where the tree is called parápará, chickens, and pigs forage for the fallen fruits in January before the rising waters have covered the ground underneath the trees. As waters rise during the flood season chickens and pigs are usually penned on platforms and locals gather the fruits to feed them. Fishermen in Cachoeira do Arari also report that guans (*Penelope* sp.), pheasant-sized game birds, are also fond of the fruits. Villagers in Macacoari along the river of the same name in Amapá feed the fruits to pigs. Also in the state of Amapá, tracajá turtles *Podocnemis unifilis* feed on the fruits during the flood season (Portal et al. 2002); these turtles are widely consumed by people in Amazonia. When flood waters subside, the tree provides welcome shade for cattle and water buffalo that have been released on pastures during the dry season, such as along the Macacoari River, affluent of the Amazon in Amapá.

Cordia tetrandra is also used for firewood and to make charcoal, such as in the vicinity of Vigia in the southern part of the Amazon estuary (Smith 2002). The Tsimane in the Bolivian Amazon also use parts of this tree for wood fuel (Martínez-Rodríguez 2009:91). Known as clammy cherry in the timber trade, the volume of *C. tetrandra* wood extracted appears to be small at the present time since other species in the genus produce valuable hardwoods, particularly *C. alliadora* and *C. goeldiana*. The fact that *C. tetrandra* is usually less than 10 m tall may save it from intense pressure from loggers, at least for a while. As more desirable woods become increasingly scarce and costly, however, attention is likely to shift to hitherto less-utilized timbers, including clammy cherry. Already, clammy cherry is featured on the website of a company based in Georgetown, Guyana, that specializes in exotic woods for crafts and custom furniture makers. Should the whaling syndrome take hold in the timber industry in the Amazon, many trees that provide fruits for fish may become targets for loggers, thereby undercutting an important food resource for both rural and urban folk.

9.4 *Cordia collococca*

Brazil Louro

Peru Sacha uva, sacha macambo, bolsa caspi, camungo tiarana

Status Wild

Louro *Cordia collococca* with immature fruits in successionary vegetation (*capoeira*). Wadrio 10 years. Inland from Carrazedo, mouth of the Xingu River, Município de Gurupá, Pará, Brazil 11-1-16

Louro *Cordia collococca* is a bush in successionary vegetation on uplands that becomes a small tree, reaching some 15 m, when fully mature. This species also occurs occasionally in floodplain forests, such as along the Amazon River near Leticia in Colombia (Prieto et al. 1995) and in the Peruvian Amazon (Dávila et al. 2015:42). The fruits turn yellow and eventually black when fully ripe and locals enjoy the sweet pulp. Coatimundis *Nasua narica*, which generally wander through mature and successionary forest in small bands, also eat the fruits and scatter the seeds (Opler et al. 1975). The fruit measures 1.5 cm by 1 cm and contains a single green seed 1.2 cm × 0.8 cm. The common name for the plant in the Peruvian Amazon is sacha uva (like the grape).

Cordia collococca ranges from Central America and several Caribbean islands south to Venezuela and the Amazon Basin. The mantled howler monkey *Alouatta palliata* has been observed feeding on this fruit in the dry tropical forest in

Guanacaste, Costa Rica (Glander 1979). Although this species does not occur in Amazonia, two related species (*Alouatta seniculus* and *A. belzebul*) are found in the Amazon Basin and *A. seniculus* is known to disperse the seeds (Rojas 2012:58). The wedge-capped capuchin monkey *Cebus olivaceus* has been feeding on this fruit in gallery forests in the savanna region of central Venezuela (Robinson 1984, 1986); this species of capuchin, which also occurs in the Guianas and Amapá and northern Pará in the Brazilian Amazon, also likely disperses the seeds. In the Magdalena River basin in northern Colombia, the rare and endangered brown spider monkey *Ateles hybridus* eats these fruit (Luna et al. 2017); related species of *Ateles* in the Amazon Basin probably also eat this fruit and disperse the seeds. Spider monkeys are widely hunted for their meat.

References

Abreu, Maria Milena de Oliveira; Mehlig, Ulf; Nascimento, Raquel do Espírito Santo Aguair do; Menezes, Moirah Paula Machado. 2006. Análise de composição florística e estrutura de um fragmento de bosque de terra firme e de um manguezal vizinhos na peninsula de Ajuruteua, Bragança, Pará. Boletim do Museu Paraense Emílio Goeldi, Ciencias Naturais 1(3):27–34

Almeida, Alexa-Sandra Farias de. 2012. Análise fitossociológica estrutural e composição florística da área de proteção ambiental margem esquerda do Rio Negro, Manaus-AM. Master's thesis, Universidade Federal do Amazonas, Manaus

Amaral, Dário Dantas do; Vieira, Ima Célia Guimarães; Almeida, Samuel Soares de; Salomão, Rafael Paiva; Silva, Antonio Sérgio Lima da; Jardim, Mário Augusto Gonçalves. 2009. Checklist da flora arbórea de remanescentes florestais da região metropolitana de Belém e valor histórico dos fragmentos, Pará, Brasil. Boletim do Museu Paraense Emílio Goeldi, Ciencias Naturais 4(3):231–289

Araujo, Henrique José Borges de. 2015. Alterações causadas por incêndio florestal na composição de espécies de uma floresta primária no sudoeste da Amazônia Brasileira. Gaia Scientia 9(1):220–231

Araújo, Maristela Machado; Oliveira, Francisco de Assis; Vieira, Ima Célia Guimarães; Barros, Paulo Luiz Contente de; Lima, Cesar Augusto Tenório de. 2001. Densidade e composição florística do banco de sementes do solo de florestas sucessionais na região do baixo Rio Guamá, Amazônia oriental. Scientia Florestalis 59:115–130

Araujo-Murakami, Alejandro; Paniagua, Narel; Cayola, L.; Jorgensen, Peter M.; Valdes, Ovidio; Macuapa, Eric; Calzadilla-Tomianovich, Miki. 2009. Diversidad y estructura florística del bosque de llanura y palmar de pantano Amazónico preandino en el sector de Chalalan, Parque Nacional Madidi, Bolivia. Kempffiana 5(1):3–27

Assis, Rafael Leandro de; Wittmann, Florian; Luize, Bruno Garcia; Haugaasen, Torbjon. 2017. Patterns of floristic diversity and composition in floodplain forests across four southern Amazon river tributaries, Brazil. Flora 229:124–140

Batista, Fábio de Jesus; Jardim, Mário Augusto Gonçalves; Medeiros, Tonny David Santiago; Lopes, Isaac Luiz Magalhães. 2011. Comparação florística e estrutural de duas florestas de várzea no estuário Amazônico, Pará, Brasil. Revista Árvore 35(2):289–298

Bennett, Bradley C.; Baker, Marc A.; Andrade, Patricia Gómez. 2002. Ethnobotany of the Shuar of Eastern Ecuador. New York Botanical Garden, Advances in Economic Botany, vol. 14.

Cain, Stanley A.; Castro, Gustavus M. de Oliveira; Pires, J. Murça; Silva, Nilo Tomâs. 1956. Application of some phytosociological techniques to Brazilian rain forest. American Journal of Botany 43(10):911–941

Campbell, David G.; Stone, Judy L.; Rosas, Jr., Arito. 1992. A comparison of the phytosociology and dynamics of three floodplain (várzea) forests of known ages, Rio Juruá, western Brazilian Amazon. Botanical Journal of the Linnean Society 108:213–237

Carim, Marcelo de Jesus Veiga. 2016. Estrutura, composição e diversidade em florestas alagáveis de várzea de maré e de igapó e suas relações com variáveis edáficas e o período de inundação no Amapá, Amazônia oriental, Brasil. Doctoral dissertation, Programa de Pós-Graduação em Botânica, INPA, Manaus

Carim, Marcelo de Jesus Veiga; Jardim, Mário Augusto Gonçalves; Medeiros, Tonny David Santiago. 2008. Composição florística e estrutura de floresta de várzea no Município de Mazagão, Estado do Amapá, Brasil. Scientia Florestalis 36(79):191–201

Carretero-Pinzón, Xyomara; Defler, Thomas R.; Ruiz-García, Manuel. 2016. How does the Colombian squirrel monkey cope with habitat fragmentation? Strategies to survive in small fragments. In: Phylogeny, Molecular Population Genetics, Evolutionary Biology and Conservation of the Neotropical Primates, Ruiz-García, M.; Shostell, J. M. (eds), Nova Science, New York, chapter 15, 1–16

Casula, Kátia Regina. 2012. Análise da similaridade florística e estrutural das formações florestais inundáveis em um trecho do alto Rio Madeira e de seus afluentes, estado de Rondônia, Brasil. Master's Thesis, Núcleo de Ciências Exatas da Terra, Universidade Federal de Rondônia, Porto Velho

Cattanio, José; Anderson, Anthony B.; Carvalho, Manoel S. 2002. Floristic composition and topographic variation in a tidal floodplain forest in the Amazon estuary. Revista Brasileira de Botanica 25(4):419–430

Cerón, Carlos Eduardo. 1995. Etnobiologia de los Cofanes de Dureno, Provincia de Sucumbíos, Ecuador. Museo Ecuatoriano de Ciencias Naturales, Herbario Nacional, Serie Monografia 10(3), Quito

Cerón, Carlos Eduardo. 1996. Etnobotanica Quichua en la vía Hollín-Loreto, Provincia del Napo. Hombre y Ambiente 25:131–170

Coelho, Roberta de Fátima Rodrigues; Miranda, Izildinha de Souza; Mitja, Danielle. 2013. Conservação das florestas do projeto de assentamento Benfica, Sudeste da Amazônia. Ciencia Florestal, Santa Maria 23(1):1–17

Coelho-Ferreira, Márlia. 2009. Medicinal knowledge and plant utilization in an Amazonian coastal community of Marudá, Pará state (Brazil). Journal of Ethnopharmacology 126:159–175

Culot, Laurence; Huynen, Marie-Claude; Gérard, Paul; Heymann, Eckhard W. 2009. Short-term post-dispersal fate of seeds defecated by two small primate species (Saguinus mystax and Saguinus fuscicollis) in the Amazonian forest of Peru. Journal of Tropical Ecology 25(3):229–238

Dantas, Mário; Rodrigues, Irenice Alves. 1980. Plantas invasoras de pastagens cultivadas na Amazônia. EMBRAPA, Centro de Pesquisa Agropecuária do Trópico Úmido, Boletim de Pesquisa, vol. 1.

Dantas, Mário; Rodrigues, Irenice Alves. 1982. Estudos fitoecológicos do trópico úmido Brasileiro: IV-levantamentos botânicos em campos do Rio Branco. EMBRAPA, Centro de Pesquisa Agropecuária do Trópico Úmido, Boletim de Pesquisa, no. 40

Dantas, Mário; Rodrigues, Irenice Alves; Müller, Nina Rosária Maradei. 1980. Estudos fitoecológicos do trópico úmido brasileiro: aspectos fitossociológicos de mata sobre latossolo amarelo em Capitão Poço, Pará. EMBRAPA-CPATU, Boletim de Pesquisa 9:1–19

Dávila, P. M.; Vega, Juan José Palacios; Vela, Marcial Martínez; Gómez, Ricardo Zárate; García, José Maco; Torres, Roger Escobedo. 2015. Cobertura y uso de la tierra de las provincias de Alto Amazonas y Ramón Castilla. Instituto de Investigaciones de la Amazonía Peruana, Iquitos, Documento Técnico No. 33

Ducke, A. 1910. Explorações scientificas no Estado do Pará. Boletim do Museu Goeldi (Museu Paraense) de Historia Natural e Ethnographia 7:100–200

Foster, Robin B.; Arce, Javier; Wachter, Tatzyana S. 1986. Dispersal and the sequential plant communities in Amazonian Peru floodplain. In: Frugivores and Seed Dispersal, Estrada, Alejandro; Fleming, Theodore H. (eds), Dr. W. Junk Publishers, Dordrecht, pp 357–370

Glander, K. E. 1979. Howling monkey feeding behavior and plant secondary compounds: a study of strategies. In: The Ecology of Arboreal Foliovores, Montgomery, G. G. (ed), Smithsonian Institution Press, Washington, DC, pp 561–574

Godoy, Judith Rosales; Petts, Geoffrey; Salo, Jukka. 1999. Riparian flooded forests of the Orinoco and Amazon basins: a comparative review. Biodiversity and Conservation 8:551–586

Gomez, Tedi Pacheco. 2011. Valoración económica de purmas de diferentes edades en la comunidad nativa de Santa Mercedes, río Putumayo, Perú. Master's thesis, UNAP, Iquitos

Gottsberger, Gerhard; Morawetz, Wilfried. 1986. Floristic, structural and phytogeographical analysis of the savannas of Humaitá (Amazonas). Flora 178:41–71

Griscom, Bronson W.; Daly, Douglas C.; Ashton, Mark S. 2007. Floristics of bamboo-dominated stands in lowland terra-firma forests of southwestern Amazonia. Journal of the Torrey Botanical Society 134(1):108–125

Guèze, Maximilien; Luz, Ana Catarina; Paneque-Gálvez, Jaime; Macía, Manuel J.; Orta-Martínez, Martí; Pino, Joan; Reyes-García, Victoria. 2014. Are ecologically important tree species the most useful? a case study from indigenous people in the Bolivian Amazon. Economic Botany 68(1):1–15

Ima, M. O. 2012. *Saberes Waorani y Parque Nacional Yasuní: Plantas, Salud y Bienestar en la Amazonía del Ecuador*. Iniciativa Yasuní ITT, Ministerio Coordinador de Patrimonio, Ministerio del Ambiente, Programa de las Naciones Unidas para el Desarrollo (PNUD) y Fondo para el Medio Ambiente Mundial (FMAM), Quito

Johnston, Mark; Colquhoun, Araminta. 1996. Preliminary ethnobotanical survey of Kurupukari: an Amerindian settlement of central Guyana. Economic Botany 50(2):182–194

Lemos, D. A. N.; Ferreira, B. G. A.; Siqueira, J. D. P.; Oliveira, M. M.; Ferreira, A. M. 2015. Floristic and phytosociology in dense "terra firme" rainforest in the Belo Monte hydroelectric plant influence area, Pará, Brazil. Brazilian Journal of Biology 75(3):257–276

Lisboa, P. L. B. 2012. *A Terra dos Aruã: Uma História Ecológica do Arquipélago do Marajó*. Museu Paraense Emílio Goeldi, Belém

Lucas, C. M. 2008. Within flood season variation in fruit consumption and seed dispersal by two characin fishes of the Amazon. *Biotropica* 40(5): 581–589.

Luize, Bruno Garcia; Venticinque, Eduardo Martins; Silva, Thiago Sanna Freire; Novo, Evlyn Marcia Leão de Moraes. 2015. A floristic survey of angiosperm species occurring at three landscapes of the central Amazon várzea, Brazil. Checklist: The Journal of Biodiversity Data 11(6):1–15

Luna, Ana Gabriela de; Link, Andrés; Montes, Andrés; Alfonso, Felipe; Mendieta, Leonardo; Di Fiore, Anthony. 2017. Increased folivory in brown spider monkeys Ateles hybridus living in a fragmented forest in Colombia. Endangered Species Research 32:123–134

Macía, Manuel J. 2011. Spatial distribution and floristic composition of trees and lianas in different forest types of an Amazonian rainforest. Plant Ecology 212:1159–1177

Maranho, Álisson Sobrinho; Salimon, Cleber Ibraim; Costa, Daniel da Silva. 2016. Woody flora of natural forest gaps in a bamboo-dominated forest remnant in southwestern Amazonia. Check List: The Journal of Biodiversity Data 12(3):1–9

Martínez-Rodríguez, María Ruth. 2009. Ethnobotanical knowledge acquisition among Tsimane' children in the Bolivian Amazon. Doctoral dissertation, University of Georgia, Athens

Medeiros, Priscilla Sanjuan de. 2016. Indicadores ambientais para funcionalidade ecológica em florestas secundárias de diferentes idades na Amazônia Oriental. Doctoral dissertation, Universidade Federal do Pará, Belém

Muniz, Francisca Helena. 2008. Padrões de floração e frutificação de árvores da Amazônia Maranhense. Acta Amazonica 38(4):617–626

Muñoz, Marianella Guevara. 2010. Evaluación de purmas con fines de manejo en la comunidad nativa de Santa Mercedes, Rio Putumayo, Perú. Master's thesis, Universidad Nacional de la Amazonia Peruana, Iquitos

Navarro, G. 2002. Provincia biogeográfica del Acre-Madre de Dios (Amazónica Suroccidental). In: Geografía Ecológica de Bolivia: Vegetación y Ambientes Acuáticos, Navarro, Gonzalo; Maldonado, Mabel (eds), Centro de Ecología Simón I. Patiño, Cochabamba, pp 51–98

Navarro, G.; Maldonado, M. 2002. Geografía Ecológica de Bolivia: Vegetación y Ambientes Acuáticos. Centro de Ecología Simón I. Patiño, Cochabamba

Nebel, Gustav; Dragsted, Jens; Vanclay, Jerome K. 2000a. Estructura y composición florística del bosque de la llanura aluvial inundable de la Amazonía Peruana: II. El sotobosque de la restinga. Folia Amazónica 10(1–2):151–181

Nebel, Gustav; Kvist, Lars Peter; Vanclay, Jerome K.; Christensen, Henning; Freitas, Luis; Ruiz, Juan. 2000b. Estructura y composición florística del bosque de la llanura aluvial en la Amazonía Peruana: I. El bosque alto. Folia Amazónica 10(1–2):91–149

Nebel, Gustav; Dragsted, Jens; Vanclay, Jerome K. 2001. Structure and floristic composition of flood plain forests in the Peruvian Amazon II. The understorey of restinga forests. Forest Ecology and Management 150: 59–77

Oliveira, S. G. de; Alfaro, Jessica W. Lynch; Veiga, Liza M. 2014. Activity budget, diet, and habitat use in the critically endangered Ka'apor capuchin monkey (*Cebus kaapori*) in Pará state, Brazil: a preliminary comparison to other capuchin monkeys. American Journal of Primatology 76:919–931

Opler, Paul A.; Baker, Herbert G.; Frankie, Gordon W. 1975. Reproductive biology of some Costa Rican Cordia species (Boraginaceae). Biotropica 7(4):234–247

Parolin, P.; Adis, J.; Silva, M. F. da; Amaral, I. L. do; Schmidt, L.; Piedade, M. T. F. 2003. Floristic composition of a floodplain forest in the Anavilhanas archipelago, Brazilian Amazonia. Amazoniana 17(3/4):399–411

Pereira, Luciano Araújo; Sobrinho, Felipe de Araújo; Costa Neto, Salustiano Vilar da. 2011. Florística e estrutura de uma mata de terra firme na Reserva de Desenvolvimento Sustentável no Rio Iratapuru, Amapá oriental, Brasil. Floresta 41(1):113–122

Pinheiro, Tatyana; Ferrari, Stephen F.; Lopes, Maria Aparecida. 2013. Activity budget, diet, and use of space by two groups of squirrel monkeys (*Saimiri sciureus*) in eastern Amazonia. Primates 54:301–308

Portal, Rubens da Rocha; Lima, Maria Auxiliadora S.; Luz, Vera Lúcia Ferreira; Bataus, Yeda Soares de Lucena; Reis, Isaías José dos. 2002. Espécies vegetais utilizadas na alimentação de Podocnemis unifilis, Troschel 1948 (Reptilia, Testudinae, Pelomedusidae) na região de Pracuúba, Amapá, Brasil. Ciencia Animal Brasileira 3(1):11–19

Porto, Maria Luiza; Longhi, Hilda Maria; Citadini, Vanilde; Ramos, Regina Feijó; Mariath, Jorge Ernesto de A. 1976. Levantamento fitossociológico em uma área de "mata-de-baixio" na Estação Experimental de Silvicultura Tropical, INPA, Manaus, Amazonas. Acta Amazonica 6(3):301–318

Prance, G. T. 1989. Economic prospects from tropical rainforest ethnobotany. In: Fragile Lands of Latin America: Strategies for Sustainable Development, Browder, John O. (Editor), pp. 61–74

Prieto, Adriana; Rangel, J. Orlando; Rudas, Agustín; Palacios, Pablo. 1995. Aspectos estructurales y tipos de vegetación de la Isla Mocagua, Rio Amazonas. Caldasia 17(82–85):463–480

Ratter, J. A. 1987. Notes on the vegetation of the Parque Nacional do Araguaia (Brazil). Notes from the Royal Botanic Garden, Edinburgh 44(2):311–342

Rech, André Rodrigo; Absy, Maria Lúcia. 2011a. Pollen storages in nests of bees of the genera Partamona, Scaura and Trigona (Hymenoptera, Apidae). Revista Brasileira de Entomologia 55(3):361–372

Rech, André Rodrigo; Absy, Maria Lúcia. 2011b. Pollen sources used by species of Meliponini (Hymenoptera: Apidae) along the Rio Negro channel in Amazonas, Brazil. Grana 50(2):150–161

Reis, Neidiane Farias Costa. 2010. Estrutura fitossociológica e seletividade de espécies em áreas inundáveis na região do baixo Rio Jaci-Paraná, Rondônia. Master's Thesis, Universidade Federal de Rondônia, Porto Velho
Robinson, J. G. 1984. Diurnal variation in foraging and diet in the wedge-capped capuchin *Cebus olivaceus*. Folia Primatologica 43:216–228
Robinson, J. G. 1986. Seasonal variation in use of time and space by the wedge-capped capuchin monkey, *Cebus olivaceus*: implications for foraging theory. Smithsonian Contributions to Zoology Number 431, Washington, DC.
Rojas, Natalia Alejandra Alvis. 2012. Patron de actividad, dieta, área de actividad y dispersion de semillas de *Alouatta seniculus* en un fragmento de Bosque en San Juan del Carare (Santander). Thesis, Universidad del Tolima, Facultad de Ciencias, Bogotá
Rondon Neto, Rubens Marques; Santos, Jeneffer Soares dos; Silva, Manoel Advíncula da; Koppe, Valdinei Cristi. 2010. Potencialidades de uso de espécies arbustivas e arbóreas em diferentes fisionomias de cerrado, em Lucas do Rio Verde/MT. Revista de Biologia e Ciências da Terra 10(2):113–126
Salomão, Rafael de Paiva; Vieira, Ima Célia Guimarães; Suemitsu, Chieno; Rosa, Nélson de Araújo; Almeira, Samuel Soares de; Amaral, Dário Dantas do; Menezes, Moirah Paulo Machado de. 2007. As florestas de Belo Monte na grande curva do rio Xingu, Amazônia Oriental. Boletim do Museu Paraense Emílio Goeldi, Ciencias Naturais 2(3):57–153
Saltos, Ricardo Vinicio Abril; Vásquez, Tomás Elías Ruiz; Lazo, Jatnel Alonso; Banguera, Derwing Viáfra; Guayasamín, Pedro Damián Ríos; Vargas, Janeth Karina Aguinda; Peñas, Ingrid Vega. 2016. The use of medicinal plants by rural populations of the Pastaza province in the Ecuadorian Amazon. Acta Amazonica 46(4):355–366
Sánchez, Judith Jacira Achong; Levy, Allison Licett Nuñez. 2013. Características de frutos consumidos por *Saguinus nigrifrons* (Primates, Callitrichidae), en la Estación Biológica Quebrado Blanco, Loreto, Perú. Master's thesis, Universidad Nacional de la Amazonía Peruana, Iquitos
Santana, José Augusto da Silva; Almeida, Wanderléa da Costa; Sousa, Luciana Karla V. dos Santos. 2004. Florística e fitossociologia em área de vegetação secundária na Amazônia oriental. Revista de Ciências Agrárias 41:105–120
Santos, Otilene dos Anjos. 2005. Fenologia reprodutiva e biologia floral de espécies do sub-bosque em uma floresta tropical úmida na região de Manaus/AM, Brasil. Master's thesis, INPA/UFAM, Manaus
Santos, Charles F.; Absy, Maria L. 2010. Polinizadores de *Bertholletia excelsa* (Lecythidales: Lecythidaceae): interações com abelhas sem ferrão (Apidade: Meliponni) e nicho trófico. Neotropical Entomology 39(6): 854–861
Schultes, R. E. and R. F. Raffauf. 1990. The Healing Forest: Medicinal and Toxic Plants of the Northwest Amazonia. Dioscorides Press, Portland, Oregon
Shepard, G. 1999. Shamanism and diversity: a Machiguenga perspective. In: Cultural and Spiritual Values of Biodiversity, D. A. Posey (ed), Intermediate Technology Publications, London, pp. 93–95
Smith, N. J. H. 1999a. *The Amazon River Forest*. Oxford University Press, New York.
Smith, N.J.H. 1999b. Biodiversity and agroforestry along the Amazon floodplain. In: *Traditional and Modern Natural Resource Management in Latin America*, F.J. Pichón, J.E. Uquillas, and J. Frechione (Editors), pp. 233–250. University of Pittsburgh Press, Pittsburgh
Smith, N. J. H. 2002. *Amazon Sweet Sea: Land, Life and Water at the River's Mouth*. University of Texas Press, Austin.
Soini, P. 1987. Ecology of the saddle-back tamarin *Saguinus fuscicollis illigeri* on the río Pacaya, northeastern Peru. Folia Primatologica 49:11–32
Solano, P. J.; Durou, S.; Corbara, B.; Quilichini, A.; Cerdan, P.; Belin-Depoux, M.; Delabie, J. H. C.; Dejean, A. 2003. Myrmecophytes of the understory of French Guianian rainforests: their distribution and their associated ants. Sociobiology 41(2):1–10

Souza, Priscila Figueira de. 2012. Florística, distribuição e diversidade beta da comunidade arbórea ao longo do gradiente hídrico em uma floresta de terra firme no interflúvio Purus-Madeira, Amazônia central. Master's thesis, INPA, Manaus.

Taurino, Tássia Cristina da Conceição Barros. 2017. Recuperação natural da diversidade de espécies de plantas em florestas secundárias na Amazônia Oriental. Master's Thesis, Universidade Federal do Pará, Belém

Thomas, Evert. 2009. New light on the floristic composition and diversity of indigenous territory and national park Isiboro-Sécure, Bolivia. Biodiversity Conservation 18:1847–1878

Valadeau, Céline; Castillo, Joaquina Alban; Sauvain, Michel; Lores, Augusto Francis; Bourdy, Geneviève. 2010. The rainbow hurts my skin: medicinal concepts and plants uses among the Yanesha (Amuesha), an Amazonian Peruvian ethnic group. Journal of Ethnopharmacology 127:175–192

Vale, Gabriel Damasco do. 2011. O efeito do solo, inundação e topografia sobre as campinaranas inundáveis na Amazônia. Master's Thesis, INPA, Manaus

Vásquez, Juan Baluarte. 1998. Composición florística de los bosques secundarios de Jenaro Herrera, Loreto. Folia Amazonica 9(1–2):195–218

Van den Eynden, Veerle. (2004). Use and management of edible non-crop plants in southern Ecuador. Ph.d. thesis. Ghent University

Vieira, Ima Célia Guimarães. 1996. Forest succession after shifting cultivation in eastern Amazonia. Doctoral dissertation, University of Stirling

Watanabe, Maurício Takashi Coutinho; Hiura, Alice Lima; Nogueira, Matheus Guimarães Cardoso. 2017. Flora das cangas da Serra dos Carajás, Pará, Brasil: Cordiaceae. Rodriguésia 68(3):955–960

Wittmann, Florian; Schöngart, Jochen; Montero, Juan Carlos; Motzer, Thomas; Junk, Wolfgang J.; Piedade, Maria T. F.; Queiroz, Helder L.; Worbes, Martin. 2006. Tree species composition and diversity gradients in white-water forests across the Amazon Basin. Journal of Biogeography 33(8):1334–1347

Wittmann, Florian; Schöngart, Jochen; Junk, Wolfgang J. 2010. Phytogeography, species diversity, community structure and dynamics of central Amazonian floodplain forests. In: Amazonian Floodplain Forests: Ecophysiology, Biodiversity and Sustainable Management, Junk, Wolfgang J.; Piedade, Maria T. F.; Wittmann, Florian; Schöngart, Jochen; Parolin, Pia (eds), Springer, Dordrecht, pp 61–102

Wittmann, Florian; Householder, Ethan; Piedade, Maria T. F.; Assis, Rafael L. de; Schöngart, Jochen; Parolin, Pia; Junk, Wolfgang J. 2013. Habitat specifity, endemism and the neotropical distribution of Amazonian white-water floodplain trees. Ecography 36:690–707

Yoneda, Masaaki. 1984. Ecological study of the saddle backed tamarin (*Saguinus fuscicollis*) in northern Bolivia. Primates 25(1):1–12

Zárate, Ricardo; Mori, Tony Jonatan; Ramirez, Fredy Francisco; Davila, Hilda Paulette; Gallardo, George Pepe; Cohello, Geancarlo. 2015. Lista actualizada y clave para la identificación de 219 especies arbóreas de los bosques sobre arena blanca de la Reserva Nacional Allpahuayo Mishana, Loreto, Peru. Acta Amazonica 45(2):133–156

Chapter 10
Bromeliaceae

10.1 *Ananas ananassoides*

Bolivia Gabaratá, piñita

Brazil Ananaí, ananás de campo, ananás de raposa, ananás bravo, ananás do miúdinho, ananás do mato, ananã pote, nananzinho, nanaí, abacaxinho, abacaxinho do mato, abacaxinho do cerrado, abacaxi do cerrado, abacaxi do campo, abacaxi do mato, abacaxi brabo, macambira; nànà'i (Tembé)

Colombia Piña de danta, piña brava

Peru Piña de zorro

Venezuela Curagua; jtuwë (Jodï)

Status Wild, cultivated

N. Smith, *Amazon Fruits: An Ethnobotanical Journey*, Ethnobiology,
https://doi.org/10.1007/978-3-031-12803-5_10

Ananaí *Ananas ananassoides* in fruit growing on the edge of a cultural forest adjacent to an open podzol with anthropogenic dark earth. Caru River, road to Silves off the Manaus-Itacoatiara highway, Amazonas, Brazil, 9-20-10

Ananaí *Ananas ananassoides* produces miniature pineapples at the end of long stems. Ananaí means the small pineapple and is aptly named because the fruits are no larger than a tennis ball. The other common name for the wild fruit, ananás do miúdinho, also means the tiny pineapple. The sweet-tasting yellow fruits are gathered for local consumption and never enter markets. These fruits are greatly appreciated because they are produced in the dry season when few native fruits are available. In the Bolivian Amazon, the fruits are used to make a drink mixed with sugar, and the fruit is eaten to expel intestinal worms (Rios and Khan 1998).

When ripe, the fruit stems lean to the ground where they are likely picked off by wild mammals which after eating the fruit disperse the tops which may then take root elsewhere depending on the habitat. The fruits also contain small seeds, some of which are dispersed by tapirs *Tapirus terrestris*, the largest terrestrial game animal in the Amazon (Barcelos et al. 2013; Herrera et al. 1999). The fondness of this fruit by the lowland tapir is also suggested by the common name for the fruit in Colombia: piña de danta (the tapir pineapple). Another possible dispersal agent is one or more species of wild dog or fox, as suggested by the common name for the fruit in Peru: piña de zorro. The fruits are sufficiently tasty to have provided an incentive for at least one indigenous group, the Jodï in the southwestern part of the Orinoco Basin, to cultivate them (Zent and Zent 2012).

This diminutive relative of pineapple ranges widely in the Amazon Basin and contiguous regions. It occurs sporadically on sandy soils in central and eastern Amazonia in vegetation called *campina* and *campinarana*, characterized by isolated stunted trees and closed canopy low forest (Braga 1977; Ferreira et al. 2013; Kinupp and Lorenzi 2014; Pires-O'Brien 1993) as well as sandstone outcrops in the Upper Xingu River Basin (Zappi et al. 2016). It occurs in the undergrowth of forests along river banks in the coastal area of the Bragantina zone in eastern Pará as well as in some inland areas of the Amazon Basin, such as in Roraima (Amaral et al. 2008; Rodrigues et al. 2017; Santos and Bastos 2014:69). This miniature pineapple is found on laterite (*canga*) plateaus in the Serra dos Carajás in southern Pará which are characterized by very thin soils (Mota et al. 2018; Santos et al. 2016a, b, c, d; Silva et al. 1996a, b). It also ranges south into the cerrado region of central Brazil, such as in the Upper Tocantins and Araguaia valleys in Pará, Mato Grosso, Goiás, and then into Minas Gerais, São Paulo, the Gran Chaco of Paraguay and northern Argentina (Baker and Collins 1939; Batalha and Martins 2004; Eiten 1972; Medeiros and Walter 2012; Medeiros et al. 2012; Pereira and Santos 2015; Silva and Neto 2017; Smith and Downs 1979:2054). It also occurs in low forest patches at the edge of savannas in the vicinity of Humaitá along the middle Madeira River (Gottsberger and Morawetz 1986). It also ranges north into the Guianas and Costa Rica. In French Guiana, it is found among granitic inselbergs rising from the inland forest (Gasc et al. 1998). In the Andean foothills, it is found up to 1210 m elevation. It is also found occasionally in home gardens in Abaetetuba and its vicinity along the lower Tocantins River in Pará where locals eat the fruits and prepare a juice from the fruits to stem bleeding (Gonçalves and Lucas 2017; Palheta et al. 2017).

Even if ananaí never makes it as a commercial fruit, it serves as an attractive garden plant in warmer climates. It is found occasionally in the ornamental trade where it is known as the cerrado pineapple. The outermost leaves turn a copper orange, making a nice contrast with the light green leaves underneath and the fruit. The flowers of this pineapple relative attract hummingbirds, another asset with respect to its ornamental properties (Machado and Oliveira 2015).

Ananaí *Ananas ananassoides* in fruit growing on the edge of the cultural forest adjacent to an open podzol with anthropogenic dark earth. Caru River, road to Silves off the Manaus-Itacoatiara highway, Amazonas, Brazil, 9-20-10

10.2 *Ananas comosus*

Bolivia Piña, ananas; cacama (Chácobo), merique, meriĝue (Tsimane)

Brazil Abacaxi, abacaxizeiro, ananás, ananá amarelo, ananá roxo, nana; nanĩ (Araweté), parawa-wɨ'a (Asurini), máviro, maawiro (Baniwa), será (Desana), nana'ĩ (Guajá), sami (Jarawara), õtõ (Juruna), nana (Ka'apor), juparaparuu, juparaparauu (Kaiabi), kãkã (Kaxinawá), akranhiti, okrañitu (Kayapó), susu (Kuikuru), kuáhlu (Nambicuara), kawak (Palikur), anana (Paumari), jacaré akãg (Sateré-Maué), nana (Tembé), chi-ná, chinü, tchinü (Ticuna), se'rá, se'ná, serã ga (Tukano), nana (Waimiri-Atroari), nãnã, rotidoro (Wajãpi)

Colombia Piña, piña de danta; txntadi, too'mi (Andoke), kitsea, kiizia (Bora), ihíbo (Cubeo), chiwilla (Inga), sẽna (Makuna), kúdzùétè (Miraña), küye, kille, kiiye (Muinane), cupuri, chuú (Nukak), tìèná (Ticuna), ofaiyi, rotidoro, rotidorai, roziyi (Witoto)

Ecuador Piña; kuish (Achuar), chiviya (Cofán), chivilla, chiwilla (Kichwa), chíu, kuish (Shuar), insí, isi (Siona-Secoya), ganabu (Waorani)

Peru Piña; tohuan (Amuesha), cúdsiha, kutsya (Bora), tivána (Campa), shim'pa (Chayahuita), chiyu (Huambisa), kankan (Kakataibo), canchi (Matsés), tsitikana, tsirianti (Matsigenka), cancan (Shipibo)

Venezuela Piña; jtineju (Jodï)

Status Planted

Several varieties of pineapple *Ananas comosus* harvested on a small farm. The elongated one on the left is called Garrafão (big bottle) and is used to make juice. The smaller variety next to it is called Mandi and is eaten fresh. Jocojó, Jocojó River, Município de Gurupá, Pará, Brazil 9-23-14

Where pineapple was domesticated has been the subject of much speculation (Smith et al. 1992:107). Some have argued that pineapple was brought into cultivation by indigenous people in Paraguay, others that it arose as a crop in the drier parts of the Brazilian northeast, and still, others that it is native to the rocky hills on the border between the Guianas and the Amazon Basin. The late Richard Evans Schultes (1979) postulates the pineapple may have originated in western Amazonia from *Ananas macrostachys,* whereas others suggest that the pineapple is a domesticated form of *Ananas ananassoides* (Clement et al. 2010). While its precise origins remain obscure, there is a good chance that the pineapple was domesticated long ago in Amazonia. The Colombian Witoto, for example, cultivate more than two dozen varieties of pineapple (Schultes 1990), far more than rural people grow outside the Amazon. Pineapple was spread by indigenous groups to the Caribbean before the arrival of Europeans (Brücher 1989:250).

To eat the fruit, the prickly skin is sliced off, or people cut off segments and hold the skin to eat the juicy, seedless pulp. The hard rind is then tossed away. But the motive for pineapple domestication may have not been for its fruit: the late Carl Sauer (1969) suggests that its earliest use was for the fiber obtained from its leaves. In the mid-nineteenth century, for example, people along the middle Tocantins used the fiber to make string for attaching arrow points to shafts for fishing (Bates 1863a:129), and this fiber was also once used to make fishing lines attached to poles along the lower Tapajós (Bates 1863b:107).

Arlinho, 11 years, eating pineapple ***Ananas comosus*** **in his family's** ***casa de forno*** **(manioc processing hut). Inland from Carrazedo, mouth of the Xingu River, Município de Gurupá, Pará, Brazil 11-2-16**

Huirapiña variety of pineapple *Ananas comosus* in the Andean foothills at 1000 m. Jepelacio, near Moyobamba, San Martin, Peru 8-18-04

Capote variety of pineapple *Ananas comosus* at a roadside stand. Comunidade Cujubim, Município de Pracuúba, Amapá, Brazil 8-30-14

Anana variety of pineapple *Ananas comosus* **with smooth leaves. A few hundred meters inland from Carrazedo, lower Rio Xingu, Pará, Brazil 9-4-14**

De la Selva variety of pineapple ***Ananas comosus*** **in a street market. Moyobamba, San Martin, Peru 8-19-04**

Pineapple was likely domesticated thousands of years ago judging by the large variety of fruit sizes and colors. The Yukuna, an Aruak group who live along the Miriti River in the watershed of the lower Caquetá, cultivate 20 varieties of pineapple (Hammen 1992:195). Some suggest that it may be one of the earliest crops in Amazonia, dating back some 8000–12,000 years (Clement et al. 2010).

Fruit size can vary from relatively small and round to large and elongated. In 1835, two British explorers reported encountering a variety of pineapple at Sarayacu along the Ucayali in the Peruvian Amazon that weighed 7.2 kg, although a Catholic priest there claimed that they can attain 11 kg (Smyth and Lowe 1836:248). One variety planted in the vicinity of Rio Branco in Acre is called Cabeça de Onça (jaguar head); the yellow fruits weigh up to 3 kg and are strongly perfumed. Another traditional variety in Acre, Gigante de Tarauacá, bears purple fruits and can weigh up to 18 kg (Silva 2011a, b:16). Some varieties are yellow when ripe, others orange, red, or green. Some varieties are grown for making juice, others for eating fresh, and still others for both eating and drinking. Pineapple is propagated by planting the leaves that sprout at the top of the fruit or around its base. Desirable mutants can thus be propagated easily.

Tuyuka woman stirring pineapple *Ananas comosus* juice mixed with tucupi (juice squeezed out of grated bitter manioc). The brew is boiled for several hours to get rid of the cyanide in the bitter manioc and is then left for several days to ferment. São Pedro, Rio Tiquié, Amazonas, Brazil 11-8-12

Coroá variety of pineapple *Ananas comosus*. The whorl of leaves around the base of the fruit and at the top can be broken off and planted. Laranjal, Igarapé Laranjal, Floresta Nacional Caxiuanã, Baia de Caxiuanã, Rio Anapú, Município de Melgaço, Pará, Brazil 5-11-15

Pineapple *Ananas comosus* tops ready for planting. Santo Antonio de Ipapucu, Lago de Tefé, Amazonas, Brazil, 8-27-12

Father and son planting pineapple *Ananas comosus* in a recently burned *roça*. A few hundred meters inland from Carrazedo, lower Rio Xingu, Pará, Brazil 9-13-14

Pineapple is eaten after cutting off the prickly skin. The pulp is somewhat crunchy and often juicy. Some indigenous cultures ferment freshly squeezed pineapple juice to make a low alcohol "beer," such as the Nambicuara in Mato Grosso and the Yukuna in the Colombian Amazon (Hammen 1992:230; Roosevelt 1926:240; Schackt 2013:129). In the eighteenth century, rural folk in the vicinity of Pozuzu in the Andean foothills of Peru used pineapples to make a beer (*chicha*) which was described as "*gustosa y delicada*" (Jaramillo-Arango 1952:297). The Tuyuka who live along the Tiquié River, an affluent of the Uaupés near the Brazilian border with Colombia, still maintain this practice. Pineapple juice is stirred in a large pan over a slow-burning fire for several hours and then left to ferment. Apinajé women who live in the interfluvial forests and scrub savanna between the Tocantins and Araguaia Rivers in eastern Pará pound pineapple fruits in a large wooden mortar (*pilão*) then mix the pulp with water and drink it without fermentation (Nimuendajú 1956). The Victorian naturalist Henry Walter Bates noted that pineapple featured prominently in one of the festivals of the Ticuna along the Solimões River. The Yukuna who live in the Miriti-Parana river basin, the last affluent of the Caquetá River before it flows into Brazil, have five pineapple festivals: lumala'a, múkapa, bejo'ola, i'iwa, and punama (Jacopin 1988). But pineapple consumption is avoided by Munduruku women for 2 weeks after they give birth so that their babies' skin does not itch and peel (Murphy and Murphy 1974:163).

Pineapple is also used in a few folk remedies. In the Peruvian Amazon, some people eat this fruit frequently in order to prevent parasite infestations in their intestines (Mejia and Rengifo 2000:160). In the watershed of the Jaú, an affluent of the middle Rio Negro in Amazonas, Brazil, river dwellers (*ribeirinhos*) eat the fruit to alleviate gastrointestinal disturbances (Rodrigues 2006) as do rural folk in the Quilombo Sangrador community in the municipality of Presidente Juscelino in Maranhão (Monteles and Pinheiro 2007). Ribeirinhos along the lower Jauaperi, another affluent of the middle Rio Negro, use the leaves to make a tea to alleviate hemorrhaging; the leaves are applied topically for the same purpose (Pedrollo et al. 2016). On Colares island in the Amazon estuary, approximately 100 km from Belém residents use the juice and macerated leaves of this fruit to expel intestinal worms from their dogs (Ritter et al. 2012). In various parts of the Brazilian Amazon, some rural use the fruit to make a syrup or juice for treating influenza, coughs, bronchitis, and anemia (Ribeiro et al. 2017; Santos et al. 2014a, b, c; Van den Berg 2010:229).

Pineapples *Ananas comosus* on the floor of a communal hut (*maloca*) during a Ticuna wedding festival. Frontispiece to vol. 2 of *The Naturalist on the River Amazons* by Henry Walter Bates, 1863b

Pineapple is known by two names in the Brazilian Amazon: abacaxi and ananas. Abacaxi usually refers to the smooth-leaved variety Cayenne Smooth, which is grown commercially since it is easier for workers to harvest. Abacaxi is an expression in Brazil used to describe difficult or problematic situations and is derived from the prickly surface of the fruits. Ananas generally refers to traditional varieties with sharp teeth along the edges of the leaves. However, the term abacaxi is sometimes applied to all varieties of pineapple. The origin of the term abacaxi is unclear but a tribe was once named after this fruit; it lived along the eastern banks of the Madeira River before moving to Itacoatiara on the northern bank of the Amazon River where it was noted among several other tribal peoples in the town in the mid-eighteenth century (Sampaio 1825:5).

In the vicinity of Boa Esperança, a small village on the western shores of Lago Amanã, a lake along the lower Juruá River in Amazonas, Brazil, ananas is sometimes planted along the edges of fields planted to manioc and other annual crops to discourage mammals, such as agouti (*Dasyprocta* spp.) from entering the fields and the crops. The Tsimane in the Bolivian Amazon also plant small patches of pineapple at the margins of their slash-and-burn fields; they also plant pineapple in fields that are being abandoned after 2–3 years of cultivating other crops such as manioc and rice (Reyes-García 2001:79). Sometimes pineapple and other fruit-bearing plants are intercropped with manioc, as is the case among the Bora who inhabit the Yaguasyacu River, an affluent of the Ampiyacu which in turn flows into the Amazon

River at Pebas, Peru (Denevan and Treacy 1987; Denevan et al. 1984) and among the Matsigenka in the southeastern Peruvian Amazon (Wezel and Ohl 2006).

Pineapple *Ananas comosus* **planted in rows along the sides of a path between two manioc fields to discourage animals from eating the tubers. Boa Esperança, Lago Amanã, Amazonas, Brazil, 9-20-12**

Pineapple is grown in home gardens in urban and rural areas, in the vicinity of manioc processing huts, and in fields (known as *roças* in the Brazilian Amazon and *chacra* in Peru). Pineapple is mostly planted on uplands, but it is grown occasionally in home gardens on floodplains, such as along the lower Solimões near its confluence with the Rio Negro (Lima and Saragoussi 2000) as well as in some upland home gardens as in the municipality of Bonito in northeastern Pará (Rosa et al. 2009), the vicinity of Marabá (Silva et al. 2014a, b, c), in the savanna region of Roraima (Pinho 2008:37), and in Comunidade Sucuri, Municipality of Cuiabá, Mato Grosso (Gonçalves and Pasa 2015). In fields, pineapple is grown both as a monocrop or intermixed with other plants such as manioc or various fruit or timber trees. Pineapple plants can survive for some 10–20 years in regrowth vegetation after manioc fields are harvested or manioc processing huts are abandoned.

Perola variety of pineapple *Ananas comosus* planted a year ago. The field has 35,000 pineapple plants. A tucumã *Astrocaryum aculeatum* palm was spared when the regrowth vegetation was cleared because it produces edible fruits. Outskirts of Santo Antonio de Ipapucu, Lago de Tefé, Amazonas, Brazil, 8-27-12

In some areas with sandy soils, pineapple can become feral such as on relic sand dunes on Marajó and Mexiana islands at the mouth of the Amazon (Meggers and Evans 1957:427; Smith 2002:130). In some areas, pineapple replicates without human intervention, although introduced water buffalo on those islands have severely trampled and in some areas destroyed the feral populations.

Pineapple *Ananas comosus* is now widely cultivated in tropical regions of the Americas, the Caribbean, Africa, Asia, and the islands in the Indian and Pacific oceans. The prickly fruit is also readily available for purchase in supermarkets in the United States, Canada, and Europe.

10.3 *Bromelia karatas*

Brazil Coatá, croatá, curaná, curauá, caroá; irawá (Aparai), karawata (Tembé), kuraiwat (Wayana)

Colombia Piñuela

Status Wild

Croatá *Bromelia karatas* among granitic rocks at the edge of a floodplain. Lajedo do Zé Maia, right bank of the middle Xingu River 1 km upstream from its confluence with the Bacajá River, Pará, Brazil, 9-10-15

Coatá or croatá *Bromelia karatas* is another wild bromeliad with edible fruits which is found in a variety of habitats ranging from sand dunes and stunted vegetation growing among granitic rocks in the Amazon Basin. It also occurs in gallery forests and scrub savanna (*cerrado, cerradão*) in the southern part of the Amazon Basin (Pereira and Santos 2015). This long-leafed bromeliad is widespread in lowland tropical America from Mexico and the Caribbean south to the Amazon Basin and the coast of Brazil south to Bahia.

Harvesting the fruits takes dexterity because the arching leaves, which can exceed 3 m, are armed with thorns shaped like cat's claws. The spines lining the edges of the sword-shaped leaves have presumably evolved to keep mammals, such as peccary, at bay. Most rodents, on the other hand, are small enough to access the fruits without being raked by the backward curving spines. Perhaps they swallow some of the seeds and later defecate them, thereby dispersing the plant. According to a cowboy on Fazenda Jilva on Marajó Island, pigs and agoutis manage to eat coatá fruits, although they likely sustain some cuts. Coatá forms virtually impenetrable thickets under the diffuse shade of the canopy or in more open areas among rocks along the shores of the Middle Xingu and on higher parts of islands in that clear water river.

Cowboy gathering coatá *Bromelia karatas* fruits. Ponto de Maguari, Fazenda Oriente, Marajó Island, Pará, Brazil 12-15-01

The fruits look like miniature red bananas, except the tart flesh is white and contains small black seeds the size of peppercorns. Bananas *Musa* arose in tropical Asia and do not contain seeds and the plants have to be propagated vegetatively. The crunchy pulp of coatá is eaten fresh and also boiled with salt, especially on northern and eastern Marajó Island which has relic sand dunes formed when sea levels were higher thus providing good habitat for *Bromelia karatas*. These ancient sand dunes, about a meter or two high, contain other useful plants and may in fact be covered with anthropogenic vegetation in some cases. For some, the coatá fruits are too acidic "*corta a boca*" so they are also boiled before being eaten. In Venezuela, the fruits are eaten and made into juice as well as jam. In Mexico, the fruits are also used to make a sauce for tacos (Hornung-Leoni 2011).

In the mid-nineteenth century, indigenous peoples along the Rio Negro extracted fiber from the leaves of caraná, as *B. karatas* is known there. The fiber was used to make bowstrings and hammocks (Spruce 1853, 1908:520).

Son of a cowboy eating coatá *Bromelia karatas* fruit. Fazenda Cajueiro, Chaves Municipality, Marajó Island, Pará, Brazil 8-18-00

References

Amaral, Dário Dantas do; Prost, Maria Thereza; Bastos, Maria de Nazaré do Carmo; Costa Neto, Salustiano Vilar; Santos, João Ubiratan Moreira dos. 2008. Restingas do litoral amazônico, estados do Pará e Amapá. Boletim do Museu Paraense Emílio Goeldi, Ciencias Naturais 3(1):35–67

Baker, Kenneth F.; Collins, J. L. 1939. Notes on the distribution and ecology of *Ananas* and *Pseudananas* in South America. American Journal of Botany 26(9):697–702

Barcelos, Adriana Renata; Bobrowiec, Paulo Estefano D.; Sanaiotti, Tânia Margarete; Gribel, Rogério. 2013. Seed germination from lowland tapir (*Tapirus terrestris*) fecal samples collected during the dry season in the northern Brazilian Amazon. Integrative Zoology 8:63–73

Batalha, Marco Antônio; Martins, Fernando Roberto. 2004. Reproductive phenology of the cerrado plant community in Emas National Park (central Brazil). Australian Journal of Botany 52:149–161

Bates, H. W. 1863a. *The Naturalist on the River Amazons*. John Murray, London, vol. 1

Bates, H. W. 1863b. *The Naturalist on the River Amazons*. John Murray, London, vol. 2

Braga, Marilene Marinho Nogueira. 1977. Anatomia foliar de Bromeliaceae da campina. Acta Amazonica 7(3) Supl.:1–74

Brücher, H. 1989. *Useful Plants of Neotropical Origin and their Wild Relatives*. Springer, Berlin

Clement, C. R., M. Cristo-Araújo, G. C. d'Echenbrugge, A. A. Pereira, and D. Picanço-Rodrigues. 2010. Origin and domestication of native Amazonian crops. *Diversity* 2:72–106.

Denevan, W. M. and J. M. Treacy. 1987. Young managed fallows at Brillo Nuevo. In: *Swidden-Fallow Agroforestry In the Peruvian Amazon*, W. M. Denevan and C. Padoch (eds.), Advances in Economic Botany vol. 5, New York Botanical Garden, pp. 8–46

Denevan, William M.; Treacy, John M.; Alcorn, Janis B.; Padoch, Christine; Denslow, Julie; Paitan, Salvador Flores. 1984. Indigenous agroforestry in the Peruvian Amazon: Bora indian management of swidden fallows. Interciencia 9(6): 346–357

de Medeiros, M. B., & Walter, B. M. T. 2012a. Composition and structure of the woody cerrado stricto sensu in northern Tocantins and southern Maranhão. Revista Árvore, 36(4), 673–683

Eiten, G. 1972. The cerrado vegetation of Brazil. Botanical Review 38(2):201–341

Ferreira, Leandro V.; Chaves, Priscilla P.; Cunha, Denise de A.; Rosário, Alessandro S. do; Parolin, Pia. 2013. A extração ilegal de areia como causa do desaparecimento de campinas e campinaranas no estado do Pará, Brasil. Pesquisas Botânica 64:157–173

Gasc, Jean-Pierre; Sarthou, Corinne; Garrouste, Romain; Villiers, Jean-François; Cremers, Georges; Thiollay, Jean-Marc. 1998. Inselbergs et savanes-roches en Guyane: biodiversité et conservation des milieux associés aux affleurements granitiques. Journal d'Agriculture Traditionnelle et de Botanique Appliquée 40(1–2):311–327

Gonçalves, Janaína Pinheiro; Lucas, Flávia Cristina Araújo. 2017. Agrobiodiversidade e etnoconhecimento em quintais de Abaetetuba, Pará, Brasil. Revista Brasileira de Biociências 15(3):119–134

Gonçalves, Karina Gondolo; Pasa, Maria Corette. 2015. A etnobotânica e as plantas medicinais na Comunidade Sucuri, Cuiabá, MT, Brasil. Interações 16(2):245–256

Gottsberger, Gerhard; Morawetz, Wilfried. 1986. Floristic, structural and phytogeographical analysis of the savannas of Humaitá (Amazonas). Flora 178:41–71

Hammen, M. C. 1992. *El Manejo del Mundo: Naturaleza y Sociedad entre los Yukuna de la Amazonia colombiana*. Tropenbos, Bogotá

Herrera, José Carlos; Taber, Andrew B.; Wallace, Robert B.; Painter R. Lilian E. 1999. Lowland tapir (Tapirus terrestris) behavioral ecology in a southern Amazonian tropical forest. Vida Silvestre Neotropical 8(1–2):31–37

Hornung-Leoni, Claudia T. 2011. Bromeliads: traditional plant foods in Latin America since Prehistoric times. Polibotánica 32:219–229

Jacopin, Pierre-Yves. 1988. Anthropological dialectics: Yukuna ritual as defensive strategy. Schweizerische Amerikanisten-Gessellschaft Bull. 52:35–46

Jaramillo-Arango, J. 1952. *Relación Historica del Viage, que Hizo a los Reynos del Peru y Chile el Botanico d. Hipolito Ruiz en el Año de 1777 hasta el de 1788, en cuya Epoca Regreso a Madrid*. Real Academia de Ciencias Exactas, Fisicas y Naturales, Madrid, vol. 1

Kinupp, V. F. and H. Lorenzi. 2014. *Plantas Alimentícas Não Convencionais (PANC) no Brasil: Guia de Identifação, Aspectos Nutricionais e Receitas Ilustradas*. Instituto Plantarum de Estudos da Flora, São Paulo.

Lima, Roberval M. B. de; Saragoussi, Muriel. 2000. Floodplain home gardens on the central Amazon in Brazil. In: The Central Amazon Floodplain: Actual Use and Options for a Sustainable Management, Junk, W. J.; Ohly, J. J.; Piedade, M. T. F.; Soares, M. G. M. (eds), Backhuys Publishers, Leiden pp 243–268

Machado, Adriana Oliveira; Oliveira, Paulo Eugênio. 2015. Diversidade beta de plantas que oferecem néctar como recurso floral aos beija-flores em cerrados do Brasil central. Rodriguésia 66(1):1–19

Medeiros, Marcelo Brilhante de; Walter, Bruno Machado Teles. 2012. Composição e estrutura de comunidades arbóreas de cerrado stricto sensu no norte do Tocantins e sul do Maranhão. Revista Árvore 36(4): 673–683

Medeiros, M., Silva, S., Walter, B., Oliveira, W., Silva, G., Gomes, B., Moser, P., Lima, I. and Cavalcanti, T., 2012. Vascular flora of the Tocantins river middle basin, Brazil. Check list, 8(5), pp. 852–885

Meggers, B. J. and C. Evans. 1957. *Archeological Investigations at the Mouth of the Amazon*. Bureau of American Ethnology, Bulletin 167, Smithsonian Institution, Washington, D. C.

Mejia, Kember; Rengifo, Elsa. 2000. Plantas Medicinales de Uso Popular en la Amazonía Peruana. Agencia Española de Cooperación Internacional, Lima

Monteles, Ricardo; Pinheiro, Claudio Urbano B. 2007. Plantas medicinais em um quilombo maranhense: uma perspectiva etnobotânica. Revista de Biologia e Ciências da Terra 7(2):1–12

Mota, Nara Furtado de Oliveira; Watanabe, Mauricio Takashi Coutinho; Zappi, Daniela Cristina; Hiura, Alice Lima; Pallos, Julieta; Viveros, Raquel Stauffer; Giulietti, Ana Maria; Viana, Pedro Lage. 2018. Cangas da Amazônia: a vegetação única de Carajás evidenciada pela lista de fanerógramas. Rodriguésia 69(3):1435–1488

Murphy, Yolanda; Murphy, Robert F. 1974. Women of the Forest. Columbia University Press, New York

Nimuendajú, C. 1956. Os Apinayé. *Boletim do Museu Paraense Emílio Goeldi* 12:1–146

Palheta, Ivanete Cardoso; Tavares-Martins, Ana Cláudia Caldeira; Lucas, Flávia Cristina Araújo; Jardim, Mário Augusto Gonçalves. 2017. Ethnobotanical study of medicinal plants in urban home gardens in the city of Abaetetuba, Pará state, Brazil. Boletín Latinoamericano y del Caribe de Plantas Medicinales y Aromáticas 16(3):206–262

Pedrollo, Camilo Tomazini; Kinupp, Valdely Ferreira; Shepard Jr., Glenn; Heinrich, Michael. 2016. Medicinal plants at Rio Jauaperi, Brazilian Amazon: ethnobotanical survey and environmental conservation. Journal of Ethnopharmacology 186:111–124

Pereira, Antonio Carlos; Santos, Eduardo Ribeiro dos. 2015. Frutas nativas do Tocantins com potencial de aproveitamento económico. Agri-Environmental Sciences 1(1):22–37

Pinho, Rachel Camargo de. 2008. Quintais agroflorestais indígenas em áreas de savana (lavrado) na Terra Indígena Araçá, Roraima. Master's thesis, INPA/UFAM, Manaus

Pires-O'Brien, M. J. 1993. Local distribution and ecology of "palha preta" – a pioneer and invasive palm in Jari, lower Amazon. *Principes* 37(4): 212–215.

Reyes-García, Victoria. 2001. Indigenous people, ethnobotanical knowledge, and market economy: a case study of the Tsimane' Amerindians in lowland Bolivia. Doctoral dissertation, University of Florida, Gainesville.

Ribeiro, Reginaldo Vicente; Bieski, Isanete Geraldini Costa; Balogun, Sikiru Olaitan; Martins, Domingos Tabajarade Oliveira. 2017. Ethnobotanical study of medicinal plants used by ribeirinhos in the north Araguaia microregion, Mato Grosso, Brazil. Journal of Ethnopharmacology 205: 69–102

Rios, Rodrigo; Khan, Bridgette. 1998. List of ethnobotanical use of Bromeliaceae. Journal of the Bromeliad Society 48:75–87

Ritter, Rhuan Amorim; Monteiro, Maria Vivina Barros; Monteiro, Frederico Ozanan Barros; Rodrigues, Silvane Tavares; Soares, Marina Lira; Silva, Jean Carlos Ramos; Palha, Maria das Dores Correia; Biondi, Germano Francisco; Rahal, Sheila Canevese; Tourinho, Manuel Malheiros. 2012. Ethnoveterinary knowledge and practices at Colares island, Pará state, eastern Amazon, Brazil. Journal of Ethnopharmacology 144:346–352

Rodrigues, Eliana. 2006. Plants and animals utilized as medicines in the Jaú National Park (JNP), Brazilian Amazon. Phytotherapy Research 20:378–391

Rodrigues, Rodrigo Schütz; Perdiz, Ricardo de Oliveira; Flores, Andréia Silva. 2017. Novas ocorrências de angiospermas para o estado de Roraima, Brasil. Rodriguésia 68(2):783–790

Roosevelt, T. 1926. *Through the Brazilian Wilderness*. Charles Scribner's Sons, New York.

Rosa, Leonilde dos Santos; Vieira, Thiago Almeida; Pires, Helaine Cristine Gonçalves. 2009. Quintais agroflorestais em comunidades rurais de Bonito, Pará. Revista Brasileira de Agroecologia 4(2):1310–1313

Sampaio, F. X. R. 1825. Diário da Viagem que em Visita, e Correição, das Povações da Capitania de S. Joze do Rio Negro fez o Ouvidor, e Intendente Geral da Mesma, no Anno de 1774 e 1775. Typographia da Academia, Lisbon

Santos, J. U. M. and M. N. Bastos. 2014. Flores e Frutas das Restingas do Estado do Pará. Universidade Federal Rural da Amazônia (UFRA), Belém

Santos, Gideão Costa; Tourinho, Manoel Malheiros; Mendes, Fernanda da Silva; Guimarães, Célia Maria Costa. 2014a. Fitossociologia e práticas de manejo tradicional em uma floresta de várzea em Santa Bárbara do Pará, Brasil. Revista de Ciências Agrárias 57(2):138–145

Santos, L. O., L. O. Santos, M. P. M. Menezes, C. R. Beasley, and U. Mehlig. 2014b. Composition and structure of a diverse tree community at the edges of a Brazilian Amazon rainforest island surrounded by marshes and mangroves. Plant Ecology 215: 1469–1481

Santos, M. R. A.; Lima, M. R.; Oliveira, C. L. L. G. 2014c. Medicinal plants used in Rondônia, western Amazon, Brazil. Revista Brasileira de Plantas Medicinais 16(3):707–720

Santos, Joielan Xipaia dos; Reis, Alisson Rodrigo Souza; Matos, Sterphane Araújo de; Leão, Fábio Miranda; Carvalho, Josiane Celerino de. 2016a. Caracterização etnobotânica de essências florestais com fins medicinais utilizadas pela etnia Xipaya, no município de Altamira-PA. Biota Amazônia 6(2):1–8

Santos, Otilene dos Anjos; Couceiro, Sheyla Regina Marques; Rezende, Alinne Costa Cavalcante; Silva, Márcia Daniela de Sousa. 2016b. Composition and richness of woody species in riparian forests in urban areas of Manaus, Amazonas, Brazil. Landscape and Urban Planning 150:70–78

Santos, R S, P G C Lima, M Coelho-Ferreira, A L Albernaz, A L Feliciano, R Scheel-Ybert. 2016c. Estudos botânicos realizados em Carajás e as perspectivas para uma abordagem etnobiológica e paleobotânica. In: Amazônia Antropogênica, M P Magalhães (ed), Museu Paraense Emílio Goeldi, Belém, pp. 199–213

Santos, Silvio Roberto Miranda dos; Kato, Osvaldo Ryohei; Tourinho, Manoel Malheiros. 2016d. Análise florística e finalidade de espécies cultivadas em sistemas agroflorestais no município de Cametá, Pará, Amazônia Oriental. Amazônia: Ciência e Desenvolvimento 13(23):9–22

Sauer, C. O. 1969. *Agricultural Origins and Dispersals: The Domestication of Animals and Foodstuffs*. M.I.T. Press, Cambridge.

Schackt, J. 2013. *A People of Stories in the Forest Myth*. Instituttet for Sammenlignende Kulturforskning, Oslo

Schultes, R. E. 1979. The Amazonia as a source of new economic plants. *Economic Botany* 33:258–266.

Schultes, R. E. 1990. Gifts of the Amazon flora to the world: Many valuable plants were originally domesticated in the now threatened rain forest of the Amazon basin. Arnoldia 50(2):21–34

Silva, Rafael Carletti Marcolino da. 2011a. Agrobiodiversidade de propriedades agrícolas familiares no município de Presidente Figueiredo, AM. Master's thesis, INPA, Manaus

Silva, Silvestre. 2011b. Fruits of the Brazilian Amazon. Metalivros, São Paulo

Silva, Regisnei Aparecido de Oliveira; Neto, Germano Guarim. 2017. O saber etnobotânico da comunidade de Retireiros do Araguaia em Luciara, Mato Grosso, Brasil. Flovet 1(9):60–85

Silva, Danilo F. da; Anunciação, Clodoaldo J. da; Nodal, Hiroshi; Reis, Odemar V. dos. 1996a. Variabilidade genética em populações naturais de cubiu da Amazônia. Horticultura Brasileira 14(1):9–14

Silva, Manoela F. F. da; Secco, Ricardo de S.; Lobo, Maria da Graça A. 1996b. Aspectos ecológicos da vegetação rupestre da Serra dos Carajás, Estado do Pará, Brasil. Acta Amazonica 26(1/2):17–44

Silva, Kátia Emidio da; Martins, Sebastião Venâncio; Fortin, Marie-José; Ribeiro, Milton Cezar; Azevedo, Celso Paulo de; Ribeiro, Carlos Antonio Álvares Soares; Santos, Nerilson Terra. 2014a. Tree species community spatial structure in a terra firme Amazon forest, Brazil. Bosque 35(3):347–355

Silva, Thiago Paixão da; Silva, Erica Micaelli de Jesus; Amorim, Ismael Alves; Aquino, Alyssandra Lima; Matos, Thaís Esllem da Silva; Rodrigues, Diego Macedo. 2014b. Levantamento de espécies vegetais e utilização em quintal agroflorestal de estabelecimento agrícola no assentamento Alegria – Marabá, Pará. Agroecossistemas 6(1):103–109

Silva, Wanuyze Adriana Silva da; Carim, Marcelo de Jesus Veiga; Guimarães, José Renan da Silva; Tostes, Luciedi de Cássia Leôncio. 2014c. Composição e diversidade florística em um trecho de floresta de terra firme no sudoeste do Estado do Amapá, Amazônia Oriental, Brasil. Biota Amazônia 4(3):31–36

Smith, N. J. H. 2002. *Amazon Sweet Sea: Land, Life and Water at the River's Mouth*. University of Texas Press, Austin.

Smith, L. B. and R. J. Downs. 1979. Bromelioideae (Bromeliaceae). New York Botanical Garden, Flora Neotropica Monograph 13, Part 3

Smith, N. J. H.; J. T. Williams; D. L. Plucknett; and J. P. Talbot. 1992. *Tropical Forests and Their Crops*. Cornell University Press, Ithaca.

Smyth, William; Lowe, Frederick. 1836. Narrative of a Journey from Lima to Pará, across the Andes and Down the Amazon: Undertaken with a view of Ascertaining the Practicality of a Navigable Communication with the Atlantic, by the Rivers Pachitea, Ucayali, and Amazon. John Murray, London

Spruce, R. 1853. Botanical objects communicated to the Kew Museum, from the Amazon River, in 1851. In: *Hooker's Journal of Botany and Kew Garden Miscellany*, W. J. Hooker (Editor), Lovell Reeve, London, vol. 5, pp. 169–177.

Spruce, R. 1908. Notes of a Botanist on the Amazon and Andes. Macmillan, London, vol. 2

Van den Berg, M. E. 2010. Plantas Medicinais na Amazônia. Museu Paraense Emílio Goeldi, Belém

Wezel, A.; Ohl, J. 2006. Homegarden plant diversity in relation to remoteness from urban centers: a case study from the Peruvian Amazon region. In: Tropical Homegardens: a time-tested example of sustainable agroforestry, Kumar, B. M.; Nair, P. K. R. (eds), pp 143–158, Advances in Agroforestry vol. 3, Springer, Dordrecht

Zappi, Daniela C.; Milliken, William; Lopes, Celia Regina A. Soares; Lucas, Eve; Piva, José Hypólito; Frisby, Sue; Biggs, Nicola; Forzza, Rafaela Campostrini. 2016. Xingu State Park vascular plant survey: filling the gaps. Brazilian Journal of Botany 39(2):751–778

Zent, S. and E. Zent. 2012. Jodï horticultural belief, knowledge and practice: incipient or integral cultivation? *Boletim do Museu Paraense Emílio Goeldi, Ciencias Humanas* 7(2): 293–338

Chapter 11
Burseraceae

11.1 *Protium heptaphyllum*

Bolivia Isigo, pacay macho, mezcla, yugua, coloradillo, huevo de perro; isigá (Chiriguano)

Brazil Breu, breu branco, breu vermelho, breu preto, breu de campina, breu da folha fina, breu da folha pequena, breu virgem, breu grande, pau de breu, areu-areu, ameixinho de cheiro, amescla de cheiro, amescla amargoso, amescla breu, amescla branca, amescla aroeira, amesclão branco, amescica, ameciqueira, amécica, amiricia, almécega, almecegueira, amesca, mesca, mescla, mescla aroeira, jauaricica, cicantá iuá; yawar'aɨ (Guajá), arakanei'ɨ (Ka'apor), yhýka, jutahýka (Piripkura), hikata-'ɨw (Tembé), warapa hi (Yanomami)

Colombia Caraña, caraño, currucay, anime, anime blanco, guacamayo, guacharaco, incienso, tachamaco; o-mo-ta' (Cubeo), hee-ta-ma-ká (Tanimuka)

Peru Copalillo, anime, caraño, currucay

Venezuela Caraña, tacamahaco, tacamajaca, catamajaca; moko jyaï (Jodï), damenu (Ye'kuana)

Status Wild, cultivated

N. Smith, *Amazon Fruits: An Ethnobotanical Journey*, Ethnobiology,
https://doi.org/10.1007/978-3-031-12803-5_11

Breu *Protium heptaphyllum* tree in fruit in upland forest. Estação Científica Ferreira Penna, Floresta Nacional Caxiuanã, Municipio de Melgaço, Pará, Brazil 4-30-15

Breu *Protium heptaphyllum* is an understory tree that reaches 15–20 m high when mature. It occurs widely in the Brazilian Amazon in upland and riverine forest and thickets of small trees and bushes on sandy soils known as *campina*, *campinarana*, and *caatinga* as well as in scrub savanna (*cerrado*) and scrub woodlands (*cerradão*) (Alencar 1990; Anderson et al. 1975; Black et al. 1950; Braga and Braga 1975; Daly et al. 2016; Ferreira et al. 2013; Fine et al. 2004, 2010; Franczak et al. 2011; Guarim Neto and Pasa 2009; Kuchla et al. 2015; Kunz et al. 2009; Lemos 2013:36; Macedo and Prance 1978; Marimon and Lima 2001; Marimon Junior and Haridasan 2005; Medeiros and Walter 2012; Medeiros et al. 2012; Prance 1996; Ratter 1987; Ratter et al. 1996; Rodrigues 1961; Rondon Neto et al. 2010; Sanaiotti and Magnusson 1995; Salomão et al. 2007a, b; Santos and Jardim 2006; Souza et al. 2006; Vicentini 2004). It also occurs in stunted vegetation on rocky outcrops with thin soils known as *cerrado rupestre*, such as in the vicinity of Nova Xavantina in Mato Grosso (Lima et al. 2009; Maracahipes et al. 2011). It is also found in some sandy soils subject to periodic flooding along the coast of northeastern Pará (Bastos et al. 1995; Costa Neto et al. 1996; Quaresma and Jardim 2017; Santos and Rosário 1988), Maranhão (Machado and Almeida 2019), and in Roraima (Vale 2011:42). It is relatively common in gallery forests in the cerrado region in the southern fringes of the Amazon forest in Mato Grosso (Corette-Pasa 2011; Kuchla et al. 2015; Lenza et al. 2015; Marimon et al. 2002, 2006, 2010; Miguel et al. 2011a, b; Oliveira Filho 1989;

Pasa et al. 2005) as well as in floodplain forest in low-lying areas in the cerrado zone of the upper Araguaia river basin in the state of Tocantins (Brito et al. 2006; Kurzatkowski et al. 2015; Martins et al. 2008). It also occurs in some floodplain forests along black water rivers in central Amazonia (Lobo 2017:59) as well as in floodplain forests in the Amazon estuary near Belém (Black et al. 1950; Jardim et al. 2013). It often gets its start in successionary vegetation (Balée 1993, 2013:64; Medeiros 2016:94; Oliveira and Jardim 1998; Rodrigues et al. 2004) and survives as an understory tree when the forest returns (Coelho et al. 2013). It also occurs on sand dunes along the northeast coast of Pará (Bastos et al. 2003:12; Santos and Bastos 2014: 71; Santos 2010:14; Silva et al. 2010a, b) as well as in forest islands surrounded by mangroves and marshes along the coast of Pará (Abreu et al. 2006; Amaral et al. 2008; Santos et al. 2014a, b, c). It also grows on patches of sandy soils on uplands in parts of the Peruvian Amazon, such as near Iquitos (Fine et al. 2005) and in the Bolivian Amazon (Araujo-Murakami et al. 2015). It is also found in riparian forest in parts of the Orinoco Basin in Venezuela (Díaz Pérez et al. 2012; Godoy et al. 1999) as well as southeastern Colombia (Cárdenas et al. 2009; Veneklaas et al. 2005). It is also found in some home gardens, such as near Macapá, the capital of Amapá (Pereira et al. 2017) and in the municipality of Cuiabá, Mato Grosso (Gonçalves and Pasa 2015). It is one of the most common trees in the forests of Amazonia, including the Bolivian Amazon (Araujo-Murakami and Ruiz 2006; Steege et al. 2013).

Breu ranges widely in the Amazon and Orinoco basins up to 900 m in the Andes and extends into the Guianas and the center-west region and Atlantic forest of Brazil. It also occurs in coastal restinga vegetation of the Brazilian northeast, such as Pernambuco (Almeida Jr. et al. 2011; Vieira et al. 2014a, b). *Protium heptaphyllum* produces fruits measuring 0.9–1.3 cm long and 2–2.7 cm in diameter. The fruits are produced in the rainy season and are also eaten by people who live in the interior. However, the fruits of *Protium heptaphyllum* do not reach urban markets, although in Iquitos, Peru, the fruits of two related species, *H. grandifolium* and *H. subserratum*, are sold in the Belén street market (Vásquez and Gentry 1989). A soft, spongy pulp surrounds the one to four black seeds of *P. heptaphyllum*. The mesocarp has a somewhat dry texture and a curious flavor, sweet but also with a hint of tartness. The white pulp is encased in red skin (pericarp) that splits open when the fruits are ripe, presumably to attract dispersal agents.

Several species of birds eat the fruits and disperse the seeds. In campina vegetation in the state of Amazonas in Brazil, several species of toucans (*Ramphastos* spp.) and araçaris (*Pteroglossus* spp.) have been observed feeding on the fruits (Macedo 1977; Macedo and Prance 1978); these birds with large bills swallow the fruits whole and thereby disperse the seeds. In deciduous forests and successionary woodlands of southern Mato Grosso and contiguous areas of Mato Grosso do Sul, Bolivia, and Paraguay, the green-cheeked parakeet *Pyrrhura molinae* eats the aril of this fruit and also possibly disperses the seeds (Ragusa-Netto 2007). Although this parakeet is not found in the Amazon Basin, other parakeets are common in many areas of Amazonia and probably also eat the fruits.

Goeldi's marmoset *Callimico goeldii*, a small monkey that lives in Amazonian forests from southern Colombia through eastern Peru and northwestern Brazil and Bolivia, is known to eat this fruit and probably disperses some of the seeds (Christen and Geissmann 1994). Although this monkey is small, it is occasionally hunted for its meat and for the pet trade. The Colombian squirrel monkey *Saimiri cassiquiarensis albigena* has also been observed feeding on this fruit and likely disperses some of the seeds (Carretero-Pinzón et al. 2016) as has the dusky titi monkey *Callicebus ornatus* which is confined to Colombia (Carretero-Pinzón and Defler 2016), the red howler monkey *Alouatta seniculus* in eastern Colombia (Stevenson et al. 2015), the bare-ear marmoset *Callithrix argentata* which is restricted to woodlands between the Tocantins River to the east, the Tapajós River to the west, and the Amazon River to the north (Albernaz and Magnusson 1999), and the ashy titi monkey *Plecturocebus cinarescens* in Rondônia (Mattos 2016:30). The gray four-eyed opossum *Philander opossum*, which occurs widely in central America and northern South America, and the bare-tailed woolly opossum *Caluromys philander* which is found in northern South America, also eat this fruit and disperses some of the seeds (Charles-Dominique et al. 1981). The red-footed tortoise *Chelonoidis carbonaria* which occurs in the Amazon Basin and some contiguous regions also eats this fruit (Wang et al. 2011) and rural people hunt this tortoise for its meat.

At least five species of stingless bees (*Melipona seminigra, M. fulva, M. interrupta, M. rufiventris, Trigona fulviventris*, and *Cephalotrigona femorata*) visit the flowers of this tree to gather pollen and rural people in the Amazon sometimes visit their nests in forests to remove some of the honey (Absy and Kerr 1977; Absy et al. 1980; Ferreira and Absy 2015, 2017; Oliveira et al. 2009a, b). This tree thus also contributes indirectly to the human diet via the much-appreciated honey produced by stingless bees.

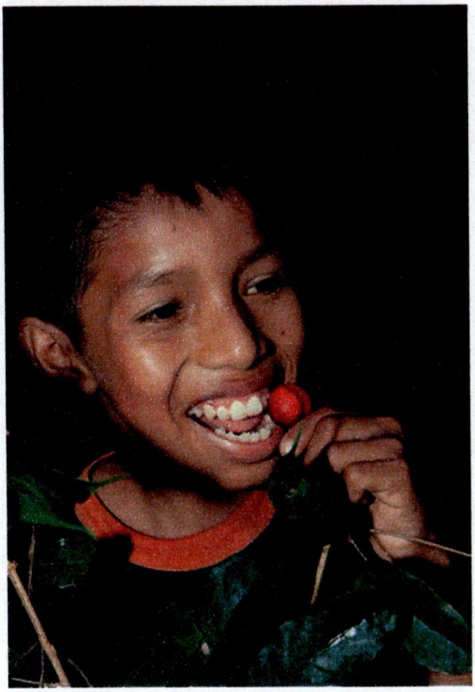

Bildoso, 12 years, eating copalillo *Protium heptaphyllum* **fruit in upland forest. Outskirts of Nauta, Rio Marañón, Loreto, Peru 4-28-06**

As with other species of *Protium*, the dark red resin of breu is used to start wooden fires for cooking and to caulk wooden boats (Brack 2003:65; Domínguez 1985:187; Pérez-Arbeláez 1956:244). The Warao of the Orinoco delta use the resin of *P. heptaphyllum* to flavor their cigars (Wilbert 1987). In the Venezuela Amazon, the Jodï burn the resin to provide lighting (Zent and Zent 2002). The Shipibo living in their village of San Francisco on the shores of Yarina Cocha near Pucallpa use the *Protium* resin to glaze some of their pots; the resin is traded from Conibo further up the Ucayali (Lathrap 1973). This resin is used for glazing pottery among indigenous groups in various parts of the Amazon (Levi-Strauss 1952). The Tanimuka who inhabit Caño Peritomé, an affluent of the Apaporis in the Colombian Amazon, burn the resin of *P. heptaphyllum* and use long tubes to blow the smoke over powdered coca leaves to impart a scent. River dwellers in the vicinity of Afuá on the northwestern coast of Marajó Island employ the resin obtained from *Protium spruceanum* to caulk their boats (Gama et al. 2003). In the early twentieth century, resin from this species was traded in some international markets where it was called elemi (Huber 1910). The wood is used in construction, fashioning furniture, and making charcoal (Aguirre-Mendoza et al. 2015; Le Cointe 1934:63; Loureiro et al. 1979a, b:86).

This plant is also used in some folk remedies. Along the Unini River, an affluent of the Rio Negro in Amazonas, Brazil, the resin is used to stem bleeding and for treating respiratory issues, stroke, and headaches (Lago et al. 2016; Santos et al. 2012). In the cerrado zone bordering the southern Amazon forest in Mato Grosso, the resin-rich bark is used to treat respiratory disorders such as colds and bronchitis, as well as headaches (Gonçalves and Pasa 2015; Guarim Neto and Morais 2003; Moreira and Guarim-Neto 2009; Violante et al. 2012). The bark of this tree is sold in the port market of Cuiabá, the capital of Mato Grosso, so that customers can make a tea from it to treat coughs (Pinto et al. 2013). Resin from the bark of this tree is also used by rural folk in the Quilombo Sangrador community in the municipality of Presidente Juscelino, Maranhão, to treat breathing disorders (Monteles and Pinheiro 2007). The Cubeo who also inhabit the Colombian Amazon use the resin to clear their nasal passages when they are suffering from bad colds (Schultes and Raffauf 1990:117). Villagers living in Marudá along the coast of Pará sometimes burn the bark to repel evil spirits and the bark is also soaked to make a bathing solution to treat rheumatism (Coelho-Ferreira 2009). The resin contains terpenoids which help cure wounds and reduce inflammation (Gilbert 2006). In the community of Bom Jardim, 60 km south of Cuiabá in Mato Grosso, locals use the leaves to make a tea to treat respiratory problems (Pasa 2011) as do rural folk in Maranhão in the eastern fringes of the Amazon Basin (Van den Berg and Silva 1986). Because of its popularity, this resin appears in some street markets in the Amazon, such as in Belém, Pará (Vieira et al. 2014a, b). The resin and leaves of this plant are used to prepare genital bathing solutions in Suriname (Van Andel et al. 2007). Resin is usually produced in trees as a protection against damage, such as attacks by animals especially insects.

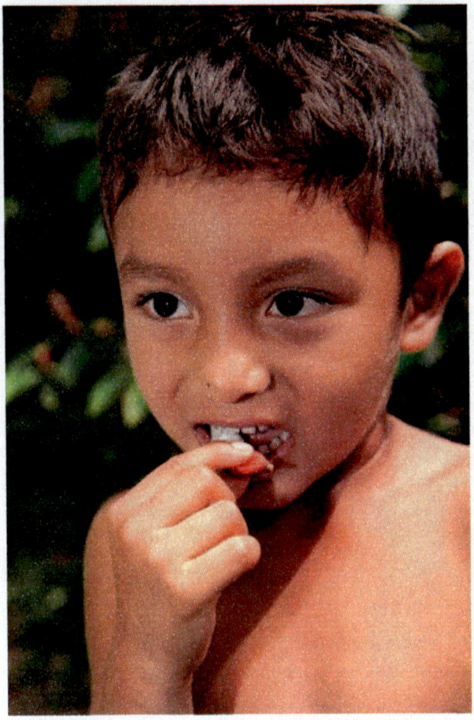

Rafael, 5 years, eating a breu *Protium heptaphyllum* fruit he gathered in regrowth forest. Pedreira, Baia de Caxiuanã, Municipio de Melgaço, Pará, Brazil 5-11-15

References

Abreu, Maria Milena de Oliveira; Mehlig, Ulf; Nascimento, Raquel do Espírito Santo Aguair do; Menezes, Moirah Paula Machado. 2006. Análise de composição florística e estrutura de um fragmento de bosque de terra firme e de um manguezal vizinhos na peninsula de Ajuruteua, Bragança, Pará. Boletim do Museu Paraense Emílio Goeldi, Ciencias Naturais 1(3):27–34

Absy, Maria Lúcia; Kerr, Warwick Estevam. 1977. Algumas plantas visitadas para obtenção de pólen por operárias de Melipona seminigra merrillae em Manaus. Acta Amazonica 7(3):309–315

Absy, Maria Lúcia; Bezerra, Eronildo Braga; Kerr, Warwick E. 1980. Plantas nectaríferas utilizadas por duas espécies de Melipona da Amazônia. Acta Amazonica 10(2):271–281

Aguirre-Mendoza, Zhofre; Loja, Ángel Fernando; Solano, Carmen; Aguirre, Nikolay. 2015. Especies Forestales más aprovechadas del Sur del Ecuador. Universidad Nacional de Loja, Ecuador

Albernaz, Ana Luisa; Magnusson, William E. 1999. Home-range size of the bare-ear marmoset (Callithrix argentata) at Alter do Chão, central Amazonia, Brazil. International Journal of Primatology 20(5):665–677

Alencar, Jurandyr da Cruz. 1990. Interpretação fenológica de espécies lenhosas de campina na Reserva Biológica de campina do INPA ao norte de Manaus. Acta Amazonica 20:145–183

Almeida Jr., Eduardo Bezerra de; Santos-Filho, Francisco Soares; Araújo, Elcida de Lima; Zickel, Carmen Sílvia. 2011. Structural characterization of the woody plants in restinga of Brazil. Journal of Ecology and the Natural Environment 3(3):95–103

Amaral, Dário Dantas do; Prost, Maria Thereza; Bastos, Maria de Nazaré do Carmo; Costa Neto, Salustiano Vilar; Santos, João Ubiratan Moreira dos. 2008. Restingas do litoral amazônico, estados do Pará e Amapá. Boletim do Museu Paraense Emílio Goeldi, Ciencias Naturais 3(1):35–67

Anderson, Anthony B.; Prance, Ghillean T.; Albuquerque, Byron W. P. de. 1975. Estudos sobre a vegetação das campinas Amazonicas-III A vegetação lenhosa da campina da reserva biológica INPA-SUFRAMA (Manaus-Caracaraí, km 62). Acta Amazonica 5(3):225–246

Araujo-Murakami, Alejandro; Ruiz, Freddy S. Zenteno. 2006. Bosques de los Andes orientales de Bolivia y sus especies útiles. In: Botánica Económica de los Andes Centrales, Moraes, M.; Ollgaard, B.; Kvist, L. P.; Borchsenius, F.; Balslev, H. (eds), Universidad Mayor de San Andrés, La Paz, pp. 146–161

Araujo-Murakami, Alejandro; Villarroel, Daniel; Pardo, Guido; Vos, Vincent A.; Parada, G. Alexander; Arroyo, Luzmila; Killeen, Timothy. 2015. Diversidad arbórea de los bosques de tierra firme de la Amazonía Boliviano. Kempffiana 11(1):1–28

Balée, W. 1993. Indigenous transformation of Amazonian forests: an example from Maranhão, Brazil. L'Homme 33(126–128):231–254

Balée, W. 2013. Cultural Forests of the Amazon: A Historical Ecology of People and Their Landscapes. University of Alabama Press, Tuscaloosa

Bastos, M. N. C.; C. S. Rosário; L. C. B. Lobato. 1995. Caracterização fitofisionômica da restinga de Algodoal, Maracanã, Pará, Brasil. Boletim do Museu Paraense Emílio Goeldi, Série Botanica 11(2): 173–197

Bastos, Maria de Nazaré do Carmo; Costa, Denise Cristina Torres; Santos, João Ubiratan Moreira dos. 2003. Vegetação de restinga: aspectos botânicos e uso medicinal. Museu Paraense Emilío Goeldi Projeto Renas, Belém

Black, G. A.; Dobzhansky, T. H.; Pavan, C. 1950. Some attempts to estimate species diversity and population density of trees in Amazonian forests. Botanical Gazette 111(4):413–425

Brack, A. 2003. *Frutas del Perú*. Universidad San Martin de Porres, Lima

Braga, Marilene Marinho Nogueira; Braga, Pedro Ivo Soares. 1975. Estudos sobre a vegetação das campinas Amazonicas IV Estudos ecológicos na campina da Reserva Biológica INPA-SUFRAMA (Manaus-Caracaraí, km 62). Acta Amazonica 5(3):247–260

Brito, Elizabeth Rodrigues; Martins, Sebastião Venâncio; Oliveira Filho, Ary Teixeira de; Silva, Elias; Silva, Alexandre Francisco da. 2006. Estrutura fitossociológica de um fragmento natural de floresta inundável em área de orizicultura irrigada, município de Lagoa da Confusão, Tocantins. Revista Árvore 30(5):829–836

Cárdenas, Dairon; Castaño, Nicolás; Sua, Sonia. 2009. Flora de la estrella fluvial de Inírida (Guianía, Colombia). Biota Colombiana 10(1–2):1–30

Carretero-Pinzón, Xyomara; Defler, Thomas R. 2016. *Callicebus ornatus*, an endemic Colombian species: demography, behavior and conservation. In: Phylogeny, Molecular Population Genetics, Evolutionary Biology and Conservation of the Neotropical Primates, Ruiz-García, M.; Shostell, J. M. (eds), Nova Science, New York, chapter 16, 1–15

Carretero-Pinzón, Xyomara; Defler, Thomas R.; Ruiz-García, Manuel. 2016. How does the Colombian squirrel monkey cope with habitat fragmentation? Strategies to survive in small fragments. In: Phylogeny, Molecular Population Genetics, Evolutionary Biology and Conservation of the Neotropical Primates, Ruiz-García, M.; Shostell, J. M. (eds), Nova Science, New York, chapter 15, 1–16

Charles-Dominique, P.; Atramentowicz, M.; Charles-Dominique, M.; Gérard, H.; Hladik, C. M.; Prévost, M. F. 1981. Les mammiferes frugivores arboricoles nocturnes d'une foret Guyanaise: inter-relations plantes-animaux. Revue d'Écologie: la Terre et la Vie 35:341–435

Christen, A.; Geissmann, T. 1994. A primate survey in northern Bolivia, with special reference to Goeldi's monkey, Callimico goeldii. International Journal of Primatology 15(2): 239–274

Coelho, Roberta de Fátima Rodrigues; Miranda, Izildinha de Souza; Mitja, Danielle. 2013. Conservação das florestas do projeto de assentamento Benfica, Sudeste da Amazônia. Ciencia Florestal, Santa Maria 23(1):1–17

Coelho-Ferreira, Márlia. 2009. Medicinal knowledge and plant utilization in an Amazonian coastal community of Marudá, Pará state (Brazil). Journal of Ethnopharmacology 126:159–175

Corette-Pasa, Maria. 2011. Abordagem etnobotânica na comunidade de Conceição-Açu, Mato Grosso, Brasil. Polibotánica 31:169–197

Costa Neto, Salustiano Vilar da; Bastos, Maria de Nazaré do Carmo; Lobato, Luiz Carlos Batista. 1996. Composição florística e fitofisionomia da restinga do Crispim, município do Marapanim, Pará. Boletim do Museu Paraense Emílio Goeldi, Série Botânica 12(2): 237–249

Daly, Douglas C.; Silveira, Marcos; Medeiros, Herison; Castro, Wenderson; Obermüller, Flávio A. 2016. The white-sand vegetation of Acre, Brazil. Biotropica 48(1):81–89

Díaz Pérez, Wilmer Antonio; Daza, Félix; Sarmiento, William. 2012. Composición florística, estructura y diversidad del bosque ribereños del Río Kakada, cuenca del Río Caura, estado Bolívar, Venezuela. Revista Científica UDO Agrícola 12(2):275–289

Domínguez, Camilo A. 1985. Amazonia Colombiana. Biblioteca Banco Popular, Bogotá

Ferreira, Marcos Gonçalves; Absy, Maria Lucia. 2015. Pollen niche and trophic interactions between colonies of Melipona (Michmelia) seminigra merrillae and Melipona (Melikerria) interrupta (Apidae: Meliponini) reared in floodplain in the Central Amazon. Arthropod-Plant Interactions 9:263–279

Ferreira, Marcos Gonçalves; Absy, Maria Lucia. 2017. Pollen analysis of honeys of Melipona (Michmelia) seminigra merrillae and Melipona (Melikerria) interrupta (Hymenoptera: Apidae) bred in Central Amazon, Brazil. Grana 56(6):436–449

Ferreira, Leandro V.; Chaves, Priscilla P.; Cunha, Denise de A.; Rosário, Alessandro S. do; Parolin, Pia. 2013. A extração ilegal de areia como causa do desaparecimento de campinas e campinaranas no estado do Pará, Brasil. Pesquisas Botânica 64:157–173

Fine, Paul V. A.; Mesones, Italo; Coley, Phyllis D. 2004. Herbivores promote habitat specialization by trees in Amazonian forests. Science 305(5684):663–665

Fine, Paul V. A.; Daly, Douglas C.; Muñoz, Gorky Villa; Mesones, Italo; Cameron, Kenneth M. 2005. The contribution of edaphic heterogeneity to the evolution and diversity of Burseraceae trees in western Amazon. Evolution 59(7):1464–1478

Fine, Paul V. A.; García-Villacorta, Roosevelt; Pitman, Nigel C. A.; Mesones, Italo; Kembel, Steven W. 2010. A floristic study of the white-sand forests of Peru. Annals of the Missouri Botanical Garden 97(3):283–305

Franczak, Daniel David; Marimon, Beatriz Schwantes; Marimon-Junior, Ben Hur; Mews, Henrique Augusto; Maracahipes, Leandro; Oliveira, Edmar Almeida de. 2011. Changes in the structure of a savanna forest over a six-year period in the Amazon-Cerrado transition, Mato Grosso state, Brazil. Rodriguésia 62(2):425–436

Gama, J. R. V., S. A. Botelho, M. Bentes-Gama, and J. R. S. Scolforo 2003. Estrutura e potencial futuro de utilização da regeneração natural de floresta de várzea alta no Município de Afuá, estado do Pará. *Ciencia Florestal, Santa Maria* 13(2):71–82

Gilbert, B. 2006. Produtos naturais industrializáveis da Amazônia. Revista Fitos 2(3):30–38

Godoy, Judith Rosales; Petts, Geoffrey; Salo, Jukka. 1999. Riparian flooded forests of the Orinoco and Amazon basins: a comparative review. Biodiversity and Conservation 8:551–586

Gonçalves, Karina Gondolo; Pasa, Maria Corette. 2015. A etnobotânica e as plantas medicinais na Comunidade Sucuri, Cuiabá, MT, Brasil. Interações 16(2):245–256

Guarim Neto, Germano; Morais, Ronan Gil de. 2003. Plantas medicinais com potencial ornamental: um estudo no cerrado de Mato Grosso. Revista Brasileira de Horticultura Ornamental 9(1):89–97

Guarim Neto, Germano; Pasa, Maria Corette. 2009. Estudo etnobotânico em uma área de cerrado no município de Acorizal, Mato Grosso. Flovet 1:5–32

Huber, J. 1910. Mattas e madeiras amazonicas. Boletim do Museu Goeldi de Historia Natural e Ethnographia 6: 91–225

Jardim, Daniel Gonçalves; Jardim, Mário Augusto G.; Quaresma, Adriano Costa; Costa, Salustiano Villar. 2013. Regeneração natural em formações florestais de uma unidade de conservação, Maracanã, Pará, Brasil. Biota Amazônia 3(2):79–87

Kuchla, Wilian José; Stepka, Thiago Floriani; Sawczuk, Alex Roberto; Lisboa, Gerson dos Santos. 2015. Florística, estrutura horizontal e distribuição diamétrica em área de transição de cerrado e floresta aluvial no município de Campos de Júlio, MT. Ambiência 11(1):13–30

Kunz, Sustanis Horn; Ivanauskas, Natália Macedo; Martins, Sebastião Venâncio. 2009. Estrutura fitossociológica de uma área de cerradão em Canarana, Estado do Mato Grosso, Brasil. Acta Scientiarum Biological Sciences 31(3):255–261

Kurzatkowski, Dariusz; Leuschner, Christoph; Homeier, Jürgen. 2015. Effects of flooding on trees in the semi-deciduous transition forests of the Araguaia floodplain, Brazil. Acta Oecologica 69:21–30

Lago, João Henrique G.; Tezoto, Jaqueline; Yazbek, Priscila B.; Cassas, Fernando; Santos, Juliana de F. L.; Rodrigues, Eliana. 2016. Exudates used as medicine by the "caboclos river-dwellers" of the Unini River, AM, Brazil--classification based in their chemical composition. Revista Brasileira de Farmacognosia 26:379–384

Lathrap, D. W. 1973. The antiquity and importance of long distance trade relationships in the moist tropics of Pre-Columbian South America. *World Archaeology* 5(2):170–186

Le Cointe, P. 1934. *Amazônia Brasileira III: Árvores e Plantas Úteis*. Livraria Classica, Belém

Lemos, Helena Lara. 2013. Vegetação arbustivo-arbórea em áreas de cerrado típico e cerrado rupestre no estado de Tocantins. Master's thesis, Departamento de Engenharia Florestal, Universidade de Brasília, Brasília

Lenza, Eddie; Santos, Josias O.; Maracahipes-Santos, Leonardo. 2015. Species composition, diversity, and vegetation structure in a gallery forest-cerrado sensu stricto transition zone in eastern Mato Grosso, Brazil. Acta Botanica Brasilica 29(3):327–338

Lévi-Strauss, C. 1952. The use of wild plants in tropical South America. *Economic Botany* 6(3):252–270

Lima, Edson de Souza; Lima, Herson Souza; Ratter, James Alexander. 2009. Mudanças pós-fogo na estrutura e composição da vegetação lenhosa, em um cerrado mesotrófico, no período de cinco anos (1997–2002) em Nova Xavantina – MT. Cerne 15(4):468–480

Lobo, Guilherme de Sousa. 2017. A alteração do regime hidrológico afeta a composição florística e estrutura de florestas de igapó? Um estudo comparativo entre um rio regulado e outro prístino na Reserva de Desenvolvimento Sustentável do Uatumã, Amazônia Central. Master's thesis, INPA, Manaus

Loureiro, Arthur A.; Silva, Marlene F. da; Alencar, Jurandyr da Cruz. 1979a. Essências Madeireiras da Amazônia. Instituto Nacional de Pesquisas da Amazônia, Manaus, vol. 1

Loureiro, Arthur A.; Silva, Marlene F. da; Alencar, Jurandyr da Cruz. 1979b. Essências Madeireiras da Amazônia. Instituto Nacional de Pesquisas da Amazônia, Manaus, vol. 2

Macedo, Miramy. 1977. Dispersão de plantas lenhosas de uma campina Amazônica. Acta Amazonica 7(1):1–69 (supl.)

Macedo, Miramy; Prance, Ghillean T. 1978. Notes on the vegetation of Amazonia II. The dispersal of plants in Amazonian white sand campinas: the campinas as functional islands. Brittonia 30(2):203–215

Machado, Monielle Alencar; Almedia Jr., Eduardo Bezerra de. 2019. Spatial structure, diversity, and edaphic factors of an area of Amazonian coast vegetation in Brazil. Journal of the Torrey Botanical Society 146(1):58–68

Maracahipes, Leandro; Lenza, Eddie; Marimon, Beatriz Schwantes; Oliveira, Edmar Almeida de; Pinto, José Roberto Rodrigues; Marimon Junior, Ben Hur. 2011. Estrutura e composição florística da vegetação lenhosa em cerrado rupestre na transição cerrado-floresta Amazônica, Mato Grosso, Brasil. Biota Neotropica 11(1):133–141

Marimon Junior, Ben Hur; Haridasan, Mundayatan. 2005. Comparação da vegetação arbórea e características edáficas de um cerradão e um cerrado sensu stricto em áreas adjacentes sobre solo distrófico no leste de Mato Grosso, Brasil. Acta Botanica Brasilica 19(4):913–926

Marimon, Beatriz Schwantes; Lima, Edson de Souza. 2001. Caracterização fitofisionômica e levantamento florístico preliminar no Pantanal dos Rio Morte-Araguaia, Cocalinho, Mato GrossoActa Botanica Brasilica 15(2):213–229

Marimon, B. S.; Felfili, J. M.; Lima, E. S. 2002. Floristics and phytosociology of the gallery forest of the Bacaba stream, Nova Xavantina, Mato Grosso, Brazil. Edinburgh Journal of Botany 59(2):303–318

Marimon, B. S.; Lima, E. S.; Duarte, T. G.; Chieregatto, L. C.; Ratter, J. A. 2006. Observations on the vegetation of northeastern Mato Grosso. IV. An analysis of the cerrado-Amazonian forest ecotone. Edinburgh Journal of Botany 63(2&3):323–341

Marimon, Beatriz Schwantes; Felfili, Jeanine Maria; Lima, Edson de Souza; Duarte, Wânia M. Gonçalves; Marimon-Júnior, Ben Hur. 2010. Environmental determinants for natural regeneration of gallery forest at the cerrado/Amazonia boundaries in Brazil. Acta Amazonica 40(1):107–118

Martins, Sebastião Venâncio; Brito, Elizabeth Rodrigues; Oliveira Filho, Ary Teixeirade; Silva, Alexandre Francisco da; Silva, Elias. 2008. Floristic composition of two wetland forests in Araguaian plain, state of Tocantins, Brazil, and comparison with other areas. Revista Árvore 32(1):129–141

Mattos, Fábio de Souza. 2016. Dieta do *Plecturocebus cinerascens* (Platyrrhini: Primates): a influência da composição química na escolha de frutos consumidos. Master's thesis, INPA, Manaus

Medeiros, Priscilla Sanjuan de. 2016. Indicadores ambientais para funcionalidade ecológica em florestas secundárias de diferentes idades na Amazônia Oriental. Doctoral dissertation, Universidade Federal do Pará, Belém

Medeiros, Marcelo Brilhante de; Walter, Bruno Machado Teles. 2012. Composição e estrutura de comunidades arbóreas de cerrado stricto sensu no norte do Tocantins e sul do Maranhão. Revista Árvore 36(4): 673–683

Medeiros, M., Silva, S., Walter, B., Oliveira, W., Silva, G., Gomes, B., Moser, P., Lima, I. and Cavalcanti, T., 2012. Vascular flora of the Tocantins river middle basin, Brazil. Check list, 8(5), pp. 852–885

Miguel, Aline; Schwantes Marimon, Beatriz; Almeida de Oliveira, Edmar; Marachaipes, Leandro; Marimon-Junior, Ben Hur. 2011a. Dinâmica da comunidade lenhosa de uma floresta de galeria na transição Cerrado-Floresta Amazônica no leste de Mato Grosso, em um período de sete anos (1999–2006). Biota Neotropica 11(1):1–9

Miguel, Aline; Marimon, Beatriz Schwantes; Oliveira, Edmar Almeida de; Marachaipes, Leandro; Marimon-Junior, Ben Hur. 2011b. Mudanças na estrutura da vegetação lenhosa em três porções da mata de galeria do Córrego Bacaba (1999–2006), Nova Xavantina-MT. Revista Árvore 35(3):725–735

Monteles, Ricardo; Pinheiro, Claudio Urbano B. 2007. Plantas medicinais em um quilombo maranhense: uma perspectiva etnobotânica. Revista de Biologia e Ciências da Terra 7(2):1–12

Moreira, Déborah Luíza; Guarim-Neto, Germano. 2009. Usos múltiplos de plantas do cerrado: um estudo etnobotânico na comunidade sítio Pindura, Rosário Oeste, Mato Grosso, Brasil. Polibotánica 27:159–190

Oliveira Filho, Ary Teixeira de. 1989. Composição florística e estrutura comunitária da floresta de galeria do Córrego da Paciência, Cuiabá (MT). Acta Botanica Brasilica 3(1):91–112

Oliveira, Francisco Plácido Magalhães; Jardim, Mário Augusto Gonçalves. 1998. Composição florística de uma floresta secundária no município de Igarapé-Açu, estado do Pará, Brasil. Boletim do Museu Paraense Emilio Goeldi, Série Botânica 14(2):127–145

Oliveira, Francisco Plácido Magalhães; Absy, Maria Lúcia; Miranda, Izildinha Souza. 2009a. Recurso polínico coletado por abelhas sem ferrão (Apidae, Meliponinae) em um fragmento de floresta na região de Manaus- Amazonas. Acta Amazonica 39(3):505–518.

Oliveira, Viviana Maria Araújo; Carneiro, Ana Lucia Basilio; Cauper, Glaucia Socorro de Barros; Pohlit, Adrian Martin. 2009b. In vitro screening of Amazonian plants for hemolytic activity and inhibition of platelet aggregation in human blood. Acta Amazonica 39(4):973–980

Pasa, Maria Corette. 2011. Saber local e medicina popular: a etnobotânica em Cuiabá, Mato Grosso, Brasil. Boletim do Museu Paraense Emílio Goeldi, Ciencias Humanas 6(1):179–196

Pasa, Maria Corette; Soares, João Juares; Guarim Neto, Germano. 2005. Estudo etnobotânico na comunidade de Conceição-Açu (alto da bacia do rio Aricá Açu, MT, Brasil). Acta Botanica Brasilica 19(2):195–207

Pereira, Luciano Araujo; Lima-Barbosa, Jackson Rodrigo de; Almeida, Maria Zélia de; Guimarães, Elsie Franklin. 2017. Diversidade de plantas em quintais quilombolas, conhecimento local sobre uso e cultivo de pimentas na Amazônia oriental, Brasil. Revista de Biologia Neotropical 14(1):56–72

Pérez-Arbeláez, Enrique. 1956. Plantas Útiles de Colombia. Liberia Colombiana, Bogotá

Pinto, Andréia Zelenski de Lara; Assis, Ana Flávia Silva de; Pereira, Arildo Gonçalo; Pasa, Maria Corette. 2013. Levantamento etnobotânico de plantas medicinais comercializados no mercado do porto em Cuiabá, Mato Grosso, Brasil. Flovet 5:51–70

Prance, G. T. 1996. Islands in Amazonia. Philosophical Transactions of the Royal Society B 351:823–833

Quaresma, Adriano Costa; Jardim, Mário Augusto G. 2017. Formations of coastal forests in the Amazon and ecological relations with vascular epiphytes. Revista Árvore 41(2):1–8

Ragusa-Netto, J. 2007. Feeding ecology of the green-cheeked parakeet (*Pyrrhura molinae*) in dry forests in western Brazil. Brazilian Journal of Biology 67(2):243–249

Ratter, J. A. 1987. Notes on the vegetation of the Parque Nacional do Araguaia (Brazil). Notes from the Royal Botanic Garden, Edinburgh 44(2):311–342

Ratter, J. A.; Bridgewater, S.; Atkinson, R.; Ribeiro, J. F. 1996. Analysis of the floristic composition of the Brazilian cerrado vegetation II: comparison of the woody vegetation of 98 areas. Edinburgh Journal of Botany 53(2):151–180

Rodrigues, William A. 1961. Aspectos fitossociológicos das caatingas do Rio Negro. Boletim do Museu Paraense Emílio Goeldi, Nova Série Botânica 15:1–41

Rodrigues, Ricardo Ribeiro; Martins, Sebastião Venâncio; Barros, Luiz Carlos de. 2004. Tropical rain forest regeneration in an area degraded by mining in Mato Grosso state, Brazil. Forest Ecology and Management 190: 323–333

Rondon Neto, Rubens Marques; Santos, Jeneffer Soares dos; Silva, Manoel Advíncula da; Koppe, Valdinei Cristi. 2010. Potencialidades de uso de espécies arbustivas e arbóreas em diferentes fisionomias de cerrado, em Lucas do Rio Verde/MT. Revista de Biologia e Ciências da Terra 10(2):113–126

Salomão, Rafael de Paiva; Vieira, Ima Célia Guimarães; Suemitsu, Chieno; Rosa, Nélson de Araújo; Almeira, Samuel Soares de; Amaral, Dário Dantas do; Menezes, Moirah Paulo Machado de. 2007a. As florestas de Belo Monte na grande curva do rio Xingu, Amazônia Oriental. Boletim do Museu Paraense Emílio Goeldi, Ciencias Naturais 2(3):57–153

Salomão, Rafael Paiva; Rosa, Nélson Araújo; Morais, Kácio Andrey Câmara. 2007b. Dinâmica da regeneração natural de árvores em áreas mineradas na Amazônia. Boletim do Museu Paraense Emílio Goeldi, Ciencias Naturais 2(2):85–139

Sanaiotti, Tânia M.; Magnusson, William E. 1995. Effects of annual fires on the production of fleshy fruits eaten by birds in a Brazilian Amazonian savanna. Journal of Tropical Ecology 11(1):53–65

Santos, Luciana Oliveira dos. 2010. Comparação da vegetação lenhosa entre borda e interior de uma mancha de mata de restinga na RESEX Mar Caeté-Taperaçu, Bragança, Pará. Master's thesis, Universidade Federal Rural da Amazônia/Museu Paraense Emílio Goeldi, Belém

Santos, J. U. M. and M. N. Bastos. 2014. Flores e Frutas das Restingas do Estado do Pará. Universidade Federal Rural da Amazônia (UFRA), Belém

Santos, Gideão Costa dos; Jardim, Mário Augusto Gonçalves. 2006. Florística e estrutura do estrato arbóreo de uma floresta de várzea no município de Santa Bárbara do Pará, Estado do Pará, Brasil. Acta Amazonica 36(4):437–446

Santos, João Ubiratan M. dos; Rosário, Carlos da S. 1988. Levantamento da vegetação fixadora das dunas de Algodoal-PA. Boletim do Museu Paraense Emílio Goeldi, Série Botânica 4(1): 133–154

Santos, Juliana de Faria Lima; Pagani, Eduardo; Ramos, José; Rodrigues, Eliana. 2012. Observations on the therapeutic practices of riverine communities of the Unini River, AM, Brazil. Journal of Ethnopharmacology 142:503–515

Santos, Gideão Costa; Tourinho, Manoel Malheiros; Mendes, Fernanda da Silva; Guimarães, Célia Maria Costa. 2014a. Fitossociologia e práticas de manejo tradicional em uma floresta de várzea em Santa Bárbara do Pará, Brasil. Revista de Ciências Agrárias 57(2):138–145

Santos, L. O., L. O. Santos, M. P. M. Menezes, C. R. Beasley, and U. Mehlig. 2014b. Composition and structure of a diverse tree community at the edges of a Brazilian Amazon rainforest island surrounded by marshes and mangroves. Plant Ecology 215: 1469–1481

Santos, M. R. A.; Lima, M. R.; Oliveira, C. L. L. G. 2014c. Medicinal plants used in Rondônia, western Amazon, Brazil. Revista Brasileira de Plantas Medicinais 16(3):707–720

Schultes, R. E. and R. F. Raffauf. 1990. The Healing Forest: Medicinal and Toxic Plants of the Northwest Amazonia. Dioscorides Press, Portland, Oregon

Silva, Marcondes Alves B. da; Melo, Laura Valdiane L.; Ribeiro, Reginaldo V.; Souza, João Paulo M. de; Lima, Joaquim Corsino S.; Martins, Domingos Tabajara de O.; Silva, Regilane M. da. 2010a. Levantamento etnobotânico de plantas utilizadas como anti-hiperlipidêmicas e anorexígenas pela população de Nova Xavantina-MT, Brasil. Revista Brasileira de Farmacognosia 20(4):549–562

Silva, Rachel Macedo da; Mehlig, Ulf; Santos, João Ubiratan Moreira dos; Menezes, Moirah Paula Machado de. 2010b. The coastal restinga vegetation of Pará, Brazilian Amazonia: a synthesis. Revista Brasileira de Botanica 33(4):563–573

Souza, Deoclides Ricardo de; Souza, Agostinho Lopes de; Leite, Helio Garcia; Yared, Jorge Alberto Gazel. 2006. Análise estrutural em floresta ombrófila densa de terra firme não explorada, Amazônia oriental. Revista Árvore 30(1):75–87

Steege, Hans ter; et al. 2013. Hyperdominance in the Amazonian tree flora. Science 342 https://doi.org/10.1126/science.1243092

Stevenson, Pablo R.; Beltrán, Marta L.; Quiñones, Marcela J.; Ahumada, Jorge A. 2015. Differences in home range, activity patterns and diet of red howler monkeys in a continuous forest and a forest fragment in Colombia. Rev. Acad. Colomb. Cienc. Ex. Fis. Nat. 39(153):503–513

Vale, Gabriel Damasco do. 2011. O efeito do solo, inundação e topografia sobre as campinaranas inundáveis na Amazônia. Master's Thesis, INPA, Manaus

Van Andel, Tinde; Behari-Ramdas, Joelaika; Havinga, Reinout; Groenendijk, Sara. 2007. The medicinal plant trade in Suriname. Ethnobotany Research & Applications 5:351–372

Van den Berg, M.E. and M.H.L. da Silva. 1986. Contribuição ao conhecimento da flora medicinal do Maranhão. In: Anais do1°Simpósio do Trópico Úmido, Belém, PA, 12 a 17 November, vol. 2, Flora e Floresta, EMBRAPA, Brasília, pp 119–125

Vásquez, R. and A. H. Gentry. 1989. Use and misuse of forest-harvested fruits in the Iquitos area. *Conservation Biology* 3(4): 350–361.

Veneklaas, E. J.; Fajarado, A.; Obregon, S.; Lozano, J. 2005. Gallery forest types and their environmental correlates in a Colombian savanna landscape. Ecography 28: 236–252

Vicentini, Alberto. 2004. A vegetação ao longo de um gradiente edáfico no Parque Nacional do Jaú. In: Janelas para a Biodiversidade no Parque Nacional do Jaú: Uma Estratégia para o Estudo da Biodiversidade na Amazônia, Borges, Sérgio Henrique; Iwanaga, Simone; Durigan, Carlos César; Pinheiro, Marcos Roberto (eds), Fundação Vitória Amazônica, Manaus, pp 105–131

Vieira, Diego dos Santos; Gama, João Ricardo Vasconcellos; Ribeiro, Renato Bezerra da Silva; Ximenes, Lucas Cunha; Corrêa, Viviane Vasconcellos; Alves, Adriana Ferreira. 2014a. Comparação estrutural entre floresta manejada e não manejada na comunidade Santo Antônio, estado do Pará. Ciência Florestal 24(4):1061–1068

Vieira, Raimundo Kennedy; Vieira, Adalena Kennedy; Kim, Jun Tae; Netravali, Anil Narayan. 2014b. Characterization of Amazonic white pitch (*Protium heptaphyllum*) for potential use as 'green' adhesive. Journal of Adhesion Science and Technology 28(10):963–974
Violante, Ivana Maria Póvoa; Hamerski, Lidilhone; Garcez, Walmir Silva; Batista, Ana Lucia; Chang, Marilene Rodrigues; Pott, Vali Joana; Garcez, Fernanda Rodrigues. 2012. Antimicrobial activity of some medicinal plants from the cerrado of the central-western region of Brazil. Brazilian Journal of Microbiology 43:1302–1308
Wang, Ellen; Donatti, Camila I.; Ferreira, Vanda L.; Raizer, Josué; Himmelstein, Jeffrey. 2011. Food habits and notes on the biology of *Chelonoidis carbonaria* (Spix 1824) (Testudinidae, Chelonia) in the southern Pantanal, Brazil. South American Journal of Herpetology 6(1):11–19
Wilbert, J. 1987. *Tobacco and Shamanism in South America*. Yale University Press, New Haven
Zent, Egleé; Zent, Stanford. 2002. Los Jodï: sabios botánicos del Amazonas venezolano. Antropologica 97–98:29–70

Chapter 12
Capparaceae

12.1 *Crataeva benthamii*

Brazil Catauarí, catauré, cataré, catoré, cabaçeira, capitari, pau d'alho, trapiá, tamara

Colombia Tamara

Peru Tamara

Status Wild

N. Smith, *Amazon Fruits: An Ethnobotanical Journey*, Ethnobiology,
https://doi.org/10.1007/978-3-031-12803-5_12

Catauarí *Crataeva benthamii* **tree in fruit on a floodplain during the high-water season. São Raimundo, Rio Urubu near Silves, Amazonas, Brazil 4-3-15**

Catauarí *Crataeva benthamii* is a small to medium-sized tree that is generally found in forests, including successionary woodlands, on floodplains of white, black, and clear water rivers throughout the Amazon Basin and never occurs on uplands (Amaral et al. 1997; Assis 2008:28;Assis et al. 2017; Casula 2012:55; Cunha and Ferreira 2012; Ferreira et al. 2013; Klinge et al. 1995; Luize et al. 2015; Myster 2018; Parolin 2000a, b, c, 2001, 2002, 2009; Parolin et al. 2002, 2004; Vieira et al. 2017; Wittmann et al. 2002, 2004, 2006, 2010; Worbes 1997). The green-grey fruits of catauarí are a favorite food of tambaqui *Colossoma macropomum*, one of the premier fishes of Amazonia (Albuquerque 2015:55; Lucas 2008). The fruits of catauarí are too big for tambaqui to swallow but they nevertheless help disperse the fruits by swimming away with them to nibble on the pulp (Kubitzki and Ziburski 1994; Parolin et al. 2010). A large tambaqui can attain 30 kg and can easily fetch the equivalent of a month's minimum wage, so farmers from Itacoatiara to Monte Alegre, and probably beyond, remember from year to year the location of catauarí trees with their orange-sized fruits in floodplain forests so that they can use them for bait. Other fruit-eating fish are caught with catauarí bait, including round, silvery pirapitinga *Colossoma bidens* which can weigh up to 20 kg and the bacu catfish

Lithodoras dorsalis (Parolin et al. 2010). River dwellers in the vicinity of Santarém also use the fruits as bait to catch tracajá *Podocnemis unifilis* turtles (Barboza et al. 2013). Matrinxã *Brycon cephalus*, a smaller fish eaten by people in the Amazon, has been documented feeding on this fruit in the Negro and Solimões watersheds in Amazonas (Pizango-Paima et al. 2001). The granulated catfish *Pterodoras granulosus* also eats this fruit and this catfish is eaten by people in the Amazon Basin (Albuquerque 2015:56). Catauarí has a long fruiting season from November, at the tail end of the low water period, to July when the flood has recently begun receding (Smith 1999:67).

Catauarí *Crataeva benthamii* fruit. São Raimundo, Rio Urubu near Silves, Amazonas, Brazil 4-3-15

The fruits, which measure 6 cm × 5.5 cm, are typically gathered from canoes, especially on the Amazon floodplain as well as along the clear water Xingu River where it is known as trapiá. Along the Amazon, people do not generally eat the fruit but ribeirinhos along the Middle Xingu do occasionally consume the fruits. The trees are spared in home gardens located on floodplains or straddling them because they provide valuable fish bait. The fruits are also used to bait suspended trotlines put out during the floodwater season to capture tracajá *Podocnemis unifilis* turtles which are much appreciated for their meat and eggs. Another much-appreciated turtle, tartaruga *Podocnemis expansa*, also eats catauarí fruit (Kubitzki and Ziburski 1994). Catauarí was noted in home gardens on the Amazon floodplain at various locations in Pará including Amador near Óbidos, several settlements near Alenquer, and along the Surubim-Açu channel near Santarém. River dwellers also spared catauarí when they cleared space for their home garden at the small village of São Raimundo along the lower Rio Urubu near Silves, Amazonas.

The wood of catauarí is used to make clubs to stun certain fish, especially piranhas and large fish, when removing them from hooks or gillnets (Smith 1981:45). The catauarí bark is used to make a bathing solution to treat *panema*, a spell that prevents people from catching fish or game. In Peru, the bark of tamara is mixed with water to treat a skin condition known as *lobosisso* which turns the skin black (Schultes and Raffauf 1990:120). In some rural parts of the Brazilian Amazon, liquid squeezed from the leaves is applied externally to alleviate rheumatism and snakebites (Le Cointe 1934:112; Moura et al. 2015).

Catauarí *Crataeva benthamii* in a village home garden. São Raimundo, Rio Urubu near Silves, Amazonas, Brazil 4-3-15

References

Albuquerque, Bianca Weiss. 2015. Frugivoria e ictiocoria em uma área de várzea na Amazônia Central brasileira. Master's thesis, INPA, Manaus

Amaral, I. L. do; Adis, J.; Prance, G. T. 1997. On the vegetation of a seasonal mixed water inundation forest near Manaus, Brazilian Amazonia. Amazoniana 14(3/4):335–347

Assis, Rafael Leandro de. 2008. Composição florística e estrutura da regeneração arbórea de florestas de várzea alta e várzea baixa na RDS Mamirauá, Amazônia central. Master's thesis, INPA, Manaus

Assis, Rafael Leandro de; Wittmann, Florian; Luize, Bruno Garcia; Haugaasen, Torbjon. 2017. Patterns of floristic diversity and composition in floodplain forests across four southern Amazon river tributaries, Brazil. Flora 229:124–140

Barboza, Roberta Sá Leitão; Barboza, Myrian Sá; Pezzuti, Juarez Carlos Brito. 2013. "Estava pescando de malhadeira, vi na praia uns cascos brilhando, era luar, abeirei a terra e fui pegar": prácticas de pesca de quelônio na várzea Amazônica (Santarém, Pará). Amazônica, Revista de Antropologia 5(3):622–653

Casula, Kátia Regina. 2012. Análise da similaridade florística e estrutural das formações florestais inundáveis em um trecho do alto Rio Madeira e de seus afluentes, estado de Rondônia, Brasil. Master's Thesis, Núcleo de Ciências Exatas da Terra, Universidade Federal de Rondônia, Porto Velho

Cunha, Denise de Andrade; Ferreira, Leandro Valle. 2012. Impacts of the Belo Monte hydroelectric dam construction on pioneer vegetation formations along the Xingu River, Pará state, Brazil. Brazilian Journal of Botany 35(2):159–167

Ferreira, Leandro Valle; Cunha, Denise A.; Chaves, Priscilla P.; Matos, Darley C. L.; Parolin, Pia. 2013. Impacts of hydroelectric dams on alluvial riparian plant communities in eastern Brazilian Amazonian. Anais da Academia Brasileira de Ciências 85(3):1013–1033

Klinge, Hans; Adis, Joachim; Worbes, Martin. 1995. The vegetation of a seasonal várzea forest in the lower Solimões River, Brazilian Amazonia. Acta Amazonica 25(3/4):201–220

Kubitzki, K. and A. Ziburski. 1994. Seed dispersal in flood plain forests of Amazonia. *Biotropica* 26(1):30–43

Le Cointe, P. 1934. *Amazônia Brasileira III: Árvores e Plantas Úteis*. Livraria Classica, Belém

Lucas, C. M. 2008. Within flood season variation in fruit consumption and seed dispersal by two characin fishes of the Amazon. *Biotropica* 40(5): 581–589.

Luize, Bruno Garcia; Venticinque, Eduardo Martins; Silva, Thiago Sanna Freire; Novo, Evlyn Marcia Leão de Moraes. 2015. A floristic survey of angiosperm species occurring at three landscapes of the central Amazon várzea, Brazil. Checklist: The Journal of Biodiversity Data 11(6):1–15

Moura, Valéria Mourão de; Sousa, Luciana A. Freitasde; Dos-Santos, Maria Cristina; Raposo, Juliana Divina Almeida; Lima, Aline Evangelista; Oliveira, Ricardo Bezerra de; Silva, Milton Nascimento da; Mourão, Rosa Helena Veras. 2015. Plants used to treat snakebites in Santarém, western Pará, Brazil: an assessment of their effectiveness in inhibiting hemorrhagic activity induced by *Bothrops jararaca* venom. Journal of Ethnopharmacology 161:224–232

Myster, Randall W. 2018. What we know about the composition and structure of igapó forests in the Amazon Basin. Botanical Review 84:394–410

Parolin, P. 2000a. Seed mass in Amazonian floodplain forests with contrasting nutrient supplies. Journal of Tropical Ecology 16(3):417–428

Parolin, P. 2000b. O uso de árvores nas florestas inundadas por água branca na Amazônia Central. Amazoniana 16(1/2):241–248

Parolin, P. 2000c. Phenology and C02-assimilation of trees in Central Amazonian floodplains. Journal of Tropical Ecology 16:465–473

Parolin, P. 2001. Seed germination and early establishment of 12 tree species from nutrient-rich and nutrient-poor central Amazonian floodplains. Aquatic Botany 70:89–103

Parolin, P. 2002. Submergence tolerance vs. escape from submergence: two strategies of seedling establishment in Amazonian floodplains. Environmental and Experimental Botany 48:177–186

Parolin, Pia. 2009. Submerged in darkness: adaptations to prolonged submergence by woody species of the Amazonian floodplains. Annals of Botany 103:359–376

Parolin, Pia; Ambruester, Nicole; Wittmann, Florian; Ferreira, Leandro; Piedade, Maria Teresa F.; Junk, Wolfgang J. 2002. A review of tree phenology in central Amazonian floodplains. Pesquisas Botânica 52:195–222

Parolin, Pia; Ferreira, Leandro Valle; Albernaz, Ana L. K. M.; Almeida, Samuel S. 2004. Tree species distribution in várzea forests of Brazilian Amazonia. Folia Geobotanica 39:371–383

Parolin, P., D. Waldhoff, and M. T. Piedade. 2010. Fruit and seed chemistry, biomass and dispersal. In: *Amazonian Floodplain Forests: Ecophysiology, Biodiversity and Sustainable Management.* W. J. Junk, M. T. Piedade, F. Wittman, and J. Schöngart (eds), Springer, Dordrecht, pp. 243–258

Pizango-Paima, Emer Gloria; Pereira-Filho, Manoel; Oliveira-Pereira, Maria Inêz de. 2001. Composição corporal e alimentar do matrinxã, *Brycon cephalus* (Günther, 1869), na Amazônia central. Acta Amazonica 31(3):509–520

Schultes, R. E. and R. F. Raffauf. 1990. The Healing Forest: Medicinal and Toxic Plants of the Northwest Amazonia. Dioscorides Press, Portland, Oregon

Smith, N. J. H. 1981. *Man, Fishes, and the Amazon.* Columbia University Press, New York

Smith, N. J. H. 1999. *The Amazon River Forest.* Oxford University Press, New York.

Vieira, Diegodos Santos; Oliveira, Marcio Leles Romarco de; Gama, João Ricardo Vasconcelos; Machado, Evandro Luiz Mendonça; Görgens, Eric Bastos; Lafetá, Bruno Oliveira; Garcia, Josinei Silva. 2017. Phytosociology of a natural fragment of the floodplain forest in the lower Tapajós River, Brazil. Bosque 38(2):357–369

Wittmann, Florian; Anhuf, Deiter; Junk, Wolfgang J. 2002. Tree species distribution and community structure of central Amazonian várzea forests by remote-sensing techniques. Journal of Tropical Ecology 18(6):805–820

Wittmann, F., W. J. Junk, and M. T. F. Piedade. 2004. The várzea forests in Amazonia: flooding and the highly dynamic geomorphology interact with natural forest succession. *Forest Ecology and Management* 196:199–212

Wittmann, Florian; Schöngart, Jochen; Montero, Juan Carlos; Motzer, Thomas; Junk, Wolfgang J.; Piedade, Maria T. F.; Queiroz, Helder L.; Worbes, Martin. 2006. Tree species composition and diversity gradients in white-water forests across the Amazon Basin. Journal of Biogeography 33(8):1334–1347

Wittmann, Florian; Schöngart, Jochen; Junk, Wolfgang J. 2010. Phytogeography, species diversity, community structure and dynamics of central Amazonian floodplain forests. In: Amazonian Floodplain Forests: Ecophysiology, Biodiversity and Sustainable Management, Junk, Wolfgang J.; Piedade, Maria T. F.; Wittmann, Florian; Schöngart, Jochen; Parolin, Pia (eds), Springer, Dordrecht, pp 61–102

Worbes, Martin. 1997. The forest ecosystem of the floodplains. Ecological Studies 126:223–265

Chapter 13
Caricaceae

13.1 *Carica papaya*

Bolivia Papaya; mapayo (Chácobo), pofi (Tsimane)

Brazil Mamão, mamão de coda, mamoeiro; managão (Arara), mãmã'i (Arawaté), yarakaci'a-ú (Asurini), baiagó (Avá-Canoeiro), iraipere (Baniwa), manu (Desana), arakačiahú (Guajá), yupã (Juruna), mãmã (Ka'apor), mõmõ (Kaiabi), barã (Kaxinawá), kàtembaré (Kayapó), moumõu (Kuikuru), maũ maũ (Sateré-Maué), mapazá (Taurepang), zarakači'a (Tembé), popayu (Ticuna), manu (Tukano), mabaia (Waimiri-Atroari), mãũ (Wajãpi), rokoari hi, xamkoro (Yanomami)

Colombia Papaya, papaya calentana, papayo, lechosa, chamburo, fruta bomba, mamona, zapote, árbol de melón; mamao (Makuna), moniya amena (Witoto)

Ecuador Papaya, papaya de mico; tsu (Achuar), papaya (Cofán), cari papaya (Kichwa), wapái (Shuar), watihiko, airo huatiji'co, huatico (Siona-Secoya), ahuaca, awaka (Waorani)

Peru Papaya, puchay, mito; wapai (Achuar), papái, papaí (Aguaruna), naimi, naimbí (Amahuaca), papue' (Amuesha), mapótsha (Campa), nampucha (Cashibo), pahpayu, pa'payo (Chayahuita), esíe (Ese Eja), wapaí (Huambisa), non puchá (Kakataibo), dectac (Matsés), tinti (Matsigenka), napucha (Pano), pucha, pocha (Shipibo), poxpayu (Ticuna), síyá (Yagua)

Venezuela Lechosa, lechosa cochinera, lechosa pajarera, lechosilla, papayo; jwaya (Jodï), shamakoro (Yanomami)

Status Planted, spontaneous in cultural settings

N. Smith, *Amazon Fruits: An Ethnobotanical Journey*, Ethnobiology,
https://doi.org/10.1007/978-3-031-12803-5_13

Female papaya *Carica papaya* in fruit. On the right is a male papaya in flower. Rio Yata, Yacuma, Beni, Bolivia, 6-2-05

Papaya *Carica papaya* is a small tree that bears green to yellow fruits along the upper portions of the trunk. Papaya is normally unbranched, but some individuals develop branches that curve upward, reminiscent of the saguaro cactus in the Sonoran Desert. Although papaya can be considered a perennial because it lives for several years, the trunk is not woody and fruits are produced within a year. The trunk is sturdy enough for this plant to reach as high as 10 m (Kempe et al. 2013) and is strong enough to support youngsters who shimmy up the single trunk to pick fruits. The Spanish donned the fruit papaya, from ababay, a Carib word. The Portuguese, on the other hand, opted for visual clues in naming the fruit: mamão, which means big breast. In the West Indies, papaya is often called pawpaw. In Venezuela, papaya is known as lechosa, derived from leche (milk) on account of the white sap that emanates when the skin of unripe fruits is cut.

Huahua variety of papaya *Carica papaya* in a market. Tingo Maria, Peru 1-19-04

Papaya is popular because of the fruit's smooth, creamy texture, and sweet taste. It is also highly nutritious, containing appreciable amounts of vitamins C and A. Some papayas produce fruits with a deep, red-orange flesh indicating particularly high levels of carotene. Papaya fruits are also rich in riboflavin, calcium, thiamin, iron, niacin, and potassium (Krishna et al. 2008). Papaya fruits vary considerably in shape and size. Some are the size of grapefruits, while others reach the dimension of a basketball. They vary in shape from round to oblong, even cylindrical. Although smaller papayas tend to travel better to markets, some large forms have relatively thick skins which minimize damage to the pulp.

Papaya *Carica papaya* harvested in a village home garden. Terra Santa, Amazon River, Pará, Brazil 6-21-94

Farmer with papaya *Carica papaya* she has just gathered in her home garden. Some varieties are green when ripe, but most are yellow. Near Igarapé-Açu, Pará, Zona Bragantina, Brazil 5-7-96

Most papaya fruits in the Amazon are eaten fresh, but in colonial times, unripe papaya fruits were cooked with squash (Daniel 1976:328). Papaya fruits are also used to make juice and smoothies in towns and cities throughout the Amazon. Papaya juice and fruit are particularly popular for breakfast, although smoothies are in demand throughout the day and into the evening. In the Peruvian Amazon, for example, a restaurant in Iquitos offers a potent smoothie dubbed the "Exterminator" (*Exterminador*) that contains papaya, plantain, guaraná *Paullinia cupana* (a vine that produces caffeine rich beans), algarrobo *Prosopis pallida* (a leguminous tree from coastal Peru that produces pods with a sweet pulp), honey, apple, ginseng, and quail eggs. In Icoaraci, a town in the Amazon estuary that has virtually merged with Belém as both urban centers burgeon, ambulatory youngsters sell homemade, frozen pushups, containing various fruits and blends, including papaya.

Papaya is a common dooryard tree in both urban and rural areas in both uplands and floodplains throughout Amazonia up to about 1500 m in the Andes. It is also intercropped in some fields, either through deliberate planting or because the trees arise spontaneously, such as in cacao plots in the municipality of São Félix do Xingu in southern Pará (Braga et al. 2019), among the Yanomamo in the Venezuelan Amazon (Harris 1971; Lizot 1980) and the Matsigenka in the Peruvian Amazon (Wezel and Ohl 2006). The Tsimane in the Bolivian Amazon plant papaya seeds in small patches at the edge of some of their slash-and-burn fields which tend to be dominated by banana, plantain, rice, sugarcane, and cacao (Reyes-García 2001:79). The Matsigenka in the Peruvian Amazon plant papaya in newly prepared fields then return during the third and fourth years to gather the fruits, as well as the fruits of other trees they have planted (Wezel and Ohl 2006). Papaya is also grown as a monocrop geared toward urban markets throughout Amazonia. Such orchards are usually found along major rivers or highways so that the fruits can reach the market in good condition.

Papaya *Carica papaya* spontaneous in a field containing several crops including manioc. Wild animals, such as certain birds, likely dispersed the seed that gave rise to this papaya. Near Terrawangã, middle Xingu River near its confluence with the Bacajá, Pará, Brazil 9-10-15

Papaya is dioecious, which means male and female flowers are borne on separate plants. However, hermaphrodites also occur and they often produce the most desirable fruit shapes. Hermaphrodite papayas are self-pollinating and 95% of the flowers self-pollinate before the petals unfurl (Fitch 2005). Male papayas do not usually produce fruit, although there are exceptions as I noted in the vicinity of Carrazedo at the mouth of the Xingu where papaya trees with male flowers and small round fruits are called *boiloa*. Male papaya trees produce large fragrant panicles of slender staminate flowers with vestigial ovaries but sometimes the panicles contain one or more perfect flowers with functioning stamens and an ovary (Fitch 2005). Commercial growers usually choose hermaphrodite papayas because space, water, and fertilizers are not taken up by male plants. Pollination is probably by wind and small flying insects. On the campus of INPA in Manaus, for example, *Scaptotrigona fulvicutis* and *Trigona williana* stingless bees have been observed visiting the flowers of this plant and carrying off some of the pollen (Marques-Souza 1996; Marques-Souza et al. 2007).

Papaya *Carica papaya* field on the Amazon floodplain. A few kilometers downstream from Anori, Rio Solimões Amazonas, Brazil 6-20-07

Papaya can grow as high as 5 m so fruits are generally harvested by either of two methods: using a pole to knock down the fruits or climbing the trunk. In the latter case, only children can perform this task because adults are too heavy and would break the stem. Children ascend the trunk with bare feet so that they have a better grip.

Dario, 8 years, harvesting papaya *Carica papaya* fruit in a village home garden. San Carlos, Rio Puinahua (a branch of the Ucayali), Reserva Nacional Pacaya-Samiria, Loreto, Peru, 4-16-06

Papaya is cultivated so widely on both uplands and floodplains in the tropics and subtropics that few people realize that it is native to tropical America. In fact, most residents in North America would probably assume that the fruit arose in Hawaii where one of the most popular commercial varieties was developed: Sunrise Solo. Known as mamão Havaií (Hawaiian papaya) in Brazil, Sunrise Solo is the most commonly encountered variety of papaya in supermarkets, including those in the Brazilian Amazon, because it is relatively compact and travels better than the larger local forms grown from unselected seed.

Papaya *Carica papaya* plantation on uplands along the lower Utcubamba River, near Bagua Grande, Peru 1-21-05

The precise origin of *Carica papaya* in tropical America is still in dispute. Papaya is apparently not found in truly wild form, although it is common as a feral plant in some areas of the humid tropics. Some scientists suggest that papaya arose in Central America including Mexico (Clement 2005; Krishna et al. 2008; Manshardt and Zee 1994; Purseglove 1974; Torke 2004), whereas some botanists and agronomists lean toward the Andean foothills in the western part of Amazonia as the hearth of papaya domestication because that is where the greatest diversity of near relatives are found (Anderson 1952:181; Brücher 1989:224; Prance 1984a, b). Nearly a century and a half ago, a Brazilian botanist, Joaquim Corrêa de Melo and his indefatigable British colleague, Richard Spruce, speculated that papaya arose in West Indies, an idea with few if any supporters today. However, they did recognize the great diversity of papaya relatives in the Andes:

> Where Papayaceae most abound is on the wooded slopes of the Andes, both on the eastern and western sides, up to 8000 feet elevation; and it is there that travelers and sedentary botanists may confidently expect to find not only materials for the more perfect elucidation of the species already partially known, but also many new species, which doubtless still remain hidden in the savage recesses of the oriental Andes (Correa de Melo and Spruce 1869)

Papaya probably had a hybrid origin involving several *Carica* species, such as *cauliflora*, *goudotiana*, *microcarpa*, *monoica*, and *pubescens* (Badillo 1971). Papaya can hybridize with several near relatives, all of which are found in the Andes. Several other species of *Carica* were domesticated at various altitudes in the Andes, including papaya de la sierra *C. pubescens*, also known as chambor, and papayita *Carica monoica* in Peru and babaco *Carica pentagona* in Ecuador. So I surmise that papaya was domesticated in the "savage recesses" of the Andean foothills. Whatever its origins, papaya was widely cultivated in tropical America when Europeans arrived in the late fifteenth century. During the colonial period, papaya was taken to other tropical regions, such as Africa, where it has become feral (Baker 1973).

Spontaneous papaya *Carica papaya* in a field cleared in floodplain forest. Near San Carlos, Rio Puinahua (a branch of the Ucayali River), Reserva Nacional Pacaya-Samiria, Loreto, Peru 8-5-03

Papaya arises spontaneously in fields and other cleared areas on floodplains and uplands and the plants are typically spared. Farmers welcome spontaneous papaya because they can take a break to snack on the fruits or take them home. Spontaneous papaya is particularly common on fertile soils, such as the floodplains of sediment-rich rivers, alfisols (*terra roxa*) derived from weathered basalt, volcanic soils, and anthropogenic black earth (*terra preta de índio*) associated with archaeological sites (Woods and McCann 1999). Papaya also arises spontaneously in irrigated fields along the desert coast of Peru, such as near Tucume in the Department of Lambayeque, where a farmer explained that papaya grows by itself ("*nasce solito*"), and he welcomes the uninvited guest in his vegetable plot. The Juruna of the middle Xingu do not shoo away birds when they feed on papaya fruits out of gratitude for their role in mythology in introducing Juruna to crops and how to plant them and prepare them before eating (Galvão 1996:361–2; Villas Boas and Villas Boas, 1975).

I recall accompanying a hunter 1 day in 1973 near Agrovila Nova Fronteira at km 80 of the Altamira-Itaituba stretch of the Transamazon Highway. João, a native of the area long before the opening of the pioneer highway in 1971, spotted some spontaneous papaya plants as we were crossing a neighbor's rice field planted on *terra roxa*. João felled one of the papayas with a single blow of his machete. The fruit had a superb taste, even though it was warm from the fierce afternoon sun. João did not feel in the slightest bit guilty about feasting on his neighbor's papaya fruits; after all, he asserted, they were a gift of nature.

Papaya *Carica papaya* fruit opened by a bird, probably a tanager. Arlinho, 11 years. Carrazedo Velho, uplands behind Carrazedo, mouth of the Xingu River, Município de Gurupá, Pará, Brazil 7-30-16

Several species of birds, particularly tanagers such as sanhaçu azul *Thraupis episcopus*, sanhaçu do coqueiro *Thraupis palmarum*, and pipira *Ramphocelus carbo*, as well as some species of oropendola (*Cacicus*) feed avidly on papaya fruits and disperse their seeds. I saw blue-gray tanagers *Thraupis episcopus* eating papaya fruits in a home garden in Agrovila Nova Fronteira at 80 km of the Altamira-Itaituba stretch of the Transamazon highway in September 1973. I also observed silver-beaked tanagers *Ramphocelus carbo* feeding on papaya fruits in the vicinity of Aripuanã in northern Mato Grosso in September 1978. The seeds are sticky, and birds pick them up while feeding on the pulp; they subsequently wipe the seeds off their beaks on branches at their perch sites. They also likely ingest some of the seeds and later defecate them in fields, home gardens, and regrowth vegetation. Opossums (*Didelphis* sp.) also eat this fruit and possibly disperse the seeds as suggested by a Wayãpi drawing (Gallois 2009:21). Moustached tamarins *Saguinus mystax* are known to disperse papaya seeds in parts of the Peruvian Amazon (Reátegui 2013:37) and two species of bats, *Sturnira lilium* and *Artibeus jamaicensis*, eat this fruit in the vicinity of Manaus and likely disperse some of the seeds (Guillaumet and Reis 1983).

Some seeds are occasionally ingested by people, especially children eager to snack on the fruits, and some of the seeds are defecated in home gardens, fields, and brush. In the eighteenth century, a Jesuit priest noted that indigenous peoples in the Orinoco valley were involved in dispersing papaya seeds when they ate the fruits (Gumilla 1791:246).

Wadrio 11 years eating papaya *Carica papaya* he gathered in successionary vegetation in an old manioc field (roça). He is ingesting some of the seeds. Near Carrazedo, mouth of the Xingu River, Município de Gurupá, Pará, Brazil 7-25-16

In addition to fresh fruit, papaya has a wide variety of other uses in Amazonia, ranging from medicinal to cosmetic and even for making musical instruments (Smith et al. 1992:161). Papaya seeds are employed as a vermifuge for livestock as well as people (Frechione et al. 1989; Saltos et al. 2016; Schultes and Raffauf 1990:121). On Colares Island in the Amazon estuary some 100 km from Belém locals make a tea from the leaves to treat hair loss among their dogs. Also on Colares Island, the dried seeds are powdered and added to the food of dogs, cats, horses, and cattle to treat worm infestations (Ritter et al. 2012) as is the case on Marajó Island in the Amazon estuary (Monteiro et al. 2011).

In Roraima, a syrup is prepared from the leaves for treating colds, while the flowers are made into a tea to aid in digestion, while the fruit is eaten on occasion as a laxative (Luz 2001). Also in Roraima, some rural people make a tea from the leaves to treat liver and stomach problems (Van den Berg 2010:204; Van den Berg and Silva 1988) as do residents of Julião, a rural community in the Reserva de Desenvolvimento Sustentável (RDS) do Tupé some 25 km west of Manaus (Veiga and Scudeller 2015). The Julião community also use the leaves and flowers of this plant to treat malaria, as do other villages in the municipality of Barcelos along the middle Rio Negro in Amazonas (Tomchinsky et al. 2017). In the municipalities of Xapuri in Acre and Pauini in southern Amazonas, rural folk make an infusion from the roots of this plant to treat malaria (Ferreira et al. 2015). Villagers living in Marudá along the coast of Pará at the mouth of the Amazon prepare a tea from the leaves or flowers to treat amoeba infections while the flowers are made into a tea to alleviate diarrhea and vomiting (Coelho-Ferreira 2009). Nearby in the municipality of Marapanim, locals boil the male flowers to treat indigestion and liver problems (Furtado et al. 1978). Rural folk in the community of Quilombo Sangrador in the municipality of Presidente Juscelino, Maranhão, use the leaves to make an infusion for treating digestive disorders (Monteles and Pinheiro 2007). Farmers in the village of Arapixuna overlooking the Amazon floodplain near Santarém feed both mature and unripe papaya fruits to their pigs so that they ingest the seeds and lighten their worm loads. Parents in rural areas in various parts of Amazonia also encourage their children to ingest the seeds if they think they are carrying roundworms.

Rural people in Colombia swallow the seeds and latex of papaya to expel intestinal worms (Pérez-Arbeláez 1956:260). In parts of the Peruvian Amazon 20 fresh seeds are mashed and mixed with honey a spoonful of which is taken once a day for no more than 4 days to get rid of intestinal worms (Lossio et al. 1995:215). Also in the Peruvian Amazon, a slice of an unripe fruit is heated in half a liter of water then a cup of this solution is drunk with each meal to treat diarrhea (Lossio et al. 1995:215). In French Guiana, papaya leaves and roots are used in combination with some other plants to treat malaria (Bertani et al. 2005; Vigneron et al. 2005).

Papaya leaves are used to wrap meat to tenderize it and to make clothes cleaner when washing them (Martina 1902:36). Rural people in Colombia cook tough meat with papaya leaves to soften it (Pérez-Arbeláez 1956:260). In the Peruvian Amazon, the fresh leaves are squeezed and applied to wounds while the fresh fruit is eaten along with some of the seeds to treat intestinal parasites (Mejia and Rengifo 2000:156).

These practices likely originated with indigenous groups. In Roraima, the Taurepang and Wapixana make an infusion using the leaves and the bitter male flowers of papaya, or a decoction of the roots, to treat malaria (Milliken 1997). Various indigenous groups in Bolivar state, Venezuela, which borders Rondônia and Pará in the Brazilian Amazon drink the juice of this fruit to treat malaria (Caraballo et al. 2004). The Guarasug'we in the Bolivian Amazon use various parts of the papaya plant to treat a variety of health issues (Hajdu and Hohmann 2012). The Chayahuita in the Peruvian Amazon make a decoction with the leaves and flowers to treat skin rashes, including acne (Odonne et al. 2013). The Amuesha who live in scattered villages in the foothills of the Andes in the central part of the Peruvian Amazon place two yellow papaya leaves which are about to fall off into boiling water. Once the water cools, half a cup is drunk three times a day to treat malaria. Latex obtained by cutting the fruit skin is applied to the skin people who are suffering from leishmaniasis (Odonne et al. 2011; Valadeau et al. 2009, 2010). In the Ecuadorian Amazon, the widespread Kichwa use the bark and leaves to make a topical application for treating leishmaniasis (Gachet et al. 2010). In the Morona Santiago and Pastaza regions of the Ecuadorian Amazon the Achuar crush papaya seeds and mix them with water which is then drunk to eliminate parasites (Giovannini 2015). The Ka'apor in eastern Amazonia make a soap for washing clothes and hammocks by using papaya leaves, a wild vine, and the inner bark of a forest tree (Balée 1993).

The leaf stems (petioles) of papaya are straight and hollow and youngsters use this anatomical feature to make wind instruments. Along the Yanayacu River in the Pacaya-Samiria National Reserve in the Peruvian Amazon, for example, boys cut off the petioles to make flutes. Few kids in rural areas of the Peruvian Amazon can afford purchased toys, so they have learned to fashion parts of plants in their environment into toys. Children on Dominica also fashion flutes from papaya petioles (James 2011:2). At the mouth of the Xingu, boys in the village of Carrazedo use dried petioles of papaya as propellers; they hook one end on to a forefinger then twirl the petiole. Over time, the custom of fashioning toys from plants in the Amazon is likely to die out as rural incomes rise and more people purchase plastic toys, often made in China.

Farm boy with a flute he fashioned from a papaya *Carica papaya* petiole. Veinte de Enero, Rio Yanayacu, Reserva Nacional Pacaya-Samiria, Loreto, Peru 3-12-04

Arlinho, 11 years, using a dried papaya *Carica papaya* petiole as a propeller. Near Carrazedo, mouth of the Xingu River, Município de Gurupá, Pará, Brazil 7-24-16

Papaya also surfaces in some myths and legends. In the Peruvian Amazon, for example, Selmira, a 67-year-old farmer who lives in San Francisco, a village in the Pacaya-Samiria National Reserve along the Marañón River, recounted the following story about chullachaqui, which is known as curupira in the Brazilian Amazon:

> Ten years ago, Dennis, a five year-old boy (*muchachito*), was working with his parents in the family field (*chacra*) near San Francisco when a chullachaqui appeared before the boy. Dennis thought it was his grandfather (*abuelo*). The chullachaqui beckoned, and Dennis followed the supernatural creature into the forest. Dennis' parents were distraught when they could not find him, so they consulted a shaman. The shaman told them that Dennis would turn up eight days later, and sure enough eight days after his disappearance, Dennis was spotted walking along the trail that leads to the family's *chacra*. Dennis' clothes were in shreds and he was holding two small papaya fruits, (*papayitas*), one in each hand. Dennis was not himself, for he did not recognize his parents. He had to be grabbed to prevent him from fleeing back into the forest. He was taken to the shaman who cured him by blowing tobacco smoke over his body.

When I visited San Francisco in 2004, Dennis was 15 years old. He did not remember anything about his experience with chullachaqui. According to Selmira, chullachaqui has one normal human foot and a smaller one, the latter like that of a tortoise. Although she has never seen the supernatural entity, Selmira claims to have seen its tracks along the shores of a nearby lake.

Papaya is featured as a symbol of forbidden fruit in *The Scent of the Green Papaya*. Billed as "a story of beauty, passion, and forbidden fruit," this intriguing film explores sexual tensions in a Vietnamese household in the 1950s.

Several artists depict papaya in their work, some going back to the colonial era, such as the Spanish painter Vicente Alban who worked in Ecuador in the eighteenth century. The French painter Jean Debret produced a beautiful canvass of the fruit in the early nineteenth century while working in Rio de Janeiro. Frida Kahlo, one of Mexico's iconic artists, produced the stunning *La novia que se espanta de ver la vida abierta* (the fiancée who is shocked to see "open" life). Painted in 1943, this provocative Mexican painter often used sensual and shocking elements in her work, and she must have been inspired by the shape and texture of papaya as well as other tropical and subtropical fruits featured in the painting.

Papaya *Carica papaya* and other fruits with a Yumbo Indian near Quito, Ecuador. Vicente Alban, *Yndio Yumbo de Maynas con su carga*. 1783. Museo de America, Madrid, 00076

Opened papaya *Carica papaya* fruit displaying seeds. Jean Debret, French artist, painted in Rio de Janeiro, Brazil, in 1818

La novia que se espanta de ver la vida abierta **by Frida Kahlo, 1943. The opened papaya** *Carica papaya* **is likely an allegory for sexuality, representing the female reproductive organ**

In addition to its fruits, papain, derived from its fruits is widely used around the world for a variety of purposes. Papain, used mostly to tenderize meat, is an enzyme obtained from the dried latex of young papaya trees and is a significant export item for Tanzania, Uganda, Zaire, and Sri Lanka (Becker 1958). Latex is normally harvested by gashing immature fruits, but recent extraction methods use papaya leaves, thereby leaving unblemished fruits for marketing. Papain is the main ingredient in meat tenderizers sold the world over. Papain's digestive properties were well known to indigenous peoples of the American tropics. Indians boil tough meat wrapped in papaya leaves. Pigs fed on papaya leaves are reputed to produce more tender pork.

Papaya *Carica papaya* fruits with longitudinal cuts to drain some of the latex; this is thought to improve the flavor. Ilha de Sumaúma, Rio Urubu, Amazonas, Brazil 4-5-15

References

Anderson, E. 1952. Plants, Man and Life. University of California Press, Berkeley

Badillo, V.M. 1971. *Monografía de la familia Caricaceae*. Associación de Profesores, Maracay, Venezuela

Baker, H. G. 1973. Evolutionary relationships between flowering plants and animals in American and African tropical forests. In: Tropical Forest Ecosystems in Africa and South America: A Comparative Review, Meggers, Betty J.; Ayensu, Edward S.; Duckworth, W. Donald (eds), Smithsonian Institution Press, Washington, DC, pp 145–159

Balée, W. 1993. *Footprints of the Forest: Ka'apor Ethnobotany, the Historical Ecology of Plant Utilization by an Amazonian People*. Columbia University Press, New York

Becker, S. 1958. The production of papain--an agricultural industry for tropical America. *Economic Botany* 12:62–79

Bertani, S.; Bourdy, G.; Landau, I.; Robinson, J. C.; Esterre, Ph.; Deharo, E. 2005. Evaluation of French Guiana traditional antimalarial remedies. Journal of Ethnopharmacology 98:45–54

Braga, Daniel P. P.; Domene, Frederico; Gandara, Flávio B. 2019. Shade trees composition and diversity in cacao agroforestry systems of southern Pará, Brazilian Amazon. Agroforestry Systems 93:1409–1421

Brücher, H. 1989. *Useful Plants of Neotropical Origin and their Wild Relatives*. Springer, Berlin

Caraballo, Alejandro; Caraballo, Brigida; Rodríguez-Acosta, Alexis. 2004. Preliminary assessment of medicinal plants used as antimalarials in the southeastern Venezuelan Amazon. Revista da Sociedade Brasileira de Medicina Tropical 37(2):185–188

Clement, C. R. 2005. Fruits. In: The Cultural History of Plants, Ghillean Prance and Mark Nesbitt (eds), Routledge, New York, pp. 77–99

Coelho-Ferreira, Márlia. 2009. Medicinal knowledge and plant utilization in an Amazonian coastal community of Marudá, Pará state (Brazil). Journal of Ethnopharmacology 126:159–175

Correa de Mello, J. and R. Spruce, 1869. Notes on Papayaceae. *Journal of Linnean Society of London, Botany* 10: 1–15

Daniel, J. 1976. *Tesouro Descoberto no Rio Amazonas*. Biblioteca Nacional, Rio de Janeiro, vol. 1

Ferreira, Almecina Balbino; Ming, Lin Chau; Haverroth, Moacir; Daly, Douglas Charles; Caballero, Javier; Ballesté, Andrea M. 2015. Plants used to treat malaria in the regions of Rio Branco-Acre state and southern Amazonas state, Brazil. International Journal of Phytocosmetics and Natural Ingredients 2(9):1–5

Fitch, M. M. M. 2005. 6.1 *Carica papaya* L. Papaya. In: *Biotechnology of Fruit and Nut Crops*, R. E. Litz (ed), CABI, Wallingford, pp. 174–207

Frechione, John; Posey, Darrell A.; Silva, Luiz Francelino da. 1989. The perception of ecological zones and natural resources in the Brazilian Amazon: an ethnoecology of Lake Coari. In: Resource Management In Amazonia: Indigenous and Folk Strategies, Posey, D. A.; Balée, W. (eds), New York Botanical Garden, Advances in Economic Botany 7:260–282

Furtado, Lourdes Gonçalves; Souza, Ruth Cortez de; Van den Berg, Maria Elizabeth. 1978. Notas sobre uso terapêutico de plantas pela população cabocla de Marapanim, Pará. Boletim do Museu Paraense Emílio Goeldi, Nova Série Antropologia 70:1–31

Gachet, María Salomé; Lecaro, Javier Salazar; Kaiser, Marcel; Brun, Reto; Navarrete, Hugo; Muñoz, Ricardo A.; Bauer, Rudolf; Schühly, Wolfgang. 2010. Assessment of anti-protozoal activity of plants traditionally used in Ecuador in the treatment of leishmaniasis. Journal of Ethnopharmacology 128:183–197

Gallois, C. 2009. *Wajãpi rena: roças, pátios e casas*. Museu do Indio-Funai, Rio de Janeiro

Galvão, Eduardo. 1996. Diários de Campo de Eduardo Galvão: Tenetehara, Kaioá, e Índios do Xingu. Editora UFRJ/Museo do Índio-FUNAI, Rio de Janeiro

Giovannini, Peter. 2015. Medicinal plants of the Achuar (Jivaro) of Amazonian Ecuador: ethnobotanical survey and comparison with other Amazonian pharmacopoeias. Journal of Ethnopharmacology 164:78–88

Guillaumet, Jean-Louis; Reis, Nélio Roberto dos. 1983. Les chauves-souris frugivores de la région de Manaus et leur rôle dans la dissémination des espèces végétales. Revue d'Écologie: La Terre et la Vie 38:147–169

Gumilla, J. 1791. Historia Natural, Civil y Geografica de las Naciones Situadas en las Riveras del Rio Orinoco. Imprenta de Carlos Gilbert y Tuto, Barcelona, vol. 2

Hajdu, Z. and J. Hohmann. 2012. An ethnopharmacological survey of the traditional medicine utilized in the community of Porvenir, Bajo Paraguá Indian Reservation, Bolivia. Journal of Ethnopharmacology 139:838–857

Harris, D.R. 1971. The ecology of swidden cultivation in the Upper Orinoco rain forest, Venezuela. *Geographical Review* 61(4):475–495

James, A. 2011. *Playing with Plants in the Nature Isle*. Forestry, Wildlife & Parks Division, Ministry of Agriculture and Forestry, Dominica, Roseau

Kempe, A.; Lautenschläger, T.; Lange, A.; Neinhuis, C. 2013. How to become a tree without wood--biochemical analysis of the stem of Carica papaya L. Plant Biology 16:264–271

Krishna, K. L.; Paridhavi, M.; Patel, Jagruti A. 2008. Review on nutritional, medicinal and pharmacological properties of papaya (Carica papaya Linn.). Natural Product Radiance 7(4):364–373

Lizot, Jacques 1980. La agricultura Yanomami. Antropológica 53:3–93

Lossio, L.C., Delgado, H.S., Ruiz, A.G., Donayre, R.A., Ruiz, J.G., Panduro, M.P., Sifuentes, T.C. 1995. Plantas Medicinales de la Amazonía Peruana. Instituto Peruano de Seguridad Social/Instituto de Medicina Tradicional, Iquitos

Luz, Francisco Joaci F. 2001. Plantas medicinais de uso popular em Boa Vista, Roraima, Brasil. Horticultura Brasileira 19(1):88–96

Manshardt, Richard M.; Zee, Francis T.-P. 1994. Papaya germplasm and breeding in Hawaii. *Fruit Varieties Journal* 48 (3): 146–152

Marques-Souza, Antonio Carlos, 1996. Fontes de pólen exploradas por *Melipona compressipes manaosensis* (Apidae: Meliponinae), abelha da Amazônia Central. Acta Amazonica 26(1/2):77–86

Marques-Souza, Antonio Carlos; Absy, Maria Lucía; Kerr, Warwick Estevam. 2007. Pollen harvest features of the central Amazonian bee Scaptotrigona fulvicutis Moure 1964 (Apidade: Meliponinae), in Brazil. Acta Botanica Brasilica 21(1):11–20
Martina, G. 1902. Estudo Chimico sobre algumas Frutas Brasileiras. Imprensa Official, Belém
Mejia, Kember; Rengifo, Elsa. 2000. Plantas Medicinales de Uso Popular en la Amazonía Peruana. Agencia Española de Cooperación Internacional, Lima
Milliken, William. 1997. Traditional anti-malarial medicine in Roraima, Brazil. Economic Botany 51(3):212–237
Monteiro, Maria Vivina Barros; Bevilaqua, Claudia Maria Leal; Palha, Maria das Dores Correia; Braga, Roberta Rocha; Schwanke, Katiane; Rodrigues, Silvane Tavares; Lameira, Osmar Alves. 2011. Ethnoveterinary knowledge of the inhabitants of Marajó Island, eastern Amazonia, Brazil. Acta Amazonica 41(2):233–242
Monteles, Ricardo; Pinheiro, Claudio Urbano B. 2007. Plantas medicinais em um quilombo maranhense: uma perspectiva etnobotânica. Revista de Biologia e Ciências da Terra 7(2):1–12
Odonne, Guillaume; Berger, Franck; Stien, Didier; Grenand, Pierre; Bourdy, Geneviève. 2011. Treatment of leishmaniasis in the Oyapock basin (French Guiana): a K.A.P. survey and analysis of the evolution of phytotherapy knowledge amongst Wayãpi Indians. Journal of Ethnopharmacology 137:1228–1239
Odonne, Guillaume; Valadeau, Celine; Alban-Castillo, Joaquina; Stien, Didier; Sauvain, Michel; Bourdy, Genevieve. 2013. Medical ethnobotany of the Chayahuita of the Paranapura basin (Peruvian Amazon). Journal of Ethnopharmacology 146:127–153
Pérez-Arbeláez, Enrique. 1956. Plantas Útiles de Colombia. Liberia Colombiana, Bogotá
Prance, G. T. 1984a. The pejibaye, *Guilielma gasipaes* (HBK) Bailey, and the papaya, *Carica papaya* L. In: Pre-Columbian Plant Migration, D. Stone (Editor), pp. 85–104. Papers of the Peabody Museum of Archeology and Ethnology, Vol. 76, Harvard University, Cambridge
Prance, G. T. 1984b. Introduction on knowledge and use of the Amazon forest. *Interciencia* 9(5):275–279
Purseglove, J.W. 1974. *Tropical Crops: Dicotyledons*. John Wiley, New York
Reátegui, Katia Wendy. 2013. Hábitos alimenticios de *Saguinus mystax* Spix "pichico barba blanca" en Padre Isla, Loreto, Peru. Master's thesis, UNAP, Iquitos
Reyes-García, Victoria. 2001. Indigenous people, ethnobotanical knowledge, and market economy: a case study of the Tsimane' Amerindians in lowland Bolivia. Doctoral dissertation, University of Florida, Gainesville.
Ritter, Rhuan Amorim; Monteiro, Maria Vivina Barros; Monteiro, Frederico Ozanan Barros; Rodrigues, Silvane Tavares; Soares, Marina Lira; Silva, Jean Carlos Ramos; Palha, Maria das Dores Correia; Biondi, Germano Francisco; Rahal, Sheila Canevese; Tourinho, Manuel Malheiros. 2012. Ethnoveterinary knowledge and practices at Colares island, Pará state, eastern Amazon, Brazil. Journal of Ethnopharmacology 144:346–352
Saltos, Ricardo Vinicio Abril; Vásquez, Tomás Elías Ruiz; Lazo, Jatnel Alonso; Banguera, Derwing Viáfra; Guayasamín, Pedro Damián Ríos; Vargas, Janeth Karina Aguinda; Peñas, Ingrid Vega. 2016. The use of medicinal plants by rural populations of the Pastaza province in the Ecuadorian Amazon. Acta Amazonica 46(4):355–366
Schultes, R. E. and R. F. Raffauf. 1990. The Healing Forest: Medicinal and Toxic Plants of the Northwest Amazonia. Dioscorides Press, Portland, Oregon
Smith, N. J. H.; J. T. Williams; D. L. Plucknett; and J. P. Talbot. 1992. *Tropical Forests and Their Crops*. Cornell University Press, Ithaca.
Tomchinsky, Bernardo; Ming, Lin Chau; Kinupp, Valdely Ferreira; Hidalgo, Ari de Freitas; Chaves, Francisco Célio Maia. 2017. Ethnobotanical study of antimalarial plants in the middle region of the Negro River, Amazonas, Brazil. Acta Amazonica 47(3):203–212
Torke, B.M. 2004. Caricaceae (Papaya Family). In: *Flowering Plants of the Neotropics*, N. Smith, S.A. Mori, A. Henderson, D.W. Stevenson, and S.V. Heald (Editors), pp. 85–86, Princeton University Press, Princeton

Valadeau, Céline; Pabon, Adriana; Deharo, Eric; Albán-Castillo, Joaquina; Estevez, Yannick; Lores, Fransis Augusto; Rojas, Rosario; Gamboa, Dionicia; Sauvain, Michel; Castillo, Denis; Bourdy, Geneviève. 2009. Medicinal plants from the Yanesha (Peru): Evaluation of the leishmanicidal and antimalarial activity of selected extracts. Journal of Ethnopharmacology 123:413–422

Valadeau, Céline; Castillo, Joaquina Alban; Sauvain, Michel; Lores, Augusto Francis; Bourdy, Geneviève. 2010. The rainbow hurts my skin: medicinal concepts and plants uses among the Yanesha (Amuesha), an Amazonian Peruvian ethnic group. Journal of Ethnopharmacology 127:175–192

Van den Berg, M. E. 2010. Plantas Medicinais na Amazônia. Museu Paraense Emílio Goeldi, Belém

Van den Berg, Maria Elisabeth; Silva, Milton Hélio Lima da. 1988. Contribuição ao conhecimento da flora medicinal de Roraima. Acta Amazonica 18(1–2) supl.:23–35

Veiga, J. B.; Scudeller, V. V. 2015. Etnobotânica e medicina popular no tratamento de malária e males asociados na comunidade ribeirinha Julião- baixo Rio Negro (Amazônia Central). Revista Brasileira de Plantas Medicinais 17(4):737–747

Vigneron, M.; Deparis, X.; Deharo, E.; Bourdy, G. 2005. Antimalarial remedies in French Guiana: a knowledge attitudes and practices study. Journal of Ethnopharmacology 98:351–360

Villas Boas, Orlando; Villas Boas, Claudio. 1975. Xingu: The Indians, Their Myths. Souvenir Press, London

Wezel, A.; Ohl, J. 2006. Homegarden plant diversity in relation to remoteness from urban centers: a case study from the Peruvian Amazon region. In: Tropical Homegardens: a time-tested example of sustainable agroforestry, Kumar, B. M.; Nair, P. K. R. (eds), pp 143–158, Advances in Agroforestry vol. 3, Springer, Dordrecht

Woods, W. I. and McCann, J. M. 1999. The anthropogenic origin and persistence of Amazonian dark earths. *Yearbook of the Conference of Latin Americanist Geographers* 25: 7–14

Chapter 14
Caryocaraceae

14.1 *Caryocar villosum*

Bolivia Castaño macho

Brazil Piquiá, piquiá eté, piquiá verdadeiro, piquiazeiro, piquí, pequi, pequirana, pequiá, suari, amêndoa de espinho; tohób' (Hupd'äh), pɨkɨ'a'ɨ (Ka'apor), pri kâ ti, pidjô-bàti (Kayapó), peki'á (Piripkura), pɨkɨ'a-'ɨw (Tembé), ehú-pu, re'upu (Tukano), ruamahi (Uaicá-Mucajaí), warky (Waimiri-Atroari), peke'a (Wajãpi), ruapa hi (Yanomami)

Colombia Barbasco

Peru Almendro, almendro de bajo, almendro pequi, almendrón

Venezuela Kumato (Yanomami)

Status Wild, spontaneous in cultural settings, planted

N. Smith, *Amazon Fruits: An Ethnobotanical Journey*, Ethnobiology,
https://doi.org/10.1007/978-3-031-12803-5_14

Bianca, 10 years, with piquiá *Caryocar villosum* fruit gathered in the forest near her village. Carrazedo, mouth of the Xingu River, Município de Gurupá, Pará, Brazil 4-15-15

Piquiá *Caryocar villosum* is a giant forest tree, reaching up to 45 m, that grows in upland forests throughout the Amazon Basin and north into the Guianas (Andrade et al. 2017; Araujo 2015; Batista et al. 2015; Bentes-Gama et al. 2009; Black et al. 1950; Carim et al. 2013, 2015; Ferreira and Prance 1999; Gama et al. 2007; Gomide 1997:111; Huber 1906; Lemos et al. 2015; Machado 2009:27; Miranda et al. 2000; Navarro 2002; Pereira et al. 2011; Pinheiro et al. 2007; Pires and Salomão 2000; Ribeiro et al. 1994; Rodrigues 1963; Silva and Rosa 1989). Because of the tree's great height, fruits are normally gathered from the ground (Scoles 2009). It also occurs occasionally in higher parts of floodplains, such as in the Amazon estuary (Amaral et al. 2009; Bentes-Gama et al. 2002; Freitas et al. 2019; Silva 2008a, b, c:15), along the clear water Jari River in eastern Amazonia (Carim 2016:87), in Bolivia (Navarro and Maldonado 2002:74), and in some depressions of the cerrado zone of the Upper Araguaia River Basin in Tocantins, Brazil (Martins et al. 2008) as well as in floodplain forests of the Upper Araguaia (Brito et al. 2006). It is also found in some successionary woodlands on uplands, such as the Belterra plateau south of Santarém in western Pará (Carvalho et al. 1986; Espírito-Santo et al. 2005), in the Rio Guamá watershed in the Amazon estuary near Belém (Scoles 2009), in the Bragantina zone of northeastern Pará (Almeida and Vieira 2001; Rayol et al. 2011; Schwartz 2007; Vieira 1996:59), in parts of eastern Pará (Medeiros 2016:91), in the Tocantins watershed (Taurino 2017:86), and in Amapá (Gomide 1997:67).

Piquiá, as the tree is known in Brazil, produces an unusual fruit. The oily yellow pulp is encased in a thick rind that is cut away or gnawed in the case of wild animals. The mesocarp can only be eaten after boiling the skinned fruits in water with salt for about half an hour to soften it. The fruits are produced in the rainy season and into the early part of the dry period (Knowles and Parrotta 1995).

The pulp normally surrounds a single seed but some fruits contain up to four seeds (Marx et al. 1997). The seed itself is protected by spines, like a miniature sea urchin hence one of its popular names in the Brazilian Amazon: amêndoa de espinho. The Victorian naturalist Henry Walter Bates captured this feature eloquently "the fruit is curious in having a hollow chamber between the pulp and the kernel, beset with hard spines which produce serious wounds if they enter the skin" (Bates 1863:48). People take great care when taking bites out of the cooked fruit for fear of pricking their lips or tongue on the short spines protruding from the nut. The pulp is usually scraped off with a spoon or knife and eaten, typically with manioc flour. Rural folk in the watershed of the Capim River in eastern Pará boil the pulp in water with salt and then eat it with manioc flour (Shanley and Rosa 2005). In the village of Marudá in Pará at the mouth of the Amazon, the spiney seeds are sometimes burned to ward off the evil eye called *mau olhado* (Coelho-Ferreira 2009), while oil from the seeds is sold in some markets, such as in Santarém, to treat muscle or joint pain (May 2016). Oil from the pulp has anti-inflammatory properties (Silva 2009:72; Xavier et al. 2011) and along the lower Tapajós is also used to treat skin burns, bruises, mycoses, and muscular pain (Alves et al. 2016). The popularity of this fruit is such that one rural community in the Tapajós, Piquiatuba, is named after the fruit (Lima et al. 2018).

Peeling piquiá *Caryocar villosum* fruits prior to boiling them. Carrazedo, mouth of the Xingu River, Município de Gurupá, Pará, Brazil 4-14-15

Cooked piquiá fruits are sometimes added to meat stews. The pulp is savory with a nutty taste and has a smooth texture, similar to avocado. Cooked piquiá fruits are also found occasionally in the street markets in the Brazilian Amazon, such as in Rio Branco, Acre (Lima 2014). Some piquiá fruits are larger than others because they contain two seeds. The Waimiri-Atroari, who live in an indigenous reserve north of Manaus, Amazonas, dry the seeds and then roast them over a fire before eating them (Milliken et al. 1992:62). Other indigenous groups gather piquiá fruits in the forest, such as the Yanomami in southern Venezuela and bordering Amazonas and Roraima in Brazil (Albert and Le Tourneau 2007). On Marajó Island and its vicinity, rural folk make a tea from the fruit and bark to treat fevers and diarrhea (Lisboa 2012:390).

Cooked piquiá *Caryocar villosum* fruits to be eaten as a snack with manioc flour. Sitio do Raimundo, Rio Curuá, Floresta Nacional Caxiuanã, Municipio de Melgaço, Pará, Brazil 5-5-15

As peeled piquiá fruits are heated in water, some of the oil rises to the surface. In the past that oil was sometimes skimmed off and used as a substitute for butter and to make soap; piquiá "butter" also entered commerce on a small scale (Kawanishi et al. 1986; Le Cointe 1922:481). Now both urban and rural residents in the Brazilian Amazon mostly use purchased soybean oil for frying and margarine made with soybean oil to spread on their crackers and bread.

Piquiá also contributes to the human diet indirectly. Piquiá fruits evolved for dispersal by animals, some of which are hunted by rural folk. When ripe during the rainy season, the cannon ball-sized fruits fall to the ground where they are sought after by a variety of mammals, including paca *Agouti paca*, a highly prized game animal, as well as agoutis (*Dasyprocta* spp.), giant armadillo *Priodontes maximus*, tapir *Tapirus terrestris*, and brocket deer (*Mazama americana*, *M. gouazoubira*) (Albert and Milliken 2009:65; Ayres and Ayres 1979). Like humans, wild animals have learned not to bite down on the fruits; rather they carefully nibble the mesocarp to avoid being pricked by spines protecting the single seed. The yellow flowers, which fall during the dry season, are also eaten by several game animals including paca and brocket deer. Hunters who live in Carrazedo at the mouth of the Xingu set up platforms (*jiraus*, *mutas*) close to piquiá trees when they are dropping flowers so that they can return at night with flashlights and shoot game. In the Caxiuanã National Forest, hunters also wait on platforms during the day close to piquiá trees to shoot diurnal agoutis when they come to feed on the fallen flowers (Lisboa et al. 2002). Rural folk in the watershed of the Capim River in eastern Pará also wait on elevated platforms close to this tree to shoot various game animals when they come to feed on the fallen flowers (Shanley and Rosa 2005). In the Tapajós extractive reserve south of Santarém in western Pará, some rural folk use seeds of this fruit to bait traps to capture some game animals (Silva et al. 2018).

The white-nosed saki monkey *Chiropotes albinasus* which occurs in upland forests south of the Amazon River between the Madeira in the west and the Xingu in the east eat the fruits of *Caryocar villosum* (Ayres 1989; Oliveira et al. 1985) as does the bearded saki *Chiropotes chiropotes* which lives in the northern part of the Brazilian Amazon and contiguous areas in Venezuela and the Guianas (Barnett et al. 2012; Boyle et al. 2012). Both monkeys may be involved in dispersing the seeds. The endangered white-nosed saki is not an important game animal but some rural people may shoot it when they are on hunts in the forest. Its survival is threatened by large-scale deforestation along the "arc of fire" stretching from eastern Amazonia south along the southern part of the basin.

Piquiá *Caryocar villosum* flowers that have fallen to the ground. The tree was spared when forest was cleared for cattle pasture. Fazenda Alvorado 2, Travessão Tiradentes, km 80 Altamira-Itaituba, Transamazônica, Município de Medicilandia, Pará, Brazil 9-5-15

Because piquiá is such a useful tree, it is typically spared when small farmers fell forest to plant their crops, such as in the vicinity of Manacapuru along the lower Solimões near its confluence with the Rio Negro (Costa and Mitja 2010). This is a tradition stemming back to indigenous cultures such as the Waimiri-Atroari in Amazonas (Milliken et al. 1992:24). Farmers living in Carrazedo at the mouth of the Xingu typically leave a buffer zone of forest or successionary vegetation around piquiá trees so that they are not killed or damaged by fires set to burn off debris in their fields. Occasionally, young piquiá trees are encountered in manioc fields near Carrazedo, evidently dispersed by game animals. When weeding manioc fields farmers spare piquiá as well as other spontaneous fruits and nut trees. However, piquiá flowers evolved to be pollinated by bats as well as some moths, bees, and beetles (Kerr et al. 1986; Martins and Gribel 2007; Prance 1985) so if deforestation becomes pronounced the pollinators will become scarce or even disappear and this will impact spared or planted piquiá trees. Conserving forests is not only important for maintaining a rich genepool of many of the Amazonian fruits but is also essential for the survival of pollinators and dispersal agents.

Piquiá *Caryocar villosum* and other fruit trees with a buffer zone of successionary vegetation to protect the trees from fires set by farmers when clearing their fields. Adjacent to an archaeological site a few hundred meters inland from Carrazedo, the mouth of the Rio Xingu, Pará, Brazil 8-1-16

Some farmers occasionally plant piquiá among their crops, such as near the village of Jocojó along a river of the same name that flows into the Amazon a few kilometers upstream from Gurupá and in the vicinity of Santa Luzia village on the shores of Lago Amanã, along the lower Rio Juruá. Along Ramal Anauera near Tomé-Açu, Pará, a farmer has planted two piquiá by his large cacao grove. One of the motives for planting the tree is that the large piquiá leaves are used to make small containers for nursery seedlings.

Young piquiá *Caryocar villosum* tree, one of nine planted 2 years ago along the edge of a field containing various crops. Jocojó, Rio Jocojó, Município de Gurupá, Pará, Brazil 4-23-15

Piquiá is also planted in some home gardens, such as near the towns of Castanhal and Breves and small communities such as Jocojó near Gurupá and Laranjal, along Igarapé which drains into the Baia de Caxiuanã, all in Pará. River dwellers along the Guamá River on the outskirts of Belém also plant this giant tree in some of their dooryard gardens (Lisboa et al. 2009) as do some Japanese-Brazilian farmers in the vicinity of Tomé-Açu, Pará (Yamada and Osaqui 2006). Also in Pará, rural people in the upland community of Santo Antônio some 125 km south of Santarém plant this tree in some of their home gardens (Almeida and Gama 2014) as do some rural folk in the community of Boa Esperança at km 43 of the Santarém to Curuá-Una highway (Garcia et al. 2017) and in the municipalities of Bragança, Igarapé-Açu, Marapanim in the zona Bragantina of northeastern Pará (Quaresma et al. 2015; Rosa et al. 2007; Silva et al. 2017a, b). This tree is also planted in home gardens of rural folk near Macapá, the capital of Amapá (Pereira et al. 2017). In São Luis, Maranhão, in the eastern fringe of the Amazon Basin, this tree is also found in some home gardens (Akinnifesi et al. 2010a, b). In Amazonas, I observed this tree in a home garden in Piraruaia, a small village along the Tefé River. Piquiá is present in some rural home gardens along Ramal do Pau Rosa, an offshoot of the BR 175 Manaus-Boa Vista highway north of Manaus (Machado 2016:30) and in the vicinity of Manaus (Guillaumet et al. 1990; Santos 2006:52; Souza 2010:26; Souza and Scudeller 2009).

Since so many rural folk have either planted or protected piquiá, it is sometimes found in cultural forests near home sites, as in the vicinity of Comunidade Nossa Senhora da Luz, Rio Tauaú, an affluent of the Parauaú River not far from Breves on Marajó Island. Piquiá is also encountered in orchards planted around some manioc processing huts (*casa de farinha*, *casa de forno*) such as at Vila Valente along the lower Tefé near its confluence with the Solimões and Jocojó along a river of the same name that flows into the lower Amazon. Ancient piquiá trees are also found near villages, such as Carariacá in the municipality of Santarém, evidently planted or spared in the past.

The Kayapó plant piquiá because they value the fruit (González-Pérez et al. 2013); they also plant it in forest patches (*apêtê*) created in scrub savanna in the Upper Xingu watershed to attract game animals (Posey 2002:202). In the Tapajós National Forest south of Santarém groves of this tree are found in the vicinity of some archaeological sites with anthropogenic dark earth suggesting that indigenous cultures have affected the distribution patterns of this species (Alves et al. 2016). A related species, pequi or piqui *Caryocar brasiliensis* is native to the cerrado region bordering southern Amazonia and is planted for its edible fruits by some indigenous groups such as the Kalapolo (Basso 1973). The distribution of *Caryocar villosum* in Amazonia has likely been influenced by indigenous cultures (Clement 2000).

Piquiá *Caryocar villosum* planted 5 years ago in a village home garden. Piraruaia, Tefé River, Amazonas, Brazil 9-8-12

Ancient piquiá *Caryocar villosum* tree at the perimeter of a village. Carariacá, Santarém Municipality, Pará, Brazil 5-15-96

Unfortunately, piquiá seems to be increasingly rare in some areas due to the timber trade. In the vicinity of Porto de Moz along the lower Xingu River in Pará local communities have long cut down some piquiá trees to make boats (Medina et al. 2009). Rural folk sometimes cut deals with loggers so that piquiá trees on their land can be felled and removed such as in the vicinity of Santarém (Costa et al. 2002; Menton et al. 2009; Silva and Bentes-Gama 2008) and in the municipalities of Oriximiná, Itupiranga, and Parauapebas, also in the state of Pará (Fiedler et al. 2012). In the vicinity of Ipixuna in Pará, loggers have drastically reduced the quantity of piquiá trees, thereby diminishing the availability of the fruits for rural people (Rist et al. 2012). Piquiá is also one of the preferred targets of loggers in the vicinity of Paragominas in eastern Pará (Veríssimo et al. 1992). Brocket deer (*Mazama americana* and *M. gouazoubira*), important game animals, are fond of these fruits and that is one reason that their numbers are diminishing.

Piquiá are felled to make planks, especially for boat construction, posts, plywood, and furniture because the wood is resistant (Huber 1910; Lisboa 2012:177; Prance 1990, 2014:83; Prance and da Silva 1973:33; Villachica 1996:17). Piquiá timber has been exported to Portugal since at least the mid-eighteenth century (Cruz 1973:280). However, care must be taken when sawing piquiá because the dust can cause skin allergies (Moura et al. 1993). In 2000, I visited a small boat building facility (*estaleiro*) at Breves on the southern coast of Marajó Island where I saw a boat under construction that had "ribs" of piquiá. In 1998, I visited Ilha dos Porcos in the Amazon estuary, where a ribeirinho showed me the small boat he was making in his backyard; its bowsprit was fashioned from piquiá wood. Piquiá is also used to stake up black pepper *Piper nigrum*, an important cash crop at Tomé-Açu, Pará. Piquiá planks were also used to construct a large round vat (*lambique*) to store cachaça at the last remaining distillery near Abaetetuba along the lower Tocantins River. Nowadays, most of the cachaça consumed in the Brazilian Amazon comes from São Paulo, Minas Gerais, or the Northeast region of Brazil.

Francisca, 58 years, with her cutting board made of piquiá *Caryocar villosum* wood. Ilha Marisol, Rio Xingu, about 45 km upstream from Altamira, Pará, Brazil 9-15-15

Piquiá has been employed for a long time by rural folk for a variety of purposes. Piquiá logs are hollowed out with axes, fire, and chisels to make dugout canoes, such as along the Upper Anajás River on Marajó Island. Piquiá is one of the more desirable trees for making dugout canoes because the wood is so durable (Shanley and Rosa 2004). An old rubber tapper in Breves showed me a knife he used to use to cut the bark of rubber trees; its curved handle was fashioned from piquiá wood. Along the middle Xingu, the wife of a fisherman uses a cutting board made with piquiá wood in her kitchen. In the village of Pedreira along the shores of Baia de Caxiuanã along the lower Anapu River, piquiá wood is employed to make bowls (*alguidar*) for soaking acai *Euterpe oleracea* fruits prior to making juice. In the nineteenth century, piquiá bark and the outer casing of the fruits was used to make ink (Baena 2004:53; Smith 1879:188). These traditional uses do not exert nearly as much pressure on piquiá populations as the timber trade that supplies planks for burgeoning urban markets (Loureiro et al. 1979:89).

Rafael, 5 years, holding an *alguidar* carved from piquiá *Caryocar villosum* wood used by his grandmother for soaking açaí *Euterpe oleracea* fruits. Pedreira, Baia de Caxiuanã, Rio Anapu, Município de Melgaço, Pará, Brazil 5-11-15

References

Akinnifesi, F. K.; Sileshi, G.; Costa, Joselene da; Moura, E. G. da; Silva, Rosele F. da; Ajayi, O. C.; Linhares, J. F. P.; Akinnifesi, A. I.; Araujo, Maria de; Rodrigues, M. A. I. 2010a. Floristic composition and canopy structure of home-gardens in São Luís city, Maranhão state, Brazil. Journal of Horticulture and Forestry 2(4):72–86

Akinnifesi, Festus Kehinde.; Sileshi, Gudeta W.; Ajayi, Oluyede C.; Akinnifesi, Adetutu I.; Moura, Emanoel G. de; Linhares, Jairo F. P.; Rodrigues, Ivanilde 2010b. Biodiversity of the urban homegardens of São Luís city, northeastern Brazil. Urban Ecosystems 13(1):129–146

Albert, Bruce; Le Tourneau, François-Michel. 2007. Ethnogeography and resource use among the Yanomami: toward a model of "reticular space". Current Anthropology 48(4):584–592

Albert, Bruce; Milliken, William. 2009. Urihi A: a terra-floresta Yanomami. Instituto Socioambiental (ISA)/Institut de Recherche pour le Devéloppement, São Paulo

Almeida, Larissa Santos de; Gama, João Ricardo Vasconcellos. 2014. Quintais agroflorestais: estrutura, composição florística e aspectos socioambientais em área de assentamento rural na Amazônia brasileira. Ciência Florestal 24(4):1041–1053
Almeida, Arlete Silva de; Vieira, Ima Célia Guimarães. 2001. Padrões florísticas e estruturais de uma cronoseqüência de florestas no munícipio de São Francisco do Pará, região Bragantina, Pará. Boletim do Museu Paraense Emílio Goeldi, Série Botânica 17(1):209–240
Alves, Rubana P.; Levis, Carolina; Clement, Charles R. 2016. Use and management of piquiá suggest in situ domestication along the lower Tapajós River, Brazilian Amazonia. Economic Botany 70(2):198–202
Amaral, Dário Dantas do; Vieira, Ima Célia Guimarães; Almeida, Samuel Soares de; Salomão, Rafael Paiva; Silva, Antonio Sérgio Lima da; Jardim, Mário Augusto Gonçalves. 2009. Checklist da flora arbórea de remanescentes florestais da região metropolitana de Belém e valor histórico dos fragmentos, Pará, Brasil. Boletim do Museu Paraense Emílio Goeldi, Ciencias Naturais 4(3):231–289
Andrade, Ricardo Teixeira Gregório de; Pansini, Susamar; Sampaio, Adeilza Felipe; Ribeiro, Maíra Silva; Cabral, Guilherme Sampaio; Manzatto, Ângelo Gilberto. 2017. Fitossociologia de uma floresta de terra firme na Amazônia sul-occidental, Rondônia, Brasil. Biota Amazônia 7(2):36–43
Araujo, Henrique José Borges de. 2015. Alterações causadas por incêndio florestal na composição de espécies de uma floresta primária no sudoeste da Amazônia Brasileira. Gaia Scientia 9(1):220–231
Ayres, J. M. 1989. Comparative feeding ecology of the uakari and bearded saki, *Cacajao* and *Chiropotes*. Journal of Human Evolution 18:697–716
Ayres, J. M.; Ayres, C. 1979. Aspectos da caça no alto Rio Aripuanã. Acta Amazonica 9(2):287–298
Baena, Antônio Ladislau Monteiro 2004. Ensaio Corográfico sobre a Província do Pará. Edições do Senado Federal, Brasilia
Barnett, Adrian A.; Boyle, Sarah A.; Pinto, Liliam P.; Lourenço, Waldete C.; Almeida, Thais; Silva, Welma Sousa; Ronchi-Teles, Beatriz; Bezerra, Bruna M.; Ross, Caroline; MacLarnon, Ann; Spironello, Wilson R. 2012. Primary seed dispersal by three Neotropical seed-predating primates (*Cacajao melanocephalus ouakary*, *Chiropotes chiropotes* and *Chiropotes albinasus*). Journal of Tropical Ecology 28(6):543–555
Basso, E. B. 1973. *The Kalapalo Indians of Central Brazil*. Holt, Rinehart and Winston, New York
Bates, H. W. 1863. *The Naturalist on the River Amazons*. John Murray, London, vol. 2
Batista, Anderson Pedro Bernardina; Aparício, Wegliane Campelo da Silva; Santos, Vanessa Silva dos; Lima, Robson Borges de; Mello, José Marcio de. 2015. Caracterização estrutural em uma floresta de terra firme no estado do Amapá, Brasil. Pesquisa Florestal Brasileira 35(81):21–33
Bentes-Gama, Michelliny de Matos; Scolforo, José Roberto Soares; Gama, João Ricardo Vasconcellos. 2002. Potencial produtivo de madeira e palmito de uma floresta secundária de várzea baixa no estuário Amazônico. Revista Árvore 26(3):311–319
Bentes-Gama, Michelliny de Matos; Leal, Guido Sanick; Barros, Joelson de Oliveira; Lopes, Raimunda Herculano; López, Giovana Fiorella Zamora; Silveira, Juliane Cardoso da. 2009. Características da estrutura de uma floresta de terra firme em Porto Velho, Rondônia. EMBRAPA Porto Velho Circular Técnica 109:1–7
Black, G. A.; Dobzhansky, T. H.; Pavan, C. 1950. Some attempts to estimate species diversity and population density of trees in Amazonian forests. Botanical Gazette 111(4):413–425
Boyle, Sarah A.; Zartman, Charles E.; Spironello, Wilson R.; Smith, Andrew T.; Beck, Harald. 2012. Implications of habitat fragmentation on the diet of bearked saki monkeys in central Amazonian forest. Journal of Mammalogy 93(4):959–976
Brito, Elizabeth Rodrigues; Martins, Sebastião Venâncio; Oliveira Filho, Ary Teixeira de; Silva, Elias; Silva, Alexandre Francisco da. 2006. Estrutura fitossociológica de um fragmento natural de floresta inundável em área de orizicultura irrigada, município de Lagoa da Confusão, Tocantins. Revista Árvore 30(5):829–836

Carim, Marcelo de Jesus Veiga. 2016. Estrutura, composição e diversidade em florestas alagáveis de várzea de maré e de igapó e suas relações com variáveis edáficas e o período de inundação no Amapá, Amazônia oriental, Brasil. Doctoral dissertation, Programa de Pós-Graduação em Botânica, INPA, Manaus

Carim, Marcelo de Jesus Veiga; Guillaumet, Jean-Louis Bernard; Guimarães, José Renan da Silva; Tostes, Luciedi Cássia Leôncio. 2013. Composição e estrutura de floresta ombrófila densa do extremo norte do Estado do Amapá, Brasil. Biota Amazônia 3(2):1–10

Carim, Marcelo de Jesus Veiga; Guimarães, José Renan da Silva; Tostes, Luciendi de Cássia Leôncio; Takiyama, Luís Roberto; Wittmann, Florian. 2015. Composition, structure and floristic diversity in dense rain forest in the eastern Amazon, Amapá, Brazil. Acta Scientiarum Biological Sciences 37(4):419–426

Carvalho, João Olegário Pereira de; Silva, José Natalino Macedo; Lopes, José do Carmo Alves; Montagner, Lise Helene; Carvalho, Manoel Sebastião Pereira de. 1986. Composição florística de uma mata secundária no planalto de Belterra no Pará. In: Anais do 1° Simpósio do Trópico Úmido, Belém, PA, 12 a 17 November, EMBRAPA, Brasília, pp 197–205

Clement, C. R. 2000. Piquiá *Caryocar villosum*. In: Biodiversidade Amazônica: Exemplos e Estratégias de Utilização, Jason W. Clay, Paulo de T. B. Sampaio, and Charles R. Clement (eds), INPA/SEBRAE, Manaus, pp 100–109

Coelho-Ferreira, Márlia. 2009. Medicinal knowledge and plant utilization in an Amazonian coastal community of Marudá, Pará state (Brazil). Journal of Ethnopharmacology 126:159–175

Costa, Joanne R.; Mitja, Danielle. 2010. Uso dos recursos vegetais por agricultores familiares de Manacapuru (AM). Acta Amazonica 40(1):49–58

Costa, Dulce Helena Martins; Carvalho, João Olegário Pereira de; Silva, José Natalino Macedo. 2002. Dinâmica da composição florística após a colheita de madeira em uma área de terra firme na Floresta Nacional do Tapajós (PA). Revista de Ciências Agrárias 38: 67–90

Cruz, Ernesto. 1973. História de Belém. Universidade Federal do Pará, Coleção Amazônica, Série José Veríssimo, Belém, vol. 1

Espírito-Santo, Fernando Del Bon; Shimabukuro, Yosio Edimir; Aragão, Luiz Eduardo Oliveira e Cruz de; Machado, Evandro Luiz Mendonça. 2005. Análise da composição florística e fitossociológica da floresta nacional do Tapajós com apoio geográfico de imagens satélites. Acta Amazonica 35(2):155–173

Ferreira, Leandro V.; Prance, Ghillean T. 1999. Ecosystem recovery in terra firme forests after cutting and burning: a comparison on species richness, floristic composition and forest structure in the Jaú National Park, Amazonia. Botanical Journal of the Linnaean Society 130:97–110

Fiedler, Nilton C.; Costa, Alexandre F. da; Soares, Thelma S.; Leite, Ângelo Márcio P. 2012. Caracterização do segmento de madeira serrada em três municípios do estado do Pará. Revista Brasileira de Ciências Agrárias 7(1):111–116

Freitas, João da Luz; Santos, Erick Silva dos; Cruz Junior, Francisco de Oliveira; Borges, César Henrique Alves; Santos, Adriano Castelo dos. 2019. Structural analysis and distribution patterns in lowland tropical forest, eastern Amazon. Journal of Agricultural Studies 7(4): 240–255

Gama, João Ricardo Vasconcellos; Souza, Agostino Lopesde Souza; Calegário, Natalino; Lana, Guilherme Campos. 2007. Fitossociologia de duas fitocenoses de floresta ombrófila aberta no município de Codó, estado do Maranhão. Revista Árvore 31(3):465–477

Garcia, Bruna Naiara Rocha; Vieira, Thiago Almeida; Oliveira, Francisco de Assis. 2017. Tree and shrub diversity in agroforestry homegardens in rural community in eastern Amazon. Floresta 47(4):543–552

Gomide, Guilherme Luís Augusto. 1997. Estrutura e dinâmica de crescimento de florestas tropicais primária e secundária no Estado do Amapá. Master's Thesis, Universidade Federal do Paraná, Brazil

González-Pérez, Sol E.; Robert, Pascale de; Coelho-Ferreira, Márlia. 2013. Seed use and socioeconomic significance in Kayapó handicrafts: a case study from Pará state, Brazil. Economic Botany 67(1):1–16

Guillaumet, J-L.; Grenand, P.; Bahri, S.; Grenand, F.; Lourd, M.; Santos, A.A. dos; Gély, A. 1990. Les jardins-vergers familiaux d'Amazonie centrale: une exemple d'utilisation de l'espace. Turrialba 40(1):63–81

Huber, Jacques. 1906. La végétation de la vallée du Rio Purus (Amazone). Bulletin de L'Herbier Boissier 6(4):249–276

Huber, J. 1910. Mattas e madeiras amazonicas. Boletim do Museu Goeldi de Historia Natural e Ethnographia 6: 91–225

Kawanishi, Kazuko; Raffauf, Robert F.; Schultes, Richard Evans. 1986. The Caryocaraceae as a source of fish poisons in the northwest Amazon. Botanical Museum Leaflets, Harvard University 30(4):247–253

Kerr, Warwick Estevam; Absy, Maria Lúcia; Souza, Antonio Carlos Marques. 1986. Espécies nectaríferas e poliniferas utilizadas pela abelha *Melipona compressipes fasciculata* (Meliponinae, Apidae) no Maranhão. Acta Amazonica 16:145–156

Knowles, Oliver H.; Parrotta, John A. 1995. Amazonian forest restoration: an innovative system for native species selection based on phenological data and field performance indices. Commonwealth Forestry Review 74(3):230–243

Le Cointe, P. 1922. *L'Amazonie Brésilienne: Le Pays-ses Habitants, ses Resources, Notes et Statistiques jusqu'en 1920*. Augustin Challamel, Paris, vol 1

Lemos, D. A. N.; Ferreira, B. G. A.; Siqueira, J. D. P.; Oliveira, M. M.; Ferreira, A. M. 2015. Floristic and phytosociology in dense "terra firme" rainforest in the Belo Monte hydroelectric plant influence area, Pará, Brazil. Brazilian Journal of Biology 75(3):257–276

Lima, Ana Carolina Barbosa de. 2014. Flavors of the city: access to regional fruit and fruit consumption in the State of Acre, Brazil. Boletim do Museu Paraense Emílio Goeldi, Ciencias Humanas 9(1):79–92

Lima, Mauriceia Colares; Couceiro, Sheyla Regina Marques; Fonseca Júnior, Élcio Meira da. 2018. Estrutura diamétrica e especial de piquiazeiro na comunidade de Piquiatuba, Floresta Nacional do Tapajós, oeste do Pará. Agroecossistemas 10(1):29–42

Lisboa, P. L. B. 2012. *A Terra dos Aruã: Uma História Ecológica do Arquipélago do Marajó*. Museu Paraense Emílio Goeldi, Belém

Lisboa, P. L. B., I. A. Gomes, R. C. Lisboa, and C. V. Urbinati. 2002. O estilo Amazônico de sobreviver: manejo dos recursos naturais. In: *Natureza, Homem e Manejo de Recursos Naturais na Região de Caxiuanã, Melgaço, Pará*, P. L. B. Lisboa (ed.), Museu Paraense Emílio Goeldi, Belém, pp. 41–170

Lisboa, Pedro L. B.; Pinon, Alerrandson; Silva, Marly Lameira da. 2009. A ocupação humana na APEG e os recursos biológicos (comunidades Porto de Ceasa e Nossa Senhora dos Navegantes). In: Aurá: Comunidades & Florestas, Lisboa, Pedro L. B. (ed), Museu Paraense Emílio Goeldi, Belém, pp 33–61

Loureiro, Arthur A.; Silva, Marlene F. da; Alencar, Jurandyr da Cruz. 1979. Essências Madeireiras da Amazônia. Instituto Nacional de Pesquisas da Amazônia, Manaus, vol. 2

Machado, Kathleen Samira da Silva. 2009. Estrutura e composição florística de uma floresta de terra firme na reserva de desenvolvimento sustentável Amanã, Amazônia Central. Master's thesis, INPA, Manaus

Machado, Danilo de Oliveira. 2016. A agrobiodiversidade de quintais agroflorestais em propriedades agrícolas familiares na BR 174, Ramal do Pau-Rosa, Manaus, AM. Master's Thesis, INPA, Manaus

Martins, Rodrigo Lemes; Gribel, Rogério. 2007. Polinização de Caryocar villosum (Aubl.) Pers. (Caryocaraceae) uma árvore emergente da Amazônia Central. Revista Brasileira de Botanica 30(1):37–45

Martins, Sebastião Venâncio; Brito, Elizabeth Rodrigues; Oliveira Filho, Ary Teixeirade; Silva, Alexandre Francisco da; Silva, Elias. 2008. Floristic composition of two wetland forests in Araguaian plain, state of Tocantins, Brazil, and comparison with other areas. Revista Árvore 32(1):129–141

Marx, Friedhelm; Andrade, Eloisa Helena A.; Maia, José Guilherme. 1997. Chemical composition of the fruit pulp of *Caryocar villosum*. Zeitschrift für Lebensmittel-Untersuchung und Forschung 204:442–444

May, Thomas. 2016. Aspectos de sostenibilidad de productos no maderables florestales con uso curativo en el oeste de Pará, Brasil. Ambiente y Desarrollo 20(38):69–84

Medeiros, Priscilla Sanjuan de. 2016. Indicadores ambientais para funcionalidade ecológica em florestas secundárias de diferentes idades na Amazônia Oriental. Doctoral dissertation, Universidade Federal do Pará, Belém

Medina, G.; Pokorny, B.; Campbell, B. M. 2009. Community forest management for timber extraction in the Amazon frontier. International Forestry Review 11(3):408–420

Menton, Mary C.S.; Merry, Frank D.; Lawrence, Anna; Brown, Nick. 2009. Company-community logging contracts in Amazonian settlements: impacts on livelihoods and NTFP harvests. Ecology and Society 14(1):39–57

Milliken, W., R. P. Miller, S. R. Pollard, and E. V. Wandelli. 1992. *The Ethnobotany of the Waimiri Atroari Indians of Brazil*. Royal Botanic Gardens, Kew

Miranda, Ivana Lobato; Carvalho, João Olegário Pereira de; Ferreira, Maria do Socorro Gonçalves; Oliveira, Lia Cunha de; Vieira, Ima Célia Guimarães; Almeida, Wanderléa da Costa; Soares, Márcio Hofman Mota; Costa, Dulce Helena Martina. 2000. Principais espécies arbóreas de valor econômico em uma floresta às margens do Lago Acapu, no Município de Oriximiná, Pará. Embrapa Amazônia Oriental Comunicado Técnico 31:1–4

Moura, Cleonice de Oliveira; Wolter Filho, Wilson; Absy, Maria Lúcia. 1993. Alergias causadas por algumas espécies madeireiras na Amazônia. In: Bases Científicas para Estratégias de Preservação e Desenvolvimento da Amazônia, Ferreira, Efrem Jorge Gondim; Santos, Geraldo Mendes dos; Leão, Elizabeth Lima Mendes; Oliveira, Luíz Antônio de Oliveira (eds), INPA, Manaus, vol. 2, pp 89–106

Navarro, G. 2002. Provincia biogeográfica del Acre-Madre de Dios (Amazónica Suroccidental). In: Geografía Ecológica de Bolivia: Vegetación y Ambientes Acuáticos, Navarro, Gonzalo; Maldonado, Mabel (eds), Centro de Ecología Simón I. Patiño, Cochabamba, pp 51–98

Navarro, G.; Maldonado, M. 2002. Geografía Ecológica de Bolivia: Vegetación y Ambientes Acuáticos. Centro de Ecología Simón I. Patiño, Cochabamba

Oliveira, J. M. S.; Lima, M. G.; Bonvincino, C.; Ayres, J. M.; Fleagle, J. G. 1985. Preliminary notes on the ecology and behavior of the Guianan saki (*Pithecia pithecia*, Linnaeus 1766; Cebidae, Primate). Acta Amazonica 15(1–2):249–263

Pereira, Luciano Araújo; Sobrinho, Felipe de Araújo; Costa Neto, Salustiano Vilar da. 2011. Florística e estrutura de uma mata de terra firme na Reserva de Desenvolvimento Sustentável no Rio Iratapuru, Amapá oriental, Brasil. Floresta 41(1):113–122

Pereira, Luciano Araujo; Lima-Barbosa, Jackson Rodrigo de; Almeida, Maria Zélia de; Guimarães, Elsie Franklin. 2017. Diversidade de plantas em quintais quilombolas, conhecimento local sobre uso e cultivo de pimentas na Amazônia oriental, Brasil. Revista de Biologia Neotropical 14(1):56–72

Pinheiro, Klewton Adriano Oliveira; Carvalho, João Olegário Pereira de; Quantz, Beatriz; Francez, Luciana Maria de Barros; Schwartz, Gustavo. 2007. Fitossociologia de uma área de preservação permanente no leste da Amazônia: indicacão de espécies para recuperação de áreas alteradas. Floresta 37(2): 175–187

Pires, João Murça; Salomão, Rafael de Paiva. 2000. Dinâmica da diversidade arbórea de um fragmento de floresta tropical primária na Amazônia oriental-1. Período: 1956 a 1992. Boletim do Museu Paraense Emílio Goeldi, Série Botânica 16(1):63–110

Posey, D. A. 2002. Kayapó Ethnoecology and Culture. Routledge, London

Prance, G. T. 1985. The pollination of Amazonian plants. In: Key Environments: Amazonia, Ghillean T. Prance and Thomas E. Lovejoy (eds), Pergamon Press, Oxford, pp. 166–191

Prance, G. T. 1990. The genus Caryocar L. (Caryocaraceae): an underexploited tropical resource. Advances In Economic Botany 8: 177–188

Prance, G. T. 2014. That Glorious Forest: Exploring the Plants and Their Indigenous Uses in Amazonia. New York Botanical Garden, New York

Prance, G.T. and M.F. da Silva. 1973. Caryocaraceae. Flora Neotropica Monograph No. 12, New York

Quaresma, Amanda Paiva; Almeida, Ruth Helena Cristo; Oliveira, Cyntia Meireles de; Kato, Osvaldo Ryohei. 2015. Composição florística e faunística de quintais agroflorestais da agricultura familiar no nordeste Paraense. Revista Verde de Agroecologia e Desenvolvimento Sustentável 10(5):76–84

Rayol, Breno Pinto; Alvino-Rayol, Fabrízia de Oliveira; Silva, Manoela Ferreira Fernandes da. 2011. Similaridade florística entre o estrato arbóreo e a regeneração natural de uma floresta secundária, no município de Bragança, nordeste do estado do Pará, Brasil. Revista Brasileira de Agroecologia 6(3):107–114

Ribeiro, José Eduardo Lahoz S.; Nelson, Bruce Walker; Silva, Marlene Freitas da; Martins, Luisa Suely S.; Hopkins, Michael. 1994. Reserva Florestal Ducke: diversidade e composição da flora vascular. Acta Amazonica 24(1–2):19–30

Rist, Lucy; Shanley, Patricia; Sunderland, Terry; Sheil, Douglas; Ndoye, Ousseynou; Liswanti, Nining; Tieguhong, Julius. 2012. The impacts of selective logging on non-timber forest products of livelihood importance. Forest Ecology and Management 268:57–69

Rodrigues, William A. 1963. Estudo de 2,6 hectares de mata de terra firme da Serra do Navio, território do Amapá. Boletim do Museu Paraense Emílio Goeldi, Nova Série Botânica 19:1–23

Rosa, Leonilde dos Santos; Silveira, Erycélede Lima; Santos, Mônica Mota dos; Modesto, Rozi da Silva; Perote, James Richard Silva; Vieira, Thiago Almeira. 2007. Os quintais agroflorestais em áreas de agricultores familiares no município de Bragança-PA: composição florística, uso de espécies e divisão de trabalho familiar. Revista Brasileira de Agroecologia 2(2):337–341

Santos, Jerffereson Lobatodos. 2006. Uso e diversidade de espécies vegetais cultivadas na Reserva de Desenvolvimento Sustentável do Tupé, Manaus, Amazonas. Master's thesis, INPA/UFAM, Manaus

Schwartz, Gustavo. 2007. Manejo sustentável de florestas secundárias: espécies potenciais no nordeste do Pará, Brasil. Amazônia: Ciência e Desenvolvimento 3(5):125–147

Scoles, Ricardo. 2009. El quintal y las frutas: recursos económicos y alimentares en la comunidad negra de Itacoã, Acará, Pará, Brasil. Acta Amazonica 39(1):1–12

Shanley, P. and N.A. Rosa. 2004. Eroding knowledge: an ethnobotanical inventory in eastern Amazonia's logging frontier. Economic Botany 58(2):135–160

Shanley, P. and N.A. Rosa. 2005. Conhecimento em erosão: um inventário etnobotânico na fronteira de exploração da Amazônia oriental. Boletim do Museu Paraense Emílio Goeldi, Série Ciências Naturais 1(1):147–171

Silva, Fabíola Andréa. 2008a. Ceramic technology of the Asurini do Xingu, Brazil: an ethnoarchaeological study of artifact variability. Journal of Archaeological Method and Theory 15:217–265

Silva, Maria Betânia Araujo da. 2008b. Composição florística e estrutura fitossociológica de uma floresta ripária situada às margens do Rio Moju, Pará. Master's thesis, Universidade Federal Rural da Amazônia, Belém

Silva, Regina Oliveira da. 2008c. Biodiversidade e políticas de conservação: o caso do parque Estadual Monte Alegre, Pará. Doctoral dissertation, Universidade de Brasília

Silva, Rosibel Rodrigues. 2009. O etnoextrativismo não-madeireiro em uma comunidade Amazônica: um estudo de caso em Santa Luzia do Buiuçuzinho-Coari/AM. Master's thesis, UFAM, Manaus

Silva, Ana Paula Ferreira Frota da; Bentes-Gama, Michelliny de Matos. 2008. Fitossociologia de uma floresta ombrófila aberta em área de assentamento rural no distrito de Jaci Paraná, Porto Velho, Rondônia. Ambiência 4(3):435–452

Silva, Manoela F. F. da; Rosa, Nelson A. 1989. Análise do estrato arbóreo da vegetação sobre jazidas de cobre na Serra dos Carajás-PA. Boletim do Museu Paraense Emílio Goeldi, Série Botânica 5(2):175–206

Silva, Hanna Kassia Machado da; Gama, João Ricardo Vasconcellos; Sousa, Randerson José de Araujo; Lameira, Mahyanny Karoline da Silva; Costa, Daniele e Lima da; Oliveira, Douglas Valente de; Rocha, Jobert Silva da; Oliveira, Thiago Gomes de Sousa. 2017a. Composição florística de quintais agroflorestais na vila Cuera, Bragança, Pará. Agroecossistemas 9(2):330–338

Silva, Silmara Bento; Sousa, Venderleia Conceição; Santos, Crissogno Mesquita; Mariano, Daiane Cinque; Okumura, Ricardo Shigueru. 2017b. Levantamento florístico do componente arbustivo-arbóreo da vegetação ciliar de fragmento no Rio Parauapebas. Agroecossistemas 9(1):99–115

Silva, Andrea Araújo da; Santos, Lizandra Elizeário; Cruz, Girlene da Silva; Ribeiro, Renato Bezerra da Silva; Gama, João Ricardo Vasconcelos. 2018. Potencial de comercialização de produtos florestais não madeireiras na área de manejo da reserva extrativista Tapajós Arapiuns Pará. Acta Tecnológica 13(1):45–63

Smith, Herbert H. 1879. Brazil, the Amazons and the Coast. Charles Scribner's Sons, New York

Souza, Célia Cristina Valero. 2010. Etnobotânica de quintais em três comunidades ribeirinhas na Amazônia central, Manaus, AM. Master's thesis, INPA, Manaus

Souza, Célia Cristina Valero; Scudeller, Veridiana Vizoni. 2009. Plantas úteis nos quintais das comunidades ribeirinhas Julião e Agrovila, Reserva de Desenvolvimento Sustentável do Tupé, Amazônia Central. Revista Brasileira de Agroecologia 4(2): 2487–2491

Taurino, Tássia Cristina da Conceição Barros. 2017. Recuperação natural da diversidade de espécies de plantas em florestas secundárias na Amazônia Oriental. Master's Thesis, Universidade Federal do Pará, Belém

Veríssimo, Adalberto; Barreto, Paulo; Mattos, Marli; Tarifa, Ricardo; Uhl, Christopher. 1992. Logging impacts and prospects for sustainable forest management in an old Amazonian frontier: the case of Paragominas. Forest Ecology and Management 55: 169–199

Vieira, Ima Célia Guimarães. 1996. Forest succession after shifting cultivation in eastern Amazonia. Doctoral dissertation, University of Stirling

Villachica, L., 1996. Frutales y hortalizas promisorios de la Amazonia. SPT-TCA; 44

Xavier, William Kathy Silva; Medeiros, Benedito Junior; Lima, Clarissa Silva; Favacho, Hugo Alexandre; Andrade, Eloisa Helena de Aguiar; Araújo, Railda Neyva Moreira; Santos, Lourivaldo Silva; Carvalho, José Carlos Tavares. 2011. Topical anti-inflammatory action of Caryocar villosum oil (Aubl) Pers. Journal of Applied Pharmaceutical Science 1(3):62–67

Yamada, M.; Osaqui, H. M. L. 2006. The role of homegardens in agroforestry development: lessons from Tomé-Açu, a Japanese-Brazilian settlement in the Amazon. In: Tropical Homegardens: a time-tested example of sustainable agroforestry, Kumar, B. M.; Nair, P. K. R. (eds), pp. 299–316, Advances in Agroforestry vol. 3, Springer, Dordrecht

Chapter 15
Celastraceae

15.1 *Peritassa laevigata*

Brazil Gulosa, gogó de guariba, cipó ovo de guariba, bampari, grão de macaco, bacupari, jatumã, cabeça de macaco; oké du'hpoa (Tukano)

Colombia Huevo de gato

Ecuador Omentacabo (Waorani)

Peru Llosna micuna, coto-y-o, murcuhuasca, parinari huasca

Venezuela Aji de paloma, bejuco de pereza, cachete de vieja, derello

Status Wild, planted

N. Smith, *Amazon Fruits: An Ethnobotanical Journey*, Ethnobiology,
https://doi.org/10.1007/978-3-031-12803-5_15

Gulosa *Peritassa laevigata* **planted in an upland manioc field by Floriano, 67 years, a former rubber tapper. Sitio Timbó, left bank of Rio Xingu about 45 km upstream from Altamira, Pará, Brazil, 9-17-15**

Gulosa *Peritassa laevigata* is usually found in floodplain forests and fruits during the rainy season. This tall but slender tree has a vine-like growth habit and is found throughout the Amazon Basin, the Atlantic forest of Brazil, the Guianas, the Upper Orinoco River in Venezuela, Tobago, and Central America as far north as Cota Rica. It occurs up to 1400 m in the Andes (Smith 1953). It also occurs in the Upper Orinoco watershed in southern Venezuela. It is also found in gallery forests south of the Amazon into northern Paraguay as well as in semi-deciduous upland forest in the transition zone between the southern Amazon and cerrado vegetation of central Brazil (Marimon and Lima 2001; Mews et al. 2012).

Rural folk gather the orange-yellow, grapefruit-sized fruits for a snack and to make juice. The fruit pulp is also used to bait hooks to attract fruit-eating fish such as pacu (several species of *Myleus*, *Metynnis*, and *Mylossoma*). Accordingly, some *ribeirinhos* plant gulosa in their home gardens, such as in Paquiçamba, a Juruna village along the middle Xingu between Altamira and Belo Monte, and at Ilha do Poção on the right bank of the Xingu some 7 km upstream from Altamira. A river dweller further up the Xingu between Altamira and the mouth of the Iriri river has planted gulosa in his upland manioc field to gather the fruits once the manioc is harvested. Several other fruit trees, both spontaneous and planted, will help turn the harvested manioc field into an orchard rather than just successionary vegetation. Although gulosa's "home" is várzea forest, it can adapt to uplands in cultural settings as with several other trees with edible fruits in Amazonia.

15.2 *Salacia cordata*

Bolivia Chuchuhuaso

Brazil Bacuri da várzea, cipó de anta

Ecuador Guasca de guitarra, comida de loro, luma blanca; saguata caspi (Kichwa)

Peru Puricho huayo, auca murco huasca, chuchuhuasi huasca, chopapuentsopar, tambor waska

Status Wild

Puricho huayo *Salacia cordata* tree in fruit floodplain forest. Manco Capac, Rio Puinahua (a branch of the Ucayali), Reserva Nacional Pacaya-Samiria, Loreto, Peru 4-15-06

Salacia cordata is an understory tree but also with the growth habit of woody vine in floodplain forests with yellow to peach-colored fruits that ripen in the rainy season. People gather the fruits by bending down branches from their canoes. In some parts of its range, *S. cordata* resembles a liana reaching some 30 m into the canopy of upland forests where the fruits are out of reach for people. Toward the canopy, the fruits are evidently eaten by parrots, as suggested by the name for the plant in Ecuador: *comida de loro*. Parrots likely disperse the seeds. Woolly monkeys *Lagothrix lagotricha*, spider monkeys *Ateles paniscus*, and red howler monkeys *Alouatta seniculus* and *A. macconnelli* also eat these fruits (Arroyo-Rodríguez et al. 2015; Defler and Defler 1996; Julliot 1996a, b) and thus contribute indirectly to the human diet because these monkeys are hunted for their meat throughout the Amazon Basin and contiguous lowland rainforest regions. They also disperse some of the seeds of this fruit. Some smaller monkeys, such as the white-faced saki *Pithecia pithecia*, bearded saki *Chiropotes santanas*, Geoffroy's saddleback tamarin *Saguinus fuscicollis nigrifrons*, and brown capuchin monkey *Cebus apella* also eat this fruit and disperse some of the seeds (Gregory 2006:52; Roble 2008:48; Sánchez and Levy 2013:28, 62).

The round fruits, 3 cm in diameter, are covered with a creamy white down (indument) and contain four seeds covered by a translucent, jelly-like pulp. Known as puricho huayo in Manco Capac, a village on the Puinahua, a branch of the Ucayali, the fruits are generally eaten in the forest and are not taken to markets. The Machiguenga and Yoras in the upper Madre de Dios also eat the fruits.

Puricho huayo fruit gathered in floodplain forest. Manco Capac, Rio Puinahua (a branch of the Ucayali), Reserva Nacional Pacaya-Samiria, Loreto, Peru 4-15-06

Salacia cordata is found below 1900 m from Mexico through Central America to northern South America, including the Amazon and Orinoco basins and the Guianas. In addition to the fruit, *S. cordata* has medicinal uses in some areas, such as in the vicinity of Chazuta along the upper Huallaga River in the Peruvian Amazon where people use the bark to make a decoction to treat rheumatism and broken bones (Sanz-Biset et al. 2009). In the Upper Juruá in Acre, Brazil, hunters wait at night when the liana is dropping fruit to shoot game.

15.3 *Salacia juruana*

Ecuador Mina caspi (Kichwa), bewee (Waorani)

Peru Bombo huayo, huasca chuchohuasi, yacu palillo; shuhshuwasha munin (Chayahuita), chuchuwasi (Mayna)

Venezuela Cachete de vieja

Status Wild

Bombo huayo *Salacia juruana* fruits in forest. 800 m north of Miraflores, Rio Tigre, Loreto, Peru, 7-1-06

The yellow fruits of *Salacia juruana* do not appear in markets but are appreciated by those traveling in upland and floodplain forests during the rainy months. The round fruits are 4–5 cm in diameter. The bumpy skin of the globular fruit is broken open to ingest the gray, translucent pulp which contains numerous orange-tan seeds. Known as bombo huayo along the lower Tigre in the Peruvian Amazon, the fruits are typically gathered from the ground. Two species of tamarin monkeys (*Saguinus fuscicollis* and *S. mystax*) feed on these fruits in upland forests in the Peruvian Amazon and likely disperse some of the seeds; the red-billed ground cuckoo *Neomorphus pucheranii* eats some of the partially eaten fruits discarded by these monkeys and may also be involved in dispersing the seeds (Siegel et al. 1989). When young *Salacia juruana*, as with other species in the genus, resembles a small tree but when it grows larger it assumes a liana-like habit when trees are nearby to offer support. In the Peruvian Amazon, the Chayahuita dry the bark to obtain powder for a poultice applied to the skin to treat leishmaniasis (Odonne et al. 2013).

Salacia juruana grows below 1600 m and is found from Panama south to northern South America including Amazonia, such as in upland forests in the vicinity of Serra dos Carajás in Pará (Lombardi and Biral 2016). *Salacia juruana* is also found in the moist Atlantic forest of Brazil, separated from the Amazon by a dry corridor of savanna-cerrado vegetation. The fruits of at least three other species of *Salacia* are eaten in Amazonia: *S. cordata*, *S. gigantea*, and *S. impressifolia*. The latter, known as uaimirapati in the Brazilian Amazon, is particularly abundant in the Rio Negro watershed (Cavalcante 2010:235).

Roy, 9 years, eating bombo huayo fruit he has gathered in forest. 800 m north of Miraflores, Rio Tigre, Loreto, Peru, 7-1-06

References

Arroyo-Rodríguez, Víctor; Andresen, Ellen; Bravo, Susana P.; Stevenson, Pablo R. 2015. Seed dispersal by howler monkeys: current knowledge, conservation implications, and future directions. In: Howler Monkeys, Developments in Primatology: Progress and Prospects, Kowalewski, M. M. et al. (eds), Springer, New York, pp 111–139

Cavalcante, P. B. 2010. *Frutas Comestíveis na Amazônia*. Museu Paraense Emílio Goeldi, Belém

Defler, Thomas R.; Defler, Sara B. 1996. Diet of a group of *Lagothrix lagothricha lagothrica* in southeastern Colombia. International Journal of Primatology 17(2):161–190

Gregory, Tremaine. 2006. Comparative socioecology of sympatric, free-ranging white-faced and bearded sakis in Brownsberg Nature Park, Suriname. Doctoral dissertation, Kent State University, Kent, Ohio

Julliot, C. 1996a. Seed dispersal by red howling monkeys (*Alouatta seniculus*) in the tropical rain forest of French Guiana. International Journal of Primatology 17(2):239–258

Julliot, C. 1996b. Fruit choice by red howler monkeys (*Alouatta seniculus*) in a tropical rain forest. American Journal of Primatology 40:261–282

Lombardi, Julio Antonio; Biral, Leonardo. 2016. Flora das cangas da Serra dos Carajás, Pará, Brasil: Celastraceae. Rodriguésia 67(5):1285–1290

Marimon, Beatriz Schwantes; Lima, Edson de Souza. 2001. Caracterização fitofisionômica e levantamento florístico preliminar no Pantanal dos Rio Morte-Araguaia, Cocalinho, Mato GrossoActa Botanica Brasilica 15(2):213–229

Mews, H. A.; Marimon, B. S.; Ratter, J. A. 2012. Observations on the vegetation of Mato Grosso, Brazil. V. Changes in the woody species diversity of a forest in the cerrado-Amazonian forest transition zone and notes on the forests of the region. Edinburgh Journal of Botany 69(2):239–253

Odonne, Guillaume; Valadeau, Celine; Alban-Castillo, Joaquina; Stien, Didier; Sauvain, Michel; Bourdy, Genevieve. 2013. Medical ethnobotany of the Chayahuita of the Paranapura basin (Peruvian Amazon). Journal of Ethnopharmacology 146:127–153

Roble, Nick. 2008. Ontogenetic resource partitioning in white faced sakis (*Pithecia pithecia*). Master's thesis, Kent State University.

Sánchez, Judith Jacira Achong; Levy, Allison Licett Nuñez. 2013. Características de frutos consumidos por *Saguinus nigrifrons* (Primates, Callitrichidae), en la Estación Biológica Quebrado Blanco, Loreto, Perú. Master's thesis, Universidad Nacional de la Amazonía Peruana, Iquitos

Sanz-Biset, Jaume; Campos-de-la-Cruz, José; Epiquién, Mirbel A.; Cañigueral, Salvador. 2009. A first survey on the medicinal plants of the Chazuta valley (Peruvian Amazon). Journal of Ethnopharmacology 122:333–362

Siegel, Charles E.; Hamilton, Joel M.; Castro, N. Rogerio. 1989. Observations of the red-billed ground-cuckoo (*Neomorphus pucheranii*) in association with tamarins (Saguinas) in northeastern Amazonian Peru. The Condor 91(3):720–722

Smith, A. C. 1953. Studies of South American Plants, XIII. Journal of the Washington Academy of Sciences 43(7):203–212

Chapter 16
Chrysobalanaceae

16.1 *Chrysobalanus icaco*

Brazil Ajuru, ajirú, ajirú roxo, abajurú, abajeru, cajurú, juru, juruba, guajurú, guajurú rosa, guajiru

Colombia Icaco, guajiru

Venezuela Cocojoru, icacillo, icaco

Status Wild, planted

Ajuru ***Chrysobalanus icaco*** **growing on a sand dune. Praia Maçarico, Salinas, Zona Bragantina, Pará, Brazil 10-8-14**

N. Smith, *Amazon Fruits: An Ethnobotanical Journey*, Ethnobiology,
https://doi.org/10.1007/978-3-031-12803-5_16

Ajuru *Chrysobalanus icaco* is a squat bush on sandy soils at the mouth of the Amazon, forming pure stands in some cases, such as on dunes near Maguary Point on Marajó Island and along the Bragantina coast of Pará (Amaral et al. 2001; Bastos 1988; Bastos et al. 1995, 2003:12; Costa Neto et al. 1996; Santos and Rosário 1988) as well as in low-lying herbaceous wetlands in the coastal zone (Amaral et al. 2008). One coastal community, Ajuruteua near Bragança, is named after the fruit. Ajuruteua means "the place of ajuru" and is a popular weekend hangout for tourists from Bragança. Bragança has a hotel with the same name as well as at least one fishing boat. Such naming is an indication of the cultural importance of ajuru fruit. Two species of stingless bees, *Melipona flavolineat*a and *Melipona fasciculata*, have been recorded carrying pollen from this plant in the eastern fringe of the Amazon rainforest in Maranhão and likely pollinate some of the flowers (Gostinski et al. 2018).

Fishing boat under repair at an *estaleiro*. The boat's name means the ice boat in the land of ajuru. Bragança, Pará, Brazil 3-28-15

In eastern Marajó where sandy beaches are common and the water is brackish, ajuru occurs up to 1 km inland. Close to the shore, sea breezes sculpt ajuru into flattened, well-rounded bushes, but further inland and better protected from the prevailing winds, the bushes can grow into a small tree as high as 5 or 6 m (Castañeda 1961:84; Prance 1972:16). Ajuru is widespread in wild stands in coastal areas of the Caribbean including Florida, Central and tropical South America, including the Atlantic forest of Brazil (Lorenzi et al. 2006; Popenoe 1920:271). This bush was widespread in the Caribbean by the time Spaniards arrived in the late fifteenth century (Patiño 1963:233). It is also found in coastal areas of West Africa; fruits must have been dispersed in the distant past by ocean currents either from Africa to the Neotropics, or the other way round. It is cultivated and naturalized in Seychelles and Fiji (Prance 1972:16).

Boys climbing a juru *Chrysobalanus icaco* tree to gather fruit in a village home garden along a black water river. The tree was planted some 50 years ago. Anjo da Guarda, Rio Pracupi, affluent of the Anapu River, Pará, Brazil 5-15-15

Jefferson, 7 years, holding a yellow-skinned variety of juru *Chrysobalanus icaco*. Anjo da Guarda, Rio Pracupi, affluent of the Anapu River, Pará, Brazil 5-15-15

The fruits vary in shape and size according to location. Some are oblong and relatively small, like olives, such as near Salvaterra on Marajó, whereas others are the shape and size of walnuts, as observed near Maguary Point on the northeast tip of the island. Many fruits are nearly round, measuring 3 cm by 2 cm or 2.5 cm × 2.2 cm. Three color forms are found: black, pink, and yellow. Along the Curuá River in the Caxiuanã National Forest, river dwellers call the yellow form "juru branco." Sometimes at least two-color forms are located in the same area.

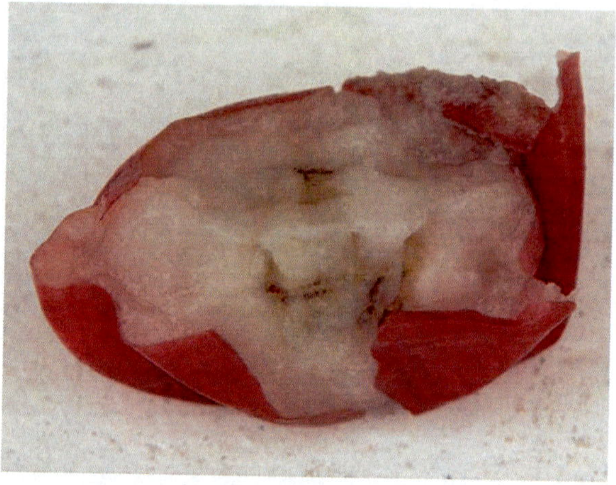

Red-skinned variety of ajuru *Chrysobalanus icaco* gathered from bushes among sand dunes. Ajuruteua, near Bragança, Pará, Brazil 3-22-15

The white, somewhat dry flesh that surrounds the single seed, which measures 2 cm × 1.5 cm, is rather insipid, so ajuru fruits are not usually dispatched to markets. Rather, they are eaten on the spot or taken home.

This bush to the small tree also has other uses, including folk medicines. In the Caxiuanã National Forest river dwellers eat ajuru fruits to control diarrhea and a tea is prepared from the leaves and petioles to treat hemorrhoids and diabetes (Rocha et al. 2013). Along the coast of Pará, locals prepare a tea from the roots to treat diabetes and urinary pain and the seeds are used to treat wounds (Bastos et al. 2003; Carneiro et al. 2010; Coelho-Ferreira 2009; Jardim and Medeiros 2006). Ajuru bears fruit in the dry season, providing a welcome treat because not many Amazonian plants bear edible fruits when the rains diminish because it is harder for the seeds to germinate. In the vicinity of Ajuruteua along the Bragantina coast of Pará, rural folk use the wood to make charcoal and directly to fuel fires (Carneiro et al. 2010).

Black-skinned variety of juru *Chrysobalanus icaco*. Mouth of the black water Rio Caxiuanã, Município de Melgaço, Pará, Brazil 5-6-15

Ajuru appears to be native to coastal areas near the sea because its fruits float and therefore can be dispersed by currents. However, indigenous people have long taken the fruits inland for planting in their communities and this interchange between different parts of eastern Amazonia continues. Along Furo dos Porcos near Macapá, for example, a river dweller obtained ajuru seeds from Maruim on the northeastern coast of Marajó and has planted them in his açaí *Euterpe oleracea* palm grove. Along the Macacoari River, an affluent of the northern branch of the Amazon near its mouth in Amapá, a farmer has planted ajuru in an agroforestry plot dominated by bitter and sweet manioc.

Juru ***Chrysobalanus icaco*** **planted in a 2-year-old field planted to a variety of crops including manioc. Adjacent to a large archaeological site. Jocojó, Rio Jocojó, Município de Gurupá, Pará, Brazil 4-23-15**

Ajuru is also planted in gardens around homes and manioc processing huts (*casa de farinha, casa de forno*) in eastern Pará, such as in the municipality of Bragança (Rosa et al. 2007) and in Abaetetuba and its vicinity along the lower Tocantins River (Gonçalves and Lucas 2017; Lobato et al. 2017). Some manioc processing huts are located away from homes and villages, such as at Jocojó in the Municipality of Gurupá, where monkeys take advantage of human absences to raid the planted ajuru trees. In the vicinity of Soure on Marajó Island in the Amazon estuary, squirrel monkeys *Saimiri sciureus* have been observed feeding on this fruit and likely disperse some of the seeds (Santos et al. 2018a, b). I have encountered ajuru in home gardens as far west as the lower Xingu and all the homeowners explained that they were planted from seeds or seedlings obtained from other locations, including a sandy beach along the lower Xingu near Porto de Moz. Ajuru is also found in some home gardens in suburbs of São Luis, Maranhão, in the eastern fringe of Amazonia (Akinnifesi et al. 2010a, b) as well as in the vicinity of Manaus, Amazonas (Souza 2010:26; Souza and Scudeller 2009; Veiga and Scudeller 2011) and in Vila Cuera, a rural community in the vicinity of Bragança in northeastern Pará (Silva et al. 2017a, b) and rural communities in the municipality of Bonito in the Bragantina zone in northeastern Pará (Vieira et al. 2012).

Juru *Chrysobalanus icaco* by the dock of a home along a black water river. Gurupá Mirim, Município de Gurupá, Pará, Brazil 7-29-16

Juru *Chrysobalanus icaco* tree in the home garden of a river dweller (*ribeirinho*). Site of a long abandoned colonial village. Icajó, Rio Amazonas, Município de Gurupá, Pará, Brazil 8-23-14

Ajuru is found in what might appear to be wild stands in the interior, such as along the shores of the lower Xingu near Porto de Moz and along the Caxiuanã River, an affluent of the lower Anapu. Such "wild" stands occur along clear and black water rivers. But they are most likely the vestiges of home gardens, long since abandoned. Interestingly, some of these "wild" stands are close to archaeological sites such as at the IBAMA station along the lower Anapu. The Xingu River basin appears to be at the western limit of the distribution of this species in the Amazon Basin (Silva et al. 1977:15).

Juru *Chrysobalanus icaco* on the floodplain of a black water river. These bushes may be vestiges of a home garden abandoned long ago. Lower Caxiuanã River, Município de Melgaço, Pará, Brazil 5-15-15

16.2 *Couepia bracteosa*

Brazil Pajurá, pajurá de racha, pajurá da mata, pajurá do mato, pajurá de anta, pajurazinho, marirana, mari-mari, umari rana, copuda miuda, batinga, caraipé, coco pau; apirahya (Piripkura)

Colombia Aguire

Peru Parinari, parinari colorado, pacharama, supay ocote

Status Wild, planted

Farmer holding a pajurá *Couepia bracteosa* fruit he gathered in a floodplain forest. Arapixuna, Santarém Municipality, Pará, Brazil 5-15-96

Pajurá *Couepia bracteosa* is a medium-sized tree reaching up to 25 m tall in mature and regrowth forests on uplands (Campbell et al. 1986; Carim et al. 2007; Carneiro 2004; Condé 2011:75; Falcão et al. 1981; Oliveira and Mori 1999; Parrotta et al. 1997; Pinheiro et al. 2007; Rayol et al. 2011; Rocha 2017:9; Scudeller et al. 2009; Silva et al. 1992; Silva et al. 1985; Spichiger et al. 1996). It also appears occasionally on some floodplains, such as near Belém, Pará (Cain et al. 1956) and along the Jaci-Paraná an affluent of the Madeira River in Rondônia (Reis 2010:31). It also occurs sometimes on the floodplains of black water rivers such as in the Jaú National Park in the Rio Negro watershed in Amazonas (Barnett et al. 2015). Pajurá bears fruit in the rainy season which measure 8 cm long and 7 cm wide (Prance 1972:227). Found mostly in central and western Amazonia, pajurá is also found further east in Amapá and north in the Guianas. In the Guianas, *Couepia bracteosa* is also found in scrub forests growing on sandy soils. Given that it is an edible fruit, and is sometimes planted in home gardens, it is possible that the "natural" distribution of pajurá, as with many other fruit trees, has been altered by human agency. Cultivated forms generally have sweeter pulp and this species was probably domesticated a long time ago by indigenous peoples (Ducke 1946).

Pajurá is gathered from an upland forest, such as in the vicinity of Arapixuna, a village overlooking the floodplain of the Amazon just upstream from its confluence with the Tapajós. The brown, rough-skinned fruits of pajurá, similar in shape and size to the Guatemalan race of avocado, are peeled to reveal a sweet, though slightly oily flesh. The pulp contains vitamin C and has a high-fiber content which helps reduce the chances of constipation (Massing et al. 2018). Hawkmoths pollinate pajurá (Prance 1984) as does a stingless bee *Melipona seminigra merrillae* (Ferreira and Absy 2017). Tapirs *Tapirus terrestris* apparently eat this fruit as suggested by one of its common names in Brazil: pajurá de anta. So, this fruit also contributes indirectly to some rural diets because tapirs are the largest terrestrial game animal in the Amazon.

Pajurá is sometimes planted in home gardens, such as in Carariacá just north of Arapixuna and along Ramal Andirobalzinho, a side road off the Santarém to Alter do Chão highway. As noted nearly a century ago, pajurá is most frequently cultivated in the vicinity of Santarém (Ducke 1922). This tree also occurs in some home gardens along the lower Purus River, an affluent of the Solimões (Machado 2018:49). Villagers obtain pajurá seedlings for their home gardens by germinating seeds from fruits gathered in the wild or from other dooryards.

Farmer with a pajurá *Couepia bracteosa* seedling and some other plants she obtained from a friend in a nearby settlement. She will plant them in her village home garden. Carariacá, Amazon River, Santarém Municipality, Pará, Brazil 7-22-96

16.3 *Couepia chrysocalyx*

Brazil Inamuirana, umarirana, parinaric wixi de quintal; yuru ohtese (Tukano)

Colombia Guarai, doroje

Ecuador Hueso, pani nari; amchichare (Awá), cacawerewe, pusaramanduru, mashca mururuya, rumi caspi (Kichwa), saka (Shuar), caya'o, juëñë, yaso sooquëë, yahi (Siona-Secoya), titecahanca (Waorani)

Peru Parinari, supay ocote, supay huayo, sacha umapu, sacha umari; apíkna, namakia apíkna, uchi apíkna (Aguaruna), tseréntsere (Jívaro), kokoti jihui (Shipibo), ngumari (Ticuna)

Venezuela Mereycura; mereycura (Baniwa)

Status Wild, planted

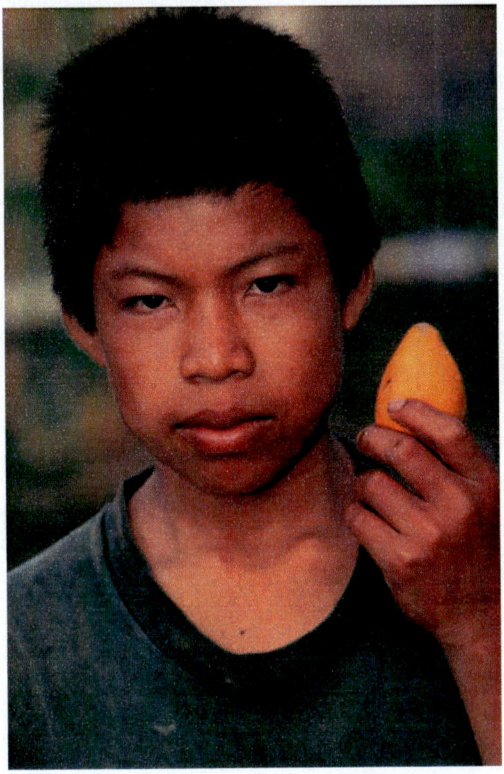

Esli, 14 years, with parinari *Couepia chrysocalyx* fruit he gathered in his home garden. Monte Bello, Reserva Nacional Pacaya-Samiria, Rio Ucayali, Loreto, Peru 5-14-03

Parinari *Couepia chrysocalyx* is native to central and western Amazonia and is widely cultivated in the basin both for its edible fruit and as an ornamental (Prance 1972:245). It occurs in the Andean foothills, such as in Ecuador (Cerón 1996). The tree grows in mature and regrowth forests on both uplands and high river banks where floods reach for only brief periods each year, such as in the village of Monte Bello along the Ucayali. The cylindrical fruits some 15 cm long ripen to an orange-yellow color toward the end of the rainy season. The pulp is orange and has a strong carrot-like smell. Parinari fruits reach some urban markets, such as Iquitos (Vasquez and Gentry 1989). In eastern Colombia, the white-fronted capuchin *Cebus albifrons* eats the fallen fruits of this tree and probably disperses some of the seeds (Defler 1979).

In the Peruvian Amazon, the bark provides a purple-red dye and the cooked fruits yield a blue-black dye which is used as an ink substitute and to stain carved gourds (Antúnez de Mayolo 1989). In the Chazuta district located in a narrow valley created by the Huallaga River in the Andean foothills of the Peruvian Amazon, the bark is used to make a decoction along with several other plants to treat tumors (Sanz-Biset and Cañigueral 2011). The Miraña who live along the Caquetá River in the Colombian Amazon burn the wood of this tree for lighting (Sánchez et al. 2005).

Although the tree can reach 20 m high in forests, it is planted in home gardens where it does not grow so tall so fruits are easier to gather. Parinari is also grown in some fields (*chacras*) alongside other perennial crops. Several other species of *Couepia* produce edible fruits in Amazonia, often endemic to relatively small areas such as pirauxi *C. paraensis*, which is confined to floodplains between Manaus and Santarém (Cavalcante 2010:210).

Parinari features in some folktales. The co-pilot of the boat I hired on one of my field trips in the Peruvian Amazon in 2004 recounted a strange experience that happened to him when he was 14 years old. William was walking along a path leading to his village called Centro America on an island in the Ucayali River in Loreto Province when he had an alarming encounter with a malicious spirit (*demonio*). It was around 9 o' clock at night when he saw a ghostly figure with red eyes (*sombra negra con ojos rojos*) in a grove of parinari trees along the path. William was terrified so he screamed. Fortunately, he was close to home and his family rushed out to ascertain what the matter was.

16.4 *Couepia subcordata*

Brazil Marirana, mari, umarirana

Ecuador Guaiji

Peru Parinari, supay ocote

Status Wild

Marirana *Couepia subcordata* fruits in floodplain forest. Fazenda Jaguatirica, near Codajás, Rio Solimões, Amazonas, Brazil 6-25-07

Marirana *Couepia subcordata* produces orange fruits, like a slim, small mango, measuring 4–9 cm long by 3–5 cm wide (Prance 1972:228). The fibrous, bright orange pulp surrounds a single seed, 2.5 cm × 1 cm. Two primates, capuchin monkeys *Cebus apella* and the monk saki *Pithecia monachus,* have been observed feeding on this fruit in the Peruvian Amazon (Aquino and Bodmer 2004) and other species of monkey likely do the same. Marirana grows on floodplains and uplands in the central and western parts of the Amazon Basin where rural people gather the fruits for a snack. This tree is considered native to the central Amazon (Lorenzi et al. 2006). The tree, which can grow as high as 35 m, has a smooth bark that sports a variety of colorful lichens.

This tree is sometimes planted in home gardens, or perhaps arises spontaneously from discarded seeds, such as along the Purus River, an affluent of the Solimões (Machado 2018:49). It is also planted occasionally along streets and public squares, especially in the Brazilian state of Amazonas, where it is generally shorter since it does not have to compete for light (Cavalcante 2010:242). Known as parinari in Peru, the fruits also turn up in the street markets of Iquitos (Vásquez and Gentry 1989). The wood is occasionally used by carpenters to provide fuelwood and charcoal (Le Cointe 1934:458; Prance and Silva 1975:82).

Paulo, 18 years, snacking on marirana ***Couepia subcordata*** **fruits. Fazenda Jaguatirica, near Codajás, Rio Solimões, Amazonas, Brazil 6-25-07**

16.5 *Couepia ulei*

Brazil Pajurá, marirana, oitirana

Peru Parinari, parinari colorado, suichi parinari, parinarillo

Status Wild

Enoc, 13 years, eating parinari *Couepia ulei* fruit he gathered from a canoe in a floodplain forest. Cocha Huylla, a few kilometers upstream from Veinte de Enero, Rio Yanayacu, Reserva Nacional Pacaya-Samiria, Loreto, Peru 1-14-06

Known as parinari in the Peruvian Amazon, *Couepia ulei* is a wild relative of the widely cultivated *Couepia chrysocalyx*. Unlike the orange-yellow fruits of the latter, however, the fruits of *C. ulei* are green with raised gray spots (lenticels). Although *C. ulei* fruits are smaller, about the size of a large chicken egg, they are nevertheless appreciated as snacks, especially by boys when out fishing. The rough-skinned fruits are 5 cm long by 2 cm wide (Prance 1972) and do not appear in markets.

Couepia ulei is a tree of sunlit riverbanks, especially in black waters, such as the Yanayacu and Samiria rivers in the Reserva Nacional Pacaya-Samiria. The tree is relatively short, usually less than 15 m, so at high water, the fruits are close to the waterline and are easily reached. Although found mostly in the wild, this fruit tree is occasionally encountered in some gardens, such as in Yagua villages along the Amazon River in Peru (Lamont et al. 1999). *Couepia ulei* is confined to the western Amazon, including the western part of the Brazilian Amazon in Acre and Amazonas, where it also grows in upland forests (Prance 1972:251). The white flowers are fragrant and attract bees, which probably serve as pollinators.

16.6 *Hirtella racemosa*

Bolivia Coloradillo, llave; chana, chanasia (Chácobo), vai'si'ma (Tsimane)

Brazil Maria pretinha, murtinha, murta de rato, murta roxa, ucuúba, ajuru, ajurúrana, ajirú do mato, ajurizinho, macucu, macucu peludo, macucu preto, macucu vermelho, macucu mirim, macucu da folha pequena, mucureiro, macucurana, cumaté, cumaterana, acarapucú, caraipé, caraipé branco, caripé, cariperana, cariperana branca, merda de rato, bosta de rato, quebra colhão, canela de cutia, uajuru do mato, azeitona roxa; mawira (Baniwa), wyra-na'ã, ira-pirã-'ɨ (Guajá), akuxi-myra, mukuku-'ɨ-wi (Ka'apor), moroya nyon pisugnu (Karitiana), pi kare ô kryre (Kayapó), wariruwai-ran (Tembé), pyrwa iepry (Waimiri-Atroari)

Peru Tamara blanca, iakui, palitza, pinsha huayo, coloradillo, apacharama, apacharana, parinari, parinarillo, huacupu; apíkna (Aguaruna), nave (Cashinahua)

Venezuela Carbonero, ceniza negra, jacazo, hicaquillo, jicaquillo, pica-pica morada, odooma üjotü; shiemo (Yanomami)

Status Wild, spontaneous in cultural settings

Maria pretinha *Hirtella racemosa* in flower and fruit in early successionary vegetation on uplands. Arlinho, 11 years. Carrazedo, mouth of the Xingu River, Município de Gurupá, Pará, Brazil 7-23-16

Maria pretinha *Hirtella racemosa* is a bush to small tree, reaching as high as 12 m, with spikelets containing some 20 small violet flowers. The small flowers produce a nutrient-rich nectar and are visited by various species of butterflies and bees that pollinate them (Kerr et al. 1986; Prance 1985). The shrub starts producing fruits when only 0.5 m high. The black fruits vary in size from 1 cm × 0.5 cm to 1.5 cm × 1 cm as I measured in the field in the vicinity of Carrazedo at the mouth of the Xingu. Most of the fruits ripen from July to September during the dry season (Alencar 1990). Sweet pink pulp surrounds a single tan seed 1.3 cm × 0.5 cm. Mostly, children eat thefruits since they are rather small and generally close to the ground. Care has to be taken when picking the fruits, however, because the plant has nectaries to attract ants. The ants in turn offer some protection (Prance 2014:191).

Squirrel monkeys *Saimiri sciureus* also eat the fruits, such as in the Peruvian Amazon, and are likely involved in dispersing the seeds along with children (Terborgh 1983:65). In Rondônia in the Brazilian Amazon, the ashy titi monkey *Plecturocebus cinerascens* has also been observed feeding on this fruit and likely disperses some of the seeds (Mattos 2016:30). In French Guiana, the grey brocket deer *Mazama gouazoubira* has been recorded eating the fruits of this plant and this deer is hunted for its meat in the Guianas and in Amazonia (Gayot et al. 2004). In Costa Rica, howler monkeys *Alouatta palliata*, spider monkeys *Ateles geoffroyi*, and the white-headed capuchin monkeys *Cebus capucinus* eat the fruits of this shrub to small trees (Chapman 1989); related species in the Amazon most likely do the same. Several bird species have been observed feeding on these fruits in the central Amazon and are probably involved in dispersing some of the seeds (Macedo 1977).

The shrub has other uses, especially among indigenous groups. The Tembé in eastern Amazonia use the trunk and branches for firewood as do the Kayapó in the Upper Xingu Basin (Posey 2002: 237). The Ka'apor, who also inhabit eastern Amazonia, use the wood to make skewers for roasting game meat (Balée 1993:73). The Taiwano of the Colombian Amazon extract oil from the flowers and seeds of this shrub to treat ear pain (Schultes and Raffauf 1990:130). Villagers in Marudá along the coast of Pará at the mouth of the Amazon make a tea from the leaves to treat diabetes (Coelho-Ferreira 2009), while the Karitiana in Rondônia rub the leaves of this plant on their legs to repel snakes (Prance 1997).

Boys eating Maria pretinha *Hirtella racemosa* fruits they have gathered in successionary vegetation. Rodrigo 11 years (on left) and Arlinho, 11 years. Carrazedo, mouth of the Xingu River, Município de Gurupá, Pará, Brazil 11-2-16

Hirtella racemosa grows in a wide range of habitats from early successionary growth to woodlands, grasslands, and sand dunes, such as along the coast of the Bragantina zone in Pará state (Amaral et al. 2008, 2015; Bastos et al. 1995; Claudino et al. 2015; Kunz et al. 2014b; Magnusson et al. 2008; Medeiros 2016:92; Sanaiotti and Magnusson 1995; Santos and Bastos 2014:76; Santos et al. 2003; Santos 2010) and in nearby Maranhão (Freire and Monteiro 1993). It is found in light gaps in upland forests, such as north of Manaus (Costa and Magnusson 2003), in Roraima and Acre (Alarcón and Peixoto 2007; Maranho and Salimon 2015), along the lower Xingu (Salomão et al. 2007), in the Bragantina zone of northeastern Pará (Jardim et al. 2013), and within woodlands dominated by a native bamboo *Guadua weberbaueri* in Acre in the southwestern portion of the Brazilian Amazon (Maranho et al. 2016). It also occurs occasionally as a weed in cattle pastures cleared in a forest, such as in eastern Pará (Mascarenhas et al. 1999). It thrives in stunted vegetation (*campina*, *campinarana*, *caatinga*) found on sandy soils that are poor in nutrients and dry out quickly, as in the Içana watershed, an affluent of the Rio Negro, and in the vicinity of Manaus (Abraão et al. 2010; Alencar 1990; Anderson et al. 1975; Braga 1979; Braga and Braga 1975; Macedo 1977).

Although it is largely confined to uplands, *Hirtella racemosa* is occasionally found on floodplains, such as along the Rio Negro, the Tapajós, and Madeira Rivers as well as some of their affluents in central Amazonia, in various parts of eastern Pará, the Upper Araguaia River Basin in the state of Tocantins and Mato Grosso, along the Rio das Pacas in the Upper Xingu watershed in northern Mato Grosso, in the Colombian Amazon, and in the Orinoco Basin in Venezuela (Brito et al. 2006, 2008; Casula 2012:55; Ferreira et al. 2013; Godoy et al. 1999; Kunz et al. 2014a, b;

Martins et al. 2008; Myster 2018; Parolin et al. 2003; Parra 2006; Reis 2010:31; Rudas and Prieto 1998; Santos and Jardim 2006; Stefanello et al. 2010). It is relatively common in the transition zone between the Amazon forest and cerrado, such as in the Upper Rio Xingu watershed (Kunz et al. 2008, 2009; Stefanello 2008:18). It also occurs in the watershed of the Caura, an affluent of the Orinoco River in southern Venezuela (Díaz Pérez et al. 2012). This species ranges widely in the Amazon Basin and is also found in other parts of the Neotropics including Suriname and some arid areas of northeastern Brazil and parts of Central America, including Mexico (Bánki et al. 2003:31; Prance 1988).

16.7 *Licania macrophylla*

Brazil Nuera, anauerá, anoerá, anoeré, anoira, pintadinha, macucú chiador, macucú roxo, macucú terra, cariperana, pajurá pedra; ana-wyra (Tembé)

Status Wild

Nuera *Licania macrophylla* **fruit used to make spinning tops** (*pião*)**. Wadrio, 10 years. Carrazedo, mouth of the Xingu River, Município de Gurupá, Pará, Brazil 7-24-16**

Nuera *Licania macrophylla* produces fruits intermediate in size between a golf ball and a tennis ball and can attain 8 cm in diameter (Prance 1972:100). They turn hard when mature and boys use them for making spinning tops (*pião*). The tree has other uses. *Licania macrophylla* is a target of loggers for its valuable timber, and the wood is also used to fuel fires.

This tree, which can reach 30 m high when fully mature, grows on floodplains, including swamp forests that remain flooded for much of the year (Bentes-Gama et al. 2002; Carim 2016:88; Casula 2012:55; Luize et al. 2015; Myster 2018; Prance 1978). However, it is also found occasionally in upland forests, such as in southern Amapá (Aparício 2011:60), north of Manaus, Amazonas (Carneiro 2004), in the vicinity of Belém in eastern Pará (Amaral et al. 2009), as well as stunted woodlands growing on patches of upland sandy soil such as in the Uatumã watershed 150 km northeast of Manaus, Amazonas (Targhetta et al. 2015). *Licania macrophylla* ranges widely in the Amazon Basin and can be found as high as 2400 m in the Andean foothills, such as in Peru. It also occurs in the Guianas.

Related species have several other uses. The bark of caraipé *Licania octandra* is burned and the ash is mixed with clay as a temper when making pottery in rural areas, although this practice is much less common nowadays with people buying aluminum pots or ceramics made in urban areas, or even imported. The wood of farinha seca *Licania micrantha* is used as fuelwood to make manioc flour in the vicinity of Manaus (Bahri 1992:67). In the vicinity of Belém, rural people use this plant to treat a variety of medical issues. A tea is made from the dry bark of this tree to treat inflammation of the reproductive systems of men and women and to treat worm infestations, amoeba infections as well as gastritis. The seeds are grated and used to treat diarrhea and to treat wounds (Gomes et al. 2006).

References

Abraão, M. B.; Shepard, Glenn H.; Nelson, Bruce W.; Baniwa, João Cláudio; Andrello, Geraldo; Yu, Douglas W. 2010. Baniwa vegetation classification in the white-sand campinarana habitat of the northwest Amazon. In: Landscape Ethnoecology: Concepts of Physical and Biotic Space, Johnson, L. M.; Hunn E. (eds), Landscape Ethnoecology: Concepts of Physical and Biotic Space, Berghahn Books, New York, pp 83–115

Akinnifesi, F. K.; Sileshi, G.; Costa, Joselene da; Moura, E. G. da; Silva, Rosele F. da; Ajayi, O. C.; Linhares, J. F. P.; Akinnifesi, A. I.; Araujo, Maria de; Rodrigues, M. A. I. 2010a. Floristic composition and canopy structure of home-gardens in São Luís city, Maranhão state, Brazil. Journal of Horticulture and Forestry 2(4):72–86

Akinnifesi, Festus Kehinde.; Sileshi, Gudeta W.; Ajayi, Oluyede C.; Akinnifesi, Adetutu I.; Moura, Emanoel G. de; Linhares, Jairo F. P.; Rodrigues, Ivanilde 2010b. Biodiversity of the urban homegardens of São Luís city, northeastern Brazil. Urban Ecosystems 13(1):129–146

Alarcón, Juan Gabriel Soler; Peixoto, Ariane Luna. 2007. Florística e fitossociologia de um trecho de um hectare de floresta de terra firme, em Caracaraí, Roraima, Basil. Boletim do Museu Paraense Emílio Goeldi, Ciencias Naturais 2(2):33–60

Alencar, Jurandyr da Cruz. 1990. Interpretação fenológica de espécies lenhosas de campina na Reserva Biológica de campina do INPA ao norte de Manaus. Acta Amazonica 20:145–183

Amaral, Dário Dantas do; Santos, João Ubiratan M. dos; Bastos, Maria de Nazaré do Carmo; Costa, Denise Cristina Torres. 2001. A vegetação da Ilha Canela, Município de Bragança – Pará, Brasil. Boletim do Museu Paraense Emílio Goeldi, Série Botânica 17(2):389–402

Amaral, Dário Dantas do; Prost, Maria Thereza; Bastos, Maria de Nazaré do Carmo; Costa Neto, Salustiano Vilar; Santos, João Ubiratan Moreira dos. 2008. Restingas do litoral amazônico, estados do Pará e Amapá. Boletim do Museu Paraense Emílio Goeldi, Ciencias Naturais 3(1):35–67

Amaral, Dário Dantas do; Vieira, Ima Célia Guimarães; Almeida, Samuel Soares de; Salomão, Rafael Paiva; Silva, Antonio Sérgio Lima da; Jardim, Mário Augusto Gonçalves. 2009. Checklist da flora arbórea de remanescentes florestais da região metropolitana de Belém e valor histórico dos fragmentos, Pará, Brasil. Boletim do Museu Paraense Emílio Goeldi, Ciencias Naturais 4(3):231–289

Amaral, Dário Dantas do; Jardim, Mário Augusto Gonçalves; Costa-Neto, Salustriano Vilar; Bastos, Maria de Nazaré do Carmo. 2015. Síndromes de dispersão de propágulos e a influência da floresta amazônica na composição de espécies lenhosas de uma restinga no litoral norte brasileiro. Biota Amazônia 5(3):28–37

Anderson, Anthony B.; Prance, Ghillean T.; Albuquerque, Byron W. P. de. 1975. Estudos sobre a vegetação das campinas Amazonicas-III A vegetação lenhosa da campina da reserva biológica INPA-SUFRAMA (Manaus-Caracaraí, km 62). Acta Amazonica 5(3):225–246

Antúnez de Mayolo, A. 1989. Peruvian natural dye plants. Economic Botany 43(2):181–191

Aparício, Wegliane Campelo da Silva. 2011. Estrutura da vegetação em diferentes ambientes na RESEX do Rio Cajari: interações solo-floresta e relações com a produção de castanha. Doctoral Dissertation, Universidade Federal Rural de Pernambuco, Recife

Aquino, Rolando; Bodmer, Richard E. 2004. Plantas útiles en la alimentación de primates en la cuenca del río Samiria, Amazonia Peruana. Neotropical Primates 12(1):1–6

Bahri, S. 1992. L'agroforesterie, une alternative pour le développement de la plaine alluviale de l'Amazone: l'exemple de l'île de Careiro. Université de Montpellier

Balée, W. 1993. *Footprints of the Forest: Ka'apor Ethnobotany, the Historical Ecology of Plant Utilization by an Amazonian People*. Columbia University Press, New York

Bánki, Olaf S.; Steege, Hans ter; Jansen-Jacobs, Marion J.; Raghoenandan, Usha P. D. 2003. Plant Diversity of the Nassau Mountains, Suriname: Report of the 2003 Expedition. National Herbarium Nederland, Utrecht

Barnett, Adrian A.; Silva, Welma Sousa; Shaw, Peter J. A.; Ramsay, Paul M. 2015. Inundation duration and vertical vegetation zonation: a preliminary description of the vegetation and structuring factors in borokotóh (hummock igapó), an overlooked high-diversity, Amazonian vegetation association. Nordic Journal of Botany 33:601–614

Bastos, Maria de Nazaré do Carmo. 1988. Levantamento florístico em restinga arenosa litorânea na Ilha de Maiandeua, Pará. Boletim do Museu Paraense Emílio Goeldi, Série Botânica 4(1): 159–169

Bastos, M. N. C.; C. S. Rosário; L. C. B. Lobato. 1995. Caracterização fitofisionômica da restinga de Algodoal, Maracanã, Pará, Brasil. Boletim do Museu Paraense Emílio Goeldi, Série Botanica 11(2): 173–197

Bastos, Maria de Nazaré do Carmo; Costa, Denise Cristina Torres; Santos, João Ubiratan Moreira dos. 2003. Vegetação de restinga: aspectos botânicos e uso medicinal. Museu Paraense Emilío Goeldi Projeto Renas, Belém

Bentes-Gama, Michelliny de Matos; Scolforo, José Roberto Soares; Gama, João Ricardo Vasconcellos; Oliveira, Antônio Donizette de. 2002. Estrutura e valoração de uma floresta de várzea alta na Amazônia. Cerne 8(1):88–102

Braga, Pedro Ivo Soares. 1979. Subdivisão fitogeográfica, tipos de vegetação, conservação e inventário florístico da floresta amazónica. Acta Amazonica 9(4) supl.:53–80

Braga, Marilene Marinho Nogueira; Braga, Pedro Ivo Soares. 1975. Estudos sobre a vegetação das campinas Amazonicas IV Estudos ecológicos na campina da Reserva Biológica INPA-SUFRAMA (Manaus-Caracaraí, km 62). Acta Amazonica 5(3):247–260

Brito, Elizabeth Rodrigues; Martins, Sebastião Venâncio; Oliveira Filho, Ary Teixeira de; Silva, Elias; Silva, Alexandre Francisco da. 2006. Estrutura fitossociológica de um fragmento natural de floresta inundável em área de orizicultura irrigada, município de Lagoa da Confusão, Tocantins. Revista Árvore 30(5):829–836

Brito, Elizabeth Rodrigues; Martins, Sebastião Venâncio; Oliveira-Filho, Ary Teixeira de; Silva, Elias; Silva, Alexandre Francisco da. 2008. Estrutura fitossociológica de um fragmento natural de floresta inundável em área de Campo Sujo, Lagoa da Confusão, Tocantins. Acta Amazonica 38(3):379–386

Cain, Stanley A.; Castro, Gustavus M. de Oliveira; Pires, J. Murça; Silva, Nilo Tomâs. 1956. Application of some phytosociological techniques to Brazilian rain forest. American Journal of Botany 43(10):911–941

Campbell, David G.; Daly, Douglas C.; Prance, Ghillean T.; Maciel, Ubirajara N. 1986. Quantitative ecological inventory of terra firme and várzea tropical forest on the Rio Xingu, Brazilian Amazon. Brittonia 38(4):369–393

Carim, Marcelo de Jesus Veiga. 2016. Estrutura, composição e diversidade em florestas alagáveis de várzea de maré e de igapó e suas relações com variáveis edáficas e o período de inundação no Amapá, Amazônia oriental, Brasil. Doctoral dissertation, Programa de Pós-Graduação em Botânica, INPA, Manaus

Carim, Samyrams; Schwartz, Gustavo; Silva, Manoela Fernandes Ferreira da. 2007. Riqueza de espécies, estrutura e composição florística de uma floresta secundária de 40 anos no leste da Amazônia. Acta Botanica Brasilica 21(2):293–308

Carneiro, Vilany Matilla Colares. 2004. Composição florística e análise estrutural da floresta primária de terra firme na bacia do Rio Cuieiras, Manaus-AM. Master's Thesis, Universidade Federal do Amazonas (UFAM)/Instituto Nacional de Pesquisas da Amazônia (INPA), Manaus

Carneiro, Diogo Borges; Barboza, Myrian Sá Leitão; Menezes, Moirah Paula. 2010. Plantas nativas úteis na Vila dos Pescadores da Reserva Extrativista Marinha Caeté-Taperaçu, Pará, Brasil. Acta Botanica Brasilica 24(4):1027–1033

Castañeda, Rafael Romero. 1961. *Frutas Silvestres de Colombia*. Bogota, vol. 1.

Casula, Kátia Regina. 2012. Análise da similaridade florística e estrutural das formações florestais inundáveis em um trecho do alto Rio Madeira e de seus afluentes, estado de Rondônia, Brasil. Master's Thesis, Núcleo de Ciências Exatas da Terra, Universidade Federal de Rondônia, Porto Velho

Cavalcante, P. B. 2010. *Frutas Comestíveis na Amazônia*. Museu Paraense Emílio Goeldi, Belém

Cerón, Carlos Eduardo. 1996. Etnobotanica Quichua en la vía Hollín-Loreto, Provincia del Napo. Hombre y Ambiente 25:131–170

Chapman, Colin A. 1989. Primate seed dispersal: the fate of dispersed seeds. Biotropica 21(2): 148–154

Claudino, Wesley Vicente; Garcia Marcos Leandro; Lopes, Célia Regina Araújo Soares, Rodrigues, Lucirene; Cabral, Fabiana Ferreira; Piva, José Hypolito; Silva, Antônio Carlos Silveiro da. 2015. Análise fitossociológica da vegetação na região do arco do desmatamento. Revista de Ciências Agroambientais, Alta Floresta, MT 13(1): 43–52

Coelho-Ferreira, Márlia. 2009. Medicinal knowledge and plant utilization in an Amazonian coastal community of Marudá, Pará state (Brazil). Journal of Ethnopharmacology 126:159–175

Condé, Tiago Monteiro. 2011. Avaliação dos impactos da vegetação após a exploração madeireira em floresta ombrófila de terra firme no Município de Caracaraí, RR. Master's thesis, Programa de Pós-Graduação em Recursos Naturais, Universidade Federal de Roraima, Boa Vista

Costa, Flávia R. C.; Magnusson, William E. 2003. Effects of selective logging on the diversity and abundance of flowering and fruiting understory plants in a central Amazonian forest. Biotropica 35(1):103–114

Costa Neto, Salustiano Vilar da; Bastos, Maria de Nazaré do Carmo; Lobato, Luiz Carlos Batista. 1996. Composição florística e fitofisionomia da restinga do Crispim, município do Marapanim, Pará. Boletim do Museu Paraense Emílio Goeldi, Série Botânica 12(2): 237–249

Defler, Thomas R. 1979. On the ecology and behaviour of *Cebus albifrons* in eastern Colombia: 1: Ecology. Primates 20(4):475–490

Díaz Pérez, Wilmer Antonio; Daza, Félix; Sarmiento, William. 2012. Composición florística, estructura y diversidad del bosque ribereños del Río Kakada, cuenca del Río Caura, estado Bolívar, Venezuela. Revista Científica UDO Agrícola 12(2):275–289

Ducke, A. 1922. Plantes nouvelles ou peu connues de la région amazonienne (II partie). Archivos do Jardim Botanico do Rio de Janeiro 3:3–269.

Ducke, A. 1946. *Plantas de Cultura Precolombiana na Amazônia brasileira. Notas Sôbre as Espécies ou Formas Espontâneas que Supostamente lhes teriam dado Origem.* Instituto Agronômico do Norte, Belém, Boletim Técnico 8, pp. 1–24

Falcão, Martha de Aguiar; Lleras, Eduardo; Kerr, Warwick Estevam. 1981. Aspectos fenológicos, ecológicos e de produtividade do pajurá (Couepia bracteosa Bentham) (Chrysobalanaceae). Acta Amazonica 11(3):473–482

Ferreira, Marcos Gonçalves; Absy, Maria Lucia. 2017. Pollen analysis of honeys of Melipona (Michmelia) seminigra merrillae and Melipona (Melikerria) interrupta (Hymenoptera: Apidae) bred in Central Amazon, Brazil. Grana 56(6):436–449

Ferreira, Leandro V.; Chaves, Priscilla P.; Cunha, Denise de A.; Matos, Darley C. L.; Parolin, Pia. 2013. Variação da riqueza e composição de espécies da comunidade de plantas entre as florestas de igapós e várzeas na Estação Científica Ferreira Penna-Caxiuanã na Amazônia oriental. Pesquisas Botânica 64:175–195

Freire, Maria Cristina C.; Monteiro, Reinaldo. 1993. Florística das praias da Ilha de São Luís, Estado do Maranhão (Brasil): diversidade de espécies e suas ocorrências no litoral Brasileiro. Acta Amazonica 23(2–3):125–140

Gayot, Marc; Henry, Olivier; Dubost, Gérard; Sabatier, Daniel. 2004. Comparative diet of the two forest cervids of the genus *Mazama* in French Guiana. Journal of Tropical Ecology 20(1):31–43

Godoy, Judith Rosales; Petts, Geoffrey; Salo, Jukka. 1999. Riparian flooded forests of the Orinoco and Amazon basins: a comparative review. Biodiversity and Conservation 8:551–586

Gomes, Marcos do Livramento; Oliveira, Jociwaldo Silva; Jardim, Mário Augusto Gonçalves; Silva, Jair Campos da. 2006. Usos medicinais e composição química das folhas de Licania macrophylla Benth. (Chrysobalanaceae). Revista Brasileira de Farmacognosia 87(1):26–29

Gonçalves, Janaína Pinheiro; Lucas, Flávia Cristina Araújo. 2017. Agrobiodiversidade e etnoconhecimento em quintais de Abaetetuba, Pará, Brasil. Revista Brasileira de Biociências 15(3):119–134

Gostinski, Luana Fontoura; Oliveira, Fabiana França; Contrera, Felipe Andrés Léon; Albuquerque, Patrícia Maia Correia de. 2018. Nicho trófico e partição de fontes polínicas por duas espécies de Melipona (Hymenoptera, Apidae) na Amazônia oriental. Oecologia Australis 22(4):449–462

Jardim, M.A.G. and Medeiros, T.D.S. 2006. Plantas oleaginosas do Estado do Pará: composição florística e usos medicinais. Revista Brasileira de Farmácia 87(4):124–127

Jardim, Daniel Gonçalves; Jardim, Mário Augusto G.; Quaresma, Adriano Costa; Costa, Salustiano Villar. 2013. Regeneração natural em formações florestais de uma unidade de conservação, Maracanã, Pará, Brasil. Biota Amazônia 3(2):79–87

Kerr, Warwick Estevam; Absy, Maria Lúcia; Souza, Antonio Carlos Marques. 1986. Espécies nectaríferas e poliniferas utilizadas pela abelha *Melipona compressipes fasciculata* (Meliponinae, Apidae) no Maranhão. Acta Amazonica 16:145–156

Kunz, Sustanis Horn; Ivanauskas, Natália Macedo; Martins, Sebastião Venâncio; Silva, Elias; Stefanello, Daniel. 2008. Aspectos florísticos e fitossociológicos de um trecho de floresta estacional perenifólia na Fazenda Triarão, Bacia do rio das Pacas, Querência, MT. Acta Amazonica 38(2):245–254

Kunz, Sustanis Horn; Ivanauskas, Natália Macedo; Martins, Sebastião Venâncio; Silva, Elias; Stefanello, Daniel. 2009. Análise da similaridade florística entre florestas do Alto Rio Xingu, da Bacia Amazônica e do Planalto Central. Revista Brasileira de Botanica 32(4):725–736

Kunz, Sustanis Horn; Moreau, Julia; Spadeto, Cristiani; Martins, Sebastião Venâncio; Stefanello, Daniel; Ivanauskas, Natália Macedo. 2014a. Estrutura da comunidade arbórea de trecho de floresta estacional sempre-verde e similaridade florística na região nordeste de Mato Grosso, Brasil. Floresta e Ambiente 21(4):429–440

Kunz, Sustanis Horn; Ivanauskas, Natália Macedo; Martins, Sebastião Venâncio; Stefanello, Daniel; Silva, Elias. 2014b. Fitossociologia do componente arbóreo de dois trechos de floresta estacional perenifólia, bacia do Rio das Pacas, Querência, MT. Ciência Florestal, Santa Maria 24(1):1–11

Lamont, Susan R.; Eshbaugh, W. Hardy; Greenberg, Adolph M. 1999. Species composition, diversity, and use of homegardens among three Amazonian villages. Economic Botany 53(3):312–326

Le Cointe, P. 1934. *Amazônia Brasileira III*: *Árvores e Plantas Úteis*. Livraria Classica, Belém

Lobato, G. de J. M.; Lucas, F. C. A.; Tavares-Martins, A. C. C.; Jardim, M. A. G.; Martorano, L. G. 2017. Diversidade de uso e aspectos socioambientais de quintais urbanos em Abaetetuba, Pará, Brasil. Revista Brasileira de Agroecologia 12(2):95–105

Lorenzi, Harri; Bacher, Luis; Lacerda, Marco; Sartori, Sergio. 2006. Frutas Brasileiras e Exóticas Cultivadas. Instituto Plantarum de Estudos da Flora Ltda, Odessa, Sao Paulo.

Luize, Bruno Garcia; Venticinque, Eduardo Martins; Silva, Thiago Sanna Freire; Novo, Evlyn Marcia Leão de Moraes. 2015. A floristic survey of angiosperm species occurring at three landscapes of the central Amazon várzea, Brazil. Checklist: The Journal of Biodiversity Data 11(6):1–15

Macedo, Miramy. 1977. Dispersão de plantas lenhosas de uma campina Amazônica. Acta Amazonica 7(1):1–69 (supl.)

Machado, Clara de Carvalho. 2018. Conhecimento e consumo de plantas alimentícias em cinco comunidades da reserva de desenvolvimento sustentável Piaguçu-Purus, Amazonas. Master's thesis, INPA, Manaus

Magnusson, William E.; Lima, Albertina P.; Albernaz, Ana L. K. M.; Sanaiotti, Tânia M.; Guillaumet, Jean-Louis. 2008. Composição florística e cobertura vegetal das savanas na região de Alter do Chão, Santarém-PA. Revista Brasileira de Botanica 31(1):165–177

Maranho, Álisson Sobrinho; Salimon, Cleber Ibraim. 2015. Maranthaceae overabundance decreases richness and abundance of regenerating woody plants in natural gaps. Neotropical Biology and Conservation 10(2):53–62

Maranho, Álisson Sobrinho; Salimon, Cleber Ibraim; Costa, Daniel da Silva. 2016. Woody flora of natural forest gaps in a bamboo-dominated forest remnant in southwestern Amazonia. Check List: The Journal of Biodiversity Data 12(3):1–9

Martins, Sebastião Venâncio; Brito, Elizabeth Rodrigues; Oliveira Filho, Ary Teixeirade; Silva, Alexandre Francisco da; Silva, Elias. 2008. Floristic composition of two wetland forests in Araguaian plain, state of Tocantins, Brazil, and comparison with other areas. Revista Árvore 32(1):129–141

Mascarenhas, Raimundo E. B.; Modesto Júnior, Moisés de S.; Dutra, Saturnino; Souza Filho, Antonio P. da S.; Teixeira Neto, José F. Teixeira. 1999. Plantas daninhas de uma pastagem cultivada de baixa produtividade no nordeste Paraense. Planta Daninha 17(3):399–418

Massing, Lais Tatiele; Mourão, Rosa Helena; Bouillet, Leoneide Érica M.; Bernardes, Regiane Sablina A.; Andrade, Eloisa Helena A.; Tremea, Agnaldo; Souza, Jesus Nazareno S.; Maia, José Guilherme S. 2018. Nutritional composition of the pulp of pajurá (Couepia bracteosa Benth.), an underutilized fruit from the Amazon. Integrative Food, Nutrition and Metabolism 5(2):1–7

Mattos, Fábio de Souza. 2016. Dieta do *Plecturocebus cinerascens* (Platyrrhini: Primates): a influência da composição química na escolha de frutos consumidos. Master's thesis, INPA, Manaus

Medeiros, Priscilla Sanjuan de. 2016. Indicadores ambientais para funcionalidade ecológica em florestas secundárias de diferentes idades na Amazônia Oriental. Doctoral dissertation, Universidade Federal do Pará, Belém

Myster, Randall W. 2018. What we know about the composition and structure of igapó forests in the Amazon Basin. Botanical Review 84:394–410

Oliveira, Alexandre A. de; Mori, Scott A. 1999. A central Amazonian terra firme forest. I. High tree species richness on poor soils. Biodiversity and Conservation 8:1219–1244

Parolin, P.; Adis, J.; Silva, M. F. da; Amaral, I. L. do; Schmidt, L.; Piedade, M. T. F. 2003. Floristic composition of a floodplain forest in the Anavilhanas archipelago, Brazilian Amazonia. Amazoniana 17(3/4):399–411

Parra, C. 2006. Estudio general de la vegetación nativa de Puerto Carreño (Vichada, Colombia). Caldasia 28(2):165–177

Parrotta, John A.; Knowles, Oliver Henry; Wunderle Jr., Joseph M. 1997. Development of floristic diversity in 10-year-old restoration forests on a bauxite mined site in Amazonia. Forest Ecology and Management 99:21–42

Patiño, V. M. 1963. Plantas Cultivadas y Animales Domesticadas en America Equinoccial. Imprensa Departamental, Cali, vol. 1

Pinheiro, Klewton Adriano Oliveira; Carvalho, João Olegário Pereira de; Quantz, Beatriz; Francez, Luciana Maria de Barros; Schwartz, Gustavo. 2007. Fitossociologia de uma área de preservação permanente no leste da Amazônia: indicacão de espécies para recuperação de áreas alteradas. Floresta 37(2): 175–187

Popenoe, W. 1920. *Manual of Tropical and Subtropical Fruits*. Macmillan, New York

Posey, D. A. 2002. Kayapó Ethnoecology and Culture. Routledge, London

Prance, G. T. 1972. *Chrysobalanaceae*. Flora Neotropica, Monograph 9, New York Botanical Garden, Bronx

Prance, G. T. 1978. The origin and evolution of the Amazon flora. Interciencia 3(4):207–222

Prance, G. T. 1984. Introduction on knowledge and use of the Amazon forest. *Interciencia* 9(5):275–279

Prance, G. T. 1985. The pollination of Amazonian plants. In: Key Environments: Amazonia, Ghillean T. Prance and Thomas E. Lovejoy (eds), Pergamon Press, Oxford, pp. 166–191

Prance, G. T. 1988. Padrões de distribuição em Chrysobalanaceae e outras famílias de plantas Amazônicas. Acta Botanica Brasilica 1(2):1–25

Prance, G. T. 1997. The ethnobotany of the Amazon indians as a tool for the conservation of biological diversity. Monografías del Jardín Botánico de Córdoba 5:135–143

Prance, G. T. 2014. That Glorious Forest: Exploring the Plants and Their Indigenous Uses in Amazonia. New York Botanical Garden, New York

Prance, G.T. and M.F. da Silva. 1975. Árvores de Manaus. INPA, Manaus

Rayol, Breno Pinto; Alvino-Rayol, Fabrízia de Oliveira; Silva, Manoela Ferreira Fernandes da. 2011. Similaridade florística entre o estrato arbóreo e a regeneração natural de uma floresta secundária, no município de Bragança, nordeste do estado do Pará, Brasil. Revista Brasileira de Agroecologia 6(3):107–114

Reis, Neidiane Farias Costa. 2010. Estrutura fitossociológica e seletividade de espécies em áreas inundáveis na região do baixo Rio Jaci-Paraná, Rondônia. Master's Thesis, Universidade Federal de Rondônia, Porto Velho

Rocha, Javier Enrique. 2017. Relación altura commercial—diámetro y abundancia—clase diamétrica en los Árboles de un bosque de colina baja, Maynas, Loreto, Perú, 2015. Master's Thesis, Facultad de Ciencias Forestales, Universidad Nacional de la Amazonía Peruana, Iquitos

Rocha, S. F. R., M. R. C. Ferreira, and M. G. Sablayrolles. 2013. Farmacopeia vegetal da comunidade Caxiuanã, na Floresta Nacional de Caxiuanã. In: Caxiuanã: Paraíso ainda Preservado, P. L. B. Lisboa (Editor), Museu Paraense Emílio Goeldi, Belém, pp. 257–285

Rosa, Leonilde dos Santos; Silveira, Erycélede Lima; Santos, Mônica Mota dos; Modesto, Rozi da Silva; Perote, James Richard Silva; Vieira, Thiago Almeira. 2007. Os quintais agroflorestais em áreas de agricultores familiares no município de Bragança-PA: composição florística, uso de espécies e divisão de trabalho familiar. Revista Brasileira de Agroecologia 2(2):337–341

Rudas, Augustín; Prieto, Adriana. 1998. Análisis florístico del parque nacional natural Amacayacu e Isla Mocagua, Amazonas (Colombia). Caldasia 20(2):142–172

Salomão, Rafael de Paiva; Vieira, Ima Célia Guimarães; Suemitsu, Chieno; Rosa, Nélson de Araújo; Almeira, Samuel Soares de; Amaral, Dário Dantas do; Menezes, Moirah Paulo Machado de. 2007. As florestas de Belo Monte na grande curva do rio Xingu, Amazônia Oriental. Boletim do Museu Paraense Emílio Goeldi, Ciencias Naturais 2(3):57–153

Sanaiotti, Tânia M.; Magnusson, William E. 1995. Effects of annual fires on the production of fleshy fruits eaten by birds in a Brazilian Amazonian savanna. Journal of Tropical Ecology 11(1):53–65

Sánchez, M.; Duivenvoorden, J. F.; Duque, A.; Miraña, P.; Cavelier, J. 2005. A stem-based ethnobotanical quantification of potential rain forest use by Mirañas in NW Amazonia. Ethnobotany Research & Applications 3:215–229

Santos, Luciana Oliveira dos. 2010. Comparação da vegetação lenhosa entre borda e interior de uma mancha de mata de restinga na RESEX Mar Caeté-Taperaçu, Bragança, Pará. Master's thesis, Universidade Federal Rural da Amazônia/Museu Paraense Emílio Goeldi, Belém

Santos, J. U. M. and M. N. Bastos. 2014. Flores e Frutas das Restingas do Estado do Pará. Universidade Federal Rural da Amazônia (UFRA), Belém

Santos, Gideão Costa dos; Jardim, Mário Augusto Gonçalves. 2006. Florística e estrutura do estrato arbóreo de uma floresta de várzea no município de Santa Bárbara do Pará, Estado do Pará, Brasil. Acta Amazonica 36(4):437–446

Santos, João Ubiratan M. dos; Rosário, Carlos da S. 1988. Levantamento da vegetação fixadora das dunas de Algodoal-PA. Boletim do Museu Paraense Emílio Goeldi, Série Botânica 4(1): 133–154

Santos, João Ubiratan M. dos; Amaral, Dário Dantas do; Gorayeb, Inocêncio de Sousa; Bastos, Maria de Nazaré do C.; Secco, Ricardo de S.; Neto, Salustiano Vilar Costa; Costa, Denise Cristina Torres. 2003. Vegetação da área de proteção ambiental Jabotitiua-Jatium, Município de Viseu, Pará, Brasil. Acta Amazonica 33(3):431–444

Santos, Deborah Cruz dos; Corrêa Neto, José de Jesus; Alvarenga, Adriano Braga Brasileiro de. 2018a. Frutos consumidos por macaco-de-cheiro *Saimiri sciureus collinsi* (Osgood, 1916) em fragmentos de florestas em uma unidade de conservação no arquipélago do Marajó, Amazônia Oriental, Brasil. Revista Brasileira de Zoociências 19(1):82–90

Santos, Lucélia Rodrigues; Ferreira, Evandro José Linhares; Silva, Cláudia Eugenio. 2018b. Florística e fitossociologia de um sub-bosque urbano em Rio Branco-AC, Brasil. Biota Amazônia 8(3):28–33

Sanz-Biset, Jaume; Cañigueral, Salvador. 2011. Plant use in the medicinal practices known as "strict diets" in Chazuta valley (Peruvian Amazon). Journal of Ethnopharmacology 137:271–288

Schultes, R. E. and R. F. Raffauf. 1990. The Healing Forest: Medicinal and Toxic Plants of the Northwest Amazonia. Dioscorides Press, Portland, Oregon

Scudeller, Veridiana Vizoni; Ramos, Renata Azevedo; Cruz, Maria Eliene Gomes da. 2009. Flora fanerogâmica da floresta de terra firme na RDS Tupé. In: Biotupé: Meio Físico, Diversidade Biológica e Sociocultural do Baixo Rio Negro, Amazônia Central, Santos-Silva, Edinaldo Nelson; Scudeller, Veridiana Vizoni (eds), pp 109–120, UEA Edições, Manaus

Silva, M. F., P. L. B. Lisbôa, R. C. L. Lisbôa 1977. Nomes Vulgares de Plantas Amazônicas. INPA, Manaus

Silva, José Natalino Macedo; Carvalho, João Olegário Pereira de; Lopes, José do Carmo Alves. 1985. Inventário florestal de uma área experimental na Floresta Nacional do Tapajós. Boletim de Pesquisa Florestal 10/11:38–110

Silva, Antônio Sérgio Lima da; Lisboa, Pedro L. B.; Maciel, Ubirajara N. 1992. Diversidade florística e estrutura em floresta densa da bacia do rio Juruá-AM. Boletim do Museu Paraense Emílio Goeldi, Série Botânica 8(2):203–258

Silva, Hanna Kassia Machado da; Gama, João Ricardo Vasconcellos; Sousa, Randerson José de Araujo; Lameira, Mahyanny Karoline da Silva; Costa, Daniele e Lima da; Oliveira, Douglas Valente de; Rocha, Jobert Silva da; Oliveira, Thiago Gomes de Sousa. 2017a. Composição florística de quintais agroflorestais na vila Cuera, Bragança, Pará. Agroecossistemas 9(2):330–338

Silva, Silmara Bento; Sousa, Venderleia Conceição; Santos, Crissogno Mesquita; Mariano, Daiane Cinque; Okumura, Ricardo Shigueru. 2017b. Levantamento florístico do componente arbustivo-arbóreo da vegetação ciliar de fragmento no Rio Parauapebas. Agroecossistemas 9(1):99–115

Souza, Célia Cristina Valero. 2010. Etnobotânica de quintais em três comunidades ribeirinhas na Amazônia central, Manaus, AM. Master's thesis, INPA, Manaus

Souza, Célia Cristina Valero; Scudeller, Veridiana Vizoni. 2009. Plantas úteis nos quintais das comunidades ribeirinhas Julião e Agrovila, Reserva de Desenvolvimento Sustentável do Tupé, Amazônia Central. Revista Brasileira de Agroecologia 4(2): 2487–2491

Spichiger, Rodolphe; Loizeau, Pierre-André; Latour, Cyrille; Barreira, Gabrielle. 1996. Tree species richness of a south-western Amazonian forest (Jenaro Herrera, Peru, 73° 40'W/4° 54'S). Candollea 5(1–2):559–577

Stefanello, Daniel. 2008. Florística, estrutura e dispersão de sementes de três trechos de floresta estacional perenifólia ribeirinha na bacia hidrográfica do Rio das Pacas, Querência, MT. Master's Thesis, Universidade Federal de Viçosa, Minas Gerais

Stefanello, Daniel; Ivanauskas, Natália Macedo; Martins, Sebastião Venâncio; Silva, Elias; Kunz, Sustanis Horn. 2010. Síndromes de dispersão de diásporos das espécies de trechos de vegetação ciliar do rio das Pacas, Querência-MT. Acta Amazonica 40(1):141–150

Targhetta, Natália; Kesselmeier, Jürgen; Wittman, Florian. 2015. Effects of the hydroedaphic gradient on tree species composition and aboveground wood biomass of oligotrophic forest ecosystems in the central Amazon basin. Folia Geobotanica 50:185–205

Terborgh, J. 1983. Five New World Primates: A Study in Comparative Ecology. Princeton University Press, Princeton

Vásquez, R. and A. H. Gentry. 1989. Use and misuse of forest-harvested fruits in the Iquitos area. *Conservation Biology* 3(4): 350–361.

Veiga, Josephina Barata da; Scudeller, Veridiana V. 2011. Quintais agroflorestais da comunidade ribeirinha São João do Tupé no baixo rio Negro, Amazonas. In: Biotupé: Meio Físico, Diversidade Biológica e Sociocultural do Baixo Rio Negro, Amazônia Central, Santos-Silva, Edinaldo Nelson; Cavalcanti, Mauro José; Scudeller, Veridiana Vizoni (eds), UEA Edições, Manaus, vol. 3, pp. 523–543

Vieira, Thiago Almeida; Rosa, Leonilde dos Santos; Santos, Maria Marly de Lourdes Silva. 2012. Agrobiodiversidade de quintais agroflorestais no município de Bonito, Estado do Pará. Revista de Ciências Agrárias 55(3):159–166

Chapter 17
Clusiaceae

17.1 *Clusia grandiflora*

Brazil Apuí, cebola brava, cebola grande da mata, ceboleira, cebola brava, manga brava, mata pau; apo'y (Ka'apor), dikada dika (Tuyuka), kypy' (Waimiri-Atroari), poriporitoto (Yanomami)

Peru Came renaco, mata palo, renaquilla

Venezuela Copei

Status Wild

Apuí *Clusia grandiflora* in flower in successionary vegetation on an abandoned Tuyuka settlement. São Pedro Antigo, Rio Tiquié, affluent of the Uaupés, Amazonas, Brazil 11-1-12

N. Smith, *Amazon Fruits: An Ethnobotanical Journey*, Ethnobiology,
https://doi.org/10.1007/978-3-031-12803-5_17

Apuí *Clusia grandiflora* is a small tree or liana found in successionary forests on uplands throughout the Amazon and Orinoco Basins as well as the Guianas. It also occurs occasionally on floodplains, such as in the Bragantina zone in northeastern Pará (Batista and Gonçalves 2013; Batista et al. 2011), along the lower Xingu (Salomão et al. 2007a), and in southeastern Colombia (Cárdenas et al. 2009). It also thrives along the sandy coast of Pará (Amaral et al. 2008; Bastos 1988; Costa Neto et al. 1996; Silva et al. 2010) and in campina and campinara vegetation growing on patches of sandy soil inland (Ferreira et al. 2013). It also grows on sand dunes along the Bragantina coast of Pará, such as on Algodoal Island in the municipality of Maracanã (Bastos et al. 1995, 2003:12). It sports large male and female flowers, reaching 16 cm in diameter, hence its species name *grandiflora*. The flowers are pollinated by various species of stingless bees including in the genus *Euglossa* as well as *Partamona pearsoni*, *Nogueirapis minor*, *Trigona williana*, *T. fulviventris*, *Scaura tenuis*, *Plebeia* sp., *Ptilotrigona lurida*, *Eufriesea pulchra*, *Ptilotrigona lurida* and *Melipona lateralis* (Bittrich and Amaral 1997; Kaminski and Absy 2006). The flowers do not produce nectar; rather the bees visit the male and female flowers to gather resin to help construct their nests (Lokvam and Braddock 1999). The resin from these flowers has antimicrobial properties which help protect the nests of the bees from bacterial infestations. After the white petals fall off, a green fruit develops that eventually reaches the proportions of a peach palm fruit. The Tuyuka, who belong to the Tukanoan linguistic family, gather the fruits for use as fish bait as explained to me by villagers along the Upper Tiquié, an affluent of the Uaupés in Amazonas, Brazil. An apuí tree bearing immature fruit was noted on the site of a long abandoned Tuyuka village along the Tiquié River and may have been planted by them in the past.

Black spider monkeys *Ateles paniscus*, which are hunted for their meat, also eat the fruits (Norconk and Kinzey 1994; Van Roosmalen 1985; Van Roosmalen and Klein 1988). Black-headed uakaris *Cacajao melanocephalus melanocephalus* likewise eat these fruits (Boubli 1999) and although these monkeys with characteristic short tails normally do not exceed 4 kg they are occasionally shot for their meat. Black-bearded sakis *Chiropotes satanas*, bearded sakis *Chiropotes chiropotes*, red-nosed bearded sakis *C. albinasus*, and white-faced sakis *Pithecia pithecia* also eat these fruits and likely disperse the seeds (Barnett et al. 2012; Boyle et al. 2012; Norconk and Kinzey 1994; Pinto 2008:75; Roble 2008:34; Van Roosmalen et al. 1981, 1988).

Residents of Caicubi, a village near the confluence of the Rio Negro and Rio Branco in Roraima, employ the latex of *Clusia grandiflora* to treat muscle sprains (Alarcón and Peixoto 2008), while the Waimiri-Atroari north of Manaus use the latex to treat fresh wounds (Milliken et al. 1992:71). Inhabitants of the village of Marudá along the coast of Pará use the fruits or flowers of this tree to make a tea to alleviate coughs (Coelho-Ferreira 2009). In Guyana, the wood of this tree is used in the furniture industry (Cummings and Read 2016).

Apuí *Clusia grandiflora* fruits are gathered by the Tuyuka for fish bait. São Pedro Antigo, Rio Tiquié, affluent of the Uaupés, Amazonas, Brazil 11-1-12

17.2 *Garcinia benthamiana*

Brazil Bacuri, bacuri coroa, bacuri selvagem, bacuri maxixe, bacurizinho, bacuripari; oruhe hi, karaka uremisihi (Yanomami)

Colombia Kói (Ticuna)

Peru Charichuelo; pengkaenum (Aguaruna)

Status Wild, spontaneous in cultural settings

Erlan, 11 years, with bacuri *Garcinia benthamiana* fruit gathered in nearby forest. Carrazedo, mouth of the Xingu River, Município de Gurupá, Pará, Brazil 4-19-15

Bacuri *Garcinia benthamiana* is a tree that can reach 25 m in floodplain and upland forests throughout the Amazon and the Guianas (Bánki et al. 2003:31; Batista et al. 2015; Berg 1979; Moser 2013:91; Santos et al. 2017b). It is also found in successionary forests and arises spontaneously from discarded seeds in some home gardens, such as along the lower Tefé River in Amazonas. The dimpled fruits, slightly smaller than tennis balls, contain three brown seeds some 3 cm long which are surrounded by a white pulp with an intriguing sweet-tart taste. The fruits are picked up from the forest floor or shaken loose by shimmying up the trunk and are eaten as a snack. Widespread capuchin monkeys *Cebus apella*, white-faced sakis *Pithecia pithecia*, and black spider monkeys *Ateles paniscus* also eat the fruits (Gregory 2006:53; Simmen and Sabatier 1996; Van Roosmalen 1985; Van Roosmalen and Klein 1988) and are likely involved in dispersing some of the seeds in forests of the Amazon and Guianas. Spider monkeys are widely hunted for their meat and so is the grey brocket deer *Mazama gouazoubira* which also eats this fruit (Gayot et al. 2004).

The thick-skinned, yellow fruits which are produced during the rainy season do not bruise easily and can be stored at home for several days. Although most bacuri fruits are consumed in the interior, they occasionally find their way to urban markets, such as in Iquitos, Peru (Vásquez and Gentry 1989). In the Peruvian Amazon, the fruits are used to produce juice and ice cream (Brack Egg 2003:86).

Wadrio, 9 years, eating bacuri *Garcinia benthamiana* fruit. Carrazedo, mouth of the Xingu River, Município de Gurupá, Pará, Brazil 4-19-15

17.3 *Garcinia brasiliensis*

Bolivia Guapomo, ocoró, achachairú, nobocorr; corivavaj, tsocoi (Tsimane)

Brazil Bacuri, bacuri liso, bacuri de anta, bacuri da várzea, bacuri selvagem, bacurizinho, bacurizeiro, bacuripari, bacuparí, bacuparí liso, bacuparí folha pequena, bacurirana, acupari, cupari, amapari, azeidinho, poroca, porocó, limão do mato, pitomba; pakuri-sõsõ-ɨ (Ka'apor)

Colombia Charichuelo, bacuri, madroño

Peru Charichuelo

Status Wild, spontaneous in cultural settings, planted

Bacuri *Garcinia brasiliensis* **in fruit on the Amazon floodplain. Near Santarém, Pará, Brazil 5-15-96**

Garcinia brasiliensis occurs on the floodplains of black, clear, and sediment-rich rivers and streams throughout the Amazon Basin, as well as the Guianas and the northeast of Brazil (Amaral et al. 2009; Assis and Wittmann 2011; Berg 1979; Campbell et al. 1986; Carim 2016:88; Carim et al. 2008; Marinho et al. 2013; Parolin 2009; Pasa et al. 2005; Ratter 1987; Salomão et al. 2007a, b; Santos et al. 2016; Wittmann et al. 2009, 2013). It also occurs in gallery forests winding through cerrado vegetation in the upper watershed of the Araguaia River in Mato Grosso (Stefanello et al. 2009). The leaves of this tree can remain submerged for several months without damage (Parolin et al. 2004). It occurs on levees (*restingas*) along the coast of Brazil from the mouth of the Amazon to Paraná in southern Brazil (Abreu et al. 2006; Lorenzi et al. 2006a, b:126). It is also found in successionary forests on uplands in the eastern fringes of the Amazon Basin in the state of Maranhão where the Ka'apor gather the fruits (Balée and Gély 1989) as well as in upland forests in the river basin of the Juruá, a southern affluent of the Solimões in Amazonas (Silva et al. 1992), in eastern Acre (Araujo 2015), and in the Bolivian Amazon (Araujo-Murakami et al. 2015). It is also found in some cacao agroforestry plots on uplands in the municipality of São Félix do Xingu in southern Pará (Braga et al. 2019). Whether *G. brasiliensis* is natural in such upland locations is not known; it is possible the tree was introduced to that environment by indigenous people. Its range also extends as far south as Paraguay and northern Argentina. It can be easily distinguished from *G. benthamiana* because *G. brasiliensis* is a shorter tree, usually under 15 m. Also, the fruits are smooth-skinned and smaller, typically three to four centimeters in diameter. The fruits of *G. brasiliensis* are rich in vitamin C (Marx and Maia 1983).

Known as bacuri liso in Brazil, *G. brasiliensis* bears lustrous fruits that are ready for gathering at high water, and locals, such as in the vicinity of Santarém, paddle up to the almost submerged trees to pluck off the Ping-Pong-sized fruits. Small motorboats will even tie up to the small trees so that passengers can pick the fruits from the deck. It is occasionally cultivated in home gardens (Cavalcante 2010:62), including the upland community of Boa Esperança at 43 km of the Santarém to Curuá-Una highway (Garcia et al. 2017), along the Ramal do Pau Rosa, an offshoot of the BR 175 Manaus-Rio Branco highway north of Manaus (Machado 2016:30), in the municipality of Rio Preto da Eva along the Manaus-Itacoatiara highway (Silva 2017:30), and the village of São Miguel in the municipality of Várzea Grande in Mato Grosso (Mamede and Pasa 2018).

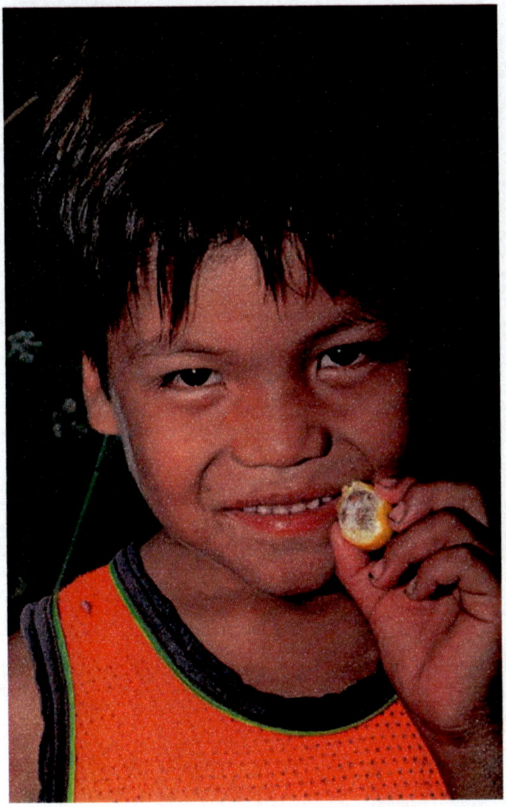

Xyam Noé, 5 years, eating a charichuelo *Garcinia brasiliensis* fruit. Yarina, Rio Yanayacu, Reserva Nacional, Pacaya-Samiria, Loreto, Peru 6-5-04

The fruits float so they are carried away by currents and are also eaten by certain species of fish such as pacu *Piaractus mesopotamicus* and *Colossoma macropomum*, the latter known as tambaqui in Brazil and gamitana in Peru (Galetti et al. 2007; Lucas 2008). Adult tambaqui are large enough to swallow the fruits whole and thus likely serve as dispersal agents. The lustrous fruits contain three elongated brown seeds, the size of a fingernail, some of which likely pass unharmed through the guts of fish as well as birds and mammals. Pacu and tambaqui are highly prized fish because their tasty flesh contains few bones. One of the common names for this tree in Rondônia in the southwestern part of the Brazilian Amazon is bacuri de anta (tapir bacuri), and tapir is the largest terrestrial game animal in Amazonia so this fruit contributes indirectly to the rural diet in some areas via this much-esteemed mammal. A species of squirrel monkey *Saimiri vanzolinii* which is confined to a small area of floodplain forest along the Solimões in the vicinity of its confluence with the Japurá also eats this fruit during the flood season (Paim et al. 2017) as does another species of squirrel monkey, *Saimiri sciureus cassiquiarensis*, in floodplain forests of the Solimões near Tefé, Amazonas (Araujo 2014:36). Both monkeys likely disperse some of the seeds.

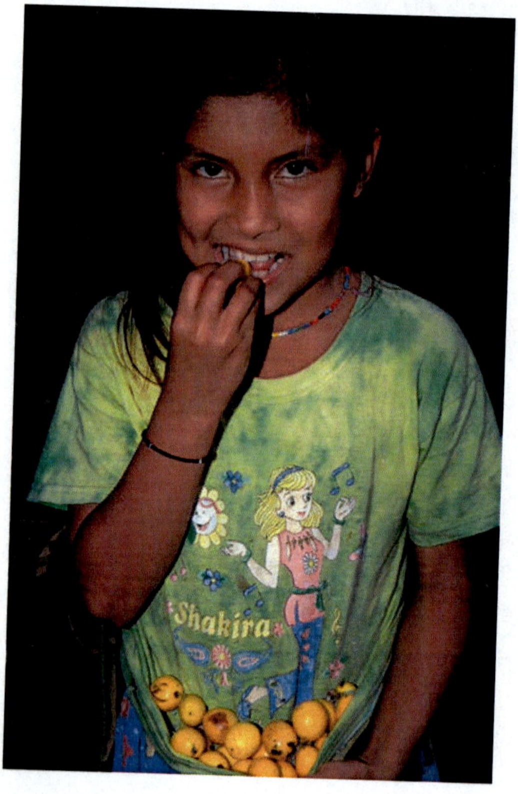

Nicivia, 9 years, with charichuelo *Garcinia brasiliensis* fruits she has picked out of the water under a tree in her village. Yarina, Rio Yanayacu, Reserva Nacional, Pacaya-Samiria, Loreto, Peru 6-5-04

Kids in Yarina, a village along the Upper Yanayacu in the Pacaya-Samiria area of the Peruvian Amazon, shimmy up the tree to gather the fruits which they place in a large pocket created by pulling up their T-shirts. Boys and girls also wade into the river to scoop up the floating fruits, known locally as charichuelo. Because of the fruit's popularity, *G. brasiliensis* is typically spared when clearing home sites along rivers and the tree is also sometimes planted, or arises spontaneously, in home gardens, such as at Porto Valente along the lower Tefé River and along the road from Santa Rosa de Yacuma to Riberalta in the Bolivian Amazon. A hooked pole is often used to dislodge the fruits in home gardens.

Charichuelo *Garcinia brasiliensis* fruits in a street market. Nauta, Rio Marañón, Loreto, Peru 4-20-06

The sweet, smooth-skinned fruits are a delicacy in rural areas and occasionally appear in markets, such as in Nauta along the lower Marañón. When the fruits do reach markets they are a bargain. On 13 March 2004, in Nauta, for example, the fruits were being sold by the bowlful containing about 40 fruits for 50 centimos, the equivalent of 280 fruits for US$ 1. Bacuri liso fruits also appear in the markets of Manaus (Kinupp and Lorenzi 2014:310). A mature *Garcinia brasiliensis* tree produces between 1000 and 1200 fruits during the rainy season (Souza et al. 1996:52).

This tree also has other uses, especially in folk medicine. In the Bolivian Amazon, the Guarasug'we drink the juice squeezed from the fruits to treat stomachache and diarrhea (Hajdu and Hohmann 2012). In the Brazilian Amazon, the leaves are used to treat tumors, inflammation of the urinary tract, as well as to alleviate pain, such as caused by arthritis (Santa-Cecília et al. 2011). In the state of Maranhão along the eastern fringe of Amazonia, the leaves are used to make a tea to treat coughs (Van den Berg and da Silva 1986). Because of its esteem as a fruit and value in herbal remedies, *Garcinia brasiliensis* is sometimes planted in home gardens, such as in the Bolivian Amazon (Murakami et al. 2016:39). The wood is used for fuel and for construction purposes in some areas, such as in Saracura, a quilombola community on a floodplain island along the middle Amazon near Santarém, Pará (Oliveira and Cavalcante 2017) and by ribeirinhos on the Solimões floodplain near its confluence with the Japurá River (Wittmann et al. 2009).

17.4 *Garcinia gardneriana*

Bolivia Aichaicheru, achachairú, achachairucillo, ocoró, bacuri; biwei' (Mosetene), pozokyryza (Rikbaktsa), hashau (Tacana), ibijqui (Tsimane')

Brazil Bacuri mirim, bacuripari, bacurizinho, bacurirana, bacuparí, jatarí mirim, guajará, coparí, coparí grande

Ecuador Aengue madaño

Peru Charichuelo, charichuelo de hoja menuda

Venezuela Naranjito, cozoiba rebalsera

Status Wild, planted

Charichuelo *Garcinia gardneriana* **fruit. Rio Puinahua (a branch of the Ucayali), 2 km upstream from San Carlos, Reserva Nacional Pacaya-Samiria, Loreto, Peru 4-18-06**

Charichuelo *Garcinia gardneriana* is an understory tree in both upland and seasonally inundated forests in the Amazon Basin. It is also occasionally found in successionary vegetation, such as in Mato Grosso in the southern part of the Brazilian Amazon (Rodrigues et al. 2004) and in the vicinity of Jenaro Herrera along the Ucayali River in the Peruvian Amazon (Vásquez 1998). It also occurs in other parts of Brazil as far south as the Rio Grande do Sul (Lorenzi 2008:99). *Garcinia gardneriana* is relatively short, typically 5–10 m tall, and therefore a shade-tolerant tree (Smith et al. 2007:98). It is particularly common along water courses (Grogan and Galvão 2006; Machado and Pinheiro 2016). The round fruits of *Garcinia gardneriana* are only 2 cm in diameter and are the smallest fruits in the genus hence one of its names in Brazil: bacuri mirim, which means the small bacuri. Despite their diminutive size, the fruits are greatly appreciated for the sweet yet tangy pulp surrounding the two to three seeds. The Tsimane in the Bolivian Amazon head into forests to gather these fruits (Reyes-García 2001:101); they also use the wood for construction and fuel (Guèze et al. 2014). Also in the Bolivian Amazon, the Mosetene boil the bark and leaves of this tree, and the solution is then drunk to treat fever (Muñoz et al. 2000).

As is the case with many wild fruit trees, people spare *Garcinia gardneriana* when clearing fields or home sites, such as in the vicinity of San Carlos along the Puinahua, an arm of the Ucayali. It is also planted in parts of the Bolivian Amazon, such as among the Yuracaré and Trinitario indigenous people (Thomas and Van Damme 2010) and in some home gardens in Mato Grosso in the Brazilian Amazon (Santos and Guarim Neto 2017). Although most *Garcinia gardneriana* fruits are eaten in rural areas, they are occasionally taken to urban markets such as Iquitos (Vásquez and Gentry 1989).

17.5 *Garcinia macrophylla*

Bolivia Achachairú, achachairú grande, putuy, madroño; buvei' (Tsimane)

Brazil Bacuri parí, bacuri casca lisa, bacuri liso, bacuri coroa, bacuri da várzea, bacuri da mata, bacurirana, bacurizinho, goiabão; turupa (Juruna), borokáa (Suruí), kotakɨ axihi (Yanomami)

Colombia Madroño, madroño amazónico, bacuri, charichuelo; gueneta manio (Nui), akaká (Nukak), pókuri (Ticuna)

Ecuador Madroño pequeño, peladera, amarillo, cafetillo, leche de perro, chora, shora; penká (Achuar), winul (Awavi), sji pistcha ka bocaca (Cayapa), pevicho, randepevicho (Cofán), pungara, pungara muyo, atun pungara, ungura yura (Kichwa), wushia penká (Shuar), majáro (Siona-Secoya), ontocayabo, untokawen (Waorani)

Peru Charichuelo, charichuelo de altura, charichuelo grande, charichuelo liso, charichuelo rugoso, charichuelo amarillo, chullachaqui caspi, ituchi caspi, lagarto caspi, durazno amarillo del monte; pegkáenum, shíig pegkáenum (Aguaruna), awimuin (Chayahuita), ino mequen (Shipibo)

Venezuela Kaanada, aicharikudek, cozoiba; tawe yaɨ (Hotï), basho kë moka hi, basho ehu hi (Yanomami)

Status Wild, planted

Arlete, 50 years, with bacuri casca lisa *Garcinia macrophylla* **fruits gathered in successionary forest. Gurupá-Mirim, Municipio de Gurupá, Pará, Brazil 4-26-15**

The smooth, yellow fruits of *Garcinia macrophylla* are a treat throughout the Amazon as well as many other tropical and subtropical parts of central and South America. Native to Amazonia, the tree grows in mature and regrowth forest on uplands as well as floodplains, including the tidal zone at the mouth of the Amazon (Almeida and Jardim 2011, 2012; Amaral et al. 2009; Araujo-Murakami et al. 2012; Batista and Gonçalves 2013; Bentes-Gama et al. 2002; Braga and Jardim 2019; Braga et al. 2015; Carim 2016:88; Cattanio et al. 2002; Gama et al. 2002, 2003; Nebel et al. 2001; Queiroz and Machado 2007, 2008; Queiroz et al. 2005, 2006, 2007; Lau and Jardim 2013; Maués et al. 2011; Miranda 2000; Muñoz 2010:22; Nebel et al. 2000a, b; Parolin 2000a, b, c; Salomão et al. 2007a, b; Silva 2008:27; Souza et al. 1996). This tree occurs up to 2000 m in the Andean foothills of the Bolivian, Peruvian and Ecuadorian Amazon (Aguirre-Mendoza et al. 2015:45; Arellano 2013:174; Macía and Fuertes 2008; Mendoza and Guerrero 2010). It also occurs in seasonal forests in the southern fringes of the Amazon Basin in Mato Grosso (Lopes et al. 2014). This tree is prolific, producing some 700 fruits a year which have a diameter of 3.8–5.6 cm and weigh around 50 g. The fruits are not too difficult to reach by climbing or using a pole since the tree is usually under 15 m tall (Silva 2011:76). This tree grows on a wide variety of soils, including sandy soils (Zárate et al. 2015).

The creamy pulp of *Garcinia macrophylla* is sufficiently sweet to entice farmers in many parts of the Amazon to cultivate the tree in their home gardens (Berg 1979; Cavalcante 2010:62; Fraser et al. 2011; Gonçalves and Lucas 2017; Lisboa et al. 2009; Lorenzi et al. 2006a, b:129; Machado 2016:30; Silva 2006:127; Yamada and Osaqui 2006). It is also grown in house yards in urban areas such as Belem (Huber 1904). One such tree planted in a home garden 26 years ago along the lower Tigre in Peru had reached 20 m high and was festooned with at least two thousand fruits at the end of the rainy season in 2006. The tree was likely so productive because it was growing on fertile alluvial soils enriched by the annual flood of the sediment-rich Tigre. River dwellers have also planted bacuripari in home gardens in the Caxiuanã National Forest in eastern Pará. In the Peruvian Amazon, the Chayahuita apply the milky latex obtained from the fruit's skin to human skin wounds (Odonne et al. 2013).

Ronaldo, 9 years, eating charichuelo *Garcinia macrophylla* fruit. Manco Capac, Rio Puinahua (a branch of the Ucayali), Reserva Nacional Pacaya-Samiria, Loreto, Peru 4-15-06

The shiny, round fruits, known as charichuelo in Peru, are usually between 6 and 8 cm in diameter and on some trees have reddish-brown streaks on the skin. Rural and urban folk alike pry open the leathery skin, which is almost a centimeter thick, with their fingers to access the moist, white pulp that surrounds the one to four brown seeds. After relishing the sweet pulp, the tan seeds, some 3.5–3.8 cm long and 1.8–3 cm in diameter, are spat out. The fruits are gathered by climbing the tree, picking up fruits that have fallen to the ground, or are plucked from canoes during high water, such as in the vicinity of Manco Capac on the Rio Puinahua in the Peruvian Amazon. On one ranch on Marajó Island in the Municipality of Ponta de Pedras, cowboy children pick the fruits from canoes while they are still green because they enjoy the tart taste and because they want to make sure they beat others to the treat (Smith 2002:118).

Boy picking unripe fruits of bacuri pari *Garcinia macrophylla* in flooded savanna (*campo*). Retiro Graças a Deus, Lavrado, watershed divide between the Arari and Anabiju, Municipality of Ponta de Pedras, Marajó Island, Para, Brazil 4-24-00

Game animals also appreciate the fruits, including the bulldog-sized paca *Agouti paca* which feeds on the ground. Woolly monkeys *Lagothrix lagotricha*, howler monkeys *Alouatta seniculus*, and black spider monkeys *Ateles paniscus*, also hunted for their meat, feed on the pulp while in trees and disperse some of the seeds (Defler and Defler 1996; Dew 2005; Kinzey and Norconk 1990; Neves and Rylands 1991; Stevenson 2000, 2005, 2007; Stevenson et al. 1994; Van Roosmalen 1985; Van Roosmalen and Klein 1988). Long-haired spider monkeys have been observed swallowing the seeds and are thus likely involved in dispersing this tree (Dew 2008). In the Peruvian Amazon, the equatorial saki *Pithecia aequatorialis* has been observed feeding on this fruit (Charpentier et al. 2015; Montalván 2015:38). The red titi monkey *Callicebus discolor* which occurs in northwestern Amazonia has also been documented eating this fruit (Montalván 2015:39). Although the saki and titi monkeys are rather small and not usually hunted for their meat, they may be involved in dispersing some of the seeds. Similarly, the red-bellied tamarin *Saguinus labiatus* has been observed feeding on this fruit in the Bolivian Amazon (Buchanan-Smith 1991). In southern Venezuela tapirs *Tapirus terrestris*, the largest terrestrial game animal in the Amazon and Orinoco Basins, also eat the fruit (Salas and Fuller 1996). The seeds are eventually defecated, and the resulting progeny thus escape some of the pathogens and pests that congregate around parental trees (Stevenson et al. 2005). In the Colombian Amazon, a game bird, paujil *Mitu silvini*, has been observed feeding on this fruit and likely disperses some of the seeds (Franco and Santamaria 1997). Tambaqui *Colossoma macropomum*, known as gamitana in Peru, is one of the most prized fish in Amazonia which eats this fruit during the high-water period and probably disperses some of the seeds (Baca 2005).

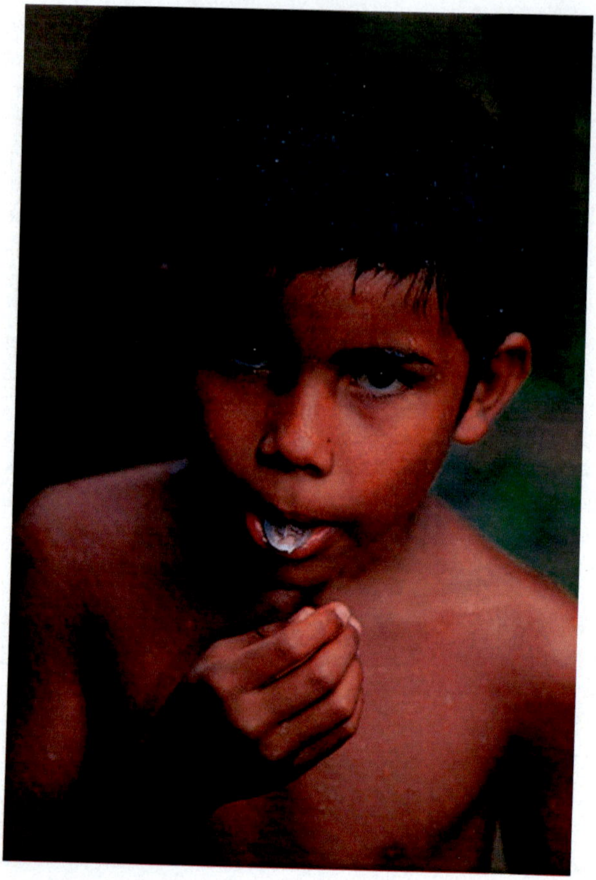

Boy ejecting seed of bacuri pari *Garcinia macrophylla* on the veranda of his home. Discarded seeds of *G. macrophylla* often sprout in home gardens. Retiro Graças a Deus, Lavrado, watershed divide between the Arari and Anabiju, Municipality of Ponta de Pedras, Marajó Island, Para, Brazil 4-24-00

Fruits of *Garcinia macrophylla* sometimes appear in urban markets during the flood season such as in Iquitos, Peru, and Leticia, Colombia (Pulido and Cavelier 2001). The tough skin protects the pulp from damage, thus facilitating transport. Although it is not a high-value timber, the wood is also used occasionally for local construction and firewood in the Brazilian Amazon (Le Cointe 1934:47), the Ecuadorian Amazon (Aguirre-Mendoza et al. 2015:45; Van den Eynden et al. 1999:42; Cerón and Ayala 1998:61), and by the Palikur in French Guiana (Ogeron et al. 2018).

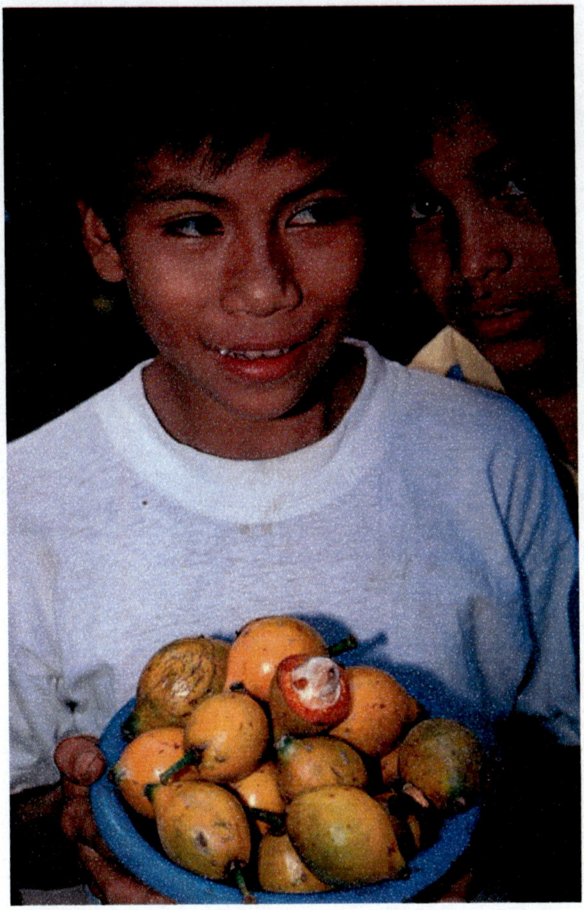

Charichuelo de altura ***Garcinia macrophylla*** **fruits with vendor in Belén market, Iquitos, Loreto, Peru 6-17-04**

In the old days at Belém's Ver-o-Peso market, the fruit was served as a chaser for potent drinks, such as brandy or cachaça. Stands selling alcoholic drinks by the glass were located by the old fort, built by the Portuguese in the early seventeenth century. Some customers appreciated the tart taste of bacuripari after downing "two fingers" (*dois dedos*) of straight cachaça. Ver-o-Peso has undergone some major reforms, and the drink stands are now gone. Fruits of *G. macrophylla* also appear sporadically throughout the rainy season from December to June in the Belén market in Iquitos, the largest city in the Peruvian Amazon.

Bacuri pari ***Garcinia macrophylla*** **fruit offered as a chaser for customers buying a shot of sugarcane alcohol (cachaça) at a stand adjacent to the Feira do Açaí. Condensed milk is also provided to counter the harsh taste of straight cachaça. Ver-o-Peso market, Belém, Pará, Brazil 12-24-01**

17.6 *Garcinia madruno*

Bolivia Ocoró, ocorosillo, camaruru, camoruro, camururo, achachairú, chapi simayu; shiquishi, tahuatë (Chácobo), antágasajo (Leco), camururu (Tacana)

Brazil Bacuri, bacuri de espinho, bacuri coroa, bacuri azedo, bacuri de anta, bacuri caraquento, bacuri de igapó, bacuri cascudinho, bacuripari, bacurizinho, limãozinho; pãxo ãhuhi, pãxoate mohi (Yanomami)

Colombia Madroño, madrona grande, bacuri, ungüento maría, tierra amarillo; podá, pada (Andoke), córop (Nukak)

Ecuador Madroño, madroño pequeño, pungara; pevicho, chipire pevicho (Cofán), pungaramuyo, pungara (Kichwa), penká (Shuar), majáro (Siona-Secoya), öntaka (Waorani)

Peru Charichuelo, charichuelo menudo, charichuelo chico, charichuelo rugoso, charichuela quillo wiqui, durazno, durazno amarillo, duraznillo calato, pariqui; pengkaenum, pegkainum, wayámpainim (Aguaruna), wayámpainim (Awajún), chomaeki (Campa)

Venezuela Doro, naranjilla, peraman de agua, cozoiba negra, pata de danto,

madroño

Status Wild, planted

Charichuelo *Garcinia madruno* in fruit in floodplain forest along the black water Rio Pucate, Reserva Nacional Pacaya-Samiria, Loreto, Peru 4-21-06

The yellow, bumpy fruits of *Garcinia madruno* are widely appreciated in the Amazon Basin and other parts of its range which include Costa Rica, Panama, the Guianas, and Venezuela (Berg 1979; Van Andel et al. 2012). Known as charichuelo in Peru and bacuri in Brazil, *Garcinia madruno* reaches some 30 m tall in forests, including successionary vegetation, on floodplains and uplands up to 1500 m in the foothills of the Andes (Barbosa et al. 2011; Mendoza and Guerrero 2010; Luize et al. 2015;

Nebel et al. 2000a, b; Paim et al. 2018; Robles 2017:47; Scudeller and Souza 2009; Silva 2017:30). This species is widespread in Amazonia and occurs on forest islands surrounded by mangroves and marshes along the coast of Pará (Rudas and Prieto 1998; Santos 2010:42; Santos et al. 2014) and in gallery forests in the transition zone to cerrado in the southern fringe of the Amazon Basin (Kuchla et al. 2015). It is found in some wetlands dominated by the tall *Mauritia flexuosa* palm, such as along the Iquitos-Nauta highway in the Peruvian Amazon (Zárate et al. 2013). *Garcinia madruno* can thrive for many years in the deep shade of the forest and starts producing fruits by the time it is 3 m tall, so the fruits are sometimes within easy reach. Indigenous groups, such as the Nukak in the Colombian Amazon, gather the fruits (Politis 1996:177), as do rural folk in other parts of the Amazon Basin. The Waorani in the Ecuadorian Amazon also plants the tree around their settlements (Ima 2012:84).

Woolly monkeys *Lagothrix lagotricha* and spider monkeys *Ateles belzebuth* also consume the fruits of *G. madruno* (Defler and Defler 1996; Van Roosmalen and Klein 1988) and these monkeys are hunted for their meat throughout the Amazon Basin. A related species of spider monkey, *Ateles hybridus*, which lives in the Magdalena River basin in northern Colombia, also eats this fruit and is also hunted for its meat (Luna et al. 2017). The Colombian squirrel monkey *Saimiri cassiquiarensis albigena* and the dusky titi monkey *Callicebus ornatus*, which is confined to eastern Colombia, also eat this fruit (Carretero-Pinzón and Defler 2016; Carretero-Pinzón et al. 2016). Although these monkeys are not normally hunted for their meat because of their small size they likely disperse some of the seeds of this fruit tree. One of the common names for this tree in the Brazilian Amazon is bacuri de anta which suggests that tapirs *Tapirus terrestris*, the large terrestrial game animal in Amazonia, also eats this fruit as has been recorded in the Upper Orinoco watershed in eastern Colombia (Alviz and Pérez-Albarracín 2015:41). Tambaqui *Colossoma macropomum*, known as gamitana in Peru, is one of the most prized fish in the Amazon also eats this fruit and thus likely disperses some of the seeds (Baca 2005).

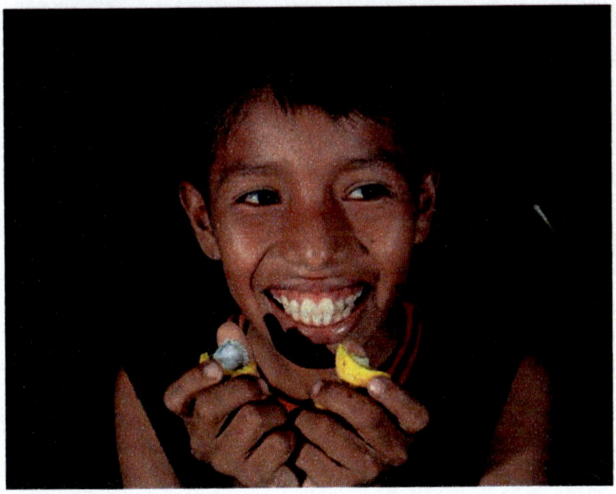

Bildoso, 12 years, eating charichuelo *Garcinia madruno* **fruit in floodplain forest. Rio Pucate, Reserva Nacional Pacaya-Samiria, Loreto, Peru 4-21-06**

The rough skin of the round fruits, some 3.5–6 cm long and 2.5–4.5 cm wide, is pried open with teeth or fingers to access the light gray pulp surrounding one to four brown seeds, measuring 2.4 cm by 0.8 cm. Pulp accounts for about one-third of the fruit and is rich in potassium (Cury et al. 2012). One youngster along the Rio Pucate in the Reserva Nacional Pacaya-Samiria broke open the skin by holding the fruit in one hand and thumbing it with the base of the thumb on the other hand. Fruits start maturing in the latter part of the dry season and continue bearing throughout the rainy season.

Charichuelo *Garcinia madruno* fruits in a street market. Mercado Belén, Iquitos, Loreto, Peru 4-30-06

The pulp has an intriguing flavor, sweet yet slightly tangy, and accounts for about one-third of the fruit's size. The fruits are eaten locally and also taken to markets, such as in Iquitos, where they are sold relatively cheap and therefore easily affordable for the majority of consumers. In the sprawling Belém market in Iquitos in November 2003, for example, customers could buy a 40 kg sack of the fruits for as little as US $4 (12 soles). For the cash-strapped, vendors had small plastic bags (*bolsitas*) containing about 25 fruits for the equivalent of US$ 0.30 (1 sole). As the rainy season progresses and more fruits become available, charichuelo prices drop slightly; in March 2004, for example, fruits of *Garcinia madruno* were selling for the equivalent of one U.S. cent each (30 for 1 sole) in the Belém market of Iquitos.

The unusual and highly agreeable flavor of *Garcinia madruno*, as well as other wild fruits in the genus, warrants attention for future selection and crop development (Pennington et al. 2004:233). In addition, the wood is a high-grade timber useful for a variety of construction purposes (Reynel et al. 2003:130). Senescent trees no longer producing appreciable quantities of fruit could therefore be harvested for timber.

17.7 *Platonia insignis*

Brazil Bacuri, bacuri açú, bacuri pari, bacuri liso, bacuri de serra, bacuri de anta, bacuri de obra, bacuri do Pará, bacuri doce, bacurizeiro, bacupari açu, bacuxiuba; mukur-'ɨ (Guajá), pakuri'ɨ (Ka'apor), pi pannê ka tire (Kayapó), pozokyry (Rikbaktsa)

Colombia Bacuri, charichuela, charichuelo, olha de mono; Ko-mê-peé-go-rê (Desana), gö-hê-reé-ka (Makuna), kõ (Puinave), oo-ká-o-ree-kê (Tanimuka), ko-paí-yo-ree (Tukano)

Peru Bacuri, matazama, matazona

Status Wild, planted

Wife of a cowboy holding bacuri *Platonia insignis* **fruit. Fazenda Anjos, Marajó Island, Pará, Brazil 12-16-01**

Bacuri *Platonia insignis* is easily recognized by its yellow, grapefruit-sized fruits that appear frequently in urban markets and along roadside stands at the mouth of the Amazon during the latter part of the dry season and the early part of the rainy period. Bacuri fruits from December to the height of the rainy season in April (Batista and Jardim 2006). The fruits are usually 7–15 cm in diameter and can weigh up to 1 kg (Ferreira and Melo 2007; Medina and Ferreira 2004; Rogez et al. 2004; Silva 2011:75). In some years, a mature bacuri tree can produce over 2000 fruits, although few fruits are produced the following year (Silva 2011:75). A mature bacuri tree produces an average of 500 fruits a year (Ferreira and Melo 2007). The thick skin (epicarp) accounts for some two-thirds of the fruit's weight but this helps protect the edible pulp (mesocarp) which only accounts for about a fifth of the fruit's weight. The green-thighed parrot *Pionites leucogaster*, the golden-winged parakeet *Brotogeris chrysoptera*, and the white-eyed parakeet *Aratinga leucophthalmus* have been observed visiting the flowers of this tree in the vicinity of Belém to obtain nectar and are involved in pollination because the pollen is sticky and cannot be transported by air currents (Maués and Venturieri 1996).

This is the best-known bacuri in eastern Amazonia because, in addition to providing copious amounts of delicious pulp, the fruit is protected by a relatively thick skin that facilitates transport to markets. Indeed, bacuri fruits have been sent to urban markets in the Brazilian Amazon for a long time. In the early nineteenth century, for example, bacuri fruits were transported by boat from Marajó Island for sale in Belém (Baena 2004:275). The fruits are opened by smashing them together, or against a hard object, to access the cream-colored pulp. The pulp is eaten raw, made into juice, jam, and in urban areas, a delicious ice cream (Shanley and Rosa 2005; Shanley et al. 2002). In the mid-nineteenth century, bacuri jam was exported from Belém to Europe (Avé-Lallemant 1961:39). The fruit is a pleasure to eat because it is largely unencumbered by seeds; only one to five large black seeds are encased in the fruit, and these are easily spat out. In the past, oil was expressed from the seeds to make soap (Schurz et al. 1925:182).

Some game animals also eat this fruit. Grey brocket deer *Mazama gouazoubira* have been recorded eating this fruit in French Guiana (Gayot et al. 2004) and this deer also occurs in Amazonia. Tapir *Tapirus terrestris*, the largest terrestrial game animal in Amazonia, also eat this fruit as suggested by one of the common names for this fruit: bacuri de anta. Black spider monkeys *Ateles paniscus* and the northern bearded saki *Chiropotes satanas chiropotes*, both of which are hunted for their meat, are known to eat the pulp (Van Roosmalen 1985; Van Roosmalen and Klein 1988; Van Roosmalen et al. 1981). Goeldi's marmoset *Callimico goeldii* has also been observed eating this fruit in the Bolivian Amazon (Pook and Pook 1981), but this small monkey is not usually hunted for its meat because of its small size.

Bacuri can grow as high as 37 m and grows in the forest as well as successionary vegetation and savannas, especially on Marajó Island. On this large island, rural folk sometimes scar the trunk to identify which trees belong to whom; they also build camps in the forest during the rainy season when the fruits are falling (Lisboa 2012:451). Bacuri is sometimes concentrated on forest islands on elevated portions of the grasslands of Marajó, some of which are anthropogenic (Amaral et al. 2019;

Schaan 2012:180). On Fazenda Jilva, a ranch in the northeastern part of Marajó, a wooded island in savanna is called Teso do Bacuri Alto (Hill of the Tall Bacuri), and indigenous pots (*gaçabas*) have been found on this relic sand dune. Bacuri thrives in successionary forest in other upland areas of Pará, such as in the municipalities of Bragança and Capitão Poço (Ferreira and Melo 2007; Rayol et al. 2011) as well as in neighboring Maranhão (Balée 1993b). Bacuri also occurs in some forests and successionary vegetation on floodplains, such as in the Amazon estuary (Aparício 2011:92; Carim 2016:88; Silva 2008:27), and in seasonally flooded stunted woodlands (*campinarana*) in Roraima (Vale 2011:44). Bacuri is particularly common in the Amazon estuary (Nascimento et al. 2007).

In Maranhão on the eastern fringes of the Amazon Basin, bacuri occurs in high density in cultural forests created by the Ka'apor (Balée 1993a, 2010) as well as in scrub savanna (*cerrado*) and stunted forest (*cerradão*), such as in the Parque Estadual do Mirador (Rodrigues and Conceição 2014), the Chapada Limpa Extractive Reserve (Santos et al. 2017a), and in the municipality of Carolina, Maranhão (Medeiros and Walter 2012). Ancient bacuri trees in open areas resemble the baobab tree of Africa, creating an almost surreal landscape reminiscent of the plains of East Africa. Old bacuri trees have thick, light-colored trunks topped by drooping umbrella-like branches. In the early part of the dry season, when bacuri is in flower, the trees put on a flush of bright, green leaves that glint pleasingly against the backdrop of a bright blue sky, so typical of the dry season on Marajó.

Bacuri *Platonia insignis* trees on an ancient sand dune in savanna on a ranch with water buffalo during the dry season. Teso do Bacuri, Fazenda Jilva, Marajó Island, Pará, Brazil 12-15-01

Bacuri is particularly common in eastern Amazonia and is probably native to Pará (Cavalcante 2010:58; Silva et al. 2011:72). Along the Capim River in eastern Pará, bacuri is one of the most commonly collected wild fruits (Shanley and Rosa 2004). Bacuri also occurs in the wild north of the Amazon River in Amapá extending into the Guianas (Pereira et al. 2011). Bacuri is also occasionally encountered in central and northwestern Amazonia including the Rio Negro and Japurá watersheds extending into the Colombian Amazon (Major et al. 2005; Pulido and Cavelier 2001; Silva et al. 2016). The "natural" distribution of bacuri is hard to define given the long history of human migrations and trade in Amazonia, but it was probably taken to parts of western Amazonia in precontact times.

Although most bacuri fruits are gathered in the wild, bacuri is also planted occasionally in home gardens on uplands and floodplains, such as on Marajó and Guajará Islands, in the Caxiuanã National Forest along the Anapu River, along the Xingu River, in the vicinity of Abaetetuba along the lower Tocantins, in the municipalities of Bonito and Capitão Poço in northeastern Pará, in the Municipality of Alenquer in western Pará, along the lower Madeira River and near Manaus, Amazonas (Gonçalves and Lucas 2017; Kawa et al. 2015; Rayol et al. 2013; Rosa et al. 2007, 2009; Souza et al. 2017; Souza 2010:26). Bacuri occurs in some home gardens on the floodplain of the Solimões/Amazon near its confluence with the Rio Negro (Castro et al. 2009; Lima and Saragoussi 2000) as well as in the Upper Tocantins watershed in the southeastern part of the Brazilian Amazon (Pereira and Santos 2015). Some villagers plant this tree in their upland home gardens in the municipalities of Igarapé-Açu and Marapanim in the zona Bragantina of northeastern Pará (Quaresma et al. 2015). Some Japanese-Brazilian farmers in the vicinity of Tomé-Açu in Pará also plant this tree in their home gardens (Yamada and Osaqui 2006). Bacuri is also planted in some home gardens in the suburbs of São Luis, Maranhão, on the eastern fringe of the Amazon Basin (Akinnifesi et al. 2010a, b).

Bacuri is also occasionally planted in agroforestry fields, such as near Tomé-Açu, where a Japanese-Brazilian farmer tends a plot containing black pepper intercropped with several species of fruit trees. The Kayapó in the interfluvial area between the Tocantins and Xingu plant bacuri in some of their manioc fields so that they can return to those areas long after the tubers have been harvested to gather fruits (Posey 1985). Some small-scale farmers on the outskirts of Rio Branco in Acre also plant this tree in their agroforestry plots (Slinger 2000).

Bacuri's intriguing flavor has inspired people in Pará to name crop varieties, street names, and even entire communities after the fruit. A mango variety cultivated widely in Pará, from Marajó to the vicinities of Tome-Açu and Itupiranga, for example, is called Bacuri because the smooth, rich-tasting fruit has no discernible fiber, unlike most mangoes, which leave their aficionados with stringy teeth. The round, yellow Bacuri variety of mango also appears in the Ver-o-Peso street market of Belém at the mouth of the Amazon. A bitter manioc cultivated in the vicinity of Caxiuanã along the Anapú is named after bacuri. A community along the Arapiuns River near Santarém is called Bacuri and bacuri is held in such esteem in Óbidos, a town along the middle Amazon, that one of the principal streets is named after the fruit. A village near Bragança in Pará is called Bacuriteua, which means the land of bacuri.

Although bacuri's flavor is sublime—sweet and creamy, yet with a hint of savory—its fame has not spread as far as some other Amazonian fruits. Bacuri is hardly known in contiguous Amazonas state, for example, although it has reached parts of the Northeast and Mato Grosso along the southern fringes of Amazonia. Bacuri is one of those Cinderella fruits that may 1 day burst onto the scene in Paris, New York, or Tokyo in the form of an exotic mousse, enticing smoothie, or trendy ice cream.

For the time being, however, only residents or visitors to large urban centers at the mouth of the Amazon, such as Belém, can savor the delights of a bacuri ice cream or milkshake. A popular microbrewery in Belém, Amazon Beer, flavors one of its ales with bacuri. In rural areas, bacuri is mostly eaten fresh, or in the case of families living on ranches, mixed with warm milk to make a room temperature milkshake.

The yellow wood is durable and has been used in boat construction going back to the early nineteenth century (Baena 2004:43; Le Cointe 1934:48; Loureiro et al. 1979:78). Bacuri continues to be the target of loggers, thus reducing its genetic diversity and option for crop improvement down the road (Silva et al. 1977:31). Bacuri wood is also used for firewood in manioc processing huts and to make charcoal for cooking such as in the Vila de Algodoal in the municipality of Maracanã along the northeast coast of Pará (Coelho-Ferreira and Jardim 2005). Indigenous groups along the Uaupés River in northwestern Amazonia use latex to treat skin infections (Schultes and Raffauf 1990:209), while on Marajó Island the latex is also used in folk remedies, such as for treating wounds (Lisboa 2012:451). The resinous latex obtained from the trunk and fruit is applied to a variety of skin issues including eczema and herpes (Van den Berg 2010:131).

References

Abreu, Maria Milena de Oliveira; Mehlig, Ulf; Nascimento, Raquel do Espírito Santo Aguair do; Menezes, Moirah Paula Machado. 2006. Análise de composição florística e estrutura de um fragmento de bosque de terra firme e de um manguezal vizinhos na peninsula de Ajuruteua, Bragança, Pará. Boletim do Museu Paraense Emílio Goeldi, Ciencias Naturais 1(3):27–34

Aguirre-Mendoza, Zhofre; Loja, Ángel Fernando; Solano, Carmen; Aguirre, Nikolay. 2015. Especies Forestales más aprovechadas del Sur del Ecuador. Universidad Nacional de Loja, Ecuador

Akinnifesi, F. K.; Sileshi, G.; Costa, Joselene da; Moura, E. G. da; Silva, Rosele F. da; Ajayi, O. C.; Linhares, J. F. P.; Akinnifesi, A. I.; Araujo, Maria de; Rodrigues, M. A. I. 2010a. Floristic composition and canopy structure of home-gardens in São Luís city, Maranhão state, Brazil. Journal of Horticulture and Forestry 2(4):72–86

Akinnifesi, Festus Kehinde.; Sileshi, Gudeta W.; Ajayi, Oluyede C.; Akinnifesi, Adetutu I.; Moura, Emanoel G. de; Linhares, Jairo F. P.; Rodrigues, Ivanilde 2010b. Biodiversity of the urban homegardens of São Luís city, northeastern Brazil. Urban Ecosystems 13(1):129–146

Alarcón, J. G. S. and A. L. Peixoto. 2008. Use of terra firme forest by Caicubi caboclos, middle Rio Negro, Amazonas, Brazil: A quantitative study. Economic Botany 62(1):60–73

Almeida, Adrielson Furtado; Jardim, Mário Augusto Gonçalves. 2011. Florística e estrutura da comunidade arbórea de uma floresta de várzea na Ilha de Sororoca, Ananindeua, Pará, Brasil. Scientia Florestalis 39(90):191–198

Almeida, Adrielson Furtado; Jardim, Mário Augusto Gonçalves. 2012. A utilização das espécies arbóreas da floresta de várzea da Ilha de Sororoca, Ananindeua, Pará, Brasil por moradores locais. Revista Brasileira de Ciências Ambientais 23:48–54

Alviz, Ángela; Pérez-Albarracín, Karen. 2015. Plan para la conservación de la danta de tierras bajas (Tapirus terrestris) en los departamentos de Casanare, Arauca y Vichada. Corporinoquia-Fundación Orinoquia Biodiversa, Yopal, Colombia

Amaral, Dário Dantas do; Prost, Maria Thereza; Bastos, Maria de Nazaré do Carmo; Costa Neto, Salustiano Vilar; Santos, João Ubiratan Moreira dos. 2008. Restingas do litoral amazônico, estados do Pará e Amapá. Boletim do Museu Paraense Emílio Goeldi, Ciencias Naturais 3(1):35–67

Amaral, Dário Dantas do; Vieira, Ima Célia Guimarães; Almeida, Samuel Soares de; Salomão, Rafael Paiva; Silva, Antonio Sérgio Lima da; Jardim, Mário Augusto Gonçalves. 2009. Checklist da flora arbórea de remanescentes florestais da região metropolitana de Belém e valor histórico dos fragmentos, Pará, Brasil. Boletim do Museu Paraense Emílio Goeldi, Ciencias Naturais 4(3):231–289

Amaral, Dário Dantas do; Rocha, Antônio Elielson; Pereira, Jorge Luis Gavina; Costa Neto, Salustiano Vilar. 2019. Identificação dos subtipos de savanas na Amazônia oriental (Pará e Amapá, Brasil) com uma chave dicotômica de individualização. Boletim do Museu Paraense Emílio Goeldi, Ciencias Humanas Naturais 14(2):183–195

Aparício, Wegliane Campelo da Silva. 2011. Estrutura da vegetação em diferentes ambientes na RESEX do Rio Cajari: interações solo-floresta e relações com a produção de castanha. Doctoral Dissertation, Universidade Federal Rural de Pernambuco, Recife

Araujo, Michele. 2014. Ecologia alimentar de *Saimiri sciureus cassiquiarensis* (Lesson, 1840) (Primates, Cebidae) em florestas de várzea da Amazônia Central Brasileira. Master's thesis, MPEG/UFPa, Belém

Araujo, Henrique José Borges de. 2015. Alterações causadas por incêndio florestal na composição de espécies de uma floresta primária no sudoeste da Amazônia Brasileira. Gaia Scientia 9(1):220–231

Araujo-Murakami, Alejandro; Poma-Chura, Abraham; Palabral, Arely; Salvatierra, Rolando; Hurtado, Fidel. 2012. Composición florística de los bosques Amazónicos de tierra firme e inundable en las proximidades de las Pampas del Sonene (río Heath), Parque Nacional Madidi, Bolivia. Kempffiana 8(1):2–21

Araujo-Murakami, Alejandro; Villarroel, Daniel; Pardo, Guido; Vos, Vincent A.; Parada, G. Alexander; Arroyo, Luzmila; Killeen, Timothy. 2015. Diversidad arbórea de los bosques de tierra firme de la Amazonía Boliviano. Kempffiana 11(1):1–28

Arellano, Gabriel. 2013. Patrones de diversidad, distribución y dominancia de plantas leñosas en los bosques montanos del Parque Nacional Madidi, Bolivia. Doctoral dissertation, Departamento de Biologia, Facultad de Ciencias, Universidad Autónoma de Madrid

Assis, Rafael Leandro de; Wittmann, Florian 2011. Forest structure and tree species composition of the understory of two central Amazonian várzea forests of contrasting flood heights. Flora 206:251–260

Avé-Lallemant, Robert. 1961. Viagem pelo Norte do Brasil no ano de 1859. Instituto Nacional do Livro, Rio de Janeiro, Vol. 2

Baca, Luis Campo. 2005. Algunos parámetros físicos, químicos y bioecológicos que influyen el comportamiento migratorio de la gamitana *Colossoma macropomum* en el Río Ucayali. In: Biología de la Poblaciones de Peces de la Amazonía y Piscicultura, Renno, Jean François; García, Carmen; Duponchelle, Fabrice; Nuñez, Jésus (eds), IIAP/IRD, Iquitos, pp 58–64

Baena, Antônio Ladislau Monteiro 2004. Ensaio Corográfico sobre a Província do Pará. Edições do Senado Federal, Brasilia

Balée, W. 1993a. Biodiversidade e os índios Amazônicos. In: *Amazônia: Etnologia e História Indígena*, E. Viveiros de Castro and M. C. da Cunha (Editors), Núcleo de História Indígena e do Indigenismo da USP/FAPESP, São Paulo, pp. 385–393

Balée, W. 1993b. Indigenous transformation of Amazonian forests: an example from Maranhão, Brazil. L'Homme 33(126–128):231–254

Balée, W. 2010. Contingent diversity on anthropic landscapes. *Diversity* 2:163–181

Balée, W. and A. Gély, 1989. Managed forest succession in Amazonia: the Ka'apor case. In: *Resource Management in Amazonia: Indigenous and Folk Strategies*, Posey, D. A. and W. Balée (Eds), pp. 129–158, Advances in Economic Botany, vol. 7, New York Botanical Garden, Bronx, New York.

Bánki, Olaf S.; Steege, Hans ter; Jansen-Jacobs, Marion J.; Raghoenandan, Usha P. D. 2003. Plant Diversity of the Nassau Mountains, Suriname: Report of the 2003 Expedition. National Herbarium Nederland, Utrecht

Barbosa, Danielle Cristine de Figueiredo; Marimon, Beatriz Schwantes; Lenza, Eddie; Marimon Junior, Ben Hur; Oliveira, Edmar Almeida De; Maracahipes, Leandro. 2011. Estrutura da vegetação lenhosa em dois fragmentos naturais de florestas inundáveis (impucas) no Parque Estadual do Araguaia, Mato Grosso. Revista Árvore 35(3):457–471

Barnett, Adrian A.; Boyle, Sarah A.; Pinto, Liliam P.; Lourenço, Waldete C.; Almeida, Thais; Silva, Welma Sousa; Ronchi-Teles, Beatriz; Bezerra, Bruna M.; Ross, Caroline; MacLarnon, Ann; Spironello, Wilson R. 2012. Primary seed dispersal by three Neotropical seed-predating primates (*Cacajao melanocephalus ouakary*, *Chiropotes chiropotes* and *Chiropotes albinasus*). Journal of Tropical Ecology 28(6):543–555

Bastos, Maria de Nazaré do Carmo. 1988. Levantamento florístico em restinga arenosa litorânea na Ilha de Maiandeua, Pará. Boletim do Museu Paraense Emílio Goeldi, Série Botânica 4(1): 159–169

Bastos, M. N. C.; C. S. Rosário; L. C. B. Lobato. 1995. Caracterização fitofisionômica da restinga de Algodoal, Maracanã, Pará, Brasil. Boletim do Museu Paraense Emílio Goeldi, Série Botanica 11(2): 173–197

Bastos, Maria de Nazaré do Carmo; Costa, Denise Cristina Torres; Santos, João Ubiratan Moreira dos. 2003. Vegetação de restinga: aspectos botânicos e uso medicinal. Museu Paraense Emilío Goeldi Projeto Renas, Belém

Batista, Fábio de Jesus; Gonçalves, Mário Augusto. 2013. Florística e estrutura da regeneração natural arbórea de uma floresta de várzea na Reserva Extrativista Chocoaré-Mato Grosso, Pará, Brasil. Biota Amazônia 3(3):139–145

Batista, Fábio de Jesus; Jardim, Mário Augusto Gonçalves. 2006. Notas sobre a morfologia floral e a fenologia do bacurizeiro (Platonia insignis Mart.) Clusiaceae, no município de Bragança, estado do Pará. Boletim do Museu Paraense Emílio Goeldi, Ciencias Naturais 1(1):183–186

Batista, Fábio de Jesus; Jardim, Mário Augusto Gonçalves; Medeiros, Tonny David Santiago; Lopes, Isaac Luiz Magalhães. 2011. Comparação florística e estrutural de duas florestas de várzea no estuário Amazônico, Pará, Brasil. Revista Árvore 35(2):289–298

Batista, Anderson Pedro Bernardina; Aparício, Wegliane Campelo da Silva; Santos, Vanessa Silva dos; Lima, Robson Borges de; Mello, José Marcio de. 2015. Caracterização estrutural em uma floresta de terra firme no estado do Amapá, Brasil. Pesquisa Florestal Brasileira 35(81):21–33

Bentes-Gama, Michelliny de Matos; Scolforo, José Roberto Soares; Gama, João Ricardo Vasconcellos. 2002. Potencial produtivo de madeira e palmito de uma floresta secundária de várzea baixa no estuário Amazônico. Revista Árvore 26(3):311–319

Berg, M. E. 1979. Revisão das espécies do gênero *Rheedia* L. (Guttiferae). Acta Amazonica 9(1):43–74

Bittrich, V.; Amaral, Maria C. E. 1997. Floral biology of some Clusia species from Central Amazonia. Kew Bulletin 52(3):617–635

Boubli, J. P. 1999. Feeding ecology of black-headed uacaris (*Cacajao melanocephalus melanocephalus*) in Pico da Neblina National Park, Brazil. International Journal of Primatology 20(5):719–749

Boyle, Sarah A.; Zartman, Charles E.; Spironello, Wilson R.; Smith, Andrew T.; Beck, Harald. 2012. Implications of habitat fragmentation on the diet of bearked saki monkeys in central Amazonian forest. Journal of Mammalogy 93(4):959–976

Brack Egg, Antonio. 2003. Frutas del Perú. Universidad San Martin de Porres, Lima

Braga, Elayne Oliveira; Jardim, Mário Augusto Gonçalves. 2019. Florística, estrutura e formas de vida do estrato inferior de uma floresta ombrófila densa aluvial, Pará, Brasil. Ciencia Florestal, Santa Maria 29(3): 1048–1059

Braga, Elayne Oliveira; Silva, Julie Andrews de França e; Pantoja, Milenna Vasconcelos; Jardim, Mário Augusto Gonçalves. 2015. Florística, estrutura fitossociológica e formas de vida do estrato inferior em um floresta de várzea Amazônica. Biota Amazônia 5(3):59–65

Braga, Daniel P. P.; Domene, Frederico; Gandara, Flávio B. 2019. Shade trees composition and diversity in cacao agroforestry systems of southern Pará, Brazilian Amazon. Agroforestry Systems 93:1409–1421

Buchanan-Smith, H. 1991. A field study on the red-bellied tamarin, Saguinus l. labiatus, in Bolivia. International Journal of Primatology 12(3):259–276

Campbell, David G.; Daly, Douglas C.; Prance, Ghillean T.; Maciel, Ubirajara N. 1986. Quantitative ecological inventory of terra firme and várzea tropical forest on the Rio Xingu, Brazilian Amazon. Brittonia 38(4):369–393

Cárdenas, Dairon; Castaño, Nicolás; Sua, Sonia. 2009. Flora de la estrella fluvial de Inírida (Guianía, Colombia). Biota Colombiana 10(1–2):1–30

Carim, Marcelo de Jesus Veiga. 2016. Estrutura, composição e diversidade em florestas alagáveis de várzea de maré e de igapó e suas relações com variáveis edáficas e o período de inundação no Amapá, Amazônia oriental, Brasil. Doctoral dissertation, Programa de Pós-Graduação em Botânica, INPA, Manaus

Carim, Marcelo de Jesus Veiga; Jardim, Mário Augusto Gonçalves; Medeiros, Tonny David Santiago. 2008. Composição florística e estrutura de floresta de várzea no Município de Mazagão, Estado do Amapá, Brasil. Scientia Florestalis 36(79):191–201

Carretero-Pinzón, Xyomara; Defler, Thomas R. 2016. *Callicebus ornatus*, an endemic Colombian species: demography, behavior and conservation. In: Phylogeny, Molecular Population Genetics, Evolutionary Biology and Conservation of the Neotropical Primates, Ruiz-García, M.; Shostell, J. M. (eds), Nova Science, New York, chapter 16, 1–15

Carretero-Pinzón, Xyomara; Defler, Thomas R.; Ruiz-García, Manuel. 2016. How does the Colombian squirrel monkey cope with habitat fragmentation? Strategies to survive in small fragments. In: Phylogeny, Molecular Population Genetics, Evolutionary Biology and Conservation of the Neotropical Primates, Ruiz-García, M.; Shostell, J. M. (eds), Nova Science, New York, chapter 15, 1–16

Castro, Albejamere Pereira de; Fraxe, Therezinha de Jesus Pinto; Santiago, Jozane Lima; Matos, Raimundo Barbosa; Pinto, Ilzon Castro. 2009. Os sistemas agroflorestais como alternativa de sustentabilidade em ecosistemas de várzea no Amazonas. Acta Amazonica 39(2):279–288

Cattanio, José; Anderson, Anthony B.; Carvalho, Manoel S. 2002. Floristic composition and topographic variation in a tidal floodplain forest in the Amazon estuary. Revista Brasileira de Botanica 25(4):419–430

Cavalcante, P. B. 2010. *Frutas Comestíveis na Amazônia*. Museu Paraense Emílio Goeldi, Belém

Cerón, Carlos Eduardo; Ayala, Consuelo Montalvo. 1998. Etnobotanica de los Huaorani de Quehueiri-ono Napo, Ecuador. Abya-Yala, Quito

Charpentier, Elvis J.; García, Gabriel; Aquino, Rolando. 2015. Uso y competición por plantas alimenticias entre *Pithecia aequatorialis* (Primates: Pitheciidae) y otros animales en la Amazonía peruana. Revista Peruana de Biología 22(2):225–232

Coelho-Ferreira, Márlia. 2009. Medicinal knowledge and plant utilization in an Amazonian coastal community of Marudá, Pará state (Brazil). Journal of Ethnopharmacology 126:159–175

Coelho-Ferreira, Márlia; Jardim, Mário Augusto G. 2005. Algumas espécies vegetais usadas pelos moradores da Ilha de Algodoal, Maiandeua, município de Maracanã, Pará. Boletim do Museu Paraense Emílio Goeldi, Série Ciências Naturais 1(2): 45–51

Costa Neto, Salustiano Vilar da; Bastos, Maria de Nazaré do Carmo; Lobato, Luiz Carlos Batista. 1996. Composição florística e fitofisionomia da restinga do Crispim, município do Marapanim, Pará. Boletim do Museu Paraense Emílio Goeldi, Série Botânica 12(2): 237–249

Cummings, Anthony R.; Read, Jane M. 2016. Drawing on traditional knowledge to identify and describe ecosystem services associated with northern Amazon's multiple-use plants. International Journal of Biodiversity Science, Ecosystem Services & Management 12(1–2):39–56

Cury, Galia Chávez; Gisbert, María del Carmen Abela; Bravo, José A.; Peñarrieta, J. Maurico; Porcel, Willy José Rendón. 2012. Estudio del fruto comestible de la especies vegetal Garcinia madruno. Revista Boliviana de Química 29(1):87–93

Defler, Thomas R.; Defler, Sara B. 1996. Diet of a group of *Lagothrix lagothricha lagothrica* in southeastern Colombia. International Journal of Primatology 17(2):161–190

Dew, J.L. 2005. Foraging, food choice, and food processing by sympatric ripe-fruit specialists: *Lagothrix lagotricha poeppigii* and *Ateles belzebuth belzebuth*. International Journal of Primatology 26(5):1107–1135

Dew, J. Lawrence 2008. Spider monkeys as seed dispersers. In: Spider Monkeys: Behavior, Ecology and Evolution of the Genus *Ateles*, Campbell, Christina J. (ed), Cambridge University Press, Cambridge, pp 155–182

Ferreira, Mariado Socorro; Melo, Marcelo. 2007. Platonia insignis Mart. species richness in secondary forests of north-eastern Pará, Brazil. Bois et Forêts des Tropiques 294(4):21–28

Ferreira, Leandro V.; Chaves, Priscilla P.; Cunha, Denise de A.; Rosário, Alessandro S. do; Parolin, Pia. 2013. A extração ilegal de areia como causa do desaparecimento de campinas e campinaranas no estado do Pará, Brasil. Pesquisas Botânica 64:157–173

Franco, Ana María; Santamaria, Marcela. 1997. Area vital, habitos alimenticios y otros aspectos de historia natural de paujil *Mitu silvini*. In: Manejo de Fauna Silvestre en la Amazonía, Fang, T.G.; Bodmer, R.E.; Aquino, R.; Valqui, M.H. (eds), Instituto de Ecologia, La Paz, Bolivia, pp 267–282

Fraser, J. A., A. B. Junqueira, and C. R. Clement. 2011. Homegardens on Amazonian Darth Earths, non-anthropogenic upland, and floodplain soils along the Brazilian middle Madeira River exhibit diverging agrobiodiversity. *Economic Botany* 65(1):1–12

Galetti, M., C. I. Donatti, M. A. Pizo, and H. C. Giacomini. 2007. Big fish are the best: seed dispersal of Bactris glaucescens by the pacu fish (Piaractus mesopotamicus) in the Pantanal, Brazil. *Biotropica* 40(3):386–389

Gama, João Ricardo Vasconcellos; Botelho, Soraya Alvarenga; Bentes-Gama, Michellinyde Matos. 2002. Composição florística e estrutura da regeneração natural de floresta secundária de várzea baixo no estuário Amazônico. Revista Árvore 26(5):559–566

Gama, J. R. V., S. A. Botelho, M. Bentes-Gama, and J. R. S. Scolforo 2003. Estrutura e potencial futuro de utilização da regeneração natural de floresta de várzea alta no Município de Afuá, estado do Pará. *Ciencia Florestal, Santa Maria* 13(2):71–82

Garcia, Bruna Naiara Rocha; Vieira, Thiago Almeida; Oliveira, Francisco de Assis. 2017. Tree and shrub diversity in agroforestry homegardens in rural community in eastern Amazon. Floresta 47(4):543–552

Gayot, Marc; Henry, Olivier; Dubost, Gérard; Sabatier, Daniel. 2004. Comparative diet of the two forest cervids of the genus *Mazama* in French Guiana. Journal of Tropical Ecology 20(1):31–43

Gonçalves, Janaína Pinheiro; Lucas, Flávia Cristina Araújo. 2017. Agrobiodiversidade e etnoconhecimento em quintais de Abaetetuba, Pará, Brasil. Revista Brasileira de Biociências 15(3):119–134

Gregory, Tremaine. 2006. Comparative socioecology of sympatric, free-ranging white-faced and bearded sakis in Brownsberg Nature Park, Suriname. Doctoral dissertation, Kent State University, Kent, Ohio

Grogan, J. and J. Galvão. 2006. Physiographic and floristic gradients across topography in transitional seasonally dry evergreen forests of southeast Pará, Brazil. *Acta Amazonica* 36(4):483–496
Guèze, Maximilien; Luz, Ana Catarina; Paneque-Gálvez, Jaime; Macía, Manuel J.; Orta-Martínez, Martí; Pino, Joan; Reyes-García, Victoria. 2014. Are ecologically important tree species the most useful? a case study from indigenous people in the Bolivian Amazon. Economic Botany 68(1):1–15
Hajdu, Z. and J. Hohmann. 2012. An ethnopharmacological survey of the traditional medicine utilized in the community of Porvenir, Bajo Paraguá Indian Reservation, Bolivia. Journal of Ethnopharmacology 139:838–857
Huber, J. 1904. Notas sobre a patria e distribuição geographica das arvores fructiferas do Pará. *Boletim do Museu Goeldi (Museu Paraense) de Historia Natural e Ethnographia* 4 (2–3):375–406
Ima, M. O. 2012. *Saberes Waorani y Parque Nacional Yasuní: Plantas, Salud y Bienestar en la Amazonía del Ecuador*. Iniciativa Yasuní ITT, Ministerio Coordinador de Patrimonio, Ministerio del Ambiente, Programa de las Naciones Unidas para el Desarrollo (PNUD) y Fondo para el Medio Ambiente Mundial (FMAM), Quito
Kaminski, Ana Claudia; Absy, Maria Lúcia. 2006. Bees visitors of three species of Clusia (Clusiaceae) flowers in central Amazonia. Acta Amazonica 36(2):259–264
Kawa, Nicholas C.; Michelangeli, José A. Clavijo; Clement, Charles R. 2015. Household agrobiodiversity management on Amazonian dark earths, oxisols, and floodplain soils on the lower Madeira River, Brazil. Human Ecology 43:339–353
Kinupp, V. F. and H. Lorenzi. 2014. *Plantas Alimentícas Não Convencionais (PANC) no Brasil: Guia de Identifação, Aspectos Nutricionais e Receitas Ilustradas*. Instituto Plantarum de Estudos da Flora, São Paulo.
Kinzey, Warren G.; Norconk, Marilyn A. 1990. Hardness as a basis of fruit choice in two sympatric primates. American Journal of Physical Anthropology 81:5–15
Kuchla, Wilian José; Stepka, Thiago Floriani; Sawczuk, Alex Roberto; Lisboa, Gerson dos Santos. 2015. Florística, estrutura horizontal e distribuição diamétrica em área de transição de cerrado e floresta aluvial no município de Campos de Júlio, MT. Ambiência 11(1):13–30
Lau, Arthur Viana; Jardim, Mário Augusto Gonçalves. 2013. Florística e estrutura da comunidade arbórea em uma floresta de várzea na área proteção ambiental, Ilha do Combu, Belém, Pará. Biota Amazônia 3(2):88–93
Le Cointe, P. 1934. *Amazônia Brasileira III: Árvores e Plantas Úteis*. Livraria Classica, Belém
Lima, Roberval M. B. de; Saragoussi, Muriel. 2000. Floodplain home gardens on the central Amazon in Brazil. In: The Central Amazon Floodplain: Actual Use and Options for a Sustainable Management, Junk, W. J.; Ohly, J. J.; Piedade, M. T. F.; Soares, M. G. M. (eds), Backhuys Publishers, Leiden pp 243–268
Lisboa, P. L. B. 2012. *A Terra dos Aruã: Uma História Ecológica do Arquipélago do Marajó*. Museu Paraense Emílio Goeldi, Belém
Lisboa, Pedro L. B.; Pinon, Alerrandson; Silva, Marly Lameira da. 2009. A ocupação humana na APEG e os recursos biológicos (comunidades Porto de Ceasa e Nossa Senhora dos Navegantes). In: Aurá: Comunidades & Florestas, Lisboa, Pedro L. B. (ed), Museu Paraense Emílio Goeldi, Belém, pp 33–61
Lokvam, John; Braddock, Joan F. 1999. Anti-bacterial function in the sexually dimorphic pollinator rewards of Clusia grandiflora (Clusiaceae). Oecologia 119:534–540
Lopes, Célia Regina Araujo Soares; Ribeiro, Ricardo da Silva; Rodrigues, Lucirene; Cabral, Fabiana Ferreira; Silva, Dennis Rodrigues. 2014. Checklist de angiospermas da região de influência da UHE SINOP, médio Teles Pires, Mato Grosso. Enciclopédia Biosfera 10(19):2016–2048
Lorenzi, H. 2008. Árvores Brasileiras: Manual de Identificação e Cultivo de Plantas Arbóreas Nativas do Brasil. Instituto Plantarum, Nova Odessa, São Paulo, vol. 1
Lorenzi, Harri; Bacher, Luis; Lacerda, Marco; Sartori, Sergio. 2006a. Frutas Brasileiras e Exóticas Cultivadas. Instituto Plantarum de Estudos da Flora Ltda, Odessa, Sao Paulo.

Lorenzi, H., L. Bacher, M. Lacerda, S. Sartori 2006b. Brazilian Fruits and Cultivated Exotics (for Consuming in Natura). Instituto Plantarum de Estudos da Flora Ltda, Odessa, São Paulo

Loureiro, Arthur A.; Silva, Marlene F. da; Alencar, Jurandyr da Cruz. 1979. Essências Madeireiras da Amazônia. Instituto Nacional de Pesquisas da Amazônia, Manaus, vol. 1

Lucas, C. M. 2008. Within flood season variation in fruit consumption and seed dispersal by two characin fishes of the Amazon. *Biotropica* 40(5): 581–589.

Luize, Bruno Garcia; Venticinque, Eduardo Martins; Silva, Thiago Sanna Freire; Novo, Evlyn Marcia Leão de Moraes. 2015. A floristic survey of angiosperm species occurring at three landscapes of the central Amazon várzea, Brazil. Checklist: The Journal of Biodiversity Data 11(6):1–15

Luna, Ana Gabriela de; Link, Andrés; Montes, Andrés; Alfonso, Felipe; Mendieta, Leonardo; Di Fiore, Anthony. 2017. Increased folivory in brown spider monkeys Ateles hybridus living in a fragmented forest in Colombia. Endangered Species Research 32:123–134

Machado, Danilo de Oliveira. 2016. A agrobiodiversidade de quintais agroflorestais em propriedades agrícolas familiares na BR 174, Ramal do Pau-Rosa, Manaus, AM. Master's Thesis, INPA, Manaus

Machado, Monielle Alencar; Pinheiro, Claudio Urbano B. 2016. Da água doce à água salgada: mudanças na vegetação de igapó em margens de lagos, rios e canais do baixo curso do rio Pindaré, Baixada Maranhense. Revista Brasileira de Geografia Física 9(5): 1410–1427

Macía, Manuel J.; Fuertes, Juan. 2008. Composición florística y estructura de los árboles en un bosque tropical montano de la Cordillera Mosetenes, Bolivia. Revista Boliviana de Ecología y Conservación Ambiental 23:1–14

Major, Julie; Clement, Charles R.; DiTommaso, Antonio. 2005. Influence of market orientation on food plant diversity of farms located on Amazonian Dark Earth in the region of Manaus, Amazonas, Brazil. Economic Botany 59(1):77–86

Mamede, Jeneffer Soares dos Santos; Pasa, Maria Corette. 2018. Os quintais agroflorestais na comunidade rural em Várzea Grande, Mato Grosso, Brazil. Biodiversidade 17(2):54–70

Marinho, Tatiana Andreza da Silva; Lopes, Aline; Assis, Rafael Leandro de; Ramos, Santiago Linorio Ferreyra; Gomes, Liene Rocha Picanço; Wittmann, Florian; Schöngart. 2013. Distribuição e crescimento de Garcinia brasiliensis Mart. e Hevea spruceana (Benth.) Müll. Arg. em um floresta inundável em Manaus, Amazonas. Ciência Florestal 23(1):223–232

Marx, F.; Maia, J. G. S. 1983. Vitamins in fruits and vegetables of the Amazon. 1. Methods for the determination of B-carotene, tocopherol and ascorbic acid with high performance liquid chromatography (HPLC). Acta Amazonia 13(5–6):823–830

Maués, Márcia Motta; Venturieri, Giorgio C. 1996. Ecologia da colonização do bacurizeiro (Platonia insignis Mart.) Clusiaceae. EMBRAPA-CATU, Boletim de Pesquisa, Belém.

Maués, Bernardo Antônio Rodrigues; Jardim, Mário Augusto Gonçalves; Batista, Fábio de Jesus; Medeiros, Tonny David Santiago; Quaresma, Adriano da Costa. 2011. Composição florística e estrutura do estrato inferior da floresta de várzea na área de proteção ambiental Ilha do Combu, município de Belém, estado do Pará. Revista Árvore 35(3):669–677

Medeiros, Marcelo Brilhante de; Walter, Bruno Machado Teles. 2012. Composição e estrutura de comunidades arbóreas de cerrado stricto sensu no norte do Tocantins e sul do Maranhão. Revista Árvore 36(4): 673–683

Medina, G. and S. Ferreira. 2004. Bacuri (*Platonia insignis* Martius), the Amazonian fruit that has become gold. In: Forest Products, Livelihoods and Conservation: Case Studies of Non-Timber Product Systems, Alexiades, Miguel N.; Shanley, Patricia (eds), CIFOR, Jakarta, vol. 3, pp. 195–210

Mendoza, Abel L. Monteagudo; Guerrero, Marleni Huamán. 2010. Catálogo de los arboles y afines de la selva central del Perú. Arnaldoa 17(2):202–242

Milliken, W., R. P. Miller, S. R. Pollard, and E. V. Wandelli. 1992. *The Ethnobotany of the Waimiri Atroari Indians of Brazil*. Royal Botanic Gardens, Kew

Miranda, Izildinha S. 2000. Análise florística e estrutural da vegetação lenhosa do Rio Comemoração, Pimenta Bueno, Rondônia, Brasil. Acta Amazonica 30(3):393–422

Montalván, Nadiuska Giannina Achong. 2015. Plantas alimenticias usadas por mamíferos mayores en la cuenca alta del Río Itaya, Loreto, Perú. Bachelor thesis, UNAP, Facultad de Ciencias Biológicas, Iquitos

Moser, Pamela. 2013. Vegetação Arbórea e sua Relação com Fatores Ambientais e Especiais em Florestas de Terra Firme no Noroeste de Rondônia, Brasil. Master's thesis, Faculdade de Tecnologia, Departamento de Engenharia Florestal, Programa de Pós-Graduação em Ciências Florestais, Universidade de Brasília

Muñoz, Marianella Guevara. 2010. Evaluación de purmas con fines de manejo en la comunidad nativa de Santa Mercedes, Rio Putumayo, Perú. Master's thesis, Universidad Nacional de la Amazonia Peruana, Iquitos

Muñoz, V.; Sauvain, M.; Bourdy, G.; Callapa, J.; Rojas, I.; Vargas, L.; Deharo, E. 2000. The search for natural bioactive compounds through a multidisciplinary approach in Bolivia. Part II. Antimalarial activity of some plants used by Mosetene indians. Journal of Ethnopharmacology 69:139–155

Murakami, Alejandro Araujo; Reyes, Juan Fermandon; Milliken, William. 2016. Frutales Silvestres y Promisorios de Pando. Herencia/Museo de Historia Natural Noel Kempff Mercado, Cobija

Nascimento, Walnice Maria Oliveira do; Carvalho, José Edmar Urano de; Müller, Carlos Hans. 2007. Ocorrência e distribuição geográfica do bacurizeiros. Revista Brasileira de Fruticultura 29(3):657–660

Nebel, Gustav; Dragsted, Jens; Vanclay, Jerome K. 2000a. Estructura y composición florística del bosque de la llanura aluvial inundable de la Amazonía Peruana: II. El sotobosque de la restinga. Folia Amazónica 10(1–2):151–181

Nebel, Gustav; Kvist, Lars Peter; Vanclay, Jerome K.; Christensen, Henning; Freitas, Luis; Ruiz, Juan. 2000b. Estructura y composición florística del bosque de la llanura aluvial en la Amazonía Peruana: I. El bosque alto. Folia Amazónica 10(1–2):91–149

Nebel, Gustav; Dragsted, Jens; Vanclay, Jerome K. 2001. Structure and floristic composition of flood plain forests in the Peruvian Amazon II. The understorey of restinga forests. Forest Ecology and Management 150: 59–77

Neves, Ana M. S.; Rylands, Anthony B. 1991. Diet of a group of howling monkeys, *Alouatta seniculus*, in an isolated forest patch in Central Amazonia. In: *A Primatologia no Brasil – 3*, Rylands, Anthony B.; Bernardes, Aline T. (eds), Fundação Biodiversitas para a Conservação da Diversidade Biológica, Belo Horizonte, pp 263–274

Norconk, Marilyn A.; Kinzey, Warren G. 1994. Challenge of Neotropical frugivory: travel patterns of spider monkeys and bearded sakis. American Journal of Primatology 34: 171–183

Odonne, Guillaume; Valadeau, Celine; Alban-Castillo, Joaquina; Stien, Didier; Sauvain, Michel; Bourdy, Genevieve. 2013. Medical ethnobotany of the Chayahuita of the Paranapura basin (Peruvian Amazon). Journal of Ethnopharmacology 146:127–153

Ogeron, Clémence; Odonne, Guillaume; Cristinoi, Antonia; Engel, Julien; Grenand, Pierre; Beauchêne, Jacques; Clair, Bruno; Davy, Damien. 2018. Palikur traditional roundwood construction in eastern French Guiana: ethnobotanical and cultural perspectives. Journal of Ethnobiology and Ethnomedicine 14:1–18

Oliveira, Patricia Chaves de; Cavalcante, Suellen. 2017. Ethnobotany in the Amazon floodplain ecosystem: a case study, Quilombo Saracura, Pará, Brazil. International Journal of Botany Studies 2(1):89–99

Paim, Fernanda P.; Chapman, Colin A.; Queiroz, Helder L. de; Paglia, Adriano P. 2017. Does resource availability affect the diet and behavior of the vulnerable squirrel monkey, *Saimiri vanzolinii*? International Journal of Primatology 38(3):572–587

Paim, Fernanda Pozzam; Valenta, Kim; Chapman, Colin A.; Paglia, Adriano Pereira; Queiroz, Helder Lima de. 2018. Tree community structure reflects niche segregation of three parapatric squirrel monkey species (Saimiri spp.). Primates https://doi.org/10.1007/s10329-018-0659-6

Parolin, P. 2000a. Seed mass in Amazonian floodplain forests with contrasting nutrient supplies. Journal of Tropical Ecology 16(3):417–428

Parolin, P. 2000b. O uso de árvores nas florestas inundadas por água branca na Amazônia Central. Amazoniana 16(1/2):241–248

Parolin, P. 2000c. Phenology and C02-assimilation of trees in Central Amazonian floodplains. Journal of Tropical Ecology 16:465–473

Parolin, Pia. 2009. Submerged in darkness: adaptations to prolonged submergence by woody species of the Amazonian floodplains. Annals of Botany 103:359–376

Parolin, P.; Simone, O. De; Haase, K.; Waldhoff, D.; Rottenberger, S.; Kuhn, U.; Kesselmeier, J.; Kleiss, B.; Schmidt, W.; Piedade, M. T. F.; Junk, W. J. 2004. Central Amazonian floodplain forests: tree adaptations in a pulsing system. The Botanical Review 70(3):357–380

Pasa, Maria Corette; Soares, João Juares; Guarim Neto, Germano. 2005. Estudo etnobotânico na comunidade de Conceição-Açu (alto da bacia do rio Aricá Açu, MT, Brasil). Acta Botanica Brasilica 19(2):195–207

Pennington, T. D., C. Reynel, and A. Daza. 2004. *Illustrated Guide to the Trees of Peru.* David Hunt, Sherborne, England

Pereira, Antonio Carlos; Santos, Eduardo Ribeiro dos. 2015. Frutas nativas do Tocantins com potencial de aproveitamento econômico. Agri-Environmental Sciences 1(1):22–37

Pereira, Luciano Araújo; Sobrinho, Felipe de Araújo; Costa Neto, Salustiano Vilar da. 2011. Florística e estrutura de uma mata de terra firme na Reserva de Desenvolvimento Sustentável no Rio Iratapuru, Amapá oriental, Brasil. Floresta 41(1):113–122

Pinto, L. P. 2008. Ecologia alimentar do cuxiú-de-nariz vermelho *Chiropotes albinasus* (Primates: Pitheciidae) na Floresta Nacional do Tapajós, Pará. Doctoral dissertation, Universidade Estadual de Campinas, São Paulo

Politis, G. G. 1996. *Nukak.* Instituto Amazónico de Investigaciones Científicas, Bogotá

Pook, A.G.; Pook, G. 1981. A field study of the socio-ecology of the Goeldi's monkey (Callimico goeldii) in northern Bolivia. Folia Primatologica 36:288–312

Posey, D. A. 1985. Native and indigenous guidelines for new Amazonian development strategies: understanding biological diversity through ethnoecology. In: Change in the Amazon Basin Volume 1: Man's Impact on Forests and Rivers, Hemming, John (ed), Manchester University Press, Manchester, pp 156–181

Pulido, María T.; Cavelier, Jaime. 2001. Comercialización de productos vegetales en los mercados de Leticia y Florencia, Amazonía colombiana. In: Evaluación de Recursos Vegetables no Maderables en la Amazonía Noroccidental, Duivenvoorden, Joost F.; Balslev, Henrik; Cavelier, Jaime; Grandez, César; Tuomisto, Hanna; Valencia, Renato (eds), Institute for Biodiversity and Ecosystem Dynamics, Amsterdam, pp. 265–310

Quaresma, Amanda Paiva; Almeida, Ruth Helena Cristo; Oliveira, Cyntia Meireles de; Kato, Osvaldo Ryohei. 2015. Composição florística e faunística de quintais agroflorestais da agricultura familiar no nordeste Paraense. Revista Verde de Agroecologia e Desenvolvimento Sustentável 10(5):76–84

Queiroz, José Antonio Leite de; Machado, Sebastião do Amaral. 2007. Potencial de utilização madeireiras de espécies florestais de várzea no município de Mazagão no estado do Amapá. Floresta 37(2):293–302

Queiroz, José Antonio Leite; Machado, Sebastião do Amaral. 2008. Fitossociologia em floresta de várzea do estuário Amazônico no estado do Amapá. Pesquisa Florestal Brasileira 57:5–20

Queiroz, José Antonio Leite de; Mochiutti, Silas; Machado, Sebastião do Amaral; Galvão, Franklin. 2005. Composição florística e estrutura de floresta em várzea alta estuarina Amazônica. Floresta 35(1):41–55

Queiroz, José Antonio Leite de; Mochiutti, Silas; Machado, Sebastião do Amaral. 2006. Distribuição diamétrica em floresta de várzea do braço norte do Rio Amazonas. Revista de Ciência Agrária 46:151–170

Queiroz, José Antonio Leite de; Machado, Sebastiã do Amaral; Hosokawa, Roberto Tuyoshi; Silva, Ivan Crespo da. 2007. Estrutura e dinâmica de floresta de várzea no estuário Amazônico no estado do Amapá. Floresta 37(3):339–352

Ratter, J. A. 1987. Notes on the vegetation of the Parque Nacional do Araguaia (Brazil). Notes from the Royal Botanic Garden, Edinburgh 44(2):311–342

Rayol, Breno Pinto; Alvino-Rayol, Fabrízia de Oliveira; Silva, Manoela Ferreira Fernandes da. 2011. Similaridade florística entre o estrato arbóreo e a regeneração natural de uma floresta secundária, no município de Bragança, nordeste do estado do Pará, Brasil. Revista Brasileira de Agroecologia 6(3):107–114

Rayol, Breno Pinto; Alvino-Rayol, Fabrízia de Oliveira; Silva, Maria da Conceição Pereira da. 2013. Caracterização de quintais agroflorestais desenvolvidos por agricultores familiares do município de Alenquer, Pará. Cadernos de Agroecologia 8(2):1–5

Reyes-García, Victoria. 2001. Indigenous people, ethnobotanical knowledge, and market economy: a case study of the Tsimane' Amerindians in lowland Bolivia. Doctoral dissertation, University of Florida, Gainesville.

Reynel, C; Pennington, T.D.; Pennington, R.T.; Flores, C.; Daza, A. 2003. *Árboles Útiles de la Amazonía Peruana y sus Usos: Un Manual con Apuntes de Identificación, Ecología y Propagación de las Especies*. Tarea Gráfica Educativa, Lima

Roble, Nick. 2008. Ontogenetic resource partitioning in white faced sakis (*Pithecia pithecia*). Master's thesis, Kent State University.

Robles, Rembrandt Ramiro Canahuire. 2017. Composición florística y estructura de la recuperación natural en un área degradado por minería en la comunidad nativa de Tres Islas, Tambopata, Madre de Dios. Master's thesis, Escuela Profesional de Ingeniería Forestal y Medio Ambiente, Puerto Maldonado, Perú

Rodrigues, Maira dos Santos; Conceição, Gonçalo Mendes da. 2014. Diversidade florística das diferentes fisionomias de cerrado do Parque Estadual do Mirador, Maranhão, Brasil. Brazilian Geographical Journal: Geosciences and Humanities research medium, Ituiutaba 5(1):139–156

Rodrigues, Ricardo Ribeiro; Martins, Sebastião Venâncio; Barros, Luiz Carlos de. 2004. Tropical rain forest regeneration in an area degraded by mining in Mato Grosso state, Brazil. Forest Ecology and Management 190: 323–333

Rogez, Hervé; Buxant, Raphaële; Mignolet, Eric; Souza, Jesus N. S.; Silva, Evaldo M.; Larondelle, Yvan. 2004. Chemical composition of the pulp of three typical Amazonian fruits: araça-boi (Eugenia stipitata), bacuri (Platonia insignis) and cupuaçu (Theobroma grandiflorum). European Food Research and Technology 218:380–384

Rosa, Leonilde dos Santos; Silveira, Erycélede Lima; Santos, Mônica Mota dos; Modesto, Rozi da Silva; Perote, James Richard Silva; Vieira, Thiago Almeira. 2007. Os quintais agroflorestais em áreas de agricultores familiares no município de Bragança-PA: composição florística, uso de espécies e divisão de trabalho familiar. Revista Brasileira de Agroecologia 2(2):337–341

Rosa, Leonilde dos Santos; Vieira, Thiago Almeida; Pires, Helaine Cristine Gonçalves. 2009. Quintais agroflorestais em comunidades rurais de Bonito, Pará. Revista Brasileira de Agroecologia 4(2):1310–1313

Rudas, Augustín; Prieto, Adriana. 1998. Análisis florístico del parque nacional natural Amacayacu e Isla Mocagua, Amazonas (Colombia). Caldasia 20(2):142–172

Salas, Leonardo A.; Fuller, Todd K. 1996. Diet of the lowland tapir (Tapirus terrestris L.) in the Tabaro River valley, southern Venezuela. Canadian Journal of Zoology 74: 1444–1451

Salomão, Rafael de Paiva; Vieira, Ima Célia Guimarães; Suemitsu, Chieno; Rosa, Nélson de Araújo; Almeira, Samuel Soares de; Amaral, Dário Dantas do; Menezes, Moirah Paulo Machado de. 2007a. As florestas de Belo Monte na grande curva do rio Xingu, Amazônia Oriental. Boletim do Museu Paraense Emílio Goeldi, Ciencias Naturais 2(3):57–153

Salomão, Rafael Paiva; Rosa, Nélson Araújo; Morais, Kácio Andrey Câmara. 2007b. Dinâmica da regeneração natural de árvores em áreas mineradas na Amazônia. Boletim do Museu Paraense Emílio Goeldi, Ciencias Naturais 2(2):85–139

Santa-Cecília, Flávia V.; Vilela, Fabiana C.; Rocha, Cláudia Q. da; Dias, Danielle F.; Cavalcante, Gustavo P.; Freitas, Lissara A. S.; Santos, Marcelo H. dos; Giusti-Paiva, Alexandre. 2011. Anti-inflammatory and antinociceptive effects of *Garcinia brasiliensis*. Journal of Ethnopharmacology 133:467–473

Santos, Luciana Oliveira dos. 2010. Comparação da vegetação lenhosa entre borda e interior de uma mancha de mata de restinga na RESEX Mar Caeté-Taperaçu, Bragança, Pará. Master's thesis, Universidade Federal Rural da Amazônia/Museu Paraense Emílio Goeldi, Belém

Santos, Josefa Silva dos; Guarim Neto, Germano. 2017. Plantas de quintais da área urbana de Comodoro, Mato Grosso, Brasil. Flovet 1(9):31–57

Santos, L. O., L. O. Santos, M. P. M. Menezes, C. R. Beasley, and U. Mehlig. 2014. Composition and structure of a diverse tree community at the edges of a Brazilian Amazon rainforest island surrounded by marshes and mangroves. Plant Ecology 215: 1469–1481

Santos, Otilene dos Anjos; Couceiro, Sheyla Regina Marques; Rezende, Alinne Costa Cavalcante; Silva, Márcia Daniela de Sousa. 2016. Composition and richness of woody species in riparian forests in urban areas of Manaus, Amazonas, Brazil. Landscape and Urban Planning 150:70–78

Santos, Edyane Moraes dos; Monteiro Neto, Carlos Alberto de Sampaio; Nascimento, Claudio Adriano de Jesus; Muniz, Francisca Helena; Barros, José de Ribamar Silva. 2017a. Genetic and population diversity of bacuri (Platonia insignis Mart.) in Chapada Limpa extractive reserve, Maranhão State, Brazil. African Journal of Biotechnology 16(50): 2317–2325

Santos, Ronaldo Oliveira dos; Lima, Robson Carmo; Lima, Robson Borges de; Aparício, Perseu da Silva; Abreu, Jadson Coelho de. 2017b. Florística e estrutura de uma comunidade arbórea na floresta estadual do Amapá, Amazônia Oriental, Brasil. Nativa Sinop 5:529–539

Schaan, D. 2012. Sacred Geographies of Ancient Amazonia: Historical Ecology of Social Complexity. Left Coast Press, Walnut Creek, California

Schultes, R. E. and R. F. Raffauf. 1990. The Healing Forest: Medicinal and Toxic Plants of the Northwest Amazonia. Dioscorides Press, Portland, Oregon

Schurz, W.L.; Hargis, O. D.; Marbut, C. F.; Manifold, C. B. 1925. Rubber production in the Amazon valley. Government Printing Office, Washington, D. C.

Scudeller, Veridiana Vizoni; Souza, Adriana Mota Gomes de. 2009. Florística da mata de igapó na Amazônia Central. In: Biotupé: Meio Físico, Diversidade Biológica e Sociocultural do Baixo Rio Negro, Amazônia Central, Santos-Silva, Edinaldo Nelson; Scudeller, Veridiana Vizoni (eds), pp 97–108, UEA Edições, Manaus

Shanley, P. and N.A. Rosa. 2004. Eroding knowledge: an ethnobotanical inventory in eastern Amazonia's logging frontier. Economic Botany 58(2):135–160

Shanley, P. and N.A. Rosa. 2005. Conhecimento em erosão: um inventário etnobotânico na fronteira de exploração da Amazônia oriental. Boletim do Museu Paraense Emílio Goeldi, Série Ciências Naturais 1(1):147–171

Shanley, Patricia; Luz, Leda; Swingland, Ian R. 2002. The faint promise of a distant market: a survey of Belém's trade in non-timber forest products. Biodiversity and Conservation 11:615–636

Silva, Suzy Cristina Pedroza da. 2006. Os sistemas agroflorestais na Reserva de Desenvolvimento Sustentável Piagaçu-Purus: uma alternativa para a agricultura familiar. Master's thesis, Universidade Federal do Amazonas, Manaus

Silva, Maria Betânia Araujo da. 2008. Composição florística e estrutura fitossociológica de uma floresta ripária situada às margens do Rio Moju, Pará. Master's thesis, Universidade Federal Rural da Amazônia, Belém

Silva, Silvestre. 2011. Fruits of the Brazilian Amazon. Metalivros, São Paulo

Silva, Edinei Santos da. 2017. Plantas alimentícias em comunidades agrícolas no município de Rio Preto da Eva-AM. Master's thesis, INPA, Manaus

Silva, M. F., P. L. B. Lisbôa, R. C. L. Lisbôa 1977. Nomes Vulgares de Plantas Amazônicas. INPA, Manaus

Silva, Antônio Sérgio Lima da; Lisboa, Pedro L. B.; Maciel, Ubirajara N. 1992. Diversidade florística e estrutura em floresta densa da bacia do rio Juruá-AM. Boletim do Museu Paraense Emílio Goeldi, Série Botânica 8(2):203–258

Silva, Rachel Macedo da; Mehlig, Ulf; Santos, João Ubiratan Moreira dos; Menezes, Moirah Paula Machado de. 2010. The coastal restinga vegetation of Pará, Brazilian Amazonia: a synthesis. Revista Brasileira de Botanica 33(4):563–573

Silva, C. R. S.; Albuquerque, PSB; Ervedosa, FR; Mota, JWS; Figueira, A; Sebbenn, AM. 2011. Understanding the genetic diversity, spatial genetic structure and mating system at the hierarchical levels of fruits and individuals of a continuous *Theobroma cacao* population from the Brazilian Amazon. Heredity 106: 973–985

Silva, Katia Emidio da; Martins, Sebastião Venâncio; Ribeiro, Carlos Antônio Alvares Soares; Santos, Nerilson Terra; Azevedo, Celso Paulo de. 2016. Structure of 15 hectares permanent plots of terra firme dense forest in central Amazon. Revista Árvore 40(4):603–615

Simmen, Bruno; Sabatier, Daniel. 1996. Diets of some French Guianan primates: food composition and food choices. International Journal of Primatology 17(5):661–693

Slinger, Vanessa A. V. 2000. Peri-urban agroforestry in the Brazilian Amazon. Geographical Review 90(2):177–190

Smith, N. J. H. 2002. *Amazon Sweet Sea: Land, Life and Water at the River's Mouth*. University of Texas Press, Austin.

Smith, N., R. Vásquez, and W. H. Wust. 2007. *Amazon River Fruits: Flavors for Conservation*. Amazon Conservation Association/Missouri Botanical Garden, Lima

Souza, Célia Cristina Valero. 2010. Etnobotânica de quintais em três comunidades ribeirinhas na Amazônia central, Manaus, AM. Master's thesis, INPA, Manaus

Souza, A. G. C., N. R. Sousa, S. E. L. Silva, C. D. M. Nunes, A. C. C. Canto, L. A. A. Cruz, 1996. *Fruteiras da Amazônia*. EMBRAPA (Empresa Brasileira de Agropecuária), Brasília

Souza, Antonio Maricélio Borges de; Alves, Keila de Nazaré; Lobato, Whesley Thiago dos Santos; Leal, Ayla Julliane da Silva; Almeida, Gabriela Mourão de; Souza, Antonio Alan Santos de; Mota, Adriano Vitti. 2017. Aspectos da segurança alimentar com base em quintais agroflorestais na comunidade rural de Santa Luzia do Induá no município de Capitão Poço, PA. Agroecosistemas 9(2): 275–287

Stefanello, Daniel; Fernandes-Bulhão, Clarissa; Martins, Sebastião Venâncio. 2009. Síndromes de dispersão de sementes em três trechos de vegetação ciliar (nascente, meio e foz) ao longo do Rio Pindaíba, MT. Revista Árvore 33(6):1051–1061

Stevenson, Pablo R. 2000. Seed dispersal by woolly monkeys (*Lagothrix lagothricha*) at Tinigua National Park, Colombia: dispersal distance, germination rates, and dispersal quantity. American Journal of Primatology 50: 275–289

Stevenson, Pablo R. 2005. Potential keystone plant species for the frugivore community at Tinigua Park, Colombia. In: Tropical Fruits and Frugivores: The Search for Strong Interactors, Dew, J. Lawrence; Boubli, Jean Philippe (eds), Springer, Dordrecht, pp 37–57

Stevenson, Pablo R. 2007. Estimates of the number of seeds dispersed by a population of primates in a lowland forest in western Amazonia. In: Seed Dispersal: Theory and its Application in a Changing World, Dennis, Andrew J.; Schupp, Eugene W.; Green, Ronda J.; Westcott, David A. (eds), CABI, Wallingford, pp 340–362

Stevenson, Pablo R.; Quiñones, Marcela J.; Ahumada, Jorge A. 1994. Ecological strategies of woolly monkeys (*Lagothrix lagotricha*) at Tinigua National Park, Colombia. American Journal of Primatology 32:123–140

Stevenson, P.R.; Pineda, M.; Samper, T. 2005. Influence of seed size on dispersal patterns of woolly monkeys (*Lagothrix lagotricha*) at Tiningua Park, Colombia. *Oikos* 110: 435–440

Thomas, Evert; Van Damme, Patrick. 2010. Plant use and management in homegardens and swiddens: evidence from the Bolivian Amazon. Agroforestry Systems 80:131–152

Vale, Gabriel Damasco do. 2011. O efeito do solo, inundação e topografia sobre as campinaranas inundáveis na Amazônia. Master's Thesis, INPA, Manaus

Van Andel, Tinde; Veldman, Sarina; Maas, Paul; Thijsse, Gerard; Eurlings, Marcel. 2012. The forgotten Hermann herbarium: A 17th century collection of useful plants from Suriname. Taxon 61(6):1296–1304

Van den Berg, M. E. 2010. Plantas Medicinais na Amazônia. Museu Paraense Emílio Goeldi, Belém

Van den Berg, M.E. and M.H.L. da Silva. 1986. Contribuição ao conhecimento da flora medicinal do Maranhão. In: Anais do1°Simpósio do Trópico Úmido, Belém, PA, 12 a 17 November, vol. 2, Flora e Floresta, EMBRAPA, Brasília, pp 119–125

Van den Eynden, Veerle; Cueva, Eduardo; Cabrera, Omar. 1999. Plantas Silvestres Comestibles del Sur del Ecuador. Ediciones Abya-Yala, Quito

Van Roosmalen, M. G. M. 1985. Habitat preferences, diet, feeding strategy and social organization of the black spider monkey (*Ateles paniscus paniscus* Linnaeus 1758) in Surinam. Acta Amazonica 15(3/4 suplemento) 1–238

Van Roosmalen, Marc G. M.; Klein, Lewis L. 1988. The spider monkeys, genus *Ateles*. In: Ecology and Behavior of Neotropical Primates, Mittermeier, Russell A.; Rylands, Anthony B.; Coimbra-Filho, Adelmar F.; Fonseca, Gustavo A.B. da (eds), Ecology and Behavior of Neotropical Primates, World Wildlife Fund, Washington, DC, vol. 2, pp 455–537

Van Roosmalen, Marc G. M.; Mittermeier, Russel A.; Milton, Katherine. 1981. The bearded sakis, genus *Chiropotes*. In: Ecology and Behavior of Neotropical Primates, Coimbra-Filho, Adelmar F.; Mittermeier, Russell A. (eds), Academia Brasileira de Ciências, Rio de Janeiro, vol. 1, pp 419–441

Van Roosmalen, Marc G. M.; Mittermeier, Russel A.; Fleagle, John G. 1988. Diet of the northern bearded saki (*Chiropotes satanas chiropotes*): a Neotropical seed predator. American Journal of Primatology 14:11–35

Vásquez, Juan Baluarte. 1998. Composición florística de los bosques secundarios de Jenaro Herrera, Loreto. Folia Amazonica 9(1–2):195–218

Vásquez, R. and A. H. Gentry. 1989. Use and misuse of forest-harvested fruits in the Iquitos area. *Conservation Biology* 3(4): 350–361.

Wittmann, Florian; Schöngart, Jochen; Queiroz, Helder L.; Wittmann, Astrid de Oliveira; Conserva, Auristela dos Santos; Piedade, Maria T. F.; Kesselmeier, Jürgen; Junk, Wolfgang J. 2009. The Amazon floodplain demonstration site: sustainable timber production and management of central Amazonian white-water floodplains. Ecohydrology & Hydrobiology 9(1):41–54

Wittmann, Florian; Householder, Ethan; Piedade, Maria T. F.; Assis, Rafael L. de; Schöngart, Jochen; Parolin, Pia; Junk, Wolfgang J. 2013. Habitat specifity, endemism and the neotropical distribution of Amazonian white-water floodplain trees. Ecography 36:690–707

Yamada, M.; Osaqui, H. M. L. 2006. The role of homegardens in agroforestry development: lessons from Tomé-Açu, a Japanese-Brazilian settlement in the Amazon. In: Tropical Homegardens: a time-tested example of sustainable agroforestry, Kumar, B. M.; Nair, P. K. R. (eds), pp. 299–316, Advances in Agroforestry vol. 3, Springer, Dordrecht

Zárate, Ricardo; Mori, Tony J.; Maco, José T. 2013. Estructura y composición florística de las comunidades vegetales del ámbito de la Carretera Iquitos-Nauta, Loreto-Perú. Folia Amazónica 22(1–2):77–89

Zárate, Ricardo; Mori, Tony Jonatan; Ramirez, Fredy Francisco; Davila, Hilda Paulette; Gallardo, George Pepe; Cohello, Geancarlo. 2015. Lista actualizada y clave para la identificación de 219 especies arbóreas de los bosques sobre arena blanca de la Reserva Nacional Allpahuayo Mishana, Loreto, Peru. Acta Amazonica 45(2):133–156

Chapter 18
Combretaceae

18.1 *Buchenavia grandis*

Brazil Tanimbuca, tanibuca da folha grande, cuiarana, cuiarana de caroço, cuiraneiras, marindiba, mirindiba, mirindiba amarela, mirindiba grande, mirindiba folha miúda; piriquiteira; katxirmia (Waimiri-Atroari)

Ecuador Torro orito caspi (Kichwa), kawe yakawee (Waorani)

Peru Ushun, yacushapana, tacho, lagartillo de altura, porotillo, chimicua

Venezuela Apamate amarillo rebalsero

Status Wild

N. Smith, *Amazon Fruits: An Ethnobotanical Journey*, Ethnobiology,
https://doi.org/10.1007/978-3-031-12803-5_18

Ushun *Buchenavia grandis* tree in floodplain forest along a black water river during the low water season. 800 m east of San Martin Tipishca, Rio Samiria, Reserva Nacional Pacaya-Samiria, Loreto, Peru, 6-30-06

Supported by impressive buttress roots, *Buchenavia grandis* is a canopy emergent in floodplain and upland forests of the Amazon Basin, the Guianas, and the Atlantic Forest of Brazil (Almeida et al. 1993; Amaral et al. 2009; Assis et al. 2017; Carneiro 2004; Condé and Tonini 2013; Corrêa 2006:128; Gama et al. 2007; Lemos et al. 2015; Lima et al. 2012; Machado 2009:28; Santos et al. 2017; Silva et al. 1986, 1992; Souza 2012:49; Vásquez 1997:232). It is also found in upland successionary forest, such as in southern Roraima (Condé 2011:75) and in the Tapajós National Forest in western Pará (Gualberto et al. 2014). It occurs on the floodplains of various water types, including clear water rivers such as the lower Xingu Jari (Carim 2016:88; Cunha 2009:48; Cunha and Ferreira 2012), sediment-rich rivers such as the Upper Madeira (Casula 2012:56), as well as black water rivers such as in the Jaú National Park in the Rio Negro watershed and in the Uatumã watershed, both in the Brazilian state of Amazonas (Barnett et al. 2015; Lobo 2017:61). It also grows in stunted woodlands (*campinarana*) on sandy soils in Roraima (Vale 2011:44) and along the Iquitos-Nauta highway in Loreto, Peru. Its range extends up to 850 m in the Andean foothills of the Peruvian Amazon (Mendoza and Guerrero 2010).

Known as ushun along the Ucayali and Marañón Rivers, this forest giant bears tan-colored fruits that fall to the ground during the rainy season. The oblong fruits, some 2.5 cm long by 1.5 cm wide, have a texture of fresh figs except they are exceptionally juicy. Locals, including indigenous groups such as the Waimiri Atroari in Amazonas, Brazil (Milliken et al. 1992:25), peel the thin skin to savor the yellow pulp that surrounds the single seed. The mesocarp is refreshing and has a unique aroma like an exotic perfume. In addition to eating the fruit fresh, they are made into juice (*refresco* in Peru). Ushun fruits stain the hands black, but unlike genipap *Genipa americana*, the dye washes off easily.

Percy and Walmer, both 12 years, gathering ushun *Buchenavia grandis* fruits in floodplain forest close to their village. Near Yarina, Rio Yanayacu, Reserva Nacional Pacaya-Samiria, Loreto, Peru 3-7-04

Buchenavia grandis also contributes to the human diet indirectly through fish and some game animals. Tambaqui *Colossoma macropomum*, for example, swallow the fruits whole and then disperse the seed (Silva et al. 2003). Tapirs *Tapirus terrestris*, the largest game animal in Amazonia, also allegedly eat the fruits and disperse the seeds in their feces. Other game animals, such as howler monkeys (*Alouatta* spp.), woolly monkeys *Lagothrix lagotricha*, and spider monkeys (*Ateles* spp.) also eat the fruits (Defler and Defler 1996; Milliken et al. 1992:66).

Ushun would be a promising candidate for domestication, except that it is a large tree that can soar for 50 m and may take a decade or so before it starts producing fruit. Given the fruit's unusual flavor, ushun might still be a worthwhile addition to agroforestry systems as long as other quicker maturing crops were interplanted. Another favorable aspect of ushun for domestication is that the wood provides valuable timber so older trees that no longer produce good fruit crops could be sold to sawmills.

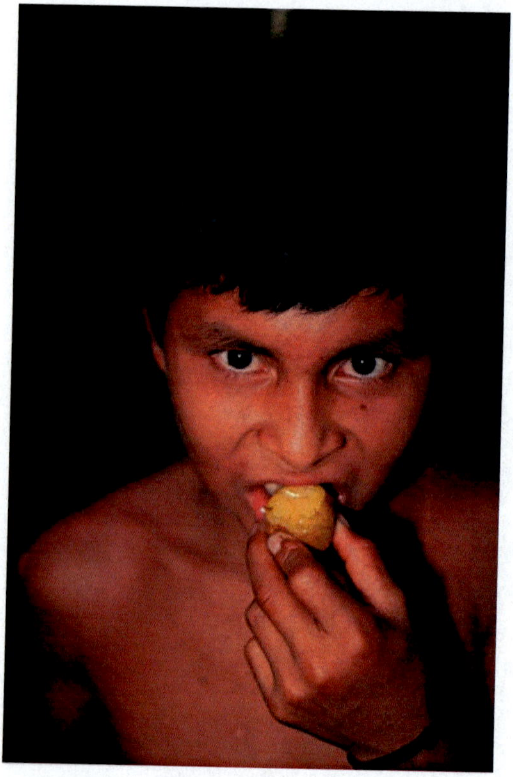

Nerio, 13 years, eating ushun *Buchenavia grandis* fruit he gathered in floodplain forest. Near Yarina, Rio Yanayacu, Reserva Nacional Pacaya-Samiria, Loreto, Peru 3-7-04

References

Almeida, Samuel S.; Lisboa, Pedro Luiz B.; Silva, Antônio Sérgio. 1993. Diversidade florística de uma comunidade arbórea na estação científica "Ferreira Penna", em Caxiuanã (Pará). Boletim do Museu Paraense Emilio Goeldi, Série Botanica 9(1):93–128

Amaral, Dário Dantas do; Vieira, Ima Célia Guimarães; Almeida, Samuel Soares de; Salomão, Rafael Paiva; Silva, Antonio Sérgio Lima da; Jardim, Mário Augusto Gonçalves. 2009. Checklist da flora arbórea de remanescentes florestais da região metropolitana de Belém e valor histórico dos fragmentos, Pará, Brasil. Boletim do Museu Paraense Emílio Goeldi, Ciencias Naturais 4(3):231–289

Assis, Rafael Leandro de; Wittmann, Florian; Luize, Bruno Garcia; Haugaasen, Torbjon. 2017. Patterns of floristic diversity and composition in floodplain forests across four southern Amazon river tributaries, Brazil. Flora 229:124–140

Barnett, Adrian A.; Silva, Welma Sousa; Shaw, Peter J. A.; Ramsay, Paul M. 2015. Inundation duration and vertical vegetation zonation: a preliminary description of the vegetation and structuring factors in borokotóh (hummock igapó), an overlooked high-diversity, Amazonian vegetation association. Nordic Journal of Botany 33:601–614

Carim, Marcelo de Jesus Veiga. 2016. Estrutura, composição e diversidade em florestas alagáveis de várzea de maré e de igapó e suas relações com variáveis edáficas e o período de inundação no Amapá, Amazônia oriental, Brasil. Doctoral dissertation, Programa de Pós-Graduação em Botânica, INPA, Manaus

Carneiro, Vilany Matilla Colares. 2004. Composição florística e análise estrutural da floresta primária de terra firme na bacia do Rio Cuieiras, Manaus-AM. Master's Thesis, Universidade Federal do Amazonas (UFAM)/Instituto Nacional de Pesquisas da Amazônia (INPA), Manaus

Casula, Kátia Regina. 2012. Análise da similaridade florística e estrutural das formações florestais inundáveis em um trecho do alto Rio Madeira e de seus afluentes, estado de Rondônia, Brasil. Master's Thesis, Núcleo de Ciências Exatas da Terra, Universidade Federal de Rondônia, Porto Velho

Condé, Tiago Monteiro. 2011. Avaliação dos impactos da vegetação após a exploração madeireira em floresta ombrófila de terra firme no Município de Caracaraí, RR. Master's thesis, Programa de Pós-Graduação em Recursos Naturais, Universidade Federal de Roraima, Boa Vista

Condé, Tiago Monteiro; Tonini, Helio. 2013. Fitossociologia de uma floresta ombrófila densa na Amazônia setentrional, Roraima, Brasil. Acta Amazonica 43(3):247–260

Corrêa, Honorly Kátia Mestre. 2006. Ecologia de dois grupos de sagüis-brancos, *Mico argentatus* (Linnaeus 1771) em um fragmento florestal natural, Santarém-Pará. Doctoral dissertation, Universidade Federal do Pará/Museu Paraense Emílio Goeldi, Belém

Cunha, D. 2009. Análise da paisagem e estrutura das comunidades vegetais das formações pioneiras do baixo Xingu, Pará. Master's thesis, Universidade Federal Rural da Amazônia/Museu Paraense Emílio Goeldi, Belém

Cunha, Denise de Andrade; Ferreira, Leandro Valle. 2012. Impacts of the Belo Monte hydroelectric dam construction on pioneer vegetation formations along the Xingu River, Pará state, Brazil. Brazilian Journal of Botany 35(2):159–167

Defler, Thomas R.; Defler, Sara B. 1996. Diet of a group of *Lagothrix lagothricha lagothrica* in southeastern Colombia. International Journal of Primatology 17(2):161–190

Gama, João Ricardo Vasconcellos; Souza, Agostino Lopesde Souza; Calegário, Natalino; Lana, Guilherme Campos. 2007. Fitossociologia de duas fitocenoses de floresta ombrófila aberta no município de Codó, estado do Maranhão. Revista Árvore 31(3):465–477

Gualberto, Midiã Larine Colares; Ribeiro, Renato Bezerra da Silva; Gama, João Ricardo Vaconcellos; Vieira, Diego dos Santos. 2014. Fitossociologia e potencial de espécies arbóreas em ecossistema sucessional na Floresta Nacional do Tapajós, Pará. Agroecossistemas 6(1):42–57

Lemos, D. A. N.; Ferreira, B. G. A.; Siqueira, J. D. P.; Oliveira, M. M.; Ferreira, A. M. 2015. Floristic and phytosociology in dense "terra firme" rainforest in the Belo Monte hydroelectric plant influence area, Pará, Brazil. Brazilian Journal of Biology 75(3):257–276
Lima, R. B. A.; Silva, J. A. A.; Marangon, L. C.; Ferreira, R. L. C.; Silva, R. K. S. 2012. Fitossociologia de um trecho de floresta ombrófila densa na Reserva de Desenvolvimento Sustentável Uacari, Carauari, Amazonas. Scientia Plena 8(1):1–12
Lobo, Guilherme de Sousa. 2017. A alteração do regime hidrológico afeta a composição florística e estrutura de florestas de igapó? Um estudo comparativo entre um rio regulado e outro prístino na Reserva de Desenvolvimento Sustentável do Uatumã, Amazônia Central. Master's thesis, INPA, Manaus
Machado, Kathleen Samira da Silva. 2009. Estrutura e composição florística de uma floresta de terra firme na reserva de desenvolvimento sustentável Amanã, Amazônia Central. Master's thesis, INPA, Manaus
Mendoza, Abel L. Monteagudo; Guerrero, Marleni Huamán. 2010. Catálogo de los arboles y afines de la selva central del Perú. Arnaldoa 17(2):202–242
Milliken, W., R. P. Miller, S. R. Pollard, and E. V. Wandelli. 1992. *The Ethnobotany of the Waimiri Atroari Indians of Brazil*. Royal Botanic Gardens, Kew
Santos, Ronaldo Oliveira dos; Lima, Robson Carmo; Lima, Robson Borges de; Aparício, Perseu da Silva; Abreu, Jadson Coelho de. 2017. Florística e estrutura de uma comunidade arbórea na floresta estadual do Amapá, Amazônia Oriental, Brasil. Nativa Sinop 5:529–539
Silva, Manoela F. F. da; Rosa, Nelson A.; Salomão, Rafael de P. 1986. Estudos botânicos na área do Projeto Ferro Carajás. 3. Aspectos florísticos da mata do aeroporto de Serra Norte – PA. Boletim do Museu Paraense Emílio Goeldi, Série Botânica 2(2):169–187
Silva, Antônio Sérgio Lima da; Lisboa, Pedro L. B.; Maciel, Ubirajara N. 1992. Diversidade florística e estrutura em floresta densa da bacia do rio Juruá-AM. Boletim do Museu Paraense Emílio Goeldi, Série Botânica 8(2):203–258
Silva, J. M., M. Pereira-Filho, and M. I. Oliveira-Pereira. 2003. Valor nutricional e energético de espécies vegetais importantes na alimentação do tambaqui. *Acta Amazonica* 33(4):687–700
Souza, Priscila Figueira de. 2012. Florística, distribuição e diversidade beta da comunidade arbórea ao longo do gradiente hídrico em uma floresta de terra firme no interflúvio Purus-Madeira, Amazônia central. Master's thesis, INPA, Manaus.
Vale, Gabriel Damasco do. 2011. O efeito do solo, inundação e topografia sobre as campinaranas inundáveis na Amazônia. Master's Thesis, INPA, Manaus
Vásquez, R. 1997. *Flórula de las Reservas Biológicas de Iquitos, Perú*. Monographs in Systematic Botany Vol. 63, Missouri Botanical Garden, St. Louis.

Chapter 19
Convolvulaceae

19.1 *Dicranostyles scandens*

Brazil Tatapunha, gogozinho

Peru Motelo huayo

Venezuela Melón orillero, yotáripê; mapaglidún (Curipaco), Meláo (Baniwa), mamánga (Yeral)

Status Wild

Tatapunha *Dicranostyles scandens* vine in fruit along a black water river. Rio Curuá, Caxiuanã National Forest, Município de Melgaço, Pará, Brazil 5-4-15

N. Smith, *Amazon Fruits: An Ethnobotanical Journey*, Ethnobiology,
https://doi.org/10.1007/978-3-031-12803-5_19

Tatapunha *Dicranostyles scandens* is a woody liana that grows along river banks and in upland forests in many parts of the Amazon Basin from Brazil to Venezuela, Colombia, Ecuador, and Bolivia. Tatapunha produces edible fruits that are gathered from canoes. The olive-shaped fruits measuring 2 cm by 1.5 cm have a sweet, black pulp surrounding a single dark-brown seed, 1.5 cm × 1 cm. The dark purple fruits are produced during the rainy season and are likely dispersed by birds, mammals, as well as fish when the river levels rise and cover the fruits.

Tatapunha vine in fruit along a black water river. Rio Curuá, Caxiuanã National Forest, Município de Melgaço, Pará, Brazil 5-4-15

19.2 *Ipomoea squamosa*

Bolivia Flor de camote

Brazil Zitik-ran (Tembé)

Peru Camutillo, camote huasca; inchinchi (Achuar), uchi inchinchi (Aguaruna)

Status Wild

Percy, 11 years, gathering camutillo *Ipomoea squamosa* **fruit for fish bait. Near Yarina, Rio Yanayacu, Reserva Nacional Pacaya-Samiria, Loreto, Peru 3-7-04**

In the Peruvian Amazon, river dwellers pick the fruits of camutillo *Ipomoea squamosa* to bait their hooks when pole or handline fishing. The baited hooks are periodically pulled in and dropped again into the water so that the plop sound imitates falling fruit to attract fish, such as palometa, known as pacu in Brazil (species of *Mylossoma*, *Myleus*, and *Metynnis*). Camutillo is a bushy vine found along water courses, including black and clear water rivers, as well as in successionary growth on uplands. People generally do not eat the pea-sized fruits which are light green or pale yellow when they ripen during the rainy season. *Ipomoea squamosa* has a wide distribution from Central America south to the Guianas, Venezuela, the Amazon Basin, the Pantanal along the upper Paraguay River, and central Brazil.

19.3 *Maripa axilliflora*

Peru Azúcar huayo, brea huayo, sacha abuta blanca, paujil micuna; natém (Achuar)

Status Wild

Azúcar huayo *Maripa axilliflora* vine with immature fruit on a small floodplain tree. Cocha Grande, near San Carlos, Rio Puinahua (a branch of the Ucayali), Reserva Nacional Pacaya-Samiria, Loreto, Peru 4-18-06

Vines are common along the banks of rivers in the Amazon, and many of them have edible fruits. One such vine in the sweet potato family, *Maripa axilliflora*, produces black fruits some 1.8–2.5 cm long and 1.2–1.4 cm wide with a single tan seed 1.5 cm by 1 cm. Along the Puinahua, an arm of the Ucayali River, the fruit is known as azúcar huayo, and as the common name suggests, the pulp is sweet and much appreciated by fishermen while they work along the shores of lakes and channels. According to fishermen, several fish much appreciated for their fine flavor also eat the fruits, including lisa (*Leporinus*) and palometa (*Mylossoma*). A dozen or so fruits are produced on each bunch, and a vine typically contains numerous such infructescences, so azúcar huayo provides a welcome and satisfying snack.

In spite of its fine flavor, however, the fruits never appear in markets. The pulp of azúcar huayo is black, an unusual color for mesocarp, but could prove of interest to chefs looking for novel combinations of color and texture for desserts. *Maripa axilliflora* is confined to western Amazonia below 1000 m from Ecuador south through Peru to Bolivia.

Azúcar huayo *Maripa axilliflora* vine with mostly immature fruit. Cocha Grande, near San Carlos, Rio Puinahua (a branch of the Ucayali), Reserva Nacional Pacaya-Samiria, Loreto, Peru 4-18-06

Chapter 20
Costaceae

20.1 *Dimerocostus strobilaceus*

Bolivia Po'o sísi (Ese Eja)

Brazil Cana de macaco

Colombia Cañahuate, caña agria, caña pangala

Ecuador Pepa dulce, untuntupi; gola chi (Cayapa), mishqui muyu, urcu huiru (Kichwa), pat, untuntu (Shuar), mimijico (Siona-Secoya), ñoneguimonkagui, wamonka (Waorani)

Peru Situlli, cañagre; shinauakush, shiwanuk, untumú kumpari (Aguaruna), yaix xin piti (Cashinahua)

Status Wild, spontaneous in cultural settings, planted

N. Smith, *Amazon Fruits: An Ethnobotanical Journey*, Ethnobiology,
https://doi.org/10.1007/978-3-031-12803-5_20

Children gathering cañagre *Dimerocostus strobilaceus* in successionary vegetation to snack on the thin pulp surrounding the seeds. Manco Capac, Rio Puinahua (a branch of the Ucayali), Reserva Nacional Pacaya-Samiria, Loreto, Peru 4-14-06

Dimerocostus strobilaceus is a conspicuous member of successionary vegetation because of its tall cane-like stems that can reach 5 m high, long fleshy leaves, and showy yellow flowers. The ginger-like plant adorns trails in the western part of Amazonia as well as Central America as far north as Honduras. Known by various names even within the confines of the Reserva Nacional Pacaya-Samiria, the perennial herb can reach up to 5 m high and sometimes forms small clumps because of its rhizomes. *Dimerocostus strobilaceus* is found as high as 900 m in the Andes.

In the Peruvian Amazon, the flowers are largely ignored, but children in the Pacaya-Samiria National Reserve gather the slender fruits. Kids use their teeth to peel off the tan-colored skin and bite off the numerous black seeds which they suck to remove the thin, translucent pulp surround the seeds. The seeds are attached to the seed pod by a white aril which is not eaten. Although the pulp surrounding the seeds is very sparse, it is sweet with nutty overtones and therefore attracts children. The seeds are spat out after sucking on them. Children in the Ecuadorian Amazon also suck on the seeds.

The trumpet-like flowers bloom in sequence along a stem, such that mature fruits, measuring 4 cm long and 1.5 cm wide, are located further down the stem. Kids often break off the infructescence and snack on the seeds as they are walking along a path through regrowth. Each pod contains about 30 seeds. Flowers and fruits are produced in both the dry and wet seasons, and the attractive flowers of *Dimerocostus strobilaceus* have caught the attention of nurseries specializing in exotic plants. The Shuar in the southern part of the Ecuadorian Amazon squeeze juice from the stem of this plant and then let it dry before placing it on a leaf, which is then applied to tumors to reduce swelling (Bennett et al. 2002:154).

Cristian, 10 years, sucking the sweet film that covers the seeds of cañagre. Near Yarina, Rio Yanayacu, Reserva Nacional Pacaya-Samiria, Loreto, Peru 6-23-06

Reference

Bennett, Bradley C.; Baker, Marc A.; Andrade, Patricia Gómez. 2002. Ethnobotany of the Shuar of Eastern Ecuador. New York Botanical Garden, Advances in Economic Botany, vol. 14.

Chapter 21
Cucurbitaceae

21.1 *Melothria pendula*

Brazil Pepino da roça, melão de São Caetano

Peru Pepinillo, cruhisma, puro purito; yuwish (Aguaruna)

Status Wild, spontaneous in cultural settings

Pepino da roça *Melothria pendula* adjacent to a cacao orchard. Gleba 23, lote 2, km 75 Altamira-Itaituba, Transamazonica, Município de Medicilandia, Pará, Brazil 9-7-15

Pepino da roça *Melothria pendula* is a weed found mostly in cultivated fields hence its name in the Brazilian Amazon: cucumber of the field. This is a wide-ranging

N. Smith, *Amazon Fruits: An Ethnobotanical Journey*, Ethnobiology,
https://doi.org/10.1007/978-3-031-12803-5_21

annual plant in open and regrowth areas from Canada south through the Caribbean and Central America and down to Argentina in South America. It occurs on uplands and as floating mats along rivers and is found as high as 1500 m in the Andean foothills. In the Serra dos Carajás, Pará, it is found in marshy grasslands on the tops of hills of metallic ore (Gomes-Klein et al. 2016).

People snack on the small, cucumber-like fruits, such as in the vicinity of Agrovila Nova Fronteira at 80 km of the Altamira-Itaituba stretch of the Transamazon Highway. Black when fully ripe but also eaten while still green, the fruits measure only 1.6 cm by 1 cm. They contain numerous light tan seeds, 0.35 cm by 0.2 cm, which are swallowed along with the pulp. The fruits are easy to gather because pepino da roça is a low climbing creeper and is a welcome weed because it does not take up much space and fruits during the dry season when far fewer wild edible fruits are available. During the nineteenth century, the German naturalist Carl von Martius noted that some rural people in the Brazilian Amazon used the fruit as a purgative (Breitbach et al. 2013).

Pepino da roça *Melothria pendula* fruits in the hands of a farmer. Gleba 23, lote 2, 75 km Altamira-Itaituba, Transamazonica, Município de Medicilandia, Pará, Brazil 9-7-15

21.2 *Sicana odorifera*

Bolivia Pabi

Brazil Melão croá, melão de caroá, melão caboclo, cruá, croá

Colombia Melocotón

Ecuador casabanana; namú (Shuar)

Peru Sicana, curua, carua, curaba, curuba, casabanana; namúk (Achuar), cencak namúk, namúk, shuwin namúk, yuwich (Aguaruna), pama (Campa), na'tun (Chayahuita), secana (Kichwa), uksha (Shuar)

Venezuela Cajuba, cajúa

Status Wild, planted

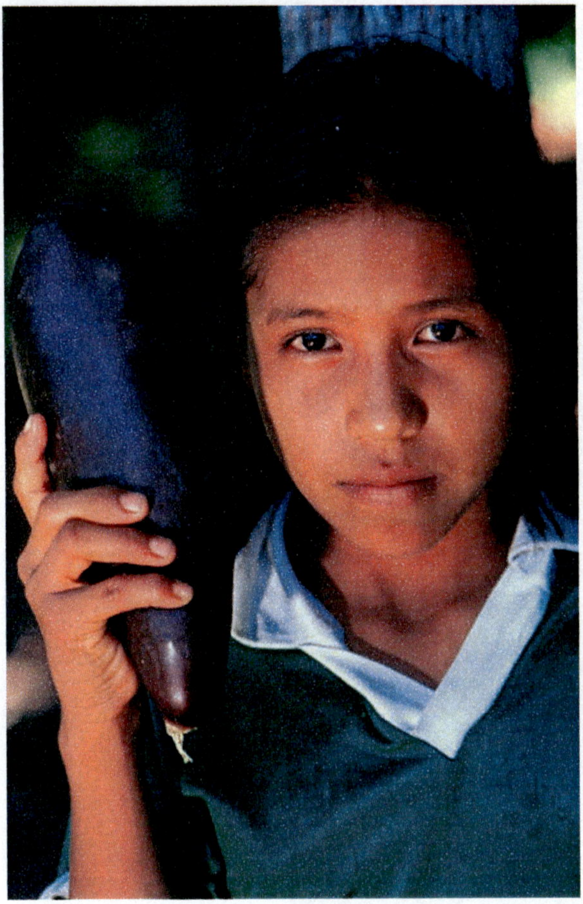

Maria, 11 years, with a sicana *Sicana odorifera* fruit she has harvested in her home garden. Near San Francisco, Rio Marañón, Reserva Nacional Pacaya-Samiria, Loreto, Peru 11-9-03

The origins of the musk cucumber *Sicana odorifera* are unclear but this perennial vine is thought to be native to southern Brazil (Lorenzi et al. 2006a, b:132). It is then spread to neighboring regions as a crop plant in pre-colonial times. It is now cultivated in lowland tropical America and parts of the Caribbean for its fruit and as an ornamental plant. It also occurs in successionary vegetation (Vásquez et al. 2010:574).

Sicana, as it is known in Peru, is a climbing vine that can grow as long as 15 m and requires a tree or sizeable trellis for support. It also takes about 6 months before it starts producing fruit. The dark purple fruits are the size and shape of a large eggplant. In the Peruvian Amazon, the fruits are boiled or roasted and are much appreciated for the sweet-tasting pulp which has an appetizing aroma. Because of their popularity, the fruits are taken to some urban markets, such as Iquitos (Vásquez and Gentry 1989). The fruits are also used in folk medicine to treat fever, constipation, and to expel intestinal worms (Schultes 1990a, b). In the Peruvian Amazon, the Chayahuita prepare a decoction of grated fruits and leaves which is drunk to take care of diarrhea (Odonne et al. 2013).

Farm boy with a sicana fruit he has harvested in his home garden. Near San Francisco, Rio Marañón, Reserva Nacional Pacaya-Samiria, Loreto, Peru 11-9-03

References

Breitbach, Ulrike B.; Niehues, Michael; Lopes, Norberto P.; Faria, Jair E. Q.; Brandão, Maria G. L. 2013. Amazonian Brazilian medicinal plants described by C.F.P. von Martius in the 19th century. Journal of Ethnopharmacology 147:180–189

Gomes-Klein, Vera Lúcia; Hall, Climbiê Ferreira; Gil, André dos Santos Bragança. 2016. Flora das cangas da Serra dos Carajás, Pará, Brasil: Cucurbitaceae. Rodriguésia 67(5): 1319–1327

Lorenzi, Harri; Bacher, Luis; Lacerda, Marco; Sartori, Sergio. 2006a. Frutas Brasileiras e Exóticas Cultivadas. Instituto Plantarum de Estudos da Flora Ltda, Odessa, Sao Paulo.

Lorenzi, H., L. Bacher, M. Lacerda, S. Sartori 2006b. Brazilian Fruits and Cultivated Exotics (for Consuming in Natura). Instituto Plantarum de Estudos da Flora Ltda, Odessa, São Paulo

Odonne, Guillaume; Valadeau, Celine; Alban-Castillo, Joaquina; Stien, Didier; Sauvain, Michel; Bourdy, Genevieve. 2013. Medical ethnobotany of the Chayahuita of the Paranapura basin (Peruvian Amazon). Journal of Ethnopharmacology 146:127–153

Schultes, R. E. 1990a. Biodynamic cucurbits in the New World Tropics. In: *Biology and Utilization of the Cucurbitaceae*, D. M. Bates, R. W. Robinson, and C. E. Jeffrey (eds.), pp. 307–317, Cornell University Press, Ithaca

Schultes, R. E. 1990b. Gifts of the Amazon flora to the world: Many valuable plants were originally domesticated in the now threatened rain forest of the Amazon basin. Arnoldia 50(2):21–34

Vásquez, R. and A. H. Gentry. 1989. Use and misuse of forest-harvested fruits in the Iquitos area. *Conservation Biology* 3(4): 350–361.

Vásquez, R., R. Rojas, H. van der Werff. 2010. *Flora del río Cenepa, Amazonas, Perú*. Missouri Botanical Garden, St. Louis, vol. 1

Chapter 22
Dichapetalaceae

22.1 *Tapura amazonica*

Bolivia Motecillo

Brazil Quina, veludo da mata, morrão, pau de bicho, tapurá, tapurá peluda, bruto, manguito, louro capitiu, grão de galo; pi kàre (Kayapó), itsui ro (Xavante)

Ecuador Aguen catohue (Waorani)

Peru Zorro caspi, guisador caspi, yutubanco

Status Wild

N. Smith, *Amazon Fruits: An Ethnobotanical Journey*, Ethnobiology,
https://doi.org/10.1007/978-3-031-12803-5_22

Zorro caspi ***Tapura amazonica*** **small tree in fruit. Quebrada Huirurui, Rio Samiria, near San Martin Tipishca, Reserva Nacional Pacaya-Samiria, Loreto, Peru 6-30-06**

The tan fruits of *Tapura amazonica* ripen in the rainy months and early part of the dry season. Measuring some 2–3 cm long and 1.5 cm wide, the fruits contain a single seed surrounded by a thin layer of jelly-like pulp. After sucking off the mesocarp, people discard the tan-colored seed which is about 1.8 cm long and 1 cm in diameter. Several indigenous groups eat the fruit, including the Kayapó of the Upper Xingu (Posey 2002:238) and the Waorani in the Ecuadorian Amazon (Mondragón and Smith 1997:30). This tree can reach 30 m tall (Prance 2007), but it starts producing fruit when only 1.5 m tall so they can be picked easily for a snack.

This bush to small tree is found in a diverse range of habitats from floodplains to upland forests and scrub savanna. For example, it occurs in floodplain forests along the black water Jauaperi River, an affluent of the Rio Negro, with separates the states of Amazonas and Roraima (Ammann 2014:17). This tree is encountered in tidal forests in the Amazon estuary and upland forests in the states of Amazonas,

Pará, and Rondônia in the Brazilian Amazon (Absy et al. 1986; Amaral et al. 2009; Lau et al. 2020; Lobo 2017:61). It also occurs in successionary vegetation as young as 5-years old as in the vicinity of Santa Mercedes, a community along the Rio Putumayo in the northern Peruvian Amazon bordering Colombia (Muñoz 2010:17). *Tapura amazonica* ranges from the Guianas through the Amazon Basin to central Brazil and the Atlantic forest. It occurs in scrub savanna (*cerradão*) in the transition zone between the Amazon forest and cerrado vegetation in Mato Grosso (Franczak et al. 2011) and in the gallery and seasonal forests in this transition zone in Tocantins (Haidar et al. 2013; Silva et al. 2016).

The common name for *Tapura amazonica* along the Samiria River an affluent of the lower Marañón is zorro caspi (fox fruit) which suggests that wild canids may serve as dispersal agents. Monkeys may also be involved in spreading the seeds. In the scrub woodlands (*cerradão*) of central Brazil, for example, the black-pincelled marmoset (*Callithrix penicillata*) feeds on the fruits (Miranda and Faria 2001).

References

Absy, Maria Lúcia; Prance, Ghillean T.; Barbosa, Edelcílio Marques. 1986. Inventário florístico de floresta natural na área da estrada Cuiabá-Porto Velho (BR 364). Acta Amazonica 16/17:85–121

Amaral, Dário Dantas do; Vieira, Ima Célia Guimarães; Almeida, Samuel Soares de; Salomão, Rafael Paiva; Silva, Antonio Sérgio Lima da; Jardim, Mário Augusto Gonçalves. 2009. Checklist da flora arbórea de remanescentes florestais da região metropolitana de Belém e valor histórico dos fragmentos, Pará, Brasil. Boletim do Museu Paraense Emílio Goeldi, Ciencias Naturais 4(3):231–289

Ammann, Stefan. 2014. Etnobotânica de árvores e palmeiras em três comunidades ribeirinhas do rio Jauaperi, na divisa entre Roraima e Amazonas. Master's Thesis, INPA, Manaus

Franczak, Daniel David; Marimon, Beatriz Schwantes; Marimon-Junior, Ben Hur; Mews, Henrique Augusto; Maracahipes, Leandro; Oliveira, Edmar Almeida de. 2011. Changes in the structure of a savanna forest over a six-year period in the Amazon-Cerrado transition, Mato Grosso state, Brazil. Rodriguésia 62(2):425–436

Haidar, Ricardo Flores; Fagg, Jeanine Maria Felfili; Pinto, José Roberto Rodrigues; Dias, Ricardo Ribeiro; Damasco, Gabriel; Silva, Lucasde Carvalho Ramos; Fagg, Christopher William. 2013. Florestas estacionais e áreas de ecótono no estado do Tocantins, Brasil: parâmetros estruturais, classificação das fitofisionomias florestais e subsídios para conservação. Acta Amazonica 43(3):261–290

Lau, Arthur Viana; Ferreira Gracialda Costa; Jardim, Mário Augusto Gonçalves. 2020. Fitossociologia e aspectos ecológicos da comunidade arbórea do Bosque Rodrigues Alves – Jardim Botânico da Amazônia, Belém, Pará, Brasil. Revista Brasileira de Geografia Física 13(2): 510–526

Lobo, Guilherme de Sousa. 2017. A alteração do regime hidrológico afeta a composição florística e estrutura de florestas de igapó? Um estudo comparativo entre um rio regulado e outro prístino na Reserva de Desenvolvimento Sustentável do Uatumã, Amazônia Central. Master's thesis, INPA, Manaus

Miranda, G.H.B., and D. S. Faria. 2001. Ecological aspects of black-pincelled marmoset (*Callithrix penicillata*) in the cerradão and dense cerrado of the Brazilian central plateau. *Brazilian Journal of Zoology* 61 (3): 397–404

Mondragón, Martha L.; Smith, Randall. 1997. Bete Quiwiguimamo Salvando el bosque para vivir sano: algunas plantas y árboles utilizadas por la Nacionalidad Huaorani de la Amazonía Ecuatoriana. Abya Yala, Quito

Muñoz, Marianella Guevara. 2010. Evaluación de purmas con fines de manejo en la comunidad nativa de Santa Mercedes, Rio Putumayo, Perú. Master's thesis, Universidad Nacional de la Amazonia Peruana, Iquitos

Posey, D. A. 2002. Kayapó Ethnoecology and Culture. Routledge, London

Prance, G. T. 2007. Flora Reserva Ducke, Amazonas, Brasil: Dichapetalaceae. Rodriguésia 58(3):487–492

Silva, Wagner Matos da; Lolis, Solange e Fátima; Viana, Rodney Haulien Oliveira. 2016. Composition and structure of the gallery forest in the Taquaruçu Grande sub-basin, Municipality of Palmas, Tocantins State. Acta Scientiarum Biological Sciences 38(1):17–24

Chapter 23
Euphorbiaceae

23.1 *Alchornea latifolia*

Bolivia Arraignon

Brazil Kasadika (Tuyuka)

Ecuador Tyunucechi; nagakamenewe (Huaorani), amarun caspi (Kichwa)

Peru Ipururo colorado, yacochihua de altura

Status Wild

N. Smith, *Amazon Fruits: An Ethnobotanical Journey*, Ethnobiology,
https://doi.org/10.1007/978-3-031-12803-5_23

A Tuyuka teenager gathering fruits of kasadika *Alchornea latifolia* **in regrowth vegetation on an abandoned village site. His face is painted from a dye obtained from the leaves of warasoa vine. São Pedro Antigo, Rio Tiquié, affluent of the Uaupés, Amazonas, Brazil 11-1-12**

Alchornea latifolia is found in mature forest and successionary vegetation on uplands and floodplains. This tree, which can reach 25 m high, has a wide distribution in the tropical lowlands of Central and South America as well as parts of the Caribbean. In the foothills of the Andes, it can be found as high up as 2260 m. It produces tiny round fruits, 0.4 cm in diameter, that are too small to be of much interest to most rural peoples in its range. In the Upper Rio Negro region of Brazil, however, the Tuyuka gather the red fruits to mix into the manioc dough when making flat bread (*beiju*). The Huaorani in the Napo Watershed also consume the fruits.

Manioc flat bread (*beiju*) with kasadika fruits prepared by a Tuyuka woman. São Pedro, Rio Tiquié, affluent of the Uaupés, Amazonas, Brazil 11-2-12

23.2 *Hevea benthamiana*

Brazil Seringa barriguda, seringa do igapó, seringa fogo, seringueira branca, seringueira casca roxa, seringueira torrada, seringueira chicote

Colombia Siringa blanca, seringa fina, seringa de rebalse

Venezuela Caucho, goma, jacia, shiringa, seringa

Status Wild, planted

Seringa barriguda *Hevea benthamiana* tree with immature fruit on the floodplain of a stream. The tree has been felled to plant crops. Enseada at Flechinha, Amazon River a few km downstream from the mouth of the Xingu, Município de Gurupá, Pará, Brazil 10-27-16

In the Brazilian Amazon, the two to three nuts inside the fruit capsules of *Hevea benthamiana* are used to bait hooks to catch tambaqui *Colossoma macropomum*, one of the most esteemed fishes of the Amazon. The fruits are also used to bait trotlines to catch tartaruga *Podocnemis expansa*, the largest turtle in the Amazon and much appreciated for its meat. The fruits of one of its near relatives, rubber *Hevea brasiliensis*, are also used to bait hooks to catch fish, especially tambaqui. *Hevea benthamiana* has been crossed with rubber *Hevea brasiliensis* to transfer resistance to South American leaf blight *Microcyclus ulei* (Hashim 1983). This fungus has devastated rubber plantations in the Amazon where rubber is native. Although

Hevea brasiliensis is the main source of rubber, wild *Hevea benthamiana* has been tapped occasionally for its latex in the past (Dean 1987:38; Ducke 1910; Schultes 1966).

Hevea benthamiana is mostly confined to floodplain forests and grows along rivers with diverse waters from black, such as the Rio Negro and its affluent the Uaupés and the Uatumã, to sediment-rich rivers such as the Amazon (Lobo 2017:61; Matos and Amaral 1999; Prance 1989; Schultes 1956). It is found in some stunted woodlands (*campinarana*) growing on sandy soils subject to seasonal flooding, such as in Roraima (Vale 2011:45). It also occurs in the Upper Orinoco in Venezuela. When mature, the tree can reach 30 m high. It does occur in some upland forests, such as in the southern fringe of the Amazon forest in the vicinity of Alta Floresta in Mato Grosso (Almeida et al. 2014). It is also planted occasionally, mostly for rubber breeding experiments. It occurs throughout Amazonia and into parts of the Upper Orinoco in Venezuela (Le Cointe 1922:316).

23.3 *Hevea brasiliensis*

Bolivia Siringa, goma, caucho; carama jihui (Chácobo)

Brazil Seringa, seringa branca, seringa barriguda, seringueira, borracha; konoli, dzeeka (Baniwa), tàk-tëg (Hupd'äh), cauuchú (Kambeba), pi ok-y (Kayapó), walíkikatsu (Nambicuara)

Colombia Caucho, siringa; isire, jitirai (Witoto)

Ecuador Caucho, siringa

Peru Shiringa, shiringa fino, siringa, chiringa, jebe, jebe fino, caucho, balata; shijig (Aguaruna), mááкiñeco (Bora), chirinka (Cocama), sekatsipini, toturopini (Matsigenka)

Venezuela Caucho banera

Status Wild, spontaneous in cultural settings, planted

Farm boy with rubber *Hevea brasiliensis* seeds he gathered from the ground by a rubber tree planted in his home garden. The tree is being tapped for its latex. Comunidade Cipoal, km 15 Santarém-Rurópolis highway, Pará, Brazil 5-11-96

Rubber *Hevea brasiliensis* is best known for its latex which is transformed into a variety of industrialized products from radial tires to latex gloves. Indeed, rubber is one of the most valuable crops to have come out of Amazonia; it is widely planted in the humid tropics, especially in Southeast Asia. The rubber boom based on the extraction of latex from wild rubber trees in the Amazon from 1870 to 1910 had a profound impact on the region's economy and culture, from the extirpation of many indigenous groups to the influx of rubber tappers from other regions, especially the northeastern region of Brazil. Attempts to establish rubber plantations in the Amazon have met with limited success due mainly to the South American leaf blight disease caused by the *Microcyclus ulei* fungus. Henry Ford's ambitious attempts to establish vast rubber plantations at Fordlandia and Belterra along the Tapajós River in the 1930s failed. Limited amounts of rubber are produced in parts of the Amazon from small operations which typically interplant rubber with other trees to reduce disease pressure (Schroth et al. 2003). The tapping of wild rubber trees has essentially ceased, at least as an economic enterprise. Some are still tapped by ribeirinhos, however, to obtain latex to fix leaks in their canoes and wooden boats. Most of the world's rubber is now produced on plantations in Southeast Asia, especially in Malaysia and Indonesia.

While latex is the main economic value of rubber trees, few people know that rubber seeds have long served as a snack in Amazonia (Wheeler 1977; Schultes 1956, 1977). Rubber seeds are highly nutritious since they contain 30% protein when the oil is expressed out (Purseglove 1974:154). The seeds are produced during the rainy season and into the early part of the dry period.

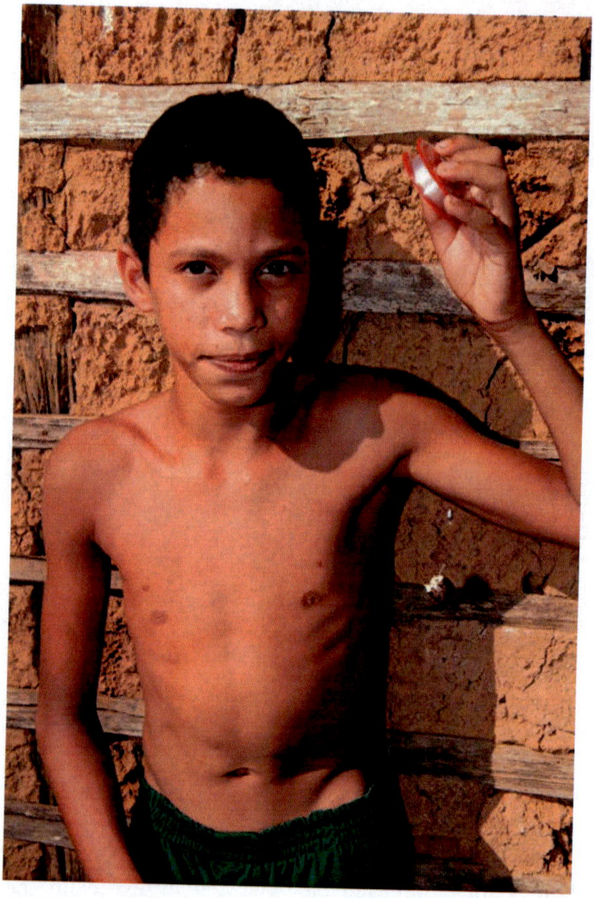

Roberto, 13 years, holding a handline baited with a rubber *Hevea brasiliensis* seed that he gathered in his home garden. Chácara do Paulo & Lu, Pedral, left bank of the middle Xingu, a few kilometers upstream from Altamira, Pará, Brazil 8-31-15

Rubber seeds are also widely used in the Amazon Basin for fish bait (Smith 1981:63, 1999). Rubber seeds can float in water for up to 2 months hence their attractiveness for several species of fish (Kubitzki 1985). Shiny rubber seeds are gathered from the ground in large quantities for local use and to sell to urban-based fishermen. Rubber seeds are relished by tambaqui *Colossoma macropomum*, one of the region's most prized fish (Goulding 1980:79; Honda 1974). Tambaqui are also fond of the seeds of a closely related species, seringa barriguda *Hevea spruceana*. Tambaqui does not occur above the Belo Monte rapids along the Xingu (except for a few areas where tambaqui have escaped from fish culture ponds); its place is taken by pacu seringa *Myleus rhomboidalis*, arguably the most valuable fish along the middle Xingu. Pacu seringa, with their distinctive silvery sides, are caught in gill-nets and on handlines baited with rubber seeds. The Cocama and Cocamilla in the Peruvian Amazon also use rubber seeds for fish bait (Ruiz 2004:105). Matrinxã *Brycon cephalus*, a widely consumed fish along the Solimões and Negro rivers in Amazonas, also eat these seeds (Pizango-Paima et al. 2001).

The largest turtle in the Amazon, tartaruga *Podocnemis expansa*, also eat the fruits (Almeida et al. 1986). People eat tartaruga turtles as well as their eggs which they dig up on sandy beaches during the later water season (Best 1984). Woolly monkeys *Lagothrix lagotricha*, bald uakaris *Cacajao calvus ucayalii*, and black-headed uakaris *Cacajao melanocephalus melanocephalus*, *C. hosomi* also eat rubber seeds (Aquino and Encarnación 1999; Boubli 1999; Boubli and Tokuda 2008; Peres 1994) and these monkeys are sometimes shot for their meat.

Pacu seringa *Myleus rhomboidalis* is one of the most prized food fishes along the middle Xingu which eats rubber seeds. Sitio São José, left bank of the middle Xingu downstream from Altamira, Pará, Brazil 8-19-15

A few indigenous groups use, or once employed rubber latex for a variety of purposes. The now extinct Omagua of the Upper Amazon once made several articles from rubber, such as water bottles (Métraux 1963). A few other groups in western Amazonia made rubber balls for games (Steward 1963). For example, the Pareci of Mato Grosso used to play a game in which they butted the homemade rubber ball with their heads (Rondon 1916:272). The Chácobo in the Bolivian Amazon dip their feet in the latex that has coagulated to make shoes (Boom 1987:23, 1989). Some English explorers took rubber bouncing balls made by native cultures back home where they were used for erasing pencil marks, hence the English name "rubber" (Imle 1978). In the nineteenth century, latex was rubbed on the skin to reduce inflammation (Edwards 1847:157).

Latex from rubber trees has been used to make rain capes in the Amazon since at least the early nineteenth century. However, as noted by the German naturalists, Johann Baptist von Spix and Carl Friedrich Philipp von Martius, such capes were extremely hot (Spix and Martius 1976:29). In some parts of the Bolivian Amazon, heavy capes are still made using this tree's latex. Some river dwellers along the Solimões prepare a tea from the bark to treat rheumatism, coughs, colds, and inflammation (Silva et al. 2007).

Rubber *Hevea brasiliensis* tree (on left) planted in the home garden of a river dweller along with other fruit and nut trees. Sitio Jatobá, Ilha do Costa Junior, middle Xingu River upstream from Altamira, Pará, Brazil 9-1-15

The food value of the seeds probably spurred many tribes in the Amazon to plant rubber and some of its near relatives (Seibert 1948). This tradition has continued. Along the middle Xingu, for example, rubber trees were found in 15 of 57 home gardens I surveyed in 2015. The home gardens were established on upland slopes bordering the river or on islands that flood for part of the rainy season. Rubber was the 15th most common species among a total of 81 fruit or nut-bearing trees and bushes I noted in those 57 home gardens. The rubber trees are no longer tapped, but *ribeirinhos* gather the seeds when they fall to the ground for fish bait. Rubber trees are also found in some home gardens in the municipalities of Igarapé-Açu and Marapanim in the zona Bragantina of northeastern Pará (Quaresma et al. 2015), near Abaetetuba along the lower Tocantins River (Gonçalves and Lucas 2017), in the vicinity of Manaus (Guillaumet et al. 1990), and in parts of Acre in the southwestern part of the Brazilian Amazon (Martins et al. 2012). Rubber trees occasionally sprout spontaneously in cultivated fields, such as near Afuá in the northwestern part of Marajó Island.

Irlene, 8 years, holding rubber *Hevea brasiliensis* seeds by a rubber tree with tapping scars. The rubber tree, which is no longer tapped, is on an archaeological site presently occupied by a village. Pontão, Rio Urubu near Silves, Amazonas, Brazil 4-4-15

With the exception of a small area west of Manaus, rubber is generally thought to be native to forests in the Amazon Basin south of the Amazon River (Kuchla et al. 2015; Purseglove 1974:148; Wycherley 1976). But given the extensive history of people using and planting rubber in the Amazon, it is difficult to discern precisely the tree's natural distribution. Forests in some areas were likely enriched with rubber long before the rubber boom. Rubber trees are sometimes encountered in anthropogenic forests on or adjacent to archaeological sites, such as along the lower Caxiuanã River, an affluent of the lower Anapú in eastern Pará, and at Sitio Santo Antonio I overlooking the Xingu River near the Belo Monte dam. Wild rubber is found throughout the Amazon in both upland forests, including woodlands dominated by native bamboo *Guadua weberbaueri* in Acre, and on floodplains of muddy, clear, and black water rivers (Almeida et al. 2004; Batista et al. 2011; Cattanio et al. 2002; Carim 2016:26; Ferreira et al. 2005; Gama et al. 2003, 2005; Maranho et al. 2016; Montagnini and Muñiz-Miret 1999; Montero and Latrubesse 2013; Montero et al. 2014; Myster 2018; Navarro and Maldonado 2002:72; Reis 2010:32; Salomão et al. 2007; Santos and Jardim 2006; Seibert 1947; Vieira et al. 2017). It also occurs in floodplain forests in the Amazon estuary and along the coast of northeastern Pará which are flooded daily at high tide (Almeida and Jardim 2012; Bentes-Gama et al. 2002; Braga and Jardim 2019; Freitas et al. 2019; Lau and Jardim 2013; Maués et al. 2011; Moegenburg 2002).

Rubber tree *Hevea brasiliensis* with old tapping scars in a village home garden. Carariacá, Igarapé do Jari, Pará, Brazil 5-15-96

Rubber trees *Hevea brasiliensis* planted on a floodplain. Fishermen are waiting in boats with seine nets for schools of fish to pass by. Rio Solimões near Manacapuru, Amazonas, Brazil 6-20-07

Rubber trees have featured in some folktales in the Amazon. A *ribeirinho* interviewed on the waterfront of Altamira in 1991 recounted an incident that happened to one of his friends (*companheiros*) in 1950. His friend was gashing rubber trees on Ilha do Farol along the middle Xingu 1 day when a thin hag suddenly appeared and warned him to be careful when tapping rubber trees. He left immediately and never returned to that island (Smith 1996:133). The mother of rubber trees (*mãe de seringa*) typically has gashes on her arms and legs similar to the cuts made on rubber trees and she will steal a person's shadow if she is not obeyed. Curupira, a stunted boy-like figure known as chullachaqui in Peru, can also confront rubber tappers. In 1994, I interviewed Bocinho, an 80-year-old farmer and former rubber tapper on Combú Island, near Belém. Bocinho recalled walking along a trail in the forest on Combú Island that led to rubber trees when he saw a small, dark man. Curupira asked for a cigarette. Bocinho did not have any so curupira hit him hard (*baque*) and told him to leave.

23.4 *Omphalea diandra*

Bolivia Tutumillo, bejuco blanco, bejuco cabeza de mono; chimo (Chácobo)

Brazil Cipó castanha, castanha de cutia, castanha de peixe, castanha de caiaté, castanha purgativa, comadre de azeite, compadre do azeite, caiaté, ouabé, maracujá de anta; ana-wɨpo (Guajá)

Ecuador Aya tullu panga (Kichwa), quenegapoga (Waorani)

Peru Habilla, sapo huasca, sapote de liana, sacha mani; kencha muxabin bimi (Cashinahua)

Status Wild

Artemio, 14 years, with habilla *Omphalea diandra* liana, he has pulled down in floodplain forest to feed on the nuts embedded in the fruit. 800 m south of Veinte de Enero, Rio Yanayacu, Reserva Nacional Pacaya-Samiria, Loreto, Peru, 6-28-06

Many lianas in western Amazonia produce edible fruits and at least one, *Omphalea diandra*, also provides tasty nuts. *Omphalea diandra* is a forest liana found in both uplands and floodplains below 600 m from Costa Rica south to northern South America, including the Amazon Basin. It also occurs in the tidal zone at the mouth of the Amazon (Amaral et al. 2009; Cattanio et al. 2002) and in floodplain forests of the black water Anapu River in eastern Pará (Ferreira et al. 2013). It is also encountered sporadically in the Caribbean, such as in Trinidad and Cuba. How this vine reached islands in the Caribbean is a mystery; the seeds are too large to be swallowed whole by most Neotropical mammals. The fruits may have drifted to some islands or Arawaks or Caribs may have taken the nuts when they dispersed from mainland South America into the Caribbean.

Along the Yanayacu River in the Reserva Nacional Pacaya-Samiria, rural folk snack on the round, 3-centimeter nuts which are called habilla. In Acre in the southwestern part of the Brazilian Amazon, *Omphalea diandra* is aptly named the nut vine (cipó castanha). In Acre, locals eat the nuts and make a soap using the nut's abundant oil. Hunters also wait near the liana at night for paca *Agouti paca* to come and feed on the fallen fruits. Agoutis *Dasyprocta* spp. also feed on these fallen fruits as suggested by one of its common names in the Brazilian Amazon: castanha de cutia. In the Peruvian Amazon, Geoffroy's saddleback tamarins *Saguinus fuscicollis nigrifrons* have been observed feeding on these fruits (Sánchez and Levy 2013:29, 71). In Surinam which borders northern Pará state in the Brazilian Amazon, the Boni once used the nuts to extract oil for cooking and to provide luster for their hair (Crévaux 1883:285).

Three to four black-skinned nuts are embedded in a dry, pale green pulp. The mesocarp and nuts are encased in a thick, olive-green peel that is broken open with the fingers to access the nuts. The grapefruit-shaped fruits are about 10–15 cm in diameter and drop to the forest floor when ripe. The fruits are thus encountered opportunistically on the ground and the nuts are usually eaten on the spot. The savory nuts do not enter commerce. Formerly, the Chácobo used sections of the liana to make bands for their wrists and ankles, and they still use bark scrapings to reduce swellings associated with skin infections (Boom 1987:23).

Artemio, 14 years, eating habilla *Omphalea diandra* nuts in floodplain forest. 800 m south of Veinte de Enero, Rio Yanayacu, Reserva Nacional Pacaya-Samiria, Loreto, Peru, 6-28-06

23.5 *Plukenetia volubilis*

Brazil Sacha inchi, amendoim inca, amêndoa lopo

Colombia Maní estrella, laurel, sacha inchi, almendro; amiai (Muinane), amo (Witoto)

Ecuador Maní de árbol, sacha inchi; estrella ticasu, tikasu, inshi (Kichwa), nuse (Shuar), komitawemo, tikaso (Waorani)

Peru Sacha inchi, sacha maní, maní silvestre, maní de monte, maní del inca, manisacha, correa; sapátar (Shuar), amui-o (Witoto)

Status Planted

Carlos, 12 years, by sacha inchi *Plukenetia volubilis* planted in his home garden for the edible seed. Sauce, San Martin, Peru 8-24-04

Another liana in the Euphorbiaceae, sacha inchi *Plukenetia volubilis*, is native to the lower montane forests of the Andes and was domesticated there in pre-contact times. It is now being widely planted in the Peruvian Amazon and increasingly in neighboring countries, such as among indigenous groups in Colombia along the Caquetá River (Veléz 1991) and Kichwa communities in the Upper Napo watershed in the Ecuadorian Amazon (Friedman et al. 1993). The seeds contain around 27% protein and are rich in Omega 3 oil, one of the "healthy" oils good for lowering LDL cholesterol and thus helps to prevent cardiovascular and neuromuscular diseases (Saavedra et al. 2012; Valente et al. 2017). Oil from the seeds is also used in the cosmetics industry. In the small town of Sauce in the foothills of the Peruvian Andes in the Department of San Martin, locals roast the seeds of sacha inchi and then grind them to make a dough which is then fried and eaten. Because of its health benefits, *Plukenetia volubilis*, sometimes known as the Andean peanut, has taken off as a commercial crop in Peru, Ecuador, Bolivia, and increasingly Brazil (Kinupp and Lorenzi 2014:390). In Peru, 38 companies export oil compressed from the seeds of this plant with Amazon Health Products being the lead exporter (Bartra 2017:37). Toasted and ground seeds are also consumed locally and exported. Sacha inchi is now planted in parts of Southeast Asia as well, particularly in Thailand. However, habitat fragmentation of wild populations, such as in the San Martin province of the Peruvian Amazon, is likely leading to genetic erosion of this valuable plant which could jeopardize the future of this valuable crop (Corazon-Guivin et al. 2008).

Seed and fruits of sachi inchi *Plukenetia volubilis* harvested in a home garden. Manaus, Amazonas, Brazil 4-1-15

References

Almeida, Adrielson Furtado; Jardim, Mário Augusto Gonçalves. 2012. A utilização das espécies arbóreas da floresta de várzea da Ilha de Sororoca, Ananindeua, Pará, Brasil por moradores locais. Revista Brasileira de Ciências Ambientais 23:48–54

Almeida, Samuel S.; Sá, Paulo G. S.; Garcia, Alvaro. 1986. Vegetais utilizadas como alimento por Podocnemis (Chelonia) na região do baixo Rio Xingu (Brasil-Pará). Boletim do Museu Paraense Emílio Goeldi, Série Botânica 2(2):199–211

Almeida, Samuel Soares de; Amaral, Dário Dantas do; Silva, Antonio Sérgio Lima da. 2004. Análise florística e estrutura de florestas de várzea no estuário amazônico. Acta Amazonica 34(4):513–524

Almeida, Eder Alves de; Lopes, Célia Regina Araujo Soares; Rodrigues, Lucirene; Simão, Silmaria Soares; Fernandes, José Martins. 2014. Estrutura fitossociológica de floresta estacional decidual submontana e floresta ombrófila aberta submontana em Alta Floresta, Mato Grosso. Enciclopédia Biosfera 10(19):1058–1078

Amaral, Dário Dantas do; Vieira, Ima Célia Guimarães; Almeida, Samuel Soares de; Salomão, Rafael Paiva; Silva, Antonio Sérgio Lima da; Jardim, Mário Augusto Gonçalves. 2009. Checklist da flora arbórea de remanescentes florestais da região metropolitana de Belém e valor histórico dos fragmentos, Pará, Brasil. Boletim do Museu Paraense Emílio Goeldi, Ciencias Naturais 4(3):231–289

Aquino, Rolando; Encarnación, Filomeno. 1999. Observaciones preliminares sobre la dieta de *Cacajao calvus ucayalii* en el nor-oriente Peruana. Neotropical Primates 7(1):1–5

Bartra, Alfonso Sánchez. 2017. Diagnóstico de la cadena de valor del sachi inchi (Plukenetia volubilis L.) en la región San Martín. Master's Thesis, Universidad Nacional de San Martín, Tarapoto

Batista, Fábio de Jesus; Jardim, Mário Augusto Gonçalves; Medeiros, Tonny David Santiago; Lopes, Isaac Luiz Magalhães. 2011. Comparação florística e estrutural de duas florestas de várzea no estuário Amazônico, Pará, Brasil. Revista Árvore 35(2):289–298

Bentes-Gama, Michelliny de Matos; Scolforo, José Roberto Soares; Gama, João Ricardo Vasconcellos. 2002. Potencial produtivo de madeira e palmito de uma floresta secundária de várzea baixa no estuário Amazônico. Revista Árvore 26(3):311–319

Best, R.C. 1984. The aquatic mammals and reptiles of the Amazon. In: The Amazon: Limnology and Landscape Ecology of a Mighty Tropical River and its Basin, H. Sioli (ed), W. Junk, Dordrecht, pp 371–412

Boom, B. M. 1987. Ethnobotany of the Chácobo Indians, Beni, Bolivia. The New York Botanical garden, Advances in Economic Botany, vol. 4

Boubli, J. P. 1999. Feeding ecology of black-headed uacaris (*Cacajao melanocephalus melanocephalus*) in Pico da Neblina National Park, Brazil. International Journal of Primatology 20(5):719–749

Boubli, J. P.; Tokuda, M. 2008. Socioecology of black uakari monkeys, Cacajao hosomi, in Pico da Neblina National Park, Brazil: the role of the peculiar spatial-temporal distribution of resources in the Neblina forests. Primate Report 75:3–10

Braga, Elayne Oliveira; Jardim, Mário Augusto Gonçalves. 2019. Florística, estrutura e formas de vida do estrato inferior de uma floresta ombrófila densa aluvial, Pará, Brasil. Ciencia Florestal, Santa Maria 29(3): 1048–1059

Carim, Marcelo de Jesus Veiga. 2016. Estrutura, composição e diversidade em florestas alagáveis de várzea de maré e de igapó e suas relações com variáveis edáficas e o período de inundação no Amapá, Amazônia oriental, Brasil. Doctoral dissertation, Programa de Pós-Graduação em Botânica, INPA, Manaus

Cattanio, José; Anderson, Anthony B.; Carvalho, Manoel S. 2002. Floristic composition and topographic variation in a tidal floodplain forest in the Amazon estuary. Revista Brasileira de Botanica 25(4):419–430

Corazon-Guivin, Mike; Rodríguez, Ángel; Cachique, Danter; Chota, Werner; Vásquez, Guillermo; Del-Castillo, Dennis; Renno, Jean-François; García-Dávila, Carmen. 2008. Diversidad genética en poblaciones naturales de sacha inchi Plukenetia volubilis L. (Euphorbiaceae) en el Departamento de San Martín (Perú). Folia Amazónica 17(1–2):83–90

Crévaux, J. 1883. Voyages dans l'Amérique du Sud. Libraire Hachette, Paris

Dean, W. 1987. Brazil and the Struggle for Rubber. Cambridge University Press, Cambridge

Ducke, A. 1910. Explorações scientificas no Estado do Pará. Boletim do Museu Goeldi (Museu Paraense) de Historia Natural e Ethnographia 7:100–200

Edwards, W. H. 1847. *A Voyage up the River Amazon, including a Residence at Pará*. John Murray, London

Ferreira, Leandro V.; Almeida, Samuel S.; Amaral, Dário D.; Parolin, Pia. 2005. Riqueza e composição de espécies da floresta de igapó e várzea da Estação Científica Ferreira Penna: subsídios para o plano de manejo da Floresta Nacional de Caxiuanã. Pesquisas Botânica 56:103–116

Ferreira, Leandro V.; Chaves, Priscilla P.; Cunha, Denise de A.; Matos, Darley C. L.; Parolin, Pia. 2013. Variação da riqueza e composição de espécies da comunidade de plantas entre as florestas de igapós e várzeas na Estação Científica Ferreira Penna-Caxiuanã na Amazônia oriental. Pesquisas Botânica 64:175–195

Freitas, João da Luz; Santos, Erick Silva dos; Cruz Junior, Francisco de Oliveira; Borges, César Henrique Alves; Santos, Adriano Castelo dos. 2019. Structural analysis and distribution patterns in lowland tropical forest, eastern Amazon. Journal of Agricultural Studies 7(4): 240–255

Friedman, J.; Bolotin, D.; Rios, M.; Mendosa, P.; Cohen, Y.; Balick, M. J. 1993. A novel method for identification and domestication of indigenous useful plants in Amazonian Ecuador. In: New Crops, Janick, J.; Simon, J. E. (eds), John Wiley, New York, pp 167–174

Gama, J. R. V., S. A. Botelho, M. Bentes-Gama, and J. R. S. Scolforo 2003. Estrutura e potencial futuro de utilização da regeneração natural de floresta de várzea alta no Município de Afuá, estado do Pará. *Ciencia Florestal, Santa Maria* 13(2):71–82

Gama, João Ricardo Vasconcellos; Souza, Agostinho Lopes de; Martins, Sebastião Venâncio; Souza, Deoclides Ricardo de. 2005. Comparação entre florestas de várzea e de terra firme do estado do Pará. Revista Árvore 29(4):607–616

Gonçalves, Janaína Pinheiro; Lucas, Flávia Cristina Araújo. 2017. Agrobiodiversidade e etnoconhecimento em quintais de Abaetetuba, Pará, Brasil. Revista Brasileira de Biociências 15(3):119–134

Goulding, Michael. 1980. *The Fishes and the Forest: Explorations in Amazonian Natural History*. University of California Press, Berkeley

Guillaumet, J-L.; Grenand, P.; Bahri, S.; Grenand, F.; Lourd, M.; Santos, A.A. dos; Gély, A. 1990. Les jardins-vergers familiaux d'Amazonie centrale: une exemple d'utilisation de l'espace. Turrialba 40(1):63–81

Hashim, I. 1983. Biology and economic importance of South American leaf blight of Hevea rubber. In: Exotic Plant Quarantine Pests and Procedures For Introduction of Plant Materials, K. G. Singh (editor), ASEAN Plant Quarantine Centre and Training Institute, Serdang, Selangor, Malaysia, pp. 27–34

Honda, E. M. S. 1974. Contribuição ao conhecimento da biologia de peixes do Amazonas II. Alimentação de tambaqui, *Colossoma bidens* (Spix). *Acta Amazonica* 4(2):47–53

Imle, E. P. 1978. Hevea rubber--past and future. *Economic Botany* 32:264–277

Kinupp, V. F. and H. Lorenzi. 2014. *Plantas Alimentícas Não Convencionais (PANC) no Brasil: Guia de Identifação, Aspectos Nutricionais e Receitas Ilustradas*. Instituto Plantarum de Estudos da Flora, São Paulo.

Kubitzki, K. 1985. Dispersal of forest plants. In: Key Environments: Amazonia, G. T. Prance and T. E. Lovejoy (eds.), Pergamon Press, Oxford, pp. 192–206

Kuchla, Wilian José; Stepka, Thiago Floriani; Sawczuk, Alex Roberto; Lisboa, Gerson dos Santos. 2015. Florística, estrutura horizontal e distribuição diamétrica em área de transição de cerrado e floresta aluvial no município de Campos de Júlio, MT. Ambiência 11(1):13–30

Lau, Arthur Viana; Jardim, Mário Augusto Gonçalves. 2013. Florística e estrutura da comunidade arbórea em uma floresta de várzea na área proteção ambiental, Ilha do Combu, Belém, Pará. Biota Amazônia 3(2):88–93

Le Cointe, P. 1922. *L'Amazonie Brésilienne: Le Pays-ses Habitants, ses Resources, Notes et Statistiques jusqu'en 1920*. Augustin Challamel, Paris, vol 2

Lobo, Guilherme de Sousa. 2017. A alteração do regime hidrológico afeta a composição florística e estrutura de florestas de igapó? Um estudo comparativo entre um rio regulado e outro prístino na Reserva de Desenvolvimento Sustentável do Uatumã, Amazônia Central. Master's thesis, INPA, Manaus

Maranho, Álisson Sobrinho; Salimon, Cleber Ibraim; Costa, Daniel da Silva. 2016. Woody flora of natural forest gaps in a bamboo-dominated forest remnant in southwestern Amazonia. Check List: The Journal of Biodiversity Data 12(3):1–9

Martins, Williane Maria de Oliveira; Martins, Lilliane Maria de Oliveira; Paiva, Fabiano Silveira; Martins, Wilton José de Oliveira; Lima Júnior, Sebastião Ferreira. 2012. Agrobiodiversidade nos quintais e roçados ribeirinhos na comunidade Boca do Môa – Acre. Biotemas 25(3):111–120

Matos, Francisca Dionízia de A.; Amaral, Iêda Leão do. 1999. Análise ecológica de um hectare em floresta ombrófila densa de terra-firme, estrada da várzea, Amazonas, Brasil. Acta Amazonica 29(3):365–379

Maués, Bernardo Antônio Rodrigues; Jardim, Mário Augusto Gonçalves; Batista, Fábio de Jesus; Medeiros, Tonny David Santiago; Quaresma, Adriano da Costa. 2011. Composição florística e estrutura do estrato inferior da floresta de várzea na área de proteção ambiental Ilha do Combu, município de Belém, estado do Pará. Revista Árvore 35(3):669–677

Métraux, A. 1963. Tribes of the Middle and Upper Amazon River. In: *Handbook of South American Indians*, J. H. Steward (Editor), Cooper Square Publishers, New York, vol. 3, pp. 687–712

Moegenburg, Susan M. 2002. Spatial and temporal variation in hydrochory in Amazonian floodplain forest. Biotropica 34(4):606–612

Montagnini, Florencia; Muñiz-Miret, Nuria. 1999. Vegetation and soils of tidal floodplains of the Amazon estuary: a comparison of várzea and terra firme forests in Pará, Brazil. Journal of Tropical Forest Science 11(2):420–437

Montero, Juan Carlos; Latrubesse, Edgardo M. 2013. The igapó of the Negro River in central Amazonia: linking late-successional inundation forest with fluvial geomorphology. Journal of South American Earth Sciences 46:137–149

Montero, Juan Carlos; Teresa, Maria; Piedade, Fernandez; Wittmann, Florian. 2014. Floristic variation across 600 km of inundation forests (igapó) along the Negro River, central Amazonia. Hydrobiologia 729:229–246

Myster, Randall W. 2018. What we know about the composition and structure of igapó forests in the Amazon Basin. Botanical Review 84:394–410

Navarro, G.; Maldonado, M. 2002. Geografía Ecológica de Bolivia: Vegetación y Ambientes Acuáticos. Centro de Ecología Simón I. Patiño, Cochabamba

Peres, C.A. 1994. Diet and feeding ecology of gray woolly monkeys (*Lagothrix lagotricha cana*) in central Amazonia: comparisons with other Atelines. International Journal of Primatology 15(3):333–372

Pizango-Paima, Emer Gloria; Pereira-Filho, Manoel; Oliveira-Pereira, Maria Inêz de. 2001. Composição corporal e alimentar do matrinxã, *Brycon cephalus* (Günther, 1869), na Amazônia central. Acta Amazonica 31(3):509–520

Prance, G. T. 1989. Economic prospects from tropical rainforest ethnobotany. In: Fragile Lands of Latin America: Strategies for Sustainable Development, Browder, John O. (Editor), pp. 61–74

Purseglove, J.W. 1974. *Tropical Crops: Dicotyledons*. John Wiley, New York

Quaresma, Amanda Paiva; Almeida, Ruth Helena Cristo; Oliveira, Cyntia Meireles de; Kato, Osvaldo Ryohei. 2015. Composição florística e faunística de quintais agroflorestais da agricultura familiar no nordeste Paraense. Revista Verde de Agroecologia e Desenvolvimento Sustentável 10(5):76–84

Reis, Neidiane Farias Costa. 2010. Estrutura fitossociológica e seletividade de espécies em áreas inundáveis na região do baixo Rio Jaci-Paraná, Rondônia. Master's Thesis, Universidade Federal de Rondônia, Porto Velho

Rondon, Candido Mariano da Silva. 1916. Missão Rondon: Apontamentos sobre os Trabalhos Realizados pela Commissão de Linhas Telegraphicas Estrategicas de Matto-Grosso ao Amazonas sob a Direcção do Coronel de Engenharia Candido Mariano da Silva Rondon de 1907 a 1915. Typ. Do Jornal do Commercio, Rio de Janeiro

Ruiz, Roxani Rivas 2004. El Gran Pescador: Técnicas de Pesca entre los Cocama-Cocamillas de la Amazonia Peruana. Pontifícia Universidad Católica del Perú, Lima

Saavedra, Juan Jorge Huamán; Silva, Boris Eltsin Fogel; Pairazamán, Patricia Isabel Escobar; Minaya, Karen Yanet Castillo. 2012. Efectos de la ingesta de Plukenetia volubilis Linneo o "Sacha inchi" en el perfil lipídico de aldutos jóvenes. Acta Medica Peruana 29(3):155–160

Salomão, Rafael de Paiva; Vieira, Ima Célia Guimarães; Suemitsu, Chieno; Rosa, Nélson de Araújo; Almeira, Samuel Soares de; Amaral, Dário Dantas do; Menezes, Moirah Paulo Machado de. 2007. As florestas de Belo Monte na grande curva do rio Xingu, Amazônia Oriental. Boletim do Museu Paraense Emílio Goeldi, Ciencias Naturais 2(3):57–153

Sánchez, Judith Jacira Achong; Levy, Allison Licett Nuñez. 2013. Características de frutos consumidos por *Saguinus nigrifrons* (Primates, Callitrichidae), en la Estación Biológica Quebrado Blanco, Loreto, Perú. Master's thesis, Universidad Nacional de la Amazonía Peruana, Iquitos

Santos, Gideão Costa dos; Jardim, Mário Augusto Gonçalves. 2006. Florística e estrutura do estrato arbóreo de uma floresta de várzea no município de Santa Bárbara do Pará, Estado do Pará, Brasil. Acta Amazonica 36(4):437–446

Schroth, Götz; Coutinho, Paulo; Moraes, Vicente H. F.; Albernaz, Ana Luisa. 2003. Rubber agroforests at the Tapajós river, Brazilian Amazon—environmentally benign land use systems in an old forest frontier region. Agriculture, Ecosystems & Environment 97:151–165

Schultes, R. E. 1956. The Amazon Indian and evolution in *Hevea* and related genera. *Journal of the Arnold Arboretum* 37:123–148

Schultes, R.E. 1966. Lactiferous plants of the Karaparaná-Igaraparaná region of Colombia. Acta Botanica Neerlandica 15: 178–189 1969,
Schultes, R. E. 1977. The odyssey of the cultivated rubber tree. *Endeavour* 1 (3/4):133–138
Seibert, R. J. 1947. A study of *Hevea* (with its economic aspects) in the Republic of Peru. Annals of the Missouri Botanical Garden 34(3):261–352
Seibert, R. J. 1948. The uses of *Hevea* for food in relation to its domestication. *Annals of the Missouri Botanical Garden* 35: 117–121
Silva, Suzy Cristina Pedroza da; Pereira, Cloves Farias; Fraxe, Therezinha de Jesus Pinto; Witkoski, Antônio Carlos; Silva, Michele Andreza Pedroza da. 2007. In: Comunidades Ribeirinhas Amazônicas: Modos de Vida e Uso dos Recursos Naturais, Fraxe, Therezinha de Jesus Pinto; Pereira, Henrique dos Santos; Witkoski, Antônio Carlos (eds), EDUA, Manaus, pp 141–154
Smith, N. J. H. 1981. *Man, Fishes, and the Amazon*. Columbia University Press, New York
Smith, N. J. H. 1996. *The Enchanted Amazon Rain Forest: Stories from a Vanishing World*. University Press of Florida, Gainesville.
Smith, N.J.H. 1999. Biodiversity and agroforestry along the Amazon floodplain. In: *Traditional and Modern Natural Resource Management in Latin America*, F.J. Pichón, J.E. Uquillas, and J. Frechione (Editors), pp. 233–250. University of Pittsburgh Press, Pittsburgh
Spix, J. B. and C. F. P. Martius, 1976. *Viagem pelo Brasil, 1817–1820*. Edições Melhoramentos, São Paulo, Vol.3
Steward, Julian H. 1963. Culture areas of the tropical forest. In: *Handbook of South American Indians*, J. H. Steward (Editor), Cooper Square Publishers, New York, vol. 3, pp. 883–899
Vale, Gabriel Damasco do. 2011. O efeito do solo, inundação e topografia sobre as campinaranas inundáveis na Amazônia. Master's Thesis, INPA, Manaus
Valente, Mágno Sávio Ferreira; Chaves, Francisco Célio Maia; Lopes, Maria Teresa Gomes; Oka, Jaisson Myiosi; Rodrigues, Rodney Alexandre Ferreira. 2017. Crop yield, genetic parameter estimation and selection of sacha inchi in central Amazon. Pesquisa Agropecuária Tropical, Goiânia 47(2):226–236
Veléz, Germán Alonso. 1991. Los frutales amazónicos cultivados por las comunidades indígenas de la región del Medio Caquetá. Colombia Amazonica 5(2): 135–183
Vieira, Diegodos Santos; Oliveira, Marcio Leles Romarco de; Gama, João Ricardo Vasconcelos; Machado, Evandro Luiz Mendonça; Görgens, Eric Bastos; Lafetá, Bruno Oliveira; Garcia, Josinei Silva. 2017. Phytosociology of a natural fragment of the floodplain forest in the lower Tapajós River, Brazil. Bosque 38(2):357–369
Wheeler, Louis C. 1977. Sementes de *Hevea* para alimentação humana. *Acta Amazonica* 7 (1): 139–143
Wycherley, P. R. 1976. Rubber Hevea brasiliensis (Euphorbiaceae). In: *Evolution of Crop Plants*, N.W. Simmonds (ed), Longman, London, pp. 77–81

Chapter 24
Fabaceae

Numerous trees and bushes in the pea or bean family (Fabaceae) produce fruits eaten by people, turtles, and fish. Some 300 species of *Inga* are found scattered throughout the American tropics and many of them bear fruits that are eaten locally as well as taken to markets. The fruits of at least a dozen species of *Inga* are gathered in the wild in the Peruvian Amazon and several are also cultivated. Promoters of agroforestry are often keen on *Inga* because, in addition to edible fruit, the trees enrich the soil with nitrogen because they are legumes (Fabaceae). The branches of *Inga* trees can be lopped off to provide mulch to suppress weeds and build up organic matter in the soil.[1] Inhabitants of the Peruvian Amazon have long gathered fruits of several wild species, often lumping them under the name shimbillo. In the Brazilian Amazon, however, all useful species of *Inga* are called ingá with a suffix for each species. In some cases, a separate name is given to species of *Inga* in the Peruvian Amazon, depending on location. Along the Ucayali, for example, the fruits of shimbillo (also called paihuaro; *Inga nobilis*) and shimbillo (also called pairafo, *Inga ruiziana*) are gathered in the higher portions of floodplain forests for domestic consumption. Other species are gathered in the wild as well as planted in home gardens, such as vaca paleta (*Inga macrophylla*), which is more commonly known as shimbillo in markets. Fruits of another shimbillo, *Inga cinnamomea*, are also gathered in floodplain forests and eaten at home as well as taken to markets.

The pulp of *Inga* fruits is protected by woody pods, hence the ease with which they can be shipped. The pods are generally opened by twisting them so that the seam splits revealing seeds surrounded by white pulp. Guaba is easily distinguished from the other species in the vast genus by its slender, deeply furrowed pods. Green when ripe, the pods are typically a meter to a meter and a half long and contain a dozen or so black seeds surrounded by a spongy white pulp. The pulp can thus be extracted conveniently in bite-size sections. A little dry and mealy in texture, the flesh has little nutritious value but is nevertheless sweet, hence its popularity. A fruit

[1] Pennington 1997; Reynel and Pennington 1997.

N. Smith, *Amazon Fruits: An Ethnobotanical Journey*, Ethnobiology,
https://doi.org/10.1007/978-3-031-12803-5_24

may not be especially nutritious but is still important culturally because people enjoy its flavor. The seeds, about 3 cm long, are spat on the ground and easily sprout. Thus it is often difficult to tell whether guaba is wild or naturalized. Guaba fruits are eaten fresh and are not made into puddings, ice creams, or milkshakes.

24.1 The Inga Fruits

Some 300 species of *Inga* are found scattered throughout the American tropics and many of them bear fruits that are eaten locally as well as taken to markets. The fruits of at least a dozen species of *Inga* are gathered in the wild in the Peruvian Amazon and several are also cultivated. Promoters of agroforestry are often keen on *Inga* because, in addition to edible fruit, the trees enrich the soil with nitrogen because they are legumes (Fabaceae). The branches of *Inga* trees can be lopped off to provide mulch to suppress weeds and build up organic matter in the soil.[2] Inhabitants of the Peruvian Amazon have long gathered fruits of several wild species, often lumping them under the name shimbillo. In the Brazilian Amazon, however, all useful species of *Inga* are called ingá with a suffix for each species. In some cases, a separate name is given to species of *Inga* in the Peruvian Amazon, depending on location. Along the Ucayali, for example, the fruits of shimbillo (also called paihuaro; *Inga nobilis*) and shimbillo (also called pairafo, *Inga ruiziana*) are gathered in the higher portions of floodplain forests for domestic consumption. Other species are gathered in the wild as well as planted in home gardens, such as vaca paleta (*Inga macrophylla*), which is more commonly known as shimbillo in markets. Fruits of another shimbillo, *Inga cinnamomea*, are also gathered in floodplain forests and eaten at home as well as taken to markets.

The pulp of *Inga* fruits is protected by woody pods, hence the ease with which they can be shipped. The pods are generally opened by twisting them so that the seam splits revealing seeds surrounded by white pulp. Guaba is easily distinguished from the other species in the vast genus by its slender, deeply furrowed pods. Green when ripe, the pods are typically a meter to a meter and a half long and contain a dozen or so black seeds surrounded by a spongy white pulp. The pulp can thus be extracted conveniently in bite-size sections. A little dry and mealy in texture, the flesh has little nutritious value but is nevertheless sweet, hence its popularity. A fruit may not be especially nutritious but is still important culturally because people enjoy its flavor. The seeds, about 3 cm long, are spat on the ground and easily sprout. Thus it is often difficult to tell whether guaba is wild or naturalized. Guaba fruits are eaten fresh and are not made into puddings, ice creams, or milkshakes.

[2] Pennington 1997; Reynel and Pennington 1997.

Trees in the genus *Inga* produce pods with pulp surrounding the seeds. Confined to Tropical America, trees in this genus are dispersed by animals, particularly monkeys. When humans arrived in warmer climes of the New World, they soon learned to appreciate the pulp, obtained by splitting open the pods. Some species thus became camp followers, sprouting from discarded seeds around hunting and fishing camps, as well as more permanent settlements. When people started clearing forests for planting crops, many species of ingá, as the trees are known in Brazil, were favored because they thrive in second growth. According to Terry Pennington of the Royal Botanic Gardens, Kew, the genus embraces some 300 species, with the greatest concentration of species in the Andean foothills of western Amazonia and southern Central America. The fruits of dozens of species are consumed in the Amazon, and several ingá are planted. Only a relatively small sample of the edible ingá are examined here, ones most likely to be encountered in markets and home gardens.

Ingá fruits travel well because the pulp is protected with a woody encasing. With growing interest in promoting agroforestry as a means to achieve "sustainable" agricultural development, ingá has received some attention because ingá are dual-purpose trees; produce useful fruits and as members of the legume family, help to enrich the soil with nitrogen.

24.2 *Cassia leiandra*

Bolivia Cañafistola sabanera

Brazil Marimari, marimari da várzea, marizeiro, mari, marí marí, ingá mari, fava marimari, canafístula do igapó, taxi preto da folha miúda, seruaia

Peru Jacaranda, marimari, copiiña

Status Wild, planted

Marimari *Cassia leiandra* **tree in fruit along a black water river. Taperebatuba, Rio Urubu, Amazonas, Brazil 4-3-15**

Marimari *Cassia leiandra* forms giant cylindrical pods, often reaching nearly a meter, which turns yellow when ripe. Marimari fruits during the annual flood and a tree can produce as many as 100 pods (Arkcoll 1984). It is easy for people to twist the pods off the partially submerged trees from canoes and boats at high water because the tree is medium-sized, usually less than 20 m tall, and heavily branched. The annual flood in the middle Amazon is typically in the 10- to 15-m range. The pods are split open to reveal a sweet, white pulp surrounding the numerous round seeds with a diameter of 2 cm.

Gathering marimari *Cassia leiandra* fruits from a boat on the Amazon floodplain. Near Santarém, Pará, Brazil 5-10-96

When the pods are still green, the pulp is also green but still edible; the mesocarp has a sweet-tart taste, reminiscent of kiwi fruit. The fruits are often picked while still green because if a person waits until the pods are fully ripe someone else, or a monkey, may beat them to it. Several species of fish that are caught for human consumption also feed on this fruit including tambaqui *Colossoma macropomum*, pirapitinga *Piaractus brachypomus*, bacu *Lithodoras dorsalis*, and matrinxã *Brycon cephalus* (Maia and Chalco 2002).

Marimari is so popular that some farmers on the Amazon floodplain have planted the tree in their backyards, such as on Ilha do Carmo in the Municipality of Alenquer and on the outskirts of Juriti, both in Pará. Near the mouth of the Amazon in Amapá, some river folk have also planted the tree in their home gardens (Steward 2013) as do some rural folk on the floodplain of the lower Madeira River in Amazonas (Kawa et al. 2015).

Suelane, 10 years, with marimari *Cassia leiandra* fruits she has picked in her home garden by the shore of the black water Rio Urubu. Taperebatuba, Amazonas, Brazil 4-3-15

This leguminous tree, which sports bright yellow flowers before setting fruit, grows in floodplain forests along sediment-rich, clear, and black water rivers in the middle and lower Amazon, and several of its affluents including the Juruá, Purus, Madeira, and Negro (Assis et al. 2017; Irwin and Barneby 1982:18; Klinge et al. 1995; Lorenzi et al. 2006a:138; Scoles 2009). The state of Amazonas, Brazil, has been postulated as the center of origin of this species because it is more common there (Silva 2011:196). It also occurs in the Atlantic forest of Brazil in Bahia (Kinupp and Lorenzi 2014:396). *Cassia leiandra* has also been collected in the Bolivian and Peruvian Amazon, though the tree appears to be much less common there. It is possible that seeds or seedlings of *C. leiandra* were introduced to the western and southwestern parts of the Amazon Basin.

Suelane, 10 years, eating marimari *Cassia leiandra* fruit. Taperebatuba, Rio Urubu, Amazonas, Brazil 4-3-15

Marimari finds its way to markets in several towns and cities along the middle Amazon, such as Manacapuru, Manaus, Itacoatiara, Juriti, Óbidos, and Santarém. In June 1994, marimari fruits were selling for the equivalent of $0.68/kg (they are usually sold in bundles, rather than weight) in Santarém, a relatively high price considering the income levels in the region. Upstream at Juriti, the fruits were even more expensive at $0.80/kg, but at Óbidos they were selling for $0.50/kg during the same month. The differences in price are not pronounced, although marimari may be more common in the Óbidos area, accounting for the lower price there.

Vendor with marimari *Cassia leiandra* fruits in a street market. Santarém, Pará, Brazil 5-10-96

This tree also has other uses. In Saracura, a quilombola community on a floodplain island along the middle Amazon near Santarém, in addition to eating the fruit locals also use the wood for fuel (Oliveira and Cavalcante 2017). Along the Jaú River and its affluents, a tributary of the Rio Negro in Amazonas, Brazil, river folk obtain an extract from the leaves to treat bruises (Rodrigues et al. 2010).

24.3 *Dialium guianense*

Bolivia Tamarindo del monte, tamarindillo

Brazil Pororoca, pororoqueira, jutaí, jutaí café, jutaí cica, jutaí mirim, jutaí peba, jutaí poca, jutaí pororoca, jutaí preto, jutaízinho, jutaícica, casca de jutaí, jatobazinho, tamarina, tamarindo, cururú, roxinho, roxinhorana, sucupira amarela, parajuba, itu, coração de negro, ferrinho, angelim preto, azedinha, tetéu; janîta'i (Arawaté), ayuru-mope-'ɨwa (Asurini), yeta-'ɨ-pipiru-'ɨ (Guajá), roromi (Ikpeng), mboimira, pu'ipiraphiu, yurupepe'ɨ (Ka'apor), okre ny (Kayapó), kâpó (Panará), juta'ía (Piripkura), ɨwa-popok-'ɨw (Tembé), deteicge, manjacjke (Ticuna), mükakö iyê (Tiriyó), de'té (Tukano), wila-takulu (Wajãpi), muyapi (Waurá), paoko xihi, paro koxihi (Yanomami), kuãxihã (Yudjá)

Colombia Puinave, moradilla, corazón negro, tres tablas, angolito, tamarindo, tamarindo de monte, brazo de pelejo, pepa agria, pichico, algarrobillo, granadillo, hueso; jicka (Karijona), kumaricama (Makuna), wo pee (Puinave), deteicge (Ticuna), llano nokairo (Witoto)

Ecuador Piedrita, tocota, cafesillo, cafetillo; teéshnumi (Achuar), machcu muyu (Kichwa), dicarimo, dikaremo, meñemo (Waorani)

Peru Palo de sangre, palo sangre, palisangre, sangre sangre, palo santo, navaha shimbillo, charapilla, charapillo, huitillo, azúcar huaillo, achapaqui, tamarindo, pelejo, polvorilla; patyujke (Bora), mbaki (Shuar), yaiño amena (Witoto)

Venezuela Tamarindo, tamarindo rebalsero, barbasco, cedro dulce, azúcar huaillo, ebanillo, dede, bergatin, cacho; ogchiri (Barí), motua aji, motuwa kë hi, sipararimi kofi (Yanomami), dede (Ye'kuana)

Status Wild, spontaneous in cultural settings, planted

Pororoca *Dialium guianense* fruits gathered from the ground in a home garden. Arapixuna, Igarapé Jari, Município de Santarém, Pará, Brazil 5-15-96

Pororoca *Dialium guianense* is a tree that can reach 35 m and drops its brown fruits, the size of large olive pits, during the rainy season and into the early part of the dry period. Children often pick them up to snack on the thin brown pulp surrounding the single seed. Pororoca is found in the forest as well as successionary vegetation close to settlements on both uplands and floodplains in Amazonia from the Amazon estuary and Maranhão in the eastern fringes of the Amazon Basin to the Andean foothills (Araujo-Murakami et al. 2015; Balée 1993b; Boubli 2002; Cain et al. 1956; Campbell et al. 1986; Carim 2016:89; Carvalho et al. 1986; Cattanio et al. 2002; Coelho et al. 2012; Condé 2011:75; Condé and Tonini 2013; Gama et al. 2002, 2003; Giraldo-Cañas 1999; Gomez 2011:77; Lemos et al. 2015; Lisboa et al. 1991; Macedo 2014:55; Muñoz 2010:23; Navarro 2002; Pires and Salomão 2000; Reis 2010:33; Salomão et al. 1988, 2007a, b; Silva and Rosa 1989; Silva et al. 2013; Taurino 2017:93), including islands on the floodplain of black water rivers (Souza and Silva 2003). It is also found in light gaps and within woodlands dominated by native bamboo, *Guadua weberbaueri* in Acre in the southwestern portion of the Brazilian Amazon (Maranho et al. 2016). In the cerrado region in the southern fringe of the Amazon Basin, this tree occurs in gallery forest (Campos Filho 2012:220; Kuchla et al. 2015). In the eastern edge of the Amazon rainforest, this species is rare in the mature forest but common in fallows (Balée 1993a:120). It also grows in the Atlantic forest of Brazil as far south as Espírito Santo and inland in Minas Gerais (Lorenzi 2009:107). The range of *D. guianense* extends north into the Guianas, Venezuela, and in Central America as far north as Mexico where the fruits are also eaten (Díaz Pérez et al. 2012; Gentry 1992; Godoy et al. 1999).

Girl with pororoca *Dialium guianense* fruits she picked from a low hanging branch in her village. Arapixuna, Igarapé Jari, Município de Santarém, Pará, Brazil 5-15-96

Pororoca fruits resemble miniature tamarind *Tamarindus indica* fruits, hence one of the local names for the fruit in Bolivia, tamarindo del monte (tamarind of the forest) and tamarina and tamarindo in parts of the Brazilian Amazon. Tamarind is also in the legume family and is native to Africa; it is widely cultivated in tropical America and Asia. The Huaorani in the Napo watershed of eastern Ecuador eat both green and ripe fruits of pororoca, either fresh or after soaking them in water (Davis and Yost 1983). The Shuar in the Peruvian Amazon eat the fruits as well as boil them and then drink the juice.

Boys picking up fruits under a pororoca *Dialium guianense* tree in their village. Arapixuna, Igarapé Jari, Município de Santarém, Pará, Brazil 7-22-96

Pororoca is found in some rural home gardens, such as in the vicinity of Abaetetuba along the lower Tocantins River in eastern Pará (Gonçalves and Lucas 2017) and at Arapixuna along Igarapé Jari near Santarém. The Ka'apor in Maranhão on the eastern fringes of the Amazon forest plant *Dialium guianense* in some of their home gardens (Balée 1993b, 2013:68). Pororoca is also found in anthropogenic forests that have been deliberately enriched with fruit and nut trees, such as on the outskirts of Comunidade Silencio along Igarapé do Mamauru, near Óbidos, Pará.

The wood of pororoca is used for a variety of construction purposes because it is durable, as for example in the vicinity of Jaci Paraná in Rondônia (Silva and Bentes-Gama 2008) and along the Putumayo River in the Peruvian Amazon (Gomez 2011:77). It is one of the preferred timber trees in the Colombian Amazon because of its durability and is used for a variety of construction purposes including bridges and flooring (Duarte 1979:55; Quevedo and Zuluaga 1994; Loureiro et al. 1979:200). In southern Ecuador, the wood from this tree is used for making furniture (Aguirre-Mendoza et al. 2015:43). One of the reasons that the wood of this tree is popular is because of its durability (Loureiro and Lisboa 1979). Rural folk along the Jauaperi River that forms the boundary between the states of Amazonas and Roraima in northern Brazil use the inner bark and resin of this tree to caulk their boats (Ammann 2014:38). Over time, this tree may become increasingly scarce as the timber trade continues expanding in the Amazon.

Dialium guianense fruits evolved for dispersal by mammals and birds, such as Spix's guan *Penelope jacquacu* (Pinto and Setz 2004). During the 6 months that pororoca is in fruit it is the most important source of fruits for at least one species of howler monkey *Alouatta belzebul* (Pinto 2002). In Mexico, another species of howler monkey *Alouatta palliata* also eats these fruits and defecates the seeds intact (Estrada and Coates-Estrada 1984). Woolly monkeys *Lagothrix lagotricha*, spider monkeys *Ateles belzebuth* and *A. paniscus*, and howler monkeys *Alouatta belzebul*, which are hunted widely in the Amazon Basin for their meat, also eat the fruits of *D. guianense* (Defler and Defler 1996; Ghilardi and Alho 1990; Nunes 1998; Peres 1994a; Stevenson 2005; Van Roosmalen 1985; Van Roosmalen and Klein 1988). Pororoca thus contributes indirectly to the rural diet via game meat. The endangered black-bearded saki *Chiropotes satanas* also eats this fruit but is not normally hunted for its meat because of its relatively small size and rarity (Veiga 2006:97).

This tree also has other uses. Pororoca wood is highly durable, and the Yanomami use it when building their communal houses (Milliken and Albert 1997a). The Waorani in the Ecuadorian Amazon also use the wood when constructing their homes and for fuel (Cerón and Ayala 1998:49). Rural people in the Tapajós-Arapiuns extractive reserve in western Pará use the wood to make furniture (McGrath et al. 2004). The Tiriyó who live in the border zone between northern Pará and Surinam squeeze liquid from the inner bark onto wounds to promote healing (Cavalcante 1973:98).

Its common name in Pará means tidal bore which only occurs in the Amazon estuary. Pororoca tidal bores occur mostly in March and April and are increasingly popular among surfers because they can ride them for extended periods, unlike waves. Why *Dialium guianense* acquired the common name of pororoca in many parts of the Amazon is a mystery. A variety of bitter manioc cultivated by farmers living in the village of Flexal near Óbidos and at Juriti is called Pororoca.

24.4 *Hymenaea courbaril*

Bolivia Paquió, paquió fruta grande, paquiocillo, papuró, carpoteca, pata de buey, algarrobo, jacojaco, jatobá, isigo, pito, copal; corama, cura pisi (Chácobo), nobori-piaquiorr (Chiquitano), bejqui' (Tsimane)

Brazil Jatobá, jatobá de fava, jatobá da mata, jatobá do cerrado, jatobá mirim, jatobá grande, jutaí, jutaí açu, jutaí bravo, jutaí grande, jutaícica, jataí, jataúba, farinheira; yɨta'i (Arawaté), itawa-'ɨ (Guajá), kóh (Hupd'äh), tarapa'ɨ, trapaɨ-pitag (Ka'apor), môj, moi, môtx têtê (Kayapó), uagi (Kuikuru), maze (Rikbaktsa), ma'ate, woaté (Sateré-Maué), mandéa (Suruí), zuta-'ɨwa-ɨw (Tembé), ke'rõ (Tukano), kero (Tuyuka), tirikir (Wapixana), a' ō wede, a'rõ (Xavante), hatohato koxihi (Yanomami)

Colombia Algarroba, algarrobo, algarrobillo, guapinol, guapinol, nazareno, pecueca, copa, copal, jatuba, nazareno

Ecuador Copal, courbaril

Peru Azúcar huayo, azúcar huayo blanco, ámbar blanco, courbaril, copal, copal americano, copal caspi, copalhuallo, huarana, algarrobo, lacre colorado; zumpa (Amahuaca), mííjíllehe (Bora), koveni (Matsigenka), yuru (Shuar)

Venezuela Algarrobo, corobore; muyeja jyaï, bũye yaɨ (Jodï), arõ kofi, arõ kotho (Yanomami), wakuda (Ye'kuana)

Status Wild, planted

Marianne, 8 years, with jutaí *Hymenaea courbaril* fruits she gathered on the floor of an anthropogenic forest on an archaeological site at the perimeter of her community. Vila Associação, mouth of the Caxiuanã River, Município de Melgaço, Pará, Brazil, 5-18-15

Jatobá *Hymenaea courbaril* is a towering tree, soaring up to 50 m, much sought after by loggers because the reddish-brown wood is highly durable (Le Cointe 1922:28; Medina et al. 2009; Miranda et al. 2000; Pérez-Arbeláez 1956:274). Also known as jutaícica, this member of the legume family is more common in upland forest, but it is also occasionally found on floodplains (Absy et al. 1986; Amaral et al. 2009; Araujo-Murakami et al. 2009; Brito et al. 2006; Cain et al. 1956; Ducke 1949:97; Martins et al. 2008; Medeiros et al. 2012; Pinheiro et al. 2007; Salomão et al. 2007a; Santos et al. 2016b), including the tidal zone in the lower Amazon region (Almeida and Jardim 2011, 2012; Carim 2016:89; Carim et al. 2008; Lau and Jardim 2013; Scoles 2009). Because some of the trees are found on floodplains, tambaqui *Colossoma macropomum*, known as gamitana in Peru, feed on some of the fruits during the high-water season (Baca 2005). Tambaqui is one of the most prized fish in the Amazon. This tree is also found in some successionary forests (Balée 1993b; Carvalho et al. 1986; Condé 2011:76; Dionisio et al. 2016; Ebert et al. 2014; Espírito-Santo et al. 2005; Rayol et al. 2006, 2011).

Jatobá is widespread in Amazonia, including the southern and eastern fringes of the basin as it transitions to scrub savanna (Abreu et al. 2014; Araujo-Murakami et al. 2015; Ferreira Júnior et al. 2008; Furley et al. 1978; Haidar et al. 2013; Ivanauskas et al. 2003, 2004; Kunz et al. 2008; Lemos 2013:37; Mews et al. 2011a, b, 2012; Muniz et al. 1994; Nascimento et al. 1997; Navarro 2002; Ratter 1987; Rondon Neto et al. 2010) in the Andean foothills, such as in Bolivia (Araujo-Murakami and Ruiz 2006) and in the highlands of Surinam (Bánki et al. 2003:23) as well as woody savannas in other parts of the Amazon Basin such as Amapá and central Pará (Amaral et al. 2019). This tree can grow to at least 300 years old (Locosselli et al. 2017).

Bats appear to be the main pollinators, although hummingbirds and moths are also involved (Lacerda et al. 2008; Lee and Langenheim 1975:38). Only a few flowers mature each day over several weeks thus encouraging pollinators to seek other jatobá trees as they feed thus spreading pollen around in what is known as a trapline syndrome. If the forests continue to shrink in Amazonia, the long-term survival of jatobá may be imperiled.

Jutaícica *Hymenaea courbaril* tree in fruit in floodplain forest along a black water river. Anapu River, Caxiuanã National Forest, Município de Melgaço, Pará, Brazil 7-7-16

This giant tree is found throughout Amazonia up to about 1300 m in the Andean foothills as well as the Guianas, some Caribbean islands, and Central America. It also occurs in gallery forest in the scrub savanna region of the Araguaia and Tocantins river basins in the southeastern fringe of the Amazon Basin as well as central and northeastern Brazil (Cabacinha and Fontes 2014; Kern et al. 1992; Marimon et al. 2002; Silva et al. 2016b; Stefanello et al. 2009). Although the fruit pods float it is possible that indigenous people took *Hymenaea courbaril* from northern South America to some of the islands in the Caribbean, such as Puerto Rico (Janzen 1975). Indigenous people in the Amazon have been gathering the pods of this giant tree for thousands of years as indicated by remnants of seeds in kitchen middens in a sandstone cave near Monte Alegre along the lower Amazon (Roosevelt 2000).

Rural folk in the Amazon gather the dark brown fruit pods, which range in size from 7 to 17 cm in length and between 4 and5 cm wide, to snack on the mesocarp surrounding the reddish-brown seeds which measure 2 cm × 1.3 cm. The pods can reach 21 cm in length (Langenheim et al. 1973). The seeds vary in number depending on the length of the pod. The tree is far too tall to climb when mature, so young and old alike gather the fallen pods from the forest floor in the dry season to snack on the dry, powdery light tan pulp surrounding the seeds. The pods are also gathered along rivers because they float, such as along the black water Rio Curuá in the Caxiuanã National Forest. The pods of younger trees on floodplain can also be reached from canoes. Although most of the fruit is produced in the rainy season, the hard pods protect the powdery mesocarp surrounding the seeds well into the dry season.

Jutaícica ***Hymenaea courbaril*** **pod floating in a black water river. Curuá River, Caxiuanã National Forest, Município de Melgaço, Pará, Brazil 7-4-16**

Jatobá can be quite prolific after it reaches maturity between 8 and 10 years old; as many as 3000 fruits can be produced annually (Souza et al. 1996). Several game animals feed on the fruits including peccaries (*Pecari tajacu* and *Tayassu pecari*); peccaries eat the pulp and discard most of the seeds onto the ground (Terborgh et al. 1993) although some of the seeds are also eaten (Beck 2005a, b). Other much-appreciated game animals, such as tapir *Tapirus terrestris*, and paca *Agouti paca*, also break open the hard wooden pods to eat the sweet mesocarp. Another game animal, agouti *Dasyprocta* spp., is the main dispersal agent for jatobá because the dachshund-sized rodent carries away some of the seeds, not all of which are retrieved (Asquith et al. 1999; Cummings et al. 2018; Gorchov et al. 2004; Hallwachs 1986; Janzen 1983). In Costa Rica, agoutis *Dasyprocta punctata* disperse the seeds of *Hymenaea courbaril* up to 200 m from the parent tree (Hallwachs 1986), a pattern likely repeated in Amazonia. Spider monkeys *Ateles belzebuth, A. paniscus*, Geoffroy's woolly monkey *Lagothrix cana*, northern bearded sakis *Chiropotes satanas chiropotes*, and bald uakaris *Cacajao calvus ucayalii* eat the fruits while red howler monkeys *Alouatta seniculus* eat the young leaves of this forest giant (Aquino and Encarnación 1999; Iwanaga and Ferrari 2001; Nunes 1998; Simmen and Sabatier 1996; Van Roosmalen and Klein 1988; Van Roosmalen et al. 1988). In the Colombian Amazon, woolly monkeys *Lagothrix lagotricha* have been reported eating these fruits when they are not yet ripe, presumably because it is easier to access the mesocarp surrounding the seeds (Stevenson 2005). These animals are all hunted for their meat by rural people.

Jatobá *Hymenaea courbaril* pod. Bacuri, Lago de Tefé, Amazonas, Brazil, 8-27-12

A farmer living along the lower Tefé River told me that the pulp is so dry it sometimes makes you choke (*entala*). The savory pulp surrounds the one to four reddish black seeds that measure 3 cm by 2 cm. In the nineteenth century, a British explorer traveling in what was then British Guiana described the mesocarp thus: "a dry fluffy substance of a light, yellowish color, which is good to eat, tasting not unlike a dry cake, being sweet, and melting in the mouth" (Brown 1876:180). Perhaps because the pulp is so powdery the Chácobo in the Bolivian Amazon use it to make a beverage (Boom 1987:24).

Vanessa, 15 years, eating jatobá *Hymenaea courbaril* **fruit purchased in a street market. Itacoatiara, Amazonas, Brazil 11-15-02**

With a piece of wood serving as a mallet, kids crack open the hard pods to reveal the prize: pale orange flesh which is so crumbly that it leaves smudges on one's lips, like confectionery sugar. Adults open the pods by stomping on them while wearing shoes or boots or using a machete to cut open the pods. Enough aficionados of the sweet, mealy pulp live in towns and cities to warrant taking some of the rear-view mirror-sized pods to market, such as Santarém and the Ver-o-Peso in Belém (Rego et al. 2011).

Boy cracking open pods of jatobá *Hymenaea courbaril* **in his backyard. Itapiranema, Lago de Tefé, Amazonas, Brazil, 8-25-12**

Jatobá is also used in various folk remedies. Jatobá's light gray bark is smooth and often partially covered with colorful patches of yellow lichen and locals collect the bark to make a tea to treat diabetes, as in the Amazon estuary. Residents of Quilombo Sangrador, a village in the municipality of Presidente Juscelino in Maranhão, use this bark to make a solution to treat blood disorders (Monteles and Pinheiro 2007). Some sawmills in the vicinity of Belém sell the bark of this tree to vendors who then offer it in street markets specializing in medicinal plants; sometimes strips of the bark are removed before dispatching the logs to sawmills (Shanley and Lux 2003). A tea from the bark is sold in the main street market of Santarém at the junction of the Amazon and Tapajós rivers to treat respiratory problems (May 2016). Inhabitants of Caruarú, a rural community on Mosqueiro Island some 70 km northeast of Belém, make a tea from the bark exudate to treat coughs (Mesquita and Tavares-Martins 2018). The Xipaya indigenous people along the lower Iriri River, an affluent of the Xingu, make a tea from the bark to treat colds and anemia (Santos et al. 2016a). In several parts of Rondônia rural folk use the inner bark to make a decoction to treat influenza (Santos et al. 2014b). According to Paul Le Cointe, a French agronomist who lived much of his professional life in the Brazilian Amazon in the early twentieth century, jatobá is a diuretic, and thus helpful in treating one of the symptoms of diabetes: excessive fluid retention (Le Cointe 1934:226). In the Municipality of Ariquemes in Rondônia, the inner bark is used to make a decoction to treat colds, anemia, inflammation, and malaria (Santos et al. 2008). The Wapixana in Roraima make a decoction from the bark which is drunk to treat the after effects of malaria (Milliken 1997). Also in Roraima, some rural folk use the bark and hard fruit case to make a tea to treat colds, coughs, and catarrh (Van den Berg and Silva 1988). The inner bark is also used to treat helminthic infestations and the resinous sap is employed to alleviate a variety of health issues from enlarged prostrate to bronchitis (Moura et al. 1993). Residents of Santa Luzia do Buiuçuzinho, a small village some 20 km west of Coari, Amazonas, make a tea from the bark to treat a variety of medical issues including diarrhea, stomachache, coughs, and cancer (Silva 2009:72). A fruit vendor in Rio Preto da Eva, a village near Manaus, explained that the emptied fruit capsules of jatobá also have medicinal value: the fruit cases are soaked overnight, and the infusion is swallowed to treat sore throat and hepatitis. Along the Unini River, an affluent of the Rio Negro in Amazonas, Brazil, rural folk use the resin to prepare remedies for flu, sore throat, and coughs (Lago et al. 2016).

Several communities in the Municipality of Manacapuru along the lower Solimões near its confluence with the Rio Negro employ the bark of this giant tree to make a tea or syrup to treat coughs, enlarged prostrate, discomfort caused by a hernia, malaria, and anemia (Vásquez 2014). In the municipalities of Xapuri in Acre and Pauini in southern Amazonas state, rural folk use the bark of this tree to make a decoction for treating liver issues, fever, and anemia (Ferreira et al. 2015). In the Bolivian Amazon, the Guarasug'we employ the bark to make a decoction for treating coughs, colds, and anemia, while the seeds are used to make a decoction for alleviating dysentery (Hajdu and Hohmann 2012). In Amapá, the inner bank is used to make a decoction which is mixed with syrup to treat coughs and bronchitis (Tenório et al. 1991). In Roraima, the inner bark is used to make a tea for treating anemia and colds (Luz 2001). In that same state in the northern part of the Brazilian Amazon, rural people also use the leaves to make a tea for treating colds and hemorrhoids. Along the Caxiuanã River, an affluent of the Anapu in Pará, rural folk use the bark to make a syrup (*xarope*) to treat sore throat. The resinous sap (*seiva*) is mixed with condensed milk and red wine and drunk to alleviate body aches and to help put on weight. In the Peruvian Amazon, the fresh bark is boiled in water and after it cools is drunk every 9 hours to treat worm infestations and diarrhea and to alleviate rheumatism (Mejia and Rengifo 2000:40). In Suriname, the resin and bark of this tree are sold in some street markets for treating hypertension and diarrhea (Van Andel et al. 2007).

Jatobá *Hymenaea courbaril* trunk with some of the bark removed to treat sore throat. The bore hole below was made to extract sap (*seiva*) to treat body aches. David Lucas 7 yrs. Vila da Associação, Caxiuanã River, FLONA Caxiuanã, Municipio de Melgaço, Pará, Brazil 7-16-16

The Arara use the resin to start fires and to caulk canoes and boats at Terrawangã village on the right bank of the Xingu River a little upstream from its confluence with the Bacajá. The Assurini who live in various villages inland from the right bank of the middle Xingu employ the resin to coat pottery, especially if it has been painted, and as a glue when putting together arrows used for hunting (Lukesch 1976:77; Silva 2008a). River dwellers use the resin to coat some of their homemade pottery in the Brazilian Amazon, such as on Marajó Island (Prance 2014:192) and in an around the Caxiuanã National Forest in the Anapu watershed in Pará (Silveira et al. 1997). The Chácobo in the Bolivian Amazon burn this resin to create light (Boom 1987:24, 1989). When the resin of jatobá crystallizes after long exposure to the air it sparkles and the Kuikuru of the Upper Xingu believe that jaguar eyes, which shine brilliantly at night, are made of it (Carneiro 1978).

A myth of the Apinajé recounts how a jaguar rescued a boy on a cliff face and took him to his abode, a jatobá trunk that was burning at one end. The Indians were unacquainted with fire at that time and ate only meat dried in the sun. The boy asked what is smoking and the jaguar explained it was a fire that would keep them warm at night. The boy eventually returned home to his village and the Indians set out to find the fire he talked about and the jaguar gave them the fire (Lévi-Strauss 1975:69).

Jatobá *Hymenaea courbaril* log containing a hive of uruçú *Melipona* sp. bees. The hive containing the stingless bees is suspended from a mango tree in a rural home garden. Raimundo Roxa, 38 yrs. Ilha dos Arara, middle Xingu River upstream from Altamira, Pará, Brazil 9-2-15

The wood and bark of jatobá are also used by indigenous peoples and river dwellers. A *ribeirinho* on Ilha dos Arara, along the middle Xingu upstream from Altamira, has a hive for native stingless bees fashioned from a jatobá log. A section of the log is suspended under a mango tree in his home garden. The bees, called uruçú *Melipona* sp., are not aggressive but will swarm and nibble anyone who disturbs their nest. Resin is tapped from the trunk to make a wax and the bark is used to make a tea for treating coughs (Parrotta et al. 1995:162). The wood of this giant tree is used for a variety of construction purposes by rural folk including for making posts, house flooring, and furniture (Loureiro and Lisboa 1979; Soares et al. 2018). The Taurepang and Sapará indigenous communities in the savanna region of Roraima use the wood of this tree to make charcoal (Perez 2011:28).

Indigenous people across the Amazon Basin also use this wood for a variety of purposes. The Deni who live in the interfluvial forest between the Juruá and Purus rivers in Amazonas, Brazil, employ jatobá wood to fashion house posts (Pezzuti and Chaves 2009), as do the Chácobo and Tsimane in the Bolivian Amazon (Reyes-García 2001:245; Zambrana et al. 2017). The Ka'apor who live in 12 villages in the eastern fringes of the Amazon forest in the state of Maranhão use the wood to make a trough to hold manioc dough (Balée 1993a:69). Several indigenous groups make canoes from the bark of jatobá including the Chácobo in the Bolivian Amazon, the Kuikuru, Mehinánaku, and Kalapálo in the Upper Xingu, the Tenetehara along the Pinaré River in Maranhão, the Mura-Pirahã along the lower Madeira, the Pirahã along the Marmelos and Maici rivers in the Madeira watershed, and the Wapishana in northern Roraima and neighboring Guyana (Boom 1987:24; Carneiro 1956, 1978; Cunha 1960:72; Farabee 1918:74; Gonçalves 2001:115; Lima 1950; Rodrigues and Oliveira 1977; Indios Kuikuru 2007:112; Wagley and Galvão 1961:218). The Kalapálo do not cut down the giant forest tree; rather they erect a platform to cut out a large section of the bark. Formerly, they used stone axes to undertake this task, but now they use steel axes obtained by trading (Cunha 1960:73).

Jutaí *Hymenaea courbaril* tree at the edge of a community. Vila Associação, Rio Caxiuanã, FLONA Caxiuanã, Pará, Brazil 7-3-16

Jatobá is occasionally found at the margins of home gardens, especially on archaeological sites such as at Sitio São Gabriel, Rio Aneba, an affluent of the lower Urubu in Amazonas state and at Vila Associação at the mouth of Caxiuanã River in eastern Pará. River dwellers take care to locate their houses a safe distance from the tree, however, in case it is struck by lightning which can break off a branch. An 80-year-old ribeirinho who lives on an island in the middle Xingu explained that he planted the jatobá on the perimeter of his home garden some 50 years ago when he worked as a rubber tapper. His property is named after the tree: Sitio Jatobá, which is located on Ilha da Costa. In the savanna region of Roraima, some indigenous people plant this tree in their home gardens (Pinho 2008; Pinho et al. 2011). Several residents of the rural community of Colônia Central in the Reserva de Desenvolvimento Sustentável Tupé, a short distance up the Rio Negro from Manaus, Amazonas, maintain this tree in their home gardens (Santos 2006:52). Several rural folk living in the community of São Miguel in the municipality of Várzea Grande in Mato Grosso also plant this tree in their home gardens (Mamede and Pasa 2018), as do some residents of the São Gonçalo Beira Rio district in Cuiabá, the capital of Mato Grosso (Mamede et al. 2015), the community of Sucuri in the municipality of Cuiabá (Gonçalves and Pasa 2015), and rural folk in the community of Boca do Môa, municipality of Cruzeiro do Sul in Acre (Martins et al. 2012). Some colonists in the vicinity of km 107 of the Altamira-Itaituba stretch of the Transamazon Highway in the municipality of Medicilândia in Pará save this tree in their cacao orchards because of its many uses (Leão et al. 2017). In French Guiana, this giant tree is sometimes planted in parks (Hoff and Cremers 2005).

The Jodï who live in the southwestern portion of the Orinoco Basin protect *Hymenaea courbaril* and other useful plants in fallow vegetation (Zent and Zent 2012). Farmers living in Carrazedo at the mouth of the Xingu typically leave a belt of successionary forest around jatobá trees to protect them from fires set to clear fields for planting manioc.

Jatobá *Hymenaea courbaril* tree (center) surrounded by a belt of successionary forest to protect it from fire. Manioc field in foreground. Inland from Carrazedo, mouth of the Xingu River, Município de Gurupá, Pará, Brazil 10-25-16

Commercial logging, though, has reduced the availability of this fruit. Logging has greatly reduced the number of reproductive trees of *Hymenaea courbaril* in the Brazilian Amazon, thereby shrinking the gene pool (Carneiro et al. 2011). Jatobá is one of the main targets of loggers in the vicinity of Paragominas in eastern Amazonia (Veríssimo et al. 1992). In addition to furniture, highly durable jatobá wood is used to stake black pepper *Piper nigrum* plants on farms in the vicinity of Tomé-Açu, Pará. Along the Rio Jacaré-Puru, a right bank affluent of the Pacajá, a family was observed using a manually operated wheel made from jatobá wood to grate manioc tubers in 2000. However more and more rural people are now using gasoline-powered graters to make manioc flour.

24.5 *Hymenaea oblongifolia*

Brazil Jutaí, jutaí mirim, jutaí da folha grande, jutaí da folha fina, jutaí da várzea, jutaí açu, jatobá, jatobaí

Colombia Algarroba, algarrobo, algarrobillo, jutaki, tamarindo, vaca paleta; kev nui (Barasana), caurú (Karijona), karápa (Nukak), wa ka ma (Puinave), makurina (Witoto)

Ecuador Caimitillo pequeño, azúcar huayo, azúcar muyo; yurú (Achuar), sojo (Siona-Secoya), guiramoncahue, wuitamoncahue, minamocawe (Waorani)

Peru Azúcar huayo, azúcar huayo chico, alfaro negro, jutahy, yutobanco, goma huayo; koveni (Matsigenka), yúru (Shuar)

Status Wild, planted

Jutaí *Hymenaea oblongifolia* fruits gathered in forest. Aninduba, Amazon River, Santarém Municipality, Pará, Brazil 7-23-96

Hymenaea oblongifolia is a forest tree along the banks of sediment-rich and clear water rivers in widely scattered locations throughout Amazonia from the estuary to around 1000 m in the Andean foothills in Colombia, Ecuador, Peru, and Bolivia (Gama et al. 2003). The rough, gray-barked tree also grows in forests in the Guianas. Capable of soaring to 50 m, *Hymenaea oblongifolia* also grows in mature upland forest, such as in the vicinity of Belém (Pires and Salomão 2000), along the lower Xingu (Salomão et al. 2007a), in the vicinity of Lábrea in the state of Amazonas bordering Acre and Rondônia (Pereira et al. 2005), and is thus another fruit tree in the region that spans both wetlands and non-flooded environments. When fully mature, *Hymenaea oblongifolia* towers above the canopy making it an attractive nesting site for scarlet macaws *Ara macao*, iconic birds with their distinctive loud squawk and brilliant red plumage (Brightsmith 2005). Jutaí also occurs in regrowth forests, such as in the vicinity of Iquitos, Peru (Macedo 2014:55) and in 5-year-old successionary vegetation in the Peruvian Amazon bordering Colombia (Gomez 2011:77; Muñoz 2010:17). Rural folk appreciate and often protect this tree because of the fruits and wood used for making poles and posts. Ribeirinhos also use the wood to fashion wheels for operating manioc graters, such as in the Caxiuanã National Forest in eastern Pará.

Farmer gathering jutaí *Hymenaea oblongifolia* fruits at the edge of his manioc field. Near Porto Novo, Rio Tapajós, Pará, Brazil 11-21-02

This giant legume produces pods the size of an adult thumb that encase a sweet green mesocarp surrounding the one or two seeds. The pulp is eaten after first cracking open the hard-wooden pods with a machete or club. Aptly named the "sugar fruit" (azúcar huayo) in Peru, the reddish-brown pods are flecked with a tan-colored resin, presumably to discourage insect predators, particularly beetles. The pulp tastes like avocado but has a dry and powdery texture. The mesocarp is also eaten by woolly monkeys *Lagothrix lagotricha* which are hunted throughout the Amazon Basin for their meat (Defler and Defler 1996; Rimachi-Taricuarima et al. 2019; Stevenson 2007) and this species of monkey likely disperses some of the seeds. Bald uakaris *Cacajao calvus ucayalii* also eat the fruits and this subspecies is hunted for its meat in the Peruvian Amazon (Aquino and Encarnación 1999). In the Peruvian Amazon, a game bird pava de monte *Pipile pipile* is known to eat this fruit and disperse some of the seeds (Begazo 1997).

In the vicinity of Aninduba, a village overlooking the Amazon floodplain a little upstream from Santarém in Brazil, children search for the avocado-shaped pods on the floor of both upland and wetland forests in the dry season. People along this stretch of the Amazon have been eating fruits of jutaí for a long time; the archaeologist Anna Roosevelt has found carbonized fruits of *Hymenaea oblongifolia* in caves and rock shelters at Serra do Pilão on the opposite side of the river to Aninduba (Roosevelt et al. 1996). Hunters and gatherers occupied Serra do Pilão over 10,000 years ago and left behind some intriguing geometric and animistic rock art.

Locals gash the bark of jutaí so that it exudes a translucent yellow resin, which is employed by a few old timers to varnish pottery. In Itapuã, a small village near Vigia at the mouth of the Amazon, a 72-year-old widow uses a stick to spread the resin on the inside of jars and bowls when they came out of the kiln. The ball of resin at the end of the stick melts because the pottery is still hot; when it cools, the resin dries to an impervious glaze. The kiln is a backyard affair, dug into the ground and fired with charcoal. Rural folk on nearby Marajó Island also use jutaí resin to glaze pots (Beck and Prance 1991). The pots are glazed when they are still hot rubbing a lump of the resin over the pot (Prance 2014:190). The use of jutaí resin for varnishing pots is dying out as people increasingly buy plastic, aluminum, and ceramic dishware made in factories. The Yukuna in the Colombian Amazon use the resin to treat fungal infections of the feet (Schultes and Raffauf 1990:242). Along the lower Ucayali in the Peruvian Amazon, some rural folk use the bark to prepare a folk remedy for treating diarrhea (Tananta 2014:65).

As with several other fruit trees, jutaí is the target of loggers on account of its fine grain which is highly suitable for hardwood flooring. Rural people along the lower Tocantins use this timber for construction purposes, including boat building as well as for some crafts (Santos et al. 2004). This wood is also used for a variety of construction purposes in the Colombian Amazon (Duarte 1979:59) and the Peruvian Amazon (Tananta 2014:65). Although the amount of *Hymenaea oblongifolia* timber that is currently traded on world markets is limited, the wood is advertised on the internet and demand is likely to grow for the rich, reddish-brown wood.

24.6 *Hymenaea parvifolia*

Bolivia Paquiosillo, paquiocillo, paquió fruta pequeña

Brazil Jutaí, jutaí mirim, jutaí pororoca, jutaí cica, jutaí pequeno, jutaí curuba, jutaí vermelho, jutaí do campo, jutaí café, jatobá, jatobá mirim, jatobá curuba, jatobazinho, louro, comer de arara; yúta-‘i, (Araweté), yeta-‘iwa (Asurini), wɨta-i-‘ɨ (Guajá), jeta’i’y, yeta-‘ɨ (Ka’apor), zuta-‘i-ɨw (Tembé), yɨta-‘i (Wajãpi), ãro kohi (Yanomami)

Colombia Algarrobo, copal; pokopá (Andoke)

Peru Azúcar huayo

Status Wild, planted

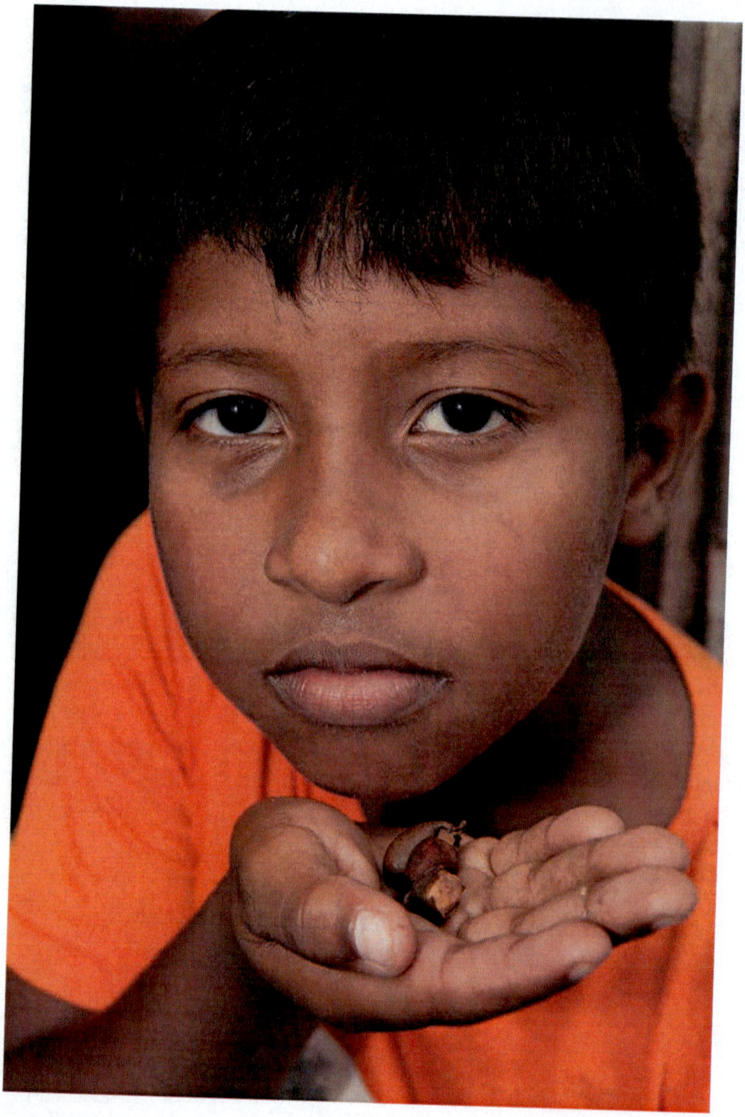

Paulo André, 11 years, with jutaí mirim *Hymenaea parvifolia* fruits gathered in upland forest. Near Caiçara, Boca do Rio Curuá, Caxiuanã National Forest, Município de Melgaço, Pará, Brazil 5-7-15

Hymenaea parvifolia is found throughout the Amazon Basin in mature upland forest as well as forest regenerating on abandoned agricultural plots, such as in the Bragantina zone in northeastern Pará (Almeida and Vieira 2001), in the Tocantins watershed in eastern Pará (Taurino 2017:89), and on the Belterra plateau behind Santarém (Alvino et al. 2005; Amaral et al. 2017; Araujo-Murakami et al. 2015; Baar et al. 2004; Carim et al. 2007; Carvalho et al. 1986; Lemos et al. 2015; Moser 2013:84; Navarro 2002; Pinheiro et al. 2007; Rayol et al. 2006; Silva et al. 2016a; Souza 2012:50). It is also found in forests of indigenous territories in Pará and neighboring Maranhão (Balée 1993b, 2013:65). This tree also regenerates in some areas that were formerly mined, such as near Porto Trombetas in Pará (Salomão et al. 2007b).

It is found in forest light gaps and in woodlands dominated by native bamboo *Guadua weberbaueri* in Acre in the southwestern portion of the Brazilian Amazon (Maranho and Salimon 2015; Maranho et al. 2016). It is also found in some grasslands with scattered trees in the lower Amazon, hence one of its common names: jutaí do campo (Huber 1910). It grows on a wide variety of soils, from clay to sandy areas covered with low forest (*campinarana*) and scrub savanna (*cerrado*, *cerradão*) in rocky areas (Langenheim et al. 1973; Lima 1958; Rayol 2006:35; Silva 2008c:94). It is also occasionally found in forests on the higher parts of floodplains, such as along the upper Madeira River and some of its affluents in Rondônia (Casula 2012:59; Reis 2010:33), along the black water Anapu River in eastern Pará (Ferreira et al. 2013b), and in the Amazon estuary near Belém (Amaral et al. 2008; Black et al. 1950). This tree can soar to 45 m, such as on a floodplain in the vicinity of Manacapuru, Amazonas, and because of the tree's height it is used by harpy eagles *Harpia harpyja* to build their nets (Luz 2004:11,35). This tree also occurs in the eastern fringe of the Amazon forest in Maranhão (Muniz et al. 1994) and is sometimes a short tree, reaching some 20 m when fully mature (Lee and Langenheim 1975:62). This tree is occasionally found in some home gardens, such as in the São Raimundo district of Manaus, Amazonas (Barbosa 2018:42), in some rural communities in the vicinity of Manaus (Souza 2010:27), and in the watershed of the lower Purus River (Silva 2006:131). This tree can live for at least 371 years (Locosselli et al. 2017).

One of the common names for *Hymenaea parvifolia* in the Brazilian Amazon, jutaí mirim, means the small jutaí and is so named because it bears the smallest fruits of the three species of *Hymenaea* that are eaten in the region. The brown fruits of jutaí mirim are only 2–4 cm long by 1.1–2.5 cm but the thin, pale orange mesocarp surrounding the single seed is sufficiently tasty to warrant gathering the fruits in upland forest. The powdery pulp is accessed by cracking open the crisp skin (epicarp), using one's teeth. The tree can soar to 45 m so is too tall to climb and gather the fruits; rural people, therefore, wait for them to drop to the ground in the rainy season. According to the Yanomami, an indigenous group in Roraima and contiguous forest in southern Venezuela, the fruits are also eaten by some game animals, such as white collared peccary *Pecari tajacu*, white-lipped peccary *Tayassu pecari* and woolly monkeys *Lagothrix lagotricha* (Albert and Milliken 2009:65; Defler and Defler 1996). Woolly monkeys have also been seen eating the fruits of *Hymenaea parvifolia* in the Tefé watershed, an affluent of the Solimões in Amazonas,

Brazil (Peres 1994a). As with many Amazonian fruits, *Hymenaea parvifolia* thus contributes indirectly to the human diet via wild meat. The red-nosed bearded saki *Chiropotes albinasus* has been recorded eating the pulp surrounding the seeds of this tree in the Tapajós National Forest south of Santarém (Pinto 2008:76). Although this monkey is not an important game animal it likely disperses some of the seeds.

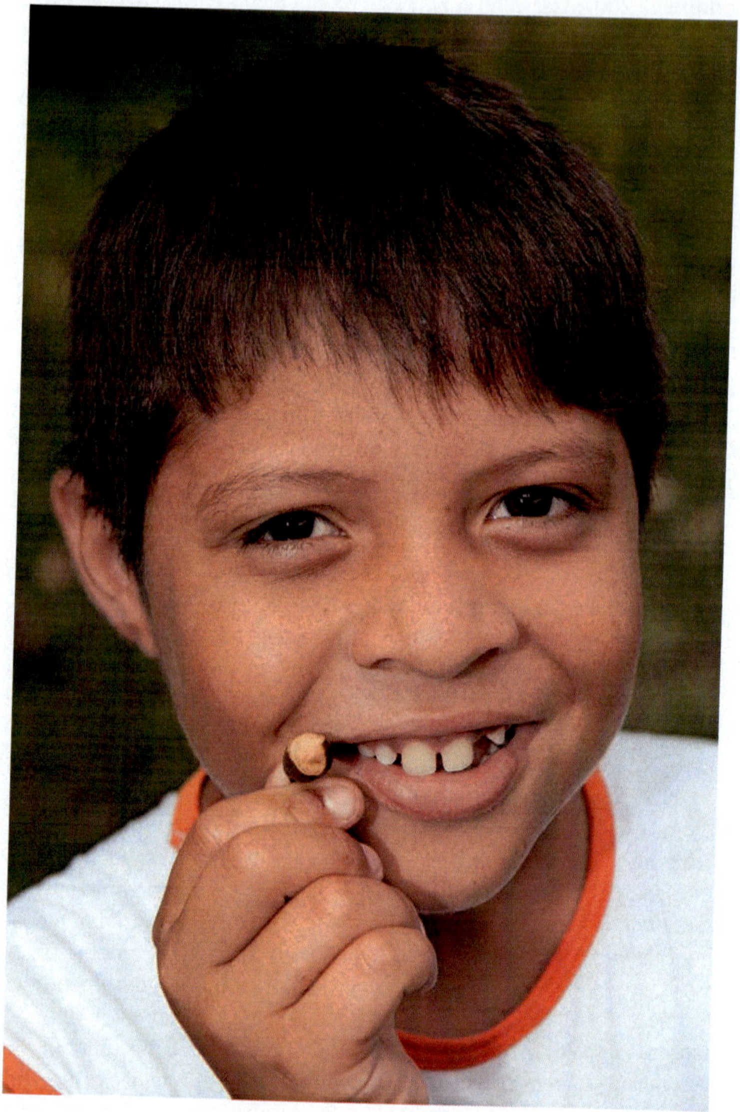

Salamão, 8 years, eating Jutaí mirim *Hymenaea parvifolia* fruit gathered in nearby forest. Near Caiçara, Boca do Rio Curuá, Caxiuanã National Forest, Município de Melgaço, Pará, Brazil 5-7-15

Along the Capim River in eastern Pará, *Hymenaea parvifolia* fruits are among the most important wild fruits gathered by rural folk (Shanley and Rosa 2004). The fruits of *Hymenaea parvifolia* have evidently been collected for a long time. Fragments of jutaí mirim fruits have been found at an archaeological site in sandstone hills near Monte Alegre, Pará, dating back 9800 to 11,200 years (Roosevelt 1999; Roosevelt et al. 1996). The fruits are also gathered by rural folk in other parts of the Amazon Basin such as the Madre de Dios department in the Peru (Phillips 1993).

In some rural areas, such as the quilombo community of Jocojó in the Municipality of Gurupá, the resin is occasionally used to glaze pottery and for illumination. The clear resin is heated and then smeared on pottery to produce a shine. This is an ancient practice as it was also recorded at Itupanema near Vila de Conde along the lower Tocantins in the 1930s (Lustosa 1976:220) and at Ereré village near Monte Alegre, Pará, in the nineteenth century (Smith 1879:378).

The resin is also used in some folk remedies, such as among the Ka'apor who burn the resin and mix the ashes with water to expel worms (Balée 1993a:109). Yanomami shamans summon the tree spirit (*utupë*), which dwells in *Hymenaea parvifolia* and some other forest trees, to expel malevolent spirits that have invaded people causing them to be ill (Albert and Milliken 2009:137). People living in and around the village of Marudá along the coast of Pará at the mouth of the Amazon obtain a syrup from the bark for treating coughs, congestion, and flu (Coelho-Ferreira 2009) as do river dwellers in the watershed of the Jaú, an affluent of the middle Rio Negro in Amazonas (Rodrigues 2006).

Resin oozing from the trunk of a jutaí cica *Hymenaea parvifolia* tree. Near Jocojó, Rio Jocojó, Município de Gurupá, Pará, Brazil 10-22-16

Jutaí mirim, as with other some species in the genus, is under pressure from the timber trade (Costa et al. 2002; Guariguata et al. 2009; Rockwell et al. 2007). It has durable wood which is in demand in rural and urban areas. The hardwood, sometimes laced with patches of resin, was in demand for railroad ties in the early twentieth century (Le Cointe 1922:29). In the mid-twentieth century, the wood was used to fashion ax handles in the municipality of Almerim along the lower Amazon (Ducke 1949:98). Currently, rural people in the Tapajós-Arapiuns extractive reserve in western Pará use the wood to make furniture (McGrath et al. 2004). Forest clearing for cattle pasture, as with many other Amazon fruit trees, is another threat to this tree.

24.7 *Inga acuminata*

Brazil Ingá; rĩa moxiririma hi (Yanomami)

Ecuador Guaba de monte; pilingas, sacha pilanga, chunda pacai, nina pacai (Kichwa), mimotan, mimuntan, mimuntahue (Waorani)

Peru Shimbillo rabo de mono, shimbillo pichico de mono, shimbillo delgado

Venezuela Guamo

Status Wild

Shimbillo delgado *Inga acuminata* in fruit along margin of a floodplain lake. Cocha Grande, near San Carlos, Rio Puinahua (a branch of the Ucayali), Reserva Nacional Pacaya-Samiria, Loreto, Peru 4-18-06

The green, elongated pods of *Inga acuminata* are typically gathered from canoes during high water because the tree grows on floodplains and is only 15–20 m tall when mature. An understory tree of floodplain forests below 1200 m, *Inga acuminata* ranges from Costa Rica south to the western and north-central part of the Amazon Basin and the Orinoco drainage. It also occurs on St. Vincent in the Caribbean (Pennington 1997:500). The white pulp surrounding the seeds is sucked off and the seeds discarded. The fruits are also consumed by wildlife, including brocket deer (*Mazama* spp.) which help disperse the seeds and supply game meat in the case of the deer (Cerón and Ayala 1998:107).

This species of *Inga* is also used in some folk remedies. The Yanomami in Roraima, Brazil, and contiguous Venezuela, mash the dried leaves of this species, mix the powder with saliva or sap from green plantains, and apply the paste to nipples of mothers whose babies are infected with the oral fungus *Candida albicans* (Albert and Milliken 2009:128; Milliken and Albert 1996; Milliken and Albert 1997b). The Waorani in the Ecuadorian Amazon use the wood in home construction and for fuel (Cerón and Ayala 1998:107).

Known as shimbillo rabo de mono (shimbillo monkey tail) along the lower Tigre in the Peruvian Amazon on account of the fruit's length and slender shape (up to 13.5 cm long and 2 cm wide), a sweet pulp surrounds the seeds. The relatively thin-skinned pod bulges conspicuously at each seed in contrast to some of the thicker, woody pods of other species of *Inga* that conceal the presence of seeds, such as *Inga cinnamomea*, which is also found along river banks. The other name for *Inga acuminata* along the lower Tigre is shimbillo pichico de mono (the monkey's penis) because the legumes are highly variable in size and are sometimes as small as 6.5 cm long and 1 cm in width.

24.8 *Inga alba*

Bolivia Pacay

Brazil Ingá vermelha, ingá vermelho, ingá ferrugem, ingá ferro, ingá pereba, ingá titica, ingá xixi, ingá xixica, ingá chichica, ingá turi, ingá tururi, ingá mole, ingá branca, ingá cordão, ingá de rosca, ingá peludo, ingá sapo, ingá casca vermelha, ingá da mata, ingá chato, ingaí, ingarana, jarandeua da folha graúda; ɨnašiši'ɨ (Araweté), iga-i-úna (Asurini), wiritáa kantsa (Baniwa), bi mené (Desana), čičipe'ɨ (Guajá), siširupe-'ɨ, ɨnaŝiŝɨ (Ka'apor), olakabapojo (Suruí), iga-pi-'iw (Tembé), bi mené (Tukano), moxima hi (Yanomami)

Colombia Chi itá, guamas, guamo churimo, guamo silvestre, guamilla; chitá (Nukak)

Ecuador Guaba de monte; sacha pacai (Kichwa), airo yapenë (Siona-Secoya), anganahue, ahuetahue, ebenbahue, hanganangue, ganahuan (Waorani)

Peru Shimbillo, shimbillo colorado, shimbillo enano, poroto shimbillo, pacae

Venezuela Guamo, guamito, paatê; bãyekalõ luwe yaɨ (Hotï)

Status Wild, spontaneous in cultural settings, planted

Ingá *Inga alba* fruit gathered by Erlan, 11 years, in regrowth forest on an archaeological site. Arapijó, inland from Carrazedo, confluence of the Xingu and Amazon Rivers, Município de Gurupá, Pará, Brazil 4-20-15

Inga alba can grow 50 m tall and occurs widely in forests in the Amazon Basin, Venezuela, the Guianas, and into Central America as far north as southern Mexico (Bánki et al. 2003:37; Pennington 1997:204). The tree is also found in central and southern Brazil. *Inga alba* grows mostly on uplands up to 1500 m in the Andean foothills, such as in the interfluvial area between the Purus and Madeira Rivers in Amazonas (Souza 2012:51), in the watershed of the Coari River, an affluent of the Solimões in Amazonas (Sousa et al. 2011), in the watershed of the lower Rio Branco in Roraima (Alarcón and Peixoto 2007), in Rondônia (Absy et al. 1986; Andrade et al. 2017; Salomão and Lisboa 1988), in the vicinity of Manaus, Amazonas (Silva et al. 2014), in upland forest south of Santarém, Pará (Espírito-Santo et al. 2005; Vieira et al. 2014), in Amapá (Carim et al. 2013; Gomide 1997:68), on Marajó Island at the mouth of the Amazon (Silva et al. 2013), and in eastern Pará (Almeida et al. 1993; Mackensen et al. 2000; Pinheiro et al. 2007; Pires and Salomão 2000; Salomão and Rosa 1989; Salomão et al. 1988, 2002, 2007a, b; Silva and Rosa 1989). It is also found in successionary vegetation of various ages on uplands such as in the vicinity of Belém and the Bragantina zone of northeastern Pará (Almeida and Vieira 2001; Oliveira and Jardim 1998; Santana et al. 2004; Souza 2019:12), in parts of eastern Pará (Medeiros 2016:92) and western Pará (Parrotta et al. 2002), 80 km north of Manaus (Fernandes Neto 2015:23), in the southern fringe of the Amazon

Basin in Mato Grosso (Medeiros 2004:14), in the Caquetá watershed in the Colombian Amazon (Vester and Cleef 1998), in the Tapajós National Forest in western Pará (Espírito-Santo et al. 2005), and as I have observed in successionary woodlands inland from Carrazedo at the confluence of the Xingu and Amazon Rivers. It is found in successionary vegetation as young as 4 years old (Coelho et al. 2003b) and grows on a wide variety of soils, including on iron-rich sedimentary rock known as *canga* in Brazil (Mattos et al. 2018; Mota et al. 2018).

This tree grows mainly on uplands, such as in Roraima (Lewis and Owen 1989), but also thrives occasionally on the higher parts of forested floodplains, such as along the coast of Pará, in tidal forest in the Amazon estuary, on Marajó Island, on the banks of the Upper Madeira River and some of its affluents in Rondônia, in the Bolivian Amazon, and in the Orinoco Basin in Venezuela (Almeida and Jardim 2011; Amaral et al. 2008, 2009; Araujo-Murakami et al. 2012; Arellano 2013:178; Batista and Gonçalves 2013; Bentes-Gama et al. 2002a, b; Braga and Jardim 2016, 2019; Carim et al. 2008; Casula 2012:24; Gama et al. 2002, 2003, 2005; Godoy et al. 1999; Jardim et al. 2013; Lau and Jardim 2013; Maués et al. 2011; Rabelo et al. 2000; Reis 2010:33; Santos and Jardim 2006; Santos et al. 2014a; Thomas 2009). It occurs in floodplain forests of various water types, including black water (Lobo 2017:55). Several species of fish sought after by river dwellers, such as pacu (various species of *Mylossoma*, *Myleus,* and *Metynnis*) and aracu (several species of *Leporinus* and *Rhytiodus*), are fond of the fruits (Ribeiro 1990).

In southwest Amazonia, this species of *Inga* is also found in bamboo vegetation dominated by *Guadua sarcocarpa* and *G. weberbaueri* (Griscom et al. 2007) and in the Uatumã watershed northeast of Manaus it is found in some stunted woods on patches of white sand called *campinarana* (Targhetta et al. 2015). It is also found in *campina* vegetation growing on sandy soils in parts of Pará (Ferreira 2009:90). It also occurs in some successionary forests, such as in southern Roraima (Condé 2011:76), in the Bragantina zone of northeastern Pará (Carim et al. 2007; Schwartz 2007; Vieira 1996:59), in the Tapajós National Forest in western Pará (Gualberto et al. 2014), in Amapá (Gomide 1997:111), and in Guyana (Van Andel 2001).

Known by various names in the Brazilian Amazon, it is one of those many fruit trees that is difficult to classify as truly wild or semi-domesticated. That is because it is often found in landscapes that have been altered by human activity, such as in successionary vegetation after cultivated plots or cattle pastures that have been abandoned (Araújo et al. 2005; Baar et al. 2004; Coelho et al. 2012; Ducke 1910; Lisboa 1989; Miller and Proctor 1998; Prata et al. 2010; Rayol et al. 2006, 2011; Uhl et al. 1988b; Vásquez 1998) and in regrowth forest on archaeological sites, such as on artificial mounds built by extinct indigenous cultures on Marajó Island (Lisboa 2012:413). Some botanists have found *Inga alba* in "virgin terra firme forest," such as near Castanhal in the Bragantina Zone (Murça-Pires et al. 1953). Such forests may be mature, but it can be hard to ascertain for sure that they are truly virgin. For example, this species of *Inga* has been found in 4- and 8-year-old successionary vegetation in the vicinity of Castanhal (Coelho et al. 2003b). This species gets a start in regrowth areas and survives in mature forest. In the vicinity of Paragominas, about half of the trees of this species resprout when fires penetrate fragmented forest (Kauffman 1991).

It is also planted, or arises spontaneously, in home gardens such as along the middle Xingu. Sometimes locals bring the pods back home and after snacking on the mesocarp spit out the seeds, some of which may later sprout in the yard. The tree starts producing fruit when relatively small so the branches can be pulled down to reach the fruit. Alternatively, youngsters climb young trees to harvest the pods, which normally range in length from 6 to 15 cm, although they can reach 25 cm in some cases (Pennington 1997:204; Reynel and Pennington 1997:56). Rural people appreciate the white pulp surrounding the black seeds, 1.5 cm by 0.8 cm. *Inga alba* is one of the most common fruits gathered by the Ka'apor in the eastern fringes of Amazonia (Balée 1993a:67).

Several species of monkey also eat the fruits and some of them disperse the seeds (Peres 1994b). The black-handed tamarin *Saguinus midas niger* feeds on the fruits of this tree and is known to ingest the seeds, thereby dispersing them (Oliveira and Ferrari 2000). In upland forests along the Upper Urucu, an affluent of the lower Coari River in Amazonas, Brazil, saddle-back tamarins *Saguinus fuscicollis* and the moustached tamarin *Saguinus mystax* have been observed feeding on the fruits of this tree (Peres 1993) and probably disperse the seeds as well. The moustached tamarin has been observed swallowing the pulp and seeds of *Inga alba* in other parts of its range in western Amazonia (Garber and Kitron 1997). The pied bare-faced tamarin *Saguinus bicolor bicolor* also feeds on these fruits in the vicinity of Manaus along the lower Rio Negro in Amazonas, Brazil (Egler 1992) as does the silvery marmoset *Mico argentatus* in upland forest in the vicinity of Alter do Chão along the lower Tapajós River (Corrêa 2006:57).

The fruits are also eaten by other monkeys, such as black spider monkeys *Ateles paniscus*, tufted capuchins *Cebus apella*, white-faced sakis *Pithecia pithecia*, golden-faced sakis *Pithecia pithecia chrysocephala*, bearded sakis *Chiropotes satanas*, Uta Hick's bearded saki *Chiropotes utahickae*, the northern bearded saki *Chiropotes sagulatus*, the red-nosed bearded sakis *Chiropotes albinasus* and ashy titi monkeys *Plecturocebus cinerascens* (Buchanan et al. 1981; Cummings et al. 2018; Gómez-Posada 2012; Kinzey and Norconk 1990; Mattos 2016:30; Norconk and Kinzey 1994; Pinto 2008:77; Port-Carvalho and Ferrari 2004; Santos et al. 2013; Setz et al. 1999; Shaffer 2013; Van Roosmalen et al. 1981, 1988), all of which are also likely involved in seed dispersal. The highly endangered Ka'apor capuchin monkey *Cebus kaapori*, which is restricted to eastern Pará and neighboring parts of Maranhão, eats the pulp around the seeds of this fruit in the dry season (Oliveira et al. 2014). Woolly monkeys *Lagothrix lagotricha* and spider monkeys *Ateles paniscus* and *A. belzebuth*, widely hunted for their meat throughout the Amazon Basin, also eat the fruits (Defler and Defler 1996; Norconk and Kinzey 1994; Nunes 1998; Peres 1994a; Simmen and Sabatier 1996; Stevenson 2000, 2004, 2005, 2007; Stevenson et al. 1994, 1997, 2002; Van Roosmalen 1985; Van Roosmalen and Klein 1988). Bearded sakis *Chiropotes chiropotes*, which live in upland forests in the northern part of the Brazilian Amazon and contiguous areas of Venezuela and the Guianas, reportedly eat the seeds of this fruit (Boyle et al. 2012) while the white-faced saki *Pithecia pithecia*, which inhabits the same region, eats the pulp and drops the seeds, thereby scattering them (Norconk and Veres 2011). The endangered black-bearded saki *Chiropotes satanas* in eastern Pará also eats this fruit and likely disperses some of the seeds (Veiga 2006:98). In eastern Colombia, squirrel monkeys *Saimiri cassiquiarensis albigena* also eat these fruits and likely disperse some of the seeds (Carretero-Pinzón et al. 2016) as does *Saimiri sciureus* in Colombia (Astwood et al. 2014) and in eastern Pará (Lima and Ferrari 2003). The bare-tailed woolly opossum *Caluromys philander*, which occurs in northern South America, is not normally hunted for its meat but also eats this fruit and likely spreads some of the seeds (Charles-Dominique et al. 1981). Parrots, which are not hunted for their meat but are sometimes kept as pets, also reportedly eat the fruits.

The flowers of *Inga alba* are visited by at least three species of stingless bees (*Melipona seminigra merrillae*, *Trigona fulviventris*, and *Cephalotrigona femorata*) in both the wet and dry seasons (Oliveira et al. 2009). Rural people gather honey from stingless bees in the Amazon and sometimes provide homes for them in segments of hollowed out logs which they suspend from the roof rafters around the outside of their homes.

As with many other fruits in Amazonia, *Inga alba* has other uses. The trunk and branches are used for fuelwood and to make charcoal, such as in the Maguari community in the Tapajós National Forest in western Pará (Gualberto et al. 2014). In Roraima, the bark of this tree is used to dye species of *Ischnosiphon* and *Philodendron* which in turn are woven into baskets and mats (Alarcón and Peixoto 2008). Along the Jauaperi River that forms the boundary between the states of Amazonas and Roraima in the Brazilian Amazon some rural folk scrape the bark of this tree into latex tapped from rubber trees *Hevea brasiliensis* to thicken it (Ammann 2014:41). The Tembé in eastern Amazonia use the bark to make a black dye for painting the gourd rattler used by shamans (Prance et al. 1987). In Suriname, rural people of indigenous and mixed African ancestry treat wounds with the sap (Hoffman 2013), and the bark of this tree is sold in street markets to treat both wounds and sores (Van Andel et al. 2007). Along the Oyapock River in French Guiana that forms the border with Amapá in the Brazilian Amazon, the Wayãpi prepare a juice from the bark of this tree to treat leishmaniasis (Odonne et al. 2011). In the Peruvian Amazon, wood is used in construction (Reynel and Pennington 1997:58). In the Ecuadorian Amazon, the Waorani use wood when building or repairing their houses and for fuel (Cerón and Ayala 1998:107).

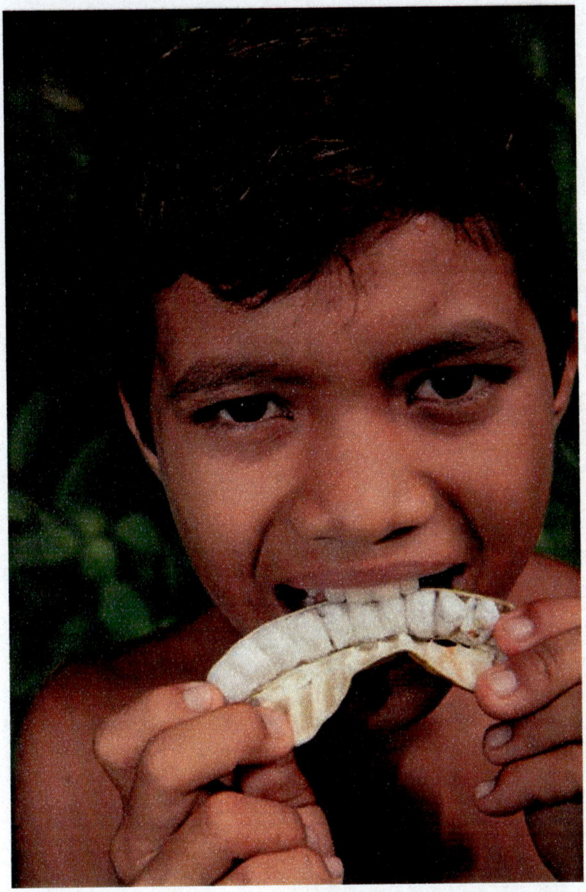

Diogo, 11 years, eating *Inga alba* fruit. Regrowth forest on an archaeological site. Arapijó, inland from Carrazedo, confluence of the Xingu and Amazon Rivers, Município de Gurupá, Pará, Brazil 4-20-15

24.9 *Inga auristellae*

Brazil Ingá, ingá xixica, ingá vermelho, ingá chinelo; iñá-paka (Arawaté), yurupi-rana-igá (Asurini), čičipe'ɨ (Guajá), ɨnapere'ɨ (Ka'apor), iga-pi'ɨw-him (Tembé), iga-sili (Wajãpi)

Colombia Guamo

Ecuador Pilingas, quina pacai (Kichwa), sisi pene, son'yio (Siona-Secoya), momontan, mimontan, mimoncahue, noiguagan (Waorani)

Peru Shimbillo, shimbillo de altura, sapo shimbillo, pacaecillo; juichani sampi, napurak sampi, sampi numi (Aguaruna)

Status Wild

Segundo, 12 years, eating sapo shimbillo *Inga auristellae* fruit. The tree resprouted when regrowth forest was cleared to plant crops. Yarina, Rio Yanayacu, Reserva Nacional Pacaya-Samiria, Loreto, Peru, 6-23-06

Inga auristellae is an understory tree of mature and regrowth forests on higher parts of floodplains and uplands in the Amazon Basin and the Guianas up to 1500 m in the Andean foothills (Amaral et al. 2009; Araújo et al. 2005; Batista et al. 2015; Cárdenas et al. 2009; Coelho et al. 2012; Lemos et al. 2015; Macía 2011; Mackensen et al. 2000; Mendoza and Guerrero 2010; Oliveira and Amaral 2005; Pennington 1997:206; Salomão et al. 2007a, b; Taurino 2017:94; Valencia et al. 2004; Vieira et al. 2017). It is also found in some 20- and 40-year-old successionary woodlands in the Bragantina zone of northeastern Pará (Almeida and Vieira 2001). In Loreto province of the Peruvian Amazon, *Morpho marcus intermedia* butterflies have been observed visiting the flowers of this plant thereby possibly pollinating them (Bardales et al. 2017).

Known as sapo shimbillo (the frog inga) along the Yanayacu, an affluent of the lower Marañón in the Peruvian Amazon, the flattened pods are gathered during the rainy season. The green pods, some 7–15 cm long and 1.5–2.3 cm wide, are easy to reach because even though the tree can reach 35 m, it starts producing fruit when only 3–4 m tall. People eat the white pulp surrounding the nine or so seeds that are ejected before the pulp is swallowed. The Waorani in the Ecuadorian Amazon also use the wood for house construction and for fuel (Cerón and Ayala 1998:108). *Inga auristellae* is sometimes cleared when preparing a field but subsequently resprouts and fruits. This arboreal legume also occurs in some cacao orchards, such as at km 107 of the Altamira-Itaituba stretch of the Transamazon Highway (Leão et al. 2017). The arils surrounding the seeds of this fruit are eaten by the Ka'apor capuchin monkey *Cebus kaapori* in eastern Pará and these highly endangered monkeys most likely disperse some of the seeds (Oliveira et al. 2014). The seeds of this fruit are also dispersed by saddleback tamarins *Saguinus fuscicollis*. In the Ecuadorian Amazon, the red titi monkey *Callicebus discolor* also eats this fruit and likely disperses some of the seeds (Carrillo-Bilbao et al. 2005).

24.10 *Inga bourgonii*

Bolivia Pacay

Brazil Ingá, ingá ferro, ingá xixica

Colombia Guamo macheto, churimbo

Ecuador Sacha pacai, guabo; chunda pacai, yana cara pacai (Kichwa), buguiahue, ebenbahue, behuetempoye (Waorani)

Peru Shimbillo, yacu shimbillo, pacae, pacae amarillo; sampi (Aguaruna)

Venezuela Paatê; hãlĩ luwe yaɨ, jani luwë jyaï (Hotï)

Status Wild

Inga bourgonii **in fruit in regrowth forest. A few 100 m inland from Carrazedo, confluence of the Xingu and Amazon Rivers, Município de Gurupá, Pará, Brazil 4-20-15**

Inga bourgonii can reach as high as 40 m high in forests and ranges throughout the Amazon, the Guianas, and the Orinoco Basin in Venezuela. It grows mostly on floodplains but also occasionally on uplands, such as in various parts of the Bolivian Amazon (Araujo-Murakami et al. 2015; Lewis and Owen 1989; Thomas 2009). In the Andean foothills it can be found as high as 1500 m (Arellano 2013:178). The pods of *Inga bourgonii* are gathered from canoes at high water in the Peruvian Amazon. The pods, some 15–17 cm long by 2.5–3.5 wide (Pennington 1997:209), are split open to chew on the spongy white pulp surrounding the seeds. The green seeds, 1.5 cm by 0.7 cm, are spat out. The fruits of *I. bourgonii* are eaten locally and are not taken to markets, probably because the trees are sparsely distributed. Along the Oyapock River in French Guiana that forms the border with Amapá in the Brazilian Amazon, the Wayãpi prepare a juice from the bark of this tree to treat leishmaniasis (Odonne et al. 2011).

Jon, 9 years, gathering *Inga bourgonii* fruits to snack on the pulp. Regrowth forest on an archaeological site. A few 100 m inland from Carrazedo, confluence of the Xingu and Amazon Rivers, Município de Gurupá, Pará, Brazil 4-20-15

At the mouth of the Xingu River a few 100 m inland from the village of Carrazedo, *Inga bourgonii* was encountered in regrowth forest on an upland archaeological site. This extensive site which has deep anthropogenic black earth (*terra preta de índio*) and ceramic material from at least one indigenous culture as well as from colonial times has a number of other fruit trees normally found on floodplains. It is possible that *Inga bourgonii* occurs naturally on uplands. But it is conceivable that people have introduced *Inga bourgonii* to some upland sites. In the case of the archaeological site inland from Carrazedo, this species may have been brought there in the past by people and the tree has managed to thrive, perhaps because of the fertile anthrosol. Residents of Carrazedo, especially children, like to visit the cultural forest on the archaeological site to gather the fruits. This species occurs in some upland cacao agroforestry plots in the municipality São Félix do Xingu in southern Pará (Braga et al. 2019).

Red howler monkeys *Alouatta seniculus* and black spider monkeys *Ateles paniscus* also feed on the fruits (Julliot and Sabatier 1993; Van Roosmalen 1985; Van Roosmalen and Klein 1988) and since these monkeys are shot for their meat the fruits also contribute indirectly to the human diet. The smaller northern bearded saki *Chiropotes satanas chiropotes* also eats the pulp surrounding the seeds; although these monkeys are less frequently shot for their meat, they probably help disperse the seeds (Van Roosmalen et al. 1988).

24.11 *Inga brachystachys*

Brazil Ingá ferrujo, ingá amarelinha, ingá de macaco, ingá jarandeua

Colombia Guamilla. Guamo

Peru Poroto shimbillo

Venezuela Siembra potro

Status Wild, spontaneous in cultural settings, planted

Wadrio, 11 years, gathering fruits of ingá ferrujo ***Inga brachystachys*** **in a floodplain forest. Carrazedo, mouth of the Xingu River, Município de Gurupá, Pará, Brazil 7-24-16**

Inga brachystachys is an understory tree in upland and floodplain forests ranging from the eastern and central part of the Brazilian Amazon into Venezuela and the Guianas. It often occurs in regrowth forest, such as in uplands in southwestern Amapá (Guimarães et al. 2016) and as I noted on an upland archaeological site inland from Carrazedo at the mouth of the Xingu. Villagers from that community often tarry in the successionary forest to gather the fruits of *Inga brachystachys* on the way to or from their cultivated fields. This small tree has also been recorded in campinarana vegetation on patches of sandy soil in the vicinity of Belém, Pará (Ferreira et al. 2013a).

The yellow-green pods vary from 12 to15 cm long by 2.8–3 cm wide, and although very little pulp surrounds the seeds, children in particular eagerly scavenge for the pods. The tree starts producing fruits when only 3 or 4 m high, so they are easy to pick. The green seeds, 2 cm × 1 cm, are spat out after ingesting the white mesocarp, some of which later sprout. *Inga brachystachys* was also found in cattle pasture at Mata 3, an archaeological site with anthropogenic dark earth on a hill overlooking the middle Xingu.

Arlinho, 11 years, eating ingá ferrujo *Inga brachystachys* fruit he has picked in a floodplain forest. Carrazedo, mouth of the Xingu River, Município de Gurupá, Pará, Brazil 7-24-16

This relatively short tree, which reaches no more than 25 m when fully mature, is also planted occasionally in home gardens in rural areas, such as along the middle Xingu where it is known as ingá ferrujo. This species is also known as ingá de macaco along the middle Xingu suggesting that the fruits are also procured by monkeys.

24.12 *Inga cinnamomea*

Bolivia Pacay

Brazil Ingá açú, ingá batelão, ingá grande, ingá baú, ingá pracuúba, ingá pracuxi, ingá chinela, ingá grossa, ingá branco, ingá boi, ingá chato, ingazeiro açú; peiu mené (Desana), ynahu'y (Ka'apor), u mené (Tukano)

Colombia Guama, guamo; tôú (Ticuna)

Peru Shimbillo, vaca paleta, vaca shimbillo, pacae, pacae amarillo, pacae Pozuzo; apash sampi (Aguaruna)

Status Wild, planted

Ingá açu *Inga cinnamomea* partially submerged during the annual flood. Portal do Miwá, Lago Miwá near Codajás, Rio Solimões, Amazonas, Brazil 6-24-07

Inga cinnamomea **pods vary in size. Arlinho, 10 years, with fruits gathered in a home garden in his village on stilts. Carrazedo, mouth of the Xingu River, Município de Gurupá, Pará, Brazil 4-21-15**

Inga cinnamomea is a tree that grows as tall as 30 m on floodplains of various river types in the Amazon from the western portion of the basin in Peru, Ecuador, and Colombia to the Amazon estuary (Cain et al. 1956; Carim 2016:89; Carim et al. 2008; Ducke 1925, 1949:24; Lobo 2017:62; Luize et al. 2015; Nebel et al. 2000a, b, 2001; Pennington 1997:343; Queiroz et al. 2007; Silva 2008b:16; Silva et al. 2013; Wittmann et al. 2013). In the Andean foothills, it can be found as high as 1750 m. *Inga cinnamomea* is also encountered in upland successionary woods (Rodrigues 1986; Uhl et al. 1988a, b), but in some cases that may be a vestige of plantings in cultivated fields.

Rodrigo, 10 years, snacking on ingá açu *Inga cinnamomea* fruit gathered in a home garden in his floodplain village. Carrazedo, mouth of the Xingu River, Município de Gurupá, Pará, Brazil 4-17-15

Known by various names throughout its range in the Amazon basin, several names reflect the relatively large fruits size such as ingá açu and ingá grande in Brazil. Both of those names mean the large ingá, a reflection of the tree's thick, cylindrical pods. Ingá batelão refers to the wide berth and shallow draft of a certain kind of wooden cargo boat, traditionally used in the estuary for hauling sugarcane or heart-of-palm obtained from açaí *Euterpe oleracea*. Light green to yellow when mature, the pods vary in length from a stubby 15 cm, in which case they contain only two or three seeds, to 25 cm in length with as many as fifteen seeds. The pods are usually 4.5–5 cm wide. When fully ripe the pods turn yellow, but they are often picked when they are still green or green turning yellow. The green to brown seeds are typically 3.8 cm by 2 cm and are surrounded by a spongy pulp, technically known as a sarcotesta, which is exceptionally sweet. Hence it is one of the more popular species of *Inga* in the Amazon.

A Shipibo girl, Neyra 10 years, selling fruits of pacae *Inga cinnamomea* and necklaces made from beads. Yarina Cocha, Pucallpa, Peru, 1-20-04

River folk gather the pods of ingá açu in the wild for home consumption or for sale in street markets, such as Óbidos, Breves, and Belém in Pará, Macapá in Amapá, and in Iquitos and Pucallpa in Peru. In 1994, a vendor in Óbidos was selling ingá baú for the equivalent of US 10 cents/pod and in 2007 the pods were fetching US 8 cents each in the market of Codajás along the Solimões. In the Mercado Belén in Iquitos, the pods sell for as little as US 4 cents each, as in 2006. The fruits are typically harvested by climbing the tree and plucking off the fruits or lopping off some of the smaller branches. Such pruning is not detrimental to future fruit production because branches resprout and it makes the crown more compact. *Inga cinnamomea* fruits at the end of the rainy season through the early part of the dry period; peak fruit production occurs when river levels are highest.

Ingá batelão *Inga cinnamomea* fruits in the Ver-o-Peso market, Belém, Pará, Brazil 11-13-98

Inga cinnamomea is occasionally planted in home gardens on both floodplains and uplands. I noted this species of *Inga* in several riverine home gardens, such as in the vicinity of Santarém, along the middle Xingu, in the village of Miraflores at the mouth of the Rio Tigre, and at San Martin Tipishca along the lower Rio Samiria in the Pacaya Samiria National Reserve in Peru. I have also seen *Inga cinnamomea* in home gardens on uplands at various locations including along Ramal Andirobalzinho, a side-road leading south from the Santarém to Alter do Chão highway and in Vila Baixa along the lower Tefé. At Taperebatuba along the lower Rio Urubu in Amazonas, Brazil, a farmer has an *Inga cinnamomea* seedling by a raised vegetable and spice bed that he plans to plant in his home garden. Ingá açu is also planted in home gardens in Belém and Manaus (Ducke 1946), in the yards of rural homes along Ramal do Pau Rosa, an offshoot of the BR 174 Manaus-Rio Branco highway north of Manaus (Machado 2016:32), in rural home gardens in the lower Purus watershed (Silva 2006:12), and along the Amazon downstream from Iquitos in Peru (Lamont et al. 1999).

Shimbillo *Inga cinnamomea* in a village home garden by a black water river. San Martin Tipishca, Rio Samiria, Reserva Nacional Pacaya-Samiria, Loreto, Peru 6-29-06

Inga cinnamomea was also encountered in various cultivated fields, such as an agroforestry plot containing a variety of fruit trees on the banks of the black water Yanayacu River near the village of Arequipa in the Pacaya-Samiria National Reserve. Shimbillo, as the tree is known locally, was planted or spared when the floodplain forest was cleared to plant crops or possibly resprouted after having been felled. In the vicinity of Santo Antonio de Ipapucu along the lower Tefé River, a farmer has planted a variety of fruit trees including ingá açu, cupuaçu *Theobroma grandiflorum*, pineapple, and mango after harvesting his manioc crop in an upland field. A river dweller at Morro dos Arara along the middle Xingu has intercropped a cacao grove with several fruit trees including *Inga cinnamomea*. This cacao grove is on an archaeological site, and *Inga cinnamomea* may have been introduced to this hill in pre-contact times.

Turtles, such as the giant river turtle *Podocnemis expansa*, and certain fruit-eating fish, particularly tambaqui *Colossoma macropomum*, may also be involved in seed dispersal. Aracu (several species of *Leporinus* and *Rhytiodus*) are fond of the fruits of ingá açú but they are likely too small to swallow the seeds (Ribeiro 1990). Nevertheless, the pulp of this fruit contributes indirectly to the diet of river dwellers because they are fond of catching aracu (Smith 1981:140). In the early nineteenth century, the German naturalists Johann Spix and Carl Martius noted that indigenous groups along the Amazon fed ingá fruits to corralled turtles (Spix and Martius 1976:83). Early explorers to the Amazon remarked that Indian villages often had thousands of giant river turtles penned in shallow water along the Amazon. Europeans, including Portuguese garrisons and missionaries, continued this practice during the colonial period (Smith 1974). For this reason, tartaruga, as the turtle is known in Brazil, was referred to as "river cattle." Today, populations of giant river turtles are greatly reduced and introduced Old World cattle and water buffalo now dominate many floodplain landscapes of the Amazon floodplain, resulting in massive destruction of forests that once provided food for abundant turtle populations as well as certain fish.

Several species of monkey also eat the fruits and are likely dispersal agents for the seeds. Woolly monkeys *Lagothrix lagotricha*, black spider monkeys *Ateles paniscus*, and tufted capuchin monkeys *Cebus apella* feed on the fruits of *Inga cinnamomea* (Defler and Defler 1996; Gómez-Posada 2012; Van Roosmalen 1985; Van Roosmalen and Klein 1988) and may be involved in dispersing some of the seeds. These monkeys are widely hunted in the Amazon Basin for their meat. Bearded sakis *Chiropotes satanas* and *C. sagulatus* also eat the fruits (Kinzey and Norconk 1990; Norconk and Kinzey 1994; Shaffer 2013), but these monkeys are not important game animals.

24.13 *Inga edulis*

Bolivia Pacay, pacay cola de mono, pacay peludo, pacae, ingá, guama; xënanë (Chácobo), nasa turudo bechu tida (Tacana), marajitity (Tsimane)

Brazil Ingá cipó, ingá de metro, ingá de corda, ingá de braço, ingá de macaco, ingá da grande, ingá comprida, ingá açu, ingá peluda, ingá xixi, ingá espada, ingá da folha miúda, ingá pé de galinha, ingazeira, ingazeiro; dapi hawadza (Baniwa), mené pagá (Desana), ykangaká (Piripkura), mokiu, mukiu (Sateré-Maué), moridia (Suruí), iga wypo (Tembé), mené, mené pahká, mihsĩ mere, mere paka, yao se (Tukano), akai (Waimiri-Atroari), mene (Wanano), krepu uhi (Yanomami)

Colombia Guama, guama churimba, guama churino, guamo largo, guamo rabo de mono, guamo rabo de mico, guamo cola de mico, guamo bejuco, guamo macheto, guamo santafereño, guaba, chimbillo, santafereño, chiruma; sidá (Andoke), tisi, tisio, áróhú'è (Miraña), mené (Tukano), jizaiko, jitaño debozairai, jisaño, jitairai (Witoto), guiro (Yukuna)

Ecuador Guaba, guaba bejuca, guaba de bejuco, guaba vejuco, guaba grande, guaba musga, guabo, guama, bariza pácai, alli pacai; wampa (Achuar), pichiche, bo sji chi, sichi jaguie, shiichi (Cayapa), onga fiño (Cofán), coto pacai, turo pacai, alli pacai, pakai, pakay waska, pacai yura, machitona (Kichwa), sampi, wámpi, wámpa, guampa (Shuar), otá bené, ho'ya bené, jo'yapene (Siona-Secoya), ago, an, gontocaohue, ontocan, nawa ägowë (Waorani)

Peru Guaba, guaba de monte, guabilla, huaba, huaba ancha, shimbillo tablacho, shimbillo alado, rujino shimbillo, pacay, pacay de monte, pacae colorado, pacae soga; wampa (Achuar), ikamia wampa, yakúm sampi, wámpa (Aguaruna), wámpa (Awajún), túútsihe (Bora), intsipa (Campa), naha (Ese Eja), pakay (Kichwa), intsipa, inchipa (Matsigenka), wambá, wampuá, sampi (Mayna), shinan, shenan, bana xenán, wishkonti xenán, poapoari xenán (Shipibo)

Venezuela Guamo rabo de mono, guamo bejuco, guama; luwë, uli luwë jyaï (Jodï), krebo u si, krebo ühi, irosisi (Yanomami)

Status Wild, planted

Inga edulis **in fruit, km 5 Anori-Mato Grosso road, Amazonas, Brazil 6-22-07**

Inga edulis is found throughout lowland tropical South America as well as in Central America where it has been introduced (Pennington 1997:744). It also grows in the cloud forests of the Andes up to 2800 m. Ted Pennington of the Royal Botanic Gardens, Kew, and a world authority on the genus *Inga*, suggests that *Inga edulis* was domesticated in the Amazon. The northwest part of the Amazon Basin has been identified as the most likely area where *Inga edulis* was brought into cultivation (Clement 1999). The tree likely radiated out from the basin with migrating indigenous groups and through trade thousands of years ago.

Guaba *Inga edulis* in fruit in front of a village church. Veinte de Enero, Yanayacu River, Pacaya-Samiria, Loreto, Peru, 7-12-10

This medium sized tree with an umbrella like crown normally reaches up to 15 m tall (Lorenzi et al. 2006b:143), but in some cases as high as 25 m (Sousa et al. 2011), and is most commonly encountered on uplands in home gardens and agroforestry fields, but is also found in successionary woodlands such as in the Zona Bragantina zone, in the vicinity of Belém, and the Tocantins watershed in Pará, southwestern Amapá, in Rondônia, in the Peruvian Amazon near Iquitos and Puerto Maldonado, and in the Putumayo watershed bordering Colombia, and in Napo province in the Ecuadorian Amazon (Alvino et al. 2005; Carim et al. 2007; Coelho et al. 2003a, b, 2012; Garcia et al. 2017; Gomez 2011:80; Guimarães et al. 2016; Irvine 1987; Lisboa 1989; Lisboa et al. 1991; Macedo 2014:52; Oliveira and Jardim 1998; Prata et al. 2010; Rayol et al. 2006, 2011; Robles 2017:47; Souza 2019:12; Taurino 2017:66; Vieira 1996:57, 59). It also grows in floodplain forest including on islands on seasonally flooded savannas, such as on Marajó Island, in the eastern fringes of the Amazon Basin, and in the tidal zone of the Amazon estuary (Almeida and Jardim

2011, 2012; Amaral et al. 2009; Batista and Gonçalves 2013; Batista et al. 2011; Bentes-Gama et al. 2002a, b; Black et al. 1950; Braga and Jardim 2019; Braga et al. 2015; Cain et al. 1956; Carim 2016:89; Carim et al. 2008; Cattanio et al. 2002; David et al. 2015; Gama et al. 2002, 2003, 2005; Jardim et al. 2013; Lau and Jardim 2013; Machado and Pinheiro 2016; Maués et al. 2011; Santos et al. 2016b; Silva 2008b:16). It also grows in swamps dominated by the tall *Mauritia flexuosa* palm in the Peruvian Amazon (Dávila et al. 2015:43). It is also found in gallery forests along water courses in cerrado vegetation in the southern part of the Amazon Basin (Corette-Pasa 2011). It is also cultivated in some home gardens on floodplains, such as along the lower Solimões near its confluence with the Rio Negro (Lima and Saragoussi 2000). Seeds of *Inga edulis* sprout in manioc fields of the Kichwa in Ecuador after these indigenous people snack on the pulp (Irvine 1989).

Several species of monkey are involved in dispersing this species as suggested by one of the common names for this tree in Brazil: ingá de macaco. At the property of a small farmer near Maués, for example, a pet Zogue-zogue *Callicebus moloch* monkey is fond of eating *I. edulis* fruits. The black-handed tamarin *Saguinus midas niger* and the white-lipped tamarin *Saguinus labiatus* feed on the fruits of this tree and are known to ingest the seeds, thereby dispersing them (Ique and Aquino 2015; Oliveira and Ferrari 2000). In the Caquetá river basin in the Colombian Amazon, the mottled-face tamarin *Saguinus inustus* also feeds on this fruit (Palacios et al. 2004) as does the equatorial saki *Pithecia aequatorialis* which lives in northeastern Peru and the Ecuadorian Amazon (Charpentier et al. 2015) and the Colombian squirrel monkey *Saimiri cassiquiarensis albigena* (Carretero-Pinzón et al. 2016). Tamarins and sakis are not usually hunted for their meat but are likely involved in dispersing the seeds of this widely consumed fruit. Widespread capuchin monkeys *Cebus apella* also feed on the fruits (Simmen and Sabatier 1996). Spider monkeys *Ateles belzebuth*, *A. chamek*, *A. paniscus* and woolly monkeys *Lagothrix lagotricha*, which are widely hunted for their meat, are known to feed on the fruits of *Inga edulis* and disperse some of the seeds (Felton et al. 2008, 2009; Palma and Stevenson 2010; Stevenson 2000, 2007; Stevenson et al. 2002; Van Roosmalen and Klein 1988). Black spider monkeys *Ateles paniscus paniscus* have been observed feeding on these fruits in Surinam (Van Roosmalen 1985). Also known as the red-faced spider monkey, *Ateles paniscus* is widespread in forested areas of northern South America. In the Colombian Amazon a game bird, paujil *Mitu silvini*, has been observed feeding on this fruit and likely disperses some of the seeds (Franco and Santamaria 1997). The widely consumed tracajá turtle *Podocnemis unifilis* also feeds on these fruits and possibly disperses some of the seeds (Portal et al. 2002). Primates, including humans, and some turtles have likely been involved in the density and distribution of this popular fruit tree for millennia.

Inga edulis is the most encountered species of *Inga* in markets in the Amazon and is also the most widely cultivated in both home gardens and fields. In a survey of 150 home gardens in upland and floodplain environments in the Brazilian Amazon, I encountered a total of 98 fruit and nut species of which 71 species are native to Amazonia. *Inga edulis* was encountered in 17 of those home gardens, by far the most common species of *Inga*. Ingá açu (*I. cinnamomea*) was the second most encountered representative of the genus in those home gardens, with 5 recordings. Along the middle Xingu, my survey of 57 home gardens on upland slopes along the river recorded 81 fruit and nut species, including 14 species of *Inga*. Again, *Inga edulis* was the most common representative of the genus, noted in 11 home gardens.

Inga edulis is easily distinguished from the other species in this large genus by its long, thin, and deeply furrowed pods. Green when ripe, the pods are typically a meter long, reaching up to 1.5 m in some cases. The Miraña who inhabit the middle Caquetá in the Colombian Amazon believe that spider monkeys (*Ateles* sp.) are a transformation of the fruits of *Inga edulis* on account of those monkeys long and thin arms and legs (Karadimas 2005:347). The baton-like fruits of *Inga edulis* have inspired a number of descriptive names for the tree in the Brazilian Amazon, such as the meter ingá (ingá de metro), the rope ingá (ingá de corda), the long ingá (ingá comprida), and the arm ingá (ingá de braço). One of the more common names for the tree in the Brazilian Amazon is the vine ingá (ingá cipó). In Bolivia and Venezuela, the fruit is known as monkey tail (pacay cola de mono, guamo rabo de mono). One of the popular names for this tree in the Ecuadorian Amazon is guaba grande which refers to its long fruit.

Pedro Luis, 13 years, gathering guaba *Inga edulis* fruits in a village home garden. San Carlos, Rio Puinahua (a branch of the Ucayali River), Reserva Nacional Pacaya-Samiria, Loreto, Peru 4-17-06

On lower branches, the pods can be easily twisted off and kids enjoy climbing the tree to reach pods on higher branches. Adults play it safe by employing a pole with a noose on the end which they use to snag the higher fruits and twist them off. Because the pods can reach over a meter and a half long, their weight varies from 250 grams to 600 grams, and the pulp accounts for about a fifth of the fruit by weight. In some areas, *Inga edulis* fruits during 9 months of the year, and annual yields can be as high as 1700 kg of fruit per tree (Falcão and Clement 2000). *Inga edulis* is especially productive in the dry season, when fewer other fruits tend to be available, so the fruits are particularly a welcome addition to the diet.

Guaba *Inga edulis* pod opened to eat the mesocarp. One of the black seeds is already sprouting. San Carlos, Rio Puinahua (a branch of the Ucayali), Reserva Nacional Pacaya-Samiria, Loreto, Peru 4-17-06

A spongy white pulp surrounds the shiny, black seeds, a dozen or more of which are nestled in each pod. The pulp can thus be extracted conveniently in bite-size sections. A little dry and mealy in texture, the pulp is nevertheless sweet. The English name for the tree, the ice cream bean, is derived from the perception that the pulp has hint of vanilla flavor. But in the Amazon at least, the mesocarp is not incorporated into ice cream, sherbet, or juice. Rather, the fruits are eaten fresh. The black seeds, some 3–4.5 cm long by 1.5 cm, are eaten fresh or roasted or boiled by the Andoke in the Colombian (La Rotta 1983:53). The seeds are usually spat on the ground and where they easily sprout. Thus it is difficult to tell where *Inga edulis* is truly wild over its vast range. The tree can persist in abandoned fields for at least 20 years (Descola 1994:137). Although the tree grows quickly, it is also long-lived, with some trees thriving for 50 years or so, thus enabling them to compete with other successionary plants but also to survive when woodlands take over.

School children snacking on guaba *Inga edulis* fruit in their village. Centro America near Tamshiyacu, Rio Amazonas, Loreto, Peru 6-15-04

Ingá cipó *Inga edulis* for sale in a public market. Mercado Central, Tefé, Amazonas, Brazil, 8-13-12

Fruits of *Inga edulis* are relatively cheap and thus accessible to a large proportion of urban populations in the region with limited buying power. In a street market in Requena along the Ucayali, for example, the equivalent of one US dollar could buy fifteen pods in May 2003. The fruit pods are sold individually or bundled; bunches of ten pods were selling for US$1.70 along the BR 316 highway near Castanhal in

Pará in August 1988. The cost of living is higher in Brazil, and fruits generally cost more there. Even so, the fruits of ingá cipó are relatively cheap.

Guaba *Inga edulis* fruits for sale in a peque-peque canoe tied up by a public market. Requena, Rio Tapiche, Loreto, Peru 4-13-06

Wild populations of *Inga edulis* are found in a variety of upland and floodplain sites in the Amazon. Most of the wild *Inga edulis* trees are found in regrowth forests rather than mature woodlands. Along the banks of the Rio Puinahua, a branch of the Ucayali in the Peruvian Amazon, *Inga edulis* grows in some old successionary forests and in the vicinity of San Carlos where cattle wander into such forests and feed on the fallen pods. *Inga edulis* is a weedy tree thus bringing it into close contact with people and rendering it easy to domesticate. On Ilha das Onças in the Amazon estuary near Belém, for example, farmers do not plant the tree but look after seedlings when they arise spontaneously from discarded seeds (Anderson et al. 1995). This propensity for naturalization is a concern in areas where the tree has been introduced, such as in Australia, where it appears on weed watch lists.

In the Brazilian Amazon, ingá cipó is found in regrowth forests in widely scattered locations, including near Barcarena at the mouth of the Tocantins. According to river dwellers in the vicinity of Santarém, ingá cipó grows wild in floodplain forests along the Amazon. On the banks of Surubim-Açu, a channel cutting across an island in the Amazon between Santarém and Alenquer, one river dweller collected fruits of ingá cipó in várzea forest to plant the seeds in his home garden. In northern Marajó, such as on Cajueiro and Mangueiras ranches in the Municipality of Chaves, and on Mexiana at the mouth of the Amazon, ingá cipó occurs on forest islands in savanna, including relict sand dunes now dominated by economic trees possibly planted long ago by indigenous cultures.

Inga edulis **(center) planted in a manioc field along with various other fruit trees. Sitio Timbó, left bank of Rio Xingu about 45 km upstream from Altamira, Pará, Brazil, 9-17-15**

In the Brazilian Amazon most of the ingá cipó fruits consumed by rural and urban folk come from trees in home gardens or agroforestry fields on floodplains as well as uplands. In home gardens, ingá cipó arises from deliberate plantings and sprouting from discarded seeds. In a field on an upland bluff along the Rio dos Macacos near Breves on Marajó Island, ingá cipó was interplanted with nine perennial and annual crops. In the Del Rey community near Paragominas, one farmer from Bahia has interplanted his cashew *Anacardium occidentale* grove with ingá cipó and eleven other perennial crops. Two hundred kilometers further west near Tailândia, a farmer has intercropped his black pepper field with ingá cipó. On the floodplain of the Rio Maracá in Amapá, a farmer has planted a polycultural field containing ingá cipó and eleven other fruit and nut crops.

Rural folk along the Tamshiyacu, a tributary of the Amazon River some 50 km southeast of Iquitos in the Peruvian Amazon, plant this tree in some fields (*chacras*) after the rice has been harvested (Hiraoka 1989). The Matsigenka in the southeastern part of the Peruvian Amazon also plant this tree in some of their fields (Wezel and Ohl 2006) as do some rural folk in their cacao *Theobroma cacao* orchards in the Alto Beni in the Andean foothills of Bolivia (López and Somarriba 2005). Occasionally, ingá cipó turns up spontaneously in agroforestry fields and home gardens, and farmers recognize the seedlings and tend to them as with their other crops. The practice of planting *Inga edulis* is undoubtedly ancient; along the Caquetá River in the Colombian Amazon, for example, the Andoke and Witoto plant this fruit tree in agroforestry plots (Eden 1988; Veléz and Veléz 1992), as do the Shuar in the Ecuadorian Amazon (Cerón 1991) and the Yuracaré and Trinitario in the Bolivian Amazon (Thomas and Van Damme 2010).

Inga edulis also has some non-food uses, including folk remedies. River dwellers in the Caxiuanã National Forest in eastern Pará, for example, use the bark to make a tea which they pour on wounds to promote healing (Lisboa et al. 2002). Also in Pará, some rural folk use the leaves and bark of this tree to make a tea which is taken to alleviate inflammation and diarrhea (Silva et al. 2007). The Kichwa in the Ecuadorian Amazon use the leaves to make a topical application for treating leishmaniasis (Gachet et al. 2010) while along the Oyapock River in French Guiana that forms the border with Amapá in the Brazilian Amazon, the Wayãpi prepare a juice from the bark of this tree to treat leishmaniasis (Odonne et al. 2011).

In Acre, this tree is used as a living fence in some cattle pastures (Franke 1999).

The trunk and branches are also used for fuelwood, such as among the Chácobo, Gwarayo, and Tsimane in the Bolivian Amazon (Reyes-García 2001:246; Toledo and Salick 2006; Zambrana et al. 2017), the Cofán and Waorani in the Ecuadorian Amazon (Cerón et al. 1994:54; Cerón and Ayala 1998:109) and rural folk along the Amazon River and in the Andean foothills in Peru (Balbin and Samaniego 1996; Marquardt et al. 2013). The Chácobo also use the wood in house construction and to make arrows for hunting and fishing (Zambrana et al. 2017). The Waorani also use the wood in house construction. The Shuar in the southern part of the Ecuadorian Amazon prefer this wood for cooking (Bennett et al. 2002:68). Some non-indigenous rural people in the Amazon also use this wood for house construction and to make charcoal, such as in the Comunidade Rural do Bacuri, 25 km southeast of Abaetetuba, Pará (Gonçalves and Lucas 2017).

Inga edulis is planted as a shade tree for coffee, cacao, and tea in Bolivia and other parts of Latin America (Léon 1998; Murakami et al. 2016:67). The relatively short tree has the additional advantages of providing a rewarding snack to workers and helps enrich the soil through nitrogen fixation in its root nodules. At least three species of stingless bees (*Melipona rufiventris*, *M. seminigra*, and *Trigona fulviventris*) visit the flowers of this tree in the central Amazon and probably help with pollination (Absy and Kerr 1977; Absy et al. 1980; Oliveira et al. 2009). Rural people gather the honey of several species of stingless bees in the nearby forests and sometimes raise them in hollowed out logs that hang under the eaves of their homes.

24.14 *Inga grandiflora*

Brazil Ingá peludo, ingá felpudo, ingá amarelo, ingá do mato, ingarana

Status Wild, planted

Ingá peludo *Inga grandiflora* **in fruit in a home garden on an archaeological site. Base de ICMBio/IBAMA, Baia de Caxiuanã, Rio Anapú, Pará, Brazil 4-30-15**

One of the common names for *Inga grandiflora* in the Brazilian Amazon, ingá peludo, reflects the pod's furry texture. The "hairy" pods are 15–25 cm long by 2–3 cm wide and contain dark olive seeds, 1.5 cm by 1 cm, which are surrounded by a spongey white mesocarp. Rural folk like to snack on these fruits because of their sweet taste. The fruits are produced in the rainy season and into the early part of the dry period. The species is called *grandiflora* because of the relatively large flowers of this genus.

Ingá peludo *Inga grandiflora* fruit. Base de ICMBio/IBAMA, Baia de Caxiuanã, Rio Anapú, Pará, Brazil 5-14-15

Inga grandiflora is a small tree in upland forests and is less common along streams in Amazonia (Alarcón and Peixoto 2007; Lemos et al. 2015; Pennington 1997:651; Souza 2012:50). It is also found occasionally in floodplain forest of various water types including the sediment-rich river Solimões in Amazonas, Brazil (Luize et al. 2015), muddy waters in the Amazon estuary (Amaral et al. 2009), and clear waters such as Rio Jari in Amapá (Carim 2016:89) and in Mato Grosso (David et al. 2015). In the Andean foothills, this tree is not found above 200 m (Reynel and Pennington 1997:178). Ingá peludo is cultivated in some upland home gardens, particularly along the middle Xingu. Several of those home gardens are on archaeological sites with fertile anthropogenic dark earth. Ingá peludo is also found on *terra preta de índio* further east in the Caxiuanã National Forest.

Ingá peludo *Inga grandiflora* in flower. Base de ICMBio/IBAMA, Baia de Caxiuanã, Rio Anapú, Pará, Brazil 5-14-15

24.15 *Inga heterophylla*

Bolivia Pacay, pacay rosado

Brazil Ingaí, ingá mirim, ingá xixica, ingá xixi, ingá xixi vermelho, ingá vermelho, ingá macaco, ingá de macaco, ingá copaíba, ingá tururi, ingá pretinho, ingá de sapo, ingá miúdo, ingá ferradura, ingá brabo, ingá do mato, ingarana, ingazinho, pacu-ingá; mahá mené (Desana), čičipe'ɨ (Guajá), ɨnapere'ɨ (Ka'apor), paku inga (Tembé), mahá mené (Tukano), 'renho (Xavante)

Colombia Guamo de loro; merecño (Karijona)

Ecuador Muranya, sampi, imik sampi (Shuar), aween (Waorani)

Peru Shimbillo, shimbillo colorado; sugkamat sampi (Aguaruna), tónoyeco (Bora), kañarigaviki (Matsigenka)

Venezuela Guamo; poorara yo (Panaré)

Status Wild, planted

Ingaí *Inga heterophylla* bush in fruit in upland successionary vegetation. Jon, 10 years, is gathering the pods to snack on the mesocarp. Inland from Carrazedo, mouth of the Xingu River, Município de Gurupá, Pará, Brazil 7-22-16

Inga heterophylla is a small to medium-sized tree found in the understory of upland forest and in successionary vegetation throughout the Amazon up to about 2000 m in the Andes (Alvino et al. 2005; Arellano 2013:178; Baar et al. 2004; Brienza Júnior 1999:26; Carim et al. 2007; Coelho et al. 2012; Espírito-Santo et al. 2005; Gomide 1997:68, 111; Ducke 1925, 1949:22; Gama et al. 2005; Ivanauskas et al. 2003; Lemos et al. 2015; Lisboa 1989; Mackensen et al. 2000; Medeiros 2016:93; Mendoza and Guerrero 2010; Oliveira and Mori 1999; Rayol et al. 2006, 2011; Rodrigues et al. 2007a, b; Salomão et al. 1988; Schwartz 2007; Silva and Rosa 1989; Taurino 2017:32). It also occurs in Venezuela, the Guianas, and southern Brazil (Pennington 1997:253). It grows up to 12 m tall (Guarim Neto 1984). In the lower Xingu, this species of *Inga* is found in upland forests dominated by various palms, such as babaçu *Attalea speciosa* and lianas (Salomão et al. 2007a). It is also found occasionally in some floodplain forests, such as along the lower clear-water Tapajós River (Vieira et al. 2017), the sediment-rich Upper Madeira River in Rondônia (Casula 2012:59), the Rio das Pacas in the upper Xingu River basin in

Mato Grosso (Kunz et al. 2014a, b; Stefanello et al. 2010), and in the Amazon estuary (Amaral et al. 2008; Carim et al. 2008; Silva et al. 2013). It also grows in some successionary vegetation on floodplains, such as in the watershed of the Mojú River in eastern Pará (Silva 2008b:27) and thrives in gallery forests snaking through cerrado vegetation and seasonal forests in the southern and southeastern fringes of the Amazon Basin (David et al. 2015; Haidar et al. 2013; Ivanauskas et al. 2004; Kunz et al. 2014b; Lenza et al. 2015; Marimon et al. 2002, 2006; Miguel et al. 2011; Oliveira Filho 1989; Stefanello 2008:18). In addition to forests, *I. heterophylla* also grows in stunted woodland (*campina, cerradão*) on sandy soils, such as near Manaus, Amazonas, Brazil, in the Colombian Amazon, in the vicinity of Iquitos in Peru, and on iron-rich hills at Serra dos Carajás in Pará state (Duque et al. 2001; Gómez et al. 2013; Mattos et al. 2018; Zárate et al. 2015). It also occurs in the transition zone between Amazonian forest and scrub savanna (*cerrado*) in Mato Grosso along the southern fringe of the Amazon Basin (Marimon et al. 2001; Mews et al. 2011a, b; Rondon Neto et al. 2010). This species is also a component of woods on old sand dunes along the northeastern coast of Pará (Amaral et al. 2015; Bastos et al. 1995; Santos 2010:42; Sousa et al. 2009). It is also found occasionally on archaeological sites with rich anthropogenic black earth. This species is clearly adapted to a wide range of habitats.

Inga heterophylla starts producing fruit when it reaches the size of a bush, around 1.5 m. Its slender pods, usually curved in a half moon shape, vary in length from 5 to 15 cm and are 1–1.5 cm wide (Reynel and Pennington 1997:70). Young and old alike gather the yellow pods to snack on the white pulp surrounding the round to slightly oblong brown seeds that measure 0.7 cm by 1 cm or 1 cm in diameter. Some youngsters also pick the fruits when they are still green, to avoid others, including monkeys, beating them to it.

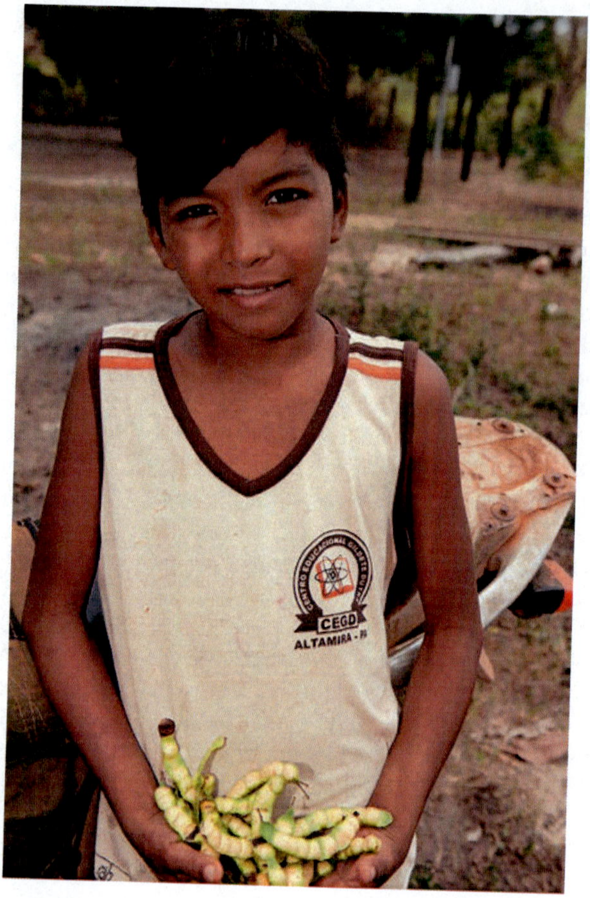

Jean, 11 years, with ingá mirim *Inga heterophylla* fruits he gathered in nearby forest. He traces part of his ancestry to the Kayapó indigenous tribe. Travessão Tiradentes, km 80 Altamira-Itaituba, Transamazonica, Município de Medicilandia, Pará, Brazil 9-5-15

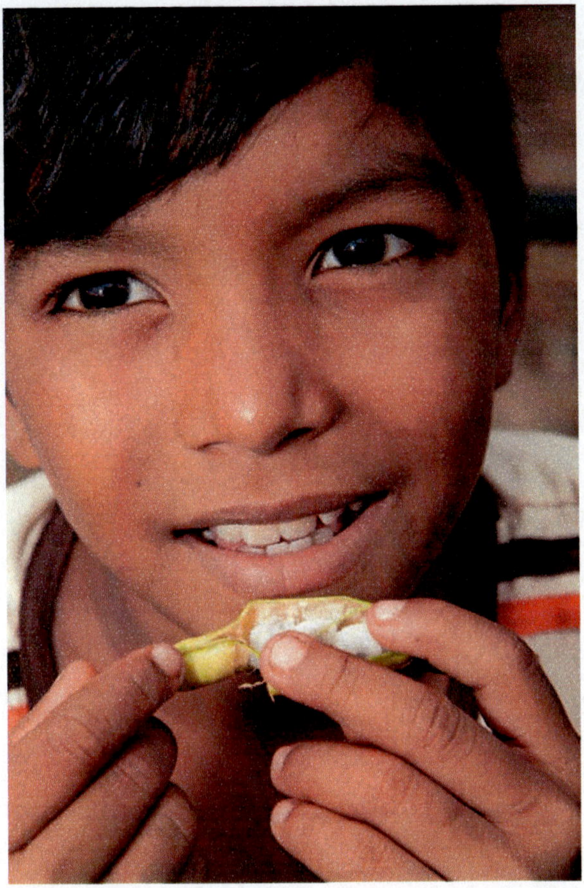

Jean, 11 years, eating an ingá mirim *Inga heterophylla* fruit. Travessão Tiradentes, km 80 Altamira-Itaituba, Transamazonica, Município de Medicilandia, Pará, Brazil 9-5-15

Inga heterophylla is occasionally planted in home gardens on uplands and floodplains, such as along the lower Solimões near its confluence with the Rio Negro (Lima and Saragoussi 2000). I found this tree in an abandoned home garden along the middle Xingu; it could have been planted or perhaps arose spontaneously from discarded seeds. The Tukano in the Upper Rio Negro watershed plant this tree in some of their manioc fields as well as in their villages (Ribeiro 1990).

The fruit is popular with some monkeys, especially howler monkeys *Alouatta* spp. (Pinto et al. 2003) and the red-nosed bearded saki *Chiropotes albinasus* (Pinto 2008:77), hence one of the local names for the tree in the Brazilian Amazon: ingá macaco. *Inga heterophylla* is also occasionally found along river banks and the fruits are eaten by fish, such as pacu (various species of *Myleus*, *Mylossoma*, and *Metynnis*). The Tembé in eastern Amazonia recognize this, hence their name for the fruit: paku inga. In the vicinity of Santa Cruz, Bolivia, cattle ranchers protect this tree in their pastures to provide shade for their cattle and because the livestock like to eat the leaves (Reynel and Pennington 1997:72).

Inga heterophylla has uses other than for fruit. In the Caxiuanã national forest in eastern Pará, for example, locals burn the bark of this tree and mix it with ash from lamps (*lamparinas*) to make a black color for dyeing handicrafts (Lisboa et al. 2002).

24.16 *Inga ingoides*

Bolivia Pacai, pacay, ingá, pacay cola de mono, pacay macho

Brazil Ingá cipó, ingá rabo de guariba, ingá pequena, ingá xixica, ingá branco, ingá de corda, ingá da beira; dia bu'i kãhse mere (Tukano), wahunama (Ye'kuana), kurumiirek (Uaicá-Mucajaí)

Colombia Guamo cafetero, guamo de Popayán

Peru Guabilla, guabo, sacha guaba, coto chupa, coto shimbillo, shimbillo, sambi, tsopi; shenan (Shipibo)

Venezuela Guamo rabo de mono

Status Wild, planted

Ingá cipó *Inga ingoides* in fruit in a home garden. Ilha Marisol, middle Xingu River upstream from Altamira, Pará, Brazil 9-3-15

The fruits of *Inga ingoides* are similar to those of the widely cultivated *Inga edulis* except that they are a little shorter and more slender. Furthermore, the furrowed pods of *Inga ingoides* are yellow-green when ripe as opposed to olive-green in the case of *Inga edulis*. Guabilla, the local name for *Inga ingoides* in the Reserva Nacional Pacaya-Samiria in Peru, means the little guaba *Inga edulis*. The slender pods of *Inga ingoides*, typically 20–25 cm long, contain between 10 and 15 black seeds, the size of a small adult fingernail, surrounded by a white, spongy pulp. After ingesting the sweet pulp, people spit out the seeds. *Inga ingoides* fruits in both the rainy and dry seasons.

Rister, 5 years, with guabilla *Inga ingoides* fruits he has gathered in a home garden. Manco Capac, Rio Puinahua (a branch of the Ucayali River), Reserva Nacional Pacaya-Samiria, Loreto, Peru 4-14-06

Another name for *Inga ingoides* in the Peruvian Amazon is coto chupa (the howler monkey sucks), which indicates that monkeys in the genus *Alouatta* also eat the fruits and probably disperse the seeds. Spider monkeys *Ateles paniscus* are known to disperse the seeds in the Bolivian Amazon (Pacheco and Simonetti 1998, 2000; Roldán and Simonetti 2001). Woolly monkeys *Lagothrix lagotricha* and tufted capuchin monkeys *Cebus apella* also eat the fruits of *I. ingoides* (Defler and Defler 1996; Gómez-Posada 2012). Woolly monkeys, spider monkeys, howler monkeys, and tufted capuchin monkeys are hunted throughout the Amazon Basin for their meat. In the Peruvian Amazon, the saddle-back tamarin *Saguinus fuscicollis illigeri* has been observed eating the fruits of this tree (Soini 1987); although these monkeys are too small to be hunted regularly for their meat they are sometimes kept as pets and are likely involved in dispersing the seeds. In the vicinity of Manaus, a related species, the pied bare-faced tamarin *Saguinus bicolor bicolor* also feeds on these fruits and likely defecate the seeds (Egler 1992). Pygmy marmosets *Cebuella pygmaea* pierce the bark of this tree with their teeth to lick up the exudate (Soini 1988); these monkeys are too small to be hunted for their meat since they generally weigh under 140 g. The slender pods of *Inga ingoides* have inspired some rural folk in Bolivia, Brazil, and Venezuela to name the fruit monkey tail. According to farmers in the village of Veinte de Enero along the Yanayacu, an affluent of the lower Marañón, parrots also break open the pods to eat the pulp.

A denizen of successionary forests on the higher parts of floodplains throughout the humid lowlands of tropical South America and the Caribbean it is found up to 1000 m in mountain foothills (Pennington 1997:723). It also occurs in upland successionary vegetation in some abandoned agricultural plots, such in the vicinity of Santa Mercedes along the Rio Putumayo in northern Peru bordering Colombia (Gomez 2011:80; Muñoz 2010:17, 23, 35) and in the Zona Bragantina region of northeastern Pará (Vieira 1996:59). The tree is normally 10–15 m high, but can reach 30 m, and is also found occasionally in the mature upland forest such as in the Peruvian Amazon and along the lower Xingu River in Pará (Salomão et al. 2007a; Vásquez 1997:342). It is also found in some gallery forests in the Upper Araguaia watershed in the southern fringe of the Amazon Basin (Cabacinha and Fontes 2014). It also occurs in thick scrub vegetation (*cerradão*) in the southern part of the Amazon Basin in the watershed of the middle Tocantins (Medeiros et al. 2012).

Inga ingoides. **Colored stipple engraving by P.J.F. Turpin, inscribed** *Inga ornata*. C.S. Kunth, Mimoses, t. 14, Paris 1820. Library, Botanischer Garten und Botanisches Museum Berlin-Dahlem, Freie Universität, Berlin. Published in: H. Walter Lack. 2009. *Alexander Von Humboldt and the Botanical Exploration of the Americas*, Prestel, Munich

Inga ingoides is also planted in home gardens, such as in the village of Manco Capac along the Puinahua, an arm of the Ucayali River and along the Tefé and Xingu rivers in Brazil. It is also cultivated occasionally in fields intercropped with other perennial fruit and nut trees. The fruits of *Inga ingoides* sometimes turn up in markets, such as Nauta on the lower Marañón. The wood is also used in construction and to make charcoal (Reynel and Pennington 1997:192).

Ingá cipó *Inga ingoides* in fruit. Planted in a home garden that floods in some years. Ilha Marisol, an island in the middle Xingu River upstream from Altamira, Pará, Brazil 9-3-15

24.17 *Inga laurina*

Bolivia Pacay, pacay chico, pacaycillo

Brazil Ingá xixi, ingá xixica, ingá de macaco, ingá cururú, ingá capuchino, ingá da mata, ingá do cerrado, ingá mirim, ingá ferrugem, ingá chichica, ingá pequeno, ingá pé de galinha, ingá banana, ingá branco, ingarana, ingaí, pitomba da mata; ynga'ía (Piripkura)

Colombia Guamo

Ecuador Guaba vainilla

Peru Guabilla, tamarindo, guama, shimbillo, shimbillo de monte

Venezuela Guamo, guamo pata de morrocoy, adukuni

Status Wild, spontaneous in cultural settings, planted

Ingá ferrugem *Inga laurina* tree in fruit in floodplain forest along a black water river. Rio Ipixuna, affluent of the lower Xingu near Carrazedo, Município de Gurupá, Pará, Brazil 4-19-15

Inga laurina has the widest distribution in the genus, ranging from northwestern Mexico, parts of the Caribbean, south through Central America to northern Argentina (Pennington 1997:165). Given its enormous range, it is difficult to pinpoint its origin because the tree was already widespread by the time Europeans arrived in the Americas. Well adapted to dry as well as humid environments, *Inga laurina* occurs as high as 1500 m on the sides of mountains, such as in the Andean foothills of Bolivia (Arellano 2013:178). It grows in upland and floodplain forests (Camargo 2018:8; Silva et al. 2017), including tidal forests in the Amazon estuary (Braga and

Jardim 2019; Braga et al. 2015), regrowth woodlands, early successionary vegetation, and dry scrub forest (*cerrado*).

Manuel, 14 years, gathering ingá ferrugem *Inga laurina* **fruits in floodplain forest along a black water river. Rio Ipixuna, affluent of the lower Xingu near Carrazedo, Município de Gurupá, Pará, Brazil 4-19-15**

The tree can reach 30 m and produces slender pods about the length of an adult's hand (Vásquez 1997:342). The pods are yellow when ripe and usually vary in length from 7 to 13 cm, although they can reach 20 cm in some cases (Reynel and

Pennington 1997:42). A thin, white pulp surrounds the green seeds, 2.5 cm by 1 cm. Although the pulp is meager it is very sweet and much appreciated by locals. On floodplains, people use canoes to maneuver up to the trees and pluck the fruits during the rainy season. Some trees bear fruits well into the dry season. Monkeys, such as the golden-backed uakari *Cacajao melanocephalus oukary* and squirrel monkeys *Saimiri sciureus cassiquiarensis*, like to feed on the fruits as well (Araujo 2014:37; Barnett and Castilho 2000; Barnett et al. 2005) as suggested by one of the common names for this tree in the Brazilian Amazon: ingá de macaco. These monkeys likely disperse some of the seeds.

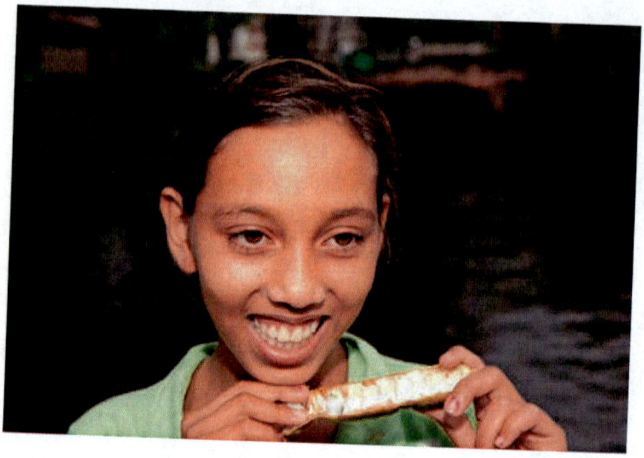

Auindisen, 14 years, eating ingá ferrugem *Inga laurina* fruit gathered by her brother in floodplain forest. Carrazedo, confluence of the Xingu and Amazon Rivers, Município de Gurupá, Pará, Brazil 4-19-15

Inga laurina occurs in home gardens on the Amazon floodplain, such as near Santarém and Oriximiná, as well as on uplands such as along the middle Xingu. Sometimes it is planted around rural homes, other times the trees sprout from discarded seeds. In the early twentieth century, Adolfo Ducke, the first resident botanist in the Brazilian Amazon, noted that ingá cururú was cultivated in the vicinity of Gurupá in the lower Amazon. Along the Rio Tauaú, an affluent of the Parauaú near Breves on Marajó Island, ingá cururú has also been planted in a home garden. In the municipality of São Félix do Xingu in southern Pará this species is found in some cacao agroforestry plots (Braga et al. 2019). In the Peruvian Andes, *Inga laurina* is employed as shade in tea plantations, whereas in Colombia and the Antilles, it is planted to provide partial shade in coffee groves (Léon 1998). The fruits turn up occasionally in markets, such as the Ver-o-Peso in Belém in November 1998.

24.18 *Inga leiocalycina*

Brazil Ingá, ingá vermelha, ingaí

Ecuador Guaba; liutipa, tuihuish pacai, sacha pacai (Kichwa), main sámpi (Shuar), aague, ananawen, qui, goihwagahue, noyhagahue (Waorani)

Peru Shimbillo, shimbillo de la cocha, shimbillo del rosario, shimbillo delgado, pacae colorado; yuwícham sampi (Aguaruna)

Venezuela Guamo, guamo cagaruta de chivo, moshima; tosha (Yanomami)

Status Wild

Shimbillo *Inga leiocalycina* tree in flower and fruit along the margin of a floodplain lake.

Cocha Grande, near San Carlos, Rio Puinahua (a branch of the Ucayali), Reserva Nacional Pacaya-Samiria, Loreto, Peru, 4-18-06

Normally found in uplands, *Inga leiocalycina* is also occasionally found on floodplains where its fruits are gathered at high water, such as in the Anavilhanas archipelago along the lower Rio Negro upstream from Manaus (Souza and Silva 2003). Although the tree can reach 35 m, it starts producing fruit when relatively short, around 3 or 4 m tall. Known as shimbillo de la cocha (the lake *Inga*) in the vicinity of San Carlos along the Puinahua River, an arm of the Ucayali, the sizeable pods can reach 40 cm in length. People twist the pods off the branches and split them open to access the white pulp. The legume bulges at the site of each seed hence another common name for *Inga leiocalycina* in the Peruvian Amazon: shimbillo del rosario (the rosary bead). The fruits are consumed locally and are not sold in markets. The Waorani in the Ecuadorian Amazon also use this wood for house construction and for fuel (Cerón and Ayala 1998:110).

Spider monkeys *Ateles paniscus* and the northern bearded saki *Chiropotes satanas chiropotes* are known to feed on the mesocarp of these fruits and are likely involved in dispersing some of the seeds (Van Roosmalen and Klein 1988). These monkeys are shot for their meat so *Inga leiocalycina* also contributes indirectly to the diet of some rural people.

A wide-ranging species from southern Mexico to the Guianas and the Amazon Basin, *Inga leiocalycina* is found in a broad range of forest environments below 1000 m from areas with a pronounced dry season to regions with heavy, almost year-round rainfall (Pennington 1997:443).

24.19 *Inga macrophylla*

Bolivia Pacay, pacay blanco

Brazil Ingá facão, ingá açu, ingá chata, ingá chinelo, ingá folha grande, ingá peluda, ingá peludo, inga péua; kawiapali (Baniwa), kai mené (Desana), mokiu tirit'i (Sateré-Maué), pa-má (Ticuna), kai mené (Tukano)

Colombia Guamo machete, guamo hembra; r-gnotairai (Witoto)

Ecuador Guambushi (Shuar), goɨ bené (Siona-Secoya)

Peru Shimbillo, shimbillo tablacho, coto shimbillo, vaca paleta, pacay, pacae, pacae coto, pacae soga, pacae de playa, pacae amarillo, pairahu, guaba, guaba paltacha, guaba peluda, guaba lanuda; kashúpa (Achuar), ahiva, aajɨvahe (Bora)

Venezuela Guamo verde; adawata aküdü (Ye'kuana)

Status Wild, spontaneous in cultural settings, planted

Vaca paleta *Inga macrophylla* fruit in a village home garden. Nueva Esperanza, Rio Marañón, Reserva Nacional Pacaya-Samiria, Loreto, Peru 11-8-03

Inga macrophylla has one of the largest fruits among the three hundred or so species in its genus with sword-like pods that can reach 1.5 m long and some 4 cm thick. No wonder, then, that one of the local names for the fruit in Brazil is the machete ingá (ingá facão) and guamo machete in Colombia. Each of the slightly furry pods contains between 15 and 30 seeds, so a single pod makes a satisfying snack. The sweet and spongy pulp is enjoyed by adults and children alike. *Inga macrophylla* bears ripe fruit toward the end of the rainy period and into the first few months of the dry season. The hardy fruits turn up in some street markets, especially in Peru, as well as along roadside stands, such as near Cajamarca and Tarapoto. The Ticuna who live along the Solimões in Brazil and the Amazon and its affluents in Colombia employ the fruit pulp to wipe their eyelids when a yellowish accretion forms (Schultes and Raffauf 1990:243). Rural folk along the Jauaperi River that forms the divide between the states of Amazonas and Roraima in the northern part of the Brazilian Amazon use these fruits to bait hooks suspended in water for catching turtles (Ammann 2014:37).

Anita, 13 years, selling pacae *Inga macrophylla* **fruits. Tabalosus, Moyobamba-Tarapoto road, San Martin, Peru, 8-21-04**

Widely distributed in forests and disturbed sites in the humid lowlands of northern South America up to 1400 m in the Andean foothills, *Inga macrophylla* occurs mostly on uplands but is also found occasionally on floodplains, such as at the confluence of the Purus and Solimões rivers in Amazon, Brazil (Luize et al. 2015; Macía and Fuertes 2008; Pennington 1997:771). This medium-sized tree is frequently cultivated in home gardens throughout its range and some of them likely arose from discarded seeds. A mature tree produces 100–150 fruits a year (Souza et al. 1996:124), one of the reasons the nitrogen-fixing tree is so popular with farmers and fishermen. It is particularly common in yards along the banks of the Ucayali and Marañón rivers in Peru. *Inga macrophylla* is also occasionally interplanted with other crops in fields, such as in the vicinity of Anori along the Solimões River. The Witoto along the Caquetá River in the Colombian Amazon plant this fruit tree in agroforestry plots, undoubtedly an ancient practice (Veléz and Veléz 1992).

Farm girl spitting out an ingá chata *Inga macrophylla* seed after eating the pulp. She gathered the fruit in her home garden. Boa Esperança near Itacoatiara, Amazonas, Brazil 12-9-06

24.20 *Inga marginata*

Bolivia Pacae pequeño, pacay, pacay chico, pacay mono, pacay rosario, pacai, siquile; shipi xënanë (Chácobo)

Brazil Ingá, ingá branco, ingá amarela, ingá mirim, ingá da mata, ingá chinelo, ingá bainha, ingá feijão, ingá de macaco, ingá facão, ingá verdadeira, ingá miúdo, ingá de leite, ingá dedo, ingazinho; ñã poko'i (Arawaté), ɨna-'i-una-'ɨwa (Asurini), čičipe'ɨ (Guajá), ɨnapere'ɨ (Ka'apor),

Colombia Guama del río, guamo churimo, guamo negrito; tóko (Andoke), mená (Tatuyo)

Ecuador Guaba, guaba del río, guaba agria, guabo, guabilla, pilingos pacay, chunda pacay, uno yutzo, wampa, sampi; tserempush (Achuar), pishillo (Cayapa), inza fino, siri fiño (Cofán), cachic, sacha pacai, turu pacai, pilinga (Kichwa), zambi, tserempach, tserempush (Shuar), siri bené (Siona-Secoya), mimoton (Waorani)

Peru Shimbillo, shimbillo del rio, poroto shimbillo, shimbillo colorado, guabilla, huaro, pacae del monte, pacae alverjita, pacae maní, moena colorada; sampi, see jempash, sejempách, sijimpak, tuish sampi (Aguaruna), shipin xenán, moka xenán (Shipibo)

Status wild, planted

Jonathan, 9 years, holding shimbillo del rio *Inga marginata* fruits that his mother has for sale in a street market. Mercado Bella Vista, Pucallpa, Ucayali, Peru 1-22-04

Inga marginata is one of the most widespread members of the genus, ranging from Mexico to Bolivia (Pennington 1997:183). A denizen of forests and disturbed sites on floodplains the tree is found as high as 2500 m in mountain valleys. *Inga marginata* also occurs in a variety of other habitats such as *terra firme* forests in Amazonian Ecuador (Navarrete 2016:31; Valencia et al. 1994) and along the lower Xingu (Salomão et al. 2007a), on abandoned pastures in the vicinity of Paragominas in eastern Pará, successionary vegetation near Belém and other parts of eastern Pará as well as in the vicinity of Puerto Maldonado in the southeastern part of the Peruvian Amazon, stunted forest (*campinarana*), in the eastern fringe of the Amazon rainforest in Maranhão, and on sandy soils with stunted vegetation, such as in Amapá (Cain et al. 1956; Gama et al. 2007; Medeiros 2016:93; Robles 2017:47; Sanaiotti et al. 1997; Santana et al. 2004; Thomas 2009; Uhl et al. 1988b).

It can reach 30 m high, especially in forests where it is competing for light. Its relatively small, slender pods vary from 7 cm long by 1–1.5 cm wide to some 20 cm by 2 cm. They are produced during the rainy season and have distinctive bulges around the seeds. The pods have up to a dozen or so seeds surrounded by a sweet, white pulp which is relished by rural and urban folk.

This fruit is eaten by certain fish that are caught by people for food, such as sabalo cola roja *Brycon erypthropterum* in the Peruvian Amazon (Murrieta and Arévalo 2004:52). Several species of monkey, including the spider monkey *Ateles paniscus chamek*, capuchin monkey *Cebus apella*, the white-fronted capuchin *Cebus albifrons*, squirrel monkeys *Saimiri sciureus* and *S. cassiquiarensis albigena*, the emperor tamarin *Saguinus imperator*, the bearded saki *Chiropotes satanas*, and the saddle-back tamarin *Saguinus fuscicollis illigeri*, and the red titi monkey *Callicebus discolor*, are also fond of the fruits and are probably involved in dispersing some of the seeds (Carretero-Pinzón et al. 2016; Carrillo-Bilbao et al. 2005; Soini 1987; Terborgh 1983:67; White 1986). In the northern part of the Colombian Amazon, woolly monkeys *Lagothrix lagotricha* eat this fruit and disperse some of the seeds (Stevenson 2000). The fondness of several species of monkey for this fruit accounts for one of its common names in the Brazilian Amazon: ingá de macaco. Pygmy marmosets *Cebuella pygmaea* pierce the bark of this tree with their teeth to lick up the exudate (Soini 1988); these monkeys are too small to be hunted for their meat since they generally weigh under 140 g.

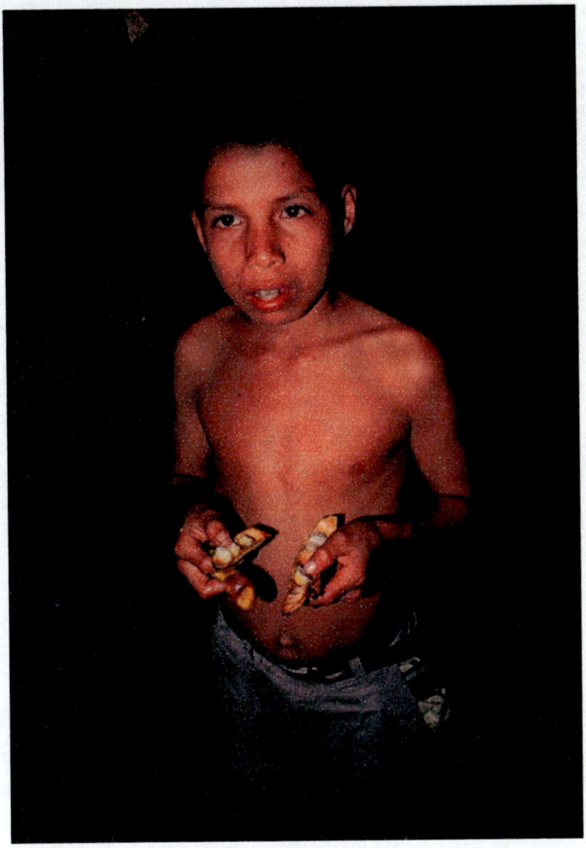

Enoc, 11 years, eating shimbillo *Inga marginata* fruits in floodplain forest near his village. Veinte de Enero, Rio Yanayacu, Reserva Nacional Pacaya-Samiria, Loreto, Peru 3-8-04

Although not common in Amazonian markets, the fruits appear occasionally for sale, such as in Macapá, the capital of Amapá (Adaime et al. 2014) and as I have observed in Pucallpa along the Ucayali River. In eastern Amazonia, the Ka'apor and Assurini gather the fruits for consumption in their villages (Balée 1993a:66). *Inga marginata* occurs occasionally in home gardens, such as along the middle Xingu. The wood of *Inga marginata*, as with several other species in the genus, is highly regarded for fuelwood and the tree is occasionally planted to shade coffee (Reynel and Pennington 1997:50).

24.21 *Inga nobilis*

Bolivia Pacay, pacay colorado, pacay pildorita; nasha turu (Tacana)

Brazil Ingá, ingá peludo, ingá branco, ingá preto, ingá canela, ingá azeda, ingá xixica, ingá amarelinha, ingá de sapo, ingá vermelho, ingá ferrugem, ingá bravo, ingaí, ingazinha; ñã poko'i (Arawaté), yurupi-rana-igá (Asurini), iga-howi-'i (Ka'apor), moridikabedá (Suruí), wãkama, rerokoakasihi (Yanomami)

Colombia Guamo, guamo rosario, guamo cerindo, guamilla; e keou wy (Maku), ta-roo'-boo-ta-re-mê, mĕ-ñĕ-ra' (Makuna), me ve (Tukano)

Ecuador Guaba, guaba negra, guabilla; wampushik (Achuar), barbasco pacai (Kichwa), wampukish, muauwa sampi, imiuk sampi, imik sámpi, samik, kunkuin sámpi (Shuar), pene (Siona-Secoya), guiyuetempuyo (Waorani)

Peru Shimbillo, sapo shimbillo, yacu shimbillo, shimbillo amargo, shimbillo de monte, incha paloma, pacaecillo del canto del rio, pacae de monte; wampukish, wampushik (Achuar), imik sampi, shagkuina, wámpushik (Aguaruna), sampi (Kichwa), shenan, koman xenán (Shipibo), unawachu cauré (Ticuna)

Venezuela Guamo caraota; wãkarha, wekama (Yanomami)

Status Wild

Geni, 8 years, gathering ingá peludo *Inga nobilis* fruit in regrowth forest on an archaeological site at the edge of his village. Vila Castanha, Rio Cariá, a left bank affluent of the lower Pracupi, in turn, an affluent of the Anapu River in the Baia de Botos, Caxiuanã National Forest, Pará, Brazil 5-16-15

Inga nobilis is a medium-sized tree encompassing several subspecies that thrives as an understory tree in floodplain and upland forests from Mexico to Bolivia, including the Amazon and Orinoco basins and central Brazil (Abreu 2012:24; Abreu et al. 2014; Araujo-Murakami et al. 2015; David et al. 2015; Godoy et al. 1999; Jardim et al. 2013; Lemos et al. 2015; Macía 2011; Nebel et al. 2000a, b, 2001; Pennington 1997:381; Pinto and Hay 2005; Salomão et al. 2007a; Santos and Jardim 2006; Thomas 2009). This tree grows in the tidal zone at the mouth of the Amazon (Almeida and Jardim 2012; Amaral et al. 2009; Cattanio et al. 2002; Jardim and Vieira 2001), including sand dunes along the coast of Pará (Amaral et al. 2015), and is found as high as 2000 m in mountain valleys, such as in the foothills of the Andes in Bolivia, Ecuador, and Peru (Arellano 2013:178; Mendoza and Guerrero 2010; Toledo 2016:30). It also occurs in regrowth forest on uplands, including on archaeological sites (Alvino et al. 2005; Rayol et al. 2006; Robles 2017:47; Rodrigues et al. 2004). *Inga nobilis* starts producing fruits when it is as little as 2 m high, so the pods are often within easy reach. The yellow-green pods are between 10 and 20 cm long, although they can reach 30 cm in length (Pérez-Arbeláez 1956:492), and contain seeds surrounded by a sweet white pulp. The cylindrical pods ripen during the rainy season and are also gathered from the forest floor when the tree grows bigger, such as along the Yanayacu River in the Reserva Nacional Pacaya-Samiria.

Several species of monkeys in the Amazon, including the capuchin monkey *Cebus apella*, howler monkeys (*Alouatta* spp.), and black-headed uakaris *Cacajao melanocephalus melanocephalus* are also fond of this fruit and are probably involved in dispersing the seeds (Boubli 1999; Terborgh 1983:67). These monkeys are hunted for their meat so *Inga nobilis* also contributes indirectly to the diet of rural people. The black-bearded saki *Chiropotes satanas satanas* has also been observed feeding on the mesocarp of these fruits in the eastern fringes of the Amazon Basin in Maranhão (Port-Carvalho and Ferrari 2004), and although this monkey is not usually hunted for its meat, it likely also disperses some of the seeds.

Indigenous groups are also fond of the fruits and probably played a role in moving this species around, particularly increasing its density around settlements. The Makuna in the Colombian Amazon, for example, enjoy eating the pulp which they believe helps preserve their teeth (Schultes and Raffauf 1990:243). The fruits of this species figure as claws of Uyush, a legendary figure in Achuar mythology who cultivated manioc and donated some of the tubers to Achuar women who did not practice agriculture at that time (Descola 1988:266).

Geni, 8 years, eating ingá peludo *Inga nobilis* fruit he gathered in regrowth forest. Vila Castanha, Rio Cariá, a left bank affluent of the lower Pracupi, in turn, an affluent of the Anapu River in the Baia de Botos, Caxiuanã National Forest, Pará, Brazil 5-16-15

This relatively short tree is often spared when clearing sites for homes and fields and thrives in successionary woods near settlements, such as in the municipality of Bragança in northeastern Pará (Rayol et al. 2011) and in the Tocantins watershed in eastern Pará (Taurino 2017:88), so villagers often have access to the fruits. The trunk and branches are also used for firewood, such as among the Ticuna, an indigenous group that lives in scattered villages along the Amazon in Peru and the Solimões in Brazil (Rengifo-Salgado et al. 2017), and for house construction among the Shuar in the Andean foothills of the southeastern part of the Ecuadorian Amazon (Pohle and Reinhardt 2004).

24.22 *Inga pilosula*

Bolivia Pacay

Brazil Ingá, ingá dos indios; virine (Tukano), poataa hi (Yanomami)

Colombia Guamo; gûiro (Yukuna)

Peru Shimbillo, shimbillo verano, shimbillo enano, purma shimbillo

Venezuela Guamo, guamo amarillo, guama de araguato, jaguatá; adawata janadü (Ye'kuana)

Status Wild, planted

Tukano boy, Jilmar, 9 years, by an *Inga pilosula* tree in his home garden. Cunuri Ponta, Uaupés River, Amazonas, Brazil, 11-12-12

Inga pilosula is shrub to small tree, usually under 15 m high, that is widespread in Amazonia, Venezuela, and the Guianas. It is mostly found in successionary vegetation on higher parts of floodplains, especially along clear and black water rivers. However, it also sometimes occurs on uplands, such as in stunted woods (*caatinga*, *campina*) growing on sandy soils (Pennington 1997:517), on parts of Marajó Island at the mouth of the Amazon (Silva et al. 2013), and on sand dunes along the coast of Pará (Santos and Bastos 2014:134; Sousa et al. 2009). Its range also extends into the scrub savanna zone of central Brazil (David et al. 2015). Found as high as 1500 m in the foothills of the Andes, *Inga pilosula* is also cultivated occasionally in some gardens, such as along the Uaupés and middle Xingu Rivers in the Brazilian Amazon, although it is not nearly as common around houses as some other species in this genus, such as *Inga edulis* and *Inga cinnamomea*. In the Caquetá river basin in the Colombian Amazon, the mottled-face tamarin *Saguinus inustus* feeds on this fruit (Palacios et al. 2004) and may be involved in dispersing some of the seeds as does the Colombian squirrel monkey *Saimiri cassiquiarensis albigena* (Carretero-Pinzón et al. 2016).

Rural folk break open the relatively flat pods, normally 10–20 cm in length (Reynel and Pennington 1997:152), to snack on the white mesocarp, but the lustrous fruits are usually not taken to urban markets. The Cubeo in the Colombian Amazon clean off sticky accretions around their eyelids with the pulp that surrounds the seeds (Schultes and Raffauf 1990:243).

24.23 *Inga rubiginosa*

Brazil Ingá, ingá pracuúba, ingá vermelho, ingá de pêlo, ingá guariba, ingá peludo, ingá peluda, ingá da folha peluda, ingá folha grossa, ingá ferrugem; iñá-paka-‘i (Araweté), iga-pe-‘iwi (Asurini), čičipe’ɨ (Guajá), tapi’irɨna’ɨ, kagwaruhu ynga (Ka’apor), arawyta pana (Waimiri-Atroari), akiki-iga (Wajãpi)

Status Wild, planted

Bento, 57 years, holding an ingá pracuúba *Inga rubiginosa* fruit gathered in regrowth forest at the perimeter of his manioc field. Pedreira, Floresta Nacional Caxiuanã, Baia de Caxiuanã, Rio Anapu, Município de Melgaço, Pará, Brazil 5-11-15

Inga rubiginosa is a medium sized tree, reaching up to 30 m in upland forests, including those in succession, throughout the Amazon Basin (Almeida and Vieira 2001; Almeida et al. 1993; Alvino et al. 2005; Amaral et al. 2009; Aparício 2011:61; Araújo et al. 2005; Carim et al. 2013; Coelho et al. 2003a, b, 2012; Ducke 1949:29; Fernandes Neto 2015:23; Gomide 1997:68, 112; Lemos et al. 2015; Lisboa 1989; Mackensen et al. 2000; Prata et al. 2010; Rayol et al. 2006, 2011; Rodrigues 1986; Rodrigues 1963; Salomão et al. 1988; Santana et al. 2004; Silva 2010:81; Silva and Rosa 1989; Taurino 2017:64; Vieira 1996:59). It is also found in some riparian forests, such as in southern Rondônia, along the Solimões River in Amazonas, the clear water Jari River in eastern Amazonia, and in the Amazon estuary (Carim 2016: 89; Luize et al. 2015; Miranda 2000; Silva 2008b:16, 27; Silva et al. 2013). It is found from Panama into the Guianas, Venezuela, and the Brazilian part of the Amazon Basin (Abreu et al. 2006; Bánki et al. 2003:37; Cummings and Read 2016; Pennington 1997:631; Salomão et al. 2007a, b; Silva et al. 1987). The sickle-shaped pods of *Inga rubiginosa*, 15–30 cm long and around 3 cm wide, are produced during the rainy season. The yellow-green pods with rusty-brown splotches contain dark olive seeds surrounded by a white mesocarp.

Black spider monkeys *Ateles paniscus*, which are widely hunted for their meat in Amazonia and the Guianas, are known to eat the mesocarp and possibly disseminate some of the seeds (Rylands 1979, 1981; Van Roosmalen 1985; Van Roosmalen and Klein 1988). The Santarém marmoset *Callithrix humeralifera* also eats these fruits and is likely involved in dispersing the seeds within its distribution from the Tapajós River in the east to the Madeira River in the west (Stevenson and Rylands 1988). This marmoset normally weighs under half a kilogram so is not usually hunted for its meat. Further west, a subspecies of this monkey, *Callithrix humeralifera intermedius* which inhabits the interfluvial forest and successionary vegetation between the Aripuanã and Roosevelt rivers in the southern part of Amazonas near the border with Mato Grosso, also eats this fruit and likely disperses some of the seeds (Rylands 1979, 1981). The northern bearded saki *Chiropotes satanas chiropotes* also feeds on the mesocarp surrounding the seeds of this fruit (Van Roosmalen et al. 1981, 1988). Other monkeys observed feeding on this fruit in the Brazilian Amazon and French Guiana include *Callicebus moloch*, *Cebus apella*, and *Ateles paniscus* and these monkeys disperse some of the seeds (Regan et al. 2001; Rylands 1981).White-faced sakis *Pithecia pithecia* which range from eastern Venezuela through the Guianas and into the northern part of the Brazilian Amazon, have been observed feeding on these fruits in Suriname and may also be involved in dispersing some of the seeds (Norconk and Veres 2011). One of the common names for this fruit in the Brazilian Amazon is ingá guariba which suggests that howler monkeys (*Alouatta* spp.) may also be involved in dispersing some of the seeds. As forests become more fragmented in Amazonia and the Guianas, the populations of these seed dispersers will likely continue to decline. In Guyana, tapir *Tapirus terrestris* and peccaries (*Pecari tajacu and Tayassu pecari*) are known to eat this fruit and likely disperse some of the seeds (Cummings and Read 2016).

Adults and youngsters alike enjoy snacking on the pulp and often taken some pods home with them to eat later. The Cubeo in the Colombian Amazon dry the flowers of this species then crunch them into a powder to be inhaled when a person is suffering from nasal congestion (Schultes and Raffauf 1990:243). Rural folk also sometimes plant the seeds in their home gardens, as near Manaus (Souza 2010:27) and as I have observed at Pedreira on the margins of Baia de Caxiuanã along the lower Anapu in eastern Pará. Agoutis (*Dasyprocta* spp.) are also fond of the fruits of *Inga rubiginosa*, as I witnessed at the Serra dos Carajás, and are likely involved in dispersing the seeds. Agoutis, known as cutia in Brazil, are popular game animals and often thrive near human habitations in rural areas.

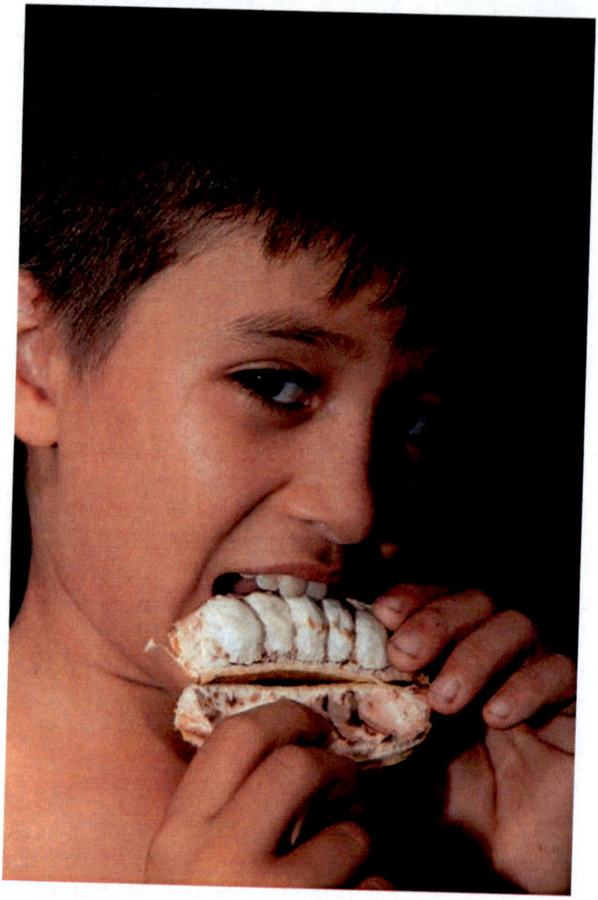

Abraão, 10 years, eating ingá pracuúba *Inga rubiginosa* fruit he gathered in regrowth forest. Pedreira, Floresta Nacional Caxiuanã, Baia de Caxiuanã, Rio Anapu, Município de Melgaço, Pará, Brazil 5-11-15

24.24 *Inga ruiziana*

Bolivia Pacay; ña'me (Tsimane)

Brazil Ingá costela de vaca, ingá peba, ingá péua

Colombia Guama de gallineta, guamo; aagüahe-güachaavogüa (Miraña)

Ecuador Guaba, guaba de río, guaba pequeña amarilla; avunbico fiño, sesenga fiño (Cofán), estihuan pacai, sacha pacai, yacu cachi, yacupacai, vaca pacai, quillu pacai (Kichwa), noca pena, huigosa (Siona-Secoya), guerequeo, huerequego, viriquiu (Waorani)

Peru Shimbillo, shimbillo carambaso, millero shimbillo, yacu shimbillo, puca shimbillo, rufintis shimbillo, rufindi shimbillo, rujindi, pacay cuero, pacae playa, ucsha guiro, rojiende; nakaar (Achuar), naji kumpari, naji, wampukish, shagkuina, sampi, múun sámpi (Aguaruna), intsipa (Campa), pairafo, shenan (Shipibo)

Venezuela Dapunia, guama

Status Wild, spontaneous in cultural settings, planted

Raul, 10 years, gathering shimbillo *Inga ruiziana* fruits in his home garden. Manco Capac, Rio Puinahua (a branch of the Ucayali River), Reserva Nacional Pacaya-Samiria, Loreto, Peru 4-14-06

Although *Inga ruiziana* is widely distributed from Nicaragua and Costa Rica south to Bolivia and in Amazonia it is confined to the western part of the Amazon basin including the Andean foothills up to 1600 m (Araujo-Murakami et al. 2015; Calzadilla-Tomianovich and Cayola 2006; Hartshorn and Poveda 1983; Mendoza and Guerrero 2010; Pennington 1997:373; Robles 2017:47; Thomas 2009). In the mid-twentieth century, the Museu Goeldi-based botanist Adolpho Ducke noted the presence of *Inga ruiziana* along the Solimões River and speculated that it was probably introduced from Peru (Ducke 1946). It grows in forests and successionary vegetation on both uplands and floodplains up to 2000 m in the Andes. In the Reserva Nacional Pacaya-Samiria, the tree is found on higher parts of the floodplains of sediment-rich rivers. It is also encountered on the floodplains of small rivers and streams along the Iquitos-Nauta highway in the Peruvian Amazon (Zárate et al.

2013). *Inga ruiziana* is typically spared when clearing sites for fields and homes to provide ready snacks.

Inga ruiziana **fruits are highly variable in shape and size. San Carlos, Rio Puinahua (a branch of the Ucayali River), Reserva Nacional Pacaya-Samiria, Loreto, Peru, 4-18-06**

Yellow when fully ripe during the rainy season, the flat pods of *I. ruiziana* are highly variable in size and shape, ranging from 7 to 22 cm long and 2 to 4 cm in diameter. Some pods are curved, like a quarter moon, while others are almost straight. People open the pods by twisting them so that they can scoop out the black seeds surrounded by a white mesocarp. The flat, elongated seeds are spat out on the ground after ingesting the sweet pulp. Some game animals, such as paca *Agouti paca* and several monkeys, as well as fish also consume the fruits (Cerón 1996).

Roy, 5 years, eating shimbillo *Inga ruiziana* fruits in his village home. He gathered the fruits while they were still green in his family's field (*chacra*) in order to beat others, including monkeys, to the prized fruit. Maypuco, Rio Marañón, Reserva Nacional Pacaya-Samiria, Loreto, Peru 3-10-04

In forests, *Inga ruiziana* can reach 35 m, but in open areas such as in fields and home gardens, the trees are generally 10 to 15 m high and form broad crowns with multiple branches thus making them easier to climb. In such locations, children clamber up the branches to obtain the fruits. The pods are sometimes gathered while still green because if they are left until they turn yellow, someone else may take the fruits. When not quite ripe, the pulp is a little dry but still sweet enough to entice young and old alike to gather the fruits.

The Shipibo spare this tree when it arises spontaneously from discarded seeds in some of their home gardens, such as in the village of Nueva Cajamarca along the middle Ucayali. Farmers in the community of Santa Rosa on the outskirts of Tingo Maria sometimes plant the tree in their fields (*chacras*). Similarly, residents of

Maypuco along the Puinahua, a branch of the Ucayali River, welcome this addition to their home gardens when it sprouts from tossed seeds. At Maypuco in the Reserva Nacional Pacaya-Samiria along the Marañón River, farmers protect shimbillo, as the tree is known there when it sprouts spontaneously in their manioc fields. In the Andean foothills, *Inga ruiziana* is sometimes spared when clearing forest to provide shade for coffee plantations.

This species of *Inga* has a variety of other uses. The wood provides fuel and the bark is used to prepare an infusion for treating diarrhea and to alleviate pain, especially those created by rheumatism and broken bones (Reynel and Pennington 1997:106). In the Bolivian Amazon, the Tsimane use the wood to make charcoal (Reyes-García 2001:246). In the vicinity of Chazuta along the upper Huallaga River in the Peruvian Amazon, people boil the bark and after it cools use the liquid as a mouthwash to treat infections caused by the yeast *Candida albicans*, and this fluid is also used to disinfect wounds (Sanz-Biset et al. 2009). Also, in the Peruvian Amazon, the Aguaruna use the sap of this tree to treat skin sores (Jernigan 2006:193) and rural people use the wood for construction in the vicinity of San Martín in the Andean foothills of southeastern Peru (Marquardt et al. 2013).

24.25 *Inga splendens*

Brazil Ingaí, ingá ferrujo, ingá ferro, ingá peludo, ingá duro, ingá cobra, ingá facão, ingá alado, ingá chato, ingá açu da mata, ingá vermelha, ingá branco, ingá boi; jatsîhupape, ñatsîoho (Araweté)

Ecuador Tindigas pacai (Kichwa)

Peru Shimbillo, paleta shimbillo

Status Wild, spontaneous in cultural settings

Ingaí *Inga splendens* in flower and fruit. Small tree on the floodplain at the mouth of the Xingu River. Artur 8 yrs. Carrazedo, mouth of the Xingu River, Município de Gurupá, Pará, Brazil 7-31-16

Ingaí *Inga splendens* is small tree that can reach 20 m high on floodplains of clear and sediment-rich rivers such as at the mouth of the Xingu, along the Jari River in eastern Amazonia, in the Amazon estuary, at the confluence of the Purus and Solimões rivers, the confluence of the Japurá and Solimões, and on floodplains of the Maracá island in Roraima (Almeida et al. 2004; Amaral et al. 2009; Carim 2016; Cattanio et al. 2002; Ducke 1949:29; Jardim and Vieira 2001; Lewis and Owen 1989; Luize et al. 2015; Paim et al. 2018; Silva et al. 2013). This species, which can reach 30 m high when fully grown is also found in upland forests and successionary vegetation and ranges widely in the Amazon Basin as well as in Venezuela and the Guianas (Holm et al. 2014; Lemos et al. 2015; Santana et al. 2004). This small to the medium-sized tree also grows in stunted woodlands (*campinarana*) on sandy soils subject to seasonal floodings, such as in Roraima (Vale 2011:48), and on coastal sand dunes in the Amazon estuary (Sousa et al. 2009).

The pods are yellow-green when ripe and normally range in size from 8 cm × 2 cm to 15 cm × 3 cm. However, the pods can reach 25 cm long in some cases (Pennington 1997:472). Rural people eat the white mesocarp and the seeds that measure 2 cm by 1 cm are spat out. Even when the fruits are not fully ripe, people still snack on the pulp though it is less sweet. Woolly monkeys *Lagothrix lagotricha*, spider monkeys *Ateles belzebuth* and tufted capuchin monkeys *Cebus apella*, widely hunted in the Amazon Basin for their meat, also eat this fruit (Defler and Defler 1996; Gómez-Posada 2012; Nunes 1998).

Children gather the fruits by climbing the tree or picking them from the ground when the trees are still young. Because of the edible fruit, people often spare the tree when clearing vegetation to build their homes along rivers. Fruits of *Inga splendens* have been gathered for a long time since botanical remains of this species have been found at archaeological sites in the Upano Valley in eastern Ecuador dating back thousands of years (Rostain 1999). The Araweté along the Ipixuna, an affluent of the middle Xingu, also use the wood for fuel.

Ingá ferrujo *Inga splendens* tree in fruit. José, 8 years, picked the fruit and is snacking on the pulp. Carrazedo, mouth of the Xingu River, Município de Gurupá, Pará, Brazil 7-23-16.

24.26 *Inga stenoptera*

Bolivia Pacay

Brazil Ingá, ingá da várzea, ingaí

Colombia Guamo, guamo copero; to-to'-ko (Cubeo), meen (Maku)

Ecuador Guaba

Peru Shimbillo, charapa shimbillo, tahuampa shimbillo, yacu shimbillo, pairajo; kashai ampuj (Achuar)

Venezuela Echepeñ yo', pihkuankae (Panaré)

Status Wild

Percy Rodrigo, 12 years, gathering fruits of charapa shimbillo *Inga stenoptera* along a black water river. The pods are 15 cm long and contain ten dark brown seeds covered by a sweet white pulp. Near Yarina, Rio Yanayacu, Reserva Nacional Pacaya-Samiria, Loreto, Peru 3-7-04

Mostly confined to river banks and the margins of lakes in the humid lowlands of northern South America (Ducke 1949:30; Nebel et al. 2001; Pennington 1997:431; Silva et al. 2013), *Inga stenoptera* occasionally crops up in upland forests, such as in the Isiboro-Sécure National Park in the Bolivian Amazon (Thomas 2009) and in the Department of Guaviare in the Colombian Amazon (Stevenson and Rodríguez 2008). It is a medium-sized tree reaching up to 25 m in seasonally flooded forests, hence one of its common names in Acre, Brazil: ingá da várzea. Known as charapa shimbillo along the Yanayacu River, an affluent of the lower Marañón in Peru, this local name refers to charapa, the giant South American river turtle *Podocnemis expansa* which is eagerly sought after for its meat. Known as tartaruga in Brazil, this turtle allegedly eats, and possibly disperses, the seeds of *Inga stenoptera*. Woolly monkeys *Lagothrix lagotricha*, also hunted for their meat, are known to eat the fruits and possibly disperse some of the seeds (Stevenson 2004, 2005, 2007). This broad crowned tree occurs along rivers with diverse sediment loads and acidity from white water to clear, and black.

The green pods, which bulge at the seeds, ripen during the flood stage, and some of them are submerged and thus within easy reach of fish and turtles. People gather the slender pods, which range from 9 to 30 cm long by 2 to 3 cm wide, from canoes to eat the sweet, white pulp surrounding the ten or so dark brown seeds. The pods contain dark brown seeds covered by a sweet white pulp. The fruits are eaten locally and do not appear in urban markets. *Inga stenoptera*, along with *I. cinnamomea* and *I. marginata*, is the best-adapted species in the genus with edible fruits capable of withstanding prolonged flooding.

Percy Rodrigo, 12 years, eating charapa shimbillo *Inga stenoptera* fruit he gathered from a canoe. Near Yarina, Rio Yanayacu, Reserva Nacional Pacaya-Samiria, Loreto, Peru 3-7-04

24.27 *Inga thibaudiana*

Bolivia Pacay colorado, pacay pequeño verde; xename (Chácobo)

Brazil Ingá, ingá do mato, ingá da mata, ingá amarela, ingá branco, ingá preta, ingá vermelha, ingá pintado, ingá barata, ingá roceiro, ingá facão, ingá de macaco, ingá rabo de macaco, ingá mococona, ingá xixica, ingá miudo, ingá de porco, ingá cipó, ingá folha pequena, ingazeiro, ingaí, ingarana; iñá-paka-i (Araweté), muruwi-yawa-iga (Asurini), wiritáa panali (Baniwa), bëhpë mené (Desana), čičipe-'ɨ (Guajá), xixirupe, inga-'ɨ-pɨtag (Ka'apor), ɨga-wi-zu-'ɨw (Tembé), bëhpë mené (Ticuna), alãkwã-iga (Wajãpi), 'renho watsédé (Xavante), shititi, tastiik (Yanomami)

Colombia Guamo, guamo cerindo, guamilla; gûiro (Yukuna)

Ecuador Guaba de monte; pakairuya (Achuar), onga fiño, oma fiño (Cofán), bariza pacai, sacha pacai (Kichwa), juicham sampi, kunkuin sampi, kashai sampi, napúrak (Shuar), noka bené (Siona-Secoya), aaun, huamuncahue, quemotahue, mimontahue, momontan, tecanamue (Waorani)

Peru Shimbillo, shimbillo blanco, shimbillo colorado, shimbillo rujente, shimbillo rufin, rufinde shimbillo, coto shimbillo, poroto shimbillo, pacae, pacae blanco, pacae lanudo, pacaecillo amarillo; napujuk sampi, dapujúk sampi (Aguaruna), naméwacháábowa (Bora), intsipa (Campa), napurak sampi (Huambisa), xopon xenán (Shipibo)

Venezuela Ichipaté, wirrita, guamo; shititi kë hi (Yanomami)

Status Wild, planted

Felipe, 14 years, gathering fruits of ingá do mato *Inga thibaudiana* in successionary vegetation at the edge of a manioc field. Pedreira, Baia de Caxiuanã, Município de Melgaço, Pará, Brazil 5-8-15

Inga thibaudiana occurs in successionary growth and light gaps in mature forests on uplands and floodplains throughout the Amazon Basin (Araujo-Murakami et al. 2009; Batista et al. 2011; Casula 2012:59; Coelho et al. 2003a, b, 2013; Dulmen 2001; Ivanauskas et al. 2003; Lobo 2017:62; Longworth et al. 2014; Oliveira and Jardim 1998; Peña-Claros 2003; Pennington 1997:488; Pitman et al. 2001; Rayol et al. 2006; Santos and Jardim 2006; Silva 2010:88; Silva 2008b:16; Taurino 2017:74; Uhl et al. 1988b; Vásquez 1998). It also grows in some stunted woodlands (*campinarana*) subject to seasonal flooding, such as in Roraima (Vale 2011:48) and on hills capped with an iron-rich sedimentary rock known as *canga*, such as on the Serra dos Carajás in Pará (Mota et al. 2018) as well as estuarine vegetation in

Maranhão (Freire and Monteiro 1993) and coastal Pará (Sousa et al. 2009). It is found as high as 1500 m in the Andean foothills, such as in central Peru (Marcelo-Peña and Rodriguez 2014; Mendoza et al. 2014). It is found in successionary vegetation as young as 4, 5, and 6 years on abandoned agricultural plots, such as in the vicinity of Santa Mercedes along the Rio Putumayo in the northern Peruvian Amazon bordering Colombia (Gomez 2011:85; Muñoz 2010:18), in the municipality of Castanhal some 60 km east of Belém (Coelho et al. 2003b), in the municipality of São Francisco do Pará in the Bragantina zone of northeastern Pará (Almeida and Vieira 2001), and 80 km north of Manaus (Fernandes Neto 2015:23). It can survive in regrowth woodlands for at least 20 years (Vieira 1996:59). It also occurs in successionary vegetation on uplands and floodplains in the Amazon delta, hence one of its common names: ingá roceiro (*roça* is the name for cleared fields in the Brazilian Amazon).

This tree grows in some floodplain forests in the southern fringes of the Amazon Basin, such as in the upper Araguaia watershed (Stefanello et al. 2010). It also occurs in successionary woodlands on uplands, such as in Rondônia (Lisboa 1989), northeastern Pará (Prata et al. 2010), southwestern Amapá (Guimarães et al. 2016), and stunted woods (*campina*) that grow on sandy soils and in gallery forest in scrub savanna (*cerrado*) in the southern fringe of the Amazon Basin (Araújo et al. 2005; Ivanauskas et al. 2004; Lenza et al. 2015; Marimon et al. 2002; Miguel et al. 2011; Stefanello 2008:18). It is found as high as 1500 m in the Andean foothills of Bolivia, Ecuador and Peru (Arellano 2013:179; Mendoza and Guerrero 2010; Van den Eynden et al. 1999:164). It also thrives on some archaeological sites with fertile anthropogenic dark earth, such as at Palmeiras on an upland slope along the left bank of the middle Xingu River upstream from Altamira and in the vicinity of Inca ruins in the Madidi National Park in the Bolivian Amazon (Macía 2008). *Inga thibaudiana* has a wide distribution in lowlands from southern Mexico through Central America and into South America as far south as Bolivia, central and southern Brazil, and the Atlantic forest of Brazil.

This medium-sized tree which can reach 35 m when fully mature is also planted in some upland home gardens, such as along the middle Xingu. The Siona-Secoya who live in scattered villages along the Upper Aguarico in the Ecuadorian Amazon also plant this tree (Vickers 1994). In addition to eating the pulp surrounding the seeds, the Shuar and the Cofán in the Ecuadorian Amazon use the trunk for fuelwood (Bennett et al. 2002:175; Cerón et al. 1994:56), while the Waorani in the central part of the Ecuadorian Amazon use this wood for house construction as well as for fuel (Cerón and Ayala 1998:111). In Acre, this tree is used for living fences surrounding cattle pastures and locals use the wood for fuel and to make charcoal (Franke 1999).

The green pods are produced in the rainy season and are 8–30 cm long by 1.8–2.5 cm wide and contain seven green seeds, measuring 2 cm by 1 cm, when the pod is 8.5 cm long. Rural folk often tarry to snack on the fruits in regrowth forest while traveling to or from their fields, or while hunting. Black spider monkeys *Ateles paniscus*, the northern bearded saki *Chiropotes satanas chiropotes*, the mottled-face tamarin *Saguinus inustus*, Geoffroy's saddleback tamarin *Saguinus fuscicollis nigrifrons*, Goeldi's marmoset *Callimico goeldii*, Santarém marmosets *Callithrix humeralifera*, white-handed titi monkeys *Callicebus lugens*, red titi monkeys *Callicebus discolor*, and squirrel monkeys *Saimiri sciureus* are also fond of the fruits and are probably involved in dispersing some of the seeds in the Amazon Basin (Alvarez and Heymann 2012; Carrillo-Bilbao et al. 2005; Lima and Ferrari 2003; Palacios et al. 2004; Porter 2001; Sánchez and Levy 2013:29; Stevenson and Rylands 1988; Van Roosmalen 1985; Van Roosmalen and Klein 1988; Van Roosmalen et al. 1981). Woolly monkeys *Lagothrix lagotricha* eat this fruit and disperse some of the seeds in the northern part of the Colombian Amazon (Stevenson 2000). While marmosets, titi monkeys, and tamarins are not normally hunted for their meat because of their small size, spider monkeys are widely hunted game animals in the Amazon. In neighboring Surinam, the northern bearded saki *Chiropotes satanas chiropotes* has been observed feeding on the mesocarp and is likely involved in dispersing some of the seeds (Van Roosmalen et al. 1988). Rural people have noted that several monkeys eat this fruit, hence one of its common names in the Brazilian Amazon: ingá de macaco. The bare-tailed woolly opossum *Caluromys philander*, which occurs in northern South America, and the kinkajou *Potos flavus* which is native to Central and South America, are not normally hunted for their meat but these marsupials also eat this fruit and likely spread some of the seeds (Charles-Dominique et al. 1981).

Rafael, 5 years, eating ingá do mato *Inga thibaudiana* fruit he gathered in successionary vegetation at the edge of a manioc field. Pedreira, Baia de Caxiuanã, Município de Melgaço, Pará, Brazil 5-8-15

24.28 *Inga velutina*

Brazil Ingá peludo, ingá peluda, ingá folha peluda, ingádefogo`, ingá vermelho

Ecuador Guaba, guabilla, guaba del mono, cutu cachic, cutupacay, cutapacai; acho fino (Cofán), emubenë, airo benë (Siona-Secoya)

Peru Pacae, pacae peludo, rosca shimbillo, shimbillo peludo, coto shimbillo, shimbillo de monte, shimbillo roscapacae

Status Wild, planted

Luz Heloisa, 11 years, eating pacae *Inga velutina* fruit gathered from a tree planted in her home garden in the foothills of the Andes at 1500 m. Oxapampa, Pasco, Peru 11-6-04

Inga velutina is widespread in the Amazon Basin up to 2000 m in the Andean foothills, such as in the central Peruvian Amazon (Mendoza and Guerrero 2010; Mendoza et al. 2014). It is a denizen of successionary vegetation on floodplains and uplands (Bentes-Gama et al. 2002a, b; Gama et al. 2002, 2003; Gualberto et al. 2014; Pennington 1997:655; Santana et al. 2003, 2004; Souza 2019:12; Vieira 1996:59). It also occurs in some riparian forests, such as in southern Rondônia, at the confluence of the Purus and Solimões river in Amazonas, and in tidal forests in the Amazon estuary in Amapá (Carim 2016:89; Luize et al. 2015; Queiroz and Machado 2008; Queiroz et al. 2006; Miranda 2000). A relatively short tree, usually no higher than 25 m, it is also occasionally encountered in the mature forest but perhaps this relatively short tree is on its way out in such environments because of the deep shade. The slender pods vary in length from 10 to 30 cm and both rural and urban folk twist the pods open to eat the sweet white pulp surrounding the seeds. The pods are slightly pubescent, hence one of the popular names for this tree in Brazil: ingá peludo. The fruits are evidently eaten by monkeys as suggested by one of its names in Pastaza, Ecuador: guaba de mono.

Inga velutina is sometimes planted in home gardens as in Oxapampa and along the middle Xingu River. It is also planted around some manioc processing huts (*casas de forno*), such as in the vicinity of Carrazedo at the mouth of the Xingu. In Acre, this tree is sometimes used for living fences around cattle pastures, and the wood is used for fuel and to make charcoal (Franke 1999).

References

Abreu, Thiago Ayres Lazzarotti. 2012. Dinâmica da vegetação arbustivo-arbórea no período de 14 anos em uma floresta de vale, Mato Grosso, Brasil. Master's thesis, Universidade de Brasília

Abreu, Maria Milena de Oliveira; Mehlig, Ulf; Nascimento, Raquel do Espírito Santo Aguair do; Menezes, Moirah Paula Machado. 2006. Análise de composição florística e estrutura de um fragmento de bosque de terra firme e de um manguezal vizinhos na peninsula de Ajuruteua, Bragança, Pará. Boletim do Museu Paraense Emílio Goeldi, Ciencias Naturais 1(3):27–34

Abreu, Thiago Ayres Lazzarotti; Pinto, José Roberto Rodrigues; Mews, Henrique Augusto. 2014. Variações na riqueza e na diversidade de espécies arbustivas e arbóreas no período de 14 anos em uma Floresta de Vale, Mato Grosso, Brasil. Rodriguésia 65(1):73–88

Absy, Maria Lúcia; Kerr, Warwick Estevam. 1977. Algumas plantas visitadas para obtenção de pólen por operárias de Melipona seminigra merrillae em Manaus. Acta Amazonica 7(3):309–315

Absy, Maria Lúcia; Bezerra, Eronildo Braga; Kerr, Warwick E. 1980. Plantas nectaríferas utilizadas por duas espécies de Melipona da Amazônia. Acta Amazonica 10(2):271–281

Absy, Maria Lúcia; Prance, Ghillean T.; Barbosa, Edelcílio Marques. 1986. Inventário florístico de floresta natural na área da estrada Cuiabá-Porto Velho (BR 364). Acta Amazonica 16/17:85–121

Adaime, Ricardo; Jesus-Barros, Cristiane Ramos de; Souza-Filho, Miguel Francisco de. 2014. Moscas-das-frutas (Diptera: Tephritidae) obtidas de frutos comercializados em feiras públicas de Macapá, Amapá. Boletim de Pesquisa e Desenvolvimento, EMBRAPA 85:1–17

Aguirre-Mendoza, Zhofre; Loja, Ángel Fernando; Solano, Carmen; Aguirre, Nikolay. 2015. Especies Forestales más aprovechadas del Sur del Ecuador. Universidad Nacional de Loja, Ecuador

Alarcón, Juan Gabriel Soler; Peixoto, Ariane Luna. 2007. Florística e fitossociologia de um trecho de um hectare de floresta de terra firme, em Caracaraí, Roraima, Basil. Boletim do Museu Paraense Emílio Goeldi, Ciencias Naturais 2(2):33–60

Alarcón, J. G. S. and A. L. Peixoto. 2008. Use of terra firme forest by Caicubi caboclos, middle Rio Negro, Amazonas, Brazil: A quantitative study. Economic Botany 62(1):60–73

Albert, Bruce; Milliken, William. 2009. Urihi A: a terra-floresta Yanomami. Instituto Socioambiental (ISA)/Institut de Recherche pour le Devéloppement, São Paulo

Almeida, Adrielson Furtado; Jardim, Mário Augusto Gonçalves. 2011. Florística e estrutura da comunidade arbórea de uma floresta de várzea na Ilha de Sororoca, Ananindeua, Pará, Brasil. Scientia Florestalis 39(90):191–198

Almeida, Adrielson Furtado; Jardim, Mário Augusto Gonçalves. 2012. A utilização das espécies arbóreas da floresta de várzea da Ilha de Sororoca, Ananindeua, Pará, Brasil por moradores locais. Revista Brasileira de Ciências Ambientais 23:48–54

Almeida, Arlete Silva de; Vieira, Ima Célia Guimarães. 2001. Padrões florísticas e estruturais de uma cronoseqüência de florestas no munícipio de São Francisco do Pará, região Bragantina, Pará. Boletim do Museu Paraense Emílio Goeldi, Série Botânica 17(1):209–240

Almeida, Samuel S.; Lisboa, Pedro Luiz B.; Silva, Antônio Sérgio. 1993. Diversidade florística de uma comunidade arbórea na estação científica "Ferreira Penna", em Caxiuanã (Pará). Boletim do Museu Paraense Emilio Goeldi, Série Botanica 9(1):93–128

Almeida, Samuel Soares de; Amaral, Dário Dantas do; Silva, Antonio Sérgio Lima da. 2004. Análise florística e estrutura de florestas de várzea no estuário amazônico. Acta Amazonica 34(4):513–524

Alvarez, Silvia J.; Heymann, Eckhard W. 2012. Brief communication: a preliminary study on the influence of physical fruit traits on fruit handling and seed fate by white-handed titi monkeys (*Callicebus lugens*). American Journal of Physical Anthropology 147:482–488

Alvino, Fabrízia de Oliveira; Silva, Manoela Ferreira F. da; Rayol, Breno Pinto. 2005. Potencial de uso das espécies arbóreas de uma floresta secundária, na Zona Bragantina, Pará, Brasil. Acta Amazonica 35(4):413–420

Amaral, Dário Dantas do; Prost, Maria Thereza; Bastos, Maria de Nazaré do Carmo; Costa Neto, Salustiano Vilar; Santos, João Ubiratan Moreira dos. 2008. Restingas do litoral amazônico, estados do Pará e Amapá. Boletim do Museu Paraense Emílio Goeldi, Ciencias Naturais 3(1):35–67

Amaral, Dário Dantas do; Vieira, Ima Célia Guimarães; Almeida, Samuel Soares de; Salomão, Rafael Paiva; Silva, Antonio Sérgio Lima da; Jardim, Mário Augusto Gonçalves. 2009. Checklist da flora arbórea de remanescentes florestais da região metropolitana de Belém e valor histórico dos fragmentos, Pará, Brasil. Boletim do Museu Paraense Emílio Goeldi, Ciencias Naturais 4(3):231–289

Amaral, Dário Dantas do; Jardim, Mário Augusto Gonçalves; Costa-Neto, Salustriano Vilar; Bastos, Maria de Nazaré do Carmo. 2015. Síndromes de dispersão de propágulos e a influência da floresta amazônica na composição de espécies lenhosas de uma restinga no litoral norte brasileiro. Biota Amazônia 5(3):28–37

Amaral, Iêda Leão do; Magnusson, William Ernest; Matos, Francisca Dionízia de Almeida; Albernaz, Ana Luisa K.; Feitosa, Yuri Oliveira; Guillaumet, Jean-Louis. 2017. Disentangling structural patterns of natural forest fragments in a savanna matrix in the eastern Brazilian Amazon. Acta Amazonica 47(2):111–122

Amaral, Dário Dantas do; Rocha, Antônio Elielson; Pereira, Jorge Luis Gavina; Costa Neto, Salustiano Vilar. 2019. Identificação dos subtipos de savanas na Amazônia oriental (Pará e Amapá, Brasil) com uma chave dicotômica de individualização. Boletim do Museu Paraense Emílio Goeldi, Ciencias Humanas Naturais 14(2):183–195

Ammann, Stefan. 2014. Etnobotânica de árvores e palmeiras em três comunidades ribeirinhas do rio Jauaperi, na divisa entre Roraima e Amazonas. Master's Thesis, INPA, Manaus

Anderson, A. B., P. Magee, A. Gély, and M. A. G. Jardim. 1995. Forest management patterns in the floodplain of the Amazon estuary. *Conservation Biology* 9(1):47–61

Andrade, Ricardo Teixeira Gregório de; Pansini, Susamar; Sampaio, Adeilza Felipe; Ribeiro, Maíra Silva; Cabral, Guilherme Sampaio; Manzatto, Ângelo Gilberto. 2017. Fitossociologia de uma floresta de terra firme na Amazônia sul-occidental, Rondônia, Brasil. Biota Amazônia 7(2):36–43

Aparício, Wegliane Campelo da Silva. 2011. Estrutura da vegetação em diferentes ambientes na RESEX do Rio Cajari: interações solo-floresta e relações com a produção de castanha. Doctoral Dissertation, Universidade Federal Rural de Pernambuco, Recife

Aquino, Rolando; Encarnación, Filomeno. 1999. Observaciones preliminares sobre la dieta de *Cacajao calvus ucayalii* en el nor-oriente Peruana. Neotropical Primates 7(1):1–5

Araujo, Michele. 2014. Ecologia alimentar de *Saimiri sciureus cassiquiarensis* (Lesson, 1840) (Primates, Cebidae) em florestas de várzea da Amazônia Central Brasileira. Master's thesis, MPEG/UFPa, Belém

Araújo, Maristela Machado; Tucker, Joanna Marie; Vasconcellos, Joanna Marie; Zarin, Daniel Jacob; Oliveira, Wilson; Sampaio, Patrícia Delamônica; Rangel-Vasconcelos, Lívia Gabrig; Oliveira, Francisco de Assis; Coelho, Roberta de Fatima Rodrigues; Aragão, Débora Veiga; Miranda, Izildinha. 2005. Padrão e processo sucessionais em florestas secundárias de diferentes idades na Amazônia oriental. Ciencia Florestal, Santa Maria 15(4):343–357

Araujo-Murakami, Alejandro; Ruiz, Freddy S. Zenteno. 2006. Bosques de los Andes orientales de Bolivia y sus especies útiles. In: Botánica Económica de los Andes Centrales, Moraes, M.; Ollgaard, B.; Kvist, L. P.; Borchsenius, F.; Balslev, H. (eds), Universidad Mayor de San Andrés, La Paz, pp. 146–161

Araujo-Murakami, Alejandro; Paniagua, Narel; Cayola, L.; Jorgensen, Peter M.; Valdes, Ovidio; Macuapa, Eric; Calzadilla-Tomianovich, Miki. 2009. Diversidad y estructura florística del bosque de llanura y palmar de pantano Amazónico preandino en el sector de Chalalan, Parque Nacional Madidi, Bolivia. Kempffiana 5(1):3–27

Araujo-Murakami, Alejandro; Poma-Chura, Abraham; Palabral, Arely; Salvatierra, Rolando; Hurtado, Fidel. 2012. Composición florística de los bosques Amazónicos de tierra firme e inundable en las proximidades de las Pampas del Sonene (río Heath), Parque Nacional Madidi, Bolivia. Kempffiana 8(1):2–21

Araujo-Murakami, Alejandro; Villarroel, Daniel; Pardo, Guido; Vos, Vincent A.; Parada, G. Alexander; Arroyo, Luzmila; Killeen, Timothy. 2015. Diversidad arbórea de los bosques de tierra firme de la Amazonía Boliviano. Kempffiana 11(1):1–28

Arellano, Gabriel. 2013. Patrones de diversidad, distribución y dominancia de plantas leñosas en los bosques montanos del Parque Nacional Madidi, Bolivia. Doctoral dissertation, Departamento de Biologia, Facultad de Ciencias, Universidad Autónoma de Madrid

Arkcoll, David B. 1984. Some leguminous trees providing useful fruits in the north of Brazil. Pesquisa Agropecuária Brasileira 19:235–239

Asquith, N. M., J. A. Terborgh, A. E. Arnold, and C. M. Riveros. 1999. The fruits the agouti ate: *Hymenaea courbaril* seed fate when its disperser is absent. *Journal of Tropical Ecology* 15:229–235

Assis, Rafael Leandro de; Wittmann, Florian; Luize, Bruno Garcia; Haugaasen, Torbjon. 2017. Patterns of floristic diversity and composition in floodplain forests across four southern Amazon river tributaries, Brazil. Flora 229:124–140

Astwood, Jorge; Rodríguez, José; Rodríguez, Karen. 2014. Foraging density for squirrel monkey *Saimiri sciureus* in two forests in Puerto Lopez, Colombia. Revista MVZ Córdoba 19(2):4158–4167

Baar, R.; M. Cordeiro, M. Denrich; and H. Fölster. 2004. Floristic inventory of secondary vegetation in agricultural systems of East-Amazonia. *Biodiversity and Conservation* 13:501–528

Baca, Luis Campo. 2005. Algunos parámetros físicos, químicos y bioecológicos que influyen el comportamiento migratorio de la gamitana *Colossoma macropomum* en el Río Ucayali. In: Biología de la Poblaciones de Peces de la Amazonía y Piscicultura, Renno, Jean François; García, Carmen; Duponchelle, Fabrice; Nuñez, Jésus (eds), IIAP/IRD, Iquitos, pp 58–64

Balbin, Isabel Oré; Samaniego, Dedy Llapapasca. 1996. Huertas domesticas como sistema tradicional de cultivo en Moena Caño, Rio Amazonas, Iquitos, Peru. Folia Amazonica 8(1):91–110

Balée, W. 1993a. *Footprints of the Forest: Ka'apor Ethnobotany, the Historical Ecology of Plant Utilization by an Amazonian People*. Columbia University Press, New York

Balée, W. 1993b. Indigenous transformation of Amazonian forests: an example from Maranhão, Brazil. L'Homme 33(126–128):231–254

Balée, W. 2013. Cultural Forests of the Amazon: A Historical Ecology of People and Their Landscapes. University of Alabama Press, Tuscaloosa

Bánki, Olaf S.; Steege, Hans ter; Jansen-Jacobs, Marion J.; Raghoenandan, Usha P. D. 2003. Plant Diversity of the Nassau Mountains, Suriname: Report of the 2003 Expedition. National Herbarium Nederland, Utrecht

Barbosa, Cristiano de Souza. 2018. Diversidade e uso de plantas úteis nas quintais do bairro de São Raimundo, zona oeste de Manaus, AM. Master's thesis, INPA, Manaus

Bardales, Joel Vásquez; Gómez, Ricardo Zárate; Canaquiri, Percy Huiñapi; Jiménez, Julio Pinedo; Hernández, Juan José Ramírez; Lamas, Gerardo; García, Pedro Vela. 2017. Plantas alimentícias de 19 especies de mariposas diurnas (Lepidoptera) en Loreto, Perú. Revista Peruana de Biología 24(1): 35–42

Barnett, A. A. and C. V. Castilho. 2000. Report on a short study of the dry season feeding ecology and habitat preferences of the golden-backed uacari or bicó, *Cacajao melanocephalus ouakary* (Cebidae: Pitheciinae), on the lower Rio Jaú, Amazonas, Brazil. Akodon Ecological Consulting, Concord, California.

Barnett, A. A., C. V. Castilho, R. L. Shapely, and A. Anicácio. 2005. Diet, habitat selection and natural history of *Cacajao melanocephalus ouakary* in Jaú National Park, Brazil. *International Journal of Primatology* 26(4):949–969

Bastos, M. N. C.; C. S. Rosário; L. C. B. Lobato. 1995. Caracterização fitofisionômica da restinga de Algodoal, Maracanã, Pará, Brasil. Boletim do Museu Paraense Emílio Goeldi, Série Botanica 11(2): 173–197

Batista, Fábio de Jesus; Gonçalves, Mário Augusto. 2013. Florística e estrutura da regeneração natural arbórea de uma floresta de várzea na Reserva Extrativista Chocoaré-Mato Grosso, Pará, Brasil. Biota Amazônia 3(3):139–145

Batista, Fábio de Jesus; Jardim, Mário Augusto Gonçalves; Medeiros, Tonny David Santiago; Lopes, Isaac Luiz Magalhães. 2011. Comparação florística e estrutural de duas florestas de várzea no estuário Amazônico, Pará, Brasil. Revista Árvore 35(2):289–298

Batista, Anderson Pedro Bernardina; Aparício, Wegliane Campelo da Silva; Santos, Vanessa Silva dos; Lima, Robson Borges de; Mello, José Marcio de. 2015. Caracterização estrutural em uma floresta de terra firme no estado do Amapá, Brasil. Pesquisa Florestal Brasileira 35(81):21–33

Beck, H. 2005a. Seed predation and dispersal by peccaries throughout the Neotropics and its consequences: a review and synthesis. In: Seed Fate: Predation, Dispersal and Seedling Establishment, Forget, Pierre-Michel; Lambert, Joanna E.; Hulme, Philip E.; Wall, Stephen B. Vander (eds), CABI, Wallingford, pp 77–115

Beck, H. T. 2005b. Caffeine, alcohol, and sweeteners. In: *The Cultural History of Plants*. Prance, G.T. and Nesbitt, M (eds), Routledge, New York, 173–190

Beck, H. T. and G.T. Prance. 1991. Ethnobotanical notes on Marajó ceramic pottery utilizing two Amazonian trees. *Boletim do Museu Paraense Emílio Goeldi, Série Antropologia, Série Botanica*, Vol 7 (2) 269–275

Begazo, A. J. 1997. Uso de las poblaciones de paujiles, pucacungas, pavas de monte y manajaracos (Cracidae, Galliformes) en el noroeste Peruano. In: Manejo de Fauna Silvestre en la Amazonía, Fang, T.G.; Bodmer, R.E.; Aquino, R.; Valqui, M.H. (eds), Instituto de Ecologia, La Paz, Bolivia, pp 257–266

Bennett, Bradley C.; Baker, Marc A.; Andrade, Patricia Gómez. 2002. Ethnobotany of the Shuar of Eastern Ecuador. New York Botanical Garden, Advances in Economic Botany, vol. 14.

Bentes-Gama, Michelliny de Matos; Scolforo, José Roberto Soares; Gama, João Ricardo Vasconcellos; Oliveira, Antônio Donizette de. 2002a. Estrutura e valoração de uma floresta de várzea alta na Amazônia. Cerne 8(1):88–102

Bentes-Gama, Michelliny de Matos; Scolforo, José Roberto Soares; Gama, João Ricardo Vasconcellos. 2002b. Potencial produtivo de madeira e palmito de uma floresta secundária de várzea baixa no estuário Amazônico. Revista Árvore 26(3):311–319

Black, G. A.; Dobzhansky, T. H.; Pavan, C. 1950. Some attempts to estimate species diversity and population density of trees in Amazonian forests. Botanical Gazette 111(4):413–425

Boom, B. M. 1987. Ethnobotany of the Chácobo Indians, Beni, Bolivia. The New York Botanical garden, Advances in Economic Botany, vol. 4

Boubli, J. P. 1999. Feeding ecology of black-headed uacaris (*Cacajao melanocephalus melanocephalus*) in Pico da Neblina National Park, Brazil. International Journal of Primatology 20(5):719–749

Boubli, J. P. 2002. Lowland floristic assessment of Pico da Neblina National Park, Brazil. Plant Ecology 160:149–167

Boyle, Sarah A.; Zartman, Charles E.; Spironello, Wilson R.; Smith, Andrew T.; Beck, Harald. 2012. Implications of habitat fragmentation on the diet of bearked saki monkeys in central Amazonian forest. Journal of Mammalogy 93(4):959–976

Braga, Elayne Oliveira; Jardim, Mário Augusto Gonçalves. 2016. Edaphic and climatic relation and its influence on the composition floristic lower stratum in a floodplain forest, Pará, Brazil. Revista Árvore 40(5):901–910

Braga, Elayne Oliveira; Jardim, Mário Augusto Gonçalves. 2019. Florística, estrutura e formas de vida do estrato inferior de uma floresta ombrófila densa aluvial, Pará, Brasil. Ciencia Florestal, Santa Maria 29(3): 1048–1059

Braga, Elayne Oliveira; Silva, Julie Andrews de França e; Pantoja, Milenna Vasconcelos; Jardim, Mário Augusto Gonçalves. 2015. Florística, estrutura fitossociológica e formas de vida do estrato inferior em um floresta de várzea Amazônica. Biota Amazônia 5(3):59–65

Braga, Daniel P. P.; Domene, Frederico; Gandara, Flávio B. 2019. Shade trees composition and diversity in cacao agroforestry systems of southern Pará, Brazilian Amazon. Agroforestry Systems 93:1409–1421

Brienza Júnior, Silvio. 1999. Biomass dynamics of fallow vegetation enriched with leguminous trees in the eastern Amazon of Brazil. Doctoral dissertation, Georg August University, Göttingen, Germany

Brightsmith, D. J. 2005. Parrot nesting in southeastern Peru: seasonal patterns and keystone trees. *Wilson Bulletin* 117(3):296–305

Brito, Elizabeth Rodrigues; Martins, Sebastião Venâncio; Oliveira Filho, Ary Teixeira de; Silva, Elias; Silva, Alexandre Francisco da. 2006. Estrutura fitossociológica de um fragmento natural de floresta inundável em área de orizicultura irrigada, município de Lagoa da Confusão, Tocantins. Revista Árvore 30(5):829–836

Brown, C. B. 1876. *Canoe and Camp Life in British Guiana*. Edward Stanford, London

Buchanan, D.B.; Mittermeier, R.A.; van Roosmalen, M.G.M. 1981. The saki monkeys, genus *Pithecia*. In: Ecology and Behavior of Neotropical Primates, Coimbra-Filho, Adelmar F.; Mittermeier, Russell A. (eds), Academia Brasileira de Ciências, Rio de Janeiro, vol. 1, pp 391–417

Cabacinha, Christian Dias; Fontes, Marco Aurélio Leite. 2014. Caracterização florística e estrutural de fragmentos de matas de galeria da bacia do Alto Araguaia. Ciência Florestal, Santa Maria 24(2):379–390

Cain, Stanley A.; Castro, Gustavus M. de Oliveira; Pires, J. Murça; Silva, Nilo Tomâs. 1956. Application of some phytosociological techniques to Brazilian rain forest. American Journal of Botany 43(10):911–941

Calzadilla-Tomianovich, Miki Hernan; Cayola, Leslie. 2006. Estructura y composición florística de un bosque amazónico de pie de monte, Área Natural de Manejo Integrado Madidi, La Paz-Bolivia. Ecología en Bolivia 41(2):117–129

Camargo, Miguel Ângelo Branco. 2018. Variação em diâmetro do tronco de 30 espécies de árvores florestais em função de variáveis ambientais em floresta de terra firme na Amazônia central. Doctoral dissertation, INPA, Manaus

Campbell, David G.; Daly, Douglas C.; Prance, Ghillean T.; Maciel, Ubirajara N. 1986. Quantitative ecological inventory of terra firme and várzea tropical forest on the Rio Xingu, Brazilian Amazon. Brittonia 38(4):369–393

Campos Filho, Eduardo Malta. 2012. Plante as Árvores do Xingu e Araguaia. Instituto Socioambiental, São Paulo

Cárdenas, Dairon; Castaño, Nicolás; Sua, Sonia. 2009. Flora de la estrella fluvial de Inírida (Guianía, Colombia). Biota Colombiana 10(1–2):1–30

Carim, Marcelo de Jesus Veiga. 2016. Estrutura, composição e diversidade em florestas alagáveis de várzea de maré e de igapó e suas relações com variáveis edáficas e o período de inundação no Amapá, Amazônia oriental, Brasil. Doctoral dissertation, Programa de Pós-Graduação em Botânica, INPA, Manaus

Carim, Samyrams; Schwartz, Gustavo; Silva, Manoela Fernandes Ferreira da. 2007. Riqueza de espécies, estrutura e composição florística de uma floresta secundária de 40 anos no leste da Amazônia. Acta Botanica Brasilica 21(2):293–308

Carim, Marcelo de Jesus Veiga; Jardim, Mário Augusto Gonçalves; Medeiros, Tonny David Santiago. 2008. Composição florística e estrutura de floresta de várzea no Município de Mazagão, Estado do Amapá, Brasil. Scientia Florestalis 36(79):191–201

Carim, Marcelo de Jesus Veiga; Guillaumet, Jean-Louis Bernard; Guimarães, José Renan da Silva; Tostes, Luciedi Cássia Leôncio. 2013. Composição e estrutura de floresta ombrófila densa do extremo norte do Estado do Amapá, Brasil. Biota Amazônia 3(2):1–10

Carneiro, R. L. 1956. La cultura de los indios Kuikurus del Brasil central. *Runa Archivo para las Ciencias del Hombre* 8:169–202

Carneiro, R. L. 1978. The knowledge and use of rain forest trees by the Kuikuru Indians of central Brazil. In: *The Nature and Status of Ethnobotany*, R. Ford (ed.), Anthropological Paper Series, Museum of Anthropology, University of Michigan, Ann Arbor, vol. 67:208–216

Carneiro, F. S.; Lacerda, A. E. B.; Lemes, M. R.; Gribel, R.; Kanashiro, M.; Wadt, L. H. O.; Sebbenn, A. M. 2011. Effects of selective logging on the mating systems and pollen dispersal of Hymenaea courbaril L. (Leguminosae) in the eastern Brazilian Amazon as revealed by microsatellite analysis. Forest Ecology and Management 262:1758–1765

Carretero-Pinzón, Xyomara; Defler, Thomas R.; Ruiz-García, Manuel. 2016. How does the Colombian squirrel monkey cope with habitat fragmentation? Strategies to survive in small fragments. In: Phylogeny, Molecular Population Genetics, Evolutionary Biology and Conservation of the Neotropical Primates, Ruiz-García, M.; Shostell, J. M. (eds), Nova Science, New York, chapter 15, 1–16

Carrillo-Bilbao, Gabriel; Fiore, Anthony Di; Fernández-Duque, Eduardo. 2005. Dieta, forrajeo y presupuesto de tiempo en cotoncillos (*Callicebus discolor*) del Parque Nacional de Yasuní en la Amazonia Ecuatoriana. Neotropical Primates 13(2):7–11

Carvalho, João Olegário Pereira de; Silva, José Natalino Macedo; Lopes, José do Carmo Alves; Montagner, Lise Helene; Carvalho, Manoel Sebastião Pereira de. 1986. Composição florística de uma mata secundária no planalto de Belterra no Pará. In: Anais do 1° Simpósio do Trópico Úmido, Belém, PA, 12 a 17 November, EMBRAPA, Brasília, pp 197–205

Casula, Kátia Regina. 2012. Análise da similaridade florística e estrutural das formações florestais inundáveis em um trecho do alto Rio Madeira e de seus afluentes, estado de Rondônia, Brasil. Master's Thesis, Núcleo de Ciências Exatas da Terra, Universidade Federal de Rondônia, Porto Velho

Cattanio, José; Anderson, Anthony B.; Carvalho, Manoel S. 2002. Floristic composition and topographic variation in a tidal floodplain forest in the Amazon estuary. Revista Brasileira de Botanica 25(4):419–430

Cavalcante, P. B. 1973. A Farmacopéia Tiriyó: Estudo Étno-Botânico. Museu Paraense Emílio Goeldi, Publicações Avulsas No. 24

Cerón, Carlos Eduardo. 1991. Manejo florístico Shuar-Achuar (Jibaro) del ecosistema Amazónico en el Ecuador. Hombre y Ambiente 5(17):7–31

Cerón, Carlos Eduardo. 1996. Etnobotanica Quichua en la vía Hollín-Loreto, Provincia del Napo. Hombre y Ambiente 25:131–170

Cerón, Carlos Eduardo; Ayala, Consuelo Montalvo. 1998. Etnobotanica de los Huaorani de Quehueiri-ono Napo, Ecuador. Abya-Yala, Quito

Cerón, C. E.; Montalvo, Consuelo G.; Umenda, José; Umenda Enma Chica. 1994. Etnobotánica y Notas sobre la Diversidad Vegetal en la Comunidad Cofán de Sinangüé, Sucumbíos, Ecuador. EcoCiencia, Quito

Charles-Dominique, P.; Atramentowicz, M.; Charles-Dominique, M.; Gérard, H.; Hladik, C. M.; Prévost, M. F. 1981. Les mammiferes frugivores arboricoles nocturnes d'une foret Guyanaise: inter-relations plantes-animaux. Revue d'Écologie: la Terre et la Vie 35:341–435

Charpentier, Elvis J.; García, Gabriel; Aquino, Rolando. 2015. Uso y competición por plantas alimenticias entre *Pithecia aequatorialis* (Primates: Pitheciidae) y otros animales en la Amazonía peruana. Revista Peruana de Biología 22(2):225–232

Clement, C. R. 1999. 1492 and the loss of Amazonian crop genetic resources. II. Crop biogeography at contact. *Economic Botany* 53(2):203–216

Coelho, Roberta de Fátima Rodrigues; Zarin, Daniel Jacob; Miranda, Izildinha Souza; Tucker, Joanna Marie. 2003a. Análise florística e estrutural de uma floresta em diferentes estágios sucessionais no município de Castanhal, Pará. Acta Amazonica 33(4):563–582

Coelho, Roberta de Fátima Rodrigues; Zarin, Daniel Jacob; Miranda, Izildinha Souza; Tucker, Joanna Marie. 2003b. Ingresso e mortalidade em uma floresta em diferentes estágios sucessionais no município de Castanhal, Pará. Acta Amazonica 33(4):619–630

Coelho, Roberta de Fátima Rodrigues; Miranda, Izildinha Souza; Mitja, Danielle. 2012. Caracterização do processo sucessional no Projeto de Assentamento Benfica, sudeste do estado do Pará, Amazônia oriental. Boletim do Museu Paraense Emílio Goeldi, Ciencias Naturais 7(3):251–282

Coelho, Roberta de Fátima Rodrigues; Miranda, Izildinha de Souza; Mitja, Danielle. 2013. Conservação das florestas do projeto de assentamento Benfica, Sudeste da Amazônia. Ciencia Florestal, Santa Maria 23(1):1–17

Coelho-Ferreira, Márlia. 2009. Medicinal knowledge and plant utilization in an Amazonian coastal community of Marudá, Pará state (Brazil). Journal of Ethnopharmacology 126:159–175

Condé, Tiago Monteiro. 2011. Avaliação dos impactos da vegetação após a exploração madeireira em floresta ombrófila de terra firme no Município de Caracaraí, RR. Master's thesis, Programa de Pós-Graduação em Recursos Naturais, Universidade Federal de Roraima, Boa Vista

Condé, Tiago Monteiro; Tonini, Helio. 2013. Fitossociologia de uma floresta ombrófila densa na Amazônia setentrional, Roraima, Brasil. Acta Amazonica 43(3):247–260

Corette-Pasa, Maria. 2011. Abordagem etnobotânica na comunidade de Conceição-Açu, Mato Grosso, Brasil. Polibotánica 31:169–197

Corrêa, Honorly Kátia Mestre. 2006. Ecologia de dois grupos de sagüis-brancos, *Mico argentatus* (Linnaeus 1771) em um fragmento florestal natural, Santarém-Pará. Doctoral dissertation, Universidade Federal do Pará/Museu Paraense Emílio Goeldi, Belém

Costa, Dulce Helena Martins; Carvalho, João Olegário Pereira de; Silva, José Natalino Macedo. 2002. Dinâmica da composição florística após a colheita de madeira em uma área de terra firme na Floresta Nacional do Tapajós (PA). Revista de Ciências Agrárias 38: 67–90

Cummings, Anthony R.; Read, Jane M. 2016. Drawing on traditional knowledge to identify and describe ecosystem services associated with northern Amazon's multiple-use plants. International Journal of Biodiversity Science, Ecosystem Services & Management 12(1–2):39–56

Cummings, Anthony R.; Read, Jane M.; Fragoso, Jose M. V. 2018. Implications of forest type and land tenure diversity for the sustainability of ecosystem services provided by northern Amazonia's multiple-use tree species. Landscape Ecology 33:423–438

Cunha, A. C. 1960. Entre os Índios do Xingu: A Verdadeira História de Diacuí. Livraria Exposição do Livro, São Paulo

David, Margô De; Gonçalves, Karina Gondolo; Guarim Neto, Germano. 2015. A subfamília Mimosoidae (Fabaceae) para a flora de Mato Grosso, Brasil. Biodiversidade 14(3):10–34

Dávila, P. M.; Vega, Juan José Palacios; Vela, Marcial Martínez; Gómez, Ricardo Zárate; García, José Maco; Torres, Roger Escobedo. 2015. Cobertura y uso de la tierra de las provincias de Alto Amazonas y Ramón Castilla. Instituto de Investigaciones de la Amazonía Peruana, Iquitos, Documento Técnico No. 33

Davis, W. E. and J. J. Yost. 1983. The ethnobotany of the Waorani of Eastern Ecuador. *Harvard Botanical Museum Leaflets* 29(3):159–217

Defler, Thomas R.; Defler, Sara B. 1996. Diet of a group of *Lagothrix lagothricha lagothrica* in southeastern Colombia. International Journal of Primatology 17(2):161–190

Descola, P. 1988. La Selva Culta: Simbolismo y Praxis en la Ecología de los Achuar. Ediciones Abya Yala, Quito

Descola, P. 1994. *In the Society of Nature: A Native Ecology in Amazonia.* Cambridge University Press, Cambridge

Díaz Pérez, Wilmer Antonio; Daza, Félix; Sarmiento, William. 2012. Composición florística, estructura y diversidad del bosque ribereños del Río Kakada, cuenca del Río Caura, estado Bolívar, Venezuela. Revista Científica UDO Agrícola 12(2):275–289

Dionisio, Luiz Fernandes Silva; Bonfim Filho, Osmar Serra; Crivelli, Bruno Robertode Souza; Gomes, Jefferson Peixoto; Oliveira, Marcelo Henrique Silva de; Carvalho, João Olegário Pereira de. 2016. Importância fitossociológica de um fragmento de floresta ombrófila densa no estado de Roraima, Brasil. Agroambiente 10(3):243–252

Duarte, Luis Enrique Acero. 1979. Principales Plantas Utiles de la Amazonia Colombiana. Instituto Geografico "Augustin Codazzi", Bogotá

Ducke, A. 1910. Explorações scientificas no Estado do Pará. Boletim do Museu Goeldi (Museu Paraense) de Historia Natural e Ethnographia 7:100–200

Ducke, A. 1925. Plantes nouvelles ou peu connues de la région amazonienne (III partie). Archivos do Jardim Botanico do Rio de Janeiro 4:1–341.

Ducke, A. 1946. *Plantas de Cultura Precolombiana na Amazônia brasileira. Notas Sôbre as Espécies ou Formas Espontâneas que Supostamente lhes teriam dado Origem.* Instituto Agronômico do Norte, Belém, Boletim Técnico 8, pp. 1–24

Ducke, A. 1949. Notas sôbre a Flora Neotrópica- II: As Leguminosas da Amazônia Brasileira. Instituto Agronômico do Norte, Boletim Técnico 18, Belém

Dulmen, Arthurvan. 2001. Pollination and phenology of flowers in the canopy to two contrasting rain forest types in Amazonia, Colombia. Plant Ecology 153(1/2):73–85

Duque, Alvaro; Sánchez, Mauricio; Cavelier, Jaime; Duivenvoorden, Joost F.; Miraña, Petei; Miraña, Juvenal; Matapí, Anibal. 2001. Relación bosque-ambiente en el medio Caquetá, Amazonía colombiana. In: Evaluación de Recursos Vegetables no Maderables en la Amazonía Noroccidental, Duivenvoorden, Joost F.; Balslev, Henrik; Cavelier, Jaime; Grandez, César; Tuomisto, Hanna; Valencia, Renato (eds), Institute for Biodiversity and Ecosystem Dynamics, Amsterdam, pp 99–129

Ebert, Alexandre; Teixeira, Leandro Ribeiro; Silva, Adriana Zanirato Contini da; Costa, Reginaldo Brito da. 2014. Natural regeneration in tropical secondary forest in southern Amazonia, Brazil. Open Journal of Forestry 4(2):151–160

Eden, Michael J. 1988. Crop diversity in tropical swidden cultivation: comparative data from Colombia and Papua New Guinea. Agriculture, Ecosystems and Environment 20:127–136

Egler, S. G. 1992. Feeding ecology of *Saguinus bicolor bicolor* (Callitrichidade: Primates) in a relict forest in Manaus, Brazilian Amazonia. Folia Primatologica 59:61–76

Espírito-Santo, Fernando Del Bon; Shimabukuro, Yosio Edimir; Aragão, Luiz Eduardo Oliveira e Cruz de; Machado, Evandro Luiz Mendonça. 2005. Análise da composição florística e fitossociológica da floresta nacional do Tapajós com apoio geográfico de imagens satélites. Acta Amazonica 35(2):155–173

Estrada, Alejandro; Coates-Estrada, Rosamond. 1984. Fruit eating and seed dispersal by howling monkeys (*Alouatta palliata*) in the tropical rain forest of Los Tuxtlas, Mexico. American Journal of Primatology 6:77–91

Falcão, M. and C. R. Clement, 2000. Fenologia e produtividade do ingá-cipó (*Inga edulis*) na Amazônia Central. Acta Amazonica 30 (2): 173–180

Farabee, W. C. 1918. *The Central Arawaks*. The University Museum, University of Pennsylvania, Philadelphia, vol. 10

Felton, Annika M.; Felton, Adam; Wood, Jeff T.; Lindenmayer, David B. 2008. Diet and feeding ecology of *Ateles chamek* in a Bolivian semihumid forest: the importance of Ficus as a staple food resource. International Journal of Primatology 29:379–403

Felton, Annika M.; Felton, Adam; Wood, Jeff T.; Foley, William J.; Raubenheimer, David; Wallis, Ian R.; Lindenmayer, David B. 2009. Nutritional ecology of Ateles chamek in lowland Bolivia: how macronutrient balancing influences food choices. International Journal of Primatology 30:675–696

Fernandes Neto, José Guedes. 2015. Estratégias funcionais das espécies dominantes na sucessão florestal na Amazônia central. Master's thesis, INPA, Manaus

Ferreira, Carlos Alberto Cid. 2009. Análise comparativa de vegetação lenhosa do ecosistema campina na Amazonia Brasileira. Doctoral dissertation, INPA/UFAM, Manaus

Ferreira Júnior, Elbert Viana; Soares, Thelma Shirlen; Costa, Marcus Filipe Fernandesda; Silva, Versides Sebastião Moraes. 2008. Composição, diversidade e similaridade florística de uma floresta tropical semidecídua de uma floresta tropical semidecídua submontana em Marcelândia, MT. Acta Amazonia 38(4):673–680

Ferreira, Leandro V.; Chaves, Priscilla P.; Cunha, Denise de A.; Rosário, Alessandro S. do; Parolin, Pia. 2013a. A extração ilegal de areia como causa do desaparecimento de campinas e campinaranas no estado do Pará, Brasil. Pesquisas Botânica 64:157–173

Ferreira, Leandro V.; Chaves, Priscilla P.; Cunha, Denise de A.; Matos, Darley C. L.; Parolin, Pia. 2013b. Variação da riqueza e composição de espécies da comunidade de plantas entre as florestas de igapós e várzeas na Estação Científica Ferreira Penna-Caxiuanã na Amazônia oriental. Pesquisas Botânica 64:175–195

Ferreira, Almecina Balbino; Ming, Lin Chau; Haverroth, Moacir; Daly, Douglas Charles; Caballero, Javier; Ballesté, Andrea M. 2015. Plants used to treat malaria in the regions of Rio Branco-Acre state and southern Amazonas state, Brazil. International Journal of Phytocosmetics and Natural Ingredients 2(9):1–5

Franco, Ana María; Santamaria, Marcela. 1997. Area vital, habitos alimenticios y otros aspectos de historia natural de paujil *Mitu silvini*. In: Manejo de Fauna Silvestre en la Amazonía, Fang, T.G.; Bodmer, R.E.; Aquino, R.; Valqui, M.H. (eds), Instituto de Ecologia, La Paz, Bolivia, pp 267–282

Franke, Idésio Luis. 1999. Principais usos e serviços de árvores e arbustos promissores que ocorrem em pastagens no Estado do Acre. EMBRAPA Acre Comunicado Técnico 106:1–6

Freire, Maria Cristina C.; Monteiro, Reinaldo. 1993. Florística das praias da Ilha de São Luís, Estado do Maranhão (Brasil): diversidade de espécies e suas ocorrências no litoral Brasileiro. Acta Amazonica 23(2–3):125–140

Furley, P. A.; Ratter, J. A.; Gifford, D. R. 1978. Observations on the vegetation of eastern Mato Grosso, Brazil. III. The woody vegetation and soils of the Morro de Fumaça, Torixoreu. Proceedings of the Royal Society of London, B 203:191–208

Gachet, María Salomé; Lecaro, Javier Salazar; Kaiser, Marcel; Brun, Reto; Navarrete, Hugo; Muñoz, Ricardo A.; Bauer, Rudolf; Schühly, Wolfgang. 2010. Assessment of anti-protozoal activity of plants traditionally used in Ecuador in the treatment of leishmaniasis. Journal of Ethnopharmacology 128:183–197

Gama, João Ricardo Vasconcellos; Botelho, Soraya Alvarenga; Bentes-Gama, Michellinyde Matos. 2002. Composição florística e estrutura da regeneração natural de floresta secundária de várzea baixo no estuário Amazônico. Revista Árvore 26(5):559–566

Gama, J. R. V., S. A. Botelho, M. Bentes-Gama, and J. R. S. Scolforo 2003. Estrutura e potencial futuro de utilização da regeneração natural de floresta de várzea alta no Município de Afuá, estado do Pará. *Ciencia Florestal, Santa Maria* 13(2):71–82

Gama, João Ricardo Vasconcellos; Souza, Agostinho Lopes de; Martins, Sebastião Venâncio; Souza, Deoclides Ricardo de. 2005. Comparação entre florestas de várzea e de terra firme do estado do Pará. Revista Árvore 29(4):607–616

Gama, João Ricardo Vasconcellos; Souza, Agostino Lopesde Souza; Calegário, Natalino; Lana, Guilherme Campos. 2007. Fitossociologia de duas fitocenoses de floresta ombrófila aberta no município de Codó, estado do Maranhão. Revista Árvore 31(3):465–477

Garber, P.A.; Kitron, U.1997. Seed swallowing in tamarins: evidence of a curative function or enhanced foraging efficiency? International Journal of Primatology 18(4): 523–538

Garcia, Bruna Naiara Rocha; Vieira, Thiago Almeida; Oliveira, Francisco de Assis. 2017. Tree and shrub diversity in agroforestry homegardens in rural community in eastern Amazon. Floresta 47(4):543–552

Gentry, A. 1992. New nontimber forest products from western South America. In: *Sustainable Harvest and Marketing of Rain Forest Products*, M. Plotkin and L. Famolare (eds), Island Press, Washington, DC, pp. 125–136

Ghilardi, R.; Alho, C.J.R. 1990. Produtividade sazonal da floresta e atividade de forrageamento animal em habitat de terra firme da Amazônia. Acta Amazonica 20:61–76

Giraldo-Cañas, Diego. 1999. Riqueza, composición y distribución florística de los paisajes fisiográficos del eje de los ríos Apaporis y Amazonas, Amazonia Colombiana. Darwiniana 37(1–2):25–35

Godoy, Judith Rosales; Petts, Geoffrey; Salo, Jukka. 1999. Riparian flooded forests of the Orinoco and Amazon basins: a comparative review. Biodiversity and Conservation 8:551–586

Gomez, Tedi Pacheco. 2011. Valoración económica de purmas de diferentes edades en la comunidad nativa de Santa Mercedes, río Putumayo, Perú. Master's thesis, UNAP, Iquitos

Gómez, Ricardo Zárate; Vargas, Tony Jonatan Mori; Pérez, Luis Andres Valles. 2013. Composición florística, diversidad y estructura de los bosques sobre arena blanca de la Reserva Nacional Allpahuayo-Mishana, Loreto, Perú. Arnaldoa 19(2):237–247

Gómez-Posada, C. 2012. Dieta y comportamiento alimentario de un grupo de mico maicero *Cebus apella* de acuerdo a la variación en la oferta de frutos y artrópodos, en la Amazonía colombiana. Acta Amazonica 42(3):363–372

Gomide, Guilherme Luís Augusto. 1997. Estrutura e dinâmica de crescimento de florestas tropicais primária e secundária no Estado do Amapá. Master's Thesis, Universidade Federal do Paraná, Brazil

Gonçalves, Marco Antonio. 2001. O Mundo Inacabado: Ação e Criação em uma Cosmologia Amazônica, Etnografia Pirahã. Editora UFRJ, Rio de Janeiro

Gonçalves, Janaína Pinheiro; Lucas, Flávia Cristina Araújo. 2017. Agrobiodiversidade e etnoconhecimento em quintais de Abaetetuba, Pará, Brasil. Revista Brasileira de Biociências 15(3):119–134

Gonçalves, Karina Gondolo; Pasa, Maria Corette. 2015. A etnobotânica e as plantas medicinais na Comunidade Sucuri, Cuiabá, MT, Brasil. Interações 16(2):245–256

Gorchov, David L.; Palmeirim, Jorge M.; Jaramillo, Margarita; Ascorra, Cesar F. 2004. Dispersal of seeds of *Hymenaea courbaril* (Fabaceae) in a logged rain forest in the Peruvian Amazon. Acta Amazonica 34(2):251–259

Griscom, Bronson W.; Daly, Douglas C.; Ashton, Mark S. 2007. Floristics of bamboo-dominated stands in lowland terra-firma forests of southwestern Amazonia. Journal of the Torrey Botanical Society 134(1):108–125

Gualberto, Midiã Larine Colares; Ribeiro, Renato Bezerra da Silva; Gama, João Ricardo Vaconcellos; Vieira, Diego dos Santos. 2014. Fitossociologia e potencial de espécies arbóreas em ecossistema sucessional na Floresta Nacional do Tapajós, Pará. Agroecossistemas 6(1):42–57

Guariguata, Manuel R.; Licona, Juan Carlos; Mostacedo, Bonifacio; Cronkleton, Peter. 2009. Damage to Brazil nut trees (*Bertholletia excelsa*) during selective timber harvesting in northern Bolivia. Forest Ecology and Management 258:788–793

Guarim Neto, Germano. 1984. Plantas do Brasil: Angiospermas do Estado de Mato Grosso- 1. Rodriguésia 36(59):105–121

Guimarães, José Renan da Silva; Carim, Marcelo de Jesus Veiga; Costa Neto, Salustiano Vilard a; Tostes, Luciedide Cássia Leôncio. 2016. Floristic diversity of secondary forest in the eastern Amazon, state of Amapá. Floresta 46(3):343–351

Haidar, Ricardo Flores; Fagg, Jeanine Maria Felfili; Pinto, José Roberto Rodrigues; Dias, Ricardo Ribeiro; Damasco, Gabriel; Silva, Lucasde Carvalho Ramos; Fagg, Christopher William. 2013. Florestas estacionais e áreas de ecótono no estado do Tocantins, Brasil: parâmetros estruturais, classificação das fitofisionomias florestais e subsídios para conservação. Acta Amazonica 43(3):261–290

Hajdu, Z. and J. Hohmann. 2012. An ethnopharmacological survey of the traditional medicine utilized in the community of Porvenir, Bajo Paraguá Indian Reservation, Bolivia. Journal of Ethnopharmacology 139:838–857

Hallwachs, W. 1986. Agoutis (Dasyprocta punctata): the inheritors of guapinol (*Hymenaea courbaril*: Leguminosae). In: Frugivores and Seed Dispersal, Estrada, Alejandro; Fleming, Theodore H. (eds), Dr. W. Junk Publishers, Dordrecht, pp 285–304

Hartshorn, G. S. and L. J. Poveda. 1983. Checklist of trees. In: *Costa Rican Natural History*, D. H. Janzen (ed.), University of Chicago Press, Chicago, pp. 158–183

Hiraoka, M. 1989. Ribereño's changing economic patterns in the Peruvian Amazon. Journal of Cultural Geography 9(2):103–119

Hoff, Michel; Cremers, Georges. 2005. Le jardin Guyanais: inventaire des plantes cultivées et des adventices des jardins de Guyane française. Journal de Botanique de la Société de France 29:3–40

Hoffman, B. 2013. Exploring Biocultural Contexts: Comparative Woody Plant Knowledge of an Indigenous and Afro-American Maroon Community in Suriname, South America. In: *African Ethnobotany in the Americas*, R. Voeks and J. Rashford (eds.), Springer, New York, pp. 335–393

Holm, J. A.; Chambers, J. Q.; Collins, W. D.; Higuchi, N. 2014. Forest response to increased disturbance in the central Amazon and comparison to western Amazonian forests. Biogeosciences 11:5773–6794

Huber, J. 1910. Mattas e madeiras amazonicas. Boletim do Museu Goeldi de Historia Natural e Ethnographia 6: 91–225

Indios Kuikuru. 2007. Kungatagohoha Igei Ngongoi: Esta é a Terra Que Nós Plantamos. Parque Indígena do Xingu, Aika e ISA, FALE/UFMG e SECAD/MEC, Belo Horizonte

Ique, Carlos A.; Aquino, Rolando. 2015. Principales especies florísticas utilizadas y dispersión de semillas por *Saguinus labiatus* E. Geoffroy, 1812 en isla Muyuy, Loreto, Perú. Ciencia Amazónica 5(1):81–88

Irvine, Dominique. 1987. El manejo del bosque secundario por los Quichua amazonicos. Hombre y Ambiente 2:121–142

Irvine, Dominique. 1989. Succession management and resource distribution in an Amazonian rain forest. Resource Management In Amazonia: Indigenous and Folk Strategies, Posey, D. A.; Balée, W. (Editors), pp. 223–237. Advances in Economic Botany, vol. 7, New York Botanical Garden.

Irwin, H. S. and R. C. Barneby. 1982. *The American Cassiinae: A Synoptical Revision of Leguminosae Tribe Cassieae Subtribe Cassiinae in the New World.* Memoirs of the New York Botanical Garden, Bronx, Vol. 35, Part 1

Ivanauskas, Natália Macedo; Monteiro, Reinaldo; Rodrigues, Ricardo Ribeiro. 2003. Alterations following a fire in a forest community of Alto Rio Xingu. Forest Ecology and Management 184:239–250

Ivanauskas, Natália Macedo; Monteiro, Reinaldo; Rodrigues, Ricardo Ribeiro. 2004. Estrutura de um trecho de floresta Amazônica na bacia do alto rio Xingu. Acta Amazonica 34(3):275–299

Iwanaga, Simone; Ferrari, Stephen F. 2001. Party size and diet of synoptic Atelids (*Ateles chamek* and *Lagothrix cana*) in southwestern Brazilian Amazonia. Folia Primatologica 72:217–227

Janzen, D. H. 1975. Behavior of *Hymenaea courbaril* when its predispersal seed predator is absent. Science 189(4197):145–147

Janzen, D. H. 1983. *Hymenaea courbaril* (Guapinol, stinking toe). In: Costa Rican Natural History, Janzen, Daniel H. (ed), University of Chicago Press, Chicago, pp. 253–256

Jardim, Mário Augusto G.; Vieira, Ima Célia Guimarães. 2001. Composição florística e estrutura de uma floresta de várzea do estuário Amazônico, Ilha do Combu, Estado do Pará, Brasil. Boletim do Museu Paraense Emílio Goeldi, Série Botânica 17(2):333–354

Jardim, Daniel Gonçalves; Jardim, Mário Augusto G.; Quaresma, Adriano Costa; Costa, Salustiano Villar. 2013. Regeneração natural em formações florestais de uma unidade de conservação, Maracanã, Pará, Brasil. Biota Amazônia 3(2):79–87

Jernigan, Kevin Arthur. 2006. An ethnobiological exploration of sensory and ecological aspects of tree identification among the Aguaruna Jívaro. Doctoral dissertation, University of Georgia, Athens

Julliot, C.; Sabatier, D. 1993. Diet of the red howler monkey (*Alouatta seniculus*) in French Guiana. International Journal of Primatology 14: 527–550

Karadimas, D. 2005. La Raison du Corps: Idéologie du Corps et Représentations de l'Environnement Chez les Miraña d'Amazonie Colombienne. Peeters, Paris

Kauffman, J. Boone. 1991. Survival by sprouting following fire in tropical forests of the eastern Amazon. Biotropica 23(3):219–224

Kawa, Nicholas C.; Michelangeli, José A. Clavijo; Clement, Charles R. 2015. Household agrobiodiversity management on Amazonian dark earths, oxisols, and floodplain soils on the lower Madeira River, Brazil. Human Ecology 43:339–353

Kern, Dirse C.; Marques, Fernando L. T.; Maurity, Clóvis W.; Atzingen, Noé von. 1992. O potencial espeleoarqueológico da região de São Geraldo do Araguaia-PA. Boletim do Museu Paraense Emílio Goeldi, Série Antropologia 8(2):157–183

Kinupp, V. F. and H. Lorenzi. 2014. *Plantas Alimentícas Não Convencionais (PANC) no Brasil: Guia de Identifação, Aspectos Nutricionais e Receitas Ilustradas.* Instituto Plantarum de Estudos da Flora, São Paulo.

Kinzey, Warren G.; Norconk, Marilyn A. 1990. Hardness as a basis of fruit choice in two sympatric primates. American Journal of Physical Anthropology 81:5–15

Klinge, Hans; Adis, Joachim; Worbes, Martin. 1995. The vegetation of a seasonal várzea forest in the lower Solimões River, Brazilian Amazonia. Acta Amazonica 25(3/4):201–220

Kuchla, Wilian José; Stepka, Thiago Floriani; Sawczuk, Alex Roberto; Lisboa, Gerson dos Santos. 2015. Florística, estrutura horizontal e distribuição diamétrica em área de transição de cerrado e floresta aluvial no município de Campos de Júlio, MT. Ambiência 11(1):13–30

Kunz, Sustanis Horn; Ivanauskas, Natália Macedo; Martins, Sebastião Venâncio; Silva, Elias; Stefanello, Daniel. 2008. Aspectos florísticos e fitossociológicos de um trecho de floresta estacional perenifólia na Fazenda Triarão, Bacia do rio das Pacas, Querência, MT. Acta Amazonica 38(2):245–254

Kunz, Sustanis Horn; Moreau, Julia; Spadeto, Cristiani; Martins, Sebastião Venâncio; Stefanello, Daniel; Ivanauskas, Natália Macedo. 2014a. Estrutura da comunidade arbórea de trecho de floresta estacional sempre-verde e similaridade florística na região nordeste de Mato Grosso, Brasil. Floresta e Ambiente 21(4):429–440

Kunz, Sustanis Horn; Ivanauskas, Natália Macedo; Martins, Sebastião Venâncio; Stefanello, Daniel; Silva, Elias. 2014b. Fitossociologia do componente arbóreo de dois trechos de floresta estacional perenifólia, bacia do Rio das Pacas, Querência, MT. Ciência Florestal, Santa Maria 24(1):1–11

La Rotta, C. 1983. Observaciones Etnobotanicas sobre algunas Especies Utilizadas por la Comunidad Indigena Andoque (Amazonas, Colombia). Universidad Nacional de Colombia, Bogotá

Lacerda, André Eduardo Biscala de; Kanashiro, Milton; Sebbenn, Alexandre Magno. 2008. Long-pollen movement and deviation of random mating in a low-density continuous population of a tropical tree Hymenaea courbaril in the Brazilian Amazon. Biotropica 40(4):462–470

Lago, João Henrique G.; Tezoto, Jaqueline; Yazbek, Priscila B.; Cassas, Fernando; Santos, Juliana de F. L.; Rodrigues, Eliana. 2016. Exudates used as medicine by the "caboclos river-dwellers" of the Unini River, AM, Brazil--classification based in their chemical composition. Revista Brasileira de Farmacognosia 26:379–384

Lamont, Susan R.; Eshbaugh, W. Hardy; Greenberg, Adolph M. 1999. Species composition, diversity, and use of homegardens among three Amazonian villages. Economic Botany 53(3):312–326

Langenheim, Jean H.; Lee, Yin-Tse; Martin, Susan S. 1973. An evolutionary and ecological perspective of Amazonian Hylaea species of Hymeneae (Leguminosae: Caesalpinioideae). Acta Amazonica 3:5–28

Lau, Arthur Viana; Jardim, Mário Augusto Gonçalves. 2013. Florística e estrutura da comunidade arbórea em uma floresta de várzea na área proteção ambiental, Ilha do Combu, Belém, Pará. Biota Amazônia 3(2):88–93

Le Cointe, P. 1922. *L'Amazonie Brésilienne: Le Pays-ses Habitants, ses Resources, Notes et Statistiques jusqu'en 1920*. Augustin Challamel, Paris, vol 2

Le Cointe, P. 1934. *Amazônia Brasileira III: Árvores e Plantas Úteis*. Livraria Classica, Belém

Leão, Fábio Miranda; Dionisio, Luiz Fernandes Silva; Silva, Nágilla Gabriella Euzébio da; Oliveira, Marcelo Henrique Silva de; D'arace, Larissa Martins Barbosa; Neves, Raphael Lobato Prado. 2017. Fitossociologia em sistemas agroflorestais com diferentes idades de implantação no município de Medicilândia, PA. Revista Agroambiente 11(1):71–81

Lee, Yin-Tse; Langenheim, Jean H. 1975. Systematics of the genus *Hymenaea* L. (Leguminosae, Caesalpinioideae, Detarieae). University of California Publications in Botany, Berkeley, vol. 69

Lemos, Helena Lara. 2013. Vegetação arbustivo-arbórea em áreas de cerrado típico e cerrado rupestre no estado de Tocantins. Master's thesis, Departamento de Engenharia Florestal, Universidade de Brasília, Brasília

Lemos, D. A. N.; Ferreira, B. G. A.; Siqueira, J. D. P.; Oliveira, M. M.; Ferreira, A. M. 2015. Floristic and phytosociology in dense "terra firme" rainforest in the Belo Monte hydroelectric plant influence area, Pará, Brazil. Brazilian Journal of Biology 75(3):257–276

Lenza, Eddie; Santos, Josias O.; Maracahipes-Santos, Leonardo. 2015. Species composition, diversity, and vegetation structure in a gallery forest-cerrado sensu stricto transition zone in eastern Mato Grosso, Brazil. Acta Botanica Brasilica 29(3):327–338
Léon, J. 1998. Inga as shade for coffee, cacao and tea: historical aspects and present day utilization. In: *The genus Inga: Utilization*, T.D. Pennington and E.C.M. Fernandes (Editors), Royal Botanic Gardens, Kew, pp. 101–115
Lévi-Strauss, C. 1975. The Raw and the Cooked, Introduction to a Science of Mythology: I. Harper Colophon Books, New York
Lewis, G. P.; Owen, P. E. 1989. Legumes of the Ilha de Maracá. Royal Botanic Gardens, Kew
Lima, Pedro E. 1950. A canoa de casca de jatobá entre os indios do Xingu. Revista do Museu Paulista 4:369–380
Lima, D. A. 1958. Viagem aos campos de Monte Alegre, Pará. Instituto Agronômico do Norte, *Boletim Técnico* 36: 99–150
Lima, Eldianne M.; Ferrari, Stephen F. 2003. Diet of a free-ranging group of squirrel monkeys (*Saimiri sciureus*) in eastern Brazilian Amazonia. Folia Primatologica 74:150–158
Lima, Roberval M. B. de; Saragoussi, Muriel. 2000. Floodplain home gardens on the central Amazon in Brazil. In: The Central Amazon Floodplain: Actual Use and Options for a Sustainable Management, Junk, W. J.; Ohly, J. J.; Piedade, M. T. F.; Soares, M. G. M. (eds), Backhuys Publishers, Leiden pp 243–268
Lisboa, P. L. B. 1989. Estudo florístico da vegetação arbórea de uma floresta secundária em Rondônia. Boletim do Museu Paraense Emílio Goeldi, Série Botânica 5(2):145–162
Lisboa, P. L. B. 2012. *A Terra dos Aruã: Uma História Ecológica do Arquipélago do Marajó*. Museu Paraense Emílio Goeldi, Belém
Lisboa, Pedro L. B.; Maciel, Ubirajara N.; Prance, Ghillean T. 1991. Some effects of colonization on the tropical flora of Amazonia: a case study from Rondônia. Kew Bulletin 46(2):187–204
Lisboa, P. L. B., I. A. Gomes, R. C. Lisboa, and C. V. Urbinati. 2002. O estilo Amazônico de sobreviver: manejo dos recursos naturais. In: *Natureza, Homem e Manejo de Recursos Naturais na Região de Caxiuanã, Melgaço, Pará*, P. L. B. Lisboa (ed.), Museu Paraense Emílio Goeldi, Belém, pp. 41–170
Lobo, Guilherme de Sousa. 2017. A alteração do regime hidrológico afeta a composição florística e estrutura de florestas de igapó? Um estudo comparativo entre um rio regulado e outro prístino na Reserva de Desenvolvimento Sustentável do Uatumã, Amazônia Central. Master's thesis, INPA, Manaus
Locosselli, Giuliano M.; Krottenthaler, Stefan; Pitsch, Philipp; Anhuf, Dieter; Ceccanti, Gregório. 2017. Age and growth rate of congeneric tree species (Hymenaea spp., Leguminosae) inhabiting different tropical biomes. Erdkunde 71(1):1–13
Longworth, J. Benjamin; Mesquita, Rita C.; Bentos, Tony V.; Moreira, Marcelo P.; Massoca, Paulo E.; Williamson, G. Bruce. 2014. Shifts in dominance and species assemblages over two decades in alternative successions in central Amazonia. Biotropica 46(5):529–537
López, Arlene; Somarriba, Eduardo. 2005. Árboles frutales en fincas de cacao orgánico del Alto Beni, Bolivia. Agroforestería en las Américas 43:38–45
Lorenzi, H. 2009. Árvores Brasileiras: Manual de Identificação e Cultivo de Plantas Arbóreas Nativas do Brasil. Instituto Plantarum, Nova Odessa, São Paulo, vol. 2
Lorenzi, Harri; Bacher, Luis; Lacerda, Marco; Sartori, Sergio. 2006a. Frutas Brasileiras e Exóticas Cultivadas. Instituto Plantarum de Estudos da Flora Ltda, Odessa, Sao Paulo.
Lorenzi, H., L. Bacher, M. Lacerda, S. Sartori 2006b. Brazilian Fruits and Cultivated Exotics (for Consuming in Natura). Instituto Plantarum de Estudos da Flora Ltda, Odessa, São Paulo
Loureiro, Arthur A.; Lisboa, Pedro L. Braga. 1979. Madeiras do município de Aripuanã e suas utilidades (Mato Grosso). Acta Amazonica 9(1: supl.):1–88
Loureiro, Arthur A.; Silva, Marlene F. da; Alencar, Jurandyr da Cruz. 1979. Essências Madeireiras da Amazônia. Instituto Nacional de Pesquisas da Amazônia, Manaus, vol. 1
Luize, Bruno Garcia; Venticinque, Eduardo Martins; Silva, Thiago Sanna Freire; Novo, Evlyn Marcia Leão de Moraes. 2015. A floristic survey of angiosperm species occurring at three

landscapes of the central Amazon várzea, Brazil. Checklist: The Journal of Biodiversity Data 11(6):1–15

Lukesch, A. 1976. Bearded Indians of the Tropical Forest: The Asuriní of the Ipiaçaba, Notes and Observations on the First Contact and Living Together. Akademische Druck und Verlagsanstalt, Graz, Austria

Lustosa, Dom Antônio de Almeida. 1976. No Estuário Amazônico "À Margem da Visita Pastoral". Conselho Estadual de Cultura do Pará, Belém.

Luz, Francisco Joaci F. 2001. Plantas medicinais de uso popular em Boa Vista, Roraima, Brasil. Horticultura Brasileira 19(1):88–96

Luz, Benjamin Bordallo da. 2004. Características de árvores emergentes utilizadas por gavião-real (*Harpia harpyja*) para nidificação no centro e leste da Amazônia Brasileira. Master's thesis, INPA/UFAM, Manaus

Macedo, Erick Joe Dossantos. 2014. Almacenamiento de carbono en la biomasa aérea del bosque primario y bosque secundario de la parcela "Muro Huayra" en la Reserva Nacional Allpahuayo Mishana, Iquitos, Perú. Master's Thesis, UNAP, Facultad de Ciencias Forestales, Iquitos

Machado, Danilo de Oliveira. 2016. A agrobiodiversidade de quintais agroflorestais em propriedades agrícolas familiares na BR 174, Ramal do Pau-Rosa, Manaus, AM. Master's Thesis, INPA, Manaus

Machado, Monielle Alencar; Pinheiro, Claudio Urbano B. 2016. Da água doce à água salgada: mudanças na vegetação de igapó em margens de lagos, rios e canais do baixo curso do rio Pindaré, Baixada Maranhense. Revista Brasileira de Geografia Física 9(5): 1410–1427

Macía, Manuel J. 2008. Woody plants diversity, floristic composition and land use history in the Amazonian rain forests of Madidi National Park, Bolivia. Biodiversity Conservation 17: 2671–2690

Macía, Manuel J. 2011. Spatial distribution and floristic composition of trees and lianas in different forest types of an Amazonian rainforest. Plant Ecology 212:1159–1177

Macía, Manuel J.; Fuertes, Juan. 2008. Composición florística y estructura de los árboles en un bosque tropical montano de la Cordillera Mosetenes, Bolivia. Revista Boliviana de Ecología y Conservación Ambiental 23:1–14

Mackensen, Jens; Tillery-Stevens, Madeleine; Klinge, Rudolf; Fölster, Horst. 2000. Site parameters, species composition, phytomass structure and element stores of a terra-firme forest in East-Amazonia, Brazil. Plant Ecology 151:101–109

Maia, Lúcia Alencar; Chalco, Fiorella Perotti. 2002. Produção de frutos de espécies da floresta de várzea da Amazônia central importantes na alimentação de peixes. Acta Amazonica 32(1):45–54

Mamede, Jeneffer Soares dos Santos; Pasa, Maria Corette. 2018. Os quintais agroflorestais na comunidade rural em Várzea Grande, Mato Grosso, Brazil. Biodiversidade 17(2):54–70

Mamede, Jeneffer Soares dos Santos; David, Margô De; Tsukamoto Filho, Antônio de Arruda; Pasa, Maria Corette. 2015. Os quintais e as manifestações culturais da comunidade de São Gonçalo Beira Rio, Cuiabá – MT. Biodiversidade 14(1):168–182

Maranho, Álisson Sobrinho; Salimon, Cleber Ibraim. 2015. Maranthaceae overabundance decreases richness and abundance of regenerating woody plants in natural gaps. Neotropical Biology and Conservation 10(2):53–62

Maranho, Álisson Sobrinho; Salimon, Cleber Ibraim; Costa, Daniel da Silva. 2016. Woody flora of natural forest gaps in a bamboo-dominated forest remnant in southwestern Amazonia. Check List: The Journal of Biodiversity Data 12(3):1–9

Marcelo-Peña, José Luis; Rodriguez, Carlos Reynel. 2014. Patrones de diversidad y composición florística de parcelas de evaluación permanente en la selva central de Perú. Rodriguésia 65(1):35–47

Marimon, B. S.; Felfili, J. M.; Haridasan, M. 2001. Studies in monodominant forests in eastern Mato Grosso, Brazil: II. A forest in the Areões Xavante Indian Reserve. Edinburgh Journal of Botany 58(3):483–497

Marimon, B. S.; Felfili, J. M.; Lima, E. S. 2002. Floristics and phytosociology of the gallery forest of the Bacaba stream, Nova Xavantina, Mato Grosso, Brazil. Edinburgh Journal of Botany 59(2):303–318

Marimon, B. S.; Lima, E. S.; Duarte, T. G.; Chieregatto, L. C.; Ratter, J. A. 2006. Observations on the vegetation of northeastern Mato Grosso. IV. An analysis of the cerrado-Amazonian forest ecotone. Edinburgh Journal of Botany 63(2&3):323–341

Marquardt, Kristina; Milestad, Rebecka; Salomonsson, Lennart. 2013. Improved fallows: a case study of an adaptive response in Amazonian swidden farming systems. Agriculture and Human Values 30:417–428

Martins, Sebastião Venâncio; Brito, Elizabeth Rodrigues; Oliveira Filho, Ary Teixeirade; Silva, Alexandre Francisco da; Silva, Elias. 2008. Floristic composition of two wetland forests in Araguaian plain, state of Tocantins, Brazil, and comparison with other areas. Revista Árvore 32(1):129–141

Martins, Williane Maria de Oliveira; Martins, Lilliane Maria de Oliveira; Paiva, Fabiano Silveira; Martins, Wilton José de Oliveira; Lima Júnior, Sebastião Ferreira. 2012. Agrobiodiversidade nos quintais e roçados ribeirinhos na comunidade Boca do Môa – Acre. Biotemas 25(3):111–120

Mattos, Fábio de Souza. 2016. Dieta do *Plecturocebus cinerascens* (Platyrrhini: Primates): a influência da composição química na escolha de frutos consumidos. Master's thesis, INPA, Manaus

Mattos, Cilene Mara Jordão de; Silva, Wanderson Luis da Silva e; Carvalho, Catarina Silvad e; Lima, Andressa Novaes; Faria, Sérgio Miana de; Lima, Haroldo Cavalcante de. 2018. Flora das cangas da Serra dos Carajás, Pará, Brasil: Leguminosae. Rodriguésia 69(3):1147–1220

Maués, Bernardo Antônio Rodrigues; Jardim, Mário Augusto Gonçalves; Batista, Fábio de Jesus; Medeiros, Tonny David Santiago; Quaresma, Adriano da Costa. 2011. Composição florística e estrutura do estrato inferior da floresta de várzea na área de proteção ambiental Ilha do Combu, município de Belém, estado do Pará. Revista Árvore 35(3):669–677

May, Thomas. 2016. Aspectos de sostenibilidad de productos no maderables florestales con uso curativo en el oeste de Pará, Brasil. Ambiente y Desarrollo 20(38):69–84

McGrath, David G.; Peters, Charles M.; Bentes, Antônio José Mota. 2004. Community forestry for small-scale furniture production in the Brazilian Amazon. In: Working Forests in the Neotropics: Conservation through Sustainable Management, Zarin, Daniel J.; Alavalapati, Janaki R. R.; Putz, Francis E.; Schmink, Marianne (eds), Columbia University Press, New York, pp. 200–220

Medeiros, Reginaldo Antonio. 2004. Dinâmica de sucessão secundária em floresta de transição na Amazônia meridional. Master's thesis, Universidade Federal de Mato Grosso, Cuiabá

Medeiros, Priscilla Sanjuan de. 2016. Indicadores ambientais para funcionalidade ecológica em florestas secundárias de diferentes idades na Amazônia Oriental. Doctoral dissertation, Universidade Federal do Pará, Belém

Medeiros, Marcelo Brilhante de; Walter, Bruno Machado Teles; Silva, Glocimar Pereira da; Gomes, Beatriz Machado; Lima, Isabel Lustz Portela; Silva, Suelma Ribeiro; Moser, Pamela; Washington, Luís Oliveira; Cavalcanti, Taciana Barbosa. 2012. Vascular flora of the Tocantins River middle basin, Brazil. Check List 8(5):852–885

Medina, G.; Pokorny, B.; Campbell, B. M. 2009. Community forest management for timber extraction in the Amazon frontier. International Forestry Review 11(3):408–420

Mejia, Kember; Rengifo, Elsa. 2000. Plantas Medicinales de Uso Popular en la Amazonía Peruana. Agencia Española de Cooperación Internacional, Lima

Mendoza, Abel L. Monteagudo; Guerrero, Marleni Huamán. 2010. Catálogo de los arboles y afines de la selva central del Perú. Arnaldoa 17(2):202–242

Mendoza, Abel L. Monteagudo; Gamarra, Luis Valenzuela; Vásquez, Rodolfo; Gonzáles, Martínez & Rocío del Pilar Rojas; Phillips, Oliver L.; Lopez-Gonzalez, Gabriela. 2014. Primer catálogo de los árboles y afines de la Reserva Comunal El Sira, Perú. Arnoldoa 21(1): 127–164

Mesquita, Ulliane de O.; Tavares-Martins, Ana Cláudia. 2018. Etnobotánica de plantas medicinales en la comunidad de Caruarú, Isla de Mosqueiro, Belém-PA, Brasil. Boletín Latinoamericano y del Caribe de Plantas Medicinales y Aromáticas 17(2):130–159

Mews, Henrique Augusto; Marimon, Beatriz Schwantes; Pinto, José Roberto Rodrigues; Silvério, Divino Vicente. 2011a. Dinâmica estrutural da comunidade lenhosa em floresta estacional semidecidual na transição cerrado-floresta Amazônica, Matos Grosso, Brasil. Acta Botanica Brasilica 25(4):845–857

Mews, Henrique Augusto; Marimon, Beatriz Schwantes; Maracahipes, Leandro; Franczak, Daniel David; Mamonon Junior, Ben Hur. 2011b. Dinâmica da comunidade lenhosa de um cerrado típico na região Nordeste do Estado de Mato Grosso, Brasil. Biota Neotropica 11(1):1–10

Mews, H. A.; Marimon, B. S.; Ratter, J. A. 2012. Observations on the vegetation of Mato Grosso, Brazil. V. Changes in the woody species diversity of a forest in the cerrado-Amazonian forest transition zone and notes on the forests of the region. Edinburgh Journal of Botany 69(2):239–253

Miguel, Aline; Marimon, Beatriz Schwantes; Oliveira, Edmar Almeida de; Marachaipes, Leandro; Marimon-Junior, Ben Hur. 2011. Mudanças na estrutura da vegetação lenhosa em três porções da mata de galeria do Córrego Bacaba (1999–2006), Nova Xavantina-MT. Revista Árvore 35(3):725–735

Miller, Robert P.; Proctor, John. 1998. A small area of young secondary forest on the Ilha de Maracá: structure and floristics. In: Maracá: The Biodiversity and Environment of an Amazonian Rainforest, Milliken, William; Ratter, James A. (eds), John Wiley, Chichester, pp 130–134

Milliken, William. 1997. Traditional anti-malarial medicine in Roraima, Brazil. Economic Botany 51(3):212–237

Milliken, W. and B. Albert. 1996. The use of medicinal plants by the Yanomami indians of Brazil. Economic Botany 50(1):10–25

Milliken, W. and B. Albert. 1997a. The construction of a new Yanomami round-house. *Journal of Ethnobiology* 17(2):215–233

Milliken, W. and B. Albert. 1997b. Plantas medicinais dos Yanomami: uma nova visão dentro do etnobotânica de Roraima. In: Homem, Ambiente e Ecologia no Estado de Roraima, Barbosa, Reinaldo I.; Ferreira, Efrem G.; Castellón, Eloy G. (eds), INPA, Manaus, 85–110

Miranda, Izildinha S. 2000. Análise florística e estrutural da vegetação lenhosa do Rio Comemoração, Pimenta Bueno, Rondônia, Brasil. Acta Amazonica 30(3):393–422

Miranda, Ivana Lobato; Carvalho, João Olegário Pereira de; Ferreira, Maria do Socorro Gonçalves; Oliveira, Lia Cunha de; Vieira, Ima Célia Guimarães; Almeida, Wanderléa da Costa; Soares, Márcio Hofman Mota; Costa, Dulce Helena Martina. 2000. Principais espécies arbóreas de valor econômico em uma floresta às margens do Lago Acapu, no Município de Oriximiná, Pará. Embrapa Amazônia Oriental Comunicado Técnico 31:1–4

Monteles, Ricardo; Pinheiro, Claudio Urbano B. 2007. Plantas medicinais em um quilombo maranhense: uma perspectiva etnobotânica. Revista de Biologia e Ciências da Terra 7(2):1–12

Moser, Pamela. 2013. Vegetação Arbórea e sua Relação com Fatores Ambientais e Especiais em Florestas de Terra Firme no Noroeste de Rondônia, Brasil. Master's thesis, Faculdade de Tecnologia, Departamento de Engenharia Florestal, Programa de Pós-Graduação em Ciências Florestais, Universidade de Brasília

Mota, Nara Furtado de Oliveira; Watanabe, Mauricio Takashi Coutinho; Zappi, Daniela Cristina; Hiura, Alice Lima; Pallos, Julieta; Viveros, Raquel Stauffer; Giulietti, Ana Maria; Viana, Pedro Lage. 2018. Cangas da Amazônia: a vegetação única de Carajás evidenciada pela lista de fanerógramas. Rodriguésia 69(3):1435–1488

Moura, Cleonice de Oliveira; Wolter Filho, Wilson; Absy, Maria Lúcia. 1993. Alergias causadas por algumas espécies madeireiras na Amazônia. In: Bases Científicas para Estratégias de Preservação e Desenvolvimento da Amazônia, Ferreira, Efrem Jorge Gondim; Santos, Geraldo Mendes dos; Leão, Elizabeth Lima Mendes; Oliveira, Luíz Antônio de Oliveira (eds), INPA, Manaus, vol. 2, pp 89–106

Muniz, Francisca Helena; Cesar, Oswaldo; Monteiro, Reinaldo. 1994. Fitossociologia da vegetação arbórea da reserva florestal do Sacavém, São Luís, Maranhão (Brasil). Acta Amazonica 24 (3/4):219–236

Muñoz, Marianella Guevara. 2010. Evaluación de purmas con fines de manejo en la comunidad nativa de Santa Mercedes, Rio Putumayo, Perú. Master's thesis, Universidad Nacional de la Amazonia Peruana, Iquitos
Murakami, Alejandro Araujo; Reyes, Juan Fermandon; Milliken, William. 2016. Frutales Silvestres y Promisorios de Pando. Herencia/Museo de Historia Natural Noel Kempff Mercado, Cobija
Murça-Pires, J.; Dobzhansky Th.; Black, G. A. 1953. An estimate of the number of species of trees in an Amazonian forest community. Botanical Gazette 114(4): 467–477
Murrieta, Javier Noriega; Arévalo, Jorge Augusto Gil. 2004. Frutos y semillas utilizados como alimento por peces de consumo y de valor comerical, en las cochas Capirona y Quincha. Universidad Nacional de la Amazonía Peruana, Iquitos
Nascimento, M. T.; Proctor, J.; Villela, D. M. 1997. Floristic structure, floristic composition and soils of an Amazonian monodominant forest on Maracá Island, Roraima, Brazil. Edinburgh Journal of Botany 54(1):1–38
Navarrete, Víctor Alfredo Torres. 2016. Diversidad, estructura y concentración de carbono en un bosque siempreverde piedmontano de 800 a 900 msnm, em la Amazonía Ecuatoriana. Master's thesis, Universidad Técnica Estatal de Queve do Unidad de Posgrado, Qevedo, Ecuador
Navarro, G. 2002. Provincia biogeográfica del Acre-Madre de Dios (Amazónica Suroccidental). In: Geografía Ecológica de Bolivia: Vegetación y Ambientes Acuáticos, Navarro, Gonzalo; Maldonado, Mabel (eds), Centro de Ecología Simón I. Patiño, Cochabamba, pp 51–98
Nebel, Gustav; Dragsted, Jens; Vanclay, Jerome K. 2000a. Estructura y composición florística del bosque de la llanura aluvial inundable de la Amazonía Peruana: II. El sotobosque de la restinga. Folia Amazónica 10(1–2):151–181
Nebel, Gustav; Kvist, Lars Peter; Vanclay, Jerome K.; Christensen, Henning; Freitas, Luis; Ruiz, Juan. 2000b. Estructura y composición florística del bosque de la llanura aluvial en la Amazonía Peruana: I. El bosque alto. Folia Amazónica 10(1–2):91–149
Nebel, Gustav; Dragsted, Jens; Vanclay, Jerome K. 2001. Structure and floristic composition of flood plain forests in the Peruvian Amazon II. The understorey of restinga forests. Forest Ecology and Management 150: 59–77
Norconk, Marilyn A.; Kinzey, Warren G. 1994. Challenge of Neotropical frugivory: travel patterns of spider monkeys and bearded sakis. American Journal of Primatology 34: 171–183
Norconk, Marilyn A.; Veres, Michael. 2011. Physical properties of fruit and seeds ingested by primate seed predators with emphasis on sakis and bearded sakis. The Anatomical Record 294:2092–2111
Nunes, Andréa. 1998. Diet and feeding ecology of *Ateles belzebuth belzebuth* at Maracá Ecological Station, Roraima, Brazil. Folia Primatologica 69:61–76
Odonne, Guillaume; Berger, Franck; Stien, Didier; Grenand, Pierre; Bourdy, Geneviève. 2011. Treatment of leishmaniasis in the Oyapock basin (French Guiana): a K.A.P. survey and analysis of the evolution of phytotherapy knowledge amongst Wayãpi Indians. Journal of Ethnopharmacology 137:1228–1239
Oliveira, Arlem Nascimento de; Amaral, Iêda Leão. 2005. Aspectos florísticos, fitossociológicos e ecológicos de um sub-bosque de terra firme na Amazônia Central, Amazonas, Brasil. Acta Amazonica 35(1):1–16
Oliveira, Patricia Chaves de; Cavalcante, Suellen. 2017. Ethnobotany in the Amazon floodplain ecosystem: a case study, Quilombo Saracura, Pará, Brazil. International Journal of Botany Studies 2(1):89–99
Oliveira, Ana Cristina M.; Ferrari, Stephen F. 2000. Seed dispersal by black-handed tamarins, Saguinus midas niger (Callitrichinae, Primates): implications for the regeneration of degraded forest habitats in eastern Amazonia. Journal of Tropical Ecology 16: 709–716
Oliveira Filho, Ary Teixeira de. 1989. Composição florística e estrutura comunitária da floresta de galeria do Córrego da Paciência, Cuiabá (MT). Acta Botanica Brasilica 3(1):91–112
Oliveira, Francisco Plácido Magalhães; Jardim, Mário Augusto Gonçalves. 1998. Composição florística de uma floresta secundária no município de Igarapé-Açu, estado do Pará, Brasil. Boletim do Museu Paraense Emilio Goeldi, Série Botânica 14(2):127–145

Oliveira, Alexandre A. de; Mori, Scott A. 1999. A central Amazonian terra firme forest. I. High tree species richness on poor soils. Biodiversity and Conservation 8:1219–1244

Oliveira, Francisco Plácido Magalhães; Absy, Maria Lúcia; Miranda, Izildinha Souza. 2009. Recurso polínico coletado por abelhas sem ferrão (Apidae, Meliponinae) em um fragmento de floresta na região de Manaus- Amazonas. Acta Amazonica 39(3):505–518.

Oliveira, S. G. de; Alfaro, Jessica W. Lynch; Veiga, Liza M. 2014. Activity budget, diet, and habitat use in the critically endangered Ka'apor capuchin monkey (*Cebus kaapori*) in Pará state, Brazil: a preliminary comparison to other capuchin monkeys. American Journal of Primatology 76:919–931

Pacheco, L.F.; Simonetti, J.A. 1998. Consecuencias demográficas para *Inga ingoides* (Mimosoideae) para la pérdida de Ateles paniscus (Cebidae), uno de sus dispersores de semillas. Ecología en Bolivia 31: 67–90

Pacheco, L.F.; Simonetti, J.A. 2000. Genetic structure of a Mimosoid tree deprived of its seed disperser, the spider monkey. *Conservation Biology* 14 (6):1766

Paim, Fernanda Pozzam; Valenta, Kim; Chapman, Colin A.; Paglia, Adriano Pereira; Queiroz, Helder Lima de. 2018. Tree community structure reflects niche segregation of three parapatric squirrel monkey species (Saimiri spp.). Primates https://doi.org/10.1007/s10329-018-0659-6

Palacios, Erwin; Rodríguez, Adriana; Castillo, Claudia. 2004. Preliminary observations on the mottled-face tamarin (*Saguinus inustus*) on the lower Rio Caquetá, Colombian Amazon. Neotropical Primates 12 (3): 123–126

Palma, Ana Cristina; Stevenson, Pablo R. 2010. Dispersión de semillas por monos araña en la Estación Biológica Cocha Cashu, Perú. In: Primatología en Colombia: Avances al Principio del Milenio, Pereira-Bengoa, Victoria; Stevenson, Pablo R.; Bueno, Marta Lucía; Nassar-Montoya, Fernando (eds), Fundación Universitaria San Martín, Bogotá, pp 19–35

Parrotta, John A.; Francis, John K.; Almeida, Rionaldo Rolo de. 1995. Trees of the Tapajós: A Photographic Field Guide. International Institute of Tropical Forestry, General Technical Report IITF-1, Rio Piedras, Puerto Rico

Parrotta, John A.; Francis, John K.; Knowles, Oliver H. 2002. Harvesting intensity affects forest structure and composition in an upland Amazonian forest. Forest Ecology and Management 169:243–255

Peña-Claros, Marielos. 2003. Changes in forest structure and species composition during secondary forest succession in the Bolivian Amazon. Biotropica 35(4):450–461

Pennington, T. D. 1997. *The Genus Inga*. Royal Botanical Gardens, Kew

Pereira, Nádia Waleska Valentim; Venturin, Nelson; Machado, Evandro Luiz Mendonça; Scolforo, José Roberto Soares; Macedo, Renato Luiz Grisi; d'Oliveira, Marcus Viníco Neves. 2005. Análise das variações temporais na florística e estrutura da comunidade arbórea de uma floresta explorada com plano de manejo. Cerne 11(3):263–282

Peres, C.A. 1993. Diet and feeding ecology of saddle-back (*Saguinus fuscicollis*) and moustached (S. mystax) tamarins in an Amazonian terra firme forest. Journal of Zoology 230:567–592

Peres, C.A. 1994a. Diet and feeding ecology of gray woolly monkeys (*Lagothrix lagotricha cana*) in central Amazonia: comparisons with other Atelines. International Journal of Primatology 15(3):333–372

Peres, C. A. 1994b. Primate responses to phenological changes in an Amazonian terra firme forest. Biotropica 26(1): 98–112

Perez, Inayê Uliana. 2011. Uso dos recursos naturais vegetais na comunidade indígena Araçá, Roraima. Master's thesis, Universidade Federal de Roraima, Boa Vista

Pérez-Arbeláez, Enrique. 1956. Plantas Útiles de Colombia. Liberia Colombiana, Bogotá

Pezzuti, J. and R. P. Chaves. 2009. Etnografia e manejo de recursos naturais pelos índios Deni, Amazonas, Brasil. *Acta Amazonica* 39(1):121–138

Phillips, O. 1993. The potential for harvesting fruits in tropical rainforests: new data from Amazonian Peru. Biodiversity and Conservation 2:18–38

Pinheiro, Klewton Adriano Oliveira; Carvalho, João Olegário Pereira de; Quantz, Beatriz; Francez, Luciana Maria de Barros; Schwartz, Gustavo. 2007. Fitossociologia de uma área de preservação permanente no leste da Amazônia: indicacão de espécies para recuperação de áreas alteradas. Floresta 37(2): 175–187

Pinho, Rachel Camargo de. 2008. Quintais agroflorestais indígenas em áreas de savana (lavrado) na Terra Indígena Araçá, Roraima. Master's thesis, INPA/UFAM, Manaus

Pinho, Rachel C.; Aflaia, Sonia S.; Miller, Robert Pritchard; Uguen, Katell; Magalhães, Leovone D.; Ayres, Marta; Freitas, Viviane; Trancoso, Ralph. 2011. Islands of fertility: soil improvement under indigenous homegardens in the savannas of Roraima, Brazil. Agroforestry Systems 81:235–247

Pinto, L. P. 2002. Diet, activity and home range of *Alouatta belzebul discolor* (Primates, Atelidae) in Paranaíta, northern Mato Grosso. Neotropical Primates 10(2):98–99

Pinto, L. P. 2008. Ecologia alimentar do cuxiú-de-nariz vermelho *Chiropotes albinasus* (Primates: Pitheciidae) na Floresta Nacional do Tapajós, Pará. Doctoral dissertation, Universidade Estadual de Campinas, São Paulo

Pinto, José Roberto R.; Hay, John du Vall. 2005. Mudanças florísticas e estruturais na comunidade arbórea de uma floresta de vale no Parque Nacional da Chapada dos Guimarães, Mato Gross, Brasil. Revista Brasileira de Botanica 28(3):523–539

Pinto, L. P. and E. Z. F. Setz. 2004. Diet of *Alouatta belzebul discolor* in an Amazonian Rain Forest of Northern Mato Grosso State, Brazil. *International Journal of Primatology* 25(6):1197–1211

Pinto, A. C. B., C. Azevedo-Ramos, and O. Carvalho. 2003. Activity patterns and diet of the howler monkey *Alouatta belzebul* in areas of logged and unlogged forest in eastern Amazonia. *Animal Biodiversity and Conservation* 26(2):39–49

Pires, João Murça; Salomão, Rafael de Paiva. 2000. Dinâmica da diversidade arbórea de um fragmento de floresta tropical primária na Amazônia oriental-1. Período: 1956 a 1992. Boletim do Museu Paraense Emílio Goeldi, Série Botânica 16(1):63–110

Pitman, Nigel C. A.; Terborgh, John W.; Silman, Miles R.; Núñez, Percy; Neill, David A.; Cerón, Carlos E.; Palacios, Walter A.; Aulestia, Milton. 2001. Dominance and distribution of tree species in upper Amazonian terra firme forests. Ecology 82(8):2101–2117

Pohle, Perdita; Reinhardt, Sylvia. 2004. Indigenous knowledge of plants and their utilization among the Shuar of the lower tropical mountain forest in southern Ecuador. Lyonia 7(2):133–149

Portal, Rubens da Rocha; Lima, Maria Auxiliadora S.; Luz, Vera Lúcia Ferreira; Bataus, Yeda Soares de Lucena; Reis, Isaías José dos. 2002. Espécies vegetais utilizadas na alimentação de Podocnemis unifilis, Troschel 1948 (Reptilia, Testudinae, Pelomedusidae) na região de Pracuúba, Amapá, Brasil. Ciencia Animal Brasileira 3(1):11–19

Port-Carvalho, Márcio; Ferrari, Stephen F. 2004. Occurrence and diet of the black bearded saki (*Chiropotes satanas satanas*) in the fragmented landscape of western Maranhão, Brazil. Neotropical Primates 12(1):17–20

Porter, L. M. 2001. Dietary differences among sympatric Callitrichinae in northern Bolivia: *Callimico goeldii, Saguinus fuscicollis* and *S. labiatus*. International Journal of Primatology 22(6):961–992

Prance, G. T. 2014. That Glorious Forest: Exploring the Plants and Their Indigenous Uses in Amazonia. New York Botanical Garden, New York

Prance, G. T.; Balée, W.; Boom, B. M. 1987. Quantitative ethnobotany and the case for conservation in Amazonia. Conservation Biology 1(4):296–310

Prata, Shirley Soares; Miranda, Izildinha de Souza; Alves, Sérgio Augusto Oliveira; Farias, Flavia Cardoso; Jardim, Fernando Cristovam da Silva. 2010. Gradiente florístico das florestas secundárias do Nordeste Paraense. Acta Amazonica 40(3):523–534

Queiroz, José Antonio Leite; Machado, Sebastião do Amaral. 2008. Fitossociologia em floresta de várzea do estuário Amazônico no estado do Amapá. Pesquisa Florestal Brasileira 57:5–20

Queiroz, José Antonio Leite de; Mochiutti, Silas; Machado, Sebastião do Amaral. 2006. Distribuição diamétrica em floresta de várzea do braço norte do Rio Amazonas. Revista de Ciência Agrária 46:151–170

Queiroz, José Antonio Leite de; Machado, Sebastiã do Amaral; Hosokawa, Roberto Tuyoshi; Silva, Ivan Crespo da. 2007. Estrutura e dinâmica de floresta de várzea no estuário Amazônico no estado do Amapá. Floresta 37(3):339–352

Quevedo, Enrique; Zuluaga, Luis. 1994. Un aproximación al proceso de domesticación de vegetales con referencia especial a los frutales en la Amazonía Colombiana. Agronomia Colombiana 11(2):190–205

Rabelo, Fernando Galvão; Zarin, Daniel Jacob; Oliveira, Francisco de Assis; Jardim, Fernando Cristóvam da Silva. 2000. Regeneração natural de florestas estuarinas na região do rio Amazonas, Amapá, Brasil. Revista Ciência Agrária 34:129–137

Ratter, J. A. 1987. Notes on the vegetation of the Parque Nacional do Araguaia (Brazil). Notes from the Royal Botanic Garden, Edinburgh 44(2):311–342

Rayol, Breno Pinto. 2006. Análise florística e estrutural da vegetação xerofítica das savannas metalófilas na Floresta Nacional de Carajás: subsídios à conservação. Master's thesis, Universidade Federal Rural da Amazônia/Museu Paraense Emílio Goeldi, Belém.

Rayol, Breno Pinto; Silva, Manoela Ferreira Fernandes da; Alvino, Fabrízia de Oliveira. 2006. Dinâmica da diversidade florística da regeneração natural de florestas secundárias, no município de Bragança, Pará, Brasil. Boletim do Museu Paraense Emílio Goeldi, Ciencias Naturais 1(2):9–27

Rayol, Breno Pinto; Alvino-Rayol, Fabrízia de Oliveira; Silva, Manoela Ferreira Fernandes da. 2011. Similaridade florística entre o estrato arbóreo e a regeneração natural de uma floresta secundária, no município de Bragança, nordeste do estado do Pará, Brasil. Revista Brasileira de Agroecologia 6(3):107–114

Regan, B. C.; Julliot, C.; Simmen, B.; Viénot, F.; Charles-Dominique, P.; Mollon, J. D. 2001. Fruits, foliage and the evolution of primate colour vision. Philosophical Transactions of the Royal Society London B 356: 229–283

Rego, Lyvia Julienne Sousa; Elly, Márcio Piero; Pereira, Jéssica Lira; Garcia, Bruna Naiara; Barbosa, Humberto Figueira; Vieira, Thiago Almeida. 2011. Produtos florestais não-madeireiros comercializados em feiras de Santarém, Pará. Cadernos de Agroecologia 6(2):1–6

Reis, Neidiane Farias Costa. 2010. Estrutura fitossociológica e seletividade de espécies em áreas inundáveis na região do baixo Rio Jaci-Paraná, Rondônia. Master's Thesis, Universidade Federal de Rondônia, Porto Velho

Rengifo-Salgado, Elsa; Rios-Torres, Sandra; Malaverri, Lizardo Fachín; Vargas-Arana, Gabriel. 2017. Saberes ancestrales sobre el uso de flora y fauna en la comunidad indígena Tikuna de Cushillo Cocha, zona fronteriza Perú-Colombia-Brasil. Revista Peruana de Biología 24(1):67–78

Reyes-García, Victoria. 2001. Indigenous people, ethnobotanical knowledge, and market economy: a case study of the Tsimane' Amerindians in lowland Bolivia. Doctoral dissertation, University of Florida, Gainesville.

Reynel, C.; Pennington, T.D. 1997. El Género *Inga* en el Perú: Morfología, Distribución y Usos. Royal Botanic Gardens, Kew

Ribeiro, Berta G. 1990. Classificação dos solos e horticultura Desâna. In: Ethnobiology: Implications and Applications, Proceedings of the First International Congress of Ethnobiology, Posey, Darrell A.; Overal, William Leslie; Clement, Charles R.; Plotkin, Mark; Elisabetsky, Elaine; Mota, Clarice Novaes da; Barros, José Flávio Pessôa (eds), Museu Paraense Emílio Goeldi, Belém, pp. 25–49

Rimachi-Taricuarima, Milagros Natalia; Pérez-Gardini, Juan Jesús; Tirado-Herrera, Emérita Rosabel; Zárate-Gómez, Ricardo; Mozombite-Pinto, Linder Felipe. 2019. Plantas consumidas por *Lagothrix lagotricha lagotricha* Humboldt, 1812 en la Amazonía peruana. Ciencia Amazónica 7(1):93–110

Robles, Rembrandt Ramiro Canahuire. 2017. Composición florística y estructura de la recuperación natural en un área degradado por minería en la comunidad nativa de Tres Islas, Tambopata, Madre de Dios. Master's thesis, Escuela Profesional de Ingeniería Forestal y Medio Ambiente, Puerto Maldonado, Perú

Rockwell, Cara A.; Kainer, Karen A.; Staudhammer, Christina L.; Baraloto, Christopher. 2007. Future crop tree damage in a certified community forest in southwestern Amazonia. Forest Ecology and Management 242:108–118

Rodrigues, William A. 1963. Estudo de 2,6 hectares de mata de terra firme da Serra do Navio, território do Amapá. Boletim do Museu Paraense Emílio Goeldi, Nova Série Botânica 19:1–23

Rodrigues, I. A. 1986. Inventário florístico em áreas do projeto Albrás-Alunorte, Barcarena-Pa. *Anais do Primeiro Simpósio do Trópico Umido, 12 a 17 novembro 1984*, EMPRABA-CPATU, Belém, vol. 2, pp. 153–166

Rodrigues, Eliana. 2006. Plants and animals utilized as medicines in the Jaú National Park (JNP), Brazilian Amazon. Phytotherapy Research 20:378–391

Rodrigues, I. and A. E. Oliveira. 1977. Alguns aspectos da ergologia Mura-Pirahã. *Boletim do Museu Paraense Emílio Goeldi, Nova Série Antropologia* 65:1–47

Rodrigues, Ricardo Ribeiro; Martins, Sebastião Venâncio; Barros, Luiz Carlos de. 2004. Tropical rain forest regeneration in an area degraded by mining in Mato Grosso state, Brazil. Forest Ecology and Management 190: 323–333

Rodrigues, Mauro Antônio Cavaleiro de Macedo; Miranda, Izildinha Souza; Kato, Maria do Socorro Andrade. 2007a. Estrutura de florestas secundárias após dois diferentes sistemas agrícolas no nordeste do estado do Pará, Amazônia Oriental. Acta Amazonica 37(4):591–598

Rodrigues, Mauro Antônio Cavaleiro Macedo; Miranda, Izildinha Souza; Kato, Maria do Socorro Andrade. 2007b. Flora e estrutura da vegetação secundária após o uso de diferentes trituradores florestais. Pesquia Agropecúaria Brasileira 42(4):459–465

Rodrigues, Eliana; Duarte-Almeida, Joaquim M.; Pires, Júlia Movilla. 2010. Perfil farmacológico e fitoquímico de plantas indicadas pelos caboclos do Parque Nacional do Jaú (AM) como potenciais analgésicas. Part I. Revista Brasileira de Farmacognosia 20(6):981–991

Roldán, Alejandra I.; Simonetti, Javier A. 2001. Plant-mammal interactions in tropical Bolivian forests with different hunting pressures. Conservation Biology 15(3):617–623

Rondon Neto, Rubens Marques; Santos, Jeneffer Soares dos; Silva, Manoel Advíncula da; Koppe, Valdinei Cristi. 2010. Potencialidades de uso de espécies arbustivas e arbóreas em diferentes fisionomias de cerrado, em Lucas do Rio Verde/MT. Revista de Biologia e Ciências da Terra 10(2):113–126

Roosevelt, A. C. 1999. Twelve thousand years of human-environment interaction in the Amazon floodplain. Advances in Economic Botany 13:371–392

Roosevelt, A. C. 2000. The lower Amazon: a dynamic human habitat. In: Imperfect Balance: Landscape Transformations in the Precolumbian Americas, Lentz, David (Editor), Columbia University Press, New York, pp 455–491

Roosevelt, A. C.; Costa, M. Lima da; Machado, C. Lopes; Michab, M.; Mercier, N.; Valladas, H.; Feathers, J.; Barnett, W.; Silveira, M. Imazio da; Henderson, A.; Silva, J.; Chernoff, B.; Reese, D. S.; Holman, J. A.; Toth, N.; Schick, K. 1996. Paleoindian cave dwellers in the Amazon: the peopling of the Americas. *Science* 272:373–384

Rostain, S. 1999. Occupations humaines et fonction domestique de monticules Préhistorique d'Amazonie équatorienne. Société Suisse des Americanistes 63:71–95

Rylands, Anthony B. 1979. Observações preliminares sobre o sagüi, *Callithrix humeralifera intermedius* (Hershkovitz, 1977) em Dardanelos, rio Aripuanã, Mato Grosso. Acta Amazonica 9(3):589–602

Rylands, Anthony B. 1981. Preliminary field observations on the marmoset, *Callithrix humeralifera intermedius* (Hershkovitz, 1977) at Dardanelos, Rio Aripuanã, Mato Grosso. Primates 22(1):46–59

Salomão, Rafael de Paiva; Lisboa, Pedro L. B. 1988. Análise ecológica da vegetação de uma floresta pluvial tropical de terra firme, Rondônia. Boletim do Museu Paraense Emílio Goeldi, Série Botânica 4(2):195–233.

Salomão, Rafael de Paiva; Rosa, Nelson de Araújo. 1989. Análise da vegetação de floresta pluvial tropical de terra firme, pelo método dos quadrantes: Serra Norte, Carajás, PA. Acta Botanica Brasilica 2(1):27–42

Salomão, Rafael de Paiva; Silva, Manoela Ferreira Fernandes da; Rosa, Nelson Araújo. 1988. Inventário ecológico em floresta pluvial tropical de terra firme, Serra Norte, Carajás, Pará. Boletim do Museu Paraense Emílio Goeldi, Série Botânica 4(1):1–46

Salomão, Rafael de Paiva; Matos, Aires Henriques de; Rosa, Nélson de Araújo. 2002. Dinâmica do sub-bosque e do estado arbóreo de floresta tropical primária fragmentada na Amazônia oriental. Acta Amazonica 32(3):387–419

Salomão, Rafael de Paiva; Vieira, Ima Célia Guimarães; Suemitsu, Chieno; Rosa, Nélson de Araújo; Almeira, Samuel Soares de; Amaral, Dário Dantas do; Menezes, Moirah Paulo Machado de. 2007a. As florestas de Belo Monte na grande curva do rio Xingu, Amazônia Oriental. Boletim do Museu Paraense Emílio Goeldi, Ciencias Naturais 2(3):57–153

Salomão, Rafael Paiva; Rosa, Nélson Araújo; Morais, Kácio Andrey Câmara. 2007b. Dinâmica da regeneração natural de árvores em áreas mineradas na Amazônia. Boletim do Museu Paraense Emílio Goeldi, Ciencias Naturais 2(2):85–139

Sanaiotti, Tânia M.; Bridgewater, Samuel; Ratter, James A. 1997. A floristic study of the savanna vegetation of the state of Amapá, Brazil, and suggestions for its conservation. Boletim do Museu Paraense Emílio Goeldi, Série Botânica 13(1):3–24

Sánchez, Judith Jacira Achong; Levy, Allison Licett Nuñez. 2013. Características de frutos consumidos por *Saguinus nigrifrons* (Primates, Callitrichidae), en la Estación Biológica Quebrado Blanco, Loreto, Perú. Master's thesis, Universidad Nacional de la Amazonía Peruana, Iquitos

Santana, José Augusto da Silva; Sousa, Luciana Karla V. dos Santos; Almeida, Wanderléa da Costa. 2003. Produção anual de serapilheira em floresta secundária na Amazônia oriental. Revista de Ciências Agrárias 40:119–232

Santana, José Augusto da Silva; Almeida, Wanderléa da Costa; Sousa, Luciana Karla V. dos Santos. 2004. Florística e fitossociologia em área de vegetação secundária na Amazônia oriental. Revista de Ciências Agrárias 41:105–120

Santos, Jerffereson Lobatodos. 2006. Uso e diversidade de espécies vegetais cultivadas na Reserva de Desenvolvimento Sustentável do Tupé, Manaus, Amazonas. Master's thesis, INPA/UFAM, Manaus

Santos, Luciana Oliveira dos. 2010. Comparação da vegetação lenhosa entre borda e interior de uma mancha de mata de restinga na RESEX Mar Caeté-Taperaçu, Bragança, Pará. Master's thesis, Universidade Federal Rural da Amazônia/Museu Paraense Emílio Goeldi, Belém

Santos, J. U. M. and M. N. Bastos. 2014. Flores e Frutas das Restingas do Estado do Pará. Universidade Federal Rural da Amazônia (UFRA), Belém

Santos, Gideão Costa dos; Jardim, Mário Augusto Gonçalves. 2006. Florística e estrutura do estrato arbóreo de uma floresta de várzea no município de Santa Bárbara do Pará, Estado do Pará, Brasil. Acta Amazonica 36(4):437–446

Santos, Silvio Roberto Miranda dos; Miranda, Izildinha de Souza; Tourinho, Manoel Malheiros. 2004. Análise florística e estrutural de sistemas agroflorestais das várzeas do rio Juba, Cametá, Pará. Acta Amazonica 34(2):251–263

Santos, Maurício Reginaldo A. dos; Lima, Maria Railda de; Ferreira, Maria das Graças R. 2008. Uso de plantas medicinais pela população de Ariquemes, Rondônia. Horticultura Brasileira 26:244–250

Santos, Ricardo R.; Vieira, Tatiana M.; Ferrari, Stephen F. 2013. Feeding ecology of Uta Hick's bearded saki (Chiropotes utahickae) on a man-made island in southeastern Brazilian Amazonia: seasonal and longitudinal variation. In: *Evolutionary Biology and Conservation of Titis, Sakis, and Uacaris*, Veiga, Liza M.; Barnett, Adrian A.; Ferrari, Stephen F.; Norconk, Marilyn A. (eds), Cambridge University Press, pp 250–254

Santos, Gideão Costa; Tourinho, Manoel Malheiros; Mendes, Fernanda da Silva; Guimarães, Célia Maria Costa. 2014a. Fitossociologia e práticas de manejo tradicional em uma floresta de várzea em Santa Bárbara do Pará, Brasil. Revista de Ciências Agrárias 57(2):138–145

Santos, M. R. A.; Lima, M. R.; Oliveira, C. L. L. G. 2014b. Medicinal plants used in Rondônia, western Amazon, Brazil. Revista Brasileira de Plantas Medicinais 16(3):707–720

Santos, Joielan Xipaia dos; Reis, Alisson Rodrigo Souza; Matos, Sterphane Araújo de; Leão, Fábio Miranda; Carvalho, Josiane Celerino de. 2016a. Caracterização etnobotânica de essências florestais com fins medicinais utilizadas pela etnia Xipaya, no município de Altamira-PA. Biota Amazônia 6(2):1–8

Santos, Otilene dos Anjos; Couceiro, Sheyla Regina Marques; Rezende, Alinne Costa Cavalcante; Silva, Márcia Daniela de Sousa. 2016b. Composition and richness of woody species in riparian forests in urban areas of Manaus, Amazonas, Brazil. Landscape and Urban Planning 150:70–78

Sanz-Biset, Jaume; Campos-de-la-Cruz, José; Epiquién, Mirbel A.; Cañigueral, Salvador. 2009. A first survey on the medicinal plants of the Chazuta valley (Peruvian Amazon). Journal of Ethnopharmacology 122:333–362

Schultes, R. E. and R. F. Raffauf. 1990. The Healing Forest: Medicinal and Toxic Plants of the Northwest Amazonia. Dioscorides Press, Portland, Oregon

Schwartz, Gustavo. 2007. Manejo sustentável de florestas secundárias: espécies potenciais no nordeste do Pará, Brasil. Amazônia: Ciência e Desenvolvimento 3(5):125–147

Scoles, Ricardo. 2009. El quintal y las frutas: recursos económicos y alimentares en la comunidad negra de Itacoã, Acará, Pará, Brasil. Acta Amazonica 39(1):1–12

Setz, E. Z. F.; Enzweiler, J.; Solferini, V. N.; Amêndola; Berton, R. S. 1999. Geophagy in the golden-faced saki monkey (*Pithecia pithecia chrysocephala*) in the Central Amazon. Journal of the Zoological Society of London 247:91–103

Shaffer, Christopher A. 2013. Feeding ecology of northern bearded sakis (*Chiropotes sagulatus*) in Guyana. American Journal of Primatology 75:568–580

Shanley, Patricia; Lux, Leda. 2003. The impacts of forest degradation on medicinal plant use and implications for health care in eastern Amazonia. Bioscience 53(6):573–584

Shanley, P. and N.A. Rosa. 2004. Eroding knowledge: an ethnobotanical inventory in eastern Amazonia's logging frontier. Economic Botany 58(2):135–160

Silva, Suzy Cristina Pedroza da. 2006. Os sistemas agroflorestais na Reserva de Desenvolvimento Sustentável Piagaçu-Purus: uma alternativa para a agricultura familiar. Master's thesis, Universidade Federal do Amazonas, Manaus

Silva, Fabíola Andréa. 2008a. Ceramic technology of the Asurini do Xingu, Brazil: an ethnoarchaeological study of artifact variability. Journal of Archaeological Method and Theory 15:217–265

Silva, Maria Betânia Araujo da. 2008b. Composição florística e estrutura fitossociológica de uma floresta ripária situada às margens do Rio Moju, Pará. Master's thesis, Universidade Federal Rural da Amazônia, Belém

Silva, Regina Oliveira da. 2008c. Biodiversidade e políticas de conservação: o caso do parque Estadual Monte Alegre, Pará. Doctoral dissertation, Universidade de Brasília

Silva, Rosibel Rodrigues. 2009. O etnoextrativismo não-madeireiro em uma comunidade Amazônica: um estudo de caso em Santa Luzia do Buiuçuzinho-Coari/AM. Master's thesis, UFAM, Manaus

Silva, Francismeire Bonadeu da. 2010. A tribo Ingeae Bentham (Leguminosae – Mimosoideae) na Floresta Nacional de Caxiuanã, Pará, Brasil. Master's thesis, Universidade Federal Rural da Amazônia/Museu Paraense Emílio Goeldi, Belém

Silva, Silvestre. 2011. Fruits of the Brazilian Amazon. Metalivros, São Paulo

Silva, Ana Paula Ferreira Frota da; Bentes-Gama, Michelliny de Matos. 2008. Fitossociologia de uma floresta ombrófila aberta em área de assentamento rural no distrito de Jaci Paraná, Porto Velho, Rondônia. Ambiência 4(3):435–452

Silva, Manoela F. F. da; Rosa, Nelson A. 1989. Análise do estrato arbóreo da vegetação sobre jazidas de cobre na Serra dos Carajás-PA. Boletim do Museu Paraense Emílio Goeldi, Série Botânica 5(2):175–206

Silva, Manoela Ferreira; Rosa, Nelson A.; Oliveira, Jorge. 1987. Estudos botânicos na área do projeto Ferro Carajás. Aspectos florísticos da mata do rio Gelado, Pará. Boletim do Museu Paraense Emílio Goeldi, Série Botânica 3(1):1–20

Silva, E. M.; Souza, J. N. S.; Rogez, H.; Rees, J. F.; Larondelle, Y. 2007. Antioxidant activities and polyphenolic contents of fifteen selected plant species from Amazonian region. Food Chemistry 101: 1012–1018

Silva, Wanderson Luis da Silva e; Gurgel, Ely Simone Cajueiro; Santos, João Ubiratan Moreira dos; Silva, Mônica Falcão da. 2013. Inventário e distribuição geográfica de Leguminosae no arquipélago de Marajó, PA, Brasil. Hoehnea 40(4):627–647

Silva, Kátia Emidio da; Martins, Sebastião Venâncio; Fortin, Marie-José; Ribeiro, Milton Cezar; Azevedo, Celso Paulo de; Ribeiro, Carlos Antonio Álvares Soares; Santos, Nerilson Terra. 2014. Tree species community spatial structure in a terra firme Amazon forest, Brazil. Bosque 35(3):347–355

Silva, Katia Emidio da; Martins, Sebastião Venâncio; Ribeiro, Carlos Antônio Alvares Soares; Santos, Nerilson Terra; Azevedo, Celso Paulo de. 2016a. Structure of 15 hectares permanent plots of terra firme dense forest in central Amazon. Revista Árvore 40(4):603–615

Silva, Wagner Matos da; Lolis, Solange e Fátima; Viana, Rodney Haulien Oliveira. 2016b. Composition and structure of the gallery forest in the Taquaruçu Grande sub-basin, Municipality of Palmas, Tocantins State. Acta Scientiarum Biological Sciences 38(1):17–24

Silva, Silmara Bento; Sousa, Venderleia Conceição; Santos, Crissogno Mesquita; Mariano, Daiane Cinque; Okumura, Ricardo Shigueru. 2017. Levantamento florístico do componente arbustivo-arbóreo da vegetação ciliar de fragmento no Rio Parauapebas. Agroecossistemas 9(1):99–115

Silveira, Isolda Maciel da; Quaresma, Helena Doris Barbosa; Guapindaia, Vera Lúcia; Machado, Ana Lúcia Machado. 1997. As populações pré-históricas e atuais. In: Caxiuanã, Lisboa, Pedro L. B. (ed), Museu Paraense Emílio Goeldi, Belém, pp 53–91

Simmen, Bruno; Sabatier, Daniel. 1996. Diets of some French Guianan primates: food composition and food choices. International Journal of Primatology 17(5):661–693

Smith, Herbert H. 1879. Brazil, the Amazons and the Coast. Charles Scribner's Sons, New York

Smith, N. J. H. 1974. Destructive exploitation of the South American river turtle. *Yearbook of the Pacific Coast Geographers* 36:85–102

Smith, N. J. H. 1981. *Man, Fishes, and the Amazon*. Columbia University Press, New York

Soares, Caio Rodrigo Alves; Sá, Heloíse Rodrigues Alves de; Rodrigues, Marcos; Goulart, Selma Lopes. 2018. Gestão e competividade: análise de um aglomerado produtiva moveleiro na Amazônia. Revista de Estudos Sociais 20(41):144–159

Soini, P. 1987. Ecology of the saddle-back tamarin *Saguinus fuscicollis illigeri* on the río Pacaya, northeastern Peru. Folia Primatologica 49:11–32

Soini, P. 1988. The pygmy marmoset, genus *Cebuella*. In: Ecology and Behavior of Neotropical Primates, Mittermeier, Russell A.; Rylands, Anthony B.; Coimbra-Filho, Adelmar F.; Fonseca, Gustavo A.B. da (eds), Ecology and Behavior of Neotropical Primates, World Wildlife Fund, Washington, DC, vol. 2, pp 79–129

Sousa, Julio dos Santos de; Bastos, Maria de Nazaré do Carmo; Rocha, Antônio Elielson Sousa. 2009. Mimosoideae (Leguminosae) do litoral Paraense. Acta Amazonica 39(4):799–812

Sousa, Julio dos Santos de; Bastos, Maria de Nazaré do Carmo; Gurgel, Ely Simone Cajueiro. 2011. O gênero *Inga* (Leguminosae-Mimosoideae) na província petrolífera de Urucu, Coari, Amazonas, Brasil. Rodriguésia 62(2):283–297

Souza, Célia Cristina Valero. 2010. Etnobotânica de quintais em três comunidades ribeirinhas na Amazônia central, Manaus, AM. Master's thesis, INPA, Manaus

Souza, Priscila Figueira de. 2012. Florística, distribuição e diversidade beta da comunidade arbórea ao longo do gradiente hídrico em uma floresta de terra firme no interflúvio Purus-Madeira, Amazônia central. Master's thesis, INPA, Manaus.

Souza, Victória Cristina Oliveira de. 2019. Avaliação do estágio sucessional utilizando a classificação em grupos ecológicos de uma floresta urbana na região metropolitana de Belém-Pará. Bachelor's thesis, Universidade Federal Rural da Amazônia (Ufra), Belém

Souza, Luiz Augusto Gomes de; Silva, Marlene Freitas da. 2003. Bioeconomical potential of Leguminosae from the lower Negro River, Amazon, Brazil. Lyonia 5(1):15–23

Souza, A. G. C., N. R. Sousa, S. E. L. Silva, C. D. M. Nunes, A. C. C. Canto, L. A. A. Cruz, 1996. *Fruteiras da Amazônia*. EMBRAPA (Empresa Brasileira de Agropecuária), Brasília

Spix, J. B. and C. F. P. Martius, 1976. *Viagem pelo Brasil, 1817–1820*. Edições Melhoramentos, São Paulo, Vol.3

Stefanello, Daniel. 2008. Florística, estrutura e dispersão de sementes de três trechos de floresta estacional perenifólia ribeirinha na bacia hidrográfica do Rio das Pacas, Querência, MT. Master's Thesis, Universidade Federal de Viçosa, Minas Gerais

Stefanello, Daniel; Fernandes-Bulhão, Clarissa; Martins, Sebastião Venâncio. 2009. Síndromes de dispersão de sementes em três trechos de vegetação ciliar (nascente, meio e foz) ao longo do Rio Pindaíba, MT. Revista Árvore 33(6):1051–1061

Stefanello, Daniel; Ivanauskas, Natália Macedo; Martins, Sebastião Venâncio; Silva, Elias; Kunz, Sustanis Horn. 2010. Síndromes de dispersão de diásporos das espécies de trechos de vegetação ciliar do rio das Pacas, Querência-MT. Acta Amazonica 40(1):141–150

Stevenson, Pablo R. 2000. Seed dispersal by woolly monkeys (*Lagothrix lagothricha*) at Tinigua National Park, Colombia: dispersal distance, germination rates, and dispersal quantity. American Journal of Primatology 50: 275–289

Stevenson, Pablo R. 2004. Fruit choice by woolly monkeys in Tinigua National Park, Colombia. International Journal of Primatology 25(2): 367–381

Stevenson, Pablo R. 2005. Potential keystone plant species for the frugivore community at Tinigua Park, Colombia. In: Tropical Fruits and Frugivores: The Search for Strong Interactors, Dew, J. Lawrence; Boubli, Jean Philippe (eds), Springer, Dordrecht, pp 37–57

Stevenson, Pablo R. 2007. Estimates of the number of seeds dispersed by a population of primates in a lowland forest in western Amazonia. In: Seed Dispersal: Theory and its Application in a Changing World, Dennis, Andrew J.; Schupp, Eugene W.; Green, Ronda J.; Westcott, David A. (eds), CABI, Wallingford, pp 340–362

Stevenson, Pablo R.; Rodríguez, María Elizabeth. 2008. Determinantes de la composición florística y efecto de borde en un fragmento de bosque en el Guaviare, Amazonia Colombiana. Revista Colombia Florestal 11:5–17

Stevenson, Miranda F.; Rylands, Anthony B. 1988. The marmosets, genus *Callithrix*. In: Ecology and Behavior of Neotropical Primates, Mittermeier, Russell A.; Rylands, Anthony B.; Coimbra-Filho, Adelmar F.; Fonseca, Gustavo A.B. da (eds), Ecology and Behavior of Neotropical Primates, World Wildlife Fund, Washington, DC, vol. 2, pp 131–222

Stevenson, Pablo R.; Quiñones, Marcela J.; Ahumada, Jorge A. 1994. Ecological strategies of woolly monkeys (*Lagothrix lagotricha*) at Tinigua National Park, Colombia. American Journal of Primatology 32:123–140

Stevenson, Pablo R.; Castellanos, María Clara; Barreto, Lina María. 1997. Estudio preliminar de dispersión de semillas por micos churucos (Lagothrix lagotricha) en el Parque Nacional Tinigua, Colombia. Universitas Scientiarum 41(1):23–30

Stevenson, Pablo R.; Castellanos, Maria Clara; Pizarro, Juan Carlos; Garavito, Mariaximena. 2002. Effects of seed dispersal by three Ateline monkey species on seed germination at Tinigua National Park, Colombia. International Journal of Primatology 23(6): 1187–1204

Steward, A. 2013. Reconfiguring Agrobiodiversity in the Amazon Estuary: Market Integration, the Açaí Trade and Smallholders' Management Practices in Amapá, Brazil. Human Ecology 41:827–840

Tananta, Leslie Virginia. 2014. Análisis del conocimiento tradicional del uso de especies vegetales en tres comunidades de la cuenca baja del río Ucayali, Loreto, Peru. Master's thesis, UNAP, Iquitos

Targhetta, Natália; Kesselmeier, Jürgen; Wittman, Florian. 2015. Effects of the hydroedaphic gradient on tree species composition and aboveground wood biomass of oligotrophic forest ecosystems in the central Amazon basin. Folia Geobotanica 50:185–205

Taurino, Tássia Cristina da Conceição Barros. 2017. Recuperação natural da diversidade de espécies de plantas em florestas secundárias na Amazônia Oriental. Master's Thesis, Universidade Federal do Pará, Belém

Tenório, Maria Alice Ramalho de Oliveira; Van den Berg, Maria Elisabeth; Menezes, Ordoenha Furtado de; Salles, Pedro. 1991. Fitoterapia: uma estratégica terapêuticas natural do Amapá. In: Medicinas Tradicionais e Medicina Occidental na Amazônia, Buchillet, Dominique (ed), Museu Paraense Emílio Goeldi, Belém

Terborgh, J. 1983. Five New World Primates: A Study in Comparative Ecology. Princeton University Press, Princeton

Terborgh, J.; Losos, E.; Riley, M.P.; Riles, M.B. 1993. Predation by vertebrates and invertebrates on the seeds of five canopy trees species of an Amazonian forest. Vegetatio 107/8:375–386

Thomas, Evert. 2009. New light on the floristic composition and diversity of indigenous territory and national park Isiboro-Sécure, Bolivia. Biodiversity Conservation 18:1847–1878

Thomas, Evert; Van Damme, Patrick. 2010. Plant use and management in homegardens and swiddens: evidence from the Bolivian Amazon. Agroforestry Systems 80:131–152

Toledo, Narcisa Noemí Cartuche. 2016. Identificación de tipos de bosques y análisis de riqueza, diversidad y productividad en la provincia de Zamora Chinchipe. Master's thesis, Facultad de Ciencias Agropecuarias, Universidad de Cuenca, Ecuador

Toledo, Marisol; Salick, Jan. 2006. Secondary succession and indigenous management in semideciduous fallows of the Amazon basin. Biotropica 38(2):161–170

Uhl, Christopher; Clark, Kathleen; Maquirino, Pedro. 1988a. Vegetation dynamics in Amazonian treefall gaps. Ecology 69: 751–763

Uhl, C.; Buschbacher, R.; Serrão, E. A. S. 1988b. Abandoned pastures in eastern Amazonia. I. Patterns of plant succession. Journal of Ecology 76:663–681

Vale, Gabriel Damasco do. 2011. O efeito do solo, inundação e topografia sobre as campinaranas inundáveis na Amazônia. Master's Thesis, INPA, Manaus

Valencia, Renato; Balslev, Henrik; Paz y Miño, Guillermo. 1994. High tree alpha-diversity in Amazonian Ecuador. Biodiversity and Conservation 3:21–28

Valencia, Renato; Foster, Robin B.; Villa, Gorky; Condit, Richard; Svenning, Jens-Christian; Hernández, Consuelo; Romoleroux, Katya; Losos, Elizabeth; Magard, Else; Balslev, Henrik. 2004. Tree species distributions and local habitat variation in the Amazon: large forest plot in eastern Ecuador. Journal of Ecology 92(2): 214–229

Van Andel, Tinde. 2001. Floristic composition and diversity of mixed primary and secondary forests in northwest Guyana. Biodiversity and Conservation 10:1645–1682

Van Andel, Tinde; Behari-Ramdas, Joelaika; Havinga, Reinout; Groenendijk, Sara. 2007. The medicinal plant trade in Suriname. Ethnobotany Research & Applications 5:351–372

Van den Berg, Maria Elisabeth; Silva, Milton Hélio Lima da. 1988. Contribuição ao conhecimento da flora medicinal de Roraima. Acta Amazonica 18(1–2) supl.:23–35

Van den Eynden, Veerle; Cueva, Eduardo; Cabrera, Omar. 1999. Plantas Silvestres Comestibles del Sur del Ecuador. Ediciones Abya-Yala, Quito

Van Roosmalen, M. G. M. 1985. Habitat preferences, diet, feeding strategy and social organization of the black spider monkey (*Ateles paniscus paniscus* Linnaeus 1758) in Surinam. Acta Amazonica 15(3/4 suplemento) 1–238

Van Roosmalen, Marc G. M.; Klein, Lewis L. 1988. The spider monkeys, genus *Ateles*. In: Ecology and Behavior of Neotropical Primates, Mittermeier, Russell A.; Rylands, Anthony B.; Coimbra-Filho, Adelmar F.; Fonseca, Gustavo A.B. da (eds), Ecology and Behavior of Neotropical Primates, World Wildlife Fund, Washington, DC, vol. 2, pp 455–537

Van Roosmalen, Marc G. M.; Mittermeier, Russel A.; Milton, Katherine. 1981. The bearded sakis, genus *Chiropotes*. In: Ecology and Behavior of Neotropical Primates, Coimbra-Filho, Adelmar F.; Mittermeier, Russell A. (eds), Academia Brasileira de Ciências, Rio de Janeiro, vol. 1, pp 419–441

Van Roosmalen, Marc G. M.; Mittermeier, Russel A.; Fleagle, John G. 1988. Diet of the northern bearded saki (*Chiropotes satanas chiropotes*): a Neotropical seed predator. American Journal of Primatology 14:11–35

Vásquez, R. 1997. *Flórula de las Reservas Biológicas de Iquitos, Perú.* Monographs in Systematic Botany Vol. 63, Missouri Botanical Garden, St. Louis.

Vásquez, Juan Baluarte. 1998. Composición florística de los bosques secundarios de Jenaro Herrera, Loreto. Folia Amazonica 9(1–2):195–218
Vásquez, Silvia Patricia Flores. 2014. Conhecimento, uso e conservação da diversidade vegetal em quatro comunidades ribeirinhas no Município Manacapuru, Amazonas. Doctoral dissertation, Programa de Pós-Graduação em Botânica, INPA, Manaus
Veiga, Liza Maria. 2006. Ecologia e comportamento do cuxiú-preto (*Chiropotes satanas*) na paisagem fragmentada da Amazônia oriental. Doctoral Dissertation, Universidade Federal do Pará, Belém
Veléz, Germán Alonso; Veléz, Antonio José. 1992. Sistema agroflorestal de "chagras" utilizado por las comunidades indígenas del medio Caquetá (Amazonia Colombiana). Colombia Amazonica 6(1): 101–134
Veríssimo, Adalberto; Barreto, Paulo; Mattos, Marli; Tarifa, Ricardo; Uhl, Christopher. 1992. Logging impacts and prospects for sustainable forest management in an old Amazonian frontier: the case of Paragominas. Forest Ecology and Management 55: 169–199
Vester, Hans F. M.; Cleef, Antoine M. 1998. Tree architecture and secondary tropical rain forest development: a case study in Araracuara, Colombian Amazon. Flora 193:75–97
Vickers, W. T. 1994. The health significance of wild plants for the Siona and Secoya. In: Eating on the Wild Side: The Pharmalogic, Ecologic, and Social Implications of Using Noncultures, N. L. Etkin (ed.), University of Arizona Press, Tucson, pp 143–165
Vieira, Ima Célia Guimarães. 1996. Forest succession after shifting cultivation in eastern Amazonia. Doctoral dissertation, University of Stirling
Vieira, Diego dos Santos; Gama, João Ricardo Vasconcellos; Ribeiro, Renato Bezerra da Silva; Ximenes, Lucas Cunha; Corrêa, Viviane Vasconcellos; Alves, Adriana Ferreira. 2014. Comparação estrutural entre floresta manejada e não manejada na comunidade Santo Antônio, estado do Pará. Ciência Florestal 24(4):1061–1068
Vieira, Diegodos Santos; Oliveira, Marcio Leles Romarco de; Gama, João Ricardo Vasconcelos; Machado, Evandro Luiz Mendonça; Görgens, Eric Bastos; Lafetá, Bruno Oliveira; Garcia, Josinei Silva. 2017. Phytosociology of a natural fragment of the floodplain forest in the lower Tapajós River, Brazil. Bosque 38(2):357–369
Wagley, C. and E. Galvão. 1961. *Os Índios Tenetehara: Uma Cultura em Transição*. Ministério da Educação e Cultura, Serviço de Documentação, Rio de Janeiro
Wezel, A.; Ohl, J. 2006. Homegarden plant diversity in relation to remoteness from urban centers: a case study from the Peruvian Amazon region. In: Tropical Homegardens: a time-tested example of sustainable agroforestry, Kumar, B. M.; Nair, P. K. R. (eds), pp 143–158, Advances in Agroforestry vol. 3, Springer, Dordrecht
White, F. 1986. Census and preliminary observations on the ecology of the black-faced black spider monkey (*Ateles paniscus chamek*) in Manu National Park, Peru. American Journal of Primatology 11:125–132
Wittmann, Florian; Householder, Ethan; Piedade, Maria T. F.; Assis, Rafael L. de; Schöngart, Jochen; Parolin, Pia; Junk, Wolfgang J. 2013. Habitat specifity, endemism and the neotropical distribution of Amazonian white-water floodplain trees. Ecography 36:690–707
Zambrana, Narel Y. Paniagua; Bussmann, Rainer; Hart, Robbie E.; Huanca, Araceli L. Moya; Soria, Gere Ortiz; Vaca, Milton Ortiz; Álvarez, David Ortiz; Morán, Jorge Soria; Morán, María Soria; Chávez, Saúl; Moreno, Bertha Chávez; Moreno, Gualberto Chávez; Roca, Oscar; Siripi, Erlin. 2017. Traditional knowledge hiding in plain sight – twenty-first century ethnobotany of the Chácobo in Beni, Bolivia. Journal of Ethnobiology and Ethnomedicine 13:1–47
Zárate, Ricardo; Mori, Tony J.; Maco, José T. 2013. Estructura y composición florística de las comunidades vegetales del ámbito de la Carretera Iquitos-Nauta, Loreto-Perú. Folia Amazónica 22(1–2):77–89
Zárate, Ricardo; Mori, Tony Jonatan; Ramirez, Fredy Francisco; Davila, Hilda Paulette; Gallardo, George Pepe; Cohello, Geancarlo. 2015. Lista actualizada y clave para la identificación de 219 especies arbóreas de los bosques sobre arena blanca de la Reserva Nacional Allpahuayo Mishana, Loreto, Peru. Acta Amazonica 45(2):133–156
Zent, S. and E. Zent. 2012. Jodï horticultural belief, knowledge and practice: incipient or integral cultivation? *Boletim do Museu Paraense Emílio Goeldi, Ciencias Humanas* 7(2): 293–338

Chapter 25
Gnetaceae

25.1 *Gnetum leyboldii*

Brazil Ituá; coro coro (Tukano)

Colombia Aarigkeiu (Miraña)

Ecuador Washik (Shuar)

Peru Paujillo, paujil ruro, hambre huayo

Status Wild

N. Smith, *Amazon Fruits: An Ethnobotanical Journey*, Ethnobiology,
https://doi.org/10.1007/978-3-031-12803-5_25

Paujillo ***Gnetum leyboldii*** **in fruit in floodplain forest along a black water stream. Quebrada Huirurui, Rio Samiria, near San Martin Tipishca, Reserva Nacional Pacaya-Samiria, Loreto, Peru 6-30-06**

Gnetum leyboldii is a liana of floodplain and upland forests with edible seeds. It is also found in stunted forests (*caatinga*) growing on sandy soils, such as in the Upper Rio Negro watershed. Known as paujillo along the Samiria River, an affluent of the lower Marañón, the orange-red fruits contain seeds that are roasted on embers and eaten by locals in the Pacaya-Samiria National Reserve as well as in the vicinity of Jenaro Herrera along the Ucayali.

The liana extends for 30 m or more into the forest canopy, so the shiny, oblong fruits, which measure some 4.5–5 cm long by 2–3 cm in diameter, are gathered from the ground. The serpentine liana fruits during the rainy season and early part of the dry period so the fruits sometimes fall into the water, where they are likely eaten by fish and turtles. Little is known about the natural history of *Gnetum leyboldii*, but another common name for the liana in Peru is paujil ruro. Paujil is the common name for guans and curassows (several species in the genera *Crax* and *Mitu*), suggesting that those popular game birds eat the fruits in the canopy and possibly disperse the seeds.

The liana is found below 1300 m from Costa Rica south to the Orinoco and Amazon basins. Known as ituá in Brazil, the seeds of *Gnetum leyboldii* and *G. venosum* are used to bait hooks when fishing for bacu *Lithodorus dorsalis*, an armorplated catfish that feeds on fruits. In the Brazilian Amazon, flour (*farinha de ituá*) is also made from the seeds for human consumption. Two additional species of *Gnetum* in the Peruvian Amazon, *G. nodiflorum* and *G. schwackeanum*, also have edible seeds.

The fruits of two near relatives in Asia, *Gnetum gnemon* and *G. ula*, are also roasted and eaten in their native areas, and that has spurred the cultivation of those Gymnosperms (Cavalcante 1978). One day, perhaps, *Gnetum leyboldii* or one of the other five species in the genus found in the Amazon Basin may also emerge as a new crop.

Reference

Cavalcante, P.B. 1978. Contribuição ao conhecimento das Gnetáceas da Amazonia (Gimnospermas). *Acta Amazônica* 8(2): 201–215

Chapter 26
Humiriaceae

26.1 *Endopleura uchi*

Bolivia Aceitillo

Brazil Uxi, uxi amarelo, uxi preto, uxi liso, uxi pucu, uxi torrado, uxizeiro, uchi, uchi liso, uchi louro, uchi pucu, uchi torrado, fruta do pobre; kremp, pidjô-kamrek (Kayapó), muntu'úba (Piripkura), iypymp ywa (Sateré-Maué), makina hi (Yanomami)

Peru Lagarto moena, uchi

Status Wild, planted

N. Smith, *Amazon Fruits: An Ethnobotanical Journey*, Ethnobiology,
https://doi.org/10.1007/978-3-031-12803-5_26

Uxi *Endopleura uchi* in upland forest. Caxiuanã National Forest, Pará, Brazil 11-8-95

Uxi *Endopleura uchi* is a giant tree with pale gray bark that soars up to 40 m on forested uplands, sometimes becoming a canopy emergent (Almeida et al. 1993). Juveniles are also found in successionary vegetation, such as in the Trombetas watershed in Pará (Parrotta et al. 1997; Salomão et al. 2007b), on the Belterra plateau south of Santarém, Pará (Carvalho et al. 1986; Espírito-Santo et al. 2005; Gualberto et al. 2014), in the Bragantina zone in northeastern Pará (Schwartz 2007), and on Fazenda Aruanã, municipality of Itacoatiara, Amazonas (Santos and Absy 2010). This tree also occurs occasionally on floodplains, such as along the Jaci-Paraná River, an affluent of the Madeira in Rondônia (Reis 2010:34).

Uxi is found in Acre, Amazonas, Pará, and northern Mato Grosso in Brazil but is more common in Pará (Cavalcante 2010:244; Corrêa 2006:126; Claudino et al. 2015; Condé 2011:76; Gama and Pinheiro 2010; Jardim 1988; Lemos et al. 2015; Lorenzi et al. 2006:149; Oliveira and Amaral 2005; Pinheiro et al. 2007; Salomão and Rosa 1989; Salomão et al. 1988, 2007a, b, 2012; Silva et al. 1985, 1986, 2014, 2016; Silva and Rosa 1989; Souza 2012:52). A few *Endopleura uchi* trees are also found in the Bolivian Amazon (Araujo-Murakami et al. 2015) and along the Upper Essequibo in Guyana. In spite of its wide distribution, populations are declining in many areas as deforestation and selective logging accelerate (Homma 2014a, b; Silva and Bentes-Gama 2008). Uxi is the target of loggers because of its durable wood, used especially for posts and beams in buildings (Parrotta et al. 1995:114).

Uxi fruits during the rainy season and into the early part of the dry period. People gather the brown-skinned fruits, typically measuring 5–7 cm long and 3–5 cm wide, from the ground because the tree is too tall to climb and there are no low-hanging branches. Annual fruit production ranges from 300 to 4000 per tree and unlike some other wild fruit trees, mature uxi trees produce fruit every year (Shanley and Gaia 2004; Shanley et al. 2012). Some trees produce fruit during the dry season (July and August) rather than during the rainy period, while others bear two crops of fruit, one in the rainy season and the other during the dry months. The fruits are nutritious, containing high levels of carotene and vitamin C, and are also high in fiber (Magalhães et al. 2007; Marx et al. 2002; Shanley et al. 2012).

Basket made with strips of jacitara *Desmoncus* sp. palm stems containing uxi *Endopleura uchi* fruits gathered in the nearby forest. Manioc processing hut inland from Carrazedo, mouth of the Xingu River, Município de Gurupá, Pará, Brazil 4-18-15

The damson-sized fruits are peeled to consume the creamy pulp surrounding a single seed. Sometimes people bite into the skin and eat it as well because it is so thin. The light tan-colored mesocarp is gnawed off the seed or mixed with manioc flour. When unripe fruits fall to the ground, rural folk sometimes place them in a shallow ditch and cover them with twigs and earth for a few days to ripen (Shanley and Gaia 2004). River dwellers along the Rio Capim in eastern Pará also make juice from the fruits and obtain oil by boiling the fruits and skin which is used for frying food (Shanley and Rosa 2004, 2005). In the early 2000s, frozen popsicles of uxi juice were popular in Belém (Shanley et al. 2002), but locally produced popsicles and ice cream in Belém are now less common as those products are increasingly dominated by large corporations using temperate zone fruits. In the mid-nineteenth century, several indigenous groups in the region around Tefé (then called Ega) held joint festivals in honor of the uxi fruit (Bates 1863:207). But Henry Walter Bates noted that the participation in the festival was declining each year and "will no doubt soon be discontinued altogether." He was right.

Rayssa, 10 years, eating *Endopleura uchi* uxi fruit gathered in nearby forest. Carrazedo, mouth of the Xingu River, Município de Gurupá, Pará, Brazil 4-19-15

These popular fruits are sometimes taken to markets, such as Portel, Belém, and Santarém in Pará (Rego et al. 2011). In 2004, sales of uxi fruits in Belém generated the equivalent of US$1.61 million (Shanley et al. 2002). In June 1994, uxi fruits were selling for the equivalent of US$ 0.23/dozen along the waterfront at Santarém. In May 1999, a street vendor in Vigia was selling 100 uxi fruits for US$ 2.42. The selling price appears to have climbed in recent years perhaps because of the declining availability of the fruit. At Barcarena in March 2015, for example, 10 fruits were fetching US$ 0.75.

Some game animals also feast on uxi fruits, including tortoises (*Chelonoidis carbonaria* and *C. denticulata*), agoutis (*Dasyprocta* spp.), paca *Agouti paca*, and tapir *Tapirus terrestris* (Barboza 2008:56–59; Ghilardi and Alho 1990), the latter the largest terrestrial animal hunted for its meat in Amazonia. The bearded saki *Chiropotes chiropotes*, the red-nosed bearded saki *Chiropotes albinasus*, and squirrel monkeys *Saimiri sciureus* eat the immature and mature fruit of this tree, and although these monkeys are not important game animals, they likely disperse some of the seeds (Boyle et al. 2012; Pinto 2008:76; Santos et al. 2018).

Rural people use the bark to make a tea to treat arthritis, diabetes, and lower cholesterol levels (Shanley and Gaia 2004). Residents of Tiningú, a quilombo community 47 km southeast of Santarém, make a tea from the bark which is drunk to treat inflammation issues (Carvalho 2015:62). Residents of the rural community of Caruarú on Mosqueiro Island some 70 km northeast of Belém use the bark of this tree for the same purpose and also to treat diabetes (Mesquita and Tavares-Martins 2018). Bark from the trunk of this tree is sold in the river port market of Cuiabá, Mato Grosso, to make a tea for women with uterine and ovarian issues (Pinto et al. 2013). Bergenin obtained from the bark of this tree has antimicrobial activity, anti-inflammatory properties, and has been shown to inhibit the growth of infectious yeasts such as *Candida albicans*, *C. tropicalis*, and *C. guilliermondii* which are a particular nuisance to women (Borges et al. 2017; Silva et al. 2009). The endocarps have traditionally been used in some areas to fashion amulets and necklaces (Shanley and Gaia 2004). In the watershed of the Jaú, an affluent of the middle Rio Negro in Amazonas, river dwellers use the bark to make an infusion to reduce fever (Rodrigues 2006). In the Brazilian Amazon, the fruit is used to make a tea to treat inflammation of the uterus (Van den Berg 2010:211). Inhabitants of Colares island in the Amazon estuary some 100 km from Belém scrape the endocarps and add the powder to the eater to treat diarrhea among their dogs (Ritter et al. 2012).

Because of the popularity of the fruits, the tree is sometimes planted in home gardens and cultivated fields. Care is taken, however, to plant the seeds along the perimeter of the house yard so that there is less likelihood of a branch, or the tree itself, crashing down on the house in the event of a severe storm. I noted uxi in various rural home gardens in Pará such as at Laranjal, a small community along a stream of the same name that flows into the Baia de Caxiuanã, Municipio de Melgaço; Sitio Andiroba near Castanhal; Comunidade Nossa Senhora da Luz, Rio Tauaú on Marajó Island; and along Ramal Anauera near Tomé-Açu. Uxi is planted in some home gardens on several Japanese-Brazilian farms in the vicinity of Tomé-Açu, Pará (Yamada and Osaqui 2006) as well as in some rural communities in the vicinity of Manaus (Souza 2010:28) and in the municipality of Rio Preta da Eva along the Manaus-Itacoatiara highway in Amazonas (Silva 2017:30). Uxi is also planted in rural home gardens in Amazonas State, such as along the lower Maués River and along Ramal do Pau Rosa, an offshoot of the BR-175 highway north of Manaus that connects Manaus with Boa Vista in Roraima (Machado 2016:33) and in the São Raimundo district of Manaus (Barbosa 2018:42).

Uxi is also planted in some agroforestry fields, such as along Ramal Anauera near Tomé-Açu. In the vicinity of Belém, some rural folk plant uxi in forest light gaps resulting from trees that have fallen because of strong market demand for the fruit (Shanley and Gaia 2004). They also transplant seedlings in the forest, clear out some of the competing vegetation, and scatter seeds from damaged fruits onto the forest floor. The density of uxi in some forests is thus a result of human intervention. The Kayapó in the interfluvial area between the Tocantins and Xingu plant *Endopleura uchi* in some of their manioc fields so that they can return to those areas long after the tubers have been harvested to gather fruits (Posey 1985).

Izalian, 9 years, next to a young uxi *Endopleura uchi* tree planted 5 years ago in a home garden. Laranjal, Igarapé Laranjal, Baia de Caxiuanã, Município de Melgaço, Pará, Brazil 5-8-15

The distribution and density of uxi, as with many other edible fruits in Amazonia, are not entirely "natural." Uxi occurs in regrowth forests on some archaeological sites with fertile black earth. The fruits of uxi are noticeably larger than trees growing on *terra preta de índio*, as noted along the Cariá River in the Caxiuanã National Forest. On the outskirts of Comunidade Nossa Senhora da Luz along Rio Tauaú, an affluent of the Parauaú near Breves on Marajó Island, uxi is found in a cultural forest containing several other fruit and nut trees, such as bacaba *Oenocarpus distichus* and Brazil nut *Bertholletia excelsa*. Similarly, uxi is growing in a cultural forest with numerous fruit and nut trees on the outskirts of Paquiçamba, a Juruna village along the middle Xingu between Altamira and Belo Monte.

Jaiane, 9 years, with uxi *Endopleura uchi* fruits. The fruit on the right is noticeably larger (7 cm x 5 cm); it came from a tree growing on fertile anthropogenic dark earth. Vila Cedro, Rio Cariá, a left bank affluent of the lower Rio Pracupi, in turn, an affluent of the Anapu River in the Baia de Botos, Caxiuanã national forest, Pará, Brazil 5-16-15

Uxi is featured in some folktales and myths. Lucas, an 81-year-old ribeirinho who lives along the lower Tefé River, recounted an incident that happened to his grandmother when he was 9 years old. His grandmother, 65 years old at the time, went into the forest near Missão, a Catholic church along the lower Tefé, with a 20-year-old woman to gather fruits of uxi and piquiá *Caryocar villosum*. By nightfall, they had not returned so a search party was organized to look for them using flashlights, but to no avail. The next day, the search party set out again and finally located the women. The women explained that they had become disoriented. Lucas claimed that this was the work of curupira "*foi trabalho de curupira*." Curupira is a legendary, boy-like figure that patrols some forests and can lead people astray.

26.2 *Humiria balsamifera*

Brazil Meri, umiri, umiri de casca cheirosa, umiribuiuçu, humiri, uruamari, juá preta, pau mirim, mirim, mirí, mucucu, uxirana; maporotti (Baniwa), uẽgühi (Kalapalo), bà-rerek (Kayapó), muréua (Tiriyó), wehti (Tukano), asoka hi (Yanomami)

Colombia Weé-tee-go, ta-ta-weé-tee-go, ma-sá-ne-tê (Barasana), mapuluzi (Piapoco), wa-toó-moo-ko (Yukuna)

Peru Payama, humiri, umiry, quinilla negra, apacharana, camisa, loro shungo, parinarillo, manchari caspi, tangarana de altura; tsiyaroki (Campa)

Venezuela Tau ate yaɨ, jtau ade (Hotï)

Status Wild

Mother and her children gathering fruits of meri *Humiria balsamifera* on the sandy shores of the Arapiuns River, a black water affluent of the lower Tapajós River. They have walked from the nearby village of Villa Franca. Pará, Brazil 8-3-93

Humiria balsamifera is widespread in the humid lowlands of tropical South America reaching up to 1600 m in the Andean foothills in the Zamora Chinchipe province of Ecuador (Toledo 2016:29) and 1800 m in the highlands of Roraima. This tree is common in some 15- to 40-year-old successionary woodlands, such as in the Upper Rio Negro watershed in Colombia and Venezuela (Saldarriaga and Uhl 1991) and in the vicinity of Igarapé-Açu in the Zona Bragantina of northeastern Pará (Oliveira and Jardim 1998). This tree, which can reach 30 m tall, grows on the sandy banks of clear, black, and sediment-rich water courses in Amazonia, including gallery forests extending into the cerrado in the southern part of the Amazon Basin (Amaral et al. 2009; Ferreira and Prance 1998; Parolin et al. 2002; Ratter et al. 1978; Schultes 1979). It also thrives on sandy soils not subject to flooding such as along the coast of northeastern Pará (Amaral et al. 2008, 2015; Bastos et al. 2003:13; Costa Neto et al. 1996; Quaresma and Jardim 2017; Santos 2010:42) as well as inland patches of sand with *campina* or *campinarana* vegetation (Barbosa and Ferreira 2004a; Demarchi et al. 2018; Ferreira 2009:44, 55, 57, 72, 75, 78, 84, 87, 90; Ferreira et al. 2013, 2014; Gómez et al. 2013; Guimarães et al. 2016; Mendonça et al. 2014, 2017; Prance and Johnson 1992; Prance and Schubart 1978; Sanaiotti et al. 1997; Sousa 2012:35; Zappi et al. 2016). Several indigenous groups in the savanna region of Roraima plant this tree in some of their home gardens on sandy soils (Pinho 2008:38). It also grows in stunted woodlands (*cerradão*) in the southern fringes of the Amazon Basin, such as in the watersheds of the Upper Xingu and Araguaia

Rivers in Mato Grosso (Ratter et al. 1973) as well as in savannas in the southern part of Rondônia (Miranda 2000; Miranda et al. 2002), in Roraima (Dantas and Rodrigues 1982), as well as on Marajó Island at the mouth of the Amazon (Amaral et al. 2019). South of the Amazon basin, *Humiria balsamifera* is found in Bahia and southeastern Brazil (Lorenzi 2009:186). Several species of bees pollinate the flowers, such as in Roraima (Holanda 2013:58).

It produces fruit when it reaches the size of a shrub and also flourishes in patches of sandy soil dominated by stunted vegetation in upland areas away from rivers, such as in the vicinity of Manaus, in the Içana watershed, an affluent of the Upper Rio Negro, in the Upper Juruá River basin in Acre in southwestern Amazonia, and in the Trombetas watershed in Pará (Abraão et al. 2008; Alencar 1990; Anderson et al. 1975; Daly et al. 2016; Lima Filho et al. 2004; Macedo 1977; Prance 1989, 1996; Ribeiro et al. 1994). On uplands, sandy patches are endowed with stunted vegetation known in Brazil as *campina* and *campinarana* (Barbosa and Ferreira 2004a, b; Mendonça 2011:86, 99). Such stands of *Humiria balsamifera* are known as umirizal in the Rio Negro watershed (Spruce 1853). *Humiria balsamifera* also grows among rocks on hills in Roraima and in contiguous areas of Venezuela (Zent and Zent 2002a, b) and in scrub savanna (*cerrado*). This tree is also found along the sandy coast of Pará (Silva et al. 2010) as well as in some successionary vegetation on uplands, such as abandoned fields in Kayapó territory in southern Pará (Posey et al. 1984).

During the dry season, when the soft, sandy beaches of the Arapiuns, an affluent of the lower Tapajós in Pará, are exposed, mothers and their children from Vila Franca at the river's mouth wander upstream and gather the fruits from the ground and bushes. Known locally are meri, the small, black fruits, about the size of a date seed, are easy to spot on the white sand. Although the fruits are quite small, *Humiria balsamifera* is a prolific fruiter. Sometimes meri stands are dense enough in floodplain forest to be called groves (*merizal*), such as along streams flowing into the lower Arapiuns. In meri stands, the fruits are harder to find in the carpet of leaves. The fruits are born in clusters at the end of branches and when ripe fall to the ground in profusion. Meri starts producing fruit when it is a small bush and youngsters can then pick them from low branches or climb them safely. A bag or basket is typically brought along to take some of the fruits home for a later snack. The fruits of meri have been gathered in large quantities in other parts of the lower Tapajós watershed for a long time (Bates 1863:150). The Hotï in the Upper Orinoco watershed in Venezuela make seasonal camps to feast on this fruit (Zent and Zent 2002a, b).

Along the Arapiuns, most of the fruits ripen in the dry season from July to December, so birds or terrestrial mammals are likely to be the most important seed dispersal agents rather than fish. Two species of game birds, the great tinamou *Tinamus major* and the grey-winged trumpeter *Psophia crepitans*, are known to eat this fruit in French Guiana (Erard et al. 2007) and these birds also occur in the Amazon Basin. In the northern part of the Amazon Basin in Guyana brocket deer, peccaries, tortoises, macaws, and other birds as well as some monkeys eat this fruit and likely disperse some of the seeds (Cummings et al. 2018). This small tree also produces mature fruit during the dry season on sandy upland soils, such as in campina vegetation north of Manaus (Alencar 1990). Rural folk claim that esteemed game animals such as deer (*Mazama americana* and *M. gouazoubira*), collared peccary *Pecari tajacu*, and tapir *Tapirus terrestris* are fond of the fruit. The grey brocket deer *Mazama gouazoubira* has been documented eating this fruit in French Guiana (Gayot et al. 2004). Capuchin monkeys *Cebus apella*, the Santarém marmoset *Callithrix humeralifera*, the Geoffroy's saddleback tamarin *Saguinus fuscicollis nigrifrons*, and the silvery marmoset *Mico argentatus* also eat these fruits and likely disperse some of the seeds (Corrêa 2006:57; Sánchez and Levy 2013:29, 72; Stevenson and Rylands 1988; Zhang 1995). A subspecies of the marmoset, *Callithrix humeralifer intermedius* which is restricted to a relatively small area between the Aripuanã and Roosevelt rivers in Mato Grosso, are known to eat this fruit and probably disperse some of the seeds (Rylands 1979, 1981). In campina vegetation in the vicinity of Manaus, the great fruit-eating bat *Artibeus lituratus* eats the fruits of this small tree and ingests the seeds, thereby dispersing them (Macedo and Prance 1978). In French Guiana, the beautiful cock of the rock *Rupia rupicola* also disperses the seeds of this fruit (Thery and Larpin 1993). This cock of the rock also occurs in the northern part of Amazonia in Brazil, Surinam, Guyana, southern Venezuela, and eastern Colombia.

In the vicinity of Puerto Ayacucho in the Upper Orinoco of Venezuela, *Humiria balsamifera* fruits are gathered in February which is the dry season there because it is north of the equator. Most wild fruits in the Amazon ripen during the rainy season, so meri is a welcome exception. No wonder, then, that families are willing to trek some distance to obtain their fruit reward. Stands of meri are well known to locals who return year after year to the same spots to relish the fruits.

Girl gathering fruits of meri *Humiria balsamifera* on the sandy shores of the Arapiuns, a black water affluent of the lower Tapajós River. She has walked with her mother and siblings from the nearby village of Villa Franca. Pará, Brazil 8-3-93

The flesh is meager, but the skin is thin and the fruit contains only one green seed, so meri fruits are not much trouble to eat. The fruits are eaten whole and the green seed is quickly spat out so that the creamy pulp can be savored. The taste and texture of meri fruits are similar to avocado, only sweeter.

Known as umiri in the coastal areas of the Bragantina Zone in eastern Pará, the fruits were so abundant on sandy soils in the 1970s that people were able to fill their canoes with the fruit (Cavalcante 2010:243). Unfortunately, many of the *campinas* in the "salty zone" (Zona Salgada) have since been scraped barren to provide building sand for the booming cities of eastern Pará, especially Belém. Numerous roads have been built or improved in that zone in the last few decades, thereby facilitating the destruction of *campinas* and associated stands of umiri as well as other botanical resources.

Humiria balsamifera also has medicinal uses. Mature trees sometimes seep resin from the bark that is used by some rural folk as an expectorant which promotes the secretion of sputum that enhances the expulsion of mucus from air passages in the lungs (Cavalcante 2010:243). Residents of Marudá, a village along the coast of Pará at the mouth of the Amazon, prepare a tea from the bark to treat hepatitis (Coelho-Ferreira 2009) and this tea is also drunk along the coast of Pará to alleviate diarrhea

(Bastos et al. 2003:21). Also along the coast of Pará, rural folk in the vicinity of Ajuruteua use the bark to make a bathing solution to treat uterine infections (Carneiro et al. 2010). In the watershed of the Jaú, an affluent of the middle Rio Negro in Amazonas, river dwellers use the bark to make an infusion for treating gastrointestinal disturbances (Rodrigues 2006). The Barasana who live along the Piraparaná River in the Colombian Amazon dry and pulverize the bark and apply the powder to wounds to promote healing (Schultes and Raffauf 1990:214).

In the vicinity of Ajuruteua along the Bragantina coast of Pará, rural folk use the wood of this small tree to construct fishing corrals (Carneiro et al. 2010). In the rural communities of Marudá, Camapá, and Bacuriteua in the Marapanim municipality of the Zona Bragantina region of northeastern Pará, this wood is used to make roof tiles (Bastos 1995). The hard wood of this tree is also used by the Kuikuro of the Upper Xingu to make the chief's house (Fausto 2017), while the Kalapalo, who also live in the Upper Xingu, use sections of the tree trunk to honor the spirits of departed tribal leaders (Guerreiro 2011). The wood of *Humiria balsamifera* is known in the timber trade as Santos mahogany. While shipments appear to be small compared to other more valuable timber species, such as tropical mahogany *Swietenia macrophylla* and tropical cedar *Cedrela odorata*, Santos mahogany could become the target of loggers in some areas as more desirable species are taken out.

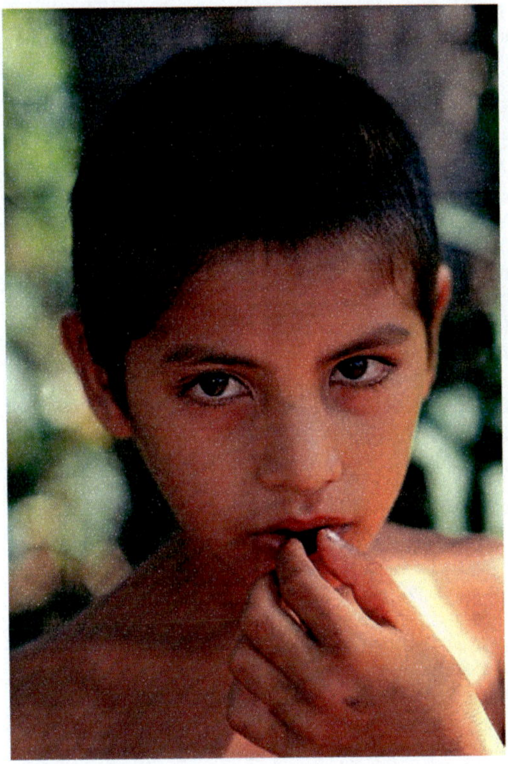

Marcelo, 9 years, eating a meri *Humiria balsamifera* fruit he gathered in a nearby floodplain forest. Near Vila Franca, lower Rio Tapajós, Pará, Brazil 11-20-02

References

Abraão, Marcia Barbosa; Nelson, Bruce W.; Baniwa, João Cláudio; Yu, Douglas W.; Shepard, Glenn H. 2008. Ethnobotanical ground-truthing: indigenous knowledge, floristic inventories and satellite imagery in the Upper Rio Negro, Brazil. Journal of Biogeography 35: 2237–2248

Alencar, Jurandyr da Cruz. 1990. Interpretação fenológica de espécies lenhosas de campina na Reserva Biológica de campina do INPA ao norte de Manaus. Acta Amazonica 20:145–183

Almeida, Samuel S.; Lisboa, Pedro Luiz B.; Silva, Antônio Sérgio. 1993. Diversidade florística de uma comunidade arbórea na estação científica "Ferreira Penna", em Caxiuanã (Pará). Boletim do Museu Paraense Emilio Goeldi, Série Botanica 9(1):93–128

Amaral, Dário Dantas do; Prost, Maria Thereza; Bastos, Maria de Nazaré do Carmo; Costa Neto, Salustiano Vilar; Santos, João Ubiratan Moreira dos. 2008. Restingas do litoral amazônico, estados do Pará e Amapá. Boletim do Museu Paraense Emílio Goeldi, Ciencias Naturais 3(1):35–67

Amaral, Dário Dantas do; Vieira, Ima Célia Guimarães; Almeida, Samuel Soares de; Salomão, Rafael Paiva; Silva, Antonio Sérgio Lima da; Jardim, Mário Augusto Gonçalves. 2009. Checklist da flora arbórea de remanescentes florestais da região metropolitana de Belém e valor histórico dos fragmentos, Pará, Brasil. Boletim do Museu Paraense Emílio Goeldi, Ciencias Naturais 4(3):231–289

Amaral, Dário Dantas do; Jardim, Mário Augusto Gonçalves; Costa-Neto, Salustriano Vilar; Bastos, Maria de Nazaré do Carmo. 2015. Síndromes de dispersão de propágulos e a influência da floresta amazônica na composição de espécies lenhosas de uma restinga no litoral norte brasileiro. Biota Amazônia 5(3):28–37

Amaral, Dário Dantas do; Rocha, Antônio Elielson; Pereira, Jorge Luis Gavina; Costa Neto, Salustiano Vilar. 2019. Identificação dos subtipos de savanas na Amazônia oriental (Pará e Amapá, Brasil) com uma chave dicotômica de individualização. Boletim do Museu Paraense Emílio Goeldi, Ciencias Humanas Naturais 14(2):183–195

Anderson, Anthony B.; Prance, Ghillean T.; Albuquerque, Byron W. P. de. 1975. Estudos sobre a vegetação das campinas Amazonicas-III A vegetação lenhosa da campina da reserva biológica INPA-SUFRAMA (Manaus-Caracaraí, km 62). Acta Amazonica 5(3):225–246

Araujo-Murakami, Alejandro; Villarroel, Daniel; Pardo, Guido; Vos, Vincent A.; Parada, G. Alexander; Arroyo, Luzmila; Killeen, Timothy. 2015. Diversidad arbórea de los bosques de tierra firme de la Amazonía Boliviano. Kempffiana 11(1):1–28

Barbosa, Cristiano de Souza. 2018. Diversidade e uso de plantas úteis nas quintais do bairro de São Raimundo, zona oeste de Manaus, AM. Master's thesis, INPA, Manaus

Barbosa, R. I. and C. A. C. Ferreira. 2004a. Densidade básica da madeira de um ecossistema de campina em Roraima, Amazônia brasileira. *Acta Amazonica* 34(4): 587–591

Barbosa, R. I. and C. A. C. Ferreira 2004b. Biomassa acima do solo de um ecossistema de "campina" em Roraima, norte da Amazônia Brasileira. Acta Amazonica 34(4):577–586

Barboza, M. S. L. 2008. O preço que a natureza pagou e os efeitos colaterais que sofremos para hoje se ter energia: uso dos recursos animais e percepção dos impactors entre os ribeirinhos do Lago de Tucuruí (PA). Master's thesis, Universidade Federal do Pará, Nucleo de Altos Estudos Amazônicas, Programa de Pós-Graduação em Desenvolvimento Sustentável do Trópico Úmido, Belém.

Bastos, Maria de Nazaré do C. 1995. A importância das formações vegetais da restinga e do manguezal para as comunidades pesqueiras. Boletim do Museu Paraense Emílio Goeldi, Série Antropologia 11(1): 41–56

Bastos, Maria de Nazaré do Carmo; Costa, Denise Cristina Torres; Santos, João Ubiratan Moreira dos. 2003. Vegetação de restinga: aspectos botânicos e uso medicinal. Museu Paraense Emilío Goeldi Projeto Renas, Belém

Bates, H. W. 1863. *The Naturalist on the River Amazons*. John Murray, London, vol. 2

Borges, Jaqueline C. Moreira; Aguiar, Raimundo W. de Sousa; Sousa, José F. de; Ripardo Filho, Haroldo S.; Guilhom, Giselle S. Pinheiro; Santos, S. Lourivaldo S.; Sousa, Pergentino

J. Cunha; Pinheiro, Bruno G.; Favacho, Hugo A. S.; Carvalho, Jose C. Tavares. 2017. Anti-inflammatory and non ulcerogenic activities of acetylbergenin. African Journal of Pharmacy and Pharmacology 11(33):402–410

Boyle, Sarah A.; Zartman, Charles E.; Spironello, Wilson R.; Smith, Andrew T.; Beck, Harald. 2012. Implications of habitat fragmentation on the diet of bearked saki monkeys in central Amazonian forest. Journal of Mammalogy 93(4):959–976

Carneiro, Diogo Borges; Barboza, Myrian Sá Leitão; Menezes, Moirah Paula. 2010. Plantas nativas úteis na Vila dos Pescadores da Reserva Extrativista Marinha Caeté-Taperaçu, Pará, Brasil. Acta Botanica Brasilica 24(4):1027–1033

Carvalho, Telma Lélia Gonçalves Schultz de. 2015. Etnofarmacologia e fisiologia de plantas medicinais do quilombo Tiningú, Santarém, Pará, Brasil. Master's thesis, Universidade Federal do Oeste do Pará, Santarém.

Carvalho, João Olegário Pereira de; Silva, José Natalino Macedo; Lopes, José do Carmo Alves; Montagner, Lise Helene; Carvalho, Manoel Sebastião Pereira de. 1986. Composição florística de uma mata secundária no planalto de Belterra no Pará. In: Anais do 1° Simpósio do Trópico Úmido, Belém, PA, 12 a 17 November, EMBRAPA, Brasília, pp 197–205

Cavalcante, P. B. 2010. *Frutas Comestíveis na Amazônia.* Museu Paraense Emílio Goeldi, Belém

Claudino, Wesley Vicente; Garcia Marcos Leandro; Lopes, Célia Regina Araújo Soares, Rodrigues, Lucirene; Cabral, Fabiana Ferreira; Piva, José Hypolito; Silva, Antônio Carlos Silveiro da. 2015. Análise fitossociológica da vegetação na região do arco do desmatamento. Revista de Ciências Agroambientais, Alta Floresta, MT 13(1): 43–52

Coelho-Ferreira, Márlia. 2009. Medicinal knowledge and plant utilization in an Amazonian coastal community of Marudá, Pará state (Brazil). Journal of Ethnopharmacology 126:159–175

Condé, Tiago Monteiro. 2011. Avaliação dos impactos da vegetação após a exploração madeireira em floresta ombrófila de terra firme no Município de Caracaraí, RR. Master's thesis, Programa de Pós-Graduação em Recursos Naturais, Universidade Federal de Roraima, Boa Vista

Corrêa, Honorly Kátia Mestre. 2006. Ecologia de dois grupos de sagüis-brancos, *Mico argentatus* (Linnaeus 1771) em um fragmento florestal natural, Santarém-Pará. Doctoral dissertation, Universidade Federal do Pará/Museu Paraense Emílio Goeldi, Belém

Costa Neto, Salustiano Vilar da; Bastos, Maria de Nazaré do Carmo; Lobato, Luiz Carlos Batista. 1996. Composição florística e fitofisionomia da restinga do Crispim, município do Marapanim, Pará. Boletim do Museu Paraense Emílio Goeldi, Série Botânica 12(2): 237–249

Cummings, Anthony R.; Read, Jane M.; Fragoso, Jose M. V. 2018. Implications of forest type and land tenure diversity for the sustainability of ecosystem services provided by northern Amazonia's multiple-use tree species. Landscape Ecology 33:423–438

Daly, Douglas C.; Silveira, Marcos; Medeiros, Herison; Castro, Wenderson; Obermüller, Flávio A. 2016. The white-sand vegetation of Acre, Brazil. Biotropica 48(1):81–89

Dantas, Mário; Rodrigues, Irenice Alves. 1982. Estudos fitoecológicos do trópico úmido Brasileiro: IV-levantamentos botânicos em campos do Rio Branco. EMBRAPA, Centro de Pesquisa Agropecuária do Trópico Úmido, Boletim de Pesquisa, no. 40

Demarchi, Layon Oreste; Scudeller, Veridiana Vizoni; Moura, Livia Carvalho; Dias-Terceiro, Randolpho Gonçalves; Lopes, Aline; Wittmann, Florian Karl; Piedade, Maria Tereza Fernandez. 2018. Floristic composition, structure and soil-vegetation relations in three white-sand soil patches in central Amazonia. Acta Amazonica 48(1):46–56

Erard, Christian; Théry, Marc; Sabatier, Daniel. 2007. Fruit characters in the diet of syntopic large frugivorous forest bird species in French Guiana. Revue d'Écologie: La Terre et la Vie 62:323–350

Espírito-Santo, Fernando Del Bon; Shimabukuro, Yosio Edimir; Aragão, Luiz Eduardo Oliveira e Cruz de; Machado, Evandro Luiz Mendonça. 2005. Análise da composição florística e fitossociológica da floresta nacional do Tapajós com apoio geográfico de imagens satélites. Acta Amazonica 35(2):155–173

Fausto, Carlos. 2017. Chefe jaguar, chefe árvore: afinidade, ancestralidade e memória no Alto Xingu. Mana 23(3):653–676

Ferreira, Carlos Alberto Cid. 2009. Análise comparativa de vegetação lenhosa do ecosistema campina na Amazonia Brasileira. Doctoral dissertation, INPA/UFAM, Manaus
Ferreira, Leandro V. and G. T. Prance. 1998. Structure and species richness of low-diversity floodplain forest on the Rio Tapajós, eastern Amazonia, Brazil. Biodiversity and Conservation 7:585–596
Ferreira, Leandro V.; Chaves, Priscilla P.; Cunha, Denise de A.; Rosário, Alessandro S. do; Parolin, Pia. 2013. A extração ilegal de areia como causa do desaparecimento de campinas e campinaranas no estado do Pará, Brasil. Pesquisas Botânica 64:157–173
Ferreira, Leandro V.; Chaves, Priscilla P.; Cunha, Denise de A.; Parolin, Pia. 2014. Florística e estrutura das campinaranas do baixo Rio Tocantins como subsídio para a criação de novas unidades de conservação no estado do Pará. Pesquisas Botânica 65:169–182
Gama, João Ricardo Vasconcellos; Pinheiro, Josiel Carneiro. 2010. Inventário florestal para adequação ambiental da Fazenda Santa Rita, Município de Santarém, Estado do Pará. Floresta 40(3):585–592
Gayot, Marc; Henry, Olivier; Dubost, Gérard; Sabatier, Daniel. 2004. Comparative diet of the two forest cervids of the genus *Mazama* in French Guiana. Journal of Tropical Ecology 20(1):31–43
Ghilardi, R.; Alho, C.J.R. 1990. Produtividade sazonal da floresta e atividade de forrageamento animal em habitat de terra firme da Amazônia. Acta Amazonica 20:61–76
Gómez, Ricardo Zárate; Vargas, Tony Jonatan Mori; Pérez, Luis Andres Valles. 2013. Composición florística, diversidad y estructura de los bosques sobre arena blanca de la Reserva Nacional Allpahuayo-Mishana, Loreto, Perú. Arnaldoa 19(2):237–247
Gualberto, Midiã Larine Colares; Ribeiro, Renato Bezerra da Silva; Gama, João Ricardo Vaconcellos; Vieira, Diego dos Santos. 2014. Fitossociologia e potencial de espécies arbóreas em ecossistema sucessional na Floresta Nacional do Tapajós, Pará. Agroecossistemas 6(1):42–57
Guerreiro, Antonio. 2011. Refazendo corpos para os mortos: as effigies mortuárias Kalapolo (Alto Xingu, Brasil). Tipití: Journal of the Society for the Anthropology of Lowland South America 9(1):1–29
Guimarães, Felipe Silva; Bueno, Guilherme Taitson; Mendes, Débora de Sena Oliveira; Souza, Jorge Batista; Diniz, Alisson Duarte. 2016. Mapeamento das unidades de paisagem natural na bacia do Rio Demini, afluente do Rio Negro – AM. Acta Geográfica 10(24):18–33
Holanda, Ana Sofia Sousa de. 2013. Estudo taxonômico de Humiriaceae no Parque Nacional do Viruá e biologia reprodutiva de duas variedades de Humiria balsamifera Aubl. Master's thesis, INPA, Manaus
Homma, A. K. O. 2014a. Agricultura na Amazônia: oportunidades de mercados e desenvolvimento regional. In: *Mercado Cadeia Produtiva e Desenvolvimento Rural na Amazônia*, A. C. Santana (ed.), Universidade Federal Rural da Amazônia, Belém, pp. 55–78
Homma, A. K. O. 2014b. Cemitério das castanheiras. In: Extrativismo Vegetal na Amazônia: História, Ecologia e Domesticação, A. K. O. Homma (ed.), EMBRAPA, Brasília, pp. 193–199
Jardim, Fernando C. da Silva. 1988. Índice de expansão florística na floresta equatorial de terra-firme. Acta Amazonica 18(3–4):211–220
Lemos, D. A. N.; Ferreira, B. G. A.; Siqueira, J. D. P.; Oliveira, M. M.; Ferreira, A. M. 2015. Floristic and phytosociology in dense "terra firme" rainforest in the Belo Monte hydroelectric plant influence area, Pará, Brazil. Brazilian Journal of Biology 75(3):257–276
Lima Filho, Diógenesde Andrade' Revilla, Juan; Amaral, Iêda L. do; Matos, Francisca Dionizia de A.; Coêlho, Luiz de Souza; Ramos, José Ferreira; Silva, Gláucio Belém da; Guedes, José de Oliveira. 2004. Aspectos florísticos de 13 hectares da área de Cachoeira Porteira-PA. Acta Amazonica 34(3):415–423
Lorenzi, H. 2009. Árvores Brasileiras: Manual de Identificação e Cultivo de Plantas Arbóreas Nativas do Brasil. Instituto Plantarum, Nova Odessa, São Paulo, vol. 2
Lorenzi, H., L. Bacher, M. Lacerda, S. Sartori 2006. Brazilian Fruits and Cultivated Exotics (for Consuming in Natura). Instituto Plantarum de Estudos da Flora Ltda, Odessa, São Paulo
Macedo, Miramy. 1977. Dispersão de plantas lenhosas de uma campina Amazônica. Acta Amazonica 7(1):1–69 (supl.)

Macedo, Miramy; Prance, Ghillean T. 1978. Notes on the vegetation of Amazonia II. The dispersal of plants in Amazonian white sand campinas: the campinas as functional islands. Brittonia 30(2):203–215

Machado, Danilo de Oliveira. 2016. A agrobiodiversidade de quintais agroflorestais em propriedades agrícolas familiares na BR 174, Ramal do Pau-Rosa, Manaus, AM. Master's Thesis, INPA, Manaus

Magalhães, Lyege Amazonas M.; Lima, Maria da Paz; Marinho, Helyde Albuquerque; Ferreira, Antonio Gilberto. 2007. Identificação de bergenina e carotenoids no fruto de uchi (*Endopleura uchi*, Humiriaceae). Acta Amazonica 37(3):447–450

Marx, Friedhelm; Andrade, Eloisa Helena A.; Zoghbi, Marias das Graças B.; Maia, José Guilherme S. 2002. Studies of edible Amazonian plants. Part 5: chemical characterization of Amazonian *Endopleura uchi* fruits. European Food Research and Technology 214:331–334

Mendonça, B. A. F. 2011. Campinaranas Amazônicas: pedogênese e relações solo-vegetação. Doctoral dissertation, Universidade de Viçosa, Viçosa

Mendonça, Bruno Araujo Furtado de; Simas, Felipe Nogueira Bello; Schaefer, Carlos Ernesto Gonçalves Reynaud; Fernandes Filho, Elpidio Inácio; Vale Júnior; José Frutuoso do Vale; Mendonça, Júlia Gaio Furtado de. 2014. Podzolized soils and paleoenvironmental implications of white-sand vegetation (campinarana) in the Viruá National Park, Brazil. Geoderma Regional 2(3):9–20

Mendonça, Bruno A. F. de; Fernandes Filho, Elpídio I.; Schaffer, Carlos E. G. R.; Mendonça, Júlia G. F. de; Vasconcelos, Bruno N. F. 2017. Soil-vegetation relationships and community structure in a "terra-firme"-white-sand vegetation gradient in Viruá National Park, northern Amazon, Brazil. Anais da Academia Brasileira de Ciências 89(2):1269–1293

Mesquita, Ulliane de O.; Tavares-Martins, Ana Cláudia. 2018. Etnobotánica de plantas medicinales en la comunidad de Caruarú, Isla de Mosqueiro, Belém-PA, Brasil. Boletín Latinoamericano y del Caribe de Plantas Medicinales y Aromáticas 17(2):130–159

Miranda, Izildinha S. 2000. Análise florística e estrutural da vegetação lenhosa do Rio Comemoração, Pimenta Bueno, Rondônia, Brasil. Acta Amazonica 30(3):393–422

Miranda, Izildinha Souza; Absy, Maria Lúcia; Rebêlo, George Henrique. 2002. Community structure of woody plants of Roraima savannahs, Brazil. Plant Ecology 164:109–123

Oliveira, Arlem Nascimento de; Amaral, Iêda Leão. 2005. Aspectos florísticos, fitossociológicos e ecológicos de um sub-bosque de terra firme na Amazônia Central, Amazonas, Brasil. Acta Amazonica 35(1):1–16

Oliveira, Francisco Plácido Magalhães; Jardim, Mário Augusto Gonçalves. 1998. Composição florística de uma floresta secundária no município de Igarapé-Açu, estado do Pará, Brasil. Boletim do Museu Paraense Emilio Goeldi, Série Botânica 14(2):127–145

Parolin, Pia; Ambruester, Nicole; Wittmann, Florian; Ferreira, Leandro; Piedade, Maria Teresa F.; Junk, Wolfgang J. 2002. A review of tree phenology in central Amazonian floodplains. Pesquisas Botânica 52:195–222

Parrotta, John A.; Francis, John K.; Almeida, Rionaldo Rolo de. 1995. Trees of the Tapajós: A Photographic Field Guide. International Institute of Tropical Forestry, General Technical Report IITF-1, Rio Piedras, Puerto Rico

Parrotta, John A.; Knowles, Oliver Henry; Wunderle Jr., Joseph M. 1997. Development of floristic diversity in 10-year-old restoration forests on a bauxite mined site in Amazonia. Forest Ecology and Management 99:21–42

Pinheiro, Klewton Adriano Oliveira; Carvalho, João Olegário Pereira de; Quantz, Beatriz; Francez, Luciana Maria de Barros; Schwartz, Gustavo. 2007. Fitossociologia de uma área de preservação permanente no leste da Amazônia: indicacão de espécies para recuperação de áreas alteradas. Floresta 37(2): 175–187

Pinho, Rachel Camargo de. 2008. Quintais agroflorestais indígenas em áreas de savana (lavrado) na Terra Indígena Araçá, Roraima. Master's thesis, INPA/UFAM, Manaus

Pinto, L. P. 2008. Ecologia alimentar do cuxiú-de-nariz vermelho *Chiropotes albinasus* (Primates: Pitheciidae) na Floresta Nacional do Tapajós, Pará. Doctoral dissertation, Universidade Estadual de Campinas, São Paulo

Pinto, Andréia Zelenski de Lara; Assis, Ana Flávia Silva de; Pereira, Arildo Gonçalo; Pasa, Maria Corette. 2013. Levantamento etnobotânico de plantas medicinais comercializados no mercado do porto em Cuiabá, Mato Grosso, Brasil. Flovet 5:51–70
Posey, D. A. 1985. Native and indigenous guidelines for new Amazonian development strategies: understanding biological diversity through ethnoecology. In: Change in the Amazon Basin Volume 1: Man's Impact on Forests and Rivers, Hemming, John (ed), Manchester University Press, Manchester, pp 156–181
Posey, Darrell A.; Frechione, John; Eddins, John; Silva, Luiz Francelinoda; Myers, Debbie; Case, Diane; MacBeath, Peter. 1984. Ethnoecology as applied anthropology in Amazonian development. Human Organization 43(2):95–107
Prance, G. T. 1989. Economic prospects from tropical rainforest ethnobotany. In: Fragile Lands of Latin America: Strategies for Sustainable Development, Browder, John O. (Editor), pp. 61–74
Prance, G. T. 1996. Islands in Amazonia. Philosophical Transactions of the Royal Society B 351:823–833
Prance, Ghillean T.; Johnson, David M. 1992. Plant collections from the plateau of Serra do Aracá (Amazonas, Brazil) and their phytogeographic affinities. Kew Bulletin 47(1):1–24
Prance, Ghillean T.; Schubart, H. 1978. Notes on the vegetation of Amazonia 1. A preliminary note on the origin of the open white sand campinas of the lower Rio Negro. Brittonia 30(1):60–63
Quaresma, Adriano Costa; Jardim, Mário Augusto G. 2017. Formations of coastal forests in the Amazon and ecological relations with vascular epiphytes. Revista Árvore 41(2):1–8
Ratter, J. A.; Richards, P. W.; Argent, G.; Gifford, D. R. 1973. Observations on the vegetation of northeastern Mato Grosso: I. The woody vegetation types of the Xavantina-Cachimbo expedition area. Philosophical Transactions of the Royal Society of London 266(880):449–492
Ratter, J. A.; G. P. Askew; R. F. Montgomery; and D. R. Gifford. 1978. Observations on the vegetation of northeastern Mato Grosso II. Forests and soils of the Rio Suiá-Missu area. *Proceedings of the Royal Society of London, Series B, Biological Sciences* 203 (1151):191–208
Rego, Lyvia Julienne Sousa; Elly, Márcio Piero; Pereira, Jéssica Lira; Garcia, Bruna Naiara; Barbosa, Humberto Figueira; Vieira, Thiago Almeida. 2011. Produtos florestais não-madeireiros comercializados em feiras de Santarém, Pará. Cadernos de Agroecologia 6(2):1–6
Reis, Neidiane Farias Costa. 2010. Estrutura fitossociológica e seletividade de espécies em áreas inundáveis na região do baixo Rio Jaci-Paraná, Rondônia. Master's Thesis, Universidade Federal de Rondônia, Porto Velho
Ribeiro, José Eduardo Lahoz S.; Nelson, Bruce Walker; Silva, Marlene Freitas da; Martins, Luisa Suely S.; Hopkins, Michael. 1994. Reserva Florestal Ducke: diversidade e composição da flora vascular. Acta Amazonica 24(1–2):19–30
Ritter, Rhuan Amorim; Monteiro, Maria Vivina Barros; Monteiro, Frederico Ozanan Barros; Rodrigues, Silvane Tavares; Soares, Marina Lira; Silva, Jean Carlos Ramos; Palha, Maria das Dores Correia; Biondi, Germano Francisco; Rahal, Sheila Canevese; Tourinho, Manuel Malheiros. 2012. Ethnoveterinary knowledge and practices at Colares island, Pará state, eastern Amazon, Brazil. Journal of Ethnopharmacology 144:346–352
Rodrigues, Eliana. 2006. Plants and animals utilized as medicines in the Jaú National Park (JNP), Brazilian Amazon. Phytotherapy Research 20:378–391
Rylands, Anthony B. 1979. Observações preliminares sobre o sagüi, *Callithrix humeralifera intermedius* (Hershkovitz, 1977) em Dardanelos, rio Aripuanã, Mato Grosso. Acta Amazonica 9(3):589–602
Rylands, Anthony B. 1981. Preliminary field observations on the marmoset, *Callithrix humeralifera intermedius* (Hershkovitz, 1977) at Dardanelos, Rio Aripuanã, Mato Grosso. Primates 22(1):46–59
Saldarriaga, J.G.; Uhl, C. 1991. Recovery of forest vegetation following slash-and-burn agriculture in the Upper Rio Negro. In: Rainforest Regeneration and Management, Gómez-Pompa, A.; Whitmore, T. C.; Hadley, M. (eds), UNESCO/Parthenon Publishing Group, Paris, pp 303–312
Salomão, Rafael de Paiva; Rosa, Nelson de Araújo. 1989. Análise da vegetação de floresta pluvial tropical de terra firme, pelo método dos quadrantes: Serra Norte, Carajás, PA. Acta Botanica Brasilica 2(1):27–42

Salomão, Rafael de Paiva; Silva, Manoela Ferreira Fernandes da; Rosa, Nelson Araújo. 1988. Inventário ecológico em floresta pluvial tropical de terra firme, Serra Norte, Carajás, Pará. Boletim do Museu Paraense Emílio Goeldi, Série Botânica 4(1):1–46

Salomão, Rafael de Paiva; Vieira, Ima Célia Guimarães; Suemitsu, Chieno; Rosa, Nélson de Araújo; Almeira, Samuel Soares de; Amaral, Dário Dantas do; Menezes, Moirah Paulo Machado de. 2007a. As florestas de Belo Monte na grande curva do rio Xingu, Amazônia Oriental. Boletim do Museu Paraense Emílio Goeldi, Ciencias Naturais 2(3):57–153

Salomão, Rafael Paiva; Rosa, Nélson Araújo; Morais, Kácio Andrey Câmara. 2007b. Dinâmica da regeneração natural de árvores em áreas mineradas na Amazônia. Boletim do Museu Paraense Emílio Goeldi, Ciencias Naturais 2(2):85–139

Salomão, Rafael de Paiva; Santana, Antônio Cordeiro; Brienza Júnior, Silvio; Gomes, Vitor Hugo Freitas. 2012. Análise fitossociológica de floresta ombrófila densa e determinação de espécies-chave para recuperação de área degradada da adequação do índice de valor de importância. Boletim do Museu Paraense Emílio Goeldi, Ciencias Naturais 7(1):57–102

Sanaiotti, Tânia M.; Bridgewater, Samuel; Ratter, James A. 1997. A floristic study of the savanna vegetation of the state of Amapá, Brazil, and suggestions for its conservation. Boletim do Museu Paraense Emílio Goeldi, Série Botânica 13(1):3–24

Sánchez, Judith Jacira Achong; Levy, Allison Licett Nuñez. 2013. Características de frutos consumidos por *Saguinus nigrifrons* (Primates, Callitrichidae), en la Estación Biológica Quebrado Blanco, Loreto, Perú. Master's thesis, Universidad Nacional de la Amazonía Peruana, Iquitos

Santos, Luciana Oliveira dos. 2010. Comparação da vegetação lenhosa entre borda e interior de uma mancha de mata de restinga na RESEX Mar Caeté-Taperaçu, Bragança, Pará. Master's thesis, Universidade Federal Rural da Amazônia/Museu Paraense Emílio Goeldi, Belém

Santos, Charles F.; Absy, Maria L. 2010. Polinizadores de *Bertholletia excelsa* (Lecythidales: Lecythidaceae): interações com abelhas sem ferrão (Apidade: Meliponni) e nicho trófico. Neotropical Entomology 39(6): 854–861

Santos, Deborah Cruz dos; Corrêa Neto, José de Jesus; Alvarenga, Adriano Braga Brasileiro de. 2018. Frutos consumidos por macaco-de-cheiro *Saimiri sciureus collinsi* (Osgood, 1916) em fragmentos de florestas em uma unidade de conservação no arquipélago do Marajó, Amazônia Oriental, Brasil. Revista Brasileira de Zoociências 19(1):82–90

Schultes, R. E. 1979. De plantis toxicariis e mundo novo tropicale commentationes. XXI. Interesting native uses of the Humiriaceae in the northwest Amazon. Journal of Ethnopharmacology 1:89–94

Schultes, R. E. and R. F. Raffauf. 1990. The Healing Forest: Medicinal and Toxic Plants of the Northwest Amazonia. Dioscorides Press, Portland, Oregon

Schwartz, Gustavo. 2007. Manejo sustentável de florestas secundárias: espécies potenciais no nordeste do Pará, Brasil. Amazônia: Ciência e Desenvolvimento 3(5):125–147

Shanley, Patricia; Gaia, Gloria 2004. Poor man's fruit turns profitable: *Endopleura uchi* in managed groves near Belem, Brazil. In: Forest Products, Livelihoods and Conservation: Case Studies of Non-Timber Product Systems, Alexiades, Miguel N.; Shanley, Patricia (eds), CIFOR, Jakarta, vol. 3, pp. 211–231

Shanley, P. and N.A. Rosa. 2004. Eroding knowledge: an ethnobotanical inventory in eastern Amazonia's logging frontier. Economic Botany 58(2):135–160

Shanley, P. and N.A. Rosa. 2005. Conhecimento em erosão: um inventário etnobotânico na fronteira de exploração da Amazônia oriental. Boletim do Museu Paraense Emílio Goeldi, Série Ciências Naturais 1(1):147–171

Shanley, Patricia; Luz, Leda; Swingland, Ian R. 2002. The faint promise of a distant market: a survey of Belém's trade in non-timber forest products. Biodiversity and Conservation 11:615–636

Shanley, Patricia; Silva, Murilo da Serra; Melo, Tadeu; Carmenta, Rachel; Nasi, Robert. 2012. From conflict of use to multiple use: forest management innovations by small holders in Amazonian logging frontiers. Forest Ecology and Management 268:70–80

Silva, Edinei Santos da. 2017. Plantas alimentícias em comunidades agrícolas no município de Rio Preto da Eva-AM. Master's thesis, INPA, Manaus

Silva, Ana Paula Ferreira Frota da; Bentes-Gama, Michelliny de Matos. 2008. Fitossociologia de uma floresta ombrófila aberta em área de assentamento rural no distrito de Jaci Paraná, Porto Velho, Rondônia. Ambiência 4(3):435–452

Silva, Manoela F. F. da; Rosa, Nelson A. 1989. Análise do estrato arbóreo da vegetação sobre jazidas de cobre na Serra dos Carajás-PA. Boletim do Museu Paraense Emílio Goeldi, Série Botânica 5(2):175–206

Silva, José Natalino Macedo; Carvalho, João Olegário Pereira de; Lopes, José do Carmo Alves. 1985. Inventário florestal de uma área experimental na Floresta Nacional do Tapajós. Boletim de Pesquisa Florestal 10/11:38–110

Silva, Manoela F. F. da; Rosa, Nelson A.; Salomão, Rafael de P. 1986. Estudos botânicos na área do Projeto Ferro Carajás. 3. Aspectos florísticos da mata do aeroporto de Serra Norte – PA. Boletim do Museu Paraense Emílio Goeldi, Série Botânica 2(2):169–187

Silva, Saulo L. da; Oliveira, Viviane Guedes de; Yano, Tomomasa; Nunomura, Ritade Cássia Saraiva. 2009. Antimicrobial activity of bergenin from *Endopleura uchi* (Huber) Cautrec. Acta Amazonica 39(1):187–192

Silva, Rachel Macedo da; Mehlig, Ulf; Santos, João Ubiratan Moreira dos; Menezes, Moirah Paula Machado de. 2010. The coastal restinga vegetation of Pará, Brazilian Amazonia: a synthesis. Revista Brasileira de Botanica 33(4):563–573

Silva, Kátia Emidio da; Martins, Sebastião Venâncio; Fortin, Marie-José; Ribeiro, Milton Cezar; Azevedo, Celso Paulo de; Ribeiro, Carlos Antonio Álvares Soares; Santos, Nerilson Terra. 2014. Tree species community spatial structure in a terra firme Amazon forest, Brazil. Bosque 35(3):347–355

Silva, Katia Emidio da; Martins, Sebastião Venâncio; Ribeiro, Carlos Antônio Alvares Soares; Santos, Nerilson Terra; Azevedo, Celso Paulo de. 2016. Structure of 15 hectares permanent plots of terra firme dense forest in central Amazon. Revista Árvore 40(4):603–615

Sousa, Maihyra Marina Pombo de. 2012. Estrutura e composição das comunidades de árvores na bacia do alto rio Negro. Master's thesis, INPA, Manaus

Souza, Célia Cristina Valero. 2010. Etnobotânica de quintais em três comunidades ribeirinhas na Amazônia central, Manaus, AM. Master's thesis, INPA, Manaus

Souza, Priscila Figueira de. 2012. Florística, distribuição e diversidade beta da comunidade arbórea ao longo do gradiente hídrico em uma floresta de terra firme no interflúvio Purus-Madeira, Amazônia central. Master's thesis, INPA, Manaus.

Spruce, R. 1853. Journal of a voyage up the Amazon and Rio Negro. *Hooker's Journal of Botany and Kew Garden Miscellan*y 5:207–215

Stevenson, Miranda F.; Rylands, Anthony B. 1988. The marmosets, genus *Callithrix*. In: Ecology and Behavior of Neotropical Primates, Mittermeier, Russell A.; Rylands, Anthony B.; Coimbra-Filho, Adelmar F.; Fonseca, Gustavo A.B. da (eds), Ecology and Behavior of Neotropical Primates, World Wildlife Fund, Washington, DC, vol. 2, pp 131–222

Thery, Marc; Larpin, Denis. 1993. Seed dispersal and vegetation dynamics at a cock-of-the-rock's lek in the tropical forest of French Guiana. Journal of Tropical Ecology 9:109–116

Toledo, Narcisa Noemí Cartuche. 2016. Identificación de tipos de bosques y análisis de riqueza, diversidad y productividad en la provincia de Zamora Chinchipe. Master's thesis, Facultad de Ciencias Agropecuarias, Universidad de Cuenca, Ecuador

Van den Berg, M. E. 2010. Plantas Medicinais na Amazônia. Museu Paraense Emílio Goeldi, Belém

Yamada, M.; Osaqui, H. M. L. 2006. The role of homegardens in agroforestry development: lessons from Tomé-Açu, a Japanese-Brazilian settlement in the Amazon. In: Tropical Homegardens: a time-tested example of sustainable agroforestry, Kumar, B. M.; Nair, P. K. R. (eds), pp. 299–316, Advances in Agroforestry vol. 3, Springer, Dordrecht

Zappi, Daniela C.; Milliken, William; Lopes, Celia Regina A. Soares; Lucas, Eve; Piva, José Hypólito; Frisby, Sue; Biggs, Nicola; Forzza, Rafaela Campostrini. 2016. Xingu State Park vascular plant survey: filling the gaps. Brazilian Journal of Botany 39(2):751–778

Zent, Egleé; Zent, Stanford. 2002a. Impactos ambientales generadores de biodiversidade: conductas ecológicas de los Hotï de la Sierra Maigualida, Amazonas Venezolano. Interciencia 27(1):9–20

Zent, Egleé; Zent, Stanford. 2002b. Los Jodï: sabios botánicos del Amazonas venezolano. Antropologica 97–98:29–70

Zhang, S. 1995. Activity and ranging patterns in relation to fruit utilization by brown capuchins (*Cebus apella*) in French Guiana. International Journal of Primatology 16(3):489–507

Chapter 27
Icacinaceae

27.1 *Poraqueiba paraensis*

Brazil Umari, umari amarela, mari; uamã (Tukano)

Colombia Umari, guacari; nemote (Witoto)

Ecuador Huacu; hua curi (Siona-Secoya)

Peru Humari, umari; nimuhe (Bora)

Venezuela Yuri

Status Wild, planted

Umari ***Poraqueiba paraensis*** **fruit from a tree planted in a home garden in a Tukano village. Cunuri Ponta, Uaupés River, Amazonas, Brazil, 11-12-12**

N. Smith, *Amazon Fruits: An Ethnobotanical Journey*, Ethnobiology,
https://doi.org/10.1007/978-3-031-12803-5_27

Umari *Poraqueiba paraensis* is a medium-sized tree in upland forests and along high banks of rivers that fruits during the rainy season (Altman et al. 1964). It mostly occurs in the state of Pará and is usually replaced further up the Amazon by a similar species, *P. sericea* (Cavalcante 2010:241; Cavalcante and Carvalho 1971; Falcão and Lleras 1980a, b). However, a few botanical specimens of *P. paraensis* have been collected in the Colombian, Ecuadorian, Peruvian, and Venezuela Amazon, where it may have been introduced by indigenous migrations or trade. I have seen *P. paraensis* cultivated in a home garden in Cunuri Ponta, a Tuyuka village along the lower Uaupés in northwestern Amazonas state in Brazil. Although umari is mostly valued for its fruit, in the early nineteenth century, the wood was used to build boats (Baena 2004:59).

Arlinho, 10 years, eating mari ***Poraqueiba paraensis*** **fruit gathered in nearby forest. Carrazedo, mouth of the Xingu River, Município de Gurupá, Pará, Brazil 4-17-15**

The fruits of *Poraqueiba paraensis*, measuring 6–8 cm long by around 4 cm in diameter, usually have an orange skin but some are yellow-orange. The fruits are much sought after by rural folk for the creamy orange pulp surrounding the single seed. The savory pulp has an oily texture reminiscent of tucumã *Astrocaryum vulgare* palm fruits. The oil content of the pulp is around 50% (Altman et al. 1964). After gnawing off the thin layer of pulp, the large, light gray seed, which occupies about 85% of the fruit by volume, is typically tossed away.

Small boat (*catraia*) tied up at a village with mari fruits for sale. Carrazedo, mouth of the Xingu River, Município de Gurupá, Pará, Brazil 4-18-15

Occasionally, however, some of the seeds are planted in raised vegetable beds (*canteiro suspenso*) so that when they sprout they can be transplanted in the home garden or in the vicinity of manioc processing huts, as is the case near Carrazedo at the mouth of the Xingu. The pulp is also scraped off and mixed with manioc flour. A Tukano lady in Manaus explained to me that in addition to eating the pulp, her tribe also soak the seed for about 2 weeks in water, mash it, and then squeeze the ground-up kernel in a sleeve-like manioc press (*tipití*). Then, the mass is mixed with manioc starch (*goma*) to make pancakes (*beiju*).

José das Graças "Poliodoro," 55 years, next to an umari tree he planted 5 years ago in his home garden. Laranjal, Igarapé Laranjal, Baia de Caxiuanã, Municipio de Melgaço, Pará, Brazil 5-8-15

Umari in a raised vegetable and spice bed (*canteiro suspenso*). Sitio Boa União, lower Rio dos Macacos, near Breves, Marajó Island, Pará, Brazil, 5-17-00

The fruits are taken to villages and towns for sale. *Poraqueiba paraensis* is particularly common in eastern Pará, and several urban localities are named after it. A district in Belém, Umarizal, is named after it but the original stands of the fruit tree there have long since given way to buildings and streets. Umarizal is the name of a small village along the Camará River in the southeastern part of Marajó Island. Marituba, a town on the outskirts of Belém, the capital of Pará, means the place of many mari, another common name for *P. paraensis* in Brazil (Venturieri 2013). Both Belém and Marituba have grown spectacularly over the last 40 years, with the urban sprawl essentially blending the cities together, and so few mari trees are left standing in that area.

Vendor with umari fruits in a street market. Vigia, Pará, Brazil 5-10-99

Game animals are also fond of umari fruits and hunters take advantage of this. They set up rudimentary platforms (*jirau*) near umari trees and wait at night with shotgun and flashlight for wild mammals to approach the fallen fruit. Villagers living in Carrazedo at the mouth of the Xingu River shoot a variety of game animals in this manner, including paca *Agouti paca*, armadillos, tapir *Tapirus terrestris*, red brocket deer known locally as veado vermelho *Mazama americana*, and the gray brock deer, called veado branco *Mazama gouazoubira*. Farmers at Carrazedo thus take care to safeguard wild fruit trees whenever possible when choosing areas to clear for planting crops.

27.2 *Poraqueiba sericea*

Brazil Umari, umari roxo, umari preto, mari, mari preto; yumári (Baniwa), pèj (Hupd'äh)

Colombia Umari, umari colorado, umari morado; niimui-e, guacure (Miraña), wama (Tatuyo)

Peru Umari; nimuhe (Bora)

Venezuela Yuri, yuri amarillo, yuri negro; dúmari (Curripaco), umari (Yeral)

Status Wild, planted

Umari *Poraqueiba sericea* fruits in a street market. Mercado Belén, Iquitos, Loreto, Peru 3-5-04

Umari *Poraqueiba sericea* is cultivated in groves, agroforestry fields, and home gardens on uplands in central and western Amazonia (Ducke 1925). It is found mostly in cultural settings, but is also encountered occasionally in the forest on uplands and as well as on high banks along rivers, successionary vegetation, and *campinarana* on sandy soils. Umari is encountered wild in upland forests, such as in the vicinity of the confluence of the Tigre and Marañón Rivers in Loreto, Peru, but over a century ago a botanist pointed out that it is hard to tell whether umari is truly wild when encountered in the forest (Huber 1904). Indeed, umari is considered by some to be an indicator of anthrosols, known as *terra preta de índio* in Brazil (Mora 2001). The distinction between truly wild populations and those that are a vestige of human interventions on the landscape is thus blurred.

Jessica, 12 years, eating a black-skinned umari fruit harvested from a cultivated tree near Tamshiyacu along the Amazon River. Arequipa, Rio Yanayacu, Reserva Nacional Pacaya-Samiria, Loreto, Peru 3-8-04

The tree, which can reach 30 m high, fruits in the rainy season. The fruits are similar in size and shape to its close relative *P. paraensis* except the skin color varies from yellow to red and black (Pennington et al. 2004:468; Reynel et al. 2003:164). Sometimes, the fruits of *P. sericea* are a little more cylindrical in shape and longer, reaching some 10 cm (Brack Egg, 1999:405). A farmer along the Iquitos-Nauta highway cultivates a large 10 cm long variety of the fruit which has a rust-colored skin. This variety is known locally as Huitoto, possibly selected by the Witoto indigenous peoples who inhabit the forest along the Amazon straddling both Peru and Colombia.

Liz Karen, 9 years, with umari fruit gathered in her home garden. Mishana, Rio Nanay, Loreto, Peru 1-8-06

The pulp surrounds a large tan-colored seed which is discarded after gnawing off the mesocarp. The Hupd'äh, also known as Nukak, are predominately hunters and gatherers in the forest in northwestern Amazonas and contiguous Colombia and call areas with lots of umari tree pèj purúb (Ramirez 2006). It is possible that these concentrations of umari in the forest may be vestiges from discarded seeds around Hupd'äh camps. The Tukano in the Colombian Amazon make a juice from the pulp which they ferment (Reichel-Dolmatoff 1996:106). Because of its cultural importance, umari has become entwined in folklore, such as among the Ticuna (Nimuendajú 1952:127). The Tukano believe that a mythical tapir is the master of umari.

This fruit is popular in both rural and urban areas. Umari fruits are common in street markets in the Peruvian Amazon, such as Iquitos, Pucallpa, and Huanaco in the Andean foothills. The fruits are relatively cheap; at Nauta along the lower Marañón River, for example, umari fruits were selling for the equivalent of US$ 1 for 126 fruits in March 2004. Umari is even featured in desserts at some high-end restaurants in Lima. *Guargueros rellenos con manjar y mousse de umari*, for example, is a funnel-shaped pastry filled with cooked condensed milk and umari mousse (Gutsche 2008).

Umari is frequently encountered in home gardens, such as at Santa Rosa along the lower Ucayali River in Peru (Padoch and de Jong 1991). When planted as a monocrop, as on the outskirts of Tamshiyacu along the Amazon River a little upstream from Iquitos, the fruits are mostly destined for urban markets. Plantations of this fruit tree are known as umaral in the Peruvian Amazon and the trees start producing fruit after only 4 years. They continue to be productive for some 70 years or more. Umari is also interplanted with other perennial crops in fields, such as near Miraflores at the mouth of the Tigre River in the Peruvian Amazon. Villagers residing in Tamshiyacu along the Amazon River upstream from Iquitos intercrop *Poraqueiba sericea* along with other fruit and nut trees that keep some agroforestry plots productive for several decades (Padoch et al. 1985). The Bora in the Peruvian Amazon cultivate *P. sericea* in their fields such that it often dominates the fallow vegetation (Denevan and Treacy 1987), a similar pattern noted in the vicinity of Iquitos with farmers of mixed ancestry (Hiraoka 1986). The Tuyuka along the Tiquié, an affluent of the Uaupés in the northwestern part of the Brazilian Amazon, also intercrop umari in some of their agricultural plots (Schmidt et al. 2010). The Ticuna along the Upper Solimões also cultivate umari in their fields (Acosta 2001).

References

Acosta, L. E. 2001. Los sistemas de producción de la etnia Ticuna del resguardo de Puerto Nariño, sur del Trapecio Amazónico: una aproximación socioeconómica. *Cadernos de Desarrollo Rural* 46:101–132

Altman, R. F. A., P. C. Oliveira, G. Enilson, M. L. Silva, and M. C. Neves. 1964. Estudos sobre o fruto de "umari" (*Poraqueiba paraensis*, Ducke e *P. sericea*, Tul.). Instituto Nacional de Pesquisas da Amazônia, Manaus, Publicação 8

Baena, Antônio Ladislau Monteiro 2004. Ensaio Corográfico sobre a Província do Pará. Edições do Senado Federal, Brasilia
Brack Egg, A. 1999. Diccionario enciclopédico de plantas útiles del Perú.
Cavalcante, P. B. 2010. *Frutas Comestíveis na Amazônia*. Museu Paraense Emílio Goeldi, Belém
Cavalcante, P. B. and M. J. C. Carvalho. 1971. O gênero *Poraqueiba* (Icacinaceae) na Amazônia. *Boletim do Museu Paraense Emílio Goeldi*, Nova Série Botânica 39:1–10
Denevan, W. M. and J. M. Treacy. 1987. Young managed fallows at Brillo Nuevo. In: *Swidden-Fallow Agroforestry In the Peruvian Amazon*, W. M. Denevan and C. Padoch (eds.), Advances in Economic Botany vol. 5, New York Botanical Garden, pp. 8–46
Ducke, A. 1925. Plantes nouvelles ou peu connues de la région amazonienne (III partie). Archivos do Jardim Botanico do Rio de Janeiro 4:1–341.
Falcão, M. and E. Lleras. 1980a. Aspectos fenológicos, ecológicos e da produtividade do umari (*Poraqueiba sericea* Tulasne). Acta Amazonica 10(3):445–462
Falcão, M. and E. Lleras. 1980b. Aspectos fenológicos, ecológicos e de produtividade do mapati (*Pourouma cecropiifolia* Mart.). Acta Amazonica 10(4):711–724
Gutsche, A. 2008. *Frutas Amazonicas: Postres Peruanas de Vanguardia*. Wust Ediciones, Lima
Hiraoka, M. 1986. Zonation of mestizo riverine farming systems in northeast Peru. *National Geographic Research* 2(3):354–371
Huber, J. 1904. Notas sobre a patria e distribuição geographica das arvores fructiferas do Pará. *Boletim do Museu Goeldi (Museu Paraense) de Historia Natural e Ethnographia* 4 (2–3):375–406
Mora, S. 2001. Suelos negros y sociedad: un sistema agrícola de entonces, un sistema agrícola de ahora? In: *Desarrollo Sostenible en La Amazonía: ¿Mito o Realidad?*, M. Hiraoka and S. Mora (eds.), Ediciones Abya Yala, Quito, pp. 31–45
Nimuendajú, C. 1952. *The Tukuna*. University of California Press, Berkeley
Padoch, C. and W. de Jong. 1991. The house gardens of Santa Rosa: diversity and variability in an Amazonian agricultural system. *Economic Botany* 45(2):166–175
Padoch, C. J.; J. Chota Inuma; W. de Jong; and J. Unruh. 1985. Amazonian agroforestry: a market-oriented system in Peru. Agroforestry Systems 3:47–58
Pennington, T. D., C. Reynel, and A. Daza. 2004. *Illustrated Guide to the Trees of Peru*. David Hunt, Sherborne, England
Ramirez, H. 2006. *A Língua dos Hupd'äh do Alto Rio Negro*. Saúde Sem Limites, São Paulo
Reichel-Dolmatoff, Gerardo. 1996. The Forest Within: The World-View of the Tukano Amazonian Indians. Themis Books, Darlington
Reynel, C; Pennington, T.D.; Pennington, R.T.; Flores, C.; Daza, A. 2003. *Árboles Útiles de la Amazonía Peruana y sus Usos: Un Manual con Apuntes de Identificación, Ecología y Propagación de las Especies*. Tarea Gráfica Educativa, Lima
Schmidt, M. V. C.; Vander Veld, Pieter; Ramos, José Barreto; Rezende, João Bosco Azevedo; Rezende, Odilon Barreto; Rezende, Renato Barreto; Resende, Alexandre Azevedo; Cabalzar, Aloisio. 2010. Capoeiras Tuyuka: processos de restauração e recuperação de terras degradadas. In: *Manejo do Mundo: Conhecimentos e Práticas dos Povos Indígenas do Rio Negro*, A. Cabalzar (ed.), Instituto Socioambiental, São Paulo, pp. 156–167
Venturieri, G.A. 2013. Indigenous strategies used to domesticate plants in Brazilian Amazon. In: Encyclopedia of Biodiversity, Levin, S. A. (ed), Academic Press, Waltham, pp. 279–292

Chapter 28
Lamiaceae

28.1 *Vitex orinocensis*

Brazil Tarumã preto, tarumã do igapó, tarumã frondoso, aquariquari branca

Colombia Guayacancito, flor morado, aceituna, aceituno, guarataro, nocuito

Peru Cormiñón, aceituna del monte, piñaraño, sacha requia, pali perro

Venezuela Guarataro; braki hi, mamo uhe (Yanomami)

Status Wild, spontaneous in cultural settings, planted

Leaves and fruits of cormiñón *Vitex orinocensis*. **San Carlos, Rio Puinahua (a branch of the Ucayali), Reserva Nacional Pacaya-Samiria, Loreto, Peru, 1-12-06**

N. Smith, *Amazon Fruits: An Ethnobotanical Journey*, Ethnobiology,
https://doi.org/10.1007/978-3-031-12803-5_28

Known as cormiñón along the lower Ucayali in Peru, *Vitex orinocensis* is a medium-sized tree, reaching 20 m high when fully mature, that grows in successionary forests on floodplains and occasionally on uplands as well, such as in the Bragantina zone in northeastern Pará in Brazil (Rayol et al. 2006, 2011). This tree is found in many tropical parts of northern South America from the Orinoco and Amazon basins to the Atlantic rainforest of Brazil. Abundant crops of shiny black olive-sized fruits appear in the latter part of the dry season and the early part of the rainy period. At such times, the ground beneath *V. orinocensis* is carpeted with thousands of fruits, and people, especially children, pick them up eagerly.

On floodplains, at least two highly esteemed fish, tambaqui *Colossoma macropomum* and bacu catfish *Lithodoras dorsalis*, eat the fruit and disperse the seeds (Parolin et al. 2010). The largest turtle in the Amazon, tartaruga *Podocnemis expansa*, also eats the fruits. People eat tartaruga turtles as well as their eggs which they dig up on sandy beaches during the later rainy season (Best 1984). Collared peccaries *Pecari tajacu*, a popular game animal, also eats this fruit (Beck 2005a). Wedge-capped capuchin monkeys *Cebus olivaceus*, which are found in Venezuela, the Guianas, and Amapá and northern Pará in the Brazilian Amazon, also eat this fruit (Robinson 1984, 1986). Although capuchin monkeys are rather small and not usually hunted for their meat, they likely disperse the seeds of this useful fruit.

Children picking up fallen fruits of cormiñón *Vitex orinocensis* **in a village home garden. San Carlos, Rio Puinahua (a branch of the Ucayali), Reserva Nacional Pacaya-Samiria, Loreto, Peru, 1-12-06**

The prune-colored pulp of *Vitex orinocensis* is savory rather than sweet, reminiscent of *Byrsonima crassifolia*, a widely planted fruit tree in tropical America and the Caribbean. The single seed is typically discarded on the ground after ingesting the pulp, and some of them sprout spontaneously. Several other species of *Vitex* produce edible fruits in the Peruvian Amazon (Brack Egg 2003:525) and the fruits of tarumã *Vitex cymosa* are used for fish bait in the Brazilian Amazon.

The juicy, sweet-tasting fruits are widely appreciated, so cormiñón is often spared when clearing forests for fields and home sites. *Vitex orinocensis* is planted and also sprouts spontaneously in home gardens from discarded seeds. It is thus found in many home gardens in villages in the Peruvian Amazon, including San Carlos along the Rio Puinahua, an arm of the Ucayali; Monte Bello at the southern edge of the Pacaya-Samiria reserve along the Ucayali; and San Francisco, a Shipibo village on the shores of Yarina Cocha near Pucallpa. The fruits occasionally turn up in markets, such as in Pucallpa where they were selling at the bargain price of 100 for US $1 in January 2004. In the Brazilian Amazon, some residents of Imperatriz in Maranhão close to the border of eastern Pará prepare an infusion with the leaves of this plant to treat diarrhea (Penido et al. 2016).

Cormiñón *Vitex orinocensis* **fruits for sale in a street market. Pucallpa, Rio Ucayali, Peru 1-22-04**

The wood of *Vitex orinocensis* is very durable and some rural folk employ it for posts to support their houses and for fencing, such as in the village of San Carlos along the Rio Puinahua. Rural folk in the Brazilian Amazon also use the timber from this tree for a variety of purposes (Le Cointe 1934:429). The tree is also sought out by commercial loggers.

References

Beck, H. 2005a. Seed predation and dispersal by peccaries throughout the Neotropics and its consequences: a review and synthesis. In: Seed Fate: Predation, Dispersal and Seedling Establishment, Forget, Pierre-Michel; Lambert, Joanna E.; Hulme, Philip E.; Wall, Stephen B. Vander (eds), CABI, Wallingford, pp 77–115

Beck, H. T. 2005b. Caffeine, alcohol, and sweeteners. In: *The Cultural History of Plants*. Prance, G.T. and Nesbitt, M (eds), Routledge, New York, 173–190

Best, R.C. 1984. The aquatic mammals and reptiles of the Amazon. In: The Amazon: Limnology and Landscape Ecology of a Mighty Tropical River and its Basin, H. Sioli (ed), W. Junk, Dordrecht, pp 371–412

Brack Egg, Antonio. 2003. Frutas del Perú. Universidad San Martin de Porres, Lima

Le Cointe, P. 1934. *Amazônia Brasileira III: Árvores e Plantas Úteis*. Livraria Classica, Belém

Parolin, P., D. Waldhoff, and M. T. Piedade. 2010. Fruit and seed chemistry, biomass and dispersal. In: *Amazonian Floodplain Forests: Ecophysiology, Biodiversity and Sustainable Management*. W. J. Junk, M. T. Piedade, F. Wittman, and J. Schöngart (eds), Springer, Dordrecht, pp. 243–258

Penido, Alexandre Batista; Morais, Selene Maia de; Ribeiro, Alan Bezerra; Silva, Ana Zélia. 2016. Ethnobotanical study of medicinal plants in Imperatriz, state of Maranhão, northeastern Brazil. Acta Amazonica 46(4):345–354

Rayol, Breno Pinto; Silva, Manoela Ferreira Fernandes da; Alvino, Fabrízia de Oliveira. 2006. Dinâmica da diversidade florística da regeneração natural de florestas secundárias, no município de Bragança, Pará, Brasil. Boletim do Museu Paraense Emílio Goeldi, Ciencias Naturais 1(2):9–27

Rayol, Breno Pinto; Alvino-Rayol, Fabrízia de Oliveira; Silva, Manoela Ferreira Fernandes da. 2011. Similaridade florística entre o estrato arbóreo e a regeneração natural de uma floresta secundária, no município de Bragança, nordeste do estado do Pará, Brasil. Revista Brasileira de Agroecologia 6(3):107–114

Robinson, J. G. 1984. Diurnal variation in foraging and diet in the wedge-capped capuchin *Cebus olivaceus*. Folia Primatologica 43:216–228

Robinson, J. G. 1986. Seasonal variation in use of time and space by the wedge-capped capuchin monkey, *Cebus olivaceus*: implications for foraging theory. Smithsonian Contributions to Zoology Number 431, Washington, DC.

Chapter 29
Lauraceae

29.1 *Nectandra amazonum*

Bolivia Laurel negro, negrillo

Brazil Louro, louro da várzea, louro do igapó, louro da beira do rio, louro da beirada, louro da praia, louro abacate, louro aritu, louro chumbo, louro branco, louro sucena, louro pimenta, louro bosta, lourinho

Status Wild

N. Smith, *Amazon Fruits: An Ethnobotanical Journey*, Ethnobiology,
https://doi.org/10.1007/978-3-031-12803-5_29

Louro *Nectandra amazonum* fruits in a calabash *Crescentia cujete* bowl. The farm boy gathered the fruits in a nearby floodplain forest along the Amazon for use as fish bait. Paraná Cachoeri near Oriximiná, Pará, Brazil 6-20-94

Nectandra amazonum is a tree of floodplain forests widely distributed in the Amazon Basin and the Guianas. This relatively short tree usually reaches no higher than 24 m and grows along sediment-rich, clear, and black water rivers, such as in the watershed of the Jaú River, a black water affluent of the middle Rio Negro in Amazonas (Parolin and Worbes 2000), in floodplain forests along the lower Rio Negro (Piedade et al. 2005; Scudeller and Vegas-Vilarrúbia 2018), and in floodplain forests along sediment-rich rivers (Parolin 2000a, c, 2002). It is also found in some estuarine forests, such as in Amapá (Freitas et al. 2019). The leaves of this tree can remain submerged for several months without damage (Parolin et al. 2004a, b). It starts producing fruit when only 3 or 4 m high. Known as louro in Brazil, the fruits are easy to collect and are used to bait hooks to fish. A wide variety of prized fish eat louro fruits including pacu (species of *Mylossoma*, *Metynnis*, and *Myleus*), aracu (*Leporinus* spp.), sardinha *Triportheus elongatus*, matrinxã *Brycon amazonicus*, and tambaqui *Colossoma macropomum* (Albuquerque 2015:55; Goulding et al. 1988:54; Gottsberger 1978; Lucas 2008; Silva et al. 2003). Each fruit contains a single seed and certain species of catfish are thought to be involved in dispersing louro seeds (Waldhoff et al. 1996).

The tree is well adapted to wetlands because saplings can maintain photosynthesis even while submerged (Parolin et al. 2010). Louro is thus a tree of floodplain forests and the fruits are usually gathered by cutting some of the smaller branches with a machete or large knife and taking them home. The fruits are typically collected from canoes. The mostly green fruits are colored yellow around the area they are attached to the tree; at full maturity, the fruits turn black. About the size of a small olive, people do not eat the fruits. The wood, however, is used for a variety of construction purposes (Loureiro and Lisboa 1979; Wittmann et al. 2009, 2010:80).

Farm boy on the front porch of his home stripping louro *Nectandra amazonum* fruits from branches he collected in nearby floodplain forest along the Amazon. He will use the fruits for fish bait. Paraná Cachoeri near Oriximiná, Pará, Brazil 6-20-94

References

Albuquerque, Bianca Weiss. 2015. Frugivoria e ictiocoria em uma área de várzea na Amazônia Central brasileira. Master's thesis, INPA, Manaus

Freitas, João da Luz; Santos, Erick Silva dos; Cruz Junior, Francisco de Oliveira; Borges, César Henrique Alves; Santos, Adriano Castelo dos. 2019. Structural analysis and distribution patterns in lowland tropical forest, eastern Amazon. Journal of Agricultural Studies 7(4): 240–255

Gottsberger, Gerhard. 1978. Seed dispersal by fish in the inundated regions of Humaitá, Amazonia. *Biotropica* 10 (3): 170–183

Goulding, Michael; Carvalho, Miriam Leal; Ferreira, Efrem G. 1988. Rio Negro, Rich Life in Poor Water: Amazonian Diversity and Foodchain Ecology as Seen through Fish Communities. SPB Academic Publishing, The Hague

Loureiro, Arthur A.; Lisboa, Pedro L. Braga. 1979. Madeiras do município de Aripuanã e suas utilidades (Mato Grosso). Acta Amazonica 9(1: supl.):1–88

Lucas, C. M. 2008. Within flood season variation in fruit consumption and seed dispersal by two characin fishes of the Amazon. *Biotropica* 40(5): 581–589.

Parolin, P. 2000a. O uso de árvores nas florestas inundadas por água branca na Amazônia Central. Amazoniana 16(1/2):241–248

Parolin, P. 2000b. Phenology and C02-assimilation of trees in Central Amazonian floodplains. Journal of Tropical Ecology 16:465–473

Parolin, P. 2002. Submergence tolerance vs. escape from submergence: two strategies of seedling establishment in Amazonian floodplains. Environmental and Experimental Botany 48:177–186

Parolin, Pia; Worbes, Martin. 2000. Wood density of trees in black water floodplains of Rio Jaú National Park, Amazonia, Brazil. Acta Amazonica 30(3):441–448

Parolin, P.; Simone, O. De; Haase, K.; Waldhoff, D.; Rottenberger, S.; Kuhn, U.; Kesselmeier, J.; Kleiss, B.; Schmidt, W.; Piedade, M. T. F.; Junk, W. J. 2004a. Central Amazonian floodplain forests: tree adaptations in a pulsing system. The Botanical Review 70(3):357–380

Parolin, Pia; Ferreira, Leandro Valle; Albernaz, Ana L. K. M.; Almeida, Samuel S. 2004b. Tree species distribution in várzea forests of Brazilian Amazonia. Folia Geobotanica 39:371–383

Parolin, P., D. Waldhoff, and M. Zerm. 2010. Photochemical capacity after submersion in darkness: How Amazonian floodplain trees cope with extreme flooding. *Aquatic Botany* 93:83–88

Piedade, M. T. F.; Junk, W. J.; Adis, J.; Parolin, P. 2005. Ecologia, zonação e colonização da vegetação arbórea das ilhas Anavilhanas. Pesquisas Botânica 56:117–144

Scudeller, Veridiana Vizoni; Vegas-Vilarrúbia, Teresa. 2018. Distribution and B-diversity of trees species in igapó forests (Negro River basin, Brazilian Amazon). Journal of Vegetation Science 29:1052–1064

Silva, J. M., M. Pereira-Filho, and M. I. Oliveira-Pereira. 2003. Valor nutricional e energético de espécies vegetais importantes na alimentação do tambaqui. *Acta Amazonica* 33(4):687–700

Waldhoff, D., U. Saint-Paich, and F. Bodo. 1996. Value of fruits and seeds from the floodplain forests of central Amazonia as food resource for fish. *Ecotropica* 2:143–156

Wittmann, Florian; Schöngart, Jochen; Queiroz, Helder L.; Wittmann, Astrid de Oliveira; Conserva, Auristela dos Santos; Piedade, Maria T. F.; Kesselmeier, Jürgen; Junk, Wolfgang J. 2009. The Amazon floodplain demonstration site: sustainable timber production and management of central Amazonian white-water floodplains. Ecohydrology & Hydrobiology 9(1):41–54

Wittmann, F. W.; Schongart, Jochen; Brito, Joneide M. de; Wittman, Astride de Oliveira; Piedade, Maria Teresa 2010. Manual of Trees from Central Amazon Várzea Floodplains: Taxonomy, Ecology, and Use. INPA, Manaus

Chapter 30
Lecythidaceae

30.1 *Allantoma lineata*

Brazil Cerú, ceruzeiro, serú, seruzeiro, tauarí, churú, xurú, cachimbeira, curanáou tauri

Colombia Badamimeenio

Venezuela Tabari

Status Wild

N. Smith, *Amazon Fruits: An Ethnobotanical Journey*, Ethnobiology,
https://doi.org/10.1007/978-3-031-12803-5_30

Cerú *Allantoma lineata* tree in fruit. The lids of the capsules have come off and the black seeds have fallen into the water. Rio Caxiuanã, Floresta Nacional Caxiuanã, Município de Melgaço, Pará, Brazil 5-9-15

Allantoma lineata is one of the wetland representatives of the Brazil nut family (Lecythidaceae) with edible nuts. Known as cerú or tauarí in the Brazilian Amazon, the taste of the slender black nuts is similar to almonds, but the nuts are only 5 cm long and are covered by a blackish-brown skin that has to be peeled off before eating. Cerú is a medium-sized tree of floodplain forests that are inundated annually upstream from the mouth of the Amazon or daily in the estuary itself (Almeida and Jardim 2011; Almeida et al. 2004; Amaral et al. 2009; Bentes-Gama et al. 2002a, b; Black et al. 1950; Freitas et al. 2019; Queiroz et al. 2006, 2007; Prance 1978; Santos et al. 2016b). This tree is found on the floodplains of various water types including acidic black water (Ferreira et al. 2005, 2013; Lobo 2017:64). This tree can reach 35 m high and is found in Amazonas and Pará in Brazil as well as the Upper Rio Negro and Orinoco watersheds in Venezuela (Chih-Hua and Mori 2002; Mori and Lepsch-Cunha 1995). It is particularly common in the Amazon estuary. The tree occasionally occurs in upland forests, usually close to rivers, and in *campinaranas* on sandy soils, but it is never common in *terra firme* environments.

Jucineide, 12 years, holding cerú ***Allantoma lineata*** **capsules. The lids have fallen off and the seeds were scattered in the floodplain forest. Baia de Caxiuanã, Anapú River, Pará, Brazil 5-6-15**

Cerú produces fruits, technically capsules, with a lid that falls off when the nuts ripen during the rainy period. Fruit maturation coincides with the high water season, thus facilitating the dispersal of the buoyant seeds by currents (Matta and Scudeller 2012; Prance and Mori 1978). The 10- to 15-cm-long capsule looks like a narrow bell with a lid, known technically as an operculum and as *embigo* in the Brazilian Amazon, on the bottom. When ripe, the wooden lid falls off scattering the nuts, usually four per capsule, into the water. If the nuts, which are 5–6 cm long, fall on the ground and manage to escape detection by predators such as feral pigs, peccaries, agoutis, and opossums, they are picked up by the next high tide and are carried off. Some of the nuts are eaten by fish, such as puffer fish (known locally as maiacu or baiacu) before they can be deposited by the outgoing tide.

On Marajó Island at the mouth of the Amazon, kids harvest cerú by one of two means. The most common method is to scoop the nuts out of the water with a pole. A plastic bottle is cut in half and the bottom part is attached to the end of a pole so that the nuts can be fished out of the water from a dock. Cerú nuts are also picked out of the water by hand from canoes; the nuts often accumulate in flotsam along the water's edge. If the children have been diligent enough, or have enough left over after snacking on the job, some of the nuts are stored at home for the rest of the family.

Girl from a cowboy family eating a cerú *Allantoma lineata* nut. Rio Flechal, Marajó Island, Pará, Brazil 5-21-99

Cerú nuts are good sources of dietary iron, zinc, and manganese (Marx et al. 2000). The nuts are also high in selenium, with just 100 g containing the recommended daily intake for the mineral. Some people, especially children, eat considerably more than 100 g of ceru nuts in a day, but "excessive" consumption of the nuts is not considered dangerous (Andrade et al. 1999b). Furthermore, as in the case with all nuts, cerú nuts are also high in protein. The high oil content of ceru nuts and the corky coating helps them float for up to 3 months (Ducke 1949; Mori et al. 2017; Prance and Mori 1978, 1979), and also provides energy. Although the nuts are small, youngsters and adults do not tire in peeling off the skin and eating them; some rural families keep large calabash bowls full of cerú nuts in their homes in the vicinity of Muaná on Marajó Island. Cerú nuts may not make it to market, but they nevertheless enhance rural diets. In the Upper Rio Orinoco watershed, rural people use the bark to make cigarette wrappers (Prance and Mori 1979).

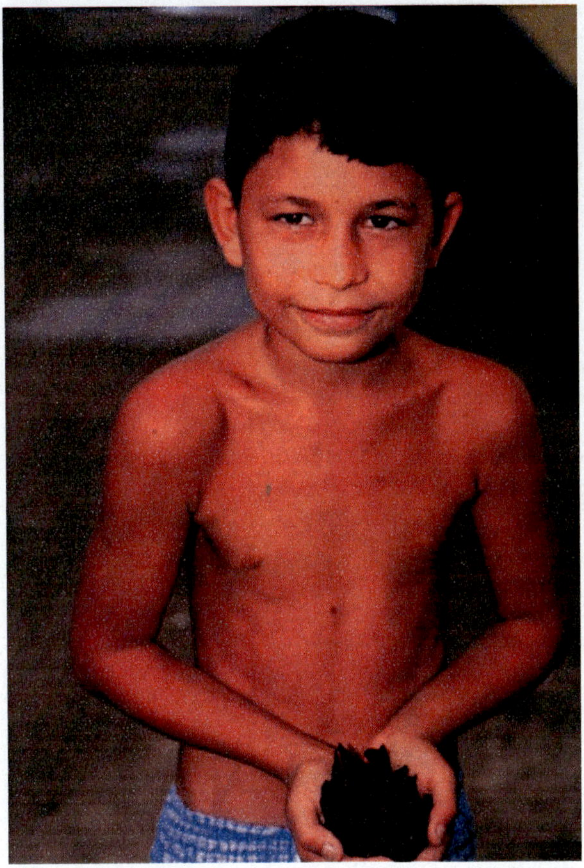

Boy with cerú *Allantoma lineata* nuts he gathered using a canoe. Muaná, Marajó Island, Pará, Brazil 5-20-99

30.2 *Bertholletia excelsa*

Bolivia Castaña, castaño, almendra, almendro, nuez amazónica, árbol de castaña; tapa, tapa ristí, tsixo (Chácobo), xiwiwi (Ese Eja)

Brazil Castanheira, castanha do Pará, castanha do Brasil, castanha da Amazônia, castanha de ouriço, maranhota, tocari; kol-re (Apinajé), toimo (Arara), yá'i (Arawaté), ja'ɨwa (Asurini), maneéh (Hupd'äh), kãtãi-ɨ (Ka'apor), minata (Kamayurá), mijo (Karitiana), ï'wa-ete (Kayabí), pi'y, pidjô-ti (Kayapó), erái (Kuruaya), arao (Makurap), txipii (Matanawi), mam (Mondé), wenïjl wenã (Munduruku), tihii (Mura), wana', wanakka (Nambikwara), txu (Parakanã), ña'i

(Parintintin), jã (Piripkura), pitsi (Rikbaktsa), we'euja, weenha (Sateré-Maué), magab (Suruí), teko-ingwer, zapuka-z-a'i (Tembé), ñoó, nho'õ (Ticuna), buhkú ba'a se (Tukano), tetky (Waimiri-Atroari) katã'e, yã (Wajãpi), iniá (Xipaya), hawari hi, hawari koko (Yanomami)

Colombia Castaña, castaña de monte, castaña del Brasil, castaño, nuez del Brasil, mano de tigre; nojá (Muinane), katâia (Ticuna), ipak na, ifákina, ifaki (Witoto)

Peru Castaña, nuez de Brasil, nuez Amazónica; chomaeki, pajotziki (Campa), muni-watsu (Cocama), shiwiwi (Ese Eja), inke (Matsigenka), tama wan (Sharanahua), ano pinti (Shipibo), taxü (Ticuna)

Venezuela Yumi, tuca, almendra; hawari koko (Yanomami), wufia (Ye'kuana)

Status Wild, spontaneous in cultural settings, planted

Brazil nut *Bertholletia excelsa* tree in old regrowth forest. Near Vila Valente, lower Rio Tefé, Amazonas, Brazil, 8-15-12

Brazil nut *Bertholletia excelsa* is a giant tree, soaring up to 60 m in upland forests of Amazonia, the Upper Orinoco, and into the Guianas. Brazil nut trees are among the largest trees in the Amazon Basin with a circumference that can attain 16.5 m (Salomão 1991). Because they are canopy emergents when fully mature, majestic harpy eagles *Harpia harpyja* sometimes choose them to build their nests (Luz 2004:36) as do some of the large jabiru storks *Jabiru mycteria* in cleared areas near water courses (Lees et al. 2013). Their straight boles terminate in a broccoli or cauliflower-like display of large branches which produce flowers that are pollinated by bees followed by mature capsules with edible nuts some 14 months later (Le Cointe 1922:455). Brazil nut is one of the longest-living trees in Amazonia, attaining 900 years in some cases (Salomão 2009).

Brazil nuts entered world trade in the nineteenth century and possibly earlier; in England, they are most popular around Christmas time when they are cracked open along with other kinds of nuts, such as hazelnuts, walnuts, and almonds. In the United States and many other countries, Brazil nuts are found in canned and salted mixed nuts. At the beginning of the twenty-first century, some 45,000 tons of Brazil nuts were gathered in the Brazilian Amazon with sales over US$33 million (Peres et al. 2004). However, Brazil nuts represent less than 2% of the world trade in edible nuts (Clay 1997), and if rampant deforestation continues in Amazonia this proportion is likely to decline.

Brazil nut *Bertholletia excelsa* **capsules (*ouriços*) picked up from the ground by Arlinho, 11 years, in an old successionary forest near an archaeological site. The Brazil nut capsules have bite marks from macaws. Rio Inajá, affluent of the lower Xingu just upstream from Carrazedo, Município de Gurupá, Pará, Brazil 10-26-16**

The 10–25 nuts are encased in a hard wooden capsule, technically called a pyxidium and known as *ouriço* in Brazil (Mori 1992). In Acre, Brazil nut trees produce an average of 65–72 capsules per year, but the number produced diminishes when lianas are draped from the tree's canopy (Kainer et al. 2006). This capsule, which can weigh up to 2.5 kg, evolved to protect the nuts while they are maturing.

Nevertheless, several species of macaws including the hyacinth macaw *Anodorhynchus hyacinthinus*, the red-and-green macaw *Ara chloropterus*, the scarlet macaw *Ara macao*, and the blue-and-yellow macaw *Ara ararauna*, nevertheless, damage a few of the capsules as they are developing when they are less hard. While these macaws can cause severe damage to the capsules on some Brazil nut trees, overall their impact on Brazil nut production is minimal (Trivedi et al. 2004). Most of the damaged capsules continue to form, however, eventually dropping to the ground when the nuts are ripe. Nut production often varies from year to year but a mature Brazil nut tree usually produces between 100 and 269 capsules per year (Ribeiro et al. 2014a). In the municipalities of São João da Baliza and Caracaraí in Roraima some Brazil nut trees produce few or no nuts in some years, but most show a peak in production every 4 years (Tonini and Pedrozo 2014).

Almost all Brazil nuts entering the trade come from trees in forests, including anthropogenic woodlands, rather than plantations. Even though Brazil nut was introduced to Sri Lanka and Southeast Asia in the late nineteenth century (Macmillan 1935:230), it never took off there as a plantation crop unlike rubber *Hevea brasiliensis*.

Brazil nut trees are self-incompatible and must be pollinated by various bees when they flower in the latter part of the dry season. The relatively large stingless bees are strong enough to force their way into the hooded flowers to feed on the nectar (Mori and Prance 1990a, b; Prance 1985). In the Manaus area, six species of bees in three genera pollinate Brazil nut flowers (Nelson et al. 1985). Further east in the municipality of Itacoatiara some 270 km from Manaus, 17 species of bees have been observed visiting Brazil nut flowers on the Aruanã property including *Eulaema mocsaryi*, *E. bombiformis*, *E. cingulata*, *Xylocopa frontalis*, *X. muscaria*, *Bombus transversalis*, *Centris flavifrons*, *C. flavilabris*, *Epicharis zonata*, and *Euglossa intersecta* (Cavalcante et al. 2012; Santos and Absy 2010, 2012). Other native bees observed visiting Brazil nut flowers on this property include *Centris denudans*, *C. americana*, *C. ferruginea*, *Eulaema meriana*, *Epicharis conica*, *E. flava*, and *Eufriesea flaviventris* (Cavalcante et al. 2012). In various parts of the Brazilian Amazon, the *Trigona cilipes* stingless bee also pollinates the flowers of this gigantic tree (Absy et al. 1984).

In Pará, nine species of native bees have been observed pollinating these flowers including *Xylocopa frontalis*, *X. aurelenta*, *Epicharis rustica*, *E. affinis*, *E. similis*, *Eulaema nigrita*, *E. cingulata*, *Bombus brevivillus*, and *B. transversalis* (Maués 2002). Pollen from the flowers is brushed off on the bodies of the euglossine bees and when they fly to another Brazil nut tree the flowers are pollinated. Some of the bees can only survive if the surrounding forest is maintained, whereas others survive in successionary vegetation (Nelson et al. 1985). In parts of Rondônia Brazil nut trees isolated from forest produce few or no nuts because pollinators cannot reach them (Lisboa et al. 1991b). Male euglossine bees do not pollinate the flowers; however, they gather the scent from orchids and form a lek to attract female bees for mating (Prance 1984). Without forest habitat for the orchids, the large euglossine bees disappear. Two hummingbird species, *Phaethornis malaris* and *P. hispidus*, have been observed visiting the flowers of this tree near Leticia in the Colombian

Amazon and may also pollinate them (Rodríguez-Flores et al. 2012). The nuts mature in about 14 months and the capsules then fall to the ground in the rainy season (Prance and Mori 1979:80).

Boys gathering Brazil nut *Bertholletia excelsa* capsules in regrowth forest on an archaeological site. They are standing next to a Brazil nut tree. Near Carrazedo, mouth of the Xingu River, Município de Gurupá, Pará, Brazil 4-15-15

Pollinated and unpollinated flowers fall to the ground in the latter part of the dry season and early part of the rainy period and are eaten by various game animals.

Hunters take advantage of the fallen flowers by setting up platforms near the trees at night to wait for brocket deer (*Mazama americana* and *M. gouazoubira*) and paca *Agouti paca* (Smith 1976).

Another reason that Brazil nut has not emerged as a major plantation crop is that it takes 8–10 years before the trees start producing nuts. Nevertheless, a Brazil nut plantation has been established on degraded pasture at km 215 of the Manaus-Itacoatiara highway (AM 010). At Fazenda Aruanã many of the Brazil nut trees are grafted and the owner has corridors of regrowth forest that are 10 m wide every 50 m. This allows for pollinators to survive and thus helps boost production. This plantation started in the late 1970s and by 2012 some 1.3 million Brazil nut trees had been planted on 4000 hectares. The nuts are sold in cans and vacuum-paced plastic bags under the Econut brand name. Brazil nut trees have been used to recoup some deforested areas, such as in bauxite mining areas in the Trombetas River watershed in Pará (Salomão et al. 2006).

The international trade in Brazil nuts spurred the construction of a railroad in 1927 linking Tucuruí (then known as Alcobaça) and Jatobal, small towns on the left margin of the Tocantins River. Steam locomotives imported from the United States plied the 117 km railroad to bypass rapids along the Tocantins which slowed the movement of Brazil nuts and other goods as well as causing many wrecks. The building and maintenance of this railroad triggered conflicts with the Parakanã tribe (Laraia and Da Matta 1978:67). Some Brazil nut gatherers also triggered hostile encounters with indigenous groups, particularly in the Tocantins watershed where Brazil nut populations have historically been larger than other parts of the Amazon Basin. With the construction of the Transamazon Highway in 1970 followed by a network of other roads, the railroad was deactivated and now most of the Brazil nuts gathered in the Tocantins watershed that are destined for export are taken by trucks to Belém.

The Tocantins valley in Pará accounts for more than half of the Brazil nuts produced in Amazonia. Marabá, at the confluence of the Tocantins and Itacaiúnas rivers, grew up from a few huts to a small town at the beginning of the twentieth century largely in response to the growing demand for Brazil nuts (Silva 1973). In 1950, Marabá dispatched 8 million kilogram of Brazil nuts to Belém, up substantially from the 269,800 kg in 1919 when rubber *Hevea brasiliensis* was far more important economically in that area (Barruel de Lagenest 1958:51). The Trombetas River, a northern affluent of the Amazon in Pará is also rich in Brazil nut groves (Ferreira Penna 1869:20). Large concentrations of Brazil nut trees, known as *castanhais* in Brazil and *castañales* in Bolivia, are also found in Acre, along the Purus River in Amazonas, and the Bolivian Amazon (Cronkleton et al. 2012; Veríssimo 1963). Smaller concentrations of Brazil nut, most likely the result of human agency, are scattered widely in other parts of the Amazon Basin. The Ecuadorian Amazon, however, seems to have few if any Brazil nut groves, most likely because the tree requires a pronounced dry season to produce flowers and the Ecuadorian Amazon has especially high rainfall and a reduced dry season.

When the capsules drop to the ground during the latter part of the dry season early part of the rainy period, agoutis (*Dasyprocta* spp.) gnaw open the capsules and scatter hoard some of the nuts, a few of which later sprout (Haugaasen et al. 2010; Kubitzki 1985; Malu and Peres 2005; Mori and Prance 1987; Peres and Baider 1997; Prance and Mori 1978). Agoutis sometimes move the capsules (*ouriços*) before gnawing them open to access the nuts (Haugaasen et al. 2012). Agoutis usually disperse the nuts up to 40 m from the parent tree (Peres et al. 1997) although in some cases the nuts are buried as far as 100 m from the parent tree (Cotta et al. 2008). Agoutis disperse some of the nuts into fallow vegetation and along the edges of the forest (Cotta et al. 2008). Such environments have more light and thus help Brazil nut seedlings get a head start (Scoles et al. 2011). In some areas, as the number of shifting cultivation cycles increases so does the density of Brazil nut trees reaching as high as 104 trees/ha in some cases (Paiva et al. 2011).

Acouchis *Myoprocta* spp., similar in shape and habits to agoutis although smaller, also gnaw open Brazil nut pyxidiums to access the nuts some of which they bury (Haugaasen et al. 2010). However, acouchis are much less common than agoutis and therefore do not play as important a role as agoutis in seed dispersal. River dwellers in the Caxiuanã National Forest in Pará told me that squirrels also gnaw open Brazil nut capsules and disperse some of the nuts. According to rural folk near Tucuruí on the Tocantins River paca *Agouti paca* also gnaws open the capsules to access the nuts. However, this bulldog-sized rodent does not bury any of the nuts, so it is not involved in seed dispersal. Either way, Brazil nut trees further enhance rural diets by providing food for game animals.

Brazil nut *Bertholletia excelsa* capsule gnawed open by an agouti. Seringalzinho, Rio Jaú, Amazonas, Brazil, 10-3-12

The nuts have long been appreciated by indigenous groups and other rural peoples in the interior. In the Upper Orinoco and Negro Rivers, for example, indigenous groups once celebrated the gathering of Brazil nuts with festivities during which considerable amounts of fermented drinks were consumed (Humboldt and Bonpland 1821:532). The Kaiabi in Mato Grosso store Brazil nuts for months as a standby in case of food shortages (Grünberg 2004). The Kayapó in southern Pará state chew Brazil nuts and feed them to their pet parakeets (Banner 1961). Brazil nuts are an important source of food for the Toromona, an indigenous group belonging to the Tacanan linguistic family, who live in the Upper Madidi and Beni Rivers in the lowlands of Bolivia. The Toromona have also long traded Brazil nuts with indigenous peoples in the Andes (Métraux 1963). As long as Brazil nuts are kept dry, they are still good to eat for at least a year.

Wadrio, 9 years, chopping open a Brazil nut *Bertholletia excelsa* capsule with a machete. Near Carrazedo, mouth of the Xingu River, Município de Gurupá, Pará, Brazil 4-15-15

Both tribal peoples and rural folk of mixed ancestry have established groves of Brazil nut trees in the vicinity of their settlements, going back thousands of years in the case of indigenous groups (Andrade et al. 2019; Levis et al. 2012; Venturieri 2001). Along the Manicoré River, an affluent of the Madeira in Amazonas, Brazil nut groves can reach densities of 20 trees/ha, likely a result of human activities (Scoles and Gribel 2015; Thomas et al. 2015). More genetic variation among Brazil nut trees is found within groves than between them. Studies of chloroplast sequencing and microsatellite markers suggest a relatively recent and rapid spread of Brazil nut trees, coinciding with the arrival of humans in Amazonia (Shepard and Ramirez 2011).

Recently planted grove of Brazil nut *Bertholletia excelsa* trees also containing mango *Mangifera indica* trees. Adjacent to a home garden abandoned some 5 years ago. Sitio do Jacinto, lower Rio Caxiuanã, upstream from Vila da Associação, FLONA Caxiuanã, Pará, Brazil 7-17-16

In a survey of 150 home gardens in various parts of the Brazilian Amazon carried out over several decades, I encountered Brazil nut trees in 24 of them. During a separate inventory of 57 home gardens along the middle Xingu in 2015, I found Brazil nut trees in 7 of them. This tree is also planted in some rural home gardens in the vicinity of Manacapuru along the lower Solimões River in Amazonas (Klüppel 2006:24), in the municipality of Rio Preto da Eva along the Manaus-Itacoatiara highway in Amazonas (Silva 2017:31), and in Rondônia (Summers et al. 2004). In the Municipality of Gurupá along the lower Amazon, some rural people plant Brazil nuts in the vicinity of their manioc processing huts (*casas de forno*) as well as their home gardens. Some farmers living along the eastern bank of the lower Tapajós

River plant Brazil nuts, as well as some other useful trees, in their manioc fields so that the abandoned *roças* will turn into productive agroforests (Schroth et al. 2003). Some farmers in the vicinity of Tomé-Açu, Pará, also plant Brazil nuts in their agroforestry plots (Bolfe and Batistella 2011).

This is likely an ancient practice. The Ticuna, for example, plant Brazil nut trees near their homes (Lima 2006:171; Nimuendajú 1963), likely a tradition that goes back many generations. The Mekranoti of the Upper Xingu plant Brazil nuts in the forest near their villages (Werner 1984:213). The Kayapó in the interfluvial area between the Tocantins and Xingu plant Brazil nut in their manioc fields because they value the nuts (Kerr and Posey 1984; Posey 1985); they also plant them on the outskirts of their villages (Ribeiro et al. 2014b). Brazil nut seedlings also become established along trails outside Brazil nut groves used by the Kayapó when some of the nuts fall from the baskets used to carry the nuts back to their villages (Ribeiro et al. 2014b). During the colonial period, Jesuits had Brazil nut trees planted near their mission at Tauá near the confluence of Acará and Mojú rivers south of Belém (Spruce 1908:16). Both the density and distribution of Brazil nut trees have thus been influenced by humans for a long time.

Brazil nut *Bertholletia excelsa* tree planted at the edge of a home garden. Mouth of the Caxiuanã River, FLONA Caxiuanã, Municipio de Melgaço, Pará, Brazil 7-1-16

Brazil nut trees are sometimes found in cultural forests on the outskirts of settlements, especially if they are on Anthrosols, known *as terra preta de índio* in Brazil. In these anthropogenic forests, patches of undergrowth are sometimes cleared to help find fallen capsules. These more open spots are occasionally planted for crops, such as the New World yam *Dioscorea trifida* and banana. Cleared patches in the undergrowth of *castanhais* that had been planted to those crops were observed along the Curuá and Caxiuanã rivers in eastern Pará. Cleared undergrowth also facilitates the sprouting of Brazil nuts that have been buried by agoutis. Along the upper Jari River, the undergrowth of some of the Brazil nut groves undergoes controlled burns to release ash to fertilize the soil and to get rid of large grasshoppers that allegedly attack the leaves of Brazil nut trees (Lins 1991:67).

Brazil nuts can sprout in the shady interior of forests but for them to proceed to the sapling stage they need increased light (Myers et al. 2000). The survival and growth rate of Brazil nut seedlings improves with more light (Peña-Claros et al. 2002). Thus people have facilitated the regeneration of Brazil nut trees in some areas by clearing some of the undergrowth in Brazil nut stands, creating light gaps along paths, and by clearing a mosaic of successionary vegetation as a result of slash-and-burn agriculture. Brazil nut seedlings grow better when planted in slash-and-burn fields due to better light conditions compared to those planted in forest gaps (Kainer et al. 1998). Brazil nut stands appear to be largely anthropogenic (Neves 2016; Scoles 2011; Scoles and Gribel 2011).

A grove of Brazil nut *Bertholletia excelsa* trees at the perimeter of a river dweller's home garden. Açaí *Euterpe oleracea* palms have been planted closer to the houses since they pose less of a risk. Furo Jacundaí, Rio Anapú, Pará, Brazil 6-29-16

A grove of Brazil nut *Bertholletia excelsa* **trees and** *castanheiros*. **Percy Lau 1949,** *Tipos e Aspectos do Brasil*, **Rio de Janeiro**

Indeed, the presence of Brazil nut trees is considered a disturbance indicator (Balée and Campbell 1990) and a sign for archaeological sites with anthropogenic dark earth (Clement et al. 2003; Woods and McCann 1999). The boundary between home gardens and cultural forests near villages and river dweller homes is often blurred. For safety reasons, Brazil nut trees are typically planted at the perimeter of home gardens to avoid damage to homes and the danger of capsules falling on people and livestock, such as chickens and pigs. Sometimes rural people cut down Brazil nut trees in their home gardens for safety reasons or to obtain timber. But the trees usually resprout. When clearing forest or successionary vegetation to plant crops or establish pasture, Brazil nut trees, including immature ones, are typically spared. Farmers along the Tefé River and its affluents refrain from weeding Brazil nut seedlings when they sprout in their manioc fields because they provide valuable nuts (Jakovac et al. 2016).

Anthropogenic forest with Brazil nut *Bertholletia excelsa* **trees intermingled with inajá** *Attalea maripa* **and mumbaca** *Astrocaryum gynacanthum* **palms. Sitio Santo Antonio, lower Rio Caxiuanã, Município de Melgaço, Pará, Brazil 5-6-15**

Farmers also interplant Brazil nut trees in some of their fields, sometimes with just one or two annuals or perennials, other times with numerous crops. The idea is to transition fields to productive orchards rather than simply abandoning them to successionary vegetation. Cacao *Theobroma cacao*, for example, is often interplanted with Brazil nut trees, or they are spared when they arise spontaneously, with other perennial crops, as in the vicinity of Tomé-Açu and Agrovila Nova Fronteira at km 80 of the Altamira-Itaituba stretch of the Transamazon Highway. At km 95 and 107 of the same stretch of the highway, two other farmers have protected Brazil nut trees that arose spontaneously in their cacao orchards. Along a side road at Km 165 of that highway, a farmer has spared a Brazil nut tree that sprouted in an agroforestry field that contains orange, tangerine, lime, avocado, maize, and mahogany *Swietenia macrophylla*. Farmers living in Nova Califórnia at km 155 of the Rio Branco-Porto Velho stretch of BR 364 have interplanted Brazil nut with cupuaçu *Theobroma grandiflorum*, peach palm *Bactris gasipaes*, banana, and manioc in some fields that were previously planted to maize and beans. Along a side-road leading from km 35 of the Marabá-Estreito stretch of the Transamazon Highway, one farmer has interplanted Brazil nut trees, one of which was grafted, with bananas. A nearby farmer has intercropped Brazil nuts with banana and cupuaçu. At km 13 of the Santarém-Belterra highway (FR 163), a farmer has intercropped black pepper *Piper nigrum* with orange and Brazil nut trees. That way when the black pepper finally succumbs to disease, especially Fusarium wilt, other useful trees will provide an income source.

Elido Trevisan by a 30-year-old Brazil nut *Bertholletia excelsa* tree in his cacao orchard. The Brazil nut tree sprouted spontaneously in a field of annual crops and subsequently, the field was planted with cacao. Gleba 23, lote 2, km 75 Altamira-Itaituba, Transamazonica, Município de Medicilandia, Pará, Brazil 9-7-15

A farmer near Machadinho in Rondônia has welcomed some spontaneous Brazil nut seedlings that sprouted among rubber *Hevea brasiliensis* and robusta coffee he planted in one field. In another field nearby he has spared spontaneous Brazil nut trees sprouting among rubber and guaraná *Paullinia cupana*. Near Ouro Preto in Rondônia, a farmer has intercropped coconut *Cocos nucifera* with cupuaçu *Theobroma grandiflorum*, soursop *Annona muricata*, banana, and Brazil nut. The field also contains spontaneous guava *Psidium guajava*. At Quatro Marcos near Cuiabá in Mato Grosso, a farmer has interplanted robusta coffee and Brazil nut. Along the Estrada do Bacabal near Maués, Amazonas, a guaraná *Paullinia cupana* field is planted with numerous annual and perennial crops including Brazil nut. Small-scale farmers, in contrast to ranchers and most plantation owners, have thus long been involved in increasing the density of these Brazil nut trees in many areas of the Amazon Basin.

Owners of land with large concentrations of Brazil nut trees grubstake Brazil nut gatherers, known as *castanheiros* in Brazil, with food and other supplies so that they can concentrate on gathering the fallen nuts. *Castanheiros* construct makeshift huts in the forest, where after gathering fallen capsules, they sit and chop them open with machetes to release the nuts (Velho 1972:63). On a typical day, a Brazil nut gatherer chops open 700–800 capsules a day thus obtaining around 100 kg of nuts (Valverde and Dias 1967:150). Mules or the *castanheiros* themselves carry the nuts to the nearest stream, river, or road where they are loaded onto canoes, boats, or trucks and taken to the owner for payment. After the owner has subtracted the cost of the gatherers' supplies, *castanheiros* typically leave with meager earnings. When the 2- to 3-month collecting season is over, *castanheiros* disperse to look for other employment, such as gold mining, tapping rubber, or cutting forest on cattle ranches. Before the Tucuruí dam flooded substantial portions of the Tocantins River in 1984, many *castanheiros* spent the hot, dry season panning for diamonds along the sandy islands and banks of the clear-water river. In the case of rural people with concentrations of Brazil nut trees near their settlements, they walk into the forest with backpack baskets to collect the capsules, cut them open, and pour the nuts into their baskets lined with broad leaves so that they do not spill out. In 1873, scientists on an expedition organized by the British Amazon Steam Navigation Company noted that rural folk living along the banks of the lower Trombetas River in Pará created mounds of Brazil nut capsules in the yards of their dwellings evidently for trade and local consumption: "all the houses along the banks had great heaps of them, in their hard outer cases, piled up in front or alongside, looking like heaps of cannon balls, and suggesting the idea that the different places were prepared for a siege" (Brown and Lidstone 1878:228).

Farmer removing the seed coat (testa) of a Brazil nut *Bertholletia excelsa* with a capsule (woody pericarp) serving as the chopping block. Pedreira, Baia de Caxiuanã, Município de Melgaço, Pará, Brazil 5-11-15

Each nut has a hard wooden shell (technically called an endocarp or testa) that is typically chipped off using a machete or large knife. In pre-contact times, indigenous peoples in the Amazon did not have metal tools so they must have cracked open the wooden seed coat with hardwood or stone. An indentation in a melon-sized rock found at an archaeological site along the Trombetas River in Para suggests it was probably used to hold Brazil nuts in their shell that were then struck with a stone or wooden club (Rodet et al. 2010). The nutshells are discarded and enrich the soil in home gardens. Some indigenous groups, such as the Arara, also carefully remove nuts so that the endocarps can be used as ankle rattles in ceremonial dances. The Kayapó of the Upper Xingu attach Brazil nut shells as pendants on belts made from fiber obtained from the buriti *Mauritia flexuosa* palm (Verswijver 1995:274). Discarded Brazil nut shells are also used to fuel dryers for Brazil nuts at some factories, such as at Óbidos along the Amazon River, which prepares Brazil nuts in laminated plastic bags for export to New York. In the Bolivian Amazon, Brazil nut shells are used to fuel some power plants (Stoian 2004).

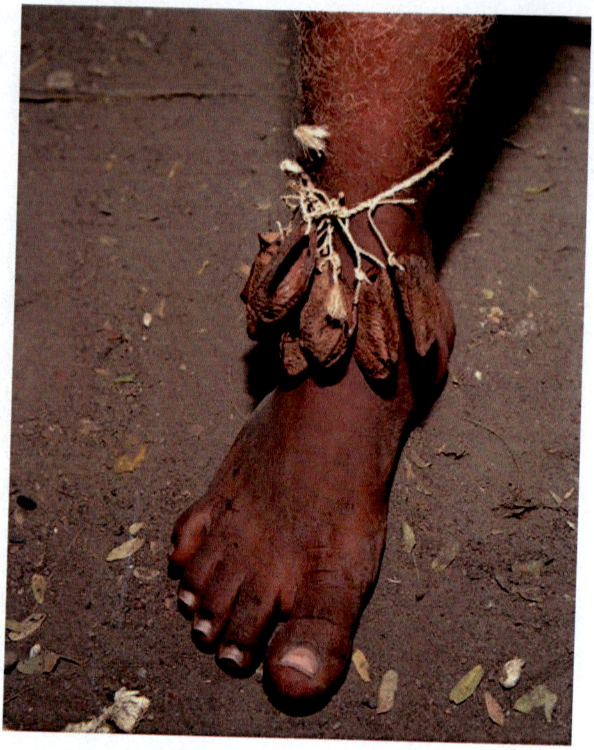

Anklet rattle made with Brazil nut *Bertholletia excelsa* endocarps. The string holding the hollowed out endocarps is fashioned from young, unopened fronds (*olho*) of inajá *Attalea maripa* palm. Terrawangã, Middle Xingu River near its confluence with the Bacajá, Terra Indígena Arara de Volta Grande, Município Senador José Porfírio, Pará, Brazil 9-10-15

Brazil nuts in their shells are used in arts and crafts in small towns. In Altamira along the Xingu, for example, Brazil nuts in their shells are used to represent sunflowers for wall hangings. At Alenquer along the Amazon River in Pará, one family of Lebanese descent makes souvenirs with Brazil nut shells that are painted and assembled to represent butterflies.

Francisca Teles da Silva, 58 years, with wall hanging made in Altamira featuring Brazil nuts *Bertholletia excelsa* and a capsule (*ouriço*) which serves as the bowl. Large black seeds of mucuna *Dioclea* sp. are in the center of the flowers. Ilha Marisol, middle Xingu River about 45 km upstream from Altamira, Pará, Brazil 9-17-15

The nuts are not only tasty but high in protein (14–17%) as well as oil, which accounts for their smooth texture. The oil content is sufficiently high (63–69%) that they can be lit and serve as miniature candles. The content of unsaturated fat in Brazil nuts is higher than any other nuts (Kluczkovski et al. 2015). In the 18th and 19th centuries, oil was extracted from the nuts to be used in lighting, cooking, and as an ointment (Daniel 1976; Hartt 1898; Herndon and Gibbon 1853:298). In the nineteenth century and early twentieth century, Brazil nut oil was also sometimes added to the fruit juice of cupuaçu *Theobroma grandiflorum* and taperebá *Spondias mombin* along the Tocantins River to reduce its acidity (Moura 1989:151). In the mid-nineteenth century, the American naturalist William Edwards noted that Brazil nuts were sometimes crushed and added to coffee "When the castanha-nuts are fresh they much resemble in taste the cocoa-nut, and the white milk, easily expressed, is no bad substitute for milk in coffee" (Edwards 1847:55). The Suruí in Rondônia rub Brazil nut oil on their bows to make them lustrous (Coimbra 1985). In the 1990s, Body Shop marketed a hair conditioner that contained Brazil nut oil. The back label explained that the oil was purchased from a Kayapó village and that customers were helping these indigenous people to protect their land. However, this venture between Body Shop and the Kayapó in the 1990s eventually fell apart, in part because it did not increase local cash income among the indigenous groups to any great extent (Ros-Tonen et al. 2008). Brazil nut oil is still used occasionally by river dwellers to cook, especially when frying game birds (Lisboa et al. 2002).

Brazil nuts are the highest known food source of selenium which helps protect against various cancers (Chunhieng et al. 2004; Yang 2009). Brazil nuts are also a good dietary source of selenium, magnesium, phosphorus, thiamin, niacin, vitamins E and B6, calcium, iron, potassium, zinc, and copper. Brazil nuts also contain substantial amounts of antioxidants which provide health benefits (John and Shahidi 2010).

Roselia, 48 years, with Brazil nuts *Bertholletia excelsa* she has coated with chocolate and sugar. Vila Valente, lower Tefé River, Amazonas, Brazil, 8-30-12

The nuts are eaten raw and also used to make homemade candies by covering them with chocolate and then coating the candies with sprinkled sugar so that they do not stick together and further enhance the flavor. These Brazil nut candies are typically made for sale in towns, such as Tefé in Amazonas, Brazil. Near the mouth of the Tefé River, a river dweller makes *castanha cristalizada* by chopping up Brazil nuts and blending them with chocolate. These are eaten at home and also sold in the nearby town of Tefé. In Riberalta in the Bolivia Amazon, Brazil nuts coated with caramelized condensed milk and sprinkled with grated coconut are produced locally.

The Gavião in eastern Pará grate Brazil nuts and mix them with cupuaçu *Theobroma grandiflorum* pulp (Iara Ferraz, personal communication). Brazil nut gatherers camped out in the forest during the harvest season would grate Brazil nuts and mix the "milk" (*leite*) with *farinha puba*, a type of manioc flour obtained after first soaking the tubers for several days (Silva 1973). In 2012, a taxi driver in Tefé recalled that when he was a boy living in the interior where there was no bread available so he would cut up pieces of tucumã *Astrocaryum aculeatum* palm fruit or Brazil nuts and add them to coffee mixed with manioc flour for breakfast.

Mari Leny, 19 years, selling Brazil nuts *Bertholletia excelsa* covered with caramelized condensed milk and grated coconut. Riberalta airport, Beni, Bolivia 5-31-05

Other recipes feature Brazil nuts in the Brazilian Amazon. Pé de moleque is a kind of cake made with manioc flour (farinha puba), chopped-up Brazil nuts, butter, and sugar. It is wrapped in banana leaves and baked on a large metal pan in an oven (*forno*) normally used for toasting manioc flour. Known as manzapí among the Juruna of the middle Xingu, this tasty dessert and snack appear to be a blend of indigenous and colonial cuisine since manioc was domesticated long before the arrival of Europeans, but the latter introduced cattle and sugarcane.

Maria Feliques de Almeida, 73 years, eating Pé de moleque made with manioc flour (*farinha puba*), chopped up Brazil nuts *Bertholletia excelsa*, butter, and sugar. Paquiçamba, Rio Xingu between Altamira and Belo Monte, Pará, Brazil 9-9-15

A flat, crisp biscuit (*beiju*) is made with manioc flour and chopped up Brazil nuts, such as along the Tocantins River. River dwellers in the Caxiuanã National Forest roast Brazil nuts in their shells among the embers of a fire to make a dish called paçoca. The charred shells are removed and the cooked nuts are pounded in a wooden mortar along with manioc flour and salt. Sometimes paçoca is prepared with boiled rather than roasted Brazil nuts. The Wapishana of the Upper Essequibo in Guyana and contiguous Roraima in Brazil pound Brazil nuts in a mortar, squeeze out the oil in a tipití, and mix the mass with manioc or sweet potato (Farabee 1918:37).

Luciano, 13 years, eating a *beiju* of manioc flour mixed with Brazil nuts *Bertholletia excelsa*. Cametá, Tocantins River, Pará, Brazil 7-1-07

In 1979 at Marabá along the Tocantins River, I witnessed a fisherman using Brazil nuts for bait. He was after a fish called pacu dente seco *Mylesinus paucisquamatus* which migrates upstream during the flood season after laying eggs to spend the rest of the year among rapids. After tossing cooked rice into the water to get the attention of fish, he cast his baited hook using a bamboo pole and then waited as the line drifted downstream. *Mylesinus paucisquamatus*, which can attain 20 cm in length, is confined to the Tocantins watershed and its populations have likely decreased with dam construction and rampant deforestation. Pirapitinga *Colossoma bidens* are also caught on hooks baited with Brazil nut.

An empty Brazil nut *Bertholletia excelsa* capsule used as a stopper at the open end of a tipiti to keep the manioc dough from spilling out. The tipiti is made from strips torn from the stem of jacitara *Desmoncus* sp. palm. Santo Antonio de Ipapucu, Lago de Tefé, Amazonas, Brazil, 8-27-12

Discarded capsules are put to a variety of uses. They serve as miniature bowls for grinding black pepper and for carrying smoky fires to dispel black flies (*Simulium* spp.) that appear in the rainy season near rapids along clear and black water rivers. Empty capsules are used as stoppers to keep manioc dough inside sleeve presses (*tipitis*) when the dough is being squeezed. Agugó is a musical instrument made with two empty Brazil nut *ouriços* attached to a piece of wood and tapped with a wooden stick to make a drum sound. This musical instrument appears to have arisen from the Afro-Brazilian musical tradition. The capsules are also used for fuelwood, especially for roasting fish, such as in the vicinity of Óbidos along the middle Amazon River, along the Tefé River, and along the lower Purus River (Silva 2006:40). In the watershed of the Purus River in Amazonas, several rural communities burn the capsules in a hole covered with some earth and leaves to create charcoal for cooking (Bentes 2007:65). During the rubber boom, the capsules were also burned to produce smoke for coagulating latex gathered from rubber trees (Le Cointe 1922:321). Some indigenous groups have also found uses for the capsules. The Paumari of the Purus, for example, use them to mix the ashes of toasted tobacco leaves with the toasted shells of cacao beans to make snuff. The snuff is inhaled using a pair of hollowed-out bird legs with round lumps of beeswax at one end to place in the nostrils (Steere 1903). The Yanomami in northern Roraima and southern Venezuela use the capsules as bowls for the preparation of hallucinogenic snuff (Albert and Milliken 2009:97). The Curuaya of the Curuá River, an affluent of the

Iriri which in turn flows into the middle Xingu, use Brazil nut capsules to gather honey from bee nests in the forest. The top of the capsule is sealed from wax obtained from the bee nest to prevent the honey from spilling (Snethlage 1910).

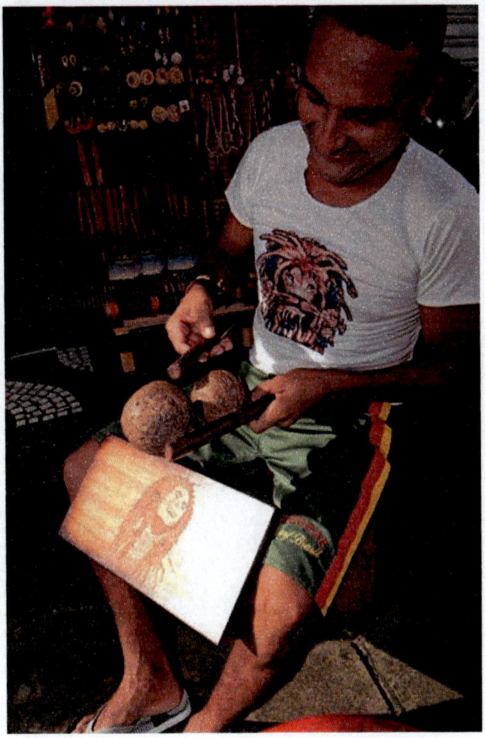

Paixão, 45 years, playing an agugó made with Brazil nut capsules *Bertholletia excelsa*. He is selling arts and crafts from a stand. Feira do Produtor, Bragança, Pará, Brazil 5-28-15

Valdizia, 52 years, grinding black pepper kernels in a Brazil nut capsule *Bertholletia excelsa* on the floor of her kitchen. Boa Esperança, Lago Amanã, Amazonas, Brazil, 9-4-12

The capsules are also used in folk remedies. A vendor in the public market at Barcarena in eastern Pará explained that the capsules are used to treat anemia. Water is placed in an *ouriço* after the nuts are emptied out, and the capsule is then placed in a refrigerator overnight. The next morning the liquid is drunk. The vendor reports that buyers come from as far as Brasília and São Paulo to obtain the Brazil nut *ouriços* for such purposes. Residents of Tiningú, a quilombo community 47 km southeast of Santarém, use the *ouriços* to make a tea for treating cancer (Carvalho 2015:64). In the Municipality of Itupiranga along the Tocantins River, peasants (*caboclos*) use the *embigo* (belly button) of Brazil nut capsules to treat hemorrhages. The *embigo*, a protrusion on the capsule where it was attached to the tree, is burned and the ashes mixed with water which is then drunk.

Brazil nut capsules *Bertholletia excelsa* at a stand containing other medicinal plants. Barcarena, Pará, Brazil 3-19-15

Brazil nut wood has been used traditionally for a variety of purposes such as making dugout canoes, as observed along the Tefé River in Amazonas and in the watershed of the Anapu River in Pará. The wood is especially durable hence its popularity for such purposes.

River dweller paddling a dugout canoe fashioned from the trunk of a Brazil nut *Bertholletia excelsa* tree. Lower Caxiuanã River, affluent of the Anapu, Municipio de Melgaço, Pará, Brazil 7-2-16

Dugout canoe made with Brazil nut *Bertholletia excelsa* wood. The seats and flooring are made with louro-namui wood. Bacuri, Lago de Tefé, Amazonas, Brazil, 8-27-12

Brazil nut wood is also used to make shingles for rural homes, along the Middle Xingu, the Tocantins River, and several affluents of the Anapú River. However, Brazil nut shingles are much less common now that asbestos and metallic roofing have been widely adopted in the Brazilian Amazon. The Suruí in Rondônia make large containers from Brazil nut wood to ferment some drinks (Coimbra 1985). Logging of Brazil nut trees has been prohibited in Brazil, Peru, and Bolivia for several decades, but the law is not always respected. One way to tell if a truck has any Brazil nut trunks is to check the cross-section for dark red resinous rings. The wood has a deep red-brown color. Sometimes the planks are sawn inside the forest to help avoid being caught. Even if Brazil nut trees are not cut down for their timber, they are sometimes damaged by logging operations targeting other trees such as in the Madre de Dios region of the Peruvian Amazon and the Pando Department in northern Bolivia (Duchelle et al. 2012).

Wattle and daub house with roof shingles made from Brazil nut *Bertholletia excelsa* wood. A charred Brazil nut tree has been cut down near the house. Farm of Elias Magres, São João Batista, Gleba Cajazeiras, Transamazon Highway, Itupiranga Municipality, Pará, Brazil 11-11-92

The inner bark of Brazil nut trees also has several uses. During the colonial period, harpoon lines and fishing nets were made with the inner bark of young Brazil nut trees to capture pirarucu *Arapaima gigas*, one of the largest fish in Amazonia (Ferreira 1972:15). Nowadays fishing nets and lines are mostly made from monofilament or multifilament nylon. In the mid-twentieth century, some cast nets on Marajó Island were still being made from twine obtained from the inner bark of Brazil nut trees (Pinto 1956). The inner bark of Brazil nut trees has been used for centuries to caulk boats and canoes in the Brazilian and Bolivian Amazon as well as the Guianas (Crévaux 1883:44; Daniel 1976:335; Hartt 1898; Huber 1910; Keller 1874:62). This is one of the preferred caulking materials because of its durability (Avé-Lallemant 1961:144). The inner bark was also used in the nineteenth century as padding for saddles in the middle Amazon (Smith 1879:313). Near Itacoatiara, Amazonas, the inner bark was also used to roll up cured tobacco leaves in the nineteenth century (Spix and Martius 1976:114).

In the Peruvian Amazon, small branches of young Brazil nut trees are cut to strip off the inner bark to make cordage to tie up sacks of charcoal. In the Caxiuanã National Forest, river dwellers employ the inner bark of Brazil nut trees to fashion head straps for baskets they carry on their backs to take manioc tubers to processing huts. Along the Rio Negro in Amazonas, the bark of young Brazil nut trees is also used for cordage (Alarcón and Peixoto 2008). The Waimiri-Atroari north of Manaus use the inner bark to fashion straps for carrying babies (Milliken et al. 1992:32). Along the Rio Maracá in Amapá river dwellers use the inner bark of Brazil nut trees to treat diarrhea.

Charcoal worker carrying Brazil nut *Bertholletia excelsa* twigs he has cut in regrowth forest. The bark will be torn into strips to serve as cordage for tying up bags of charcoal. Near Tamshiyacu, Rio Amazonas, Loreto, Peru 6-16-04

Brazil nut also has medicinal uses. The inner bark is pounded and then put in cold water to soak for 1 or 2 hours before drinking. Rural folk in the vicinity of Maués use this tea as a remedy for diarrhea, as well as for body aches and malaria. Several villages in the vicinity of Barcelos along the middle Rio Negro, Amazonas, prepare a decoction of the bark of this tree to treat malaria (Tomchinsky et al. 2017). Some rural people in Rondônia also macerate the inner bark of this tree to treat malaria and diarrhea (Santos et al. 2014b). In the Bolivian Amazon, some indigenous people boil the bark in water and then add sugar to treat malaria and anemia (Ponz et al. 2005:51). Along the lower Rio Negro near Manaus, some rural folk prepare a decoction with the bark of this tree to treat inflammation (Scudeller et al. 2009). Inhabitants of Santa Luzia do Buiuçuzinho, a small rural community some 20 km west of Coari, Amazonas, use the bark to make an infusion to treat inflammations and stomachache (Silva 2009:72). Residents of Marudá, a village along the coast of Pará at the mouth of the Amazon, prepare syrup from the resin-rich inner bark to treat flu and lung congestion as evidenced by coughing (Coelho-Ferreira 2009). River dwellers along the lower Jauaperi, an affluent of the middle Rio Negro, use the bark to make a tea for treating anemia, stomachache, malaria, dysentery, urine infections, and swelling (Pedrollo et al. 2016). The Ticuna along the Solimões use the bark to make a tree for treating diarrhea and to pour on wounds. They also prepare a bathing solution using the bark to treat rheumatism (Lima 2006:171). Brazil nut trees produce a dark red resin when the bark is attacked by burrowing insects, especially certain beetles. Sometimes the resin is produced in such profusion that it drips down the trunk. Locals gather the resin to burn and for use in folk remedies. Rural folk along the Unini River, an affluent of the Rio Negro in Amazonas use the water-soluble resin to treat diarrhea and hemorrhoids (Lago et al. 2016). At Terrawangã, an Arara village on the right bank of the Xingu River near its confluence with the Bacajá, the resin is burned and the smoke is inhaled to treat headaches, sinusitis, and heart conditions. In Roraima, in the northern part of the Brazilian Amazon, some people extract oil from the nuts which are applied to wounds to stem the bleeding (Luz 2001). In the Peruvian Amazon, some 50 g of the bark is boiled in 1 l of water and when the solution cools one cup is drunk three times a day to treat rheumatism (Lossio et al. 1995:89). In the Purus watershed in Amazonas, some rural folk use Brazil nut leaves to make a bathing solution to treat inflammation among women while the *ouriços* are used for treating hepatitis and hemorrhages (Bentes 2007:67). These folk medicines have not drastically reduced Brazil nut populations; rather it provides an incentive to plant even more of the trees.

Edinaldo Arara, 47 years, inspecting resin seeping from the trunk of a Brazil nut *Bertholletia excelsa* tree. The resin is used for a variety of folk remedies. Near Terrawangã, Middle Xingu River near its confluence with the Bacajá, Terra Indígena Arara de Volta Grande, Município Senador José Porfírio, Pará, Brazil 9-10-15

Brazil nut trees are entwined in some indigenous myths. The Araweté, a Tupi group of the middle Xingu watershed, believe that a spirit called koropi lives in Brazil nut trees and this supernatural entity is a hunter and cannibal (Viveiros de Castro 1992:80). The Suruí in Rondônia believe a mythical figure made Brazil nut trees grow exceptionally tall so that they could not be climbed by people (Mindlin 1995:52). The great height of Brazil nut trees is also featured in a legend of the

Arara in Rondônia. According to the Arara, God lifted an Indian into a Brazil nut tree because he was gathering up ouriços. God used to live on earth and gather ouriços. Originally, Brazil nut trees were quite short, about the height of ingá, but he made Brazil nut trees grow tall so that he could put prisoners in the canopy of the Brazil nut trees. God no longer lives on earth (Gabas and Arara 2009:87). Brazil nut features in a Gavião-Ikolen legend:

> A harpy eagle had a nest in a Brazil nut tree. Two male Indians set off to steal the harpy eagle nestlings. One was married, and the other was passionately in love with his companion's wife.
>
> "Let's go get the harpy eagle nestlings!"
>
> They were friends, not brothers. They got a pole and used trees surrounding the Brazil nut tree to climb higher. Eventually one of the men reached the nest, but his companion tossed the pole away so that he would not be able to get down. When he noticed that pole was hanging from another tree, he complained furiously
>
> "Why did you do this? Don't do this!"
>
> His "friend" left for home and told the brother of his abandoned companion that he did not want to come home right away. So the brother went to check on what was going on. He noticed the hanging pole and realized that his brother had been tricked.
>
> "Why did your friend do this rotten thing to you?"
>
> "He took away the pole, I can't come down!"
>
> "I will get another one so that you can return!"
>
> "No, I don't want to return, I want to stay up here"
>
> The brother tried to convince him to come down, but he was too scared, and he said he would never come back home. As the sun went down, the brother returned home, distraught. When he got home, he told others what had happened. He asked them to help rescue his brother. When they entered the forest they were horrified; the young brother was making sounds like a harpy eagle. His body was already sprouting feathers and down.
>
> "I am not coming down!"
>
> So the man transformed into a harpy eagle. Later the man-harpy eagle attacked the man who had taken up with his wife and killed him. (Mindlin 2001:57–60)

Brazil nut also surfaces in the lore of rural folk of mixed ancestry. At Taperinha, a ranch along the Rio Maicá near Santarém, a ghost of a black man with a gray beard allegedly appears in a Brazil nut grove near the port of the property. Two archaeological sites are located on this ranch which was a sugarcane plantation during the colonial period. The current owner of the property thinks the ghostly apparition indicates that some gold is likely buried nearby. The ghost could be the spirit of a slave on the former sugarcane plantation.

At Arapaxi, a small village along a river of the same name in the Municipality of Chaves on Marajó Island, a villager named Raimundo recounted a strange incident that happened to him when he went on a fishing trip to nearby Mexiana Island. He was 25 years old at the time and was ascending Igarapé Pau Mulato with a companion when they came abreast of two large Brazil nut trees on a high embankment along a curve in the stream. A loud noise was heard, similar to the sound of a buttress root being struck. Brazil nut trees do not produce buttress roots, known locally as *sapopemas*, and Raimundo interpreted the unexpected sound as a warning from a supernatural entity "*dando aviso que tem mãe aqui*." Raimundo felt that the warning was from the "mother of the forest"; he explained that when you venture into unfamiliar territory you should not mess with things that do not belong to you (*a gente não meixe com que não pertence a gente*). So Raimundo and his companion left promptly.

At the village of Carrazedo at the mouth of the Xingu River, Marcos recounted a strange experience that his cousin, Valdir, had in the forest while gathering Brazil nuts. Valdir became lost in the forest in the Jari River watershed when he came upon a forest spirit called João da Mata. João da Mata asked Valdir what he was doing. Valdir explained that he was lost, so João da Mata pointed out the direction where his co-workers were in a temporary forest shelter. João da Mata asked for a bottle of cachaça and some tobacco as a gift. Valdir asked where he should leave the presents and he was told anywhere he liked. So when Valdir reached the *barraca* (hut) he got a bottle of cachaça and some tobacco and went back with the presents. The next day Valdir checked the spot where he had left the presents and they were gone.

Brazil nut *Bertholletia excelsa* tree protected in a farming landscape by successionary vegetation left around the trunk to protect it from fires set by farmers when clearing their fields. Manioc field in the foreground. Carrazedo, mouth of the Xingu River, Município de Gurupá, Pará, Brazil 8-1-16

In spite of a long tradition of enriching home gardens and cultural forests with Brazil nuts, as well as providing a buffer zone of successionary vegetation around the trees so that they are not burned when farmers clear their fields, this giant tree is under threat in many areas. Commercial loggers are diminishing populations of Brazil nuts, even though cutting down Brazil nut trees to sell the wood is illegal in Brazil (Homma et al. 2014). However, sawmills have to be careful because the reddish-brown wood of Brazil nut sometimes has dark resin rings that help reveal its identity.

Another threat to Brazil nut populations is posed by dam construction to generate hydroelectric power. The Tucuruí dam, which closed in 1984, flooded 2000 sq. km of forest, mostly on uplands. Many Brazil nut trees were drowned (Fearnside 2001). Another large dam at Belo Monte along the middle Xingu closed in 2016 and more are planned along other major tributaries of the Amazon, including the Tapajós.

Dead Brazil nut *Bertholletia excelsa* tree in a cattle pasture invaded by babaçu *Attalea speciosa* palms. This upland bluff overlooking the middle Xingu River was formerly covered by forest. Downstream from Altamira, Pará, Brazil 8-17-15

An even larger threat to Brazil nut trees is posed by cattle ranchers (Homma 2014a; Woodroof 1979:182). In the vicinity of Xapuri, Acre, for example, some 80,000 Brazil nut trees have succumbed to clearing to create cattle pasture (Dwyer 1990:21). Although ranchers typically spare Brazil nut trees when clearing forests to avoid fines and to provide nuts for their employees and shade for cattle, the trees become vulnerable to lightning strikes and strong wind gusts. Furthermore, pollinators of Brazil nut trees become scarce or disappear when large tracts of forest are cleared to establish settlements and pastures (Lisboa et al. 1991a). So even when Brazil nut trees are left standing by ranchers their nut production typically declines drastically. Over time they slowly die off. They can only return if the pasture is abandoned to successionary vegetation. Small farmers who clear mature or regrowth forest for cattle pasture also generally leave Brazil nut trees standing but because the area of pasture is much smaller, many of the Brazil nut trees thrive.

In the 1950s, 30,000–35,000 tons of Brazil nuts were exported annually from Amazonia and were once the second most important export item from the region (Dias 1959). In 1973, 52,095 tons of Brazil nuts were marketed in the Brazilian

Amazon as follows: 37,675 tons in Pará, 8193 tons in Amazonas, 2162 tons in Acre, and 2050 tons in Rondônia (Benchimol 1977:672). However, the forests of eastern and southern Pará as well as Rondônia and Acre have shrunk dramatically over the last several decades due to colonization and various development projects (RAISG 2015). The "arc of fire" that extends from eastern Amazonia across the southern part of the basin has led to many populations of Brazil nut trees being decimated. An estimated 70% of the Brazil nut trees in southeastern Pará fell victim to loggers, cattle ranchers, and small farmers by 1997 (Homma 2014b). Indigenous reserves are some of the last strongholds for wild Brazil nut trees because deforestation is much less pronounced there. Because of rampant deforestation along the arc of fire, Brazil now accounts for only 37% of Brazil nut exports; Bolivia produces 50% of Brazil nut exports because forest clearing is less pronounced there (Stoian 2004). Brazil nut stands in the vicinity of Marabá in eastern Pará have been devastated by rampant deforestation in the last few decades (Salomão 2014). Bolivia surpassed Brazil as an exporter of Brazil nuts in the mid-1990s (Zuidema 2003:10). Brazil nut exports from Bolivia have risen from US$16 million in 1990 to US$192 million in 2015 (Murakami et al. 2016:12). In 2007, Bolivia exported some 55,000 tons of shelled Brazil nuts (Cronkleton et al. 2012). Most of the Brazil nut processing plants in that country are located in Riberalta (Soriano et al. 2012; Stoian 2000). In 1996, Brazil nut processing plants in Riberalta at the junction of the Beni and Madre de Dios rivers in the Bolivian Amazon provided employment for 5499 people, and 2814 of those were full-time jobs (Stoian 2005).

30.3 *Couroupita guianensis*

Bolivia Tutumillo de bajura

Brazil Castanha de macaco, castanha de galinha, coité de macaco, coité brava, abricó de macaco, cuia de macaco, cuieira da mata, macacarecuia, cabeça de macaco, curupita

Colombia Coco de mono, cocuelo, maraco

Ecuador Bala de cañón, caucho; lostundu muyu, lustunda, supay mate, lispungo (Kichwa), huatisa sa (Siona-Secoya), pancabocahue, pankabukagu, pankabukabu, pankabokahue (Waorani)

Peru Ayahuma, ayaúma, misa colorado; wishí (Achuar), shishim (Aguaruna), shishim (Shuar)

Venezuela Coco de mono, taparochuco

Status Wild, planted

Ayahuma ***Couroupita guianensis*** **tree in fruit and flower in a home garden. Near San Francisco, Rio Marañón, Reserva Nacional Pacaya-Samiria, Loreto, Peru 11-9-03**

Couroupita guianensis is a large tree of floodplain forests, reaching up to 40 m, and its fruits are used for livestock feed and folk remedies. People never eat the pulp or seeds. It also occurs in some upland forests, such as in the Amazon estuary near Belém (Amaral et al. 2009). The Waorani in the Ecuadorian Amazon also use the wood for home construction and for fuel (Cerón and Ayala 1998:88). In the rural community of Santa Rita along the Solimões River near Benjamin Constant, this tree was planted and has become spontaneous (Moura 2018:49).

Couroupita guianensis ranges from the Amazon estuary to the edge of the Andes, Venezuela, the Guianas, and into Central America as far north as Costa Rica. The cannonball-like fruits are born on the edge of long stalks that extend from the trunk and main branches of the tree. The fruits are gathered by climbing the tree and twisting them off their stalks or by picking them off the forest floor. The tree fruits in the rainy season so in low-lying areas the indehiscent capsules fall into the water where they float and can be retrieved from a canoe or by wading (Smith 1999:105).

Couroupita guianensis is highly variable and at one time was divided into a number of different species (Mori et al. 1990:100). A similar species, *Couroupita subsessilis*, is also called ayahuma in Peru and castanha de macaco in Brazil, and its fruits are fed to livestock. The latter species is taller than *C. guianensis*, reaching some 55 m, but the fruits are slightly smaller and have a shorter fruit stalk. *Couroupita subsessilis* is confined to sediment-rich rivers of the Amazon, whereas *C. guianensis* also occurs in the Guianas, Venezuela, northern Colombia, and Panama. *Couroupita guianensis* appears to be more common than *C. subsessilis* and observations here pertain mostly to the former species.

The hard-wooden shell is split open with a machete so that chickens, Muscovy ducks, turkeys, and pigs can access the pink-red pulp that surrounds numerous seeds. Ayahuma is typically spared when clearing home sites, such as near the village of San Francisco along the Marañón River. In this manner, livestock feed is readily at hand. This tree has also been encountered in home gardens belonging to the Yagua indigenous group along the Amazon River in Peru (Lamont et al. 1999). However, a house is never built adjacent to the towering tree because falling fruits would damage the house and might cause injury. *Couroupita guianensis* is also occasionally planted in home gardens on floodplains, such as along the lower Solimões near its confluence with the Rio Negro (Lima and Saragoussi 2000) and as I observed in the small settlement of Urucurituba on an island in the Amazon River near Alenquer, Pará. On the outskirts of Juriti, a small town on the right bank of the Amazon upstream from Santarém, a home garden straddles uplands and the floodplain and a castanha de macaco was planted long ago on the part of the home garden that floods annually. River dwellers also spare *Couroupita guianensis* when it sprouts spontaneously in their home gardens, such as in Comunidade de São Pedro, Paraná do Baixo, in the municipality of Alenquer. This tree is also encountered occasionally in upland home gardens of rural people along Ramal do Pau Rosa, an offshoot of the BR 174 highway that links Manaus with Boa Vista in Amazonas (Machado 2016:34).

Muscovy ducks, chickens and a turkey feeding on a castanha de macaco *Couroupita guianensis* fruit in a village home garden. Arapixuna, Santarém municipality, Pará, Brazil 5-15-96

The common name for this tree in Brazil suggests that the seeds are dispersed by monkeys. Farmers in the village of San Carlos along the Puinahua, a branch of the Ucayali River, report that monkeys feed on the ayahuma fruits. Peccaries (*Pecari tajacu* and *Tayassu pecari*), forest deer (*Mazama americana* and *M. gouazoubira*), and tapir *Tapirus terrestris* also consume the fruits and may pass some of the seeds undamaged (Cerón and Ayala 1998:88; Mori 2004). Peccaries, deer, and tapir are widely hunted in Amazonia for their meat, so this tree contributes indirectly to the diet of rural people.

Rural folk use ayahuma fruits in various home remedies, including the treatment of supernatural afflictions (Luna and Amaringo 1991:13). In the village of Yarina along the Yanayacu, an affluent of the Marañón, men prepare a bathing solution from the fruits of *Couroupita guianensis* to cure stubborn cases of "bad luck" (*mala suerte*). This condition, called *panema* in Brazil, results from a spell or transgression, such as killing too many game animals. The ayahuma bath is taken three mornings in a row, presumably to invoke the Trinity. At San Carlos along the Rio Puinahua, a branch of the Ucayali, locals prepare a bathing solution from the bark of ayahuma to treat rheumatism and fever "*para sacar la calentura del cuerpo*." In the medicinal plant section of the Mercado Belén in Iquitos, ayahuma capsules were available in several stalls in 2006 to treat bad luck and skin allergies. In some rural areas of the Peruvian Amazon, the flower buds are crushed and applied to aching teeth (Mejia and Rengifo 2000:36), while in the vicinity of Chazuta along the Upper Huallaga River people apply the fruit pulp on the infected skin of their dogs (Sanz-Biset et al. 2009). Also in the Peruvian Amazon, the bark of *Couroupita guianensis* is sometimes added to hallucinogenic brews consisting of crushed stems of ayahuasca *Banisteriopsis caapi* with leaves of chacruna *Psychotria viridis* (Luna 1984). In Pastaza province in the Ecuadorian Amazon the bark, leaves, and fruits are used to treat infections and pain (Saltos et al. 2016) and in other parts of the Ecuadorian Amazon, the Waorani apply the pulp of ripe fruits to the skin to treat fungal infections (Mondragón and Smith 1997:137).

Ayahuma *Couroupita guianensis* fruit in the medicinal plant section of a street market. Mercado Belén, Iquitos, Peru 1-7-06

30.4 *Eschweilera coriacea*

Bolivia Charque, charqui, blanquillo, matamata, pancho, pancho amargo, pancho blanco, charque, tamarindo, estopa

Brazil Matamatá, matamatá branco, matamatá roxo, matamatá vermelho, matamatá preto, matamatá amarelo, matamatá jibóia, matamatá folha graúda, matamatá casca grossa, maturi, morrão, tiriba branca, toari, estopeiro, ripeira, quiriba, gema de ovo; iwitiri-oho, iwitiri-'iwa (Arawaté), iwɨritiri-'ɨwa (Asurini), wiri'ɨ (Guajá), parawa'yhu, iwiri'y, parawa'ɨ (Ka'apor), awawiha-'ɨw-pihun (Tembé), dwoma, doma (Waimiri-Atroari), hõtokomãri, hõkotomari, hokoto uhi (Yanomami)

Colombia Machimango, carguero, carguero negro, carguero de gavilán, fono, fono blanco, turú, popai, matamatá, matamatá blanco, matamata negro, guasca, guasco, coco majagua; ká-ma-kö (Cubeo)

Ecuador Coshnua, cupo, cashnum, sabroso, matamatá; unguapi (Cofán), quebrachi (Kichwa), sacha iñako (Shuar), toa, asao (Siona-Secoya), mencawue, buencabuhue, buencabo (Waorani)

Peru Machimango, machimango blanco, machimango colorado, machimango negro, machimanguillo, almendrillo, aletón, palo misho, misa, misa blanca, pínush, huacapu; ati, pínush, shuwát (Aguaruna), tohjúhe (Bora), shuwat (Huambisa), ison chomo, joxo ison chomo, shinon rao (Shipibo)

Venezuela Coco de mono, totuma de mono, puepuyan; jocorouji (Yanomamo)

Status Wild

Matamatá *Eschweilera coriacea* flower gathered from the floor of an anthropogenic forest adjacent to an archaeological site. David Lucas 7 years, Vila da Associação, Caxiuanã River, FLONA Caxiuanã, Municipio de Melgaço, Pará, Brazil 7-16-16

Matamatá *Eschweilera coriacea* is a large tree reaching 37 m in upland and floodplain forests throughout the Amazon Basin (Assis et al. 2017; Bentes-Gama et al. 2002b; Carim 2016:26, 91; Cattanio et al. 2002; Dávila et al. 2015:32, 41; Gama et al. 2002, 2003; Korning et al. 1991; Lima et al. 2007; Macía 2011; Mackensen et al. 2000; Mori and Prance 1990a, b:205; Pires and Salomão 2000; Reynafarje 2014:11; Salomão and Lisboa 1988; Salomão et al. 1988, 2002; Santos and Jardim 2006; Silva and Rosa 1989; Spichiger et al. 1996; Vieira et al. 2014), including woodlands of an anthropogenic nature. This tree occurs on floodplains of various water types, including acidic black waters such as in the Apaporis Basin in the Colombian Amazon and along the lower Rio Negro, and in the Uatumã in Amazonas, Brazil (Cano and Stevenson 2009; Lobo 2017:65; Parolin et al. 2002) as well as the Arapiuns River, an affluent of the lower Tapajos near its confluence with the Amazon (Kubitzki 1989). It also occurs in some tidal forests in the Amazon estuary (Almeida and Jardim 2011, 2012; Braga and Jardim 2019; Braga et al. 2015; Silva 2008).

It also occurs in successionary forests, including 20-, 30-, and 40-year-old woods regenerating in the Bragantina zone in northeastern Pará, 10-year-old successionary growth in the Tapajós National Forest, 20-year-old successionary forest in the Tocantins watershed, as well as tidal flats in the Amazon estuary (Almeida and Vieira 2001; Alvino et al. 2005; Aparicio et al. 2014; Balée 1993b; Bentes-Gama et al. 2002b; Coelho et al. 2012; Espírito-Santo et al. 2005; Gomide 1997:111; Gualberto et al. 2014; Medeiros 2016:92; Oliveira and Jardim 1998; Prata et al. 2010; Rayol et al. 2006, 2011; Schwartz 2007; Silva 2008:28, 2015:63). It can be found in successionary vegetation as young as 5 years, such as abandoned agricultural plots in the vicinity of Santa Mercedes along the Putumayo River in the northern Peruvian Amazon bordering Colombia (Gomez 2011:81; Muñoz 2010:18). It thrives in light gaps and woods dominated by a native bamboo *Guadua weberbaueri* in Acre in the southwestern portion of the Brazilian Amazon (Maranho et al. 2016). This tree is found in some cacao orchards at km 107 of the Altamira-Itaituba stretch of the Transamazon Highway (Leão et al. 2017). It is also found in the stunted forest (*campinarana*) which grows in areas with deep sandy soils such as in the vicinity of Manaus, Amazonas (Almeida 2012:58).

Eschweilera coriacea ranges widely in the Amazon Basin up to about 1700 m in the Andean foothills and also occurs in the Guianas, Venezuela, and Panama (Bánki et al. 2003:35; Godoy et al. 1999; Korning and Balslev 1994b; Larue et al. 2002; Matta and Scudeller 2012; Milliken 1998; Mori and Lepsch-Cunha 1995; Rudas and Prieto 1998; Thomas 2009; Valencia et al. 2004). *Eschweilera coriacea* is categorized as a hyperdominant and is one of the most common species in most forest types (Almeida et al. 2012; Carneiro 2004; Duque et al. 2009; Giraldo-Cañas 1999; Lima Filho et al. 2001; Souza et al. 2006; Steege et al. 2013). This tree can endure as long as a thousand years (Laurance et al. 2004). At least some of the hyperdominant forest species are likely the result of human agency such as planting, concentrating, or dispersing their propagules, clearing vegetation around the desirable trees, and/or fertilizing them with refuse (Roosevelt 2014). Such may well be the case with *Eschweilera coriacea*.

This tree produces nut pods (*ouriços*) similar to those of sapucaia *Lecythis pisonis*, only smaller. Rural people remove the pod lids to access the nuts which are attached by fleshy arils. The nuts are eaten locally but they never reach markets. Macaws are also fond of the nuts which they access using their powerful beaks. Some game animals also eat the fallen pink and yellow flowers, such as paca *Agouti paca*, while certain monkeys, including the black-headed uakari *Cacajao melanocephalus melanocephalus*, the red uakari *Cacajao calvus ucayalii*, and the endangered black-bearded saki *Chiropotes satanas* consume the nuts (Boubli 1999; Bowler and Bodmer 2011; Veiga 2006:97). Although uakaris and sakis are not important game animals, they are nevertheless hunted occasionally for their meat. Woolly monkeys *Lagothrix lagotricha lagotricha* have been recorded eating the leaves and flowers of this tree in the Peruvian Amazon (Rimachi-Taricuarima et al. 2019); woolly monkeys are hunted for their meat, so this plant contributes indirectly to the diet of rural people in the Amazon.

Eschweilera coriacea has other uses, especially among indigenous groups. The Waimiri-Atroari who live north of Manaus, Amazonas, use the inner bark to fashion head straps to bring home game killed in the forest; they also use the wood to fashion posts on the outer ring of their communal houses (Milliken and Albert 1997; Milliken et al. 1992:32, 39). The Asurini in the Xingu watershed also use the inner bark for straps and the wood for house posts (Balée 1989). The Deni who live along the Cunhuá River in the Purus watershed and the Arawaté who live along the Ipixuna, an affluent of the middle Xingu, use bark strips torn from young trees for tying up items, such as on backpacks, as do the Yanomami who live in a large area straddling the Brazilian state of Roraima and Venezuela (Shepard 2006). Rural folk in the Peruvian Amazon also employ strips torn from the bark for straps (Vásquez 1997:420). The Ka'apor in eastern Amazonia employ the bark to coagulate blood.

The Arawaté and the Yanomami also use the tree for firewood. The Tembé in the eastern fringes of the Amazon forest use the wood for fuel when making manioc flour and also for home construction (Balée 1987), as do river dwellers in other parts of the Amazon Basin, such as in the vicinity of Leticia in Colombia. The trunk is also used to fashion canoes, such as in the Peruvian Amazon, because the wood is hard and durable. Matamatá wood is highly resistant to water and shipworm (*Neoteredo* sp.), known as turu in Brazil, so is highly desired to provide posts for bridges and elevated walkways on floodplains (Huber 1910). Some residents of Jericó community, Garrafão do Norte, in northeastern Pará use this durable wood to make planks, posts, and furniture (Vieira 2005:105). The Palikur in French Guiana use this wood for a variety of construction purposes (Ogeron et al. 2018). These traditional uses of *Eschweilera coriacea* are under threat since matamatá produces valuable timber so the trees are under threat from commercial loggers, as in the vicinity of Jaci Paraná in Rondônia (Silva and Bentes-Gama 2008).

30.5 *Grias neuberthii*

Colombia Cocoro, cocora, kokora, pepe guara, mortecino, mula muerta, chupa; té-té-ku-chu (Cofán)

Ecuador Pitón, pitón panga, sacha pitón, pitun ruya, passo, membrillo, huevos de burro, cocora, aguacatillo, bola de chivo; apai (Achuar), ttette'ccu, t'cucho (Cofán), pitón, sacha pitón, pitun, jahuiyo (Kichwa), kãsi, cansüi soquëñë, casë (Siona-Secoya), benhahue, huengahue, nuencahue, noguincahue, wegaka, wenaka, wenawe, wingaka (Waorani)

Peru Sachamangua, mangua, sachamango, chope, sacha chope, membrillo; apái, apia, kuapiu (Aguaruna), pitun (Kichwa), apái (Mayna), jihuin huarana, ni mancoa (Shipibo), mangua chicü (Ticuna)

Status Wild, planted

Sachamangua *Grias neuberthii* in flower and fruit in a home garden. Soledad, Rio Itaya, Loreto, Peru 1-7-06

The Brazil nut family has at least five wetland representatives with edible fruits or nuts. Three of them—*Grias neuberthii*, *Gustavia longifolia*, and *Couroupita guianensis*—produce pulp eaten by people and livestock. The floodplain members of the Brazil nut family with edible nuts, *Lecythis pisonis* and *Allantoma lineata*, are confined to central and eastern Amazonia. *Grias neuberthii* also grows in some upland forests, such as in the Ecuadorian Amazon (Grimes et al. 1994; Grubb et al. 1963;

Korning and Balslev 1994b; Macía 2011; Marcalla 2012:17; Navarrete 2016:10; Pitman et al. 2005).

Known as sachamangua ("mango-like" or "wild mango") in Peru, *Grias neuberthii* is both wild and cultivated in western Amazonia. The range of *Grias neuberthii* extends from Peru to the Ecuadorian and Colombian Amazon (Friedman et al. 1993; Korning and Balslev 1994a, b; Korning et al. 1994; Prance and Mori 1979). *Grias neuberthii* occurs up to 1000 m in the Andes and may overlap with a similar species, *Grias peruviana*, also called sachamangua, in the high selva (Pennington et al. 2004).

Enoc, 13 years, picking sachamangua *Grias neuberthii* fruit in a home garden. Buenos Aires, Rio Pucate, Reserva Nacional Pacaya-Samiria, Loreto, Peru, 1-14-06

Sachamangua is an understory tree, usually less than 20 m high, that occurs in floodplain forests as well as uplands. Sachamangua sometimes forms stands in the forest; in the vicinity of Jenaro Herrera along the Ucayali, for example, densities of mature trees can reach 248 individuals/ha (Peters 1990). In other areas of the

Peruvian Amazon, stands of this species can reach 500 individuals per hectare (Villachica et al. 1996:245). It is also commonly planted in home gardens, such as along rivers in the Ucayali and Marañón watersheds. The Yagua plant this tree in some of their home gardens along the Amazon downstream from Iquitos (Lamont et al. 1999). Some farmers intercrop *Grias neuberthii* in their fields, as in the vicinity of the confluence of the Napo and Coca Rivers in Ecuador (Uquillas et al. 1992), and Victoria along the Puinahua, a branch of the Ucayali River. The Cofán in the northern part of the Ecuadorian Amazon bordering Colombia also plant this tree in their fields (Cerón 1995:109). Some concentrations of sachamangua in forests may thus be the result of human activities.

Grias neuberthii flowers in the dry season and most of the fruits mature during the rainy months. The elongated fruits of sachamangua vary in size from 10 to 18 cm long and 5 to 10 cm in diameter and are born along the trunk (technically called cauliflory). Many of the fruits are within easy reach. The olive skin is peeled to reveal an orange-yellow pulp with a crisp texture similar to carrots. The pulp surrounding the single seed is relatively thin, comprising only about 15% of the fruit by volume. People either gnaw or cut off the mesocarp. The firm pulp has a nutty flavor, similar to almonds, and is rich in vitamin A since it contains 1.2 mg of carotene per 100 g of pulp (Peters and Hammond 1990). The pulp is also boiled to extract the oil which contains appreciable amounts of the omega 3 fatty acid, good for lowering bad cholesterol (Alva et al. 2002). The Siona-Secoya in the Ecuadorian Amazon bake the fruits over fire embers (Vickers 1994) and they also ferment the pulp in water to make an alcoholic drink called *chicha* in Spanish (Marcalla 2012:100). The highly nutritious fruits appear in markets all year-round but are more common in the rainy season.

The fruits, seeds, and inner bark of this tree also have various medicinal uses. In the Ecuadorian Amazon, the Siona-Secoya grate the fruit and mix it with water which is then drunk to serve as a purgative (Schultes and Raffauf 1990:227). The Cofán who live in the northern part of the Ecuadorian Amazon boil the single seed found in each fruit in water and then when the liquid cools drink it to provoke vomiting which cleans out the stomach (Cerón et al. 1994:50). The Achuar who inhabit the Morona Santiago and Pastaza provinces of the Ecuadorian Amazon treat malaria by scraping the bark of this tree and mixing it with water which is then boiled; after cooling the infusion is drunk to provoke vomiting (Giovannini 2015). Kichwa women along the Napo River in the Ecuadorian Amazon undertake a ritual 1 week after giving birth by drinking a concoction of the fruit and bark of this tree which provokes vomiting. This act is thought to ward off evil spirits and fortify the baby (Iglesias 1991). The Kichwa also use the seeds to make a paste which is applied to skin ulcers caused by leishmaniasis (Gachet et al. 2010). In the Peruvian Amazon, the Kichwa prepare an infusion from the stem or decoction from the bark of this tree to treat intestinal parasites and malaria (Roumy et al. 2007). Rural folk in the Peruvian Amazon also use the fruits in a similar fashion for the same purpose (Flores 1997:251). Natives along the Chico River use the cambium to induce vomiting in people suffering from malaria. The seed is employed in an enema for treating dysentery (Schultes and Raffauf 1990:228).

Pedro Luis, 13 years, eating a sachamangua *Grias neuberthii* fruit gathered in a home garden. San Carlos, Rio Puinahua (a branch of the Ucayali River), Reserva Nacional Pacaya-Samiria, Loreto, Peru 4-17-06

In Ecuador, the Kichwa protect the tree because of its edible fruits and its various medicinal uses. Furthermore, the fruits are considered food for Sacha Runa, a forest spirit (Bennett 1992). The Siona-Secoya mix the grated fruits with water for use as a purgative (Vickers 1994; Vickers and Plowman 1984). Various other parts of the tree are also used medicinally in the Peruvian Amazon (Castner et al. 1998:57). The Waorani in the Ecuadorian Amazon eat this fruit and use the wood in house construction and for fuel and they fashion forked poles to support various trees near their homes that are laden with fruit (Cerón and Ayala 1998:89). The Waorani use this wood to make smoke for blackening their arrows to make them more durable and the wood is also used to dig holes in their fields to plant manioc and plantain (Mondragón and Smith 1997:76).

Peeled sachamangua *Grias neuberthii* **fruit in a public market. Requena, Rio Tapiche, Loreto, Peru 4-13-06**

Sachamangua is an ideal candidate for further development as a crop and for the management of wild and anthropogenic stands. The mesocarp is highly nutritious, the oil has medicinal uses, and the tree is in fruit much of the year thus avoiding a labor bottleneck at harvest time. Furthermore, the fruits travel well and sell briskly in markets. Eventually, more distant markets for the fruit could be developed for the hardy fruit, perhaps pitched as the "Amazon tree carrot."

30.6 *Gustavia augusta*

Bolivia Coquito

Brazil Jenipaparana, jeniparana, genipaparana, geniparana, geniporana, murrão, mucurão, pau general, mata fome, matamatá, matamatá do igapó, orelha de burro, canela de velha, castanharana, castanha fedorenta, periquito castanha, pau fedorento, cachimbo juruparana, janipara, envira de cutia, ripeira, tauarí liso; janiparani'i (Arawaté), yanipa-ran-ɨwa (Asurini), tikõtara-'ɨ (Guajá), mitupusu'ɨ (Ka'apor), zanipa-ran-iw (Tembé), a'i-walipi (Wajãpi), pooko sihi (Yanomami)

Colombia Castaña, matamatá, matamatá pequeña, tifaicugao

Peru Chopé, sacha chopé, sacha mango, copa de mono, castaña, mangua, barbacoa caspi

Venezuela Koyamo, guatero, guatoso, baraboro, velero, cacao cimarrón, malabar de agua, coco de mono, rosa de muerto, kudumadashi; hõnokore (Yanomami)

Status Wild

Jenipaparana *Gustavia augusta* in flower and fruit on a floodplain island of a clear water river. Ilha do Palhal, Rio Xingu between Altamira and Belo Monte, Pará, Brazil 8-13-15

One of several members of the Brazil nut family with edible fruits or nuts, *Gustavia augusta* grows in a wide range of habitats and environments but the fist-sized fruits (technically a dehiscent pyxidium) are only eaten sporadically. It is found in riverine forests along black, clear, and muddy rivers throughout the Amazon, Venezuela, and the Guianas (Amaral et al. 1997; Andel 2003; Batista et al. 2011; Bentes-Gama et al. 2002a, b; Carim 2016:26; Casula 2012:62; Díaz Pérez et al. 2012; Fernandes-Corrêa and Furch 1992; Gama et al. 2002; Godoy et al. 1999; Irmler 1977; Klinge et al. 1995; Luize et al. 2015; Montero and Latrubesse 2013; Montero et al. 2014;

Myster 2018; Navarro 2002; Navarro and Maldonado 2002:69; Parolin 2000a, b, c; Prance 1978, 1979, 1980; Reis 2010:34; Ratter 1987; Santos and Jardim 2006; Scudeller and Vegas-Vilarrúbia 2018; Wittmann et al. 2002, 2010). It also occurs in the tidal zone of the Amazon estuary (Almeida and Jardim 2011, 2012; Amaral et al. 2009; Batista and Gonçalves 2013; Bentes-Gama et al. 2002b; Carim et al. 2008; Cattanio et al. 2002; Freitas et al. 2019; Maués et al. 2011; Queiroz et al. 2005, 2006, 2007; Santos et al. 2014a; Vale et al. 2014) and neighboring Maranhão (Freire and Monteiro 1993). This tree is also found on some sand dunes along the coast of the Bragantina zone in northeastern Pará (Santos and Rosário 1988; Santos 2010:43). The leaves of this tree can remain submerged for several months without damage (Parolin et al. 2004a, b).

It also occurs occasionally in some upland forests, such as along the Iquitos-Nauta highway in the Peruvian Amazon (Zárate et al. 2013), in the Floresta Nacional do Tapajós 69 km south of Santarém, Pará (Sandel and Carvalho 2000), in the watershed of the lower Rio Branco in Roraima (Alarcón and Peixoto 2007), on Maracá Island along the Uraricoera River, a tributary of the Rio Branco in Roraima (Nascimento et al. 1997), the lower Xingu in Pará (Salomão et al. 2007a), in Amapá (Santos et al. 2017), and in the vicinity of Belém (Mackensen et al. 2000; Pires and Salomão 2000) and Manaus (Camargo 2018:7). This bush to small tree, which can reach 22 m high when fully mature (Prance and Mori 1979:157), occasionally occurs on uplands, including forest and successionary vegetation (Absy et al. 1986; Balée 1993b, 2013:65; Campbell et al. 1986; Carim et al. 2007; Claudino et al. 2015; Espírito-Santo et al. 2005; Gomez 2011:86; Guimarães et al. 2016; Lemos et al. 2015; Moser 2013:27; Muñoz 2010:25, 37; Rayol et al. 2006, 2011; Salomão et al. 2002; Schwartz 2007; Souza 2019:12). In the municipality of Bragança in northeastern Pará, this tree has been recorded in 44-year-old successionary woodlands (Muniz et al. 2007). It also occurs in light gaps and within woodlands dominated by native bamboo *Guadua weberbaueri* in Acre in the southwestern part of the Brazilian Amazon (Maranho and Salimon 2015; Maranho et al. 2016). It also grows on sand dunes in the coastal region of Pará state (Lisboa 2012:461; Santos and Bastos 2014:123) as well as on some of the higher grasslands on Marajo Island at the mouth of the Amazon (Miranda 1909). The flowers are pollinated by at least one species of stingless bee, *Melipona fasciculata* (Martins et al. 2011).

Along the Yanayacu River in the Reserva Nacional Pacaya-Samiria, rural folk generally eschew the fruits whereas in a nearby watershed, the Samiria River, locals appreciate the pulp. Jenipaparana, one of the many names for this fruit in Brazil, is also consumed by river folk along the Xingu River. In a survey of 57 home gardens I conducted along the middle Xingu in 2015, *Gustavia augusta* was found in five of them. Many of the home gardens surveyed had been abandoned because they will be permanently flooded by the Belo Monte reservoir, so it was not possible to ascertain whether the jenipaparana had been planted or arose spontaneously in those house yards. The Yagua indigenous group along the Amazon River downstream from Iquitos also plant this tree in some of their home gardens (Lamont et al. 1999). In the municipality of São Félix do Xingu in southern Pará this tree is encountered in some cacao agroforestry plots (Braga et al. 2019).

The "pulp" is the placenta attached to the inside of the capsule and the funicles, which is a yellow fleshy material attached to the black seeds. The edible portion of the fruit is accessed by prying off the lid at the bottom of the capsule. The capsules vary from 3 to 7 cm long and 3 to 8 cm wide and are born in the rainy season to coincide with the annual flood (Schöngart et al. 2002). Young trees can survive being submerged and can even maintain photosynthesis under water (Parolin et al. 2010). The fruits are typically gathered for local consumption by canoeing up to the seasonally flooded trees. *Gustavia augusta* starts producing fruits when it reaches the size of a 2-m shrub and although a mature tree can reach 20 m, many fruits are easily accessible at high water given the wide amplitude of floods in most parts of Amazonia.

The fruits are also eaten by fish and game animals, thus indirectly contributing to the diet of locals. Currents likely disperse the seeds as well as certain fish, such as gamitana *Colossoma macropomum*, known as tambaqui in Brazil, pacu (several genera), and sabalo (*Brycon*) which feed on the fruits according to riverine folk along the Yanayacu River and may occasionally ingest entire seeds which are later passed. In the Brazilian Amazon, tambaqui has been documented eating the fruits of *Gustavia augusta* (Honda 1974). The largest turtle in the Amazon, tartaruga *Podocnemis expansa*, also eats the fruits. People eat tartaruga turtles as well as their eggs which they dig up on sandy beaches during the later water season (Best 1984).

Monkeys are also possible dispersal agents, as suggested by the common name for *Gustavia augusta* in Yarina along the Yanayacu River: copa de mono (monkey's cup). Spider monkeys (*Ateles* spp.), capuchin monkeys (*Cebus apella*, *C. kaapori*), squirrel monkeys *Saimiri sciureus*, and bearded sakis *Chiropotes satanas* are known to feed on the fruits and rodents scatter hoard some of the nuts when they drop to the ground (Ghilardi and Alho 1990; Oliveira et al. 2014). Spider monkeys are hunted for their meat but capuchin monkeys, squirrel monkeys, and bearded sakis are relatively small primates and are not usually shot. The white-faced saki monkey *Pithecia pithecia*, which inhabits eastern Venezuela, the Guianas, and the northern part of the Brazilian Amazon, scrapes off the aril attached to the seeds and eats this fleshy appendage (Norconk and Veres 2011; Roble 2008:34); the seeds are then dropped and some of them are likely spread away from the tree. In French Guiana, the grey brocket deer *Mazama gouazoubira* has been documented eating the seeds of this fruit (Gayot et al. 2004); this is a much procured game animal in the Guianas and the Amazon Basin.

Gustavia augusta is also employed in a variety of folk remedies, including the treatment of leishmaniasis (Souza et al. 2001). The Ka'apor in the eastern fringe of the Amazon rainforest prepare a decoction using the inner bark of this tree which is swallowed over several days to treat venomous snake bites; shavings of the inner bark are also soaked and applied to the sites of snake bites (Balée 1993a:111; Vieira 2005:103). In the vicinity of Belém in the Amazon estuary, the nuts are used to treat inflammation (Jardim and Medeiros 2006). The pulp of this fruit has antioxidant properties (García-Torres et al. 2009).

The tree has other uses as well. In Roraima, a thin strip of wood (lath or slat) is obtained from the trunk to finish off the tops of woven baskets (Alarcón and Peixoto 2008). In French Guiana, the Palikur use the wood of this tree for a variety of construction purposes (Ogeron et al. 2018) as do some rural people in the Amazon estuary in the state of Amapá (Pinedo-Vasquez et al. 2001). In some rural areas of the Brazilian Amazon, the roots are used to prepare a laxative and the leaves are used to make a tea to treat jaundice (Le Cointe 1934:177). This species has not been domesticated for fruit production, but its attractive flowers and capsules have won it a place in some botanical gardens (Prance and Mori 1979). Nurseries that specialize in tropical plants also occasionally stock seedlings of *Gustavia augusta*. The tree's large, showy pink and white flowers have also inspired artists, such as the late Margaret Mee, and painters decorating pottery for sale.

Gustavia augusta. **Painting by Antonio Cortes Alcocer, eighteenth century. Archivo del Real Jardin Botanico (CSIC), Madrid, III, 2673**

30.7 *Gustavia longifolia*

Brazil Chope, castanha fedorenta

Colombia Cocora, fono, matamatá, mula muerta; jameda ime ñukuri (Witoto)

Ecuador Pacora, paso silvestre, iñaco, tripa de pollo; ttette'ccu, aycopi (Cofán), ayampaso, sacha paso muyo, allan pasu (Kichwa), apái (Shuar), jo'co (Siona-Secoya), doquinca, moinka, noginka, noguincahue, noguiwe, noinga, nugicahue (Waorani)

Peru Chopé, sacha chopé, chopé masha, chopé sachavaca, chope amarillo, sacha papa, sacha manga, sacha vaya, mangua, tripa de pollo, rugoso, paso silvestre; kuap, apái (Achuar), pamáusuki, apái (Mayna)

Status Wild, planted

Chope *Gustavia longifolia* in fruit in the understory of a floodplain forest along a black water river. Rio Pucate, Reserva Nacional Pacaya-Samiria, Loreto, Peru 4-21-06

Chope *Gustavia longifolia* is an understory tree that can reach 30 m high in the central and western part of the Amazon from roughly the Rio Negro and Madeira rivers in Brazil west to Colombia, Ecuador, and Peru (Korning and Balslev 1994a, b; Korning et al. 1991, 1994; Macía 2011; Macía and Svenning 2005; Navarrete 2016:28; Prance and Mori 1979; Santos et al. 2018; Valencia et al. 2004). Normally found in upland forests, the tree is also found on the higher parts of floodplains, as in the Reserva Nacional Pacaya-Samiria in Peru. As the species name suggests, the tree has exceptionally long leaves, similar to those of sachamangua (*Grias neuberthii*), another member of the Brazil nut family with edible fruits. The reddish-brown capsules ripen toward the end of the rainy season.

The round fruits are about 8 cm in diameter and have leathery rather than woody skins with green lids. The skin is cut open to access the pulp inside the reddish-brown fruits. Chope fruits contain a yellow pulp surrounding about nine black seeds, similar in shape and size to those of breadfruits or chestnuts. The savory mesocarp tastes similar to the nuts of Brazil nut *Bertholletia excelsa* and sapucaia *Lecythis pisonis*. Rural folk in parts of the Peruvian Amazon extract oil from the pulp for domestic consumption (Flores 1997:106).

Indigenous peoples and mestizos use this plant for a variety of other purposes, including the treatment of various health issues. For example, in the vicinity of Chazuta along the upper Huallaga River in the Peruvian Amazon, rural folk make a decoction from the leaves to treat malaria and relieve the discomfort of colon contractions (Sanz-Biset et al. 2009). The Cofán in the northern part of the Ecuadorian Amazon grate the bark into water and then boil; after cooling, the liquid is applied to treat a variety of skin problems (Cerón 1995:109). In addition to eating the fruits, the Shuar in the southern Ecuadorian Amazon use the hardwood for posts and for fuel (Bennett et al. 2002:193). Likewise, in the Ecuadorian Amazon, the Waorani eat the fruits and use the wood for fuel (Cerón and Ayala 1998:90).

Alessandro, 8 years, with a chope *Gustavia longifolia* fruit he gathered in the forest on the steep bank of a black water river. Rio Pucate, Reserva Nacional Pacaya-Samiria, Loreto, Peru 6-4-04

Although the *Gustavia longifolia* can become a sizeable tree, it starts producing fruits when it is the size of a short bush, around 1 m high. When it becomes larger, the tree is sometimes cut down to gather the fruits, such as in parts of the Peruvian Amazon. In the vicinity of Buenos Aires, a village at the confluence of the Pucate and Yanayacu rivers in the Reserva Nacional Pacaya-Samiria, the fruits on one chope tree were defended by large black ants. The ants inflict a painful bite but at least have no stingers. Several trees in the Amazon, as in other tropical lands, have evolved commensal relationships with ants; in the case of *Gustavia longifolia* some of the twigs are hollow to provide shelter for the ants.

The Bora who live about 150 km northeast of Iquitos in the Peruvian Amazon cultivate *Gustavia longifolia* in managed fallows. Rural folk in the departments of Loreto and San Martín in the Peruvian Amazon also plant this tree as do the Waorani of the Ecuadorian Amazon (Flores 1997:105).

Squirrel monkeys *Saimiri sciureus* have been observed eating chope fruits in floodplains forests of the Yarapa River in Loreto Province of the Peruvian Amazon. It is not known, however, if monkeys are involved in dispersing the seeds. Chope fruits appear occasionally in markets, such as Iquitos (Vásquez and Gentry 1989). Black-headed uakaris *Cacajao melanocephalus melanocephalus* have been observed feeding on the nuts in forests near Pico da Neblina in Amazonas, Brazil, close to the border with Venezuela (Boubli 1999). Squirrel monkeys are generally not hunted because of their small size but it is likely that rural people hunt uakaris, especially indigenous groups.

30.8 *Lecythis pisonis*

Colombia Lairo jero (Witoto)

Brazil Sapucaia, sapucaia vermelha, macaco de cuia, cumbaca de macaco, marmita de macaco, cumbaca de macaco, castanha de macaco, caçamba do mato, jabucai, jatereua, jarana, castanharana, sabucaié; kol-ti (Apinajé), japukai (Arawaté), ya-makai-'ɨ (Guajá), japukwai'ɨ (Ka'apor), pi'y tê krê ti (Kayapó), iwyri'yw (Tembé), awanma (Waimiri-Atroari)

Colombia Nuez del paraíso, sapucaia; iairo jero (Witoto)

Peru Castaña de monte, castaña silvestre, sacha castaña, machimango, machín mango, olla de mono, nuez del paraíso, sapucaya

Status Wild, planted

Sapucaia *Lecythis pisonis* in fruit in a home garden on the Amazon floodplain. Fazenda São Lourenço, Paraná Miri, Alenquer Municipality, Pará, Brazil 3-20-93

Sapucaia *Lecythis pisonis* is a giant tree with edible nuts, soaring to 50 m in the upland forest (Almeida et al. 1993; Espírito-Santo et al. 2005; Mori and Prance 1990b:297; Silva et al. 2014), but tends to be smaller with a broad crown, like a spreading oak tree, on floodplains, especially in areas that have been cleared, such as for cattle pasture. This tree occurs in some woodlands in the tidal zone along the coast of northeastern Pará (Almeida and Jardim 2012). Sapucaia is more common in central and eastern Amazonia, but also occurs sporadically in the western and southern part of the basin as well as the Atlantic forest of Brazil (Amaral et al. 2017; Barnett et al. 2015; Batista et al. 2015; Gama and Pinheiro 2010; Lemos et al. 2015; Lorenzi et al. 2006a, b:156; Mori 1990; Mori et al. 1990:291; Salomão et al. 2002, 2007b; Silva et al. 2018). Sapucaia also thrives in successionary forests and on archaeological sites with anthropogenic dark earth and is thus associated with some cultural forests (Almeida and Vieira 2001; Alvino et al. 2005; Balée 1993b, 2010; Gualberto et al. 2014; Hiraoka et al. 2003; Prata et al. 2010; Rayol et al. 2011; Santana et al. 2004; Schwartz 2007; Taurino 2017:102; Vieira 1996:59). It also occurs in some cacao orchards at km 107 of the Altamira-Itaituba stretch of the Transamazon Highway (Leão et al. 2017).

Also called castanha in the vicinity of Santarém, the 10–50 nuts are encased in a large, bowl-like shell (technically a capsule) that hangs upside down. Some nuts studied in Mato Grosso varied in length from 3.48 to 4.92 cm (Braga et al. 2007). The capsules can reach 30 cm in width and weigh up to 7 kg, such as in the vicinity of Afuá, a small town on stilts in northwestern Marajó Island near the mouth of the Amazon (Silva 2011:238). The tasty nuts are secured by a fleshy appendage called an aril and are tamped inside the inverted "bowl" by a saucer-shaped lid. The nuts are eaten fresh and in the Peruvian Amazon also boiled and roasted (Brack 2003:126). When the nuts are ripe, the lid, called an operculum, falls off and bats, such as the greater spear-nosed bat *Phyllostomus hastatus*, often abscond with the nuts to feed on the attached succulent arils (Mori 2001; Mori et al. 2017). After the bats have finished dining on the fleshy attachments, the nuts are dropped to the ground where some eventually germinate (Prance 2014:142). When the lid falls off monkeys sometimes beat the bats and eat the nuts. The capsules of the better-known Brazil nut (*Bertholletia excelsa*), which is confined to upland forests, are indehiscent; the lid remains shut until it is gnawed open by agoutis and squirrels, the main dispersal agents for Brazil nuts. Even though the nuts of sapucaia taste better than those of Brazil nut trees the former are rarely encountered in markets because most of the nuts are dispersed rapidly or eaten on the spot by wild animals (Sampaio 2000).

Sapucaia *Lecythis pisonis* nuts with attached arils. Near Santarém, Pará, Brazil 5-10-96

Pollinated by carpenter bees (*Xylocopa frontalis*) which are strong enough to push through the hooded flowers (Mori et al. 1990:298, 2017), sapucaia blooms in the dry season and the nuts are harvested during the rainy season from January to May. To subvert the bats or prevent the nuts from spilling on the ground, people frequently harvest the nuts just before they are fully ripe. Sapucaia nuts are between 4 and 6 cm long and 3 cm wide and weigh about 5.5 g. A mature tree can produce close to 80 kg of nuts annually (Andrade et al. 1999b). Sapucaia flowers fall to the ground, whether pollinated or not, and are eaten by game animals such as paca, while collared peccary, and brocket deer (Barboza 2008:55). A matrilineal moiety among the Apinayé who inhabit interfluvial uplands between the Tocantins and Araguaia rivers is named after this tree (Nimuendajú 1939:21).

Sapucaia is occasionally planted in home gardens. In my survey of 150 home gardens in the Brazilian Amazon conducted over several decades I noted sapucaia planted in six of them, including the home yards of river dwellers on Ilha do Carmo, a floodplain island along the Amazon in the Municipality of Alenquer, near Arapixuna in the Municipality of Santarém, and on the outskirts of Juriti up the Amazon from Santarém. In a separate survey of 57 home gardens along the middle Xingu in 2015, I found sapucaia trees in three of them. Rural folk living along Ramal do Pau Rosa a side-road leading from the BR 175 highway linking Manaus with Boa Vista in Roraima also plant this tree in some of their home gardens (Machado 2016:34) as do some rural folk in the vicinity of Abaetetuba along the lower Tocantins River in eastern Pará (Gonçalves and Lucas 2017), in the Municipality of Alenquer in western Pará (Rayol et al. 2013), and in Comodo community in Mato Grosso (Santos and Guarim Neto 2017). Some rural communities near Macapá, the capital of Amapá, also plant this tree in their home gardens (Pereira et al. 2017).

Sapucaia *Lecythis pisonis* capsule (*ouriço*) in an anthropogenic forest adjacent to an archaeological site. David Lucas 7 years. Vila da Associação, Caxiuanã River, FLONA Caxiuanã, Municipio de Melgaço, Pará, Brazil 7-16-16

Sapucaia nuts are consumed locally and sold in nearby markets, such as Santarém. In the latter part of the nineteenth century, small quantities of sapucaia nuts made their way to the London market where they fetched three times as much as Brazil nuts (Hooker 1849; Smith 1879:313). In the mid-nineteenth century, the botanist Richard Spruce remarked that sapucaia nuts are "delicious, resembling in taste the sweetest of filberts, and are much preferred by everybody to the common Castanha, the fruit of Bertholletia excelsa" (Spruce 1853). In the early 1900s, some six metric tons of sapucaia nuts were exported annually; in the United States they were called paradise nuts, a savvy marketing ploy (Akers 1912:69). In the middle of the nineteenth century, sapucaia capsules were sold in Covent Garden, London, under the name "monkey's drinking-cup" (Bates 1863a:67). Today, sapucaia capsules, which can reach 25 cm across, are sometimes used in the Amazon as flowerpots, such as in Juriti, Pará, and are also sold in urban centers to serve as ashtrays.

Abiu ***Pouteria caimito*** **seedling in a sapucaia** ***Lecythis pisonis*** **capsule in a home garden. The abiu will be transplanted when it is a little larger. Ilha dos Porcos, near Afuá, Amazon estuary, Pará, Brazil 8-17-98**

Curiously, sapucaia nuts are apparently no longer exported. Known in English as paradise nuts, they were able to penetrate overseas markets a century ago because regular steamships plied the Atlantic between the Amazon the United States and Europe. Paradise nuts were so named not because eating them ensures entry into heaven but because of their sublime taste. Indeed, some consider it the best-tasting nut in the world, and I agree (Rosengarten 1984:306). Steamships first started making their way up the Amazon in 1853, and as the rubber boom took off, more lines and crossings were added to Europe and the United States. Sapucaia nuts, which contain 20% protein may make a comeback, however, because they taste even better than Brazil nuts, which are already well accepted in North American and European markets, but they have half the fat content, an important consideration in markets with many weight-conscious consumers. But that would entail protecting forests and keeping out illegal loggers. Plantations could be established, but perhaps not in the Amazon where diseases and pests could trigger epidemics if the trees are planted close together.

Sapucaia and Brazil nuts are high in barium and selenium, a potential health concern because of those elements' carcinogenic properties when eaten in high doses, but it appears that those minerals are mostly bound in the nuts in a form that is hard for the body to absorb, and thus pass largely undigested. Sapucaia nuts are high in protein, especially in methionine; levels of that amino acid are twice the levels found in whole egg protein (Andrade et al. 1999a). Methionine boosts the absorption of iron; additionally, sapucaia nuts are high in iron. The consumption of sapucaia nuts would thus help alleviate anemia. The Ka'apor who inhabit the eastern fringes of Amazonia in the state of Maranhão do not permit their children to eat sapucaia nuts so that they do not become lazy (Balée 1993a:67). Sapucaia nuts are sometimes ground up and fed to backyard chickens, such as along Igarapé Jari near Santarém.

Sapucaia also has a few medicinal uses. In the Municipality of Bragança in northeastern Pará, for example, a juice is squeezed from the leaves and rubbed on the skin to treat itching, especially when triggered by tick bites (Rios et al. 2001; Silva et al. 2008:116). In the northeast part of Mato Grosso the bark of this tree is used to make a decoction to treat venereal disease and itching (Ribeiro et al. 2017). In Nova Xavantina in Mato Grosso, the leaves are used to make a decoction to treat anorexia and hypolipidemic conditions, such as caused by high levels of cholesterol in the blood (Silva et al. 2010). In Saracura, a quilombola community on a floodplain island along the middle Amazon near Santarém, the leaves and bark are used to prepare tea and bathing solutions to treat muscle pain and rheumatism (Oliveira and Cavalcante 2017). The Xipaya indigenous people along the lower Iriri, an affluent of the Xingu, use the bark and leaves of this tree to make a bathing solution and a tea to treat mycosis (Santos et al. 2016a). In the early nineteenth century, the empty capsules of sapucaia were filled with water and after they had soaked for a while and were swallowed by people suffering from urine retention and/or edema (Baena 2004:69; Martina 1902:115). Residents of Colares Island some 100 km from Belém in the Amazon estuary macerate the leaves of this tree to treat dogs and water buffalo suffering from myiasis (Ritter et al. 2012). More recently, antimicrobial peptides in the seeds of this tree have been shown to inhibit the fungal pathogen *Candida albicans* which can infect people (Vieira et al. 2015).

The inner bark of sapucaia is used to make twine, such as on the Amazon floodplain near Alenquer, Pará, where rural folk use sapucaia string to tie up fish. The Waimiri-Atroari north of Manaus, Amazonas, use the bark as one of their fish poisons (Milliken et al. 1992:78). The Caishána (Cayuishána), an Arawak tribe that once inhabited the forest in the vicinity of Tonantins on the Upper Solimões River before becoming extinct in the early twentieth century, used to make a small apron from the inner bark of sapucaia Lecythis pisonis to cover the private parts of men and women (Bates 1863b:375).

Sapucaia timber is durable and harvested in the Amazon for a variety of construction uses (Parrotta et al. 1995:146; Veríssimo et al. 1992; Wittmann et al. 2009). In the watershed of the Capim River in eastern Pará, rural folk use this wood to make roof tiles and fence posts (Shanley and Rosa 2005) and in the Jericó community, Garrafão do Norte, in northeastern Pará rural folk also use the wood to make furniture (Vieira 2005:105). The dark reddish-brown wood has distinctive resin canals, like its distant cousin the Brazil nut tree. In the early nineteenth century, sapucaia wood was used in boat building (Baena 2004:56). Sapucaia would thus make an ideal candidate for floodplain agroforestry with seed production in the early years followed by harvesting the trees for timber once they have reached a suitable size. In the Comunidade Rural do Bacuri, 25 km southeast of Abaetetuba, Pará, the community of small farmers uses this wood in house construction and to make charcoal (Gonçalves and Lucas 2017).

References

Absy, Maria Lúcia; Camargo, João M. F.; Kerr, Warwick E.; Miranda, Ires Paula de Andrade. 1984. Espécies de plantas visitadas por Meliponinae (Hymenoptera: Apoidea) para coleta de pólen na região do Médio Amazonas. Revista Brasileira de Biologia 44(2):227–237

Absy, Maria Lúcia; Prance, Ghillean T.; Barbosa, Edelcílio Marques. 1986. Inventário florístico de floresta natural na área da estrada Cuiabá-Porto Velho (BR 364). Acta Amazonica 16/17:85–121

Akers, C. E. 1912. The Amazon Valley: Its Rubber Industry and Other Resources. Waterlow and Sons, London

Alarcón, Juan Gabriel Soler; Peixoto, Ariane Luna. 2007. Florística e fitossociologia de um trecho de um hectare de floresta de terra firme, em Caracaraí, Roraima, Basil. Boletim do Museu Paraense Emílio Goeldi, Ciencias Naturais 2(2):33–60

Alarcón, J. G. S. and A. L. Peixoto. 2008. Use of terra firme forest by Caicubi caboclos, middle Rio Negro, Amazonas, Brazil: A quantitative study. Economic Botany 62(1):60–73

Albert, Bruce; Milliken, William. 2009. Urihi A: a terra-floresta Yanomami. Instituto Socioambiental (ISA)/Institut de Recherche pour le Devéloppement, São Paulo

Almeida, Alexa-Sandra Farias de. 2012. Análise fitossociológica estrutural e composição florística da área de proteção ambiental margem esquerda do Rio Negro, Manaus-AM. Master's thesis, Universidade Federal do Amazonas, Manaus

Almeida, Adrielson Furtado; Jardim, Mário Augusto Gonçalves. 2011. Florística e estrutura da comunidade arbórea de uma floresta de várzea na Ilha de Sororoca, Ananindeua, Pará, Brasil. Scientia Florestalis 39(90):191–198

Almeida, Adrielson Furtado; Jardim, Mário Augusto Gonçalves. 2012. A utilização das espécies arbóreas da floresta de várzea da Ilha de Sororoca, Ananindeua, Pará, Brasil por moradores locais. Revista Brasileira de Ciências Ambientais 23:48–54

Almeida, Arlete Silva de; Vieira, Ima Célia Guimarães. 2001. Padrões florísticas e estruturais de uma cronoseqüência de florestas no munícipio de São Francisco do Pará, região Bragantina, Pará. Boletim do Museu Paraense Emílio Goeldi, Série Botânica 17(1):209–240

Almeida, Samuel S.; Lisboa, Pedro Luiz B.; Silva, Antônio Sérgio. 1993. Diversidade florística de uma comunidade arbórea na estação científica "Ferreira Penna", em Caxiuanã (Pará). Boletim do Museu Paraense Emilio Goeldi, Série Botanica 9(1):93–128

Almeida, Samuel Soares de; Amaral, Dário Dantas do; Silva, Antonio Sérgio Lima da. 2004. Análise florística e estrutura de florestas de várzea no estuário amazônico. Acta Amazonica 34(4):513–524

Almeida, Larissa Santos de; Gama, João Ricardo Vasconcellos; Oliveira, Francisco de Assis; Carvalho, João Olegário Pereira de; Gonçalves, Danielly Caroline Miléo; Araújo, Giovânia Carvalho. 2012. Fitossociologia e uso múltiplo de espécies arbóreas em floresta manejada, Comunidade Santo Antônio, município de Santarém, Estado do Pará. Acta Amazonica 42(2):185–194

Alva, Alenguer; Vásquez, Oscar; Cunibertti, Roberto; Castillo, Abel del; Guerra, Wilson. 2002. Extracción y caracterización de ácidos grasos de la especie *Grias neuberthii* Macht (Sachamango). Revista Amazónica de Investigación 2: 103–106

Alvino, Fabrízia de Oliveira; Silva, Manoela Ferreira F. da; Rayol, Breno Pinto. 2005. Potencial de uso das espécies arbóreas de uma floresta secundária, na Zona Bragantina, Pará, Brasil. Acta Amazonica 35(4):413–420

Amaral, I. L. do; Adis, J.; Prance, G. T. 1997. On the vegetation of a seasonal mixed water inundation forest near Manaus, Brazilian Amazonia. Amazoniana 14(3/4):335–347

Amaral, Dário Dantas do; Vieira, Ima Célia Guimarães; Almeida, Samuel Soares de; Salomão, Rafael Paiva; Silva, Antonio Sérgio Lima da; Jardim, Mário Augusto Gonçalves. 2009. Checklist da flora arbórea de remanescentes florestais da região metropolitana de Belém e valor histórico dos fragmentos, Pará, Brasil. Boletim do Museu Paraense Emílio Goeldi, Ciencias Naturais 4(3):231–289

Amaral, Iêda Leão do; Magnusson, William Ernest; Matos, Francisca Dionízia de Almeida; Albernaz, Ana Luisa K.; Feitosa, Yuri Oliveira; Guillaumet, Jean-Louis. 2017. Disentangling structural patterns of natural forest fragments in a savanna matrix in the eastern Brazilian Amazon. Acta Amazonica 47(2):111–122

Andel, T. R. van. 2003. Floristic composition and diversity of three swamp forests in northwest Guyana. Plant Ecology 167:293–317

Andrade, E.H.A.; Maia, J.G.S.; Streich, R.; Marx, F. 1999a. Seed composition of Amazonian Lecythidaceae species: Part 3 in the Series "Studies of edible Amazonian plants". Journal of Food Composition and Analysis 12: 37–51

Andrade, A.C.S.; Ramos, F.N.; Souza, A.F.; Loureiro, M.B.; Bastos, R. 1999b. Flooding effects in seedlings of *Cytharexyllum myrianthum* Cham. and *Genipa americana* L.: responses of two Neotropical lowland tree species. Revista Brasilieira de Botanica 22 (2): 281–285

Andrade, Victor L. Caetano; Flores, Bernardo M.; Levis, Carolina; Clement, Charles R.; Roberts, Patrick; Schöngart, Jochen. 2019. Growth rings of Brazil nut trees (*Bertholletia excelsa*) as a living record of historical human disturbance in Central Amazonia. PLOS One 14(4): e0214128

Aparicio, Perseu Silva; Sotta, Eleneide Doff; Guedes, Marcelino Carneiro; Aparício, Wegliane Campelo da Silva; Oliveira, Lana Patricia; Souza, Raianny Nayara. 2014. Níveis de regeneração natural em floresta de terra firme no Amapá Brasil. Revista Árvore 38(4):699–710

Assis, Rafael Leandro de; Wittmann, Florian; Luize, Bruno Garcia; Haugaasen, Torbjon. 2017. Patterns of floristic diversity and composition in floodplain forests across four southern Amazon river tributaries, Brazil. Flora 229:124–140

Avé-Lallemant, Robert. 1961. Viagem pelo Norte do Brasil no ano de 1859. Instituto Nacional do Livro, Rio de Janeiro, Vol. 2

Baena, Antônio Ladislau Monteiro 2004. Ensaio Corográfico sobre a Província do Pará. Edições do Senado Federal, Brasilia

Balée, W. 1987. A etnobotânica quantitativa dos índios Tembé (Rio Gurupi, Pará). Boletim do Museu Paraense Emílio Goeldi, Série Botânica 3(1):29–50

Balée, W. 1989. The culture of Amazonian forests. In: R*esource Management in Amazonia: Indigenous and Folk Strategies*, Posey, D. A. and W. Balée (Eds), pp. 1–21, Advances in Economic Botany, vol. 7, New York Botanical Garden, Bronx, New York.

Balée, W. 1993a. *Footprints of the Forest: Ka'apor Ethnobotany, the Historical Ecology of Plant Utilization by an Amazonian People*. Columbia University Press, New York

Balée, W. 1993b. Indigenous transformation of Amazonian forests: an example from Maranhão, Brazil. L'Homme 33(126–128):231–254

Balée, W. 2010. Contingent diversity on anthropic landscapes. *Diversity* 2:163–181

Balée, W. 2013. Cultural Forests of the Amazon: A Historical Ecology of People and Their Landscapes. University of Alabama Press, Tuscaloosa

Balée, W. and D. G. Campbell 1990. Evidence for the successional status of liana forest (Xingu River Basin, Amazonian Brazil). Biotropica 22(1):36–47

Bánki, Olaf S.; Steege, Hans ter; Jansen-Jacobs, Marion J.; Raghoenandan, Usha P. D. 2003. Plant Diversity of the Nassau Mountains, Suriname: Report of the 2003 Expedition. National Herbarium Nederland, Utrecht

Banner, Horace. 1961. O índio Kayapó em seu acampamento. Boletim do Museu Paraense Emílio Goeldi, Nova Série, Antropologia, vol. 13: 1–51

Barboza, M. S. L. 2008. O preço que a natureza pagou e os efeitos colaterais que sofremos para hoje se ter energia: uso dos recursos animais e percepção dos impactors entre os ribeirinhos do Lago de Tucuruí (PA). Master's thesis, Universidade Federal do Pará, Nucleo de Altos Estudos Amazônicas, Programa de Pós-Graduação em Desenvolvimento Sustentável do Trópico Úmido, Belém.

Barnett, Adrian A.; Silva, Welma Sousa; Shaw, Peter J. A.; Ramsay, Paul M. 2015. Inundation duration and vertical vegetation zonation: a preliminary description of the vegetation and structuring factors in borokotóh (hummock igapó), an overlooked high-diversity, Amazonian vegetation association. Nordic Journal of Botany 33:601–614

Barruel de Lagenest, H. D. 1958. Marabá: Cidade do Diamante e da Castanha. Editôra Anhambi, São Paulo

Bates, H. W. 1863a. *The Naturalist on the River Amazons*. John Murray, London, vol. 1

Bates, H. W. 1863b. *The Naturalist on the River Amazons*. John Murray, London, vol. 2

Batista, Fábio de Jesus; Gonçalves, Mário Augusto. 2013. Florística e estrutura da regeneração natural arbórea de uma floresta de várzea na Reserva Extrativista Chocoaré-Mato Grosso, Pará, Brasil. Biota Amazônia 3(3):139–145

Batista, Fábio de Jesus; Jardim, Mário Augusto Gonçalves; Medeiros, Tonny David Santiago; Lopes, Isaac Luiz Magalhães. 2011. Comparação florística e estrutural de duas florestas de várzea no estuário Amazônico, Pará, Brasil. Revista Árvore 35(2):289–298

Batista, Anderson Pedro Bernardina; Aparício, Wegliane Campelo da Silva; Santos, Vanessa Silva dos; Lima, Robson Borges de; Mello, José Marcio de. 2015. Caracterização estrutural em uma floresta de terra firme no estado do Amapá, Brasil. Pesquisa Florestal Brasileira 35(81):21–33

Benchimol, Samuel. 1977. Amazônia: Um Pouco-Antes e Além-Depois. Editora Umberto Calderaro, Manaus

Bennett, Bradley C. 1992. Plants and people of the Amazonian rainforests: the role of ethnobotany in sustainable development. Bioscience 42(8):599–607

Bennett, Bradley C.; Baker, Marc A.; Andrade, Patricia Gómez. 2002. Ethnobotany of the Shuar of Eastern Ecuador. New York Botanical Garden, Advances in Economic Botany, vol. 14.

Bentes, Evely Sevalho. 2007. Extrativismo da castanha-do-Brasil (*Bertholletia excelsa* Hubl.) na Reserva de Desenvolvimento Sustentável Piaguçu-Purus. Master's thesis, Universidade Federal do Amazonas, Manaus

Bentes-Gama, Michelliny de Matos; Scolforo, José Roberto Soares; Gama, João Ricardo Vasconcellos; Oliveira, Antônio Donizette de. 2002a. Estrutura e valoração de uma floresta de várzea alta na Amazônia. Cerne 8(1):88–102

Bentes-Gama, Michelliny de Matos; Scolforo, José Roberto Soares; Gama, João Ricardo Vasconcellos. 2002b. Potencial produtivo de madeira e palmito de uma floresta secundária de várzea baixa no estuário Amazônico. Revista Árvore 26(3):311–319

Best, R.C. 1984. The aquatic mammals and reptiles of the Amazon. In: The Amazon: Limnology and Landscape Ecology of a Mighty Tropical River and its Basin, H. Sioli (ed), W. Junk, Dordrecht, pp 371–412

Black, G. A.; Dobzhansky, T. H.; Pavan, C. 1950. Some attempts to estimate species diversity and population density of trees in Amazonian forests. Botanical Gazette 111(4):413–425

Bolfe, Édson Luis; Batistella, Mateus. 2011. Análise florística e estrutural de sistemas silviagrícolas em Tomé-Açu, Pará. Pesquisa Agropecuária Brasileira 46(10):1139–1147

Boubli, J. P. 1999. Feeding ecology of black-headed uacaris (*Cacajao melanocephalus melanocephalus*) in Pico da Neblina National Park, Brazil. International Journal of Primatology 20(5):719–749

Bowler, Mark; Bodmer, Richard E. 2011. Diet and food choice in Peruvian red uakaris (*Cacajao calvus ucayalii*): selective or opportunistic seed predation? International Journal of Primatology 32:1109–1122

Brack, A. 2003. *Frutas del Perú*. Universidad San Martin de Porres, Lima

Braga, Elayne Oliveira; Jardim, Mário Augusto Gonçalves. 2019. Florística, estrutura e formas de vida do estrato inferior de uma floresta ombrófila densa aluvial, Pará, Brasil. Ciencia Florestal, Santa Maria 29(3): 1048–1059

Braga, Lúcia Filgueiras; Sousa, Marcílio Pereira; Gilberti, Sirlene; Carvalho, Marco Antônio Camillo de. 2007. Caracterização morfométrica de sementes de castanha de sapucaia (*Lecythis pisonis* Cambess, Lecythidaceae). Revista de Ciências Agro-Ambientais, Alta Floresta 5(1):111–116

Braga, Elayne Oliveira; Silva, Julie Andrews de França e; Pantoja, Milenna Vasconcelos; Jardim, Mário Augusto Gonçalves. 2015. Florística, estrutura fitossociológica e formas de vida do estrato inferior em um floresta de várzea Amazônica. Biota Amazônia 5(3):59–65

Braga, Daniel P. P.; Domene, Frederico; Gandara, Flávio B. 2019. Shade trees composition and diversity in cacao agroforestry systems of southern Pará, Brazilian Amazon. Agroforestry Systems 93:1409–1421

Brown, C.B. and W. Lidstone. 1878. *Fifteen Thousand Miles on the Amazon and its Tributaries*. Edward Stanford, London

Camargo, Miguel Ângelo Branco. 2018. Variação em diâmetro do tronco de 30 espécies de árvores florestais em função de variáveis ambientais em floresta de terra firme na Amazônia central. Doctoral dissertation, INPA, Manaus

Campbell, David G.; Daly, Douglas C.; Prance, Ghillean T.; Maciel, Ubirajara N. 1986. Quantitative ecological inventory of terra firme and várzea tropical forest on the Rio Xingu, Brazilian Amazon. Brittonia 38(4):369–393

Cano, Ángela; Stevenson, Pablo R. 2009. Diversidad y composición florística de tres tipos de bosque en la Estación Biológica Caparú, Vaupés. Revista Colombia Florestal 12:63–80

Carim, Marcelo de Jesus Veiga. 2016. Estrutura, composição e diversidade em florestas alagáveis de várzea de maré e de igapó e suas relações com variáveis edáficas e o período de inundação no Amapá, Amazônia oriental, Brasil. Doctoral dissertation, Programa de Pós-Graduação em Botânica, INPA, Manaus

Carim, Samyrams; Schwartz, Gustavo; Silva, Manoela Fernandes Ferreira da. 2007. Riqueza de espécies, estrutura e composição florística de uma floresta secundária de 40 anos no leste da Amazônia. Acta Botanica Brasilica 21(2):293–308

Carim, Marcelo de Jesus Veiga; Jardim, Mário Augusto Gonçalves; Medeiros, Tonny David Santiago. 2008. Composição florística e estrutura de floresta de várzea no Município de Mazagão, Estado do Amapá, Brasil. Scientia Florestalis 36(79):191–201

Carneiro, Vilany Matilla Colares. 2004. Composição florística e análise estrutural da floresta primária de terra firme na bacia do Rio Cuieiras, Manaus-AM. Master's Thesis, Universidade Federal do Amazonas (UFAM)/Instituto Nacional de Pesquisas da Amazônia (INPA), Manaus

Carvalho, Telma Lélia Gonçalves Schultz de. 2015. Etnofarmacologia e fisiologia de plantas medicinais do quilombo Tiningú, Santarém, Pará, Brasil. Master's thesis, Universidade Federal do Oeste do Pará, Santarém.

Castner, J. L.; Timme, S. L.; Duke, J. A. 1998. A Field Guide to Medicinal and Useful Plants of the Upper Amazon. Feline Press, Gainesville

Casula, Kátia Regina. 2012. Análise da similaridade florística e estrutural das formações florestais inundáveis em um trecho do alto Rio Madeira e de seus afluentes, estado de Rondônia, Brasil. Master's Thesis, Núcleo de Ciências Exatas da Terra, Universidade Federal de Rondônia, Porto Velho

Cattanio, José; Anderson, Anthony B.; Carvalho, Manoel S. 2002. Floristic composition and topographic variation in a tidal floodplain forest in the Amazon estuary. Revista Brasileira de Botanica 25(4):419–430

Cavalcante, M. C.; Oliveira, F. F.; Maués, M. M.; Freitas, B. M. 2012. Pollination requirements and the foraging behavior of potential pollinators of cultivated Brazil nut (*Bertholletia excelsa* Bonpl.) trees in central Amazonian rainforest. Psyche https://doi.org/10.1155/2012/978019

Cerón, Carlos Eduardo. 1995. Etnobiologia de los Cofanes de Dureno, Provincia de Sucumbíos, Ecuador. Museo Ecuatoriano de Ciencias Naturales, Herbario Nacional, Serie Monografia 10(3), Quito

Cerón, Carlos Eduardo; Ayala, Consuelo Montalvo. 1998. Etnobotanica de los Huaorani de Quehueiri-ono Napo, Ecuador. Abya-Yala, Quito

Cerón, C. E.; Montalvo, Consuelo G.; Umenda, José; Umenda Enma Chica. 1994. Etnobotánica y Notas sobre la Diversidad Vegetal en la Comunidad Cofán de Sinangüé, Sucumbíos, Ecuador. EcoCiencia, Quito

Chih-Hua, Tsou; Scott A. Mori. 2002. Seed coat anatomy and its relationship to seed dispersal in subfamily Lecythidoideae of the Lecythidaceae (The Brazil Nut Family). Bot. Bull. Acad. Sin. 43: 37–56

Chunhieng, Thavarith; Goli, Thierry; Piombo, Georges; Pioch, Daniel; Brochier, José; Montet, Didier. 2004. Recent analysis of the composition of Brazil nut Bertholletia excelsa. Bois et Forêts des Tropiques 280(2): 91–98

Claudino, Wesley Vicente; Garcia Marcos Leandro; Lopes, Célia Regina Araújo Soares, Rodrigues, Lucirene; Cabral, Fabiana Ferreira; Piva, José Hypolito; Silva, Antônio Carlos Silveiro da. 2015. Análise fitossociológica da vegetação na região do arco do desmatamento. Revista de Ciências Agroambientais, Alta Floresta, MT 13(1): 43–52

Clay, Jason W. 1997. Brazil nuts: the use of a keystone species for conservation and development. In: Harvesting Wild Species: Implications for Biodiversity Conservation, Johns Hopkins University Press, Baltimore, pp 246–282

Clement, C. R., J. M. McCann, and N. J. H. Smith. 2003. Agrobiodiversity in Amazonia and its relationship with dark earths. In: *Amazonian Dark Earths: Origin, Properties, Management*, J. Lehmann, D. C. Kern, B. Glaser, W. I. Woods (Editors), Kluwer, Dordrecht, pp. 150–178

Coelho, Roberta de Fátima Rodrigues; Miranda, Izildinha Souza; Mitja, Danielle. 2012. Caracterização do processo sucessional no Projeto de Assentamento Benfica, sudeste do estado do Pará, Amazônia oriental. Boletim do Museu Paraense Emílio Goeldi, Ciencias Naturais 7(3):251–282

Coelho-Ferreira, Márlia. 2009. Medicinal knowledge and plant utilization in an Amazonian coastal community of Marudá, Pará state (Brazil). Journal of Ethnopharmacology 126:159–175

Coimbra, C. E. A. 1985. Estudos de ecologia humana entre os Suruí do Parque Indígena Aripuanã, Rondônia: Elementos de etnozoologia. Boletim do Museu Paraense Emílio Goeldi, Série Antropologia 2(1):9–36

Cotta, Jamie N.; Kainer, Karen A.; Wadt, Lúcia H.O.; Staudhammer, Christina L. 2008. Shifting cultivation effects on Brazil nut (*Bertholletia excelsa*) regeneration. Forest Ecology and Management 256:28–35

Crévaux, J. 1883. Voyages dans l'Amérique du Sud. Libraire Hachette, Paris

Cronkleton, Peter; Guariguata, Manuel R.; Albornoz, Marco A. 2012. Multiple use forestry planning: timber and Brazil nut management in the community forests of northern Bolivia. Forest Ecology and Management 268:49–56

Daniel, J. 1976. *Tesouro Descoberto no Rio Amazonas*. Biblioteca Nacional, Rio de Janeiro, vol. 1

Dávila, P. M.; Vega, Juan José Palacios; Vela, Marcial Martínez; Gómez, Ricardo Zárate; García, José Maco; Torres, Roger Escobedo. 2015. Cobertura y uso de la tierra de las provincias de Alto Amazonas y Ramón Castilla. Instituto de Investigaciones de la Amazonía Peruana, Iquitos, Documento Técnico No. 33

Dias, C. V. 1959. Aspectos geográficos do comércio da castanha no médio Tocantins. Revista Brasileira de Geografia 21(4):77–91

Díaz Pérez, Wilmer Antonio; Daza, Félix; Sarmiento, William. 2012. Composición florística, estructura y diversidad del bosque ribereños del Río Kakada, cuenca del Río Caura, estado Bolívar, Venezuela. Revista Científica UDO Agrícola 12(2):275–289

Duchelle, Amy E.; Guariguata, Manuel R.; Less, Giuliano; Albornoz, Marco Antonio; Chavez, Andrea; Melo, Tadeu. 2012. Evaluating the opportunities and limitations to multiple use of Brazil nuts and timber in western Amazonia. Forest Ecology and Management 268:39–48

Ducke, Adolfo. 1949. Árvores amazônicas e sua propagação: adaptação de frutos ou sementes de árvores amazônicas a diversos meios de propagação em espécies de estreita afinidade botanica, porém em condições mesológicas diferentes. Boletim do Museu Paraense Emílio Goeldi 10: 81–92

Duque, Álvaro; Phillips, Juan Fernando; Hildebrand, Patricio von; Posada, Carlos Alberto; Prieto, Adriana; Rudas, Augustín; Suescún, Mabel; Stevenson, Pablo. 2009. Distance decay of tree species similarity in protected areas on terra firme forests in Colombian Amazonia. Biotropica 41(5):599–607

Dwyer, Augusta. 1990. Into the Amazon: The Struggle for the Rain Forest. Sierra Club Books, San Francisco

Edwards, W. H. 1847. *A Voyage up the River Amazon, including a Residence at Pará*. John Murray, London

Espírito-Santo, Fernando Del Bon; Shimabukuro, Yosio Edimir; Aragão, Luiz Eduardo Oliveira e Cruz de; Machado, Evandro Luiz Mendonça. 2005. Análise da composição florística e fitossociológica da floresta nacional do Tapajós com apoio geográfico de imagens satélites. Acta Amazonica 35(2):155–173

Farabee, W. C. 1918. *The Central Arawaks*. The University Museum, University of Pennsylvania, Philadelphia, vol. 10

Fearnside, P. M. 2001. Environmental impacts of Brazil's Tucuruí dam: unlearned lessons for hydroelectric development in Amazonia. Environmental Management 27(3):377–396

Fernandes-Corrêa, Ana Francisca; Furch, Bodo. 1992. Investigations on the tolerance of several trees to submergence in blackwater (Igapó) and whitewater (Várzea) inundation forests near Manaus, Central Amazonia. Amazoniana 12(1):71–84

Ferreira, A. R. 1972. Descrição do peixe pirarucu (1787). In: Viagem Filosófica pelas Capitanias do Grão-Pará, Rio Negro, Mato Grosso e Cuiabá, A. R. Ferreira, Conselho Federal de Cultura, Rio de Janeiro

Ferreira Penna, D. S. 1869. *A Região Occidental da Provincia do Pará*. Typographia do Diario, Belém

Ferreira, Leandro V.; Almeida, Samuel S.; Amaral, Dário D.; Parolin, Pia. 2005. Riqueza e composição de espécies da floresta de igapó e várzea da Estação Científica Ferreira Penna: subsídios para o plano de manejo da Floresta Nacional de Caxiuanã. Pesquisas Botânica 56:103–116

Ferreira, Leandro V.; Chaves, Priscilla P.; Cunha, Denise de A.; Matos, Darley C. L.; Parolin, Pia. 2013. Variação da riqueza e composição de espécies da comunidade de plantas entre as florestas de igapós e várzeas na Estação Científica Ferreira Penna-Caxiuanã na Amazônia oriental. Pesquisas Botânica 64:175–195

Flores, Salvador. 1997. *Cultivo de Frutales Nativos Amazonicos: Manual para el Extensionista*. Tratado de Cooperación Amazónica, Lima

Freire, Maria Cristina C.; Monteiro, Reinaldo. 1993. Florística das praias da Ilha de São Luís, Estado do Maranhão (Brasil): diversidade de espécies e suas ocorrências no litoral Brasileiro. Acta Amazonica 23(2–3):125–140

Freitas, João da Luz; Santos, Erick Silva dos; Cruz Junior, Francisco de Oliveira; Borges, César Henrique Alves; Santos, Adriano Castelo dos. 2019. Structural analysis and distribution patterns in lowland tropical forest, eastern Amazon. Journal of Agricultural Studies 7(4): 240–255

Friedman, J.; Bolotin, D.; Rios, M.; Mendosa, P.; Cohen, Y.; Balick, M. J. 1993. A novel method for identification and domestication of indigenous useful plants in Amazonian Ecuador. In: New Crops, Janick, J.; Simon, J. E. (eds), John Wiley, New York, pp 167–174

Gabas, N. and S. K. P. Arara. 2009. Mitos Arara. Museu Paraense Emílio Goeldi, Belém

Gachet, María Salomé; Lecaro, Javier Salazar; Kaiser, Marcel; Brun, Reto; Navarrete, Hugo; Muñoz, Ricardo A.; Bauer, Rudolf; Schühly, Wolfgang. 2010. Assessment of anti-protozoal activity of plants traditionally used in Ecuador in the treatment of leishmaniasis. Journal of Ethnopharmacology 128:183–197

Gama, João Ricardo Vasconcellos; Pinheiro, Josiel Carneiro. 2010. Inventário florestal para adequação ambiental da Fazenda Santa Rita, Município de Santarém, Estado do Pará. Floresta 40(3):585–592

Gama, João Ricardo Vasconcellos; Botelho, Soraya Alvarenga; Bentes-Gama, Michellinyde Matos. 2002. Composição florística e estrutura da regeneração natural de floresta secundária de várzea baixo no estuário Amazônico. Revista Árvore 26(5):559–566

Gama, J. R. V., S. A. Botelho, M. Bentes-Gama, and J. R. S. Scolforo 2003. Estrutura e potencial futuro de utilização da regeneração natural de floresta de várzea alta no Município de Afuá, estado do Pará. *Ciencia Florestal, Santa Maria* 13(2):71–82

García-Torres, D. E.; Solís, V. Sotero; Mancini, D. A. P.; Torres, R. P.; Mancini-Filho, J. 2009. Actividad antioxidante de los extractos del chopé (*Gustavia augusta* L.). Revista de la Sociedad Química del Perú 75(3):374–381

Gayot, Marc; Henry, Olivier; Dubost, Gérard; Sabatier, Daniel. 2004. Comparative diet of the two forest cervids of the genus *Mazama* in French Guiana. Journal of Tropical Ecology 20(1):31–43

Ghilardi, R.; Alho, C.J.R. 1990. Produtividade sazonal da floresta e atividade de forrageamento animal em habitat de terra firme da Amazônia. Acta Amazonica 20:61–76

Giovannini, Peter. 2015. Medicinal plants of the Achuar (Jivaro) of Amazonian Ecuador: ethnobotanical survey and comparison with other Amazonian pharmacopoeias. Journal of Ethnopharmacology 164:78–88

Giraldo-Cañas, Diego. 1999. Riqueza, composición y distribución florística de los paisajes fisiográficos del eje de los ríos Apaporis y Amazonas, Amazonia Colombiana. Darwiniana 37(1–2):25–35

Godoy, Judith Rosales; Petts, Geoffrey; Salo, Jukka. 1999. Riparian flooded forests of the Orinoco and Amazon basins: a comparative review. Biodiversity and Conservation 8:551–586

Gomez, Tedi Pacheco. 2011. Valoración económica de purmas de diferentes edades en la comunidad nativa de Santa Mercedes, río Putumayo, Perú. Master's thesis, UNAP, Iquitos

Gomide, Guilherme Luís Augusto. 1997. Estrutura e dinâmica de crescimento de florestas tropicais primária e secundária no Estado do Amapá. Master's Thesis, Universidade Federal do Paraná, Brazil

Gonçalves, Janaína Pinheiro; Lucas, Flávia Cristina Araújo. 2017. Agrobiodiversidade e etnoconhecimento em quintais de Abaetetuba, Pará, Brasil. Revista Brasileira de Biociências 15(3):119–134

Grimes, Alicia; Loomis, Sally; Jahnige, Paul; Burnham, Margo; Onthank, Karen; Alarcón, Rocio; Cuenca, Walter Palacios; Martinez, Carlos Cerón; Neill, David; Balick, Michael; Bennett, Brad; Mendelsohn, Robert. 1994. Valuing the rain forest: the economic value of nontimber forest products in Ecuador. Ambio 23(7):405–410

Grubb, P. J.; Lloyd, J. R.; Pennington, T. D.; Whitmore, T. C. 1963. A Comparison of Montane and Lowland Rain Forest in Ecuador I. The Forest Structure, Physiognomy, and Floristics. Journal of Ecology 51(3): 567–601

Grünberg, G. 2004. Os Kaiabi do Brasil Central: História e Etnografia. Instituto Socioambiental, São Paulo

Gualberto, Midiã Larine Colares; Ribeiro, Renato Bezerra da Silva; Gama, João Ricardo Vaconcellos; Vieira, Diego dos Santos. 2014. Fitossociologia e potencial de espécies arbóreas em ecossistema sucessional na Floresta Nacional do Tapajós, Pará. Agroecossistemas 6(1):42–57

Guimarães, José Renan da Silva; Carim, Marcelo de Jesus Veiga; Costa Neto, Salustiano Vilard a; Tostes, Luciedide Cássia Leôncio. 2016. Floristic diversity of secondary forest in the eastern Amazon, state of Amapá. Floresta 46(3):343–351

Hartt, C. F. 1898. O Rio Tocantins. Boletim do Museu Paraense Emílio Goeldi 2:181–191

Haugaasen, Joanne M. Tuck; Haugaasen, Torbjorn; Peres, Carlos A.; Gribel, Rogerio; Wegge, Per. 2010. Seed dispersal of the Brazil nut tree (*Bertholletia excelsa*) by scatter-hoarding rodents in a central Amazonian forest. Journal of Tropical Ecology 26:251–262

Haugaasen, Joanne M. Tuck; Haugaasen, Torbjorn; Peres, Carlos A.; Gribel, Rogerio; Wegge, Per. 2012. Fruit removal and natural seed dispersal of the Brazil nut (*Bertholletia excelsa*) in central Amazonia, Brazil. Biotropica 44(2):205–210

Herndon, W. L. and L. Gibbon. 1853. Exploration of the Valley of the Amazon, made under the Direction of the Navy Department. Rober Armstrong, Washington, D.C., vol. 1

Hiraoka, Mario; Yamamoto, Shozo; Matsumoto, Eiji; Nakamura, Satoshi; Falesi, Italo C.; Baena, Antonio Ronaldo Camacho. 2003. Contemporary use and management of Amazonian Dark

Earths. In: Amazonian Dark Earths: Origin, Properties, Management, Lehmann, Johannes; Kern, Dirse C; Glaser, Bruno; Woods, William I. (eds.), Kluwer, Dordrecht, pp. 387–406

Homma, A. K. O. 2014a. Agricultura na Amazônia: oportunidades de mercados e desenvolvimento regional. In: *Mercado Cadeia Produtiva e Desenvolvimento Rural na Amazônia*, A. C. Santana (ed.), Universidade Federal Rural da Amazônia, Belém, pp. 55–78

Homma, A. K. O. 2014b. Cemitério das castanheiras. In: Extrativismo Vegetal na Amazônia: História, Ecologia e Domesticação, A. K. O. Homma (ed.), EMBRAPA, Brasília, pp. 193–199

Homma, A. K. O., R. T. Walker, R. A. Carvalho, J. A. Conto, C. A. P. Ferreira. 2014. Políticas agrícolas e econômicas para a conservação de recursos naturais: o caso de castanhais em lotes de colonos no sul do Pará. In: Extrativismo Vegetal na Amazônia: História, Ecologia e Domesticação, A. K. O. Homma (ed.), EMBRAPA, Brasília, pp. 177–191

Honda, E. M. S. 1974. Contribuição ao conhecimento da biologia de peixes do Amazonas II. Alimentação de tambaqui, *Colossoma bidens* (Spix). *Acta Amazonica* 4(2):47–53

Hooker, W. J. 1849. Zabucajo, a new excellent nut, recently imported. Hooker's Journal of Botany and Kew Garden Miscellany 1:21–22

Huber, J. 1910. Mattas e madeiras amazonicas. Boletim do Museu Goeldi de Historia Natural e Ethnographia 6: 91–225

Humboldt, A. and A. Bonpland. 1821. Personal Narrative of Travels to the Equinoctial Regions of the New Continent during the Years 1799–1804. Longman, Hurst, Rees, Orme, and Brown, London, vol. 5

Iglesias, G. 1991. Medicina herbolaria de los Quichuas del Napo: la cultura fitoterapeútica de las mujeres. In: Las Plantas y El Hombre: Memorias del Primer Simposio Ecuatoriano de Etnobotánica Ecónomica, Ríos, M.; Pedersen, H. Borgtoft (eds), Ediciones Abya-Yala, Quito, pp 119–127

Irmler, U. 1977. Inundation-forest types in the vicinity of Manaus. Biogeographica 8:17–29

Jakovac, Catarina C.; Peña-Claros, Marielos; Mesquita, Rita C.G.; Bongers, Frans; Kuyper, Thomas W. 2016. Swiddens under transition: consequences of agricultural intensification in the Amazon. Agriculture, Ecosystems and Environment 218:116–125

Jardim, M.A.G. and Medeiros, T.D.S. 2006. Plantas oleaginosas do Estado do Pará: composição florística e usos medicinais. Revista Brasileira de Farmácia 87(4):124–127

John, Jenny A.; Shahidi, Fereidoon. 2010. Phenolic compounds and antioxidant activity of Brazil nut (*Bertholletia excelsa*). Journal of Functional Foods 2:196–209

Kainer, Karen A.; Duryea, Mary L.; Macêdo, Nazaré Costa de; Williams, Kimberlyn. 1998. Brazil nut seedling establishment and autoecology in extractive reserves of Acre, Brazil. Ecological Applications 8(2):397–410

Kainer, Karen A.; Wadt, Lúcia H. O.; Gomes-Silva, Daisy A. P.; Capanu, Marinela. 2006. Liana loads and their association with Bertholletia excelsa fruit and nut production, diameter growth and crown attributes. Journal of Tropical Ecology 22(2):147–154

Keller, F. 1874. The Amazon and Madeira Rivers, Sketches and Descriptions from the Note-Book of an Explorer. Chapman and Hall, London

Kerr, Warwick Estevam; Posey, Darrell Addison. 1984. Informações adicionais sobre a agricultura dos Kayapó. Interciencia 9(6): 392–400

Klinge, Hans; Adis, Joachim; Worbes, Martin. 1995. The vegetation of a seasonal várzea forest in the lower Solimões River, Brazilian Amazonia. Acta Amazonica 25(3/4):201–220

Kluczkovski, Ariane Mendonça; Martine, Maristela; Mundim, Silmara Miranda; Simões, Renata Hanan; Nascimento, Karine Silveira; Marinho, Helyde Albuquerque; Kluczkovski Junior, Augusto. 2015. Properties of Brazil nuts: a review. African Journal of Biotechnology 14(8):642–648

Klüppel, Marina Pinheiro. 2006. Sistemas agrícolas e plantas medicinais em terra preta de índio da Amazônia Central. Master's thesis, INPA/UFAM, Manaus

Korning, Jorgen; Balslev, Henrik. 1994a. Growth rates and mortality patterns of tropical lowland tree species and the relation to forest structure in Amazonian Ecuador. Journal of Tropical Ecology 10:151–166

Korning, Jorgen; Balslev, Henrik. 1994b. Growth and mortality of trees in Amazonian tropical rain forest in Ecuador. Journal of Vegetation Science 4:77–86

Korning, Jorgen; Thomsen, Karsten; Ollgaard, Benjamin. 1991. Composition and structure of a species rich Amazonian rain forest obtained by two different sample methods. Nordic Journal of Botany 11(1):103–110

Korning, Jorgen; Thomsen, Karsten; Dalsgaard, Kristian; Nornberg, Per. 1994. Characters of three Udults and their relevance to the composition and structure of virgin rain forest of Amazonian Ecuador. Geoderma 63:145–164

Kubitzki, K. 1985. Dispersal of forest plants. In: Key Environments: Amazonia, G. T. Prance and T. E. Lovejoy (eds.), Pergamon Press, Oxford, pp. 192–206

Kubitzki, K. 1989. The ecogeographical differentiation of Amazonian inundation forests. Plant Systematics and Evolution 162:285–304

Lago, João Henrique G.; Tezoto, Jaqueline; Yazbek, Priscila B.; Cassas, Fernando; Santos, Juliana de F. L.; Rodrigues, Eliana. 2016. Exudates used as medicine by the "caboclos river-dwellers" of the Unini River, AM, Brazil--classification based in their chemical composition. Revista Brasileira de Farmacognosia 26:379–384

Lamont, Susan R.; Eshbaugh, W. Hardy; Greenberg, Adolph M. 1999. Species composition, diversity, and use of homegardens among three Amazonian villages. Economic Botany 53(3):312–326

Laraia, Roque de Barros; Da Matta, Roberto. 1978. Indios e Castanheiros: A Empresa Extrativa e os Índios no Médio Tocantins. Paz e Terra, São Paulo

Larue, Muriel; Ringuet, Stéphane; Sabatier, Daniel; Forget, Pierre-Michel. 2002. Fruit richness and seasonality in a fragmented landscape of French Guiana. Revue d'Écologie: La Terre et la Vie 57:39–57

Laurance, William F.; Nasimento, Henrique E. M.; Laurance, Susan G.; Condit, Richard; D'Angelo, Sammya; Andrade, Ana. 2004. Inferred longevity of Amazonian rainforest trees based on a long-term demographic study. Forest Ecology and Management 190:131–143

Le Cointe, P. 1922. *L'Amazonie Brésilienne: Le Pays-ses Habitants, ses Resources, Notes et Statistiques jusqu'en 1920*. Augustin Challamel, Paris, vol 1

Le Cointe, P. 1934. *Amazônia Brasileira III*: *Árvores e Plantas Úteis*. Livraria Classica, Belém

Leão, Fábio Miranda; Dionisio, Luiz Fernandes Silva; Silva, Nágilla Gabriella Euzébio da; Oliveira, Marcelo Henrique Silva de; D'arace, Larissa Martins Barbosa; Neves, Raphael Lobato Prado. 2017. Fitossociologia em sistemas agroflorestais com diferentes idades de implantação no município de Medicilândia, PA. Revista Agroambiente 11(1):71–81

Lees, Alexander C.; Zimmer, Kevin J.; Marantz, Curtis A.; Whittaker, Andrew; Davis, Bradley J. W.; Whitney, Bret M. 2013. Alta Floresta revisited: an updated review of the avifauna of the most intensively surveyed locality in south-central Amazonia. Bulletin of the British Ornithologists' Club 133(3):178–239

Lemos, D. A. N.; Ferreira, B. G. A.; Siqueira, J. D. P.; Oliveira, M. M.; Ferreira, A. M. 2015. Floristic and phytosociology in dense "terra firme" rainforest in the Belo Monte hydroelectric plant influence area, Pará, Brazil. Brazilian Journal of Biology 75(3):257–276

Levis, Carolina; Souza, Priscilla Figueira de; Schietti, Juliana; Thaise, Emilio; Pinto, José Luiz Purri da Veiga; Clement, Charles R.; Costa, Flavia R. C. 2012. Historical human footprint on modern tree species composition in the Purus-Madeira interfluve, Central Amazonia. PLOS One 7(11):1–10

Lima, D. 2006. Ngiã nüna tadaugü i torü naãne (vamos cuidar da nossa terra). Editora UFMG, Belo Horizonte

Lima, Roberval M. B. de; Saragoussi, Muriel. 2000. Floodplain home gardens on the central Amazon in Brazil. In: The Central Amazon Floodplain: Actual Use and Options for a Sustainable Management, Junk, W. J.; Ohly, J. J.; Piedade, M. T. F.; Soares, M. G. M. (eds), Backhuys Publishers, Leiden pp 243–268

Lima, Adriano José Nogueira; Teixeira, Liliane Marins; Carneiro, Vilany Matilla Colares; Santos, Joaquim dos; Higuchi, Niro. 2007. Análise da estrutura e do estoque de fitomassa de uma floresta secundária da região de Manaus AM, dez anos após corte raso seguido de fogo. Acta Amazonica 37(1):49–54

Lima Filho, Diógenes de Andrade; Matos, Francisca Dionizia de Almeida; Amaral, Iêda L. do; Revilla, Juan; Coêlho, Luiz de Souza; Ramos, José Ferreira; Santos, José Lima dos. 2001. Inventário florístico de floresta ombrófila densa de terra firme, na região do Rio Urucu, Amazonas, Brasil. Acta Amazonica 31(4):565–579

Lins, Cristóvão. 1991. Jari: 70 Anos de História. Dataforma, Rio de Janeiro

Lisboa, P. L. B. 2012. *A Terra dos Aruã: Uma História Ecológica do Arquipélago do Marajó*. Museu Paraense Emílio Goeldi, Belém

Lisboa, P. L. B., U.N. Maciel, and G. T. Prance. 1991a. Perdendo Rondônia. In: Amazônia, Ciência Hoje, Bloch Editors, Rio de Janeiro, pp. 74–83

Lisboa, Pedro L. B.; Maciel, Ubirajara N.; Prance, Ghillean T. 1991b. Some effects of colonization on the tropical flora of Amazonia: a case study from Rondônia. Kew Bulletin 46(2):187–204

Lisboa, P. L. B., I. A. Gomes, R. C. Lisboa, and C. V. Urbinati. 2002. O estilo Amazônico de sobreviver: manejo dos recursos naturais. In: *Natureza, Homem e Manejo de Recursos Naturais na Região de Caxiuanã, Melgaço, Pará*, P. L. B. Lisboa (ed.), Museu Paraense Emílio Goeldi, Belém, pp. 41–170

Lobo, Guilherme de Sousa. 2017. A alteração do regime hidrológico afeta a composição florística e estrutura de florestas de igapó? Um estudo comparativo entre um rio regulado e outro prístino na Reserva de Desenvolvimento Sustentável do Uatumã, Amazônia Central. Master's thesis, INPA, Manaus

Lorenzi, Harri; Bacher, Luis; Lacerda, Marco; Sartori, Sergio. 2006a. Frutas Brasileiras e Exóticas Cultivadas. Instituto Plantarum de Estudos da Flora Ltda, Odessa, Sao Paulo.

Lorenzi, H., L. Bacher, M. Lacerda, S. Sartori 2006b. Brazilian Fruits and Cultivated Exotics (for Consuming in Natura). Instituto Plantarum de Estudos da Flora Ltda, Odessa, São Paulo

Lossio, L.C., Delgado, H.S., Ruiz, A.G., Donayre, R.A., Ruiz, J.G., Panduro, M.P., Sifuentes, T.C. 1995. Plantas Medicinales de la Amazonía Peruana. Instituto Peruano de Seguridad Social/Instituto de Medicina Tradicional, Iquitos

Luize, Bruno Garcia; Venticinque, Eduardo Martins; Silva, Thiago Sanna Freire; Novo, Evlyn Marcia Leão de Moraes. 2015. A floristic survey of angiosperm species occurring at three landscapes of the central Amazon várzea, Brazil. Checklist: The Journal of Biodiversity Data 11(6):1–15

Luna, L. E. 1984. The healing practices of a Peruvian shaman. Journal of Ethnopharmacology 11:123–133

Luna, Luis Eduardo; Amaringo, Pablo. 1991. Ayahuasca Visions: The Religious Iconography of a Peruvian Shaman. North Atlantic Books, Berkeley

Luz, Francisco Joaci F. 2001. Plantas medicinais de uso popular em Boa Vista, Roraima, Brasil. Horticultura Brasileira 19(1):88–96

Luz, Benjamin Bordallo da. 2004. Características de árvores emergentes utilizadas por gavião-real (*Harpia harpyja*) para nidificação no centro e leste da Amazônia Brasileira. Master's thesis, INPA/UFAM, Manaus

Machado, Danilo de Oliveira. 2016. A agrobiodiversidade de quintais agroflorestais em propriedades agrícolas familiares na BR 174, Ramal do Pau-Rosa, Manaus, AM. Master's Thesis, INPA, Manaus

Macía, Manuel J. 2011. Spatial distribution and floristic composition of trees and lianas in different forest types of an Amazonian rainforest. Plant Ecology 212:1159–1177

Macía, Manuel J.; Svenning, Jens-Christian. 2005. Oligarchic dominance in western Amazonian plant communities. Journal of Tropical Ecology 21(6):613–626

Mackensen, Jens; Tillery-Stevens, Madeleine; Klinge, Rudolf; Fölster, Horst. 2000. Site parameters, species composition, phytomass structure and element stores of a terra-firme forest in East-Amazonia, Brazil. Plant Ecology 151:101–109

Macmillan, H. F. 1935. Tropical Planting and Gardening with Special Reference to Ceylon. Macmillan, London

Malu, S.P.J, and C. A. Peres. 2005. Population density and home range size of red-rumped agoutis (*Dasyprocta leporina*) within and outside a natural Brazil nut stand in Southeastern Amazonia. Biotropica 37(2):317–321

Maranho, Álisson Sobrinho; Salimon, Cleber Ibraim. 2015. Maranthaceae overabundance decreases richness and abundance of regenerating woody plants in natural gaps. Neotropical Biology and Conservation 10(2):53–62

Maranho, Álisson Sobrinho; Salimon, Cleber Ibraim; Costa, Daniel da Silva. 2016. Woody flora of natural forest gaps in a bamboo-dominated forest remnant in southwestern Amazonia. Check List: The Journal of Biodiversity Data 12(3):1–9

Marcalla, Gabriela Fernanda Moya. 2012. Etnobotánica de las comunidades de Puerto Bolívar, Tarapuya, Aboquëhuira y Sototsiaya de la nacionalidad Siona, Provincia Sucumbios, Ecuador. Bachelors thesis, Universidad Central del Ecuador, Quito

Martina, G. 1902. Estudo Chimico sobre algumas Frutas Brasileiras. Imprensa Official, Belém

Martins, Adriana Cecília Lima; Rêgo, Márcia Maria Corrêa; Carreira, Léa Maria Medeiros; Albuquerque, Patricia Maia Correia de. 2011. Espectro polínico de mel de tiúba (*Melipona fasciculata* Smith, 1854, Hymenoptera, Apidae). Acta Amazonica 41(2):183–190

Marx, F.; Maia, J.G.S.; Andrade, E.H.A. 2000. Mineral Contents in Amazonian Lecythidaceae Shell Fruits. Acta Horticulturae 531: 241–244

Matta, Luisa Brasil Viana; Scudeller, Veridiana Vizoni. 2012. Lecythidaceae Poit. in the Tupé Sustainable Development Reserve, Manaus, Brazil. Brazilian Journal of Botany 35(2):195–217

Maués, Márcia Motta. 2002. Reproductive phenology and pollination of the Brazil nut tree (*Bertholletia excelsa* Humb. & Bonpl. Lecythidaceae) in eastern Amazonia. In: Pollinating Bees: The Conservation Link between Agriculture and Nature, Kevan, P.; Imperatriz Fonseca, V. L. (eds), pp. 245–254, Ministry of Environment, Brasilia

Maués, Bernardo Antônio Rodrigues; Jardim, Mário Augusto Gonçalves; Batista, Fábio de Jesus; Medeiros, Tonny David Santiago; Quaresma, Adriano da Costa. 2011. Composição florística e estrutura do estrato inferior da floresta de várzea na área de proteção ambiental Ilha do Combu, município de Belém, estado do Pará. Revista Árvore 35(3):669–677

Medeiros, Priscilla Sanjuan de. 2016. Indicadores ambientais para funcionalidade ecológica em florestas secundárias de diferentes idades na Amazônia Oriental. Doctoral dissertation, Universidade Federal do Pará, Belém

Mejia, Kember; Rengifo, Elsa. 2000. Plantas Medicinales de Uso Popular en la Amazonía Peruana. Agencia Española de Cooperación Internacional, Lima

Métraux, A. 1963. Tribes of eastern Bolivia and the Madeira headwaters. In: *Handbook of South American Indians*, J. H. Steward (Editor), Cooper Square Publishers, New York, vol. 3, pp 381–454

Milliken, William. 1998. Structure and composition of one hectare of central Amazonian terra firme forest. Biotropica 30(4):530–537

Milliken, William; Albert, Bruce. 1997. The use of medicinal plants by the Yanomami indians of Brazil, Part II. Economic Botany 51(3):264–278

Milliken, W., R. P. Miller, S. R. Pollard, and E. V. Wandelli. 1992. *The Ethnobotany of the Waimiri Atroari Indians of Brazil*. Royal Botanic Gardens, Kew

Mindlin, B. 1995. Unwritten Stories of the Suruí Indians of Rondônia. Institute of Latin American Studies, University of Texas, Austin

Mindlin, B. 2001. Couro dos Espíritos: Namoro, Pajés e Cura entre os Índios Gavião-Ikolen de Rondônia. Editora Senac/Terceiro Nome, São Paulo

Miranda, V. C. 1909. Os campos de Marajó e a sua flora considerados sob o ponto de vista pastoril. *Boletim do Museu Goeldi (Museu Paraense) de Historia Natural e Ethnographia* 5(1): 96–151

Mondragón, Martha L.; Smith, Randall. 1997. Bete Quiwiguimamo Salvando el bosque para vivir sano: algunas plantas y árboles utilizadas por la Nacionalidad Huaorani de la Amazonía Ecuatoriana. Abya Yala, Quito

Montero, Juan Carlos; Latrubesse, Edgardo M. 2013. The igapó of the Negro River in central Amazonia: linking late-successional inundation forest with fluvial geomorphology. Journal of South American Earth Sciences 46:137–149

Montero, Juan Carlos; Teresa, Maria; Piedade, Fernandez; Wittmann, Florian. 2014. Floristic variation across 600 km of inundation forests (igapó) along the Negro River, central Amazonia. Hydrobiologia 729:229–246

Mori, S. A. 1990. Diversificação e conservação das Lecythidaceae Neotropicais. Acta Bot. Bras. 4(1):45–68

Mori, S.A. 1992. The Brazil nut industry–past, present, and future. In: Sustainable Harvest and Marketing of Rain Forest Products, Plotkin, Mark; Famolare, Lisa (eds), Island Press, Washington, D.C.

Mori, Scott A. 2001. A familia da castanha-do-Pará: símbolo do Rio Negro. In: Florestas do Rio Negro, Oliveira, Alexandre Adalardo; Daly, Douglas C. (eds), pp. 119–141, Companhia das Letras/UNIP, São Paulo

Mori, S. A. 2004. Lecythidaceae (Brazil Nut Family). In: Flowering Plants of the Neotropics, Smith, Nathan; Mori, Scott A.; Henderson, Andrew; Stevenson, Dennis W.; Heald, Scott V. (Editors), pp. 207–208. Princeton University Press, Princeton

Mori, S. A. and N. Lepsch-Cunha. 1995. The Lecythidaceae of a central Amazonian moist forest. Memoirs of the New York Botanical Garden 75:1–55

Mori, S. A. and G. T. Prance. 1987. Species diversity, phenology, plant-animal interactions, and their correlation with climate, as illustrated by the Brazil nut family (Lecythidaceae). In: The Geophysiology of Amazonia: Vegetation and Climate Interactions, R. E. Dickinson (ed.), J. Wiley, New York, pp. 69–89

Mori, S. A. and G. T. Prance. 1990a. Taxonomy, ecology, and economic botany of the Brazil nut (*Bertholletia excelsa*) Humb. & Bonpl.: Lecythidaceae. Advances in Economic Botany 8: 130–150

Mori, S. A. and G. T. Prance. 1990b. Lecythidaceae-Part II: The Zygomorphic-flowered New World Genera (Couroupita, Corythophora, Bertholletia, Couratari, Eschweilera, & Lecythis). New York Botanical Garden, Flora Neotropica, Monograph 21 (II)

Mori, Scott A.; Prance, Ghillean T.; de Zeeuw, Carl H. 1990. Lecythidaceae--Part II. Flora Neotropica, Monograph 21 (II), New York Botanical Garden

Mori, Scott A.; Kiernan, Elizabeth A.; Smith, Nathan P.; Kelly, Lawrence M.; Huang, Ya-Yi; Prance, Ghillean T.; Thiers, Barbara. 2017. Observations on the phytogeography of the Lecythidaceae clade (Brazil nut family). Phytoneuron 30:1–85

Moser, Pamela. 2013. Vegetação Arbórea e sua Relação com Fatores Ambientais e Especiais em Florestas de Terra Firme no Noroeste de Rondônia, Brasil. Master's thesis, Faculdade de Tecnologia, Departamento de Engenharia Florestal, Programa de Pós-Graduação em Ciências Florestais, Universidade de Brasília

Moura, Ignácio Baptista de. 1989. De Belém a S. João do Araguaia, Vale do Rio Tocantins. Fundação Cultural do Pará Tancredo Neves (FCPTN)/Secretaria de Estado da Cultura (SECULT), Belém

Moura, Anna Carolinedos Santos. 2018. As plantas e a qualidade dos agroecossistemas de Santa Rita, Benjamin Constant, Amazonas. Master's thesis, Universidade Federal do Amazonas, Tabatinga

Muniz, Ana Lúcia Vilhena; Silva, Manoela Ferreira Fernandes da; Araújo, Érika Luiza Souza; Alvino, Fabrizia de Oliveira. 2007. Dinâmica do estrato arbóreo de florestas secundárias no nordeste do Pará (Bragança). Revista Brasileira de Biociências 5(1):603–605

Muñoz, Marianella Guevara. 2010. Evaluación de purmas con fines de manejo en la comunidad nativa de Santa Mercedes, Rio Putumayo, Perú. Master's thesis, Universidad Nacional de la Amazonia Peruana, Iquitos

Murakami, Alejandro Araujo; Reyes, Juan Fermandon; Milliken, William. 2016. Frutales Silvestres y Promisorios de Pando. Herencia/Museo de Historia Natural Noel Kempff Mercado, Cobija

Myers, Gillian P.; Newton, Adrian C.; Melgarejo, Orlando. 2000. The influence of canopy gap size on natural regeneration of Brazil nut (*Bertholletia excelsa*) in Bolivia. Forest Ecology and Management 127:119–129

Myster, Randall W. 2018. What we know about the composition and structure of igapó forests in the Amazon Basin. Botanical Review 84:394–410

Nascimento, M. T.; Proctor, J.; Villela, D. M. 1997. Floristic structure, floristic composition and soils of an Amazonian monodominant forest on Maracá Island, Roraima, Brazil. Edinburgh Journal of Botany 54(1):1–38

Navarrete, Víctor Alfredo Torres. 2016. Diversidad, estructura y concentración de carbono en un bosque siempreverde piedmontano de 800 a 900 msnm, em la Amazonía Ecuatoriana. Master's thesis, Universidad Técnica Estatal de Queve do Unidad de Posgrado, Qevedo, Ecuador

Navarro, G. 2002. Provincia biogeográfica del Acre-Madre de Dios (Amazónica Suroccidental). In: Geografía Ecológica de Bolivia: Vegetación y Ambientes Acuáticos, Navarro, Gonzalo; Maldonado, Mabel (eds), Centro de Ecología Simón I. Patiño, Cochabamba, pp 51–98

Navarro, G.; Maldonado, M. 2002. Geografía Ecológica de Bolivia: Vegetación y Ambientes Acuáticos. Centro de Ecología Simón I. Patiño, Cochabamba

Nelson, B.W.; Absy, M.L.; Barbosa, E.M.; Prance, G.T. 1985. Observations on flower visitors to *Bertholletia excelsa* H.B.K. and *Couratari tenuicarpa* A.C. SM. (Lecythidaceae). Acta Amazonica 15(1–2):225–234

Neves, Eduardo G. 2016. A tale of three species or the ancient soul of tropical forests. In: Tropical Forest Conservation: Long-Term Processes of Human Evolution, Cultural Adaptations and Consumption Patterns, N. Sanz (ed.), UNESCO, Mexico, pp. 228–244

Nimuendajú, C. 1939. The Apinayé. Catholic University of America Press, Washington, D.C.

Nimuendajú, C. 1963. The Tucuna. In: *Handbook of South American Indians*, J. H. Steward (Editor), Cooper Square Publishers, New York, vol. 3, pp 713–725

Norconk, Marilyn A.; Veres, Michael. 2011. Physical properties of fruit and seeds ingested by primate seed predators with emphasis on sakis and bearded sakis. The Anatomical Record 294:2092–2111

Ogeron, Clémence; Odonne, Guillaume; Cristinoi, Antonia; Engel, Julien; Grenand, Pierre; Beauchêne, Jacques; Clair, Bruno; Davy, Damien. 2018. Palikur traditional roundwood construction in eastern French Guiana: ethnobotanical and cultural perspectives. Journal of Ethnobiology and Ethnomedicine 14:1–18

Oliveira, Patricia Chaves de; Cavalcante, Suellen. 2017. Ethnobotany in the Amazon floodplain ecosystem: a case study, Quilombo Saracura, Pará, Brazil. International Journal of Botany Studies 2(1):89–99

Oliveira, Francisco Plácido Magalhães; Jardim, Mário Augusto Gonçalves. 1998. Composição florística de uma floresta secundária no município de Igarapé-Açu, estado do Pará, Brasil. Boletim do Museu Paraense Emilio Goeldi, Série Botânica 14(2):127–145

Oliveira, S. G. de; Alfaro, Jessica W. Lynch; Veiga, Liza M. 2014. Activity budget, diet, and habitat use in the critically endangered Ka'apor capuchin monkey (*Cebus kaapori*) in Pará state, Brazil: a preliminary comparison to other capuchin monkeys. American Journal of Primatology 76:919–931

Paiva, Paulo Marcelos; Guedes, Marcelino Carneiro; Funi, Claudia. 2011. Brazil nut conservation through shifting cultivation. Forest Ecology and Management 261:508–514

Parolin, P. 2000a. Seed mass in Amazonian floodplain forests with contrasting nutrient supplies. Journal of Tropical Ecology 16(3):417–428

Parolin, P. 2000b. O uso de árvores nas florestas inundadas por água branca na Amazônia Central. Amazoniana 16(1/2):241–248

Parolin, P. 2000c. Phenology and C02-assimilation of trees in Central Amazonian floodplains. Journal of Tropical Ecology 16:465–473

Parolin, Pia; Ambruester, Nicole; Wittmann, Florian; Ferreira, Leandro; Piedade, Maria Teresa F.; Junk, Wolfgang J. 2002. A review of tree phenology in central Amazonian floodplains. Pesquisas Botânica 52:195–222

Parolin, P.; Simone, O. De; Haase, K.; Waldhoff, D.; Rottenberger, S.; Kuhn, U.; Kesselmeier, J.; Kleiss, B.; Schmidt, W.; Piedade, M. T. F.; Junk, W. J. 2004a. Central Amazonian floodplain forests: tree adaptations in a pulsing system. The Botanical Review 70(3):357–380

Parolin, Pia; Ferreira, Leandro Valle; Albernaz, Ana L. K. M.; Almeida, Samuel S. 2004b. Tree species distribution in várzea forests of Brazilian Amazonia. Folia Geobotanica 39:371–383

Parolin, P., D. Waldhoff, and M. Zerm. 2010. Photochemical capacity after submersion in darkness: How Amazonian floodplain trees cope with extreme flooding. *Aquatic Botany* 93:83–88

Parrotta, John A.; Francis, John K.; Almeida, Rionaldo Rolo de. 1995. Trees of the Tapajós: A Photographic Field Guide. International Institute of Tropical Forestry, General Technical Report IITF-1, Rio Piedras, Puerto Rico

Pedrollo, Camilo Tomazini; Kinupp, Valdely Ferreira; Shepard Jr., Glenn; Heinrich, Michael. 2016. Medicinal plants at Rio Jauaperi, Brazilian Amazon: ethnobotanical survey and environmental conservation. Journal of Ethnopharmacology 186:111–124

Peña-Claros, Marielos; Boot, René G.A.; Dorado-Lora, Jhovania; Zonta, Armelinda. 2002. Enrichment planting of *Bertholletia excelsa* in secondary forest in the Bolivian Amazon: effect of cutting line width on survival, growth and crown traits. Forest Ecology and Management 161: 159–168

Pennington, T. D., C. Reynel, and A. Daza. 2004. *Illustrated Guide to the Trees of Peru*. David Hunt, Sherborne, England

Pereira, Luciano Araujo; Lima-Barbosa, Jackson Rodrigo de; Almeida, Maria Zélia de; Guimarães, Elsie Franklin. 2017. Diversidade de plantas em quintais quilombolas, conhecimento local sobre uso e cultivo de pimentas na Amazônia oriental, Brasil. Revista de Biologia Neotropical 14(1):56–72

Peres, C.A. and C. Baider. 1997. Seed dispersal, spatial distribution and population structure of Brazilnut trees (*Bertholletia excelsa*) in southeastern Amazonia. Journal of Tropical Ecology 13:595–616

Peres, C. A., L. C. Schiesari, and C. L. Dias-Leme. 1997. Vertebrate predation of Brazil-nuts (*Bertholletia excelsa*, Lecythidaceae), an agouti-dispersed Amazonia seed crop: a test of the escape hypothesis. Journal of Tropical Ecology 13:69–79

Peres, C. A.; Baider, Claudia; Zuidema, Pieter A.; Wadt, Lúcia H. O.; Kainer, Karen A.; Gomes-Silva, Daisy A. P.; Salomão, Rafael P.; Simões, Luciana L.; Franciosi, Eduardo R. N.; Valverde, Fernando Conejo; Gribel, Rogério; Shepard Jr., Glenn H.; Ka. 2004. Demographic threats to the sustainability of Brazil nut exploitation. Science 302:2112–2114

Peters, Charles M. 1990. Population ecology and management of forest fruit trees in Peruvian Amazonia. In: *Alternatives to Deforestation. Steps toward Sustainable Use of the Amazon Rainforest*, Anderson, Anthony B. (Editor), pp. 86–98. Columbia University Press, New York

Peters, Charles M.; Hammond, Elysa J. 1990. Fruits from the flooded forests of Peruvian Amazon: yield estimates for natural populations of three promising species. Advances In Economic Botany 8: 159–176

Pinedo-Vasquez, Miguel; Zarin, Daniel J.; Coffey, Kevin; Padoch, Christine; Rabelo, Fernando. 2001. Post-boom logging in Amazonia. Human Ecology 29(2):219–239

Pinto, M. M. 1956. Contribuição ao estudo da pesca na região do Rio Arari (Ilha de Marajó). Revista Brasileira de Geografia 18(3):373–407

Pires, João Murça; Salomão, Rafael de Paiva. 2000. Dinâmica da diversidade arbórea de um fragmento de floresta tropical primária na Amazônia oriental-1. Período: 1956 a 1992. Boletim do Museu Paraense Emílio Goeldi, Série Botânica 16(1):63–110

Pitman, Nigel C. A.; Cerón, Carlos E.; Reyes, Carmita I.; Thurber, Mark; Arellano, Jorge. 2005. Catastrophic natural origin of a species-poor tree community in the world's richest forest. Journal of Tropical Ecology 21:559–568

Ponz, Elizabeth; Carpio, Griceldo; Meo, Severo. 2005. La Medicina Tradicional de los Tacana y Machineri: Conocimientos Prácticos de las Plantas Medicinales. Investigaciones Regionales Pando, La Paz

Posey, D. A. 1985. Native and indigenous guidelines for new Amazonian development strategies: understanding biological diversity through ethnoecology. In: Change in the Amazon Basin Volume 1: Man's Impact on Forests and Rivers, Hemming, John (ed), Manchester University Press, Manchester, pp 156–181

Prance, G. T. 1978. The origin and evolution of the Amazon flora. Interciencia 3(4):207–222

Prance, G. T. 1979. Notes on the vegetation of Amazonia III. The terminology of Amazonian forest types. Brittonia 31(1):26–38

Prance, G. T. 1980. A terminologia dos tipos de florestas amazônicas sujeitas a inundação. Acta Amazonica 10(3):495–504

Prance, G. T. 1984. Introduction on knowledge and use of the Amazon forest. *Interciencia* 9(5):275–279

Prance, G. T. 1985. The pollination of Amazonian plants. In: Key Environments: Amazonia, Ghillean T. Prance and Thomas E. Lovejoy (eds), Pergamon Press, Oxford, pp. 166–191

Prance, G. T. 2014. That Glorious Forest: Exploring the Plants and Their Indigenous Uses in Amazonia. New York Botanical Garden, New York

Prance, Ghillean T.; Mori, Scott A. 1978. Observations on the fruits and seeds of Neotropical Lecythidaceae. Brittonia 30(1):21–33

Prance, Ghillean T.; Mori, Scott A. 1979. Lecythidaceae-Part I: The Actinomorphic-flowered New World Lecythidaceae (Asteranthos, Gustavia, Grias, Allantoma, & Cariniana). New York Botanical Garden, Flora Neotropica Monograph No. 21, New York.

Prata, Shirley Soares; Miranda, Izildinha de Souza; Alves, Sérgio Augusto Oliveira; Farias, Flavia Cardoso; Jardim, Fernando Cristovam da Silva. 2010. Gradiente florístico das florestas secundárias do Nordeste Paraense. Acta Amazonica 40(3):523–534

Queiroz, José Antonio Leite de; Mochiutti, Silas; Machado, Sebastião do Amaral; Galvão, Franklin. 2005. Composição florística e estrutura de floresta em várzea alta estuarina Amazônica. Floresta 35(1):41–55

Queiroz, José Antonio Leite de; Mochiutti, Silas; Machado, Sebastião do Amaral. 2006. Distribuição diamétrica em floresta de várzea do braço norte do Rio Amazonas. Revista de Ciência Agrária 46:151–170

Queiroz, José Antonio Leite de; Machado, Sebastiã do Amaral; Hosokawa, Roberto Tuyoshi; Silva, Ivan Crespo da. 2007. Estrutura e dinâmica de floresta de várzea no estuário Amazônico no estado do Amapá. Floresta 37(3):339–352

RAISG. 2015. Deforestation in the Amazon (1970–2013). RAISG Amazonian Network of Georeferenced Socio-Environmental Information, Instituto Sociambiental, São Paulo

Ratter, J. A. 1987. Notes on the vegetation of the Parque Nacional do Araguaia (Brazil). Notes from the Royal Botanic Garden, Edinburgh 44(2):311–342

Rayol, Breno Pinto; Silva, Manoela Ferreira Fernandes da; Alvino, Fabrízia de Oliveira. 2006. Dinâmica da diversidade florística da regeneração natural de florestas secundárias, no município de Bragança, Pará, Brasil. Boletim do Museu Paraense Emílio Goeldi, Ciencias Naturais 1(2):9–27

Rayol, Breno Pinto; Alvino-Rayol, Fabrízia de Oliveira; Silva, Manoela Ferreira Fernandes da. 2011. Similaridade florística entre o estrato arbóreo e a regeneração natural de uma floresta secundária, no município de Bragança, nordeste do estado do Pará, Brasil. Revista Brasileira de Agroecologia 6(3):107–114

Rayol, Breno Pinto; Alvino-Rayol, Fabrízia de Oliveira; Silva, Maria da Conceição Pereira da. 2013. Caracterização de quintais agroflorestais desenvolvidos por agricultores familiares do município de Alenquer, Pará. Cadernos de Agroecologia 8(2):1–5

Reis, Neidiane Farias Costa. 2010. Estrutura fitossociológica e seletividade de espécies em áreas inundáveis na região do baixo Rio Jaci-Paraná, Rondônia. Master's Thesis, Universidade Federal de Rondônia, Porto Velho

Reynafarje, Carlos Alfredo. 2014. Relación entre la estructura diamétrica y la abundancia en tres tipos de bosque en el distrito del Alto Nanay, Loreto, Perú, 2014. Facultad de Ciencias Forestales, Universidad Nacional de la Amazonía Peruana (UNAP), Iquitos

Ribeiro, Maria Beatriz N.; Jerozolimski, Adriano; Robert, Pascale de; Magnusson, William E. 2014a. Brazil nut stock and harvesting at different spatial scales in southeastern Amazonia. Forest Ecology and Management 319:67–74

Ribeiro, Maria Beatriz N.; Jerozolimski, Adriano; Robert, Pascale de; Salles, Nilson V.; Kayapó, Biribiri; Pimentel, Tania P.; Magnusson, William E. 2014b. Anthropogenic landscape in south-eastern Amazon: contemporary impacts of low-intensity harvesting and dispersal of Brazil nuts by the Kayapó indigenous People. PLOS One 9(7) e102187

Ribeiro, Reginaldo Vicente; Bieski, Isanete Geraldini Costa; Balogun, Sikiru Olaitan; Martins, Domingos Tabajarade Oliveira. 2017. Ethnobotanical study of medicinal plants used by ribeirinhos in the north Araguaia microregion, Mato Grosso, Brazil. Journal of Ethnopharmacology 205: 69–102

Rimachi-Taricuarima, Milagros Natalia; Pérez-Gardini, Juan Jesús; Tirado-Herrera, Emérita Rosabel; Zárate-Gómez, Ricardo; Mozombite-Pinto, Linder Felipe. 2019. Plantas consumidas por *Lagothrix lagotricha lagotricha* Humboldt, 1812 en la Amazonía peruana. Ciencia Amazónica 7(1):93–110

Rios, M.; Silva, R.C.V.; Sabogal, C.; Martins, J.; Silva, R.N.; Brito, R.R.; Brito, I.M.; Brito, M.F.; Silva, J.R.; Ribeiro, R.T. 2001. Benefícios das Plantas da Capoeira para a Comunidade de Benjamin Constant, Pará, Amazônia Brasileira. CIFOR, Belém

Ritter, Rhuan Amorim; Monteiro, Maria Vivina Barros; Monteiro, Frederico Ozanan Barros; Rodrigues, Silvane Tavares; Soares, Marina Lira; Silva, Jean Carlos Ramos; Palha, Maria das Dores Correia; Biondi, Germano Francisco; Rahal, Sheila Canevese; Tourinho, Manuel Malheiros. 2012. Ethnoveterinary knowledge and practices at Colares island, Pará state, eastern Amazon, Brazil. Journal of Ethnopharmacology 144:346–352

Roble, Nick. 2008. Ontogenetic resource partitioning in white faced sakis (*Pithecia pithecia*). Master's thesis, Kent State University.

Rodet, Maria Jacqueline; Guapindaia, Vera; Matos, Amauri. 2010. Análise tecnológica e cadeia operatória: uma nova proposta para a indústria lítica lascada das culturas ceramistas da Amazônia. In: Arqueologia Amazônica vol. 2, Edithe Pereira and Vera Guapindaia (eds), MPEG/IPHAN/SECULT, Belém, pp 681–711

Rodríguez-Flores, Claudia; Stiles, F. Gary; Arizmendi, María del Coro. 2012. Pollination network of a hermit hummingbird community (Trochilidae, Phaethornithinae) and their nectar resources in the Colombian Amazon. Ornitologia Neotropical 23:85–100

Roosevelt, A. C. 2014. The Amazon and the Anthropocene: 13,000 years of human influence in a tropical rainforest. Anthropocene 4:69–87

Rosengarten, Frederic. 1984. The Book of Edible Nuts. Walker and Company, New York

Ros-Tonen, Mirjam A. F.; Van Andel, Tinde; Morsello, Carla; Otsuki, Kei; Rosendo, Sergio; Scholz, Imme. 2008. Forest-related partnerships in Brazilian Amazonia: there is more to sustainable forest management than reduced impact logging. Forest Ecology and Management 256: 1482–1497

Roumy, V.; Garcia-Pizango, G.; Gutierrez-Choquevilca, A.-L; Ruiz, L.; Jullian, V.; Winterton, P.; Fabre, N.; Moulis, C.; Valentin, A. 2007. Amazonian plants from Peru used by Quechua and Mestizo to treat malaria with evaluation of their activity. Journal of Ethnopharmacology 112:482–489

Rudas, Augustín; Prieto, Adriana. 1998. Análisis florístico del parque nacional natural Amacayacu e Isla Mocagua, Amazonas (Colombia). Caldasia 20(2):142–172

Salomão, Rafael de Paiva. 1991. Estrutura e densidade de *Bertholletia excelsa* H. & B. ("castanheira") nas regiões de Carajás e Marabá, Estado do Pará. Boletim do Museu Paraense Emílio Goeldi, Série Botanica 7(1): 47–68

Salomão, Rafael de Paiva. 2009. Densidade, estrutura e distribuição especial de castanheira-do-brasil (*Bertholletia excelsa* H. & B.) em dois platôs de floresta ombrófolia densa na Amazônia setentorial brasileira. Boletim do Museu Paraense Emílio Goeldi, Ciências Naturais 4(1): 11–25

Salomão, Rafael de Paiva. 2014. A castanheira: história natural e importância socioeconómica. Boletim do Museu Paraense Emílio Goeldi, Ciencias Naturais 9(2):259–266

Salomão, Rafael de Paiva; Lisboa, Pedro L. B. 1988. Análise ecológica da vegetação de uma floresta pluvial tropical de terra firme, Rondônia. Boletim do Museu Paraense Emílio Goeldi, Série Botânica 4(2):195–233.

Salomão, Rafael de Paiva; Silva, Manoela Ferreira Fernandes da; Rosa, Nelson Araújo. 1988. Inventário ecológico em floresta pluvial tropical de terra firme, Serra Norte, Carajás, Pará. Boletim do Museu Paraense Emílio Goeldi, Série Botânica 4(1):1–46

Salomão, Rafael de Paiva; Matos, Aires Henriques de; Rosa, Nélson de Araújo. 2002. Dinâmica do sub-bosque e do estado arbóreo de floresta tropical primária fragmentada na Amazônia oriental. Acta Amazonica 32(3):387–419

Salomão, Rafael de Paiva; Rosa, Nélson Araújo; Castilho, Alexandre; Morais, Kácio Andrey Câmara. 2006. Castanheira-do-brasil recuperando áreas degradadas e provendo alimento e renda para comunidades da Amazônia setentrional. Boletim do Museu Paraense Emílio Goeldi, Série Ciencias Naturais 1(2):65–78

Salomão, Rafael de Paiva; Vieira, Ima Célia Guimarães; Suemitsu, Chieno; Rosa, Nélson de Araújo; Almeira, Samuel Soares de; Amaral, Dário Dantas do; Menezes, Moirah Paulo Machado de. 2007a. As florestas de Belo Monte na grande curva do rio Xingu, Amazônia Oriental. Boletim do Museu Paraense Emílio Goeldi, Ciencias Naturais 2(3):57–153

Salomão, Rafael Paiva; Rosa, Nélson Araújo; Morais, Kácio Andrey Câmara. 2007b. Dinâmica da regeneração natural de árvores em áreas mineradas na Amazônia. Boletim do Museu Paraense Emílio Goeldi, Ciencias Naturais 2(2):85–139

Saltos, Ricardo Vinicio Abril; Vásquez, Tomás Elías Ruiz; Lazo, Jatnel Alonso; Banguera, Derwing Viáfra; Guayasamín, Pedro Damián Ríos; Vargas, Janeth Karina Aguinda; Peñas, Ingrid Vega. 2016. The use of medicinal plants by rural populations of the Pastaza province in the Ecuadorian Amazon. Acta Amazonica 46(4):355–366

Sampaio, P. 2000. Castanha-sapucaia *Lecythis pisonis*. In: Biodiversidade Amazônica: Exemplos e Estratégias de Utilização, Jason W. Clay, Paulo de T. B. Sampaio, and Charles R. Clement (eds), INPA/SEBRAE, Manaus, pp 140–147

Sandel, Maureen Peggy; Carvalho, João Olegário Pereira de. 2000. Composição florística e estrutura de uma área de cinco hectares de mata alta sem babaçu na floresta nacional do Tapajós. Embrapa Amazônia Oriental, Belém, Documentos 63, pp. 1–19

Santana, José Augusto da Silva; Almeida, Wanderléa da Costa; Sousa, Luciana Karla V. dos Santos. 2004. Florística e fitossociologia em área de vegetação secundária na Amazônia oriental. Revista de Ciências Agrárias 41:105–120

Santos, Luciana Oliveira dos. 2010. Comparação da vegetação lenhosa entre borda e interior de uma mancha de mata de restinga na RESEX Mar Caeté-Taperaçu, Bragança, Pará. Master's thesis, Universidade Federal Rural da Amazônia/Museu Paraense Emílio Goeldi, Belém

Santos, Charles F.; Absy, Maria L. 2010. Polinizadores de *Bertholletia excelsa* (Lecythidales: Lecythidaceae): interações com abelhas sem ferrão (Apidade: Meliponni) e nicho trófico. Neotropical Entomology 39(6): 854–861

Santos, Charles Fernando dos; Absy, Maria Lúcia. 2012. Interactions between carpenter bees and orchid bees (Hymenoptera: Apidae) in flowers of *Bertholletia excelsa* Bonpl. (Lecythidaceae). Acta Amazonica 42(1):89–94

Santos, J. U. M. and M. N. Bastos. 2014. Flores e Frutas das Restingas do Estado do Pará. Universidade Federal Rural da Amazônia (UFRA), Belém

Santos, Josefa Silva dos; Guarim Neto, Germano. 2017. Plantas de quintais da área urbana de Comodoro, Mato Grosso, Brasil. Flovet 1(9):31–57

Santos, Gideão Costa dos; Jardim, Mário Augusto Gonçalves. 2006. Florística e estrutura do estrato arbóreo de uma floresta de várzea no município de Santa Bárbara do Pará, Estado do Pará, Brasil. Acta Amazonica 36(4):437–446

Santos, João Ubiratan M. dos; Rosário, Carlos da S. 1988. Levantamento da vegetação fixadora das dunas de Algodoal-PA. Boletim do Museu Paraense Emílio Goeldi, Série Botânica 4(1): 133–154

Santos, Gideão Costa; Tourinho, Manoel Malheiros; Mendes, Fernanda da Silva; Guimarães, Célia Maria Costa. 2014a. Fitossociologia e práticas de manejo tradicional em uma floresta de várzea em Santa Bárbara do Pará, Brasil. Revista de Ciências Agrárias 57(2):138–145

Santos, M. R. A.; Lima, M. R.; Oliveira, C. L. L. G. 2014b. Medicinal plants used in Rondônia, western Amazon, Brazil. Revista Brasileira de Plantas Medicinais 16(3):707–720

Santos, Joielan Xipaia dos; Reis, Alisson Rodrigo Souza; Matos, Sterphane Araújo de; Leão, Fábio Miranda; Carvalho, Josiane Celerino de. 2016a. Caracterização etnobotânica de essências florestais com fins medicinais utilizadas pela etnia Xipaya, no município de Altamira-PA. Biota Amazônia 6(2):1–8

Santos, Otilene dos Anjos; Couceiro, Sheyla Regina Marques; Rezende, Alinne Costa Cavalcante; Silva, Márcia Daniela de Sousa. 2016b. Composition and richness of woody species in riparian forests in urban areas of Manaus, Amazonas, Brazil. Landscape and Urban Planning 150:70–78

Santos, Ronaldo Oliveira dos; Lima, Robson Carmo; Lima, Robson Borges de; Aparício, Perseu da Silva; Abreu, Jadson Coelho de. 2017. Florística e estrutura de uma comunidade arbórea na floresta estadual do Amapá, Amazônia Oriental, Brasil. Nativa Sinop 5:529–539

Santos, Lucélia Rodrigues; Ferreira, Evandro José Linhares; Silva, Cláudia Eugenio. 2018. Florística e fitossociologia de um sub-bosque urbano em Rio Branco-AC, Brasil. Biota Amazônia 8(3):28–33

Sanz-Biset, Jaume; Campos-de-la-Cruz, José; Epiquién, Mirbel A.; Cañigueral, Salvador. 2009. A first survey on the medicinal plants of the Chazuta valley (Peruvian Amazon). Journal of Ethnopharmacology 122:333–362

Schöngart, Jochen; Piedade, Maria Teresa F.; Ludwigshausen, Sabine; Horna, Viviana; Worbes, Martin. 2002. Phenology and stem-growth periodicity of tree species in Amazonian floodplain forests. Journal of Tropical Ecology 18:581–597

Schroth, Götz; Coutinho, Paulo; Moraes, Vicente H. F.; Albernaz, Ana Luisa. 2003. Rubber agroforests at the Tapajós river, Brazilian Amazon—environmentally benign land use systems in an old forest frontier region. Agriculture, Ecosystems & Environment 97:151–165

Schultes, R. E. and R. F. Raffauf. 1990. The Healing Forest: Medicinal and Toxic Plants of the Northwest Amazonia. Dioscorides Press, Portland, Oregon

Schwartz, Gustavo. 2007. Manejo sustentável de florestas secundárias: espécies potenciais no nordeste do Pará, Brasil. Amazônia: Ciência e Desenvolvimento 3(5):125–147

Scoles, Ricardo. 2011. Do rio Madeira ao rio Trombetas: novas evidências ecológicas e históricas da origem antrópica dos castanhas amazônicos. Novos Cadernos NAEA 14(2):265–282

Scoles, Ricardo; Gribel, Rogério. 2011. Population structure of Brazil nut (*Bertholletia excelsa*, Lecythidaceae) stands in two areas with different occupation histories in the Brazilian Amazon. Human Ecology 39: 455–464

Scoles, Ricardo; Gribel, Rogério. 2015. Human influence on the regeneration of the Brazil nut tree (*Bertholletia excelsa* Bonpl., Lecythidaceae) at Capanã Grande lake, Manicoré, Amazonas, Brazil. Human Ecology 43:843–854

Scoles, Ricardo; Gribel, Rogério; Klein, Gilmar Nicolau. 2011. Crescimento e sobrevivência de castanheira (*Bertholletia excelsa* Bonpl.) em diferentes condições ambientais na região do rio Trombetas, Oriximiná, Pará. Boletim do Museu Paraense Emílio Goeldi, Série Ciências Naturais 6(3):273–293

Scudeller, Veridiana Vizoni; Vegas-Vilarrúbia, Teresa. 2018. Distribution and Β-diversity of trees species in igapó forests (Negro River basin, Brazilian Amazon). Journal of Vegetation Science 29:1052–1064

Scudeller, Veridiana Vizoni; Veiga, Josephina Barata da; Araújo-Jorge, Lúcia Helena de. 2009. Etnoconhecimento de plantas de uso medicinal nas comunidades São João do Tupé e Central (Reserva de Desenvolvimento Sustentável do Tupé). In: Biotupé: Meio Físico, Diversidade Biológica e Sociocultural do Baixo Rio Negro, Amazônia Central, Santos-Silva, Edinaldo Nelson; Scudeller, Veridiana Vizoni (eds), UEA Edições, Manaus, vol. 2, pp 185–199

Shanley, P. and N.A. Rosa. 2005. Conhecimento em erosão: um inventário etnobotânico na fronteira de exploração da Amazônia oriental. Boletim do Museu Paraense Emílio Goeldi, Série Ciências Naturais 1(1):147–171

Shepard, Glenn. 2006. Etnobiologia Yanomami. In: Projeto de Conservação e Utilização Sustentável da Diversidade Biológica Brasileira-PROBIO, Boubli, Jean Philippe; Pontual, Francisco Boavista (editors), Projeto de Conservação e Utilização Sustentável da Diversidade Biológica Brasileira-PROBIO, MMA/BIRD/GEF/CNPq, Brasilia, pp. 102–128

Shepard, Glenn H.; Ramirez, Henri. 2011. "Made in Brazil": human dispersal of the Brazil nut (*Bertholletia excelsa*, Lecythidaceae) in ancient Amazonia. Economic Botany 65(1):44–65

Silva, D. 1973. A castanha do Pará como fator inicial de desenvolvimento de Marabá: perspectivas atuais. Instituto de Geografia, Geografia Econômica, Universidade de São Paulo, vol. 12

Silva, Suzy Cristina Pedroza da. 2006. Os sistemas agroflorestais na Reserva de Desenvolvimento Sustentável Piagaçu-Purus: uma alternativa para a agricultura familiar. Master's thesis, Universidade Federal do Amazonas, Manaus

Silva, Maria Betânia Araujo da. 2008. Composição florística e estrutura fitossociológica de uma floresta ripária situada às margens do Rio Moju, Pará. Master's thesis, Universidade Federal Rural da Amazônia, Belém

Silva, Rosibel Rodrigues. 2009. O etnoextrativismo não-madeireiro em uma comunidade Amazônica: um estudo de caso em Santa Luzia do Buiuçuzinho-Coari/AM. Master's thesis, UFAM, Manaus

Silva, Silvestre. 2011. Fruits of the Brazilian Amazon. Metalivros, São Paulo

Silva, Wheriton Fernando Moreira da. 2015. Florística, dinâmica e estrutura da regeneração natural em floresta explorada na FLONA do Tapajós, PA. Master's Thesis, INPA, Manaus

Silva, Edinei Santos da. 2017. Plantas alimentícias em comunidades agrícolas no município de Rio Preto da Eva-AM. Master's thesis, INPA, Manaus

Silva, Ana Paula Ferreira Frota da; Bentes-Gama, Michelliny de Matos. 2008. Fitossociologia de uma floresta ombrófila aberta em área de assentamento rural no distrito de Jaci Paraná, Porto Velho, Rondônia. Ambiência 4(3):435–452

Silva, Manoela F. F. da; Rosa, Nelson A. 1989. Análise do estrato arbóreo da vegetação sobre jazidas de cobre na Serra dos Carajás-PA. Boletim do Museu Paraense Emílio Goeldi, Série Botânica 5(2):175–206

Silva, Aglair Gomes da; et al. 2008. Aldeias indígenas Mura. Editora da Universidade Federal do Amazonas, Manaus

Silva, Marcondes Alves B. da; Melo, Laura Valdiane L.; Ribeiro, Reginaldo V.; Souza, João Paulo M. de; Lima, Joaquim Corsino S.; Martins, Domingos Tabajara de O.; Silva, Regilane M. da. 2010. Levantamento etnobotânico de plantas utilizadas como anti-hiperlipidêmicas e anorexígenas pela população de Nova Xavantina-MT, Brasil. Revista Brasileira de Farmacognosia 20(4):549–562

Silva, Kátia Emidio da; Martins, Sebastião Venâncio; Fortin, Marie-José; Ribeiro, Milton Cezar; Azevedo, Celso Paulo de; Ribeiro, Carlos Antonio Álvares Soares; Santos, Nerilson Terra. 2014. Tree species community spatial structure in a terra firme Amazon forest, Brazil. Bosque 35(3):347–355

Silva, Andrea Araújo da; Santos, Lizandra Elizeário; Cruz, Girlene da Silva; Ribeiro, Renato Bezerra da Silva; Gama, João Ricardo Vasconcelos. 2018. Potencial de comercialização de produtos florestais não madeireiras na área de manejo da reserva extrativista Tapajós Arapiuns Pará. Acta Tecnológica 13(1):45–63

Smith, Herbert H. 1879. Brazil, the Amazons and the Coast. Charles Scribner's Sons, New York

Smith, N. J. H. 1976. Utilization of game along Brazil's Transamazon Highway. *Acta Amazonica* 6(4):455–466

Smith, N. J. H. 1999. *The Amazon River Forest*. Oxford University Press, New York.

Snethlage, E. 1910. A travessia entre o Xingú e o Tapajoz. Boletim do Museu Goeldi (Museu Paraense) de Historia Natural e Ethnographia 7:49–92

Soriano, Marlene; Kainer, Karen A.; Staudhammer, Christina L.; Soriano, Eduardo. 2012. Implementing multiple forest management in Brazil nut-rich community forests: effects of logging on natural regeneration and forest disturbance. Forest Ecology and Management 268:92–102

Souza, Victória Cristina Oliveira de. 2019. Avaliação do estágio sucessional utilizando a classificação em grupos ecológicos de uma floresta urbana na região metropolitana de Belém-Pará. Bachelor's thesis, Universidade Federal Rural da Amazônia (Ufra), Belém

Souza, Afonso Duarte Leão de; Rocha, Arnaldo F.; Imbiriba da; Pinheiro, Maria Lúcia Belém et al. 2001. Chemical constituents of *Gustavia augusta* l. (Lecythidaceae). *Química. Nova* 24(4):439–442

Souza, Deoclides Ricardo de; Souza, Agostinho Lopes de; Leite, Helio Garcia; Yared, Jorge Alberto Gazel. 2006. Análise estrutural em floresta ombrófila densa de terra firme não explorada, Amazônia oriental. Revista Árvore 30(1):75–87

Spichiger, Rodolphe; Loizeau, Pierre-André; Latour, Cyrille; Barreira, Gabrielle. 1996. Tree species richness of a south-western Amazonian forest (Jenaro Herrera, Peru, 73° 40'W/4° 54'S). Candollea 5(1–2):559–577

Spix, J. B. and C. F. P. Martius, 1976. *Viagem pelo Brasil, 1817–1820*. Edições Melhoramentos, São Paulo, Vol.3

Spruce, R. 1853. Botanical objects communicated to the Kew Museum, from the Amazon River, in 1851. In: *Hooker's Journal of Botany and Kew Garden Miscellany*, W. J. Hooker (Editor), Lovell Reeve, London, vol. 5, pp. 169–177.

Spruce, R. 1908. Notes of a Botanist on the Amazon and Andes. Macmillan, London, vol. 1

Steege, Hans ter; et al. 2013. Hyperdominance in the Amazonian tree flora. Science 342 https://doi.org/10.1126/science.1243092

Steere, J. B. 1903 A narrative of a visit to Indian tribes of the Purús River, Brazil. Report of the U.S. National Museum for the Year Ending June 30, 1901, Government Printing Office, Washington, D. C., pp. 363–393

Stoian, D. 2000. Shifts in forest production extraction: the post-rubber era in the Bolivian Amazon. International Tree Crops Journal 10:277–297

Stoian, D. 2004. Harvesting windfalls: the Brazil nut (*Bertholletia excelsa*) economy in the Bolivian Amazon. In: Forest Products, Livelihoods and Conservation: Case Studies of Non-Timber Product Systems, Alexiades, Miguel N.; Shanley, Patricia (eds), CIFOR, Jakarta, vol. 3, pp. 85–110

Stoian, D. 2005. Making the best of two worlds: rural and peri-urban livelihood options sustained by nontimber forest products from the Bolivian Amazon. World Development 33(9):1473–1490

Summers, Percy M.; Browder, John O.; Pedlowski, Marcos A. 2004. Tropical forest management and silvicultural practices by small farmers in the Brazilian Amazon: recent farm-level evidence from Rondônia. Forest Ecology and Management 192:161–177

Taurino, Tássia Cristina da Conceição Barros. 2017. Recuperação natural da diversidade de espécies de plantas em florestas secundárias na Amazônia Oriental. Master's Thesis, Universidade Federal do Pará, Belém

Thomas, Evert. 2009. New light on the floristic composition and diversity of indigenous territory and national park Isiboro-Sécure, Bolivia. Biodiversity Conservation 18:1847–1878

Thomas, Evert; Caicedo, Carolina Alcázar; McMichael, Crystal H.; Corvera, Ronald; Loo, Judy. 2015. Uncovering spatial patterns in the natural and human history of Brazil nut (*Bertholletia excelsa*) across the Amazon Basin. Journal of Biogeography 42: 1367–1382

Tomchinsky, Bernardo; Ming, Lin Chau; Kinupp, Valdely Ferreira; Hidalgo, Ari de Freitas; Chaves, Francisco Célio Maia. 2017. Ethnobotanical study of antimalarial plants in the middle region of the Negro River, Amazonas, Brazil. Acta Amazonica 47(3):203–212

Tonini, Helio; Pedrozo, Cássia Ângelo. 2014. Variações anuais na produção de frutos e sementes de castanheira-do-Brasil (*Bertholletia excelsa* Bonpl., Lecythidaceae) em florestas nativas de Roraima. Revista Árvore 38(1):133–143

Trivedi, M. R., Cornejo, F. H. and Watkinson, A. R. 2004. Seed predation on Brazil nuts (Bertholletia excelsa) by macaws (Psittacidae) in Madre de Dios, Peru. Biotropica, 36(1), pp.118–122

Uquillas, J., A. Ramírez, and C. Seré 1992. Are modern agroforestry practices economically viable? a case study in the Ecuadorian Amazon. In: Financial and Economic Analysis of Agroforestry Systems, G. Sullivan, S. Huke, and J. Fox (eds.), NFTA-FSP, East-West Center, Honolulu, pp. 273–292

Vale, Igor do; Costa, Luiz Gonzaga Silva; Miranda, Izildinha Souza. 2014. Espécies indicadas para a recomposição da floresta ciliar da sub-bacia do Rio Peixe-Boi, Pará. Ciência Florestal 24(3):573–582

Valencia, Renato; Foster, Robin B.; Villa, Gorky; Condit, Richard; Svenning, Jens-Christian; Hernández, Consuelo; Romoleroux, Katya; Losos, Elizabeth; Magard, Else; Balslev, Henrik.

2004. Tree species distributions and local habitat variation in the Amazon: large forest plot in eastern Ecuador. Journal of Ecology 92(2): 214–229
Valverde, Orlando; Dias, Catharina. 1967. A Rodovia Belém-Brasília: Estudo de Geografia Regional. IBGE, Rio de Janeiro
Vásquez, R. 1997. *Flórula de las Reservas Biológicas de Iquitos, Perú*. Monographs in Systematic Botany Vol. 63, Missouri Botanical Garden, St. Louis.
Vásquez, R. and A. H. Gentry. 1989. Use and misuse of forest-harvested fruits in the Iquitos area. *Conservation Biology* 3(4): 350–361.
Veiga, Liza Maria. 2006. Ecologia e comportamento do cuxiú-preto (*Chiropotes satanas*) na paisagem fragmentada da Amazônia oriental. Doctoral Dissertation, Universidade Federal do Pará, Belém
Velho, Otávio G. 1972. Frentes de Expansão e Estrutura Agrária: Estudo do Processo de Penetração numa Área da Transamazônica. Zahar Editores, Rio de Janeiro
Venturieri, G. A. 2001. Crop improvement and biodiversity. Encyclopedia of Biodiversity 1:897–91
Veríssimo, José. 1963. Castanhais. In: Amazônia: Tipos e Aspectos, Superintendência do Plano de Valorização Econômica da Amazônia (SPVEA), Rio de Janeiro, pp 28–31
Veríssimo, Adalberto; Barreto, Paulo; Mattos, Marli; Tarifa, Ricardo; Uhl, Christopher. 1992. Logging impacts and prospects for sustainable forest management in an old Amazonian frontier: the case of Paragominas. Forest Ecology and Management 55: 169–199
Verswijver, G. 1995. *Kaiapo: Material Culture-Spiritual World*. Museum für Völkerkunde, Frankfurt
Vickers, W. T. 1994. The health significance of wild plants for the Siona and Secoya. In: Eating on the Wild Side: The Pharmalogic, Ecologic, and Social Implications of Using Noncultures, N. L. Etkin (ed.), University of Arizona Press, Tucson, pp 143–165
Vickers, W. T. and T. Plowman. 1984. *Useful Plants of the Siona and Secoya Indians*. Field Museum of Natural History, Fieldiana Botany New Series, no. 15
Vieira, Ima Célia Guimarães. 1996. Forest succession after shifting cultivation in eastern Amazonia. Doctoral dissertation, University of Stirling
Vieira, Paulo R. 2005. Floresta e comunidade: cotidiano de famílias em Jericó, Garrafão do Norte, Pará. Master's thesis, Universidade Federal do Pará, Belém
Vieira, Diego dos Santos; Gama, João Ricardo Vasconcellos; Ribeiro, Renato Bezerra da Silva; Ximenes, Lucas Cunha; Corrêa, Viviane Vasconcellos; Alves, Adriana Ferreira. 2014. Comparação estrutural entre floresta manejada e não manejada na comunidade Santo Antônio, estado do Pará. Ciência Florestal 24(4):1061–1068
Vieira, Maria Eliza Brambila; Vasconcelos, Ilka Maria; Machado, Olga Lima Tavares; Gomes, Valdirene Moreira; Carvalho, André de Oliveira. 2015. Isolation, characterization and mechanism of action of an antimicrobial peptide from *Lecythis pisonis* seeds with inhibitory activity again *Candida albicans*. Acta Biochim. Biophys. Sin 47(9):716–729
Villachica, Hugo; Carvalho, José Edmar Urano de; Muller, Carlos Hans; Díaz, Camilo; Almanza, Martha; Mercado, Roberto Samanez; Izquierdo, Juan; Egg, Antônio Brack. 1996. Frutales y Hortalizas Promisorios de la Amazonia. Tratado de Cooperación Amazonica, Lima.
Viveiros de Castro, E. 1992. From the Enemy's Point of View: Humanity and Divinity in an Amazonian Society. University of Chicago Press, Chicago
Werner, D. 1984. Amazon Journey: An Anthropologist's Year among Brazil's Mekranoti Indians. Simon and Schuster, New York
Wittmann, Florian; Anhuf, Deiter; Junk, Wolfgang J. 2002. Tree species distribution and community structure of central Amazonian várzea forests by remote-sensing techniques. Journal of Tropical Ecology 18(6):805–820
Wittmann, Florian; Schöngart, Jochen; Queiroz, Helder L.; Wittmann, Astrid de Oliveira; Conserva, Auristela dos Santos; Piedade, Maria T. F.; Kesselmeier, Jürgen; Junk, Wolfgang J. 2009. The Amazon floodplain demonstration site: sustainable timber production and management of central Amazonian white-water floodplains. Ecohydrology & Hydrobiology 9(1):41–54

Wittmann, Florian; Schöngart, Jochen; Junk, Wolfgang J. 2010. Phytogeography, species diversity, community structure and dynamics of central Amazonian floodplain forests. In: Amazonian Floodplain Forests: Ecophysiology, Biodiversity and Sustainable Management, Junk, Wolfgang J.; Piedade, Maria T. F.; Wittmann, Florian; Schöngart, Jochen; Parolin, Pia (eds), Springer, Dordrecht, pp 61–102

Woodroof, Jasper Guy 1979. Tree Nuts: Production, Processing, Products. AVI Publishing Company, Westport, Connecticut

Woods, W. I. and McCann, J. M. 1999. The anthropogenic origin and persistence of Amazonian dark earths. *Yearbook of the Conference of Latin Americanist Geographers* 25: 7–14

Yang, Jun. 2009. Brazil nuts and associated health benefits: a review. LWT – Food Science and Technology 42:1573–1580

Zárate, Ricardo; Mori, Tony J.; Maco, José T. 2013. Estructura y composición florística de las comunidades vegetales del ámbito de la Carretera Iquitos-Nauta, Loreto-Perú. Folia Amazónica 22(1–2):77–89

Zuidema, Pieter A. 2003. Ecología y Manejo del Árbol de Castaña (*Bertholletia excelsa*). PROMAB Serie Científica Nro. 6, Riberalta, Bolivia

Chapter 31
Malpighiaceae

31.1 *Bunchosia armeniaca*

Bolivia Cereza, ciruela verde, ciruela de fraile

Brazil Marmelo, caferana, ciruela, siruela, ameixa braba, ameixa do Pará, café do Amazonas, cafezinho, café do Amazonas, moranga, caramelo, falso guaraná

Colombia Mamelo, mamey, ciruela de fraile

Ecuador Ciruela, ciruela verde, ciruela silvestre, ushón; ila (Kichwa), ontarecon-camo (Waorani)

Peru Cansaboca, ciruela, ciruela del fraile, ciruelo de cansa boca, indano, usuma

Venezuela Palo de colmena, ciruela de fraile

Status Wild, spontaneous in cultural settings, planted

N. Smith, *Amazon Fruits: An Ethnobotanical Journey*, Ethnobiology,
https://doi.org/10.1007/978-3-031-12803-5_31

Ciruela *Bunchosia armeniaca* bush in fruit in a home garden. San Joaquin de Omaguas, Loreto, Peru, 7-11-10

Bunchosia armeniaca is a bush to small tree, normally under 10 m high but attaining 30 m in some cases, that grows in forests, successionary growth, and home gardens on both uplands and floodplains. Ciruela, one of its names in Peru, is the same name applied to red mombin *Spondias purpurea* because the reddish-orange fruits of *Bunchosia armeniaca* resemble those of red mombin. Cansaboca (tires your mouth) is another name for *Bunchosia armeniaca* in the Reserva Nacional Pacaya-Samiria in Loreto, Peru, because the mesocarp is chewy. The fruits of *Bunchosia armeniaca* are typically 3–4 cm long and are produced after the plant reaches 15–18 months old (Watson 1985:319).

Ciruela is more common in the western portion of the Amazon Basin from Colombia south to Bolivia and is considered native to the Andean foothills up to 2300 m (Arellano 2013:183; Cartagena 2014:71; Loza et al. 2010; Mendoza and Guerrero 2010; Popenoe 1924; Giacomotti 2016:32; Giacomotti and Reynel 2018; Ugent and Ochoa 2006:183). Ciruela only extends into Brazil in Acre and western Amazonas state. Ciruela is cultivated in home gardens as high as 1800 m in the Peruvian Andes, such as in Oxapampa, Peru. In the Ecuadorian Andes, it has been recorded as high as 2400 m (Popenoe 1924). The medium-sized tree is also planted in home gardens along banks of rivers, such as the Nanay, Itaya, and Ucayali because the tree can tolerate flooding for brief periods during the height of the rainy season. The red to orange fruits contain one or two seeds surrounded by a sweet pulp. Relished as a snack, the fruits occasionally turn up in markets, such as Iquitos.

Shirly, 11 years, gathering ciruela *Bunchosia armeniaca* **fruit in her home garden. Soledad, Rio Itaya, Loreto, Peru 1-7-06**

The precise origins of *Bunchosia armeniaca* are unknown, but it was domesticated in pre-contact times in tropical South America, possibly in Peru or Ecuador. Another possibility is that it came into the Amazon basin from the northern part of South America or the Caribbean (Ducke 1946). Seeds have been found in archaeological sites associated with the Inca civilization in the Andes. The much older Mochica culture along Peru's northern desert coast fashioned jars decorated with molded representations of the fruit (Towle 1961:61). *Bunchosia armeniaca* has been cultivated along the Peruvian coast since at least 2600 BP (Cohen 1977). The coast of Peru is too dry for *Bunchosia armeniaca* to have arisen there, so the tree was likely introduced by indigenous cultures and planted in irrigated orchards.

31.2 *Byrsonima chrysophylla*

Bolivia Muela

Brazil Muruci da mata, muruci do mato, muruci de pomba, muruci da capoeira, muruci do campo, muruci do miúdo, muruci grande, muruci pequeno, muruci pitanga, muruci amarelo, muruci da praia, semaneira; amiã kurupïpï (Juruna)

Peru Indano, indano de monte alto, quilla caspi

Status Wild

Muruci da mata *Byrsonima chrysophylla* **in upland forest along a black water river. Curuá River, FLONA Caxiuanã, Municipio de Melgaço, Pará, Brazil 7-3-16**

Muruci da mata *Byrsonima chrysophylla* is a medium-sized tree with glistening leaves that reaches up to 30 m high in upland forest with light gaps, such as along rivers or in successionary forest where it is not too dark for it to survive (Alvino et al. 2005; Aud and Ferraz 2012; Bentos et al. 2008; Cárdenas et al. 2009; Malheiros et al. 2009; Medeiros 2016:91; Rayol et al. 2011; Salomão and Rosa 1989). Muruci da mata starts bearing fruit when only 2 m high and also occurs in gallery forests in savanna areas, such as in Bolivia, and in sandy areas with stunted vegetation called *campina* and *campinarana* in Brazil (Ferreira 2009:67, 84, 87). In Mato Grosso, the tree also occurs occasionally in scrub savanna (*cerrado*). Muruci also grows on floodplains in forests, meadows, and on sandy beaches along sediment-rich, black and clear water rivers, such as the Amazon floodplain near Manaus, Amazonas, and along streams in the city itself as well as black water rivers in the vicinity of Manaus (Parolin 2000a, b, c; Parolin et al. 2002; Santos et al. 2016). In the coastal zone of Pará and Amapá, muruci also grows on sand dunes and in grasslands around sand dunes and tidal forests (Amaral et al. 2008; Santos and Rosário 1988). This tree also occurs in stunted vegetation growing among iron-rich sedimentary rocks on hill tops in the Serra dos Carajás in Pará (Amorim et al. 2018; Mota et al. 2018). This tree ranges widely in the Amazon Basin and north into Venezuela and the Guianas. In the Andean foothills, it can be found as high as 1200 m.

In the coastal zone of Maranhão, 16 species of bees visit the flowers which are produced in the rainy season; several species of *Centris* and *Trigona fulviventris* are the major pollinators of this medium-sized tree (Ribeiro et al. 2008). In Manaus, Amazonas, the stingless bee *Melipona rufiventris paraensis* has been observed collecting pollen from the flowers of this tree (Marques-Souza et al. 1995).

The fruits are yellow-green when ripe, 1 cm × 1.3 cm, and contain a single dark brown seed 0.8 cm × 0.5 cm. The fruiting period runs from March to the end of July, thus spanning the height of the rainy season into the dry period. According to river dwellers in the Caxiuanã National Forest in Pará capuchin monkeys *Cebus apella*, known as macaco prego in Brazil, eat the fruit. The critically endangered Ka'apor capuchin monkey *Cebus kaapo*ri is known to eat this fruit in eastern Pará (Oliveira et al. 2014). The bearded saki monkey *Chiropotes chiropotes*, which inhabits upland forests in the northern part of the Brazilian Amazon and contiguous areas of Venezuela and the Guianas, also eats this fruit and likely disperses some of the seeds in its feces (Boyle et al. 2012). Certain birds are also fond of the fruit, such as pigeons, as suggested by one of its local names in the Brazilian Amazon: muruci de pomba. Rural folk eat the fruits and also make juice from them. In Julião, a rural community in the Reserva de Desenvolvimento Sustentável (RDS) do Tupé some 25 km west of Manaus, people use the bark of this tree to make a tea for treating stomach problems (Veiga and Scudeller 2015).

Muruci da mata *Byrsonima chrysophylla* fruits gathered from the ground at the edge of an upland forest. Curuá River, FLONA Caxiuanã, Municipio de Melgaço, Pará, Brazil 6-30-16

Because its fruits are appreciated by farmers in the interior, the tree is often spared when clearing areas to build manioc processing huts, such as in the vicinity of Carrazedo at the mouth of the Xingu and Jocojó, a quilombo community along the Jocojó River, an affluent of the Amazon near Gurupá. It is also planted occasionally in home gardens such as in Manaus, Amazonas (Kew specimen K1001221058 collected by Sir Ghillean Prance), in rural communities near Manaus (Souza 2010:29), along the lower Madeira River, Amazonas (Kawa et al. 2015), in São Gabriel da Cachoeira along the Upper Rio Negro in Amazonas (Pinto 2012:71), and as I have observed in the Pracajurá village on the margins of the Baia de Caxiuanã, Rio Anapu, Municipio de Portel, Pará. In the Peruvian Amazon, the bark is used as a source of red-brown dye (Antúnez de Mayolo 1989). At least four species of stingless bees (*Melipona seminigra merrillae, Melipona fulva, Trigona fulviventris*, and *Cephalotrigona femorata*) visit the flowers of this small tree during the rainy season (Oliveira et al. 2009) and rural people sometimes gather the honey of these bees in the wild. Some stingless bees are also kept in hollow logs suspended from the outside eaves of roofs in the Amazon to provide a ready source of honey.

31.3 *Byrsonima crassifolia*

Bolivia Totaicillo del campo

Brazil Muruci, muruci do campo, muruci da praia, muruci da restinga, muruci do varjão, muruci amarelo, muruci grande, murici, muricizeiro, mirixi, mirixi miúdo, mirixi vermelho, barbatimão, nanche, simaneira, canjicão; mùy′ (Hupd'äh), manakuyu ha (Juruna), pidjô nhinhu nhikai, kutenk (Kayapó), ugitikuegü (Kuikuru), murupa (Tiriyó)

Colombia Peraleja hembra, peralejo, yuco, yaca, maache, nancé, chaparro manteca, mantequera, muruci, balsam rojo, cereza, noro

Peru Cerezo, chupi cara, indano, indano colorado, murushi, quillo sisa, nanci, nance, nancite, yoco, chaparro, chaparro manteca, chaparro de sabana, maricas, cimarron, changugo, paralejo

Venezuela Manteco, chaparro manteco; ke'cho (Panaré)

Status Wild, spontaneous in cultural settings, planted

Muruci *Byrsonima crassifolia* fruits starting to ripen in a home garden. Fazenda Jilva, Marajó Island, Pará, Brazil, 12-14-01

Muruci *Byrsonima crassifolia* is widely appreciated for its abundant yellow fruits, the size of blueberries. As with blueberries and many other fruits, the size of muruci fruits varies somewhat but they are generally slightly flattened and measure 1.3–1.4 cm by 1.8 cm or 1.5 cm by 1.9 cm. The fruits usually contain just one seed, but occasionally two or three (Gordon et al. 2011). This small tree, typically under 10 m but occasionally attaining 20 m, is common on savannas and scrub grasslands (*cerrado*) in the tropical lowlands of Central and South America, including Cuba (Cárdenas et al. 2009; Coomes and Grubb 1996; Furley 1999; Medeiros et al. 2012; Miranda 1993, 1995; Prance 1996; Ratter 1987; Ratter et al. 1996; Ross 1994). In savanna areas (*campos*) and scrub savannas (*cerradão*) of the Amazon Basin it is one of the more common trees (Albernaz and Magnusson 1999; Amaral et al. 2019; Barbosa et al. 2005; Dantas and Rodrigues 1982; Furley and Ratter 1990; Guarim Neto et al. 1994; Kubitzki 1979; Lemos 2013:38; Magnusson et al. 2008; Miranda et al. 2002, 2006; Ratter et al. 2003; Rondon Neto et al. 2010; Sanaiotti 1996:401; Sanaiotti and Magnusson 1995; Sanaiotti et al. 1997; Santos et al. 2003; Takeuchi 1960a, b). It also occurs on sand dunes along the coast of the Bragantina zone in northeastern Pará and contiguous Maranhão and patches of sandy soil inland with *campina* and *campinarana* vegetation, such as in the vicinity of Belém and in Roraima (Amaral et al. 2001, 2008, 2015; Barbosa and Ferreira 2004a, b; Barbosa et al. 2012; Bastos 1988; Bastos et al. 1995, 2003; Costa Neto et al. 1996; Ferreira 2009:53, 72, 75, 78, 84, 87, 90; Ferreira et al. 2013; Freire and Monteiro 1993; Quaresma and Jardim 2017; Santos and Rosário 1988; Silva et al. 2010). It is also found on hills with thin soils on rocks rich in iron ore at Serra dos Carajás in southern Pará.

Muruci *Byrsonima crassifolia* in flower on a granitic inselberg in campo cerrado. Pedra do Indio, Município de Ferreira Gomes, Amapá, Brazil 8-29-14

The burning of savannas by early hunters and gatherers likely increased favorable habitat for muruci by thwarting the encroachment of forest. Also, muruci is stimulated rather than killed by the relatively light fires that periodically sweep savannas (Cintra and Sanaiotti 2005). Local enrichment of muruci may also have occurred as people discarded seeds around camps and along trails. Muruci also thrives in successionary vegetation, so when people started farming and clearing the forest some 8000–10,000 years ago, populations of muruci proliferated further.

Muruci is now found from Mexico, where it is known as nance, through Central America and the Caribbean south to Paraguay. In Amazonia, it is apparently absent in the western part of the basin, probably because rainfall is higher and there are fewer open habitats where it can thrive.

In the Amazon, muruci is found on savannas (*campo* and *campo cerrado*), in rural and urban home gardens, and some cultivated fields. During the Pleistocene, the forest shrunk in some parts of the basin as the climate became cooler and drier during four major glaciations thereby probably allowing *Byrsonima crassifolia* to penetrate deeper into Amazonia from the cerrado of central Brazil. As ice sheets advanced in colder regions, savannas penetrated deep into the Amazonia basin especially in the southern and eastern portion of the basin because the climate became drier (Kipnis et al. 2005). After the last Ice Age ended some 10,000 years ago, one or more dry periods occurred again during the Holocene. From 9000 to 3700 years ago the climate became drier again and the savanna returned to several parts of the Amazon basin especially in the Tocantins and Xingu watersheds (Behling and Hooghiemstra 2000; Hermanowski et al. 2012). Today pockets of savanna are found in various parts of the central, southern, and eastern parts of Amazonia and some of them are now shrinking as the climate has become more humid. Wild muruci is found in some of these savanna pockets, such as in Amapá, Roraima, and on Marajó Island (Bastos 1984; Eiten 1972).

Muruci *Byrsonima crassifolia* **tree in flower in a home garden. Thaynna, 10 years, with genipap** *Genipa americana* **body paint. Paquiçamba, a Juruna village along the middle Xingu, Pará, Brazil 9-8-15**

Muruci trees produce abundant flowers in the dry and wet seasons which are visited by native stingless bees, such as *Melipona seminigra* and *Centris flavifrons* (Oliveira et al. 2009; Rêgo et al. 2008) as well as the introduced honeybee *Apis mellifera*. At least 25 native bees have been observed visiting the flowers of this tree in Maranhão to collect pollen or oil (Albuquerque and Rêgo 1989; Rêgo and Albuquerque 1989). Stingless bees produce honey which is much appreciated by rural people in the interior of the Amazon. In savannas, several species of ants including the aggressive fire ant, are also attracted to the flowers, presumably to obtain nectar.

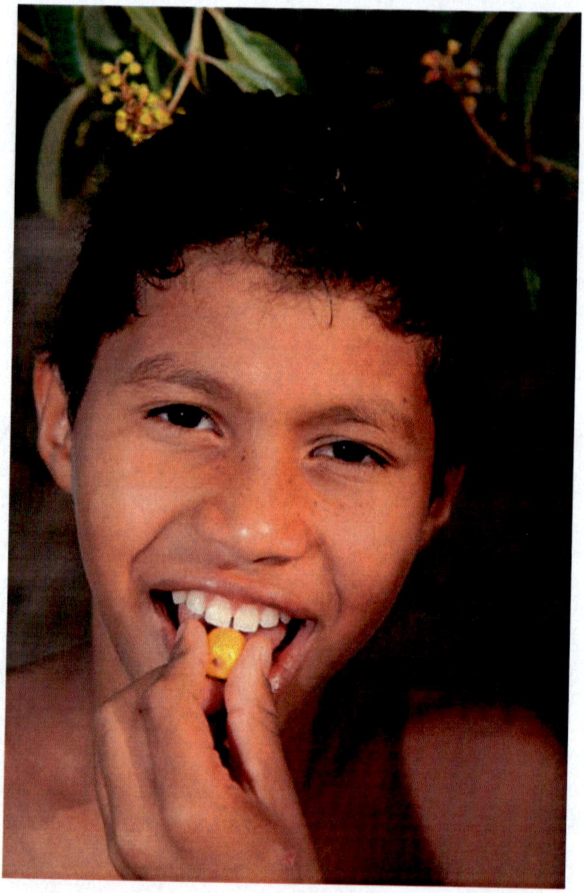

Ronaldo, 11 years, eating muruci *Byrsonima crassifolia* fruit he picked in his home garden. Chácara do Paulo & Lu, Pedral, left bank of the middle Xingu, a few kilometers upstream from Altamira, Pará, Brazil 8-31-15

The fruits are eaten fresh or made into juice in the countryside. Sometimes the juice is mixed with cachaça to make a potent alcoholic drink (*batida de muruci*), such as on Marajó Island (Lisboa 2012:196). The fruits are also sold by roadside vendors and in urban street markets. Muruci has a relatively long fruiting season in eastern Amazonia, extending from the height of the dry season in September to the latter part of the rainy season in May. In ice cream parlors, muruci is often one of the featured regional fruit flavors, while poorer urban residents typically opt for frozen muruci juice in cigar-shaped plastic bags, known locally as chopp.

Boy selling muruci *Byrsonima crassifolia* fruits along a highway. He is by a speed bump which slows traffic, thus tempting drivers and passengers to buy the fruit. Aimoré near Salinas, Pará, Brazil, 12-31-01

Frozen muruci juice is sold by ambulant vendors carrying Styrofoam containers for about a dime, such as at the Boca do Camará on Marajó where the ferry from Icoaraci docks twice a day. In Maués, Amazonas, vendors with pushcarts also sell muruci popsicles on wooden sticks.

Boy selling muruci ***Byrsonima crassifolia*** **popsicles (*chopp*) which are kept cold in a Styrofoam container. Boca do Camará, Marajó Island, Pará, Brazil, 12-16-01**

Muruci's architecture varies depending on the habitat. In old regrowth vegetation, muruci is slender and relatively tall as it competes with other plants for light. In advanced stages of succession muruci is crowded out. On sunlit savannas, muruci takes on the guise of a small, stunted tree or even a bush. Although muruci does not usually grow on floodplains, it thrives in savannas that become soggy during the rainy season, such as on Marajó Island. Sandy ridges (tesos) in the grasslands of Marajó, vestiges of old sand dunes or possibly formed by indigenous groups in the distant past, provide ideal habitat for muruci as well as other trees that provide edible fruit (Lisboa 2012:410; Miranda 1909). In the Peruvian Amazon, the Campa sometimes create pockets of savanna by burning fields cleared in the forest after the crops have been harvested, usually after 2 or 3 years. *Byrsonima crassifolia* then invades some of these grassland patches after being dispersed by birds (Sauer 1988:108). However, muruci is absent on low-lying savannas that are under a meter or so of water for several months each year.

Muruci is particularly common in the Bragantina Zone east of Belém, the most densely settled part of the Amazon. Farmers engaged in repeated slash and burn cycles in the Bragantina zone have sometimes led to the development of savannas dominated by muruci and rank sapé grass. Much of the forest in the Bragantina zone has been cleared over the last century, creating plenty of regrowth woods and grassland for muruci to proliferate, especially where the soils are sandy (Baar et al. 2004; Lima 1958). Birds, such as the blue-gray and silver-beaked tanagers (*Thraupis episcopus* and *Ramphocelus carbo*), likely disperse muruci across the largely cultural landscape. In the dry tropical forest of western Cosa Rica, the mantled howler monkey *Alouatta palliata* has been recorded eating this fruit and most likely disperses the seeds (Glander 1979, 1981). Although *Alouatta palliata* does not occur in the Amazon Basin related species do and most likely also eat the fruits. On Marajó Island in the Amazon estuary, squirrel monkeys *Saimiri sciureus* have been observed feeding on this fruit and likely disperse some of the seeds (Santos et al. 2018). Muruci fruits are picked in successionary vegetation, such as in the vicinity of Arapixuna near Santarém and Vigia, and on savannas, such as along the Tauá River on Marajó Island and in the vicinity of Villa Franca on the lower Tapajós. Muruci fruits have been gathered on the savannas of Marajó for millennia as evidenced by their remains in the middens of the Marajoara civilization (Roosevelt 1991:375).

The fruits of at least one wild relative of muruci are gathered in the forests of Amazonia. In the Caxiuanã National Forest along the lower Anapú, for example, locals gather fruits of muruci da mata ("muruci of the forest"), probably *Byrsonima crispa*, which according to Paulo Cavalcante in his authoritative work on the edible fruits of the Amazon, is endemic to the forests of Pará. Carbonized fruits of *Byrsonima crispa* have been uncovered in sandstone caves near Monte Alegre along the middle Amazon once occupied by hunters and gatherers (Roosevelt et al. 1996; Roosevelt 1998). River folk along the middle Xingu gather fruits of muruci da mata as well as planted *Byrsonima crassifolia* which they call simply muruci or muruci do campo, recognizing the latter species preference for more open habitats.

Muruci is also cultivated in home gardens throughout the Brazilian Amazon, such as in the upland community of Santo Antônio some 125 km south of Santarém, Pará (Almeida and Gama 2014), the community of São Domingos along the northern border of the Tapajós National Forest (Matos 2015:39), in various locations in the Brazilian state of Amazonas (Guillaumet et al. 1990; Silva 2017:31) and in the vicinity of Macapá, the capital of Amapá (Pereira et al. 2017). In a survey of 150 home gardens I carried out in various parts of the Amazon Basin over several decades I found muruci in 20 of them. In a separate survey of 57 home gardens along the middle Xingu in 2015, I encountered muruci in 15 of them. Muruci is more common in home gardens along the middle Xingu for two reasons: the soil tends to be more sandy and avenues of savanna likely penetrated along this river during dry climatic cycles in the past. One island along the middle Xingu upstream from Altamira has so many muruci trees that it is known locally as Murucituba (the place of muruci). At least some of the muruci in home gardens along the middle Xingu arose from seeds brought from Belém, such as at Ilha Tracoá 2 between Altamira and Belo Monte. Muruci might have also been introduced to the middle Xingu by indigenous people bringing the seeds or seedlings from the headwaters of the river. The Kayapó who live in scattered villages in the Upper Xingu watershed plant this fruit tree in forest patches (*apêtê*) created in scrub savanna to attract game animals (Posey 2002:202). Some indigenous groups also cultivate this tree in their home gardens, such as the Araçá in northern Roraima (Pinho et al. 2011).

Although muruci is largely confined to uplands, it is planted in home gardens on higher parts of floodplains such as along the middle Xingu and along Surubim Açu on the Amazon floodplain near Santarém. Near the village of Jocojó in the municipality of Gurupá and in the Caxiuanã National forest, both in eastern Pará, muruci is also planted around some manioc processing huts.

Muruci *Byrsonima crassifolia* trees (on right) in a home garden with several other fruit trees including mango and macaúba *Acrocomia aculeata* palms. Tracoal 2 archaeological site, Xingu River between Altamira and Belo Monte, Pará, Brazil 8-17-15

Some farmers intercrop muruci in their fields such as in the Murumuru community near Santarém where a river dweller has interplanted muruci in his cacao grove. In the São Pedro community along the Pararijós River near Breves on Marajó Island, a farmer has planted muruci in a one-hectare agroforestry plot containing açaí *Euterpe oleracea*, cashew *Anacardium occidentale*, cupuaçu *Theobroma grandiflorum*, lime, tangerine, and spontaneous virola *Virola surinamensis*, a valuable timber species. In the vicinity of Igarapé-Açu in the Zona Bragantina region of northeastern Pará, some farmers intercrop muruci in agroforestry fields (Vieira et al. 2007). Near Porto Novo along the lower Tapajós, a farmer has planted muruci in a field containing six varieties of bitter manioc.

Muruci also arises spontaneously in fields and home gardens, and because it is useful, it is usually allowed to grow. On the lower slopes of the scarp overlooking the Amazon floodplain at Murumuru near Santarém, for example, a farmer has let a number of spontaneous fruit trees, including muruci, tucumã *Astrocaryum vulgare*, and inajá *Attalea maripa*, grow in the family's one-hectare manioc field that has been interplanted with maize and watermelon. Once the short cycle crops are harvested, the field will be abandoned to successionary vegetation, but the family will periodically return to gather the "left over" fruits that nature provided.

Muruci provides several other useful products besides fruit for treating health issues. In the municipality of Gurupá, rural folk informed me that the bark is soaked in water for several hours and the "tea" is drunk to treat diarrhea, as is the case in the vicinity of Manacapuru along the lower Solimões near its confluence with the Rio Negro (Vásquez et al. 2014). In the Caxiuanã National Forest in eastern Pará, muruci bark is soaked in water and the solution is applied to teeth that are aching and is also used to wash wounds (Rocha et al. 2013) as is the case along the coast of Pará (Bastos et al. 2003:18). In the vicinity of Ajuruteua along the Bragantina coast of Pará the bark is also used to make a solution to treat wounds (Carneiro et al. 2010). In various parts of the Brazilian Amazon, the leaves are used to make tea for treating inflammation of ovaries and hemorrhoids (Van den Berg 2010:217; Van den Berg and da Silva 1986). The bark is also soaked in water and then drunk to assuage fevers and treat tuberculosis (Le Cointe 1934:292). In various parts of Pará in the eastern part of the Brazilian Amazon, the leaves and bark are also used to make a tea that serves as a diuretic (Silva et al. 2007). The Tiriyó who inhabit the hilly region along the border between northern Pará and Surinam scrape the bark and then squeeze the shaving so that the liquid drips onto wounds or burns to promote healing (Cavalcante 1973:102). Cotton is also used to soak up this liquid which is then applied to the wound as a band-aid.

On Colares Island in the southern part of the Amazon estuary, muruci bark is boiled and the infusion is then brushed on boats to help preserve the wood. Over a century ago, the same solution was used to dye cotton shirts yellow along the middle and lower Amazon (Bates 1863:23). Until relatively recently, muruci was used to dye sails of fishing boats in Vigia and its environs at the mouth of the Amazon. Sailing vessels are increasingly using polypropylene for their sails, however, because they do not rot. Artificial dyes and plastic are increasingly putting an end to the use of vegetable dyes in Amazonia.

Stopper made from muruci *Byrsonima crassifolia* wood used for keeping manioc dough inside tipitis which are used to squeeze tucupi juice from grated manioc prior to making flour. Arlinho, 11 years. Terra Preta, Rio Camutá an affluent of the Ipixuna River, in turn an affluent of the lower Xingu, Município de Gurupá, Pará, Brazil 10-28-16

Muruci is reputed to be an excellent fuelwood and a valued source of aromatic smoke for preserving foodstuffs and beverages. The Sateré-Maué in the Maués valley in Amazonas prefer muruci wood for their ovens used to toast guaraná *Paullinia cupana* seeds and to cure the salami-shaped guaraná sticks formed by mixing ground guaraná seeds with water (Henman 1982). In the town of Maués, the owner of one of the bakeries asserted that muruci is one of the preferred woods for making bread. At the São Raimundo community along the lower Urubu River in Amazonas, river dwellers employ muruci wood to make tapioca pancakes (*beiju*) and in the Caxiuanã National Forest in eastern Pará river dwellers like to use muruci wood as fuel to prepare manioc flour. Inhabitants of the village of Algodoal in the municipality of Maracanã along the northeastern coast of Pará also use the trunk and branches for fuel (Coelho-Ferreira and Jardim 2005). In Roraima, the Taurepang and Sapará indigenous communities use the branches of this tree for fuelwood (Perez 2011:29). The wood of this small tree is also used for fuel in rural parts of the Magdalena Valley in Colombia (Pérez-Arbeláez 1956:479).

The wood is also used to make stoppers so that manioc dough is not squeezed out of tipitis, sleeve-like presses that are used to expel the tucupi juice from grated manioc tubers. In Roraima, a state in the northern part of the Brazilian Amazon, people make a tea from the inner bark to treat diarrhea, inflammation, and malaria (Luz 2001).

31.4 *Heteropterys orinocensis*

Brazil Sarabá

Venezuela Bejuco mariposa, bejuco perro de agua, bejuco candela

Status Wild

Sarabá *Heteropterys orinocensis* on the floodplain of a black water river during the flood season. The winged seeds are used as fish bait. Pontão, Rio Urubu near Silves, Amazonas, Brazil 4-4-15

Sarabá *Heteropterys orinocensis* is a vine on floodplains that produces winged seeds during the annual flood. River folk gather these seeds from canoes to use as bait to catch fish. Unlike some fleshy fruits which are also used for fish bait, sarabá seeds can be stored for months without deteriorating. This relatively slender liana is largely confined to black and clear water rivers in central, southern, and western parts of the Brazilian Amazon as well as the Upper Orinoco in Venezuela. It is also occasionally found along the margins of lakes on the floodplains of sediment-rich rivers, such as the middle Madeira in Amazonas and along the Solimões near Manacapuru. Much of the sediment in such lakes decants out as there is little current, so the water is clear or dark.

Sarabá winged seeds gathered on the floodplain of a black water river. The wings are removed and the seeds are used to bait fish hooks. Pontão, Rio Urubu near Silves, Amazonas, Brazil 4-4-15

References

Albernaz, Ana Luisa; Magnusson, William E. 1999. Home-range size of the bare-ear marmoset (Callithrix argentata) at Alter do Chão, central Amazonia, Brazil. International Journal of Primatology 20(5):665–677

Albuquerque, Patrícia M. C. de; Rêgo, Márcia Maria C. 1989. Fenologia das abelhas visitantes de murici (Byrsonima crassifolia, Malpighiaceae). Boletim do Museu Paraense Emilio Goeldi, Série Zoologia 5(2): 163–178

Almeida, Larissa Santos de; Gama, João Ricardo Vasconcellos. 2014. Quintais agroflorestais: estrutura, composição florística e aspectos socioambientais em área de assentamento rural na Amazônia brasileira. Ciência Florestal 24(4):1041–1053

Alvino, Fabrízia de Oliveira; Silva, Manoela Ferreira F. da; Rayol, Breno Pinto. 2005. Potencial de uso das espécies arbóreas de uma floresta secundária, na Zona Bragantina, Pará, Brasil. Acta Amazonica 35(4):413–420

Amaral, Dário Dantas do; Santos, João Ubiratan M. dos; Bastos, Maria de Nazaré do Carmo; Costa, Denise Cristina Torres. 2001. A vegetação da Ilha Canela, Município de Bragança – Pará, Brasil. Boletim do Museu Paraense Emílio Goeldi, Série Botânica 17(2):389–402

Amaral, Dário Dantas do; Prost, Maria Thereza; Bastos, Maria de Nazaré do Carmo; Costa Neto, Salustiano Vilar; Santos, João Ubiratan Moreira dos. 2008. Restingas do litoral amazônico, estados do Pará e Amapá. Boletim do Museu Paraense Emílio Goeldi, Ciencias Naturais 3(1):35–67

Amaral, Dário Dantas do; Jardim, Mário Augusto Gonçalves; Costa-Neto, Salustriano Vilar; Bastos, Maria de Nazaré do Carmo. 2015. Síndromes de dispersão de propágulos e a influência da floresta amazônica na composição de espécies lenhosas de uma restinga no litoral norte brasileiro. Biota Amazônia 5(3):28–37

Amaral, Dário Dantas do; Rocha, Antônio Elielson; Pereira, Jorge Luis Gavina; Costa Neto, Salustiano Vilar. 2019. Identificação dos subtipos de savanas na Amazônia oriental (Pará e Amapá, Brasil) com uma chave dicotômica de individualização. Boletim do Museu Paraense Emílio Goeldi, Ciencias Humanas Naturais 14(2):183–195

Amorim, André M.; Vasconcellos, Liziane Vilela; Silva Júnior, Valdir S. 2018. Flora das cangas da Serra dos Carajás, Pará, Brasil: Malpighiaceae. Rodriguésia 69(3):1221–1235

Antúnez de Mayolo, A. 1989. Peruvian natural dye plants. Economic Botany 43(2):181–191

Arellano, Gabriel. 2013. Patrones de diversidad, distribución y dominancia de plantas leñosas en los bosques montanos del Parque Nacional Madidi, Bolivia. Doctoral dissertation, Departamento de Biologia, Facultad de Ciencias, Universidad Autónoma de Madrid

Aud, Fabiana F.; Ferraz, Isolde D. K. 2012. Seed size influence on germination responses to light and temperature of seven pioneer tree species from the Central Amazon. Anais da Academia Brailerira de Ciências 84(3):759–766

Baar, R.; M. Cordeiro, M. Denrich; and H. Fölster. 2004. Floristic inventory of secondary vegetation in agricultural systems of East-Amazonia. *Biodiversity and Conservation* 13:501–528

Barbosa, R. I. and C. A. C. Ferreira. 2004a. Densidade básica da madeira de um ecossistema de campina em Roraima, Amazônia brasileira. *Acta Amazonica* 34(4): 587–591

Barbosa, R. I. and C. A. C. Ferreira 2004b. Biomassa acima do solo de um ecossistema de "campina" em Roraima, norte da Amazônia Brasileira. Acta Amazonica 34(4):577–586

Barbosa, Reinaldo Imbrozio; Nascimento, Sebastião Pereira do; Amorim, Paulo Atlântico Figueiredo de; Silva, Rosemberg Ferreira da. 2005. Notas sobre a composição arbóreo-arbustiva de uma fisionomia das savanas de Roraima, Amazônia Brasileira. Acta Botanica Brasilica 19(2):323–329

Barbosa, Reinaldo Iembrozio; Mourão Jr., Moisés; Casadio, Giliola Maria Lima; Silva, Silvio José Reis da. 2012. Reproductive phenology of the main tree species in the Roraima savanna, Brazilian Amazon. Ecotropica 18(2):81–92

Bastos, Maria de Nazaré do C. 1984. Levantamento florístico dos campos do Estado do Pará. I – Campo de Joanes (Ilha de Marajó). Boletim do Museu Paraense Emílio Goeldi, Série Botânica 1(1/2): 67–86

Bastos, Maria de Nazaré do Carmo. 1988. Levantamento florístico em restinga arenosa litorânea na Ilha de Maiandeua, Pará. Boletim do Museu Paraense Emílio Goeldi, Série Botânica 4(1): 159–169

Bastos, M. N. C.; C. S. Rosário; L. C. B. Lobato. 1995. Caracterização fitofisionômica da restinga de Algodoal, Maracanã, Pará, Brasil. Boletim do Museu Paraense Emílio Goeldi, Série Botanica 11(2): 173–197

Bastos, Maria de Nazaré do Carmo; Costa, Denise Cristina Torres; Santos, João Ubiratan Moreira dos. 2003. Vegetação de restinga: aspectos botânicos e uso medicinal. Museu Paraense Emilío Goeldi Projeto Renas, Belém

Bates, H. W. 1863. *The Naturalist on the River Amazons*. John Murray, London, vol. 2

Behling, H. and H. Hooghiemstra. 2000. Holocene Manejo de Fauna Silvestre en la Amazoníae Amazon rainforest–savanna dynamics and climatic implications: high-resolution pollen record from Laguna Loma Linda in eastern Colombia. Journal of Quaternary Science 15(7):687–695

Bentos, Tony V.; Mesquita, Rita C. G.; Williamson, G. Bruce. 2008. Reproductive phenology of central Amazon pioneer trees. Tropical Conservation Science 1(3): 186–203

Boyle, Sarah A.; Zartman, Charles E.; Spironello, Wilson R.; Smith, Andrew T.; Beck, Harald. 2012. Implications of habitat fragmentation on the diet of bearked saki monkeys in central Amazonian forest. Journal of Mammalogy 93(4):959–976

Cárdenas, Dairon; Castaño, Nicolás; Sua, Sonia. 2009. Flora de la estrella fluvial de Inírida (Guianía, Colombia). Biota Colombiana 10(1–2):1–30

Carneiro, Diogo Borges; Barboza, Myrian Sá Leitão; Menezes, Moirah Paula. 2010. Plantas nativas úteis na Vila dos Pescadores da Reserva Extrativista Marinha Caeté-Taperaçu, Pará, Brasil. Acta Botanica Brasilica 24(4):1027–1033

Cartagena, Susana Esther Cotito. 2014. Diversidad y composición florística de un área de bosque ribereño premontano del valle de Chanchamayo. Master's thesis, Universidad Nacional Agraria La Molina, Lima

Cavalcante, P. B. 1973. A Farmacopéia Tiriyó: Estudo Étno-Botânico. Museu Paraense Emílio Goeldi, Publicações Avulsas No. 24
Cintra, R. and T. M. Sanaiotti. 2005. Fire effects on the composition of a bird community in an Amazonian savanna. Brazilian Journal of Biology 65(4):683–695
Coelho-Ferreira, Márlia; Jardim, Mário Augusto G. 2005. Algumas espécies vegetais usadas pelos moradores da Ilha de Algodoal, Maiandeua, município de Maracanã, Pará. Boletim do Museu Paraense Emílio Goeldi, Série Ciências Naturais 1(2): 45–51
Cohen, M. N. 1977. Population pressure and the origins of agriculture: an archaeological example from the coast of Peru. In: Origins of Agriculture, C. A. Reed (ed.), Mouton, The Hague, pp. 135–177
Coomes, David A.; Grubb, Peter J. 1996. Amazonian caatings and related communities at La Esmeralda, Venezuela: forest structure, physiognomy and floristics. Vegetatio 122(2):167–191
Costa Neto, Salustiano Vilar da; Bastos, Maria de Nazaré do Carmo; Lobato, Luiz Carlos Batista. 1996. Composição florística e fitofisionomia da restinga do Crispim, município do Marapanim, Pará. Boletim do Museu Paraense Emílio Goeldi, Série Botânica 12(2): 237–249
Dantas, Mário; Rodrigues, Irenice Alves. 1982. Estudos fitoecológicos do trópico úmido Brasileiro: IV-levantamentos botânicos em campos do Rio Branco. EMBRAPA, Centro de Pesquisa Agropecuária do Trópico Úmido, Boletim de Pesquisa, no. 40
Ducke, A. 1946. *Plantas de Cultura Precolombiana na Amazônia brasileira. Notas Sôbre as Espécies ou Formas Espontâneas que Supostamente lhes teriam dado Origem.* Instituto Agronômico do Norte, Belém, Boletim Técnico 8, pp. 1–24
Eiten, G. 1972. The cerrado vegetation of Brazil. Botanical Review 38(2):201–341
Ferreira, Carlos Alberto Cid. 2009. Análise comparativa de vegetação lenhosa do ecosistema campina na Amazonia Brasileira. Doctoral dissertation, INPA/UFAM, Manaus
Ferreira, Leandro V.; Chaves, Priscilla P.; Cunha, Denise de A.; Rosário, Alessandro S. do; Parolin, Pia. 2013. A extração ilegal de areia como causa do desaparecimento de campinas e campinaranas no estado do Pará, Brasil. Pesquisas Botânica 64:157–173
Freire, Maria Cristina C.; Monteiro, Reinaldo. 1993. Florística das praias da Ilha de São Luís, Estado do Maranhão (Brasil): diversidade de espécies e suas ocorrências no litoral Brasileiro. Acta Amazonica 23(2–3):125–140
Furley, Peter A. 1999. The nature and diversity of neotropical savanna vegetation with particular reference to the Brazilian cerrados. Global Ecology and Biogeography 8: 223–241
Furley, Peter A.; Ratter, James A. 1990. I. Pedological and botanical variations across the forest-savanna transition on Maracá Island. The Geographical Journal 156(3):251–266
Giacomotti, José Guillermo. 2016. Evaluación de la dinámica forestal en un área de bosque secundario tardío en el fundo La Génova, Chanchamayo. Master's thesis, Universidad Nacional Agraria La Molina, Lima
Giacomotti, José Guillermo; Reynel, Carlos. 2018. Mortalidad y recrutamiento de árboles en un bosque secundario tardío del valle de Chanchamayo, Perú. Revista Florestal del Perú 33(1):42–51
Glander, K. E. 1979. Howling monkey feeding behavior and plant secondary compounds: a study of strategies. In: The Ecology of Arboreal Foliovores, Montgomery, G. G. (ed), Smithsonian Institution Press, Washington, DC, pp 561–574
Glander, K. E. 1981. Feeding patterns in mantled howling monkeys. In: Foraging Behavior: Ecological, Ethological, and Psychological Approaches, Kamil, A. C.; Sargent, T. D. (eds), Garland Press, New York, pp 231–257
Gordon, André; Jungfer, Elvira; Silva, Bruno Alexandre da; Maia, José Guilherme S.; Marx, Friedhelm. 2011. Phenolic constituents and antioxidant capacity of four underutilized fruits from the Amazon region. Journal of Agricultural and Food Chemistry 59:7688–7699
Guarim Neto, Germano; Guarim, Vera Lucia M. S.; Prance, Ghillean T. 1994. Structure and floristic composition of the trees of an area of cerrado near Cuiabá, Mato Grosso, Brazil. Kew Bulletin 49(3):499–509
Guillaumet, J-L.; Grenand, P.; Bahri, S.; Grenand, F.; Lourd, M.; Santos, A.A. dos; Gély, A. 1990. Les jardins-vergers familiaux d'Amazonie centrale: une exemple d'utilisation de l'espace. Turrialba 40(1):63–81

Henman, A. R. 1982. Guaraná (*Paullinia cupana* var. *sorbilis*): Ecological and social perspectives on an economic plant of the central Amazon basin. Journal of Ethnopharmacology 6:311–338

Hermanowski, Barbara; Costa, Marcondes Lima da; Behling, Hermann. 2012. Environmental changes in southeastern Amazonia during the last 25,000 yr revealed from a paleoecological record. Quaternary Resarch 77:138–148

Kawa, Nicholas C.; Michelangeli, José A. Clavijo; Clement, Charles R. 2015. Household agro-biodiversity management on Amazonian dark earths, oxisols, and floodplain soils on the lower Madeira River, Brazil. Human Ecology 43:339–353

Kipnis, Renato; Caldarelli, Solange Bezerra; Oliveira, Wesley Charles de. 2005. Contribuição para a cronologia da colonização amazônica e suas implicações teóricas. Revista de Arqueologia 18:81–93

Kubitzki, K. 1979. Ocorrência de *Kielmeyera* nos "campos de Humaitá" e a natureza dos "campos"—flora da Amazônia. Acta Amazonica 9(2):401–404

Le Cointe, P. 1934. *Amazônia Brasileira III: Árvores e Plantas Úteis*. Livraria Classica, Belém

Lemos, Helena Lara. 2013. Vegetação arbustivo-arbórea em áreas de cerrado típico e cerrado rupestre no estado de Tocantins. Master's thesis, Departamento de Engenharia Florestal, Universidade de Brasília, Brasília

Lima, D. A. 1958. Viagem aos campos de Monte Alegre, Pará. Instituto Agronômico do Norte, *Boletim Técnico* 36: 99–150

Lisboa, P. L. B. 2012. *A Terra dos Aruã: Uma História Ecológica do Arquipélago do Marajó*. Museu Paraense Emílio Goeldi, Belém

Loza, Isabel; Moraes, Mónica; Jorgensen, Peter M. 2010. Variación de la diversidad y composición florística en Relación a la elevación en un bosque montano boliviano (PNANMI Madidi). Ecología en Bolivia 45(2):87–100

Luz, Francisco Joaci F. 2001. Plantas medicinais de uso popular em Boa Vista, Roraima, Brasil. Horticultura Brasileira 19(1):88–96

Magnusson, William E.; Lima, Albertina P.; Albernaz, Ana L. K. M.; Sanaiotti, Tânia M.; Guillaumet, Jean-Louis. 2008. Composição florística e cobertura vegetal das savanas na região de Alter do Chão, Santarém-PA. Revista Brasileira de Botanica 31(1):165–177

Malheiros, Antonio Francisco; Higuchi, Niro; Santos, Joaquim dos. 2009. Análise estrutural da floresta tropical úmida do município de Alta Floresta, Mato Grosso, Brasil. Acta Amazonica 38(3):539–548

Marques-Souza, Antonio Carlos; Absy, Maria Lúcia; Kerr, Warwick Estevam; Peralta, Francisco Javier Aguilera. 1995. Pólen coletado por duas espécies de Meliponíneos (Hymenoptera: Apidae) da Amazônia. Revista Brasileira de Biologia 55(4):855–864

Matos, Caio Barros. 2015. Diversidade e usos de quintais agroflorestais da comunidade São Domingos, Floresta Nacional do Tapajós, Belterra, PA. Master's thesis, Universidade Federal do Oeste do Pará, Santarém.

Medeiros, Priscilla Sanjuan de. 2016. Indicadores ambientais para funcionalidade ecológica em florestas secundárias de diferentes idades na Amazônia Oriental. Doctoral dissertation, Universidade Federal do Pará, Belém

Medeiros, Marcelo Brilhante de; Walter, Bruno Machado Teles; Silva, Glocimar Pereira da; Gomes, Beatriz Machado; Lima, Isabel Lustz Portela; Silva, Suelma Ribeiro; Moser, Pamela; Washington, Luís Oliveira; Cavalcanti, Taciana Barbosa. 2012. Vascular flora of the Tocantins River middle basin, Brazil. Check List 8(5):852–885

Mendoza, Abel L. Monteagudo; Guerrero, Marleni Huamán. 2010. Catálogo de los arboles y afines de la selva central del Perú. Arnaldoa 17(2):202–242

Miranda, V. C. 1909. Os campos de Marajó e a sua flora considerados sob o ponto de vista pastoril. *Boletim do Museu Goeldi (Museu Paraense) de Historia Natural e Ethnographia* 5(1): 96–151

Miranda I. S. 1993. Estrutura do estrato arbóreo do cerrado amazônico em Alter-do-Chão, Pará, Brasil. Revista Brasileira de Botânica 16(2):143–150

Miranda, I. S. 1995. Fenologia do estrato arbóreo de uma comunidade de cerrado em Alter-do-Chão, PA. Revista Brasileira de Botânica 18(2):235–240

Miranda, Izildinha Souza; Absy, Maria Lúcia; Rebêlo, George Henrique. 2002. Community structure of woody plants of Roraima savannahs, Brazil. Plant Ecology 164:109–123
Miranda, Izildinha Souza; Almeida, Samuel Soares; Dantas, Paulo Jorge. 2006. Florística e estrutura de comunidades arbóreas em cerrados de Rondônia, Brasil. Acta Amazonica 36(4):419–430
Mota, Nara Furtado de Oliveira; Watanabe, Mauricio Takashi Coutinho; Zappi, Daniela Cristina; Hiura, Alice Lima; Pallos, Julieta; Viveros, Raquel Stauffer; Giulietti, Ana Maria; Viana, Pedro Lage. 2018. Cangas da Amazônia: a vegetação única de Carajás evidenciada pela lista de fanerógramas. Rodriguésia 69(3):1435–1488
Oliveira, Francisco Plácido Magalhães; Absy, Maria Lúcia; Miranda, Izildinha Souza. 2009. Recurso polínico coletado por abelhas sem ferrão (Apidae, Meliponinae) em um fragmento de floresta na região de Manaus- Amazonas. Acta Amazonica 39(3):505–518.
Oliveira, S. G. de; Alfaro, Jessica W. Lynch; Veiga, Liza M. 2014. Activity budget, diet, and habitat use in the critically endangered Ka'apor capuchin monkey (*Cebus kaapori*) in Pará state, Brazil: a preliminary comparison to other capuchin monkeys. American Journal of Primatology 76:919–931
Parolin, P. 2000a. Seed mass in Amazonian floodplain forests with contrasting nutrient supplies. Journal of Tropical Ecology 16(3):417–428
Parolin, P. 2000b. O uso de árvores nas florestas inundadas por água branca na Amazônia Central. Amazoniana 16(1/2):241–248
Parolin, P. 2000c. Phenology and C02-assimilation of trees in Central Amazonian floodplains. Journal of Tropical Ecology 16:465–473
Parolin, Pia; Ambruester, Nicole; Wittmann, Florian; Ferreira, Leandro; Piedade, Maria Teresa F.; Junk, Wolfgang J. 2002. A review of tree phenology in central Amazonian floodplains. Pesquisas Botânica 52:195–222
Pereira, Luciano Araujo; Lima-Barbosa, Jackson Rodrigo de; Almeida, Maria Zélia de; Guimarães, Elsie Franklin. 2017. Diversidade de plantas em quintais quilombolas, conhecimento local sobre uso e cultivo de pimentas na Amazônia oriental, Brasil. Revista de Biologia Neotropical 14(1):56–72
Perez, Inayê Uliana. 2011. Uso dos recursos naturais vegetais na comunidade indígena Araçá, Roraima. Master's thesis, Universidade Federal de Roraima, Boa Vista
Pérez-Arbeláez, Enrique. 1956. Plantas Útiles de Colombia. Liberia Colombiana, Bogotá
Pinho, Rachel C.; Aflaia, Sonia S.; Miller, Robert Pritchard; Uguen, Katell; Magalhães, Leovone D.; Ayres, Marta; Freitas, Viviane; Trancoso, Ralph. 2011. Islands of fertility: soil improvement under indigenous homegardens in the savannas of Roraima, Brazil. Agroforestry Systems 81:235–247
Pinto, Ilzon Castro. 2012. Agrobiodiversidade de quintais agroflorestais urbanos e perfil social de etnias indígenas em São Gabriel da Cachoeira, AM. Doctoral dissertation, Universidade Federal de Lavras, Minas Gerais
Popenoe, W. 1924. Economic fruit-bearing plants of Ecuador. Smithsonian Institution, United States National Museum, Contributions from the United States National Herbarium 24(5): 101–134
Posey, D. A. 2002. Kayapó Ethnoecology and Culture. Routledge, London
Prance, G. T. 1996. Islands in Amazonia. Philosophical Transactions of the Royal Society B 351:823–833
Quaresma, Adriano Costa; Jardim, Mário Augusto G. 2017. Formations of coastal forests in the Amazon and ecological relations with vascular epiphytes. Revista Árvore 41(2):1–8
Ratter, J. A. 1987. Notes on the vegetation of the Parque Nacional do Araguaia (Brazil). Notes from the Royal Botanic Garden, Edinburgh 44(2):311–342
Ratter, J. A.; Bridgewater, S.; Atkinson, R.; Ribeiro, J. F. 1996. Analysis of the floristic composition of the Brazilian cerrado vegetation II: comparison of the woody vegetation of 98 areas. Edinburgh Journal of Botany 53(2):151–180
Ratter, J. A.; Bridgewater, S.; Ribeiro, J. F. 2003. Analysis of the floristic composition of the Brazilian cerrado vegetation III Comparison of the woody vegetation of 376 areas. Edinburgh Journal of Botany 60(1):57–109

Rayol, Breno Pinto; Alvino-Rayol, Fabrízia de Oliveira; Silva, Manoela Ferreira Fernandes da. 2011. Similaridade florística entre o estrato arbóreo e a regeneração natural de uma floresta secundária, no município de Bragança, nordeste do estado do Pará, Brasil. Revista Brasileira de Agroecologia 6(3):107–114

Rêgo, Márcia Maria C.; Albuquerque, Patrícia M. C. de. 1989. Comportamento das abelhas visitantes de murici, *Byrsonima crassifolia* (L.) Kunth, Malpighiaceae. Boletim do Museu Paraense Emilio Goeldi, Série Zoologia 5(2): 179–193

Rêgo, Márcia M. C.; Albuquerque, Patrícia M. C.; Ramos, Marina C.; Carreira, Léa M. 2008. Aspectos da biologia e nidificação de *Centris flavifrons* (Friese) (Hymenoptera: Apidae, Centridini), um dos principais polinizadores do murici (*Byrsonima crassifolia* L. Kunth, Malpighiaceae), no Maranhão. Neotropical Entomology 35(5):579–587

Ribeiro, Éville Karina Maciel Delgado; Rêgo, Márcia Maria Corrêa; Machado, Isabel Cristina Sobreira. 2008. Cargas polínicas de abelhas polinizadoras de Byrsonima chrysophylla Kunth. (Malpighiaceae): fidelidade e fontes alternativas de recursos florais. Acta Botânica Brasilica 22(1):165–171

Rocha, S. F. R., M. R. C. Ferreira, and M. G. Sablayrolles. 2013. Farmacopeia vegetal da comunidade Caxiuanã, na Floresta Nacional de Caxiuanã. In: Caxiuanã: Paraíso ainda Preservado, P. L. B. Lisboa (Editor), Museu Paraense Emílio Goeldi, Belém, pp. 257–285

Rondon Neto, Rubens Marques; Santos, Jeneffer Soares dos; Silva, Manoel Advíncula da; Koppe, Valdinei Cristi. 2010. Potencialidades de uso de espécies arbustivas e arbóreas em diferentes fisionomias de cerrado, em Lucas do Rio Verde/MT. Revista de Biologia e Ciências da Terra 10(2):113–126

Roosevelt, A. C. 1991. *Moundbuilders of the Amazon: Geophysical Archaeology on Marajo Island, Brazil*. Academic Press, San Diego

Roosevelt, A. C. 1998. Ancient and modern hunter-gatherers of lowland South America: an evolutionary problem. In: Advances in Historical Ecology, W. Balée (ed.), Columbia University Press, New York, pp. 190–212

Roosevelt, A. C.; Costa, M. Lima da; Machado, C. Lopes; Michab, M.; Mercier, N.; Valladas, H.; Feathers, J.; Barnett, W.; Silveira, M. Imazio da; Henderson, A.; Silva, J.; Chernoff, B.; Reese, D. S.; Holman, J. A.; Toth, N.; Schick, K. 1996. Paleoindian cave dwellers in the Amazon: the peopling of the Americas. *Science* 272:373–384

Ross, S. 1994. Soil nutrients and organic matter in forest and savanna habitats on Maracá Island. In: The Rainforest Edge: Plant and Soil Ecology of Maracá, J. Hemming (ed.), Manchester University Press, Manchester, pp. 63–91

Salomão, Rafael de Paiva; Rosa, Nelson de Araújo. 1989. Análise da vegetação de floresta pluvial tropical de terra firme, pelo método dos quadrantes: Serra Norte, Carajás, PA. Acta Botanica Brasilica 2(1):27–42

Sanaiotti, Tânia Margarete. 1996. The woody flora and soils of seven Brazilian Amazonian dry savanna areas. Doctoral dissertation, University of Stirling, Scotland

Sanaiotti, Tânia M.; Magnusson, William E. 1995. Effects of annual fires on the production of fleshy fruits eaten by birds in a Brazilian Amazonian savanna. Journal of Tropical Ecology 11(1):53–65

Sanaiotti, Tânia M.; Bridgewater, Samuel; Ratter, James A. 1997. A floristic study of the savanna vegetation of the state of Amapá, Brazil, and suggestions for its conservation. Boletim do Museu Paraense Emílio Goeldi, Série Botânica 13(1):3–24

Santos, João Ubiratan M. dos; Rosário, Carlos da S. 1988. Levantamento da vegetação fixadora das dunas de Algodoal-PA. Boletim do Museu Paraense Emílio Goeldi, Série Botânica 4(1): 133–154

Santos, João Ubiratan M. dos; Amaral, Dário Dantas do; Gorayeb, Inocêncio de Sousa; Bastos, Maria de Nazaré do C.; Secco, Ricardo de S.; Neto, Salustiano Vilar Costa; Costa, Denise Cristina Torres. 2003. Vegetação da área de proteção ambiental Jabotitiua-Jatium, Município de Viseu, Pará, Brasil. Acta Amazonica 33(3):431–444

Santos, Otilene dos Anjos; Couceiro, Sheyla Regina Marques; Rezende, Alinne Costa Cavalcante; Silva, Márcia Daniela de Sousa. 2016. Composition and richness of woody species in riparian forests in urban areas of Manaus, Amazonas, Brazil. Landscape and Urban Planning 150:70–78

Santos, Deborah Cruz dos; Corrêa Neto, José de Jesus; Alvarenga, Adriano Braga Brasileiro de. 2018. Frutos consumidos por macaco-de-cheiro *Saimiri sciureus collinsi* (Osgood, 1916) em fragmentos de florestas em uma unidade de conservação no arquipélago do Marajó, Amazônia Oriental, Brasil. Revista Brasileira de Zoociências 19(1):82–90

Sauer, J. D. 1988. *Plant Migration: The Dynamics of Geographic Patterning in Seed Plant Species*. University of California Press, Berkeley.

Silva, Edinei Santos da. 2017. Plantas alimentícias em comunidades agrícolas no município de Rio Preto da Eva-AM. Master's thesis, INPA, Manaus

Silva, E. M.; Souza, J. N. S.; Rogez, H.; Rees, J. F.; Larondelle, Y. 2007. Antioxidant activities and polyphenolic contents of fifteen selected plant species from Amazonian region. Food Chemistry 101: 1012–1018

Silva, Rachel Macedo da; Mehlig, Ulf; Santos, João Ubiratan Moreira dos; Menezes, Moirah Paula Machado de. 2010. The coastal restinga vegetation of Pará, Brazilian Amazonia: a synthesis. Revista Brasileira de Botanica 33(4):563–573

Souza, Célia Cristina Valero. 2010. Etnobotânica de quintais em três comunidades ribeirinhas na Amazônia central, Manaus, AM. Master's thesis, INPA, Manaus

Takeuchi, Masayuki. 1960a. A estrutura da vegetação na Amazônia III: a mata de campina na região do Rio Negro. Boletim do Museu Paraense Emílio Goeldi, Nova Série Botânica 8:1–13

Takeuchi, Masayuki. 1960b. A estrutura da vegetação na Amazônia II: as savanas do norte da Amazônia. Boletim do Museu Paraense Emílio Goeldi, Nova Série Botânica 7:1–14

Towle, Margaret A. 1961. *The Ethnobotany of Pre-Columbian Peru*. Aldine, Chicago

Ugent, Donald; Ochoa, Carlos M. 2006. La Etnobotánica del Perú: Desde la Prehistoria al Presente. CONCYEC (Consejo Nacional de Ciencia, Tecnología e Innovación Tecnológica), Lima

Van den Berg, M. E. 2010. Plantas Medicinais na Amazônia. Museu Paraense Emílio Goeldi, Belém

Van den Berg, M.E. and M.H.L. da Silva. 1986. Plantas medicinais do Amazonas. In: Anais do1°Simpósio do Trópico Úmido, Belém, PA, 12 a 17 November, vol. 2, Flora e Floresta, EMBRAPA, Brasília, pp. 127–133

Vásquez, S.P.F., Mendonça, M.S.D. and Noda, S.D.N. 2014. Etnobotânica de plantas medicinais em comunidades ribeirinhas do Município de Manacapuru, Amazonas, Brasil. Acta amazônica, 44, pp. 457–472

Veiga, J. B.; Scudeller, V. V. 2015. Etnobotânica e medicina popular no tratamento de malária e males asociados na comunidade ribeirinha Julião- baixo Rio Negro (Amazônia Central). Revista Brasileira de Plantas Medicinais 17(4):737–747

Vieira, Thiago Almeida; Rosa, Leonilde dos Santos; Vasconcelos, Paulo César Silva; Santos, Mônica Mota dos; Modesto, Rozi da Silva. 2007. Sistemas agroflorestais em áreas de agricultores familiares em Igarapé-Açu, Pará: caracterização florística, implantação e manejo. Acta Amazonica 37(4):549–558

Watson, E. 1985. Cultivos Tropicales Adaptados a la Selva Alta Peruana, particularmente al Alto Huallaga. Banco Agrario del Perú, Lima

Chapter 32
Malvaceae

32.1 *Guazuma ulmifolia*

Bolivia Coco, coquito, coquino, bacheche, chirimoya; ava ava (Tsimane)

Brazil Mutamba, mutamba preta, mutamba de capoeira, mutamba da praia, mutambo, mutambeira, pojó, envireira, enveira do campo, cabeça de negro, chico magro, guaxima, guaxima macho, guaxima torcida, guamaca, guazuma, fruta de macaco, embiribeira, embira, periquiteira, cacaman, camacã, pau de pomba, pau de bicho, pau pólvora, araticum bravo, buruma, ibixuna; yane-'I (Arawaté), apɨ'a-'ɨ (Ka'apor), kiri-wɨra-'ɨw (Tembé)

Colombia Guácimo, guásimo, guasmo, nacedero

Ecuador Guásimo, guázimo, guácimo, guasmo, huasi caspi

Peru Bolaina, bolaina de bajo, bolaina negra, guásimo, guácima, guázumo, huásimo, cerezo, hapayillo, papayillo, yumanasi, yumanasa, lluicho vainilla, atadijo; ampejiniroki (Campa), xeshta (Shipibo)

Venezuela Guácimo, guácimo blanco, guácima, guamo dulce; jwaejka jyaï (Jodï), ya'ra ipun yo' (Panaré), pohoroa mosi fi (Yanomami)

Status Wild, spontaneous in cultural settings

N. Smith, *Amazon Fruits: An Ethnobotanical Journey*, Ethnobiology,
https://doi.org/10.1007/978-3-031-12803-5_32

Mutamba *Guazuma ulmifolia* tree in fruit in an abandoned home garden. Adjacent to Sitio Senhor Heladio, left bank of the Rio Xingu about 50 km upstream from Altamira, Pará, Brazil 9-17-15

Guazuma ulmifolia thrives in successionary vegetation, such as in abandoned fields, on the higher parts of floodplains of black, clear, and sediment-rich rivers. It is more common, however, along muddy rivers, such as at the confluence of the Amazon and Madeira rivers (Luize et al. 2015), along the Upper Madeira in Rondônia (Casula 2012:69), along the Purus, a southern affluent of the Solimões (Huber 1906b), and in the Amazon estuary (Freitas et al. 2019). It is also found in successionary woodlands in the Amazon estuary (Alves et al. 1997; Cattanio et al. 2002; Queiroz et al. 2007) as well as on some uplands, such as in Rondônia (Lisboa 1989; Moser 2013:98), in the Tocantins watershed in eastern Pará (Coelho et al. 2012; Taurino 2017:71), in southern Amapá (Aparício 2011:64; Guimarães et al. 2016b), in the Andean foothills in the Peruvian Amazon (Giacomotti 2016:32), and in the Bolivian Amazon (Araujo-Murakami et al. 2015).

It can survive in the understory of mature floodplain forests (Huber 1910; Thomas 2009), and also occurs in grasslands such as the Pampa del Sacramento in the Peruvian Amazon (Huber 1906c) and in gallery forests along water courses in cerrado vegetation bordering the Amazon forest in the south (Corette-Pasa 2011; Ivanauskas et al. 2004; Marimon et al. 2002; Medeiros and Walter 2012; Medeiros et al. 2012). It is also occasionally found in savanna woodlands (*cerrado*) and dry forest (*cerradão*) in the Araguaia watershed (Ratter 1987), in the Upper Tocantins River Basin (Coimbra et al. 2002:155; Oliveira et al. 2018; Pinheiro et al. 2018), in Rondônia (Loverde-Oliveira et al. 2010), and in the municipalities of Acorizal and Santo Antônio do Leverger in Mato Grosso in the southern fringe of the Amazon Basin (Borges and Shepherd 2005; Guarim Neto and Pasa 2009). The tree, which can reach 30 m when fully mature, also occurs occasionally on uplands (*terra firme*) that are not subject to flooding, such as along the lower Xingu (Salomão et al. 2007a) and in some cacao agroforestry plots in the municipality of São Félix do Xingu in southern Pará (Braga et al. 2019).

This fast-growing tree is widespread in tropical America from Mexico to northern Argentina, extending up to 1800 m in the Andean foothills. It is considered native to Amazonia (Prance and Silva 1975:248). In the Brazilian Amazon, where it is generally known as mutamba or pojó, this tree grows on floodplains of the Amazon and some of its tributaries, such as the Xingu. The branches are sometimes adorned with the suspended nests of yellow-rumped caciques *Cacicus cela*, as in the village of Nuevo Esperanza along the Marañón River. In the distance, the nests might look like large fruits, but the fruits are much smaller and round, between 2 and 3 cm in diameter.

Bolaina *Guazuma ulmifolia* tree with yellow-rumped cacique *Cacicus cela* nests in successionary vegetation. Nueva Esperanza, Rio Marañón, Reserva Nacional Pacaya-Samiria, Loreto, Peru 11-8-03

At first glance, bolaina fruits would seem unlikely candidates for a snack. The skin of the cherry tomato-sized fruits is hard and knobby, and there is no succulent pulp inside. The fruits resemble the hard, woody fruits of sweetgum *Liquidambar styraciflua* in the witch hazel family (Hamamelidaceae), a successionary tree of temperate and subtropical woods in the south and the southeastern United States as well as parts of Mexico and Guatemala, only they are not quite as prickly.

Pojó *Guazuma ulmifolia* fruits gathered on the Amazon floodplain. Piracuera da Cima near Santarém, Pará, Brazil 7-25-96

It was only when I was walking on a trail from the village of Monte Bello along the Ucayali to some rice fields that I discovered that kids relish the fruits. I asked the farmer who was taking me to the rice fields the name of the dark brown fruits scattered on a section of the trail traversing some old successionary vegetation. "Bolaina," he replied. When I asked him whether people use the fruit for anything, he answered no. But when I glanced at the farmers' two young daughters and their 11-year-old friend, Iber, they were all sucking enthusiastically on the black fruits. "*Se chupa*" explained Iber, the son of another farmer. The woody fruits are not eaten; rather youngsters relish the sweet sap that exudes from the creviced skin; the flavor is somewhat reminiscent of tamarind *Tamarindus indica*, a popular exotic fruit in many parts of tropical America, except that it has a chocolate-like aroma. The fruits fall in the dry season, and once they have been on the ground for a month or more, they lose their appealing sap.

Germán, 8 years, sucking on a bolaina *Guazuma ulmifolia* fruit gathered in regrowth forest adjacent to his village. Yarina, Rio Yanayacu, Reserva Nacional Pacaya-Samiria, Loreto, Peru 7-24-10

I subsequently verified that kids in other villages in the Pacaya-Samiria National Reserve, such as Yarina along the Yanayacu and Nueva Esperanza along the Marañón, also gather the freshly fallen bolaina fruits to suck out the syrup. In the village of San Carlos, a day's journey down the Ucayali from Monte Bello, 12-year-old Richter explained that he used a machete to cut open the fruits so that the pink syrup flows more readily. Children also suck on the fruits along the Xingu and Amazon rivers in Brazil.

Because the fruits are relished, especially by children, *Guazuma ulmifolia* is typically spared while clearing home sites or when it arises spontaneously in home gardens. In a survey of fruit and nut trees in 150 rural home gardens in the Brazilian Amazon conducted over several decades, I found *Guazuma ulmifolia* in 12 of them. In a separate inventory of fruits and nut shrubs and trees in 57 riverine home gardens along the middle Xingu in 2015, I found mutamba in 7 of them. This species is also found in some home gardens in Rosário Oeste, Mato Grosso, a town in the cerrado region bordering the southern part of the Amazon forest (Guarim Neto and Amaral 2010), in the savanna region of Roraima (Pinho 2008:39), and in the vicinity of São Domingos do Araguaia, in southeastern Pará (Pereira et al. 2010). Home gardens often contain a rich assortment of fruit and nut trees containing exotic and native species. In a survey of 150 home gardens, for example, I found 98 exotic and native species of fruit and nut plants. Along the middle Xingu, a total of 81 species of fruit and nut trees were observed. The Jodï in the southwestern part of the Orinoco Basin allow *Guazuma ulmifolia*, as well as other useful trees, to grow when they arise spontaneously in their cultivated plots (Zent and Zent 2012). Mutamba is also planted in some public squares, such as in Manaus (Prance and Silva 1975:248).

Bolaina evidently evolved the ability to produce sap in its fruits to attract dispersal agents, such as monkeys, tortoises, tapirs, and peccaries. Black spider monkeys (*Ateles* spp.), red howler monkeys *Alouatta seniculus*, the white-collared peccary *Pecari tajacu*, and the white-lipped peccary *Tayassu pecari* which are all hunted for their meat, are known to eat these fruits (Beck 2005; Keuroghlian et al. 2009; Nunes 1998; Prado 2013; Rojas 2012:58; Salazar 2007; Wallace 2005; Van Roosmalen and Klein 1988). Spider monkeys *Ateles chamek* have been recorded eating this fruit in the Bolivian Amazon (Felton et al. 2008, 2009). Squirrel monkeys *Saimiri sciureus*, not normally hunted for their meat because of their small size, eat this fruit and these monkeys are sometimes kept as pets (Pontes 1997). Some dispersal agents might crunch the fruits and ingest some of the numerous gray-green seeds, similar in size to those of the tomato, which would later be defecated and sprout. This role has been suggested for the white-collared peccary and the white-lipped peccary (Beck 2005). In Costa Rica, Baird's tapir *Tapirus bairdii* and collared peccaries *Pecari tajacu* chew the fruits. The collared peccary occurs in small bands in Amazonia, but Baird's tapir is replaced by *Tapirus terrestris*. In the cerrado region of Brazil, *Tapirus terrestris* eats this fruit and disperses some of the seeds (Prado 2012:18). Daniel Janzen notes that the fruits have a rock-hard core which helps protect at least some of the seeds from being masticated; horses can ingest up to 2000 fruits a day and a horse-like mammal may have been a major dispersal agent for the tree during the Pleistocene (Janzen and Martin 1982). Other dispersal agents might carry the fruits to other locations and simply lick the sap. Either way, progeny would be dispersed.

Guazuma ulmifolia has a number of other uses. Strips of bark are removed to prepare various home remedies, as in the vicinity of Nueva Esperanza along the lower Marañón river in the Peruvian Amazon. In the Brazilian Amazon, the leaves are used to make a solution for washing wounds and hemorrhoids and to make a tea to treat high blood pressure (Van den Berg 2010:226). In Mato Grosso in the southern part of the Brazilian Amazon, the bark is used to treat skin diseases and gastric ulcers (Violante et al. 2012), while the leaves are used to prepare a tea that serves as a laxative (Guarim Neto and Morais 2003). Also in Mato Grosso, residents of the rural community of Bom Jesus some 60 km south of Cuiabá use the leaves to make a bathing solution to treat wounds and hemorrhoids (Pasa 2011). The bark of this tree is sold at the Mercado do Porto in Cuiabá, Mato Grosso, to make a tea to reduce inflammations of various kinds (Pinto et al. 2013). Residents of Quilombo Sangrador, a rural community in the municipality of Presidente Juscelino, Maranhão, use the leaves to make a tea to assist with pregnancy (Monteles and Pinheiro 2007). In the Ecuadorian Amazon, a decoction of the fruits is prepared to treat coughs and the wood is used to make fences and for firewood (Van den Eynden et al. 1999:73).

The wood of this tree is used for a variety of purposes by carpenters on the floodplain of the Solimões in the vicinity of its confluence with the Japurá River (Wittmann et al. 2009). In the Bolivian Amazon, this tree is the preferred firewood among the Tsimane (Reyes-García 2001:100). In the savanna region of Roraima, the Taurepang and Sapará indigenous communities use the wood of this tree to create tiles for their houses (Perez 2011:27). Along the middle Xingu, rural folk use the sap from mutamba fruits to rub into their hair and to treat earache. Mutamba oil has been used to make hair more lustrous for a long time (Le Cointe 1922a, b:491) and mucilage extracted from the bark is still used in hair and skin lotions as well as some shaving creams (Fryxell 2004). The inner bark is also boiled to make tea for treating syphilis and skin problems (Le Cointe 1934:206; Silva et al. 1977:147). In the cerrado zone in Mato Grosso bordering the Amazon forest in the south, the leaves are used to make a poultice which is applied to wounds and burns (Moreira and Guarim-Neto 2009). In Colombia, the tree is planted as a living fence and to provide shade for cattle. The wood is also used for fuel (Pérez-Arbeláez 1956:718).

The oil was formerly used in making perfume as well (Le Cointe 1922a:44). Although bolaina is a tree of successionary forest, the wood is remarkably hard and durable and sought after for fuelwood, such as along Paraná do Baixo on the Amazon floodplain near Alenquer. In Veinte de Enero along the lower Yanayacu, one farmer mashes up bolaina fruits in *aguardiente*, a white rum that serves as the principal ingredient in many home-spun alcoholic beverages (*tragos*) in the Peruvian Amazon. After steeping for 8 days, the pieces of bolaina are filtered out and the sweetened *trago* is stored in a glass bottle. Many different *aguardiente* concoctions are made locally with a variety of fruits and barks. Some of them are passed off as "medicinal," a convenient excuse, no doubt, for the occasional overindulgence. In the vicinity of Jenaro Herrera along the lower Ucayali, rural folk prepare an aromatic tobacco for homemade cigarettes by soaking bolaina fruits in *aguardiente* for several days then mashing them up and sprinkling the liquid on tobacco leaves that are still being cured.

32.2 *Herrania mariae*

Brazil Cacau jacaré, cacau de morcego, cacau cabeça de cutia, cacau de quina, cacau estrela, cacau nazaré, cacau embaúba, cacaurana, cacaurana de quina, cacauí, cacauzinho, cacauú; nisan paubin (Kaxinawá), sokorohaek (Rikbaktsa)

Ecuador cacao de monte; kushikiam (Achuar), tserenda (Kichwa), kushíkiam, kushínkiap (Shuar), boguinca (Waorani)

Peru Cacahuillo, sacha cacahuillo, kushi bakau; kushikami, kuaíwakam kusikam, kuaíkam (Achuar), kushikam, kushi wakam (Aguaruna)

Status Wild, spontaneous in cultural settings, planted

Wadrio, 9 years, with cacau jacaré *Herrania mariae* fruits he gathered by swimming up to a tree in floodplain forest. Carrazedo, mouth of the Xingu River, Município de Gurupá, Pará, Brazil 4-21-15

Cacau jacaré *Herrania mariae* is so named in Brazil on account of the fruit's corrugated skin similar to a caiman's back. Attractive crimson and white flowers adorn the slender trunk of this understory tree before the fruits are produced. The green pods are up to 12 cm long and typically reach maturity during the latter part of the rainy season. Green when ripe, the ribbed pods are broken open to reveal a sweet pulp that surrounds the dozen or so seeds. *Herrania mariae* fruits are eaten locally and only occasionally reach markets, such as in the Brazilian Amazon.

This relative of cacao also has medicinal properties. In the vicinity of Chazuta along the upper Huallaga River in the Peruvian Amazon, local folk crush the seeds of this relative of cacao and mix them with the leaves of *Senna hirsuta* Fabaceae in warm water; the liquid is then swallowed to alleviate acute coughs (Sanz-Biset et al. 2009). In the Ecuadorian Amazon, the Shuar use the leaves and bark to make a wash to treat snake bites (Bennett et al. 2002:269). On the Island of Combu in the Amazon estuary near Belém rural folk use the seeds to treat intestinal worms (Jardim and Medeiros 2006).

Herrania mariae **in flower in a home garden. The long white ligules probably evolved to attract pollinators. Sitio São Miguel, Ressaca do Cacauzinho, Lago Amanã, Amazonas, Brazil, 9-5-12**

A small tree found mostly on floodplains throughout the Amazon, including the estuary and southern Guyana, locals can easily gather the pods from the trunks of the diminutive trees since they start producing fruit when only 2 m tall and do not exceed 15 m in height when fully mature (Schultes 1958). *Herrania mariae* is more common in seasonally flooded forests along sediment-rich rivers. It also thrives in successionary vegetation on floodplains, such as in the Amazon estuary (Gama et al. 2003), and it occurs occasionally in upland forests, such as in Rondônia in the southwestern part of the Brazilian Amazon (Absy et al. 1986). The Shuar in the southern part of the Ecuadorian Amazon plant this cacao relative because of its useful fruit (Bennett et al. 2002:56).

Braz, 62 years, eating cacau jacaré *Herrania mariae* fruit he picked in his home garden on an archaeological site. Vila Valente, lower Tefé River, Amazonas, Brazil, 8-14-12

In some upland home gardens some *Herrania mariae* fruit in the dry season, such as along the lower Tefé River in Amazonas. I only saw *Herrania mariae* in a few upland home gardens and they were all on archaeological sites with anthropogenic dark earth, presumably because they are more fertile than most *terra firme* soils. Although not a common component of home gardens, it does turn up occasionally in the yards of rural folk in the Brazilian Amazon, such as in the vicinity of Manaus (Guillaumet et al. 1990). Cacau jacaré also occurs in groves of fruit trees around manioc processing huts near Carrazedo, a village at the mouth of the Xingu River.

Cacau jacaré *Herrania mariae* tree in a grove of fruit trees around a manioc processing hut. Jon, 9 years. Near Carrazedo, mouth of the Xingu River, Município de Gurupá, Pará, Brazil 4-15-15

Because the pulp is appreciated by young and old alike, *Herrania mariae* is usually not cut down when clearing fields or home sites. In Peru, for example, the tree is often spared when clearing forest for home gardens or fields, as in Yarina, a village along the Yanayacu River in the Pacaya-Samiria National Reserve. In Ecuador, the Achuar also avoid cutting down *Herrania mariae* when clearing forest for their fields. Should *Herrania mariae* become established as a crop, the forest will have to be preserved to safeguard its pollinators. Flowers of *Herrania* smell like decaying vegetable and animal matter which attracts phorid flies. The bodies of phorid flies are covered with bristles that pick up the sticky pollen grains and transfer them to other flowers (Young 1994:130).

32.3 *Herrania nitida*

Brazil Cacau jacaré; cha-tê-rá (Ticuna)

Colombia Cacao silvestre, cacao del monte, cacao caimán, maraca de borugo; koo-ra-tá, otonasaré, curata (Karijona), taajki aae (Miraña)

Ecuador Sacha cacao, cacao de monte; coquio'cho (Cofán), sicu patasi, sicu sambo, ilma cambi, cambig, cambia, sacha kambij (Kichwa), sënorí (Siona-Secoya), bogikawe, boinka, bonguinca (Waorani)

Peru Cacahuillo, cacahuillo lanudo, sacha cacao, cacao del monte, cacao silvestre, flor de araña; kxipnalli (Candoshi), ne sa ka te (Cashinahua), kushíikiám (Mayna)

Status Wild, planted

Cacahuillo *Herrania nitida* with flowers and fruit in a home garden. The tree was spared when the village home site was cleared. Yarina, Rio Yanayacu, Reserva Nacional Pacaya-Samiria, Loreto, Peru 1-15-06

Herrania nitida is a distant relative of cacao that thrives on the higher parts of floodplains and uplands throughout the Amazonia but is more common in the western part of the basin including Acre in Brazil and low-lying rainforest areas of Bolivia, Peru, Ecuador, and Colombia (Dávila et al. 2015:43; Friedman et al. 1993; Santos et al. 2018a, b; Schultes 1944, 1958). Confined to lowlands below 800 m, *Herrania nitida* is a short tree, usually reaching no higher than 8 m, that thrives in the deep shade of forests on floodplains and *terra firme*. This small tree also occurs in light gaps in upland forest, such as in Acre (Maranho and Salimon 2015). The elongated pods are born along the trunk and are thus easy to twist off. Measuring some 7–11 cm long by 3–5 cm wide, the fruits have a pointed tip and remain green when ripe in the rainy season. The thick skin is deeply furrowed with numerous ridges. People enjoy the sweet yet tangy white pulp surrounding the two dozen or so seeds contained in the pod. The Karijona in the Colombian Amazon roast the seeds and ingest them to treat stomachache (Schultes and Raffauf 1990:446). In the Colombian Amazon a game bird, paujil *Mitu silvini*, has been observed feeding on this fruit and likely disperses some of the seeds (Franco and Santamaria 1997).

In spite of their appealing flavor, *Herrania nitida* fruits rarely if ever reach markets, nor are the seeds fermented and roasted to make chocolate. Locals typically spare the tree when clearing forest for a home garden or field, such as in Yarina and Veinte de Enero along the Yanayacu River and San Carlos along the Ucayali River, all villages in the Pacaya-Samiria National Forest, Peru. The Cofán in the northern part of the Ecuadorian Amazon plant this tree in some of their fields (Cerón 1995:144).

Abel, 6 years, eating cacahuillo *Herrania nitida* fruit in a home garden. Veinte de Enero, Rio Yanayacu, Reserva Nacional Pacaya-Samiria, Loreto, Peru 6-28-06

32.4 *Herrania nycterodendron*

Bolivia Chocolatillo palmado

Colombia Cacao de monte, cacao de murciélago; mu sé ge ke (Witoto)

Ecuador Cambig, cambia (Kichwa), sacha erania, cacaguillo del monte, buikahue, buinca (Waorani)

Peru Cacahuillo, cacao de murciélago, palo de murciélago, cacao de chimbe, palo de chimbe; kushi wakam (Aguaruna)

Status Wild

Cacahuillo ***Herrania nycterodendron*** **tree in flower in forest. Near Veinte de Enero, Rio Yanayacu, Reserva Nacional Pacaya-Samiria, Loreto Peru, 6-28-06**

Similar in appearance to its close relatives *Herrania mariae* and *H. nitida*, *H. nycterodendron* is confined to the forests of western Amazonia up to around 900 m. *Herrania nycterodendron* occurs in upland and floodplain forests. Apparently less common that its close cousins, the fruits of *Herrania nycterodendron* are also appreciated by rural folk for a snack while in the forest hunting or gathering other fruits, such as in the Pacaya-Samiria Reserve in the Peruvian Amazon. The Waorani in the

Ecuadorian Amazon eat the creamy white aril pulp surrounding the seeds (Cerón and Ayala 1998:165). The fruits of *Herrania nycterodendron* are 10–12 cm long by 4–5 cm in diameter and the tree does not grow taller than 8 m so fruits are usually easy to pick (Schultes 1958).

32.5 *Matisia cordata*

Bolivia Zapote, sapote

Brazil Sapote, sapota, sapota do Solimões, sapota preta, sapotila; o-tê'-rre (Ticuna)

Colombia Zapote, zapote chupachupa, zapote de monte, sapote blanco, matasano, chupa chupa

Ecuador Sapote, sapote molinillo, sapote amarillo, sapote blanco, sapote de monte, sapotillo, zapote; pau (Achuar), sacha sapote, sapote yura, sapote muyo (Kichwa), sáput (Shuar), aposí, apasi, apashi, naso apashi (Siona-Secoya), meñea-hue, bokawe, bocahue (Waorani)

Peru Sapote, sapote de monte, sapote amarillo, sapotillo, zapote, sacha zapote, chupa chupa, mamey colorado; paaú (Aguaruna), panashiteki (Asháninka), atiri (Cocama), bata (Ese Eja), 'o ichibin (Kakataibo)

Venezuela Mame colorado

Status Wild, spontaneous in cultural settings, planted

Margarita, 12 years, eating a sapote *Matisia cordata* fruit gathered in a home garden. Cabo Lopez, Rio Itaya, Loreto, Peru 1-7-06

Matisia cordata is one of the more promising fruits of the Amazon on account of its nutritional value, fine flavor, and attractive appearance. The deep orange color of the pulp indicates that the fruit is a good source of vitamin A; indeed, carotene levels of 2400 ug/100 g of pulp have been recorded for the fruit (Ducke 1946; Souza et al. 1996:180). The fruit is juicy, similar in taste to mango, and also slightly fibrous. The appearance of the fruit is also appealing, with an olive-colored skin that is slightly velvety to the touch. Furthermore, sapote can be easily grafted, an advantage when propagating desirable varieties.

Matisia cordata is native to central and western Amazonia (Lorenzi et al. 2006a, b; Reynel et al. 2003:77; Schultes 1977) and occurs in the Andean foothills up to around 1800 m. The eastern limit of its range appears to be in the vicinity of Tefé along the Solimões in the Brazilian Amazon (Brack 2003:59; Ducke 1946). I saw *Matisia cordata* in several home gardens along the lower Tefé River in 2012, but not further east. A river dweller near Tefé told me that sapote was brought to the area some 70 years ago by a Peruvian farmer. However, a few sapote trees have been found in home gardens further down the Solimões in Coari and Manacapuru, the latter close to Manaus (Clement 1982).

This large-leafed tree grows in upland forests and is occasionally also found on higher parts of floodplains. Sapote is a sizeable tree, reaching as high as 45 m in successionary forests. Under cultivation, however, the tree normally does not exceed 20 m because of reduced competition for light. Sapote fruits have five seeds, about the size of a large date. Capuchin monkeys (*Cebus* sp.) may be involved in dispersing the seeds as suggested by one of the common names ascribed to sapote by the Siona-Secoya in the Ecuadorian Amazon (Vickers and Plowman 1984). Also in the Ecuadorian Amazon, woolly monkeys *Lagothrix lagotricha poeppigii* and white-bellied spider monkeys *Ateles belzebuth belzebuth* have been observed eating the pulp of this fruit after removing the thick skin (Dew 2005, 2008; Link and Fiore 2006; Suarez 2006). In the southeastern Peruvian Amazon, spider monkeys *Ateles belzebuth chamek* have been observed feeding on this fruit and are known to disperse the seeds (Palma and Stevenson 2010, 2013). Woolly monkeys and spider monkeys are hunted for their meat throughout the Amazon Basin. Pygmy marmosets *Cebuella pygmaea*, which occur in western Amazonia, use their sharp teeth to pierce the bark of this tree to feed on the exudate (Soini 1988); these monkeys are too small, normally weighing less than 140 g, to be hunted for their meat. At least one species of stingless bee, *Trigona williana*, pollinates the flowers of this tree in the central Amazon and the honey of this species is appreciated in some rural communities (Marques-Souza et al. 1996).

Sapote *Matisia cordata* tree in a village home garden on an archaeological site. Yarina, Yanayacu River, Reserva Nacional Pacaya-Samiria, Loreto, Peru 7-23-10

Sapote bears globular fruits, similar in shape to Navel oranges but usually bigger, between 5 and 15 cm long and 5 and 15 cm in diameter. Fruit weight varies from 150 g to 1.3 kg (Clement 1989; Kerr and Clement 1980). A fully mature tree can produce as many as 8000 fruits (Kerr 1986), thus sapote is one of the most productive fruit trees in the region. Farmers in San Carlos report that sapote starts producing 5 years after planting the seed. The fruits mature in the rainy season and are gathered by climbing the tree or knocking them down with a pole. The fruit's thick skin helps prevent damage when they fall to the ground or are transported to markets, such as in Iquitos, Requena, Pucallpa, and Tingo Maria in Peru and Leticia in the Colombian Amazon (Duarte 1979:47). Sapote fruits are durable enough to be taken by boat from Tefé to Manaus, a journey that takes several days. Sapote fruits are often a bargain; in January 2004, for example, one US dollar would buy 20 sapote fruits in Pucallpa. A farmer who lives in Victoria along the Puinahua, a branch of the Ucayali, claims that sapote fruits stay fresh 2 weeks after picking even without refrigeration.

This tree is also used in some folk remedies. Ticuna women along the Solimões River in Amazonas state, Brazil, use the fruit to make a decoction with the soft inner stem of *Pouzolzia formicaria* (Urticaceae) to deal with menstrual problems (Schultes and Raffauf 1990:111). In the Ecuadorian Amazon, the Kichwa use the leaves to make a topical application to treat leishmaniasis (Gachet et al. 2010).

Sapote *Matisia cordata* fruits in a street market. Nauta, Rio Marañón, Loreto, Peru 4-29-06

Sapote is commonly planted in home gardens throughout the Peruvian Amazon especially along the lower Ucayali (Huber 1906b). Some farmers in the Peruvian Amazon, including the Shipibo at Nueva Cajamarca along the Ucayali, also plant sapote in their fields (*chacras*) including agroforestry plots devoted to cacao *Theobroma cacao* production (Vebrova et al. 2014). Some of the sapote trees found in cultivated fields arose from seeds discarded after snacking on the pulp, as in the vicinity of Veinte de Enero along the lower Yanayacu in the Pacaya-Samiria National Reserve. In some areas, such as in the village of San Carlos along the Puinahua, sapote fruits are gathered in the forest as well as in home gardens and fields. The Shuar in the southern part of the Ecuadorian Amazon also plant this fruit tree (Bennett et al. 2002:57). Sapote fruits in home gardens tend to be larger than those gathered in the wild (Ducke 1946), but considerable variation is found in the size of fruits around homes and in the forest. Some "wild" sapotes may be relics of abandoned fields or home sites, or their descendants.

Abandoned field (*chacra*) with plantain and sapote *Matisia cordata* trees (light gray trunks) on an archaeological site with black earth. Salva Vida, Quebrada Wiuri, near San Martin Tipishca, Samiria River, Reserva Nacional Pacaya-Samiria, Loreto, Peru 7-16-10

This tree surfaces in indigenous lore. The Mayoruna recount a myth about an enormous flood that covered the land in remote times and only one man and a woman survived by climbing a sapote tree. They survived by eating the fruits of this tree until the flood subsided and then they started the repopulation of their world (Maroni 1988:171).

32.6 *Pseudobombax munguba*

Brazil Munguba, munguba branca, mungubeira, ponga, envireira, envira; ütchine (Ticuna)

Colombia Algodón, plao de algodón, algodón de árbol

Ecuador Charapa poto (Kichwa)

Peru Punga, punga colorada, punga roja

Status Wild, spontaneous in cultural settings, planted

Munguba *Pseudobombax munguba* tree in fruit on the Amazon floodplain. Much of the forest has been cleared for cattle pasture. Paraná do Cambixe, Careiro Island, Amazonas, Brazil 6-27-07

Munguba *Pseudobombax munguba* is a medium-sized tree reaching 30 m high on floodplains of sediment-rich, clear, and black water rivers throughout the Amazon Basin, sometimes becoming the dominant tree (Irmler 1977; Kvist and Nebel 2001; Wittmann and Parolin 1999; Worbes 1997). The flowers are pollinated by a single, large bat *Phyllostomus hastatus* (Gribel and Gibbs 2002). It produces red pods which break open while on the tree to disperse the seeds. The seeds are embedded in fluff and are thus carried away by the wind. When they land on water, some fish, such as the much-prized tambaqui *Colossoma macropomum*, feed on them (Araújo-Lima and Goulding 1997:61; Roubach and Saint-Paul 1994). The seeds are also used to bait hooks to catch pacu (various species and genera) and pirapitinga *Colossoma bidens*, such as on Ilha Grande a few kilometers down the Amazon from Óbidos. When the flowers drop into water, some fish, such as species of sardiña *Triportheus* spp., feed on the slimy, gelatinous flowers. Known as sardinha in Brazil, these fish swim in schools and are much appreciated by rural folk. Munguba thus contributes indirectly to the diet of people on Amazonia by providing food for sought-after fish.

The fluff surrounding the seeds is sometimes used to stuff pillows, such as on ranches on Marajó Island. In the vicinity of Tefé along the middle Solimões, the fluff is used to stuff safety jackets on boats (Conserva 2007:21). In the colonial period, the "cotton" from munguba and the kapok *Ceiba pentandra*, another large floodplain tree, was exported to Portugal to serve as building insulation and felt inside hats (Spix and Martius 1976:83). The seed fluff accounts for a common name for the tree in the Colombian Amazon: algodón.

Dona Tiava with her granddaughter holding a pillow stuffed with munguba *Pseudobombax munguba* fluff that surrounds the seeds. Boa Vista, near Santa Cruz do Arari, Marajó Island, Pará, Brazil 5-14-99

In the colonial period, the fluff around kapok *Ceiba pentandra* seeds was used to stuff mattresses and pillows but that practice has been largely replaced by artificial materials. Even chicken feathers are no longer used to stuff pillows. *Ceiba pentandra* was even taken to European colonies in Asia and the Pacific to provide stuffing for pillows and mattresses. Some indigenous groups used to use the fluff for their blowgun darts, but very few indigenous peoples still use blowguns.

The inner bark of *Pseudobombax munguba* was once used by the Mura along the Amazon near Parintins to weave their hammocks (Bates 1863:305). In the eighteenth century, twine was made from the inner bank of munguba in various parts of the Brazilian Amazon (Sampaio 1825:66). The tree also has medicinal properties. In Peru, for example, the bark is used to prepare a decoction for treating snake bites (Schultes 1983). Along the lower Ucayali River in the Peruvian Amazon, the bark of this tree is grated into sugarcane alcohol (*aguardiente*) and drunk to treat rheumatism (Mendoza 2014). In some rural communities along the shores of Lake Amanã situated in the watershed of the lower Japurá, a tributary of the Solimões, a tea is prepared from the seeds of this fruit which is taken by women who have just given birth to help "close the uterus" (Câmpera 2017:102).

Because munguba has several uses it is typically spared by river folk when clearing their home sites. It thus turns up in some home gardens, such as Comunidade Januaria on Ilha do Carmo in Alenquer Municipality, Pará, and on the outskirts of Juriti along the Amazon River upstream from Santarém. It also sprouts spontaneously in some home gardens, such as at Sitio Camanova, Piracauera da Cima, on the Amazon floodplain near Santarém. It is also planted in some home gardens, such as at the residence of the head cowboy at Fazenda Santa Helena, north of Santa Cruz do Arari, Lago Arari, on Marajó Island.

Munguba *Pseudobombax munguba* trees along the waterfront of a village on an archaeological site. The houses are on stilts because this community floods during the rainy season. Bom Jesus de Baré, Lago Amanã, Amazonas, Brazil 9-2-12

The soft wood of munguba is used as mulch for vegetable beds such as at Carariacá, Municipality of Santarém, Pará; Mamauru along the Amazon River 10 km downstream from Óbidos; and along Igarapé Jari and on Ilha Taperá, both near Santarém, Pará. The rotting wood is collected from trees that have fallen.

Vendor with tobacco roll wrapped with the inner bark of munguba *Pseudobombax munguba*. **Mercado Central, Tefé, Amazonas, Brazil 8-13-12**

The inner bark is also sometimes used to make cordage, as at São Cirico de Urucurituba near Santarém. On Marajó Island, munguba cordage is used to sling hammocks and to tie the feet of pigs destined for butcher shops in urban centers. In the vicinity of Tefé the inner bark, called *envira de munguba*, is used to wrap tobacco that has been pressed into salami-shaped rolls. Locals shave off the locally prepared tobacco on to cigarette paper. Munguba is felled sometimes by loggers, such as on Marajó Island at the mouth of the Amazon and in the Anavilhanas archipelago along the lower Rio Negro, although it is not a high-value timber tree (Scabin et al. 2011).

32.7 *Theobroma bicolor*

Bolivia Chocolatillo, papaste, pataste

Brazil Cacaurana, cacau banana, cacau de índio, cacau do Perú, cacau peruano, cacau bafu, cacau melão, cacau da mata, cacau bravo, cacau quadrado, cacauí, cacauti, cupuaçu do Solimões, baraturi, macambo; ngu (Ticuna), baraturi (Tuyuka), himara amohi, himoro amohi (Yanomami)

Colombia Maracá, maracas, maraco, cacao maracó, cacao silvestre, macambo, bacao, copoazu, pataxte; pai', po'oi (Andoke), aao (Bora), heé-a (Maku), eo, ao (Makuna), aae (Miraña), jajo'o (Muinane), há-ha (Tanimuka), mizeko, mïzéna, mizeyi, musena, ma-ra-ká, mizena (Witoto)

Ecuador Cacao blanco, cacao de monte, cacao silvestre, maraca, patas, patas muyo, macambo; kuchiwakamp (Achuar), mak'av ɨ-, macavu (Cofán), kila, quila, patas, patas culas, patas yura, cula, challigua cacao (Kichwa), patasi (Runa), wakámpe, wakam (Shuar), iu, musejei (Siona-Secoya), capemoenca, cupehuenca, cupemuenca, koräu (Waorani)

Peru Macambo, cacahuillo; wakám (Aguaruna), ááhe (Bora), macambo pura (Cocama), nukanëmo (Kakataibo), machin tsau, nḯi torampi (Shipibo), ngu (Ticuna), maracá, mïzéna (Witoto)

Venezuela Himare (Yanomami)

Status Wild, planted

Macambo *Theobroma bicolor* fruit with vendor in a street market. Puerto Nanay, Iquitos, Loreto, Peru 4-13-06

Macambo *Theobroma bicolor* bears fruits that are distinguished by the corrugate-like veins that protrude from the skin in contrast to cacao *Theobroma cacao* which usually has a relatively smooth skin. Yellow when ripe, the pods are the shape of a rugby ball and are gathered in the wild as well as from planted trees in home gardens and fields (*chacras*) during the rainy season. Macambo pods are typically larger than those of cacao, normally between 15 and 20 cm long and 10 and 12 cm wide. The other singular characteristic of macambo is that the fruits are gathered mainly for the flat, round seeds which are roasted and eaten like chestnuts. This occurred in western Amazonia in pre-contact times (Bletter and Daly 2006) and continues to this day, such as among the Shuar in the Ecuadorian Amazon (Van den Eynden 2004:75). Before the arrival of Europeans, cacao *Theobroma cacao* seeds were also roasted but they were then ground and made into chocolate in Central America. In Central America, however, a chocolate beverage is made with seeds of *Theobroma bicolor* and it is considered superior to cacao because the drink is less oily (Gordon 1982; Léon 1984).

Macambo *Theobroma bicolor* seeds and plantain roasting on a brazier in a street market. Puerto Nanay, Iquitos, Loreto, Peru 1-8-06

Some 30 seeds are typically found in *Theobroma bicolor* fruits, but pod size varies and so does the number of seeds. The sticky, yellow pulp of macambo fruits is also eaten, but the flavor is rather bland and slightly savory, rather than sweet and tangy like the pulp surrounding cacao "beans." According to a villager at Yarina along the Yanayacu River, an affluent of the lower Marañón, several game animals are fond of macambo fruits including white-collared peccary *Pecari tajacu*, agouti *Dasyprocta* spp., and paca *Agouti paca*.

Opened macambo *Theobroma bicolor* pod showing the edible pulp surrounding the seeds. Veinte de Enero, Rio Yanayacu, Reserva Nacional Pacaya-Samiria, Loreto, Peru 4-21-06

Theobroma bicolor is a small understory tree reaching up to 20 m in forests on uplands and higher parts of floodplains in western Amazonia occurring up to 800 m in the Andean foothills (Castañeda 1961:147; Ducke 1946; Mendoza and Guerrero 2010). It is also found in upland forest in the southern fringes of the Amazon forest in the border area between Amazonas and neighboring Mato Grosso in Brazil (Claudino et al. 2015). It is also found in successionary vegetation, such as in the Colombian Amazon (La Rotta 1983:82) and the Ecuadorian Amazon (Irvine 1987). The shade-tolerant tree can be found as high as 800 m in the Andean foothills. Wild *Theobroma bicolor* in the Peruvian Amazon bears pods that are smaller than cultivated forms (Miller and Nair 2006). It is also found on the banks (*restingas*) of rivers of diverse watercolors and acidity. This relative of cacao was present in Central America in pre-contact times (Kufer and McNeil 2006). How it got there, whether carried by humans from South America or by natural dispersal, is still the subject of debate. Like cacao, *Theobroma bicolor* was likely introduced into Central America in precontact times by indigenous people (Young 1994:26).

Victor, 7 years, snacking on macambo *Theobroma bicolor* pulp. Manco Capac, Rio Puinahua (a branch of the Ucayali), Reserva Nacional Pacaya-Samiria, Loreto, Peru 4-14-06

This relative of cacao is widely cultivated in western Amazonia, such as among various indigenous groups in the Colombian Amazon (Gasché 1971; Veléz 1991) and in the western part of the Brazilian Amazon, such as along the Solimões River and in the vicinity of Manaus (Lima and Saragoussi 2000; Souza 2010:29) and in the municipality of Rio Preto da Eva along the Manaus-Itacoatiara highway (Silva 2017:31). A century ago, the Swiss botanist Jacques Huber noted that indigenous folk gathered the fruits of wild *Theobroma bicolor* on the floodplain of the sediment-rich Purus (Huber 1906a, b). The Ticuna cultivate the tree along the Upper Solimões in Brazil and the Amazon River across the border in Peru and it is also grown by other indigenous groups along parts of the Rio Negro (Clement et al. 2001; Ducke 1953; Rengifo-Salgado et al. 2017). Along the Tiquié, an affluent of the Uaupés in the northwestern part of the Brazilian Amazon, *Theobroma bicolor* is only found in home gardens in villages of the Tukano and Tuyuka; these indigenous groups told me that the tree is not found in the surrounding forest. The Tuyuka also plant *Theobroma bicolor* in some of their fields (Schmidt et al. 2010). The Bora who live along the Yaguasyacu River, an affluent of the Ampiyacu which flows into the Amazon River at Pebas in Colombia, plant macambo and other fruit trees in some of their manioc fields so that they become orchards once the root crop has been harvested (Denevan and Treacy 1987). The Witoto along the Caquetá in the Colombian Amazon also interplant *Theobroma bicolor* in the middle of their manioc fields so that they can gather fruits some 2–3 years after the root crop has been harvested (Veléz and Veléz 1992).

The Coreguaje, Kamejaya, and Witoto who inhabit parts of the Caquetá watershed in the Colombian Amazon also plant the tree (Hammen 1992; Trujillo and Correa-Múnera 2010; Veléz and Veléz 1999:51). Some argue that *Theobroma bicolor* is not native to the Colombian Amazon (Baker et al. 1953). *Theobroma bicolor* has been cultivated for at least 4700 years along the middle Caquetá and its affluents (Herrera et al. 1992; Mora et al. 1991; Walschburger and Hildebrand 1988). It is also grown outside of the Amazon in the Chocó bordering Colombia and Panama and in the Esmeraldas region of coastal Ecuador (Gordon 1982).

It is particularly abundant in the Iquitos area of the Peruvian Amazon (Pound 1938). Macambo is frequently cultivated in home gardens of river folk and indigenous groups in Loreto Province, of which Iquitos is the capital. Villagers in the vicinity of Iquitos sometimes plant *Theobroma bicolor* along with several other fruit trees in their fields after harvesting short-cycle crops in order to form orchards (Coomes et al. 2000; Coomes and Ban 2004; Hiraoka 1989). I saw macambo interplanted with 14 other kinds of fruit trees as well as manioc in a field near Pucallpa along the Ucayali. On the outskirts of San Pedro de Cumbaza near Tarapoto, macambo was interplanted in an orchard containing cacao, oranges, and sapote *Matisia cordata*. Macambo thus becomes part of successionary forests in some areas of western Amazonia.

Further north in the Ecuadorian Amazon, *Theobroma bicolor* is also planted in home gardens (Allen 1987). The Runa also plant *Theobroma bicolor* and consider the tree an indicator of former human occupation of the area (Irvine 1989). The

Aguarana, Achuar, and Kichwa also cultivate this tree in the Ecuadorian Amazon (Boster 1983; Descola 1994; Sirén 2007).

Vendors in street markets, such as Iquitos and Nauta, skewer the poker chip-sized seeds with long, slender sticks and roast them over charcoal. The tan-colored seeds are roasted for about half an hour until crunchy. The toasted seeds taste like peanuts, and at only 1 US cent apiece, sell briskly. PERSA, a small company in Iquitos specializing in the processing of local fruits and other produce, sells toasted macambo seeds in plastic bags to local customers. One day, macambo seeds may be a big hit in the party circuit in Europe and North America as a novelty snack. On a farm in the outskirts of San Pedro de Cumbaza near Tarapoto, macambo seeds are fried along with eggs for breakfast. Although fruits of *Theobroma bicolor* rarely reach markets in the Brazilian Amazon, they are occasionally found in street markets of Rio Branco, the capital of Acre (Lima 2014).

Farmers bringing macambo *Theobroma bicolor* **and guaba** *Inga edulis* **fruits to market. Requena, Rio Trapiche, Loreto, Peru 4-13-06**

Macambo seeds also feature in *chilcano de pescado,* a light fish broth eaten at breakfast, lunch, or dinner in the Peruvian Amazon. The seeds add texture and another savory flavor to the dish, which is especially popular in rural homes. *Chilcano* with macambo seeds is not served in restaurants and is only occasionally eaten in towns. Walter, a driver for one of the better hotels in Iquitos, recounted with a sense of nostalgia the *chilcano* with macambo seeds his grandmother used to prepare at Timareo, a rural community a couple of hours upstream by small boat from Iquitos. When Walter was a kid, he used to spend a lot of time on his grandmother's small farm and he remembers that she boiled macambo seeds and sliced green plantains, seasoned with garlic and cilantro. Walter, now in his 40s, noted with some sadness that his children are city kids, and that they do not know country cooking. His grandparents have died, and the land has been sold to a stranger.

Moises, 11 years, with a small-fruited form (*menudo*) of macambo *Theobroma bicolor* gathered from a tree in his home garden. The tree sprouted from the seed of a wild macambo. Yarina, Rio Yanayacu, Reserva Nacional Pacaya-Samiria, Loreto, Peru 7-24-10

Rural folk along the Ucayali in the vicinity of the Pacaya-Samiria National Reserve recognize two fruit forms of macambo. One form is relatively large (*grande*), about the size of a cantaloupe, and is always found in cultivation in home gardens or fields. The other, called *menudo*, is about a third smaller and is found in the forest. Macambo thus appears to have been brought into cultivation a long time ago, since larger fruit forms have been selected by locals for propagation. However, small-fruited forms (*menudo*) are also found in home gardens, such as at Monte Bello along the Ucayali and Yarina along the Yanayacu River, an affluent of the lower Marañón. In some cases, home gardens contain both the *grande* and *menudo* varieties of macambo, such as at San Carlos along the Puinahua, an arm of the Ucayali River. Seeds from wild types occasionally sprout spontaneously near homes after being discarded such as in Yarina along the Yanayacu River.

One farmer in Buenos Aires, a village on the bank of the Pucate River, a black water affluent of the Yanayacu in Pacaya-Samiria, posited another hypothesis concerning the large and small fruit forms of macambo. José Santos, who still speaks a few words of his ancestral tongue, Cocama, suggested to me that when macambo trees are young they produce larger fruits, but as the trees age, the fruits become progressively smaller. Another farmer in Yarina, a village along the middle Yanayacu, agreed that macambo fruits range from small to large, but asserted that the smaller forms are from wild trees. The latter farmer distinguishes three forms of the fruit: *grande* (large), *mediano* (medium), and *menudo* (small), suggesting a gradient. The distinction between wild and cultivated macambo is thus somewhat blurred, with medium sized fruits possibly representing variability within wild populations or domesticated forms that have become feral after home sites are abandoned.

In addition to fruit and edible seeds, macambo wood is also employed in house construction. At Veinte de Enero along the lower Yanayacu, for example, macambo poles are used for the main support posts for some of the village houses. Once the opened pods have dried they serve as bowls, as at Lago do Jacu, near Bella Vista do Sapa along the Tefé River in Amazonas, Brazil. The Yanomami in northern Roraima, Brazil, and contiguous forest in southern Venezuela use the inner bark to fashion straps for carrying baskets (Albert and Milliken 2009:100). The Ticuna who live in widely scattered villages along the Solimões River in Amazonas, Brazil, and contiguous Amazon River in Peru, use the leaves to make an infusion which is drunk to treat headaches and fever (Rengifo-Salgado et al. 2017).

Rosinei, 13 years, holding a bowl fashioned from a cacaurana *Theobroma bicolor* pod in his home. Lago do Jacu, near Bella Vista do Sapa, Tefé River, Amazonas, Brazil 9-9-12

32.8 *Theobroma cacao*

Bolivia Cacao, cacao criollo, chocolate, chocolatillo; mora, cuashe (Tacana), chocorate, chocorati (Tsimane)

Brazil Cacau, cacau da mata, cacaueiro; áka-'i (Araweté), yuára poro (Aruá), aka-ɨwa (Asurini), kakawa (Baniwa), ako'o-'ɨ (Guajá), kaka-ɨ, kaka wira (Ka'apor), kuben krã ti (Kayapó), ñumi (Parintintin), waijak (Rikbaktsa), waworẽ (Sateré-Maué), akôptia (Suruí), aka-ɨw-ete (Tembé), tcharere (Ticuna), kakawa (Tukano), arapuru, walapulu (Wajãpi), pohoroa nahi, poroa unahi, poro unahi, waiporounahi (Yanomami)

Colombia Cacao; bai-uc' (Maku), joaeõ, a-ba-ca-ra-á (Makuna), aso-ya-ee (Piratapuya), há-ha (Tanimuka), tchaperé (Ticuna), muselle, uam teg ma, riama misena, miegie (Witoto)

Ecuador Cacao, cacao platanito, cacao amarillo, cacao rojo, cacao de monte, kakáu; kawarunch (Achuar), cacao (Colorado), runa cacao, sacha cacao, killu kakáu (Kichwa), chukuchúk (Shuar), iu, tsié (Siona-Secoya), bikinka, bogïnka, cacabohue biquinea, cupehuenca, cupemuenca (Waorani)

Peru Cacao; kúchukuch, ikiammaya (Achuar), bakáu, kakaú, wakam (Aguaruna), vakao (Awajún), chaxon runxan (Amahuaca), kemitaki (Campa), nja xun de xan (Cashinahua), kuahe (Ese Eja), nukan (Kakataibo), cacahu caspi, cacahuillo (Kichwa), dadan dëso (Matses), sarigemini, sariyemineki, kakao (Matsigenka), cacahua (Pano), torampi, turanqui (Shipibo), chaxpere (Ticuna), musena (Witoto)

Venezuela Cacao; dairukbáa (Barí), pohoroa mosi hi, bohoroa anahi, pojoroa, shiekõmi natha hi (Yanomami)

Status Wild, planted

Rayone, 9 years, eating cacao *Theobroma cacao* pulp. She picked the fruit in her home garden on a riverbank. Porto Val, lower Tefé River, Amazonas, Brazil 8-30-12

Cacao *Theobroma cacao* is an understory tree that produces fruit pods on its trunks and main branches. Most of the pods are produced in the rainy season, but cacao planted in home gardens and orchards sometimes fruits throughout the year. The elongated pods resemble an American football, but some pods are smaller. They also vary in color when ripe from green to yellow, pale orange, red, or magenta. Some wild cacao trees along the Ucayali bear orange fruits. This diversity is most pronounced in the wild and in home gardens. Commercial groves tend to be dominated by cacao with yellow pods, known in the Brazilian Amazon as the Hibrido variety, but not all yellow fruits in home gardens are of this variety.

Juan, 68 years, opening pods of diverse varieties of cacao *Theobroma cacao* on his farm. San Pedro de Cumbaza near Tarapoto, San Martin, Peru 8-23-04

Roxo variety of cacao *Theobroma cacao* in an orchard planted by a river dweller. Morro do Arara, right margin of the middle Xingu River upstream from Altamira, Pará, Brazil 9-3-15

Wild animals, especially monkeys and squirrels, carry off the pods to access the sweet white pulp surrounding the seeds, a snapshot of natural history captured in ancient Mayan vessel paintings and carvings (Aguilar-Moreno, 2006; Ogata et al. 2006). Capuchin monkeys (*Cebus apella* and *C. albifrons*) smash open the pods against branches to access the mesocarp (Terborgh 1983:76). In the mid-nineteenth century, the famous English naturalist Henry Walter Bates and the American zoologist Herbert Smith noted that capuchin monkeys and squirrel monkeys *Saimiri sciureus* raid cacao groves to smash open the pods and eat the pulp, thereby scattering the seeds (Smith 1879:104). Capuchin monkeys were noted carrying off cacao pods, thus likely dispersing the seeds (Bates 1863:323). Farmers tell me that spider monkeys (*Ateles* spp.) do the same. In parts of the Bolivian and Peruvian Amazon Goeldi's marmoset *Callimico goeldii*, brown-mantled tamarin *Saguinus fuscicollis*, and white-lipped tamarin *Saguinus labiatus* have been observed feeding on the pulp of cacao and may disperse some of the seeds (Porter 2001; Sánchez and Levy 2013:30). However, in some areas deforestation is so advanced that monkeys are increasingly rare. In the Colombian Amazon a game bird, paujil *Mitu silvini*, has been observed feeding on this fruit and likely disperses some of the seeds (Franco and Santamaria 1997).

A farmer at km 75 of the Altamira-Medicilandia stretch of the Transamazon remarked that paca *Agouti paca* and agoutis *Dasyprocta* spp. are also fond of cacao pulp but can only access pods close to the ground since these rodents cannot climb trees. Paca have been recorded eating cacao pulp from planted trees near a rural village in central French Guiana (Gély 1986). Pacas and agoutis, which are much appreciated game animals, only eat the mesocarp and leave the beans on the ground. While the mesocarp is a tempting treat, the seed itself is bitter tasting, thus discouraging dispersal agents from biting down on "beans" (Young 1994). When people arrived in the Amazon Basin they continued this tradition and in some cases made juice from the pulp (Heriarte 1964).

Anderson, 10 years, with cacao *Theobroma cacao* he gathered in forest on the outskirts of his village. San Carlos, Rio Puinahua (a branch of the Ucayali), Reserva Nacional Pacaya-Samiria, Loreto, Peru 1-11-06

Cacao is native to forests on uplands and higher parts of floodplains in the Amazon Basin, the southern part of the Orinoco watershed, and contiguous areas of the Guianas. It occurs in montane forest up to 900 m in the Andean foothills. It is found wild in the Upper Amazon where it can be abundant in forests as noticed by British naturalists in the middle nineteenth century (Bates 1863:270; Spruce 1908b:4). The late Richard Evans Schultes has suggested that cacao is native to northwest Amazonia in Ecuador and Colombia (Schultes 1977). Wild stands of cacao can reach up to 142 trees per hectare in parts of Rondônia (Lisboa et al. 1991).

The shifting course of many rivers in the Amazon lowlands, such as the Ucayali in Peru and the Purus in Brazil, has contributed to the genetic diversity of the crop by undercutting banks with older trees and depositing fresh alluvium on the opposite side to be colonized by young cacao and other plants. The highly dynamic nature of some floodplains thus creates opportunities for genetic recombination in cacao populations. Cacao populations along narrow rivers are generally more homogeneous (Bartley 2005:2, 51–52).

Chocolate is adored the world over, but few people have any idea what kind of plant is the source of chocolate, nor where it is cultivated. Fewer still know that the cacao *Theobroma cacao* tree is native to Amazonia and that the fate of genes for possibly enriching the crop rests in the hands of those who live in the Amazon, or have plans to develop it. The price of one's favorite chocolate bar may rest ultimately on the fate of forests in the Amazon Basin.

There is more to chocolate than just the sensual texture and pleasing taste. It turns out there is a physiological basis to why so many people crave chocolate. Chocolate contains compounds that mimic cannabinoid drugs and contains serotonin, a neurotransmitter than acts as an antidepressant. Chocolate also contains theobromine, a stimulant in the xanthine group similar to caffeine. Caffeine is mildly addictive, and so is chocolate. Furthermore cocoa "butter" is rich in stearic acid which does not raise LDL cholesterol, the bad kind of cholesterol that can lead to strokes and heart attacks (Bonanome and Grundy 1988).

Cacao is grown in many lowland tropical regions to produce the seeds or beans that are fermented, dried, and then ground to make chocolate. This practice arose in Mexico where cacao was introduced in precontact times. The Aztecs and Mayans also used cacao beans for currency (Alden 1976; Stone 1984). Cacao has been cultivated in the moist lowlands of Central America for at least 3000 years (Cheesman 1944).

How cacao arrived in Central America is unknown, but it happened thousands of years ago (Smith 2009). Wild cacao in Mexico is genetically distinct from that found in South America, suggesting that it has been in Central America long enough to evolve distinct forms (Cruz et al. 1995). Specialized ceramic bowls used for drinking chocolate, similar to ones still in use, have been uncovered at an archaeological site in the Ulua River valley of Honduras that dates back 4600 years. Cacao could have migrated naturally into Central America before the arrival of people (Ogata et al. 2006), but that would be difficult considering the barrier of the high Andes. Cacao cannot grow in the cold, bleak meadows of upper mountain slopes and valleys. Cacao likely reached Central America with the help of people by various routes.

Hunters and gatherers brought cacao up the Rio Negro and down the Orinoco at least 16,000 years ago (Lathrap 1977). At that time, just the pulp surrounding the beans inside pods was likely consumed. Chocolate making was discovered later in Mexico. From the Orinoco, cacao was carried as seed or seedlings over some low passes in the Venezuelan cordilleras into the Maracaibo basin, known today for a fine cacao called Porcelano. The Guajira peninsula, at the mouth of Lake Maracaibo, is too dry for cacao, so again people took seeds or seedlings to more humid locations along the Caribbean coast of Colombia near its border with Panama from whence it could have been taken into Central America.

Cacao was likely taken across the Andes to the Pacific coast of South America in early times (Cobley 1956). The lowest route across the Andes is Porcuya Pass in northern Peru at 2200 m, but the environment on both sides of that watershed divide is too arid to grow cacao. If cacao seeds or seedlings were carried through the Porcuya pass in ancient times, they would have had to be taken several hundred kilometers north along the Pacific coast to northern Peru or Ecuador where conditions are sufficiently humid and warm year-round to grow the crop. Cacao may have been taken across the Andes by indigenous people further north in Ecuador. The extinct Mayo Chinchipe culture which flourished from 5500 to 2500 BP in the foothills of the Andes in southern Ecuador once fashioned stone grinders to match the shape of cacao pods (Valdez 2013). Remnants of cacao have been found in circular archaeological mounds held in place by walls of stone that date back more than 5000 years (Rostain 2017:93, 144). The Valdivia culture, which thrived along the coast of Ecuador from 5500 years BP to 3800 years BP, likely took cacao to Mexico by large dugout canoes and rafts (Wolters 1999, 2001).

The late Richard Evans Schultes suggested that the motive for spreading cacao may have been its early use in curing and supernatural rituals. Even today, the Witoto of the Colombian Amazon use cacao pods cut in half as cups to hold potent tobacco juice which is taken by shamans and village elders to solemnize agreements (Schultes and Raffauf 1992:97). According to Maria Van den Berg, a retired botanist who worked at the Goeldi Museum in Belém, rural and urban folk in the Brazilian Amazon employ cocoa butter to treat bruises (Van den Berg and Silva 1986).

The precise distribution of truly wild cacao in Amazonia is unclear because people have moved the plant around for a long time. In the mid-nineteenth century, for example, one farmer along the Tocantins River had planted some 60,000 cacao trees in the forest (Wallace 1889:53). The boundaries between wild and planted cacao are a bit blurred. People in Amazonia have always appreciated snacking on the creamy sweet pulp surrounding the seeds inside the pods which are produced along the trunk and main branches; the pulp is also made into juice. Cacao may have been confined originally to the central and western parts of the Amazon up to the foothills of the Andes before humans arrived in Amazonia some 15–20,000 years ago (Cheesman 1944). The highest levels of genetic diversity in *Theobroma cacao* are found in the Upper Amazon from Acre in Brazil and southern Peru north to Ecuador and Colombia (Sereno et al. 2006; Thomas et al. 2012; Zhang et al. 2006).

Humans not only extended cacao's range they also increased its density by discarding seeds at camps and settlements, as well as planting the seeds. Along the lower Tocantins, for example, rural folk in the nineteenth century planted cacao in the understory of forest on some of the islands in the river (Bates 1863:145). Indigenous groups brought pods back to temporary camps and villages as a snack. After sucking off the sweet pulp, the seeds were discarded and some eventually sprouted. Some of the spontaneous seedlings were probably nurtured. Two centuries ago, the German naturalist Alexander von Humboldt noted that indigenous groups along the Orinoco River left piles of cacao seeds around their campsites (Humboldt and Bonpland, 1818). Cacao, then, is an example of the anthropogenic nature of forests in certain parts of the Amazon Basin.

Along the Purus, the once numerous tribe of the same name relied on wild fruits, especially cacao, for sustenance. Father João Daniel noted that the industrious Purus planted orchards and the forest along their river was rich in cacao (Daniel 1976:264; Daniel 2004:26). Thus the density of "wild" cacao in at least some parts of the forest in western and central Amazonia is a cultural artifact. In the mid-nineteenth century, Paul Marcoy, a French explorer, remarked on the abundance of cacao along the Amazon in the vicinity of Tefé and noted that the shores of lakes and river channels were bordered with "forests of wild cacao" (Marcoy 1873:455). The Tefé River has an abundance of archaeological sites and numerous tribes lived in that area in the colonial period, so many of the forests are in fact anthropogenic.

In the eighteenth century, Francisco Xavier Sampaio reported large concentrations of cacao along certain portions of the middle Amazon between Manacapuru and Coari. An astute observer of the landscape and culture of the region, the traveling judge speculated that the cacao was "planted by nature." In describing Aramaça Island along the Upper Amazon near the mouth of the Javari in 1774, Sampaio noted that cacao was so abundant that the entire island was practically a plantation (Sampaio 1825:69). It seems likely that indigenous groups planted cacao on Aramaça and other islands along the middle Amazon. The Madeira, a southern affluent of the Amazon, has long had a reputation for the density of cacao and the quality of the beans. In the second half of the seventeenth century, the Jesuit João Felipe Bettendorf noted that the Portuguese made trips to the Madeira River because of the abundance of cacao groves and the fine beans which were exported (Bettendorf 1910:355). Although that stretch of the Amazon and the Madeira falls within the natural distribution of cacao, its density was likely the result of human agency.

How far east in Amazonia cacao was found before humans arrived is a matter of conjecture. Over a century ago, the Swiss Botanist Jacques Huber remarked on how difficult it often is to differentiate truly wild cacao from feral populations (Huber 1906a, b, c). When a cacao plantation is abandoned and the forest returns at least some of the trees can survive as noted by Richard Spruce along the Amazon near its confluence with the Rio Negro in the mid-nineteenth century (Spruce 1908a, b, vol. 1:233). Cacao is generally considered native in forests west of the Tapajós River. North of the Amazon River, Adolpho Ducke, an Austrian entomologist turned botanist who spent most of his professional life in the Brazilian Amazon, suggested that cacao is native to forests west of Óbidos (Ducke 1946).

Still, numerous travelers have remarked on what they consider to be wild cacao along the lower Amazon, including the estuary. In the middle of the seventeenth century, for example, João Daniel, a Jesuit missionary, remarked on the abundance of wild cacao ("*da natureza*") on some floodplain islands around Marajó. Toward the close of the seventeenth century, another Jesuit priest, João Felipe Bettendorf, described stands of cacao on Marajó as "a gift of nature, much appreciated by the inhabitants" (Bettendorf 1910:26). Indigenous people likely brought cacao to the estuary long before the arrival of Europeans, and when they disappeared, the cacao became feral. For example, the late Betty Meggers, an archaeologist with the Smithsonian Institution, found cacao on an indigenous mound along the Anajás River on Marajó and suggested that indigenous people brought the widely appreciated plant to the island (Meggers and Evans 1957).

Homemade chocolate made from *Theobroma cacao* which is shaved and mixed with hot water and drunk with added sugar. The farmer, Raimundo Jordão 83 years, made the chocolate from cacao he grows in his home garden and around his manioc processing hut. Terra preta, Rio Camutá, affluent of the Ipixuna, in turn an affluent of the lower Xingu, Município de Gurupá, Pará, Brazil 7-27-16

Amazonians never made chocolate, although the Purus nation along the river of the same name apparently ate raw, unfermented cacao seeds during the colonial period even though they are bitter (Daniel 1976:264). The art of making chocolate was introduced to the Amazon during the colonial period by Spaniards who had learned of the seductive beverage from the Aztecs in Mexico. A rather greasy chocolate drink is made from cacao today by riverine folk in parts of western Amazonia; it appears to have been drunk sparingly during the colonial period. Folk chocolate for drinking is made into the firm and slightly slippery cylinders or patties, which are melted in boiling water. The cocoa butter rises to the surface and can be quite unpalatable to an outsider accustomed to processed chocolate drinks with sugar and milk powder in which much of the cocoa butter has been removed.

Franco, 10 years, scraping chocolate made by his mother from cacao *Theobroma cacao* grown in their home garden. The shavings will be added to hot water to make a drink. Vila Valente, lower Tefé River, Amazonas, Brazil, 8-30-12

Along the Madeira and Purús rivers, for example, locals still fashion oblong blocks of ground cacao for making chocolate drinks and desserts. Cocoa butter is not expressed from homemade chocolate, and the resulting oily texture is an acquired taste. Round, waxy tablets of artisanal chocolate are also prepared in some other parts of Amazonia, such as the Mayo valley in the foothills of the Andes in Peru. The chocolate tablets, destined for making beverages rather than eating, are sometimes flavored with cinnamon and clove and appear in regional markets such as Tarapoto and Iquitos. Tabalosos, a small town on the asphalted highway connecting Tarapoto with Moyobamba, is a center for artisanal chocolate in the Peruvian Amazon and several vendors typically sell smooth, oblong to rounded chocolate bars along the roadside.

Brigida, 35 years, with her son Monjer, 11 years, selling homemade chocolate made from *Theobroma cacao* **by a roadside. Tabalosos, Tarapoto-Moyobamba Highway, San Martin, Peru 8-21-04**

The King of Portugal spurred wider growing of cacao in the lower Amazon after a 1677 edict encouraging its cultivation. Jesuits had cacao planted around their missions in the Amazon by 1760 from seeds gathered from the forest, especially along the Madeira River. Cacao plantations were established around Alenquer along the middle Amazon by the close of the seventeenth century (Bartley et al. 1987), as well as in the vicinity of Cametá along the lower Tocantins. Cacao dominated exports from the Brazilian Amazon from the early 1730s for some 150 years (Alden 1976; Gomes 2018). Most of the cacao was sent to France. Between 1730 and 1822, cacao accounted for at least half of the value of exports from Belém to Portugal (Anderson 1999:58). Cacao exports from the Amazon peaked in 1888 (Le Cointe 1922a:294).

In the nineteenth century, the central Amazon between Santarém and Óbidos had the densest plantations of cacao (Spruce 1908a:78). The concentration of the cacao plantations was such that one property seemed to blend into the next as captured in a nineteenth-century novel, *O Cacaulista*, by the Brazilian writer Inglês de Souza. The romance is set on a cacao farm a little upstream from Óbidos:

> The shady cacao grove linked the farm with neighboring properties, and behind the house was a black lagoon, formed by an arm of the Amazon River, and ringed with dense thickets of the arum plant (Souza 1973)

During the colonial period, cacao plantations, some quite extensive, were established in various locations along the Amazon from Parintins to the estuary and along the lower Tocantins and lower Tapajós. One such plantation, Cacao Imperial, with 40,000 trees was established on the right bank of the Amazon a little downstream from Óbidos in the early 1700s and was still in production in 1831 when it was abandoned, probably because of the scarcity of labor as introduced diseases ravaged the indigenous population. Along the banks of the Amazon, cacao was planted as a monocrop in a cleared floodplain forest. In 1846, the American traveler William Edwards noted the abundance of planted cacao in the vicinity of Santarém: "Sometimes the plantations are of vast extent, and one can walk for miles along the river, from one to the other, as freely as through an orchard" (Edwards 1847:106). In the late 1800s, the banks of the Amazon between Juriti and Alenquer were described as being continuously lined with cacao orchards varying in width from 50 to 200 m. A little further downstream near Prainha, a couple of English surveyors for the Amazon Steam Navigation Company described the scenery in 1873 thus:

> These narrow bands of cacoa trees bordering the edges of the Lower Amazon form one of the most marked and characteristic features of its scenery. Their light-green leaves, tinged with brown, contrast in a pleasing manner with the tall dark wall of forest foliage behind. (Brown and Lidstone 1878:98)

Farmer paddling into his cacao *Theobroma cacao* grove interplanted with rubber *Hevea brasiliensis* trees on the Amazon floodplain. Paraná do Cambixe, Careiro Island, Amazonas, Brazil 6-27-07

Although great floods in 1859, 1892, 1895, and 1900 destroyed a great deal of cacao along the middle Amazon, many growers replanted (Le Cointe 1903). But after the massive flood of 1953 many cacao groves were subsequently planted to jute, a fiber crop introduced to the Amazon by Japanese-Brazilians in the 1930s (Harris 2000; Smith 1999). Cacao does well on floodplains as long as the inundation period is brief. In describing the floodplain forests on islands of the lower Tocantins in the 1870s, Charles Hartt, a geologist from Cornell University, noted that the cacao trees "were in their element" (Hartt 1898). One of the reasons that farmers in the lower Amazon planted cacao on floodplains was to avoid damage by leaf cutter ants (Lustosa 1976:299); there ants (several species of *Atta*) strip certain trees of their leaves to provide nourishment for fungi in the large underground nests. River dwellers along the lower Tocantins still harvest cacao in the vestiges of these ancient orchards on floodplain islands a little upstream from Cametá. These relict orchards, which self-seed according to locals, blend with the floodplain forest that is re-asserting itself, thus forming a cultural forest. Cacautuba, a village along the Tefé River, means the place of cacao and is so named because of the large number of cacao trees in the forest surrounding the settlement. Cacautuba is on an archaeological site with anthropogenic dark earth and the concentration of cacao trees in the area is likely the result of human activities over an extended period.

Cacao *Theobroma cacao* planted among babaçu *Attalea speciosa* palms and taperebá *Spondias mombin*. Babaçu and taperebá are spared because they produce useful fruits. Near Bela Vista 1 archaeological site, Xingu River upstream from Belo Monte, Pará, Brazil 8-26-15

In the Amazon, cacao is still planted in small orchards, both on higher parts of floodplains and on uplands. Along the Ucayali River, for example, cacao is incorporated in some plantain fields so that it takes over once the plantain is no longer productive after a few years. Cacao is also planted following maize harvests. These cacao orchards often have other useful trees, such as fruit and timber species, which help reduce disease pressure on the cacao. Cacao is also planted in home gardens for its pulp which is eaten as a snack, to make juice, and to make chocolate. Many rural folk in the Peruvian and Brazilian Amazon farmers skip the bean fermentation step when preparing chocolate.

Cacao *Theobroma cacao* intercropped with açaí *Euterpe oleracea* palms, banana, and other fruit and nut trees. Km 88 Altamira-Itaituba, Transamazon Highway, Pará, Brazil 8-23-15

In my survey of 150 home gardens carried out over several decades in the Brazilian Amazon, I found cacao in 23% of them. In a separate study of 57 home gardens along the middle Xingu in 2015, cacao was found in 44% of them. If homes are abandoned cacao trees in the yard can still thrive if the forest returns. In some communities, farmers build manioc processing huts (*casa de forno, casa da farinha*) close to their fields so that they do not have to carry the heavy manioc tubers long distances to their settlements. Many of these manioc processing huts have small orchards established around them, sometimes containing cacao, as in the vicinity of Carrazedo at the mouth of the Xingu and further east at Jocojó, also in the Municipality of Gurupá.

Manuel, 75 years, harvesting cacao *Theobroma cacao* with a machete in floodplain forest. Rio Panité on an island in the Tocantins River, Pará, Brazil 7-2-07

Cacao pods are harvested by various means: chopping them off the trunk and lower branches with a machete, using a pole (*podão*) with a small sickle-like metal blade at one end to cut them free, or climbing the tree. Only agile youngsters can safely climb cacao trees since grownups are likely to break the slender branches if they get up in the trees which are normally under 10 m high but occasionally reach 17 m.

Adenor, 14 years, harvesting cacao *Theobroma cacao* in his home garden using a pole with attached blade (*podão*). Rio Joroca on a floodplain island in the Tocantins River, Pará, Brazil 7-2-07

Arlinho, 11 years, gathering a cacao *Theobroma cacao* pod from a tree planted in floodplain forest. Carrazedo, mouth of the Xingu River, Município de Gurupá, Pará, Brazil 7-24-16

Worker drying cacao *Theobroma cacao* beans on a retractable drawer on a small farm. Km 95 Altamira-Itaituba, Gleba 29, lote 10, Transamazon Highway, Pará, Brazil 7-12-96

Since the pods have a thick skin, they are not damaged when they fall to the ground. After the pods are gathered up, they are broken open, and the beans with attached pulp are emptied into bins or baskets lined with large leaves. In the nineteenth century, river dwellers along the Tocantins River in Pará often emptied the pods into the shells of tartaruga *Podocnemis expansa* turtles (Moura 1989:66), but this the largest turtle in Amazonia has declined dramatically due to the widespread gathering of its eggs on sandy beaches (Smith 1974). The beans ferment for a few days before being spread out on large drawers that can be pushed under a roof of a shed (known as *tendão* in the Brazilian Amazon) in the event of rain. On some farms, the cacao beans are protected from rain by a clear plastic awning, or they are simply covered by a tarp if it rains. These days in the Altamira area farmers skip the bean fermentation process to save time and money. Instead, the beans are spread out on drawers for immediate drying in the sun. Periodically workers, or the farmers themselves, walk up and down the enormous drawers turning over the beans with their feet.

Farmer raking cacao *Theobroma cacao* beans on a mat made from arumã *Ischnosiphon obliquus*. The blue plastic tarp is used to cover the beans when it rains. Combú Island, near Belém, Pará, Brazil 12-15-94

José "Zezão" Ascelino, 58 years, turning over cacao *Theobroma cacao* beans with his feet. This farmer has opted for a shed with walls and ceiling of plastic to keep the beans dry rather than one with a retractable drawer. Morro dos Arara, right margin of the middle Xingu River upstream from Altamira, Pará, Brazil 9-3-15

Along the Transamazon near Altamira and Medicilandia, farmers usually hire workers to gather cacao pods and undertake pruning and weeding. But in the vicinity of Santarem, cacao growers would sometimes help each other as well as hire a few workers to prune and weed their orchards. The reward was a festive occasion called *puxirum* during which the orchard owner would provide some 60 people with food and drink after they had completed the task. The drink was potent *cachaça* poured from 20-l glass flagons (*frasceiras*) produced at distilleries near Abaetetuba along the Tocantins River. A cow or bullock was butchered and the volunteers and workers were served a BBQ (*churrasco*) along with rice and manioc flour. This tradition appears to have ceased in the mid-1960s and while cachaça continues to be consumed in copious quantities in urban and rural areas it is now mostly produced in the northeastern and central-southern parts of Brazil. In the mid-nineteenth century, a German explorer noted that in the vicinity of Cametá along the lower Tocantins River families growing cacao would team up to help each other with tending their groves thus providing opportunities to strengthen friendships and boost their income (Avé-Lallemant 1961:39).

Cacao has been used traditionally for various purposes in addition to juice, eating the pulp, and making chocolate. During the nineteenth and early twentieth centuries, villagers and rural people along the Tocantins, such as at Cametá and Baião, burned cacao pods and mixed the ashes with cocoa butter or oil from andiroba nuts *Carapa guianensis* to make soap, known as *sabão de cacau* (Bates 1863:128;

Ferreira Penna 1864:46; Lustosa 1976:302; Souza 1873:215). Soap was also made with burned cacao pods in the nineteenth century along the Amazon between Óbidos and Alenquer (Herndon and Gibbon 1853:298). The Puinave who inhabit the Inírida River, an affluent of the Orinoco in Colombia, combine the ashes obtained from burning the bark of wild cacao with an infusion made with the bark of *Virola calophylla*, a powerful entheogen (Davis 2004). The mixture is then sifted and put into a snail shell with two bird-bone tubes; one is inserted in the shaman's nostril while an assistant blows the hallucinogenic compound into the shaman's nose using the other straw. The concoction puts the shaman into a deep sleep and the shouts he emits are interpreted by an assistant. In the past, the Yanomami who inhabit northern Roraima, Brazil, and contiguous forest in southern Venezuela, used to make sticks from cacao wood which they would rub between their hands into a wooden base to generate hot shavings to start fires while traveling away from their communal houses (Albert and Milliken 2009:90; Lizot 1978).

Cacao is also used in folk medicine throughout Amazonia. In the Colombian Amazon, an infusion of the leaves is taken as a heart tonic as well as a diuretic and the Ingano tribe employs cacao bark to make a decoction for washing a skin disorder known locally as *sarna* (Schultes and Raffauf 1990:447). In the Peruvian Amazon, the outer skin of the cacao bean is cooked to obtain a thick chocolate one cup of which is drunk three times a day to treat dry cough (Mejia and Rengifo 2000:50). Also in the Peruvian Amazon, approximately 30 g of the thick pod shell is boiled in 2 l of water and when it cools one cup of the liquid is drunk with meals and to reduce inflammation seven leaves are boiled in 1 l of water and when the solution cools it is used to wash the swollen area (Lossio et al. 1995:75). In Roraima in the northern part of the Brazilian Amazon, some rural folk prepare a drink from the beans to treat liver problems (Van den Berg and Silva 1988).

In the Amazon estuary and along the lower Tocantins, cacao has always been planted in the forest and in agroforestry systems rather than as a monoculture. In the mid- nineteenth century, for example, the English naturalist Alfred Russell Wallace noted that cacao was planted in partially cleared forest so that the remaining trees provided shade; no wonder that Charles Frederick Hartt, a geologist from Cornell University, described cacao along the Tocantins in the late 1800s as being "in its element." Along the lower Tocantins and at the mouth of the Amazon, cacao is still gathered on a small scale in cultural forests to prepare the beans for the market. One reason that cacao has thrived in those places is that it has always been grown in the forest alongside other economic plants and is therefore but one of a basket of plants used for income and subsistence.

Although cattle have been grazed on the floodplain of the middle Amazon for close to two centuries, the rapidly expanding market for beef in the region has accelerated the removal of remaining cacao plantings and floodplain forest. Low cacao prices since the late 1980s have also encouraged farmers to switch to other land uses, especially cattle ranching. Floodplain landscapes of the middle Amazon have thus witnessed the waxing and waning of many economic activities, from subsistence farming in pre-contact times, to cacao estates, then jute farms, and now vast expanses of cleared areas for cattle and water buffalo production.

The second boom in cacao planting occurred on the uplands in the 1970s and early 1980s, when the Brazilian government decided to open up the Amazon for settlement and development by bulldozing highways through the vast interfluvial rainforests. But not all of the thousands of settlers that poured into the Amazon along the Transamazon and other pioneer highways had access to credit to plant cacao. Only for those farmers with fertile alfisols (*terra roxa*) were eligible for financing and technical support to establish cacao plantings. Between 1976 and 1988, some 90,000 hectares or cacao were planted in the Amazon, mainly along the BR 364 highway and associated side roads in Rondônia and in the Altamira area of the Transamazon in Pará. In the early 1990s, cacao covered some 30,000 hectares within a 250-km radius of Altamira in the early 1990s, mostly concentrated between km 75 and km 140 of the Altamira-Uruará stretch of the highway and its side roads.

But the concentration of cacao in a few zones along pioneer highways only made it easier for witches' broom, caused by the pathogenic fungus *Crinipellis perniciosa*, to reach all the groves (Stewart 1994:81). Within a few years, many of the cacao groves along the Transamazon and in Rondônia became infected with the disease. When cacao prices are high, such as occurred in the early 1980s, the disease is managed by rigorous pruning of infected branches. But when prices dip, as has happened since the late 1980s the groves are neglected or converted to other land uses, such as cattle pasture or maize production.

Neglected cacao orchards nevertheless served a useful purpose: it taught many farmers the benefits of intercropping. While the cacao orchards were idled, farmers were no longer weeding. So seeds from forest timber and fruit trees germinated and flourished, thereby diversifying cacao orchards spontaneously. One farmer near Ouro Preto in Rondônia is so pleased with the timber trees, especially pinho cuiabano *Schizolobium parahyba* var. *amazonicum*, that have become established in his cacao orchard that he plans to save the remaining forest on his land as a future source of the seed of valuable timber species. A Transamazon colonist some 75 km west of Altamira has spared a mature ipê roxo *Tabebuia impetiginosa* tree when felling forest for his cacao plantation because it produces a highly durable timber. Now, numerous seedlings from the "mother" tree have become established in the orchard. Timber trees will eventually dominate that agroforestry plot as the cacao becomes senescent. Over a dozen timber species have established themselves in cacao groves in the Amazon.

The spontaneous establishment of timber species in cacao groves has underscored the importance of maintaining forest patches nearby. The connection between the number and diversity of valuable timber species in cacao and the proximity and size of the undisturbed forest has not escaped the attention of farmers. Many growers are now consciously protecting remaining tracts of forest on their property not only to provide a source of timber seeds for spontaneous regeneration in cacao but also to sell the seed to buyers. The buying and selling of timber seeds and seedlings is a growing business in the Amazon, and some growers allow collectors to comb through their woodlots for a price.

Some wild fruit trees have also become established in cacao orchards in the Amazon, thereby providing welcome snacks to families and workers. Yellow mombin *Spondias mombin*, used to make a refreshing juice, ingá (several species of *Inga*) which produces a sweet pulp surrounding its seeds, and guava *Psidium guajava*, have arisen spontaneously in cacao groves along the Transamazon. Some farmers get rid of guava when they turn up in their cacao groves because they say that guava has an aggressive root system that competes with cacao and also takes up a lot of water in addition to nutrients. Other fruit trees, both native and exotic, sometimes planted in cacao groves include açaí *Euterpe oleracea*, buriti *Mauritia flexuosa*, bacaba *Oenocarpus mapora*, peach palm *Bactris gasipaes*, coconut *Cocos nucifera*, biribá *Annona mucosa*, sweetsop *Annona squamosa*, conde *Annona glabra*, cupuaçu *Theobroma grandiflorum*, cupuí *Theobroma speciosum*, cashew *Anacardium occidentale*, juru *Chrysobalanus icaco*, piquiá *Caryocar villosum*, tuturuba *Pouteria macrophylla*, pineapple *Ananas comosus*, papaya *Carica papaya*, passionfruit *Passiflora edulis*, carambola *Averrhoa carambola*, avocado *Persea americana*, Barbados cherry *Malpighia emarginata*, banana, jackfruit *Artocarpus heterophyllus*, lime, tangerine, orange, and mango *Mangifera indica*. Sometimes cacao groves contain only one other fruit tree, other times as many as seventeen.

Brazil nut *Bertholletia excelsa* also arises spontaneously in some cacao orchards, and the trees start producing protein-rich nuts a decade or so after germinating. Several timber trees, such as ipê (two species of *Tabebuia*), tropical cedar *Cedrela odorata*, and mahogany *Swietenia macrophylla*, also arise spontaneously or are planted in some cacao orchards, such as in the municipality of Medicilandia along the Transamazon and in the vicinity of Tomé-Açu, a colony started by Japanese immigrants in the 1930s. Cacao is also sometimes intercropped with rubber *Hevea brasiliensis* and black pepper *Piper nigrum*, such as on small farms near Tomé-Açu in eastern Pará. At Sitio Ihumá near Machadinho in Rondônia a farmer has intercropped cacao with robusta coffee; that way he does not have all the eggs in one basket when it comes to generating income. Agroforestry rather than monocropping thus characterizes most cacao orchards in the Amazon.

Cacao is attacked by several diseases, depending on the region, but witches' broom is one of the most serious disease threats to cacao production. The pathogen that causes the disease, *Crinipellis perniciosa*, is native to the Amazon but the fungus does not cause catastrophic damage to wild trees because they are genetically diverse and tend to be dispersed through the forest. Only when monocultures of genetically similar cacao are planted in the cleared forest does the disease trigger massive outbreaks. That is why the Amazon has never been a major source of cacao. Virtually all of the chocolate consumed in towns and cities in the Amazon comes from cacao grown in other regions.

Cacao breeders have been aware of the perilous genetic foundation of the world's chocolate industry for a long time. In 1938, an agronomist from Trinidad's Department of Agriculture, F. J. Pound, set out to comb the western Amazon for sources of resistance to cacao diseases, especially witches' broom. Two of his collections, SCA 6 and SCA 8, from the Scavina farm near Iquitos in the Peruvian Amazon, proved resistant to witches' broom and have been bred into much of the cacao grown in areas where the disease occurs. Both selections produce beans that are too small for commercial production, so they were crossed with more desirable types. But new sources of resistance are needed because the pathogen has evolved into numerous races or pathotypes. Single gene resistance rarely holds up for long in all areas, and such is the case with Scavina and witches' broom. Scavina resistance has broken down in the Brazilian Amazon, Ecuador, and Trinidad. Even in Bahia, where the pathogen has only been around for a little over a dozen years, several new races have apparently evolved.

Witches' broom spread out of the Amazon over a century ago, putting major cacao-producing regions at risk. Witches' broom was reported in Surinam in 1895, Trinidad in 1928, the delta of the Orinoco in 1937, Grenada in 1948, and Panama in 1986 (Smith et al. 1992:58). Should witches' broom reach West Africa or Southeast Asia, the price of chocolate would likely soar; West Africa currently produces close to two-thirds of the world's supply of cacao. The arrival of witches' broom in Bahia, Brazil, in 1989, could be a harbinger of things to come in West Africa and Asia unless steps are taken to diversify the genetic base of cacao and the cropping systems in which it is planted.

Cacao was introduced to Bahia in the northeast of Brazil in the eighteenth century because the state had a hot, humid climate near the coast ideally suited to cacao, but without the host of diseases that plague the crop in the Amazon. Bahia is separated from the Amazon by a broad corridor of savanna, which is too dry to harbor hosts of witches' broom. Thus liberated from the intense disease pressure that characterizes cacao plantations in the Amazon, the crop flourished in Bahia. By 1870, Bahia was the leading producer of cacao in Brazil. By 1890, cacao accounted for about one-fifth of the state's revenue and after 1900, its proportion of the state's income oscillated between one-third and one-half. Cacao had become thoroughly entwined in Bahia's economy and culture, replacing sugarcane that dominated the state's economy during the colonial period. Cacao figures prominently in several novels of Jorge Amado, one of Brazil's better-known writers. Jorge Amado was born in Ilhéus, Bahia's cacao port, in 1912, and his first novel, *Cacau: Romance* had gone through 34 editions by 1980 (Amado 1980). It describes life, lore, and lust on a cacao estate in Bahia with an edgy, class struggle tone.

Then in 1989, disaster struck. Witches' broom was detected on some cacao estates in Bahia and escaped before it could be stamped out. Precisely how the pathogen reached Bahia is not known, but it did not disperse naturally. The strain of witches' broom introduced to Bahia has been traced to Rondônia; people either brought the pathogen with planting material or some spores hitched a ride on peoples' clothes or farm implements. By 2000, the combined effects of witches' broom and low cacao prices had shrunk cacao production in Bahia by 75%. In spite of the ravages caused by witches' broom, Bahia still accounts for 78% of the cacao planted in Brazil (Mendes 2014).

Forest conservation is essential for the long-term vibrancy of cacao and agroforestry in Bahia and other regions for several other reasons. Forests sometimes contain wild populations of many crops and tropical forests also contain a rich array of candidates for future crop domestication. Finally, they provide a habitat for pollinators, seed dispersal agents, and biocontrol agents. One farmer at km 75 of the Altamira-Uruará stretch of the Transamazon told me that army ants provide a valuable service by cleaning out insect pests in his cacao grove and he is aware that if he destroys the nearby forest, the habitat for the army ants will be eliminated. The army ants also attack scorpions which can sting workers when they are pickup up cacao pods from the ground.

Outside of a relatively small circle of plant breeders and specialists in plant genetic resources, few people appreciate that most of the genetic diversity of cacao is found in the forests of Amazonia. The entire foundation of the world's chocolate industry rests upon only a handful of cacao parents, a precarious footing for such a global enterprise. Annual chocolate sales in the Unite States alone exceed $3 billion. Wild populations of cacao are increasingly under threat from widespread deforestation (Silva et al. 2011) so the gene pool of this important crop plant is shrinking.

In order to sustain cacao production over the long haul, the crop needs to be diversified on two fronts: genetically, and in the way, it is cultivated. In many areas outside of the Amazon, cacao is a monocrop, which makes it particularly vulnerable to disease and pest attacks. Cacao is more resilient when interplanted with other trees and bushes. Polyculture and upgrading the genetic resistance of cacao hinge on forest conservation, because the forests of Amazonia contain most of the genetic diversity of cacao.

32.9 *Theobroma grandiflorum*

Bolivia Copuasú, copoazú, cupuazú

Brazil Cupuaçu, cupuaçuzeiro, cupuí, cupú, pupu; kupi-'í, kopi'i (Araweté), kopowaso, mawildoa makadalipe (Baniwa), kɨ-pɨ-'ɨ, ako'o'ɨ (Guajá), kupihu'y, kupyhu'ytawa, kɨpɨhu'ɨ (Ka'apor), bàri-djô, ñejaka (Kayapó), ñumitahɨm (Parintintin), jumitahýma (Piripkura), kupu (Sateré-Maué), kupy, kupi'a'ɨw (Tembé), wahpékara, wahpékara sawë (Tukano), aka' (Waimiri-Atroari), kapɨai (Wajãpi)

Colombia Copuasú, copoazú, copoasú, cupuassú, maraca; taraira (Barasana), bayañaju (Makuna), us-per' (Piratapuya), ñee-au (Tanimuka), kôpú (Ticuna)

Ecuador Cola blanco, copoasu, copoazú

Peru Copoasú, copoazú, cupasu, sacha cacao; kemito, shamomoyaki (Campa)

Venezuela Copoazú, cupuazú; mamakuke (Ye'kuana)

Status Wild, spontaneous in cultural settings, planted

Maria, 9 years, by a cupuaçu *Theobroma grandiflorum* fruit in her home garden. Urubu River near Caretas, Amazonas, Brazil, 10-16-12

Like cacao, cupuaçu *Theobroma grandiflorum* is an understory tree but with lighter colored leaves and is native to the upland forests of eastern Amazonia, rather than the western part of the basin. It is mostly confined to uplands, whereas wild cacao grows on both floodplains and well-drained sites. However, cupuaçu is found occasionally in home gardens on floodplains, such as along the lower Solimões River and Careiro Island along the Amazon just after its confluence with the Solimões near Manaus (Castro et al. 2009; Lima and Saragoussi 2000), in floodplain forests along the lower Xingu (Salomão et al. 2007a), and in the tidal forest on Combu Island at the mouth of Guamá River near Belém (Cattanio et al. 2002). It is possible that the cupuaçu trees found in the floodplain and estuarine forests are a result of past human agency. In the wild, cupuaçu trees can grow as high as 16 m as they compete for light, but in plantations and home gardens, they are more compact and the fruits are harvested by hand. Trees can produce commercial quantities of fruit for as long as two decades (Yamada and Gholz 2002). Cupuaçu flowers during the tail end of the rainy season and into the dry period and they are pollinated by several species of bees and fruits mature during the rainy period (Falcão and Lleras 1983). The following native stingless bees are known to pollinate this tree in the state of Pará: abelha mosquito *Plebeia minima*, jataí *Tetragonisca angustula*, *Trigona fluviventris*, and *Aparatrigona impuctata* (Maués et al. 2000:12, 13) and several species of *Trigona* bees also pollinate this tree in the neighboring state of Amazonas (Rech and Absy 2011). A small species of beetle, *Antityphona thoa*, is also known to pollinate the flowers of this tree (Venturieri et al. 1997).

Wild cupuaçu is particularly common in the Tocantins watershed with outlying populations found as far west as the plateau overlooking the majestic Tapajós River as well as east of the Tocantins in the forests of northeastern Maranhão (Ducke 1946). The Tocantins is the most developed tributary of the Amazon, which means the valley has undergone the most deforestation. Numerous wild cupuaçu trees have succumbed to clearing for cattle ranches or farms, have been crushed by logging crews after valuable hardwoods, or have rotted under the placid waters of the reservoir that formed in 1984 when the Tucuruí dam closed. Wild cupuaçu is widely dispersed in the forest, but the gene pool is shrinking.

Cupuaçu is not generally found wild north of the Amazon River, apparently because the river is too broad for natural seed dispersal. Wild cupuaçu has been reported from non-flooded sites in the Venezuelan state of Amazonas (Rondón and Cumana, 2005). As in the case of some reports of "wild" cacao, it is possible that wild cupuaçu in Venezuela is actually feral.

Rodrigo, 10 years, with opened cupuaçu *Theobroma grandiflorum* pod. Carrazedo, mouth of the Xingu River, Município de Gurupá, Pará, Brazil 4-21-15

According to a woodsman (*mateiro*) who lives in Tucuruí on the lower Tocantins, capuchin monkeys (*Cebus* spp.) disperse cupuaçu seeds when they smash open the fruit pods to access the creamy pulp; the seeds are discarded, sometimes at some distance from the tree. Agoutis (*Dasyprocta* spp.) apparently bury some of the seeds that have been discarded by monkeys or obtained from ripe fruits that have fallen to the ground. Paca *Agouti paca*, a highly prized game animal, is apparently fond of cupuaçu pulp when the ripe fruits fall to the ground, but it is unlikely they bury any of the seeds. Rural folk in the vicinity of Altamira along the Xingu River wait at night by cupuaçu trees in the forest to shoot paca, tapir *Tapirus terrestris,* and brocket deer *Mazama americana* and *M. gouazoubira* (Moran 1979). The red-footed tortoise *Chelonoidis carbonaria*, the yellow-footed tortoise *C. denticulata*, and the twist-neck turtle *Platemys platycephala* also eat these fruits and rural people gather up these reptiles in Amazonian upland forests for their meat (Ghilardi and Alho 1990). The twist-neck turtle lives along shallow streams but wanders into upland forest in search of fallen fruit.

Game is not alone in being partial to the seductive flavor of cupuaçu pulp. Long before Europeans arrived, indigenous folk gathered the fruits in the forest and brought them back to their camps and villages. Some of the discarded seeds sprouted, thereby enriching the vegetation surrounding their abodes.

Adenora, 35 years, separating cupuaçu *Theobroma grandiflorum* **pulp from the seeds at her home. She will use the pulp to make juice. Carrazedo, mouth of the Xingu River, Município de Gurupá, Pará, Brazil 4-19-15**

Cupuaçu pods encase seeds covered with a delicious, sticky pulp. Although the pulp has little nutritional value, its rich, tangy overtones quickly win converts. When in season during the wetter months, women break open the velvety pods and snip off the pulp with scissors. Sugar and water are then added to make a refreshing drink. To save time when making drinks, the pale cream pulp and attached seeds are sometimes stirred vigorously in water. The remaining seeds with some of the pulp still attached are then discarded after finishing the drink.

Cupuaçu is also used in some folk remedies. In the Caxiuanã National Forest, river dwellers use cupuaçu sap to treat wounds (Lisboa et al. 2002). Residents in Jocojó, a quilombo community in the Municipality of Gurupá in the lower Amazon, soak cupuaçu bark in water for a few hours before drinking the solution to stem diarrhea. In the vicinity of Manacapuru along the lower Solimões near its confluence with the Rio Negro, rural folk use the bark to make a tea to treat hemorrhoids and gastritis (Vásquez 2014:18; Vásquez et al. 2014). In various parts of Rondônia, cupuaçu leaves are used to make an infusion to lower blood pressure (Santos et al. 2008, 2014a, b, c) while in Rio Branco, Acre, this same infusion is used to lower cholesterol levels (Siveiro et al. 2012). The Ticuna who live along the Upper Solimões and some of its affluents as well as contiguous parts of the Colombian Amazon make a tea from the crushed seeds to relieve abdominal pain (Schultes and Raffauf 1990:448). Juice is squeezed from the leaves to treat bronchitis and kidney infections (Stasi et al. 1989).

Kailani, 6 years, holding a round cupuaçu *Theobroma grandiflorum* fruit gathered in her home garden. Laranjal, Igarapé Laranjal, Baia de Caxiuanã, Município de Melgaço, Pará, Brazil 5-8-15

Unlike cacao pods which sport a variety of colors and textures, cupuaçu pods are always brown and are covered by fine hairs. This pubescence has evolved to help protect the fruits against insect attacks. But as in the case of cacao, the shape of pods can vary. Cupuaçu pods range from round to oblong. Cupuaçu fruits are between 12 and 25 cm long and 10 and 12 cm in diameter and can weigh as much as 2 kg (Cavalcante 2010:103; Rogez et al. 2004). Also, cupuaçu fruits are born along branches and not along the trunk, as is the case with cacao.

Max, 6 years, by a cupuaçu *Theobroma grandiflorum* tree in his home garden. Silves, Amazonas, Brazil 4-2-15

Cupuaçu ice cream is a perennial favorite in the cities of the Brazilian Amazon and the pulp also flavors some locally produced yogurts and frozen pushups, known as *chopp* or *dindin*. The Shambo factory in Iquitos also produces copoazú ice cream for local ice cream parlors, restaurants, and food stores. Frozen pushups are a cottage industry throughout Amazonia and many other tropical parts of Latin America and are typically sold by youngsters, especially boys. Mothers purchase small, tubular plastic bags and fill them with juices which are then placed in freezers in their kitchens. The women then send out some of their children, or those of neighbors or relatives, to sell the pushups. Although not expensive, sales from pushups are valuable supplements to family income in poorer neighborhoods. Children in the Amazon typically attend school either in the morning, afternoon, or night, so they have time to do chores.

A chilled dessert (*creme de cupuaçu*) made with cupuaçu pulp, condensed milk, and cream is always a favorite at festive meals, such as birthdays and Christmas. *Creme de Cupuaçu* may be a waistline expander and a no-no for those afflicted with elevated levels of "bad" cholesterol, nevertheless it is one of the most tempting Amazonian desserts. The artery clogging recipe calls for mixing cupuaçu pulp, one can of condensed milk, and two cans of cream in a blender, then pouring the pasty mixture onto a bed of cookies in a glass bowl or dish and placing it in the freezer. After 3 hours in the freezer, the dessert is placed in a refrigerator until ready to serve. Some *creme de cupuaçu* recipes include spices or flavorings, such as nutmeg or vanilla.

Cupuaçu jam (*geleia*) is also a cottage industry and typically sold in glass containers in stores. Cupuaçu jam is not spread on bread or toast as is the case with most fruit preserves in Europe and North America; rather it is employed as a filling in cakes (*bolo de cupuaçu*) and to fill chocolate-covered candies (*bombom de cupuaçu*). Cupuaçu-filled chocolates, almost as big as chicken eggs, are another small-scale industry in Belém and its vicinity, and are made to order or can be purchased in certain stores and gift shops, especially those catering to tourists.

When cupuaçu was domesticated is not known, but some indigenous groups in eastern Amazonia plant cupuaçu (Balée and Moore 1991; Posey 1985). Cupuaçu has likely been cultivated by several indigenous peoples in the eastern part of Amazonia for a long time. The fame of cupuaçu apparently spread, and indigenous groups traded seeds or seedlings. In remote times, cupuaçu was taken up the Rio Negro and into the Upper Orinoco, a similar route taken by cacao and guaraná *Paullinia cupana*. Known as copoazú in Venezuela, it is cultivated in backyards in San Carlos along the upper reaches of the Negro (Sanchez et al. 1988). Only unlike cacao, cupuaçu never reached Central America in pre-contact times. Cupuaçu has been in the Upper Rio Negro and Orinoco long enough to be incorporated into the healing practices of indigenous folk. Desana shamans, for example, bless cupuaçu juice to facilitate women experiencing difficult births; the idea here is that the viscous, slippery pulp of cupuaçu will lubricate the birth canal (Buchillet 1988).

Cupuaçu was also taken up the Amazon beyond its juncture with the Negro in pre-contact times. The Ticuna who inhabit the banks of the Amazon near the border between Brazil and Colombia, for example, plant cupuaçu in managed fallows. Cupuaçu is cultivated along the Caquetá and Uaupés Rivers in the Colombian Amazon (Allen 1987; Baker et al. 1953). Cupuaçu is also planted as far west as the lower Ucayali in Peru. In the Peruvian Amazon, the novelty fruit is known as copoazú, and it appears to be catching on. The fruits occasionally turn up in markets, and several hotels and stores are named Copoazú in Iquitos and Requena.

Although not native to wetlands, cupuaçu thrives in home gardens on the higher portions of floodplains, such as on the banks of the Amazon in the vicinity of Santarém and along some of its tributaries, including the Maracá in Amapá. Some ribeirinhos along the lower Madeira River in Amazonas plant this tree in their home gardens as do rural folk in the tidal zone near Belém (Anderson 1991; Kawa et al. 2015). Backyard gardens are repositories for much genetic diversity of cupuaçu, especially so in the light of massive deforestation of wild gene pools in eastern Amazonia. One home garden along the Xingu River near Altamira, for example, had a cupuaçu tree bearing fruit in July, well after the normal fruiting season. Cupuaçu fruits fetch between 50 cents and two dollars in urban markets, depending on the time of the year and the size of the fruit. Prices tend to be higher at the beginning of the season.

Cupuaçu *Theobroma grandiflorum* **planted with other fruit crops including banana, açaí** *Euterpe oleracea* **palm, and pineapple in a home garden on a riverbank. Rio Limoeiro, an affluent of the lower Tocantins, Pará, Brazil 7-5-07**

In a survey of 150 home gardens in the Brazilian Amazon conducted over several decades, I found cupuaçu in 37% of them. In a separate inventory of 57 home gardens along the middle Xingu in 2015, cupuaçu was present in 42% of them. In a survey of 40 home gardens along the lower Urubu River in Amazonas, Brazil, cupuaçu was found in 65% of them (Lins et al. 2015). Cupuaçu is so common in the home gardens of one community that it is named after the fruit: Cuputeua (the place of cupu). Cuputeua is on an archaeological site with terra preta along the Rio Uruaí, in the watershed of the Ipixuna River, an affluent of the lower Xingu. Cupuaçu is also planted occasionally in the vicinity of manioc processing huts, as near Carrazedo at the mouth of the Xingu.

Cupuaçu *Theobroma grandiflorum* seedling planted by an açaí *Euterpe oleracea* palm in a home garden. Joanderson 10 years. Jocojó, Rio Jocojó, Município de Gurupá, Pará, Brazil 8-4-16

Cupuaçu is a popular dooryard tree not only in Pará where it is native but also in Amazonas, Rondônia, Acre, and Amapá. In the Caxiuanã National Forest and along the middle Xingu, some of the cupuaçu in home gardens established by river dwellers arose spontaneously from seeds tossed after removing the pulp from fruits obtained in the forest. In other home gardens, the seeds of wild cupuaçu were deliberately planted. The gene pool of cultivated cupuaçu is thereby enriched in eastern Amazonia.

For a long time, cupuaçu lingered as a backyard tree. But starting in the 1970s, cupuaçu emerged as a valuable cash crop for some farmers in the Brazilian Amazon with good road links to urban centers. A few growers cultivate cupuaçu as a monocrop, such as the 100-hectare orchard on Fazenda Buriti near Castanhal in Pará, but most commercial cupuaçu growers are smallholders who usually grow cupuaçu in mixed cropping systems, such as in the vicinity of Igarapé-Açu in the Zona Bragantina area of Pará (Vieira et al. 2007). Farmers have learned that intercropping reduces disease pressure and helps spread income throughout the year.

Manioc interplanted with cupuaçu *Theobroma grandiflorum* and banana so that the field will remain productive after the root crop is harvested. Igarapé Ubim, Lago Amanã, Amazonas, Brazil 9-17-12

Cupuaçu is grown in many different configurations in agroforestry fields. More than two dozen crops can be found in some agroforestry plots involving cupuaçu. Most cupuaçu orchards contain at least four other intercropped perennials, thus forming rich pockets of agrobiodiversity on farmlands. Bananas, orange, black pepper, robusta coffee, mango, coconut, and soursop *Annona muricata* are some of the fruit, spice, and beverage crops interplanted with cupuaçu. Papaya and guava often sprout spontaneously from seeds dropped by birds.

Cupuaçu ***Theobroma grandiflorum*** **in an orchard interplanted with peach palm** ***Bactris gasipaes*** **and açaí** ***Euterpe oleracea*****. Outskirts of Gurupá-Mirim, Município de Gurupá, Pará, Brazil 4-25-15**

One farmer in the vicinity of Agrovila Nova Fronteira at km 80 of the Altamira-Medicilandia stretch of the Transamazon has interplanted mahogany *Swietenia macrophylla* in one of his agroforestry plots that contain cupuaçu, passionfruit, robusta coffee, orange, lime, and tangerine. In that way, the field will generate income for several decades even as some of the perennial crops become senescent and no longer productive. Similarly, a Japanese-Brazilian farmer near Tomé-Açu has planted two timber species, macaúba *Acrocomia aculeata* and mahogany, in his cupuaçu orchard. Freijó *Cordia alliodora* is another timber tree planted in some cupuaçu orchards, such as in the vicinity of Tomé-Açu. Manioc generally produces from 6 to 24 months after planting; thereafter the field is abandoned to successionary vegetation or planted with perennials, such as cupuaçu.

Cupuaçu *Theobroma grandiflorum* **interplanted with açaí** *Euterpe oleracea* **palms in a former manioc field. Boa Esperança, Lago Amanã, Amazonas Brazil, 9-18-12**

Cupuaçu fruits are highly perishable and demand for the fresh fruit in markets is easily saturated, so much of the pulp is frozen. The greatest profits are realized during the dry season, provided that sufficient freezer space is available. But few farmers have freezers, and even freezer space at cooperatives is sometimes overwhelmed before the harvest is over, as happened with some communities along the Rio Branco-Porto Velho highway in Acre in the mid-1990s. Large, well-organized cooperatives, such as CAMTA (Cooperative Agrícola Mixta de Tomé-Açu) operated by Japanese-Brazilian farmers in the vicinity of Tomé-Açu, have invested in a fruit processing plant so that spoilage is minimal. Along the Transamazon, a local company, DICACAU, began processing cupuaçu pulp at its plants in Altamira and km 140 of the Altamira-Uruará stretch of the highway in 1996.

Two hundred and fifty years ago, a Jesuit missionary, Father João Daniel, pondered why no one had made chocolate from cupuaçu seeds (Daniel 1976). Only in the latter part of the twentieth century were experiments begun on how to make *cupulate* from cupuaçu. One company in Manaus, Chocam, marketed cupuaçu "chocolate" in powdered form for a few years in the 1990s before going out of business. One problem is that candy made from the fermented seeds of cupuaçu cannot be legally called chocolate. Brazilians have called the product *cupulate*; whether that name would attract curious consumers outside of the Amazon is unclear. Some rural folk living along parts of the Solimões, Madeira, and Tocantins Rivers used to make *cupulate,* but this practice has largely disappeared (Venturieri and Aguiar 1988). I have not seen any *cupulate* in stores in the Brazilian Amazon for many years.

Efforts to launch cupuaçu juice in North America have met limited success, in part because it is expensive to import frozen pulp. The Knudsen Company in California marketed a cupuaçu beverage for a few years in the late 1980s, but soon withdrew the product. Knudsen's cupuaçu drink had little resemblance to that derived from pure cupuaçu pulp, either fresh or frozen. In the mid-1990s, Snapple launched *Samoan Splash*, a blend of orange juice, strawberries, and cupuaçu, none of which have anything to do with cupuaçu. As in the case of the Knudsen product, *Samoan Splash* only lasted a few years, possibly because a dried extract of cupuaçu pulp was used, rather than frozen pulp. The now defunct Life Plus Company marketed *Jungle Ade* online for a brief period in the 1990s: like the other products containing cupuaçu, it was a blend, this one consisting of camu camu *Myrciaria dubia*, passionfruit, cashew fruit, guava, and cupuaçu. An 8-ounce serving of *Jungle Ade* was priced at $16.

Cupuaçu will probably never become popular in North America, Europe, or Southeast Asia until the juice is available as a frozen concentrate in supermarkets, or is sold in chilled containers as a stand-alone juice reconstituted from frozen pulp rather than powder. Cupuaçu has a distinctive flowery aroma, an attribute that could be used in promoting ice cream as well as juice. Currently, orange, lemon, apple, grape, and cranberry dominate the concentrated frozen juice section of supermarkets in the United States. Only in markets with a lot of ethnic diversity and purchasing power, such as Honolulu, Miami, or New York, does one find frozen concentrates of tropical fruits, such as passionfruit or guava. Cupuaçu might make a sustained "breakthrough" in the US or European markets first via smoothies, rather than as frozen pulp in supermarkets. Customers are generally unwilling to buy a fruit or frozen pulp without some prior exposure.

Cupuaçu *Theobroma grandiflorum* **infected with witches' broom** *Moniliophthora perniciosa*, **known locally as vassoura de bruxa. Gleba 23, lote 2, km 75 Altamira-Medicilandia, Transamazon highway, Pará, Brazil 9-7-15**

In agroforestry systems, cupuaçu is not plagued with excessive diseases or pests, but should plantings become more extensive, especially in monocrop stands, disease and pest resistance will become an issue. Like cacao, cupuaçu is susceptible to witches' broom caused by the fungus *Moniliophthora perniciosa*, so those crops should not be planted together. Thus far, little systematic screening has been done with cupuaçu for resistance to diseases or pests.

Another issue for the future development of cupuaçu as a major commercial crop is the labor cost in removing the pulp from the seeds. At present, the pulp is removed by hand, using scissors. Although labor costs in the Brazilian Amazon are lower than those in North America and Europe, they are rising. Mechanical removal of the pulp can damage the seeds, thus altering the taste of the pulp. Some clones of cupuaçu have been located that do not produce seeds, but yields are low.

As with most fruits in the Amazon, cupuaçu bears in the rainy season to facilitate seed germination. Most tropical fruits have recalcitrant seeds so they quickly lose their viability and cannot be dried and frozen in seed gene banks. In order to avoid harvesting bottlenecks, highly productive clones are needed that reach peak harvest time in different months during the rainy season. A few trees even produce fruit at the onset of the dry season. A mosaic of clones with different fruiting periods could thus be planted, as has happened with the avocado industry in Florida.

32.10 *Theobroma obovatum*

Bolivia Chocolatillo macho

Brazil Cabeça de urubu, cupuarana, cupurana, cupuaçu bravo, cupuaçu da mata, cupuí, cupuí bravo, cacaurana, cacau de macaco, urubuacanga, urubuí

Colombia Cacao silvestre, cupu pequeño, copoazú pequeño; ígi (Nukak), win-cheek' (Puinave)

Peru Cacahuillo, ushapa cacao, cacao del monte, cacao colorado, sacha cacao, sacha cacahuillo, huamanzamana; me chocotu ñi, me sino ñi (Maijuna)

Status Wild

Theobroma obovatum **fruit, 5 cm tall by 4 cm wide. Sitio São Miguel, Ressaca do Cacauzinho, Lago Amanã, Amazonas, Brazil 9-5-12**

Theobroma obovatum is found in the central and western parts of Amazonia from Bolivia north to Peru, Ecuador, and Colombia, including the states of Amazonas,

Acre, Rondônia, and Mato Grosso in Brazil. Confined to lowlands below 400 m, the tree is also found in the Amazonian portion of southern Venezuela (Ducke 1953; Rondón and Cumana 2005).

A relative of cacao, fruits of *Theobroma obovatum* are gathered during the rainy season for the sweet, creamy pulp that surrounds the seeds. Known as cacahuillo in the Peruvian Amazon, *Theobroma obovatum* is a small understory tree, usually under 25 m tall, in forests on uplands and along the banks of rivers and streams. It occurs on the higher parts of floodplains along black, clear, and sediment-rich rivers, such as along the creamed coffee-colored Caquetá in the Colombian Amazon (Duivenvoorden and Lips 1995:49) and the Tefé River in Amazonas, Brazil (Assis et al. 2017). *Theobroma obovatum* occurs in both mature and successionary forests, including light gaps in upland forest such as in Acre (Maranho and Salimon 2015). The slender tree produces fruits when only 4 m high tall so they can be gathered from the lower branches and youngsters can easily climb into the branches.

Theobroma obovatum **with gray fruits that will turn yellow when ripe. The texture of the skin is reminiscent of the head and neck of the black vulture** ***Coragyps atratus*****, hence one of the common names for the fruit in Brazil: cabeça de urubu. Regrowth forest on the outskirts of Vila Valente, lower Tefé River, Amazonas, Brazil 8-15-12**

Unlike cacao *Theobroma cacao*, cacahuillo fruits are born on the ends of branches rather than along the trunk and main branches. The globular fruits, some 5–11 cm long by 3–8 cm wide, have a thick, leathery yellow skin that is broken open with the fingers to reveal the thin pulp which surrounds about 15 seeds. The slippery pulp has an intriguing flavor because it is exceptionally sweet yet with a hint of tartness. People suck the thin layers of pulp from the seeds, which are the size of plump currants, and then eject the "beans." Cacahuillo fruits are mostly eaten in the forest or brought home for snacks. However, they occasionally turn up in urban markets, such as Iquitos (Vásquez and Gentry 1989). In upland forest along Quebrado Blanco, a tributary stream of the Rio Tahuayo which flows into the Amazon River in the Peruvian Amazon, Geoffroy's saddleback tamarins *Saguinus fuscicollis nigrifrons* have been observed eating the pulp of this fruit and likely disperse some of the seeds since they are ingested (Sánchez and Levy 2013:30, 76) as is the case with white-handed titi monkeys *Callicebus lugens* in the Colombian Amazon (Alvarez and Heymann 2012).

Bildoso, 8 years, with cacahuillo *Theobroma obovatum* fruit he gathered in a forest. Near Veinte de Enero, Rio Yanayacu, Reserva Nacional Pacaya-Samiria, Loreto, Peru 3-8-04

Cacahuillo is susceptible to witches' broom (Baker et al. 1953), so if the tree is ever domesticated it will have to be planted in diverse agroforestry systems to reduce disease pressure, or highly resistant material will have to be located if it is to be planted in monocrop orchards. *Theobroma obovatum* is found in some home gardens of the Yagua ethnic group along the Amazon downstream from Iquitos (Lamont et al. 1999), but it is not clear whether those trees were planted or arose from discarded seeds. This relative of cacao thus appears to be on the threshold of domestication.

32.11 *Theobroma speciosum*

Bolivia Chocolate, chocolatillo, cacao de macaco, cacao silvestre; nohotë (Chácobo)

Brazil Cupuí, cacauú, cacauí, cacaúba, cacau da mata, cacau do mato, cacau de macaco, cacau de mico, cacau bravo, cacau jacaré, cacau folha miúda, cacau amarelo, cacau azul, cacau coroa, cacauhy, cacaurana; aka'á-wi-'i (Arawaté), aka-'ɨwa (Asurini), ako'o-'ɨ (Guajá), kaka-'iran-'ɨ (Ka'apor), kubenkrã, Kubenkrà (Kayapó), kupu'ía (Piripkura), akôhba (Suruí), aka'u-'ɨw (Tembé), aka-'ɨ-w (Wajãpi)

Colombia kú (Ticuna)

Ecuador Cacao silvestre; chukochok (Achuar), challua cacao (Kichwa), potaque poma (Siona-Secoya), tubenacahue, tubenaca (Waorani),

Peru Cacao de monte, cacao chancho, cacahuillo, sacha cacao, macambo; tájwahe (Bora), tangaroki (Matsigenka), makampo (Shipibo)

Status Wild, planted

Hilane, 8 years, with cacaurana *Theobroma speciosum* fruit she picked in regrowth forest close to her rural home. Boa Esperança near Itacoatiara, Amazonas, Brazil 12-9-06

Another wild relative of cacao widely appreciated for its fruits is the diminutive *Theobroma speciosum*, known by various names in the Brazilian Amazon including the small cacao (cacauí), like cacao (cacaurana), cacao of the forest (cacau da mata), and wild cacao (cacau bravo). *Theobroma speciosum* has a slender trunk that reaches up to 20 m in the understory of upland forest in Amazonia (Amaral et al. 2000). The small deep red flowers occur in clusters along the trunk and are pollinated by several species of flies in the genus *Drosophila* which are active from dawn until dusk (Silva and Martins 2009). If deforestation continues unchecked in the Amazon these pollinators, as well as others, are likely to be drastically reduced or even disappear.

Theobroma speciosum **in flower in a cacao grove. Gleba 23, lote 2, km 75 Altamira-Medicilandia, Transamazon Highway, Pará, Brazil 9-7-15**

Theobroma speciosum occurs widely in the Amazon and Orinoco Basins, the Guianas, and into Central America. It occurs in upland forests, such as the Noel Kempff National Park and in the Department of Beni in Bolivia (Boom 1986; Wallace et al. 1998), in Rondônia (Lisboa et al. 1991; Salomão and Lisboa 1988), in northern Mato Grosso (Perigolo et al. 2017), in the Tapajós National Forest in western Pará (Carvalho 2002; Espírito-Santo et al. 2005; Silva et al. 1985), is relatively common in upland forests along the middle and lower Xingu River in Pará (Lemos et al. 2015; Salomão et al. 2007a), in the Serra dos Carajás in southeastern Pará (Salomão and Rosa 1989), in the forested vicinity of the Caxiuanã Scientific Station in eastern Pará (Almeida et al. 1993), and in the vicinity of Belém (Pires and Salomão 2000) and Manaus (Scudeller et al. 2009). It is also found in the forest-cerrado transition zone in southern Pará (Salm 2004).

Although confined mainly to uplands, cacauí is also found occasionally in floodplain forests, such as along the Purús, in southern and western Rondônia, along the lower Xingu, the Parauapebas River, the Amazon estuary, the Ucayali in Peru, and the Madre de Dios watershed and in Parque Nacional Madidi in the Bolivian Amazon (Amaral et al. 2009; Araujo-Murakami et al. 2012; Cain et al. 1956; Casula 2012:69; Freitas et al. 2019; Huber 1906a, b, 1910; Miranda 2000; Navarro and Maldonado 2002:69; Salomão et al. 2007a, b; Silva et al. 2017a, b). It is also found in floodplain forests, such as in the tidal zone at the mouth of the Amazon (Cattanio et al. 2002). It thrives in forest as well as successionary vegetation (Balée 1993b; Balée and Campbell 1990; Carvalho et al. 1986; Gualberto et al. 2014; Lisboa 1989; Macedo 2014:56; Rodrigues 1986; Silva 2015:63; Taurino 2017:80; Uhl et al.

1988). It is also found in bamboo vegetation dominated by *Guadua sarcocarpa* and *G. weberbaueri* in the southeast Peruvian Amazon (Griscom et al. 2007). In the Brazilian Amazon, *Theobroma speciosum* flowers in the dry season and produces fruit from the end of the dry period and into the early part of the rainy season.

These fruits evolved to encourage animals to disperse the seeds, such as the Santarém marmoset *Callithrix humeralifera*, the saddleback tamarin *Saguinus fuscicollis*, and the Ka'apor capuchin monkey *Cebus kaapori* (Oliveira et al. 2014; Rylands 1979; Stevenson and Rylands 1988; Yoneda 1984). Other monkeys observed feeding on this fruit in the Brazilian Amazon include marmosets *Callithrix humeralifer intermedius*, capuchin monkeys *Cebus apella*, and spider monkeys *Ateles paniscus* (Rylands 1981).

Rodrigo, 10 years, gathering cacauí *Theobroma speciosum* fruit in regrowth forest at the edge of an archaeological site with anthropogenic dark earth. Outskirts of Carrazedo, mouth of the Xingu River, Município de Gurupá, Pará, Brazil 4-14-15

The dimpled orange fruits are born on the trunk and are relatively easy to pick because the tree rarely surpasses 15 m in height and many of the fruits are close to the ground (Silva 2011a, b:104). Many fruits can thus be picked while standing, whereas others can be reached by shimmying up the trunk. Locals break open the grapefruit-sized fruits to relish the sweet, white pulp surround the seeds. The round to elliptical fruits are 9–11.5 cm long by 7–8 cm in diameter and contain 20–40 light tan seeds measuring 2.5 cm by 1.5 cm. This relative of cacao fruits almost year-round, such as in Rondônia (Rodrigues et al. 2007). Only rural folk have the privilege of savoring cacaui fruits for they never enter markets. Among the Ka'apor in the eastern fringes of Amazonia, males and postmenopausal women relish the fruit but younger women avoid eating it because of concerns that it might provoke excessive menstruation (Balée 1993a). Over 60 years ago, the French agronomist Paul Le Cointe (1934:91) noted that cacauí seeds produce an excellent "chocolate," but no entrepreneur seems to have followed up on that intriguing observation.

Wadrio, 9 years, eating cacauí *Theobroma speciosum* fruit in regrowth forest at the edge of an archaeological site with anthropogenic dark earth. Outskirts of Carrazedo, mouth of the Xingu River, Município de Gurupá, Pará, Brazil 4-14-15

In some parts of its extensive range, *Theobroma speciosum* is cultivated, as noted over a century ago by the Baron Santa-Anna Nery (1885), a Brazilian historian and leading intellectual during the rubber boom. Although not common in home gardens, *Theobroma speciosum* is sometimes planted in dooryards, such as in the Municipality of Santarém, along the Xingu River, and along the Pracupi and Cariá River in the Caxiuanã National Forest in eastern Pará. Some Japanese-Brazilian farmers in the vicinity of Tomé-Açu, Pará, also plant this small tree in their home gardens (Yamada and Osaqui 2006) as do some rural folk along the Ramal do Pau Rosa, a side road leading from the BR 174 Manaus-Boa Vista highway in Amazonas (Machado 2016:35), in the vicinity of Manaus (Guillaumet et al. 1990), in Itacoã an Afro-Brazilian community in the Rio Guamá watershed near Belém (Scoles 2009), and in the vicinity of São Domingos do Araguaia in southeastern Pará (Pereira et al. 2010). This relative of cacao is also found in some home gardens in Manaus (Barbosa 2018:43) and along the lower Madeira River, Amazonas (Kawa et al. 2015). In the vicinity of Agrovila Nova Fronteira at km 80 of the Altamira-Medicilandia stretch of the Transamazon Highway, several *Theobroma speciosum* have sprouted spontaneously in a cacao grove and the farmer has spared them so that workers can snack on the fruits. Even when farmers cut down cacauí when clearing vegetation to plant crops, they often resprout such as in the Xingu watershed.

***Theobroma speciosum*, known locally as cacau bravo, reprouting at the edge of a cultivated field (*roça*) that was cleared in regrowth forest containing babaçu *Attalea speciosa* palms. The cacau bravo was cut down when the successionary vegetation was cleared but it has resprouted. About 3 km inland from the village of Terrawangã, Terra Indígena Arara de Volta Grande, Município Senador José Porfírio, Pará, Brazil 9-10-15**

In the eastern fringes of Amazonia, the Ka'apor deliberately swallow the seeds and defecate them, thus increasing the density of this useful fruit tree in their managed fallows (Balée 1989a, b; Balée and Gély 1989). Indeed, the tree seems to be more common on disturbed sites than in mature woods and is considered an indicator of cultural forests (Balée 2010). Some indigenous people in the Amazon plant this tree in their fields and home gardens, such as the Achuar in the Ecuadorian Amazon (Cerón 1991). In the Bolivian Amazon, the Chácobo also cultivate this tree (Boom 1989), while the Tacana mash the leaves of this tree and apply them to the forehead to relieve headaches (Bourdy et al. 2000).

32.12 *Theobroma subincanum*

Bolivia Chocolatillo de gajo, chocolatillo de copa, chocolate del monte, cacao silvestre

Brazil Cupuí, cupuí do igapó, cupuí da mata, cupuaí, cupurana, cupu da mata, cupu do mato, cupu brabo, cupuzinho, cupuaçu do mato, cacau cupu, cacau peludo, cacauí, kara; mawiloda (Baniwa), shina (Jamandi), kɨpɨ'a'ɨ (Ka'apor), wahpékara, pohpe kara (Tukano), tusu (Tuyuka), aka' (Waimiri-Atroari), waiporo unahi (Yanomami)

Colombia Cacao de monte, cacao de mico, maraca de monte, macambo, copoazú de monte; bauk (Maku), abe-ka-rá (Makuna), buué-aae (Miraña), ígi baká (Nukak)

Ecuador Cacao de monte, cacao de mono, cacao de venado, sacha cacao, cacao rojo, wakáme, tepenca; wakamp (Achuar), shan'cco coquio'cho (Cofán), cushillo cambiaj, cambi, sacha cambia, sacha cacao, chonta cacao, puca cacao, puka kambi (Kichwa), wakáme, wakam, wakamp (Shuar), tsie (Siona-Secoya), tobenaca, tepënka, tepencahue (Waorani)

Peru Macambillo, cacahuillo, sacha cacao, cacao del monte, cacao blanco, cacao colorado, cacao amarillo, machinga; akágnum, uchi wakám, (Aguaruna), akágnum (Awajún), nësa nukan (Kakataibo), chocotu ñi (Maijuna), yuwiikia (Mayna), meené-ro (Witoto)

Venezuela Cacao montero, cacao de monte, maveni; joa (Yanomami), maveni (Ye'kuana)

Status Wild, planted

Wildy, 9 years, with macambillo *Theobroma subincanum* fruits. He shimmied up a 15 m tree (behind him) in upland forest to pick the fruits. Outskirts of Miraflores, Rio Tigre, Loreto, Peru 4-26-06

Theobroma subincanum is another wild relative of cacao with edible fruits. This tree is found in lowlands and in mountains and hills up to 1500 m throughout the Amazon and Orinoco Basins as well as the Guianas (Aparício 2011:64; Araujo-Murakami et al. 2015; Carim et al. 2015; Claudino et al. 2015; Korning et al. 1991, 1994; Lemos et al. 2015; Lima et al. 2012; Machado 2009:30; Mendoza and Guerrero 2010; Moser 2013:90; Navarro 2002; Pereira et al. 2011; Pires and Salomão 2000; Rodrigues 1963; Rudas and Prieto 1998). It is the most widespread and abundant species of *Theobroma* in the Colombian Amazon (Baker et al. 1953). This straight-boled tree, which can soar to 40 m when fully mature, grows in forests and successionary vegetation on uplands and higher parts of floodplains (Amaral et al. 2009; Bentes-Gama et al. 2002a, b; Black et al. 1950; Cain et al. 1956; Campos 2009:31; Carim 2016:92; Carim et al. 2008; Casula 2012:69; Coelho et al. 2013; Espírito-Santo et al. 2005; Ferreira and Prance 1999; Gama et al. 2002, 2003; Gonzáles et al. 2011; Navarro and Maldonado 2002:67; Macedo and Anderson 1993; Pitman et al. 2014; Salomão et al. 2007a; Santana et al. 2004; Silva 2008a, b, c:16, 28; Silva et al. 2014a, b, c). It is found in successionary vegetation as young as 5 years on some uplands, such as in the watershed of the lower Tefé River in Amazonas (Jakovac et al. 2016). It is more common on the banks of forest streams (Cavalcante 2010:106; Santos et al. 2016a, b, c, d).

The Tuyuka who live along the Upper Tiquié, an affluent of the Uaupés in the northwestern part of the Brazilian Amazon, typically spare the tree when clearing land to plant crops because the fruits are edible. The Achuar in the Ecuadorian Amazon cultivate this tree which persists for many years in successionary vegetation; they even plant two varieties of this tree which is not normally considered a "domesticated" species (Descola 1994:137).

A farm family living in the watershed of the Rio Ipixuna, an affluent of the lower Xingu, has planted cupuí in an orchard around their manioc processing hut. Some Japanese-Brazilian farmers in the vicinity of Tomé-Açu, Pará, also plant this tree in their home gardens (Yamada and Osaqui 2006) as do some rural folk living along the Ramal do Pau Rosa, a side road that connects with the BR 175 Manaus-Boa Vista highway north of Manaus in Amazonas (Machado 2016:35), in the municipality of Rio Preto da Eva along the Manaus-Itacoatiara highway (Silva 2017:31), and near Abaetetuba along the lower Tocantins in eastern Pará (Gonçalves and Lucas 2017).

Arlinho, 10 years, eating cupuí *Theobroma subincanum* fruit gathered from a tree planted near a manioc processing hut. The *casa de forno* is on an archaeological site with anthropogenic dark earth. Comunidade Terra Preta, Igarapé Camutá, affluent of the Rio Ipixuna, tributary of the lower Xingu, Município de Gurupá, Pará, Brazil 4-16-15

The fruits are produced in the rainy season and are born on the branches, never on the trunk. Because of its height and relatively slender branches, only youngsters, particularly boys, venture up to the canopy to pick the fruits. Adults may be lucky enough to encounter freshly fallen fruits before some forest animals have relished the pulp. The orange-brown pods, covered in short, fine hair, are cracked open on a tree or log to access the slippery pulp. The wooden-shelled fruits vary somewhat in size and shape but are usually oblong and measure 7–14 cm long by 5–9 cm wide.

Various species of native monkeys including woolly monkeys *Lagothrix lagotricha*, spider monkeys *Ateles paniscus*, red howler monkeys *Alouatta seniculus*, tufted capuchin monkeys *Cebus apella*, collared titis *Callicebus torquatus*, equatorial sakis *Pithecia aequatorialis*, golden-faced sakis *Pithecia pithecia chrysocephala*, black-headed uakaris *Cacajao melanocephalus melanocephalus*, and saddle-back tamarins *Saguinus fuscicollis* also eat the fruits and probably disperse some of the seeds (Boubli 1999; Charpentier et al. 2015; Defler and Defler 1996; Gómez-Posada 2012; Guillotin et al. 1994; Montalván 2015:39; Palacios et al. 1997; Peres 1994; Regan et al. 2001; Setz et al. 1999; Simmen and Sabatier 1996). Woolly monkeys, spider monkeys, and howler monkeys are widespread in Amazonia and are the targets of hunters after game. The smaller white-lipped tamarin *Saguinus labiatus* eats the pulp of this fruit and disperses some of the seeds (Ique and Aquino 2015); although this monkey is not normally hunted for its meat it is sometimes kept as a pet. One of the common names for this tree in the Ecuadorian Amazon is cacao de venado (deer cacao), suggesting that one or both species of *Mazama* feed on the pulp.

Boys climbing a macambillo *Theobroma subincanum* tree in upland forest to pick the fruits. Outskirts of Miraflores, Rio Tigre, Loreto, Peru 6-11-04

The pale, tan-colored mesocarp of *Theobroma subincanum* is sublime, perhaps the best tasting wild fruit from the Amazon. The pulp surrounds some 15 seeds, each measuring 1.5 cm by 1 cm, and is sweet and tart, reminiscent of a blend between cacao and cupuaçu *Theobroma grandiflorum*. Indeed, one of the names for *Theobroma subincanum* in the Brazilian state of Amazonas is cupuarana, which means like cupuaçu. Only rural folk know about this delectable fruit, however, because they do not enter markets.

Not only could *Theobroma subincanum* provide an interesting novel flavor for ice creams, sherbets, yogurts, and puddings, the beans could be used to make a new type of "chocolate." In the Carchi region of northern Ecuador, for example, locals already prepare a beverage from the "beans" of *Theobroma subincanum*.

In addition to fresh fruit and "chocolate" beverage, *Theobroma subincanum* is employed for a variety of other purposes in the Amazon. The wood is occasionally used in house construction, such as the outer ring of posts for communal houses among the Waimiri-Atroari who live north of Manaus, Amazonas (Milliken et al. 1992:39) and individual houses among the Waorani in the Ecuadorian Amazon (Cerón and Ayala 1998:168). The Waorani also use the wood of this tree to fuel their fires. In the past, the Yanomami who inhabit northern Roraima, Brazil, and contiguous forest in southern Venezuela, used to make sticks from *Theobroma subincanum* wood which they would rub between their hands into a wooden base to generate hot shavings to start fires while traveling away from their communal houses (Albert and Milliken 2009:90). In the interfluvial forest between the Napo and Huambuno in Ecuador, ash obtained by burning the bark is mixed with tobacco juice to treat malaria, yellow fever, and fungal infections of the skin. The Dení and Jamandi, who inhabit the Juruá and Upper Purus watersheds respectively, make an intoxicating snuff by mixing crushed coca leaves, tobacco, and ashes obtained by burning the bark of *Theobroma subincanum* (Prance 1972a, c). The Witoto and Bora in the Colombian and Peruvian Amazon make a recreational snuff by mixing the ash of *Theobroma subincanum* bark with tobacco (Schultes and Raffauf 1991). The Waiká indigenous people who inhabit the Amazon rainforest in the border zone between Brazil and Venezuela add the ash from the bark of this tree to an hallucinogenic snuff (Schultes 1969).

References

Absy, Maria Lúcia; Prance, Ghillean T.; Barbosa, Edelcílio Marques. 1986. Inventário florístico de floresta natural na área da estrada Cuiabá-Porto Velho (BR 364). Acta Amazonica 16/17:85–121

Aguilar-Moreno, M. 2006. The good and evil of chocolate in colonial Mexico. In: *Chocolate in Mesoamerica: A Cultural History of Cacao*, C.L. McNeil (Ed.), pp. 273–288. University Press of Florida, Gainesville.

Albert, Bruce; Milliken, William. 2009. Urihi A: a terra-floresta Yanomami. Instituto Socioambiental (ISA)/Institut de Recherche pour le Devéloppement, São Paulo

Alden, D. 1976. The significance of cacao production in the Amazon region during the late colonial period: an essay in comparative economic history. *Proceedings of the American Philosophical Society* 120(2):103–135.

Allen, J. B. 1987. London Cocoa Trade Amazon Project: Final Report Phase 2. Cocoa Growers' Bulletin 39:1–4

Almeida, Samuel S.; Lisboa, Pedro Luiz B.; Silva, Antônio Sérgio. 1993. Diversidade florística de uma comunidade arbórea na estação científica "Ferreira Penna", em Caxiuanã (Pará). Boletim do Museu Paraense Emilio Goeldi, Série Botanica 9(1):93–128

Alvarez, Silvia J.; Heymann, Eckhard W. 2012. Brief communication: a preliminary study on the influence of physical fruit traits on fruit handling and seed fate by white-handed titi monkeys (*Callicebus lugens*). American Journal of Physical Anthropology 147:482–488

Alves, Diógenes S.; Soares, João Vianei; Amaral, Silvana; Mello, Eliana M. K.; Almeida, Sérgio A. S.; Silva, Osman Fernandes da; Silveira, Andrea M. 1997. Biomass of primary and secondary vegetation in Rondônia, Western Brazilian Amazon. Global Change Biology 3: 451–461

Amado, J. 1980, *Cacau: Romance*. Editora Record, Rio de Janeiro

Amaral, Iêda Leão do; Matos, F. Dionizia A.; Lima, José. 2000. Composição florística e parâmetros estruturais de um hectare de floresta densa de terra firme no Rio Uatumã, Amazônia, Brasil. Acta Amazonica 30(3):377–392

Amaral, Dário Dantas do; Vieira, Ima Célia Guimarães; Almeida, Samuel Soares de; Salomão, Rafael Paiva; Silva, Antonio Sérgio Lima da; Jardim, Mário Augusto Gonçalves. 2009. Checklist da flora arbórea de remanescentes florestais da região metropolitana de Belém e valor histórico dos fragmentos, Pará, Brasil. Boletim do Museu Paraense Emílio Goeldi, Ciencias Naturais 4(3):231–289

Anderson, A. 1991. Forest management strategies by rural inhabitants in the Amazon estuary. In: Rainforest Regeneration and Management, Gómez-Pompa, A.; Whitmore, T. C.; Hadley, M. (eds), UNESCO, Paris, pp. 351–360

Anderson, R. L. 1999. Colonization as Exploitation in the Amazon Rain Forest, 1758–1911. University Press of Florida, Gainesville

Aparício, Wegliane Campelo da Silva. 2011. Estrutura da vegetação em diferentes ambientes na RESEX do Rio Cajari: interações solo-floresta e relações com a produção de castanha. Doctoral Dissertation, Universidade Federal Rural de Pernambuco, Recife

Araujó-Lima, C. and M. Goulding. 1997. So Fruitful a Fish: Ecology, Conservation, and Aquaculture of the Amazon's Tambaqui. Columbia University Press, New York

Araujo-Murakami, Alejandro; Poma-Chura, Abraham; Palabral, Arely; Salvatierra, Rolando; Hurtado, Fidel. 2012. Composición florística de los bosques Amazónicos de tierra firme e inundable en las proximidades de las Pampas del Sonene (río Heath), Parque Nacional Madidi, Bolivia. Kempffiana 8(1):2–21

Araujo-Murakami, Alejandro; Villarroel, Daniel; Pardo, Guido; Vos, Vincent A.; Parada, G. Alexander; Arroyo, Luzmila; Killeen, Timothy. 2015. Diversidad arbórea de los bosques de tierra firme de la Amazonía Boliviano. Kempffiana 11(1):1–28

Assis, Rafael Leandro de; Wittmann, Florian; Luize, Bruno Garcia; Haugaasen, Torbjon. 2017. Patterns of floristic diversity and composition in floodplain forests across four southern Amazon river tributaries, Brazil. Flora 229:124–140

Avé-Lallemant, Robert. 1961. Viagem pelo Norte do Brasil no ano de 1859. Instituto Nacional do Livro, Rio de Janeiro, Vol. 2

Baker, R.E.D.; Cope, F.W.; Holliday, P.C.; Bartley, B.G.; Taylor, D.J. 1953. The Anglo-Colombian cacao collecting expedition. In: A Report on Cacao Research, Imperial College of Tropical Agriculture, Trinidad, pp. 8–29

Balée, W. 1989a. Nomenclatural patterns in Ka'apor ethnobotany. *Journal of Ethnobiology* 9(1):1–24

Balée, W. 1989b. The culture of Amazonian forests. In: R*esource Management in Amazonia: Indigenous and Folk Strategies*, Posey, D. A. and W. Balée (Eds), pp. 1–21, Advances in Economic Botany, vol. 7, New York Botanical Garden, Bronx, New York.

Balée, W. 1993a. *Footprints of the Forest: Ka'apor Ethnobotany, the Historical Ecology of Plant Utilization by an Amazonian People*. Columbia University Press, New York

Balée, W. 1993b. Indigenous transformation of Amazonian forests: an example from Maranhão, Brazil. L'Homme 33(126–128):231–254

Balée, W. 2010. Contingent diversity on anthropic landscapes. *Diversity* 2:163–181

Balée, W. and D. G. Campbell 1990. Evidence for the successional status of liana forest (Xingu River Basin, Amazonian Brazil). Biotropica 22(1):36–47

Balée, W. and A. Gély, 1989. Managed forest succession in Amazonia: the Ka'apor case. In: R*esource Management in Amazonia: Indigenous and Folk Strategies*, Posey, D. A. and W. Balée (Eds), pp. 129–158, Advances in Economic Botany, vol. 7, New York Botanical Garden, Bronx, New York.

Balée, W. and D. Moore. 1991. Similarity and variation in plant names in five Tupi-Guarani languages (eastern Amazonia). *Bulletin of the Florida Museum of Natural History, Biological Sciences* 35 (4): 209–262

Barbosa, Cristiano de Souza. 2018. Diversidade e uso de plantas úteis nas quintais do bairro de São Raimundo, zona oeste de Manaus, AM. Master's thesis, INPA, Manaus

Bartley, B. G. D. 2005. *The Genetic Diversity of Cacao and its Utilization*. CAB International, Wallingford

Bartley, B.G.D., P.F.R. Machado, D. Ahnert, J.P. Barriga, and C.M.V.C. de Almeida. 1987. Descrição de populações de cacau da Amazônia brasileira, 1: observações preliminares sobre populações de Alenquer, Para. In: *Anais da 10a. Conferencia Internacional de Pesquisas em Cacau, Santo Domingo, 17–23 maio*, pp. 665–672.

Bates, H. W. 1863. *The Naturalist on the River Amazons*. John Murray, London, vol. 1

Beck, H. 2005. Seed predation and dispersal by peccaries throughout the Neotropics and its consequences: a review and synthesis. In: Seed Fate: Predation, Dispersal and Seedling Establishment, Forget, Pierre-Michel; Lambert, Joanna E.; Hulme, Philip E.; Wall, Stephen B. Vander (eds), CABI, Wallingford, pp 77–115

Bennett, Bradley C.; Baker, Marc A.; Andrade, Patricia Gómez. 2002. Ethnobotany of the Shuar of Eastern Ecuador. New York Botanical Garden, Advances in Economic Botany, vol. 14.

Bentes-Gama, Michelliny de Matos; Scolforo, José Roberto Soares; Gama, João Ricardo Vasconcellos; Oliveira, Antônio Donizette de. 2002a. Estrutura e valoração de uma floresta de várzea alta na Amazônia. Cerne 8(1):88–102

Bentes-Gama, Michelliny de Matos; Scolforo, José Roberto Soares; Gama, João Ricardo Vasconcellos. 2002b. Potencial produtivo de madeira e palmito de uma floresta secundária de várzea baixa no estuário Amazônico. Revista Árvore 26(3):311–319

Betendorf, J.F. 1910. Chronica da missão dos padres da Companhia de Jesus no Estado do Maranhão. *Revista do Instituto Historico e Geographico Brazileiro* 72(1): 1–682

Black, G. A.; Dobzhansky, T. H.; Pavan, C. 1950. Some attempts to estimate species diversity and population density of trees in Amazonian forests. Botanical Gazette 111(4):413–425

Bletter, Nathaniel; Daly, Douglas C. 2006. Cacao and its relatives in South America: an overview of taxonomy, ecology, biogeography, chemistry, and ethnobotany. In: Chocolate in Mesoamerica: A Cultural History of Cacao, McNeil, Cameron L. (ed), University Press of Florida, Gainesville, pp 31–68

Bonanome, Andrea; Grundy, Scott M. 1988. Effect of dietary stearic acid on plasma cholesterol and lipoprotein levels. New England Journal of Medicine 318(19):1244–1248

Boom, B. M. 1986. A forest inventory in Amazonian Bolivia. Biotropica 18(4):287–294

Boom, B. M. 1989. Use of plant resources by the Chácobo. In: *Resource Management in Amazonia: Indigenous and Folk Strategies*, D. A. Posey and W. Balée (Editors), Advances in Economic Botany, vol. 7, New York Botanical Garden, Bronx, pp. 78–96

Borges, Hélida B. Nogueira; Shepherd, George J. 2005. Flora e estrutura do estrato lenhoso numa comunidade de Cerrado em Santo Antônio do Leverger, MT, Brasil. Revista Brasileira de Botânica 28(1):61–74

Boster, James. 1983. A comparison of the diversity of Jivaroan gardens of the tropical forest. Human Ecology 11 (1): 47–68

Boubli, J. P. 1999. Feeding ecology of black-headed uacaris (*Cacajao melanocephalus melanocephalus*) in Pico da Neblina National Park, Brazil. International Journal of Primatology 20(5):719–749

Bourdy, G.; DeWalt, S. J.; Chávez de Michel, L. R.; Roca, A.; Deharo, E.; Muñoz, V.; Balderrama, L.; Quenevo, C.; Gimenez, A. 2000. Medicinal plant uses of the Tacana, an Amazonian Bolivian ethnic group. Journal of Ethnopharmacology 70:87–109

Brack, A. 2003. *Frutas del Perú*. Universidad San Martin de Porres, Lima

Braga, Daniel P. P.; Domene, Frederico; Gandara, Flávio B. 2019. Shade trees composition and diversity in cacao agroforestry systems of southern Pará, Brazilian Amazon. Agroforestry Systems 93:1409–1421

Brown, C.B. and W. Lidstone. 1878. *Fifteen Thousand Miles on the Amazon and its Tributaries*. Edward Stanford, London

Buchillet, D. 1988. Interpretação da doença e simbolismo ecológico entre os índios Desana. Boletim to Museu Paraense Emílio Goeldi, Série Antropologia 4 (1): 27–42

Cain, Stanley A.; Castro, Gustavus M. de Oliveira; Pires, J. Murça; Silva, Nilo Tomâs. 1956. Application of some phytosociological techniques to Brazilian rain forest. American Journal of Botany 43(10):911–941

Câmpera, Luiza Maria Fonseca. 2017. O lago encantado e o caminho da chuva: noções de corpo, cura e cosmologia no Médio Solimões. Master's thesis, Universidade Federal do Amazonas, Manaus

Campos, Luis Enrique. 2009. Evaluación de la regeneración natural de los claros en el bosque de la llanura aluvial del río Nanay, Puerto Almendra, Loreto. Master's thesis, Universidad Nacional de la Amazonía Peruana, Iquitos

Carim, Marcelo de Jesus Veiga. 2016. Estrutura, composição e diversidade em florestas alagáveis de várzea de maré e de igapó e suas relações com variáveis edáficas e o período de inundação no Amapá, Amazônia oriental, Brasil. Doctoral dissertation, Programa de Pós-Graduação em Botânica, INPA, Manaus

Carim, Marcelo de Jesus Veiga; Jardim, Mário Augusto Gonçalves; Medeiros, Tonny David Santiago. 2008. Composição florística e estrutura de floresta de várzea no Município de Mazagão, Estado do Amapá, Brasil. Scientia Florestalis 36(79):191–201

Carim, Marcelo de Jesus Veiga; Guimarães, José Renan da Silva; Tostes, Luciendi de Cássia Leôncio; Takiyama, Luís Roberto; Wittmann, Florian. 2015. Composition, structure and floristic diversity in dense rain forest in the eastern Amazon, Amapá, Brazil. Acta Scientiarum Biological Sciences 37(4):419–426

Carvalho, João Olegário Pereira de 2002. Changes in the spatial distribution of tree species in a terra firme rain forest in Brazilian Amazonia after logging. Revista de Ciência Agrária 37:53–70

Carvalho, João Olegário Pereira de; Silva, José Natalino Macedo; Lopes, José do Carmo Alves; Montagner, Lise Helene; Carvalho, Manoel Sebastião Pereira de. 1986. Composição florística de uma mata secundária no planalto de Belterra no Pará. In: Anais do 1° Simpósio do Trópico Úmido, Belém, PA, 12 a 17 November, EMBRAPA, Brasília, pp 197–205

Castañeda, Rafael Romero. 1961. *Frutas Silvestres de Colombia*. Bogota, vol. 1.

Castro, Albejamere Pereira de; Fraxe, Therezinha de Jesus Pinto; Santiago, Jozane Lima; Matos, Raimundo Barbosa; Pinto, Ilzon Castro. 2009. Os sistemas agroflorestais como alternativa de sustentabilidade em ecosistemas de várzea no Amazonas. Acta Amazonica 39(2):279–288

Casula, Kátia Regina. 2012. Análise da similaridade florística e estrutural das formações florestais inundáveis em um trecho do alto Rio Madeira e de seus afluentes, estado de Rondônia, Brasil. Master's Thesis, Núcleo de Ciências Exatas da Terra, Universidade Federal de Rondônia, Porto Velho

Cattanio, José; Anderson, Anthony B.; Carvalho, Manoel S. 2002. Floristic composition and topographic variation in a tidal floodplain forest in the Amazon estuary. Revista Brasileira de Botanica 25(4):419–430

Cavalcante, P. B. 2010. *Frutas Comestíveis na Amazônia*. Museu Paraense Emílio Goeldi, Belém

Cerón, Carlos Eduardo. 1991. Manejo florístico Shuar-Achuar (Jibaro) del ecosistema Amazónico en el Ecuador. Hombre y Ambiente 5(17):7–31

Cerón, Carlos Eduardo. 1995. Etnobiologia de los Cofanes de Dureno, Provincia de Sucumbíos, Ecuador. Museo Ecuatoriano de Ciencias Naturales, Herbario Nacional, Serie Monografia 10(3), Quito

Cerón, Carlos Eduardo; Ayala, Consuelo Montalvo. 1998. Etnobotanica de los Huaorani de Quehueiri-ono Napo, Ecuador. Abya-Yala, Quito

Charpentier, Elvis J.; García, Gabriel; Aquino, Rolando. 2015. Uso y competición por plantas alimenticias entre *Pithecia aequatorialis* (Primates: Pitheciidae) y otros animales en la Amazonía peruana. Revista Peruana de Biología 22(2):225–232

Cheesman, E.E. 1944. Notes on the nomenclature, classification and possible relationships of cacao populations. *Tropical Agriculture* 21(8): 144–159

Claudino, Wesley Vicente; Garcia Marcos Leandro; Lopes, Célia Regina Araújo Soares, Rodrigues, Lucirene; Cabral, Fabiana Ferreira; Piva, José Hypolito; Silva, Antônio Carlos Silveiro da. 2015. Análise fitossociológica da vegetação na região do arco do desmatamento. Revista de Ciências Agroambientais, Alta Floresta, MT 13(1): 43–52

Clement, C. R. 1982. Considerações sobre a sapota da America do Sul. Proceedings of the Tropical Region, American Society of Horticultural Science 25:427–432

Clement, C.R. 1989. A center of crop genetic diversity in western Amazonia: A new hypothesis of indigenous fruit-crop distribution. *Biotropica* 39(9): 624–631

Clement, Charles R.; Noda, Hiroshi; Noda, Sandra do N.; Martins, Ayrton L. Urizzi; Silva, Gleissimar Campelo da. 2001. Recursos frutícolas na várzea e na terra firme em onze comunidades rurais do Alto Solimões, Amazonas, Brasil. Acta Amazonica 31 (3): 521–527

Cobley, L.S. 1956. *An Introduction to the Botany of Tropical Crops*. Longmans, Green and Company, London.

Coelho, Roberta de Fátima Rodrigues; Miranda, Izildinha Souza; Mitja, Danielle. 2012. Caracterização do processo sucessional no Projeto de Assentamento Benfica, sudeste do estado do Pará, Amazônia oriental. Boletim do Museu Paraense Emílio Goeldi, Ciencias Naturais 7(3):251–282

Coelho, Roberta de Fátima Rodrigues; Miranda, Izildinha de Souza; Mitja, Danielle. 2013. Conservação das florestas do projeto de assentamento Benfica, Sudeste da Amazônia. Ciencia Florestal, Santa Maria 23(1):1–17

Coimbra, C. E. A.; Flowers, Nancy N.; Salzano, Francisco M.; Santos, Ricardo V. 2002. The Xavánte in Transition: Health, Ecology, and Bioanthropology in Central Brazil. University of Michigan Press, Ann Arbor.

Conserva, Auristella dos Santos. 2007. Germinação de sementes, emergência e recrutamento de plântulas de dez espécies arbóreas das várzeas das Reservas de Desenvolvimento Sustentável Amanã e Mamirauá, Amazônia central. Doctoral dissertation, INPA, Manaus

Coomes, Oliver T.; Ban, Natalie. 2004. Cultivated plant species diversity in home gardens of Amazonian peasant village in northeastern Peru. Economic Botany 58(3):420–434

Coomes, Oliver T.; Grimard, Franque; Burt, Graeme J. 2000. Tropical forests and shifting cultivation: secondary forest fallow dynamics among traditional farmers of the Peruvian Amazon. Ecological Economics 32:109–124

Corette-Pasa, Maria. 2011. Abordagem etnobotânica na comunidade de Conceição-Açu, Mato Grosso, Brasil. Polibotánica 31:169–197

Cruz, M., R. Whitkus, A. Gómez-Pompa, and L. Mota-Bravo, 1995. Origins of cacao cultivation. *Nature* 375: 542–543

Daniel, J. 1976. *Tesouro Descoberto no Rio Amazonas*. Biblioteca Nacional, Rio de Janeiro, vol. 1

Daniel, J. 2004. Tesouro Descoberto no Máximo Rio Amazonas. Contraponto, Rio de Janeiro, vol. 2

Dávila, P. M.; Vega, Juan José Palacios; Vela, Marcial Martínez; Gómez, Ricardo Zárate; García, José Maco; Torres, Roger Escobedo. 2015. Cobertura y uso de la tierra de las provincias de Alto Amazonas y Ramón Castilla. Instituto de Investigaciones de la Amazonía Peruana, Iquitos, Documento Técnico No. 33

Davis, W. 2004. *The Lost Amazon: The Photographic Journey of Richard Evans Schultes*. Chronicle Books, San Francisco.

Defler, Thomas R.; Defler, Sara B. 1996. Diet of a group of *Lagothrix lagothricha lagothrica* in southeastern Colombia. International Journal of Primatology 17(2):161–190

Denevan, W. M. and J. M. Treacy. 1987. Young managed fallows at Brillo Nuevo. In: *Swidden-Fallow Agroforestry In the Peruvian Amazon*, W. M. Denevan and C. Padoch (eds.), Advances in Economic Botany vol. 5, New York Botanical Garden, pp. 8–46

Descola, P. 1994. *In the Society of Nature: A Native Ecology in Amazonia*. Cambridge University Press, Cambridge

Dew, J.L. 2005. Foraging, food choice, and food processing by sympatric ripe-fruit specialists: *Lagothrix lagotricha poeppigii* and *Ateles belzebuth belzebuth*. International Journal of Primatology 26(5):1107–1135

Dew, J. Lawrence 2008. Spider monkeys as seed dispersers. In: Spider Monkeys: Behavior, Ecology and Evolution of the Genus *Ateles*, Campbell, Christina J. (ed), Cambridge University Press, Cambridge, pp 155–182

Duarte, Luis Enrique Acero. 1979. Principales Plantas Utiles de la Amazonia Colombiana. Instituto Geografico "Augustin Codazzi", Bogotá

Ducke, A. 1946. *Plantas de Cultura Precolombiana na Amazônia brasileira. Notas Sôbre as Espécies ou Formas Espontâneas que Supostamente lhes teriam dado Origem*. Instituto Agronômico do Norte, Belém, Boletim Técnico 8, pp. 1–24

Ducke, A. 1953. As espécies brasileiras do gênero *Theobroma* L. Boletim Técnico do Instituto Agronômico do Norte 28: 3–20.

Duivenvoorden, J.F. and Lips, J.M. 1995. A land-ecological study of soils, vegetation, and plant diversity in Colombian Amazonia. Tropenbos Foundation, Wageningen

Edwards, W. H. 1847. *A Voyage up the River Amazon, including a Residence at Pará*. John Murray, London

Espírito-Santo, Fernando Del Bon; Shimabukuro, Yosio Edimir; Aragão, Luiz Eduardo Oliveira e Cruz de; Machado, Evandro Luiz Mendonça. 2005. Análise da composição florística e fitossociológica da floresta nacional do Tapajós com apoio geográfico de imagens satélites. Acta Amazonica 35(2):155–173

Falcão, M. and E. Lleras. 1983. Aspectos fenológicos, ecológicos e de produtividade do cupuaçu *Theobroma grandiflorum* (Willd. ex Spreng.) Schum. Acta Amazonica 13(5–6):725–735

Felton, Annika M.; Felton, Adam; Wood, Jeff T.; Lindenmayer, David B. 2008. Diet and feeding ecology of *Ateles chamek* in a Bolivian semihumid forest: the importance of Ficus as a staple food resource. International Journal of Primatology 29:379–403

Felton, Annika M.; Felton, Adam; Wood, Jeff T.; Foley, William J.; Raubenheimer, David; Wallis, Ian R.; Lindenmayer, David B. 2009. Nutritional ecology of Ateles chamek in lowland Bolivia: how macronutrient balancing influences food choices. International Journal of Primatology 30:675–696

Ferreira Penna, D.S. 1864. *O Tocantins e o Anapú*. Tip. de Federico Rhossard, Belém

Ferreira, Leandro V.; Prance, Ghillean T. 1999. Ecosystem recovery in terra firme forests after cutting and burning: a comparison on species richness, floristic composition and forest structure in the Jaú National Park, Amazonia. Botanical Journal of the Linnaean Society 130:97–110

Franco, Ana María; Santamaria, Marcela. 1997. Area vital, habitos alimenticios y otros aspectos de historia natural de paujil *Mitu silvini*. In: Manejo de Fauna Silvestre en la Amazonía, Fang, T.G.; Bodmer, R.E.; Aquino, R.; Valqui, M.H. (eds), Instituto de Ecologia, La Paz, Bolivia, pp 267–282

Freitas, João da Luz; Santos, Erick Silva dos; Cruz Junior, Francisco de Oliveira; Borges, César Henrique Alves; Santos, Adriano Castelo dos. 2019. Structural analysis and distribution patterns in lowland tropical forest, eastern Amazon. Journal of Agricultural Studies 7(4): 240–255

Friedman, J.; Bolotin, D.; Rios, M.; Mendosa, P.; Cohen, Y.; Balick, M. J. 1993. A novel method for identification and domestication of indigenous useful plants in Amazonian Ecuador. In: New Crops, Janick, J.; Simon, J. E. (eds), John Wiley, New York, pp 167–174

Fryxell, 2004. 'Sterculiaceae (Chocolate Family)' in, Smith, N., Mori, S.A., Henderson, A., Stevenson, D.W. and Heald, S.V., 2004. Flowering plants of the Neotropics. Princeton University Press

Gachet, María Salomé; Lecaro, Javier Salazar; Kaiser, Marcel; Brun, Reto; Navarrete, Hugo; Muñoz, Ricardo A.; Bauer, Rudolf; Schühly, Wolfgang. 2010. Assessment of anti-protozoal activity of plants traditionally used in Ecuador in the treatment of leishmaniasis. Journal of Ethnopharmacology 128:183–197

Gama, João Ricardo Vasconcellos; Botelho, Soraya Alvarenga; Bentes-Gama, Michellinyde Matos. 2002. Composição florística e estrutura da regeneração natural de floresta secundária de várzea baixo no estuário Amazônico. Revista Árvore 26(5):559–566

Gama, J. R. V., S. A. Botelho, M. Bentes-Gama, and J. R. S. Scolforo 2003. Estrutura e potencial futuro de utilização da regeneração natural de floresta de várzea alta no Município de Afuá, estado do Pará. *Ciencia Florestal, Santa Maria* 13(2):71–82

Gasché, Jürg. 1971. Quelques prolongments sociaux des pratiques horticoles et culinares chez les Indians Witoto. Journal de la Société des Américanistes (Paris, N.S.) 60(1):317–327

Gély, Anne. 1986. Agriculture créole; une étude de cas: Saül (Guyane Française). Journal d'Agriculture Traditionnelle et de Botanique Appliquée 33:143–192

Ghilardi, R.; Alho, C.J.R. 1990. Produtividade sazonal da floresta e atividade de forrageamento animal em habitat de terra firme da Amazônia. Acta Amazonica 20:61–76

Giacomotti, José Guillermo. 2016. Evaluación de la dinámica forestal en un área de bosque secundario tardío en el fundo La Génova, Chanchamayo. Master's thesis, Universidad Nacional Agraria La Molina, Lima

Gomes, Carlos Valério Aguiar. 2018. Ciclos económicos do extrativismo na Amazônia na visão dos viajantes naturalistas. Boletim do Museu Paraense Emílio Goeldi, Ciencias Humanas 13(1):129–146

Gómez-Posada, C. 2012. Dieta y comportamiento alimentario de un grupo de mico maicero *Cebus apella* de acuerdo a la variación en la oferta de frutos y artrópodos, en la Amazonía colombiana. Acta Amazonica 42(3):363–372

Gonçalves, Janaína Pinheiro; Lucas, Flávia Cristina Araújo. 2017. Agrobiodiversidade e etnoconhecimento em quintais de Abaetetuba, Pará, Brasil. Revista Brasileira de Biociências 15(3):119–134

Gonzáles, Augustín; Mejía, Kember Mateo; Torres, Guiuseppe Melecio; Sotero, Victor Erasmo; Vásquez, Joel; Vásquez, Armando; Imán, Sixto Alfredo; Fernández, Andrés. 2011. Contribuciones al Conocimiento de *Theobroma subincanum* "macambillo" (Sterculiaceae) frutal nativo de gran potencial en la Amazonía peruana. In: Contribuciones al Conocimiento de Frutales Amazónicos, IIAP (Instituto de Investigación de la Amazonía Peruana), Iquitos, pp. 61–83

Gordon, B. L. 1982. A Panama Forest and Shore: Natural History and Amerindian Culture in Bocas del Toro Province, Panama. Boxwood Press, Pacific Grove, California

Gribel, Rogério; Gibbs, Peter E. 2002. High outbreeding as a consequence of selfed ovule mortality and single vector bat pollination in the Amazonian tree Pseudobombax munguba (Bombacaceae). International Journal of Plant Sciences 163(6):1035–1043

Griscom, Bronson W.; Daly, Douglas C.; Ashton, Mark S. 2007. Floristics of bamboo-dominated stands in lowland terra-firma forests of southwestern Amazonia. Journal of the Torrey Botanical Society 134(1):108–125

Gualberto, Midiã Larine Colares; Ribeiro, Renato Bezerra da Silva; Gama, João Ricardo Vaconcellos; Vieira, Diego dos Santos. 2014. Fitossociologia e potencial de espécies arbóreas em ecossistema sucessional na Floresta Nacional do Tapajós, Pará. Agroecossistemas 6(1):42–57

Guarim Neto, G. and Amaral, C.N.D. 2010. Aspectos etnobotânicos de quintais tradicionais dos moradores de Rosário Oeste, Mato Grosso, Brasil. Polibotánica, (29), pp. 191–212

Guarim Neto, Germano; Morais, Ronan Gil de. 2003. Plantas medicinais com potencial ornamental: um estudo no cerrado de Mato Grosso. Revista Brasileira de Horticultura Ornamental 9(1):89–97

Guarim Neto, Germano; Pasa, Maria Corette. 2009. Estudo etnobotânico em uma área de cerrado no município de Acorizal, Mato Grosso. Flovet 1:5–32

Guillaumet, J-L.; Grenand, P.; Bahri, S.; Grenand, F.; Lourd, M.; Santos, A.A. dos; Gély, A. 1990. Les jardins-vergers familiaux d'Amazonie centrale: une exemple d'utilisation de l'espace. Turrialba 40(1):63–81

Guillotin, M.; Dubost, G.; Sabatier, D. 1994. Food choice and food competition among the three major primate species of French Guiana. Journal of Zoology 233:551–579

Guimarães, Felipe Silva; Bueno, Guilherme Taitson; Mendes, Débora de Sena Oliveira; Souza, Jorge Batista; Diniz, Alisson Duarte. 2016a. Mapeamento das unidades de paisagem natural na bacia do Rio Demini, afluente do Rio Negro – AM. Acta Geográfica 10(24):18–33

Guimarães, José Renan da Silva; Carim, Marcelo de Jesus Veiga; Costa Neto, Salustiano Vilard a; Tostes, Luciedide Cássia Leôncio. 2016b. Floristic diversity of secondary forest in the eastern Amazon, state of Amapá. Floresta 46(3):343–351

Hammen, M. C. 1992. *El Manejo del Mundo: Naturaleza y Sociedad entre los Yukuna de la Amazonia colombiana*. Tropenbos, Bogotá

Harris, M. 2000. *Life on the Amazon: The Anthropology of a Brazilian Peasant Village*. Oxford University Press, Oxford.

Hartt, C. F. 1898. O Rio Tocantins. Boletim do Museu Paraense Emílio Goeldi 2:181–191

Heriarte, Mauricio de, 1964. *Descriçam do Estado do Maranham, Para, Corupa, Rio das Amazonas (1662)*. Akademische Druck und Verlagsanstalt, Graz, Austria

Herndon, W. L. and L. Gibbon. 1853. Exploration of the Valley of the Amazon, made under the Direction of the Navy Department. Rober Armstrong, Washington, D.C., vol. 1

Herrera, Luisa Fernanda; Cavelier, Inés; Rodríguez, Camilo; Mora, Santiago. 1992. The technical transformation of an agricultural system in the Colombian Amazon. World Archaeology 24 (1):98–113

Hiraoka, M. 1989. Ribereño's changing economic patterns in the Peruvian Amazon. Journal of Cultural Geography 9(2):103–119

Huber, Jacques. 1906a. Sur l'indigénat du *Theobroma cacao* dans les alluvions du Purus et sur quelques autres espèces du genre *Theobroma*. Bulletin de L'Herbier Boissier 6:272–274

Huber, Jacques. 1906b. La végétation de la vallée du Rio Purus (Amazone). Bulletin de L'Herbier Boissier 6(4):249–276

Huber, Jacques. 1906c. Materiaes para a flora Amazonica VI. Plantas vasculares colligidas e observados no baixo Ucayali e no Pampa del Sacramento, nos mezes de outubro a dezembro de 1898. Boletim do Museu Goeldi de Historia Natural e Ethnographia 4:510–619

Huber, J. 1910. Mattas e madeiras amazonicas. Boletim do Museu Goeldi de Historia Natural e Ethnographia 6: 91–225

Humboldt, A. and A. Bonpland. 1818. *Personal Narrative of Travels to the Equinoctial Regions of the New Continent during the Years 1799–1804, Vol. 4*. Longman, Hurst, Rees, Orme, and Brown, London.

Ique, Carlos A.; Aquino, Rolando. 2015. Principales especies florísticas utilizadas y dispersión de semillas por *Saguinus labiatus* E. Geoffroy, 1812 en isla Muyuy, Loreto, Perú. Ciencia Amazónica 5(1):81–88

Irmler, U. 1977. Inundation-forest types in the vicinity of Manaus. Biogeographica 8:17–29

Irvine, Dominique. 1987. El manejo del bosque secundario por los Quichua amazonicos. Hombre y Ambiente 2:121–142

Irvine, Dominique. 1989. Succession management and resource distribution in an Amazonian rain forest. Resource Management In Amazonia: Indigenous and Folk Strategies, Posey, D. A.; Balée, W. (Editors), pp. 223–237. Advances in Economic Botany, vol. 7, New York Botanical Garden.

Ivanauskas, Natália Macedo; Monteiro, Reinaldo; Rodrigues, Ricardo Ribeiro. 2004. Estrutura de um trecho de floresta Amazônica na bacia do alto rio Xingu. Acta Amazonica 34(3):275–299

Jakovac, Catarina C.; Bongers, Frans; Kuyper, Thomas W.; Mesquita, Rita C. G.; Peña-Carlos, Marielos. 2016. Land use as a filter for species composition in Amazonian secondary forests. Journal of Vegetation Science 27:1104–1116

Janzen, D. H. and P. S. Martin. 1982. Neotropical anachronisms: The fruits the gomphotheres ate. Science 215(4528):19–27

Jardim, M.A.G. and Medeiros, T.D.S. 2006. Plantas oleaginosas do Estado do Pará: composição florística e usos medicinais. Revista Brasileira de Farmácia 87(4):124–127

Kawa, Nicholas C.; Michelangeli, José A. Clavijo; Clement, Charles R. 2015. Household agrobiodiversity management on Amazonian dark earths, oxisols, and floodplain soils on the lower Madeira River, Brazil. Human Ecology 43:339–353

Kerr, W. E. 1986. Agricultura e seleções genéticas de plantas. In: Suma Etnológica Brasileira, Vozes, Petrópolis, Part 1 E, pp. 159–171

Kerr, W. E. and C. R. Clement. 1980. Prácticas agrícolas de consequências genéticas que possibilitaram aos índios da Amazônia uma melhor adaptaçao às condiçoes ecológicas da região. Acta Amazonica 10(2):251–261

Keuroghlian, Alexine; Eaton, Donald P.; Desbiez, Arnaud L. J. 2009. The response of a landscape species, white-lipped peccaries, to seasonal resource fluctuations in a tropical wetland, the Brazilian pantanal. International Journal of Biodiversity and Conservation 1(4):87–97

Korning, Jorgen; Thomsen, Karsten; Ollgaard, Benjamin. 1991. Composition and structure of a species rich Amazonian rain forest obtained by two different sample methods. Nordic Journal of Botany 11(1):103–110

Korning, Jorgen; Thomsen, Karsten; Dalsgaard, Kristian; Nornberg, Per. 1994. Characters of three Udults and their relevance to the composition and structure of virgin rain forest of Amazonian Ecuador. Geoderma 63:145–164

Kufer, Johanna; McNeil, Cameron L. 2006. The jaguar tree (Theobroma bicolor Bonpl.). In: Chocolate in Mesoamerica: A Cultural History of Cacao, McNeil, Cameron L. (ed), University Press of Florida, Gainesville, pp. 94–104

Kvist, L. P. and G. Nebel. 2001. A review of Peruvian flood plain forests: ecosystems, inhabitants and resource use. Forest Ecology and Management 150:3–26

La Rotta, C. 1983. Observaciones Etnobotanicas sobre algunas Especies Utilizadas por la Comunidad Indigena Andoque (Amazonas, Colombia). Universidad Nacional de Colombia, Bogotá

Lamont, Susan R.; Eshbaugh, W. Hardy; Greenberg, Adolph M. 1999. Species composition, diversity, and use of homegardens among three Amazonian villages. Economic Botany 53(3):312–326

Lathrap, D.W. 1977. Our father the cayman, our mother the gourd: Spinden revisited, or a unitary model for the emergence of agriculture in the New World. In: *Origins of Agriculture*, C.A. Reed (Editor), pp. 713–751. Mouton, The Hague.

Le Cointe, P. 1903. Le Bas Amazone. *Annales de Géographie* 12:54–66.

Le Cointe, P. 1922a. *L'Amazonie Brésilienne: Le Pays-ses Habitants, ses Resources, Notes et Statistiques jusqu'en 1920.* Augustin Challamel, Paris, vol 1

Le Cointe, P. 1922b. *L'Amazonie Brésilienne: Le Pays-ses Habitants, ses Resources, Notes et Statistiques jusqu'en 1920.* Augustin Challamel, Paris, vol 2

Le Cointe, P. 1934. *Amazônia Brasileira III: Árvores e Plantas Úteis*. Livraria Classica, Belém

Lemos, D. A. N.; Ferreira, B. G. A.; Siqueira, J. D. P.; Oliveira, M. M.; Ferreira, A. M. 2015. Floristic and phytosociology in dense "terra firme" rainforest in the Belo Monte hydroelectric plant influence area, Pará, Brazil. Brazilian Journal of Biology 75(3):257–276

Léon, J. 1984. The spread of Amazonian crops in Mesoamerica: the botanical evidence. In: Pre-Columbian Plant Migration, D. Stone (ed), Peabody Museum of Archaeology and Ethnology, Harvard University, Cambridge pp 165–173

Lima, Ana Carolina Barbosa de. 2014. Flavors of the city: access to regional fruit and fruit consumption in the State of Acre, Brazil. Boletim do Museu Paraense Emílio Goeldi, Ciencias Humanas 9(1):79–92

Lima, Roberval M. B. de; Saragoussi, Muriel. 2000. Floodplain home gardens on the central Amazon in Brazil. In: The Central Amazon Floodplain: Actual Use and Options for a Sustainable Management, Junk, W. J.; Ohly, J. J.; Piedade, M. T. F.; Soares, M. G. M. (eds), Backhuys Publishers, Leiden pp 243–268

Lima, R. B. A.; Silva, J. A. A.; Marangon, L. C.; Ferreira, R. L. C.; Silva, R. K. S. 2012. Fitossociologia de um trecho de floresta ombrófila densa na Reserva de Desenvolvimento Sustentável Uacari, Carauari, Amazonas. Scientia Plena 8(1):1–12

Link, Andres; Fiore, Anthony Di. 2006. Seed dispersal by spider monkeys and its importance in the maintenance of neotropical rain-forest diversity. Journal of Tropical Ecology 22:235–246

Lins, Juliana; Lima, Helena P.; Baccaro, Fabricio B.; Kinupp, Valdely; Shepard Glenn H.; Clement, Charles R. 2015. Pre-Columbian floristic legacies in modern homegardens of Central Amazonia. PLOS One https://doi.org/10.1371/journal.pone.0127067

Lisboa, P. L. B. 1989. Estudo florístico da vegetação arbórea de uma floresta secundária em Rondônia. Boletim do Museu Paraense Emílio Goeldi, Série Botânica 5(2):145–162

Lisboa, Pedro L. B.; Maciel, Ubirajara N.; Prance, Ghillean T. 1991. Some effects of colonization on the tropical flora of Amazonia: a case study from Rondônia. Kew Bulletin 46(2):187–204

Lisboa, P. L. B., I. A. Gomes, R. C. Lisboa, and C. V. Urbinati. 2002. O estilo Amazônico de sobreviver: manejo dos recursos naturais. In: *Natureza, Homem e Manejo de Recursos Naturais na Região de Caxiuanã, Melgaço, Pará*, P. L. B. Lisboa (ed.), Museu Paraense Emílio Goeldi, Belém, pp. 41–170

Lizot, Jacques 1978. Connaissance et usage des plantes sauvages chez les Yanomami. In: Unidad y Variedad: Ensayos Antropológicos en Homenaje a José M. Cruxent, Wagner, Erika; Zucchi, Alberta (eds), IVIC, Caracas, pp 129–171

Lorenzi, Harri; Bacher, Luis; Lacerda, Marco; Sartori, Sergio. 2006a. Frutas Brasileiras e Exóticas Cultivadas. Instituto Plantarum de Estudos da Flora Ltda, Odessa, Sao Paulo.

Lorenzi, H., L. Bacher, M. Lacerda, S. Sartori 2006b. Brazilian Fruits and Cultivated Exotics (for Consuming in Natura). Instituto Plantarum de Estudos da Flora Ltda, Odessa, São Paulo

Lossio, L.C., Delgado, H.S., Ruiz, A.G., Donayre, R.A., Ruiz, J.G., Panduro, M.P., Sifuentes, T.C. 1995. Plantas Medicinales de la Amazonía Peruana. Instituto Peruano de Seguridad Social/Instituto de Medicina Tradicional, Iquitos

Loverde-Oliveira, Simoni Maria; Freitas, Maria Niely; Araújo, Patrícia Karina Barbosa; Costa, Ineilian Bruna Correa. 2010. Fragmento de cerrado urbano da Universidade Federal de Mato Grosso, campus de Rondonópolis, Mato Grosso. Revista Biodiversidade 9(1):74–90

Luize, Bruno Garcia; Venticinque, Eduardo Martins; Silva, Thiago Sanna Freire; Novo, Evlyn Marcia Leão de Moraes. 2015. A floristic survey of angiosperm species occurring at three landscapes of the central Amazon várzea, Brazil. Checklist: The Journal of Biodiversity Data 11(6):1–15

Lustosa, Dom Antônio de Almeida. 1976. No Estuário Amazônico "À Margem da Visita Pastoral". Conselho Estadual de Cultura do Pará, Belém.

Macedo, Erick Joe Dossantos. 2014. Almacenamiento de carbono en la biomasa aérea del bosque primario y bosque secundario de la parcela "Muro Huayra" en la Reserva Nacional Allpahuayo Mishana, Iquitos, Perú. Master's Thesis, UNAP, Facultad de Ciencias Forestales, Iquitos

Macedo, Domingo S.; Anderson, Anthony B. 1993. Early ecological changes associated with logging in an Amazon floodplain. Biotropica 25(2):151–163

Machado, Kathleen Samira da Silva. 2009. Estrutura e composição florística de uma floresta de terra firme na reserva de desenvolvimento sustentável Amanã, Amazônia Central. Master's thesis, INPA, Manaus

Machado, Danilo de Oliveira. 2016. A agrobiodiversidade de quintais agroflorestais em propriedades agrícolas familiares na BR 174, Ramal do Pau-Rosa, Manaus, AM. Master's Thesis, INPA, Manaus

Maranho, Álisson Sobrinho; Salimon, Cleber Ibraim. 2015. Maranthaceae overabundance decreases richness and abundance of regenerating woody plants in natural gaps. Neotropical Biology and Conservation 10(2):53–62

Marcoy, P. 1873. *A Journey across South America from the Pacific Ocean to the Atlantic Ocean*. Blackie and Son, London

Marimon, B. S.; Felfili, J. M.; Lima, E. S. 2002. Floristics and phytosociology of the gallery forest of the Bacaba stream, Nova Xavantina, Mato Grosso, Brazil. Edinburgh Journal of Botany 59(2):303–318

Maroni, P. 1988. *Noticias Autenticas del Famoso rio Marañón y Misión Apostólica de la Compañia de Jesús de la Provincia de Quito en los Dilatados Bosques de dicho Rio, Escribialas por los Años de 1738*. IIAP, Iquitos.

Marques-Souza, Antonio Carlos; Moura, Cleonice de Oliveira; Nelson, Bruce Walker. 1996. Pollen collected by *Trigona williana* (Hymenoptera: Apidae) in central Amazonia. Revista de Biologia Tropical 44(2):567–573

Maués, Márcia Motta; Sousa, Lindaurea Alves de; Miyanaga, Ryoichi. 2000. Insetos polinizadores do cupuaçuzeiro (*Theobroma grandiflorum* (Willd. Ex Sprengel) Schum, Sterculiaceae) no estado do Pará, Brasil. Embrapa Amazônia Oriental, Belém, Circular Técnica 12

Medeiros, Marcelo Brilhante de; Walter, Bruno Machado Teles. 2012. Composição e estrutura de comunidades arbóreas de cerrado stricto sensu no norte do Tocantins e sul do Maranhão. Revista Árvore 36(4): 673–683

Medeiros, M., Silva, S., Walter, B., Oliveira, W., Silva, G., Gomes, B., Moser, P., Lima, I. and Cavalcanti, T., 2012. Vascular flora of the Tocantins river middle basin, Brazil. Check list, 8(5), pp. 852–885

Meggers, B. J. and C. Evans. 1957. *Archeological Investigations at the Mouth of the Amazon.* Bureau of American Ethnology, Bulletin 167, Smithsonian Institution, Washington, D. C.

Mejia, Kember; Rengifo, Elsa. 2000. Plantas Medicinales de Uso Popular en la Amazonía Peruana. Agencia Española de Cooperación Internacional, Lima

Mendes F. A. T. 2014. Descrição da cadeia produtiva do cacau na Amazônia. In: Mercado Cadeia Produtiva e Desenvolvimento Rural na Amazônia, A. C. Santana (ed.), Universidade Federal Rural da Amazônia (UFRA), Belém, pp. 113–139

Mendoza, Napoleón Vela. 2014. Las manifestaciones naturales como indicadores del calendario bosquesino. Mundo Amazónico 5: 377–423

Mendoza, Abel L. Monteagudo; Guerrero, Marleni Huamán. 2010. Catálogo de los arboles y afines de la selva central del Perú. Arnaldoa 17(2):202–242

Miller, Robert Pritchard; Nair, P. K. R. 2006. Indigenous agroforestry systems in Amazonia: from prehistory to today. Agroforestry Systems 66:151–164

Milliken, W., R. P. Miller, S. R. Pollard, and E. V. Wandelli. 1992. *The Ethnobotany of the Waimiri Atroari Indians of Brazil.* Royal Botanic Gardens, Kew

Miranda, Izildinha S. 2000. Análise florística e estrutural da vegetação lenhosa do Rio Comemoração, Pimenta Bueno, Rondônia, Brasil. Acta Amazonica 30(3):393–422

Montalván, Nadiuska Giannina Achong. 2015. Plantas alimenticias usadas por mamíferos mayores en la cuenca alta del Río Itaya, Loreto, Perú. Bachelor thesis, UNAP, Facultad de Ciencias Biológicas, Iquitos

Monteles, Ricardo; Pinheiro, Claudio Urbano B. 2007. Plantas medicinais em um quilombo maranhense: uma perspectiva etnobotânica. Revista de Biologia e Ciências da Terra 7(2):1–12

Mora, Santiago; Herrera, Luisa Fernanda; Cavelier, Inés; Rodríguez, Camilo. 1991. Cultivars, Anthropic Soils and Stability: A Preliminary Report of Archaeological Research in Araracuara, Colombian Amazon. University of Pittsburgh Latin American Archaeology Reports No. 2, Pittsburgh

Moran, Emilio F. 1979. The Transamazonica: coping with a new environment. In: Brazil Anthropological Perspectives: Essays in Honor of Charles Wagley, Margolis, Maxine L.; Carter, William E. (eds), Columbia University Press, New York, pp 133–159

Moreira, Déborah Luíza; Guarim-Neto, Germano. 2009. Usos múltiplos de plantas do cerrado: um estudo etnobotânico na comunidade sítio Pindura, Rosário Oeste, Mato Grosso, Brasil. Polibotánica 27:159–190

Moser, Pamela. 2013. Vegetação Arbórea e sua Relação com Fatores Ambientais e Especiais em Florestas de Terra Firme no Noroeste de Rondônia, Brasil. Master's thesis, Faculdade de Tecnologia, Departamento de Engenharia Florestal, Programa de Pós-Graduação em Ciências Florestais, Universidade de Brasília

Moura, Ignácio Baptista de. 1989. De Belém a S. João do Araguaia, Vale do Rio Tocantins. Fundação Cultural do Pará Tancredo Neves (FCPTN)/Secretaria de Estado da Cultura (SECULT), Belém

Navarro, G. 2002. Provincia biogeográfica del Acre-Madre de Dios (Amazónica Suroccidental). In: Geografía Ecológica de Bolivia: Vegetación y Ambientes Acuáticos, Navarro, Gonzalo; Maldonado, Mabel (eds), Centro de Ecología Simón I. Patiño, Cochabamba, pp 51–98

Navarro, G.; Maldonado, M. 2002. Geografía Ecológica de Bolivia: Vegetación y Ambientes Acuáticos. Centro de Ecología Simón I. Patiño, Cochabamba

Nunes, Andréa. 1998. Diet and feeding ecology of *Ateles belzebuth belzebuth* at Maracá Ecological Station, Roraima, Brazil. Folia Primatologica 69:61–76

Ogata, N., A. Gómez-Pompa, and K.A. Taube, 2006. The domestication and distribution of *Theobroma cacao* L. in the Neotropics. In: *Chocolate in Mesoamerica: A Cultural History of Cacao*, C.L. McNeil (Ed.), pp. 69–89. University Press of Florida, Gainesville.

Oliveira, S. G. de; Alfaro, Jessica W. Lynch; Veiga, Liza M. 2014. Activity budget, diet, and habitat use in the critically endangered Ka'apor capuchin monkey (*Cebus kaapori*) in Pará state,

Brazil: a preliminary comparison to other capuchin monkeys. American Journal of Primatology 76:919–931

Oliveira, Lucicléia Mendes de; Sousa, Rhonan Martins de; Correa, Norma Elena Reynosa; Santos, André Ferreira dos; Giongo, Marcos. 2018. Florística e síndromes de dispersão de um fragmento de cerrado ao sul do estado do Tocantins. Scientia Agraria Paranaensis 17(1): 104–111

Palacios, Erwin; Rodríguez, Adriana; Defler, Thomas R. 1997. Diet of a group of *Callicebus torquatus lugens* (Humboldt, 1812) during the annual resource bottleneck in Amazonian Colombia. International Journal of Primatology 18(4):503–522

Palma, Ana Cristina; Stevenson, Pablo R. 2010. Dispersión de semillas por monos araña en la Estación Biológica Cocha Cashu, Perú. In: Primatología en Colombia: Avances al Principio del Milenio, Pereira-Bengoa, Victoria; Stevenson, Pablo R.; Bueno, Marta Lucía; Nassar-Montoya, Fernando (eds), Fundación Universitaria San Martín, Bogotá, pp 19–35

Palma, Ana Cristina; Stevenson, Pablo R. 2013. Dispersión de semillas por monos araña en las Estación Biológica Cocha Cashu, Parque Nacional Manu, Perú. In: ReporteManu: Pasión por la Investigación en la Amazonía Peruana, Groenendijk, Trovar; Wust, Walter (eds), San Diego Zoo Global Peru, Cusco, pp. 178–195

Pasa, Maria Corette. 2011. Saber local e medicina popular: a etnobotânica em Cuiabá, Mato Grosso, Brasil. Boletim do Museu Paraense Emílio Goeldi, Ciencias Humanas 6(1):179–196

Pereira, Carmen Nazário; Maneschy, Rosana Quaresma; Oliveira, Pâmela Dias; Oliveira, Ilmaione Keiza de Souza. 2010. Caracterização de quintais agroflorestais no projeto de assentamento Belo Horizonte I, São Domingos do Araguaia, Pará. Agroecossistemas 2(1):73–81

Pereira, Luciano Araújo; Sobrinho, Felipe de Araújo; Costa Neto, Salustiano Vilar da. 2011. Florística e estrutura de uma mata de terra firme na Reserva de Desenvolvimento Sustentável no Rio Iratapuru, Amapá oriental, Brasil. Floresta 41(1):113–122

Peres, C.A. 1994. Diet and feeding ecology of gray woolly monkeys (*Lagothrix lagotricha cana*) in central Amazonia: comparisons with other Atelines. International Journal of Primatology 15(3):333–372

Perez, Inayê Uliana. 2011. Uso dos recursos naturais vegetais na comunidade indígena Araçá, Roraima. Master's thesis, Universidade Federal de Roraima, Boa Vista

Pérez-Arbeláez, Enrique. 1956. Plantas Útiles de Colombia. Liberia Colombiana, Bogotá

Perigolo, Natália Alves; Medeiros, Marcelo Brilhante de; Simon, Marcelo Fragomeni. 2017. Vegetation types of the upper Madeira River in Rondônia, Brazil. Brittonia https://doi.org/10.1007/s12228-017-9505-1

Pinheiro, Renato Torres; Marcelino, Dianes Gomes; Moura, Dieyson Rodrigues. 2018. Espécies arbóreas de uso múltiplo e sua importância na conservação da biodiversidade nas áreas verdes urbanas de Palmas, Tocantins. Desenvolvimento e Meio Ambiente 49: 264–282

Pinho, Rachel Camargo de. 2008. Quintais agroflorestais indígenas em áreas de savana (lavrado) na Terra Indígena Araçá, Roraima. Master's thesis, INPA/UFAM, Manaus

Pinto, Andréia Zelenski de Lara; Assis, Ana Flávia Silva de; Pereira, Arildo Gonçalo; Pasa, Maria Corette. 2013. Levantamento etnobotânico de plantas medicinais comercializados no mercado do porto em Cuiabá, Mato Grosso, Brasil. Flovet 5:51–70

Pires, João Murça; Salomão, Rafael de Paiva. 2000. Dinâmica da diversidade arbórea de um fragmento de floresta tropical primária na Amazônia oriental-1. Período: 1956 a 1992. Boletim do Museu Paraense Emílio Goeldi, Série Botânica 16(1):63–110

Pitman, Nigel C. A.; Andino, Juan Ernesto Guevara; Aulestia, Milton; Cerón, Carlos E.; Neill, David A.; Palacios, Walter; Rivas-Torres, Gonzalo; Silman, Miles R.; Terborgh, John F. 2014. Distribution and abundance of tree species in swamp forests of Amazonian Ecuador. Ecography 37: 902–915

Pontes, Antonio Rossano Mendes. 1997. Habitat partitioning among primates in Maracá Island, Roraima, northern Brazilian Amazonia. International Journal of Primatology 18(2):131–157

Porter, L. M. 2001. Dietary differences among sympatric Callitrichinae in northern Bolivia: *Callimico goeldii*, *Saguinus fuscicollis* and *S. labiatus*. International Journal of Primatology 22(6):961–992

Posey, D. A. 1985. Native and indigenous guidelines for new Amazonian development strategies: understanding biological diversity through ethnoecology. In: Change in the Amazon Basin Volume 1: Man's Impact on Forests and Rivers, Hemming, John (ed), Manchester University Press, Manchester, pp 156–181

Pound, F. J. 1938. Cacao and Witchbroom Disease (*Marasmius perniciosus*) of South America with Notes on other Species of Theobroma. Yuille's Printerie, Port-of-Spain, Trinidad and Tobago

Prado, Helena Alves do. 2012. Frugivoria e dispersão de sementes por anta (*Tapirus terrestris*) em uma área de plantio florestal no cerrado, MS. Master's thesis, Universidade Estadual Paulista, Rio Claro, São Paulo

Prado, Helbert Medeiros. 2013. Feeding ecology of five Neotropical ungulates: a critical review. Oecologia Australis 17(4): 459–473

Prance, G. T. 1972a. An ethnobotanical comparison of four tribes of Amazonian Indians. *Acta Amazonica* 2(2): 7–27

Prance, G. T. 1972b. Ethnobotanical notes from Amazonian Brazil. Economic Botany 26(3):221–237

Prance, G.T. and M.F. da Silva. 1975. Árvores de Manaus. INPA, Manaus

Queiroz, José Antonio Leite de; Machado, Sebastiã do Amaral; Hosokawa, Roberto Tuyoshi; Silva, Ivan Crespo da. 2007. Estrutura e dinâmica de floresta de várzea no estuário Amazônico no estado do Amapá. Floresta 37(3):339–352

Ratter, J. A. 1987. Notes on the vegetation of the Parque Nacional do Araguaia (Brazil). Notes from the Royal Botanic Garden, Edinburgh 44(2):311–342

Rech, André Rodrigo; Absy, Maria Lúcia. 2011. Pollen sources used by species of Meliponini (Hymenoptera: Apidae) along the Rio Negro channel in Amazonas, Brazil. Grana 50(2):150–161

Regan, B. C.; Julliot, C.; Simmen, B.; Viénot, F.; Charles-Dominique, P.; Mollon, J. D. 2001. Fruits, foliage and the evolution of primate colour vision. Philosophical Transactions of the Royal Society London B 356: 229–283

Rengifo-Salgado, Elsa; Rios-Torres, Sandra; Malaverri, Lizardo Fachín; Vargas-Arana, Gabriel. 2017. Saberes ancestrales sobre el uso de flora y fauna en la comunidad indígena Tikuna de Cushillo Cocha, zona fronteriza Perú-Colombia-Brasil. Revista Peruana de Biología 24(1):67–78

Reyes-García, Victoria. 2001. Indigenous people, ethnobotanical knowledge, and market economy: a case study of the Tsimane' Amerindians in lowland Bolivia. Doctoral dissertation, University of Florida, Gainesville.

Reynel, C; Pennington, T.D.; Pennington, R.T.; Flores, C.; Daza, A. 2003. *Árboles Útiles de la Amazonía Peruana y sus Usos: Un Manual con Apuntes de Identificación, Ecología y Propagación de las Especies*. Tarea Gráfica Educativa, Lima

Rodrigues, William A. 1963. Estudo de 2,6 hectares de mata de terra firme da Serra do Navio, território do Amapá. Boletim do Museu Paraense Emílio Goeldi, Nova Série Botânica 19:1–23

Rodrigues, I. A. 1986. Inventário florístico em áreas do projeto Albrás-Alunorte, Barcarena-Pa. *Anais do Primeiro Simpósio do Trópico Umido, 12 a 17 novembro 1984*, EMPRABA-CPATU, Belém, vol. 2, pp. 153–166

Rodrigues, Santina; Caetano, Diego G.; Caetano, Creucí Maria. 2007. Especies frutíferas do centro-sul do Estado de Rondônia, Amazônia brasileira. Acta Agronómica 56(2):69–74

Rogez, Hervé; Buxant, Raphaële; Mignolet, Eric; Souza, Jesus N. S.; Silva, Evaldo M.; Larondelle, Yvan. 2004. Chemical composition of the pulp of three typical Amazonian fruits: araça-boi (Eugenia stipitata), bacuri (Platonia insignis) and cupuaçu (Theobroma grandiflorum). European Food Research and Technology 218:380–384

Rojas, Natalia Alejandra Alvis. 2012. Patron de actividad, dieta, área de actividad y dispersion de semillas de *Alouatta seniculus* en un fragmento de Bosque en San Juan del Carare (Santander). Thesis, Universidad del Tolima, Facultad de Ciencias, Bogotá

Rondón, J.B. and L.J. Cumana. 2005. Revisión taxonómica del género *Theobroma* (Sterculiaceae) en Venezuela. *Acta Botanica Venezuelica* 28(1):113–134

Rostain, S. 2017. Amazonie: Les 12 Travaux des Civilisations Précolombiens. Belin, Paris

Roubach, R. and U. Saint-Paul. 1994. Use of fruits and seeds from Amazonian inundated forests in feeding trials with Colossoma macropomum (Cuvier, 1818) (Pisces, Characidae). Journal of Applied Ichthyology 10:134–140

Rudas, Augustín; Prieto, Adriana. 1998. Análisis florístico del parque nacional natural Amacayacu e Isla Mocagua, Amazonas (Colombia). Caldasia 20(2):142–172

Rylands, Anthony B. 1979. Observações preliminares sobre o sagüi, *Callithrix humeralifera intermedius* (Hershkovitz, 1977) em Dardanelos, rio Aripuanã, Mato Grosso. Acta Amazonica 9(3):589–602

Rylands, Anthony B. 1981. Preliminary field observations on the marmoset, *Callithrix humeralifera intermedius* (Hershkovitz, 1977) at Dardanelos, Rio Aripuanã, Mato Grosso. Primates 22(1):46–59

Salazar, Romer Salvador Miserendino. 2007. Dieta de los chanchos de monte: taitetu (*Tayassu tajacu*) y tropero (*Tayassu pecari*) en el territorio indigena Siriono, Beni, Bolivia. Revista Boliviana de Ecología y Conservación Ambiental 21:43–56

Salm, Rodolfo. 2004. Tree species diversity in a seasonally-dry forest: the case of the Pinkaití site, in the Kayapó Indigenous Area, southern limits of the Amazon. Acta Amazonica 34(3):435–443

Salomão, Rafael de Paiva; Lisboa, Pedro L. B. 1988. Análise ecológica da vegetação de uma floresta pluvial tropical de terra firme, Rondônia. Boletim do Museu Paraense Emílio Goeldi, Série Botânica 4(2):195–233.

Salomão, Rafael de Paiva; Rosa, Nelson de Araújo. 1989. Análise da vegetação de floresta pluvial tropical de terra firme, pelo método dos quadrantes: Serra Norte, Carajás, PA. Acta Botanica Brasilica 2(1):27–42

Salomão, Rafael de Paiva; Vieira, Ima Célia Guimarães; Suemitsu, Chieno; Rosa, Nélson de Araújo; Almeira, Samuel Soares de; Amaral, Dário Dantas do; Menezes, Moirah Paulo Machado de. 2007a. As florestas de Belo Monte na grande curva do rio Xingu, Amazônia Oriental. Boletim do Museu Paraense Emílio Goeldi, Ciencias Naturais 2(3):57–153

Salomão, Rafael Paiva; Rosa, Nélson Araújo; Morais, Kácio Andrey Câmara. 2007b. Dinâmica da regeneração natural de árvores em áreas mineradas na Amazônia. Boletim do Museu Paraense Emílio Goeldi, Ciencias Naturais 2(2):85–139

Sampaio, F. X. R. 1825. Diário da Viagem que em Visita, e Correição, das Povações da Capitania de S. Joze do Rio Negro fez o Ouvidor, e Intendente Geral da Mesma, no Anno de 1774 e 1775. Typographia da Academia, Lisbon

Sánchez, Judith Jacira Achong; Levy, Allison Licett Nuñez. 2013. Características de frutos consumidos por *Saguinus nigrifrons* (Primates, Callitrichidae), en la Estación Biológica Quebrado Blanco, Loreto, Perú. Master's thesis, Universidad Nacional de la Amazonía Peruana, Iquitos

Sanchez, P., K. Jaffe, R. Gonzalez, I. Garcia, M.C. Muller, and J.J. Tortolero. 1988. *Proyecto Colección de Cacao Silvestre en el Amazonas*. International Board for Plant Genetic Resources, Rome, Informe Nr 3-PR3/11 IBPGR-Cocoa

Santa-Anna Nery, F. J. 1885. *Le Pays des Amazones, l'El-Dorado, les Terres a Caoutchouc*. Frinzine, Paris

Santana, José Augusto da Silva; Almeida, Wanderléa da Costa; Sousa, Luciana Karla V. dos Santos. 2004. Florística e fitossociologia em área de vegetação secundária na Amazônia oriental. Revista de Ciências Agrárias 41:105–120

Santos, Maurício Reginaldo A. dos; Lima, Maria Railda de; Ferreira, Maria das Graças R. 2008. Uso de plantas medicinais pela população de Ariquemes, Rondônia. Horticultura Brasileira 26:244–250

Santos, Gideão Costa; Tourinho, Manoel Malheiros; Mendes, Fernanda da Silva; Guimarães, Célia Maria Costa. 2014a. Fitossociologia e práticas de manejo tradicional em uma floresta de várzea em Santa Bárbara do Pará, Brasil. Revista de Ciências Agrárias 57(2):138–145

Santos, L. O., L. O. Santos, M. P. M. Menezes, C. R. Beasley, and U. Mehlig. 2014b. Composition and structure of a diverse tree community at the edges of a Brazilian Amazon rainforest island surrounded by marshes and mangroves. Plant Ecology 215: 1469–1481

Santos, M. R. A.; Lima, M. R.; Oliveira, C. L. L. G. 2014c. Medicinal plants used in Rondônia, western Amazon, Brazil. Revista Brasileira de Plantas Medicinais 16(3):707–720

Santos, Joielan Xipaia dos; Reis, Alisson Rodrigo Souza; Matos, Sterphane Araújo de; Leão, Fábio Miranda; Carvalho, Josiane Celerino de. 2016a. Caracterização etnobotânica de essências florestais com fins medicinais utilizadas pela etnia Xipaya, no município de Altamira-PA. Biota Amazônia 6(2):1–8

Santos, Otilene dos Anjos; Couceiro, Sheyla Regina Marques; Rezende, Alinne Costa Cavalcante; Silva, Márcia Daniela de Sousa. 2016b. Composition and richness of woody species in riparian forests in urban areas of Manaus, Amazonas, Brazil. Landscape and Urban Planning 150:70–78

Santos, R S, P G C Lima, M Coelho-Ferreira, A L Albernaz, A L Feliciano, R Scheel-Ybert. 2016c. Estudos botânicos realizados em Carajás e as perspectivas para uma abordagem etnobiológica e paleobotânica. In: Amazônia Antropogênica, M P Magalhães (ed), Museu Paraense Emílio Goeldi, Belém, pp. 199–213

Santos, Silvio Roberto Miranda dos; Kato, Osvaldo Ryohei; Tourinho, Manoel Malheiros. 2016d. Análise florística e finalidade de espécies cultivadas em sistemas agroflorestais no município de Cametá, Pará, Amazônia Oriental. Amazônia: Ciência e Desenvolvimento 13(23):9–22

Santos, Deborah Cruz dos; Corrêa Neto, José de Jesus; Alvarenga, Adriano Braga Brasileiro de. 2018a. Frutos consumidos por macaco-de-cheiro *Saimiri sciureus collinsi* (Osgood, 1916) em fragmentos de florestas em uma unidade de conservação no arquipélago do Marajó, Amazônia Oriental, Brasil. Revista Brasileira de Zoociências 19(1):82–90

Santos, Lucélia Rodrigues; Ferreira, Evandro José Linhares; Silva, Cláudia Eugenio. 2018b. Florística e fitossociologia de um sub-bosque urbano em Rio Branco-AC, Brasil. Biota Amazônia 8(3):28–33

Sanz-Biset, Jaume; Campos-de-la-Cruz, José; Epiquién, Mirbel A.; Cañigueral, Salvador. 2009. A first survey on the medicinal plants of the Chazuta valley (Peruvian Amazon). Journal of Ethnopharmacology 122:333–362

Scabin, Andressa Bárbara; Costa, Flávia Regina Capellotto; Schöngart, Jochen. 2011. The spatial distribution of illegal logging in the Anavilhanas archipelago (Central Amazonia) and logging impacts on species. Environmental Conservation 39(2): 111–121

Schmidt, M. V. C.; Vander Veld, Pieter; Ramos, José Barreto; Rezende, João Bosco Azevedo; Rezende, Odilon Barreto; Rezende, Renato Barreto; Resende, Alexandre Azevedo; Cabalzar, Aloisio. 2010. Capoeiras Tuyuka: processos de restauração e recuperação de terras degradadas. In: *Manejo do Mundo: Conhecimentos e Práticas dos Povos Indígenas do Rio Negro*, A. Cabalzar (ed.), Instituto Socioambiental, São Paulo, pp. 156–167

Schultes, R. E. 1944. Plantae Austro-Americanae, II: De investigationibus generis Herrania diversae observations. Caldasia 2(9):325–336

Schultes, R.E. 1958. A synopsis of the genus *Herrania. Journal of the Arnold Arboretum* 39:216–278

Schultes, R.E. 1969. Hallucinogens of plant origin. Science 163(3864):245–254

Schultes, R. E. 1977. Diversas plantas comestíveis nativas do noroeste da Amazonia. Acta Amazonica 7(3):317–327

Schultes, R. E. 1983. De plantis toxicariis e Mundo Novo tropicale commentationes XXXII: Notes, primarily of field tests and native nomenclature, on biodynamic plants of the northwest Amazon. Harvard University Botanical Museum Leaflets 29(3):251–272

Schultes, R. E. and R. F. Raffauf. 1990. The Healing Forest: Medicinal and Toxic Plants of the Northwest Amazonia. Dioscorides Press, Portland, Oregon

Schultes, R. E. and R. F. Raffauf. 1991. De plantis toxicariis e Mundo Novo Tropicale Commentationes XXXVI: phytochemical and Ethnopharmacological notes on the Solanaceae of the Northwest Amazon. In: Solanaceae III: Taxonomy, Chemistry, Evolution, Hawkes, J. G.; Lester, R. N.; Nee, M.; Estrada, N. (eds.), Royal Botanic Gardens, Kew, pp. 25–49

Schultes, R. E. and R. F. Raffauf. 1992. *Vine of the Soul: Medicine Men, their Plants and Rituals in the Colombian Amazonia*. Synergetic Press, Oracle, Arizona

Scoles, Ricardo. 2009. El quintal y las frutas: recursos económicos y alimentares en la comunidad negra de Itacoã, Acará, Pará, Brasil. Acta Amazonica 39(1):1–12
Scudeller, Veridiana Vizoni; Ramos, Renata Azevedo; Cruz, Maria Eliene Gomes da. 2009. Flora fanerogâmica da floresta de terra firme na RDS Tupé. In: Biotupé: Meio Físico, Diversidade Biológica e Sociocultural do Baixo Rio Negro, Amazônia Central, Santos-Silva, Edinaldo Nelson; Scudeller, Veridiana Vizoni (eds), pp 109–120, UEA Edições, Manaus
Sereno, Maria L.; Albuquerque, Paulo S.B.; Vencovsky, Roland; Figueira, Antonio. 2006. Genetic diversity and natural population structure of cacao (*Theobroma cacao* L.) from the Brazilian Amazon evaluated by microsatellite markers. Conservation Genetics 7:13–24
Setz, E. Z. F.; Enzweiler, J.; Solferini, V. N.; Amêndola; Berton, R. S. 1999. Geophagy in the golden-faced saki monkey (*Pithecia pithecia chrysocephala*) in the Central Amazon. Journal of the Zoological Society of London 247:91–103
Silva, Fabíola Andréa. 2008a. Ceramic technology of the Asurini do Xingu, Brazil: an ethnoarchaeological study of artifact variability. Journal of Archaeological Method and Theory 15:217–265
Silva, Maria Betânia Araujo da. 2008b. Composição florística e estrutura fitossociológica de uma floresta ripária situada às margens do Rio Moju, Pará. Master's thesis, Universidade Federal Rural da Amazônia, Belém
Silva, Regina Oliveira da. 2008c. Biodiversidade e políticas de conservação: o caso do parque Estadual Monte Alegre, Pará. Doctoral dissertation, Universidade de Brasília
Silva, Rafael Carletti Marcolino da. 2011a. Agrobiodiversidade de propriedades agrícolas familiares no município de Presidente Figueiredo, AM. Master's thesis, INPA, Manaus
Silva, Silvestre. 2011b. Fruits of the Brazilian Amazon. Metalivros, São Paulo
Silva, Wheriton Fernando Moreira da. 2015. Florística, dinâmica e estrutura da regeneração natural em floresta explorada na FLONA do Tapajós, PA. Master's Thesis, INPA, Manaus
Silva, Edinei Santos da. 2017. Plantas alimentícias em comunidades agrícolas no município de Rio Preto da Eva-AM. Master's thesis, INPA, Manaus
Silva, Alessandra de Azevedo Rodrigues da; Martins, Marlúcia Bonifácio. 2009. Polinização de cacauí *Theobroma speciosum* Willd. Ex Spreng. (Sterculiaceae) por drosofilídeos em área de occorrência natural. In: Caxiuanã: Desafios para a Conservação de uma Floresta Nacional na Amazônia, P. L. B. Lisboa (ed.), Museu Paraense Emílio Goeldi, Belém, pp. 431–440
Silva, M. F., P. L. B. Lisbôa, R. C. L. Lisbôa 1977. Nomes Vulgares de Plantas Amazônicas. INPA, Manaus
Silva, José Natalino Macedo; Carvalho, João Olegário Pereira de; Lopes, José do Carmo Alves. 1985. Inventário florestal de uma área experimental na Floresta Nacional do Tapajós. Boletim de Pesquisa Florestal 10/11:38–110
Silva, C. R. S.; Albuquerque, PSB; Ervedosa, FR; Mota, JWS; Figueira, A; Sebbenn, AM. 2011. Understanding the genetic diversity, spatial genetic structure and mating system at the hierarchical levels of fruits and individuals of a continuous *Theobroma cacao* population from the Brazilian Amazon. Heredity 106: 973–985
Silva, Kátia Emidio da; Martins, Sebastião Venâncio; Fortin, Marie-José; Ribeiro, Milton Cezar; Azevedo, Celso Paulo de; Ribeiro, Carlos Antonio Álvares Soares; Santos, Nerilson Terra. 2014a. Tree species community spatial structure in a terra firme Amazon forest, Brazil. Bosque 35(3):347–355
Silva, Thiago Paixão da; Silva, Erica Micaelli de Jesus; Amorim, Ismael Alves; Aquino, Alyssandra Lima; Matos, Thaís Esllem da Silva; Rodrigues, Diego Macedo. 2014b. Levantamento de espécies vegetais e utilização em quintal agroflorestal de estabelecimento agrícola no assentamento Alegria – Marabá, Pará. Agroecossistemas 6(1):103–109
Silva, Wanuyze Adriana Silva da; Carim, Marcelo de Jesus Veiga; Guimarães, José Renan da Silva; Tostes, Luciedi de Cássia Leôncio. 2014c. Composição e diversidade florística em um trecho de floresta de terra firme no sudoeste do Estado do Amapá, Amazônia Oriental, Brasil. Biota Amazônia 4(3):31–36
Silva, Hanna Kassia Machado da; Gama, João Ricardo Vasconcellos; Sousa, Randerson José de Araujo; Lameira, Mahyanny Karoline da Silva; Costa, Daniele e Lima da; Oliveira, Douglas

Valente de; Rocha, Jobert Silva da; Oliveira, Thiago Gomes de Sousa. 2017a. Composição florística de quintais agroflorestais na vila Cuera, Bragança, Pará. Agroecossistemas 9(2):330–338
Silva, Silmara Bento; Sousa, Venderleia Conceição; Santos, Crissogno Mesquita; Mariano, Daiane Cinque; Okumura, Ricardo Shigueru. 2017b. Levantamento florístico do componente arbustivo-arbóreo da vegetação ciliar de fragmento no Rio Parauapebas. Agroecossistemas 9(1):99–115
Simmen, Bruno; Sabatier, Daniel. 1996. Diets of some French Guianan primates: food composition and food choices. International Journal of Primatology 17(5):661–693
Sirén, A. H. 2007. Population growth and land use intensification in a subsistence-based indigenous community in the Amazon. Human Ecology 35:669–680
Siveiro, A.; Delunardo, T. A.; Haverroth, M.; Oliveira, L. C.; Mendonça, A. M. S. 2012. Plantas medicinais em quintais urbanos de Rio Branco, Acre. Revista Brasileira de Plantas Medicinais 14(4):598–610
Smith, Herbert H. 1879. Brazil, the Amazons and the Coast. Charles Scribner's Sons, New York
Smith, N. J. H. 1974. Destructive exploitation of the South American river turtle. *Yearbook of the Pacific Coast Geographers* 36:85–102
Smith, N.J.H. 1999. Biodiversity and agroforestry along the Amazon floodplain. In: *Traditional and Modern Natural Resource Management in Latin America*, F.J. Pichón, J.E. Uquillas, and J. Frechione (Editors), pp. 233–250. University of Pittsburgh Press, Pittsburgh
Smith, N.J.H. 2009. Jewels of the forest. In: From Peru to the World: Chocolate, A,. Gutsche et al. (eds), Wust Ediciones, Lima, pp. 31–37
Smith, N. J. H.; J. T. Williams; D. L. Plucknett; and J. P. Talbot. 1992. *Tropical Forests and Their Crops*. Cornell University Press, Ithaca.
Soini, P. 1988. The pygmy marmoset, genus *Cebuella*. In: Ecology and Behavior of Neotropical Primates, Mittermeier, Russell A.; Rylands, Anthony B.; Coimbra-Filho, Adelmar F.; Fonseca, Gustavo A.B. da (eds), Ecology and Behavior of Neotropical Primates, World Wildlife Fund, Washington, DC, vol. 2, pp 79–129
Souza, F. B. 1873. *Lembranças e Curiosidades do Valle do Amazonas*. Typografia do Futuro, Belém.
Souza, I. 1973. *O Cacaulista*. Universidade Federal do Pará, Belém.
Souza, Célia Cristina Valero. 2010. Etnobotânica de quintais em três comunidades ribeirinhas na Amazônia central, Manaus, AM. Master's thesis, INPA, Manaus
Souza, A. G. C., N. R. Sousa, S. E. L. Silva, C. D. M. Nunes, A. C. C. Canto, L. A. A. Cruz, 1996. *Fruteiras da Amazônia*. EMBRAPA (Empresa Brasileira de Agropecuária), Brasília
Spix, J. B. and C. F. P. Martius, 1976. *Viagem pelo Brasil, 1817–1820*. Edições Melhoramentos, São Paulo, Vol.3
Spruce, R. 1908a. Notes of a Botanist on the Amazon and Andes. Macmillan, London, vol. 1
Spruce, R. 1908b. Notes of a Botanist on the Amazon and Andes. Macmillan, London, vol. 2
Stasi, Luiz Claudio di; Santos, Elza Maria Guimarães; Santos, Claudenice Moreira dos; Hiruma, Clélia Akiko. 1989. Plantas Medicinais na Amazônia. Editora UNESP, São Paulo
Stevenson, Miranda F.; Rylands, Anthony B. 1988. The marmosets, genus *Callithrix*. In: Ecology and Behavior of Neotropical Primates, Mittermeier, Russell A.; Rylands, Anthony B.; Coimbra-Filho, Adelmar F.; Fonseca, Gustavo A.B. da (eds), Ecology and Behavior of Neotropical Primates, World Wildlife Fund, Washington, DC, vol. 2, pp 131–222
Stewart, Douglas Ian. 1994. After the Trees: Living on the Transamazon Highway. University of Texas Press, Austin
Stone, D. 1984. Pre-Columbian migration of *Theobroma cacao* Linnaeus and *Manihot esculenta* Crantz from northern South America into Mesoamerica: a partially hypothetical view. In: Pre Columbian Plant Migration, D. Stone (ed), Peabody Museum of Archaeology and Ethnology, Harvard University, Cambridge, pp 67–83
Suarez, Scott A. 2006. Diet and travel costs for spider monkeys in a nonseasonal, hyperdiverse environment. International Journal of Primatology 27(2):411–436
Taurino, Tássia Cristina da Conceição Barros. 2017. Recuperação natural da diversidade de espécies de plantas em florestas secundárias na Amazônia Oriental. Master's Thesis, Universidade Federal do Pará, Belém

Terborgh, J. 1983. Five New World Primates: A Study in Comparative Ecology. Princeton University Press, Princeton

Thomas, Evert. 2009. New light on the floristic composition and diversity of indigenous territory and national park Isiboro-Sécure, Bolivia. Biodiversity Conservation 18:1847–1878

Thomas, Evert; Zonneveld, Maarten van; Loo, Judy; Hodgkin, Toby; Galluzzi, Gea; Etten, Jacob van. 2012. Present spatial diversity patterns of Theobroma cacao L. in the Neotropics reflect genetic differentiation in Pleistocene refugia followed by human-influenced dispersal. PLOS One 7(10) e47676

Trujillo, William; Correa-Múnera, Marco 2010. Plantas usadas por una comunidad indígena Coreguaje en la Amazonía Colombiana. Caldasia 32(1):1–20

Uhl, C.; Buschbacher, R.; Serrão, E. A. S. 1988. Abandoned pastures in eastern Amazonia. I. Patterns of plant succession. Journal of Ecology 76:663–681

Valdez, F. 2013. Mayo Chinchipe: hacia un replanteamiento del origin de las sociedades complejas en la civilizaci6n Andina. In: Arqueología Amazónica: Las Civilizaciones Ocultas del Bosque Tropical, F. Valdez (ed.), Ediciones Abya Yala, Quito, pp. 99–145

Van den Berg, M. E. 2010. Plantas Medicinais na Amazônia. Museu Paraense Emílio Goeldi, Belém

Van den Berg, M.E. and M.H.L. da Silva. 1986. Plantas medicinais do Amazonas. In: Anais do1°Simpósio do Trópico Úmido, Belém, PA, 12 a 17 November, vol. 2, Flora e Floresta, EMBRAPA, Brasília, pp. 127–133

Van den Berg, Maria Elisabeth; Silva, Milton Hélio Lima da. 1988. Contribuição ao conhecimento da flora medicinal de Roraima. Acta Amazonica 18(1–2) supl.:23–35

Van den Eynden, Veerle. 2004. Use and management of edible non-crop plants in southern Ecuador. Doctoral dissertation, Universiteit Gent, Belgium

Van den Eynden, Veerle; Cueva, Eduardo; Cabrera, Omar. 1999. Plantas Silvestres Comestibles del Sur del Ecuador. Ediciones Abya-Yala, Quito

Van Roosmalen, Marc G. M.; Klein, Lewis L. 1988. The spider monkeys, genus *Ateles*. In: Ecology and Behavior of Neotropical Primates, Mittermeier, Russell A.; Rylands, Anthony B.; Coimbra-Filho, Adelmar F.; Fonseca, Gustavo A.B. da (eds), Ecology and Behavior of Neotropical Primates, World Wildlife Fund, Washington, DC, vol. 2, pp 455–537

Vásquez, Silvia Patricia Flores. 2014. Conhecimento, uso e conservação da diversidade vegetal em quatro comunidades ribeirinhas no Município Manacapuru, Amazonas. Doctoral dissertation, Programa de Pós-Graduação em Botânica, INPA, Manaus

Vásquez, R. and A. H. Gentry. 1989. Use and misuse of forest-harvested fruits in the Iquitos area. *Conservation Biology* 3(4): 350–361.

Vásquez, Silvia Patricia Flores; Mendonça, Maria Silvia de; Noda, Sandra do Nascimento. 2014. Etnobotânica de plantas medicinais em comunidades ribeirinhas do Município de Manacapuru, Amazons, Brasil. Acta Amazonica 44(4):457–472

Vebrova, Hana; Lojka, Bohdan; Husband, Thomas P.; Zans, Maria Elena Chuspe; Van Damme, Patrick; Rollo, Alexandr; Kalousova, Marie. 2014. Tree diversity in cacao agroforests in San Alejandro, Peruvian Amazon. Agroforestry Systems 88:1101–1115

Veléz, Germán Alonso. 1991. Los frutales amazónicos cultivados por las comunidades indígenas de la región del Medio Caquetá. Colombia Amazonica 5(2): 135–183

Veléz, Germán Alonso; Veléz, Antonio José. 1992. Sistema agroflorestal de "chagras" utilizado por las comunidades indígenas del medio Caquetá (Amazonia Colombiana). Colombia Amazonica 6(1): 101–134

Veléz, G. A. and A. J. Veléz. 1999. Sistema Agroflorestal de las Chagras Indígenas del Medio Caquetá. Tropenbos, Estudios en la Amazonia Colombiana, Bogotá, vol. 17

Venturieri, Giorgini Augusto; Aguiar, Jaime Paiva Lopes. 1988. Composição do chocolate caseiro de amêndoas de cupuaçu (Theobroma grandiflorum Willd ex Spreng Schum). Acta Amazonica 18(1–2):3–8

Venturieri, Giorgio Cristino; Maués, Márcia Motta; Miyanaga, Ryoichi. 1997. Polonização do cupuaçuzeiro (*Theobroma grandiflorum*, Sterculiaceae): um caso de cantarofilia em uma fruteira Amazônica. Seminário Internacional sobre Pimenta-do-reino e Cupuaçu, Embrapa Amazônia Oriental, Documentos 89, Belém, Pará, pp. 341–350

Vickers, W. T. and T. Plowman. 1984. *Useful Plants of the Siona and Secoya Indians*. Field Museum of Natural History, Fieldiana Botany New Series, no. 15
Vieira, Thiago Almeida; Rosa, Leonilde dos Santos; Vasconcelos, Paulo César Silva; Santos, Mônica Mota dos; Modesto, Rozi da Silva. 2007. Sistemas agroflorestais em áreas de agricultores familiares em Igarapé-Açu, Pará: caracterização florística, implantação e manejo. Acta Amazonica 37(4):549–558
Violante, Ivana Maria Póvoa; Hamerski, Lidilhone; Garcez, Walmir Silva; Batista, Ana Lucia; Chang, Marilene Rodrigues; Pott, Vali Joana; Garcez, Fernanda Rodrigues. 2012. Antimicrobial activity of some medicinal plants from the cerrado of the central-western region of Brazil. Brazilian Journal of Microbiology 43:1302–1308
Wallace, Alfred Russel. 1889. A Narrative of Travels on the Amazon and Rio Negro, with an Account of the Native Tribes, and Observations on the Climate, Geology, and Natural History of the Amazon Valley. Ward, Lock & Co., London
Wallace, R. B. 2005. Seasonal variations in diet and foraging behavior of Ateles chamek in a southern Amazonian tropical forest. International Journal of Primatology 26(5): 1053–1075
Wallace, Robert B.; Painter, R. Lilian E.; Taber, Andrew B. 1998. Primate diversity, habitat preferences, and population density estimates in Noel Kempff Mercado National Park, Santa Cruz Department, Bolivia. American Journal of Primatology 46:197–211
Walschburger, Thomás; Hildebrand, Patriciovon. 1988. Observaciones sobre la utilización eastacional del bosque humedo tropical por los indígenas del Río Mirití (Amazonas, Colombia). Colombia Amazónica 3 (1):51–74
Wittmann, Florian; Parolin, Pia. 1999. Phenology of six tree species from central Amazonian várzea. Ecotropica 5(1):51–57
Wittmann, Florian; Schöngart, Jochen; Queiroz, Helder L.; Wittmann, Astrid de Oliveira; Conserva, Auristela dos Santos; Piedade, Maria T. F.; Kesselmeier, Jürgen; Junk, Wolfgang J. 2009. The Amazon floodplain demonstration site: sustainable timber production and management of central Amazonian white-water floodplains. Ecohydrology & Hydrobiology 9(1):41–54
Wolters, B. 1999. Zur verbreitungsgeschichte und ethnobotanik indianischer kulturpflanzen, insbesondere des kakaobaumes. *Journal of Applied Botany* 73: 128–137.
Wolters, B 2001. Dissemination of American economic plants on pre-Columbian sea routes by Amerindians. Migration and Diffusion 1(7):40–58
Worbes, Martin. 1997. The forest ecosystem of the floodplains. Ecological Studies 126:223–265
Yamada, M. and H.L. Gholz. 2002. An evaluation of agroforestry systems as a rural development option for the Brazilian Amazon. *Agroforestry Systems* 55:81–87.
Yamada, M.; Osaqui, H. M. L. 2006. The role of homegardens in agroforestry development: lessons from Tomé-Açu, a Japanese-Brazilian settlement in the Amazon. In: Tropical Homegardens: a time-tested example of sustainable agroforestry, Kumar, B. M.; Nair, P. K. R. (eds), pp. 299–316, Advances in Agroforestry vol. 3, Springer, Dordrecht
Yoneda, Masaaki. 1984. Ecological study of the saddle backed tamarin (*Saguinus fuscicollis*) in northern Bolivia. Primates 25(1):1–12
Young, A.H. 1994. *The Chocolate Tree: A Natural History of Cacao*. Smithsonian Institution Press, Washington, DC
Zent, S. and E. Zent. 2012. Jodï horticultural belief, knowledge and practice: incipient or integral cultivation? *Boletim do Museu Paraense Emílio Goeldi, Ciencias Humanas* 7(2): 293–338
Zhang, Dapeng; Arevalo-Gardini, Enrique; Mischke, Sue; Zúñiga-Cernades, Luis; Barreto-Chavez, Alejandro; Aguila, Jorge Adriazola del. 2006. Genetic diversity and structure of managed and semi-natural populations of cocoa (*Theobroma cacao*) in the Huallaga and Ucayali valleys of Peru. Annals of Botany 98: 647–655

Chapter 33
Melastomataceae

33.1 *Bellucia grossularioides*

Bolivia Guayabochi, guayabilla; huicama (Chácobo)

Brazil Goiaba de anta, goiaba de veado, goiaba de peixe, araçá de anta, muúba, muúba da mata, jambo da mata, jambo miúdo, jambo selvagem, jambo roxo da selva, jambinho, papaterra, tintarana; toporoda (Baniwa), mu-'ɨ (Ka'apor), hemu'y'ýwa (Piripkura), pakui'aran (Sateré-Maué), papater (Tembé), pisuiu (Wajãpi), makasi hi, pɨtɨma hi (Yanomami)

Colombia Níspero, guayabo de pava, guayabo silvestre, guayabo anselmo, guayabo de danta, guayabo de mico, guayabo de monte, guayabo cimarrón, guayabo silvestre, guayabilla, guayabito, pumarroso, pomarroso, pomarrosa, pomo silvestre, pera de agua, tuno negro, coronillo, guacury de danta; we-keé-to (Barasana), tacarique (Cubeo), tse-djoo'-ka, cherd-yoó-ka (Puinave), wach-kee'-to (Tukano)

Peru Níspero, níspero del monte, disper, fruto estrella

Venezuela Guayaba de danta, camburito, pumarroso, makasi, manzana de corona; cará-ré-yek (Pemón), makasi (Yanomami), kadade enano (Ye'kuana)

Status Wild

N. Smith, *Amazon Fruits: An Ethnobotanical Journey*, Ethnobiology,
https://doi.org/10.1007/978-3-031-12803-5_33

Jambo da mata *Bellucia grossularioides* fruits from a tree in successionary forest. Fazenda Alvorado 2, Travessão Tiradentes, km 80 Altamira-Medicilandia, Transamazon Highway, Pará, Brazil 9-5-15

Bellucia grossularioides is found in upland successionary vegetation throughout the Amazon Basin (Alvino et al. 2005; Araújo et al. 2005; Barnett et al. 2015; Bentos et al. 2008; Bustamante 2009:36; Carim et al. 2007; Coelho et al. 2003a, b, 2012; Ebert et al. 2014; Espírito-Santo et al. 2005; Fernandes Neto 2015:22; Gualberto et al. 2014; Ivanauskas et al. 2003; Lima Filho et al. 2004; Lorenzi et al. 2006:170; Medeiros 2016:91; Rayol et al. 2006, 2011; Rocha et al. 2016; Saldarriaga and Uhl 1991; Salomão et al. 2007b; Santos and Absy 2010; Taurino 2017:32; Uhl 1987). It also grows in successionary forests in Guyana (Van Andel 2001), French Guiana (Foresta et al. 1984) and in Amazonas, Venezuela (Villarreal et al. 2003). It can survive in successionary forest for at least 20 years, as in the case of an abandoned cattle pasture 80 km north of Manaus, Amazonas (Longworth et al. 2014). It also occurs in single or multiple treefall gaps in mature forest, such as in the Upper Rio Negro in Venezuela (Uhl et al. 1988).

Although mostly found on uplands, this tree also grows occasionally on floodplains, such as in the Upper Araguaia basin in the southern fringes of the Amazon

Basin in Brazil (Brito et al. 2008; Kurzatkowski et al. 2015; Martins et al. 2008), in gallery forest in the transition zone between the Amazon forest and savannas in southern Rondônia (Miranda 2000), on the floodplain of the Upper Madeira in western Rondônia (Casula 2012:62), in low-lying woods on savannas subject to periodic flooding in the Madidi region of Bolivia (Fuentes 2005), in the Xingu watershed (Salomão et al. 2007a; Zappi et al. 2016), and in the Upper Orinoco watershed in Colombia and Venezuela (Cabrera-Amaya and Rivera-Díaz 2016; Godoy et al. 1999).

Although not usually encountered in backyard gardens, it has also been recorded in some rural home gardens along Ramal do Pau Rosa, a side road leading off BR 175 north of Manaus that connects Manaus with Boa Vista in Roraima (Machado 2016:35) and in other rural communities in the vicinity of Manaus (Souza 2010:29). Its range extends into the Guianas, Central America, and as far north as Mexico.

The medium-sized tree, which can reach 30 m when fully mature, branches close to the ground so people can easily reach the fruits. Also, it starts producing fruit when only 2 m tall. Unlike most fruit-bearing trees in Amazonia, *Bellucia grossularioides* produces fruits all year-round. The majority of plants in the Amazon fruit during the rainy season since it is easier for seeds to germinate in moist soil. Others only produce in the drier months. The flowers are pollinated by several genera of stingless bees and the "reward" is pollen since no nectar is produced (Absy et al. 1980; Marques-Souza 1996; Marques-Souza et al. 1995, 1996, 2002, 2007; Oliveira et al. 2009a, b; Renner 1987). Rural people in Amazonia sometimes visit the nests of stingless bees in the forest to remove some of the honey. The distinction between *Bellucia grossularioides* and its close relative *B. pentamera* is sometimes blurry because they are known to hybridize when in close proximity (Antei 1996).

Yellow when ripe, *Bellucia grossularioides* fruits are the size of a small plum and contain numerous tiny seeds which are ingested along with the pulp. The fruits are widely consumed by rural folk in the Amazon including indigenous groups such as the Chácobo in Bolivia (Boom 1989). *Bellucia grossularioides* fruits occasionally reach urban markets, such as Iquitos (Vásquez and Gentry 1989). The Kurupukari in central Guyana also consume these fruits (Johnston and Colquhoun 1996).

Nispero *Bellucia grossularioides* fruits that Xiam, 7 years, is eating in regrowth forest. Outskirts of Nauta, Rio Marañón, Loreto, Peru 4-28-06

Several game animals are also fond of the fruits. Paca *Agouti paca* feed on them at night when they fall to the ground. Tapir *Tapirus terrestris* also eat the fruits as suggested by some of the common names for this fruit: goiaba de anta (tapir guava) and araçá da anta (tapir araçá) in Brazil and guayaba de danta in Colombia and Venezuela. This species of tapir has been confirmed eating the fruits of *Bellucia grossularioides* in Roraima in the northern Brazilian Amazon, in French Guiana which borders Amapá in northeastern Amazonia, and in the Upper Orinoco watershed in eastern Colombia (Alviz and Pérez-Albarracín 2015:40; Barcelos 2011:37; Barcelos et al. 2013; Henry et al. 2000). The Yanomami also report that tapirs eat this fruit (Albert and Milliken 2009:65). In the vicinity of Tefé, the tree is known as goiaba de veado (deer guava). In Suriname and Amazonia, red brocket deer *Mazama americana* and grey brocket deer *Mazama gouazoubira* are known to eat this fruit (Branan et al. 1985; Prado 2013).

Black spider monkeys *Ateles paniscus* and howler monkeys *Alouatta seniculus*, which are widely hunted for their meat in Amazonia and the Guianas, are also fond of the fruits and probably disperse some of the seeds (Neves and Rylands 1991; Renner 1989; Stevenson et al. 2015; Van Roosmalen 1985; Van Roosmalen and Klein 1988). The white-faced saki *Pithecia pithecia*, the brown capuchin *Cebus apella*, the highly endangered Ka'apor capuchin monkey *Cebus kaapori*, the squirrel monkey *Saimiri sciureus*, the Colombian squirrel monkey *Saimiri cassiquiarensis albigena*, the dusky titi monkey *Callicebus ornatus* and the ashy titi monkey *Plecturocebus cinarescens* also eat this fruit (Astwood et al. 2014; Carretero-Pinzón and Defler 2016; Carretero-Pinzón et al. 2016; Gregory 2006:53; Mattos 2016:30; Oliveira et al. 2014; Pinheiro et al. 2013; Roble 2008:48); although these monkeys are not usually hunted for their meat because of their small size, they probably disperse some of the seeds. In eastern Colombia, white-fronted capuchins *Cebus albifrons* have been recorded eating this fruit and likely disperse some of the seeds (Defler 1979). Game animals help disperse the seeds of *Bellucia grossularioides* when they defecate them. The fruits resemble guava in size and color and both are full of seeds. *Bellucia grossularioides* is thus one of the many Amazonian fruits which contribute indirectly to the human diet by feeding game animals. This fruit is also consumed by some wild mammals not normally hunted for food: Linnaeus's mouse opossum *Marmosa marina* and the bare-tailed woolly opossum *Caluromys philander* nocturnal animals that are widespread in forested areas of the tropical lowlands of South America also consume these fruits and disperses the seeds (Charles-Dominique et al. 1981).

The "tapir guava" is occasionally also found on the higher parts of floodplains and some fish are fond of the fruits. This aspect of the region's natural history did not escape the attention of a small-scale rancher who has a property at km 46 of the Manaus-Caracari highway. His property has a spring so in 1972 he created a fish pond and planted several fruit and nut trees around the margin to provide food for the fish. He stocked the pond with several fruit-eating fish including tambaqui *Colossoma macropomum* and pacu (*Metynnis* sp.). Unfortunately, most people who raise tambaqui use commercial fish food which alters the taste of this premier fish.

Goiaba de anta has other uses in addition to fruit. Along the Jau, a tributary of the Rio Negro in Amazonas, locals obtain a red dye from the bark for coloring some of their basketry. Resin from the bark is used to dye various plant fibers in the Rio Negro watershed (Schneider et al. 2016). The Tembé in eastern Pará also obtain a red dye from the bark. In the vicinity of Manacapuru along the lower Solimões near its confluence with the Rio Negro, locals use the bark to make a tea to treat gastritis and the wood is used for fuel (Costa and Mitja 2010; Vásquez 2014:18; Vásquez et al. 2014). In Suriname, the leaves are sold in some street markets to make a genital bathing solution (Van Andel et al. 2007) while in parts of the Colombian Amazon rural people use the fruit to treat fungal infections (Lizcano et al. 2010). In Amapá, some rural people tend to this tree in their fallows, fields, and home gardens for its timber (Pinedo-Vasquez et al. 2001).

33.2 *Bellucia pentamera*

Bolivia Guayabilla

Brazil Jambo branco, jambo do mato, goiaba de anta, araçá de anta

Colombia Sacha guayaba, guayaba de mico, guayaba de pava, guayabo coronillo, guayabo de mico, guayabo de monte, guayabilla, guayavita limón, coronillo, coronel, tuno, manzana de corona; túkare (Nukak)

Ecuador Sacha manzana, manzana de monte; cuengiaupatticho (Cofán), payas, huagra payas, quisa muyu yura (Kichwa), tuntia, túnkia, tuunki, tuinka (Shuar), titepencahue, eweyimo, ewegikabo (Waorani)

Peru Níspero, sacha níspero, níspero del monte, disper, bispera, maitoki, cola de caballo, fruto estrella, estrellita, manzanita de la selva, mansanita tropical, manzanay, guayaba de monte, guayabilla, campana huayo, verrugosa; chanchák (Achuar), yujash, yujách, tugkia (Aguaruna), maetoki (Campa), parúm (Mayna)

Venezuela Manzanita, manzana de corona; kadade (Ye'kuana)

Status Wild, spontaneous in cultural settings

Eliseo, 9 years, with nispero *Bellucia pentamera* fruits he gathered from a tree in successionary vegetation. Trail from Veinte de Enero to Buenos Aires, Rio Yanayacu, Reserva Nacional Pacaya-Samiria, Loreto, Peru 1-14-06

Bellucia pentamera is common in many disturbed sites from Mexico to Bolivia. In Amazonia, it is confined mostly to the western and northern part of the basin, including Roraima and Amapá in Brazil (Araujo-Murakami et al. 2015; Phillips 1993; Renner 1987; Tuomisto et al. 2002). In the Andes, this bee-pollinated tree is found as high as 1700 m. *Bellucia pentamera* was introduced as a fruit tree to various parts of Africa, Malaysia, Borneo, Sumatra, and Java during the colonial period where it has become naturalized (Antei 1996; Karsten 1861). This medium-sized tree which can reach 20 m grows in successionary vegetation on higher parts of floodplains as well as on uplands, such as in the watershed of the Putumayo River which borders Peru and Colombia (Gomez 2011:89; Myster 2019).

Bellucia pentamera bears copious crops of walnut-sized fruits in the rainy season. Hundreds of tiny brown seeds are embedded in the cream-colored pulp, ideal for animal dispersal (Renner 1987). Some monkeys, bats, tapirs, tortoises, birds, and ants are fond of the yellow fruits and are likely involved in seed dispersal. For example, woolly monkeys *Lagothrix lagotricha* and the critically endangered Caquetá titi monkey *Plecturocebus caquetensis* have been recorded eating these fruits in the Colombian Amazon (Acero-Murcia et al. 2018; Stevenson 2007) as has the collared titi monkey *Callicebus torquatus* and the saddle-back tamarin *Saguinus fuscicollis* in the Peruvian Amazon (Montalván 2015:39). In the Ecuadorian Amazon, the red titi monkey *Callicebus discolor* also eats this fruit and likely disperses some of the seeds (Carrillo-Bilbao et al. 2005). Geoffroy's saddleback tamarin *Saguinus fuscicollis nigrifrons* also eats the fruits of this tree and likely disperses some of the seeds (Sánchez and Levy 2013:30, 77). The fondness of some monkeys for this fruit is encapsulated in one of the common names for this tree in Colombia: guayabo de mico. In the Colombian Amazon a game bird, paujil *Mitu silvini*, has been observed feeding on this fruit and likely disperses some of the seeds (Franco and Santamaria 1997).

Farm boys snacking on nispero *Bellucia pentamera* fruits in a tree spared when the vegetation was cleared for a field (*chacra*). Outskirts of Miraflores, Rio Tigre, Loreto, Peru 3-9-04

In exposed locations, such as along the borders of fields, this tree has a broad, umbrella-shaped crown. The crown is narrower, however, in regrowth forests where they can grow as tall as 25 m in search of light. People pick the fruits from the low-hanging branches or off the ground, while youngsters clamber up the trunk and along the branches to pluck the fruits. Nispero fruits are eaten fresh at home or in the field and occasionally enter markets such as in Iquitos (Vásquez and Gentry

1989). Because the fruits are valued, *Bellucia pentamera* is typically spared when clearing vegetation to plant crops or when it arises spontaneously in fields, such as in the vicinity of Miraflores at the mouth of the Tigre River in the Peruvian Amazon.

This tree also has other uses. In the Ecuadorian Amazon, the fruit peel is used in a concoction for treating diarrhea. In the Peruvian Amazon, rural folk use the leaves and branches to prepare decoctions for treating malaria, leishmaniasis, and bacterial infections (Rojas et al. 2009). In rural parts of the Department of Loreto in the Peruvian Amazon, the leaves are used to make a brown dye (Yauri 2016:39). In the Ecuadorian Amazon, the Huaorani use the fruits to bait fish hooks while the Shuar use the inner bark to make cordage (Bennett et al. 2002: 202).

33.3 *Clidemia capitellata*

Brazil Buxixu, buyuyu, olho de gato; kana (Tukano)

Colombia Niguito, mora blanca; popuyo (Yukuna)

Peru Mullaca lanuda, mullaca morada, mullaquilla

Venezuela Tepuru' kechañe tipeñ men (Panaré)

Status Wild, spontaneous in cultural settings

Buxixu *Clidemia capitellata* in a harvested manioc field. Seringalzinho, Rio Jau, Amazonas, Brazil 10-2-12

Clidemia capitellata is one of at least eight species in the genus with edible fruits. This herbaceous plant is found in early successionary vegetation such as harvested manioc fields and along paths and roads. It also occurs as a weed in some guaraná *Paullinia cupana* plantations in the vicinity of Manaus, Amazonas (Miléo et al. 2007), and in some cattle pastures in the vicinity of Santana do Araguaia in southern Pará (Dantas and Rodrigues 1980). Usually, under 3 m tall, this pubescent shrub often occurs in dense formations and is found from Vera Cruz in Mexico to northern South America (Gleason 1939). This species occurs up to 2300 m in the Andean foothills, such as in Bolivia (Arellano 2013:184). The spindly shrub grows on uplands as well as seasonally flooded savannas and river banks throughout Amazonia as well as the Guianas and the Northeast and Atlantic forests in Brazil. It is also found in stunted woodlands growing on iron-rich sedimentary rock (*canga*) on hill-tops in the Serra dos Carajás, Pará (Mota et al. 2018; Rocha et al. 2017). It is also occasionally found in some light gaps in the understorey of upland forest, such as in the Reserva de Desenvolvimento Sustentável do Tupé along the left bank of the Rio Negro some 25 km west of Manaus (Corrêa et al. 2017). The "hairiness" of *Clidemia capitellata* explains one of the common names for the plant in Peru: mullaca lanuda (woolly mullaca).

The fruits measure 1 cm by 0.75 cm and are born in clusters year-round. The purple fruits turn black when fully ripe and contain numerous tiny seeds. Although the mesocarp is a little insipid, rural folk, such as in the Rio Negro watershed, often snack on them while walking to or from their fields. Known as buxixu or buyuyu in Brazil, the black currant-sized fruits are a welcome snack because some of them are produced in the dry season when few other fruits are available. On Colares Island some 100 km from Belém in the Amazon estuary locals prepare a poultice from the leaves to treat wounds on their dogs (Ritter et al. 2012).

Lamara, 8 years, picking buxixu *Clidemia capitellata* fruits in a harvested manioc field now being taken over by successionary vegetation. Seringalzinho, Rio Jau, Amazonas, Brazil 10-3-12

33.4 *Clidemia hirta*

Brazil Uvinha, catininga, buxixú, buxixú preto, buxuxú, buiuiú, capança branca, peixinha preta, pixirica, abranda fogo; tangara-ran-wyra (Tembé), ka'a rã, ka'nã (Tukano)

Colombia Azulejo

Ecuador Mora, chinchango; chimchiak (Shuar), obogimo (Waorani)

Peru Mullaca morada, mullaca de altura, mullaca negra, mullaca amarilla, mullaca peluda, pajar mullaca, yana mullaca, mullaquilla, matapasto, papailla, huayito, sacha mulalce, rifarillo; yurán kmis (Mayna), moca shimon (Shipibo), tesiguiño-ey (Witoto)

Venezuela Bullubullu

Status Wild, spontaneous in cultural settings

Buxixú *Clidemia hirta* in successionary vegetation. Seringalzinho, Rio Jau, Amazonas, Brazil 10-2-12

A "weed" in one region, can be a resource in another. Or even in the same area. Such is the case of *Clidemia hirta*, a shrub native to the humid lowlands of Central and South America and parts of the Caribbean (Gleason 1939). In the Andean foothills it can be found as high as 1400 m. It occurs as an understorey plant in forest on uplands as well as floodplains, including tidal forests in the Amazon estuary, as well as disturbed sites (Cárdenas et al. 2009; Corrêa et al. 2017; Lau and Jardim 2014a, b; Maués et al. 2011). It is also found in some savannas, such as in the vicinity of

Humaitá along the Madeira River in Amazonas, Brazil (Gottsberger and Morawetz 1986). It occurs on a wide variety of soil types, including patches of sandy soil with stunted vegetation known as *campina* and *campinarana* in Brazil and *varillal* in Peru (Ramírez 2015:23). It is also found in open vegetation and stunted woodlands on iron-rich sedimentary plateaus (*canga*) in the Serra dos Carajás, Pará (Mota et al. 2018; Rocha et al. 2017). It also occurs as a weed in some guaraná *Paullina cupana* plantations in the vicinity of Manaus, Amazonas (Miléo et al. 2007) as well as cattle pastures such as at the Fazenda Aruanã, km 232 of the AM-010 Manaus-Itacoatiara highway (Dantas 1979). The flowers are pollinated by stingless bees, such as *Bombus transversalis* and *Melipona fulva* near Manaus, Amazonas (Ferreira et al. 1994) and *Tetragonisca angustula* in the vicinity of Santarém, Pará (Novais and Absy 2013; Novais et al. 2014).

Children in various parts of the Peruvian Amazon, such as in the vicinity of Jenaro Herrera on the banks of the Ucayali, come home with purple-stained mouths and hands after they have been snacking on the juicy, pubescent fruits. The dark purple fruits provide convenient snacks for those walking through successionary growth on their way to fields or the forest. The fruits are easy to reach because the plants are usually under 2 m and start producing fruit when only 50 cm high. Sometimes they can reach 5 m, but the flexible stems can be easily pulled over to obtain the fruits. This plant contributes indirectly to the diet of rural people via game birds such as guans and tinamous which are fond of the fruit. The largest game animal in Amazonia, the lowland tapir *Tapirus terrestris*, also eats this fruit and disperses the seeds (Barcelos 2011:37; Barcelos et al. 2013). However, the pubescent shrub can be a nuisance when it competes with crops and pasture, hence one of its names in Peru: mata pasto (pasture killer).

Girl snacking on papailla *Clidemia hirta* fruits in successionary vegetation at the edge of her village. Yarina, Rio Yanayacu, Reserva Nacional Pacaya-Samira, Loreto, Peru 6-5-04

Clidemia hirta fruits in both the dry and rainy seasons, thus providing a welcome treat to youngsters all year-round. Although fruits are small and rather insipid, the shrubs are often abundant in early successionary growth on floodplains and uplands, especially by trails, along the edges of cultivated fields, and in sun-dappled orchards. It has been found as a weed in some manioc fields, such as along the Manaus-Itacoatiara highway (Albuquerque 1980). The shrubs are also prolific, producing some 500–1000 berries each year (Heywood 2005). The seeds are dispersed by birds (Renner 1989). Humans are also likely dispersal agents because the seeds are tiny, with as many as 100 per fruit (Wester and Wood 1977), and are ingested along with the pulp. Few people in rural areas of the Amazon have flushing toilets linked to sewage systems. In the municipality of Barcarena on the right bank of the lower Tocantins as it enters Marajó locals use this plant to treat wounds and burns (Almorozo and Gély 1988).

In regions where the plant has been introduced by accident, *Clidemia hirta* is a major weed in fields and even invades forests, thereby overpowering some of the native plants. One of the English names for this shrub, Koster's curse, attests to the plant's nefarious reputation outside its native range (Heywood 2005). Introduced to Fiji from the Guianas prior to 1890, the plant reached Java in the late nineteenth century. It became established in Samoa, British Solomon Islands, and Tonga by 1934 and was first noted on Oahu, Hawaii, in 1941 (DeWalt et al. 2004). By 1971, Koster's curse had turned up in Palau, and is now also naturalized in Sabah, the Malay Peninsula, Sri Lanka, India, Tanzania, and Madagascar (Peters 2001).

33.5 *Clidemia japurensis*

Brazil Buyuyu, buiuiú azul; ka'a, ka'nã (Tukano)

Peru Mullaca sapu, mullaca morada, papailla, matapasto, huayito, rifarillo, mullaquilla; cáâu (Bora)

Status Wild, spontaneous in cultural settings

Buyuyu ***Clidemia japurensis*** **in flower and fruit in successionary vegetation on an abandoned village site. São Pedro Antigo, Rio Tiquié, affluent of the Uaupés, Amazonas, Brazil 11-1-12**

Clidemia japurensis is another weedy member of the Melastomataceae with edible fruits. It is a widespread shrub below 700 m on uplands in the Amazon and Orinoco Basins, the Guianas, and Central America as far north as Costa Rica. In the Brazilian Amazon, it appears to be largely absent from the state of Para; I have not seen it in that state, nor have any herbarium specimens been collected in the southeastern part of the Amazon Basin. In Brazil it is confined to Amazonas, Acre, Rondônia, Roraima, and Amapá. It also occasionally occurs on higher parts of floodplains, although it is much more common on well-drained sites that are never inundated.

Segundo, 12 years, picking mata pasto *Clidemia japurensis* fruits in successionary vegetation at the edge of his village. Yarina, Rio Yanayacu, Reserva Nacional Pacaya-Samiria, Loreto, Peru 4-23-06

In the vicinity of Jenaro Herrera along the lower Ucayali, *Clidemia japurensis* is known as mullaca sapo (the frog mullaca) to distinguish it from its more common relative, mullaca morada (*Clidemia hirta*). Along the Yanayacu, an affluent of the lower Marañón, *Clidemia japurensis* is known as mata pasto (the pasture killer), aptly named because it is a vigorous invader of cleared areas. Children in Yarina along the Upper Yanayacu also refer to the plant as papailla.

Clidemia japurensis can be distinguished from *C. hirta* by its more pointed leaves that are slightly hairier. Also, the fruits of *Clidemia japurensis* are light blue, rather than dark purple, and are juicier. Another difference is that mullaca sapu can grow into a small tree and thus survive in regrowth vegetation longer than mullaca morada. In the upland forest near Jenaro Herrera, for example, a farmer gathered fruits from the twigs of a 4-m tall mullaca sapu slowly being shaded out by taller trees taking over an abandoned field.

The fruits are small, measuring only 0.8 cm long by 0.5 cm wide, yet kids especially take breaks when helping their parents with weeding or harvesting in fields to pluck the fruits from the chest-high bushes. Boys also enjoy the fruits as they pass through fields on their way to hunt in the forest with their fathers, and both boys and girls eat the soft-skinned, hairy fruits when going to or from the forest with their parents to gather fruits or medicinal plants. Adults rarely if ever eat the fruits.

33.6 *Clidemia juruensis*

Bolivia Guayabilla; tantitewe (Yuracaré)

Peru Pucacucrosacha, rifarillo

Status Wild

Clidemia juruensis **shrub in old regrowth forest. Outskirts of Nauta, Rio Marañón, Loreto Peru 4-28-06**

Clidemia juruensis is a relatively short shrub, reaching no higher than 2.5 m, with spectacularly large leaves for the genus. The pubescent leaves are 25 cm long and 12 cm wide. The pale blue fruits, 1.5 cm by 1 cm, are clearly discernible in the dark understory of forests and thus attract the attention of dispersal agents. Both the leaves and fruits have a velvety texture. Although not common, the fruits of this shrub are appreciated by rural folk in the Peruvian Amazon. However, people have to be careful when picking the fruits because some of the petioles are inhabited by ants that will attack if disturbed. This wild fruit is surely a candidate for eventual domestication as an ornamental plant, if not for the fruit.

Clidemia juruensis is found in western Amazonia from Ecuador south to Bolivia as well as Acre and Rondônia in the southwestern part of the Brazilian Amazon. It also occurs in the Upper Rio Negro in Venezuela and in Guyana. Although more common on uplands, it also grows in some floodplain forests.

Clidemia juruensis **shrub in old regrowth forest. Outskirts of Nauta, Rio Marañón, Loreto Peru 4-28-06**

33.7 *Clidemia novemnervia*

Brazil Buyuyu, buxixú, buxuxú; kara (Tukano), to (Tuyuka)

Peru Huayito, mullaca, mullaca morada, mullaquilla, guayabilla

Venezuela Boyuboyu; o'mochahcho, tëhpe puru këmën (Panaré)

Status Wild, spontaneous in cultural settings

Buyuyu *Clidemia novemnervia* in successionary vegetation on the site of an abandoned Tuyuka village. São Pedro Antigo, Rio Tiquié, affluent of the Uaupés, Amazonas, Brazil 11-1-12

Buyuyu *Clidemia novemnervia* has distinctive spear-point-like leaves with a velvety texture. The stem, petioles, and fruits have a fine red pubescence. The fruits turn dark purple when ripe and are picked for snacks by rural people, such as the Tukano and Tuyuka along the Tiquié, an affluent of the Uaupés near the border between Brazil and Colombia. The fruits are similar in size and shape to black currants and measure 1.5 cm by 1 cm. They are produced in the late part of the rainy season through the dry period.

Clidemia novemnervia, like most species in the genus, flourishes in early successionary vegetation. The shrub can reach 4 m high but starts producing fruit when only half a meter tall so most fruits are within easy reach. It is found on uplands and occasionally on floodplains in Amazonia and its range also extends into Venezuela, the Guianas, and Central America as far north as Mexico (Gleason 1939). On floodplains, this plant seems to be largely confined to sandy areas that are periodically inundated by black or clear water rivers and streams. It also thrives on some patches of upland sandy soil with vegetation known as *campinarana* in Brazil and *varillal* in Peru, such as in the Allpahuayo-Mishana National Reserve near Iquitos in the Peruvian Amazon (Ramírez 2015:23). It is also found along the margins of some savannas, such as in Roraima.

33.8 *Clidemia octona*

Ecuador Sau chinchak (Shuar)

Peru Capulí, mullaquilla; chinchak, dupi (Aguaruna), suwir chanchak (Shuar)

Status Wild, spontaneous in cultural settings

Clidemia octona. **Colored stipple engraving by L. Boucquet after P.J.F. Turpin inscribed** *Melastoma octona*. **A. Bonpland, Monographia Melastomacearum, vol. 1, t. 4, Paris 1806. Published in: H. Walter Lack. 2009. Alexander Von Humboldt and the Botanical Exploration of the Americas, Prestel, Munich**

Children snack on another weedy *Clidemia*, known as capulí *Clidemia octona* in the foothills of the Peruvian Andes. Capulí is similar to *Clidemia hirta* except that it has larger leaves and can tolerate more shade, such as in fruit orchards. Kids eagerly ingest the hairy plant's juicy fruits in part because they often do not have access to purchased snacks. The common name comes from mountain black cherry *Prunus serotina*, known as capulí in Spanish-speaking countries in Latin America. Mountain black cherry was introduced to the Andes from Central America in pre-contact times. Mountain black cherry also bears small, dark purple fruits, albeit larger than those of *Clidemia hirta*. Mountain black cherry is a planted tree whereas *Clidemia octona* is a spontaneous herb in forests and disturbed sites. Most fruits of *Clidemia octona* are eaten in rural areas, but they sometimes appear in street markets, such as Tingo Maria in the foothills of the Peruvian Andes.

Saul, 11 years, with capulí *Clidemia octona* fruits he gathered near his rural home at 700 m in the Andean foothills. Santa Rosa near Tingo Maria, Peru 1-23-04

Clidemia octona is found in uplands and along river banks in Amazonia, central Brazil, the Guianas, and Central America as far north as Mexico. In the Andean foothills it can be found up to about 1000 m. The shrub is normally 1–2 m high but can reach 4 m on occasions. The fruits are thus easy to reach, or the shrub can be easily bent over to gather them.

33.9 *Clidemia rubra*

Brazil Maria pretinha, frutinha, buiuiú, quaresmão; ka'a rã, ka'nã (Tukano)

Colombia Azulejo, mortiño

Peru Mullaca, mullaquilla

Status Wild, spontaneous in cultural settings

Wadrio, 11 years, gathering maria pretinha *Clidemia rubra* fruits in upland successionary vegetation. Carrazedo, mouth of the Xingu River, Município de Gurupá, Pará, Brazil 7-23-16

Maria pretinha *Clidemia rubra* produces purple fruits when fully ripe. The oval-shaped fruits are 4–7 mm in diameter and are clustered in twos and threes at the juncture of the leaf petioles with the stem. Youngsters, such as in the vicinity of Carrazedo at the mouth of the Xingu, are fond of the juicy fruits which are easy to reach because the herb starts fruiting when only 40 cm high and reaches no higher than 3 m. These fruits are a good source of dietary fiber and contain several minerals helpful for human health such as calcium, manganese, and zinc (Gordon et al. 2011a, b). The pubescent fruits appear in both the dry and rainy seasons and contain numerous tiny seeds that are dispersed by birds and people.

Wadrio, 8 years, gathering fruits of Maria pretinha *Clidemia rubra* along a trail in regrowth forest. A few hundred meters inland from Carrazedo, mouth of the Rio Xingu, Município de Gurupá, Pará, Brazil 9-5-14

Clidemia rubra is a shrub that is widely distributed in the lowlands of Latin America from Mexico to Bolivia including most of Brazil (Gleason 1939). In the Andean foothills, it occurs up to 1600 m. *Clidemia rubra* is found in successionary vegetation on uplands, such as along paths and in cultivated fields, as well as on savannas as in Amapá and Roraima. It is also found in the understorey of some upland forests, such as in the Reserva de Desenvolvimento Sustentável do Tupé along the left bank of the Rio Negro some 25 km west of Manaus (Corrêa et al. 2017) and in Mato Grosso (Guarim Neto 1984). It also occurs on some sandy soils with low, stunted vegetation called *campina* and *campinarana* in Brazil and *varillal* in Peru (Ramírez 2015:23). It has become a weed in some guaraná *Paullina cupana* plantations in the vicinity in Amazonas, Brazil (Miléo et al. 2007; Silva et al. 2007a, b, c, d).

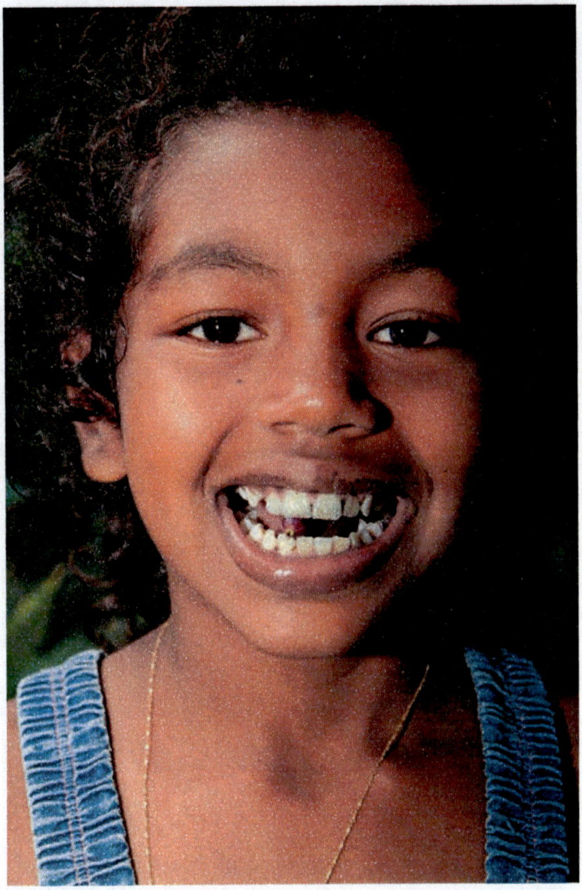

Rayssa, 9 years, eating Maria pretinha *Clidemia rubra* fruit she has gathered in successionary vegetation near her village. Igarapé Divisa, Carrazedo, mouth of the Rio Xingu, Município de Gurupá, Pará, Brazil 8-24-14

33.10 *Clidemia septuplinervia*

Ecuador Empujari, sik (Shuar), gomequenean, emegiwe, nuigui, abahue (Waorani)

Peru Matapasto, mullaca morada; chinchák (Humabisa)

Status Wild

Matapasto ***Clidemia septuplinervia*** **in regrowth forest. Near Yarina, Rio Yanayacu, Reserva Nacional Pacaya-Samiria, Loreto, Peru 4-23-06**

Clidemia is replete with shrubs containing small, edible berries, and the western Amazon, especially around the confluence of the Marañón and Ucayali Rivers, seems to be particularly rich in edible *Clidemia*. People along the lower Marañón and its affluents eat the fruits of at least half a dozen species, and three of them occur in wetlands: *Clidemia hirta*, *C. japurensis*, and *C. septuplinervia*. With its bright, cobalt blue fruits the latter is particularly striking. Another unusual feature of *Clidemia septuplinervia* is that the fruits are cauliflorous; that is they grow along the trunk and branches, rather than at the end of the stems.

The fruits are so bright the bush almost looks like it has neon lighting inside the dim interior of forests. *Clidemia septuplinervia* may eventually make it as an ornamental plant, rather than as a commercial fruit. This attractive shrub flowers and fruits simultaneously, displaying four colors: small white flowers, fruits that are pink at the beginning turning pale blue then cobalt blue when ripe. Flowers and fruits at various stages of maturation are found on the same bush, making for a particularly colorful array.

Whereas most *Clidemia* fruits are barely 1 cm in diameter, the round fruits of *Clidemia septuplinervia* are 1.2–1.5 cm wide. The shrub is also prolific; children along the Yanayacu River in the Reserva Nacional Pacaya-Samiria eagerly pluck a hundred or more of the shiny fruits from a single shrub. The iridescent fruits are easily reached because the bush fruits by the time it reaches half a meter. Even when fully grown at 4 m, the stem can be easily bent over to reach the pubescent fruits. In the Ecuadorian Amazon, the Huaorani also consume the fruits, while the Kichwa mash the fruits and squeeze the juice into the eye to alleviate pain.

Boys picking matapasto fruits in regrowth forest. Near Yarina, Rio Yanayacu, Reserva Nacional Pacaya-Samiria, Loreto, Peru 4-23-06

Clidemia septuplinervia is widespread in mature and old regrowth forests in tropical America, ranging from Central America to northern South America as far south as Bolivia and Brazil. It is found in uplands and in seasonally wet or swampy areas up to 1000 m in the Andes. In the Reserva Nacional Pacaya-Samiria, the shrub grows in regrowth forests on formerly cultivated sites on the higher parts of floodplains.

33.11 *Leandra dichotoma*

Peru Matapasto, mullaquilla

Status Wild

Leandra dichotoma **in successionary vegetation. Near Nauta, Rio Marañón, Loreto, Peru 4-28-06**

Leandra dichotoma grows in light gaps in upland forests as well as successionary vegetation in abandoned fields and along trails and is widely distributed from Belize to Argentina. In Amazonia, it is more common in the western part of the basin and in the Andean foothills can be found as high as 1500 m. This understory shrub, usually less than 4 m tall although it can reach 10 m, starts producing fruit when only 40 cm high so they are mostly within easy reach. It produces fruits in candelabra fashion which turn black when ripe. The fruits are attached to the end of stalks (technically called a peduncle). The round fruits are 0.8–0.9 cm in diameter and are eaten by rural folk, especially children. The fruits are consumed on the spot and do not enter markets. The pubescent, drip-tip leaves are quite large, measuring 11 cm × 7 cm.

33.12 *Maieta guianensis*

Brazil Buxixu, remela de gato; iatwa ia (Waimiri-Atroari)

Colombia Morillo, palo de majiña

Ecuador Yuca de quinde, chanchák chichi, uchan; payas (Achuar), shallipu, uchan, pischu-micuna-muyu-yura, puca muyo, shigshi panga (Kichwa), jempe aja, chinchak (Shuar), ewueico, egueyi, guiya enegui, emeguimo, obeguimo, dikaromo-hue (Waorani)

Peru Rifarillo, pucaquito caspi, mullaquilla, añallo; chinchak, jempe umpuágbau, numkaya chinchak, suir chinchak (Aguaruna)

Venezuela Boyuyo; ademutu (Ye'kuana)

Status Wild

Maieta guianensis **in upland successionary forest. Near Nauta, Rio Marañón, Loreto, Peru 4-28-06**

Maieta guianensis is an understory shrub, usually under 2 m high, in upland and floodplain forests, including old successionary growth. It is widespread in Amazonia and occurs up to about 1000 m in the Andean foothills. This short herb reaches no more than 3 m and also occurs in the Guianas and Venezuela (Bánki et al. 2003:36).

Maieta guianensis starts producing fruit when only 30 cm high, so they are easy to pick. The red fruits turn black when ripe. Rural people stop to snack on the fruits when they encounter them. The oblong to ovate fruits vary in size from 0.9 to 1.25 cm long by 0.5 cm to 0.75 in diameter. Adults and children alike have to take care when picking the fruits, however, because ants shelter in hollows (technically called domatia) at the base of some of the leaf stems (petioles) and may attack them although *Pheidole minutula* which has been recorded living on this plant 70 km north of Manaus is not particularly aggressive (Vasconcellos 1991, 1993). Another species of ant, *Crematogaster laevis*, also helps protect this plant from herbivores

(Izzo et al. 2009; Vasconcellos 2000). These ants live in domatia on their host plant and help protect it against herbivores (Solano et al. 2003). Several other fruit-bearing plants in the Melastomataceae as well as other botanical families are myrmecophytes, which means they evolved a mutually beneficial relationship with several species of ants. Juice squeezed from the fruits is used to treat sore eyes in the Ecuadorian Amazon while the Shuar sometimes eat the ants living on this plant to relieve thirst (Bennett et al. 2002:204).

33.13 *Miconia ceramicarpa*

Brazil Uvinha, tinteiro vermelho, papa terra; ju'y ka'a (Ka'apor)

Status Wild

Uvinha *Miconia ceramicarpa* in successionary vegetation. Pedreira, Floresta Nacional Caxiuanã, Baia de Caxiuanã, Rio Anapu, Municipio de Melgaço, Pará, Brazil 5-11-15

Uvinha (small grape) is a shrub in successionary vegetation including old regrowth forests. It occurs mostly on uplands though it is also found occasionally on floodplains, such as on and near Marajó Island at the mouth of the Amazon (Bentes-Gama et al. 2002a, b; Gama et al. 2002; Queiroz et al. 2005). Uvinha *Miconia ceramicarpa* produces blue fruits measuring 0.8 cm × 0.6 cm when only 50 cm tall so they are within easy reach. The shrub is relatively short, rarely exceeding 2 m. Children especially are fond of snacking on the pubescent fruits. This understory shrub is confined to central and eastern parts of the Amazon Basin, the Guianas, and Venezuela.

Rafael, 5 years, gathering uvinha *Miconia ceramicarpa* fruits in successionary forest near a manioc field. Pedreira, Floresta Nacional Caxiuanã, Baia de Caxiuanã, Rio Anapu, Municipio de Melgaço, Pará, Brazil 5-11-15

33.14 *Miconia longifolia*

Bolivia Piton requenito

Brazil Malva, tinteiro, erva de passarinho, jacatirão

Colombia Quiebra hueso

Peru Rifari, rifarillo, vinho huayo; chanchák (Achuar), bin kun (Cashinahua), echepa-ey (Witoto)

Status Wild

Genjis, 10 years, with rifari *Miconia longifolia* fruits gathered along Quebrada Huirurui, a stream near his village. San Martin Tipishca, Rio Samiria, Reserva Nacional Pacaya-Samiria, Loreto, Peru 6-30-06

Miconia longifolia is one of at least four species of *Miconia* with edible fruits in the Amazon Basin. An understory tree of forests on uplands as well as along the banks of rivers and streams, *Miconia longifolia* can reach 20 m high but starts producing fruit by the time it reaches 2 m, so the fruits are often easily accessible or can be reached by bending the lower branches. This long-leafed *Miconia* is confined to lowlands below 1000 m throughout the Amazon Basin and the neighboring Guianas and also occurs in Central America as far north as Mexico.

Known as vino huayo along the Samiria River, a name also used for an unrelated wild fruit (*Coccoloba densifrons*, Polygonaceae), *Miconia longifolia* produces numerous black fruits at the end of its branches. The shiny black fruits are born on pink pedicels and measure 1 cm long by 0.8 cm wide. The juicy fruits contain numerous seeds which are swallowed along with the pulp. As with many species of *Miconia* and *Clidemia*, the seeds are likely dispersed by birds as well as people. Erva de passarinho, one of the common names for *Miconia longifolia* in the Brazilian Amazon, indicates that locals are aware that birds feast on the fruits.

33.15 *Miconia nervosa*

Bolivia Guayabilla; hocótsëmo (Chácobo)

Brazil Buyuyu; jacatirão; wararaha-ka'a (Guajá), tangaraita'y (Ka'apor), xãri hi (Yanomami)

Colombia Tuno

Ecuador Pischcu-micuna-muyu-yura, puca muyo (Kichwa), aíkiam sésa (Mayna), suir chinchak, amtunar chinchak, chanchák (Shuar), doye hue'oco (Siona-Secoya), Enhueyiyamo, enhueyi (Waorani)

Peru Matapasto, mullaca, mullaca morada, mullaquilla, pucacurosacha, rifarillo

Status Wild

Matapasto *Miconia nervosa* in flower and fruit in a forest. Near Yarina, Rio Yanayacu, Reserva Nacional Pacaya-Samiria, Loreto, Peru 6-26-06

Miconia nervosa thrives as an understory shrub in old successionary vegetation and mature forests on both uplands and floodplains. It also occurs in stunted forest growing on rocks rich in iron ore (canga) on hills in the Serra dos Carajás, Pará (Rocha et al. 2017). In Costa Rica, it is more common in treefall gaps than in the forest understory (Ellison et al. 1993), but in the Peruvian Amazon, I encountered this small tree in the forest understory. This plant is also found in the Nassau Mountains of Suriname (Bánki et al. 2003:37).

This shrub to small tree, which can reach 4 m tall, produces slightly flattened dark purple fruits measuring some 0.7 cm long by 1 cm wide. People eat the pubescent fruits which are produced year-round and are full of tiny seeds which are swallowed along with the sweet, juicy pulp. Along the Yanayacu River in the Reserva Nacional Pacaya-Samiria, children typically tarry when the shrub is in fruit to pick the berries.

Cristian, 10 years, and Donaldo, 12 years, picking matapasto *Miconia nervosa* **fruits in a forest. Near Yarina, Rio Yanayacu, Reserva Nacional Pacaya-Samiria, Loreto, Peru 6-26-06**

Miconia nervosa can attain 6 m but produces fruit when only 50 cm high, so they are usually within easy reach of youngsters. Although *Miconia nervosa* does not form dense stands, a 1.8-m-high bush can provide several dozen fruits. The Chácobo, who live along the Upper Ivón, a tributary of the Beni in the Bolivian Amazon, also eat the fruits as do the Huaorani in the Ecuadorian Amazon (Boom 1989). The Siona-Secoya in the Ecuadorian Amazon use this fruit for fish bait (Marcalla 2012:102). This plant is also used in folk medicine. The Kichwa macerate the fruits to make drops to relieve irritated eyes and the Chácobo drink a leaf decoction to treat stomach disorders (Boom 1987:29).

Native to the Neotropics below 2400 m, *Miconia nervosa* is found from Belize in Central America to northern South America, including the Amazon and Orinoco Basins as well as Trinidad. It also occurs outside of Amazonia in parts of the north-eastern, central, and eastern parts of Brazil. *Miconia nervosa* is dispersed by people, birds, including guans (*Penelope* spp.) and curassows (*Crax* spp.). Guans and curassows are esteemed for their meat, so the fruits contribute indirectly to the human diet. The Kichwa name for this plant means "bird eats fruit tree." Leaf-litter ants (*Pheidole* spp.) extract the seeds from the droppings of fruit-eating animals and take them to their nests and refuse piles, some of which survive and later sprout (Levey and Byrne 1993). Saddle-backed tamarins *Saguinus fuscicollis* also feed on the fruits.

33.16 *Miconia serrulata*

Brazil Buxixu, canela de velho; uru-wɨra (Tembé), xãri hi (Yanomami)

Colombia Tabaida

Ecuador Chanchák (Shuar), garaguibe (Waorani)

Peru Matapasto, rifari, rifarillo, caracha caspi, calsoncillo, palo estrella blanco; chanchák (Achuar, Mayna), chijáwe, ukuínmanch (Aguaruna)

Status Wild

Matapasto *Miconia serrulata* in successionary forest. Near Yarina, Rio Yanayacu, Reserva Nacional Pacaya-Samiria, Loreto, Peru 6-23-06

Miconia serrulata is one of the numerous shrubby melastomes with edible fruits found in both uplands and floodplains in the Amazon such as in Acre, Brazil (Medeiros et al. 2014). The shrub ranges from Belize in Central America east to the Caribbean and south to northern South America. A denizen of light gaps in mature forest and regrowth below 1800 m, this shrub reaches the dimensions of a small tree at 20 m, though it starts bearing fruits when only 1–2 m high, so they are often easy to reach.

Along the Yanayacu River in the Reserva Nacional Pacaya-Samiria, *Miconia serrulata* is known as matapasto, a name used for several species of weedy *Clidemia* as well as some species of *Miconia*. Youngsters are especially fond of gathering the fruits in the understory of floodplain forests. The Huaorani of the Ecuadorian Amazon also consume the fruits. People have to be careful when picking the fruits because the shrub is sometimes inhabited by small but aggressive ants.

The dark blue fruits are pear-shaped, only upside down, measuring 1.2 cm long and 1 cm wide. The soft-skinned fruits contain a purple pulp full of tiny seeds. The seeds are simply ingested along with the juicy mesocarp; people are thus likely involved in spreading the seeds along with other dispersal agents, especially birds. In upland forests of the Ecuadorian Amazon, six understory birds have been recorded eating these fruits and dispersing the seeds: the striped manakin *Machaeropterus regulus*, blue-crowned manakin *Lepidothrix coronata*, white-crowned manakin *Pipra pipra*, white-tailed manakin *Pipra filicauda*, golden-headed manakin *Pipra erythrocephala*, and the white-necked thrush *Turdus albicollis* (Blendinger et al. 2008, 2011). The Achuar in the Peruvian Amazon chew the leaves and apply them to cuts.

Miconia serrulata is classified as "noxious" in the Global Compendium of Weeds, while in the Peruvian Amazon it is a food resource. This negative image of *Miconia serrulata* by agriculture departments and quarantine services arises from the trouble other species of *Miconia* and *Clidemia hirta*, another member of the Melastomataceae, have wrought on native vegetation and farmland where they have been introduced, such as Hawaii and Australia.

33.17 *Mouriri apiranga*

Brazil Piranga, apiranga, uapiringa, criolí, araçá branco, murici, mirauba, meraúba, meralba, cereja vermelha, canela de cutia, pimentão laranja

Status Wild

Girls gathering piranga *Mouriri apiranga* fruits along the sandy bank of a black water river. They have walked from the nearby village of Vila Franca with their mother. Rio Arapiuns, affluent of the lower Tapajós, Pará, Brazil 8-30-93

Confined to the Brazilian and Bolivian Amazon, *Mouriri apiranga* is a shrub to small tree, reaching as high as 11 m, along the banks of rivers and streams that produces small orange fruits containing one to three seeds (Araujo-Murakami et al. 2015; Cavalcante 2010:43; Kunz et al. 2014a, b; Miranda 2000; Morley 1976:217; Ratter et al. 1978; Stefanello 2008:18; Stefanello et al. 2010; Wallace et al. 1998). It thrives in tidal forests in the Amazon estuary and is also found in gallery forest in the cerrado area of Mato Grosso in the southern fringe of Amazonia such as in the Upper Araguaia watershed (Amaral et al. 2009; Stefanello et al. 2009). It is also found in a seasonal upland forest in the southern fringes of the Amazon Basin in northeastern Mato Grosso as well as in scrub savanna (Kunz et al. 2014b; Rondon Neto et al. 2010).

Known locally as piranga or apiranga, the fruits vary in size even on the same tree; some are as small as a black currant whereas others are as large as the fruit of acerola, also known as Barbados cherry. The flesh under the think skin is translucent and sweet, hence its popularity. Piranga fruits during the dry season and is therefore likely dispersed by birds and mammals rather than fish. A river dweller along the lower Tapajós mentioned to me that some bats and pigeons are fond of the fruits.

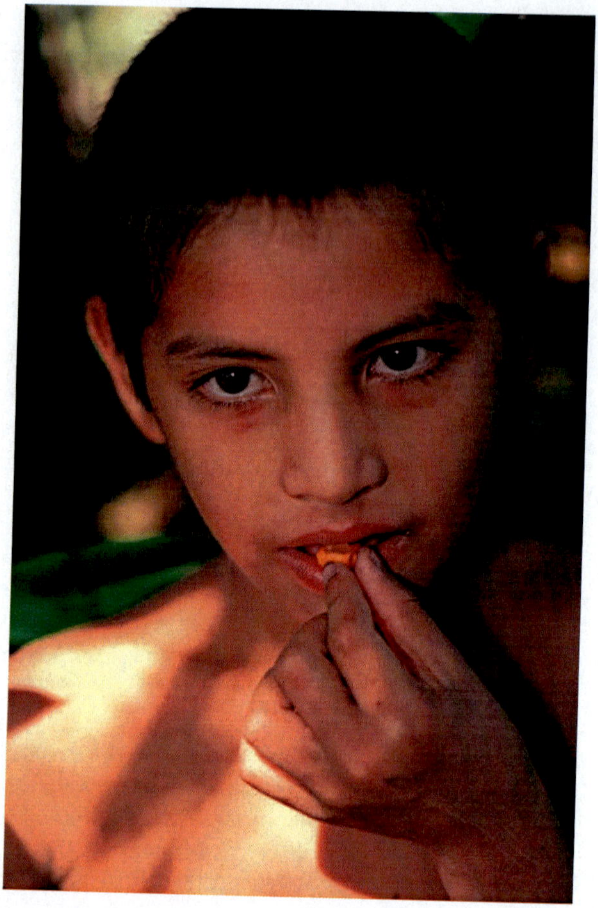

Marcelo, 9 years, eating a piranga *Mouriri apiranga* fruit he gathered in floodplain forest along the Rio Arapiuns. Lower Tapajós River near Vila Franca, Pará, Brazil 11-20-02

People have likely also been involved in dispersing the seeds of *Mouriri apiranga*. People spit out the seeds after ingesting the mesocarp, and some land far from the tree from which they were born. Carbonized *Mouriri apiranga* seeds dating back some 10,000 years have been found in a sandstone cave at Pedra Pintada near Monte Alegre (Roosevelt 1998). Hunters and gatherers evidently took fruits of piranga and several other fruit and nut trees to their rock shelters and caves along the middle Amazon.

Piranga is particularly common on the sandy shores of black water rivers (Myster 2018). Along the Arapiuns River, an affluent of the lower Tapajós, for example, the shrub-like tree forms small stands. In late August when water levels are low, villagers from Vila Franca at the mouth of the Arapiuns walk for an hour or two along the exposed beaches of the river to reach the stands and gather the fruits. Many fruits are within arm's reach because piranga starts producing when only 2–3 m high, but others can be picked by youngsters who scramble up the multi-stemmed and branched tree that normally does not exceed 19 m. Some of the fruits are relished on the spot, while others are taken home. Piranga is also common in patches along the black water Maués River, an affluent of the middle Amazon, where it also fruits at the height of the dry season. The orange color of piranga fruits has inspired a local name for an upland palm; along the Maués the tucumã palm (*Astrocaryum vulgare*) is called tucumã piranga, on account of its deep orange fruit.

Boy climbing a piranga *Mouriri apiranga* tree to pick fruits along the bank of a black water river. Rio Arapiuns, affluent of the lower Tapajós, Pará, Brazil 8-30-93

33.18 *Mouriri grandiflora*

Bolivia Guapomocillo

Brazil Camutin, camutim, remela de cachorro, arpoeira, meraúba, miraúba, gurguí, araçá, daicu, olho de tucunaré, tucunaré mereçá, guajaraí

Colombia Arecheche; socodó (Andoke)

Ecuador Mulchi; ajinumi (Achuar), sharimiat, uwínim (Shuar), aiquë'jeo (Siona-Secoya)

Peru Guayabilla, navidad caspi, charichuelo, lanza caspi, lanza huayo, arenilla; uwínim (Achuar)

Venezuela Dawicú hoja gruesa, dereyo, yúkürruda; morokoto ke hi, morokoi (Yanomami)

Status Wild

Boy swimming up to a *Mouriri grandiflora* tree along a black water river to gather fruits. Rio Ipixuna, affluent of the lower Xingu near Carrazedo, Município de Gurupá, Pará, Brazil 4-19-15

Mouriri grandiflora is a tree that grows in forests along river and stream banks throughout the Amazon Basin and Upper Orinoco, including the Guianas (Araujo-Murakami et al. 2015; Assis et al. 2017; Cain et al. 1956; Duque et al. 2005; Luize et al. 2015; Morley 1976:99; Nebel et al. 2000a, b, 2001a, b; Santos and Jardim 2006; Thomas 2009; Wittmann et al. 2013). The tree appears on the floodplains of diverse water types, including black water rivers (Myster 2018), and is found in tidal forests in the Amazon estuary (Amaral et al. 2009; Bentes-Gama et al. 2002b; Cattanio et al. 2002; Gama et al. 2003). Although mostly confined to floodplains, it is occasionally found in upland forests, such as in the vicinity of Iquitos in the Peruvian Amazon (Vásquez 1997:470) and in 5- and 10-year-old successionary woodlands such as in the vicinity of Santa Mercedes along the Rio Putumayo in the northern Peruvian Amazon bordering Colombia (Gomez 2011:81; Muñoz 2010:23).

Wadrio, 9 years, with *Mouriri grandiflora* fruits he gathered in a riverbank forest. Rio Ipixuna, affluent of the lower Xingu near Carrazedo, Município de Gurupá, Pará, Brazil 4-19-15

Like most Amazon trees, *Mouriri grandiflora* fruits in the rainy season. Although the tree can attain 30 m, it starts fruiting when only 3 m high so people can paddle up to the partially submerged branches and easily pick the orange fruits. The weak branches tend to hang (Morley 1976:96), thus making it easier to gather the fruits. The globular fruits, 3.5 cm by 3 cm, contain two to five black seeds embedded in a sweet white pulp. Some of the fruits are smaller with a diameter of 1.5–2.5 cm. The seeds usually measure 1.5 cm by 1 cm; smaller fruits have fewer of them.

Jon, 9 years, with *Mouriri grandiflora* fruit he gathered in a floodplain forest near his village. Carrazedo, mouth of the Rio Xingu, Município de Gurupá, Pará, Brazil 4-19-15

Camutim, as the tree is known in the Amazon estuary, is one of those Cinderella fruits that never appear in markets but is nevertheless appreciated by rural folk. Along the Maracá River in Amapá, for example, boys clamber up the trees and along the branches to pluck the fruits before monkeys and toucans beat them to it. The round, orange-yellow fruits are produced in clusters at the ends of branches. Ripe fruits that have not been taken by birds or mammals fall to the ground, or into the water, where people pick them up.

The tree has other uses. In French Guiana and along the Upper Orinoco, rural people use the fruits as fish bait, while in Peru the Achuar use the wood to build their homes as do river dwellers along the Solimões in the vicinity of its confluence with the Japurá (Wittmann et al. 2009). In the Ecuadorian Amazon, the tree provides wood for construction and fuel (Bennett et al. 2002:207; Van den Eynden et al. 1999:154). The Ticuna who live along the Solimões and some of its affluents in Brazil as well as rivers that flow into the Amazon in contiguous Colombia dry the leathery leaves then pound them and use the resulting powder to clean their teeth (Schultes and Raffauf 1990:299).

33.19 *Mouriri guianensis*

Brazil Socoró, criviri, capão, miriúba, gurguri, creoli, creoula, criurí, quiriri, murta, ururi, goiabarana, camutim, roncador, puçá, criviri, gurguri; davicú (Tukano), tɨhɨ hi (Yanomami)

Colombia Cometure

Ecuador Sacha pilansi (Kichwa)

Peru Lanza caspi, cedro macho

Venezuela Cameture, cometure, cometura, cascarito

Status Wild, spontaneous in cultural settings, planted

Fisherman baiting a trotline with socoró *Mouriri guianensis* fruit on an Amazon floodplain lake. Lago Curuçá, a few kilometers downstream from Itacoatiara, Amazonas Brazil 6-2-77

Mouriri guianensis (synonym *M. ulei*) is a tree in mature forests, grasslands, and successionary vegetation throughout the Amazon Basin into the Northeast and Atlantic Forest of Brazil, the Guianas, Venezuela, and Trinidad (Balée 1993; Bastos et al. 1995; Casula 2012:63; Coelho-Ferreira and Jardim 2005; Godoy et al. 1999; Machado and Pinheiro 2016; Maués et al. 2011; Morley 1976:170, 1983; Santos et al. 2003). Although mostly confined to floodplains, it is also occasionally found on uplands as well as on sandy soils along the coast of the Bragantina zone in northeastern Pará (Lima et al. 2014). It is also found in some home gardens in French Guiana (Hoff and Cremers 2005). Several species of stingless bees pollinate the flowers (Gostinski et al. 2018; Oliveira et al. 2016).

This widely distributed tree starts producing fruit when only 4 m high and can eventually reach 35 m, although it is usually shorter than that. Known as socoró in Brazil, riverine inhabitants gather the orange-red fruits as an occasional snack but mainly to be used to bait fish hooks. Socoró fruits are used to bait single hooks attached by a line to the end of a pole as well as trotlines with several hooks. Much esteemed tambaqui *Colossoma macropomum* are fond of the fruits and are the main target when using socoró bait. Socoró has a relatively brief fruiting season, and most of the trees have finished bearing their cherry-like fruits by the time the flood crests. People gather socoró fruits as the water rise and can keep them at home for several days before they start deteriorating.

Along Igarapé Jari, a channel that connects the lower Tapajós with the Amazon River, the home garden of a farmer and small-scale rancher contains several nut and fruit trees including socoró. A farmer along Surubim-Açu on the Amazon floodplain near Santarém safeguards a spontaneous socoró in his home garden because he and his family enjoy eating the fruits. At Amador on a floodplain island along the Amazon between Carariacá and Óbidos, a family has planted socoró in their home garden which is inundated each year because the children are fond of eating the fruit. In Venezuela, the wood is used to fashion beams for houses (Morley 1976:171).

33.20 *Mouriri tessmannii*

Peru Guayaba de restinga

Status Wild

Guayaba de restinga *Mouriri tessmannii* in fruit in floodplain forest. Quebrada Huirurui, affluent of the Samiria River, near San Martin Tipishca, Reserva Nacional Pacaya-Samiria, Loreto Peru 4-25-06

An understory tree in forests on higher parts of floodplains, *Mouriri tessmannii* produces fruits that vary in size even on the same tree from 4 cm long and 4 cm wide to 5 cm long by 6 cm in width. The smooth-skinned fruits turn orange-yellow when ripe and contain a single, tan-colored seed some 1.2 cm by 1.8 cm. The skin is relatively thin so a good deal of the fruit is edible pulp.

Guayaba de restinga *Mouriri tessmannii* fruits in floodplain forest. Quebrada Huirurui, affluent of the Samiria River, near San Martin Tipishca, Reserva Nacional Pacaya-Samiria, Loreto Peru 4-25-06

Known as guayaba de restinga (river bank guava) along the lower Samiria, an affluent of the Marañón River, locals appreciate the yellow pulp even though it is a bit stringy and mealy. Residents of the village of San Martin Tipishca along the lower Samiria report that deer (*Mazama* sp.) and paca *Agouti paca*, both esteemed game animals, also feed on the fruits when they drop to the ground. Although this understory tree can reach 22 m it produces fruit by the time it reaches 3 or 4 m so some of the lower branches can be pulled down to pick fruits. *Mouriri tessmannii* occurs occasionally in upland forests as well and is confined to the Amazon portions of Ecuador and Peru.

References

Absy, Maria Lúcia; Bezerra, Eronildo Braga; Kerr, Warwick E. 1980. Plantas nectaríferas utilizadas por duas espécies de Melipona da Amazônia. Acta Amazonica 10(2):271–281

Acero-Murcia, Adriana; Almario, Leidy J.; García, Javier; Defler, Thomas R.; López, René. 2018. Diet of the Caquetá titi (*Plecturocebus caquetensis*) in a disturbed forest fragment in Caquetá, Colombia. Primate Conservation 32:1–17

Albert, Bruce; Milliken, William. 2009. Urihi A: a terra-floresta Yanomami. Instituto Socioambiental (ISA)/Institut de Recherche pour le Devéloppement, São Paulo

Albuquerque, José Maria de. 1980. Identificação de plantas invasoras de cultura da região de Manaus. Acta Amazonica 10(1):47–95

Almorozo, Maria Christina de Mello; Gély, Anne. 1988. Uso de plantas medicinais por caboclos do baixo Amazonas, Barcarena, PA, Brasil. Boletim do Museu Paraense Emílio Goeldi, Série Botânica 4(1):47–131

Alvino, Fabrízia de Oliveira; Silva, Manoela Ferreira F. da; Rayol, Breno Pinto. 2005. Potencial de uso das espécies arbóreas de uma floresta secundária, na Zona Bragantina, Pará, Brasil. Acta Amazonica 35(4):413–420

Alviz, Ángela; Pérez-Albarracín, Karen. 2015. Plan para la conservación de la danta de tierras bajas (Tapirus terrestris) en los departamentos de Casanare, Arauca y Vichada. Corporinoquia-Fundación Orinoquia Biodiversa, Yopal, Colombia

Amaral, Dário Dantas do; Vieira, Ima Célia Guimarães; Almeida, Samuel Soares de; Salomão, Rafael Paiva; Silva, Antonio Sérgio Lima da; Jardim, Mário Augusto Gonçalves. 2009. Checklist da flora arbórea de remanescentes florestais da região metropolitana de Belém e valor histórico dos fragmentos, Pará, Brasil. Boletim do Museu Paraense Emílio Goeldi, Ciencias Naturais 4(3):231–289

Antei, G. 1996. Karsten Flora Columbiae. Seguros Bolivar, Bogotá

Araújo, Maristela Machado; Tucker, Joanna Marie; Vasconcellos, Joanna Marie; Zarin, Daniel Jacob; Oliveira, Wilson; Sampaio, Patrícia Delamônica; Rangel-Vasconcelos, Lívia Gabrig; Oliveira, Francisco de Assis; Coelho, Roberta de Fatima Rodrigues; Aragão, Débora Veiga; Miranda, Izildinha. 2005. Padrão e processo sucessionais em florestas secundárias de diferentes idades na Amazônia oriental. Ciencia Florestal, Santa Maria 15(4):343–357

Araujo-Murakami, Alejandro; Villarroel, Daniel; Pardo, Guido; Vos, Vincent A.; Parada, G. Alexander; Arroyo, Luzmila; Killeen, Timothy. 2015. Diversidad arbórea de los bosques de tierra firme de la Amazonía Boliviano. Kempffiana 11(1):1–28

Arellano, Gabriel. 2013. Patrones de diversidad, distribución y dominancia de plantas leñosas en los bosques montanos del Parque Nacional Madidi, Bolivia. Doctoral dissertation, Departamento de Biologia, Facultad de Ciencias, Universidad Autónoma de Madrid

Assis, Rafael Leandro de; Wittmann, Florian; Luize, Bruno Garcia; Haugaasen, Torbjon. 2017. Patterns of floristic diversity and composition in floodplain forests across four southern Amazon river tributaries, Brazil. Flora 229:124–140

Astwood, Jorge; Rodríguez, José; Rodríguez, Karen. 2014. Foraging density for squirrel monkey *Saimiri sciureus* in two forests in Puerto Lopez, Colombia. Revista MVZ Córdoba 19(2):4158–4167

Balée, W. 1993. Indigenous transformation of Amazonian forests: an example from Maranhão, Brazil. L'Homme 33(126–128):231–254

Bánki, Olaf S.; Steege, Hans ter; Jansen-Jacobs, Marion J.; Raghoenandan, Usha P. D. 2003. Plant Diversity of the Nassau Mountains, Suriname: Report of the 2003 Expedition. National Herbarium Nederland, Utrecht

Barcelos, Adriana Renata Camila Wisniewski. 2011. Dieta frugívora e potencial de dispersão de sementes de *Tapirus terrestris* (Mammalia, Perissodactyla) em um moasaico de floresta/campinarana, Amazônia, Brasil. Master's thesis, INPA, Manaus

Barcelos, Adriana Renata; Bobrowiec, Paulo Estefano D.; Sanaiotti, Tânia Margarete; Gribel, Rogério. 2013. Seed germination from lowland tapir (*Tapirus terrestris*) fecal samples collected during the dry season in the northern Brazilian Amazon. Integrative Zoology 8:63–73

Barnett, Adrian A.; Silva, Welma Sousa; Shaw, Peter J. A.; Ramsay, Paul M. 2015. Inundation duration and vertical vegetation zonation: a preliminary description of the vegetation and structuring factors in borokotóh (hummock igapó), an overlooked high-diversity, Amazonian vegetation association. Nordic Journal of Botany 33:601–614

Bastos, M. N. C.; C. S. Rosário; L. C. B. Lobato. 1995. Caracterização fitofisionômica da restinga de Algodoal, Maracanã, Pará, Brasil. Boletim do Museu Paraense Emílio Goeldi, Série Botanica 11(2): 173–197

Bennett, Bradley C.; Baker, Marc A.; Andrade, Patricia Gómez. 2002. Ethnobotany of the Shuar of Eastern Ecuador. New York Botanical Garden, Advances in Economic Botany, vol. 14.

Bentes-Gama, Michelliny de Matos; Scolforo, José Roberto Soares; Gama, João Ricardo Vasconcellos; Oliveira, Antônio Donizette de. 2002a. Estrutura e valoração de uma floresta de várzea alta na Amazônia. Cerne 8(1):88–102

Bentes-Gama, Michelliny de Matos; Scolforo, José Roberto Soares; Gama, João Ricardo Vasconcellos. 2002b. Potencial produtivo de madeira e palmito de uma floresta secundária de várzea baixa no estuário Amazônico. Revista Árvore 26(3):311–319

Bentos, Tony V.; Mesquita, Rita C. G.; Williamson, G. Bruce. 2008. Reproductive phenology of central Amazon pioneer trees. Tropical Conservation Science 1(3): 186–203

Blendinger, Pedro G.; Loiselle, Bette A.; Blake, John G. 2008. Crop size, plant aggregation, and microhabitat type affect fruit removal by birds from individual melastome plants in the Upper Amazon. Oecologia 158:273–283

Blendinger, Pedro G.; Blake, John G.; Loiselle, Bette A. 2011. Connecting fruit production to seedling establishment in two co-occurring Miconia species: consequences of seed dispersal by birds in upper Amazonia. Oecologia https://doi.org/10.1007/s00442-011-1956-6

Boom, B. M. 1987. Ethnobotany of the Chácobo Indians, Beni, Bolivia. The New York Botanical garden, Advances in Economic Botany, vol. 4

Boom, B. M. 1989. Use of plant resources by the Chácobo. In: *Resource Management in Amazonia: Indigenous and Folk Strategies*, D. A. Posey and W. Balée (Editors), Advances in Economic Botany, vol. 7, New York Botanical Garden, Bronx, pp. 78–96

Branan, William V.; Werkhoven, Marga C. M.; Marchinton, R. Larry. 1985. Food habits of brocket and white-tailed deer in Suriname. Journal of Wildlife Management 49(4):972–976

Brito, Elizabeth Rodrigues; Martins, Sebastião Venâncio; Oliveira-Filho, Ary Teixeira de; Silva, Elias; Silva, Alexandre Francisco da. 2008. Estrutura fitossociológica de um fragmento natural de floresta inundável em área de Campo Sujo, Lagoa da Confusão, Tocantins. Acta Amazonica 38(3):379–386

Bustamante, Gina G. Frausin. 2009. Frutos, sementes e órgãos tuberosos na alimentação da etnia Sateré-Mawé dos rios Marau e Urupadi (Maués-Amazonas). Master's thesis, Universidade Federal do Amazonas, Manaus

Cabrera-Amaya, Diego Muaricio; Rivera-Díaz, Orlando. 2016. Composición florística y estructura de los bosques ribereños de la cuenca baja del río Pauto, Casanare, Colombia. Caldasia 38(1):53–85

Cain, Stanley A.; Castro, Gustavus M. de Oliveira; Pires, J. Murça; Silva, Nilo Tomâs. 1956. Application of some phytosociological techniques to Brazilian rain forest. American Journal of Botany 43(10):911–941

Cárdenas, Dairon; Castaño, Nicolás; Sua, Sonia. 2009. Flora de la estrella fluvial de Inírida (Guianía, Colombia). Biota Colombiana 10(1–2):1–30

Carim, Samyrams; Schwartz, Gustavo; Silva, Manoela Fernandes Ferreira da. 2007. Riqueza de espécies, estrutura e composição florística de uma floresta secundária de 40 anos no leste da Amazônia. Acta Botanica Brasilica 21(2):293–308

Carretero-Pinzón, Xyomara; Defler, Thomas R. 2016. *Callicebus ornatus*, an endemic Colombian species: demography, behavior and conservation. In: Phylogeny, Molecular Population Genetics, Evolutionary Biology and Conservation of the Neotropical Primates, Ruiz-García, M.; Shostell, J. M. (eds), Nova Science, New York, chapter 16, 1–15

Carretero-Pinzón, Xyomara; Defler, Thomas R.; Ruiz-García, Manuel. 2016. How does the Colombian squirrel monkey cope with habitat fragmentation? Strategies to survive in small fragments. In: Phylogeny, Molecular Population Genetics, Evolutionary Biology and Conservation of the Neotropical Primates, Ruiz-García, M.; Shostell, J. M. (eds), Nova Science, New York, chapter 15, 1–16

Carrillo-Bilbao, Gabriel; Fiore, Anthony Di; Fernández-Duque, Eduardo. 2005. Dieta, forrajeo y presupuesto de tiempo en cotoncillos (*Callicebus discolor*) del Parque Nacional de Yasuní en la Amazonia Ecuatoriana. Neotropical Primates 13(2):7–11

Casula, Kátia Regina. 2012. Análise da similaridade florística e estrutural das formações florestais inundáveis em um trecho do alto Rio Madeira e de seus afluentes, estado de Rondônia, Brasil. Master's Thesis, Núcleo de Ciências Exatas da Terra, Universidade Federal de Rondônia, Porto Velho

Cattanio, José; Anderson, Anthony B.; Carvalho, Manoel S. 2002. Floristic composition and topographic variation in a tidal floodplain forest in the Amazon estuary. Revista Brasileira de Botanica 25(4):419–430

Cavalcante, P. B. 2010. *Frutas Comestíveis na Amazônia*. Museu Paraense Emílio Goeldi, Belém

Charles-Dominique, P.; Atramentowicz, M.; Charles-Dominique, M.; Gérard, H.; Hladik, C. M.; Prévost, M. F. 1981. Les mammiferes frugivores arboricoles nocturnes d'une foret Guyanaise: inter-relations plantes-animaux. Revue d'Écologie: la Terre et la Vie 35:341–435

Coelho, Roberta de Fátima Rodrigues; Zarin, Daniel Jacob; Miranda, Izildinha Souza; Tucker, Joanna Marie. 2003a. Análise florística e estrutural de uma floresta em diferentes estágios sucessionais no município de Castanhal, Pará. Acta Amazonica 33(4):563–582

Coelho, Roberta de Fátima Rodrigues; Zarin, Daniel Jacob; Miranda, Izildinha Souza; Tucker, Joanna Marie. 2003b. Ingresso e mortalidade em uma floresta em diferentes estágios sucessionais no município de Castanhal, Pará. Acta Amazonica 33(4):619–630

Coelho, Roberta de Fátima Rodrigues; Miranda, Izildinha Souza; Mitja, Danielle. 2012. Caracterização do processo sucessional no Projeto de Assentamento Benfica, sudeste do estado do Pará, Amazônia oriental. Boletim do Museu Paraense Emílio Goeldi, Ciencias Naturais 7(3):251–282

Coelho-Ferreira, Márlia; Jardim, Mário Augusto G. 2005. Algumas espécies vegetais usadas pelos moradores da Ilha de Algodoal, Maiandeua, município de Maracanã, Pará. Boletim do Museu Paraense Emílio Goeldi, Série Ciências Naturais 1(2): 45–51

Corrêa, André Luis; Scudeller, Veridiana Vizoni; Goldenberg, Renato. 2017. Melastomataceae in the Reserva de Desenvolvimento Sustentável do Tupé, Amazonas, Brazil. Phytotaxa 323(2):101–127

Costa, Joanne R.; Mitja, Danielle. 2010. Uso dos recursos vegetais por agricultores familiares de Manacapuru (AM). Acta Amazonica 40(1):49–58

Dantas, Mário. 1979. Pastagens da Amazônia Central: Ecologia e fauna do solo. Acta Amazonica 9(2) suplemento.

Dantas, Mário; Rodrigues, Irenice Alves. 1980. Plantas invasoras de pastagens cultivadas na Amazônia. EMBRAPA, Centro de Pesquisa Agropecuária do Trópico Úmido, Boletim de Pesquisa, vol. 1.

Defler, Thomas R. 1979. On the ecology and behaviour of *Cebus albifrons* in eastern Colombia: 1: Ecology. Primates 20(4):475–490

DeWalt, Saara J.; Denslow, Julie S.; Ickes, Kalan. 2004. Enemy release and habitat expansion of the invasive shrub *Clidemia hirta* (Melastomataceae). *Ecology* 85(2):471–483

Duque, Alvaro J.; Duivenvoorden, Joost F.; Cavelier, Jaime; Sánchez, Mauricio; Polanía, Carolina; León, Andrea. 2005. Ferns and Melastomataceae as indicators of vascular plant composition in rain forests of Colombian Amazon. Plant Ecology 178:1–13

Ebert, Alexandre; Teixeira, Leandro Ribeiro; Silva, Adriana Zanirato Contini da; Costa, Reginaldo Brito da. 2014. Natural regeneration in tropical secondary forest in southern Amazonia, Brazil. Open Journal of Forestry 4(2):151–160

Ellison, Aaron M.; Denslow, Julie Sloan; Loiselle, Bette A.; Brenés M., Danilo. 1993. Seed and seedling ecology of Neotropical Melastomaceae. Ecology 74(6):1733–1749

Espírito-Santo, Fernando Del Bon; Shimabukuro, Yosio Edimir; Aragão, Luiz Eduardo Oliveira e Cruz de; Machado, Evandro Luiz Mendonça. 2005. Análise da composição florística e fitossociológica da floresta nacional do Tapajós com apoio geográfico de imagens satélites. Acta Amazonica 35(2):155–173

Fernandes Neto, José Guedes. 2015. Estratégias funcionais das espécies dominantes na sucessão florestal na Amazônia central. Master's thesis, INPA, Manaus

Ferreira, Sidney Alberto do N.; Cohen, Isaac; Jansen, Mauro Roberto A. 1994. Biologia reprodutiva de *Clidemia hirta* (L.) D. Don (Melastomataceae). Acta Amazonica 23(3/4):183–188

Foresta, H. de; Charles-Dominique, P.; Erard, C.; Prévost, M. F. 1984. Zoochorie et premiers stades de la regeneration naturelle après coupe en forêt Guyanaise. Revue d'Écologie: La Terre et la Vie 39(4):369–409

Franco, Ana María; Santamaria, Marcela. 1997. Area vital, habitos alimenticios y otros aspectos de historia natural de paujil *Mitu silvini*. In: Manejo de Fauna Silvestre en la Amazonía, Fang, T.G.; Bodmer, R.E.; Aquino, R.; Valqui, M.H. (eds), Instituto de Ecologia, La Paz, Bolivia, pp 267–282

Fuentes, Alfredo. 2005. Una introducción a la vegetación de la región de Madidi. Ecología en Bolivia 40(3):1–31

Gama, João Ricardo Vasconcellos; Botelho, Soraya Alvarenga; Bentes-Gama, Michellinyde Matos. 2002. Composição florística e estrutura da regeneração natural de floresta secundária de várzea baixo no estuário Amazônico. Revista Árvore 26(5):559–566

Gama, J. R. V., S. A. Botelho, M. Bentes-Gama, and J. R. S. Scolforo 2003. Estrutura e potencial futuro de utilização da regeneração natural de floresta de várzea alta no Município de Afuá, estado do Pará. *Ciencia Florestal, Santa Maria* 13(2):71–82

Gleason, H. A. 1939. The genus *Clidemia* in Mexico and Central America. Brittonia 392):97–140
Godoy, Judith Rosales; Petts, Geoffrey; Salo, Jukka. 1999. Riparian flooded forests of the Orinoco and Amazon basins: a comparative review. Biodiversity and Conservation 8:551–586
Gomez, Tedi Pacheco. 2011. Valoración económica de purmas de diferentes edades en la comunidad nativa de Santa Mercedes, río Putumayo, Perú. Master's thesis, UNAP, Iquitos
Gordon, André; Schadow, Benjamin; Quijano, Clara E.; Marx, Friedhelm. 2011a. Chemical characterization and antioxidant capacity of berries from *Clidemia rubra* (Aubl.) Mart. (Melastomataceae). Food Research International 44: 2120–2127
Gordon, André; Jungfer, Elvira; Silva, Bruno Alexandre da; Maia, José Guilherme S.; Marx, Friedhelm. 2011b. Phenolic constituents and antioxidant capacity of four underutilized fruits from the Amazon region. Journal of Agricultural and Food Chemistry 59:7688–7699
Gostinski, Luana Fontoura; Oliveira, Fabiana França; Contrera, Felipe Andrés Léon; Albuquerque, Patrícia Maia Correia de. 2018. Nicho trófico e partição de fontes polínicas por duas espécies de Melipona (Hymenoptera, Apidae) na Amazônia oriental. Oecologia Australis 22(4):449–462
Gottsberger, Gerhard; Morawetz, Wilfried. 1986. Floristic, structural and phytogeographical analysis of the savannas of Humaitá (Amazonas). Flora 178:41–71
Gregory, Tremaine. 2006. Comparative socioecology of sympatric, free-ranging white-faced and bearded sakis in Brownsberg Nature Park, Suriname. Doctoral dissertation, Kent State University, Kent, Ohio
Gualberto, Midiã Larine Colares; Ribeiro, Renato Bezerra da Silva; Gama, João Ricardo Vaconcellos; Vieira, Diego dos Santos. 2014. Fitossociologia e potencial de espécies arbóreas em ecossistema sucessional na Floresta Nacional do Tapajós, Pará. Agroecossistemas 6(1):42–57
Guarim Neto, Germano. 1984. Plantas do Brasil: Angiospermas do Estado de Mato Grosso- 1. Rodriguésia 36(59):105–121
Henry, Olivier; Feer, Francois; Sabatier, Daniel. 2000. Diet of the lowland tapir (*Tapirus terrestris* L.) in French Guiana. Biotropica 32(2):3364–368
Heywood, V. 2005. Invasives. In: The Cultural History of Plants, Ghillean Prance and Mark Nesbitt (eds), Routledge, New York, pp. 377–386
Hoff, Michel; Cremers, Georges. 2005. Le jardin Guyanais: inventaire des plantes cultivées et des adventices des jardins de Guyane française. Journal de Botanique de la Société de France 29:3–40
Ivanauskas, Natália Macedo; Monteiro, Reinaldo; Rodrigues, Ricardo Ribeiro. 2003. Alterations following a fire in a forest community of Alto Rio Xingu. Forest Ecology and Management 184:239–250
Izzo, T. J.; Bruna, E. M.; Vasconcelos, H. L.; Inouye, B. D. 2009. Cooperative colony founding alters the outcome of interspecific competition between Amazonian plant-ants. Insectes Sociaux 56:341–345
Johnston, Mark; Colquhoun, Araminta. 1996. Preliminary ethnobotanical survey of Kurupukari: an Amerindian settlement of central Guyana. Economic Botany 50(2):182–194
Karsten, G.K.W.H. 1861. Flora Columbiae. Berlin, Vol. 1, plate 87.
Kunz, Sustanis Horn; Moreau, Julia; Spadeto, Cristiani; Martins, Sebastião Venâncio; Stefanello, Daniel; Ivanauskas, Natália Macedo. 2014a. Estrutura da comunidade arbórea de trecho de floresta estacional sempre-verde e similaridade florística na região nordeste de Mato Grosso, Brasil. Floresta e Ambiente 21(4):429–440
Kunz, Sustanis Horn; Ivanauskas, Natália Macedo; Martins, Sebastião Venâncio; Stefanello, Daniel; Silva, Elias. 2014b. Fitossociologia do componente arbóreo de dois trechos de floresta estacional perenifólia, bacia do Rio das Pacas, Querência, MT. Ciência Florestal, Santa Maria 24(1):1–11
Kurzatkowski, Dariusz; Leuschner, Christoph; Homeier, Jürgen. 2015. Effects of flooding on trees in the semi-deciduous transition forests of the Araguaia floodplain, Brazil. Acta Oecologica 69:21–30
Lau, Arthur Viana; Jardim, Mário Augusto Gonçalves. 2014a. Relação entre o banco de sementes e a composição química do solo em uma floresta de várzea. Biota Amazônia 4(2):96–101

Lau, Arthur Viana; Jardim, Augusto Gonçalves. 2014b. Composição e densidade do banco de sementes em uma floresta de várzea, Ilha do Combu, Belém-PA, Brasil. Biota Amazônia 4(3):5–14

Levey, D.J.; Byrne, M.M. 1993. Complex ant-plant interactions: rain-forest ants as secondary dispersers and post-dispersal seed predators. Ecology 74 (6): 1802–1812

Lima, Laíce Fernanda Gomes de; Santos, João Ubiratan Moreira dos; Rosário, Alessandro Silva do; Baumgratz, José Fernando Andrade. 2014. Melastomataceae em formações costeiras de restingas no Pará, Brasil. Acta Amazonica 44(1):45–58

Lima Filho, Diógenesde Andrade' Revilla, Juan; Amaral, Iêda L. do; Matos, Francisca Dionizia de A.; Coêlho, Luiz de Souza; Ramos, José Ferreira; Silva, Gláucio Belém da; Guedes, José de Oliveira. 2004. Aspectos florísticos de 13 hectares da área de Cachoeira Porteira-PA. Acta Amazonica 34(3):415–423

Lizcano, Leandro J.; Bakkali, Fadil; Ruiz-Larrea, M. Begoña; Ruiz-Sanz, José Ignacio. 2010. Antioxidant activity and polyphenol content of aqueous extracts from Colombian Amazonian plants with medicinal use. Food Chemistry 119:1566–1570

Longworth, J. Benjamin; Mesquita, Rita C.; Bentos, Tony V.; Moreira, Marcelo P.; Massoca, Paulo E.; Williamson, G. Bruce. 2014. Shifts in dominance and species assemblages over two decades in alternative successions in central Amazonia. Biotropica 46(5):529–537

Lorenzi, Harri; Bacher, Luis; Lacerda, Marco; Sartori, Sergio. 2006. Frutas Brasileiras e Exóticas Cultivadas. Instituto Plantarum de Estudos da Flora Ltda, Odessa, Sao Paulo.

Luize, Bruno Garcia; Venticinque, Eduardo Martins; Silva, Thiago Sanna Freire; Novo, Evlyn Marcia Leão de Moraes. 2015. A floristic survey of angiosperm species occurring at three landscapes of the central Amazon várzea, Brazil. Checklist: The Journal of Biodiversity Data 11(6):1–15

Machado, Danilo de Oliveira. 2016. A agrobiodiversidade de quintais agroflorestais em propriedades agrícolas familiares na BR 174, Ramal do Pau-Rosa, Manaus, AM. Master's Thesis, INPA, Manaus

Machado, Monielle Alencar; Pinheiro, Claudio Urbano B. 2016. Da água doce à água salgada: mudanças na vegetação de igapó em margens de lagos, rios e canais do baixo curso do rio Pindaré, Baixada Maranhense. Revista Brasileira de Geografia Física 9(5): 1410–1427

Marcalla, Gabriela Fernanda Moya. 2012. Etnobotánica de las comunidades de Puerto Bolívar, Tarapuya, Aboquëhuira y Sototsiaya de la nacionalidad Siona, Provincia Sucumbios, Ecuador. Bachelors thesis, Universidad Central del Ecuador, Quito

Marques-Souza, Antonio Carlos, 1996. Fontes de pólen exploradas por *Melipona compressipes manaosensis* (Apidae: Meliponinae), abelha da Amazônia Central. Acta Amazonica 26(1/2):77–86

Marques-Souza, Antonio Carlos; Absy, Maria Lúcia; Kerr, Warwick Estevam; Peralta, Francisco Javier Aguilera. 1995. Pólen coletado por duas espécies de Meliponíneos (Hymenoptera: Apidae) da Amazônia. Revista Brasileira de Biologia 55(4):855–864

Marques-Souza, Antonio Carlos; Moura, Cleonice de Oliveira; Nelson, Bruce Walker. 1996. Pollen collected by *Trigona williana* (Hymenoptera: Apidae) in central Amazonia. Revista de Biologia Tropical 44(2):567–573

Marques-Souza, Antonio Carlos; Miranda, Ires Paula de Andrade; Moura, Cleonice de Oliveira; Rabelo, Afonso; Barbosa, Edelcílio Marques. 2002. Características morfológicas e bioquímicas do pólen coletado por cinco espécies de Meliponíneos da Amazônia central. Acta Amazonica 32(2):217–229

Marques-Souza, Antonio Carlos; Absy, Maria Lucía; Kerr, Warwick Estevam. 2007. Pollen harvest features of the central Amazonian bee Scaptotrigona fulvicutis Moure 1964 (Apidade: Meliponinae), in Brazil. Acta Botanica Brasilica 21(1):11–20

Martins, Sebastião Venâncio; Brito, Elizabeth Rodrigues; Oliveira Filho, Ary Teixeirade; Silva, Alexandre Francisco da; Silva, Elias. 2008. Floristic composition of two wetland forests in Araguaian plain, state of Tocantins, Brazil, and comparison with other areas. Revista Árvore 32(1):129–141

Mattos, Fábio de Souza. 2016. Dieta do *Plecturocebus cinerascens* (Platyrrhini: Primates): a influência da composição química na escolha de frutos consumidos. Master's thesis, INPA, Manaus

Maués, Bernardo Antônio Rodrigues; Jardim, Mário Augusto Gonçalves; Batista, Fábio de Jesus; Medeiros, Tonny David Santiago; Quaresma, Adriano da Costa. 2011. Composição florística e estrutura do estrato inferior da floresta de várzea na área de proteção ambiental Ilha do Combu, município de Belém, estado do Pará. Revista Árvore 35(3):669–677

Medeiros, Priscilla Sanjuan de. 2016. Indicadores ambientais para funcionalidade ecológica em florestas secundárias de diferentes idades na Amazônia Oriental. Doctoral dissertation, Universidade Federal do Pará, Belém

Medeiros, Herison; Obermüller, Flávio; Daly, Douglas C.; Silveira, Marcos; Castro, Wenderson; Forzza, Rafaela Campostrini. 2014. Botanical advances in southwestern Amazonia: the flora of Acre (Brazil) five years after the first catalogue. Phytotaxa 177(2):101–117

Miléo, L. J.; Silva, J. F.; Bentes, J. L. S.; Christoffoleti, P. J. 2007. Plantas daninhas hospedeiras alternativas de *Colletotrichum guaranicola* em cultivos de guaraná no estado do Amazonas. Planta Daninha 25(4):771–782

Miranda, Izildinha S. 2000. Análise florística e estrutural da vegetação lenhosa do Rio Comemoração, Pimenta Bueno, Rondônia, Brasil. Acta Amazonica 30(3):393–422

Montalván, Nadiuska Giannina Achong. 2015. Plantas alimenticias usadas por mamíferos mayores en la cuenca alta del Río Itaya, Loreto, Perú. Bachelor thesis, UNAP, Facultad de Ciencias Biológicas, Iquitos

Morley, T. 1976. *Memecylaceae (Melastomataceae)*. Flora Neotropical Monograph 15, New York Botanical Garden, Bronx.

Morley, Thomas. 1983. A directional cline in *Mouriri guianensis* (Melastomataceae). Acta Amazonica 13(5–6):799–811

Mota, Nara Furtado de Oliveira; Watanabe, Mauricio Takashi Coutinho; Zappi, Daniela Cristina; Hiura, Alice Lima; Pallos, Julieta; Viveros, Raquel Stauffer; Giulietti, Ana Maria; Viana, Pedro Lage. 2018. Cangas da Amazônia: a vegetação única de Carajás evidenciada pela lista de fanerógramas. Rodriguésia 69(3):1435–1488

Muñoz, Marianella Guevara. 2010. Evaluación de purmas con fines de manejo en la comunidad nativa de Santa Mercedes, Rio Putumayo, Perú. Master's thesis, Universidad Nacional de la Amazonia Peruana, Iquitos

Myster, Randall W. 2018. What we know about the composition and structure of igapó forests in the Amazon Basin. Botanical Review 84:394–410

Myster, Randall W. 2019. Primary and secondary igapó forests in the Peruvian Amazon: floristics, physical structure and the predictive value of soil bulk density. Journal of Plant Studies 8(1):20–29

Nebel, Gustav; Dragsted, Jens; Vanclay, Jerome K. 2000a. Estructura y composición florística del bosque de la llanura aluvial inundable de la Amazonía Peruana: II. El sotobosque de la restinga. Folia Amazónica 10(1–2):151–181

Nebel, Gustav; Kvist, Lars Peter; Vanclay, Jerome K.; Christensen, Henning; Freitas, Luis; Ruiz, Juan. 2000b. Estructura y composición florística del bosque de la llanura aluvial en la Amazonía Peruana: I. El bosque alto. Folia Amazónica 10(1–2):91–149

Nebel, Gustav; Kvist, Lars Peter; Vanclay, Jerome K.; Christensen, Henning; Freitas, Luis; Ruíz, Juan. 2001a. Structure and floristic composition of flood plain forests in the Peruvian Amazon I. Overstorey. Forest Ecology and Management 150:27–57

Nebel, Gustav; Dragsted, Jens; Vanclay, Jerome K. 2001b. Structure and floristic composition of flood plain forests in the Peruvian Amazon II. The understorey of restinga forests. Forest Ecology and Management 150: 59–77

Neves, Ana M. S.; Rylands, Anthony B. 1991. Diet of a group of howling monkeys, *Alouatta seniculus*, in an isolated forest patch in Central Amazonia. In: *A Primatologia no Brasil – 3*, Rylands, Anthony B.; Bernardes, Aline T. (eds), Fundação Biodiversitas para a Conservação da Diversidade Biológica, Belo Horizonte, pp 263–274

Novais, Jaílson Santos de; Absy, Maria Lúcia. 2013. Palynological examination of the pollen pots of native stingless bees from the lower Amazon region in Pará, Brazil. Palynology 37(2):218–230
Novais, Jaílson S. de; Garcêz, Ana Cristina A.; Absy, Maria Lúcia; Santos, Francisco de Assis R. dos. 2014. Comparative pollen spectra of *Tetragonisca angustula* (Apidae, Meliponni) from the lower Amazon (N Brazil) and caatingas (NE Brazil). Apidologie 46:417–431
Oliveira, Francisco Plácido Magalhães; Absy, Maria Lúcia; Miranda, Izildinha Souza. 2009a. Recurso polínico coletado por abelhas sem ferrão (Apidae, Meliponinae) em um fragmento de floresta na região de Manaus- Amazonas. Acta Amazonica 39(3):505–518.
Oliveira, Viviana Maria Araújo; Carneiro, Ana Lucia Basilio; Cauper, Glaucia Socorro de Barros; Pohlit, Adrian Martin. 2009b. In vitro screening of Amazonian plants for hemolytic activity and inhibition of platelet aggregation in human blood. Acta Amazonica 39(4):973–980
Oliveira, S. G. de; Alfaro, Jessica W. Lynch; Veiga, Liza M. 2014. Activity budget, diet, and habitat use in the critically endangered Ka'apor capuchin monkey (*Cebus kaapori*) in Pará state, Brazil: a preliminary comparison to other capuchin monkeys. American Journal of Primatology 76:919–931
Oliveira, Fabiana dos Santos; Ribeiro, Monique Hellen Martins; Nunez, Cecilia Veronica; Albuquerque, Patricia Maia Correia de. 2016. Flowering phenology of *Mouriri guianensis* (Melastomataceae) and its interaction with the crepuscular bee *Megalopta amoena* (Halictidae) in the restinga of Lençóis Maranhenses National Park, Brazil. Acta Amazonica 46(3):281–290
Peters, Halton A. 2001. *Clidemia hirta* invasion at the Pasoh Forest Reserve: an unexpected plant invasion in an undisturbed tropical forest. Biotropica 33(1):60–68
Phillips, O. 1993. The potential for harvesting fruits in tropical rainforests: new data from Amazonian Peru. Biodiversity and Conservation 2:18–38
Pinedo-Vasquez, Miguel; Zarin, Daniel J.; Coffey, Kevin; Padoch, Christine; Rabelo, Fernando. 2001. Post-boom logging in Amazonia. Human Ecology 29(2):219–239
Pinheiro, Tatyana; Ferrari, Stephen F.; Lopes, Maria Aparecida. 2013. Activity budget, diet, and use of space by two groups of squirrel monkeys (*Saimiri sciureus*) in eastern Amazonia. Primates 54:301–308
Prado, Helbert Medeiros. 2013. Feeding ecology of five Neotropical ungulates: a critical review. Oecologia Australis 17(4): 459–473
Queiroz, José Antonio Leite de; Mochiutti, Silas; Machado, Sebastião do Amaral; Galvão, Franklin. 2005. Composição florística e estrutura de floresta em várzea alta estuarina Amazônica. Floresta 35(1):41–55
Ramírez, Diego Nicolás Macedo. 2015. Evaluación cualitativa de la regeneración natural sobre arena blanca en áreas intervenidas en la carretera Iquitos-Nauta, Loreto, Perú. Biology thesis, Universidad Nacional de la Amazonía Peruana, Iquitos
Ratter, J. A.; G. P. Askew; R. F. Montgomery; and D. R. Gifford. 1978. Observations on the vegetation of northeastern Mato Grosso II. Forests and soils of the Rio Suiá-Missu area. *Proceedings of the Royal Society of London, Series B, Biological Sciences* 203 (1151):191–208
Rayol, Breno Pinto; Silva, Manoela Ferreira Fernandes da; Alvino, Fabrízia de Oliveira. 2006. Dinâmica da diversidade florística da regeneração natural de florestas secundárias, no município de Bragança, Pará, Brasil. Boletim do Museu Paraense Emílio Goeldi, Ciencias Naturais 1(2):9–27
Rayol, Breno Pinto; Alvino-Rayol, Fabrízia de Oliveira; Silva, Manoela Ferreira Fernandes da. 2011. Similaridade florística entre o estrato arbóreo e a regeneração natural de uma floresta secundária, no município de Bragança, nordeste do estado do Pará, Brasil. Revista Brasileira de Agroecologia 6(3):107–114
Renner, S. S. 1987. Reproductive biology of *Bellucia* (Melastomataceae). Acta Amazonica 16:197–208
Renner, S. S. 1989. A survey of reproductive biology in Neotropical Melastomataceae and Memecylaceae. Annals of the Missouri Botanical Garden 76:496–518
Ritter, Rhuan Amorim; Monteiro, Maria Vivina Barros; Monteiro, Frederico Ozanan Barros; Rodrigues, Silvane Tavares; Soares, Marina Lira; Silva, Jean Carlos Ramos; Palha, Maria

das Dores Correia; Biondi, Germano Francisco; Rahal, Sheila Canevese; Tourinho, Manuel Malheiros. 2012. Ethnoveterinary knowledge and practices at Colares island, Pará state, eastern Amazon, Brazil. Journal of Ethnopharmacology 144:346–352

Roble, Nick. 2008. Ontogenetic resource partitioning in white faced sakis (*Pithecia pithecia*). Master's thesis, Kent State University.

Rocha, Gustavo P. E.; Vieira, Daniel L. M.; Simon, Marcelo F. 2016. Fast natural regeneration in abandoned pastures in southern Amazonia. Forest Ecology and Management 370:93–101

Rocha, Keila Cristina de Jesus; Goldenberg, Renato; Meirelles, Julia; Viana, Pedro Lage. 2017. Flora das cangas da Serra dos Carajás, Pará, Brasil: Melastomataceae. Rodriguésia 68(3): 997–1034

Rojas, Joahan; Satalaya, Johann R.; Grandez, Maritza; Vilchez, Luis; Rengifo, Elsa; Salamanca, Efrain; Flores, Ninoska; Giménez, Alberto; Ávila, Juan Antonio; Ruiz, Grace. 2009. Actividad leismanicida de plantas medicinales de la Amazonia Peruana. Revista Boliviana de Química 26(2):43–48

Rondon Neto, Rubens Marques; Santos, Jeneffer Soares dos; Silva, Manoel Advíncula da; Koppe, Valdinei Cristi. 2010. Potencialidades de uso de espécies arbustivas e arbóreas em diferentes fisionomias de cerrado, em Lucas do Rio Verde/MT. Revista de Biologia e Ciências da Terra 10(2):113–126

Roosevelt, A. C. 1998. Ancient and modern hunter-gatherers of lowland South America: an evolutionary problem. In: Advances in Historical Ecology, W. Balée (ed.), Columbia University Press, New York, pp. 190–212

Saldarriaga, J.G.; Uhl, C. 1991. Recovery of forest vegetation following slash-and-burn agriculture in the Upper Rio Negro. In: Rainforest Regeneration and Management, Gómez-Pompa, A.; Whitmore, T. C.; Hadley, M. (eds), UNESCO/Parthenon Publishing Group, Paris, pp 303–312

Salomão, Rafael de Paiva; Vieira, Ima Célia Guimarães; Suemitsu, Chieno; Rosa, Nélson de Araújo; Almeira, Samuel Soares de; Amaral, Dário Dantas do; Menezes, Moirah Paulo Machado de. 2007a. As florestas de Belo Monte na grande curva do rio Xingu, Amazônia Oriental. Boletim do Museu Paraense Emílio Goeldi, Ciencias Naturais 2(3):57–153

Salomão, Rafael Paiva; Rosa, Nélson Araújo; Morais, Kácio Andrey Câmara. 2007b. Dinâmica da regeneração natural de árvores em áreas mineradas na Amazônia. Boletim do Museu Paraense Emílio Goeldi, Ciencias Naturais 2(2):85–139

Sánchez, Judith Jacira Achong; Levy, Allison Licett Nuñez. 2013. Características de frutos consumidos por *Saguinus nigrifrons* (Primates, Callitrichidae), en la Estación Biológica Quebrado Blanco, Loreto, Perú. Master's thesis, Universidad Nacional de la Amazonía Peruana, Iquitos

Santos, Charles F.; Absy, Maria L. 2010. Polinizadores de *Bertholletia excelsa* (Lecythidales: Lecythidaceae): interações com abelhas sem ferrão (Apidade: Meliponni) e nicho trófico. Neotropical Entomology 39(6): 854–861

Santos, Gideão Costa dos; Jardim, Mário Augusto Gonçalves. 2006. Florística e estrutura do estrato arbóreo de uma floresta de várzea no município de Santa Bárbara do Pará, Estado do Pará, Brasil. Acta Amazonica 36(4):437–446

Santos, João Ubiratan M. dos; Amaral, Dário Dantas do; Gorayeb, Inocêncio de Sousa; Bastos, Maria de Nazaré do C.; Secco, Ricardo de S.; Neto, Salustiano Vilar Costa; Costa, Denise Cristina Torres. 2003. Vegetação da área de proteção ambiental Jabotitiua-Jatium, Município de Viseu, Pará, Brasil. Acta Amazonica 33(3):431–444

Schneider, Larissa; Ferrara, Camila R.; Vogt, Richard C.; Schaffer, Chuck. 2016. Subsistence-level chelonian exploitation on the Rio Negro and one viable alternative. Chelonian Conservation and Biology 15(1):36–42

Schultes, R. E. and R. F. Raffauf. 1990. The Healing Forest: Medicinal and Toxic Plants of the Northwest Amazonia. Dioscorides Press, Portland, Oregon

Silva, Andréa Leme; Tamashiro, Jorge; Begossi, Alpina. 2007a. Ethnobotany of riverine populations from the Rio Negro, Amazonia (Brazil). Journal of Ethnobiology 27(1): 46–72

Silva, E. M.; Souza, J. N. S.; Rogez, H.; Rees, J. F.; Larondelle, Y. 2007b. Antioxidant activities and polyphenolic contents of fifteen selected plant species from Amazonian region. Food Chemistry 101: 1012–1018

Silva, Sheila Maria Garcia da; Nascimento, Keila Gardênia Silva; Fraxe, Therezinha de Jesus Pinto; Braga, Pedro Ivo Soares. 2007c. A "Saúde" nas comunidades focais do Projeto PIATAM: o etnoconhecimento e as plantas medicinais. In: Comunidades Ribeirinhas Amazônicas: Modos de Vida e Uso dos Recursos Naturais, Fraxe, Therezinha de Jesus Pinto; Pereira, Henrique dos Santos; Witkoski, Antônio Carlos (eds), EDUA, Manaus, pp 109–138

Silva, Suzy Cristina Pedroza da; Pereira, Cloves Farias; Fraxe, Therezinha de Jesus Pinto; Witkoski, Antônio Carlos; Silva, Michele Andreza Pedroza da. 2007d. In: Comunidades Ribeirinhas Amazônicas: Modos de Vida e Uso dos Recursos Naturais, Fraxe, Therezinha de Jesus Pinto; Pereira, Henrique dos Santos; Witkoski, Antônio Carlos (eds), EDUA, Manaus, pp 141–154

Solano, P. J.; Durou, S.; Corbara, B.; Quilichini, A.; Cerdan, P.; Belin-Depoux, M.; Delabie, J. H. C.; Dejean, A. 2003. Myrmecophytes of the understory of French Guianian rainforests: their distribution and their associated ants. Sociobiology 41(2):1–10

Souza, Célia Cristina Valero. 2010. Etnobotânica de quintais em três comunidades ribeirinhas na Amazônia central, Manaus, AM. Master's thesis, INPA, Manaus

Stefanello, Daniel. 2008. Florística, estrutura e dispersão de sementes de três trechos de floresta estacional perenifólia ribeirinha na bacia hidrográfica do Rio das Pacas, Querência, MT. Master's Thesis, Universidade Federal de Viçosa, Minas Gerais

Stefanello, Daniel; Fernandes-Bulhão, Clarissa; Martins, Sebastião Venâncio. 2009. Síndromes de dispersão de sementes em três trechos de vegetação ciliar (nascente, meio e foz) ao longo do Rio Pindaíba, MT. Revista Árvore 33(6):1051–1061

Stefanello, Daniel; Ivanauskas, Natália Macedo; Martins, Sebastião Venâncio; Silva, Elias; Kunz, Sustanis Horn. 2010. Síndromes de dispersão de diásporos das espécies de trechos de vegetação ciliar do rio das Pacas, Querência-MT. Acta Amazonica 40(1):141–150

Stevenson, Pablo R. 2007. Estimates of the number of seeds dispersed by a population of primates in a lowland forest in western Amazonia. In: Seed Dispersal: Theory and its Application in a Changing World, Dennis, Andrew J.; Schupp, Eugene W.; Green, Ronda J.; Westcott, David A. (eds), CABI, Wallingford, pp 340–362

Stevenson, Pablo R.; Beltrán, Marta L.; Quiñones, Marcela J.; Ahumada, Jorge A. 2015. Differences in home range, activity patterns and diet of red howler monkeys in a continuous forest and a forest fragment in Colombia. Rev. Acad. Colomb. Cienc. Ex. Fis. Nat. 39(153):503–513

Taurino, Tássia Cristina da Conceição Barros. 2017. Recuperação natural da diversidade de espécies de plantas em florestas secundárias na Amazônia Oriental. Master's Thesis, Universidade Federal do Pará, Belém

Thomas, Evert. 2009. New light on the floristic composition and diversity of indigenous territory and national park Isiboro-Sécure, Bolivia. Biodiversity Conservation 18:1847–1878

Tuomisto, Hanna; Ruokolainen, Kalle; Poulsen, Axel Dalbreg; Moran, Robbin C.; Quintana, Catalina; Cañas, Gustavo; Celi, Jorge. 2002. Distribution and diversity of Pteridophytes and Melastomataceae along edaphic gradients in Yasuní National Park, Ecuadorian Amazon. Biotropica 43(4):516–533

Uhl, Christopher. 1987. Factors controlling succession following slash-and-burn agriculture in Amazonia. Journal of Ecology 75(2):377–407

Uhl, Christopher; Clark, Kathleen; Maquirino, Pedro. 1988. Vegetation dynamics in Amazonian treefall gaps. Ecology 69: 751–763

Van Andel, Tinde. 2001. Floristic composition and diversity of mixed primary and secondary forests in northwest Guyana. Biodiversity and Conservation 10:1645–1682

Van Andel, Tinde; Behari-Ramdas, Joelaika; Havinga, Reinout; Groenendijk, Sara. 2007. The medicinal plant trade in Suriname. Ethnobotany Research & Applications 5:351–372

Van den Eynden, Veerle; Cueva, Eduardo; Cabrera, Omar. 1999. Plantas Silvestres Comestibles del Sur del Ecuador. Ediciones Abya-Yala, Quito

Van Roosmalen, M. G. M. 1985. Habitat preferences, diet, feeding strategy and social organization of the black spider monkey (*Ateles paniscus paniscus* Linnaeus 1758) in Surinam. Acta Amazonica 15(3/4 suplemento) 1–238

Van Roosmalen, Marc G. M.; Klein, Lewis L. 1988. The spider monkeys, genus *Ateles*. In: Ecology and Behavior of Neotropical Primates, Mittermeier, Russell A.; Rylands, Anthony B.; Coimbra-Filho, Adelmar F.; Fonseca, Gustavo A.B. da (eds), Ecology and Behavior of Neotropical Primates, World Wildlife Fund, Washington, DC, vol. 2, pp 455–537

Vasconcellos, Heraldo L. 1991. Mutualism between Maieta guianensis Aubl., a myrmecophytic melastome, and one of its ant inhabitants: ant protection against insect herbivores. Oecologia 87(2):295–298

Vasconcellos, Heraldo L. 1993. Ant colonization of Maieta guianensis seedlings, an Amazon ant-plant. Oecologia 95:439–443

Vasconcellos, Heraldo L. 2000. Relationship between plant size and ant associates in two Amazonian ant-plants. Biotropica 32(1):100–111

Vásquez, R. 1997. *Flórula de las Reservas Biológicas de Iquitos, Perú*. Monographs in Systematic Botany Vol. 63, Missouri Botanical Garden, St. Louis.

Vásquez, Silvia Patricia Flores. 2014. Conhecimento, uso e conservação da diversidade vegetal em quatro comunidades ribeirinhas no Município Manacapuru, Amazonas. Doctoral dissertation, Programa de Pós-Graduação em Botânica, INPA, Manaus

Vásquez, R. and A. H. Gentry. 1989. Use and misuse of forest-harvested fruits in the Iquitos area. *Conservation Biology* 3(4): 350–361.

Vásquez, Silvia Patricia Flores; Mendonça, Maria Silvia de; Noda, Sandra do Nascimento. 2014. Etnobotânica de plantas medicinais em comunidades ribeirinhas do Município de Manacapuru, Amazons, Brasil. Acta Amazonica 44(4):457–472

Villarreal, Alberto; Arends, Ernesto; Escalante, Eduardo. 2003. Caracterización estructural y florística de sistemas tradicionales Conucos-Barbechos de etnia Piaroa, Amazonas, Venezuela. Revista Forestal Venezolana 47(2): 115–124

Wallace, Robert B.; Painter, R. Lilian E.; Taber, Andrew B. 1998. Primate diversity, habitat preferences, and population density estimates in Noel Kempff Mercado National Park, Santa Cruz Department, Bolivia. American Journal of Primatology 46:197–211

Wester, L. L. and H. B. Wood. 1977. Koster's curse (*Clidemia hirta*), a weed pest in Hawaiian forests. *Environmental Conservation* 4(1): 35–41

Wittmann, Florian; Schöngart, Jochen; Queiroz, Helder L.; Wittmann, Astrid de Oliveira; Conserva, Auristela dos Santos; Piedade, Maria T. F.; Kesselmeier, Jürgen; Junk, Wolfgang J. 2009. The Amazon floodplain demonstration site: sustainable timber production and management of central Amazonian white-water floodplains. Ecohydrology & Hydrobiology 9(1):41–54

Wittmann, Florian; Householder, Ethan; Piedade, Maria T. F.; Assis, Rafael L. de; Schöngart, Jochen; Parolin, Pia; Junk, Wolfgang J. 2013. Habitat specifity, endemism and the neotropical distribution of Amazonian white-water floodplain trees. Ecography 36:690–707

Yauri, Geraldine Yesenia Espinoza. 2016. Tintes vegetales de la sierra y selva del Perú: un estudio etnobotánico en los departamentos de Ancash, Loreto y Cusco. Master's thesis, Universidad Peruana Cayetano Heredia, Lima

Zappi, Daniela C.; Milliken, William; Lopes, Celia Regina A. Soares; Lucas, Eve; Piva, José Hypólito; Frisby, Sue; Biggs, Nicola; Forzza, Rafaela Campostrini. 2016. Xingu State Park vascular plant survey: filling the gaps. Brazilian Journal of Botany 39(2):751–778

Chapter 34
Menispermaceae

34.1 *Abuta grandifolia*

Bolivia Pitón, pitón-caserana, odó-odó, fruta de ciento pie, huevo de mono, caferano, ambaibillo; huanicousa (Chácobo), nuririjaja (Tacana), odo' odo' (Tsimane')

Brazil Pitomba da folha dura, pitomba da mata, pitomba do mato, pitombinha, catuaba, catuabinha, xexuá, grão de galo, cafèrana, orelha de onça, abuta, abuta amarela, abuta branca, buta, butuá, bota vermelha, pau de ferro, pau bôta, cipó, cipó de bôta, cipó abuta, cipó pitomba, murtinha; ihipá-ra-ti (Arawaté), wari-makwa-'ɨ (Guajá), aputa (Ka'apor), warapa yokü (Tiriyó), yani-ipo (Wajãpi), itahi mosihi (Yanomami)

Colombia Calentura caspi, chilicaspi, curare, abuta, vibuajeinia, taque taque, palo amargo, aralén; ojejikaka (Andoke), taque-pura-que (Cubeo), vibua-jeiria-mirsimarika (Makuna), mamutuka (Nukak), ancabesux (Siona-Secoya)

Ecuador Motelo caspi, motelo muyo, palo de motelo, planta de tortuga, jinuni, ajinjiembre; tteteccu'cho, titicocho tsakiko (Cofán), tara tara, tara tara muyo, para tara, yahuati caspi, yahuati-panga, payanchi, payanchi yurac, sicutara, caupanga, montelomuyo, suruqui (Kichwa), tseasnúm, machap, tscas nack, sakap, payanshi (Shuar), acahuesë, sita pipi, dayawi uo, ancabesux (Siona-Secoya), bagamö, bagahue, bayabeymo, öntame, vagamue (Waorani)

Peru Breahuayo, pichico huayo, abuta caspi, abuta hembra, achuni caspi, achuni sanango, sanango, motelo sanango, palo de motelo, soga, carpinteiro sacha, trompetero sacha, parapara, chuchuasillo amargo, azúcar huayo; caimitillo, pancha muca; kuntsé-kuntsé (Achuar), tsegasnum, tseasnumi, tseasnum painkinsh (Aguaruna), ojé-ji-ka-ka (Andoke), kana pan binu (Cashinahua), tanan ishi kayura

N. Smith, *Amazon Fruits: An Ethnobotanical Journey*, Ethnobiology,
https://doi.org/10.1007/978-3-031-12803-5_34

(Chayahuita), yisakwi (Ese Eja), tseasnum (Huambisa), chichieu cunitá (Kichwa), kaposhipini, tanpiapari (Matsigenka), nishipacha, pancha muca (Shipibo)

Venezuela Cafecillo, frijol de rabipelado, parcha; hui atä daö (Piaroa), mayadu-waima-g (Ye'kuana)

Status Wild

Victor, 8 years, picking pichico huayo *Abuta grandifolia* **in regrowth forest along a stream bank 600 m north of Miraflores, Rio Tigre, Loreto, Peru 7-1-06**

A slender tree in mature forests and successionary vegetation, *Abuta grandifolia* produces olive-sized yellow fruits that are gathered from lower branches. The fruits are often within easy reach because the understory tree is rarely taller than 10 m and produces fruit by the time it is 1 m high. Only youngsters can safely climb the liana tree; however, because the trunk and branches are so thin they can break easily hence some of the common names for this tree in the Brazilian Amazon: cipó abuta, cipó de bôta, and cipó pitomba. These fruits are largely consumed by children along the Jauaperi River which separates the states of Amazonas and Roraima in the northern Brazilian Amazon (Ammann 2014:36).

It is widely distributed in the Amazon and Orinoco Basins as well as the Guianas. Usually found below 1,000 m, this liana-like treelet grows on uplands and flood-plains, including swamps dominated by *Mauritia flexuosa* palms (Araujo-Murakami et al. 2009; Cárdenas et al. 2009; Casula 2012:64; Kunz et al. 2010, 2014a, b; Lobo 2017:66; Myster 2007, 2015; Ratter et al. 1978; Sousa et al. 2014, 2018) and along streams in Manaus, Amazonas (Santos et al. 2016). This vine-like small tree grows in gallery forest in the cerrado region of Mato Grosso (Marimon et al. 2002; Stefanello 2008:19; Stefanello et al. 2009, 2010), in dense scrub savanna called *cerradão* (Marimon and Lima 2001; Ratter 1987; Rondon Neto et al. 2010), in open vegetation growing on iron-rich sedimentary rock (*canga*) in the Serra dos Carajás in Pará (Lima and Teixeira 2018; Mota et al. 2018), and in stunted vegetation known in Brazil as *campina*, *campinarana*, and *caatinga* that grows on sandy patches. This liana-like tree also grows in stunted woodlands (*campinarana*) subject to annual flooding, such as in Roraima (Vale 2011:52). It also occurs in some upland successionary forests taking over abandoned agricultural plots, such as in the Tocantins watershed in eastern Pará (Taurino 2017:87) and in the vicinity of Santa Mercedes along the Rio Putumayo River in northern Peruvian Amazon bordering Colombia (Gomez 2011:87; Muñoz 2010:30).

Rolly, 8 years, picking breahuayo ***Abuta grandifolia*** **fruits in upland forest. Near Miraflores, lower Rio Tigre, Loreto, Peru 6-11-04**

The fruits (carpels) are born singly or in clusters. Technically, the entire cluster is a fruit, but the individual carpels look like separate fruits. Each carpel is oblong when mature and measures some 2.5 cm long and 1.5 cm in diameter and contains a single light tan seed surrounded by slimy pulp. Although the mesocarp is sparse, it has a taste similar to cream cheese and is thus popular. In parts of the Peruvian Amazon, the fruit is considered an aphrodisiac (Brack Egg 2003:131). Indigenous groups, such as the Waimiri Atroari in Amazonas, Brazil, the Huaorani in the Ecuadorian Amazon, and the Tsimane in the Bolivian Amazon are also fond of the fruits (Ima 2012; Martínez-Rodríguez 2009:91; Milliken et al. 1992). The fruits are eaten locally and do not usually enter markets, although they appear occasionally in the street markets of Santarém in Pará (Rego et al. 2011).

Some monkeys are involved in dispersing *Abuta grandifolia*. Along the Tigre, an affluent of the lower Marañón in the Peruvian Amazon, one of the common names for the plant is pichico huayo which means fruit of the saddle-back tamarin *Saguinus fuscicollis*. Both the saddle-backed tamarin and the moustached tamarin *Saguinus mystax* have been documented feeding on this fruit and dispersing the seeds in the

Quebrada Blanco watershed, an affluent of the Tahuayo River in the northeastern part of the Peruvian Amazon (Culot et al. 2009; Sánchez and Levy 2013:30, 79). A related species, the mottled-face tamarin *Saguinus inustus* has been documented feeding on this fruit in the Caquetá watershed in the Colombian Amazon (Palacios et al. 2004). Bald uakaris *Cacajao calvus ucayalii* are known to eat this fruit in the Peruvian Amazon and this monkey is hunted for its meat (Aquino and Encarnación 1999). The equatorial saki *Pithecia aequatorialis*, also known as the red-bearded saki, also eats this fruit (Charpentier et al. 2015) but this monkey, which occurs in northeastern Peru and the Ecuadorian Amazon, is relatively small, usually weighing under 2.5 kg, so it is not regularly hunted for its meat. However, it is likely involved in dispersing some of the seeds. Black spider monkeys *Ateles paniscus* and woolly monkeys *Lagothrix lagotricha* also eat this fruit and are also hunted throughout the Amazon and neighboring Guianas (Stevenson 2000; Van Roosmalen 1985; Van Roosmalen and Klein 1988). Along the Caquetá in the Colombian Amazon, mottled-faced tamarins *Saguinus inustus* have been observed ingesting the fruits (Palacios et al. 2004).

Fallen fruits are eaten by a variety of other animals, including game such as tortoises (*Chelonoidis carbonaria, C. denticulata*) as suggested by common names for the fruit in the Ecuadorian Amazon: motelo caspi, motelo muyo (tortoise fruit), and palo de motelo, planta de tortuga (tortoise tree). In the Ecuadorian Amazon, hunters rub the noses of their dogs with leaves of *Abuta grandifolia* in the belief that it will help them locate tortoises. Seeds of this fruit have been found in the feces of the maned wolf *Chrysocyon brachyurus*, a long-legged predator which inhabits cerrado vegetation in the southern fringes of the Amazon Basin (Castro and Emmons 2012).

In addition to fruits, *Abuta grandifolia* has a variety of uses in folk medicine. In the watershed of the Jaú, an affluent of the middle Rio Negro in Amazonas, Brazil, the leaves are used to make an infusion for treating gastrointestinal disturbances and the bark is used to make a drink to reduce fever (Rodrigues 2006). Rural folk along the Jauaperi, another affluent of the middle Rio Negro, use the bark to make tea to treat stomachache, gastritis, and inflammation (Pedrollo et al. 2016). Several villages in the vicinity of Barcelos along the middle Rio Negro make an infusion of the leaves of this plant to treat malaria (Tomchinsky et al. 2017), as do some rural settlements in the municipality of Xapuri in Acre and the municipality of Pauini in southern Amazonas state (Ferreira et al. 2015). In the vicinity of Manaus along the lower Rio Negro, some rural folk prepare a decoction with the flowers of this tree to alleviate diarrhea, pain, and inflammation (Scudeller et al. 2009). Residents of Enfarrusca, a rural village near Bragança in northeastern Pará, make a paste with the roots or bark of this tree to treat wounds (Freitas and Fernandes 2006). Some rural people in the Colombian Amazon use the bark to treat diabetes, snake bites, and microbial infections (Lizcano et al. 2010).

Several indigenous communities along the middle Vaupés River in the Colombian Amazon use the bark and roots of this tree to treat a variety of health issues including snake bites, whooping cough, and headaches (Ramírez et al. 2017). The Shipibo who inhabit various villages in the vicinity of Pucallpa along the Ucayali River in the Peruvian Amazon use the bark to make a decoction to treat a variety of medical conditions including diabetes, malaria, stomachache, and arthritis (Polesna et al.

2011). Some rural folk along the lower Ucayali use the bark to prepare folk remedies for treating stomach infections and wounds (Tananta 2014:67), while this plant is used in the Madre de Dios watershed to reduce dental pain and to treat diabetes, fever, rheumatism, and stomach ulcers (Pumacallahui-Salcedo et al. 2016). The Ese Eja who inhabit the Madre de Dios watershed in the southeastern Peruvian Amazon use this liana to make an infusion to treat tuberculosis and lung hemorrhages (Desmarchelier et al. 1995); the Ese Eja also mash the young leaves and apply them to skin infections and a decoction is prepared from the bark to treat intestinal infections (Arévalo-Lopéz et al. 2018). Rural folk in various parts of the Peruvian Amazon use the leaves of this liana to treat tuberculosis and other lung issues (Desmarchelier et al. 1997). Kichwa women in the southern Ecuadorian Amazon drink an infusion of the bark to prevent pregnancy (Cerón 1996). The Tiriyó who inhabit the Tumucumaque hills and surrounding lowlands along the border between northern Pará and Surinam scrape the bark into water and then drink the solution to alleviate stomach pain (Cavalcante 1973:42).

Various indigenous groups and rural folk in the Ecuadorian Amazon also use this plant in folk remedies. Rural people boil the roots to treat sore throat and the bark is boiled and the liquid is drunk after it cools to treat coughs. The Kichwa chip off pieces of the bark and soak the shavings for a few hours before drinking the solution to treat diarrhea and fever. The Kichwa also boil the leaves to make a compress for infected eyes (Schultes and Raffauf 1990:303). The Waorani scrape the bark and boil the shavings which are then drunk when it cools to treat stomachache and diarrhea. The Waorani also boil the ripe fruits in water and half a cup of the solution is drunk twice a day to alleviate headaches (Mondragón and Smith 1997:31). The Siona-Secoya scrape the bark and squeeze the liquid which is then taken to treat snakebites. They also use an infusion of the leaves to reduce fever (Schultes and Raffauf 1990:303).

In the Peruvian Amazon, various parts of this plant are used in folk remedies by rural people (Mejia and Rengifo 2000:14). A decoction of the roots is drunk to treat anemia and stem bleeding. Mashed roots are mixed with sugarcane alcohol (aguardiente) a small glass of which is then drunk three times a day for 8 days to treat rheumatism and anemia (Lossio et al. 1995:31). To stem bleeding (hemorrhaging), this same solution is taken every 4 h. To treat diabetes, the bark is scraped into water and the solution is drunk after meals for 30 days. Rural people in the province of Maynas in the Peruvian Amazon use the bark to make a tea which is taken as a diuretic and to treat uterine infections and diabetes (Jovel et al. 1996). Kichwa people in the Chazuta district of the upper Huallaga River in the Peruvian Amazon macerate the stem and roots in rum which is drunk to treat rheumatism and as an aphrodisiac (Sanz-Biset et al. 2009). The Chayahuita in the Peruvian Amazon make a decoction from the bark to treat diarrhea (Odonne et al. 2013). The Ese Eja who live in parts of the Madre de Dios watershed in the Peruvian Amazon grind the stem and then boil it. After the liquid cools, it is drunk to treat tuberculosis (Desmarchelier et al. 1996).

This small understory tree has several other uses, especially among indigenous tribes. The Cofán employ the seeds of *Abuta grandifolia* in handicrafts. Also in the Colombian Amazon, the Andoke, Cubeo, Karijona, and Makuna scrape the roots and bark of *Abuta grandifolia* and in some cases mix it with scrapings from the roots of several species of *Strychnos* to make curare, a poison used to tip blowgun darts (La Rotta 1983:62; Mejía and Turbay 2009). The Cubeo in the Colombian Amazon use this strong wood to make arrows (Ramírez et al. 2017). Waorani boys in the Ecuadorian Amazon make toy bows from the slender branches and the wood is also used for fuel (Cerón and Ayala 1998:105). The Tsimane in the Bolivian Amazon also use the wood for fuel as well as for construction purposes (Guèze et al. 2014).

34.2 *Anomospermum chloranthum*

Brazil Abuta

Colombia Veneno de churuco

Ecuador Max-so-ae-o (Siona-Secoya)

Peru Coto huayo, huevo de coto, ampihuasca

Status Wild

Boys climbing a coto huayo *Anomospermum chloranthum* liana to shake loose some fruits so that they fall to the forest floor. Near Yarina, Rio Yanayacu, Reserva Nacional Pacaya-Samiria, Loreto, Peru 6-25-06

Lianas are prominent features in tropical rainforests and they provide a wide range of uses such as cordage for construction, medicines, emergency water, and fruit. *Anomospermum chloranthum* is one such liana, producing shiny orange fruits the size of a chicken's egg during the rainy season. Known as coto huayo (howler monkey fruit) along the Yanayacu and Samiria rivers in the Reserva Nacional Pacaya-Samiria, locals gather the fruits from the forest floor because the woody liana climbs high into the canopy. The widespread squirrel monkey *Saimiri sciureus* are known to eat the fruits (Terborgh 1983:68). Although howler monkeys (*Alouatta* spp.) eat mostly leaves, they also eat some fruits and may also feast on *Anomospermum chloranthum* fruits as suggested by the name for this tree in the Pacaya-Samiria National Reserve in the Peruvian Amazon: coto huayo. In French Guiana, spider monkeys *Ateles paniscus*, which also occur in the Amazon Basin, are known to eat the fruits (Guillotin et al. 1994). Howler monkeys and spider monkeys are widely hunted for their meat.

A layer of creamy, tan-colored pulp surrounds the single black seed. Fruits serve as a forest snack and do not appear in markets. This liana, some 40 cm thick and 50 m long when fully grown, ranges from Costa Rica south through the western part of the Amazon Basin reaching as far east as the Madeira River. *Anomospermum chloranthum* occurs in upland and floodplain forests up to 1000 m in the Andean foothills.

Donaldo, 12 years, holding a coto huayo *Anomospermum chloranthum* fruit. He climbed the coto huayo liana, he is grasping to shake loose some of the fruits. Near Yarina, Rio Yanayacu, Reserva Nacional Pacaya-Samiria, Loreto, Peru 6-25-06

34.3 *Orthomene schomburgkii*

Brazil Papo de mutum, meia lua, pimenta de reino braba, bosta de capivara, grão de guariba; chawn-aat' (Maku)

Colombia Wan-yo' (Puinave), ha-na-peé (Yukuna)

Peru Abuta, huano, ampihuasca morado

Venezuela Cachete de temblador, bejuco de chigüire

Status Wild

Papo de mutum *Orthomene schomburgkii*, **a vine bearing fruit in floodplain forest along a black water river. Rio Inajá, affluent of the lower Rio Xingu a few km upstream from Carrazedo, Município de Gurupá, Pará, Brazil 8-1-16**

Papo de mutum *Orthomene schomburgkii* is an elongated vine (*cipó*) reaching up to some 30 m in length that grows in forests and successionary vegetation on uplands. It is also found in floodplain forests, especially along black water rivers. It produces edible yellow-orange fruits, measuring 4.5 cm by 3 cm, that contain a single seed. The mesocarp is pale yellow and has a sweet/tart taste. The fruits are eaten by locals, such as along the lower Xingu and its affluents. However, they never reach markets. This vine is widespread in the Amazon Basin, reaching as high as 1500 m in the Andean foothills. It is also found in northeastern and southeastern Brazil, the Guianas, Venezuela, and Central America as far north as Honduras.

Boys gathering papo de mutum *Orthomene schomburgkii* fruits from a canoe to snack on the pulp. Rio Inajá, affluent of the lower Rio Xingu a few km upstream from Carrazedo, Município de Gurupá, Pará, Brazil 8-1-16

This vine also has medicinal uses. The Cofán along the Sucumbios River in the Ecuadorian Amazon make a tea from the leaves to induce sleep. Along the Pira Paraná River in the Brazilian Amazon close to its border with Colombia, the Maku wash wounds that are taking a long time to heal with a decoction of the twigs of *Orthomene schomburgkii*. The bark of this vine is used by the Ticuna and Cofán as an ingredient in curare which is used to tip blowgun darts (Schultes and Raffauf 1990:309).

References

Ammann, Stefan. 2014. Etnobotânica de árvores e palmeiras em três comunidades ribeirinhas do rio Jauaperi, na divisa entre Roraima e Amazonas. Master's Thesis, INPA, Manaus

Aquino, Rolando; Encarnación, Filomeno. 1999. Observaciones preliminares sobre la dieta de *Cacajao calvus ucayalii* en el nor-oriente Peruana. Neotropical Primates 7(1):1–5

Araujo-Murakami, Alejandro; Paniagua, Narel; Cayola, L.; Jorgensen, Peter M.; Valdes, Ovidio; Macuapa, Eric; Calzadilla-Tomianovich, Miki. 2009. Diversidad y estructura florística del bosque de llanura y palmar de pantano Amazónico preandino en el sector de Chalalan, Parque Nacional Madidi, Bolivia. Kempffiana 5(1):3–27

Arévalo-Lopéz, Diandra; Nina, Nélida; Ticona, Juan C.; Limachi, Ivan; Salamanca, Efrain; Udaeta, Enrique; Paredes, Crispin; Espinoza, Boris; Serato, Alcides; Garnica, David; Limachi, Abigail; Coaquira, Dayana; Salazar, Sarah; Flores, Ninoska; Sterner, Olov; Giménez, Alberto. 2018. Leishmanicidal and cytotoxic activity from plants used in Tacana traditional medicine (Bolivia). Journal of Ethnopharmacology 216:120–133

Brack Egg, Antonio. 2003. Frutas del Perú. Universidad San Martin de Porres, Lima

Cárdenas, Dairon; Castaño, Nicolás; Sua, Sonia. 2009. Flora de la estrella fluvial de Inírida (Guianía, Colombia). Biota Colombiana 10(1–2):1–30

Castro, José Miguel; Emmons, Louise H. 2012. Variation in diet and resources. In: The Maned Wolves of Noel Kempff Mercado National Park. Emmons, L. H. (ed.), Smithsonian Contributions to Zoology Number 639, Smithsonian Institution Scholarly Press, pp 37–55

Casula, Kátia Regina. 2012. Análise da similaridade florística e estrutural das formações florestais inundáveis em um trecho do alto Rio Madeira e de seus afluentes, estado de Rondônia, Brasil. Master's Thesis, Núcleo de Ciências Exatas da Terra, Universidade Federal de Rondônia, Porto Velho

Cavalcante, P. B. 1973. A Farmacopéia Tiriyó: Estudo Étno-Botânico. Museu Paraense Emílio Goeldi, Publicações Avulsas No. 24

Cerón, Carlos Eduardo. 1996. Etnobotanica Quichua en la vía Hollín-Loreto, Provincia del Napo. Hombre y Ambiente 25:131–170

Cerón, Carlos Eduardo; Ayala, Consuelo Montalvo. 1998. Etnobotanica de los Huaorani de Quehueiri-ono Napo, Ecuador. Abya-Yala, Quito

Charpentier, Elvis J.; García, Gabriel; Aquino, Rolando. 2015. Uso y competición por plantas alimenticias entre *Pithecia aequatorialis* (Primates: Pitheciidae) y otros animales en la Amazonía peruana. Revista Peruana de Biología 22(2):225–232

Culot, Laurence; Huynen, Marie-Claude; Gérard, Paul; Heymann, Eckhard W. 2009. Short-term post-dispersal fate of seeds defecated by two small primate species (Saguinus mystax and Saguinus fuscicollis) in the Amazonian forest of Peru. Journal of Tropical Ecology 25(3):229–238

Desmarchelier, Cristian; Mongelli, Elena; Coussio, Jorge; Giulietti, Ana; Ciccia, Graciela. 1995. Etnobotánica y bioactividad de plantas medicinales utilizadas por un grupo indígena Takana de la Amazonia Peruana. Acta Farmacéutica Bonaerense 14(3):195–208

Desmarchelier, C.; Gurni, A.; Ciccia, G.; Giulietti, A. M. 1996. Ritual and medicinal plants of the Ese'ejas of the Amazonian rainforest (Madre de Dios, Perú). Journal of Ethnopharmacology 52:45–51

Desmarchelier, C.; Repetto, M.; Coussio, J.; Llesuy, S.; Ciccia, G. 1997. Total reactive antioxidant potental (TRAP) and total antioxidant reactivity (TAR) of medicinal plants used in southwest Amazonia (Bolivia and Peru). International Journal of Pharmacognosy 35(4):288–296

Ferreira, Almecina Balbino; Ming, Lin Chau; Haverroth, Moacir; Daly, Douglas Charles; Caballero, Javier; Ballesté, Andrea M. 2015. Plants used to treat malaria in the regions of Rio Branco-Acre state and southern Amazonas state, Brazil. International Journal of Phytocosmetics and Natural Ingredients 2(9):1–5

Freitas, Jussara Costa de; Fernandes, Marcus Emanuel Barroncas. 2006. Uso de plantas medicinais pela comunidade de Enfarrusca, Bragança, Pará. Boletim do Museu Paraense Emílio Goeldi, Ciencias Humanas 1(3):11–26

Gomez, Tedi Pacheco. 2011. Valoración económica de purmas de diferentes edades en la comunidad nativa de Santa Mercedes, río Putumayo, Perú. Master's thesis, UNAP, Iquitos

Guèze, Maximilien; Luz, Ana Catarina; Paneque-Gálvez, Jaime; Macía, Manuel J.; Orta-Martínez, Martí; Pino, Joan; Reyes-García, Victoria. 2014. Are ecologically important tree species the most useful? a case study from indigenous people in the Bolivian Amazon. Economic Botany 68(1):1–15

Guillotin, M.; Dubost, G.; Sabatier, D. 1994. Food choice and food competition among the three major primate species of French Guiana. Journal of Zoology 233:551–579

Ima, M. O. 2012. *Saberes Waorani y Parque Nacional Yasuní: Plantas, Salud y Bienestar en la Amazonía del Ecuador*. Iniciativa Yasuní ITT, Ministerio Coordinador de Patrimonio, Ministerio del Ambiente, Programa de las Naciones Unidas para el Desarrollo (PNUD) y Fondo para el Medio Ambiente Mundial (FMAM), Quito

Jovel, E. M.; Cabanillas, J.; Towers, G. H. N. 1996. An ethnobotanical study of the traditional medicine of the mestizo people of Suni Miraño, Loreto, Peru. Journal of Ethnopharmacology 53: 149–156

Kunz, Sustanis Horn; Martins, Sebastião Venâncio; Ivanauskas, Natália Macedo; Silva, Elias; Stefanello, Daniel. 2010. Estrutura fitossociológica de um trecho de floresta estacional perenifólia, bacia do Rio das Pacas, Querência, MT. Cerne 16(2):115–122

Kunz, Sustanis Horn; Moreau, Julia; Spadeto, Cristiani; Martins, Sebastião Venâncio; Stefanello, Daniel; Ivanauskas, Natália Macedo. 2014a. Estrutura da comunidade arbórea de trecho de floresta estacional sempre-verde e similaridade florística na região nordeste de Mato Grosso, Brasil. Floresta e Ambiente 21(4):429–440

Kunz, Sustanis Horn; Ivanauskas, Natália Macedo; Martins, Sebastião Venâncio; Stefanello, Daniel; Silva, Elias. 2014b. Fitossociologia do componente arbóreo de dois trechos de floresta estacional perenifólia, bacia do Rio das Pacas, Querência, MT. Ciência Florestal, Santa Maria 24(1):1–11

La Rotta, C. 1983. Observaciones Etnobotanicas sobre algunas Especies Utilizadas por la Comunidad Indigena Andoque (Amazonas, Colombia). Universidad Nacional de Colombia, Bogotá

Lima, Carla Teixeira de; Teixeira, Michella Del Rei. 2018. Flora das cangas da Serra dos Carajás, Pará, Brasil: Menispermaceae. Rodriguésia 69(1):147–151

Lizcano, Leandro J.; Bakkali, Fadil; Ruiz-Larrea, M. Begoña; Ruiz-Sanz, José Ignacio. 2010. Antioxidant activity and polyphenol content of aqueous extracts from Colombian Amazonian plants with medicinal use. Food Chemistry 119:1566–1570

Lobo, Guilherme de Sousa. 2017. A alteração do regime hidrológico afeta a composição florística e estrutura de florestas de igapó? Um estudo comparativo entre um rio regulado e outro prístino na Reserva de Desenvolvimento Sustentável do Uatumã, Amazônia Central. Master's thesis, INPA, Manaus

Lossio, L.C., Delgado, H.S., Ruiz, A.G., Donayre, R.A., Ruiz, J.G., Panduro, M.P., Sifuentes, T.C. 1995. Plantas Medicinales de la Amazonía Peruana. Instituto Peruano de Seguridad Social/Instituto de Medicina Tradicional, Iquitos

Marimon, Beatriz Schwantes; Lima, Edson de Souza. 2001. Caracterização fitofisionômica e levantamento florístico preliminar no Pantanal dos Rio Morte-Araguaia, Cocalinho, Mato GrossoActa Botanica Brasilica 15(2):213–229

Marimon, B. S.; Felfili, J. M.; Lima, E. S. 2002. Floristics and phytosociology of the gallery forest of the Bacaba stream, Nova Xavantina, Mato Grosso, Brazil. Edinburgh Journal of Botany 59(2):303–318

Martínez-Rodríguez, María Ruth. 2009. Ethnobotanical knowledge acquisition among Tsimane' children in the Bolivian Amazon. Doctoral dissertation, University of Georgia, Athens

Mejia, Kember; Rengifo, Elsa. 2000. Plantas Medicinales de Uso Popular en la Amazonía Peruana. Agencia Española de Cooperación Internacional, Lima

Mejía, Luis Eduardo; Turbay, Sandra. 2009. Los venenos de cacería en la Amazonia colombiana: sustancias letales o fuente de vitalidad? Boletín de Antropología Universidad de Antioquia 23(40):129–153

Milliken, W., R. P. Miller, S. R. Pollard, and E. V. Wandelli. 1992. *The Ethnobotany of the Waimiri Atroari Indians of Brazil.* Royal Botanic Gardens, Kew

Mondragón, Martha L.; Smith, Randall. 1997. Bete Quiwiguimamo Salvando el bosque para vivir sano: algunas plantas y árboles utilizadas por la Nacionalidad Huaorani de la Amazonía Ecuatoriana. Abya Yala, Quito

Mota, Nara Furtado de Oliveira; Watanabe, Mauricio Takashi Coutinho; Zappi, Daniela Cristina; Hiura, Alice Lima; Pallos, Julieta; Viveros, Raquel Stauffer; Giulietti, Ana Maria; Viana, Pedro Lage. 2018. Cangas da Amazônia: a vegetação única de Carajás evidenciada pela lista de fanerógramas. Rodriguésia 69(3):1435–1488

Muñoz, Marianella Guevara. 2010. Evaluación de purmas con fines de manejo en la comunidad nativa de Santa Mercedes, Rio Putumayo, Perú. Master's thesis, Universidad Nacional de la Amazonia Peruana, Iquitos

Myster, Randall W. 2007. Interactive effects of flooding and forest gap formation on tree composition and abundance in the Peruvian Amazon. Folia Geobotanica 42:1–9

Myster, Randall W. 2015. Flooding x tree fall gap interactive effects on blackwater forest floristics and physical structure in the Peruvian Amazon. Journal of Plant Interactions 10(1):126–131
Odonne, Guillaume; Valadeau, Celine; Alban-Castillo, Joaquina; Stien, Didier; Sauvain, Michel; Bourdy, Genevieve. 2013. Medical ethnobotany of the Chayahuita of the Paranapura basin (Peruvian Amazon). Journal of Ethnopharmacology 146:127–153
Palacios, Erwin; Rodríguez, Adriana; Castillo, Claudia. 2004. Preliminary observations on the mottled-face tamarin (*Saguinus inustus*) on the lower Rio Caquetá, Colombian Amazon. Neotropical Primates 12 (3): 123–126
Pedrollo, Camilo Tomazini; Kinupp, Valdely Ferreira; Shepard Jr., Glenn; Heinrich, Michael. 2016. Medicinal plants at Rio Jauaperi, Brazilian Amazon: ethnobotanical survey and environmental conservation. Journal of Ethnopharmacology 186:111–124
Polesna, Lucie; Polesny, Zbynek; Clavo, Mirella Z.; Hansson, Anders; Kokoska, Ladislav. 2011. Ethnopharmacological inventory of plants used in Coronel Portillo Province of Ucayali Department, Peru. Pharmaceutical Biology 49(2):125–136
Pumacallahui-Salcedo, Eliseo; Ríos-Falcón, Víctor; Mollinedo-Chura, Richar-Marlón. 2016. Construcción de un secador solar para el secado de plantas medicinales de comercialización en la Provincia de Tambopata. El Ceprosimad 4(2):53–67
Ramírez, O.; Blair, S.; Medio M. T. V.; Cardona, F.; Pabón. 2017. Ethnobotany of medicinal plants used to treat malaria by traditional healers from ten indigenous Colombian communities located in Vaupes Medio. Biodiversity International Journal 1(2):1–17
Ratter, J. A. 1987. Notes on the vegetation of the Parque Nacional do Araguaia (Brazil). Notes from the Royal Botanic Garden, Edinburgh 44(2):311–342
Ratter, J. A.; G. P. Askew; R. F. Montgomery; and D. R. Gifford. 1978. Observations on the vegetation of northeastern Mato Grosso II. Forests and soils of the Rio Suiá-Missu area. *Proceedings of the Royal Society of London, Series B, Biological Sciences* 203 (1151):191–208
Rego, Lyvia Julienne Sousa; Elly, Márcio Piero; Pereira, Jéssica Lira; Garcia, Bruna Naiara; Barbosa, Humberto Figueira; Vieira, Thiago Almeida. 2011. Produtos florestais não-madeireiros comercializados em feiras de Santarém, Pará. Cadernos de Agroecologia 6(2):1–6
Rodrigues, Eliana. 2006. Plants and animals utilized as medicines in the Jaú National Park (JNP), Brazilian Amazon. Phytotherapy Research 20:378–391
Rondon Neto, Rubens Marques; Santos, Jeneffer Soares dos; Silva, Manoel Advíncula da; Koppe, Valdinei Cristi. 2010. Potencialidades de uso de espécies arbustivas e arbóreas em diferentes fisionomias de cerrado, em Lucas do Rio Verde/MT. Revista de Biologia e Ciências da Terra 10(2):113–126
Sánchez, Judith Jacira Achong; Levy, Allison Licett Nuñez. 2013. Características de frutos consumidos por *Saguinus nigrifrons* (Primates, Callitrichidae), en la Estación Biológica Quebrado Blanco, Loreto, Perú. Master's thesis, Universidad Nacional de la Amazonía Peruana, Iquitos
Santos, Otilene dos Anjos; Couceiro, Sheyla Regina Marques; Rezende, Alinne Costa Cavalcante; Silva, Márcia Daniela de Sousa. 2016. Composition and richness of woody species in riparian forests in urban areas of Manaus, Amazonas, Brazil. Landscape and Urban Planning 150:70–78
Sanz-Biset, Jaume; Campos-de-la-Cruz, José; Epiquién, Mirbel A.; Cañigueral, Salvador. 2009. A first survey on the medicinal plants of the Chazuta valley (Peruvian Amazon). Journal of Ethnopharmacology 122:333–362
Schultes, R. E. and R. F. Raffauf. 1990. The Healing Forest: Medicinal and Toxic Plants of the Northwest Amazonia. Dioscorides Press, Portland, Oregon
Scudeller, Veridiana Vizoni; Veiga, Josephina Barata da; Araújo-Jorge, Lúcia Helena de. 2009. Etnoconhecimento de plantas de uso medicinal nas comunidades São João do Tupé e Central (Reserva de Desenvolvimento Sustentável do Tupé). In: Biotupé: Meio Físico, Diversidade Biológica e Sociocultural do Baixo Rio Negro, Amazônia Central, Santos-Silva, Edinaldo Nelson; Scudeller, Veridiana Vizoni (eds), UEA Edições, Manaus, vol. 2, pp 185–199
Sousa, Julio dos Santos de; Bastos, Maria de Nazaré do Carmo; Gurgel, Ely Simone Cajueiro. 2014. Estudo taxonômico do gênero *Abuta* (Menispermaceae) no Estado do Pará, Brasil. Acta Amazonica 44(2):175–184

Sousa, Julio dos Santosde; Gurgel, Ely Simone Cajueiro; Bastos, Maria de Nazaré do Carmo. 2018. Estudo taxonômico de *Abuta* (Menispermaceae) no estado do Acre, Brasil. Rodriguésia 69(2):477–488

Stefanello, Daniel. 2008. Florística, estrutura e dispersão de sementes de três trechos de floresta estacional perenifólia ribeirinha na bacia hidrográfica do Rio das Pacas, Querência, MT. Master's Thesis, Universidade Federal de Viçosa, Minas Gerais

Stefanello, Daniel; Fernandes-Bulhão, Clarissa; Martins, Sebastião Venâncio. 2009. Síndromes de dispersão de sementes em três trechos de vegetação ciliar (nascente, meio e foz) ao longo do Rio Pindaíba, MT. Revista Árvore 33(6):1051–1061

Stefanello, Daniel; Ivanauskas, Natália Macedo; Martins, Sebastião Venâncio; Silva, Elias; Kunz, Sustanis Horn. 2010. Síndromes de dispersão de diásporos das espécies de trechos de vegetação ciliar do rio das Pacas, Querência-MT. Acta Amazonica 40(1):141–150

Stevenson, Pablo R. 2000. Seed dispersal by woolly monkeys (*Lagothrix lagothricha*) at Tinigua National Park, Colombia: dispersal distance, germination rates, and dispersal quantity. American Journal of Primatology 50: 275–289

Tananta, Leslie Virginia. 2014. Análisis del conocimiento tradicional del uso de especies vegetales en tres comunidades de la cuenca baja del río Ucayali, Loreto, Peru. Master's thesis, UNAP, Iquitos

Taurino, Tássia Cristina da Conceição Barros. 2017. Recuperação natural da diversidade de espécies de plantas em florestas secundárias na Amazônia Oriental. Master's Thesis, Universidade Federal do Pará, Belém

Terborgh, J. 1983. Five New World Primates: A Study in Comparative Ecology. Princeton University Press, Princeton

Tomchinsky, Bernardo; Ming, Lin Chau; Kinupp, Valdely Ferreira; Hidalgo, Ari de Freitas; Chaves, Francisco Célio Maia. 2017. Ethnobotanical study of antimalarial plants in the middle region of the Negro River, Amazonas, Brazil. Acta Amazonica 47(3):203–212

Vale, Gabriel Damasco do. 2011. O efeito do solo, inundação e topografia sobre as campinaranas inundáveis na Amazônia. Master's Thesis, INPA, Manaus

Van Roosmalen, M. G. M. 1985. Habitat preferences, diet, feeding strategy and social organization of the black spider monkey (*Ateles paniscus paniscus* Linnaeus 1758) in Surinam. Acta Amazonica 15(3/4 suplemento) 1–238

Van Roosmalen, Marc G. M.; Klein, Lewis L. 1988. The spider monkeys, genus *Ateles*. In: Ecology and Behavior of Neotropical Primates, Mittermeier, Russell A.; Rylands, Anthony B.; Coimbra-Filho, Adelmar F.; Fonseca, Gustavo A.B. da (eds), Ecology and Behavior of Neotropical Primates, World Wildlife Fund, Washington, DC, vol. 2, pp 455–537

Chapter 35
Metteniusaceae

35.1 *Poraqueiba paraensis*

Brazil Umari, umari amarela, umarizeiro, mari, mari gordo; doomali aamo (Baniwa), uamã, wamã, wamu ñisé (Tukano)

Colombia Umarí, guacari; nemote (Witoto)

Ecuador Huacu; hua curi (Siona-Secoya)

Peru Humarí, umarí; nimuhe (Bora)

Venezuela Yuri

Status Wild, planted

N. Smith, *Amazon Fruits: An Ethnobotanical Journey*, Ethnobiology,
https://doi.org/10.1007/978-3-031-12803-5_35

Umari *Poraqueiba paraensis* fruit from a tree planted in a home garden in a Tukano village. Cunuri Ponta, Uaupés River, Amazonas, Brazil, 11-12-12

Umari *Poraqueiba paraensis* is a medium-sized tree, usually around 8 m tall when mature, that grows in upland forests and along high banks of rivers. It fruits during the rainy season, as with most other regional fruits (Altman et al. 1964). *Poraqueiba paraensis* mostly occurs in the state of Pará and is usually replaced further up the Amazon by a similar species, *Poraqueiba sericea* (Cavalcante 2010:241; Cavalcante and Carvalho 1971; Falcão and Lleras 1980). However, a few botanical specimens of *Poraqueiba paraensis* have been collected in the Colombian, Ecuadorian, Peruvian, and Venezuela Amazon, where it may have been introduced by indigenous migrations or trade. I have seen *Poraqueiba paraensis* cultivated in a home garden in Cunuri Ponta, a Tuyuka village along the lower Uaupés in northwestern Amazonas state in Brazil. Although umari is mostly valued for its fruit, in the early nineteenth century, the wood was used to build boats (Baena 2004:59).

Arlinho, 10 years, eating mari *Poraqueiba paraensis* fruit gathered in nearby forest. Carrazedo, mouth of the Xingu River, Município de Gurupá, Pará, Brazil 4-17-15

The fruits of *Poraqueiba paraensis*, measuring 6–8 cm long by around 4 cm in diameter, usually have an orange skin but some are yellow-orange. The fruits are much sought after by rural folk for the creamy orange pulp surrounding the single seed. The savory pulp has an oily texture reminiscent of tucumã *Astrocaryum vulgare* palm fruits. The oil content of the pulp is around 50% (Altman et al. 1964). After gnawing off the thin layer of pulp the large, light gray seed, which occupies about 85% of the fruit by volume, is typically tossed away.

Small boat (*catraia*) tied up at a village with mari *Poraqueiba paraensis* fruits for sale. Carrazedo, mouth of the Xingu River, Município de Gurupá, Pará, Brazil 4-18-15

Occasionally, however, some of the seeds are planted in raised vegetable beds (*canteiro suspenso*) so that when they sprout they can be transplanted in the home garden or in the vicinity of manioc processing huts, as is the case near Carrazedo at the mouth of the Xingu. This tree is also planted in some upland home gardens in the Rio Guamá watershed some 13 km from Belém (Scoles 2009) and in the vicinity of Manacapuru along the lower Solimões River in Amazonas (Klüppel 2006:24). The pulp is also scraped off and mixed with manioc flour. A Tukano lady in Manaus explained to me that in addition to eating the pulp, her tribe also soak the seed for about 2 weeks in water, then mash it and squeeze the ground-up kernel in a sleeve-like manioc press (*tipití*). Then the mass is mixed with manioc starch (*goma*) to make pancakes (*beiju*).

José das Graças "Poliodoro," 55 years, next to an umari *Poraqueiba paraensis* tree, he planted 5 years ago in his home garden. Laranjal, Igarapé Laranjal, Baia de Caxiuanã, Municipio de Melgaço, Pará, Brazil 5-8-15

Umari *Poraqueiba paraensis* seedling in a raised vegetable and spice bed (*canteiro suspenso*). Sitio Boa União, lower Rio dos Macacos, near Breves, Marajó Island, Pará, Brazil, 5-17-00

The fruits are taken to villages and towns for sale. *Poraqueiba paraensis* is particularly common in eastern Pará, and several urban localities are named after it. A district in Belém, Umarizal, is named after it but the original stands of the fruit tree there have long since given way to buildings and streets. Umarizal is the name of a small village along the Camará River in the southeastern part of Marajó Island. Marituba, a town on the outskirts of Belém, the capital of Pará, means the place of many mari, another common name for *Poraqueiba paraensis* in Brazil (Venturieri 2013). Both Belém and Marituba have grown spectacularly over the last 40 years, with the urban sprawl essentially blending the cities together, and so few mari trees are left standing in that area.

Vendor with umari *Poraqueiba paraensis* fruits in a street market. Vigia, Pará, Brazil 5-10-99

Game animals are also fond of umari fruits and hunters take advantage of this. They set up rudimentary platforms (*jirau*) near umari trees and wait at night with shotgun and flashlight for wild mammals to approach the fallen fruit. Villagers living in Carrazedo at the mouth of the Xingu River shoot a variety of game animals in this manner, including paca *Agouti paca*, armadillos, tapir *Tapirus terrestris*, red brocket deer known locally as veado vermelho *Mazama americana*, and the gray brock deer, called veado branco *Mazama gouazoubira*. Farmers at Carrazedo thus take care to safeguard wild fruit trees whenever possible when choosing areas to clear for planting crops.

35.2 *Poraqueiba sericea*

Brazil Umari, umari roxo, umari preto, mari, mari amarelo, mari preto, mari gordo, mari do Amazonas, marizeiro, muiraximbé; yumári, dumarí, doomali, doomali eewaperi (Baniwa), më (Desana) pèj purúb (Hupd'äh), techi (Ticuna), wamë, wamu (Tukano)

Colombia Umarí, umarí colorado, umarí morado, umarí amarillo, umarí negro, umarí verde, mari, guacuré, guacuré de monte, guacuri plancheto, guacuri plancheto liso, guacuri rojo, guacuri verde, yurí; ko-ata' (Andoke), wa-mö (Barasana), mö'-hoy-yê' (Cubeo), mëë (Desana), wamü (Makuna). niimui-e, guacure (Miraña), matifaño'o, tudua, niimi, ji'o ifa'o (Muinane), vamë (Piratapuya), how, ja-o (Puinave), wama, wamü (Tatuyo), teechi (Ticuna), vamë (Tukano), vamë (Wanano), nemona, nemozi, mo na, illulo, necana (Witoto)

Peru Umarí, umarí amarillo, umarí rojo, umarí negro, humarí, mari negro, guacure; nímuhe (Bora), nê-ká-soó, nê-ká (Witoto)

Venezuela Yuri, yuri amarillo, yuri negro; dúmari (Curripaco), umari (Yeral)

Status Wild, planted

Umari *Poraqueiba sericea* **fruits in a street market. Mercado Belén, Iquitos, Loreto, Peru 3-5-04**

Umari *Poraqueiba sericea* is cultivated in groves, agroforestry fields, and home gardens on uplands in central and western Amazonia up to 800 m in the Andean foothills (Barbosa 2018:42; Coomes and Ban 2004; Coomes and Burt 1997; De Jong 1996; Duarte 1979:91; Ducke 1925; Fontes 2011; Lamont et al. 1999; Lourenço et al. 2009; Machado 2016:33; Mendoza and Guerrero 2010; Padoch et al. 1985; Pulido and Cavelier 2001:37; Silva 2009:50, 2011; Souza 2010:28; Souza and Scudeller 2009). It is found mostly in cultural settings but is also encountered occasionally in the Amazon basin in forests on uplands and as well as on high banks along rivers, successionary vegetation, and *campinarana* on sandy soils. Umari is encountered wild in some upland forests, such as in the vicinity of the confluence of the Tigre and Marañón Rivers in Loreto, Peru, in the interfluvial area between the Purus and Madeira Rivers in Amazonas, Brazil, the municipality of Barcelos along the middle Rio Negro and north of Manaus (Carneiro 2004; Silva et al. 2007; Souza 2012:52). Two species of bats, *Sturnira lilium* and *Artibeus jamaicensis*, eat this fruit in the vicinity of Manaus and likely disperse some of the seeds (Guillaumet and Reis 1983).

But over a century ago, a botanist who worked at the Goeldi museum in Belém pointed out that it is hard to tell whether umari is truly wild when encountered in forests (Huber 1904). This fruit tree has also been encountered in 5- and 20-year-old successionary forests on uplands in the vicinity of Santa Mercedes along the Rio Putumayo in northern Peruvian Amazon bordering Colombia (Gomez 2011:85; Muñoz 2010:35). Indeed, umari is considered by some to be an indicator of Anthrosols, known as *terra preta de índio* in Brazil (Mora 2001). The distinction between truly wild populations and those that are a vestige of human interventions on the landscape is thus blurred.

Jessica, 12 years, eating a black-skinned umari *Poraqueiba sericea* fruit harvested from a cultivated tree near Tamshiyacu along the Amazon River. Arequipa, Rio Yanayacu, Reserva Nacional Pacaya-Samiria, Loreto, Peru 3-8-04

Flowers are produced in the dry season and they are pollinated by at least ten species of bees (Falcão and Lleras 1980). This tree, which can reach 30 m high, fruits in the rainy season. In the vicinity of Manaus, at least ten species of bees visit the flowers and are presumably involved in pollinating them. The fruits are similar in size and shape to its close relative *Poraqueiba paraensis* except the skin color varies from yellow to red, green, and black, hence its various local names in the Colombian Amazon (Pennington et al. 2004:468; Reynel et al. 2003:164). Sometimes, the fruits of *Poraqueiba sericea* are a little more cylindrical in shape and longer, reaching some 10 cm. A farmer along the Iquitos-Nauta highway cultivates a large 10 cm long variety of the fruit which has a rust-colored skin. This variety is known locally as Huitoto, possibly selected by the Witoto indigenous peoples who inhabit the forest along the Amazon straddling both Peru and Colombia.

Liz Karen, 9 years, with umari *Poraqueiba sericea* **fruit gathered in her home garden. Mishana, Rio Nanay, Loreto, Peru 1-8-06**

The pulp surrounds a large tan-colored seed which is discarded after gnawing off the mesocarp. The pulp has high levels of vitamin A (Aguiar et al. 1980) so this fruit is a particularly healthy addition to the diet. In the Peruvian Amazon, rural folk mix the pulp of this fruit with liquid (*cahuana*) extracted from manioc tubers which is then drunk (Brack 2003:117). The Hupd'äh, also known as Nukak, are predominately hunters and gatherers in the forest in northwestern Amazonas, Brazil, and contiguous Colombia and gather fruits from umari groves (Ramirez 2006). It is possible that these concentrations of umari in the forest may be vestiges from discarded seeds around Hupd'äh camps. Indeed, one of the rapids along the Upper Uaupés River near the border between Brazil and Colombia is named after this fruit (Wallace 1889:238). The Tukano in the Colombian Amazon make a juice from the pulp which they ferment (Jacopin 1988; Reichel-Dolmatoff 1996:106). Because of its cultural importance, umari has become entwined in folklore, such as among the Ticuna (Nimuendajú 1952:127). The Tukano believe that a mythical tapir is the master of umari and in a Desana myth tapirs eat the fruits (Wright 1998:63).

This fruit is popular in both rural and urban areas. Umari fruits are common in street markets in the Peruvian Amazon, such as Iquitos, Pucallpa, and Huánuco in the Andean foothills. The fruits are relatively cheap; at Nauta, along the lower Marañón River, for example, umari fruits were selling for the equivalent of US$ 1 for 126 fruits in March 2004. Umari is even featured in desserts at some high-end restaurants in Lima. *Guargueros rellenos con manjar y mousse de umari*, for example, is a funnel-shaped pastry filled with cooked condensed milk and umari mousse (Gutsche 2008). The pulp is used in western Amazonia to make a fermented beverage (Schultes and Raffauf 1990:217).

Poraqueiba sericea has other uses, especially among indigenous groups. The Witoto in the Peruvian Amazon use the fruits as an ingredient in making a contraceptive and the leaves are used widely in northwestern Amazonia to make a tea for treating dysentery. In some rural communities in the municipality of Barcelos along the middle Rio Negro, Amazonas, the seeds of this fruit are cooked to treat malaria (Tomchinsky et al. 2017). In the Brazilian Amazon, an oil is extracted from the fruit to fry fish (Schultes and Raffauf 1990:217). In the Colombian Amazon, wood from this tree is used for a variety of construction purposes (Duarte 1979:91).

Umari is frequently encountered in home gardens, such as at Santa Rosa along the lower Ucayali River in Peru (Padoch and de Jong 1991) and in the municipality of Rio Preto da Eva along the Manaus-Itacoatiara highway in Amazonas, Brazil (Silva 2017:31). When planted as a monocrop, as on the outskirts of Tamshiyacu along the Amazon River a little upstream from Iquitos, the fruits are mostly destined for urban markets. Plantations of this fruit tree are known as umaral in the Peruvian Amazon and the trees start producing fruit after only 4 years. They continue to be productive for some 70 years or more. Umari is also interplanted with other perennial crops in fields, such as among the Witoto along the Caquetá River in the Colombian Amazon (Sastre and Reichel 1978; Veléz 1992; Veléz and Veléz 1992) and near Miraflores at the mouth of the Tigre River in the Peruvian Amazon. Villagers residing in Tamshiyacu along the Amazon River upstream from Iquitos intercrop *Poraqueiba sericea* along with other fruit and nut trees that keep some agroforestry plots productive for several decades (Padoch et al. 1985). The Bora in the Peruvian Amazon cultivate *Poraqueiba sericea* in their home gardens as well as in some of their fields such that it sometimes dominates the fallow vegetation (Denevan and Treacy 1987), a similar pattern noted in the vicinity of Iquitos with farmers of mixed ancestry (Hiraoka 1986). Umarí can survive for 20–30 years in fallow orchards and is appreciated by the Bora because the fallen fruits are eaten by game animals such as paca *Agouti paca* (Denevan and Treacy 1987). Likewise, the Baniwa along the Içana River, an affluent of the Upper Rio Negro, note that certain game animals feed on the fallen fruit, including tapir *Tapirus terrestris*, armadillos (various genera and species), agouti *Dasyprocta* spp., and paca (Fontes 2011).

The Tuyuka along the Tiquié, an affluent of the Uaupés in the northwestern part of the Brazilian Amazon, also intercrop umari in some of their agricultural plots (Schmidt et al. 2010). The Ticuna along the Upper Solimões also cultivate umari in their fields (Acosta 2001) as do various indigenous groups along the middle Caquetá watershed in the Colombian Amazon (Veléz 1991). This tree can produce fruits for several decades.

References

Acosta, L. E. 2001. Los sistemas de producción de la etnia Ticuna del resguardo de Puerto Nariño, sur del Trapecio Amazónico: una aproximación socioeconómica. *Cadernos de Desarrollo Rural* 46:101–132

Aguiar, Jaime Paiva Lopes; Marinho, Helyde Albuquerque; Rebêlo, Yolanda Silva; Shrimpton, Roger. 1980. Aspectos nutritivos de alguns frutos da Amazônia. Acta Amazonica 10(4):755–758

Altman, R. F. A., P. C. Oliveira, G. Enilson, M. L. Silva, and M. C. Neves. 1964. Estudos sobre o fruto de "umari" (*Poraqueiba paraensis*, Ducke e *P. sericea*, Tul.). Instituto Nacional de Pesquisas da Amazônia, Manaus, Publicação 8

Baena, Antônio Ladislau Monteiro 2004. Ensaio Corográfico sobre a Província do Pará. Edições do Senado Federal, Brasilia

Barbosa, Cristiano de Souza. 2018. Diversidade e uso de plantas úteis nas quintais do bairro de São Raimundo, zona oeste de Manaus, AM. Master's thesis, INPA, Manaus

Brack, A. 2003. *Frutas del Perú*. Universidad San Martin de Porres, Lima

Carneiro, Vilany Matilla Colares. 2004. Composição florística e análise estrutural da floresta primária de terra firme na bacia do Rio Cuieiras, Manaus-AM. Master's Thesis, Universidade Federal do Amazonas (UFAM)/Instituto Nacional de Pesquisas da Amazônia (INPA), Manaus

Cavalcante, P. B. 2010. *Frutas Comestíveis na Amazônia*. Museu Paraense Emílio Goeldi, Belém

Cavalcante, P. B. and M. J. C. Carvalho. 1971. O gênero *Poraqueiba* (Icacinaceae) na Amazônia. *Boletim do Museu Paraense Emílio Goeldi*, Nova Série Botânica 39:1–10

Coomes, Oliver T.; Ban, Natalie. 2004. Cultivated plant species diversity in home gardens of Amazonian peasant village in northeastern Peru. Economic Botany 58(3):420–434

Coomes, O. T.; Burt, G. J. 1997. Indigenous market-oriented agroforestry: dissecting local diversity in western Amazonia. Agroforestry Systems 37:27–44

De Jong, W. 1996. Swidden-fallow agroforestry in Amazonia: diversity at close distance. Agroforestry Systems 34: 277–290

Denevan, W. M. and J. M. Treacy. 1987. Young managed fallows at Brillo Nuevo. In: *Swidden-Fallow Agroforestry In the Peruvian Amazon*, W. M. Denevan and C. Padoch (eds.), Advances in Economic Botany vol. 5, New York Botanical Garden, pp. 8–46

Duarte, Luis Enrique Acero. 1979. Principales Plantas Utiles de la Amazonia Colombiana. Instituto Geografico "Augustin Codazzi", Bogotá

Ducke, A. 1925. Plantes nouvelles ou peu connues de la région amazonienne (III partie). Archivos do Jardim Botanico do Rio de Janeiro 4:1–341.

Falcão, M. and E. Lleras. 1980. Aspectos fenológicos, ecológicos e da produtividade do umari (*Poraqueiba sericea* Tulasne). Acta Amazonica 10(3):445–462

Fontes, Orlando Andrade. 2011. Umari. In: O Que a Gente Precisa para Viver e estar Bem no Mundo, Miguel, Abílio Júlio Brazão; Pacheco, Tiago (eds), Instituto Sociambental/Associação do Conselho da Escola Pamáali, São Gabriel da Cachoeira, Amazonas, Brazil

Gomez, Tedi Pacheco. 2011. Valoración económica de purmas de diferentes edades en la comunidad nativa de Santa Mercedes, río Putumayo, Perú. Master's thesis, UNAP, Iquitos

Guillaumet, Jean-Louis; Reis, Nélio Roberto dos. 1983. Les chauves-souris frugivores de la région de Manaus et leur rôle dans la dissémination des espèces végétales. Revue d'Écologie: La Terre et la Vie 38:147–169

Gutsche, A. 2008. *Frutas Amazonicas: Postres Peruanas de Vanguardia*. Wust Ediciones, Lima

Hiraoka, M. 1986. Zonation of mestizo riverine farming systems in northeast Peru. *National Geographic Research* 2(3):354–371

Huber, J. 1904. Notas sobre a patria e distribuição geographica das arvores fructiferas do Pará. *Boletim do Museu Goeldi (Museu Paraense) de Historia Natural e Ethnographia* 4 (2–3):375–406

Jacopin, Pierre-Yves. 1988. Anthropological dialectics: Yukuna ritual as defensive strategy. Schweizerische Amerikanisten-Gessellschaft Bull. 52:35–46

Klüppel, Marina Pinheiro. 2006. Sistemas agrícolas e plantas medicinais em terra preta de índio da Amazônia Central. Master's thesis, INPA/UFAM, Manaus

Lamont, Susan R.; Eshbaugh, W. Hardy; Greenberg, Adolph M. 1999. Species composition, diversity, and use of homegardens among three Amazonian villages. Economic Botany 53(3):312–326

Lourenço, José Nestor de; Sousa, Silas Garcia de; Wandelli, Elisa Vieira; Lourenço, Francisneide de Sousa; Guimarães, Rosângela dos Reis; Campos, Lian da Silva; Silva, Roberto Lima da; Martins, Vivian Fernanda Carneiro. 2009. Agrobiodiversidade nos quintais agroflorestais em três assentamentos na Amazônia Central. Revista Brasileira de Agroecologia 4(2):965–969

Machado, Danilo de Oliveira. 2016. A agrobiodiversidade de quintais agroflorestais em propriedades agrícolas familiares na BR 174, Ramal do Pau-Rosa, Manaus, AM. Master's Thesis, INPA, Manaus

Mendoza, Abel L. Monteagudo; Guerrero, Marleni Huamán. 2010. Catálogo de los arboles y afines de la selva central del Perú. Arnaldoa 17(2):202–242

Mora, S. 2001. Suelos negros y sociedad: un sistema agrícola de entonces, un sistema agrícola de ahora? In: *Desarrollo Sostenible en La Amazonía: ¿Mito o Realidad?,* M. Hiraoka and S. Mora (eds.), Ediciones Abya Yala, Quito, pp. 31–45

Muñoz, Marianella Guevara. 2010. Evaluación de purmas con fines de manejo en la comunidad nativa de Santa Mercedes, Rio Putumayo, Perú. Master's thesis, Universidad Nacional de la Amazonia Peruana, Iquitos

Nimuendajú, C. 1952. *The Tukuna*. University of California Press, Berkeley

Padoch, C. and W. de Jong. 1991. The house gardens of Santa Rosa: diversity and variability in an Amazonian agricultural system. *Economic Botany* 45(2):166–175

Padoch, C. J.; J. Chota Inuma; W. de Jong; and J. Unruh. 1985. Amazonian agroforestry: a market-oriented system in Peru. Agroforestry Systems 3:47–58

Pennington, T. D., C. Reynel, and A. Daza. 2004. *Illustrated Guide to the Trees of Peru.* David Hunt, Sherborne, England

Pulido, María T.; Cavelier, Jaime. 2001. Comercialización de productos vegetales en los mercados de Leticia y Florencia, Amazonía colombiana. In: Evaluación de Recursos Vegetables no Maderables en la Amazonía Noroccidental, Duivenvoorden, Joost F.; Balslev, Henrik; Cavelier, Jaime; Grandez, César; Tuomisto, Hanna; Valencia, Renato (eds), Institute for Biodiversity and Ecosystem Dynamics, Amsterdam, pp. 265–310

Ramirez, H. 2006. *A Língua dos Hupd'äh do Alto Rio Negro*. Saúde Sem Limites, São Paulo

Reichel-Dolmatoff, Gerardo. 1996. The Forest Within: The World-View of the Tukano Amazonian Indians. Themis Books, Darlington

Reynel, C; Pennington, T.D.; Pennington, R.T.; Flores, C.; Daza, A. 2003. *Árboles Útiles de la Amazonía Peruana y sus Usos: Un Manual con Apuntes de Identificación, Ecología y Propagación de las Especies*. Tarea Gráfica Educativa, Lima

Sastre, C.; Reichel, H. 1978. Notas botanicas sobre la region de Araracuara (Rio Caquetá, comisaría del amazonas, Colombia). Bulletin de l'Institut Français d'Études Andines 7(1–2): 105–117

Schmidt, M. V. C.; Vander Veld, Pieter; Ramos, José Barreto; Rezende, João Bosco Azevedo; Rezende, Odilon Barreto; Rezende, Renato Barreto; Resende, Alexandre Azevedo; Cabalzar, Aloisio. 2010. Capoeiras Tuyuka: processos de restauração e recuperação de terras degradadas. In: *Manejo do Mundo: Conhecimentos e Práticas dos Povos Indígenas do Rio Negro*, A. Cabalzar (ed.), Instituto Socioambiental, São Paulo, pp. 156–167

Schultes, R. E. and R. F. Raffauf. 1990. The Healing Forest: Medicinal and Toxic Plants of the Northwest Amazonia. Dioscorides Press, Portland, Oregon

Scoles, Ricardo. 2009. El quintal y las frutas: recursos económicos y alimentares en la comunidad negra de Itacoã, Acará, Pará, Brasil. Acta Amazonica 39(1):1–12

Silva, Rosibel Rodrigues. 2009. O etnoextrativismo não-madeireiro em uma comunidade Amazônica: um estudo de caso em Santa Luzia do Buiuçuzinho-Coari/AM. Master's thesis, UFAM, Manaus

Silva, Rafael Carletti Marcolino da. 2011. Agrobiodiversidade de propriedades agrícolas familiares no município de Presidente Figueiredo, AM. Master's thesis, INPA, Manaus
Silva, Edinei Santos da. 2017. Plantas alimentícias em comunidades agrícolas no município de Rio Preto da Eva-AM. Master's thesis, INPA, Manaus
Silva, Andréa Leme; Tamashiro, Jorge; Begossi, Alpina. 2007. Ethnobotany of riverine populations from the Rio Negro, Amazonia (Brazil). Journal of Ethnobiology 27(1): 46–72
Souza, Célia Cristina Valero. 2010. Etnobotânica de quintais em três comunidades ribeirinhas na Amazônia central, Manaus, AM. Master's thesis, INPA, Manaus
Souza, Priscila Figueira de. 2012. Florística, distribuição e diversidade beta da comunidade arbórea ao longo do gradiente hídrico em uma floresta de terra firme no interflúvio Purus-Madeira, Amazônia central. Master's thesis, INPA, Manaus.
Souza, Célia Cristina Valero; Scudeller, Veridiana Vizoni. 2009. Plantas úteis nos quintais das comunidades ribeirinhas Julião e Agrovila, Reserva de Desenvolvimento Sustentável do Tupé, Amazônia Central. Revista Brasileira de Agroecologia 4(2): 2487–2491
Tomchinsky, Bernardo; Ming, Lin Chau; Kinupp, Valdely Ferreira; Hidalgo, Ari de Freitas; Chaves, Francisco Célio Maia. 2017. Ethnobotanical study of antimalarial plants in the middle region of the Negro River, Amazonas, Brazil. Acta Amazonica 47(3):203–212
Veléz, Germán Alonso. 1991. Los frutales amazónicos cultivados por las comunidades indígenas de la región del Medio Caquetá. Colombia Amazonica 5(2): 135–183
Veléz, Germán Alonso. 1992. Estudio fenológico de diecinueve Frutales silvestres utilizadas por las comunidades indígenas de la región de Araracuara- Amazonia Colombiana. Colombia Amazonica 6(1): 135–186
Veléz, Germán Alonso; Veléz, Antonio José. 1992. Sistema agroflorestal de "chagras" utilizado por las comunidades indígenas del medio Caquetá (Amazonia Colombiana). Colombia Amazonica 6(1): 101–134
Venturieri, G.A. 2013. Indigenous strategies used to domesticate plants in Brazilian Amazon. In: Encyclopedia of Biodiversity, Levin, S. A. (ed), Academic Press, Waltham, pp. 279–292
Wallace, Alfred Russel. 1889. A Narrative of Travels on the Amazon and Rio Negro, with an Account of the Native Tribes, and Observations on the Climate, Geology, and Natural History of the Amazon Valley. Ward, Lock & Co., London
Wright, R. M. 1998. Cosmos, Self, and History in the Baniwa Religion. University of Texas Press, Austin

Chapter 36
Moraceae

36.1 *Bagassa guianensis*

Brazil Tatajuba, tatajuba poca, amarelão, cachacerio, bagaceira, amapárana, garrote, inharé; taráiko-'i (Araweté), tarawire-'iwa (Asurini), taraka-ɨ (Guajá), tareka'ɨ (Ka'apor), ywaputuka'ywa (Piripkura), pakasa (Wajãpi), hapakara hi (Yanomami)

Ecuador Yiak númi, yampiu (Shuar)

Peru Loro micuna

Status Wild, spontaneous in cultural settings, planted

N. Smith, *Amazon Fruits: An Ethnobotanical Journey*, Ethnobiology,
https://doi.org/10.1007/978-3-031-12803-5_36

Tatajuba *Bagassa guianensis* **interplanted in a cacao grove. Gleba 23, lote 2, km 75 Altamira-Itaituba, Transamazon Highway, Município de Medicilandia, Pará, Brazil 9-7-15**

Bagassa guianensis is a large tree reaching 40 m tall that produces grapefruit-sized fruits in the rainy season and early part of the dry period (Magalhães 1982). *Bagassa guianensis* occurs in the upland forests throughout Amazonia and into the Guianas. It appears to be more common, however, in eastern Amazonia (Amaral et al. 2009; Balée 1993b; Lemos et al. 2015; Lorenzi 2009:249; Pinheiro et al. 2007; Santos et al. 2017). It is often found in liana forests which are considered anthropogenic

(Balée and Campbell 1990; Salomão et al. 2007a, b), and occasionally in floodplain forests such as along the lower Xingu (Campbell et al. 1986). It also occurs in successionary vegetation on uplands, such as in the Tocantins watershed, in cattle pastures in the vicinity of Paragominas in eastern Pará, in the Bragantina zone in northeastern Pará, in the Tapajós National Forest in western Pará, and in Mato Grosso (Coelho et al. 2012; Dantas and Rodrigues 1980; Espírito-Santo et al. 2005; Gomide 1997:111; Guarim Neto 1984; Schwartz 2007; Uhl et al. 1988; Taurino 2017:70).

Some of the fruits appear to be eaten by bats, while others fall to the ground and are much procured by a large assortment of game animals, including brocket deer (*Mazama americana* and *M. gouazoubira*), paca *Agouti paca*, tapir *Tapirus terrestris*, peccaries (*Pecari tajacu* and *Tayassu pecari*), the yellow-footed tortoise *Chelonoidis denticulata*, and the red-footed tortoise *Chelonoidis carbonaria*, guans *Penelope* spp., curassows *Crax* spp., and trumpeters *Psophia* spp. (Albert and Milliken 2009:65; Ayres and Ayres 1979; Balée 1993a:82; Barboza 2008:56, 58; Cummings et al. 2018; Fragoso and Huffman 2000; Gayot et al. 2004; Henry et al. 2000; Moskovits 1998; Moskovits and Bjorndal 1990; Prado 2013; Smith 1976). Hunters sometimes set up small platforms near the tree and wait at night with flashlight and shotgun at the ready.

Several species of monkey that are hunted during the day also eat the fruits and disperse some of the seeds. Red howler monkeys *Alouatta seniculus* and *A. macconnelli* feed on the fruits dispersing the seeds up to 360 m from the tree (Arroyo-Rodríguez et al. 2015; Julliot 1996a, b; Julliot and Sabatier 1993; Macias 2013:20; Regan et al. 2001; Simmen 1992). Spider monkeys *Ateles belzebuth* and *A. paniscus* also eat the fruits and are widely hunted in the Amazon Basin (Guillotin et al. 1994; Kinzey and Norconk 1990; Norconk and Kinzey 1994; Nunes 1998; Pontes 1997; Regan et al. 2001; Simmen 1992; Simmen and Sabatier 1996; Van Roosmalen 1985; Van Roosmalen and Klein 1988). Capuchin monkeys *Cebus apella* also eat the fruits (Regan et al. 2001; Simmen and Sabatier 1996) and are also involved in dispersing the seeds. Capuchin monkeys are not important game animals, but they are sometimes kept as pets by rural folk.

Tatajuba *Bagassa guianensis* interplanted in a cacao grove. Gleba 23, lote 2, km 75 Altamira-Itaituba, Transamazon Highway, Município de Medicilandia, Pará, Brazil 9-7-15

Tatajuba also provides valuable timber and is under threat from loggers as well as cattle ranchers clearing forests to plant pasture grass (Fiedler et al. 2012; Veríssimo et al. 1992). During the colonial period and the early part of the twentieth century, the durable wood was used in boat and house construction (Baena 2004:57; Huber 1910; Le Cointe 1922:47) as is still the case today, such as in Mato Grosso (Loureiro and Lisboa 1979) and among the Palikur in French Guiana (Ogeron et al. 2018). This durable wood is also used to make furniture in some areas, such as Parauapebas in northeastern Pará (Soares et al. 2018). The Galibi, who inhabit the watershed of the Oiapoque River which flows along the boundary between Amapá and French Guiana, use the trunk to make large canoes (Arnaud 1989:40). The Kurupukari in central Guyana use the wood to make bows for hunting and to make certain crafts (Johnston and Colquhoun 1996). After soaking in water for several days, the inner bark is used to make bags and some indigenous groups use it to fashion masks used in ritual events (Oliveira et al. 1991). The Shuar in the southern Ecuadorian Amazon use the bark for cordage (Bennett et al. 2002:211). Some residents of the Jericó community, Garrafão do Norte, in northeastern Pará use the wood of this tree to make planks and furniture (Vieira 2005:105).

Also known as amarelão on account of the large, yellow fruits, the tree sometimes arises spontaneously in cacao groves, such as along the Transamazon Highway in the Municipality of Medicilandia. Farmers tend to spare tatajuba as well as other timber and fruit trees when they sprout in cacao plantations. That way, the land can remain productive even when the cacao trees have become senescent. Some farmers, such as at km 75 of the Altamira-Medicilandia stretch of the Transamazon Highway, plant tatajuba alongside teak, an exotic timber tree, or in their cacao groves. Some farmers in the municipality of São Félix do Xingu in southern Pará either plant or maintain this tree in their cacao groves (Braga et al. 2019). Tatajuba is also found in some agroforestry plots maintained by small farmers in the vicinity of Tomé-Açu in eastern Pará (Bolfe and Batistella 2011) as well as in the agricultural colony of Apiaú in Roraima (Ferreira et al. 2011).

The wood and fruits of tatajuba have long been valued by indigenous people in the Amazon. The Waimiri-Atroari north of Manaus spare the tree when clearing forest for their fields or take care of spontaneous seedlings when they arise among their crops (Miller 1994). The Juruna (they call themselves Yudjá) make wooden mortars (*pilão*) from the trunks of *Bagassa guianensis*, as noted at Paquiçamba, a village along the Altamira-Belo Monte stretch of the Xingu River. The wife of the Juruna chief at Paquiçamba explained that the pilão she uses was fashioned by a Juruna group in Mato Grosso. The Ka'apor in Maranhão avoid eating the fruit because it is thought to cause baldness, although they recognize that the fruits are eaten by some game animals, including tapir (Balée 1993a:82). The wood from this tree also enters the timber trade in the Brazilian Amazon (Loureiro et al. 1979:143; Parrotta et al. 1995:252; Reis et al. 2014).

36.2 *Brosimum lactescens*

Bolivia Sapito, virila negra, quecho colorado, quecho negro, quecho amarillo, quecho del bajo, quechu, chicle, leche-leche, gomaliña, mururé, maca nui; xanate, tĕco (Chácobo)

Brazil Mururé vermelho, mururé branco, muiratinga, muirapiranga, muirapiranga branca, mujijum, canela de cutia, canela de velho, caucho, inharé, inharé amarelo, inharé preto, manixí, paman, pama, pama pequena, pama de índio, amapaí, amapá roxo, amapá pequeno, janitá, pau Brasil do Araguaia, pau Rainha, pau de leite amarelo, leiteiro, leiteiro de tucano, leiteira; manitê; tamamari-'ɨ (Guajá), mirapitãrã, wari'y (Ka'apor), manitê (Kaxinawá), kãri nahi, kãrã ãshi hi (Yanomami)

Colombia Chirricle, árbol vaca, charo, caucho

Ecuador Sande; sangangasirande, sangangacho (Cofán), simona yura, toro ardilla caspi (Kichwa), supinim, yapit (Shuar), nea sati, yatio (Siona Secoya), yarasillo; cayumenehue, quemituguemo (Waorani)

Peru Tamamuri, tamamuri dulce, shimbillo, chimicua, mashonás colorado, capinuri de altura, manchinga, congona, mururé, casha huayo, café con leche, leche caspi, panguana, panguana de hoja chica, palisangre blanco, urpay manchinga

Venezuela Yuquito, charo macho, cajamin, guaimero, misionero negro; tɨtɨweka yaɨ (Hotï), keretokoparkorudek (Pemón), sajurna (Ye'kuana)

Status Wild

Tamamuri ***Brosimum lactescens*** **fruits on a forest floor. Quebrada Yarina, Rio Yanayacu, Pacaya-Samiria, Loreto, Peru, 6-22-06**

Brosimum lactescens is a giant tree of both upland and floodplain forests throughout Amazonia that can soar to 50 m, a true canopy emergent. It is found in the Andean foothills, such as at 1100 m in the Madidi National Park, Bolivia (Plata 2008:54). It occurs along rivers of various water types, including black water such as the Rio Yanayacu in the Peruvian Amazon and clear water such as the Jari and Xingu watershed in the eastern Amazon and the Rio das Pacas in the southeastern fringe of the Amazon forest (Carim 2016:92; Mostacedo et al. 2006; Stefanello et al. 2010; Zappi et al. 2016). It is also found in tidal forests in the Amazon estuary in the vicinity of Belém (Amaral et al. 2009). It also occurs in successionary forests, such as in the Trombetas watershed (Parrotta et al. 1997), in the Tapajós National Forest (Espírito-Santo et al. 2005), and in the vicinity of Belém (Souza 2019:11), as well as in the vicinity of Madre de Dios in the Peruvian Amazon (Robles 2017:48). The tree is so lofty that people only gather the fruits from the ground. Several game animals also eat the orange fruits on the forest floor, such as the nocturnal paca *Agouti paca* and the diurnal agouti *Dasyprocta* spp. (Albert and Milliken 2009:65). In the state of Acre in the Brazilian Amazon, hunters in the Upper Juruá wait at night for game to come and feed below towering *Brosimum lactescens* when it is in fruit. This tree is also found in Suriname that borders the northern part of the Amazon Basin (Bánki et al. 2003:23) and in seasonal forests in the southern part of the Amazon Basin in Mato Grosso (Lopes et al. 2014).

Many species of monkey consume the fruits of *Brosimum lactescens* in the forest canopy including black-capped capuchin monkeys *Cebus apella*, white-fronted capuchins *Cebus albifrons*, squirrel monkeys *Saimiri sciureus* and *S. vanzolinii*, emperor tamarins *Saguinus imperator*, saddle-back tamarins *Saguinus fuscicollis*, golden-mantled tamarin *Saguinus tripartitus*, howler monkeys (*Alouatta* spp.), spider monkeys (*Ateles* spp.), white uakaris *Cacajao calvus*, collared titis *Callicebus torquatus*, bearded sakis *Chiropotes satanas*, red-nosed bearded sakis *Chiropotes albinasus*, and woolly monkeys *Lagothrix lagotricha*, *L. cana* (Ayres 1989; Defler and Defler 1996; Heymann 2000; Iwanaga and Ferrari 2001; Kinzey and Norconk 1990; Norconk and Kinzey 1994; Paim et al. 2017; Palacios et al. 1997; Peres 1994; Pinto 2008:77; Stevenson 2000, 2004, 2005, 2007; Suarez 2006; Symington 1988; Terborgh 1983:69; Van Roosmalen 1985; Van Roosmalen and Klein 1988). White-bellied spider monkeys *Ateles belzebuth* are known to disperse the seeds of *Brosimum lactescens*, which are round with a diameter of 10 mm, such as in the Upper Amazon in Colombia (Stevenson and Link 2010; Stevenson et al. 2002) and in the Peruvian Amazon (Cruz 2018:53). In the Peruvian Amazon, dung beetles have been observed removing feces left by some monkeys, thereby further dispersing the seeds (Andresen 1999). In the Manu National Park in the southeastern Peruvian Amazon, the chestnut-eared aracari *Pteroglossus castanotis*, a species of toucan, eats this fruit and disperses some of the seeds (Rosado 2013). Both humans and wild animals are thus involved in dispersing the seeds. Several of those monkey species are hunted so *Brosimum lactescens* contributes indirectly to the diet of rural people via game meat. Several game birds also consume the fruit of this tree in the Peruvian Amazon including paujil *Crax mitu*, pucacunga *Penelope jacquacu*, pava de monte *Pipile pipile*, and manajaraco *Ortalis motmot* (Begazo 1997). In the Colombian Amazon another game bird, paujil *Mitu silvini*, has been observed feeding on this fruit and likely disperses some of the seeds (Franco and Santamaria 1997). In the Madre de Dios region of the Peruvian Amazon, the yellow-footed tortoise *Chelonoidis denticulata* is known the eat this fruit and disperse some of the seeds, and rural people gather this tortoise to eat (Guzmán and Stevenson 2008).

At high water, trees on floodplains can be partially inundated allowing fish important for subsistence and commerce, such as sardinha (*Triportheus*), pacu (various genera), and aracu (*Leporinus*), to also feed on the fruits when they fall into the water. Dispersal agents are attracted by the fruit's soft, mealy pulp that surrounds a single light green seed. The slightly flattened fruit is 1.7 cm long by 2.1 cm wide and contains a single light green to tan seed approximately 1 cm in diameter. However, the fruits sometimes reach 2.5 cm in diameter and can contain as many as three slightly elongated seeds (Parrotta et al. 1995:258).

The fruits are relatively small but they are abundant. Thousands of fruits are produced by a single tree at the height of the rainy season leading into the drier period as water levels drop. The yellow-speckled orange fruits are easily spotted on the forest floor. Tamamuri fruits are mostly eaten in the forest, as in the vicinity of Veinte de Enero along the lower Yanayacu in the Reserva Nacional Pacaya-Samiria, or taken back to the village to be eaten. On rare occasions, the colorful fruits appear in markets, such as in Iquitos (Vásquez and Gentry 1989).

Boys gathering tamamuri *Brosimum lactescens* fruits in forest on a streambank. The tree with buttress roots in the background is dropping the fruits. Quebrada Yarina, Rio Yanayacu, Reserva Nacional Pacaya-Samiria, Loreto, Peru 6-22-06

Brosimum lactescens is a wide-ranging species, occurring from Belize in Central America to Bolivia and southern Brazil. The tree is found below 1600 m in the Guianas as well as the Orinoco and Amazon drainage basins. It is conceivable that some of the places where *Brosimum lactescens* occurs is the result of past human activity. For example, seed remains of *Brosimum lactescens* have been found at the

Peña Roja archaeological site along the middle Caquetá in the Colombian Amazon (Morcote-Ríos et al. 2014). It is possible that some of the trees from which the fruits were gathered arose from people tossing seeds after eating the pulp. A related species, *Brosimum alicastrum*, is often found around Mayan sites in Mexico, Guatemala, and Belize (Gomez-Pompa 1987; Gomez-Pompa et al. 1987; Turner and Miksicek 1984).

The tree's copious white latex has a variety of uses, such as in the Ecuadorian Amazon where the Kichwa employ the sap to speed the healing of cuts and to fortify lactating mothers. In the Upper Juruá in Acre, Brazil, rubber tappers use the creamy sap to coagulate latex tapped from *Hevea brasiliensis* trees. Hence one of the common names for *Brosimum lactescens* in Acre: caucho. In Bolivia, some rural people make chewing gum from the latex of *Brosimum lactescens*, accounting for another name for this tree: chicle. Wood from the tree is also used to make fires for cooking.

36.3 *Maclura tinctoria*

Bolivia Mora, mora grande, mora chica, tatayiva, tataguiba, upia; qüeototi (Chácobo), cotyij'cotyij' (Tsimane)

Brazil Tatajuca, taiúva, taiuveira, taúba, tatajuba, tatajuba de tinta, tatajuba de espinho, jataíba, limãorana, limorana, jurema de espinho, amora branca, amora brava, amora do mato, amora de árvore, amoreira, amoreira branca, moreira, mora, jataíba, amarelinho, pau amarelo, pau de fogo, pau moreira, apé

Colombia Incira, mora, dinde, abrojo, cauchillo

Ecuador Mora fino, mora, moral bobo, moral fino, chiap, sota, leylapo; chíap (Shuar)

Peru Insira, insira caspi, insira amarilla, incira, morero, tosbarao, limulana, zarzamora, lanche; tsáchig (Awajún), wajtsajinu-juhillahi (Bora), coron (Shipibo)

Venezuela Mora

Status Wild

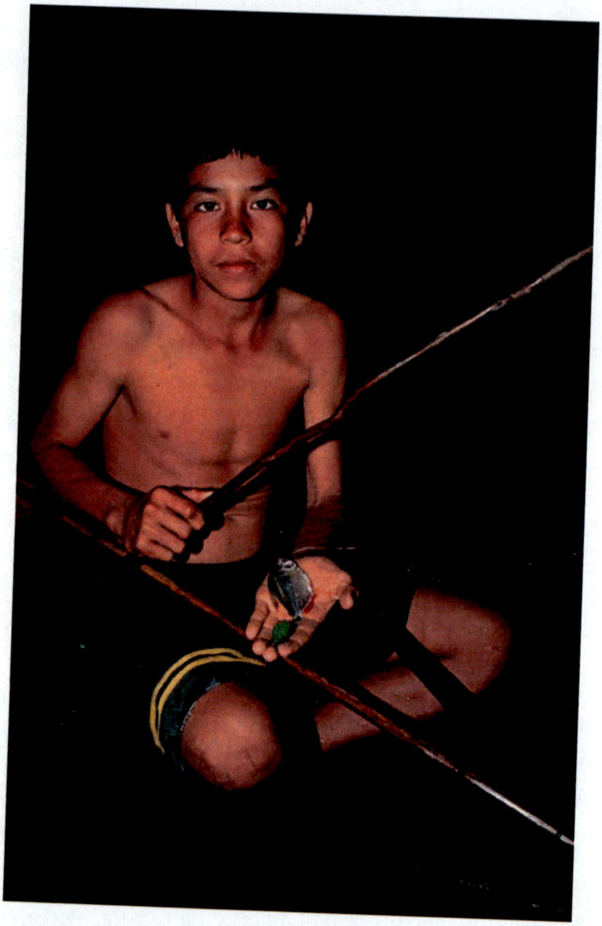

Hildebrando, 14 years, with insira *Maclura tinctoria* fruit. He has caught a sardiña (*Triportheus* sp.) using insira as bait. Rio Yanayacu near Veinte de Enero, Reserva Nacional Pacaya-Samiria, Loreto, Peru 6-14-04

Maclura tinctoria is a tree reaching 35 m along the banks of rivers and streams that produces small nobly fruits with a skin texture reminiscent of raspberries. It occurs on the floodplains of various water types, including black water, clear water, and sediment-rich rivers (Huber 1906; Kuchla et al. 2015; Lamotte 1990; Wittmann et al. 2002). In the Peruvian Amazon, the fruits are used mostly for fish bait. The tree, resembling a willow as noted by Richard Spruce over 150 years ago, occurs in forests and successionary vegetation and starts producing fruit when only 3 or 4 m high so river dwellers can paddle up to it during the annual flood and pick the fruits. People have to be careful when picking the fruits because the young branches sport spines, hence one of the names for this tree in Mato Grosso, Brazil: tatajuba de espinho. The prickles account for two of the common names for the tree in the

Brazilian Amazon: limãorana (lime-like) and jurema de espinho (spiny jurema). The fruits are yellow when fully ripe but serve as fish bait while still green. The main fish caught with this bait is sardiña (*Triportheus* spp.), known as sardinha in Brazil. Fish are likely among the dispersal agents for the numerous dark tan seeds inside the fruits.

Rare black-bearded saki *Chiropotes satanas* monkeys in eastern Amazonia and Venezuela eat the fruits (Kinzey and Norconk 1993). In seasonally dry forests of Costa Rica, howler monkeys *Alouatta palliata* and capuchin monkeys *Cebus capucinus* also eat the fruits and swallow the seeds, thus dispersing them (Wehncke et al. 2004). Related species of *Alouatta* and *Cebus* in Amazonia most likely do the same. In the Magdalena River basin in northern Colombia, the brown spider monkey *Ateles hybridus*, eats this fruit and related species in Amazonia probably do the same. On the floodplain of the Solimões in the vicinity of its confluence with the Japurá a species of squirrel monkey *Saimiri vanzolinii* with a restricted distribution also eats this fruit (Paim et al. 2017), as does another squirrel monkey *Saimiri sciureus cassiquiarensis* in floodplain forests of the Solimões near Tefé (Araujo 2014:38). In gallery and deciduous woodlands just outside the southern fringes of the Amazon Basin in Mato Grosso do Sul in Brazil the green-cheeked parakeet *Pyrrhura molinae* has been observed eating the seeds and pulp of this tree (Ragusa-Netto 2007). Although this parakeet does not occur in Amazonia, other parakeets probably do the same in the Amazon Basin.

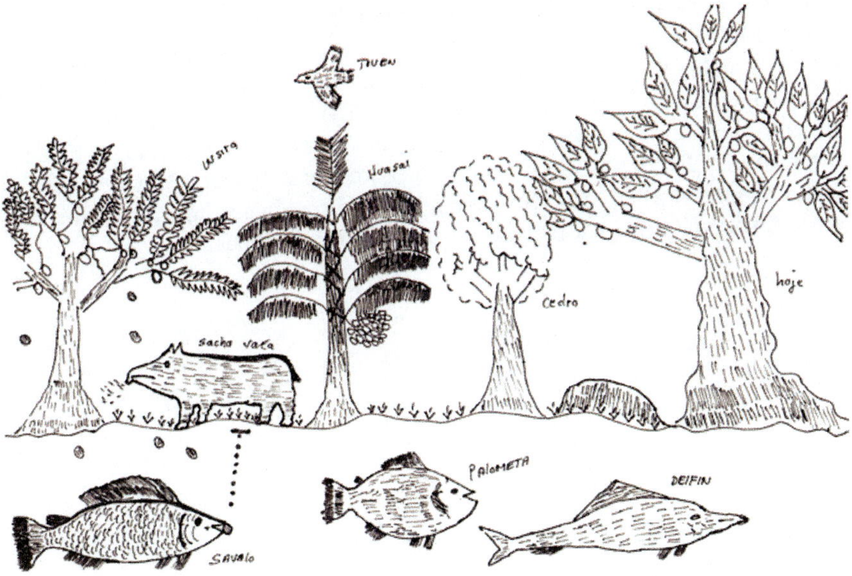

Tapir *Tapirus terrestris*, known as sacha vaca in Peru, and sabalo (*Brycon* sp.) fish-eating insira *Maclura tinctoria* fruit (on left). Drawing by Hildebrando, 14 years. Veinte de Enero, Rio Yanayacu, Reserva Nacional Pacaya-Samiria, Loreto, Peru 6-14-04

The tree has other uses. In the Peruvian Amazon and in Acre, Brazil, the latex is applied to aching teeth to relieve pain as well as to alleviate gingivitis (Kvist et al. 2001; Sanz-Biset et al. 2009). Rural folk inhabiting Quilombo Sangrador in the municipality of Presidente Juscelino, Maranhão, use the latex to treat bone and muscle disorders (Monteles and Pinheiro 2007). In Nova Xavantina in Mato Grosso, the inner bark is soaked in water and drunk to treat anorexia and hypolipidemic conditions, such as caused by high levels of cholesterol in the blood (Silva et al. 2010). The wood is used for building houses, fences, and boats in various parts of the Amazon Basin (Bennett et al. 2002:215; Kvist et al. 2001; Stagegaard et al. 2002; Van den Eynden et al. 1999:56; Wittmann et al. 2010; Wittmann et al. 2009). In Saracura, a quilombola community along the middle Amazon near Santarém, Pará, the wood is used for fuel (Oliveira and Cavalcante 2017).

Maclura tinctoria is found along riverbanks and on uplands, such as in the Xingu watershed in Pará (Lemos et al. 2015), in the forests of Mato Grosso in the southern part of the Amazon Basin (Abreu et al. 2014), and in seasonal forests and cerrado in the Upper Tocantins and Araguaia river basins in southeastern Amazonia (Haidar et al. 2013; Oliveira et al. 2018). It is found in some upland forested areas at 1100 m in the Andean foothills of Bolivia (Plata 2008:82). It occurs in some successionary vegetation on uplands, such as in the Tocantins watershed (Taurino 2017:78) and in 5-year-old and 20-year-old regrowth on abandoned agricultural plots in the vicinity of Santa Mercedes along the Rio Putumayo in the northern Peruvian Amazon bordering Colombia (Muñoz 2010:18, 36). This tree also occurs in some cacao agroforestry plots in the municipality of São Félix do Xingu in southern Pará (Braga et al. 2019). Known as taiúva in some parts of the Brazilian Amazon, it occurs on an extensive archaeological site with anthropogenic black earth at Palmeiras overlooking the middle Xingu upstream from Altamira. The Palmeiras archaeological site is currently covered with successionary vegetation. *Maclura tinctoria* is found in many parts of Brazil and the fruits are sometimes eaten locally. *Maclura tinctoria* occurs as high as 1800 m in the Andean foothills and ranges from Argentina north to the Amazon Basin, Venezuela, Colombia, and the Caribbean.

Boys picking insira *Maclura tinctoria* fruits for use as fish bait. Rio Yanayacu near Veinte de Enero, Reserva Nacional Pacaya-Samiria, Loreto, Peru 6-14-04

36.4 *Perebea longepedunculata*

Peru Manito de gato, chimicua, misho chaqui

Status Wild

Manito de gato *Perebea longepedunculata* **fruit. Near Yarina, Rio Yanayacu, Reserva Nacional Pacaya-Samiria, Loreto, Peru 4-23-06**

Aptly named manito de gato along the Yanayacu River, an affluent of the lower Marañón, on account of the fruit's segments that resemble the pads on a cat's paw, *Perebea longepedunculata* fruits during the rainy season. The bright red fruits are small, some 1.2–1.5 cm long and 1.2–2.5 cm in diameter, but greatly appreciated for their sweet and juicy pulp. The fruit's three to six tightly packed segments each contain a single brown seed, about 0.8 cm by 0.6 cm, which is ejected after relishing the pulp.

Cristian, 10 years, climbing a manito de gato tree *Perebea longepedunculata* to gather fruit. Successionary forest near Yarina, Rio Yanayacu, Reserva Nacional Pacaya-Samiria, Loreto, Peru 4-23-06

The fruits are gathered by climbing the small tree which is normally less than 14 m high and is already fruiting by the time it reaches 2 m. An understory tree in successionary forest below 200 m, *Perebea longepedunculata* is mostly found in the Peruvian Amazon although it has also been recorded in the Upper Rio Negro watershed in Venezuela. Mostly confined to areas that are seasonally flooded (Nebel et al. 2001), it also occurs occasionally in upland forests.

In the Ecuadorian Amazon, the Achuar eat the fruits of a related species, *Perebea guianensis*, which they gather in the forest (Descola 1994:73). The Wajãpi, a Tupi-Guarani group who live in Amapá and contiguous French Guiana, do the same (Grenand 1992).

Segundo, 12 years, with manito de gato *Perebea longepedunculata* fruit he gathered in successionary forest. Near Yarina, Rio Yanayacu, Reserva Nacional Pacaya-Samiria, Loreto, Peru 4-23-06

36.5 *Sorocea duckei*

Brazil Caiembé, lixeira, caraipé do igapó, jauitá

Peru Ensina negra, chimicua amarillo

Status Wild

Caiembé *Sorocea duckei* tree in fruit on the Amazon floodplain. Lago Curuçá near Itacoatiara, Amazonas, Brazil June 1977

Caiembé *Sorocea duckei* is a small floodplain tree in the lowlands of the Brazilian and Peruvian Amazon. It reaches as high as 20 m in forests, including successionary growth, but starts producing fruit when only 3 m tall. The fruits are eaten by certain fish and turtles when they drop in the water. At the height of the flood, turtles and fish can swim among the lower branches and bite off the fruit. Acará-açu *Astronotus ocellatus*, known in the aquarium trade as oscar, is one of the highly prized fish that are fond of caiembé fruits. Fishermen are keenly aware of this habit so they bait trotlines containing four to eight hooks with the fruits and suspend them by caimbé trees. The yellow-spotted Amazon turtle *Podocnemis unifilis*, known as tracajá in the Brazilian Amazon, is attracted to the blue-black fruits of caimbé. Caiembé fruits until the end of July when the floodwaters have normally been receding for about a month. Fruits of the cramuri *Gymnoluma glabrescens* tree, which are also employed as trotline bait for the yellow-spotted Amazon turtle, are only available until May when the waters are still rising (Smith 1981:65, 96). The leaves are rough, like sandpaper, hence one of the common names for the tree in Brazil: lixeira.

Tracajá *Podocnemis unifilis* turtle caught on a trotline baited with caimbé *Sorocea duckei* fruit. Another tracajá is in the canoe. Lago Curuçá near Itacoatiara, Amazonas, Brazil June 1977

References

Abreu, Thiago Ayres Lazzarotti; Pinto, José Roberto Rodrigues; Mews, Henrique Augusto. 2014. Variações na riqueza e na diversidade de espécies arbustivas e arbóreas no período de 14 anos em uma Floresta de Vale, Mato Grosso, Brasil. Rodriguésia 65(1):73–88

Albert, Bruce; Milliken, William. 2009. Urihi A: a terra-floresta Yanomami. Instituto Socioambiental (ISA)/Institut de Recherche pour le Devéloppement, São Paulo

Amaral, Dário Dantas do; Vieira, Ima Célia Guimarães; Almeida, Samuel Soares de; Salomão, Rafael Paiva; Silva, Antonio Sérgio Lima da; Jardim, Mário Augusto Gonçalves. 2009. Checklist da flora arbórea de remanescentes florestais da região metropolitana de Belém e valor histórico dos fragmentos, Pará, Brasil. Boletim do Museu Paraense Emílio Goeldi, Ciencias Naturais 4(3):231–289

Andresen, Ellen. 1999. Seed dispersal by monkeys and the fate of dispersed seeds in a Peruvian rain forest. Biotropica 31(1):145–158

Araujo, Michele. 2014. Ecologia alimentar de *Saimiri sciureus cassiquiarensis* (Lesson, 1840) (Primates, Cebidae) em florestas de várzea da Amazônia Central Brasileira. Master's thesis, MPEG/UFPa, Belém

Arnaud, Expedito. 1989. O Índio e a Expansão Nacional. Edições CEJUP, Belém

Arroyo-Rodríguez, Víctor; Andresen, Ellen; Bravo, Susana P.; Stevenson, Pablo R. 2015. Seed dispersal by howler monkeys: current knowledge, conservation implications, and future directions. In: Howler Monkeys, Developments in Primatology: Progress and Prospects, Kowalewski, M. M. et al. (eds), Springer, New York, pp 111–139

Ayres, J. M. 1989. Comparative feeding ecology of the uakari and bearded saki, *Cacajao* and *Chiropotes*. Journal of Human Evolution 18:697–716

Ayres, J. M.; Ayres, C. 1979. Aspectos da caça no alto Rio Aripuanã. Acta Amazonica 9(2):287–298

Baena, Antônio Ladislau Monteiro 2004. Ensaio Corográfico sobre a Província do Pará. Edições do Senado Federal, Brasilia

Balée, W. 1993a. *Footprints of the Forest: Ka'apor Ethnobotany, the Historical Ecology of Plant Utilization by an Amazonian People*. Columbia University Press, New York

Balée, W. 1993b. Indigenous transformation of Amazonian forests: an example from Maranhão, Brazil. L'Homme 33(126–128):231–254

Balée, W. and D. G. Campbell 1990. Evidence for the successional status of liana forest (Xingu River Basin, Amazonian Brazil). Biotropica 22(1):36–47

Bánki, Olaf S.; Steege, Hans ter; Jansen-Jacobs, Marion J.; Raghoenandan, Usha P. D. 2003. Plant Diversity of the Nassau Mountains, Suriname: Report of the 2003 Expedition. National Herbarium Nederland, Utrecht

Barboza, M. S. L. 2008. O preço que a natureza pagou e os efeitos colaterais que sofremos para hoje se ter energia: uso dos recursos animais e percepção dos impactors entre os ribeirinhos do Lago de Tucuruí (PA). Master's thesis, Universidade Federal do Pará, Nucleo de Altos Estudos Amazônicas, Programa de Pós-Graduação em Desenvolvimento Sustentável do Trópico Úmido, Belém.

Begazo, A. J. 1997. Uso de las poblaciones de paujiles, pucacungas, pavas de monte y manajaracos (Cracidae, Galliformes) en el noroeste Peruano. In: Manejo de Fauna Silvestre en la Amazonía, Fang, T.G.; Bodmer, R.E.; Aquino, R.; Valqui, M.H. (eds), Instituto de Ecologia, La Paz, Bolivia, pp 257–266

Bennett, Bradley C.; Baker, Marc A.; Andrade, Patricia Gómez. 2002. Ethnobotany of the Shuar of Eastern Ecuador. New York Botanical Garden, Advances in Economic Botany, vol. 14.

Bolfe, Édson Luis; Batistella, Mateus. 2011. Análise florística e estrutural de sistemas silviagrícolas em Tomé-Açu, Pará. Pesquisa Agropecuária Brasileira 46(10):1139–1147

Braga, Daniel P. P.; Domene, Frederico; Gandara, Flávio B. 2019. Shade trees composition and diversity in cacao agroforestry systems of southern Pará, Brazilian Amazon. Agroforestry Systems 93:1409–1421

Campbell, David G.; Daly, Douglas C.; Prance, Ghillean T.; Maciel, Ubirajara N. 1986. Quantitative ecological inventory of terra firme and várzea tropical forest on the Rio Xingu, Brazilian Amazon. Brittonia 38(4):369–393

Carim, Marcelo de Jesus Veiga. 2016. Estrutura, composição e diversidade em florestas alagáveis de várzea de maré e de igapó e suas relações com variáveis edáficas e o período de inundação no Amapá, Amazônia oriental, Brasil. Doctoral dissertation, Programa de Pós-Graduação em Botânica, INPA, Manaus

Coelho, Roberta de Fátima Rodrigues; Miranda, Izildinha Souza; Mitja, Danielle. 2012. Caracterização do processo sucessional no Projeto de Assentamento Benfica, sudeste do estado do Pará, Amazônia oriental. Boletim do Museu Paraense Emílio Goeldi, Ciencias Naturais 7(3):251–282

Cruz, Raúl Federico Bello Santa. 2018. Comportamiento de monos arañas (*Ateles chamek*) reintroducidos en el sureste de la Amazonía Peruana. Master's thesis, Universidad Nacional Agraria La Molina, Lima

Cummings, Anthony R.; Read, Jane M.; Fragoso, Jose M. V. 2018. Implications of forest type and land tenure diversity for the sustainability of ecosystem services provided by northern Amazonia's multiple-use tree species. Landscape Ecology 33:423–438

Dantas, Mário; Rodrigues, Irenice Alves. 1980. Plantas invasoras de pastagens cultivadas na Amazônia. EMBRAPA, Centro de Pesquisa Agropecuária do Trópico Úmido, Boletim de Pesquisa, vol. 1.

Defler, Thomas R.; Defler, Sara B. 1996. Diet of a group of *Lagothrix lagothricha lagothrica* in southeastern Colombia. International Journal of Primatology 17(2):161–190

Descola, P. 1994. *In the Society of Nature: A Native Ecology in Amazonia.* Cambridge University Press, Cambridge

Espírito-Santo, Fernando Del Bon; Shimabukuro, Yosio Edimir; Aragão, Luiz Eduardo Oliveira e Cruz de; Machado, Evandro Luiz Mendonça. 2005. Análise da composição florística e fitossociológica da floresta nacional do Tapajós com apoio geográfico de imagens satélites. Acta Amazonica 35(2):155–173

Ferreira, Liane Marise Moreira; Tonini, Helio; Araújo, Sílvio Levy Franco de. 2011. Viabilidade econômica de sistema agroflorestal em área de agricultor familiar na região do Apiaú, Município de Mucajaí, Roraima. EMBRAPA Roraima, Documentos 46

Fiedler, Nilton C.; Costa, Alexandre F. da; Soares, Thelma S.; Leite, Ângelo Márcio P. 2012. Caracterização do segmento de madeira serrada em três municípios do estado do Pará. Revista Brasileira de Ciências Agrárias 7(1):111–116

Fragoso, José M. V.; Huffman, Jean M. 2000. Seed-dispersal and seedling recruitment patterns by the last Neotropical megafaunal element in Amazonia, the tapir. Journal of Tropical Ecology 16:369–385

Franco, Ana María; Santamaria, Marcela. 1997. Area vital, habitos alimenticios y otros aspectos de historia natural de paujil *Mitu silvini.* In: Manejo de Fauna Silvestre en la Amazonía, Fang, T.G.; Bodmer, R.E.; Aquino, R.; Valqui, M.H. (eds), Instituto de Ecologia, La Paz, Bolivia, pp 267–282

Gayot, Marc; Henry, Olivier; Dubost, Gérard; Sabatier, Daniel. 2004. Comparative diet of the two forest cervids of the genus *Mazama* in French Guiana. Journal of Tropical Ecology 20(1):31–43

Gomez-Pompa, A. 1987. On Maya silviculture. Mexican Studies 3(1):1–17

Gomez-Pompa, Arturo; Salvador, José; Sosa, Victoria. 1987. The 'Pet Kot': a man-made tropical forest of the Maya. Interciencia 12(1):10–15

Gomide, Guilherme Luís Augusto. 1997. Estrutura e dinâmica de crescimento de florestas tropicais primária e secundária no Estado do Amapá. Master's Thesis, Universidade Federal do Paraná, Brazil

Grenand, P. 1992. The use and cultural significance of the secondary forest among the Wayapi Indians. In: *Sustainable Harvest and Marketing of Rain Forest Products*, M. Plotkin and L. Famolare (Editors), Island Press, Washington, DC, pp. 27–40

Guarim Neto, Germano. 1984. Plantas do Brasil: Angiospermas do Estado de Mato Grosso– 1. Rodriguésia 36(59):105–121

Guillotin, M.; Dubost, G.; Sabatier, D. 1994. Food choice and food competition among the three major primate species of French Guiana. Journal of Zoology 233:551–579

Guzmán, Adriana; Stevenson, Pablo R. 2008. Seed dispersal, habitat selection and movement patterns in the Amazonian tortoise, *Geochelone denticulata*. Amphibia-Reptilia 29:463–472

Haidar, Ricardo Flores; Fagg, Jeanine Maria Felfili; Pinto, José Roberto Rodrigues; Dias, Ricardo Ribeiro; Damasco, Gabriel; Silva, Lucasde Carvalho Ramos; Fagg, Christopher William. 2013. Florestas estacionais e áreas de ecótono no estado do Tocantins, Brasil: parâmetros estruturais, classificação das fitofisionomias florestais e subsídios para conservação. Acta Amazonica 43(3):261–290

Henry, Olivier; Feer, Francois; Sabatier, Daniel. 2000. Diet of the lowland tapir (*Tapirus terrestris* L.) in French Guiana. Biotropica 32(2):3364–368
Heymann, Eckhard W. 2000. Field observations of the golden-mantled tamarin, Saguinus tripartitus, on the Río Curaray, Peruvian Amazonia. Folia Primatologica 71:392–398
Huber, Jacques. 1906. La végétation de la vallée du Rio Purus (Amazone). Bulletin de L'Herbier Boissier 6(4):249–276
Huber, J. 1910. Mattas e madeiras amazonicas. Boletim do Museu Goeldi de Historia Natural e Ethnographia 6: 91–225
Iwanaga, Simone; Ferrari, Stephen F. 2001. Party size and diet of synoptic Atelids (*Ateles chamek* and *Lagothrix cana*) in southwestern Brazilian Amazonia. Folia Primatologica 72:217–227
Johnston, Mark; Colquhoun, Araminta. 1996. Preliminary ethnobotanical survey of Kurupukari: an Amerindian settlement of central Guyana. Economic Botany 50(2):182–194
Julliot, C. 1996a. Seed dispersal by red howling monkeys (*Alouatta seniculus*) in the tropical rain forest of French Guiana. International Journal of Primatology 17(2):239–258
Julliot, C. 1996b. Fruit choice by red howler monkeys (*Alouatta seniculus*) in a tropical rain forest. American Journal of Primatology 40:261–282
Julliot, C.; Sabatier, D. 1993. Diet of the red howler monkey (*Alouatta seniculus*) in French Guiana. International Journal of Primatology 14: 527–550
Kinzey, Warren G.; Norconk, Marilyn A. 1990. Hardness as a basis of fruit choice in two sympatric primates. American Journal of Physical Anthropology 81:5–15
Kinzey, Warren G.; Norconk, Marilyn A. 1993. Physical and chemical properties of fruit and seeds eaten by *Pithecia* and *Chiropotes* in Surinam and Venezuela. International Journal of Primatology 14(2):207–227
Kuchla, Wilian José; Stepka, Thiago Floriani; Sawczuk, Alex Roberto; Lisboa, Gerson dos Santos. 2015. Florística, estrutura horizontal e distribuição diamétrica em área de transição de cerrado e floresta aluvial no município de Campos de Júlio, MT. Ambiência 11(1):13–30
Kvist, Lars Peter; Andersen, Martin K.; Stagegaard, Jesper; Hesselsoe, Martin; Llapapasca, Consuelo. 2001. Extraction from woody forest plants in flood plain communities in Amazonian Peru: use, choice, evaluation and conservation status of resources. Forest Ecology and Management 150:147–174
Lamotte, Sandrine. 1990. Fluvial dynamics and succession in the lower Ucayali River basin, Peruvian Amazonia. Forest Ecology and Management 33/34: 141–156
Le Cointe, P. 1922. *L'Amazonie Brésilienne: Le Pays-ses Habitants, ses Resources, Notes et Statistiques jusqu'en 1920*. Augustin Challamel, Paris, vol 2
Lemos, D. A. N.; Ferreira, B. G. A.; Siqueira, J. D. P.; Oliveira, M. M.; Ferreira, A. M. 2015. Floristic and phytosociology in dense "terra firme" rainforest in the Belo Monte hydroelectric plant influence area, Pará, Brazil. Brazilian Journal of Biology 75(3):257–276
Lopes, Célia Regina Araujo Soares; Ribeiro, Ricardo da Silva; Rodrigues, Lucirene; Cabral, Fabiana Ferreira; Silva, Dennis Rodrigues. 2014. Checklist de angiospermas da região de influência da UHE SINOP, médio Teles Pires, Mato Grosso. Enciclopédia Biosfera 10(19):2016–2048
Lorenzi, H. 2009. Árvores Brasileiras: Manual de Identificação e Cultivo de Plantas Arbóreas Nativas do Brasil. Instituto Plantarum, Nova Odessa, São Paulo, vol. 2
Loureiro, Arthur A.; Lisboa, Pedro L. Braga. 1979. Madeiras do município de Aripuanã e suas utilidades (Mato Grosso). Acta Amazonica 9(1: supl.):1–88
Loureiro, Arthur A.; Silva, Marlene F. da; Alencar, Jurandyr da Cruz. 1979. Essências Madeireiras da Amazônia. Instituto Nacional de Pesquisas da Amazônia, Manaus, vol. 2
Macias, Javier Humberto Alvarado. 2013. Caracterizacion de la población y descripción de los hábitos alimenticios del mono aullador, Alouatta seniculus (Linnaeus, 1766), (Primate: Atelidae) en un parche de bosque, Vereda El Caucho, Nunchia, Casanare, Colombia. Master's thesis, Universidad Pedagógica y Tecnológica, Tunja, Colombia
Magalhães, Luís M. S. 1982. Produção de sementes de essências nativas em floresta primária na Amazônia. Acta Amazonica 12(2):257–262

Miller, R. P. 1994. Estudo da fruticultura tradicional dos índios Waimiri-Atroari: base para a extensão agroflorestal. In: *Anais do I Congresso Brasileiro sôbre Sistemas Agroflorestais, Porto Velho, 3–7 Julho*, L.J. Montoya and M. Medrado (Editors), EMBRAPA (Centro Nacional de Pesquisas de Florestas/Centro de Pesquisa Agroflorestal de Rondônia), vol. 2, pp. 449–462

Monteles, Ricardo; Pinheiro, Claudio Urbano B. 2007. Plantas medicinais em um quilombo maranhense: uma perspectiva etnobotânica. Revista de Biologia e Ciências da Terra 7(2):1–12

Morcote-Ríos, Gaspar; Bocanegra, Francisco Javier Aceituno; Sicard, Tomás León. 2014. Recolectores del Holoceno temprano en la floresta Amazónica Colombiana. In: Antes de Orellana: Actas del 3er Encuentro Internacional Arqueología Amazónica, S. Rostain (ed), Instituto Francés de Estudios Andinos, Quito, pp. 39–50

Moskovits, D. 1998. Population and ecology of the tortoises *Geochelone carbonaria* and *G. denticulata* on the Ilha de Maracá. In: Maracá: The Biodiversity & Environment of an Amazonian Rainforest, W. Milliken and J. Ratter (eds.), John Wiley, Chichester, pp. 263–284

Moskovits, Debra K.; Bjorndal, Karen A. 1990. Diet and food preferences of the tortoises *Geochelone carbonaria* and *G. denticulata* in northwestern Brazil. Herpetologica 46(2):207–218

Mostacedo, Bonifacio; Balcazar, Julio; Montero, Juan Carlos. 2006. Tipos de bosque, diversidad y composición florística en la Amazonia sudoeste de Bolivia. Ecología en Bolivia 41(2):99–116

Muñoz, Marianella Guevara. 2010. Evaluación de purmas con fines de manejo en la comunidad nativa de Santa Mercedes, Rio Putumayo, Perú. Master's thesis, Universidad Nacional de la Amazonia Peruana, Iquitos

Nebel, Gustav; Dragsted, Jens; Vanclay, Jerome K. 2001. Structure and floristic composition of flood plain forests in the Peruvian Amazon II. The understorey of restinga forests. Forest Ecology and Management 150: 59–77

Norconk, Marilyn A.; Kinzey, Warren G. 1994. Challenge of Neotropical frugivory: travel patterns of spider monkeys and bearded sakis. American Journal of Primatology 34: 171–183

Nunes, Andréa. 1998. Diet and feeding ecology of *Ateles belzebuth belzebuth* at Maracá Ecological Station, Roraima, Brazil. Folia Primatologica 69:61–76

Ogeron, Clémence; Odonne, Guillaume; Cristinoi, Antonia; Engel, Julien; Grenand, Pierre; Beauchêne, Jacques; Clair, Bruno; Davy, Damien. 2018. Palikur traditional roundwood construction in eastern French Guiana: ethnobotanical and cultural perspectives. Journal of Ethnobiology and Ethnomedicine 14:1–18

Oliveira, Patricia Chaves de; Cavalcante, Suellen. 2017. Ethnobotany in the Amazon floodplain ecosystem: a case study, Quilombo Saracura, Pará, Brazil. International Journal of Botany Studies 2(1):89–99

Oliveira, Jorge; Almeidas, Samuel Soares de; Potyguara, Raimunda Vilhena; Lobatos, Luiz Carlos Batista. 1991. Espécies vegetais produtoras de fibras utilizadas por comunidades Amazônicas. Boletim do Museu Paraense Emílio Goeldi, Série Botânica 7(2):393–428

Oliveira, Lucicléia Mendes de; Sousa, Rhonan Martins de; Correa, Norma Elena Reynosa; Santos, André Ferreira dos; Giongo, Marcos. 2018. Florística e síndromes de dispersão de um fragmento de cerrado ao sul do estado do Tocantins. Scientia Agraria Paranaensis 17(1): 104–111

Paim, Fernanda P.; Chapman, Colin A.; Queiroz, Helder L. de; Paglia, Adriano P. 2017. Does resource availability affect the diet and behavior of the vulnerable squirrel monkey, *Saimiri vanzolinii*? International Journal of Primatology 38(3):572–587

Palacios, Erwin; Rodríguez, Adriana; Defler, Thomas R. 1997. Diet of a group of *Callicebus torquatus lugens* (Humboldt, 1812) during the annual resource bottleneck in Amazonian Colombia. International Journal of Primatology 18(4):503–522

Parrotta, John A.; Francis, John K.; Almeida, Rionaldo Rolo de. 1995. Trees of the Tapajós: A Photographic Field Guide. International Institute of Tropical Forestry, General Technical Report IITF-1, Rio Piedras, Puerto Rico

Parrotta, John A.; Knowles, Oliver Henry; Wunderle Jr., Joseph M. 1997. Development of floristic diversity in 10-year-old restoration forests on a bauxite mined site in Amazonia. Forest Ecology and Management 99:21–42

Peres, C.A. 1994. Diet and feeding ecology of gray woolly monkeys (*Lagothrix lagotricha cana*) in central Amazonia: comparisons with other Atelines. International Journal of Primatology 15(3):333–372

Pinheiro, Klewton Adriano Oliveira; Carvalho, João Olegário Pereira de; Quantz, Beatriz; Francez, Luciana Maria de Barros; Schwartz, Gustavo. 2007. Fitossociologia de uma área de preservação permanente no leste da Amazônia: indicacão de espécies para recuperação de áreas alteradas. Floresta 37(2): 175–187

Pinto, L. P. 2008. Ecologia alimentar do cuxiú-de-nariz vermelho *Chiropotes albinasus* (Primates: Pitheciidae) na Floresta Nacional do Tapajós, Pará. Doctoral dissertation, Universidade Estadual de Campinas, São Paulo

Plata, Edwin Javier Ticona. 2008. Estructura y composición florística del bosque húmedo pluviestacional subandino de Yungas del sector noreste de Apolo, Parque Nacional Madidi, La Paz, Bolivia. Master's thesis, Universidad Mayor de San Andrés, La Paz, Bolivia

Pontes, Antonio Rossano Mendes. 1997. Habitat partitioning among primates in Maracá Island, Roraima, northern Brazilian Amazonia. International Journal of Primatology 18(2):131–157

Prado, Helbert Medeiros. 2013. Feeding ecology of five Neotropical ungulates: a critical review. Oecologia Australis 17(4): 459–473

Ragusa-Netto, J. 2007. Feeding ecology of the green-cheeked parakeet (*Pyrrhura molinae*) in dry forests in western Brazil. Brazilian Journal of Biology 67(2):243–249

Regan, B. C.; Julliot, C.; Simmen, B.; Viénot, F.; Charles-Dominique, P.; Mollon, J. D. 2001. Fruits, foliage and the evolution of primate colour vision. Philosophical Transactions of the Royal Society London B 356: 229–283

Reis, Luciane Pereira; Reis, Alisson Rodrigo S.; Carvalho, Josiane Celerino de; Silva, Érica Fernanda Rodrigues da; Silva, Jhonata Ribeiro da. 2014. Caracterização anatômica de madeiras comercializados como perna-manca nas estâncias de Altamira-PA. Enciclopédia Biosfera 10(19):463–479

Robles, Rembrandt Ramiro Canahuire. 2017. Composición florística y estructura de la recuperación natural en un área degradado por minería en la comunidad nativa de Tres Islas, Tambopata, Madre de Dios. Master's thesis, Escuela Profesional de Ingeniería Forestal y Medio Ambiente, Puerto Maldonado, Perú

Rosado, Benedicto Baca. 2013. Caracterización ecológica de Pteroglossus castanotis (Gould, 1833), Ramphastidae en Manu. In: ReporteManu: Pasión por la Investigación en la Amazonía Peruana, Groenendijk, Trovar; Wust, Walter (eds), San Diego Zoo Global Peru, Cusco, pp. 266–293

Salomão, Rafael de Paiva; Vieira, Ima Célia Guimarães; Suemitsu, Chieno; Rosa, Nélson de Araújo; Almeira, Samuel Soares de; Amaral, Dário Dantas do; Menezes, Moirah Paulo Machado de. 2007a. As florestas de Belo Monte na grande curva do rio Xingu, Amazônia Oriental. Boletim do Museu Paraense Emílio Goeldi, Ciencias Naturais 2(3):57–153

Salomão, Rafael Paiva; Rosa, Nélson Araújo; Morais, Kácio Andrey Câmara. 2007b. Dinâmica da regeneração natural de árvores em áreas mineradas na Amazônia. Boletim do Museu Paraense Emílio Goeldi, Ciencias Naturais 2(2):85–139

Santos, Ronaldo Oliveira dos; Lima, Robson Carmo; Lima, Robson Borges de; Aparício, Perseu da Silva; Abreu, Jadson Coelho de. 2017. Florística e estrutura de uma comunidade arbórea na floresta estadual do Amapá, Amazônia Oriental, Brasil. Nativa Sinop 5:529–539

Sanz-Biset, Jaume; Campos-de-la-Cruz, José; Epiquién, Mirbel A.; Cañigueral, Salvador. 2009. A first survey on the medicinal plants of the Chazuta valley (Peruvian Amazon). Journal of Ethnopharmacology 122:333–362

Schwartz, Gustavo. 2007. Manejo sustentável de florestas secundárias: espécies potenciais no nordeste do Pará, Brasil. Amazônia: Ciência e Desenvolvimento 3(5):125–147

Silva, Marcondes Alves B. da; Melo, Laura Valdiane L.; Ribeiro, Reginaldo V.; Souza, João Paulo M. de; Lima, Joaquim Corsino S.; Martins, Domingos Tabajara de O.; Silva, Regilane M. da. 2010. Levantamento etnobotânico de plantas utilizadas como anti-hiperlipidêmi-

cas e anorexígenas pela população de Nova Xavantina-MT, Brasil. Revista Brasileira de Farmacognosia 20(4):549–562
Simmen, B. 1992. Competitive utilization of *Bagassa* fruits by sympatric howler and spider monkeys. Folia Primatologica 55:155–160
Simmen, Bruno; Sabatier, Daniel. 1996. Diets of some French Guianan primates: food composition and food choices. International Journal of Primatology 17(5):661–693
Smith, N. J. H. 1976. Utilization of game along Brazil's Transamazon Highway. *Acta Amazonica* 6(4):455–466
Smith, N. J. H. 1981. *Man, Fishes, and the Amazon*. Columbia University Press, New York
Soares, Caio Rodrigo Alves; Sá, Heloíse Rodrigues Alves de; Rodrigues, Marcos; Goulart, Selma Lopes. 2018. Gestão e competividade: análise de um aglomerado produtiva moveleiro na Amazônia. Revista de Estudos Sociais 20(41):144–159
Souza, Victória Cristina Oliveira de. 2019. Avaliação do estágio sucessional utilizando a classificação em grupos ecológicos de uma floresta urbana na região metropolitana de Belém-Pará. Bachelor's thesis, Universidade Federal Rural da Amazônia (Ufra), Belém
Stagegaard, Jesper; Sorensen, Marten; Kvist, Lars Peter. 2002. Estimations of the importance of plant resources extracted by inhabitants of the Peruvian Amazon flood plains. Perspectives in Plant Ecology, Evolution and Systematics 5(2):103–122
Stefanello, Daniel; Ivanauskas, Natália Macedo; Martins, Sebastião Venâncio; Silva, Elias; Kunz, Sustanis Horn. 2010. Síndromes de dispersão de diásporos das espécies de trechos de vegetação ciliar do rio das Pacas, Querência-MT. Acta Amazonica 40(1):141–150
Stevenson, Pablo R. 2000. Seed dispersal by woolly monkeys (*Lagothrix lagothricha*) at Tinigua National Park, Colombia: dispersal distance, germination rates, and dispersal quantity. American Journal of Primatology 50: 275–289
Stevenson, Pablo R. 2004. Fruit choice by woolly monkeys in Tinigua National Park, Colombia. International Journal of Primatology 25(2): 367–381
Stevenson, Pablo R. 2005. Potential keystone plant species for the frugivore community at Tinigua Park, Colombia. In: Tropical Fruits and Frugivores: The Search for Strong Interactors, Dew, J. Lawrence; Boubli, Jean Philippe (eds), Springer, Dordrecht, pp 37–57
Stevenson, Pablo R. 2007. Estimates of the number of seeds dispersed by a population of primates in a lowland forest in western Amazonia. In: Seed Dispersal: Theory and its Application in a Changing World, Dennis, Andrew J.; Schupp, Eugene W.; Green, Ronda J.; Westcott, David A. (eds), CABI, Wallingford, pp 340–362
Stevenson, Pablo R.; Link, Andrés. 2010. Fruit preferences of *Ateles belzebuth* in Tinigua Park, northwestern Amazonia. International Journal of Primatology 31:393–407
Stevenson, Pablo R.; Castellanos, Maria Clara; Pizarro, Juan Carlos; Garavito, Mariaximena. 2002. Effects of seed dispersal by three Ateline monkey species on seed germination at Tinigua National Park, Colombia. International Journal of Primatology 23(6): 1187–1204
Suarez, Scott A. 2006. Diet and travel costs for spider monkeys in a nonseasonal, hyperdiverse environment. International Journal of Primatology 27(2):411–436
Symington, M. McFarland. 1988. Food competition and foraging party size in the black spider monkey (*Ateles paniscus chamek*). Behaviour 105(1/2): 117–134
Taurino, Tássia Cristina da Conceição Barros. 2017. Recuperação natural da diversidade de espécies de plantas em florestas secundárias na Amazônia Oriental. Master's Thesis, Universidade Federal do Pará, Belém
Terborgh, J. 1983. Five New World Primates: A Study in Comparative Ecology. Princeton University Press, Princeton
Turner, B. L. and C. H. Miksicek. 1984. Economic plant species associated with prehistoric agriculture in the Maya lowlands. Economic Botany 38(2):179–193
Uhl, C.; Buschbacher, R.; Serrão, E. A. S. 1988. Abandoned pastures in eastern Amazonia. I. Patterns of plant succession. Journal of Ecology 76:663–681
Van den Eynden, Veerle; Cueva, Eduardo; Cabrera, Omar. 1999. Plantas Silvestres Comestibles del Sur del Ecuador. Ediciones Abya-Yala, Quito

Van Roosmalen, M. G. M. 1985. Habitat preferences, diet, feeding strategy and social organization of the black spider monkey (*Ateles paniscus paniscus* Linnaeus 1758) in Surinam. Acta Amazonica 15(3/4 suplemento) 1–238

Van Roosmalen, Marc G. M.; Klein, Lewis L. 1988. The spider monkeys, genus *Ateles*. In: Ecology and Behavior of Neotropical Primates, Mittermeier, Russell A.; Rylands, Anthony B.; Coimbra-Filho, Adelmar F.; Fonseca, Gustavo A.B. da (eds), Ecology and Behavior of Neotropical Primates, World Wildlife Fund, Washington, DC, vol. 2, pp 455–537

Vásquez, R. and A. H. Gentry. 1989. Use and misuse of forest-harvested fruits in the Iquitos area. *Conservation Biology* 3(4): 350–361.

Veríssimo, Adalberto; Barreto, Paulo; Mattos, Marli; Tarifa, Ricardo; Uhl, Christopher. 1992. Logging impacts and prospects for sustainable forest management in an old Amazonian frontier: the case of Paragominas. Forest Ecology and Management 55: 169–199

Vieira, Paulo R. 2005. Floresta e comunidade: cotidiano de famílias em Jericó, Garrafão do Norte, Pará. Master's thesis, Universidade Federal do Pará, Belém

Wehncke, Elisabet V.; Valdez, Catherine Numa; Domínguez, César A. 2004. Seed dispersal and defecation patterns of *Cebus capucinus* and *Alouatta palliata*: consequences for seed dispersal effectiveness. Journal of Tropical Ecology 20:535–543

Wittmann, Florian; Anhuf, Deiter; Junk, Wolfgang J. 2002. Tree species distribution and community structure of central Amazonian várzea forests by remote-sensing techniques. Journal of Tropical Ecology 18(6):805–820

Wittmann, Florian; Schöngart, Jochen; Queiroz, Helder L.; Wittmann, Astrid de Oliveira; Conserva, Auristela dos Santos; Piedade, Maria T. F.; Kesselmeier, Jürgen; Junk, Wolfgang J. 2009. The Amazon floodplain demonstration site: sustainable timber production and management of central Amazonian white-water floodplains. Ecohydrology & Hydrobiology 9(1):41–54

Wittmann, Florian; Schöngart, Jochen; Junk, Wolfgang J. 2010. Phytogeography, species diversity, community structure and dynamics of central Amazonian floodplain forests. In: Amazonian Floodplain Forests: Ecophysiology, Biodiversity and Sustainable Management, Junk, Wolfgang J.; Piedade, Maria T. F.; Wittmann, Florian; Schöngart, Jochen; Parolin, Pia (eds), Springer, Dordrecht, pp 61–102

Zappi, Daniela C.; Milliken, William; Lopes, Celia Regina A. Soares; Lucas, Eve; Piva, José Hypólito; Frisby, Sue; Biggs, Nicola; Forzza, Rafaela Campostrini. 2016. Xingu State Park vascular plant survey: filling the gaps. Brazilian Journal of Botany 39(2):751–778

Chapter 37
Muntingiaceae

37.1 *Muntingia calabura*

Bolivia Uvilla, uvillo negro, curumín, curiu; bojno'ta (Tsimane)

Brazil Curumurí, curumi, curumín, curuminzeiro, calabura, pau de seda, montigia; tamauã (Kaxinawá)

Colombia Cerezo de restinga, chicható, chirriador, chitató, periquito, majagüito, nigua

Ecuador Niguito, frutillo, cerezo, achotillo, sapán nigüito, nigüito

Peru Yumanasa, yumanaza, tomaque, cerecillo, cerezo, cerezo caspi, cerezo silvestre, basayo blanco, tomatillo, bolaina, bolaina chata, bolaina yumanaza, mullaca huayo, atadijo, achotillo, lumanasa; xapuan (Cashinahua)

Venezuela Guacimillo, nigua, nigüite, mochocho

Status Wild, spontaneous in cultural settings, planted

N. Smith, *Amazon Fruits: An Ethnobotanical Journey*, Ethnobiology,
https://doi.org/10.1007/978-3-031-12803-5_37

Yumanasa *Muntingia calabura* fruits gathered in a home garden. Bretaña, Rio Puinahua, a branch of the Ucayali, Reserva Nacional Pacaya-Samiria, Loreto, Peru 4-16-06

Widespread in tropical America and the Caribbean, the pale scarlet fruits of *Muntingia calabura* are eaten throughout its range and also sold in some markets, such as in Mexico and in Iquitos, Peru (Morton 1987:65; Vásquez and Gentry 1989). Like guava *Psidium guajava*, its precise origins are unclear. Some think it was introduced to Brazil from Central America (Lorenzi et al. 2006a, b:441). If so, it has long become naturalized in the Amazon Basin. *Muntingia calabura* sports feathery, lime-green leaves and has been introduced to numerous islands in the Pacific and Indian oceans, typically as an ornamental. The tree, which can reach 20 m and starts producing fruit when 2 m tall, is widely cultivated in Brazil. River dwellers on Ilha Grande, a floodplain island along the Amazon near Óbidos in Brazil, protect curumurí when they sprout spontaneously in their home gardens.

A tree of successionary woodlands, it is relatively common along the banks of sediment-rich rivers, such as the Ucayali and Rio Manu in Peru, in the Amacayacu Park in the Colombian Amazon, and the Purus, Madeira and lower Amazon Rivers in Brazil (Foster et al. 1986; Huber 1906; Perigolo 2014:21; Prance 1989; Rudas and Prieto 1998). In some disturbed sites near water courses in the Peruvian Amazon, yumanasa forms dense stands known as *yumanasales*. This medium-sized tree, which reaches 15 m when fully mature, also occurs occasionally on uplands in Amazonia, though it is much more common on floodplains. This small tree is also found in cultivated areas of the Yuracaré and Trinitario indigenous people in the Bolivian Amazon (Thomas and Van Damme 2010). In the Andean foothills, this tree occurs up to 2000 m (Mendoza and Guerrero 2010).

Rister, 5 years, with yumanasa *Muntingia calabura* fruits, he gathered at the edge of his village. Manco Capac, Rio Puinahua, a branch of the Ucayali, Reserva Nacional Pacaya-Samiria, Loreto, Peru 4-14-06

The sweet, juicy fruits of *Muntingia calabura* are mainly a children's treat in the Peruvian Amazon. Kids clamber the lower branches of yumanasa to pluck the round fruits, some 1–1.5 cm in diameter, or gather them from the ground. The fruits contain numerous light tan-colored seeds which are ingested along with the pulp of the same color; children are thus involved in dispersing the tree. Several species of birds, including parakeets, some bats, and widespread coatimundi *Nasua narica* also eat the fruits and disperse the seeds (Fleming and Heithaus 1981; Fleming et al. 1985; Foster et al. 1986; Heithaus et al. 1975; Thomas et al. 1988). One of the common names for this tree in Colombia is periquito, suggesting that parakeets are fond of the fruits.

In Costa Rica, the white-faced capuchin monkey *Cebus capucinus* and spider monkey *Ateles geoffroyi* are also involved in dispersing the seeds of *Muntingia calabura* (Chapman 1988, 1989; Wehncke et al. 2004). Although the white-faced capuchin is confined to Central America and the extreme northwestern part of South America a related species, the tufted capuchin (known as macaco-prego in Brazil) *Cebus apella* is relatively common in Amazonia where they also likely eat these fruits. *Ateles geoffroyi* is confined to Central America but several other species of spider monkeys are found in the Amazon Basin and likely also feed on the fruits. The mantled howler monkey *Alouatta palliata* is known to eat this fruit in Central America (Glander 1979, 1981; Neville et al. 1988); although this species does not occur in the Amazon Basin-related species of howler monkey do and probably disperse the seeds of this plant. Many species of monkeys are considered game animals in the Amazon, so this fruit contributes indirectly to the diet of rural people. In the Peruvian Amazon, the moustached tamarin *Saguinus mystax* and the white-lipped tamarin *Saguinus labiatus* disperse the seeds of this fruit (Ique and Aquino 2015; Reátegui 2013:37); these monkeys are not normally hunted for their meat because of their small size but they are sometimes kept as a pet in rural areas. The smooth-skinned fruits of *Muntingia calabura* contain one-and-a-half times as much vitamin C as oranges. Yumanasa fruits during the rainy season, although some trees bear fruit during the drier months as well.

Yumanasa trees serve a variety of other uses. In the village of Nueva Esperanza along the Marañón River, for example, farmers remove strips of bark to bundle stalks of plantains and the wood is used in house construction. In southern Ecuador, the wood is used to make fences and as firewood (Van den Eynden et al. 1999:43). The Kaxinawá, a Pano group in Acre, use the mast fibers to lash bundles of peanuts. On Ilha Grande, a floodplain island along the Amazon near Óbidos in Brazil, fishermen use the fruits to bait hooks to catch pacu (several genera) and pirapitinga *Colossoma bidens*. In Saracura, a quilombola community on a floodplain island along the middle Amazon near Santarém, the leaves are used to make a tea to treat diarrhea (Oliveira and Cavalcante 2017).

References

Chapman, Colin. 1988. Patterns of foraging and range use by three species of Neotropical primates. Primates 29(2): 177–194

Chapman, Colin A. 1989. Primate seed dispersal: the fate of dispersed seeds. Biotropica 21(2): 148–154

Fleming, Theodore H.; Heithaus, E. Raymond. 1981. Frugivorous bats, seed shadows, and the structure of tropical forests. Biotropica 13(2):45–53

Fleming, Theodore H.; Williams, Charles F.; Bonaccorso, Frank J.; Herbst, Larry H. 1985. Phenology, seed dispersal, and colonization in Muntingia calabura, a neotropical pioneer tree. American Journal of Botany 72(3):383–391

Foster, Robin B.; Arce, Javier; Wachter, Tatzyana S. 1986. Dispersal and the sequential plant communities in Amazonian Peru floodplain. In: Frugivores and Seed Dispersal, Estrada, Alejandro; Fleming, Theodore H. (eds), Dr. W. Junk Publishers, Dordrecht, pp 357–370

Glander, K. E. 1979. Howling monkey feeding behavior and plant secondary compounds: a study of strategies. In: The Ecology of Arboreal Foliovores, Montgomery, G. G. (ed), Smithsonian Institution Press, Washington, DC, pp 561–574

Glander, K. E. 1981. Feeding patterns in mantled howling monkeys. In: Foraging Behavior: Ecological, Ethological, and Psychological Approaches, Kamil, A. C.; Sargent, T. D. (eds), Garland Press, New York, pp 231–257

Heithaus, E. Raymond; Fleming, Theodore H.; Opler, Paul A. 1975. Foraging patterns and resource utilization in seven species of bats in a seasonal tropical forest. Ecology 56(4):841–854

Huber, Jacques. 1906. La végétation de la vallée du Rio Purus (Amazone). Bulletin de L'Herbier Boissier 6(4):249–276

Ique, Carlos A.; Aquino, Rolando. 2015. Principales especies florísticas utilizadas y dispersión de semillas por *Saguinus labiatus* E. Geoffroy, 1812 en isla Muyuy, Loreto, Perú. Ciencia Amazónica 5(1):81–88

Lorenzi, Harri; Bacher, Luis; Lacerda, Marco; Sartori, Sergio. 2006a. Frutas Brasileiras e Exóticas Cultivadas. Instituto Plantarum de Estudos da Flora Ltda, Odessa, Sao Paulo.

Lorenzi, H., L. Bacher, M. Lacerda, S. Sartori 2006b. Brazilian Fruits and Cultivated Exotics (for Consuming in Natura). Instituto Plantarum de Estudos da Flora Ltda, Odessa, São Paulo

Mendoza, Abel L. Monteagudo; Guerrero, Marleni Huamán. 2010. Catálogo de los arboles y afines de la selva central del Perú. Arnaldoa 17(2):202–242

Morton, J. F. 1987. Fruits of Warm Climates. Creative Resource Systems, Winterville, North Carolina

Neville, Melvin K.; Glander, Kenneth E.; Braza, Francisco; Rylands, Anthony B. 1988. The howling monkeys, genus *Alouatta*. In: Ecology and Behavior of Neotropical Primates, Mittermeier, Russell A.; Rylands, Anthony B.; Coimbra-Filho, Adelmar F.; Fonseca, Gustavo A.B. da (eds), World Wildlife Fund, Washington, DC, vol. 2, pp 349–453

Oliveira, Patricia Chaves de; Cavalcante, Suellen. 2017. Ethnobotany in the Amazon floodplain ecosystem: a case study, Quilombo Saracura, Pará, Brazil. International Journal of Botany Studies 2(1):89–99

Perigolo, Natália Alves. 2014. Caracterização dos tipos vegetacionais do médio Rio Madeira, Rondônia. Master's thesis, Departamento de Botânica, Instituto de Ciências Biológicas, Universidade de Brasília

Prance, G. T. 1989. Economic prospects from tropical rainforest ethnobotany. In: Fragile Lands of Latin America: Strategies for Sustainable Development, Browder, John O. (Editor), pp. 61–74

Rudas, Augustín; Prieto, Adriana. 1998. Análisis florístico del del parque nacional natural amacayacu e isla mocagua, amazonas (colombia). Caldasia, pp.142–172.

Reátegui, Katia Wendy. 2013. Hábitos alimenticios de *Saguinus mystax* Spix "pichico barba blanca" en Padre Isla, Loreto, Peru. Master's thesis, UNAP, Iquitos

Thomas, Evert; Van Damme, Patrick. 2010. Plant use and management in homegardens and swiddens: evidence from the Bolivian Amazon. Agroforestry Systems 80:131–152

Thomas, Donald W.; Cloutier, Danielle; Provencher, Marc; Houle, Christian. 1988. The shape of bird- and bat-generated seed shadows around a tropical fruit tree. Biotropica 20(4):347–348

Van den Eynden, Veerle; Cueva, Eduardo; Cabrera, Omar. 1999. Plantas Silvestres Comestibles del Sur del Ecuador. Ediciones Abya-Yala, Quito

Vásquez, R. and A. H. Gentry. 1989. Use and misuse of forest-harvested fruits in the Iquitos area. *Conservation Biology* 3(4): 350–361.

Wehncke, Elisabet V.; Valdez, Catherine Numa; Domínguez, César A. 2004. Seed dispersal and defecation patterns of *Cebus capucinus* and *Alouatta palliata*: consequences for seed dispersal effectiveness. Journal of Tropical Ecology 20:535–543

Chapter 38
Myrtaceae

38.1 *Campomanesia lineatifolia*

Bolivia Sacha vilunto, pichana

Brazil Guabariba, guabiraba, guabirabeira, gabiroba, araçá lima, araçá do lago, abiu de macu

Colombia Guayaba de leche, guayaba de mono, palillo, champa, champo, palillo

Ecuador Guayabillo, hueso, palillo; masíka ma'nya, arari ma'nya (Siona-Secoya)

Peru Palillo, palillo caspi, reme; shajima (Aguaruna), reme (Kichwa)

Status Wild, spontaneous in cultural settings, planted

N. Smith, *Amazon Fruits: An Ethnobotanical Journey*, Ethnobiology,
https://doi.org/10.1007/978-3-031-12803-5_38

Palillo *Campomanesia lineatifolia* fruits harvested in a home garden. Veinte de Enero, Rio Yanayacu, Reserva Nacional Pacaya-Samiria, Loreto, Peru 1-16-06

Campomanesia lineatifolia is native to western Amazonia, from northern Bolivia to Colombia, including the western edge of Amazonas state in Brazil (Cavalcante 2010: 125; Fuentes 2018; Lorenzi et al. 2006a: 183; Villachica et al. 1996: 211). It is also found in the Upper Rio Negro watershed in Venezuela where it was likely introduced. The medium-sized tree can reach up to 25 m and is found up to 1500 m in the Andes (Brack 1999: 96). The tree grows in forests and disturbed sites on uplands and higher parts of the floodplains as well as along the margins of lakes (Vásquez 1997: 546). Piranha preta *Serrasalmus rhombeus*, the largest of some twenty species of piranha, eat the fruits when they fall into the water and likely disperse the seeds (Goulding 1980: 161). Although piranha preta is predominately a carnivore it does eat wild fruits during the annual flood stage. People catch and eat piranha preta so *Campomanesia lineatifolia* contributes indirectly to the human diet in Amazonia. In the Peruvian Amazon, the moustached tamarin *Saguinus mystax* also eats this fruit and disperses some of the seeds (Reátegui 2013: 37); although this monkey is not normally hunted for its meat by people because of its small size it is kept as a pet in some rural homes. Domesticated in the Amazon in pre-contact times, *Campomanesia lineatifolia* diffused as a crop to other parts of Peru, including the arid coast, such as Pachacamac and the Nazca Valley, where the tree has been grown under irrigation for some 2600 years (Cohen 1977; Towle 1961: 72).

Known as palillo in Peru, the tree is planted in home gardens, such as in Moyobamba in the Andean foothills (Works 1990) and along the Marañón, Ucayali, and Amazon rivers (Lamont et al. 1999). The yellow fruits are ready for eating in the latter part of the dry season and the first few months of the rainy period. The flattened fruits, about 6 cm in diameter, have a cream-colored pulp with a slight granular texture, similar to guava *Psidium guajava*, which is widely appreciated by adults and children alike. One of the common names for this fruit in Ecuador, guayabillo, reflects the similarities between the fruits of guava and *Campomanesia lineatifolia*. Each fruit, which can attain 7 cm in diameter and weigh 140 g, contains about ten flat, round seeds. The light green to tan-colored seeds are spat out after ingesting the pulp, hence the origins of some of the spontaneous trees in home gardens. The fruits are used for making jams and ice cream and the wood is also used for making fires in parts of the Peruvian Amazon (Flores 1997: 212; Ugent and Ochoa 2006: 201).

Called perfume guava in English on account of the pleasing aroma emanating from the fruits, the appealing smell was the focus of cologne industry in Peru in the nineteenth century. In 1888, Herbert Guillaume, the consul-general for Peru in Southampton, England, described palillo-scented water as "exceedingly agreeable." Alas, the bottled mixture, called *mistura*, is no longer manufactured. The Siona-Secoya in the Ecuadorian Amazon crush the leaves to extract a perfume (Vickers and Plowman 1984: 24). The wood is also useful in Peru (Jaramillo-Arango 1952: 325) and in Ecuador, the limbs are fashioned into fence posts.

The Peruvian environmentalist Antonio Brack Egg suggests that palillo has potential for the juice industry. Although virtually unknown in eastern Amazonia, the late Brazilian botanist Paulo Cavalcante also called attention to the potential of this sweet fruit for commercial juices and ice cream (Cavalcante 2010: 125). In addition to their appealing flavor, the fruits are rich in vitamin C.

Selva, 3 years, eating a palillo *Campomanesia lineatifolia* fruit gathered in her home garden. Nueva Esperanza, Rio Marañón, Reserva Nacional Pacaya-Samiria, Loreto, Peru 11-8-03

38.2 *Eugenia discreta*

Peru Lanza huayo, lanza caspi, lancetilla, guayabilla

Status Wild

Lanza huayo *Eugenia discreta* **fruiting in a forest. 800 m north of Miraflores, Rio Tigre, Loreto, Peru 7-1-06**

An understory tree reaching some 18 m high, *Eugenia discreta* produces abundant fruits during the rainy season. The yellow to orange fruits are easy to spot and provide a welcome snack for those walking in forests, such as along the Yanayacu and Tigre, affluents of the lower Marañón. The bright color of the fruit undoubtedly evolved to attract dispersal agents, such as monkeys.

Willy, 12 years, gathering Lanza huayo *Eugenia discreta* fruits in the forest near his village. 800 m north of Miraflores, Rio Tigre, Loreto, Peru 7-1-06

The slightly flattened fruits, measuring some 2 cm long by 2.2–2.5 cm wide, are attached singly or in clusters (fascicles) of two to three along the slender branches. *Eugenia discreta* produces fruit by the time it reaches 4 m high and the fruits are either gathered from the ground or the tree is cut down. If youngsters are present, they usually climb the tree and strip the fruits from the branches, thus saving the tree. People use their thumbs to pry open the thick spongy rind to reach the several seeds that are covered with a sweet white pulp. People have to be on the watch for black ants that sometimes protect the tree and bite those plucking the fruits such as in the vicinity of Veinte de Enero along the Rio Yanayacu, an affluent of the lower Marañón.

Known as lancetilla along the Yanayacu and lanza huayo along the lower Tigre, this tree is encountered on floodplains, such as old riverbanks (*restingas*) and along the margins of *Mauritia flexuosa* palm swamps known as *aguajales* in Peru. The fruits of this small tree are eaten by at least one species of fish eaten by people, sabalo cola roja *Brycon erypthropterum* in the Peruvian Amazon (Murrieta and

Arévalo 2004: 52). A denizen of mature and successionary forests below 300 m, *Eugenia discreta* is confined to western Amazonia in Peru and Bolivia.

38.3 *Eugenia inundata*

Brazil Ipixuna, araçá do igapó, araçá da várzea, goiaba araçá, goiabarana, goiabarana da várzea, goiabinha, arati, murta

Peru Sacha cereso, seresa, guayabilla, juanache, fanache, mishquina

Status Wild, spontaneous in cultural settings

Josécle "Teto" 11 years gathering Ipixuna *Eugenia inundata* fruits in successionary vegetation on uplands. Outskirts of Carrazedo, mouth of the Xingu River, Município de Gurupá, Pará, Brazil 7-23-16

Ipixuna *Eugenia inundata* is a bush that can reach 8 m in successionary vegetation (*capoeira*) and forests on floodplains and uplands throughout the Amazon. This species occurs more frequently on floodplains of white, clear, and black water rivers and streams, as indicated by its species name (Cunha 2009: 78; Cunha and Ferreira 2012; Fernandes-Corrêa and Furch 1992; Ferreira et al. 2013a; Irmler 1977; Kubitzki 1989; Leite 2017: 111; Luize et al. 2015; Machado 2018: 48, 123; Marimon and Lima 2001; Parolin 2009; Parolin et al. 2002; Piedade et al. 2005; Prance 1978, 1979, 1980; Scudeller and Souza 2009; Wittmann et al. 2010: 188; Worbes 1997). Also known as araçá do igapó in Brazil, this bush can survive protracted periods of flooding, often as long as 270 days (Junk 1989; Oliveira et al. 1991).

Locals, especially children, eat the small red fruits measuring 1 cm × 0.7 cm which contain a single tan seed measuring 0.9 cm by 0.6 cm. Although the juicy mesocarp is parsimonious, the fruits are available during the dry season when far fewer of the regional fruits are available.

This fruit also contributes indirectly to the human diet in the Amazon since it is one of nearly two dozen fruits and nuts eaten by tambaqui *Colossoma macropomum*, known as gamitana in Peru, one of the region's most prized fish (Araujó-Lima and Goulding 1997: 61; Baca 2005). Three species of turtle (*Podocnemis erythrocephala, P. expansa, and P. unifilis*) also eat this fruit and likely disperse some of the seeds (Cunha 2013: 66); these turtles are hunted for their meat. The increasingly endangered wattled curassow *Crax globulosa*, which is also hunted for its meat, eats the fruits of this bush and probably disperses some of the seeds (Leite 2017: 111).

38.4 *Eugenia lambertiana*

Brazil Murta, murta do campo, murtinha, murtinha de igapó, mortinha, araçá boi, goiaba, goiabinha, ginja, vacuna do campo

Peru Quinilla, sacha caimitillo, yacuruna, guayabilla, laja caspi, arrayán; mujaya sajimat (Aguaruna)

Status Wild

Quinilla *Eugenia lambertiana* in fruit in a floodplain forest along a black water river. Near Yarina, Rio Yanayacu, Reserva Nacional Pacaya-Samiria, Loreto, Peru 4-22-06

Eugenia lambertiana is a shrub to small tree that produces edible, yellow fruits that turn black when fully ripe. The round fruits, 1–1.5 cm in diameter, contain a sweet pulp that envelops a single round seed, typically 0.8 cm in diameter. The light tan seed is spat out after ingesting the pulp. Fruits of *Eugenia lambertiana* are eaten locally and are not taken to markets. The tree, typically under 15 m high when fully grown, is widespread in the lowlands of northern South America as well as the Lesser Antilles (Maguire 1969; Rayol et al. 2011). In the Andean foothills of the Peruvian Amazon, it occurs up to 1000 m (Mendoza and Guerrero 2010). It grows in mature and successionary forests on both floodplains and uplands but is more common along water courses. It is also found on some sand dunes along the coastal zone of northeastern Pará (Santos and Rosário 1988; Santos 2010: 43) and in scrub savanna (*campinarana*) in parts of the Upper Xingu watershed (Zappi et al. 2016).

River dweller and his son picking quinilla *Eugenia lambertiana* fruits in a floodplain forest. Near Yarina, Rio Yanayacu, Reserva Nacional Pacaya-Samiria, Loreto, Peru 4-22-06

The fruits are typically gathered from canoes when rivers and streams overflow their banks in the rainy season. The tree produces fruit by the time it reaches 2 m high, so some fruits are under water at the height of the flood. Several fish important for subsistence and commerce reportedly consume the fruits of *Eugenia lambertiana* including pacu (*Mylossoma*), sabalo (*Brycon*), gamitana (*Colossoma macropomum*), lisa (*Leporinus*), and shuyo (*Erythrinus*). This flood-tolerant tree thus also contributes indirectly to the regional diet and income. A species of the squirrel monkey, *Saimiri vanzolinii,* also eats this fruit in the floodplain forest of the Solimões in the vicinity of its confluence with the Japurá (Paim et al. 2017) but squirrel monkeys are small and not usually hunted for their meat.

38.5 *Eugenia multirimosa*

Ecuador Meyegomoe (Huaorani)

Peru Lanza caspi, guayabilla

Status Wild

Lanza caspi *Eugenia multirimosa* tree 5 m tall in fruit in floodplain forest. Rio Nahuapa, affluent of the lower Tigre, 3.8 km northwest of Miraflores, Loreto, Peru 7-2-06

Eugenia multirimosa is one of half a dozen or so species of *Eugenia* with edible fruits in Amazonia. This shrub or understory tree, which is found as high as 1800 m in Amazon headwaters, is also encountered along the humid Pacific side of Colombia and Ecuador. Known as lanza caspi along the lower Tigre, an affluent of the lower Marañón, the spindly tree thrives in forests on higher parts of the floodplain as well as on uplands.

The bright orange fruits are born in clusters (fascicles) of four to six along the branches. The tree grows as high as 25 m but fruits by the time it is 2 m high, so they can be reached from the ground or after a quick climb. The elongated fruits, some 4 cm long by 2 cm wide, contain a single seed. The relatively large seed fills most of the fruit cavity so people simply suck on the seeds to remove the thin layer of tan-colored pulp. The mesocarp is too thin to eat, so it is simply savored "*se chupa, no mas*." Although the fruits of lanza caspi are not very filling, the taste is sweet and intriguing. *Eugenia multirimosa* is typical of many wild fruits that produce just enough pulp to entice dispersal agents without wasting energy and nutrients. Only under domestication do humans select varieties that produce generous amounts of mesocarp.

Roy, 9 years, with lanza caspi *Eugenia multirimosa* fruits he gathered in floodplain forest. Rio Nahuapa, affluent of the lower Tigre, 3.8 km northwest of Miraflores, Loreto, Peru 7-2-06

38.6 *Eugenia punicifolia*

Bolivia Fichal blanco

Brazil Morta, mortinha, mortinga, murta, murtinha, raposinha, frutinha, pedra ume-caá, pedra úmida, goiabarana, pimenta de raposa, cereja do cerrado, cereja da savana, cereja da praia, mantimento de araponga, pitomba, pitanga do campo, suim vermelho, suim amarelo; suêm (Tukano)

Peru Guayabilla, araza

Venezuela Guayabita

Status Wild

Mortinha *Eugenia punicifolia* bush in successionary vegetation. Near Jocojó, Rio Jocojó, affluent of the lower Amazon, Município de Gurupá, Pará, Brazil 9-21-14

Eugenia punicifolia is a shrub reaching up to 8 m tall and widely distributed in the Amazon Basin south to Paraguay and north into the Guianas and the Orinoco Basin. It occurs up to 1700 m in the Andean foothills (Maguire 1969). It has two qualities that have earned it a place in the pantheon of Amazonian fruits: it bears during the dry season, when options for gathering wild fruits are limited, and it thrives in disturbed upland sites with degraded soils. This bush to small tree fruits year-round, such as in the Serra dos Carajás in Pará, one of the few native plants of Amazonia with edible plants to do so (Silva et al. 1996; Trindade et al. 2018). It also occurs on and near sand dunes along the northeast coast of Pará (Amaral et al. 2008; Bastos et al. 1995; Costa Neto et al. 1996; Jardim et al. 2016; Santos and Bastos 2014: 178; Santos and Rosário 1988), patches of sand inland with campina or campinarana vegetation (Ferreira 2009: 90, 92, 93, 96; Ferreira et al. 2013b; de Mendonça et al. 2017), scrub savannas (Amaral et al. 2019; Barbosa et al. 2005; Conceição and Aragão 2010; Lima et al. 2009; Magnusson et al. 2008; Maguire 1969; Medeiros and Walter 2012; Miranda and Absy 2000; Sanaiotti and Magnusson 1995; Santos et al. 2015), and open vegetation and stunted woodlands growing on iron-rich sedimentary rock (*canga*) on the Serra dos Carajás in Pará (Mota et al. 2018), so this bush is clearly adapted to soils low in nutrients. Mortinga is a colonizer of abandoned fields, especially on sites that have been farmed and left to fallow many times with relatively short intervals between slashing and burning (Baar et al. 2004). Soils in such situations, such as near villages, are thus drained of an already meager stock of nutrients. With so many areas of the Amazon degraded by cattle ranching or ever-shorter cycles of swidden farming, mortinga deserves a look as a future crop to help rehabilitate degraded landscapes.

Eilisangela, 9 years, gathering mortinha *Eugenia punicifolia* fruit along a path in successionary vegetation near her village. Jocojó, Rio Jocojó, affluent of the lower Amazon, Município de Gurupá, Pará, Brazil 9-21-14

Mortinha also grows in stunted woodlands on sandy soils, called *campina* and *campinarana* in Brazil, such as in parts of Roraima (Mendonça 2011). The bush is found on mounds built by extinct indigenous cultures on seasonally flooded grasslands on Marajó Island at the mouth of the Amazon (Lisboa 2012: 413) as well as other parts of grasslands on that island (Bastos 1984). It grows in scrub savanna, called cerrado in Brazil, found mainly in the southern part of the Amazon Basin where rainfall decreases (Gomes et al. 2011; Maracahipes et al. 2011). This shrub also occurs occasionally on floodplains, such as along the Amazon near its confluence with the Madeira River (Luize et al. 2015), in the Amazon estuary (Amaral et al. 2009), and in estuarine vegetation in adjacent Maranhão (Freire and Monteiro 1993). This bush is also found in some home gardens in the São Gonçalo Beira Rio district of Cuiabá, Mato Grosso (Mamede et al. 2015).

Mortinga bushes are only 2–3 m tall when mature, so gathering the red fruits is relatively easy. Although the round fruits are quite small, some 1.8–2.3 cm in diameter, the compact shrub bears them in profusion, so it is well worth tarrying on the way to agricultural plots, collect wood, hunt, or gather forest products. Children, such as from the village of Santana do Ituqui on the banks of the Amazon a little downstream from Santarém, scout out mortinga bushes to feast on the round fruits. Mortinga fruits taste similar to guava *Psidium guajava* and are usually eaten on the spot. The tiny seeds are ingested along with the pulp, thus people and other animals, including birds, disperse them (Foster et al. 1986). In the vicinity of Manacapuru along the lower Solimões near its confluence with the Rio Negro, people squeeze juice from the leaves and apply it to wounds to promote healing and also for treating uterine problems (Van den Berg 2010: 220; Vásquez et al. 2014: 18; Vásquez et al. 2014).

38.7 *Eugenia stipitata*

Bolivia Arazá

Brazil Araçá boi, araçá olho de boi, araçá, goiaba araçá, goiaba bahia, goiaba selvagem, goiaba brasileira; aratzé, wayawa panali (Baniwa), passakáni (Desana)

Colombia Arazá, palillo ácido

Ecuador Arazá, guayabillo, guayabo de monte, membrillo de montaña, membrillo silvestre, membrillo de oriente; kamush (Shuar), arari (Siona-Secoya)

Peru Arazá, arazá buey, guaya brasilera, guayaba brasilera, guayaba brasileña, guayabilla; akúku (Achuar), aracha (Ticuna)

Status Wild, planted

Arazá ***Eugenia stipitata*** **tree in fruit in a home garden. Lago Avispa near Requena, Rio Ucayali, Loreto, Peru 5-11-03**

Eugenia stipitata is a bush to a small tree reaching some 20 m high that produces orange or grapefruit-sized fruits which are slightly pubescent. The pulp is too tart to eat fresh; rather it is usually mixed with water and sugar and drunk. The Achuar in the Peruvian Amazon eat the fruit, but only after adding salt; they believe it helps relieve colds. The Achuar are on to something because the fruits contain twice as much vitamin C as oranges (Giacometti and Lleras 1994). Extracts from this fruit show that it could be used as a preventive agent against cancer (Neri-Numa et al. 2013). The fruits also have high antioxidant properties (Cuellar et al. 2013; Garzón et al. 2012; Vargas et al. 2005).

Eugenia stipitata has evidently been cultivated for some time since there is a generous amount of pulp surrounding the light tan seeds, the size of orange pips. Wild fruits generally weigh around 30 g whereas cultivated forms can reach 12 cm in diameter and weigh 740 g (Giacometti and Lleras 1994). When under cultivation *Eugenia stipitata* can produce 1000 fruits a year and each fruit is about three-quarters pulp (Falcão et al. 2000).

Araçá boi *Eugenia stipitata* fruit picked in a home garden on a small farm. Near Igarapé Açu, Pará, Brazil 8-14-98

The slightly flattened fruits are produced in the dry and wet seasons and are occasionally taken to markets, such as in Iquitos, Requena, Tingo Maria, and Yurimaguas in the Peruvian Amazon where they are sold to make juice and ice cream. This fruit sometimes turns up in several street markets in the Brazilian Amazon as well, as in Macapá, the capital of Macapá (Adaime et al. 2014). In Manaus, Oram Chocolates makes individually wrapped boutique chocolates with centers flavored with various fruits including araçá-boi. In Iquitos, Persa makes a jam with the fruit called mermelada de arazá. At least eight species of stingless bees (*Melipona fulva*, *M. interrupta*, *M. lateralis*, *M. pseudocentris*, *M. seminigra*, *Ptilotrigona lurida*, *Trigona fulviventris,* and *T. williana*) pollinate the flowers of this small tree year-round in the central Amazon (Falcão et al. 1988; Ferreira and Absy 2015, 2017; Marques-Souza et al. 1996). Rural people gather the honey of several species of stingless bees in the wild and in some cases hang hollowed out sections of logs from the eaves of their homes to serve as nesting sites for these bees so *Eugenia stipitata* contributes indirectly to the human diet via honey. The wood of this tree is also used for fuel to cook meals and boil water, such as along the Amazon River in Peru (Balbin and Samaniego 1996).

This understory treelet is occasionally found in forests, such as in the vicinity of Miraflores at the confluence of the Tigre and Marañón rivers. The moustached tamarin *Saguinus mystax* feeds on this fruit in floodplain forests along the Amazon River in Peru and disperses some of the seeds (Reátegui 2013: 37). Wild forms of *Eugenia stipitata* are designated as subspecies *stipitata* whereas cultivated forms are assigned to the *sororia* subspecies (Clement et al. 1997). *Eugenia stipitata* was domesticated in the western part of the Amazon and thence was spread by humans to other parts of the basin (Flores 1997: 27; Lorenzi et al. 2006b: 212). It is particularly common in the Peruvian Amazon (Villachica et al. 1996: 27).

Eugenia stipitata is planted in home gardens in many parts of the Brazilian and Peruvian Amazon. Because of market demand for the yellow-skinned fruit, small-scale farmers sometimes plant small orchards of the tree, such as in the vicinity of Extrema, a small town near the border between Acre and Rondônia in the south-western part of the Brazilian Amazon. In parts of Rondônia, some farmers have established groves of mixed fruit trees including araçá-boi, presumably to reduce disease and pest pressure. Araçá-boi is planted on uplands and higher parts of flood-plains that are only submerged for brief periods each year.

Araçá-boi ***Eugenia stipitata*** **fruits picked in a home garden. Vila Valente, lower Tefé River, Amazonas, Brazil 8-14-12**

38.8 *Myrcia bracteata*

Brazil Mortinha, murta, murta peluda, murta cabeluda, murtinha cabeluda, goiabinha, vaúna; mãkêra (Tiriyó)

Colombia Arrayán; daira (Andoke)

Ecuador Buen ruhue (Waorani)

Peru Rupiña, rupiña caspi, millua mullaca, napurshi

Venezuela Curame, arecuna, cafetillo; chipokuiyek (Pemón), kumasi (Ye'kuana)

Status Wild

Leandro, 11 years, with mortinha *Myrcia bracteata* fruits he gathered in successionary vegetation on the outskirts of a village. Anjo da Guarda, Rio Pracupi, affluent of the Anapu River which flows into the Baia de Botos along the Anapu River, FLONA Caxiuanã, Pará, Brazil 5-15-15

Myrcia bracteata is a bush to small tree in forest and successionary vegetation throughout the Amazon Basin, the Guianas, and the Orinoco watershed (Araújo et al. 2005; Brienza Júnior 1999: 26; Carvalho et al. 1986; Cavalcante 1973: 60; Coelho et al. 2003a, b, 2013; Godoy et al. 1999; Gualberto et al. 2014; Medeiros 2016: 93; Prata et al. 2010; Rayol et al. 2011; Rodrigues et al. 2007b; Schwartz 2007; Taurino 2017: 93; Vieira 1996: 134). It thrives in open forest with lianas and palms in the eastern fringes of the Amazonia in Maranhão (Gama et al. 2007), open vegetation and stunted woodlands growing on iron-rich sedimentary rock (*canga*) on the Serra dos Carajás in Pará (Mota et al. 2018; Trindade et al. 2018) and is particularly common in 15- to 40-year-old successionary vegetation in the municipality of Bragança in eastern Pará (Carim et al. 2007; Muniz et al. 2007; Oliveira and

Jardim 1998) and in 8-year-old *capoeira* in the municipality of Castanhal some 60 km east of Belém (Coelho et al. 2003b). This plant also invades some man-made cattle pastures in the Amazon, such as in eastern and northeastern Pará (Dutra et al. 2004; Hohnwald et al. 2010, 2015). It is also found along the sunlit margins of savannas and rivers (Maguire 1969; Sanaiotti and Magnusson 1995), the cerrado vegetation of central Brazil (Batalha and Martins 2004), as well as on sand dunes along the northeast coast of Pará (Santos and Bastos 2014: 179). It can survive in regrowth forest some 30–40 years old (Alvino et al. 2005; Rayol et al. 2006), but probably does not survive in fully mature forests because of the deep shade. Usually under 8 m tall, it has a wide distribution, but the bush is not particularly common. In the Andean foothills it is found as high as 2000 m.

The dark purple fruits, 1.4–1.5 cm × 1 cm, contain a single tan-colored round seed 1 cm in diameter. Children for the most part gather the fruits because they are relatively small. This short bush is encountered along paths in the vicinity of villages and cultivated fields and the fruits are eaten locally, usually on the spot. It grows on uplands and higher parts of floodplains of diverse water types from black to clear and muddy rivers. It is also found in stunted vegetation (campina) on sandy soils, such as in the Bragantina Zone in northeastern Pará, on savanna (campo, cerrado) such as on Marajó Island and the Upper Tapajós, and in forest light gaps. Squirrel monkeys *Saimiri sciureus* have been observed feeding on this fruit in the vicinity of the Tucuruí reservoir along the Tocantins River in eastern Pará and may disperse some of the seeds (Pinheiro et al. 2013).

In addition to the edible fruit, this small tree has other uses. Residents of Marudá, a village along the coast of Pará at the mouth of the Amazon, use the leaves to prepare a bathing solution to cleanse the uterus after giving birth (Coelho-Ferreira 2009). The Tiriyó who live in the vicinity of the border between northern Pará and Surinam prepare a bathing solution to treat kidney issues by crushing leaves and scraping the branches of this tree into water which is then boiled; after it cools the bathing solution is used to treat kidney issues (Cavalcante 1973: 60). Residents of Tiningú, a quilombo community 47 km southeast of Santarém, use the leaves and flowers of this plant to make a tea which is drunk to alleviate inflammation (Carvalho 2015: 66). The Andoke in the Colombian Amazon apply the leaves of this plant to the body to avoid madness and wave off evil spells (La Rotta 1983: 64). The Waorani in the Ecuadorian Amazon use the wood of this small tree in house construction and for fuel (Cerón and Ayala 1998: 132).

38.9 *Myrcia fallax*

Bolivia Pichana

Brazil Morta, murta grande, murta parida, murta branca, murtinha, murteira, cumatí, cumatê, frutinheira, fruta do mato, fruta de passarinho, araçá, goiabinha, guamirim; pi dju djure (Kayapó), warea xihi (Yanomami)

Colombia Arrayán

Ecuador Saco, saca, saca saca, saca blanca, saca colorada, arrayán

Peru Lanche chico, pitariki, puca caspi, guayabilla, huayatillo hoja grande, ahuashi yacu, yapi, rupiña, luichimapiche, capiron de altura; sáka (Aguaruna)

Venezuela Curame; mahcho'kai'mën (Panaré)

Status Wild, spontaneous in cultural settings, planted

Morta *Myrcia fallax* in successionary forest. Distrito Tarumã, Manaus, Amazonas, Brazil, 4-1-15

Morta *Myrcia fallax* is a small tree that can reach 25 m high in successionary vegetation and forests on uplands, although it is also found occasionally on floodplains of black, clear, and sediment-rich rivers such as along the Tefé and the Solimões rivers in Amazonas, Brazil, the clear-water Jari and Araguaia rivers in eastern Amazonia, and the Jaci-Paraná River, an affluent of the Upper Madeira in Rondônia (Assis and Wittmann 2011; Assis et al. 2017; Carim 2016: 93; Kurzatkowski et al. 2015; Miranda 2000; Reis 2010: 36; Santana et al. 2004) and in the watershed of the clear water Upper Araguaia River in the transition zone between scrub savanna and the Amazon rain forest (Barbosa et al. 2011; Brito et al. 2006; Martins et al. 2008). It also occurs occasionally in tidal forests in the Amazon estuary (Braga and Jardim 2019; Braga et al. 2015) including sand dunes not subject to flooding (Jardim et al. 2016). It is also found in stunted woods (*campina*, *campinarana*) on sandy soils, grasslands, and scrub savanna (Costa et al. 2014; Ferreira 2009: 90; Ferreira et al. 2013b; Magnusson et al. 2008; Ratter et al. 1996) and in the sand dune zone of coastal Pará and Amapá (Amaral et al. 2008; Bastos 1988; Santos 2010: 43). It ranges widely in tropical America from eastern Mexico, the Antilles, to the Guianas, Amazon and Orinoco basins, and southern Brazil (Bánki et al. 2003: 38; Maguire 1969; Sarthou et al. 2010). This small tree is found in stunted vegetation on some granitic inselbergs surrounded by forests in French Guiana (Sarthou et al. 2003). In the Andes, it can be found in cloud forests as high as 2500 m (Arellano 2013: 190; Mendoza and Guerrero 2010).

Morta sprouts spontaneously in some home gardens, such as on the outskirts of Manaus. The Kayapó who live in scattered settlements in the Upper Xingu plant this species for medicinal uses and because the fruits attract game animals (Posey 2002: 243). One of the most prized fish in the Amazon, tambaqui *Colossoma macropomum*, also eats the fruits (Araujó-Lima and Goulding 1997: 61). Some rural people along the Ucayali River in the Peruvian Amazon use the wood of this small tree to make furniture (Tananta 2014: 68).

The small dark purple fruits, the size of red currants, are eaten but do not appear in markets. The fruits contain numerous tiny black seeds which are ingested along with the pulp. In the vicinity of Manaus, this fruit is one of the more popular foods for the pied bare-faced tamarin *Saguinus bicolor bicolor* which likely disperses the seeds (Egler 1992). These fruits are likely dispersed by other animals, including some fruit-eating birds. The small size of the fruit is reflected in two of the common names for it in the Peruvian Amazon: lanche chico (small snack) and huayatillo hoja grande (big-leaved small fruit). In the Venezuelan Amazon, the tree's sap is painted onto the inside of gourds to make them more durable. In the Amazon Basin and contiguous forested areas of Maranhão, at least five species of stingless bees (*Melipona seminigra merrillae*, *M. fasciculata*, *M. fulva*, *Trigona fulviventris*, *Cephalotrigona femorata*) gather pollen from the flowers of *Myrcia fallax* (Martins et al. 2011; Oliveira et al. 2009) and rural people are fond of the honey produced by these wild, and in some cases, semi-domesticated bees.

38.10 *Myrcia guianensis*

Brazil Mortinha, mortinheira, murta, araçá; warda (Waimiri-Atroari)

Ecuador Saka (Shuar), buguehuemo (Waorani)

Peru Guayabilla, yacushapana blanca

Status Wild, spontaneous in cultural settings, planted

Mortinha *Myrcia guianensis* bush in fruit in upland successionary vegetation. Near Jocojó, Município de Gurupá, Pará, Brazil 9-21-14

Mortinha *Myrcia guianensis* is a bush to small tree reaching 18 m tall in successionary growth and forests on uplands and occasionally on floodplains. It occupies a broad ecological niche since it also occurs in seasonally flooded grasslands, stunted forests on sandy soils (campina, campinarana), and scrub savanna (cerrado), including the central region of Brazil (Batalha and Martins 2004; Marimon et al. 2012). The compact shrub starts bearing fruits when only 1 m high so they are often within easy reach. The edible fruits turn black when fully ripe. The round fruits are 0.5 cm in diameter and are eaten whole by rural people. In the village of Jocojó along a river of the same name that flows into the lower Amazon, a farmer obtained seeds from a mortinha bush near his manioc processing hut and planted them in his home garden in the village.

Lucio Fernandes, 55 years, by a mortinha *Myrcia guianensis* bush he planted from seed in his home garden. The mortinha is next to a juru *Chrysobalanus icaco* bush (on the right) which also produces edible fruit. Jocojó, Rio Jocojó, Município de Gurupá, Pará, Brazil 4-24-15

Myrcia guianensis occurs throughout the Amazon Basin, the northeast, central and southern Brazil, Paraguay, Bolivia, and the Guianas (Almeida Jr. et al. 2011; Araujo-Murakami et al. 2015; Maguire 1969; Sarthou et al. 2010). In the Andean foothills, it is found up to 2100 m. This bush to small tree is also found in parts of the Caribbean and Panama.

38.11 *Myrciaria dubia*

Bolivia Camu camu, gupurucillo

Brazil Camu camu, araçá da praia, araçá da várzea, araçá d'água, araçá de igapó araçá açu, sarão, sarabão, caçari, folha miuda, pitanga, crista de galo, jabuticaba, jabuticabeira preta; cici'i (Arawaté)

Colombia Guayabo, guayabito, camu camu

Ecuador Camu camu

Peru Camu camu, camocamo; moka bimi (Shipibo)

Venezuela Guayabito

Status Wild, planted

Camu camu *Myrciaria dubia* fruits from a market in Requena. Veinte de Enero, Rio Yanayacu, Loreto, Peru 4-20-06

Camu camu *Myrciaria dubia* is a bush or treelet of lake and river shores well adapted to prolonged flooding. Lustrous red fruits, the size of large marbles, are born at the ends of branches during the rainy season. The flowers are pollinated mainly by stingless bees in various genera including *Nannotrigona punctata*, *Scaptotrigona postica*, and *Trigona pallens* (Heard 1999; Maués 2002; Peters and Vasquez 1986). The shrub with numerous stems and branches can reach 15 m tall but camu camu starts producing fruit when only 1.5 m high so the fruits are often submerged during the annual flood and are eaten by fish. Other ripe fruits fall into the water where they are also eaten by fish and probably turtles which disperse the seeds. The fruits sink, so they are not dispersed by currents. In the wild, people typically gather the round fruits from canoes to make juice. *Myrciaria dubia* grows on the floodplains of clear, black, and sediment-rich rivers throughout the Amazon and Orinoco Basins and into the Guianas. In some floodplain areas, camu camu can form dense stands; along the shores of Supay lake along the Ucayali, for example, the shrub reaches densities of 27,800/ha (Gonzales 2006). Fruit production varies by year, but in 1999, camu cam fruit production in that lake reached 15,700 kg/ha.

Araçá açu *Myrciaria dubia* in fruit during the annual flood of a black water river. Rio Urubu near Pontão, Amazonas, Brazil 4-4-15

Fruit yields of wild stands range from 1.7 to 12 tons/ha/year (Riva and Gonzales 1997: 2). A single bush can yield 10 kg of fruit (Clement et al. 1997). Camu camu sometimes forms extensive stands, especially along the Putumayo, Napo, Nanay, Tigre, the lower Ucayali, and the Javari Rivers (IIAP 2001; Peters et al. 1989). In the central and eastern part of the Amazon, the largest concentrations of camu camu are found in the Xingu and Negro watersheds (Prance 1989).

Sarão *Myrciaria dubia* among granitic boulders on an island along a clear water river. The bush is in flower because it is the dry season and becomes mostly submerged during the height of the annual flood. Ilha do Mascarado, middle Xingu River upstream from Altamira, Pará, Brazil 9-1-15

Camu camu caught the attention of the juice and health food industries in the 1980s following publicity about the fruit's extraordinarily high levels of vitamin C. Camu camu fruits contain about 30–40 times as much vitamin C as oranges and 100 times as much vitamin C as lemons; they are thus used in the manufacture of some vitamin C pills (Costa et al. 2013; Peters and Hammond 1990; Prance 1989; Sanchez 1995). With an ascorbic acid content usually between 1800 and 3000 mg/100 g of fresh pulp, camu camu is the world's richest plant in terms of vitamin C, surpassing even rose hips (Aguiar and Souza 2018; Flores 1997; Genovese et al. 2008; Pinedo and de Jong 2004; Rodrigues et al. 2001; Rufino et al. 2010). In some areas, such as parts of Roraima, ascorbic acid levels of camu camu fruits reach as high as 6112 mg/100 g, the highest of any fruit in the world (Yuyama et al. 2002). The pulp of camu camu is also a good source of some minerals such as sodium, potassium, calcium, zinc, magnesium, and manganese (Akter et al. 2011). Agrícola San Juan in Pucallpa along the Ucayali converts camu camu fruits into powder for export to cosmetic companies in Japan and the United States. Founded in 1995, Agrícola San Juan is a subsidiary of the San Juan beer company. Camu camu pulp and juice is also exported from Peru to Japan and the European Union (Chirinos et al. 2010).

Fruit juice producers in Pucallpa and Iquitos purchase camu camu fruits gathered in the wild from various locations in the Peruvian Amazon, such as the lower Tigre River and El Supay Lake near Jenaro Herrera on the lower Ucayali. Camu camu fruits also appear in street markets in Iquitos for purchasers to make their own juice. In the Colombian Amazon, three factories were processing camu camu fruits in El Estrecho at the mouth of the Putumayo in the 1990s; those factories processed about 400 tons of the fruit annually.

Vendor bagging camu camu *Myrciaria dubia* fruits in a street market. Many of the fruits were picked while still green to prevent others getting them. Mercado Belén, Iquitos, Peru 1-7-06

Markets for the fruit in the Brazilian Amazon are smaller. Traditionally, the fruits have been consumed sporadically in rural areas, such as along the clear water Xingu where it is called sarão or sarabão. Longtime residents along the Maués River note that the fruits, known locally as araçá da praia (the beach araçá), are exceedingly sour even when ripe. Buyers appear occasionally along the Maués to take the fruits to Manaus, but the wild stands of camu camu along the sandy shores of this black water river are barely tapped. Because they are so tart, most camu camu fruits are used for making juice with plenty of sugar. The fruits are mashed and the two to four green, raisin-sized seeds found in each fruit are sieved out before mixing the slippery pulp with water.

In 1996, the federal government in Peru instigated a camu camu planting program to generate more fruits for juice exports to Japan. Camu camu seedlings start producing fruit 3 years after planting. Spurred by the government credit program for camu camu, some 4500 ha of the crop were planted in the Peruvian Amazon between 1997 and 1999. Some farmers opted to interplant their camu camu with other crops, such as maize, as in the Bella Vista community near Pucallpa on the Ucayali floodplain. About 500 ha of camu camu have been planted by a landowner along part of the margin of Lago Avispa on the outskirts of Requena at the confluence of the Ucayali and Tapiche Rivers. Camu camu is also planted in some home gardens along the Amazon River in Peru (Balbin and Samaniego 1996).

The phenology of cultivated camu camu is often different from that of wild stands, since the former sometimes fruit sporadically throughout the year (Penn 2006). In uplands near Pucallpa, camu camu plantations produce fruit from October, in the middle of the dry season, until May toward the end of the rainy period (Riva and Gonzales 1997: 10). Camu camu planted along the margins of Lago Avispa, as well as in the village of Miraflores at the confluence of the Tigre and Marañón rivers, fruits in the dry season whereas natural stands generally bear fruit at high water. Exceptionally severe dry seasons, such as in 2005, generally disrupt fruiting patterns of wild stands. In Cocha El Supay and associated lakes near Jenaro Herrera, for example, the extensive camu camu stands produced fruit in the dry season in 2005 instead of at the beginning of the rainy season. If economically viable plantations can be established on a sufficient scale, camu camu fruits would be available almost year-round, an advantage in keeping processing plants open.

In Iquitos, the Shambo Company makes camu camu-flavored popsicles and ice cream. Peruvian entrepreneurs are also starting to sell camu camu candy online. Camu camu is also used in a variety of folk remedies and alcoholic concoctions (*tragos*) in the Peruvian Amazon. Along the upper Yanayacu River, an affluent of the lower Marañón, for example, one ingenious farmer employs the bark of camu camu in a "medicinal" *trago* that contains a variety of barks and fruits steeped in aguardiente, known as cachaça in Brazil. Near Tingo Maria, another farmer sells camu camu liqueur from a roadside stand.

The camu camu mini-boom in the Peruvian Amazon has slowed somewhat in recent years due to a variety of factors. Japan no longer purchases significant quantities of camu camu juice because Alberto Fujimori, the ex-President of Peru who was responsible for encouraging entrepreneurs in Japan to import the fruit juice, left office in 2000. By 2003, only one factory in Iquitos was still processing camu camu fruits. Yields of planted camu camu have often been disappointing (Arkcoll and Clement 1989). In the Brazilian Amazon, customers have recently grown to like another vitamin C-rich tropical fruit, acerola *Malpighia punicifolia*, which was introduced from the Caribbean and is now widely planted in agroforestry fields and home gardens in tropical parts of Brazil.

A cargo and passenger boat named after camu camu. Distrito Belén, Rio Itaya, Iquitos, Peru 8-2-03

Nevertheless, camu camu will continue to be marketed in towns in the Amazon and cities in other regions. Around the year 2000, for example, camu camu fruits were being trucked for sale in supermarkets in Lima, such as Supermercado Metro in the San Miguel district, where the fruits were fetching US$3/kg in August 2003. Also, San Natura sells camu camu yogurt in Lima.

Camu camu is often cultivated alongside other crops in agroforestry formations, thus helping protect the land from soil erosion and increasing agrobiodiversity. In the vicinity of Veinte de Enero along the Yanayacu, an affluent of the lower Marañón, for example, a farmer has interplanted camu camu with plantain, banana, and papaya in his floodplain field. Furthermore, camu camu cultivation is compatible with efforts to sustain fisheries because the fruits are a favorite food of several fish important in commerce and subsistence, such as *Colossoma macropomum*, known as tambaqui in Brazil and gamitana in Peru (Baca 2005; Silva et al. 2003) and matrinxã *Brycon cephalus* (Pizango-Paima et al. 2001). Indeed, in some areas of the Peruvian Amazon, farmers place gillnets in their camu camu plantations at high water to catch fruit-eating fish (Penn 2006).

38.12 *Myrciaria floribunda*

Bolivia Guapurucillo, sahuinito, guayaba

Brazil Mulatinho, araçá, araçá péua, araçá do mato, murta, pele de velha, turcuetá, ginginha, ginja da mata, goiabinha, goiabinha folha fina, goiabinha vermelha, goiaba braba, goiaba araçá, goiaba de anta, goiaba da mata, goiabarana, camu camu, cambuí

Colombia Oratchá (Ticuna)

Ecuador Mulchi (Kichwa), ime ge way (Waorani)

Peru Camu camu del grande, camu camu gigante, camu camu árbol, camu camu del bajial, camu camu negro, camu camu de altura, camucamillo, usilla, guayabilla, shawinto; sáka (Aguaruna)

Venezuela Erekodek (Pemón)

Status Wild, planted

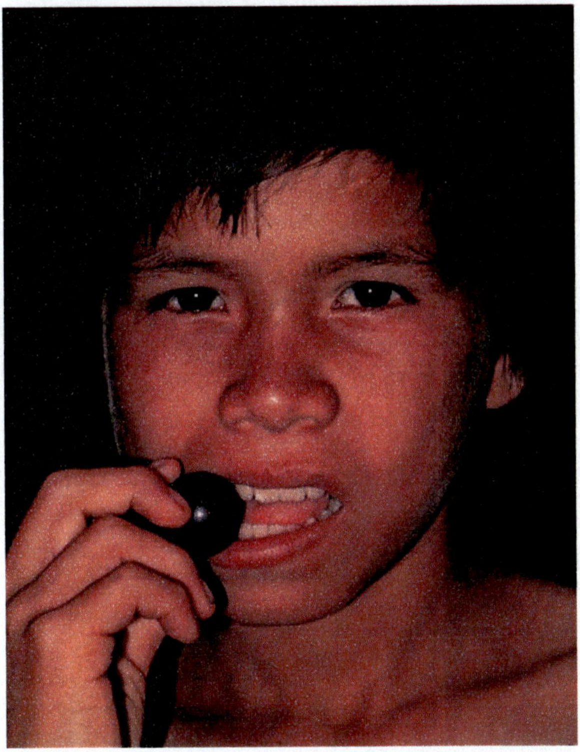

Julio, 12 years, eating *Myrciaria floribunda* fruit he gathered in floodplain forest. Cocha Huayllahua near Veinte de Enero, Rio Yanayacu, Reserva Nacional Pacaya-Samiria, Loreto, Peru 3-11-04

The fruits of *Myrciaria floribunda* are similar to those of camu camu *Myrciaria dubia*, round and dark red to purple, but slightly bigger. Furthermore, *Myrciaria floribunda* is often a tall tree rather than a bush. Indeed, *Myrciaria floribunda* can soar to 40 m (Brack Egg 2003: 137) and is also distinguished by a smooth trunk with flaking bark. *Myrciaria floribunda* is found in upland and floodplain forests as well as scrub savanna (*cerrado*) in central and South America and the Caribbean (Araujo-Murakami et al. 2009; Kimber 1988: 53; Kvist et al. 1995; Pereira et al. 2011; Thomas 2009) and occurs up to 1800 m in the Andean foothills. It is also found occasionally on canga plateaus with open vegetation in the Serra dos Carajás in Pará (Rayol 2006: 40). It is relatively common on rocky islands on the floodplain of the clear water Xingu River (Cunha 2009: 37; Cunha and Ferreira 2012), and is also found along the clear water Jari River in eastern Amazonia and the muddy Mazagão, an affluent of the Amazon in the estuary (Carim 2016: 93). It is also found on floodplains of some black water rivers (Myster 2018; Worbes 1997). It thrives in upland successionary forests, such as in the Bragantina zone in northeastern Pará, the Tocantins watershed in eastern Pará (Taurino 2017: 96), the Tapajós National Forest in western Pará (Espírito-Santo et al. 2005), and in seasonal forest in the transition zone between forest and cerrado in that state of Mato Grosso and Tocantins in Brazil (Araujo et al. 2009; Brienza Júnior 1999: 27; Carim et al. 2007; Coelho et al. 2003b; Haidar et al. 2013; Mews et al. 2012; Prata et al. 2010; Rayol et al. 2006; Rodrigues et al. 2007a). It also occurs as a weed in some cattle pastures, such as in northeastern Pará (Hohnwald et al. 2015).

In central and western Amazonia *Myrciaria floribunda* also grows along the margins of clear and black water rivers and lakes, such as along the Yanayacu River, an affluent of the lower Marañón (Vieira et al. 2017). This tree is occasionally interplanted with crops in floodplain fields, such as in the vicinity of Veinte de Enero along the lower Yanayacu in the Pacaya-Samiria National Reserve.

Myrciaria floribunda starts producing fruit when it is a small bush, but when it becomes a tree the fruits can only be harvested by climbing up the trunk and moving along the branches. In the case of trees growing on floodplains, little harm usually occurs in the event of a fall because the tree only fruits during the annual flood when the water is deep under the trees. Fewer fruits are usually gathered from upland trees because of the greater risks involved. The pale gray pulp surrounding the one to three green seeds is relished as a snack in spite of its tartness. The raisin-sized seeds are usually spat out.

This widespread tree has other uses. In the village of Yarina along the Upper Yanayacu River in the Peruvian Amazon, a resident makes a medicinal drink (*trago*) using aguardiente (sugarcane alcohol), the bark of *Myrciaria floribunda*, other wild plants, and honey. The mixture is left to soak for a month and is usually taken in the morning as a pick me up. The Chácobo in the Bolivian Amazon boil the bark and when the liquid has cooled drink it to treat vomiting, stomachache, and diarrhea. In the Ecuadorian Amazon, bark shavings are used to make an infusion to treat scorpion stings. In French Guiana, the Palikur use this wood for a variety of construction purposes (Ogeron et al. 2018).

Camu camu gigante *Myrciaria floribunda* tree with characteristic peeling bark in floodplain forest. Cocha Huayllahua near Veinte de Enero, Rio Yanayacu, Reserva Nacional Pacaya-Samiria, Loreto, Peru 1-14-06

38.13 *Psidium acutangulum*

Bolivia Guayabilla

Brazil Araçá, araçá pêra, araçá peu, araçá brava, araçá piranga, araçá do Pará, araçá goiaba, araçá boi, araçandeua, goiaba, goiaba araçá, goiaba do mato, goiaba azeda, goiaba de suco, goiaba vermelha, goiabarana, goiabinha; wayaawa (Ye'kuana)

Colombia Guayaba coronilla, arazá pera; nujoko (Cubeo)

Ecuador Guayaba de monte, guayaba del campo, guayaba agria; arari (Siona-Secoya)

Peru Guayaba del agua, sacha guayaba, guayaba silvestre, guayabilla, arazá pera; wapuk (Aguaruna), tanan wi'wi (Chayahuita), shawíntu (Kichwa), kírimp (Mayna), jenen bimpish (Shipibo)

Venezuela Guayaba morocotera, guayaba del monte, guayabo de agua, guayabita

Status Wild, spontaneous in cultural settings, planted

Manuel, 14 years, eating an unripe goiaba araçá *Psidium acutangulum* fruit in upland successionary vegetation near an archaeological site. A few hundred meters inland from Carrazedo, mouth of the Xingu River, Município de Gurupá, Pará, Brazil 8-18-14

Psidium acutangulum, a wild relative of the widely cultivated guava *Psidium guajava*, is a small tree in forests and regrowth vegetation in wetlands of the Amazon and Orinoco basins, such as in the southern part of the Colombian Amazon (Rudas and Prieto 1998), along the lower Purus River, a southern tributary of the Solimões (Machado 2018: 124) and along the Amazon/Solimões River in Brazil (Parolin 2002, 2009; Parolin et al. 2004; Wittmann and Parolin 1999; Wittmann et al. 2002). It is found in some floodplain forests of black water rivers, such as along the margins of Tupé lake near Manaus (Scudeller and Souza 2009). In seasonally flooded savannas of the Bolivian Amazon and the llanos of Venezuela, *Psidium acutangulum* grows in gallery forests (Killeen et al. 1993). It is also found in some upland forests in the Amazon, such as in municipality of Barcelos along the Rio Negro in Amazonas, Brazil (Silva et al. 2007) and along the lower Xingu (Salomão et al. 2007). It occurs as high as 1250 m in the Andean foothills of Peru (Mendoza and Guerrero 2010), and its range also extends along the Pacific coast of Ecuador.

This tree is normally under 8 m high but can reach as tall as 18 m. It grows mainly along lake margins and river banks, including along sediment-rich rivers (Worbes 1997). It bears fruit in the flood season so locals can canoe right up to the tree and pluck them off (Schöngart et al. 2002). In upland areas, the peak of fruiting in the Brazilian Amazon is from October to December in the dry season (Falcão et al. 1992). The fruits are yellow and weigh around 240 g when ripe. The fruits float so they are dispersed by currents as well as fish, turtles, and various mammals (Kubitzki and Ziburski 1994; Waldhoff et al. 1996). Some of the fish that eat the fruits are highly esteemed by riverine folk, such as tambaqui *Colossoma macropomum* and the bacu catfish *Lithodoras dorsalis* (Parolin et al. 2010). At least one species of stingless bee, *Trigona williana*, pollinates the flowers of this tree in the central Amazon and the honey of this wild bee is appreciated in rural areas (Marques-Souza et al. 1996).

Guayaba del agua *Psidium acutangulum* tree on the margin of a black water lake. A few kilometers upstream from Veinte de Enero, Rio Yanayacu, Reserva Nacional Pacaya-Samiria, Loreto, Peru, 1-14-06

Psidium acutangulum appears occasionally in upland successionary vegetation, such as adjacent to an archaeological site behind Carrazedo at the mouth of the Xingu. It is possible that people introduced the tree to this upland location centuries or thousands of years ago. It is also found in a regrowth forest near another archaeological site at Laranjal, a small settlement along a stream of the same name in the Caxiuanã National Forest in eastern Pará. *Psidium acutangulum* is considered native to western Amazonia (Lorenzi et al. 2006a: 235) so people have likely been moving this useful tree around for a long time.

It is planted occasionally in home gardens on floodplains, such as along the lower Solimões near its confluence with the Rio Negro (Lima and Saragoussi 2000) as well as some upland home gardens, such as at Vila Valente along the lower Tefé River in Amazonas, in the vicinity of Manaus and Manacapuru also in Amazonas, in Rio Branco the capital of Acre, and in the municipality of Abaetetuba along the lower Tocantins River in Pará (Gonçalves and Lucas 2017; Klüppel 2006: 24; Lobato et al. 2017; Machado 2016: 36; Major et al. 2005; Siviero et al. 2011). Near Bom Socorro, a community along Paraná Tambaqui adjacent to the floodplain of the lower Japurá, the tree is found in a small orchard around an abandoned manioc processing hut. On the margins of Lago Amanã along the lower Japurá River, cattle eat the fruits of *Psidium acutangulum* thereby dispersing them in pasture where some of them sprout. In the Ecuadorian Amazon, the Siona-Secoya plant this tree (Vickers and Plowman 1984).

The round fruits of *Psidium acutangulum* range in diameter from 3 to 8 cm and can weigh up to 250 g. Some fruits are as long as 14 cm (Ferreira 1982). The fruits contain 55–76% pulp with numerous seeds embedded in the mesocarp (Andrade et al. 1993). The seeds are swallowed along with the pulp, thereby facilitating dispersal of this tree. This tree evolved for dispersal by birds, wild mammals, and tortoise when fruits drop to the ground. The crisp fruits make a satisfying snack, and the numerous orange pip-sized seeds are generally ingested along with the pale-yellow pulp. The thin peel of *Psidium acutangulum*, which is eaten along with the pulp much like an apple, is high in fiber. The fruits are also an excellent source of vitamin C, containing 389 mg of vitamin C/100 g in the pulp (Andrade et al. 1993; Ramos et al. 2015). The fruits are also used to make juice, such as in the vicinity of Oriximiná in Pará, as well as frozen popsicles, as in the Tefé area.

The generous size, crunchy texture, and tantalizing taste of *Psidium acutangulum* fruits make it an appealing candidate for development as an orchard crop in wetlands. The intriguing flavor of *Psidium acutangulum*, which bears some resemblance to the strawberry guava *Psidium cattleianum*, has begun to catch the attention of agronomists and fruit farmers outside of the plant's native region, such as in Costa Rica, where *Psidium acutangulum* was introduced in the early 1980s. Under plantation conditions, fruit yields would likely be in the 10–44 tons/ha range (Clement and Silva 1994). Although normally found in or near water in the wild, *Psidium acutangulum* can be cultivated on uplands where it fruits in the dry season when given fertilizer in contrast to wild populations which bear fruits during the rainy months (Falcão et al. 1992).

This aquatic relative of guava also has other uses. In the Peruvian Amazon, for example, the Chayahuita, Kichwa, and the Mayna boil the bark and drink the concoction to treat diarrhea (Odonne et al. 2013). Also in the Peruvian Amazon, the Shipibo-Conibo use the leaves of this tree to treat fungal infections caused by *Candida albicans*, *Sporothrix schenckii*, and *Trichophyton mentagrophytes* (Wen et al. 2011). In Abaetetuba along the lower Tocantins River in Pará, locals prepare a tea with the twigs and leaves of *Psidium acutangulum* to treat the same condition (Palheta et al. 2017). The Ticuna along the Solimões and across the border in the Colombian Amazon use the leaves to prepare a bathing solution for treating hemorrhoids (Schultes and Raffauf 1986a, b, 1990: 339). The Cubeo in the Colombian Amazon prepare a decoction from the fresh stem bark to serve as an emetic (Ramírez et al. 2017). River dwellers along the lower Jauaperi River, an affluent of the middle Rio Negro, prepare a drinking solution with the bark to treat dysentery (Pedrollo et al. 2016). In French Guiana, the Wayana prepare a decoction and bathing solution of this plant to treat malaria (Houël et al. 2015) while the Wayãpi who live in several villages along the Oyapock River that borders French Guiana and Amapá in the Brazilian Amazon use the bark to make a juice to treat leishmaniasis (Odonne et al. 2011). Farmers who live along Paraná Cachoeiri on the Amazon floodplain near Oriximiná use the wood for fuel when cooking.

38.14 *Psidium guajava*

Bolivia Guayaba, guayabo, guayabilla; chuará-catoco, bahua quëxti (Chácobo), buhué (Tacana), coti' (Tsimane)

Brazil Goiaba, goiaba branca, goiaba vermelha, goiaba doce, goiabeira; kuyawá (Araweté), comache (Aruá), wayawa (Baniwa), waiava (Desana), waya (Guajá), kuyaba (Juruna), waiya, wajaba, waya'y (Ka'apor), pidjô krekanine'k, pidgô kamrek (Kayapó), tafoti (Kuikuru), waiawa (Sateré-Maué), wazap (Tembé), o-ra-cha (Ticuna), waiava, wayawa (Tukano), kuya (Wajãpi)

Colombia Guayaba, guayaba dulce, guayaba rosada, guayabo; tuche (Bora), oracha, taué (Ticuna)

Ecuador Guayaba; kírimp, kirim (Achuar), samindo'cho (Cofán), kírimp, wayap (Shuar), quëma (Siona-Secoya), paiganca, paigonkamo, peïngonka (Waorani)

Peru Guayaba, guayabilla, guyayabo, araza; kirím (Achuar), sháwi (Aguaruna), yocaan (Amahuaca), llom, ahuecah (Amuesha), tomáshiki (Campa), bimpish (Cashibo), wi'wi (Chayahuita), kima (Cocama), esajo (Ese Eja), wayaap (Huambisa), kana bimpish, uxu bimpish, xón bimpish (Kakataibo), shahuinto (Kichwa), komashiki, komashivi (Matsigenka), bimpishi, binysish (Shipibo), oracha (Ticuna)

Venezuela Guayaba

Status Wild, spontaneous in cultural settings, planted

Nilsiane, 5 years, eating guava *Psidium guajava* she picked in a home garden. Porto Valente, lower Tefé River, Amazonas, Brazil, 8-22-12

Guava *Psidium guajava* is native to the Neotropics and is now widely planted in the tropics and subtropics in both the Old and New World. Guava has also become feral in many tropical regions, including the Amazon, where it grows spontaneously on disturbed sites on uplands and as well as floodplains as noted over a century ago by the Swiss botanist Jacques Huber (1910). It is found as high as 2400 m in the Andean foothills. Whether it was domesticated in one area and then spread, or was already wild in many parts of Tropical America and then brought into cultivation by different indigenous groups is unclear. Guava is wild or feral from Florida to southern South America including parts of the Caribbean. Two species of stingless bees including *Trigona truculenta* visit the flowers of this small tree in the vicinity of Tefé along the Solimões River in the Brazilian Amazon and likely pollinate them (Oliveira et al. 2013: 206) as does *Melipona interrupta* in the upland forest near Manaus (Ferreira and Absy 2018).

Pasture with spontaneous guava *Psidium guajava* trees. Cattle eat the fruits thereby dispersing some of the seeds. Costa Canaboca, Rio Solimões, Amazonas, Brazil 6-20-07

Whether native or introduced, people have created favorable conditions for guava to spread in Amazonia for thousands of years by clearing forests to plant crops and establish settlements. Several species of birds, ranging from blue-gray tanagers *Thraupis episcopus* and silver-beaked tanagers *Ramphocelus carbo* to various parrots and parakeets, feed avidly on guava fruits and disperse the seeds. I observed blue-gray tanagers feeding on guava fruits in the back garden of the house I was renting in Manaus, Amazonas, in July 1976. Various monkeys also eat the fruits (Carvalho 1957; Pinheiro et al. 2013; Santos et al. 2018) as well as at least one species of bat, *Artibeus lituratus*, in the vicinity of Manaus (Guillaumet and Reis 1983). The red-footed tortoise *Chelonoidis carbonaria*, which occurs widely in the Amazon Basin and some surrounding areas, also eats this fruit and disperses some of the seeds (Wang et al. 2011). Guava fruits are perfectly "designed" for animal dispersal; it is often easier to simply swallow the BB pellet-sized seeds than to painstakingly remove the pulp and spit them out. People have thus been involved in dispersing guava seeds. Sometimes feral guava forms dense thickets, much favored by pigs in search of fallen fruit, as in a pasture on Caviana Island at the mouth of the Amazon. Other domestic animals, such as cattle, horses, and even dogs, are involved in dispersing guava in the Amazon as they are in other parts of the American tropics. In fact, guava is a common "weed" in pastures (Rocha et al. 2016), dispersed by birds and cattle which are fond of the fruit. Guava occurs in some pastures in other regions, such as Central America (Allen 1956). Large-scale ranchers do not appreciate them, but small-scale farmers don't mind spontaneous guava because they can gather the fruits. It also occurs as a weed in some guaraná *Paullinia cupana* plantations in Amazonas, Brazil (Miléo et al. 2007; Silva et al. 2007).

Sun parakeet *Aratinga solstitialis* **eating a guava** *Psidium guajava* **fruit. Alexandre Rodrigues Ferreira,** *Viagem Filosófica pelas Capitanias do Grão Pará, Rio Negro, Mato Grosso e Cuiabá, 1783–1792*, **Conselho Federal de Cultura, Rio de Janeiro, 1971, Iconografia, volume 2, plate 100**

People harvest fruits of feral and planted guava by foot, or in the case of floodplains, from canoes. People often pick guava fruits before they are fully ripe because if they wait too long birds or wild mammals might make off with them. Often the fruits are in easy reach because guava starts producing fruit when it is only 1 m high. When the tree matures it can reach 10 m so fruits higher up are brought down using a hooked pole or youngsters climb the trunk and walk along the branches. The fruits are firm, so they are not damaged when they fall to the ground.

Alex, 9 years, gathering guava *Psidium guajava* fruits in successionary vegetation by a manioc field. Outskirts of Vila Cedro, Rio Cariá, Floresta Nacional Caxiuanã, Pará, Brazil 5-16-15

Boys climbing a guava *Psidium guajava* tree in a home garden to gather fruit. Igarapé Inajá, affluent of the lower Xingu, Pará, Brazil 9-10-14

Guava fruits year-round, but yields are generally higher during the rainy season and into the early part of the dry period. The globular fruits vary in size, a reflection of the great diversity of the wild and domesticated gene pools in the Amazon. Fruit size varies from that of a golf ball to almost the dimensions of a soft ball. Wild fruits are not necessarily smaller; I have seen feral guava fruits on Caviana Island the size of baseballs. Although wild fruits have a reputation of being inferior, they are often just as tasty as domesticated forms and are not always full of seeds. Skin color when ripe ranges from lime green to yellow and flesh color varies from pale peach to deep orange-red, depending on the levels of carotene. Feral populations of guava in the Amazon and elsewhere may be an overlooked source of desirable traits for improving commercial cultivars. In the Amazon, most of the fruits are eaten fresh and are highly nutritious because they contain as much vitamin C as oranges. The fruits also contain high levels of dietary fiber (Jiménez-Escrig et al. 2001). Guava juice, popsicles, and ice cream are popular in some urban areas, such as Iquitos. Guava is a popular fruit throughout Brazil. A thick guava jam, known as *goiabada*, is consumed throughout the country as a stand-alone dessert, or with white cheese.

Hans, 10 years, eating a large variety of guava *Psidium guajava* he gathered in the home garden on his farm in the Andean foothills. Fundo Toropampa, Rio Huancabamba near Pozuzo, Pasco, Peru 11-9-04

Guava is also cultivated in home gardens on uplands and floodplains throughout the Amazon and other regions. It is grown in rural and urban home gardens, such as Belém (Madaleno 2000). Indeed, it is one of the most common fruit trees in home gardens: in a survey of 150 rural home gardens in the Brazilian Amazon carried out over several decades I found guava in a third of them. In a separate survey of 57 home gardens along the middle Xingu in 2015, I found guava in half of them. Guava sometimes arises spontaneously in home gardens and is spared, such as among some indigenous groups on the savanna of Roraima (Pinho 2008: 38) and in the village of Silencio along Igarapé do Mamauru near Óbidos. Guava is also planted close to some manioc processing huts, such as in the vicinity of Jocojó in the Municipality of Gurupá in Pará.

Guava *Psidium guajava* in the understory of a home garden containing other fruit trees including açaí *Euterpe oleracea* palms. Costa Canaboca, Rio Solimões, Amazonas, Brazil 6-20-07

Guava is sometimes planted along with other crops in fields, such as along the Transamazon Highway in the municipality of Medicilandia, in the vicinity of Itacoatiara along the Amazon River, and near Pucallpa, Peru. The Matsigenka in the Manu National Park of the southeastern part of the Peruvian Amazon also occasionally plant guava in some of their manioc fields (Wezel and Ohl 2006). Guava is also spontaneous in some agroforestry fields, such as in the vicinity of Altamira along the middle Xingu, near Santarém at the confluence of the Amazon and Tapajós, and along roads leading to Ouro Preto in Rondônia. One farmer near Agrovila Nova Fronteira along the Transamazon Highway 80 km west of Altamira considered spontaneous guava a nuisance in his cacao grove. At first, he let the guava trees grow because his large family appreciates the fruit. But he eventually resolved to cut them all down, alleging that the guava's root system is too aggressive and was "robbing" water from the more valuable cacao. The fruits are relatively firm even when ripe, so they arrive in markets in good shape. Kids are sometimes dispatched to small towns, such as Santa Cruz do Arari on Marajó Island at the mouth of the Amazon, to hawk the fruits in small net bags.

Sandra 44 years, weaving a mat using spindles of wood obtained from guava *Psidium guajava* and acapu. The mat is composed of stems of piri that are cut along streams and lakeshores. Comunidade de Aruã, Chaves Municipality, Marajó Island, Pará, Brazil 5-18-01

Guava is also used in folk medicine. In various parts of the Brazilian Amazon, rural people use the liquid squeezed from the leaves to treat diarrhea and dysentery (Lisboa et al. 2002; Luz 2001; Murakami et al. 2016: 63; Pedrollo et al. 2016; Penido et al. 2016; Rodrigues 2006; Schultes and Raffauf 1990: 339). In the Municipality of Ariquemes in Rondônia and in various parts of Roraima, a decoction is prepared from the inner bark for treating the same condition, and the leaves and buds are used to make an infusion to treat stomachache (Santos et al. 2008; Van den Berg and Silva 1988). Residents of Caruarú, a rural community on Mosqueiro Island some 70 km northeast of Belém, also prepare a tea from the bark for treating diarrhea (Mesquita and Tavares-Martins 2018) while on Marajó Island in the Amazon estuary, people make a tea from the leaves to treat diarrhea among their animals and the bark is used to make a tea to promote healing of animal wounds (Monteiro et al. 2011). In Rio Branco, Acre, the bark is used to make a tea to treat gastroenteritis (Siveiro et al. 2012). The Chácobo and rural folk in the Bolivian Amazon do the same (Boom 1987: 33). The Chácobo drink this decoction three times a day until their bowel movements return to normal.

In the Andean foothills of the central Peruvian Amazon, the Amuesha make an infusion of the leaves, or fresh leaf buds are eaten raw, to treat diarrhea that is accompanied by intestinal cramps and fever (Valadeau et al. 2010). Rural folk in the vicinity of Chazuta along the upper Huallaga River in the Peruvian Amazon use the bark to make a decoction for treating diarrhea (Sanz-Biset et al. 2009), as do the Ese

Eja in the Madre de Dios watershed (Desmarchelier et al. 1995, 1996). In various parts of the Peruvian Amazon, rural folk treat diarrhea by boiling the bark and buds and three teaspoons of the liquid are swallowed every day; also, the leaves are boiled in water for 3–5 minutes, and then when the liquid cools three glasses are drunk per day. To treat the same condition, some people in the Peruvian Amazon mix six leaves and four unripe fruits in a liter of water a cupful of which is then drunk with meals (Lossio et al. 1995: 128). To reduce swelling, the bark is boiled in water for 3–5 minutes and then two or three glasses of the liquid are drunk per day once it has cooled. To treat hemorrhages, the bark is boiled and two cups of the liquid are drunk per day, one in the morning the other in the afternoon. To treat intestinal parasites an infusion of fresh leaves is drunk three times a day. To ameliorate diarrhea the ripe fruits are eaten (Mejia and Rengifo 2000: 94). And to expel intestinal worms seven leaves are crushed in half a liter of water and left to soak for 24 hours; a cup of this solution is then drunk twice a day (Lossio et al. 1995: 128).

In the Ecuadorian Amazon, the Kichwa use the bark and leaves to make a paste which they apply to skin affected by leishmaniasis (Gachet et al. 2010). The Kichwa also used the leaves and fruits to prepare a remedy for stomachache and diarrhea (Saltos et al. 2016). Also in the Ecuadorian Amazon, the Achuar make a decoction from the leaves to treat diarrhea (Giovannini 2015). In the southern part of the Ecuadorian Amazon, the Shuar prepare a tonic from the fruits to treat diarrhea; two or three fruits are placed in 1 l of water for 10–15 minutes and the solution is drunk after each meal for two or 3 days (Bennett et al. 2002: 222). To treat children with upset stomachs, the Shuar place three to five fresh leaves of this bush to small tree in water which is then heated for 5–8 minutes, and when the solution cools it is given to the children after each meal for two to 3 days.

Along the Jari River which borders Pará and Amapá, cowboys treat cattle suffering from diarrhea with liquid squeezed from the bark of guava *Psidium guajava* trees (Lins 1991: 97). In Ecuador and Peru, the leaves, flowers, and/or the bark are boiled to make a tea to treat the same condition (Hajdu and Hohmann 2012). The Matsigenka in the Peruvian Amazon use guava leaves in a warm bath for ritual cleansing on the third day of mourning after someone has passed away. This tribe believes that such baths will prolong the lives of people because the tree is so common around settlements, often arising spontaneously and thus they never "die" (Shepard 2002). In Amapá, the leaves are used to make a decoction for alleviating stomachache. Residents of Marudá, a village along the coast of Pará at the mouth of the Amazon, prepare a bathing solution from the inner bark to cleanse the uterus after birth. Young shoots are also used to make a bathing solution to treat swollen legs prior to childbirth (Coelho-Ferreira 2009). In Surinam which borders Roraima in the Brazilian Amazon, Saramaccan maroons prepare a bathing solution from the leaves of this plant to lower fever among babies (Ruysschaert et al. 2009) and in Guyana rural people boil the leaves and dry bark of this tree to treat dysentery, diarrhea and intestinal worms (Austin and Bourne 1992).

Guava has other non-food uses as well. The trunk and branches are used for fuelwood, including charcoal, such as among the Tsimane in the Bolivian Amazon (Reyes-García 2001: 251) and in the community of Algodoal in the municipality of Maracanã along the northeastern coast of Pará (Coelho-Ferreira and Jardim 2005). The Tsimane also use the wood to fashion arrows for hunting and fishing. On Marajó Island, rural inhabitants fashion mats (*esteira*) from the stems of piri *Cyperus giganteus*. The wooden spindles used for stitching the mat together are called maçaroca and are fashioned from wood obtained from guava and acapu *Vouacapoua americana*. The twine is envira de miriti obtained from the fronds of *Mauritia flexuosa*, an aquatic palm.

Mother picking guava *Psidium guajava* leaves to make a tea to treat her son's stomachache. Former US Air Force base used during the Second World War, Amapá, Brazil 8-30-14

38.15 *Psidium guineense*

Bolivia Guayabilla, guayabilla de pampa, guayaba, churuma; coti (Tsimane)

Brazil Goiaba araçá, goiaba branca, goiaba amarela, goiaba da guiné, goiabinha araçá, goiabinha selvagem, araçá, araçá pera, araçá pedra, araçá do campo, araçá do cerrado, araçá guaçú, araçá mirim, araçá azedo, araçá mirim, araçá bravo, araçá vedaleiro, araçá verdadeiro, araçá comum, araçá goiaba, araçáí, araçázinho, araçázeiro, marmelada de cachorro; aratzé (Baniwa), kamokàtytx, idjy-kryre (Kayapó), passakáni (Tukano)

Colombia Guayaba agria, guayaba, guayabo, coronilla, arrayán

Ecuador Guayabilla, allpa guayaba

Peru Guayaba acida, guayaba brava, guayabillo

Venezuela Guayabo de monte, maraduru; maradao (Pemón)

Status Wild, spontaneous in cultural settings, planted

Goiaba araçá *Psidium guineense* fruits gathered in a home garden. Laranjal, Igarapé Laranjal, Floresta Nacional Caxiuanã, Baia de Caxiuanã, Rio Anapu, Município de Melgaço, Pará, Brazil 5-11-15

Goiaba araçá *Psidium guineense*, like its close relative guava *Psidium guajava*, occurs spontaneously in grasslands, scrub savanna, home gardens, and disturbed areas in the Amazon Basin and Andean foothills (Borges and Shepherd 2005; Penso et al. 2009; Pinho 2008: 38). This plant is an invasive weed in some pastures but is tolerated in some areas, such as in southern Ecuador, because of its edible fruits

(Van den Eynden 2004: 105). This bush to small tree, which can reach 12 m, is also found occasionally in forests such as in successionary woodlands around rocky outcrops in the Upper Orinoco watershed in eastern Colombia (Parra 2006) and in riparian vegetation along streams such as in and around Manaus, Amazonas (Santos et al. 2016; Scudeller and Souza 2009) and on the floodplains of the Mamoré and Beni rivers in the Bolivian Amazon (Navorro et al. 2010). It is widely distributed from Mexico to southern South America including some islands in the Caribbean. In the Andean foothills, it is found as high as 2800 m where the rare and endangered spectacled bear *Tremarctos ornatus* eats the fruits. In the cerrado of central Brazil, the tapir *Tapirus terrestris* eats this fruit and disperses some of the seeds (Prado 2012: 18); this tapir, the largest terrestrial game animal, also occurs in the Amazon Basin.

The fruits are generally smaller than guava, about the size of small hen eggs. Large fruits are 3.8 cm by 3.5 cm, whereas smaller ones, which are more common, are 2.8 cm by 2.5 cm. The fruits contain numerous seeds 2–3 mm in diameter (Gordon et al. 2011) which are swallowed by people along with the pulp; some birds and mammals do likewise, thus dispersing the seeds, as is the case with *Carollia perspicillata* and *Artibeus lituratus* bats in the vicinity of Manaus (Guillaumet and Reis 1983). The relatively small size of the fruit accounts for one of the common names for this species in the Brazilian Amazon: araçázinho. And guayabilla, a common name for this bush in Bolivia, means the small guava. The larger fruits tend to be found in home gardens. The fruits contain numerous light tan seeds surrounded by a light-colored mesocarp in contrast to the pink pulp of guava. The fruits are rich in calcium, potassium, phosphorus, niacin, vitamin C, and vitamin A (Silva et al. 2011: 50). The shrub starts producing fruit when under a meter tall, so they are often within easy reach of children, such as among the Tsimane in the Bolivian Amazon (Martínez-Rodríguez 2009: 63). In home gardens, *Psidium guineense* is planted and also occurs spontaneously.

Araçá, one of the common names for the fruit in the Brazilian Amazon, is brought to street markets in Belém from October to March which spans the height of the dry season into the rainy period (Cavalcante 2010: 44). In Amapá, the fruits were not seen in markets, but a goiaba araçá in a home garden along the AP 70 highway some 55 km from Macapá had ripe fruit during a visit in August 1998. The fruits are also used in folk remedies. In the Ecuadorian Amazon, a decoction of unripe fruits is drunk to treat diarrhea and a decoction of the roots is used to alleviate coughs (Van den Eynden et al. 1999: 58). The Tsimane in the Bolivian Amazon make a tea from bark or leaves to stem diarrhea (Martínez-Rodríguez 2009: 91).

Levi, 7 years, eating a goiaba araçá *Psidium guineense* fruit he gathered in his home garden. Sitio do Dominga, Igarapé Laranjal, Floresta Nacional Caxiuanã, Baia de Caxiuanã, Rio Anapu, Município de Melgaço, Pará, Brazil 5-13-15

References

Adaime, Ricardo; Jesus-Barros, Cristiane Ramos de; Souza-Filho, Miguel Francisco de. 2014. Moscas-das-frutas (Diptera: Tephritidae) obtidas de frutos comercializados em feiras públicas de Macapá, Amapá. Boletim de Pesquisa e Desenvolvimento, EMBRAPA 85:1–17

Aguiar, Jaime Paiva Lopes; Souza, Francisca das Chagas do Amaral. 2018. Antioxidant capacidant and bioactive compounds and health benefits of camu-camu puree (*Myrciaria dubia* [H.B.K.] McVaugh). International Journal of Development Research 8(6):20742–20745

Akter, Mst. Sorifa; Oh, Sejong; Eun, Jong-Bang; Ahmed, Maruf. 2011. Nutritional compositions and health promoting phytochemicals of camu-camu (*Myrciaria dubia*) fruit: a review. Food Research International 44: 1728–1732

Allen, P. H. 1956. *The Rain Forests of Golfo Dulce*. University of Florida Press, Gainesville

Almeida Jr., Eduardo Bezerra de; Santos-Filho, Francisco Soares; Araújo, Elcida de Lima; Zickel, Carmen Sílvia. 2011. Structural characterization of the woody plants in restinga of Brazil. Journal of Ecology and the Natural Environment 3(3):95–103

Alvino, Fabrízia de Oliveira; Silva, Manoela Ferreira F. da; Rayol, Breno Pinto. 2005. Potencial de uso das espécies arbóreas de uma floresta secundária, na Zona Bragantina, Pará, Brasil. Acta Amazonica 35(4):413–420

Amaral, Dário Dantas do; Prost, Maria Thereza; Bastos, Maria de Nazaré do Carmo; Costa Neto, Salustiano Vilar; Santos, João Ubiratan Moreira dos. 2008. Restingas do litoral amazônico, estados do Pará e Amapá. Boletim do Museu Paraense Emílio Goeldi, Ciencias Naturais 3(1):35–67

Amaral, Dário Dantas do; Vieira, Ima Célia Guimarães; Almeida, Samuel Soares de; Salomão, Rafael Paiva; Silva, Antonio Sérgio Lima da; Jardim, Mário Augusto Gonçalves. 2009. Checklist da flora arbórea de remanescentes florestais da região metropolitana de Belém e valor histórico dos fragmentos, Pará, Brasil. Boletim do Museu Paraense Emílio Goeldi, Ciencias Naturais 4(3):231–289

Amaral, Dário Dantas do; Rocha, Antônio Elielson; Pereira, Jorge Luis Gavina; Costa Neto, Salustiano Vilar. 2019. Identificação dos subtipos de savanas na Amazônia oriental (Pará e Amapá, Brasil) com uma chave dicotômica de individualização. Boletim do Museu Paraense Emílio Goeldi, Ciencias Humanas Naturais 14(2):183–195

Andrade, Jerusa de S.; Aragão, Carlos G.; Ferreira, Sidney A. do N. 1993. Caracterização física e química dos frutos de araçá-pera (*Psidium acutangulum* D.C.). Acta Amazonica 23: 213–217

Araújo, Maristela Machado; Tucker, Joanna Marie; Vasconcellos, Joanna Marie; Zarin, Daniel Jacob; Oliveira, Wilson; Sampaio, Patrícia Delamônica; Rangel-Vasconcelos, Lívia Gabrig; Oliveira, Francisco de Assis; Coelho, Roberta de Fatima Rodrigues; Aragão, Débora Veiga; Miranda, Izildinha. 2005. Padrão e processo sucessionais em florestas secundárias de diferentes idades na Amazônia oriental. Ciencia Florestal, Santa Maria 15(4):343–357

Araujo, Rosalia de Aguiare; Costa, Reginaldo Brito da; Felfili, Jeanine Maria; Kuntz, Ivani; Melo e Sousa, Roberto Antonio Ticle de; Dorval, Alberto. 2009. Florística e estrutura de fragmento florestal em área de transição na Amazônia Matogrossense no município de Sinop. Acta Amazonica 39(4):865–878

Araujó-Lima, C. and M. Goulding. 1997. So Fruitful a Fish: Ecology, Conservation, and Aquaculture of the Amazon's Tambaqui. Columbia University Press, New York

Araujo-Murakami, Alejandro; Paniagua, Narel; Cayola, L.; Jorgensen, Peter M.; Valdes, Ovidio; Macuapa, Eric; Calzadilla-Tomianovich, Miki. 2009. Diversidad y estructura florística del bosque de llanura y palmar de pantano Amazónico preandino en el sector de Chalalan, Parque Nacional Madidi, Bolivia. Kempffiana 5(1):3–27

Araujo-Murakami, Alejandro; Villarroel, Daniel; Pardo, Guido; Vos, Vincent A.; Parada, G. Alexander; Arroyo, Luzmila; Killeen, Timothy. 2015. Diversidad arbórea de los bosques de tierra firme de la Amazonía Boliviano. Kempffiana 11(1):1–28

Arellano, Gabriel. 2013. Patrones de diversidad, distribución y dominancia de plantas leñosas en los bosques montanos del Parque Nacional Madidi, Bolivia. Doctoral dissertation, Departamento de Biologia, Facultad de Ciencias, Universidad Autónoma de Madrid

Arkcoll, D. B.; Clement, C. R. 1989. Potential new food crops from the Amazon. New Crops for Food and Industry. Wickens, G. E.; Haq, N.; Day, P. (Editors), pp. 150–165

Assis, Rafael Leandro de; Wittmann, Florian 2011. Forest structure and tree species composition of the understory of two central Amazonian várzea forests of contrasting flood heights. Flora 206:251–260

Assis, Rafael Leandro de; Wittmann, Florian; Luize, Bruno Garcia; Haugaasen, Torbjon. 2017. Patterns of floristic diversity and composition in floodplain forests across four southern Amazon river tributaries, Brazil. Flora 229:124–140

Austin, Daniel F.; Bourne, Godfrey R. 1992. Notes on Guyana's medical ethnobotany. Economic Botany 46(3):293–298

Baar, R.; M. Cordeiro, M. Denrich; and H. Fölster. 2004. Floristic inventory of secondary vegetation in agricultural systems of East-Amazonia. *Biodiversity and Conservation* 13:501–528

Baca, Luis Campo. 2005. Algunos parámetros físicos, químicos y bioecológicos que influyen el comportamiento migratorio de la gamitana *Colossoma macropomum* en el Río Ucayali. In: Biología de la Poblaciones de Peces de la Amazonía y Piscicultura, Renno, Jean François; García, Carmen; Duponchelle, Fabrice; Nuñez, Jésus (eds), IIAP/IRD, Iquitos, pp 58–64

Balbin, Isabel Oré; Samaniego, Dedy Llapapasca. 1996. Huertas domesticas como sistema tradicional de cultivo en Moena Caño, Rio Amazonas, Iquitos, Peru. Folia Amazonica 8(1):91–110

Bánki, Olaf S.; Steege, Hans ter; Jansen-Jacobs, Marion J.; Raghoenandan, Usha P. D. 2003. Plant Diversity of the Nassau Mountains, Suriname: Report of the 2003 Expedition. National Herbarium Nederland, Utrecht

Barbosa, Reinaldo Imbrozio; Nascimento, Sebastião Pereira do; Amorim, Paulo Atlântico Figueiredo de; Silva, Rosemberg Ferreira da. 2005. Notas sobre a composição arbóreo-arbustiva de uma fisionomia das savanas de Roraima, Amazônia Brasileira. Acta Botanica Brasilica 19(2):323–329

Barbosa, Danielle Cristine de Figueiredo; Marimon, Beatriz Schwantes; Lenza, Eddie; Marimon Junior, Ben Hur; Oliveira, Edmar Almeida De; Maracahipes, Leandro. 2011. Estrutura da vegetação lenhosa em dois fragmentos naturais de florestas inundáveis (impucas) no Parque Estadual do Araguaia, Mato Grosso. Revista Árvore 35(3):457–471

Bastos, Maria de Nazaré do C. 1984. Levantamento florístico dos campos do Estado do Pará. I – Campo de Joanes (Ilha de Marajó). Boletim do Museu Paraense Emílio Goeldi, Série Botânica 1(1/2): 67–86

Bastos, Maria de Nazaré do Carmo. 1988. Levantamento florístico em restinga arenosa litorânea na Ilha de Maiandeua, Pará. Boletim do Museu Paraense Emílio Goeldi, Série Botânica 4(1): 159–169

Bastos, M. N. C.; C. S. Rosário; L. C. B. Lobato. 1995. Caracterização fitofisionômica da restinga de Algodoal, Maracanã, Pará, Brasil. Boletim do Museu Paraense Emílio Goeldi, Série Botanica 11(2): 173–197

Batalha, Marco Antônio; Martins, Fernando Roberto. 2004. Reproductive phenology of the cerrado plant community in Emas National Park (central Brazil). Australian Journal of Botany 52:149–161

Bennett, Bradley C.; Baker, Marc A.; Andrade, Patricia Gómez. 2002. Ethnobotany of the Shuar of Eastern Ecuador. New York Botanical Garden, Advances in Economic Botany, vol. 14.

Boom, B. M. 1987. Ethnobotany of the Chácobo Indians, Beni, Bolivia. The New York Botanical garden, Advances in Economic Botany, vol. 4

Borges, Hélida B. Nogueira; Shepherd, George J. 2005. Flora e estrutura do estrato lenhoso numa comunidade de Cerrado em Santo Antônio do Leverger, MT, Brasil. Revista Brasileira de Botânica 28(1):61–74

Brack, A. 1999. *Diccionario Enciclopédico de Plantas Útiles del Perú.* Centro de Estudios Regionales Andinos Bartolomé de Las Casas (CBC), Cusco

Brack Egg, Antonio. 2003. Frutas del Perú. Universidad San Martin de Porres, Lima

Braga, Elayne Oliveira; Jardim, Mário Augusto Gonçalves. 2019. Florística, estrutura e formas de vida do estrato inferior de uma floresta ombrófila densa aluvial, Pará, Brasil. Ciencia Florestal, Santa Maria 29(3): 1048–1059

Braga, Elayne Oliveira; Silva, Julie Andrews de França e; Pantoja, Milenna Vasconcelos; Jardim, Mário Augusto Gonçalves. 2015. Florística, estrutura fitossociológica e formas de vida do estrato inferior em um floresta de várzea Amazônica. Biota Amazônia 5(3):59–65

Brienza Júnior, Silvio. 1999. Biomass dynamics of fallow vegetation enriched with leguminous trees in the eastern Amazon of Brazil. Doctoral dissertation, Georg August University, Göttingen, Germany

Brito, Elizabeth Rodrigues; Martins, Sebastião Venâncio; Oliveira Filho, Ary Teixeira de; Silva, Elias; Silva, Alexandre Francisco da. 2006. Estrutura fitossociológica de um fragmento natural de floresta inundável em área de orizicultura irrigada, município de Lagoa da Confusão, Tocantins. Revista Árvore 30(5):829–836

Carim, Marcelo de Jesus Veiga. 2016. Estrutura, composição e diversidade em florestas alagáveis de várzea de maré e de igapó e suas relações com variáveis edáficas e o período de inundação no Amapá, Amazônia oriental, Brasil. Doctoral dissertation, Programa de Pós-Graduação em Botânica, INPA, Manaus

Carim, Samyrams; Schwartz, Gustavo; Silva, Manoela Fernandes Ferreira da. 2007. Riqueza de espécies, estrutura e composição florística de uma floresta secundária de 40 anos no leste da Amazônia. Acta Botanica Brasilica 21(2):293–308

Carvalho, C. T. 1957. Notas sôbre a biologia do *Ramphocelus carbo* (Passeres, Thraupidae). Boletim do Museu Paraense Emilio Goeldi, Zoologia 5:1–20

Carvalho, Telma Lélia Gonçalves Schultz de. 2015. Etnofarmacologia e fisiologia de plantas medicinais do quilombo Tiningú, Santarém, Pará, Brasil. Master's thesis, Universidade Federal do Oeste do Pará, Santarém.

Carvalho, João Olegário Pereira de; Silva, José Natalino Macedo; Lopes, José do Carmo Alves; Montagner, Lise Helene; Carvalho, Manoel Sebastião Pereira de. 1986. Composição florística de uma mata secundária no planalto de Belterra no Pará. In: Anais do 1° Simpósio do Trópico Úmido, Belém, PA, 12 a 17 November, EMBRAPA, Brasília, pp 197–205

Cavalcante, P. B. 1973. A Farmacopéia Tiriyó: Estudo Étno-Botânico. Museu Paraense Emílio Goeldi, Publicações Avulsas No. 24

Cavalcante, P. B. 2010. *Frutas Comestíveis na Amazônia*. Museu Paraense Emílio Goeldi, Belém

Cerón, Carlos Eduardo; Ayala, Consuelo Montalvo. 1998. Etnobotanica de los Huaorani de Quehueiri-ono Napo, Ecuador. Abya-Yala, Quito

Chirinos, Rosana; Galarza, Jorge; Betalleluz-Pallardel, Indira; Pedreschi, Romina; Campos, David. 2010. Antioxidant compounds and antioxidant capacity of Peruvian camu camu (*Myrciaria dubia* (H.B.K.) McVaugh) fruit at different maturity stages. Food Chemistry 120: 1019–1024

Clement, Charles R.; Silva, Danilo Fernandes da. 1994. Amazonian small fruits with commercial potential. Fruit Varieties Journal 48 (3): 152–158

Clement, C. R., S. S. Alfaia, J. H. Iriarte-Martel, K. Yuyama, J. B. M. Gomes, J. Leeuwen, L. A. G. Souza, and W. B. C. Flores. 1997. Fruteiras nativas e exóticas. In: *Duas Décadas de Contribuições do INPA à Pesquisa Agronômica no Tróprico Úmido*, Noda, Hiroshi; Souza, Luiz Augusto Gomes de (Editors), pp. 111–129, INPA, Manaus

Coelho, Roberta de Fátima Rodrigues; Zarin, Daniel Jacob; Miranda, Izildinha Souza; Tucker, Joanna Marie. 2003a. Análise florística e estrutural de uma floresta em diferentes estágios sucessionais no município de Castanhal, Pará. Acta Amazonica 33(4):563–582

Coelho, Roberta de Fátima Rodrigues; Zarin, Daniel Jacob; Miranda, Izildinha Souza; Tucker, Joanna Marie. 2003b. Ingresso e mortalidade em uma floresta em diferentes estágios sucessionais no município de Castanhal, Pará. Acta Amazonica 33(4):619–630

Coelho, Roberta de Fátima Rodrigues; Miranda, Izildinha de Souza; Mitja, Danielle. 2013. Conservação das florestas do projeto de assentamento Benfica, Sudeste da Amazônia. Ciencia Florestal, Santa Maria 23(1):1–17

Coelho-Ferreira, Márlia. 2009. Medicinal knowledge and plant utilization in an Amazonian coastal community of Marudá, Pará state (Brazil). Journal of Ethnopharmacology 126:159–175

Coelho-Ferreira, Márlia; Jardim, Mário Augusto G. 2005. Algumas espécies vegetais usadas pelos moradores da Ilha de Algodoal, Maiandeua, município de Maracanã, Pará. Boletim do Museu Paraense Emílio Goeldi, Série Ciências Naturais 1(2): 45–51

Cohen, M. N. 1977. Population pressure and the origins of agriculture: an archaeological example from the coast of Peru. In: Origins of Agriculture, C. A. Reed (ed.), Mouton, The Hague, pp. 135–177

Conceição, G. M.; Aragão, J. G. 2010. Diversidade e importância econômica das Myrtaceae do cerrado, Parque Estadual do Mirador, Maranhão. Scientia Plena 6(7):1–8

Costa, Andre Gustavo Vasconcelos; Garcia-Diaz, Diego F.; Jimenez, Paula; Silva, Pollyanna Ibrahim. 2013. Bioactive compounds and health benefits of exotic tropical red-black berries. Journal of Functional Foods 5:539–549

Costa, Reginaldo Brito da; Silveira, Ediléia Patrícia da; Souza, Thaianny Rodrigues de. 2014. Estrutura da vegetação de cerrado sensu stricto em terra indígena no noroeste do estado de Mato Grosso. Multitemas 46: 95–110
Costa Neto, Salustiano Vilar da; Bastos, Maria de Nazaré do Carmo; Lobato, Luiz Carlos Batista. 1996. Composição florística e fitofisionomia da restinga do Crispim, município do Marapanim, Pará. Boletim do Museu Paraense Emílio Goeldi, Série Botânica 12(2): 237–249
Cuellar, Fabio A.; Ariza, Edna; Anzola, Cecilia; Restrepo, Patricia. 2013. Capacidad antioxidante del arazá (*Eugenia stipitata* Mc Vaugh) durante la maduración. Revista Colombiana de Quimíca 42(2):21–28
Cunha, D. 2009. Análise da paisagem e estrutura das comunidades vegetais das formações pioneiras do baixo Xingu, Pará. Master's thesis, Universidade Federal Rural da Amazônia/Museu Paraense Emílio Goeldi, Belém
Cunha, Fernando Lima Rodrigues da. 2013. Dieta de quatro espécies do gênero *Podocnemis* (Testudines, Podocnemididae) na Reserva de Desenvolvimento Sustentável Uatumã, Amazonas, Brasil. Master's thesis, INPA, Manaus
Cunha, Denise de Andrade; Ferreira, Leandro Valle. 2012. Impacts of the Belo Monte hydroelectric dam construction on pioneer vegetation formations along the Xingu River, Pará state, Brazil. Brazilian Journal of Botany 35(2):159–167
Desmarchelier, Cristian; Mongelli, Elena; Coussio, Jorge; Giulietti, Ana; Ciccia, Graciela. 1995. Etnobotánica y bioactividad de plantas medicinales utilizadas por un grupo indígena Takana de la Amazonia Peruana. Acta Farmacéutica Bonaerense 14(3):195–208
Desmarchelier, C.; Gurni, A.; Ciccia, G.; Giulietti, A. M. 1996. Ritual and medicinal plants of the Ese'ejas of the Amazonian rainforest (Madre de Dios, Perú). Journal of Ethnopharmacology 52:45–51
Dutra, Saturnino; Silva, Manoela Ferreira Fernandes da; Queiróz, Waldenei Travassos de. 2004. Fitossociologia de plantas invasoras de pastagens cultivadas na região Nordeste Paraense. Revista de Ciência Agrária 42:143–164
Egler, S. G. 1992. Feeding ecology of *Saguinus bicolor bicolor* (Callitrichidade: Primates) in a relict forest in Manaus, Brazilian Amazonia. Folia Primatologica 59:61–76
Espírito-Santo, Fernando Del Bon; Shimabukuro, Yosio Edimir; Aragão, Luiz Eduardo Oliveira e Cruz de; Machado, Evandro Luiz Mendonça. 2005. Análise da composição florística e fitossociológica da floresta nacional do Tapajós com apoio geográfico de imagens satélites. Acta Amazonica 35(2):155–173
Falcão, Martha de Aguiar; Chavez, Wanders B.; Ferreira, Sidney A. N.; Clement, Charles R.; Barros, Maria José B.; Brito, Janete M. C. de; Santos, Tereza C. T. Dos. 1988. Aspectos fenológicos e ecológicos do "araçá-boi" (*Eugenia stipitata* McVaugh) na Amazônia central. 1. Plantas juvenis. Acta Amazonica 18(3–4):27–38
Falcão, Martha de Aguiar; Ferreira, Sidney Alberto do N.; Clement, Charles R.; Santos, Tereza Cristina T. dos; Souza, Rosana de Medeiros. 1992. Crescimento e fenologia de araçá-pera (*Psidium acutangulum* DC). Acta Amazonica 22 (3): 285–293
Falcão, Martha de Aguiar; Galvão, Rozana de M. Souza; Clement, Charles R.; Ferreira, Sidney A. do N.; Sampaio, Sebastiana das Graças. 2000. Fenologia e produtividade do araçá-boi (*Eugenia stipitata*, Myrtaceae) na Amazônia central. Acta Amazonica 30(1):9–21
Fernandes-Corrêa, Ana Francisca; Furch, Bodo. 1992. Investigations on the tolerance of several trees to submergence in blackwater (Igapó) and whitewater (Várzea) inundation forests near Manaus, Central Amazonia. Amazoniana 12(1):71–84
Ferreira, S.A.N. 1982. Observação da germinação de sementes de araçá-pêra (*Psidium acutangulum* D. C.). Acta Amazonica 12(3):503–507
Ferreira, Carlos Alberto Cid. 2009. Análise comparativa de vegetação lenhosa do ecosistema campina na Amazonia Brasileira. Doctoral dissertation, INPA/UFAM, Manaus
Ferreira, Marcos Gonçalves; Absy, Maria Lucia. 2015. Pollen niche and trophic interactions between colonies of Melipona (Michmelia) seminigra merrillae and Melipona (Melikerria) interrupta (Apidae: Meliponini) reared in floodplain in the Central Amazon. Arthropod-Plant Interactions 9:263–279

Ferreira, Marcos Gonçalves; Absy, Maria Lucia. 2017. Pollen analysis of honeys of Melipona (Michmelia) seminigra merrillae and Melipona (Melikerria) interrupta (Hymenoptera: Apidae) bred in Central Amazon, Brazil. Grana 56(6):436–449

Ferreira, Marcos Gonçalves; Absy, Maria Lucia. 2018. Pollen niche of Melipona (Melikerria) interrupta (Apidae: Meliponini) bred in a meliponary in a terra-firme forest in the central Amazon. Palynology 42(2):199–209

Ferreira, Leandro Valle; Cunha, Denise A.; Chaves, Priscilla P.; Matos, Darley C. L.; Parolin, Pia. 2013a. Impacts of hydroelectric dams on alluvial riparian plant communities in eastern Brazilian Amazonian. Anais da Academia Brasileira de Ciências 85(3):1013–1033

Ferreira, Leandro V.; Chaves, Priscilla P.; Cunha, Denise de A.; Rosário, Alessandro S. do; Parolin, Pia. 2013b. A extração ilegal de areia como causa do desaparecimento de campinas e campinaranas no estado do Pará, Brasil. Pesquisas Botânica 64:157–173

Flores, Salvador. 1997. *Cultivo de Frutales Nativos Amazonicos: Manual para el Extensionista.* Tratado de Cooperación Amazónica, Lima

Foster, Robin B.; Arce, Javier; Wachter, Tatzyana S. 1986. Dispersal and the sequential plant communities in Amazonian Peru floodplain. In: Frugivores and Seed Dispersal, Estrada, Alejandro; Fleming, Theodore H. (eds), Dr. W. Junk Publishers, Dordrecht, pp 357–370

Freire, Maria Cristina C.; Monteiro, Reinaldo. 1993. Florística das praias da Ilha de São Luís, Estado do Maranhão (Brasil): diversidade de espécies e suas ocorrências no litoral Brasileiro. Acta Amazonica 23(2–3):125–140

Fuentes, Alfredo F. 2018. Novedades florísticas de la región del Madidi: nuevos registros de plantas vasculares, adiciones al catálago de Bolivia y especies poco conocidas. Kempffiana 14(2):42–62

Gachet, María Salomé; Lecaro, Javier Salazar; Kaiser, Marcel; Brun, Reto; Navarrete, Hugo; Muñoz, Ricardo A.; Bauer, Rudolf; Schühly, Wolfgang. 2010. Assessment of anti-protozoal activity of plants traditionally used in Ecuador in the treatment of leishmaniasis. Journal of Ethnopharmacology 128:183–197

Gama, João Ricardo Vasconcellos; Souza, Agostino Lopesde Souza; Calegário, Natalino; Lana, Guilherme Campos. 2007. Fitossociologia de duas fitocenoses de floresta ombrófila aberta no município de Codó, estado do Maranhão. Revista Árvore 31(3):465–477

Garzón, G. Astrid; Narváez-Cuenca, Carlos-Eduardo; Kopec, Rachel E.; Barry, Andrew M.; Riedl, Kenneth M.; Schwartz, Steven. 2012. Determination of carotenoids, total phenolic content, and antioxidant activity of arazá (*Eugenia stipitata* McVaugh), an Amazonian fruit. Journal of Agricultural and Food Chemistry 60: 4709–4717

Genovese, M. I.; Pinto, M. Da Silva; Gonçalves, A. E. De Souza Schmidt; Lajolo, F. M. 2008. Bioactive compounds and antioxidant capacity of exotic fruits and commercial frozen pulps from Brazil. Food Science and Technology International 14(3):207–214

Giacometti, D. and E. Lleras. 1994. Subtropical Myrtaceae. In: Neglected Crops: 1492 from a Different Perspective, Bermejo, J. E.; León, Jorge (eds.), Food and Agriculture Organization, Rome, pp. 229–237

Giovannini, Peter. 2015. Medicinal plants of the Achuar (Jivaro) of Amazonian Ecuador: ethnobotanical survey and comparison with other Amazonian pharmacopoeias. Journal of Ethnopharmacology 164:78–88

Godoy, Judith Rosales; Petts, Geoffrey; Salo, Jukka. 1999. Riparian flooded forests of the Orinoco and Amazon basins: a comparative review. Biodiversity and Conservation 8:551–586

Gomes, Letícia; Lenza, Eddie; Maracahipes, Leandro; Marimon, Beatriz Schwantes; Oliveira, Edmar Almeida de. 2011. Comparações florísticas e estruturais entre duas comunidades lenhosas de cerrado típico e cerrado rupestre, Mato Grosso, Brasil. Acta Botanica Brasilica 25(4):865–875

Gonçalves, Janaína Pinheiro; Lucas, Flávia Cristina Araújo. 2017. Agrobiodiversidade e etnoconhecimento em quintais de Abaetetuba, Pará, Brasil. Revista Brasileira de Biociências 15(3):119–134

Gonzales, Augustín 2006. Caracterización biofísica de los habitats de poblaciones naturales de camu camu *Myrciaria dubia* (H.B.K) McVaugh. Folia Amazónica 15(1–2):29–38

Gordon, André; Jungfer, Elvira; Silva, Bruno Alexandre da; Maia, José Guilherme S.; Marx, Friedhelm. 2011. Phenolic constituents and antioxidant capacity of four underutilized fruits from the Amazon region. Journal of Agricultural and Food Chemistry 59:7688–7699

Goulding, Michael. 1980. *The Fishes and the Forest: Explorations in Amazonian Natural History*. University of California Press, Berkeley

Gualberto, Midiã Larine Colares; Ribeiro, Renato Bezerra da Silva; Gama, João Ricardo Vaconcellos; Vieira, Diego dos Santos. 2014. Fitossociologia e potencial de espécies arbóreas em ecossistema sucessional na Floresta Nacional do Tapajós, Pará. Agroecossistemas 6(1):42–57

Guillaumet, Jean-Louis; Reis, Nélio Roberto dos. 1983. Les chauves-souris frugivores de la région de Manaus et leur rôle dans la dissémination des espèces végétales. Revue d'Écologie: La Terre et la Vie 38:147–169

Haidar, Ricardo Flores; Fagg, Jeanine Maria Felfili; Pinto, José Roberto Rodrigues; Dias, Ricardo Ribeiro; Damasco, Gabriel; Silva, Lucasde Carvalho Ramos; Fagg, Christopher William. 2013. Florestas estacionais e áreas de ecótono no estado do Tocantins, Brasil: parâmetros estruturais, classificação das fitofisionomias florestais e subsídios para conservação. Acta Amazonica 43(3):261–290

Hajdu, Z. and J. Hohmann. 2012. An ethnopharmacological survey of the traditional medicine utilized in the community of Porvenir, Bajo Paraguá Indian Reservation, Bolivia. Journal of Ethnopharmacology 139:838–857

Heard, Tim A. 1999. The role of stingless bees in crop pollination. Annual Review of Entomology 44:183–206

Hohnwald, Stefan; Abréu, Eliana M. Acioli; Krummel, Timo; Trautwein, Julia; Veiga, Jonas Bastos da; Wollny, Clemens B. A.; Azevedo, Braga Célia M. Calandrinide; Gerold, Gerhard. 2010. Degraded pasture distribution and woody enrichment strategies for pasture fertility preservation in the Bragantina region, north-eastern Amazon. Erdkunde 64(1):17–31

Hohnwald, Stefan; Rischkowsky, Barbara; King, John M.; Camarão, Ari P.; Rodrigues Filho, José A.; Zeppenfeld, Thorsten. 2015. Intensive cattle browsing did not prevent fallow recuperation on smallholder grass-capoeira pastures in the NE-Amazon. Agroforestry Systems 89:813–828

Houël, Emeline; Fleury, Marie; Odonne, Guillaume; Nardella, Flore; Bourdy, Geneviève; Vonthron-Sénécheau, Catherine; Villa, Pascal; Obrecht, Adeline; Eparvier, Véronique; Deharo, Eric; Stien, Didier. 2015. Antiplasmodial and anti-inflammatory effects of an antimalarial remedy from the Wayana Amerindians, French Guiana: takamalaimë (*Psidium acutangulum* Mart. ex DC., Myrtaceae). Journal of Ethnopharmacology 166:279–285

Huber, J. 1910. Mattas e madeiras amazonicas. Boletim do Museu Goeldi de Historia Natural e Ethnographia 6: 91–225

IIAP. 2001. Sistema de Producción de camu camu. Instituto de Investigaciones de la Amazonia Peruana, Iquitos

Irmler, U. 1977. Inundation-forest types in the vicinity of Manaus. Biogeographica 8:17–29

Jaramillo-Arango, J. 1952. *Relación Historica del Viage, que Hizo a los Reynos del Peru y Chile el Botanico d. Hipolito Ruiz en el Año de 1777 hasta el de 1788, en cuya Epoca Regreso a Madrid*. Real Academia de Ciencias Exactas, Fisicas y Naturales, Madrid, vol. 1

Jardim, Daniel Gonçalves; Jardim, Mário Augusto Gonçalves; Quaresma, Adriano Costa; Costa Neto, Salustiano Vilar. 2016. Fatores edáficos e as variações florísticas da regeneração natural em duas formações florestais. Biota Amazônia 6(2):66–72

Jiménez-Escrig, Antonio; Rincón, Mariela; Pulido, Raquel; Saura-Calixto, Fulgencio. 2001. Guava fruit (*Psidium guajava* L.) as a new source of antioxidant dietary fiber. Journal of Agricultural and Food Chemistry 49:5489–5493

Junk, W. J. 1989. Flood tolerance and tree distribution in central Amazonian floodplains. In: Tropical Forests: Botanical Dynamics, Speciation and Diversity, Holm-Nielsen, L. B.; Nielsen, I. C.; Balslev, H. (eds), Academic Press, London, pp. 47–64

Killeen, Timothy J.; García, Emilia; Beck, Stephan G. 1993. Guía de Árboles de Bolivia. Herbario Nacional de Bolivia/Missouri Botanical Garden, La Paz.

Kimber, C. T. 1988. Martinique Revisited: The Changing Plant Geographies of a West Indian Island. Texas A&M University Press, College Station

Klüppel, Marina Pinheiro. 2006. Sistemas agrícolas e plantas medicinais em terra preta de índio da Amazônia Central. Master's thesis, INPA/UFAM, Manaus

Kubitzki, K. 1989. The ecogeographical differentiation of Amazonian inundation forests. Plant Systematics and Evolution 162:285–304

Kubitzki, K. and A. Ziburski. 1994. Seed dispersal in flood plain forests of Amazonia. *Biotropica* 26(1):30–43

Kurzatkowski, Dariusz; Leuschner, Christoph; Homeier, Jürgen. 2015. Effects of flooding on trees in the semi-deciduous transition forests of the Araguaia floodplain, Brazil. Acta Oecologica 69:21–30

Kvist, Lars P.; Andersen, Martin K.; Hesselsoe, Martin; Vanclay, Jerome K. 1995. Estimating use-values and relative importance of Amazonian flood plain trees and forests to local inhabitants. Commonwealth Forestry Review 74(4):293–300

La Rotta, C. 1983. Observaciones Etnobotanicas sobre algunas Especies Utilizadas por la Comunidad Indigena Andoque (Amazonas, Colombia). Universidad Nacional de Colombia, Bogotá

Lamont, Susan R.; Eshbaugh, W. Hardy; Greenberg, Adolph M. 1999. Species composition, diversity, and use of homegardens among three Amazonian villages. Economic Botany 53(3):312–326

Leite, Gabriel Augusto. 2017. Conhecer para conservar: *Crax globulosa* (Aves: Cracidae), da genética de populações à seleção do habitat. Doctoral dissertation, Instituto Nacional de Pesquisa da Amazônia, Manaus

Lima, Roberval M. B. de; Saragoussi, Muriel. 2000. Floodplain home gardens on the central Amazon in Brazil. In: The Central Amazon Floodplain: Actual Use and Options for a Sustainable Management, Junk, W. J.; Ohly, J. J.; Piedade, M. T. F.; Soares, M. G. M. (eds), Backhuys Publishers, Leiden pp 243–268

Lima, Edson de Souza; Lima, Herson Souza; Ratter, James Alexander. 2009. Mudanças pós-fogo na estrutura e composição da vegetação lenhosa, em um cerrado mesotrófico, no período de cinco anos (1997–2002) em Nova Xavantina – MT. Cerne 15(4):468–480

Lins, Cristóvão. 1991. Jari: 70 Anos de História. Dataforma, Rio de Janeiro

Lisboa, P. L. B. 2012. *A Terra dos Aruã: Uma História Ecológica do Arquipélago do Marajó.* Museu Paraense Emílio Goeldi, Belém

Lisboa, P. L. B., I. A. Gomes, R. C. Lisboa, and C. V. Urbinati. 2002. O estilo Amazônico de sobreviver: manejo dos recursos naturais. In: *Natureza, Homem e Manejo de Recursos Naturais na Região de Caxiuanã, Melgaço, Pará*, P. L. B. Lisboa (ed.), Museu Paraense Emílio Goeldi, Belém, pp. 41–170

Lobato, G. de J. M.; Lucas, F. C. A.; Tavares-Martins, A. C. C.; Jardim, M. A. G.; Martorano, L. G. 2017. Diversidade de uso e aspectos socioambientais de quintais urbanos em Abaetetuba, Pará, Brasil. Revista Brasileira de Agroecologia 12(2):95–105

Lorenzi, Harri; Bacher, Luis; Lacerda, Marco; Sartori, Sergio. 2006a. Frutas Brasileiras e Exóticas Cultivadas. Instituto Plantarum de Estudos da Flora Ltda, Odessa, Sao Paulo.

Lorenzi, H., L. Bacher, M. Lacerda, S. Sartori 2006b. Brazilian Fruits and Cultivated Exotics (for Consuming in Natura). Instituto Plantarum de Estudos da Flora Ltda, Odessa, São Paulo

Lossio, L.C., Delgado, H.S., Ruiz, A.G., Donayre, R.A., Ruiz, J.G., Panduro, M.P., Sifuentes, T.C. 1995. Plantas Medicinales de la Amazonía Peruana. Instituto Peruano de Seguridad Social/Instituto de Medicina Tradicional, Iquitos

Luize, Bruno Garcia; Venticinque, Eduardo Martins; Silva, Thiago Sanna Freire; Novo, Evlyn Marcia Leão de Moraes. 2015. A floristic survey of angiosperm species occurring at three landscapes of the central Amazon várzea, Brazil. Checklist: The Journal of Biodiversity Data 11(6):1–15

Luz, Francisco Joaci F. 2001. Plantas medicinais de uso popular em Boa Vista, Roraima, Brasil. Horticultura Brasileira 19(1):88–96

Machado, Danilo de Oliveira. 2016. A agrobiodiversidade de quintais agroflorestais em propriedades agrícolas familiares na BR 174, Ramal do Pau-Rosa, Manaus, AM. Master's Thesis, INPA, Manaus

Machado, Clara de Carvalho. 2018. Conhecimento e consumo de plantas alimentícias em cinco comunidades da reserva de desenvolvimento sustentável Piaguçu-Purus, Amazonas. Master's thesis, INPA, Manaus

Madaleno, Isabel. 2000. Urban agriculture in Belém, Brazil. Cities 17(1):73–77

Magnusson, William E.; Lima, Albertina P.; Albernaz, Ana L. K. M.; Sanaiotti, Tânia M.; Guillaumet, Jean-Louis. 2008. Composição florística e cobertura vegetal das savanas na região de Alter do Chão, Santarém-PA. Revista Brasileira de Botanica 31(1):165–177

Maguire, Bassett. 1969. The Botany of the Guaya Highland, Part VIII, Myrtaceae. Memoirs of the New York Botanical Garden 18(2):55–286

Major, Julie; Clement, Charles R.; DiTommaso, Antonio. 2005. Influence of market orientation on food plant diversity of farms located on Amazonian Dark Earth in the region of Manaus, Amazonas, Brazil. Economic Botany 59(1):77–86

Mamede, Jeneffer Soares dos Santos; David, Margô De; Tsukamoto Filho, Antônio de Arruda; Pasa, Maria Corette. 2015. Os quintais e as manifestações culturais da comunidade de São Gonçalo Beira Rio, Cuiabá – MT. Biodiversidade 14(1):168–182

Maracahipes, Leandro; Lenza, Eddie; Marimon, Beatriz Schwantes; Oliveira, Edmar Almeida de; Pinto, José Roberto Rodrigues; Marimon Junior, Ben Hur. 2011. Estrutura e composição florística da vegetação lenhosa em cerrado rupestre na transição cerrado-floresta Amazônica, Mato Grosso, Brasil. Biota Neotropica 11(1):133–141

Marimon, Beatriz Schwantes; Lima, Edson de Souza. 2001. Caracterização fitofisionômica e levantamento florístico preliminar no Pantanal dos Rio Morte-Araguaia, Cocalinho, Mato GrossoActa Botanica Brasilica 15(2):213–229

Marimon, Beatriz Schwantes; Marimon-Junior, Ben Hur; Mews, Henrique Augusto; Jancoski, Halina Soares; Franczak, Daniel David; Lima, Herson Souza; Lenza, Eddie; Rossete, Amintas Nazareth; Moresco, Micheli Cristina. 2012. Florística dos campos de murundus do Pantanal do Araguaia, Mato Grosso, Brasil. Acta Botanica Brasilica 26(1):181–196

Marques-Souza, Antonio Carlos; Moura, Cleonice de Oliveira; Nelson, Bruce Walker. 1996. Pollen collected by *Trigona williana* (Hymenoptera: Apidae) in central Amazonia. Revista de Biologia Tropical 44(2):567–573

Martínez-Rodríguez, María Ruth. 2009. Ethnobotanical knowledge acquisition among Tsimane' children in the Bolivian Amazon. Doctoral dissertation, University of Georgia, Athens

Martins, Sebastião Venâncio; Brito, Elizabeth Rodrigues; Oliveira Filho, Ary Teixeirade; Silva, Alexandre Francisco da; Silva, Elias. 2008. Floristic composition of two wetland forests in Araguaian plain, state of Tocantins, Brazil, and comparison with other areas. Revista Árvore 32(1):129–141

Martins, Adriana Cecília Lima; Rêgo, Márcia Maria Corrêa; Carreira, Léa Maria Medeiros; Albuquerque, Patricia Maia Correia de. 2011. Espectro polínico de mel de tiúba (*Melipona fasciculata* Smith, 1854, Hymenoptera, Apidae). Acta Amazonica 41(2):183–190

Maués, Márcia Motta. 2002. Reproductive phenology and pollination of the Brazil nut tree (*Bertholletia excelsa* Humb. & Bonpl. Lecythidaceae) in eastern Amazonia. In: Pollinating Bees: The Conservation Link between Agriculture and Nature, Kevan, P.; Imperatriz Fonseca, V. L. (eds), pp. 245–254, Ministry of Environment, Brasilia

Medeiros, Priscilla Sanjuan de. 2016. Indicadores ambientais para funcionalidade ecológica em florestas secundárias de diferentes idades na Amazônia Oriental. Doctoral dissertation, Universidade Federal do Pará, Belém

Medeiros, Marcelo Brilhante de; Walter, Bruno Machado Teles. 2012. Composição e estrutura de comunidades arbóreas de cerrado stricto sensu no norte do Tocantins e sul do Maranhão. Revista Árvore 36(4): 673–683

Mejia, Kember; Rengifo, Elsa. 2000. Plantas Medicinales de Uso Popular en la Amazonía Peruana. Agencia Española de Cooperación Internacional, Lima

Mendonça, B. A. F. 2011. Campinaranas Amazônicas: pedogênese e relações solo-vegetação. Doctoral dissertation, Universidade de Viçosa, Viçosa

Mendonça, Bruno A. F. de; Fernandes Filho, Elpídio I.; Schaffer, Carlos E. G. R.; Mendonça, Júlia G. F. de; Vasconcelos, Bruno N. F. 2017. Soil-vegetation relationships and community structure in a "terra-firme"-white-sand vegetation gradient in Viruá National Park, northern Amazon, Brazil. Anais da Academia Brasileira de Ciências 89(2):1269–1293

Mendoza, Abel L. Monteagudo; Guerrero, Marleni Huamán. 2010. Catálogo de los arboles y afines de la selva central del Perú. Arnaldoa 17(2):202–242

Mesquita, Ulliane de O.; Tavares-Martins, Ana Cláudia. 2018. Etnobotánica de plantas medicinales en la comunidad de Caruarú, Isla de Mosqueiro, Belém-PA, Brasil. Boletín Latinoamericano y del Caribe de Plantas Medicinales y Aromáticas 17(2):130–159

Mews, H. A.; Marimon, B. S.; Ratter, J. A. 2012. Observations on the vegetation of Mato Grosso, Brazil. V. Changes in the woody species diversity of a forest in the cerrado-Amazonian forest transition zone and notes on the forests of the region. Edinburgh Journal of Botany 69(2):239–253

Miléo, L. J.; Silva, J. F.; Bentes, J. L. S.; Christoffoleti, P. J. 2007. Plantas daninhas hospedeiras alternativas de *Colletotrichum guaranicola* em cultivos de guaraná no estado do Amazonas. Planta Daninha 25(4):771–782

Miranda, Izildinha S. 2000. Análise florística e estrutural da vegetação lenhosa do Rio Comemoração, Pimenta Bueno, Rondônia, Brasil. Acta Amazonica 30(3):393–422

Miranda, I. S.; Absy, M. L. 2000. Fisionomia das savanas de Roraima, Brasil. Acta Amazonica 30(3):423–440

Monteiro, Maria Vivina Barros; Bevilaqua, Claudia Maria Leal; Palha, Maria das Dores Correia; Braga, Roberta Rocha; Schwanke, Katiane; Rodrigues, Silvane Tavares; Lameira, Osmar Alves. 2011. Ethnoveterinary knowledge of the inhabitants of Marajó Island, eastern Amazonia, Brazil. Acta Amazonica 41(2):233–242

Mota, Nara Furtado de Oliveira; Watanabe, Mauricio Takashi Coutinho; Zappi, Daniela Cristina; Hiura, Alice Lima; Pallos, Julieta; Viveros, Raquel Stauffer; Giulietti, Ana Maria; Viana, Pedro Lage. 2018. Cangas da Amazônia: a vegetação única de Carajás evidenciada pela lista de fanerógramas. Rodriguésia 69(3):1435–1488

Muniz, Ana Lúcia Vilhena; Silva, Manoela Ferreira Fernandes da; Araújo, Érika Luiza Souza; Alvino, Fabrizia de Oliveira. 2007. Dinâmica do estrato arbóreo de florestas secundárias no nordeste do Pará (Bragança). Revista Brasileira de Biociências 5(1):603–605

Murakami, Alejandro Araujo; Reyes, Juan Fermandon; Milliken, William. 2016. Frutales Silvestres y Promisorios de Pando. Herencia/Museo de Historia Natural Noel Kempff Mercado, Cobija

Murrieta, Javier Noriega; Arévalo, Jorge Augusto Gil. 2004. Frutos y semillas utilizados como alimento por peces de consumo y de valor comerical, en las cochas Capirona y Quincha. Universidad Nacional de la Amazonía Peruana, Iquitos

Myster, Randall W. 2018. What we know about the composition and structure of igapó forests in the Amazon Basin. Botanical Review 84:394–410

Navorro, Gonzalo; Molina, José Antonio; Agostinelli, Emanuela; Lumbreras, Ana; Ferreira, Wanderlei. 2010. Towards an ecological classification of flooded savannas in Beni (Bolivia). Acta Botanica Gallica 157(2):265–273

Neri-Numa, Iramaia Angélica et al. 2013. Evaluation of the antioxidant, antiproliferative and antimutagenic potential of araçá-boi fruit (*Eugenia stipitata* McVaugh—Myrtaceae) of the Brazilian Amazon forest. Food Research International 50:70–76

Odonne, Guillaume; Berger, Franck; Stien, Didier; Grenand, Pierre; Bourdy, Geneviève. 2011. Treatment of leishmaniasis in the Oyapock basin (French Guiana): a K.A.P. survey and analysis of the evolution of phytotherapy knowledge amongst Wayãpi Indians. Journal of Ethnopharmacology 137:1228–1239

Odonne, Guillaume; Valadeau, Celine; Alban-Castillo, Joaquina; Stien, Didier; Sauvain, Michel; Bourdy, Genevieve. 2013. Medical ethnobotany of the Chayahuita of the Paranapura basin (Peruvian Amazon). Journal of Ethnopharmacology 146:127–153

Ogeron, Clémence; Odonne, Guillaume; Cristinoi, Antonia; Engel, Julien; Grenand, Pierre; Beauchêne, Jacques; Clair, Bruno; Davy, Damien. 2018. Palikur traditional roundwood construction in eastern French Guiana: ethnobotanical and cultural perspectives. Journal of Ethnobiology and Ethnomedicine 14:1–18

Oliveira, Francisco Plácido Magalhães; Jardim, Mário Augusto Gonçalves. 1998. Composição florística de uma floresta secundária no município de Igarapé-Açu, estado do Pará, Brasil. Boletim do Museu Paraense Emilio Goeldi, Série Botânica 14(2):127–145

Oliveira, Jorge; Almeidas, Samuel Soares de; Potyguara, Raimunda Vilhena; Lobatos, Luiz Carlos Batista. 1991. Espécies vegetais produtoras de fibras utilizadas por comunidades Amazônicas. Boletim do Museu Paraense Emílio Goeldi, Série Botânica 7(2):393–428

Oliveira, Francisco Plácido Magalhães; Absy, Maria Lúcia; Miranda, Izildinha Souza. 2009. Recurso polínico coletado por abelhas sem ferrão (Apidae, Meliponinae) em um fragmento de floresta na região de Manaus- Amazonas. Acta Amazonica 39(3):505–518.

Oliveira, Favízia Freitas de; Richers, Bárbara Tadzia Trautman; Silva, Jacson Rodrigues da; Farias, Rinéias Cunha; Matos, Tércio Alves de Lima. 2013. Guia Ilustrado das Abelhas "Sem-Ferrão" das Reservas Amanã e Mamirauá, Amazonas, Brasil (Hymenoptera, Apidae, Meliponini). IDSM, Tefé

Paim, Fernanda P.; Chapman, Colin A.; Queiroz, Helder L. de; Paglia, Adriano P. 2017. Does resource availability affect the diet and behavior of the vulnerable squirrel monkey, *Saimiri vanzolinii*? International Journal of Primatology 38(3):572–587

Palheta, Ivanete Cardoso; Tavares-Martins, Ana Cláudia Caldeira; Lucas, Flávia Cristina Araújo; Jardim, Mário Augusto Gonçalves. 2017. Ethnobotanical study of medicinal plants in urban home gardens in the city of Abaetetuba, Pará state, Brazil. Boletín Latinoamericano y del Caribe de Plantas Medicinales y Aromáticas 16(3):206–262

Parolin, P. 2002. Submergence tolerance vs. escape from submergence: two strategies of seedling establishment in Amazonian floodplains. Environmental and Experimental Botany 48:177–186

Parolin, Pia. 2009. Submerged in darkness: adaptations to prolonged submergence by woody species of the Amazonian floodplains. Annals of Botany 103:359–376

Parolin, Pia; Ambruester, Nicole; Wittmann, Florian; Ferreira, Leandro; Piedade, Maria Teresa F.; Junk, Wolfgang J. 2002. A review of tree phenology in central Amazonian floodplains. Pesquisas Botânica 52:195–222

Parolin, Pia; Ferreira, Leandro Valle; Albernaz, Ana L. K. M.; Almeida, Samuel S. 2004. Tree species distribution in várzea forests of Brazilian Amazonia. Folia Geobotanica 39:371–383

Parolin, P., D. Waldhoff, and M. T. Piedade. 2010. Fruit and seed chemistry, biomass and dispersal. In: *Amazonian Floodplain Forests: Ecophysiology, Biodiversity and Sustainable Management*. W. J. Junk, M. T. Piedade, F. Wittman, and J. Schöngart (eds), Springer, Dordrecht, pp. 243–258

Parra, C. 2006. Estudio general de la vegetación nativa de Puerto Carreño (Vichada, Colombia). Caldasia 28(2):165–177

Pedrollo, Camilo Tomazini; Kinupp, Valdely Ferreira; Shepard Jr., Glenn; Heinrich, Michael. 2016. Medicinal plants at Rio Jauaperi, Brazilian Amazon: ethnobotanical survey and environmental conservation. Journal of Ethnopharmacology 186:111–124

Penido, Alexandre Batista; Morais, Selene Maia de; Ribeiro, Alan Bezerra; Silva, Ana Zélia. 2016. Ethnobotanical study of medicinal plants in Imperatriz, state of Maranhão, northeastern Brazil. Acta Amazonica 46(4):345–354

Penn, James W. 2006. The cultivation of camu camu (*Myrciaria dubia*): a tree planting programme in the Peruvian Amazon. *Forests, Trees and Livelihoods* 16: 85–101

Penso, Sarah; Bauer, Maristela de Oliveira; Chichorro, José Franklim; Gondim, Carlos Alberto; Vasconcellos, Lívia Vieira. 2009. Caracterização estacional de uma pastagem natural do cerrado Mato-Grossense submetida ao pastejo. Ciência Animal Brasileira 10(1):124–134

Pereira, Luciano Araújo; Sobrinho, Felipe de Araújo; Costa Neto, Salustiano Vilar da. 2011. Florística e estrutura de uma mata de terra firme na Reserva de Desenvolvimento Sustentável no Rio Iratapuru, Amapá oriental, Brasil. Floresta 41(1):113–122

Peters, Charles M.; Hammond, Elysa J. 1990. Fruits from the flooded forests of Peruvian Amazon: yield estimates for natural populations of three promising species. Advances In Economic Botany 8: 159–176

Peters, Charles M.; Vasquez, A. 1986. Estudios ecológicos de camu-camu (Myrciaria dubia). I. Producción de frutas en poblaciones naturales. Acta Amazonica 16/17:161–174

Peters, Charles M.; Balick, Michael J.; Kahn, Francis; Anderson, Anthony B. 1989. Oligarchic forests of economic plants in Amazonia: utilization and conservation of an important tropical resource. Conservation Biology 3(4):341–349

Piedade, M. T. F.; Junk, W. J.; Adis, J.; Parolin, P. 2005. Ecologia, zonação e colonização da vegetação arbórea das ilhas Anavilhanas. Pesquisas Botânica 56:117–144

Pinedo, M. and W. de Jong. 2004. Camu camu *Myrciaria dubia* (HBK) McVaugh from the river plains of the Peruvian Amazon. In: Forest Products, Livelihoods and Conservation: Case Studies of Non-Timber Product Systems, Alexiades, Miguel N.; Shanley, Patricia (eds), CIFOR, Jakarta, vol. 3, pp. 263–279

Pinheiro, Tatyana; Ferrari, Stephen F.; Lopes, Maria Aparecida. 2013. Activity budget, diet, and use of space by two groups of squirrel monkeys (*Saimiri sciureus*) in eastern Amazonia. Primates 54:301–308

Pinho, Rachel Camargo de. 2008. Quintais agroflorestais indígenas em áreas de savana (lavrado) na Terra Indígena Araçá, Roraima. Master's thesis, INPA/UFAM, Manaus

Pizango-Paima, Emer Gloria; Pereira-Filho, Manoel; Oliveira-Pereira, Maria Inêz de. 2001. Composição corporal e alimentar do matrinxã, *Brycon cephalus* (Günther, 1869), na Amazônia central. Acta Amazonica 31(3):509–520

Posey, D. A. 2002. Kayapó Ethnoecology and Culture. Routledge, London

Prado, Helena Alves do. 2012. Frugivoria e dispersão de sementes por anta (*Tapirus terrestris*) em uma área de plantio florestal no cerrado, MS. Master's thesis, Universidade Estadual Paulista, Rio Claro, São Paulo

Prance, G. T. 1978. The origin and evolution of the Amazon flora. Interciencia 3(4):207–222

Prance, G. T. 1979. Notes on the vegetation of Amazonia III. The terminology of Amazonian forest types. Brittonia 31(1):26–38

Prance, G. T. 1980. A terminologia dos tipos de florestas amazônicas sujeitas a inundação. Acta Amazonica 10(3):495–504

Prance, G. T. 1989. Economic prospects from tropical rainforest ethnobotany. In: Fragile Lands of Latin America: Strategies for Sustainable Development, Browder, John O. (Editor), pp. 61–74

Prata, Shirley Soares; Miranda, Izildinha de Souza; Alves, Sérgio Augusto Oliveira; Farias, Flavia Cardoso; Jardim, Fernando Cristovam da Silva. 2010. Gradiente florístico das florestas secundárias do Nordeste Paraense. Acta Amazonica 40(3):523–534

Ramírez, O.; Blair, S.; Medio M. T. V.; Cardona, F.; Pabón. 2017. Ethnobotany of medicinal plants used to treat malaria by traditional healers from ten indigenous Colombian communities located in Vaupes Medio. Biodiversity International Journal 1(2):1–17

Ramos, Andrezza S.; Souza, Rodrigo O. S.; Boleti, Ana Paula de A.; Bruginski, Estevan R. D.; Lima, Emerson S.; Campos, Francinete R.; Machado, Marcos B. 2015. Chemical characterization and antioxidant capacity of araçá-pera (*Psidium acutangulum*): an exotic Amazon fruit. Food Research International 75: 315–327

Ratter, J. A.; Bridgewater, S.; Atkinson, R.; Ribeiro, J. F. 1996. Analysis of the floristic composition of the Brazilian cerrado vegetation II: comparison of the woody vegetation of 98 areas. Edinburgh Journal of Botany 53(2):151–180

Rayol, Breno Pinto. 2006. Análise florística e estrutural da vegetação xerofítica das savannas metalófilas na Floresta Nacional de Carajás: subsídios à conservação. Master's thesis, Universidade Federal Rural da Amazônia/Museu Paraense Emílio Goeldi, Belém.

Rayol, Breno Pinto; Silva, Manoela Ferreira Fernandes da; Alvino, Fabrízia de Oliveira. 2006. Dinâmica da diversidade florística da regeneração natural de florestas secundárias, no município de Bragança, Pará, Brasil. Boletim do Museu Paraense Emílio Goeldi, Ciencias Naturais 1(2):9–27

Rayol, Breno Pinto; Alvino-Rayol, Fabrízia de Oliveira; Silva, Manoela Ferreira Fernandes da. 2011. Similaridade florística entre o estrato arbóreo e a regeneração natural de uma floresta secundária, no município de Bragança, nordeste do estado do Pará, Brasil. Revista Brasileira de Agroecologia 6(3):107–114

Reátegui, Katia Wendy. 2013. Hábitos alimenticios de *Saguinus mystax* Spix "pichico barba blanca" en Padre Isla, Loreto, Peru. Master's thesis, UNAP, Iquitos

Reis, Neidiane Farias Costa. 2010. Estrutura fitossociológica e seletividade de espécies em áreas inundáveis na regiã̦o do baixo Rio Jaci-Paraná, Rondônia. Master's Thesis, Universidade Federal de Rondônia, Porto Velho

Reyes-García, Victoria. 2001. Indigenous people, ethnobotanical knowledge, and market economy: a case study of the Tsimane' Amerindians in lowland Bolivia. Doctoral dissertation, University of Florida, Gainesville.

Riva, Rita; Gonzales, Isaias. 1997. Tecnología del Cultivo de Camu Camu *Myrciaria dubia* HBK Mc Vaugh en la Amazonia Peruana. Instituto Nacional de Investigación Agraria, Estación Experimental Pucallpa, Pucallpa

Rocha, Gustavo P. E.; Vieira, Daniel L. M.; Simon, Marcelo F. 2016. Fast natural regeneration in abandoned pastures in southern Amazonia. Forest Ecology and Management 370:93–101

Rodrigues, Eliana. 2006. Plants and animals utilized as medicines in the Jaú National Park (JNP), Brazilian Amazon. Phytotherapy Research 20:378–391

Rodrigues, Roberta B.; Menezes, Hilary C. de; Cabral, Lourdes M. C.; Dornier, Manuel; Reynes, Max. 2001. An Amazonian fruit with a high potential as a natural source of vitamin C: the camu-camu (*Myrciaria dubia*). Fruits 56:345–354

Rodrigues, Santina; Caetano, Diego G.; Caetano, Creucí Maria. 2007a. Especies frutíferas do centro-sul do Estado de Rondônia, Amazônia brasileira. Acta Agronómica 56(2):69–74

Rodrigues, Mauro Antônio Cavaleiro Macedo; Miranda, Izildinha Souza; Kato, Maria do Socorro Andrade. 2007b. Flora e estrutura da vegetação secundária após o uso de diferentes trituradores florestais. Pesquia Agropecúaria Brasileira 42(4):459–465

Rudas, Augustín; Prieto, Adriana. 1998. Análisis florístico del parque nacional natural Amacayacu e Isla Mocagua, Amazonas (Colombia). Caldasia 20(2):142–172

Rufino, Maria do Socorro M.; Alves, Ricardo E.; Brito, Edy S. de; Pérez-Jiménez, Jara; Saura-Calixto, Fulgencio; Mancini-Filho, Jorge. 2010. Bioactive compounds and antioxidant capacities of 18 non-traditional tropical fruits from Brazil. Food Chemistry 121:996–1002

Ruysschaert, Sofie; Van Andel, Tinde; Van de Putte, Kobeke; Van Damme, Patrick. 2009. Bathe the baby to make it strong and healthy: Plant use and child care among Saramaccan Maroons in Suriname. Journal of Ethnopharmacology 121:148–170

Salomão, Rafael de Paiva; Vieira, Ima Célia Guimarães; Suemitsu, Chieno; Rosa, Nélson de Araújo; Almeira, Samuel Soares de; Amaral, Dário Dantas do; Menezes, Moirah Paulo Machado de. 2007. As florestas de Belo Monte na grande curva do rio Xingu, Amazônia Oriental. Boletim do Museu Paraense Emílio Goeldi, Ciencias Naturais 2(3):57–153

Saltos, Ricardo Vinicio Abril; Vásquez, Tomás Elías Ruiz; Lazo, Jatnel Alonso; Banguera, Derwing Viáfra; Guayasamín, Pedro Damián Ríos; Vargas, Janeth Karina Aguinda; Peñas, Ingrid Vega. 2016. The use of medicinal plants by rural populations of the Pastaza province in the Ecuadorian Amazon. Acta Amazonica 46(4):355–366

Sanaiotti, Tânia M.; Magnusson, William E. 1995. Effects of annual fires on the production of fleshy fruits eaten by birds in a Brazilian Amazonian savanna. Journal of Tropical Ecology 11(1):53–65

Sanchez, P. A. 1995. Science in agroforestry. *Agroforestry Systems* 30:5–55

Santana, José Augusto da Silva; Almeida, Wanderléa da Costa; Sousa, Luciana Karla V. dos Santos. 2004. Florística e fitossociologia em área de vegetação secundária na Amazônia oriental. Revista de Ciências Agrárias 41:105–120

Santos, Luciana Oliveira dos. 2010. Comparação da vegetação lenhosa entre borda e interior de uma mancha de mata de restinga na RESEX Mar Caeté-Taperaçu, Bragança, Pará. Master's thesis, Universidade Federal Rural da Amazônia/Museu Paraense Emílio Goeldi, Belém

Santos, J. U. M. and M. N. Bastos. 2014. Flores e Frutas das Restingas do Estado do Pará. Universidade Federal Rural da Amazônia (UFRA), Belém

Santos, João Ubiratan M. dos; Rosário, Carlos da S. 1988. Levantamento da vegetação fixadora das dunas de Algodoal-PA. Boletim do Museu Paraense Emílio Goeldi, Série Botânica 4(1): 133–154

Santos, Maurício Reginaldo A. dos; Lima, Maria Railda de; Ferreira, Maria das Graças R. 2008. Uso de plantas medicinais pela população de Ariquemes, Rondônia. Horticultura Brasileira 26:244–250

Santos, Leonardo Maracahipes; Lenza, Eddie; Santos, Josias Oliveira dos; Marimon, Beatriz Schwantes; Eisenlohr, Pedro V.; Marimon Junior, Ben Hur; Feldpausch, Ted R. 2015. Diversity, floristic composition, and structure of the woody vegetation of the cerrado in the cerrado-Amazon transition zone in Mato Grosso, Brazil. Brazilian Journal of Botany 38(4):877–887

Santos, Otilene dos Anjos; Couceiro, Sheyla Regina Marques; Rezende, Alinne Costa Cavalcante; Silva, Márcia Daniela de Sousa. 2016. Composition and richness of woody species in riparian forests in urban areas of Manaus, Amazonas, Brazil. Landscape and Urban Planning 150:70–78

Santos, Deborah Cruz dos; Corrêa Neto, José de Jesus; Alvarenga, Adriano Braga Brasileiro de. 2018. Frutos consumidos por macaco-de-cheiro *Saimiri sciureus collinsi* (Osgood, 1916) em fragmentos de florestas em uma unidade de conservação no arquipélago do Marajó, Amazônia Oriental, Brasil. Revista Brasileira de Zoociências 19(1):82–90

Sanz-Biset, Jaume; Campos-de-la-Cruz, José; Epiquién, Mirbel A.; Cañigueral, Salvador. 2009. A first survey on the medicinal plants of the Chazuta valley (Peruvian Amazon). Journal of Ethnopharmacology 122:333–362

Sarthou, Corinne; Villers, Jean-François; Ponge, Jean-François. 2003. Shrub vegetation on tropical granitic inselbergs in French Guiana. Journal of Vegetation Science 14:645–652

Sarthou, Corinne; Larpin, Denis; Fonty, Emile; Pavoine, Sandrine; Ponge, Jean-François. 2010. Stability of plant communities along a tropical inselberg ecotone in French Guiana (South America). Flora 205(10):682–694

Schöngart, Jochen; Piedade, Maria Teresa F.; Ludwigshausen, Sabine; Horna, Viviana; Worbes, Martin. 2002. Phenology and stem-growth periodicity of tree species in Amazonian floodplain forests. Journal of Tropical Ecology 18:581–597

Schultes, Richard Evans; Raffauf, Robert F. 1986a. De plantis toxicariis e Mundo Novo tropicale commentationes XXXVII. Miscellaneous notes on medicinal and toxic plants of the northwest Amazon. Botanical Museum Leaflets, Harvard University 30(4):255–285

Schultes, Richard Evans; Raffauf, Robert F. 1986b. De plantis toxicariis e mundo novo tropicale commentationes XXXVIII: ethnopharmacological and alkaloidal notes on plants of the northwest Amazon. Caldasia 15:71–75

Schultes, R. E. and R. F. Raffauf. 1990. The Healing Forest: Medicinal and Toxic Plants of the Northwest Amazonia. Dioscorides Press, Portland, Oregon

Schwartz, Gustavo. 2007. Manejo sustentável de florestas secundárias: espécies potenciais no nordeste do Pará, Brasil. Amazônia: Ciência e Desenvolvimento 3(5):125–147

Scudeller, Veridiana Vizoni; Souza, Adriana Mota Gomes de. 2009. Florística da mata de igapó na Amazônia Central. In: Biotupé: Meio Físico, Diversidade Biológica e Sociocultural do Baixo Rio Negro, Amazônia Central, Santos-Silva, Edinaldo Nelson; Scudeller, Veridiana Vizoni (eds), pp 97–108, UEA Edições, Manaus

Shepard, G. 2002. Three days of weeping: dreams, emotion, and death in the Peruvian Amazon. Medical Anthropology Quarterly 16(2):200–229

Silva, Manoela F. F. da; Secco, Ricardo de S.; Lobo, Maria da Graça A. 1996. Aspectos ecológicos da vegetação rupestre da Serra dos Carajás, Estado do Pará, Brasil. Acta Amazonica 26(1/2):17–44
Silva, J. M., M. Pereira-Filho, and M. I. Oliveira-Pereira. 2003. Valor nutricional e energético de espécies vegetais importantes na alimentação do tambaqui. *Acta Amazonica* 33(4):687–700
Silva, Andréa Leme; Tamashiro, Jorge; Begossi, Alpina. 2007. Ethnobotany of riverine populations from the Rio Negro, Amazonia (Brazil). Journal of Ethnobiology 27(1): 46–72
Silva, C. R. S.; Albuquerque, PSB; Ervedosa, FR; Mota, JWS; Figueira, A; Sebbenn, AM. 2011. Understanding the genetic diversity, spatial genetic structure and mating system at the hierarchical levels of fruits and individuals of a continuous *Theobroma cacao* population from the Brazilian Amazon. Heredity 106: 973–985
Siveiro, A.; Delunardo, T. A.; Haverroth, M.; Oliveira, L. C.; Mendonça, A. M. S. 2012. Plantas medicinais em quintais urbanos de Rio Branco, Acre. Revista Brasileira de Plantas Medicinais 14(4):598–610
Siviero, Amauri; Delunardo, Thiago Andrés; Haverroth, Moacir; Oliveira, Luis Cláudio de; Mendonça, Ângela Maria Silva. 2011. Cultivo de espécies alimentares em quintais urbanos de Rio Branco, Acre, Brasil. Acta Botanica Brasilica 25(3):549–556
Tananta, Leslie Virginia. 2014. Análisis del conocimiento tradicional del uso de especies vegetales en tres comunidades de la cuenca baja del río Ucayali, Loreto, Peru. Master's thesis, UNAP, Iquitos
Taurino, Tássia Cristina da Conceição Barros. 2017. Recuperação natural da diversidade de espécies de plantas em florestas secundárias na Amazônia Oriental. Master's Thesis, Universidade Federal do Pará, Belém
Thomas, Evert. 2009. New light on the floristic composition and diversity of indigenous territory and national park Isiboro-Sécure, Bolivia. Biodiversity Conservation 18:1847–1878
Towle, Margaret A. 1961. *The Ethnobotany of Pre-Columbian Peru*. Aldine, Chicago
Trindade, Jonilson Ribeiro; Rosário, Alessandro Silva do; Santos, João Ubiratan Moreira dos. 2018. Flora das cangas da Serra dos Carajás, Pará, Brasil: Myrtaceae. Rodriguésia 69(3): 1259–1277
Ugent, Donald; Ochoa, Carlos M. 2006. La Etnobotánica del Perú: Desde la Prehistoria al Presente. CONCYEC (Consejo Nacional de Ciencia, Tecnología e Innovación Tecnológica), Lima
Valadeau, Céline; Castillo, Joaquina Alban; Sauvain, Michel; Lores, Augusto Francis; Bourdy, Geneviève. 2010. The rainbow hurts my skin: medicinal concepts and plants uses among the Yanesha (Amuesha), an Amazonian Peruvian ethnic group. Journal of Ethnopharmacology 127:175–192
Van den Berg, M. E. 2010. Plantas Medicinais na Amazônia. Museu Paraense Emílio Goeldi, Belém
Van den Berg, Maria Elisabeth; Silva, Milton Hélio Lima da. 1988. Contribuição ao conhecimento da flora medicinal de Roraima. Acta Amazonica 18(1–2) supl.:23–35
Van den Eynden, Veerle. 2004. Use and management of edible non-crop plants in southern Ecuador. Doctoral dissertation, Universiteit Gent, Belgium
Van den Eynden, Veerle; Cueva, Eduardo; Cabrera, Omar. 1999. Plantas Silvestres Comestibles del Sur del Ecuador. Ediciones Abya-Yala, Quito
Vargas, Adrián Mauricio; Camelo, Ángela Patricia Rivera; Cuenca, Carlos Eduardo Narváez. 2005. Capacidad anatioxidante durante la maduración de arazá (*Eugenia stipitata* Mc Vaugh). Revista Colombiana de Quimíca 34(1): 57–65
Vásquez, R. 1997. *Flórula de las Reservas Biológicas de Iquitos, Perú*. Monographs in Systematic Botany Vol. 63, Missouri Botanical Garden, St. Louis.
Vásquez, Silvia Patricia Flores; Mendonça, Maria Silvia de; Noda, Sandra do Nascimento. 2014. Etnobotânica de plantas medicinais em comunidades ribeirinhas do Município de Manacapuru, Amazons, Brasil. Acta Amazonica 44(4):457–472
Vickers, W. T. and T. Plowman. 1984. *Useful Plants of the Siona and Secoya Indians*. Field Museum of Natural History, Fieldiana Botany New Series, no. 15

Vieira, Ima Célia Guimarães. 1996. Forest succession after shifting cultivation in eastern Amazonia. Doctoral dissertation, University of Stirling

Vieira, Diegodos Santos; Oliveira, Marcio Leles Romarco de; Gama, João Ricardo Vasconcelos; Machado, Evandro Luiz Mendonça; Görgens, Eric Bastos; Lafetá, Bruno Oliveira; Garcia, Josinei Silva. 2017. Phytosociology of a natural fragment of the floodplain forest in the lower Tapajós River, Brazil. Bosque 38(2):357–369

Villachica, Hugo; Carvalho, José Edmar Urano de; Muller, Carlos Hans; Díaz, Camilo; Almanza, Martha; Mercado, Roberto Samanez; Izquierdo, Juan; Egg, Antônio Brack. 1996. Frutales y Hortalizas Promisorios de la Amazonia. Tratado de Cooperación Amazonica, Lima.

Waldhoff, D., U. Saint-Paich, and F. Bodo. 1996. Value of fruits and seeds from the floodplain forests of central Amazonia as food resource for fish. *Ecotropica* 2:143–156

Wang, Ellen; Donatti, Camila I.; Ferreira, Vanda L.; Raizer, Josué; Himmelstein, Jeffrey. 2011. Food habits and notes on the biology of *Chelonoidis carbonaria* (Spix 1824) (Testudinidae, Chelonia) in the southern Pantanal, Brazil. South American Journal of Herpetology 6(1):11–19

Wen, Louise; Haddad, Mohamed; Fernández, Irma; Espinoza, Giovana; Ruiz, Candy; Neyra, Edgar; Bustamante, Beatriz; Rojas, Rosario. 2011. Actividad antifúngica de cuatro plantas usadas en la medicina tradicional Peruana: Aislamiento de 3'-formil-2',4',6'-trihidroxidihidro-chalcona, principio ativo de *Psidium acutangulum*. Revista de la Sociedad Química del Perú 77(3):199–204

Wezel, A.; Ohl, J. 2006. Homegarden plant diversity in relation to remoteness from urban centers: a case study from the Peruvian Amazon region. In: Tropical Homegardens: a time-tested example of sustainable agroforestry, Kumar, B. M.; Nair, P. K. R. (eds), pp 143–158, Advances in Agroforestry vol. 3, Springer, Dordrecht

Wittmann, Florian; Parolin, Pia. 1999. Phenology of six tree species from central Amazonian várzea. Ecotropica 5(1):51–57

Wittmann, Florian; Anhuf, Deiter; Junk, Wolfgang J. 2002. Tree species distribution and community structure of central Amazonian várzea forests by remote-sensing techniques. Journal of Tropical Ecology 18(6):805–820

Wittmann, F. W.; Schongart, Jochen; Brito, Joneide M. de; Wittman, Astride de Oliveira; Piedade, Maria Teresa 2010. Manual of Trees from Central Amazon Várzea Floodplains: Taxonomy, Ecology, and Use. INPA, Manaus

Worbes, Martin. 1997. The forest ecosystem of the floodplains. Ecological Studies 126:223–265

Works, M. A. 1990. Continuity and conversion of house gardens in western Amazonia. *Yearbook of the Association of Pacific Coast Geographers* 52:31–64

Yuyama, Kaoru; Aguiar, Jaime P. L.; Yuyama, Lucia K. O. 2002. Camu-camu: um fruto fantástico como fonte de vitamina C. Acta Amazonica 32(1):169–174

Zappi, Daniela C.; Milliken, William; Lopes, Celia Regina A. Soares; Lucas, Eve; Piva, José Hypólito; Frisby, Sue; Biggs, Nicola; Forzza, Rafaela Campostrini. 2016. Xingu State Park vascular plant survey: filling the gaps. Brazilian Journal of Botany 39(2):751–778

Chapter 39
Opiliaceae

39.1 *Agonandra brasiliensis*

Brazil Marfim, marfim de veado, marfim preto, marfim azeitona, pau marfim, muirapuamarana, agonandra, cerveja de pobre, cervejinha, quina de veado, quina doce, tatu, tinge cuia, quinze cuia; kangwaruhumyra, parei'a'y (Ka'apor)

Venezuela Aceituna, aceituno

Status Wild

Marfim *Agonandra brasiliensis* fruits. Pedra do Navio, adjacent to Paredão Valha me Deus, left margin of the Xingu River 8 km upstream from Altamira, Pará, Brazil 8-31-15

N. Smith, *Amazon Fruits: An Ethnobotanical Journey*, Ethnobiology,
https://doi.org/10.1007/978-3-031-12803-5_39

Marfim *Agonandra brasiliensis* grows up to 30 m tall in upland and floodplain forests in Amazonia, the Guianas, Venezuela, and south to Paraguay. It is also found in scrub savanna (*cerrado, cerradão*) and some grasslands in the Brazilian Amazon (Dantas and Rodrigues 1982; Goulart et al. 2013; Guarim Neto et al. 1994; Medeiros and Walter 2012; Medeiros et al. 2008, 2012; Oliveira et al. 2018; Oliveira Filho et al. 1989; Pereira et al. 2015; Silva 2011: 42) as well as on some sand dunes along the coast of northeastern Pará (Santos 2010:44). This tree is also found on some sandstone outcrops in the Upper Xingu watershed (Zappi et al. 2016). This fruit tree arises spontaneously in some home gardens, such as on sandy soils in the savanna region of Roraima (Pinho 2008:38).

Marfim *Agonandra brasiliensis* starts producing fruit when only 2 m high. The taste of this fruit is not particularly appealing, so it does neither appear in markets nor is it commonly consumed. The red-footed tortoise *Chelonoidis carbonaria* also eats this fruit (Wang et al. 2011), and some rural people in the Amazon hunt this tortoise for its meat. The pale orange fruits of marfim are gathered in Maranhão to make soap. This practice appears to be on the decline as rural incomes increase and people buy soap in stores. The tree also provides valuable timber. The Ka'apor in the eastern fringes of the Amazon forest in Maranhão prepare a remedy from the bark to treat skin infections.

References

Dantas, Mário; Rodrigues, Irenice Alves. 1982. Estudos fitoecológicos do trópico úmido Brasileiro: IV-levantamentos botânicos em campos do Rio Branco. EMBRAPA, Centro de Pesquisa Agropecuária do Trópico Úmido, Boletim de Pesquisa, no. 40

Goulart, Letícia Silveira; Teles, Helder Lopes; Mendes, Vanessa Alves; Vieira, Marizete Cavalcante de Souza; Moura, Selma Vieira de; Ramon, Júliade Lima; Souza, Jacqueline Marques; Vieira, José Cavalcante Souza; Campos, Érica Pereira de. 2013. Prospecção antifúngica em Agonandra brasiliensis. Revista Brasileira de Farmácia 94(3):289–294

Guarim Neto, Germano; Guarim, Vera Lucia M. S.; Prance, Ghillean T. 1994. Structure and floristic composition of the trees of an area of cerrado near Cuiabá, Mato Grosso, Brazil. Kew Bulletin 49(3):499–509

Medeiros, Marcelo Brilhante de; Walter, Bruno Machado Teles. 2012. Composição e estrutura de comunidades arbóreas de cerrado stricto sensu no norte do Tocantins e sul do Maranhão. Revista Árvore 36(4): 673–683

Medeiros, Marcelo Brilhante de; Walter, Bruno Machado Teles; Silva, Glocimar Pereira. 2008. Fitossociologia do cerrado stricto sensu no município de Carolina, MA, Brasil. Cerne 14(4):285–294

Medeiros, Marcelo Brilhante de; Walter, Bruno Machado Teles; Silva, Glocimar Pereira da; Gomes, Beatriz Machado; Lima, Isabel Lustz Portela; Silva, Suelma Ribeiro; Moser, Pamela; Washington, Luís Oliveira; Cavalcanti, Taciana Barbosa. 2012. Vascular flora of the Tocantins River middle basin, Brazil. Check List 8(5):852–885

Oliveira Filho, Ary Teixeira de; Shepherd, George John; Martins, Fernando Roberto. 1989. Environmental factors affecting physiognomic and floristic variation in an area of cerrado in central Brazil. Journal of Tropical Ecology 5: 413–431

Oliveira, Lucicléia Mendes de; Sousa, Rhonan Martins de; Correa, Norma Elena Reynosa; Santos, André Ferreira dos; Giongo, Marcos. 2018. Florística e síndromes de dispersão de um fragmento de cerrado ao sul do estado do Tocantins. Scientia Agraria Paranaensis 17(1): 104–111

Pereira, Firmino Cardos; Lenza, Eddie; Maracahipes-Santos, Leonardo; Mews, Henrique Augusto; Gomes, Letícia; Lima, Stefany Lorranyny; Santos, Kelyane Silva Melo. 2015. Comparação dos métodos de parcelas e pontos-quadrantes para descrever uma comunidade lenhosa de cerrado típico. Biotemas 28(2): 61–72

Pinho, Rachel Camargo de. 2008. Quintais agroflorestais indígenas em áreas de savana (lavrado) na Terra Indígena Araçá, Roraima. Master's thesis, INPA/UFAM, Manaus

Santos, Luciana Oliveira dos. 2010. Comparação da vegetação lenhosa entre borda e interior de uma mancha de mata de restinga na RESEX Mar Caeté-Taperaçu, Bragança, Pará. Master's thesis, Universidade Federal Rural da Amazônia/Museu Paraense Emílio Goeldi, Belém

Silva, Silvestre. 2011. Fruits of the Brazilian Amazon. Metalivros, São Paulo

Wang, Ellen; Donatti, Camila I.; Ferreira, Vanda L.; Raizer, Josué; Himmelstein, Jeffrey. 2011. Food habits and notes on the biology of *Chelonoidis carbonaria* (Spix 1824) (Testudinidae, Chelonia) in the southern Pantanal, Brazil. South American Journal of Herpetology 6(1):11–19

Zappi, Daniela C.; Milliken, William; Lopes, Celia Regina A. Soares; Lucas, Eve; Piva, José Hypólito; Frisby, Sue; Biggs, Nicola; Forzza, Rafaela Campostrini. 2016. Xingu State Park vascular plant survey: filling the gaps. Brazilian Journal of Botany 39(2):751–778

Chapter 40
Passifloraceae

40.1 *Passiflora costata*

Brazil Maracujá de capoeira

Venezuela Parcha

Status Wild, spontaneous in cultural settings

N. Smith, *Amazon Fruits: An Ethnobotanical Journey*, Ethnobiology,
https://doi.org/10.1007/978-3-031-12803-5_40

Maracujá de capoeira *Passiflora costata* in successionary growth in an abandoned cattle pasture. Farm of Sebastião Silva, 74 years, Ilha do Poção, middle Xingu River upstream from Altamira, Pará, Brazil 9-1-15

Maracujá de capoeira *Passiflora costata* is a vine in successionary vegetation and forests that flowers in the dry season and fruits during the rainy period. Locals eat the green fruits which are not taken to markets. *Passiflora costata* is found on the floodplains of clear, black, and sediment-rich rivers but is nowhere common. This wild relative of passionfruit is confined to the Amazon and Orinoco Basins and parts of the Guianas (Hernández and Bernal 2000; Tillett 2003).

40.2 *Passiflora foetida*

Bolivia Aguratigua, pachio, lokhoste

Brazil Maracujá da capoeira, maracujá de cheiro, maracujá fedido, maracujá do campo, maracujá do mato, maracujá de cesto, maracujá de estalo, maracujá morceguinho, maracujá de cobra, maracujá de estalo, maracujá de papoco, maracujá de pipoco, maracujá de praia, maracujá poca, maracujá fedorento, maracujázinho, maracujázinho do mato, marva, urubú-muracajá; murukuia (Tukano)

Colombia Cinco chagas, granadilla de monte; dárua (Nukak)

Ecuador Bedoca, granadilla, granadilla silvestre, berdulaga; mangu tsi tsi muyo (Kichwa), patúkmai munchi (Shuar)

Peru Campanita, bolsa mullaca, puchu puchu, puchepuche, granadilla, granadilla de culebra, granadilla del sapo, granadilla liana arbustiva, bedoque, bedoca, ñorbo cimarrón, ñorbo hediondo, puro puro, tuna; anchi múnchi, kistian múnchi (Aguaruna)

Venezuela Bolita de gato, parcha culebra, parcha sabanera, parchita; morocuyu (Baniwa), ewápa (Curripaco)

Status Wild, spontaneous in cultural settings

Maracujá da capoeira ***Passiflora foetida*** **in flower and fruit. Ubim, Lago Amanã, Amazonas, Brazil, 9-18-12**

Passiflora foetida is different in two regards from most of the edible passion fruits. The pale orange to yellow fruits are quite small, the size of a cherry tomato, some 1.5–2 cm by 2 cm. The fruits are covered by a web-like covering with small protruding appendages (technically a feathery bract). This covering is removed and the skin of the fruits is broken to swallow the slippery pulp (known technically as a gelatinous aril). The sweet, pale gray pulp is ingested along with the numerous black seeds, such as in communities along the shores of Lago Amanã along the lower Japurá in Amazonas, Brazil. In the Ecuadorian Amazon, rural people sometimes use the fruits to make juice. Not all people in the Amazon are fond of this fruit. Some colonists along the Transamazon Highway in the Municipality of Medicilandia, for example, do not consider the fruits edible, perhaps because they migrated from southern Brazil. Several indigenous groups are fond of the fruit including the Nukak in the northwestern Colombian Amazon (Politis 1996: 177). The Macuxi in Roraima in the northern Brazilian Amazon make a decoction of this vine to treat malaria (Milliken 1997). In Guyana, the whole plant is boiled to make a tea to expel intestinal nematodes and flatworms (Austin and Bourne 1992). Seeds of *Passiflora foetida* have been found at the Maicura archaeological site along the Rio Purité, an affluent of the lower Içá in the Colombian Amazon, which was occupied from AD 880 to AD 1190 (Morcote-Ríos 2006, 2008: 94).

Maracujá da capoeira *Passiflora foetida* **fruit gathered in successionary vegetation. Ubim, Lago Amanã, Amazonas, Brazil, 9-18-12**

After fruiting within a year the vine dies, unlike some other species in this genus which are perennial. The flowers are sweet smelling and open in the morning and typically close by noon; the species name refers to the foul smell that is emitted when the hairs on the leaves and fruit covering are crushed (Ulmer et al. 2004: 291). However, the late Richard Spruce attributes the species name to the smell of the flowers:

> one of the commonest of tropical weeds, and unique in a tribe whose flowers exhale such exquisite odors for its heavy narcotic smell, quite recalling that of the roosting-places of the Urubú or turkey-buzzard, whence its Indian name Urubú-muracajá (Spruce 1908: 6).

Passiflora foetida is widespread in successionary vegetation on uplands and occasionally on floodplains in the Amazon, including in the understory of floodplain forests in the estuary (Jardim et al. 2013; Machado 2018: 125; Maués et al. 2011; Silva et al. 2013). It is normally found in low-lying vegetation in the vicinity of rural dwellings and fields and takes advantage of fencing to attach itself when climbing. In the Andes, it is found as high as 2800 m. On the northeast coast of Pará and along the coast of neighboring Maranhão it thrives on sand dunes (Albuquerque et al. 2007; Santos and Bastos 2014) and it also occurs in savanna woodlands (*cerrado*) in the southern part of the Amazon Basin (Medeiros et al. 2012) as well as along the margins of patches of white sandy soils (*campinas*) in Acre in southwestern Amazonia (Daly et al. 2016). In the Orinoco watershed in eastern Colombia, this wild relative of passionfruit grows in savannas (Parra 2006).

Many species of passionfruit are pollinated by hummingbirds which are native to the Americas but some insects also pollinate some of the passion fruits; for example, the *Centris flavifrons* bee has been recorded visiting the flowers of *Passiflora foetida* in Maranhão, Brazil (Rêgo et al. 2008) and six other bee species (*Apis mellifera*, *Centris leprieuri*, *Mesoplia rufipes*, *Partamona helleri*, *Pseudaugochloropsis pandora*, *Xylocopa carbonaria*) have been observed visiting the flowers of this plant in sand dune vegetation near São Luís, Maranhão (Gottsberger et al. 1988). In the Peruvian Amazon, diurnal butterflies *Dyadula phaetusa* and *Dryas iulia alcionea* visit the flowers of this plant thus likely pollinating them (Bardales et al. 2017).

This species of passionfruit also occurs outside of the Amazon Basin such as in French Guiana (Hoff and Cremers 2005) and in irrigated farmland along the Pacific Coast of Peru where the fruits are also gathered and eaten. In Surinam which borders Roraima in the Brazilian Amazon Saramaccan maroons prepare a bathing solution using the whole plant to treat pustules on the skin of babies (Ruysschaert et al. 2009). In southern Venezuela bordering the Amazon Basin children are particularly fond of this fruit (Tillett 2003). On mountainous Dominica in the eastern Caribbean, I have seen children avidly pick the fruits along roadsides and eat them on the spot. Native to Tropical America, this species has been introduced to other regions, including tropical Asia, Africa, Australia, and the Pacific, where it has become naturalized (Fosberg 1960; Purseglove 1974: Ulmer et al. 2004: 291).

40.3 *Passiflora glandulosa*

Brazil Maracujá do mato, maracujá de paca; tayrakem-oho (Arawaté), mainumikwe-'a (Guajá)

Status Wild, spontaneous in cultural settings

Maracujá do mato ***Passiflora glandulosa*** **in successionary vegetation. Benedito 10 years. Jocojó, Rio Jocojó, Município de Gurupá, Pará, Brazil 10-22-16**

Passiflora glandulosa produces orange-yellow fruits toward the end of the dry season which are eaten by rural people, such as along the Xingu River in Pará. Some game animals also eat the fruit as suggested by one of the common names for this plant at the mouth of the Xingu: maracujá de paca. Paca *Agouti paca* is a nocturnal rodent the size of a bulldog and one of the most esteemed game animals in Amazonia. Black spider monkeys *Ateles paniscus* also eat the entire fruits (Van Roosmalen 1985; Van Roosmalen and Klein 1988) and thus likely help disperse the seeds. These monkeys are hunted for their meat throughout the Amazon and neighboring Guianas. The smaller northern bearded saki *Chiropotes satanas chiropotes* and the white-faced saki *Pithecia pithecia* have also been observed feeding on the mesocarp of this fruit in Surinam and are also likely involved in dispersing some of the seeds (Norconk and Veres 2011; Van Roosmalen et al. 1981, 1988). These sakis also occur in the northern part of the Brazilian Amazon. Similarly, the bare-faced tamarin *Saguinus bicolor bicolor*, which is also too small to be hunted for meat, also eats these fruits in successionary forests on uplands, including stunted woodland (*campinarana*) growing on sandy patches (Egler 1992). Squirrel monkeys *Saimiri sciureus*, which are not normally hunted for their meat because of their small size, also eat this fruit and likely disperse some of the seeds (Santos et al. 2018). The bare-tailed woolly opossum *Caluromys philander*, which occurs in northern South America, is not normally hunted for its meat but also eats this fruit and likely spreads some of the seeds (Charles-Dominique et al. 1981).

The red-flowered vine grows in successionary vegetation, including regrowth forest, on uplands, such as in the watershed of the Urucu River, a southern affluent of the Solimões (Cruz et al. 2013), in the vicinity of Paragominas in eastern Pará (Gerwing and Uhl 2002), and among trees planted to produce pulp on the Jari plantation in northern Pará (Pires 1992). It is also found occasionally in stunted woodlands (*campina*) on sandy soils including sand dunes along the northeast coast of Pará, savannas, and on some river banks (Santos and Bastos 2014: 199). It is also found in open vegetation growing on iron-rich sedimentary rock (*canga*) on hilltops in the Serra dos Carajás in Pará (Koch and Ilkiu-Borges 2016; Mota et al. 2018). *Passiflora glandulosa* is mostly confined to the Brazilian Amazon, the Orinoco Basin, and the Guianas (Granville 1988; Ulmer et al. 2004: 293); however, it also occurs in the Colombian Amazon (Hernández and Bernal 2000). This passionfruit species also occurs sporadically in parts of the Brazilian Northeast, such as Ceará and Pernambuco (Lima-Neto et al. 2017). This vine can attain 10 m and thus thrives in forests.

The attractive flowers have caught the attention of some nurseries specializing in ornamental plants although it is not yet widely cultivated. This species is a perennial and can last some 15 years so it may grace more gardens and greenhouses in the tropics and subtropics down the road.

40.4 *Passiflora nitida*

Bolivia Pachio

Brazil Maracujá do mato, maracujá da mata, maracujá do cerrado, maracujá do rato, maracujá da roça, maracujá da capoeira, maracujá de cheiro, maracujá suspiro, maracujá bravo, maracujazinho; mootoro akadétta (Baniwa), murukuya-pinĩ (Ka'apor), murukui'a (Sateré-Maué), marakuza (Tembé); burucuya, murukuia (Tukano), meretxa, myretxi (Waimiri-Atroari)

Colombia Granadilla, granadilla bolsosa, granadilla silvestre, granadilla de monte, burucuña; duda (Andoke), ee-djá (Makuna), murucuña (Wanano), usillao (Witoto)

Ecuador Granadilla; dayomemenica (Waorani)

Peru Granadilla, granadilla de olor, poropuer, purupuru, maracuyá; kaya munchi, munchi, tseré munchi, washi munchi (Aguaruna), kaxori (Kakataibo), poşhodi (Matses)

Venezuela Parcha rebalsera, bejuco parcha, parcha

Status Wild, spontaneous in cultural settings, planted

Maracujá da capoeira *Passiflora nitida* in successionary growth occupying a harvested manioc field. Seringalzinho, Rio Jau, Amazonas, Brazil, 10-2-12

Passiflora nitida is a vine of disturbed sites and forest edges and light gaps on floodplains and uplands throughout Amazonia, the Orinoco Basin, the Guianas, and into Central America as far north as Costa Rica (Ferreira et al. 2013a, c; Hernández and Bernal 2000; Killip 1938: 374; Muñoz 2010: 27; Silva 2011: 188; Ulmer et al. 2004: 305). It also occurs in stunted woodlands (*campina*) on sandy soils, scrub savanna (*cerrado*) south of the Amazon Basin in central Brazil as well as in gallery forests extending into Bahia and Goiás (Braga et al. 2006; Lorenzi et al. 2006a, b: 270). It also thrives in estuarine vegetation in coastal Maranhão (Freire and Monteiro 1993), on some sand dunes along the coast of northeastern Pará (Santos and Rosário 1988), and also occurs as a weed in some cattle pastures, such as in the vicinity of Paragominas in Pará (Dantas and Rodrigues 1980).

A relative of the cultivated passionfruit *Passiflora edulis* which was domesticated in southern Brazil, the fruits of *Passiflora nitida* are usually orange when ripe, although some vines produce yellow fruits. The spongy skin that encases the jelly-like sack full of seeds is much thicker than in *Passiflora edulis* which is known as maracuyá in Peru and maracujá in Brazil. Furthermore, *Passiflora nitida* fruits are slightly oblong, measuring some 5–11 cm long by 4–8 cm in diameter, in contrast to the round fruits of *Passiflora edulis* and lack the strong perfume of the latter. Other cultivated passion fruits in Peru either have much larger fruits, such as *Passiflora quadrangularis*, or are grown in the high Andes, such as *Passiflora ligularis*. The blue and white flowers of *Passiflora nitida* are attractive and this species could be a candidate for an ornamental plant someday.

Maracujá da capoeira *Passiflora nitida* **with orange fruits in early successionary growth. Seringalzinho, Rio Jau, Amazonas, Brazil, 10-2-12**

People use their thumbs or teeth to break open the skin and rupture the underlying opaque membrane that contains sweet, translucent pulp. The 50–100 gray seeds, 0.7–0.9 cm long by 0.5 cm wide, are usually swallowed along with the slippery pulp but sometimes they are spat out. Either way, people are involved in dispersing the seeds. According to inhabitants of the Urubu River in Amazonas, Brazil, monkeys eat and disperse the seeds of *Passiflora nitida*. Howler monkeys *Alouatta belzebul* are known to eat the fruits in eastern Amazonia (Pinto et al. 2003), and woolly monkeys *Lagothrix lagotricha*, bald uakaris *Cacajao calvus*, monk sakis *Pithecia monachus*, golden-faced saki monkey *Pithecia pithecia chrysocephala*, and golden and squirrel monkeys *Saimiri sciureus* have been observed feeding on these fruits in various parts of the Amazon (Setz et al. 1999; Voss and Fleck 2011). Widespread capuchin monkeys *Cebus apella* are known to eat this fruit and likely disperse some of the seeds (Gregory 2006: 55). In the Peruvian Amazon, a spider monkey *Ateles belzebuth chamek* eats this fruit and disperses some of the seeds (Palma and Stevenson 2013). Red brocket deer *Mazama americana* are also fond of the fruit (Barboza 2008: 56). The ruderal vine is also known as maracujá do rato (the rat passionfruit) in the vicinity of Santarém and Gurupá, suggesting that small rodents are also involved in dispersing the seeds. Both howler monkeys and brocket deer are hunted for their meat, so *Passiflora nitida* also contributes indirectly to the human diet.

Joelson, 11 years, gathering maracujá da roça *Passiflora nitida* **fruits in successionary vegetation. Outskirts of Jocojó village, Rio Jocojó, Município de Gurupá, Pará, Brazil 4-23-15**

Adults and children alike snack on the fruits when they encounter them along trails in successionary vegetation. This vine is also encountered clambering on manioc plants, such as in the vicinity of Jocojó, a quilombo settlement along a river of the same name in the municipality of Gurupá. Rural folk also sometimes bring the fruits home to eat them later, as in the village of Carrazedo at the mouth of the Xingu. The fruits are sometimes used for making juice, such as in the village of Yarina along the Yanayacu River in the Pacaya-Samiria National Reserve in Peru. In some areas of the Amazon Basin, this plant is used in folk medicine. For example, the Makuna in the Colombia Amazon consider this fruit helpful to the elderly when they are losing their eyesight (Schultes and Raffauf 1990: 360) and in some parts of the Brazilian Amazon, the leaves are used to make an infusion for treating gastrointestinal disorders (Vargas et al. 2016).

Mikeli, 11 years, eating a maracujá da roça ***Passiflora nitida*** **fruit she gathered in successionary vegetation. Outskirts of Jocojó village, Rio Jocojó, Município de Gurupá, Pará, Brazil 4-23-15**

A farmer in Taperebatuba, a small community along the lower Rio Urubu in Amazonas, Brazil, has transplanted a *Passiflora nitida* seedling he encountered in successionary vegetation adjacent to his manioc field to his home garden. This species of passionfruit is also encountered in some home gardens of rural folk along the Ramal do Pau Rosa, a side-road north of Manaus leading off from the BR 175 Manaus-Rio Branco highway (Machado 2016: 36). It is also cultivated in parts of the Peruvian Amazon (Flores 1997: 109; Vásquez 1997: 571) and by the Yuracaré and Trinitario indigenous people in the Bolivian Amazon (Thomas and Van Damme 2010).

Francisco, 52 years, with maracujá do mato *Passiflora nitida* bearing blue flowers in his home garden. He transplanted this vine-like shrub as a seedling from successionary vegetation. Taperebatuba, Rio Urubu, Amazonas, Brazil 4-3-15

The mesocarp of the widely cultivated *Passiflora edulis*, in contrast, is tart and therefore mixed with sugar and water to make juice. Typical of a wild fruit, the edible portion of *Passiflora nitida* is relatively small; the thick, spongy skin occupies at least half the fruit. *Passiflora nitida* fruits mainly during the dry season although it is occasionally encountered with fruit during the rainy period such as in the Pacaya-Samiria National Reserve in the Peruvian Amazon and along the lower Xingu and its affluents, such as the Ipixuna.

Vendor with maracujá do rato *Passiflora nitida* **fruits in a pile (center) and cultivated passionfruit** *Passiflora edulis* **fruits suspended in net bags (on right). Santarém, Pará, Brazil 5-13-96**

Known as granadilla in Peru and various names in Brazil such as maracujá do mato (the weedy maracujá), the fruits of *Passiflora nitida* are sold in street markets and along roads in many parts of Amazonia. In Brazil, for example, the fruits appear in the markets of Santarém in Pará, Rio Preto da Eva near Manaus, and in Rio Branco, Acre (Lima 2014). In Peru, they are sold in various towns including Nauta along the lower Marañón, Pucallpa along the Ucayali, and Iquitos along the Amazon.

Graciela, 19 years, with her baby daughter selling granadilla *Passiflora nitida* fruits at a roadside stand. Matichacra, km 35 Tarma-San Ramón highway, Junin, Peru 11-4-04

40.5 *Passiflora quadrangularis*

Bolivia Badea, quijón, tumbo

Brazil Maracujá açu, maracujá grande, maracujá graúdo, maracujá gigante, maracujá mamão, maracujá melão, maracujá do mato, maracujá de kilo; murukui'a (Sateré-Maué), murukuia pahkã (Tukano)

Colombia Badea, badeilla, estebadea, tumbo; riama usillao (Witoto)

Ecuador Badea, granadilla, bate, tumbo, parcha dulce, parcha granadina, parcha real, quijón, taxo; badea (Kichwa), tasiɨ (Siona-Secoya)

Peru Tumbo, tumbo silvestre, tumbo gigante, tumbo chino, tumbo de Chiclayo, tumbo costeño, granadilla, granadilla de monte, granadilla real, granadilla gigante, maracuyá cascudo, badea, apincoya, barbadina, ajo sacha; munchi, túmpu daek (Aguaruna), exuberu (Cashibo), non kaxori (Kakataibo), tsimoritoki (Matsigenka), túmpu, tánku (Mayna)

Venezuela Parcha granadilla, granadilla

Status Wild, spontaneous in cultural settings, planted

Tumbo *Passiflora quadrangularis* vine in fruit climbing a planted taperiba *Spondias dulcis* tree in a home garden. Bretaña, Rio Puinahua (a branch of the Ucayali), Loreto, Peru 4-16-06

Tumbo *Passiflora quadrangularis* is easily recognized because it has the largest fruits in the genus, which accounts for its common names in Brazil: maracujá açu and maracujá grande (the big passionfruit). One of the common names for the fruit in Peru is tumbo gigante, with good reason. The oblong fruits are 10–30 cm long and 10–18 cm wide (Leal 1990; Martin and Nakasone 1970), orders of magnitude larger than other cultivated or wild species of *Passiflora* encountered in Amazonia. They can weigh up to 4 kg (Ulmer et al. 2004: 68). Giant granadilla, as the fruit is known in English, reaches the dimensions of a small watermelon. In Peru, tumbo fruits are eaten fresh and made into juice and their hardy skins allow them to be transported to urban markets, such as Iquitos, with little or no damage. The vine can grow as long as 10 m and the fruits are produced during the rainy and dry seasons.

Vendor with a tumbo *Passiflora quadrangularis* fruit in a street market. Mercado Belén, Iquitos, Loreto, Peru 8-2-03

Likely domesticated in montane forests that cloak the valleys of Amazon head-waters in the central or northern Andes, tumbo was probably introduced to Brazil during colonial times (Ducke 1946; Lorenzi et al. 2006b: 272). A wild form of *Passiflora quadrangularis* with relatively small fruits is found from Colombia north to Nicaragua (Ulmer et al. 2004: 311). The vine is mostly found under cultivation but occurs occasionally in successionary forests and in light gaps within or at the edges of mature forests on uplands and floodplains. In the Ecuadorian Andes, it can be found at altitudes of 2200 m (Popenoe 1924).

Blue-flowered *Passiflora quadrangularis* is widely cultivated on trellises and fences in backyards and fields throughout the western portion of the Amazon basin from Bolivia north to Colombia. Sometimes the planted vine climbs fruit trees rather than man-made structures. Tumbo thrives as high as 1800 m in the Andes, such as in Oxapampa in the Chontabamba Valley in Peru. Farmers in the dry, irrigated valleys on the Pacific side of the Andes in Peru also grow the vine, such as along the Jequetepeque and Silencio rivers in the northern part of the country. This practice may have been developed by indigenous cultures thousands of years ago. By the colonial period, at least tumbo was grown in irrigated fields near Lima (Jaramillo-Arango 1952: 40). Giant granadilla is also grown in southern Venezuela bordering the Amazon (Tillett 2003) and in Central America and the Caribbean and was introduced to tropical Asia in the colonial period (Morton 1987: 329). In some areas where it has been introduced, such as Hawaii and other islands in the Pacific, the vine has become naturalized and is considered a weed.

In Peru, tumbo is planted on both floodplains, as along the Ucayali, and on upland sites, such as in Cumbaza, a village near Tarapoto. Tumbo produces fruit for several years, but it eventually succumbs to diseases, pests, or senescence. Some farmers take advantage of the investment in poles and wires to trellis other climbing crops. Near Lamas, a coffee-growing region in the Mayo Valley, Peru, for example, one farmer planted caihua *Cyclanthera pedata*, a climbing squash, after his tumbo yields declined. Sometimes farmers interplant tumbo with other crops, such as in the vicinity of San Pedro de Cumbaza in San Martin, Peru, rather than adopt a relay race strategy. In the Ecuadorian Amazon, the Siona-Secoya gather *Passiflora quadrangularis* fruits in the wild (Vickers 1994).

Passiflora quadrangularis has several other uses. Various parts of the plant are used in folk remedies such as to make painkilling ointments and to prepare sedatives (Ulmer et al. 2004: 66–67). In the Chimborazo district of Ecuador, the leaves of this plant are soaked for 10 minutes then the infusion is drunk three times a day to lower cholesterol and control levels of glucose in the blood. Also in the Chimborazo district, the leaves are mashed and applied to swollen areas of the body (Escalante 2016: 58). Tumbo produces attractive iris-blue flowers and is thus planted as an ornamental as well as for its fruits, as in Oxapampa. Indeed, seeds of *Passiflora quadrangularis* can be purchased from online nurseries specializing in tropical plants, which sometimes refer to the plant as "passion melon."

Bionic, 11 years, eating a tumbo *Passiflora quadrangularis* fruit he picked in his home garden. Manco Capac, Rio Puinahua (a branch of the Ucayali River), Reserva Nacional Pacaya-Samiria, Loreto, Peru 4-14-06

40.6 *Passiflora quadriglandulosa*

Brazil Maracujá do rato

Colombia Granadilla

Ecuador Anto wa yu (Siona-Secoya)

Peru Granadilla, granadilla agria, granadilla del monte; munchi (Achuar), julung (Kichwa)

Status Wild, spontaneous in cultural settings

Maracujá do rato *Passiflora quadriglandulosa* in flower with immature fruit in early successionary growth. Seringalzinho, Rio Jau, Amazonas, Brazil, 10-2-12

Passiflora quadriglandulosa is a perennial vine in successionary vegetation and forests on uplands and floodplains along the Amazon and its tributaries. It is also found in parts of Venezuela and the Guianas (Tillett 2003; Ulmer et al. 2004: 311). This plant occurs as a weed in some cattle pastures, such as Fazenda Aruanã at km 232 of AM 10 highway connecting Manaus with Itacoatiara, Amazonas (Dantas 1979). It sports brilliant red flowers that can reach 15 cm in diameter and this species may be a candidate for domestication as an ornamental plant. Not many people eat the fruits because they are sour, but some rural folk in the Peruvian Amazon do. Ripe fruits drop to the ground where agoutis (*Dasyprocta* spp.) go after the seeds embedded in the pulp. These terrestrial rodents, the size of a dachshund, are popular game animals throughout Amazonia and are hunted during the day. *Passiflora quadriglandulosa* thus contributes indirectly to the rural diet via game meat. The Achuar in the Peruvian Amazon crush the leaves, boil them in water, and then drink the tea when it cools to treat ulcers. This vine is found in Amazonia, parts of Venezuela, the Caribbean, and the Guianas.

40.7 *Passiflora riparia*

Bolivia Pachio grande

Brazil Maracujá bravo, maracujá de cheiro, maracujá do igapó, maracujá de cacho, maracujá suspiro, maracujazinho; murukujapini (Ka'apor), murukuia (Tukano)

Ecuador Muchi

Peru Granadilla, granadilla blanca, granadilla amarilla, sacha granadilla, washi munchi, lulu, puru puru, chinchorcon; tseje munchi, washi munchi (Aguaruna), yeychotsocop (Amuesha), unushupi (Chayahuita)

Status Wild

Granadilla *Passiflora riparia* along the shore of a floodplain lake. Cocha Grande, near San Carlos, Rio Puinahua (a branch of the Ucayali), Reserva Nacional Pacaya-Samiria, Loreto, Peru 4-18-06

The fruits of *Passiflora riparia* are distinguished from other edible fruits in this genus by their yellow skin with green patches and light gray spots. In addition, the skin is bumpy unlike most of the edible wild passion fruits that have smooth epicarps. Other wild species of *Passiflora* with edible fruit that sometimes occur in the same area as *Passiflora riparia* have different colors and skin textures. The fruits of *Passiflora nitida*, for example, are uniform orange or yellow-orange and smooth-skinned, whereas the fruits of *Passiflora serratodigitata* are shiny green. The globular fruits of *Passiflora riparia* are some 3–5 cm in diameter and are broken open with the fingers to access the translucent pulp surrounding the numerous dark gray seeds. The fruits are usually eaten rather than made into juice.

This riparian passionfruit is found wild throughout the Amazon and Orinoco Basins up to 2000 m. It is also found in parts of the Caribbean, likely introduced there. A vine found mostly along lake shores and the banks of rivers and streams, fruiting coincides with the annual flood. People can thus gather the fruits easily from canoes. The fruits float and are also dispersed by fish (Hoehne 1946: 62).

The fruits occasionally reach markets, such as Nauta on the lower Marañón. This vine, which can attain 7 m, is also found on seasonally flooded savannas, as in the Bolivian lowlands. It also occurs occasionally in disturbed upland sites, such as in the eastern fringes of the Amazon forest in Maranhão and parts of the Peruvian Amazon. The presence of *Passiflora riparia* in upland successionary vegetation may be the result of human introductions, either deliberate or inadvertently from discarded seeds.

Bildoso, 12 years, eating granadilla *Passiflora riparia* fruit purchased in a street market. Outskirts of Nauta, Rio Marañón, Loreto, Peru 4-28-06

40.8 *Passiflora serratodigitata*

Bolivia Vatisira

Brazil Maracujá pedra, maracujá do rato, maracujá do mato, maracujá de cobra, maracujá de urubu, maracujá de boi, maracujá de cinco pernas, tagua tagua; bo-e-ra (Ticuna)

Colombia Granadilla, granadilla de monte, cocorilla

Ecuador Granadilla, granadillo, maracuyá de monte

Peru Puropuro, granadilla; munchi (Achuar), munchimas, daék munchi (Aguaruna)

Status Wild, spontaneous in cultural settings

Puropuro *Passiflora serratodigitata* in successionary vegetation on the outskirts of a village. San Carlos, Rio Puinahua (a branch of the Ucayali), Reserva Nacional Pacaya-Samiria, Loreto, Peru 1-12-06

Puropuro *Passiflora serratodigitata* grows on floodplains and occasionally uplands, but seems to be more common in seasonally flooded areas. This vine can grow some 10 m long when mature and is generally found in successionary vegetation. *Passiflora serratodigitata* occurs sporadically in disturbed sites in the Amazon and Orinoco Basins as well as the Guianas and Panama (Parrotta et al. 1997; Ulmer et al. 2004: 318). It occurs as high as 2200 m in the Andes. This wild passionfruit is also found in the cerrado region of central Brazil (Braga et al. 2006). The vigorous vine also thrives on a few Caribbean Islands, such as Puerto Rico, Trinidad, and Martinique.

The shiny fruits are green when ripe and are born in the rainy season when locals, especially children, gather the fruits. The round fruits are 4–9 cm in diameter with a spongy skin, about 1.5 cm thick. People break open the skin with their teeth or fingers to access the sweet, gelatinous pulp. The pale orange pulp is ingested along with the numerous small gray seeds. Sometimes, though, people spit the seeds out. Most fruits are gathered and eaten in the wild, but it also arises spontaneously in some home gardens, such as at San Carlos along the Rio Puinahua, a branch of the Ucayali. Maracujá do rato, one of the common names for *Passiflora serratodigitata* in the Brazilian Amazon, suggests that rodents are also involved in dispersing the seeds. It is also occasionally planted in the Amazon (Lorenzi et al. 2006b: 273). In the cerrado region of central Brazil bordering the Amazon Basin in the south, some people cultivate this species in their home gardens, such as in the states of Mato Grosso and Tocantins (Braga et al. 2006).

César, 2 years, eating a puropuro *Passiflora serratodigitata* **fruit on the shoulders of his brother Kevin, 9 years. San Carlos, Rio Puinahua (a branch of the Ucayali), Reserva Nacional Pacaya-Samiria, Loreto, Peru 1-12-06**

As with several other species in the genus, *Passiflora serratodigitata* is planted as an ornamental outside of its native range, such as in Europe. People in colder climates appreciate the purple-mauve and white flowers, but they must be grown in glasshouses to avoid frost damage. Seeds of *Passiflora serratodigitata* are readily obtainable from online nurseries.

40.9 *Passiflora spectabilis*

Peru Granadilla, pucu-pucu

Status Wild

Kiener, 10 years, eating a granadilla *Passiflora spectabilis* **fruit he gathered in early successionary growth on sandy soil. Side-road at Km 22 Iquitos-Nauta highway, Loreto, Peru 7-5-06**

Passiflora spectabilis appears to be confined to the Peruvian Amazon where it is found as high as 1600 m in the foothills of the Andes. It grows in the forest and disturbed sites on uplands, especially in sandy areas. It is rare and the fruits do not appear in markets. However, rural people snack on the yellow fruits when they encounter them near their fields during the dry season.

References

Albuquerque, Patricia Maia Correia de; Camargo, João Maria Franco de; Mendonça, José Ângelo Cordeiro. 2007. Bee community of a beach dune ecosystem on Maranhão island, Brazil. Brazilian Archives of Biology and Technology 50(6):1005–1018

Austin, Daniel F.; Bourne, Godfrey R. 1992. Notes on Guyana's medical ethnobotany. Economic Botany 46(3):293–298

Barboza, M. S. L. 2008. O preço que a natureza pagou e os efeitos colaterais que sofremos para hoje se ter energia: uso dos recursos animais e percepção dos impactors entre os ribeirinhos do Lago de Tucuruí (PA). Master's thesis, Universidade Federal do Pará, Nucleo de Altos Estudos Amazônicas, Programa de Pós-Graduação em Desenvolvimento Sustentável do Trópico Úmido, Belém.

Bardales, Joel Vásquez; Gómez, Ricardo Zárate; Canaquiri, Percy Huiñapi; Jiménez, Julio Pinedo; Hernández, Juan José Ramírez; Lamas, Gerardo; García, Pedro Vela. 2017. Plantas alimentícias de 19 especies de mariposas diurnas (Lepidoptera) en Loreto, Perú. Revista Peruana de Biología 24(1): 35–42

Braga, Marcelo Fideles; Junqueira, Nilton Tadeu Vilela; Faleiro, Fabio Gellape; Agostini-Costa, Tania S.; Bernacci, Luis Carlos. 2006. Maracujá-do-cerrado. In: Frutas Nativas da Região Centro-Oeste do Brasil, Vieira, Roberto Fontes; Costa, Tânia da Silveira Agostini; Silva, Dijalma Barbosa da; Ferreira, Francisco Ricardo; Sano, Sueli Matiko (eds), EMBAPA Recursos Genéticos e Biotecnologia, Brasília, pp 216–233

Charles-Dominique, P.; Atramentowicz, M.; Charles-Dominique, M.; Gérard, H.; Hladik, C. M.; Prévost, M. F. 1981. Les mammiferes frugivores arboricoles nocturnes d'une foret Guyanaise: inter-relations plantes-animaux. Revue d'Écologie: la Terre et la Vie 35:341–435

Cruz, Ana Paula Oliveira; Sousa, Julio dos Santos de; Bastos, Mariade Nazaré do Carmo; Barbosa, Camilo Veríssimo de Oliveira. 2013. Passiflora (Passifloraceae) na Província Petrolífera de Urucu, Coari, Amazonas, Brasil. Rodriguésia 64(1):113–122

Daly, Douglas C.; Silveira, Marcos; Medeiros, Herison; Castro, Wenderson; Obermüller, Flávio A. 2016. The white-sand vegetation of Acre, Brazil. Biotropica 48(1):81–89

Dantas, Mário. 1979. Pastagens da Amazônia Central: Ecologia e fauna do solo. Acta Amazonica 9(2) suplemento.

Dantas, Mário; Rodrigues, Irenice Alves. 1980. Plantas invasoras de pastagens cultivadas na Amazônia. EMBRAPA, Centro de Pesquisa Agropecuária do Trópico Úmido, Boletim de Pesquisa, vol. 1.

Ducke, A. 1946. *Plantas de Cultura Precolombiana na Amazônia brasileira. Notas Sôbre as Espécies ou Formas Espontâneas que Supostamente lhes teriam dado Origem.* Instituto Agronômico do Norte, Belém, Boletim Técnico 8, pp. 1–24

Egler, S. G. 1992. Feeding ecology of *Saguinus bicolor bicolor* (Callitrichidade: Primates) in a relict forest in Manaus, Brazilian Amazonia. Folia Primatologica 59:61–76

Escalante, Luis Javier Sánchez. 2016. Estudio etnobotánico, macro y micro-morfológico del género Passiflora utilizadas como sedantes en la provincia de Chimborazo. Master's thesis, Escuela Superior Politécnica de Chimborazo, Riobamba, Ecuador

Ferreira, Leandro Valle; Cunha, Denise A.; Chaves, Priscilla P.; Matos, Darley C. L.; Parolin, Pia. 2013a. Impacts of hydroelectric dams on alluvial riparian plant communities in eastern Brazilian Amazonian. Anais da Academia Brasileira de Ciências 85(3):1013–1033

Ferreira, Leandro V.; Chaves, Priscilla P.; Cunha, Denise de A.; Matos, Darley C. L.; Parolin, Pia. 2013b. Variação da riqueza e composição de espécies da comunidade de plantas entre as florestas de igapós e várzeas na Estação Científica Ferreira Penna-Caxiuanã na Amazônia oriental. Pesquisas Botânica 64:175–195

Flores, Salvador. 1997. *Cultivo de Frutales Nativos Amazonicos: Manual para el Extensionista.* Tratado de Cooperación Amazónica, Lima

Fosberg, F. R. 1960. Vegetation of Micronesia, I. General description of the vegetation of the Marianas Islands, and a detailed consideration of the vegetation of Guam. *Bulletin of the American Museum of Natural History* 119:1–75

Freire, Maria Cristina C.; Monteiro, Reinaldo. 1993. Florística das praias da Ilha de São Luís, Estado do Maranhão (Brasil): diversidade de espécies e suas ocorrências no litoral Brasileiro. Acta Amazonica 23(2–3):125–140

Gerwing, Jeffrey J.; Uhl, Christopher. 2002. Pre-logging liana cutting reduces liana regeneration in logging gaps in the eastern Brazilian Amazon. Ecological Applications 12(6):1642–1651

Gottsberger, Gerhard; Camargo, João M. F.; Silberbauer-Gottsberger, Ilse. 1988. A bee-pollinated tropical community: The beach dune vegetation of Ilha de São Luís, Maranhão, Brazil. Botanische Jahrbücher für Systematik, Pflanzengeschichte und Pflanzengeographie 109(4):469–500

Granville, Jean-Jacques de. 1988. Phytogeographical characteristics of the Guianan Forests. Taxon 37(3):578–594

Gregory, Tremaine. 2006. Comparative socioecology of sympatric, free-ranging white-faced and bearded sakis in Brownsberg Nature Park, Suriname. Doctoral dissertation, Kent State University, Kent, Ohio
Hernández, Alexandra; Bernal, Rodrigo. 2000. Lista de especies de Passifloraceae de Colombia. Biota Colombiana 1(3):320–355
Hoehne, F. C. 1946. Frutas Indígenas. Instituto de Botânica, São Paulo
Hoff, Michel; Cremers, Georges. 2005. Le jardin Guyanais: inventaire des plantes cultivées et des adventices des jardins de Guyane française. Journal de Botanique de la Société de France 29:3–40
Jaramillo-Arango, J. 1952. *Relación Historica del Viage, que Hizo a los Reynos del Peru y Chile el Botanico d. Hipolito Ruiz en el Año de 1777 hasta el de 1788, en cuya Epoca Regreso a Madrid*. Real Academia de Ciencias Exactas, Fisicas y Naturales, Madrid, vol. 1
Jardim, Daniel Gonçalves; Jardim, Mário Augusto G.; Quaresma, Adriano Costa; Costa, Salustiano Villar. 2013. Regeneração natural em formações florestais de uma unidade de conservação, Maracanã, Pará, Brasil. Biota Amazônia 3(2):79–87
Killip, Ellsworth P. 1938. The American Species of Passifloraceae. Field Museum of Natural History, Botanical Series, Volume 19, Part 1, Publication 407, Chicago
Koch, Ana Kelly; Ilkiu-Borges, Anna Luiza. 2016. Flora das cangas da Serra dos Carajás, Pará, Brasil: Passifloraceae. Rodriguésia 67(5): 1431–1436
Leal, F. 1990. Granadilla: In: Fruits of Tropical and Subtropical Origin, Nagy, Steven; Shaw, Philip E.; Wardowski, Wilfred F.(eds), pp. 322–327, Florida Science Source, Lake Alfred, Florida
Lima, Ana Carolina Barbosa de. 2014. Flavors of the city: access to regional fruit and fruit consumption in the State of Acre, Brazil. Boletim do Museu Paraense Emílio Goeldi, Ciencias Humanas 9(1):79–92
Lima-Neto, Abelardo Barbosa Moreira; Marques, Márcia Maria Mendes; Mendes, Francisca Noélia Pereira; Vieira, Ícaro Gusmão Pinto; Diniz, Derlange Belizário; Guedes, Maria Izabel Florindo. 2017. Antioxidant activity and physicochemical analysis of passion fruit (Passiflora glandulosa Cav.) pulp native to Cariri region. Acta Scientiarum Biological Sciences 39(4):417–422
Lorenzi, Harri; Bacher, Luis; Lacerda, Marco; Sartori, Sergio. 2006a. Frutas Brasileiras e Exóticas Cultivadas. Instituto Plantarum de Estudos da Flora Ltda, Odessa, Sao Paulo.
Lorenzi, H., L. Bacher, M. Lacerda, S. Sartori 2006b. Brazilian Fruits and Cultivated Exotics (for Consuming in Natura). Instituto Plantarum de Estudos da Flora Ltda, Odessa, São Paulo
Machado, Danilo de Oliveira. 2016. A agrobiodiversidade de quintais agroflorestais em propriedades agrícolas familiares na BR 174, Ramal do Pau-Rosa, Manaus, AM. Master's Thesis, INPA, Manaus
Machado, Clara de Carvalho. 2018. Conhecimento e consumo de plantas alimentícias em cinco comunidades da reserva de desenvolvimento sustentável Piaguçu-Purus, Amazonas. Master's thesis, INPA, Manaus
Martin, Franklin W.; Nakasone, Henry Y. 1970. The edible species of Passiflora. Economic Botany 24: 333–343
Maués, Bernardo Antônio Rodrigues; Jardim, Mário Augusto Gonçalves; Batista, Fábio de Jesus; Medeiros, Tonny David Santiago; Quaresma, Adriano da Costa. 2011. Composição florística e estrutura do estrato inferior da floresta de várzea na área de proteção ambiental Ilha do Combu, município de Belém, estado do Pará. Revista Árvore 35(3):669–677
Medeiros, Marcelo Brilhante de; Walter, Bruno Machado Teles; Silva, Glocimar Pereira da; Gomes, Beatriz Machado; Lima, Isabel Lustz Portela; Silva, Suelma Ribeiro; Moser, Pamela; Washington, Luís Oliveira; Cavalcanti, Taciana Barbosa. 2012. Vascular flora of the Tocantins River middle basin, Brazil. Check List 8(5):852–885
Milliken, William. 1997. Traditional anti-malarial medicine in Roraima, Brazil. Economic Botany 51(3):212–237
Morcote-Ríos, G. 2006. Plantas y gentes antiguas en un igapó estacional del interfluvio Solimões-Içá (Amazonas-Putumayo). In: Pueblos y Paisajes Antiguos de la Selva Amazónica, Morcote-Ríos,

Gaspar; Mora, Santiago; Franky, Carlos (eds), Universidad Nacional de Colombia, Bogotá, pp 253–259

Morcote-Ríos, G. 2008. Antiguos Habitantes en Ríos de Aguas Negras: Ecosistemas y Cultivos en el Interfluvio Amazonas, Putumayo Colombia-Brasil. Instituto de Ciencias Naturales, Facultad de Ciencias, Universidad Nacional de Colombia, Biblioteca José Jerónimo Triana No. 19, Bogotá

Morton, J. F. 1987. Fruits of Warm Climates. Creative Resource Systems, Winterville, North Carolina

Mota, Nara Furtado de Oliveira; Watanabe, Mauricio Takashi Coutinho; Zappi, Daniela Cristina; Hiura, Alice Lima; Pallos, Julieta; Viveros, Raquel Stauffer; Giulietti, Ana Maria; Viana, Pedro Lage. 2018. Cangas da Amazônia: a vegetação única de Carajás evidenciada pela lista de fanerógramas. Rodriguésia 69(3):1435–1488

Muñoz, Marianella Guevara. 2010. Evaluación de purmas con fines de manejo en la comunidad nativa de Santa Mercedes, Rio Putumayo, Perú. Master's thesis, Universidad Nacional de la Amazonia Peruana, Iquitos

Norconk, Marilyn A.; Veres, Michael. 2011. Physical properties of fruit and seeds ingested by primate seed predators with emphasis on sakis and bearded sakis. The Anatomical Record 294:2092–2111

Palma, Ana Cristina; Stevenson, Pablo R. 2013. Dispersión de semillas por monos araña en las Estación Biológica Cocha Cashu, Parque Nacional Manu, Perú. In: ReporteManu: Pasión por la Investigación en la Amazonía Peruana, Groenendijk, Trovar; Wust, Walter (eds), San Diego Zoo Global Peru, Cusco, pp. 178–195

Parra, C. 2006. Estudio general de la vegetación nativa de Puerto Carreño (Vichada, Colombia). Caldasia 28(2):165–177

Parrotta, John A.; Knowles, Oliver Henry; Wunderle Jr., Joseph M. 1997. Development of floristic diversity in 10-year-old restoration forests on a bauxite mined site in Amazonia. Forest Ecology and Management 99:21–42

Pinto, A. C. B., C. Azevedo-Ramos, and O. Carvalho. 2003. Activity patterns and diet of the howler monkey *Alouatta belzebul* in areas of logged and unlogged forest in eastern Amazonia. *Animal Biodiversity and Conservation* 26(2):39–49

Pires, Maria Joaquina. 1992. A check-list of the invasive species of forestry plantation in lower Amazon, NW. Acta Amazonica 22(1):3–15

Politis, G. G. 1996. *Nukak*. Instituto Amazónico de Investigaciones Científicas, Bogotá

Popenoe, W. 1924. Economic fruit-bearing plants of Ecuador. Smithsonian Institution, United States National Museum, Contributions from the United States National Herbarium 24(5): 101–134

Purseglove, J.W. 1974. *Tropical Crops: Dicotyledons*. John Wiley, New York

Rêgo, Márcia M. C.; Albuquerque, Patrícia M. C.; Ramos, Marina C.; Carreira, Léa M. 2008. Aspectos da biologia e nidificação de *Centris flavifrons* (Friese) (Hymenoptera: Apidae, Centridini), um dos principais polinizadores do murici (*Byrsonima crassifolia* L. Kunth, Malpighiaceae), no Maranhão. Neotropical Entomology 35(5):579–587

Ruysschaert, Sofie; Van Andel, Tinde; Van de Putte, Kobeke; Van Damme, Patrick. 2009. Bathe the baby to make it strong and healthy: Plant use and child care among Saramaccan Maroons in Suriname. Journal of Ethnopharmacology 121:148–170

Santos, J. U. M. and M. N. Bastos. 2014. Flores e Frutas das Restingas do Estado do Pará. Universidade Federal Rural da Amazônia (UFRA), Belém

Santos, João Ubiratan M. dos; Rosário, Carlos da S. 1988. Levantamento da vegetação fixadora das dunas de Algodoal-PA. Boletim do Museu Paraense Emílio Goeldi, Série Botânica 4(1): 133–154

Santos, Deborah Cruz dos; Corrêa Neto, José de Jesus; Alvarenga, Adriano Braga Brasileiro de. 2018. Frutos consumidos por macaco-de-cheiro *Saimiri sciureus collinsi* (Osgood, 1916) em fragmentos de florestas em uma unidade de conservação no arquipélago do Marajó, Amazônia Oriental, Brasil. Revista Brasileira de Zoociências 19(1):82–90

Schultes, R. E. and R. F. Raffauf. 1990. The Healing Forest: Medicinal and Toxic Plants of the Northwest Amazonia. Dioscorides Press, Portland, Oregon
Setz, E. Z. F.; Enzweiler, J.; Solferini, V. N.; Amêndola; Berton, R. S. 1999. Geophagy in the golden-faced saki monkey (*Pithecia pithecia chrysocephala*) in the Central Amazon. Journal of the Zoological Society of London 247:91–103
Silva, Silvestre. 2011. Fruits of the Brazilian Amazon. Metalivros, São Paulo
Silva, Eduardo Oliveira; Santos, João Ubiratan Moreira; Dias, Ana Cristina Andrade de Aguiar. 2013. Passifloraceae na área de proteção ambiental de Belém, PA, Brasil. Rodriguésia 64(4):829–845
Spruce, R. 1908. Notes of a Botanist on the Amazon and Andes. Macmillan, London, vol. 1
Thomas, Evert; Van Damme, Patrick. 2010. Plant use and management in homegardens and swiddens: evidence from the Bolivian Amazon. Agroforestry Systems 80:131–152
Tillett, S. 2003. Passifloraceae. In: Flora of the Venezuelan Guayana, vol. 7 Myrtaceae-Plumbaginaceae, Berry, Paul E.; Yatskievych, Kay; Holst, Bruce K. (eds), Missouri Botanical Garden, St. Louis, pp 625–667
Ulmer, T., J. M. MacDougal, and B. Ulmer. 2004. Passiflora: Passionflowers of the World. Timber Press, Portland
Van Roosmalen, M. G. M. 1985. Habitat preferences, diet, feeding strategy and social organization of the black spider monkey (*Ateles paniscus paniscus* Linnaeus 1758) in Surinam. Acta Amazonica 15(3/4 suplemento) 1–238
Van Roosmalen, Marc G. M.; Klein, Lewis L. 1988. The spider monkeys, genus *Ateles*. In: Ecology and Behavior of Neotropical Primates, Mittermeier, Russell A.; Rylands, Anthony B.; Coimbra-Filho, Adelmar F.; Fonseca, Gustavo A.B. da (eds), Ecology and Behavior of Neotropical Primates, World Wildlife Fund, Washington, DC, vol. 2, pp 455–537
Van Roosmalen, Marc G. M.; Mittermeier, Russel A.; Milton, Katherine. 1981. The bearded sakis, genus *Chiropotes*. In: Ecology and Behavior of Neotropical Primates, Coimbra-Filho, Adelmar F.; Mittermeier, Russell A. (eds), Academia Brasileira de Ciências, Rio de Janeiro, vol. 1, pp 419–441
Van Roosmalen, Marc G. M.; Mittermeier, Russel A.; Fleagle, John G. 1988. Diet of the northern bearded saki (*Chiropotes satanas chiropotes*): a Neotropical seed predator. American Journal of Primatology 14:11–35
Vargas, Fabiano S. de; Almeida, Patricia D. O.; Boleti, Ana Paula A. de; Pereira, Maria M.; Souza, Tatiane P. de; Vasconcellos, Marne C. de; Nunez, Cecilia Veronica; Pohlit, Adrian M.; Lima, Emerson S. 2016. Antioxidant activity and peroxidase inhibition of Amazonian plants extracts traditionally used as anti-inflammatory. BMC Complementary and Alternative Medicine 16:83–91
Vásquez, R. 1997. *Flórula de las Reservas Biológicas de Iquitos, Perú*. Monographs in Systematic Botany Vol. 63, Missouri Botanical Garden, St. Louis.
Vickers, W. T. 1994. The health significance of wild plants for the Siona and Secoya. In: Eating on the Wild Side: The Pharmalogic, Ecologic, and Social Implications of Using Noncultures, N. L. Etkin (ed.), University of Arizona Press, Tucson, pp 143–165
Voss, Robert S.; Fleck, David W. 2011. Mammalian diversity and Matses ethnomammalogy in Amazonian Peru Part 1: Primates. Bulletin of the American Museum of Natural History 351:1–81

Chapter 41
Polygalaceae

41.1 *Diclidanthera penduliflora*

Peru Aceitunilla, coto huayo

Status Wild

Aceitunilla *Diclidanthera penduliflora* in fruit in floodplain forest. Near Manco Capac Rio Puinahua (a branch of the Ucayali), Reserva Nacional Pacaya-Samiria, Loreto, Peru 4-15-06

A liana or spindly tree of floodplain forests that can attain 7 m in length, *Diclidanthera penduliflora* produces round black fruits during the rainy season. People gather the glossy fruits from canoes during the annual flood when rivers invade the surrounding forest, often for several kilometers. A treat for those who venture into flooded forests, the juicy fruits, 1–2 cm in diameter, are generally eaten on the spot and are never taken to market.

N. Smith, *Amazon Fruits: An Ethnobotanical Journey*, Ethnobiology,
https://doi.org/10.1007/978-3-031-12803-5_41

Confined to the Amazon Basin, the "little olive" (aceitunilla) as the tree is known along the Rio Puinahua, an arm of the Ucayali, lingers as one of the numerous minor fruits that 1 day may emerge as a fashionable new fruit on the tables of urban folk in the Amazon and in other regions. Another name for the fruit in the Peruvian Amazon is coto huayo (howler monkey fruit) suggesting that the tree contributes indirectly to the human diet via game meat.

Aceitunilla in fruit in floodplain forest during the rainy season. Near Manco Capac Rio Puinahua (a branch of the Ucayali), Reserva Nacional Pacaya-Samiria, Loreto, Peru 4-15-06

41.2 *Moutabea aculeata*

Bolivia Jazmín, guapurucillo, achaichurucillo

Ecuador Omentaca, meñemo (Waorani)

Peru Meta huayo, coto huayo, caimito de monte, monte lucuma; kenúnch (Mayna)

Venezuela Wakara ãsi hi (Yanomami)

Status Wild

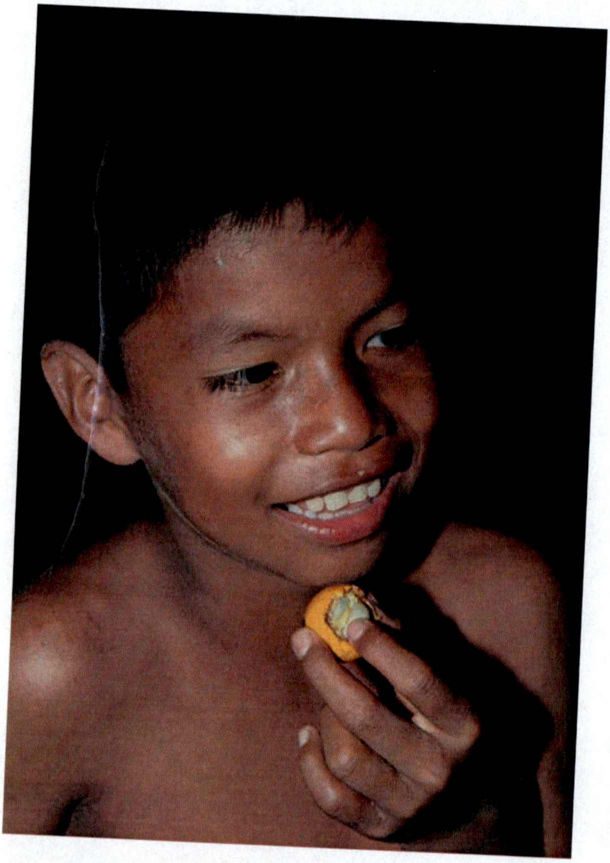

Victor, 8 years, eating a meta huayo *Moutabea aculeata* fruit he gathered from the floor of a floodplain forest. Rio Tigre, 2.5 km upstream from Miraflores, Loreto, Peru, 7-1-06

One of the several lianas with edible fruits, *Moutabea aculeata* produces ping-pong-sized fruits, measuring some 3.5 cm long and 3 cm wide, in the rainy season. The thick liana extends for 30 m or so when fully mature into the canopy, so the yellow fruits are usually gathered from the ground. People suck the slippery, translucent pulp that surrounds the four, light tan seeds and then spit the latter out. The seeds are approximately 2 cm long and 1 cm in diameter and occupy much of the fruit cavity. Typical of most wild fruits, then, the pulp is relatively small in the proportion of the fruit. Nevertheless, locals appreciate the fruits which are eaten mostly in the forest and do not enter markets.

A denizen of floodplain forests, such as along the banks of the lower Tigre in the Peruvian Amazon, this liana is also found occasionally on uplands throughout its range in western Amazonia, including Acre in Brazil. *Moutabea aculeata* is mostly confined to lowlands, but it can be found as high as 1700 m in the Andes.

Chapter 42
Polygonaceae

42.1 *Coccoloba densifrons*

Bolivia Guayabichi

Brazil Uva brava, olho de passarinho

Colombia Vino

Ecuador Licuaneo, acha caspi; rumi caspi; motilón (Kichwa), iyiknum (Shuar), nec'a cono (Siona-Secoya), bedeyowëmo, bereyahuehue, bereveyohueme, beyuguemo, biririyuem (Waorani)

Peru Vino huayo, tangarana blanca, sabalo huayo, romero; mutúpach (Achuar), peetunch (Aguaruna), motelo caspi (Kichwa), nemoño (Witoto)

Status Wild

N. Smith, *Amazon Fruits: An Ethnobotanical Journey*, Ethnobiology,
https://doi.org/10.1007/978-3-031-12803-5_42

Vino huayo *Coccoloba densifrons* on the bank of a sediment-rich river. Lower Rio Tigre, Loreto, Peru 4-26-06

Coccoloba densifrons is one of several species of tree in the genus with fruit bunches similar to grapes; hence, two of its common names: vino huayo (the wine fruit) in Peru and uva brava (wild grape) in Brazil. However, the round fruits are smaller than those of grapes, measuring only 1 cm in diameter, and contain a single seed compared to the several pips typically found in grapes. The seed in vino huayo fruits is 0.9 cm long and 0.8 cm wide, and since the black skin is relatively thick, the layer of pulp that coats the seeds is thin. The purple mesocarp is sufficiently sweet and juicy, however, to tempt adults and children alike. Some 40–50 fruits are found on each bunch, and *Coccoloba densifrons* produces dozens of infructescences during the rainy months. So although the pulp is meager, lots of fruits are produced each year.

As is often the case with wild fruits, the pulp is parsimonious, just enough to entice dispersal agents. Woolly monkeys *Lagothrix lagothricha* are known to disperse the seeds of *Coccoloba densifrons* in the Colombian Amazon (Stevenson 2000), and this wide-ranging monkey probably disperses the seeds of the tree in other parts of the western Amazon. Red howler monkeys *Alouatta seniculus*, which occur in western Amazonia, also feed on these fruits and disperse the seeds in their feces (Yumoto et al. 1999). In eastern Colombia, an endemic species of squirrel monkey *Saimiri cassiquiarensis albigena* also eats this fruit and likely disperses some of the seeds (Carretero-Pinzón et al. 2016). In the Ecuadorian Amazon, one of the Kichwa names for the fruit is motelo caspi (fruit of the tortoise), suggesting that the yellow-footed tortoise *Chelonoidis denticula* feeds on the fruits when they are ripe and fall to the ground. Other game animals, such as guans (*Penelope* spp.) and peccaries (*Pecari tajacu* and *Tayassu pecari*), also eat the fruit (Ima 2012). Some prized fish, such as species of *Prochilodus* which are known as sábalo in Peru and curimatá in Brazil, are also fond of the fruit. The tree can reach 35 m high, but starts producing fruit when only 1.5 m tall, so fish can swim up to the fruits during the flood season when the trees are young.

Coccoloba densifrons is found throughout the Amazon Basin. It is typically found in the lowlands along riverbanks and lake margins of sediment-rich, clear, and black water rivers (Cunha and Ferreira 2012; Luize et al. 2015; Macía 2011; Nebel et al. 2000a, b, 2001; Wittmann et al. 2006). It is also sometimes found in upland forests and occurs as high as 800 m in the Andean foothills. Its range also extends into the cerrado region of central Brazil (Batalha and Martins 2004). The small tree, which can reach 15 m when fully mature, is often spared when clearing for settlements, such as in the villages of Veinte de Enero and Yarina along the Yanayacu River in the Reserva Nacional Pacaya-Samiria, Loreto, Peru. It is thus incorporated in some home gardens. In cleared areas, *Coccoloba densifrons* does not grow as tall as in the forest and produces a more spreading crown making it easier for people to harvest the fruits.

Besides fruit, *Coccoloba densifrons* is useful for other purposes, including medicine. In the Peruvian Amazon, some indigenous people, such as the Mayna, boil the bark to treat dysentery. The hard and durable wood is used for making paddles, in home construction, and for firewood, such as among the Waorani in the Ecuadorian Amazon (Cerón and Ayala 1998: 142).

Franciso, 10 years, gathering vino huayo *Coccoloba densifrons* fruit in his home garden on a riverbank. Yarina, Rio Yanayacu, Reserva Nacional Pacaya-Samiria, Loreto, Peru 6-24-06

References

Batalha, Marco Antônio; Martins, Fernando Roberto. 2004. Reproductive phenology of the cerrado plant community in Emas National Park (central Brazil). Australian Journal of Botany 52:149–161

Carretero-Pinzón, Xyomara; Defler, Thomas R.; Ruiz-García, Manuel. 2016. How does the Colombian squirrel monkey cope with habitat fragmentation? Strategies to survive in small fragments. In: Phylogeny, Molecular Population Genetics, Evolutionary Biology and Conservation of the Neotropical Primates, Ruiz-García, M.; Shostell, J. M. (eds), Nova Science, New York, chapter 15, 1–16

Cerón, Carlos Eduardo; Ayala, Consuelo Montalvo. 1998. Etnobotanica de los Huaorani de Quehueiri-ono Napo, Ecuador. Abya-Yala, Quito

Cunha, Denise de Andrade; Ferreira, Leandro Valle. 2012. Impacts of the Belo Monte hydroelectric dam construction on pioneer vegetation formations along the Xingu River, Pará state, Brazil. Brazilian Journal of Botany 35(2):159–167

Ima, M. O. 2012. *Saberes Waorani y Parque Nacional Yasuní: Plantas, Salud y Bienestar en la Amazonía del Ecuador*. Iniciativa Yasuní ITT, Ministerio Coordinador de Patrimonio, Ministerio del Ambiente, Programa de las Naciones Unidas para el Desarrollo (PNUD) y Fondo para el Medio Ambiente Mundial (FMAM), Quito

Luize, Bruno Garcia; Venticinque, Eduardo Martins; Silva, Thiago Sanna Freire; Novo, Evlyn Marcia Leão de Moraes. 2015. A floristic survey of angiosperm species occurring at three landscapes of the central Amazon várzea, Brazil. Checklist: The Journal of Biodiversity Data 11(6):1–15

Macía, Manuel J. 2011. Spatial distribution and floristic composition of trees and lianas in different forest types of an Amazonian rainforest. Plant Ecology 212:1159–1177

Nebel, Gustav; Dragsted, Jens; Vanclay, Jerome K. 2000a. Estructura y composición florística del bosque de la llanura aluvial inundable de la Amazonía Peruana: II. El sotobosque de la restinga. Folia Amazónica 10(1–2):151–181

Nebel, Gustav; Kvist, Lars Peter; Vanclay, Jerome K.; Christensen, Henning; Freitas, Luis; Ruiz, Juan. 2000b. Estructura y composición florística del bosque de la llanura aluvial en la Amazonía Peruana: I. El bosque alto. Folia Amazónica 10(1–2):91–149

Nebel, Gustav; Dragsted, Jens; Vanclay, Jerome K. 2001. Structure and floristic composition of flood plain forests in the Peruvian Amazon II. The understorey of restinga forests. Forest Ecology and Management 150: 59–77

Stevenson, Pablo R. 2000. Seed dispersal by woolly monkeys (*Lagothrix lagothricha*) at Tinigua National Park, Colombia: dispersal distance, germination rates, and dispersal quantity. American Journal of Primatology 50: 275–289

Wittmann, Florian; Schöngart, Jochen; Montero, Juan Carlos; Motzer, Thomas; Junk, Wolfgang J.; Piedade, Maria T. F.; Queiroz, Helder L.; Worbes, Martin. 2006. Tree species composition and diversity gradients in white-water forests across the Amazon Basin. Journal of Biogeography 33(8):1334–1347

Yumoto, Takakazu; Kimura, Koshin; Nishimura, Akisato. 1999. Estimation of the retention times and distances of seed dispersed by two monkey species, Alouatta seniculus and Lagothrix lagotricha, in a Colombian forest. Ecological Research 14:179–191

Chapter 43
Primulaceae

43.1 *Clavija elliptica*

Peru Tucunarinari, tucunareñahui, huevos de gato, curarina sacha, gallo poroto; mas pan (Cashinahua)

Status Wild

N. Smith, *Amazon Fruits: An Ethnobotanical Journey*, Ethnobiology,
https://doi.org/10.1007/978-3-031-12803-5_43

Tucunareñahui *Clavija elliptica* **an understory shrub in a forest bearing fruit. Near Yarina, Rio Yanayacu, Reserva Nacional Pacaya-Samiria, Loreto, Peru 6-26-06**

Scarcely a meter and a half tall when fully grown, *Clavija elliptica* looks like a bonsai tree in the middle of the forest. Along the Yanayacu River in the Reserva Nacional Pacaya-Samiria, the diminutive tree is known as tucunareñahui, which means eye of the tucunaré (*Cichla* spp.), a popular game and sport fish. The "eye" in this case refers to the yellow spot surrounded by black on the fish's tail which evolved to mimic the eye of a large animal to scare off predators. Tucunareñahui fruits are bright yellow and round, similar to the ocellus on tucunaré's tail. Locals gather the lustrous fruits, which start appearing when the shrub is only half a meter high, during the rainy season and early part of the dry period and typically eat them on the spot. The fruits are about 3 cm in diameter and little pulp surrounds the three to five brown seeds in each fruit. Nevertheless, the translucent mesocarp is sufficiently sweet to entice people and seed dispersal agents. Although tucunareñahui can produce up to two dozen fruits, the shrub is usually solitary and thinly dispersed in forests, so not enough fruits are gathered to take to markets. In the vicinity of Chazuta along the Upper Huallaga River in the Peruvian Amazon locals crush the fruits and apply them to skin affected by scabies (Sanz-Biset et al. 2009). The miniature trees are widely dispersed in mature and regrowth forests on higher parts of floodplains in the Peruvian Amazon, Acre in Brazil, and south into the Bolivian lowlands.

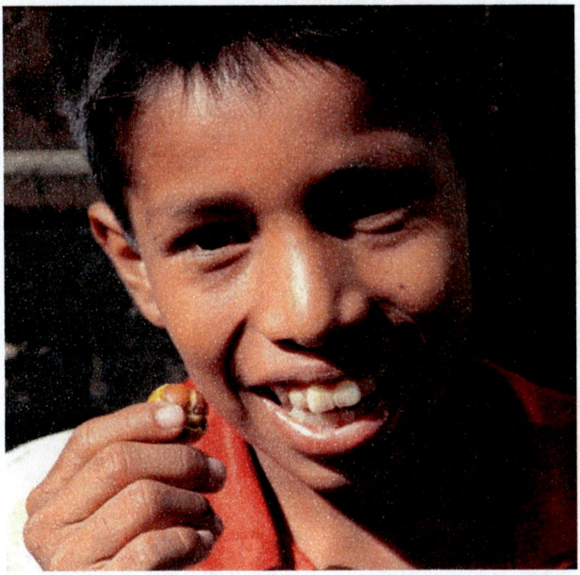

Bildoso, 10 years, with tucunareñahui *Clavija elliptica* fruits he gathered in successionary forest near his village. Veinte de Enero, Rio Yanayacu, Reserva Nacional Pacaya-Samiria, Loreto, Peru 1-14-06

43.2 *Clavija lancifolia*

Bolivia Huevo de perro, huevo de gato, huevitos de perdiz; do'opi (Ese Eja), doradyes (Tsimane)

Brazil Testiculo de guariba, olho de urubu, remela de cachorro, remela de velho, muirapuama, marapuama, pau marfim; janithópiã (Arawetê), jawcirupe'arenda

(Asurini), kamiča-apini'a-'ɨ (Guajá), karume-pitag-i (Ka'apor), zawiči-pɨta (Tembé), horikima hi (Yanomami)

Venezuela Cola de pava; yumoko, yuu mohi, yuu kë mo hi (Yanomami)

Status Wild, spontaneous in cultural settings

Clavija lancifolia **in the understory of a forest near an archaeological site. Ilha do Palhal, Rio Xingu between Altamira and Belo Monte, Pará, Brazil 8-13-15**

Clavija lancifolia is similar in size and shape to its near relative *Clavija elliptica* except that the fruits are orange when ripe rather than yellow. The round fruits of *Clavija lancifolia* are 1.4–2.7 cm in diameter and contain five to seven brown seeds which occupy most of the fruit. A thin pale orange pulp surrounds the seeds which measure 1.4 cm × 1 cm when the fruit is 2.5 cm in diameter. Sometimes the seeds are round, measuring 0.9 cm in diameter.

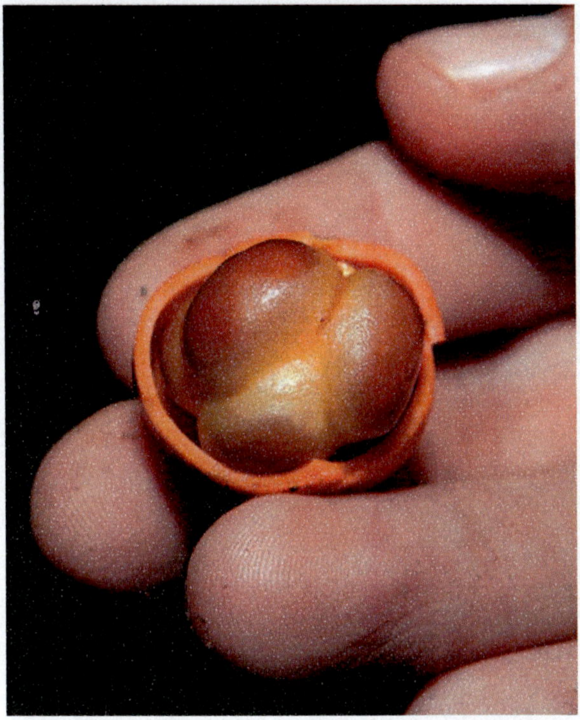

Clavija lancifolia **fruit 2.5 cm × 2.5 cm with 5 brown seeds 1.4 cm × 1 cm, covered with a thin layer of sweet translucent pulp. Ilha do Palhal, Rio Xingu between Altamira and Belo Monte, Pará, Brazil 8-13-15**

The various local names for the fruit do not sound very appetizing: howler monkey balls (*testiculo de guariba*), dog's balls (*huevo de perro*), vulture eyes (*olho de urubu*), and old man's eye boogers (*remela de velho*). Although the mesocarp is meager it is sweet and rural people suck the pulp and spit out the seeds. Also, the fruits typically occur in bunches, and as many as 15 can be found on a single plant. Fruits mature in both the dry and wet seasons. In some cases, a half dozen of the shrubs occur in close proximity, such as near the Sitio Tucunaré archaeological site on Ilha Palhal along the middle Xingu, so the harvest can be bountiful. In the community of Enfarrusca near Bragança in northeastern Pará, some rural folk make an infusion with the bark or roots of this plant which is then placed on painful areas of the body caused by rheumatism (Freitas and Fernandes 2006).

I noted the presence of *Clavija lancifolia* on or very close to several other archaeological sites along the middle Xingu including Sitio Santo Antonio I near the Belo Monte, Sitio Mamoal some 8 km downstream from Altamira, and Palmeiras and Ilha dos Arara both upstream from Altamira. *Clavija lancifolia* is also present in an archaeological site further east at Sitio do Forte, Baia de Caxiuanã, Rio Anapu. It is possible that indigenous people, rubber tappers, and other river dwellers brought the fruits to those sites or the treelets flourished there because of the rich anthropogenic black earth, and sites of ancient villages.

Clavija lancifolia stands out in the forest because of the bright orange fruits and large leaves. The fruits are easy to reach because *Clavija lancifolia* is a small bush in understory forests, especially in regrowth woodlands (Alvino et al. 2005). It starts producing fruit when only 50 cm high and reaches only 3 m when fully grown. Mature shrubs form branches and bend close to the ground when laden with fruits making them accessible to terrestrial dispersal agents. Little is known, however, about the natural history of this and many other fruit trees in Amazonia.

Clavija lancifolia **in the anthropogenic forest on an archaeological site. Sitio do Forte, Baia de Caxiuanã, Rio Anapu, Pará, Brazil 5-14-15**

Clavija lancifolia is found mostly on well-drained uplands throughout Amazonia and occurs up to about 1400 m in the foothills of the Andes. For example, it occurs in stunted woodlands growing on iron-rich sedimentary rock (*canga*) on plateaus in the Serra dos Carajás in Pará (Mota et al. 2018). It also occurs in some floodplain forests, such as at the confluence of the Solimões and Japurá rivers (Assis 2008: 39). Its range extends into the Orinoco Basin and the Guianas. It often occurs in regrowth forests and I spotted one in an abandoned ribeirinho home garden along the middle Xingu; the latter may have sprouted from a seed discarded by the former occupant or perhaps it was brought to the home garden by a wild bird or mammal.

43.3 *Stylogyne cauliflora*

Bolivia Limoero; viñaj (Tsimane)

Brazil Casca grossa

Peru Shambo huayo, vino huayo, ojo de tucunare

Status Wild

Shambo huayo *Stylogyne cauliflora* in floodplain forest. Near Miraflores, Rio Tigre, Loreto, Peru, 7-1-06

A treelet in mature and regrowth forests below 800 m on floodplains in western Amazonia from Colombia south to Bolivia, including Acre in Brazil, *Stylogyne cauliflora* produces numerous pink fruits clustered along the branches. The round fruits are only 0.5 cm in diameter, making them one of the smallest fruits eaten in wetlands in the Peruvian Amazon. However, dozens of fruits can be gathered from a single shrub or tree. *Stylogyne cauliflora* can reach 10 m, but it starts fruiting by the time it is 2 m high so fruits can often be reached from the ground. Higher fruits can be gathered by pushing against the trunk so that the tree leans. The small tree also occurs occasionally in stunted woods (campina) on sandy soils that are poorly drained.

Known as shambo huayo along the lower Tigre, the juicy fruits are eaten on the spot, or bunches are taken home for later consumption. In spite of their appealing flavor, the single-seeded fruits never reach markets, probably because the diminutive, shade-tolerant trees are sparsely distributed.

Similar in some respects to its near relative, *Stylogyne longifolia*, which is also called shambo huayo along the lower Tigre, *Stylogyne cauliflora* can be distinguished from the former by its preference for higher ground, its smaller leaves, the more clustered fruits, and fact that the fruits are pink when ripe rather than dark purple.

Elias, 11 years, gathering shambo huayo fruits in floodplain forest. Near Miraflores, Rio Tigre, Loreto, Peru, 7-1-06

43.4 *Stylogyne longifolia*

Bolivia Guapuraqui

Brazil Açaí da mata

Peru Shambo huayo, verano huayo, vino huayo

Status Wild

Cowboy with açaí da mata *Stylogyne longifolia* fruit he has gathered in the understory of a floodplain forest in the Amazon estuary. Fazenda São Luis, Rio Pacajá, Caviana Island, Pará, Brazil 5-22-01

The dark purple fruits of *Stylogyne longifolia* are a tasty treat in widely scattered parts of the Amazon Basin. The long-leafed shrub to small tree is also found in Venezuela, the Guianas, the Amazon Basin, and northern and western Colombia. Normally 2–5 m tall, *Stylogyne longifolia* can reach as high as 12 m in search of light in the dim interior of the forest. Shrubs start bearing fruits by the time they are 2 m tall so they are easily gathered from canoes or while standing on river banks. The shiny fruits appear in bunches containing about ten fruits each during the rainy season. The round fruits, about 0.8 cm in diameter, contain a single, round green seed measuring 0.6 cm across. The thin layer of green pulp is juicy. The skin is delicate so the fruits do not travel well. Only rural folk are thus familiar with this fruit.

Stylogyne longifolia grows in a variety of wetland and upland habitats. The shade-tolerant tree is found in mature and regrowth forest along the banks of sediment-rich, clear, and black water rivers, as well as in palm swamps dominated by *Mauritia flexuosa*. The tree thrives on a range of soils, from the muddy banks of tidal streams in the Amazon estuary to sandy beaches along the tannin-stained Upper Rio Negro. It is also found in disturbed sites in upland forests, such as in Acre, Brazil. *Stylogyne longifolia* thrives up to about 1000 m in montane forests in the Andes.

At the mouth of the Amazon, the fruits are called açaí da mata, the forest açaí, because its fruits are similar in shape and color to those of the açaí palm *Euterpe oleracea*, only they are smaller and with thinner skins. On Caviana Island in the Amazon estuary, one cowboy claimed that the fruits are eaten on the spot and not taken home if there are young children because they would eat too many of them and get a stomachache. A similar warning is issued to country children in England about eating too many elderberries (*Sambucus*). In spite of such concerns, however, *Stylogyne longifolia* could be an interesting topping for desserts because of its appealing flavor and lustrous color.

References

Alvino, Fabrízia de Oliveira; Silva, Manoela Ferreira F. da; Rayol, Breno Pinto. 2005. Potencial de uso das espécies arbóreas de uma floresta secundária, na Zona Bragantina, Pará, Brasil. Acta Amazonica 35(4):413–420

Assis, Rafael Leandro de. 2008. Composição florística e estrutura da regeneração arbórea de florestas de várzea alta e várzea baixa na RDS Mamirauá, Amazônia central. Master's thesis, INPA, Manaus

Freitas, Jussara Costa de; Fernandes, Marcus Emanuel Barroncas. 2006. Uso de plantas medicinais pela comunidade de Enfarrusca, Bragança, Pará. Boletim do Museu Paraense Emílio Goeldi, Ciencias Humanas 1(3):11–26

Mota, Nara Furtado de Oliveira; Watanabe, Mauricio Takashi Coutinho; Zappi, Daniela Cristina; Hiura, Alice Lima; Pallos, Julieta; Viveros, Raquel Stauffer; Giulietti, Ana Maria; Viana, Pedro Lage. 2018. Cangas da Amazônia: a vegetação única de Carajás evidenciada pela lista de fanerógramas. Rodriguésia 69(3):1435–1488

Sanz-Biset, Jaume; Campos-de-la-Cruz, José; Epiquién, Mirbel A.; Cañigueral, Salvador. 2009. A first survey on the medicinal plants of the Chazuta valley (Peruvian Amazon). Journal of Ethnopharmacology 122:333–362

Chapter 44
Rubiaceae

44.1 *Alibertia edulis*

Bolivia Conservillo, bi-chi, mermelada, marmelada de cachorro, marmelada bola, marmeleiro, guayabilla, tutumillo; tosa (Chácobo), dyincava' (Tsimane)

Brazil Puruí, puruí massa, puruí pequeno, puruízinho, apuruí, apuruí cascudo, apuruizinho, apuí, jenipapim, jenipapinho, marmelada do cerrado, marmelada de bezerro, marmelada bola, marmelada bolo, marmelada de cachorro, marmelada macho, marmelada olho de boi, marmeladinha, marmelo de cachorro, marmelo do cerrado, marmeleiro, marmelão, marmelinho, Maria preta, Maria pretinha, caneleiro, goiabinha, goiaba preta, goiaba do mato, olho de boi, cubil, furoia, furuía, canela de veado, coruba, bola; kiraru (Ka'apor), motu, roi krã ti (Kayapó), tamari-kawaw (Tembé)

Colombia Perita, pera, peralejo

Ecuador Juan colorado

Peru Caimitillo, capironilla, huitillo; ta nin ti (Cashinahua), sacha huito (Kichwa)

Status Wild, spontaneous in cultural settings, planted

N. Smith, *Amazon Fruits: An Ethnobotanical Journey*, Ethnobiology,
https://doi.org/10.1007/978-3-031-12803-5_44

Apuruí *Alibertia edulis* fruit. Maçaranduba, Lago Miwá near Codajás, Rio Solimões, Amazonas, Brazil 6-24-07

Alibertia edulis is a small tree, typically under 10 m tall but reaching as high as 22 m in some situations, with lustrous leaves and fruits that vary in color from rusty red to orange-yellow. It occurs in a wide range of habitats in Amazonia, including some upland forests such as in the Jari watershed in Amapá (Pereira et al. 2011), on hills containing copper ore in the Serra dos Carajás, Pará (Silva and Rosa 1989), in the forest transition zone with cerrado vegetation in Mato Grosso (Borges and Shepherd 2005; Ferreira Júnior et al. 2008; Lopes et al. 2014; Loverde-Oliveira et al. 2010; Oliveira Filho 1992; Oliveira Filho and Martins 1991; Pinto et al. 2014), on coastal sand dunes in northeastern Pará (Jardim et al. 2013, 2016; Santos and Rosário 1988), and in the eastern fringe of the Amazon Basin in Maranhão (Muniz et al. 1994). It is also found in successionary vegetation, such as in Mato Grosso (Rodrigues et al. 2004), in eastern and western Pará (Espírito-Santo et al. 2005; Medeiros 2016: 91), and in the Amazon estuary (Amaral et al. 2009), but it is not clear how long it can survive in the shady interior of a mature forest. This small tree invades some cattle pastures, such as in the vicinity of Santana do Araguaia in southern Pará (Dantas and Rodrigues 1980). This fruit tree has been recorded in 40-year-old successionary forests in the Bragantina zone of northeastern Pará (Rayol et al. 2006) as well as forest islands along the coast of northeastern Pará (Abreu et al. 2006; Bastos et al. 1995; Santos et al. 2003). It occurs in some scrub savannas in the southern part of the Amazon Basin (Guarim Neto 1984; Guarim Neto and Pasa 2009; Marimon Junior and Haridasan 2005; Matos et al. 2020; Medeiros et al. 2012; Penso et al. 2009; Rondon Neto et al. 2010; Santos et al. 2015), in the vicinity of Alter do Chão along the lower Tapajós River (Magnusson et al. 2008; Marimon and Lima 2001), in the Monte Alegre state park in Pará (Silva 2008: 94), and to the east of the Amazon region in south and eastern Maranhão (Rodrigues and Conceição 2014). It also grows in gallery forests in scrub savannas, such as in the Upper Tocantins and Araguaia watersheds as well as in the Municipality of Cuiabá in the southern part of the Brazilian Amazon (Corette-Pasa 2011; Homeier et al. 2017; Oliveira Filho 1989; Pasa et al. 2005; Pereira and Santos 2015).

It is also found in floodplain successionary forest in the tidal zone of the Amazon estuary (Andrade et al. 2008; Montagnini and Muñiz-Miret 1999), along the Solimões River, and in black water floodplain forest (*igapó*) in the Rio Negro watershed in Amazonas (Assis and Wittmann 2011; Barnett et al. 2015; Ferreira 1997; Parolin 2000a, b, c, 2002; Scudeller and Souza 2009). This tree occurs in some floodplain forests along the clear water Araguaia River in the southeastern part of the Amazon Basin (Kurzatkowski et al. 2015). It is also found in some floodplain forests along the sediment-rich rivers, such as the Ucayali some 10 km south of Jenaro Herrera in the Peruvian Amazon (Nebel et al. 2000a, b), along the lower Purus, an affluent of the Solimões (Machado 2018: 126) and at the confluence of the Japura and Solimões (Assis 2008: 38).

The multi-branched tree fruits in the rainy season and starts producing fruit when only 1.3 m tall, so they are often within easy reach. The slightly flattened fruits vary in size from 1.8 cm × 2.5 cm to 5 cm × 6 cm and contain numerous reddish-brown seeds (0.5 cm × 0.7 cm to 0.9 cm × 0.6 cm) embedded in a strong-tasting pulp. The texture of *Alibertia edulis* fruits is similar to that of guava *Psidium edulis*. Although the fruits thus contain little pulp, locals nevertheless enjoy sucking on the pulp and spitting out the seeds. Some just swallow the seeds along with the pulp. The pulp is often used for making juice, such as in the vicinity of Tefé and Codajás in Amazonas, Brazil. Along the Arapiuns, an affluent of the lower Tapajós, puruí fruits during the dry season when few other wild fruits are available. Along the Rio Yanayacu, an affluent of the lower Marañón, caimitillo fruits appear during the annual flood and are gathered from canoes. The fruits are also used to bait hooks to catch fish such as pacu (various species of *Mylossoma*, *Metynnis,* and *Myleus*).

Kerlem, 7 years, eating apuruí *Alibertia edulis* fruit she gathered in successionary vegetation on the outskirts of her village. Maçaranduba, Lago Miwá near Codajás, Rio Solimões, Amazonas, Brazil 6-24-07

Thought to be native to Amazonia, the tree grows in a wide range of habitats on uplands and floodplains. The tree appears to be more common on higher parts of the floodplains of water courses with various water types, from tannin-stained black waters to clear and muddy streams and rivers. It also thrives along the forest boundary with the savanna; the Kayapó, for example, plant this tree in forest islands on the savanna to attract game (Posey 1985a). They also plant it in their manioc fields so that they can return there to gather the fruits long after the manioc has been harvested (Posey 1985b). This tree is also cultivated in agroforestry plots in some other areas of the Brazilian Amazon such as in the vicinity of Tomé-Açu in eastern Pará (Bolfe and Batistella 2011). Because *Alibertia edulis* is fire tolerant it also occurs on open and scrub savannas (Ratter et al. 1996), as well as seasonally flooded grasslands. It is also found in stunted woodlands (campina) on sandy soils such as in the

Bragantina zone in northeastern Pará. The tree ranges outside of the Amazon Basin including central Brazil and the Atlantic forest, the Caribbean, and Central America as far north as Mexico.

David, 9 years, holding an unripe apuruí *Alibertia edulis* fruit in successionary vegetation on the outskirts of his village. Maçaranduba, Lago Miwá near Codajás, Rio Solimões, Amazonas, Brazil 6-24-07

Puruí occurs on indigenous mounds on Marajó Island, possibly planted there in the past (Lisboa 2012: 410; Miranda 1909). *Alibertia edulis* is also found in some home gardens usually arising from discarded seeds, such as in the vicinity of Abaetetuba along the lower Tocantins in Pará (Gonçalves and Lucas 2017), along the lower Madeira River in Amazonas (Kawa et al. 2015) and in some urban home gardens such as in Comodo city, Mato Grosso (Santos and Guarim Neto 2017). I have observed this tree along the lower Tefé River, an affluent of the Solimões, on the shores of Lago Amanã along the lower Japurá, and along the Curuá and Cariá

Rivers in the Caxiuanã National Forest in eastern Pará. Humans have likely been involved in increasing the density of this tree if not its distribution.

The tree is also found in forests and disturbed habitats in Central America. In Central America, Baird's tapir *Tapirus bairdii* disperses seeds of *Alibertia edulis*. A closely related tapir *Tapirus terrestris,* known in English as the lowland tapir or South American tapir, probably does the same in Amazonia as it has been recorded eating this fruit in the Upper Orinoco Basin in eastern Colombia (Alviz and Pérez-Albarracín 2015: 41) and in the cerrado region of central Brazil (Prado 2012: 18). Seeds of this fruit have been found in the feces of the maned wolf *Chrysocyon brachyurus*, a long-legged predator which inhabits cerrado vegetation in the southern fringes of the Amazon Basin (Castro and Emmons 2012).

This plant also has other uses. The wood is used for making posts, flooring, and walls in rural homes in southern Ecuador (Aguirre-Mendoza et al. 2015: 32, 77), and this plant is also used in some folk remedies. The fruit juice is thought to contain a lot of iron ("*conta muito ferro*") and is thus taken to relieve anemia along the lower Tefé River. In the cerrado region of Mato Grosso bordering the southern part of the Amazon rainforest the leaves are used to make a tea to treat colds (Moreira and Guarim-Neto 2009) while in the rural community of Rio dos Couros, 60 km from Cuiabá in Mato Grosso, locals make a tea from the leaves to treat heart problems (Costa et al. 2017). In the Peruvian Amazon, a decoction of crushed leaves or the fruit is taken to stem diarrhea. In the Bolivian Amazon, the Chacobo cut up the ripe fruit and boil it in water for 15 minutes then drink the solution when it cools to treat digestive problems (Muñoz et al. 2000). In the Caxiuanã national forest in eastern Pará, Brazil, the pulp of puruí fruits is plastered on the umbilical cord of babies to promote healing (Rocha et al. 2013).

44.2 *Borojoa sorbilis*

Brazil Puruí, puruí massa, puruí grande, apuruí grande

Colombia Borojó nativo, borojó de monte

Ecuador Borojó, fruto de guanta, parvi grande; chanange ichhoro'cho (Cofán)

Peru Borojó, parvi grande

Status Wild, planted

Puruí *Borojoa sorbilis* fruit gathered in a rural home garden for making juice. Vila Valente, lower Tefé River, Amazonas, Brazil, 8-22-12

Borojoa sorbilis is a small tree, typically under 8 m tall, that is native to upland forests in the river basins of the Solimões, Madeira, and middle Purus in the state of Amazonas, Brazil (Cavalcante 2010: 219). It is also found in the upland forests some 80 km north of Manaus, Amazonas (Boom and Campos 1991) as well as in the forests of Amazonian Ecuador where nocturnal pacas *Agouti paca* are known to eat this fruit (Cerón 1995: 128). This bulldog-sized terrestrial rodent is a popular game animal throughout the Amazon and is sometimes raised in captivity by rural people. This tree is also cultivated in some home gardens, such as along the Tefé River, an affluent of the Solimões, and on the shores of Lago Amanã along the lower Japurá. Rural folk living along Ramal do Pau Rosa, a side-road leading off the BR 175 highway that connects Manaus and Boa Vista, also grow this fruit tree in some of their home gardens (Machado 2016: 37) as do some ribeirinhos in the watershed of the lower Purus River (Silva 2006: 127). The fruits are similar in many respects to puruí *Alibertia edulis* except they are larger, the size of a grapefruit or small cantaloupe, and the pulp is an orange-brown color. Numerous light tan seeds are embedded in the mesocarp which has a pungent odor. The brown-skinned fruits, which are mostly produced in the dry season, are peeled and mixed with water and sugar to make juice.

44.3 *Coutarea hexandra*

Bolivia Tajibo negro; nishiraoxó (Chácobo)

Brazil Carambá, papa-terra, capirona, capança vermelha, quina, quina brava, quina doce, quina quina, quina do mato, quina do Pará, quina branca, quina roxa, quineira, murta do mato

Colombia Mediagola, chaparrito

Peru Guacamayo caspi, sanango; ngoxü arü naxi (Ticuna)

Venezuela Campanilla, cabrito negro

Status Wild

Macileia, 11 years, with carambá *Coutarea hexandra* in successionary vegetation. Near Vila Cedro, Rio Cariá, a left bank affluent of the lower Rio Pracupi, in turn an affluent of the Anapu River that flows into the Baia de Botos, Caxiuanã National Forest, Pará, Brazil 5-16-15

Carambá *Coutarea hexandra* is a shrub to a small tree reaching 27 m tall in successionary growth in areas not subject to flooding that produces pods containing winged seeds. *Coutarea hexandra* also occurs in the understory of forests, including patches of woodlands in savanna, such as in Amapá. It grows mostly in lowlands, but it can be found as high as 2800 m in the Andean foothills. It is widely distributed in South America from Argentina and Paraguay northwards to the Amazon and Orinoco Basins, the Guianas, and into Central America as far north as Mexico.

Carambá starts producing fruit (technically a capsule) when only 2 m high. Children break open the elongated pods which resemble miniature okras with their fingers and snack on the winged seeds. Kids also use them to play games. Adults do not seem to show much interest in carambá so it is one of the children's fruits.

This plant also has medicinal uses. The Chácobo in the Bolivian Amazon squeeze the stem and mix the exudate with water which is then drunk to treat diarrhea (Boom 1987: 37). The Chácobo also grate the stem bark and boil a handful of it in 2 l of water until the water is reduced to 1 l; after it cools, half a cup of the solution is drunk three times a day to treat malaria, vomiting, and fever (Muñoz et al. 2000). In rural areas of Rondônia in the southwestern part of the Brazilian Amazon, people use the bark for preparing a tea to treat malaria (Brandão et al. 1992) as well as stomachache and fever (Lima et al. 2011). In Colombia, the bark is used as a vermifuge (Pérez-Arbeláez 1956: 665).

Macileia, 11 years, with carambá *Coutarea hexandra* capsule and winged seeds. Near Vila Cedro, Rio Cariá, a left bank affluent of the lower Rio Pracupi, in turn an affluent of the Anapu River in the Baia de Botos, Caxiuanã National Forest, Pará, Brazil 5-16-15

44.4 *Genipa americana*

Bolivia Bí, bigrande, yandipa, manzana de monte, jagua dulce, huito; nané, nanë (Chácobo), nobirr (Chiquitano), ñandipa (Chiriguano), bitu (Kichwa), danipa (Paucerna), ndía (Sirionó), aquid'ru (Tacana), tyi', ki, pa'ñe (Tsimane')

Brazil Genipapo, genipapeiro, jenipapo, jenipapo da água, jenipapo manso, jenipa, jenipapeiro, jenipaporana, jenipapinho, jenipaba, jenipaca, jenipava; janypawa (Amondawa), curubajó, û'a' (Arara), yanípa-'i (Araweté), wetxo'a (Aruá), yanípa-iwa (Asurini), wesua (Cinta larga), vesoá (Gavião), yanupa-'ɨ (Guajá), d'àd (Hupd'äh), yiniro'ɨ, yenipa-'ɨ, ñandyta (Ka'apor), kida pasoj (Karitiana), mroti, mrotire, pidjô-tyk (Kayapó), anga (Kuikuru), wisute (Matis), amuwan (Palikur), nyandipáb (Parintintin), jandypáwa (Piripkura), wexoah (Suruí), zanipaw-'iw (Tembé), diawé (Tukano, uu (Tupari), janupavá (Uru-Eu-Wau-Wau), yanipa (Wajãpi), kulupë (Wayana), hoo maxihi (Yanomami), wede'rã (Xavante)

Colombia Huito, uito, uvito, jagua, jaguo, jaguarito, caruto, pepa de pintura, maluci, tinto; jigikt (Witoto)

Ecuador Huito, jagua, jagua dulce, piton; suá, sha (Achuar), shiño (Cofán), huito, huituc, guito, wituk (Kichwa), súa, sula (Shuar), aíro toa, we'e, huée (Siona-Secoya), huencahue (Waorani)

Peru Huito, jagua, jahua, yaguayagua, anna, yacuruna huito, huitillo, huitoc, uvito, wito, witu, lana, sanango; súwa, suwánim, sua (Achuar, Aguaruna), nanu (Amahuaca), ana (Campa), nanë (Cashibo), ihsa (Chayahuita), janipa (Cocama), akuisho (Ese Eja), nanë (Kakataibo), ana (Matsigenka), nandi, nanë, lana, launa (Shipibo), jidoro (Witoto)

Venezuela Caruto, jagua, xagua; tana hi (Yanomami)

Status Wild, spontaneous in cultural settings, planted

Jon, 9 years, gathering jenipapo *Genipa americana* fruits from the ground in successionary forest. Archaeological site with anthropogenic dark earth. Site also of Arapijó, a colonial era village abandoned some 50 years ago. A few hundred meters inland from Carrazedo, mouth of the Xingu River, Municipality of Gurupá, Pará, Brazil 4-20-15

Known as genipap in English, *Genipa americana* is a tree, reaching up to 35 m, that is native to floodplain forests, including early successionary vegetation, in many parts of the Amazon including the tidal zone in the estuary (Almeida and Jardim 2011, 2012; Aparício 2011: 94; Araujo-Murakami et al. 2012; Assis and Wittmann 2011; Assis et al. 2017; Batista et al. 2013; Casula 2012: 67; Freitas et al. 2019; Huber 1904; Kvist et al. 1995; Lau and Jardim 2013; Machado and Pinheiro 2016; Maués et al. 2011; Miranda 2000; Montagnini and Muñiz-Miret 1999; Navarro 2002; Queiroz et al. 2006; Robles 2017: 49; Santos et al. 2014a, 2016b; Vieira et al. 2017). It occurs on the floodplains of rivers of various water types, including clear water such as the Rio Jari, an affluent of the lower Amazon (Carim 2016: 94), the upper Xingu (Zappi et al. 2016), and the Jaci-Paraná River which is an affluent of the Upper Madeira in Rondônia (Reis 2010: 37). In the Andean foothills of the Peruvian Amazon, it is found as high as 1000 m (Mendoza and Guerrero 2010). It is also found occasionally in upland forest, such as in Ecuador (Balslev et al. 1987; Cerón 1991), in the Parque Nacional Madidi in Bolivia (Araujo-Murakami et al. 2012), and in the watershed of the Purus, an affluent of the Solimões (Machado 2018: 126). It is also found in the understory of gallery forest in the cerrado region of central Brazil bordering the southern Amazon Basin (Marimon et al. 2002; Stefanello et al. 2009) as well as in savannas in Roraima (Barbosa et al. 2005; Dantas and Rodrigues 1982; Miranda et al. 2002; Takeuchi 1960). It also ranges widely outside of Amazonia from Argentina to Belize in Central America and parts of the Caribbean. It is widely cultivated for its edible fruit which is also used to make body paint and to rub into hair to make it shiny black. The flowers are pollinated by several species of stingless bees (Marques-Souza et al. 2002).

Although it is also occasionally found in upland woods and scrub savannas (Ratter et al. 2003), genipap is predominately a wetland tree, especially along sediment-rich rivers. As with many fruits that were widespread in tropical America by the time Europeans arrived, the precise origins of genipap are unclear. Some suggest that it is native outside of the Amazon basin, such as in Trinidad and Tobago (Williams and Williams Jr. 1951: 173). But based on its intricate cultural connections, Amazonia is more likely the domestication hearth for genipap.

Genipap is wild and cultivated in many parts of the Amazon, especially on floodplains, from the river's mouth to the Andean foothills in Bolivia, Peru, Ecuador, and Colombia. Along the middle Amazon, genipap thrives in home gardens along the river banks such as at Amador near Óbidos, Urucurituba and Paraná do Baixo near Alenquer, and Piracauera de Cima near Santarém. In a survey of 150 home gardens in the Brazilian Amazon carried out over several decades, I found jenipapo in 17 of them (11%). In a separate survey of 57 home gardens undertaken along the middle Xingu in 2015, I found *Genipa americana* in 10 of them (17%). Genipap is also planted in some home gardens in urban areas, such as São Luis in Maranhão (Akinnifesi et al. 2010a, b) and Cuiabá, the capital of Mato Grosso (Mamede et al. 2015).

Some of the genipap trees in home gardens sprouted from discarded seeds, such as on floodplain farms in the vicinity of Afuá in the northwest part of Marajó, where the ground has been fertilized by kitchen residues and human and livestock feces and urine and where more light penetrates the ground than in closed-canopy floodplain forests. Anthrosols associated with archaeological sites, such as Monte Carmelo along the Upper Anajás on Marajó Island, are also fertile ground for spontaneous genipap. Rural folk also plant genipap around their homes and occasionally in agroforestry plots on floodplains, such as in the vicinity of Santarém and along the lower Samiria River, an affluent of the Marañón.

The tree can withstand several months of flooding in home gardens, orchards, and forests. Seedling growth of *Genipa americana* slows when inundated for 3 months, but they survive (Andrade et al. 1999). In the Amazon estuary, genipap grows wild in a variety of habitats from wooded streams to non-flooded relict sand dunes in the northern part of Marajó, and grasslands. Seasonally flooded savannas are widespread in the eastern half of Marajó, and genipap has colonized some of them (Huber 1904). In pockets of savanna surrounded by forest near Chaves along the northern coast of Marajó, genipap sometimes occurs as isolated trees, possibly arising from seeds discarded by people, livestock, or birds.

Suzy, 11 years, eating genipap *Genipa americana* fruit gathered in a home garden on an archaeological site. Terra Preta near Silves, Rio Urubu, Amazonas, Brazil 4-3-15

As with many trees that produce widely appreciated fruits, genipap occurs both in the wild and in cultivation. Indeed, it can be difficult to ascertain which trees in forest are truly wild as opposed to relicts from old home sites. Possibly brought into cultivation at various locations in the Amazon basin at different times, genipap is common in backyards throughout the region. In some areas, such as at Monte Bello in the Pacaya-Samiria area of the Ucayali, people gather huito fruits in both their backyards and in the surrounding wetland forests. Long before Europeans arrived in the Americas, indigenous societies most likely took *Genipa americana* as a crop into Central America.

During the end of the dry season and throughout the rainy months, genipap produces baseball to grapefruit-sized fruits that are unmistakable in markets or on the forest floor because of their powerful musty aroma and slightly fuzzy brown skin. The potent odor emitted by the fruit has evidently evolved to attract dispersal agents that range from arboreal mammals, such as tayra *Eira barbara* (Sunquist et al. 1989), to terrestrial game animals, such as tapir *Tapirus terrestris*, peccaries (*Pecari tajacu* and *Tayassu pecari*), agouti (*Dasyprocta* spp.), paca *Agouti paca*, brocket deer (*Mazama americana* and *M. gouazoubira*), the red-footed tortoise *Chelonoidis carbonaria* and the yellow-footed tortoise *Chelonoidis denticulata* (Alviz and Pérez-Albarracín 2015: 41; Ayres and Ayres 1979; Beck 2005; Branan et al. 1985; Fragoso and Huffman 2000; Jerozolimski et al. 2009; Moskovits and Bjorndal 1990; Robinson and Eisenberg 1985; Tobler et al. 2010; Wang et al. 2011). Genipap thus contributes indirectly to the rural diet when people consume the above game animals (except tayra which are not hunted for their meat). According to a farmer living near Igarapé-Miri in the southern part of the Amazon estuary, possums are fond of the genipap fruits. Seeds of this fruit have been found in the feces of the maned wolf *Chrysocyon bachyurus*, a long-legged predator which inhabits cerrado vegetation in the southern fringes of the Amazon Basin (Castro and Emmons 2012).

Several species of monkeys, such as the white-faced capuchin *Cebus capucinus*, the wedge-capped capuchin *Cebus olivaceus*, squirrel monkeys *Saimiri sciureus*, howler monkeys *Alouatta palliata*, spider monkeys *Ateles belzebuth* and *A. geoffroyi*, are also likely dispersal agents for genipap because they are known to eat the fruits (Chapman 1989; Nunes 1998; Panger et al. 2002; Pontes 1997; Robinson 1984, 1986). Seeds of this tree have been found in the feces of the red howler monkey *Alouatta seniculus* which is widespread in northern and western Amazonia as well as Venezuela and the Guianas (Rojas 2012: 58). Although *Alouatta palliata* and *Ateles geoffroyi* do not occur in the Amazon Basin related species most likely fulfill a similar function. *Genipa americana* seeds have also been found in the feces of the moustached tamarin *Saguinus mystax* and the white-lipped tamarin *Saguinus labiatus* in the Peruvian Amazon (Ique and Aquino 2015; Reátegui 2013: 37) and woolly monkeys *Lagothrix lagotricha* in the Colombian Amazon (Stevenson 2000).

Genipap fruits can float and certain fish, including the highly prized tambaqui *Colossoma macropomum*, are fond of these fruits which they eat when they fall into the water and later defecate the seeds (Araujó-Lima and Goulding 1997: 61; Goulding and Carvalho 1982). The fruits float for a while before sinking to the bottom of flooded forests (Souza et al. 1999). The Wapishana who live in the Upper Essequibo watershed straddling Guyana and Roraima, Brazil, toss genipap fruits into the water and shoot pacu *Myleus pacu* with a bow and arrow when they come to feed on them (Farabee 1918: 58). The Palikur in the Oiapoque watershed use the fruit for fish bait (Castro 2013: 78). The largest turtle in the Amazon, tartaruga *Podocnemis expansa*, also eats this fruit (Cunha 2013: 53). People eat tartaruga turtles as well as their eggs which they dig up on sandy beaches during the low water season (Best 1984). The Amazon manatee *Trichechus inunguis* is also fond of genipap fruit (Domning 1980) but this large aquatic mammal is endangered due to overhunting. Hunters use canoes or platforms in flooded grasslands to harpoon this freshwater manatee, the largest native mammal in Amazonia.

The numerous light brown seeds embedded in the cream-colored pulp are easy to swallow because they are the size of orange pips, only flatter. People throughout the Amazon relish the soft, juicy pulp, which is also made into a fermented drink in the eastern part of the basin. Along the lower Tefé River some rural folk use tipitis normally used for squeezing manioc dough to extract the juice from genipap when there are a lot of the fruits and they want to sell the juice.

Genipap bears fruit at the end of branches, so agile kids and young adults are best suited for plucking the delicate fruits from the crown. The soft, brown fruits, some 8–15 cm long and 5–12 cm in diameter, are popular throughout the region. The fruits are also knocked off lower branches with poles, although they tend to split if they are not caught before hitting the ground. When mature, genipap fruits copiously for about a month or two, producing some 400–600 fruits (Souza et al. 1996: 132). Sometimes so many fruits ripen at the same time that many fall to the ground, such as around the homes of cowboys on Anjos ranch on Marajó Island, where pigs devour them. Some trees bear ripe fruit in the rainy season, whereas others are ready for picking in the dry months. Genipap fruits thus appear sporadically in markets throughout the year. Some are even trucked long distances, such as from Pucallpa to Lima in Peru. The fruits are also eaten fresh, often mashed with sugar. One female farmer near Maués, Amazonas, suggested that jenipapo fruits are good for those suffering from anemia.

Some households also make jam from the fruits. The practice of making jam (*mermelada*) from genipap and other Amazonian fruits was likely introduced by the Portuguese and Spanish; the Jesuit priest João Daniel reports that people made genipap jam in the Brazilian Amazon in the mid-eighteenth century (Daniel 1976: 335). A relatively new chocolatier in Manaus, Oiram Chocolates, sells chocolates filled with genipap jam; the smooth, chewy chocolates, individually wrapped in green foil, are sold in boutique outlets in the city. Oiram Chocolates with genipap and other fruit centers also appear in snack baskets on the top of room refrigerators in some of the upmarket hotels in Manaus.

Rural folk in the Brazilian Amazon ferment genipap fruits in water to produce a light brown, pungent liqueur. General stores in Juriti, a riverside town along the middle Amazon in Pará State that has fallen on hard times with the collapse of the jute boom, often sell *caldo de jenipapo*, a mixture of genipap liqueur and sugarcane juice. Patrons enjoy *caldo de jenipapo* by the glass anytime during the day or night, because there are no laws governing times when liquor can be sold in the Brazilian Amazon. The practice of converting genipap fruits to an alcoholic beverage is likely ancient. Over a century ago, people in the Brazilian Amazon mashed genipap pulp in water and let it ferment to produce a mildly alcoholic beverage with a strong smell and pleasant taste (Martina 1902: 28); this may even have been a motivating force in domesticating the fruit by indigenous cultures. In the Peruvian Amazon, locals often circumvent the fermentation process by mixing huito juice with sugarcane alcohol (*aguardiente*) to form a potent drink called *huitoachado*, as in Moyobamba.

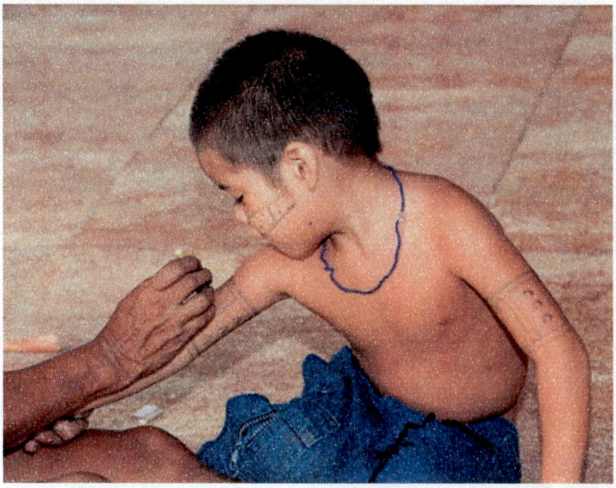

Tembé mother applying genipap *Genipa americana* body paint on her son, Mi, 4 years. They reside in an indigenous village in eastern Pará and are visiting Belém, 4-7-15

Another reason that genipap might have been domesticated is that it is used to make a black dye for body painting. In the Amazon, body painting, intricate feather work, and adornments such as lip disks and necklaces have often taken the place of clothes as personal adornment among indigenous people. The juice from freshly cut genipap fruits produces a black stain, widely employed by indigenous groups for body paint in Amazonia, especially for festive occasions. At Nueva Cajamarca along the Ucayali, for example, the Shipibo mash the green fruits in bowls fashioned from calabash gourds to tint their hair. The dye is also applied to the face, hands, and feet on festive occasions. Further upstream, the Conibo, who are now largely acculturated, used to paint themselves with genipap in the nineteenth century (Marcoy 1873, vol. 3: 27, 29). In the Huallaga valley, the Cholones painted their bodies with genipap in the nineteenth century to ward off insects, especially sandflies, as did the Panatahua (Herndon and Gibbon 1853: 138; Izaguirre 1927). In the Upper Uaupés watershed in the Colombian Amazon, pubescent Cubeo girls are painted with genipap as part of their ritual passage into adulthood (Goldman 1963: 180). In the southern Ecuadorian Amazon, the Shuar paint the figure of a serpent on the body of a person who has been bitten by a snake so that he or she will recover (Bennett et al. 2002: 245). The Shuar also use the fruit to color their hair (Pohle and Reinhardt 2004). Also in the Ecuadorian Amazon, the Waorani prepare a body paint with these fruits by grating them with the spiney stilt roots of the *Socratea exorrhiza* palm which is then mixed with the ash obtained by burning the bark of *Grias neuberthii* (which produces edible fruit), to darken the color and make it more durable (Mondragón and Smith 1997: 73).

Maricã, 10 years, a Juruna girl with genipap *Genipa americana* body paint. Paquiçamba, Rio Xingu near its confluence with the Bacajá, Pará, Brazil 9-8-15

Many tribes in the Brazilian Amazon once used, and in some cases still use, genipap for body paint. Along the Solimões, the Ticuna paint newborns with genipap dye because it is thought to help protect them against sickness. When Ticuna girls reach puberty they are again painted with genipap along with others attending the initiation ceremony (Gruber 1997: 19). The Ticuna also sometimes use genipap to dye twine obtained from the chambira palm *Astrocaryum chambira* which they use for weaving hammocks, storage bags, and string for necklaces (Glenboski 1983: 24). Genipap is such a prominent part of Ticuna culture that one of the tribal clans is named after the fruit (Gruber 1997: 19). Also along the Solimões, women belonging to the once numerous Cambeba (Omagua) tribe used to paint their bodies and hair with genipap juice in the seventeenth century (Fritz 1922: 49). In the Javari watershed in the boundary area between Amazonas, Brazil, and Peru, Matis males sometimes paint their bodies with genipap dye when hunting with their exceptionally long blowguns. They paint their bodies thus to imitate jaguars, but avoid the boy paint when hunting with bows and arrows because it is thought to scare away terrestrial game (Erikson 2001). In southwestern Amazonia, the Suruí in Rondônia are aficionados of body painting with genipap (Mindlin 1995: 21, 42). The now extinct Hypuriná of the Purus River once painted their whole bodies with genipap dye (Chandless 1866). In the northern part of the Amazon basin, the Wayapí employ genipap fruits to make body paint (Campbell 1989: 82) as do the Baniwa in the Içana watershed (Koch-Grünberg 1995: 113). Men and women of the Parintintin tribe who once inhabited the lower Madeira River used teeth of agoutis *Dasyprocta* spp. To tattoo their faces (Nimuendajú 1924).

Kawany, 6 years, with genipap *Genipa americana* body paint. Juruna village, Paquiçamba, Rio Xingu near its confluence with the Bacajá, Pará, Brazil 9-9-15

In eastern Amazonia, the Kayapó mix genipap juice with charcoal to paint their faces (Banner 1961), and the Gaviões who inhabit rapidly shrinking forests on the other side of the Tocantins River and once wore no clothes of any kind, also use genipap as well as annatto *Bixa orellana* for body paint (Arnaud 1964). Men and women in the Tapirapé tribe also use genipap for body paint, often in conjunction with annatto (Baldus 1970: 110). In the southern part of the Amazon, the Kaiabi use genipap to make body paint; this indigenous group once inhabited the Rio dos Peixes, an affluent of the Juruena in Mato Grosso, but was relocated to the Xingu indigenous park in the 1970s (Grünberg 2004: 146). The Araweté along the middle Xingu only paint their bodies with genipap when hunting and formerly when they undertook attacks on other indigenous groups (Viveiros de Castro 1992: 70). As noted by Jean de Lery, a Calvinist missionary to Brazil in the sixteenth century, genipap body paint does not wash off while swimming and lasts for about 12 days (Léry 1961: 104). In the early seventeenth century, a Capuchin priest who visited parts of Maranhão noted that genipap dye made an excellent ink for writing (Abbeville 1945: 169).

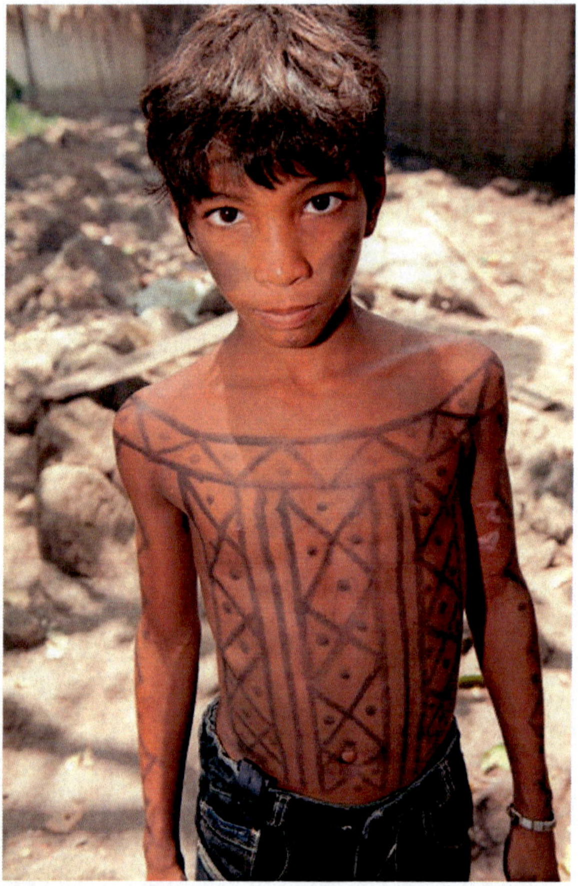

Ricardo, 10 years, an Arara boy with genipap *Genipa americana* body paint. Terrawangã, Rio Xingu 1 km upstream from the mouth of the Bacajá, Pará, Brazil 9-10-15

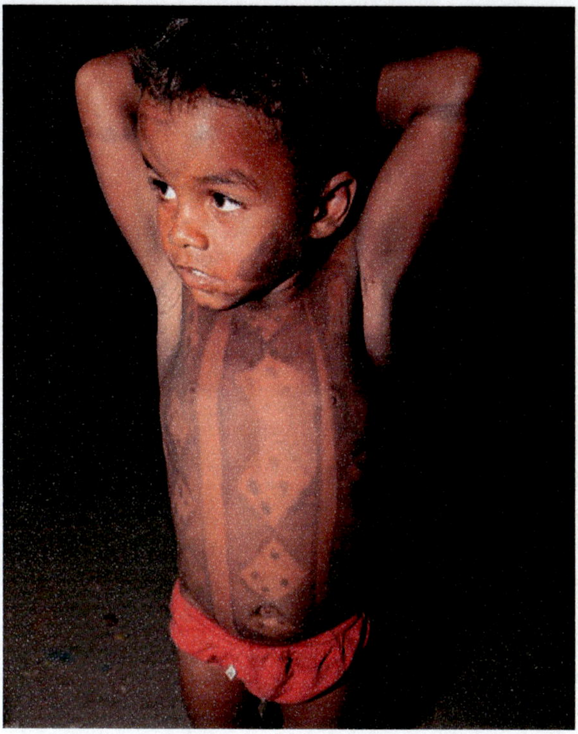

Maum, 4 years, an Arara boy with genipap *Genipa americana* body paint. Terrawangã, Rio Xingu 1 km upstream from the mouth of the Bacajá, Pará, Brazil 9-10-15

The Wayana who inhabit the Upper Rio Paru de Leste watershed in northern Pará believe that a supernatural man-caterpillar called ulupëakë taught them how to use genipap for body paint:

> In the old days, the Wayana did not use genipap paint, kulupë.
>
> A girl went to a stream, to bathe. Nearby she noticed many fallen fruits so she picked up some. She noticed that they had painted designs on them. She thought the designs were beautiful. Later she tossed the fruits away and returned home.
>
> That night a boy arrived at the village. A few people asked the boy who he was, but he did not respond. He was looking for the girl who had gone to the stream to bathe.
>
> At that time there were a lot of people around and the boy asked here and there until he met up with the girl who had gone to the stream. The boy said to the girl that his name was kulupëakë and that he lived close to the stream.
>
> He stayed around, and lay down with the girl in a hammock. In the early hours of the morning, before sunup, he left. The next day, the same thing happened.
>
> The girl's father asked her why she only had a husband at night. He suggested that next time she should not let the boy leave, so that he would remain in the village.
>
> The girl passed on the message to kulupëakë. He agreed and stayed in the village.
>
> Early the next day, they saw that he was a handsome boy, painted with genipap all over his body with various designs: maipuli axikë, akunwaiak, wakala portipë. Since the Wayana

> did not know how to paint themselves, he painted the body of the girl, other females, and eventually everyone.
>
> But the genipap was used up. The girl said that they would need to gather the fruits to be able to paint themselves. The boy agreed so he went with the girl to the place where the genipap tree was growing. When they were getting near, kulupëakë said to the girl to wait for him, to avoid her getting near the genipap tree. She was not to see him picking the fruits. But she did not obey and proceeded so that she could watch him. That was when she noticed that when he climbed up the tree he became a giant caterpillar. The girl then became angry and shouted that she no longer liked him, and that he was to go away and never return.
>
> After this, the Wayana started to use genipap for body painting as the supernatural caterpillar had taught them (Velthem and Linke 2010: 35–36)

Genipap dye is also used to paint vessels and the Cofán in the northern part of the Ecuadorian Amazon dye fiber obtained from the chambira palm *Astrocaryum chambira* (Cerón 1995: 131). The widespread use of genipap as a paint points to Amazonia, rather than other regions in tropical America, as the probable ancestral home of the tree. Over a century ago, Jacques Huber, a Swiss botanist and former director of the Goeldi Museum in Belém, noted that genipap was cultivated more for its fine black dye than for its fruit. Perhaps no single attribute can be pinpointed as the stimulus for genipap's entry into the fold of tended plants; its multiple uses have been the main incentive.

Genipap is incorporated in various folk remedies in Amazonia. In Peru, for example, the black stain reportedly promotes the healing of skin lesions. Also in Peru along the Amazon near the border with Brazil, the Ticuna use the seeds to make an infusion which is swallowed to treat parasites (Rengifo-Salgado et al. 2017). The Chayahuita heat the pulp and seeds then squeeze out the juice and apply it to skin infections (Odonne et al. 2013). The Chayahuita also grate unripe genipap fruits then heat it and squeeze out the juice which is mixed with ashes from burned banana peel and then apply it to the skin to treat cutaneous leishmaniasis (Odonne et al. 2009). The Chayahuita also decoct the ripe fruits for an hour with leaves of *Calycophyllum spruceanum* and then drink the solution to treat anemia. In various parts of the Peruvian Amazon, rural folk use this plant to treat a variety of medical conditions (Mejia and Rengifo 2000: 108). To treat hemorrhages and pellagra the bark is cooked in water and then a cup of the solution is drunk three times a day. To treat anemia and bronchial infections liquid is squeezed from the fruit and drunk and a syrup is also prepared to treat the same conditions by boiling the fruits in water containing sugar. Grated roots are cooked in water which is drunk when fasting to serve as a purgative. To treat dandruff the seeds are cooked and the liquid is poured on the scalp. To treat vaginal inflammation a mature fruit is cut into pieces and placed in a liter of water which is then heated and used as a bathing solution for the vagina (Lossio et al. 1995: 157; Ramírez and Vásquez 1998). In the vicinity of Iquitos in the Peruvian Amazon, the fruits are eaten to treat bronchitis (Ramírez and Vásquez 1998). The Achuar in the Pastaza watershed of the Peruvian Amazon use the immature pulp to facilitate the extraction of teeth (Lewis and Elvin-Lewis 1984).

In the vicinity of Chazuta along the Upper Huallaga River, locals macerate the fruit in rum and then drink it as an aphrodisiac (Sanz-Biset et al. 2009).

The Tsimane in the Bolivian Amazon use this dye to treat skin infections, such as those caused by fungi (Reyes-García 2001: 113). The Siona (Cioni) of the Upper Putumayo in the Colombian Amazon rub the leaves on the skin to relieve itching, treat bacterial skin infections, as well as to repel biting insects (Hardenburg 1913: 86).

Genipap is also used in folk remedies in the Brazilian Amazon. The Marúbo who live in the headwaters of the Ituí and Curuçá rivers, affluents of the Javari in southwest Amazonas, Brazil, rub genipap dye on sore parts of the body to promote healing, although this practice is becoming less common (Montagner 1996: 16). In the middle Amazon, the Saterê-Maué mix the juice of genipap and annatto *Bixa orellana* to ward off sickness (Pereira 1954: 75). The Makuxi in Roraima roast genipap fruits and mix the pulp with manioc dough to treat wounds caused by stingrays (Diniz 1971). Although genipap is not one of the more popular fruit juices, river dwellers sometimes drink it to treat anemia and infections, such as in the Caxiuanã National Forest in eastern Pará, in the municipality of Barcarena in northeastern Pará, along the lower Solimões near its confluence with the Rio Negro, and in the vicinity of Manaus (Amorozo 1997; Rocha et al. 2013; Vásquez 2014: 19; Vásquez et al. 2014; Veiga and Scudeller 2015). The Xipaya, an indigenous group along the lower Iriri River, an affluent of the Xingu, use the fruit and bark to make a tea to treat anemia (Santos et al. 2016a). In Roraima and Mato Grosso, people also employ the leaves to make a tea to alleviate anemia while the fruit is eaten to treat to diabetes (David and Pasa 2015; Luz 2001). Rural people in the northeast of Mato Grosso eat the fresh fruits to treat diabetes and liver problems (Ribeiro et al. 2017). In various parts of Rondônia rural folk prepare an infusion of the inner bark for treating anemia, the inner bark is placed under feet to reduce swelling, and a decoction of the leaves is swallowed to treating diabetes (Santos et al. 2014b). Some residents of Imperatriz in Maranhão close to the border with Pará prepare a decoction using the bark of this tree to treat inflammation (Penido et al. 2016). In the colonial period, the inner bark was applied to hernias, but it could take up to 2 months for them to be cured in this way (Sampaio 1969). In the early nineteenth century a German naturalist, Carl Friedrich Phillip von Martius, noted that the fruit was used to prepare bathing solutions to treat syphilitic ulcers in the vicinity of Santarém along the Amazon River near its confluence with the Tapajós (Breitbach et al. 2013).

Bottle gourd *Lagenaria siceraria* painted with genipap *Genipa americana* dye. Made by a Juruna subgroup in Mato Grosso. Held by Edilardo, 9 years, a Juruna boy. Paquiçamba, Rio Xingu near its confluence with the Bacajá, Pará, Brazil 9-9-15

Also during the colonial period, genipap wood was in demand for various construction tasks. In 1760, for example, the king of Portugal requested pau d'arco (*Tabebuia* sp.) and genipap to make carriages (Cruz 1973: 280). Also in the mid-eighteenth century, Father João Daniel, an astute observer of natural history and culture, noted that in addition to wood for carriages, the genipap timber was used for poles and gun stocks (Daniel 1976: 335). Today, the Kuikuru of the Upper Xingu use genipap wood to pierce ear lobes (Carneiro 1978). Rural people in various parts of the Amazon Basin use the wood for a variety of purposes including house posts, furniture, handles for axes and shovels, and to make canoes (Guèze et al. 2014; Kvist et al. 2001; Parrotta et al. 1995: 308; Silva et al. 1977: 113; Stagegaard et al. 2002; Wittmann et al. 2009). The bark of genipap is used to make cordage in some rural areas of the Amazon (Wittmann and Wittmann 2010).

Genipap is so thoroughly entwined in the rural and urban culture of the Brazilian Amazon that communities and properties are named after the fruit. On Marajó Island, for example, Jenipapo is a village built entirely on stilts along the low-lying banks of the Arari River near its southern exit from Arari Lake. Near Salvaterra on the same island, a ranch is named Fazenda Jenipapo. Near Abaetetuba, a bustling town on the lower Tocantins River in the southern estuary of the Amazon, one community is called Jenipa-úba, which means the place where there are a lot of genipap trees.

For a fruit, so widely consumed and used for adornment and curing, it is not surprising that genipap surfaces in the regional lore. Along the Purus River, for example, the Cashinahua ascribe the origin of ayahuasca, an entheogenic forest vine, to a tapir that tossed a genipap fruit into a lake, which then transformed into an anaconda. The aquatic serpent subsequently surfaced from the lake, changed into a beautiful woman with elaborate genipap body paint, and seduced the tapir (Lagrou 2000). This myth reflects some natural history, including the likelihood that tapirs eat genipap fruits, that both tapirs and genipap are often close to water, and male tapirs have an unusually large reproductive organ and are the subject of a number of sexual encounters involving people in folklore among various indigenous groups.

Genipap's affinity with wetlands surfaces in another myth, this one recounted by the Kamaiurá, who inhabit the Upper Xingu. In a story about the tense relationship between two ancestral brothers, a spider, a squirrel, and an opossum accompany one of the brothers to gather genipap fruits by a lagoon. The creatures decide to play with the Mother of Genipap, who resides in the water, and when the opossum tried to hide it carelessly left its tail hanging in the water where fish tore out the hair, hence the naked tails of opossums to this day (Villas Boas and Villas Boas 1975). Genipap also surfaces in a myth of the Xikrin who live in a single village along the Rio Cateté above its confluence with the Itacaiúnas in the municipality of Marabá in Pará. In this myth, two men fashion a blind in a genipap tree to kill an eagle that comes to eat the fruit (Vidal 1977: 207).

44.5 *Guettarda spruceana*

Brazil Angélica, murucututú

Status Wild, spontaneous in cultural settings

Angelica *Guettarda spruceana* in successionary vegetation with fruits not yet fully mature. Village cemetery, inland from Carrazedo, mouth of the Xingu River, Município de Gurupá, Pará, Brazil 10-24-16

Angelica *Guettarda spruceana* is found in early successionary growth (*capoeira*), grasslands (*campo*) and scrub savanna (*cerrado*) on uplands, especially in areas with sandy soils or rocky outcrops. Although *Guettarda spruceana* is largely confined to uplands it is found occasionally on floodplains. It produces edible yellow fruit, 1 cm × 0.7 cm, that are generally consumed by people while walking in rural areas. Spider monkeys *Ateles belzebuth* are also known to eat the fruits and these monkeys are hunted for the meat (Nunes 1998). The fruits never reach markets. This bush, which can reach 17 m when fully mature, is widespread but not common in Amazonia and is also found in the Guianas.

44.6 *Posoqueria latifolia*

Brazil Comida de urubu, mongoló, cabeça-rana, angelica da várzea, baga de macaco, laranja de macaco, açucena do mato, papa terra; txawa napü (Tiriyó), warë nahi, weri nahi (Yanomami)

Colombia Narciso de monte, cachaco, churumbelo; yö-kwö'-gee (Barasana)

Ecuador Wayus numi (Achuar), a tun tsicta, guayusa puru, matiri muyu, mindal (Kichwa)

Peru Lanzahuayo, huitillo, sacha huitillo, sacha huito, raya caspi, quena caspi, ucullucui, estrela ucullucui; káyuk wai, námukum (Achuar), jincagnum (Aguaruna)

Venezuela Jiomojimo; raharaha (Yanomami)

Status Wild

Kiener, 8 years, with lanzahuayo *Posoqueria latifolia* fruits he gathered in a forest. Near Yarina, Rio Yanayacu, Reserva Nacional Pacaya-Samiria, Loreto, Peru, 3-7-04

Lanzahuayo *Posoqueria latifolia* is a medium-sized tree that can reach 22 m high in the understory of forests that produces orange-yellow fruits. The globular fruits, some 4 cm in diameter, are broken open to access the pale gray pulp surrounding 2–4 brown seeds. The tree starts producing fruit when only 2 m tall and has long, spreading branches so the fruits can sometimes be accessed by pulling down on the tips of the branches. Alternatively, youngsters clamber up the trunk to pluck the fruits. The mesocarp is sweet and creamy and thus makes for a tasty snack while in the forest during the rainy season. Monkeys are known to eat the fruits and most likely serve as dispersal agents; two of the local names for this tree in Costa Rica—*guayaba de mico* and *fruta de mono*—mean monkey fruit (Beach 1983) and in parts of Brazil it is called *baga de macaco* which means monkey berry (Lorenzi et al. 2006: 289). Capuchin monkeys *Cebus apella* have been recorded eating these fruits in French Guiana (Guillotin et al. 1994); this widespread monkey undoubtedly does the same in Amazonia and may be involved in dispersing the seeds. This tree sports numerous large aromatic flowers (Le Cointe 1934: 9).

Kiener, 8 years, gathering lanzahuayo *Posoqueria latifolia* fruits in a forest. Near Yarina, Rio Yanayacu, Reserva Nacional Pacaya-Samiria, Loreto, Peru, 3-7-04

Posoqueria latifolia is also used in various folk remedies. In the Brazilian Amazon, the leaves are used to make a tea to treat liver problems such as hepatitis (Van den Berg 2010: 223; Van den Berg and da Silva 1986). The Tiriyó in northern Pará bordering Surinam grate the twigs and the shavings are then boiled in water; when the solution cools it is applied to wounds. The gratings are also plastered on the wounds to promote healing (Cavalcante 1973: 108–9). In the Ecuadorian Amazon, the Kichwa boil the bark to make a tea which is taken by women who have just given birth, and the leaves are cooked to create a bathing solution. Pounded bark is stirred into water and the liquid is dripped into the nose to treat colds. The bark is also scraped and heated in water to treat skin infections. In the Colombian Amazon, the Barasana pulverize the aromatic flowers and use the powder as a flea repellant (Schultes and Raffauf 1990: 392).

In the Amazon, *Posoqueria latifolia* is found mostly on higher parts of floodplains and old alluvial surfaces, such as swamps dominated by the *Mauritia flexuosa* palm (Amaral et al. 2009; Araujo-Murakami et al. 2009; Jardim et al. 2013; Silva 2008: 28). In flood-prone forests, it appears to be more common along black and clear water rivers (Ferreira et al. 2013). It also grows in some grasslands between sand dunes in parts of the coastal zone in the Amazon estuary (Margalho et al. 2009). It also occurs on some uplands, including scrub savannas and stunted woods (*campina*) on sandy soils, and can be found as high as 1900 m in the Andean foothills (Boom and Campos 1991; Mendoza and Guerrero 2010; Muniz et al. 1994). It also occurs on mounds (*tesos*) built by extinct indigenous cultures on Marajó Island at the mouth of the Amazon (Lisboa 2012: 415). This tree with distinctive white pendulous flowers is widely distributed in the Neotropics from Central America including Mexico, parts of the Caribbean, and South America to southern Brazil.

44.7 *Psychotria loretensis*

Ecuador Pishcu-muyu, pishcu-micuna-muya-yuri (Kichwa)

Peru Huanchaco huayo, paloma micuna; supinim (Aguaruna)

Status Wild

Genjis, 10 years, with huanchaco huayo *Psychotria loretensis* fruits gathered in a nearby floodplain forest. San Martin Tipishca, Rio Samiria, Reserva Nacional Pacaya-Samiria, Loreto, Peru, 6-30-06

Several plants in the coffee family in the Amazon have edible fruits, including the relatively well-known *Genipa americana*, but some of them are "minor" fruits known only to river dwellers because they never enter markets. *Psychotria loretensis* is one such fruit. The red fruits of *Psychotria loretensis* are gathered in floodplain forests, such as along the lower Samiria River, an affluent of the Marañón. In the Reserva Nacional Pacaya-Samiria, this shrub, which can grow 6 m tall, is known as huanchaco huayo (fruit of the huanchaco bird). Huanchaco (*Sturnella militaris*) is a red-breasted blackbird found in open and disturbed sites in Amazonia, hence the fruit's common name in Pacaya-Samiria. *Psychotria loretensis* occurs in central and western Amazonia, including occasionally in upland forests, as well as in the Guianas.

The round fruits are only 0.8 cm in diameter, but are born in closely group panicles and are sufficiently sweet to entice travelers in the forest to stop and eat them. The fruits contain three dark tan seeds, measuring some 0.5 cm by 0.3 cm, and are generally swallowed along with the pulp. The fruits are easily reached because the shrub starts producing fruit when a little over 1 m high and generally reaches no higher than 5 m.

The leaves of a related species, *Psychotria viridis*, are used in some ayahuasca preparations. Ayahuasca *Banisteriopsis caapi* is an entheogenic vine used by shamans in western Amazonia. Hence the name of the genus, *Psychotria*, which suggests mind-altering properties.

44.8 *Rudgea loretensis*

Peru Cotoruntu, sapo huayo, supinim; shantantar (Achuar), loroñahui (Kichwa), jápka, samírku sa (Mayna)

Status Wild

Cotoruntu *Rudgea loretensis* in fruit in floodplain forest along a black water river. Near Yarina, Rio Yanayacu, Reserva Nacional Pacaya-Samiria, Loreto, Peru 4-22-06

Rudgea loretensis is one of the rare plants that produces edible white fruits. The shrub or small tree is confined to the western part of the Amazon Basin in Bolivia, Peru, Ecuador, Colombia, and Acre and western Amazonas in Brazil. Although found mostly below 500 m, it is occasionally found growing as high as 1500 m in the Andean foothills. A denizen of forest on the higher parts of floodplains as well as on uplands, *Rudgea loretensis* bears round to ellipsoid fruits in the rainy season.

The fruits vary in size, from 1.6 cm × 1 cm to 3 cm long and 3 cm wide. The spongy fruits contain a single flat, light-green seed. The shade-tolerant shrub or treelet is scattered widely in the forest and does not form stands.

Cristian, 10 years, picking sapo huayo *Rudgea loretensis* fruit in a forest understory. Near Yarina, Rio Yanayacu, Reserva Nacional Pacaya-Samiria, Loreto, Peru 6-26-06

Locals gather the fruits by plucking them from the lower branches which are easily reached from the ground or canoeing up to them at the height of the annual flood. *Rudgea loretensis* bears fruit when it is 30 cm high, so even very young children pick the fruits. Mature fruits fall to the ground when ripe, so they can be picked from the forest floor. Mature trees can reach 7 m, but the trunk is slender enough to bend the tree over to obtain fruits from the higher branches. The small tree is not sturdy enough to be climbed safely.

Gilder 10 years (on left) and his brother, Donaldo 12 years, picking up cotoruntu *Rudgea loretensis* fruits from the floor of a floodplain forest. Quebrada Sacarita, affluent of the Rio Yanayacu, downstream from Yarina, Reserva Nacional Pacaya-Samiria, Loreto, Peru 6-25-06

Some of the villagers who live in Yarina along the Yanayacu, an affluent of the lower Marañón, refer to the fruit as cotoruntu; the prefix coto means howler monkey, suggesting that those monkeys (*Alouatta* spp.) are dispersal agents for the tree. At Veinte de Enero, a village along the lower Yanayacu, residents told me that frogs eat the fruits, hence one of its names in the Peruvian Amazon: sapo huayo. Game animals, such as paca *Agouti paca*, also allegedly eat the fruit. In the Ecuadorian Amazon, the bark is used to make a remedy for stomachache while the Chayahuita in the Peruvian Amazon crush the leaves and apply the poultice on wounds and ulcers caused by leishmaniasis (Odonne et al. 2009).

Cristian, 10 years, eating cotoruntu *Rudgea loretensis* fruit in the forest understory. Near Yarina, Rio Yanayacu, Reserva Nacional Pacaya-Samiria, Loreto, Peru 6-24-06

44.9 *Sabicea villosa*

Brazil Buxixu, cipó rasteira, toalha de buta, nambucaín, catiara, morango do roçado

Colombia Mortiño pipa, piña de abeja

Ecuador Yutuimias (Achuar), naék (Shuar) cogoñemuncamo (Waorani)

Peru Mullaca huasca, puchen mullaca, pulma de mullaca; cuíhja tsitsiíne, cúrɨhjau (Bora), yutuímas (Huambisa), yawar mullu (Kichwa), yuránkmis, napi (Mayna)

Status Wild, spontaneous in cultural settings

Buxixu *Sabicea villosa* in early successional growth. Ubim, Lago Amanã, Amazonas, Brazil 9-18-12

Sabicea villosa is a vine draped in early successionary vegetation that produces dark red hairy fruits that are eaten in rural areas. Although the round fruits are small, 1.2 cm by 1 cm, they are produced in small clusters and are easy to reach. Adults and children alike appreciate the spongy fruits which they most often encounter along paths to their fields and manioc processing huts. The pink to white pulp is sweet and reminiscent of some varieties of apples. Numerous light tan seeds, about 100 per fruit, are embedded in the mesocarp which are swallowed along with the pulp. In the Apaporis watershed in the Colombian Amazon, native people make a tea from the astringent leaves to treat malaria (Schultes and Raffauf 1990: 400). Also in the Colombian Amazon, the Barasana who live along the Pirá-paraná River give fruits of a related species, *Sabicea amazonensis*, to boys in an initiative rite after a shaman has blown a smoke over berries (Hugh-Jones 1979: 63). In the southern part of the Ecuadorian Amazon, the Shuar eat these fruits to treat bites from conga ants (Bennett et al. 2002: 248).

Girls picking catiara *Sabicea villosa* fruits in second growth along the side of a road leading to their village. Jocojó, Rio Jocojó, Município de Gurupá, Pará, Brazil 4-23-15

This vine, which can attain 5 m in length, is found in successionary vegetation mostly on uplands, including anthropogenic savannas, but also occurs occasionally along the banks of streams and rivers. It is widely distributed in the Amazon Basin into Central America as far north as Belize. It also occurs in parts of the Caribbean. It is found as high as 1400 m in the Andean foothills. Where it evolved is unclear, but it was most likely in Amazonia when humans arrived. Over time as forests have been cleared, the populations of this vine have increased. Along the middle Caquetá River in the Colombian Amazon, for example, *Sabicea villosa* is found in manioc fields planted by the Andoke (Eden and Andrade 1987).

Anadilson "Pretinho," 11 years, eating *Sabicea villosa* fruit he has picked in successionary vegetation on uplands. Carrazedo, mouth of the Xingu River, Município de Gurupá, Pará, Brazil 8-1-16

References

Abbeville, Claude d'. 1945. História da Missão dos Padres Capuchinhos na Ilha do Maranhão e Terras Circunvizinhas. Livraria Martins Editôra, São Paulo

Abreu, Maria Milena de Oliveira; Mehlig, Ulf; Nascimento, Raquel do Espírito Santo Aguair do; Menezes, Moirah Paula Machado. 2006. Análise de composição florística e estrutura de um fragmento de bosque de terra firme e de um manguezal vizinhos na peninsula de Ajuruteua, Bragança, Pará. Boletim do Museu Paraense Emílio Goeldi, Ciencias Naturais 1(3):27–34

Aguirre-Mendoza, Zhofre; Loja, Ángel Fernando; Solano, Carmen; Aguirre, Nikolay. 2015. Especies Forestales más aprovechadas del Sur del Ecuador. Universidad Nacional de Loja, Ecuador

Akinnifesi, F. K.; Sileshi, G.; Costa, Joselene da; Moura, E. G. da; Silva, Rosele F. da; Ajayi, O. C.; Linhares, J. F. P.; Akinnifesi, A. I.; Araujo, Maria de; Rodrigues, M. A. I. 2010a. Floristic composition and canopy structure of home-gardens in São Luís city, Maranhão state, Brazil. Journal of Horticulture and Forestry 2(4):72–86

Akinnifesi, Festus Kehinde.; Sileshi, Gudeta W.; Ajayi, Oluyede C.; Akinnifesi, Adetutu I.; Moura, Emanoel G. de; Linhares, Jairo F. P.; Rodrigues, Ivanilde 2010b. Biodiversity of the urban homegardens of São Luís city, northeastern Brazil. Urban Ecosystems 13(1):129–146

Almeida, Adrielson Furtado; Jardim, Mário Augusto Gonçalves. 2011. Florística e estrutura da comunidade arbórea de uma floresta de várzea na Ilha de Sororoca, Ananindeua, Pará, Brasil. Scientia Florestalis 39(90):191–198

Almeida, Adrielson Furtado; Jardim, Mário Augusto Gonçalves. 2012. A utilização das espécies arbóreas da floresta de várzea da Ilha de Sororoca, Ananindeua, Pará, Brasil por moradores locais. Revista Brasileira de Ciências Ambientais 23:48–54

Alviz, Ángela; Pérez-Albarracín, Karen. 2015. Plan para la conservación de la danta de tierras bajas (Tapirus terrestris) en los departamentos de Casanare, Arauca y Vichada. Corporinoquia-Fundación Orinoquia Biodiversa, Yopal, Colombia

Amaral, Dário Dantas do; Vieira, Ima Célia Guimarães; Almeida, Samuel Soares de; Salomão, Rafael Paiva; Silva, Antonio Sérgio Lima da; Jardim, Mário Augusto Gonçalves. 2009. Checklist da flora arbórea de remanescentes florestais da região metropolitana de Belém e valor histórico dos fragmentos, Pará, Brasil. Boletim do Museu Paraense Emílio Goeldi, Ciencias Naturais 4(3):231–289

Amorozo, Maria Christina de Mello 1997. Algumas notas adicionais sobre o emprego de plantas e outros produtos com fins terapêuticas pela população cabocla do Município de Barcarena, PA, Brasil. Boletim do Museu Paraense Emilio Goeldi, Série Botânica 13(2):192–213

Andrade, A.C.S.; Ramos, F.N.; Souza, A.F.; Loureiro, M.B.; Bastos, R. 1999. Flooding effects in seedlings of *Cytharexyllum myrianthum* Cham. and *Genipa americana* L.: responses of two Neotropical lowland tree species. Revista Brasilieira de Botanica 22 (2): 281–285

Andrade, Teresinha Maria de; Assis, Rafael Leandro de; Wittmann, Florian; Schöngart, Jochen; Piedade, Maria Teresa Fernandez. 2008. Padrões de regeneração em clareiras de origem antrópica na várzea da RDS Mamirauá, Amazônia central. Uakari 4(2): 19–32

Aparício, Wegliane Campelo da Silva. 2011. Estrutura da vegetação em diferentes ambientes na RESEX do Rio Cajari: interações solo-floresta e relações com a produção de castanha. Doctoral Dissertation, Universidade Federal Rural de Pernambuco, Recife

Araujó-Lima, C. and M. Goulding. 1997. So Fruitful a Fish: Ecology, Conservation, and Aquaculture of the Amazon's Tambaqui. Columbia University Press, New York

Araujo-Murakami, Alejandro; Paniagua, Narel; Cayola, L.; Jorgensen, Peter M.; Valdes, Ovidio; Macuapa, Eric; Calzadilla-Tomianovich, Miki. 2009. Diversidad y estructura florística del bosque de llanura y palmar de pantano Amazónico preandino en el sector de Chalalan, Parque Nacional Madidi, Bolivia. Kempffiana 5(1):3–27

Araujo-Murakami, Alejandro; Poma-Chura, Abraham; Palabral, Arely; Salvatierra, Rolando; Hurtado, Fidel. 2012. Composición florística de los bosques Amazónicos de tierra firme e inundable en las proximidades de las Pampas del Sonene (río Heath), Parque Nacional Madidi, Bolivia. Kempffiana 8(1):2–21

Arnaud, Expedito. 1964. Notícia sôbre os índios Gaviões de oeste, Rio Tocantins, Pará. Boletim do Museu Paraense Emílio Goeldi, Nova Série, Antropologia, vol. 20: 1–35

Assis, Rafael Leandro de. 2008. Composição florística e estrutura da regeneração arbórea de florestas de várzea alta e várzea baixa na RDS Mamirauá, Amazônia central. Master's thesis, INPA, Manaus

Assis, Rafael Leandro de; Wittmann, Florian 2011. Forest structure and tree species composition of the understory of two central Amazonian várzea forests of contrasting flood heights. Flora 206:251–260

Assis, Rafael Leandro de; Wittmann, Florian; Luize, Bruno Garcia; Haugaasen, Torbjon. 2017. Patterns of floristic diversity and composition in floodplain forests across four southern Amazon river tributaries, Brazil. Flora 229:124–140

Ayres, J. M.; Ayres, C. 1979. Aspectos da caça no alto Rio Aripuanã. Acta Amazonica 9(2):287–298

Baldus, H. 1970. Tapirapé: Tribo Tupí no Brasil Central. Companhia Editora Nacional, São Paulo

Balslev, Henrik; Luteyn, James; Ollgaard, Benjamin; Holm-Nielsen, Lauritz B. 1987. Composition and structure of adjacent unflooded and floodplain forest in Amazonian Ecuador. Opera Botanica 92: 37–57

Banner, Horace. 1961. O índio Kayapó em seu acampamento. Boletim do Museu Paraense Emílio Goeldi, Nova Série, Antropologia, vol. 13: 1–51

Barbosa, Reinaldo Imbrozio; Nascimento, Sebastião Pereira do; Amorim, Paulo Atlântico Figueiredo de; Silva, Rosemberg Ferreira da. 2005. Notas sobre a composição arbóreo-arbustiva de uma fisionomia das savanas de Roraima, Amazônia Brasileira. Acta Botanica Brasilica 19(2):323–329

Barnett, Adrian A.; Silva, Welma Sousa; Shaw, Peter J. A.; Ramsay, Paul M. 2015. Inundation duration and vertical vegetation zonation: a preliminary description of the vegetation and structuring factors in borokotóh (hummock igapó), an overlooked high-diversity, Amazonian vegetation association. Nordic Journal of Botany 33:601–614

Bastos, M. N. C.; C. S. Rosário; L. C. B. Lobato. 1995. Caracterização fitofisionômica da restinga de Algodoal, Maracanã, Pará, Brasil. Boletim do Museu Paraense Emílio Goeldi, Série Botanica 11(2): 173–197

Batista, Anderson Pedro Bernardina; Santos, Vanessa Silva dos; Aparício, Wegliane Campelo da Silva; Aparício, Perseu da Silva; Silva, Diego Armando Silva da. 2013. Similaridade e gradientes de riqueza florística em um floresta de várzea na cidade de Macapá. Revista Verde de Agroecologia e Desenvolvimento Sustentável 8(4):152–158

Beach, J. 1983. *Posoqueria latifolia* (boca de vieja, guayaba de mico, fruta de mono). In: Costa Rican Natural History, Janzen, Daniel H. (ed), University of Chicago Press, Chicago, pp. 307–308

Beck, H. 2005. Seed predation and dispersal by peccaries throughout the Neotropics and its consequences: a review and synthesis. In: Seed Fate: Predation, Dispersal and Seedling Establishment, Forget, Pierre-Michel; Lambert, Joanna E.; Hulme, Philip E.; Wall, Stephen B. Vander (eds), CABI, Wallingford, pp 77–115

Bennett, Bradley C.; Baker, Marc A.; Andrade, Patricia Gómez. 2002. Ethnobotany of the Shuar of Eastern Ecuador. New York Botanical Garden, Advances in Economic Botany, vol. 14.

Best, R.C. 1984. The aquatic mammals and reptiles of the Amazon. In: The Amazon: Limnology and Landscape Ecology of a Mighty Tropical River and its Basin, H. Sioli (ed), W. Junk, Dordrecht, pp 371–412

Bolfe, Édson Luis; Batistella, Mateus. 2011. Análise florística e estrutural de sistemas silviagrícolas em Tomé-Açu, Pará. Pesquisa Agropecuária Brasileira 46(10):1139–1147

Boom, B. M. 1987. Ethnobotany of the Chácobo Indians, Beni, Bolivia. The New York Botanical garden, Advances in Economic Botany, vol. 4

Boom, B. M.; Campos, M. T. V. 1991. A preliminary account of the Rubiaceae of a central Amazonian terra firme forest. Boletim do Museu Paraense Emílio Goeldi, Série Botânica 7(2):223–247

Borges, Hélida B. Nogueira; Shepherd, George J. 2005. Flora e estrutura do estrato lenhoso numa comunidade de Cerrado em Santo Antônio do Leverger, MT, Brasil. Revista Brasileira de Botânica 28(1):61–74

Branan, William V.; Werkhoven, Marga C. M.; Marchinton, R. Larry. 1985. Food habits of brocket and white-tailed deer in Suriname. Journal of Wildlife Management 49(4):972–976

Brandão, M. G. L.; Grandi, T. S. M.; Rocha, E. M. M.; Sawyer, D. R.; Krettli, A. U. 1992. Survey of medicinal plants used as antimalarials in the Amazon. Journal of Ethnopharmacology 36: 175–182

Breitbach, Ulrike B.; Niehues, Michael; Lopes, Norberto P.; Faria, Jair E. Q.; Brandão, Maria G. L. 2013. Amazonian Brazilian medicinal plants described by C.F.P. von Martius in the 19th century. Journal of Ethnopharmacology 147:180–189

Campbell, A. T. 1989. *To Square with Genesis: Causal Statements and Shamanic Ideas in Wayapí.* Edinburgh University Press, Edinburgh

Carim, Marcelo de Jesus Veiga. 2016. Estrutura, composição e diversidade em florestas alagáveis de várzea de maré e de igapó e suas relações com variáveis edáficas e o período de inundação no Amapá, Amazônia oriental, Brasil. Doctoral dissertation, Programa de Pós-Graduação em Botânica, INPA, Manaus

Carneiro, R. L. 1978. The knowledge and use of rain forest trees by the Kuikuru Indians of central Brazil. In: *The Nature and Status of Ethnobotany*, R. Ford (ed.), Anthropological Paper Series, Museum of Anthropology, University of Michigan, Ann Arbor, vol. 67:208–216

Castro, E. 2013. Artefatos e Matérias-Primas dos Povos Indígenas do Oiapoque. Iepé/Museu Kuahí, São Paulo

Castro, José Miguel; Emmons, Louise H. 2012. Variation in diet and resources. In: The Maned Wolves of Noel Kempff Mercado National Park. Emmons, L. H. (ed.), Smithsonian Contributions to Zoology Number 639, Smithsonian Institution Scholarly Press, pp 37–55

Casula, Kátia Regina. 2012. Análise da similaridade florística e estrutural das formações florestais inundáveis em um trecho do alto Rio Madeira e de seus afluentes, estado de Rondônia, Brasil. Master's Thesis, Núcleo de Ciências Exatas da Terra, Universidade Federal de Rondônia, Porto Velho

Cavalcante, P. B. 1973. A Farmacopéia Tiriyó: Estudo Étno-Botânico. Museu Paraense Emílio Goeldi, Publicações Avulsas No. 24

Cavalcante, P. B. 2010. *Frutas Comestíveis na Amazônia.* Museu Paraense Emílio Goeldi, Belém

Cerón, Carlos Eduardo. 1991. Manejo florístico Shuar-Achuar (Jibaro) del ecosistema Amazónico en el Ecuador. Hombre y Ambiente 5(17):7–31

Cerón, Carlos Eduardo. 1995. Etnobiologia de los Cofanes de Dureno, Provincia de Sucumbíos, Ecuador. Museo Ecuatoriano de Ciencias Naturales, Herbario Nacional, Serie Monografia 10(3), Quito

Chandless, W. 1866. Ascent of the River Purus. Journal of the Royal Geographical Society 36: 86–118

Chapman, Colin A. 1989. Primate seed dispersal: the fate of dispersed seeds. Biotropica 21(2): 148–154

Corette-Pasa, Maria. 2011. Abordagem etnobotânica na comunidade de Conceição-Açu, Mato Grosso, Brasil. Polibotánica 31:169–197

Costa, Ineilian B. C. da; Bonfim, Filipe P. G.; Pasa, Maria C.; Montero, Daniel A. V. 2017. Ethnobotanical survey of medicinal flora in the rural community Rio dos Couros, state of Mato Grosso, Brazil. Boletin Latinoamericano y del Caribe de Plantas Medicinales y Aromáticas 16(1): 53–67

Cruz, Ernesto. 1973. História de Belém. Universidade Federal do Pará, Coleção Amazônica, Série José Veríssimo, Belém, vol. 1

Cunha, Fernando Lima Rodrigues da. 2013. Dieta de quatro espécies do gênero *Podocnemis* (Testudines, Podocnemididae) na Reserva de Desenvolvimento Sustentável Uatumã, Amazonas, Brasil. Master's thesis, INPA, Manaus

Daniel, J. 1976. *Tesouro Descoberto no Rio Amazonas.* Biblioteca Nacional, Rio de Janeiro, vol. 1

Dantas, Mário; Rodrigues, Irenice Alves. 1980. Plantas invasoras de pastagens cultivadas na Amazônia. EMBRAPA, Centro de Pesquisa Agropecuária do Trópico Úmido, Boletim de Pesquisa, vol. 1.

Dantas, Mário; Rodrigues, Irenice Alves. 1982. Estudos fitoecológicos do trópico úmido Brasileiro: IV-levantamentos botânicos em campos do Rio Branco. EMBRAPA, Centro de Pesquisa Agropecuária do Trópico Úmido, Boletim de Pesquisa, no. 40

David, Margôde; Pasa, Maria Corette. 2015. As plantas medicinais e a etnobotânica em Várzea Grande, MT, Brasil. Interações 16(1):97–108

Diniz, E. S. 1971. O xamanismo dos índios Makuxi. Journal de la Société Americanistes 60:65–73

Domning, D. P. 1980. Distribution and status of manatees *Trichechus* spp. near the mouth of the Amazon River, Brazil. *Biological Conservation* 19: 85–97

Eden, Michael J.; Andrade, Angela. 1987. Ecological aspects of swidden cultivation among the Andoke and Witoto Indians of the Colombian Amazon. Human Ecology 15(3): 339–359

Erikson, P. 2001. Myth and material culture: Matis blowguns, palm trees, and ancestor spirits. In: Beyond the Visible and Material: The Amerindianization of Society in the Work of Peter Rivière, Rival, Laura; Whitehead, Neil (eds), Oxford University Press, Oxford, pp. 101–121

Espírito-Santo, Fernando Del Bon; Shimabukuro, Yosio Edimir; Aragão, Luiz Eduardo Oliveira e Cruz de; Machado, Evandro Luiz Mendonça. 2005. Análise da composição florística e fitossociológica da floresta nacional do Tapajós com apoio geográfico de imagens satélites. Acta Amazonica 35(2):155–173

Farabee, W. C. 1918. *The Central Arawaks*. The University Museum, University of Pennsylvania, Philadelphia, vol. 10

Ferreira, Leandro V. 1997. Effects of the duration of flooding on species richness and floristic composition in three hectares in the Jaú National Park in floodplain forests in central Amazonia. Biodiversity and Conservation 6:1353–1363

Ferreira Júnior, Elbert Viana; Soares, Thelma Shirlen; Costa, Marcus Filipe Fernandesda; Silva, Versides Sebastião Moraes. 2008. Composição, diversidade e similaridade florística de uma floresta tropical semidecídua de uma floresta tropical semidecídua submontana em Marcelândia, MT. Acta Amazonia 38(4):673–680

Ferreira, Leandro V.; Chaves, Priscilla P.; Cunha, Denise de A.; Matos, Darley C. L.; Parolin, Pia. 2013. Variação da riqueza e composição de espécies da comunidade de plantas entre as florestas de igapós e várzeas na Estação Científica Ferreira Penna-Caxiuanã na Amazônia oriental. Pesquisas Botânica 64:175–195

Fragoso, José M. V.; Huffman, Jean M. 2000. Seed-dispersal and seedling recruitment patterns by the last Neotropical megafaunal element in Amazonia, the tapir. Journal of Tropical Ecology 16:369–385

Freitas, João da Luz; Santos, Erick Silva dos; Cruz Junior, Francisco de Oliveira; Borges, César Henrique Alves; Santos, Adriano Castelo dos. 2019. Structural analysis and distribution patterns in lowland tropical forest, eastern Amazon. Journal of Agricultural Studies 7(4): 240–255

Fritz, S. 1922. Journal of the Travels and Labours of Father Samuel Fritz in the River of the Amazons between 1686 and 1723. Hakluyt Society, London

Glenboski, L. L. 1983. *The Ethnobotany of the Tukuna Indians, Amazonas, Colombia*. Universidad Nacional de Colombia, Museo de Historia Natural, Biblioteca Jose Jeronimo Triana, Bogotá

Goldman, I. 1963. The Cubeo: Indians of the Northwest Amazon. University of Illinois Press, Urbana

Gonçalves, Janaína Pinheiro; Lucas, Flávia Cristina Araújo. 2017. Agrobiodiversidade e etnoconhecimento em quintais de Abaetetuba, Pará, Brasil. Revista Brasileira de Biociências 15(3):119–134

Goulding, Michael; Carvalho, Mirian Leal. 1982. Life history and management of the tambaqui (*Colossoma macropomum*, Characidae): an important Amazonian fish food. Revista Brasileira de Zoologia 1(2):107–133

Gruber, Jussara Gomes. 1997. Ticuna: O Livro das Árvores. OGPTB (Organização Geral dos Professores Ticuna Bilíngües), Benjamin Constant, Amazonas, Brazil

Grünberg, G. 2004. Os Kaiabi do Brasil Central: História e Etnografia. Instituto Socioambiental, São Paulo

Guarim Neto, Germano. 1984. Plantas do Brasil: Angiospermas do Estado de Mato Grosso- 1. Rodriguésia 36(59):105–121

Guarim Neto, Germano; Pasa, Maria Corette. 2009. Estudo etnobotânico em uma área de cerrado no município de Acorizal, Mato Grosso. Flovet 1:5–32

Guèze, Maximilien; Luz, Ana Catarina; Paneque-Gálvez, Jaime; Macía, Manuel J.; Orta-Martínez, Martí; Pino, Joan; Reyes-García, Victoria. 2014. Are ecologically important tree species the most useful? a case study from indigenous people in the Bolivian Amazon. Economic Botany 68(1):1–15

Guillotin, M.; Dubost, G.; Sabatier, D. 1994. Food choice and food competition among the three major primate species of French Guiana. Journal of Zoology 233:551–579

Hardenburg, W. E. 1913. The Putumayo, The Devil's Paradise; Travels in the Peruvian Amazon Region and an Account of the Atrocities Committed upon the Indians Therein. T. Fisher Unwin, London

Herndon, W. L. and L. Gibbon. 1853. Exploration of the Valley of the Amazon, made under the Direction of the Navy Department. Rober Armstrong, Washington, D.C., vol. 1

Homeier, Jürgen; Kurzatkowski, Dariusz; Leuschner, Christoph. 2017. Stand dynamics of the drought-affected floodplain forests of Araguaia River, Brazilian Amazon. Forest Ecosystems 4:1–10

Huber, J. 1904. Notas sobre a patria e distribuição geographica das arvores fructiferas do Pará. *Boletim do Museu Goeldi (Museu Paraense) de Historia Natural e Ethnographia* 4 (2–3):375–406

Hugh-Jones, S. 1979. The Palm and the Pleiades: Initiation and Cosmology in Northwest Amazonia. Cambridge University Press, Cambridge

Ique, Carlos A.; Aquino, Rolando. 2015. Principales especies florísticas utilizadas y dispersión de semillas por *Saguinus labiatus* E. Geoffroy, 1812 en isla Muyuy, Loreto, Perú. Ciencia Amazónica 5(1):81–88

Izaguirre, B. 1927. Descripción histórico-etnográfica de algunas tribus orientales del Perú. Boletín de la Sociedad Geográfica de Lima 44:5–36

Jardim, Daniel Gonçalves; Jardim, Mário Augusto G.; Quaresma, Adriano Costa; Costa, Salustiano Villar. 2013. Regeneração natural em formações florestais de uma unidade de conservação, Maracanã, Pará, Brasil. Biota Amazônia 3(2):79–87

Jardim, Daniel Gonçalves; Jardim, Mário Augusto Gonçalves; Quaresma, Adriano Costa; Costa Neto, Salustiano Vilar. 2016. Fatores edáficos e as variações florísticas da regeneração natural em duas formações florestais. Biota Amazônia 6(2):66–72

Jerozolimski, A., M. B. Ribeiro, and M. Martins. 2009. Are tortoises important seed dispersers in Amazonian forests? *Oecologia* 161: 517–528

Kawa, Nicholas C.; Michelangeli, José A. Clavijo; Clement, Charles R. 2015. Household agrobiodiversity management on Amazonian dark earths, oxisols, and floodplain soils on the lower Madeira River, Brazil. Human Ecology 43:339–353

Koch-Grünberg, T. 1995. Dos Años entre los Indios: Viajes por el Noroeste Brasileño 1903/1905. Editorial Universidad Nacional, Bogotá

Kurzatkowski, Dariusz; Leuschner, Christoph; Homeier, Jürgen. 2015. Effects of flooding on trees in the semi-deciduous transition forests of the Araguaia floodplain, Brazil. Acta Oecologica 69:21–30

Kvist, Lars P.; Andersen, Martin K.; Hesselsoe, Martin; Vanclay, Jerome K. 1995. Estimating use-values and relative importance of Amazonian flood plain trees and forests to local inhabitants. Commonwealth Forestry Review 74(4):293–300

Kvist, Lars Peter; Andersen, Martin K.; Stagegaard, Jesper; Hesselsoe, Martin; Llapapasca, Consuelo. 2001. Extraction from woody forest plants in flood plain communities in Amazonian Peru: use, choice, evaluation and conservation status of resources. Forest Ecology and Management 150:147–174

Lagrou, Elsje Maria. 2000. Two ayahuasca myths from the Cashinahua of northwestern Brazil. In: Ayahuasca Reader: Encounters with the Amazon's Sacred Vine, Luna, L.E.; White, S.F. (Editors), pp. 31–35. Synergetic Press, Santa Fe.

Lau, Arthur Viana; Jardim, Mário Augusto Gonçalves. 2013. Florística e estrutura da comunidade arbórea em uma floresta de várzea na área proteção ambiental, Ilha do Combu, Belém, Pará. Biota Amazônia 3(2):88–93

Le Cointe, P. 1934. *Amazônia Brasileira III: Árvores e Plantas Úteis*. Livraria Classica, Belém

Léry, J. 1961.Viagem à Terra do Brasil. Biblioteca do Exercito, Rio de Janeiro

Lewis, Walter H.; Elvin-Lewis, Memory P. F. 1984. Plants and dental care among the Jívaro of the Upper Amazon Basin. Advances in Economic Botany 1: 53–61

Lima, Renato Abreu; Magalhães, Sandra Aparecida; Santos, Maurício Reginaldo Alves dos. 2011. Levantamento etnobotânico de plantas medicinais utilizadas na cidade de Vilhena, Rondônia. Revista Pesquisa & Criação 10(2):165–179

Lisboa, P. L. B. 2012. *A Terra dos Aruã: Uma História Ecológica do Arquipélago do Marajó*. Museu Paraense Emílio Goeldi, Belém

Lopes, Célia Regina Araujo Soares; Ribeiro, Ricardo da Silva; Rodrigues, Lucirene; Cabral, Fabiana Ferreira; Silva, Dennis Rodrigues. 2014. Checklist de angiospermas da região de influência da UHE SINOP, médio Teles Pires, Mato Grosso. Enciclopédia Biosfera 10(19):2016–2048

Lorenzi, H., L. Bacher, M. Lacerda, S. Sartori 2006. Brazilian Fruits and Cultivated Exotics (for Consuming in Natura). Instituto Plantarum de Estudos da Flora Ltda, Odessa, São Paulo

Lossio, L.C., Delgado, H.S., Ruiz, A.G., Donayre, R.A., Ruiz, J.G., Panduro, M.P., Sifuentes, T.C. 1995. Plantas Medicinales de la Amazonía Peruana. Instituto Peruano de Seguridad Social/Instituto de Medicina Tradicional, Iquitos

Loverde-Oliveira, Simoni Maria; Freitas, Maria Niely; Araújo, Patrícia Karina Barbosa; Costa, Ineilian Bruna Correa. 2010. Fragmento de cerrado urbano da Universidade Federal de Mato Grosso, campus de Rondonópolis, Mato Grosso. Revista Biodiversidade 9(1):74–90

Luz, Francisco Joaci F. 2001. Plantas medicinais de uso popular em Boa Vista, Roraima, Brasil. Horticultura Brasileira 19(1):88–96

Machado, Danilo de Oliveira. 2016. A agrobiodiversidade de quintais agroflorestais em propriedades agrícolas familiares na BR 174, Ramal do Pau-Rosa, Manaus, AM. Master's Thesis, INPA, Manaus

Machado, Clara de Carvalho. 2018. Conhecimento e consumo de plantas alimentícias em cinco comunidades da reserva de desenvolvimento sustentável Piaguçu-Purus, Amazonas. Master's thesis, INPA, Manaus

Machado, Monielle Alencar; Pinheiro, Claudio Urbano B. 2016. Da água doce à água salgada: mudanças na vegetação de igapó em margens de lagos, rios e canais do baixo curso do rio Pindaré, Baixada Maranhense. Revista Brasileira de Geografia Física 9(5): 1410–1427

Magnusson, William E.; Lima, Albertina P.; Albernaz, Ana L. K. M.; Sanaiotti, Tânia M.; Guillaumet, Jean-Louis. 2008. Composição florística e cobertura vegetal das savanas na região de Alter do Chão, Santarém-PA. Revista Brasileira de Botanica 31(1):165–177

Mamede, Jeneffer Soares dos Santos; David, Margô De; Tsukamoto Filho, Antônio de Arruda; Pasa, Maria Corette. 2015. Os quintais e as manifestações culturais da comunidade de São Gonçalo Beira Rio, Cuiabá – MT. Biodiversidade 14(1):168–182

Marcoy, P. 1873. *A Journey across South America from the Pacific Ocean to the Atlantic Ocean*. Blackie and Son, London

Margalho, Luciano Ferreira; Rocha, Antonio Elielson Sousa da; Secco, Ricardo de Souza. 2009. Rubiaceae Juss. da restinga da APA de Algodoal/Maiandeua, Maracanã, Pará, Brasil. Boletim do Museu Paraense Emílio Goeldi, Ciencias Naturais 4(3):303–339

Marimon Junior, Ben Hur; Haridasan, Mundayatan. 2005. Comparação da vegetação arbórea e características edáficas de um cerradão e um cerrado sensu stricto em áreas adjacentes sobre solo distrófico no leste de Mato Grosso, Brasil. Acta Botanica Brasilica 19(4):913–926

Marimon, Beatriz Schwantes; Lima, Edson de Souza. 2001. Caracterização fitofisionômica e levantamento florístico preliminar no Pantanal dos Rio Morte-Araguaia, Cocalinho, Mato GrossoActa Botanica Brasilica 15(2):213–229

Marimon, B. S.; Felfili, J. M.; Lima, E. S. 2002. Floristics and phytosociology of the gallery forest of the Bacaba stream, Nova Xavantina, Mato Grosso, Brazil. Edinburgh Journal of Botany 59(2):303–318

Marques-Souza, Antonio Carlos; Miranda, Ires Paula de Andrade; Moura, Cleonice de Oliveira; Rabelo, Afonso; Barbosa, Edelcílio Marques. 2002. Características morfológicas e bioquímicas do pólen coletado por cinco espécies de Meliponíneos da Amazônia central. Acta Amazonica 32(2):217–229

Martina, G. 1902. Estudo Chimico sobre algumas Frutas Brasileiras. Imprensa Official, Belém

Matos, Vanda Fernandes de; Lolis, Solange de Fátima; Coimbra, Ronaldo Rodrigues; Viana, Rodney Haulien Oliveira. 2020. Phytosociology in areas of the Brazilian cerrado: cerrado stricto sensu, municipality of Porto Nacional, state of Tocantins. Acta Scientiarum Biological Sciences 42: e50353

Maués, Bernardo Antônio Rodrigues; Jardim, Mário Augusto Gonçalves; Batista, Fábio de Jesus; Medeiros, Tonny David Santiago; Quaresma, Adriano da Costa. 2011. Composição florística e estrutura do estrato inferior da floresta de várzea na área de proteção ambiental Ilha do Combu, município de Belém, estado do Pará. Revista Árvore 35(3):669–677

Medeiros, Priscilla Sanjuan de. 2016. Indicadores ambientais para funcionalidade ecológica em florestas secundárias de diferentes idades na Amazônia Oriental. Doctoral dissertation, Universidade Federal do Pará, Belém

Medeiros, Marcelo Brilhante de; Walter, Bruno Machado Teles; Silva, Glocimar Pereira da; Gomes, Beatriz Machado; Lima, Isabel Lustz Portela; Silva, Suelma Ribeiro; Moser, Pamela; Washington, Luís Oliveira; Cavalcanti, Taciana Barbosa. 2012. Vascular flora of the Tocantins River middle basin, Brazil. Check List 8(5):852–885

Mejia, Kember; Rengifo, Elsa. 2000. Plantas Medicinales de Uso Popular en la Amazonía Peruana. Agencia Española de Cooperación Internacional, Lima

Mendoza, Abel L. Monteagudo; Guerrero, Marleni Huamán. 2010. Catálogo de los arboles y afines de la selva central del Perú. Arnaldoa 17(2):202–242

Mindlin, B. 1995. Unwritten Stories of the Suruí Indians of Rondônia. Institute of Latin American Studies, University of Texas, Austin

Miranda, V. C. 1909. Os campos de Marajó e a sua flora considerados sob o ponto de vista pastoril. *Boletim do Museu Goeldi (Museu Paraense) de Historia Natural e Ethnographia* 5(1): 96–151

Miranda, Izildinha S. 2000. Análise florística e estrutural da vegetação lenhosa do Rio Comemoração, Pimenta Bueno, Rondônia, Brasil. Acta Amazonica 30(3):393–422

Miranda, Izildinha Souza; Absy, Maria Lúcia; Rebêlo, George Henrique. 2002. Community structure of woody plants of Roraima savannahs, Brazil. Plant Ecology 164:109–123

Mondragón, Martha L.; Smith, Randall. 1997. Bete Quiwiguimamo Salvando el bosque para vivir sano: algunas plantas y árboles utilizadas por la Nacionalidad Huaorani de la Amazonía Ecuatoriana. Abya Yala, Quito

Montagner, Delvair. 1996. A Morada das Almas: Representações das Doenças e das Terapêuticas entre os Marúbo. Museu Paraense Emílio Goeldi, Belém

Montagnini, Florencia; Muñiz-Miret, Nuria. 1999. Vegetation and soils of tidal floodplains of the Amazon estuary: a comparison of várzea and terra firme forests in Pará, Brazil. Journal of Tropical Forest Science 11(2):420–437

Moreira, Déborah Luíza; Guarim-Neto, Germano. 2009. Usos múltiplos de plantas do cerrado: um estudo etnobotânico na comunidade sítio Pindura, Rosário Oeste, Mato Grosso, Brasil. Polibotánica 27:159–190

Moskovits, Debra K.; Bjorndal, Karen A. 1990. Diet and food preferences of the tortoises *Geochelone carbonaria* and *G. denticulata* in northwestern Brazil. Herpetologica 46(2):207–218

Muniz, Francisca Helena; Cesar, Oswaldo; Monteiro, Reinaldo. 1994. Fitossociologia da vegetação arbórea da reserva florestal do Sacavém, São Luís, Maranhão (Brasil). Acta Amazonica 24 (3/4):219–236

Muñoz, V.; Sauvain, M.; Bourdy, G.; Callapa, J.; Bergeron, S.; Rojas, I.; Bravo, J. A.; Balderrama, L.; Ortiz, B.; Gimenez, A.; Deharo, E. 2000. A search for natural bioactive compounds in Bolivia through a multidisciplinary approach Part I. Evaluation of the antimalarial activity of plants used by the Chacobo Indians. Journal of Ethnopharmacology 69:127–137

Navarro, G. 2002. Provincia biogeográfica del Acre-Madre de Dios (Amazónica Suroccidental). In: Geografía Ecológica de Bolivia: Vegetación y Ambientes Acuáticos, Navarro, Gonzalo; Maldonado, Mabel (eds), Centro de Ecología Simón I. Patiño, Cochabamba, pp 51–98

Nebel, Gustav; Dragsted, Jens; Vanclay, Jerome K. 2000a. Estructura y composición florística del bosque de la llanura aluvial inundable de la Amazonía Peruana: II. El sotobosque de la restinga. Folia Amazónica 10(1–2):151–181

Nebel, Gustav; Kvist, Lars Peter; Vanclay, Jerome K.; Christensen, Henning; Freitas, Luis; Ruiz, Juan. 2000b. Estructura y composición florística del bosque de la llanura aluvial en la Amazonía Peruana: I. El bosque alto. Folia Amazónica 10(1–2):91–149

Nimuendajú, C. 1924. Os indios Parintintin do Rio Madeira. Journal de La Société des Americanistes (Paris, N.S.) 16: 201–278

Nunes, Andréa. 1998. Diet and feeding ecology of *Ateles belzebuth belzebuth* at Maracá Ecological Station, Roraima, Brazil. Folia Primatologica 69:61–76

Odonne, G.; Bourdy, G.; Castillo, D.; Estevez, Y.; Lancha-Tangoa, A.; Alban-Castillo, J.; Deharo, E.; Rojas, R.; Stien, D.; Sauvain, M. 2009. Ta'ta', Huayani: perception of leishmaniasis and evaluation of medicinal plants used by the Chayahuita in Peru. Part II. Journal of Ethnopharmacology 126:149–158

Odonne, Guillaume; Valadeau, Celine; Alban-Castillo, Joaquina; Stien, Didier; Sauvain, Michel; Bourdy, Genevieve. 2013. Medical ethnobotany of the Chayahuita of the Paranapura basin (Peruvian Amazon). Journal of Ethnopharmacology 146:127–153

Oliveira Filho, Ary Teixeira de. 1989. Composição florística e estrutura comunitária da floresta de galeria do Córrego da Paciência, Cuiabá (MT). Acta Botanica Brasilica 3(1):91–112

Oliveira Filho, Ary Teixeira de. 1992. The vegetation of Brazilian 'murundus' – the island-effect on the plant community. Journal of Tropical Ecology 8:465–486

Oliveira Filho, A. T.; Martins, F. R. 1991. A comparative study of five cerrado areas in southern Mato Grosso, Brazil. Edinburgh Journal of Botany 48(3):307–332

Panger, M.A.; Perry, S.; Rose, L.; Gros-Louis, J.; Vogel, E.; Mackinnon, K.C.; Baker, M. 2002. Cross-site differences in foraging behavior of white-faced capuchins (*Cebus capucinus*). American Journal of Physical Anthropology 119: 52–66

Parolin, P. 2000a. Seed mass in Amazonian floodplain forests with contrasting nutrient supplies. Journal of Tropical Ecology 16(3):417–428

Parolin, P. 2000b. O uso de árvores nas florestas inundadas por água branca na Amazônia Central. Amazoniana 16(1/2):241–248

Parolin, P. 2000c. Phenology and C02-assimilation of trees in Central Amazonian floodplains. Journal of Tropical Ecology 16:465–473

Parolin, P. 2002. Submergence tolerance vs. escape from submergence: two strategies of seedling establishment in Amazonian floodplains. Environmental and Experimental Botany 48:177–186

Parrotta, John A.; Francis, John K.; Almeida, Rionaldo Rolo de. 1995. Trees of the Tapajós: A Photographic Field Guide. International Institute of Tropical Forestry, General Technical Report IITF-1, Rio Piedras, Puerto Rico

Pasa, Maria Corette; Soares, João Juares; Guarim Neto, Germano. 2005. Estudo etnobotânico na comunidade de Conceição-Açu (alto da bacia do rio Aricá Açu, MT, Brasil). Acta Botanica Brasilica 19(2):195–207

Penido, Alexandre Batista; Morais, Selene Maia de; Ribeiro, Alan Bezerra; Silva, Ana Zélia. 2016. Ethnobotanical study of medicinal plants in Imperatriz, state of Maranhão, northeastern Brazil. Acta Amazonica 46(4):345–354

Penso, Sarah; Bauer, Maristela de Oliveira; Chichorro, José Franklim; Gondim, Carlos Alberto; Vasconcellos, Lívia Vieira. 2009. Caracterização estacional de uma pastagem natural do cerrado Mato-Grossense submetida ao pastejo. Ciência Animal Brasileira 10(1):124–134

Pereira, N. 1954. Os Índios Maués. Organização Simões, Rio de Janeiro

Pereira, Antonio Carlos; Santos, Eduardo Ribeiro dos. 2015. Frutas nativas do Tocantins com potencial de aproveitamento económico. Agri-Environmental Sciences 1(1):22–37

Pereira, Luciano Araújo; Sobrinho, Felipe de Araújo; Costa Neto, Salustiano Vilar da. 2011. Florística e estrutura de uma mata de terra firme na Reserva de Desenvolvimento Sustentável no Rio Iratapuru, Amapá oriental, Brasil. Floresta 41(1):113–122

Pérez-Arbeláez, Enrique. 1956. Plantas Útiles de Colombia. Liberia Colombiana, Bogotá

Pinto, José Roberto Rodrigues; Mews, Henrique Augusto; Jancoski, Halina Soares; Marimon, Beatriz Schwantes; Bomfim, Bárbara de Oliveira. 2014. Woody vegetation in a floodplain campo de murundus in central Brazil. Acta Botanica Brasilica 28(4): 519–526

Pohle, Perdita; Reinhardt, Sylvia. 2004. Indigenous knowledge of plants and their utilization among the Shuar of the lower tropical mountain forest in southern Ecuador. Lyonia 7(2):133–149

Pontes, Antonio Rossano Mendes. 1997. Habitat partitioning among primates in Maracá Island, Roraima, northern Brazilian Amazonia. International Journal of Primatology 18(2):131–157

Posey, D. A. 1985a. Indigenous management of tropical forest ecosystems: the case of the Kayapó Indians of the Brazilian Amazon. *Agroforestry Systems* 3:139–158

Posey, D. A. 1985b. Native and indigenous guidelines for new Amazonian development strategies: understanding biological diversity through ethnoecology. In: Change in the Amazon Basin Volume 1: Man's Impact on Forests and Rivers, Hemming, John (ed), Manchester University Press, Manchester, pp 156–181

Prado, Helena Alves do. 2012. Frugivoria e dispersão de sementes por anta (*Tapirus terrestris*) em uma área de plantio florestal no cerrado, MS. Master's thesis, Universidade Estadual Paulista, Rio Claro, São Paulo

Queiroz, José Antonio Leite de; Mochiutti, Silas; Machado, Sebastião do Amaral. 2006. Distribuição diamétrica em floresta de várzea do braço norte do Rio Amazonas. Revista de Ciência Agrária 46:151–170

Ramírez, Mauro Vásquez; Vásquez, Juan Balaurte. 1998. La extracción de productos forestales diferentes de la madera en el ambito de Iquitos, Peru. Folia Amazonica 9(1–2):69–92

Ratter, J. A.; Bridgewater, S.; Atkinson, R.; Ribeiro, J. F. 1996. Analysis of the floristic composition of the Brazilian cerrado vegetation II: comparison of the woody vegetation of 98 areas. Edinburgh Journal of Botany 53(2):151–180

Ratter, J. A.; Bridgewater, S.; Ribeiro, J. F. 2003. Analysis of the floristic composition of the Brazilian cerrado vegetation III Comparison of the woody vegetation of 376 areas. Edinburgh Journal of Botany 60(1):57–109

Rayol, Breno Pinto; Silva, Manoela Ferreira Fernandes da; Alvino, Fabrízia de Oliveira. 2006. Dinâmica da diversidade florística da regeneração natural de florestas secundárias, no município de Bragança, Pará, Brasil. Boletim do Museu Paraense Emílio Goeldi, Ciencias Naturais 1(2):9–27

Reátegui, Katia Wendy. 2013. Hábitos alimenticios de *Saguinus mystax* Spix "pichico barba blanca" en Padre Isla, Loreto, Peru. Master's thesis, UNAP, Iquitos

Reis, Neidiane Farias Costa. 2010. Estrutura fitossociológica e seletividade de espécies em áreas inundáveis na região do baixo Rio Jaci-Paraná, Rondônia. Master's Thesis, Universidade Federal de Rondônia, Porto Velho

Rengifo-Salgado, Elsa; Rios-Torres, Sandra; Malaverri, Lizardo Fachín; Vargas-Arana, Gabriel. 2017. Saberes ancestrales sobre el uso de flora y fauna en la comunidad indígena Tikuna de Cushillo Cocha, zona fronteriza Perú-Colombia-Brasil. Revista Peruana de Biología 24(1):67–78

Reyes-García, Victoria. 2001. Indigenous people, ethnobotanical knowledge, and market economy: a case study of the Tsimane' Amerindians in lowland Bolivia. Doctoral dissertation, University of Florida, Gainesville.

Ribeiro, Reginaldo Vicente; Bieski, Isanete Geraldini Costa; Balogun, Sikiru Olaitan; Martins, Domingos Tabajarade Oliveira. 2017. Ethnobotanical study of medicinal plants used by ribeirinhos in the north Araguaia microregion, Mato Grosso, Brazil. Journal of Ethnopharmacology 205: 69–102

Robinson, J. G. 1984. Diurnal variation in foraging and diet in the wedge-capped capuchin *Cebus olivaceus*. Folia Primatologica 43:216–228

Robinson, J. G. 1986. Seasonal variation in use of time and space by the wedge-capped capuchin monkey, *Cebus olivaceus*: implications for foraging theory. Smithsonian Contributions to Zoology Number 431, Washington, DC.

Robinson, J. G.; Eisenberg, J. F. 1985. Group size and foraging habits of the collared peccary *Tayassu tajacu*. Journal of Mammalogy 66(1):153–155

Robles, Rembrandt Ramiro Canahuire. 2017. Composición florística y estructura de la recuperación natural en un área degradado por minería en la comunidad nativa de Tres Islas, Tambopata, Madre de Dios. Master's thesis, Escuela Profesional de Ingeniería Forestal y Medio Ambiente, Puerto Maldonado, Perú

Rocha, S. F. R., M. R. C. Ferreira, and M. G. Sablayrolles. 2013. Farmacopeia vegetal da comunidade Caxiuanã, na Floresta Nacional de Caxiuanã. In: Caxiuanã: Paraíso ainda Preservado, P. L. B. Lisboa (Editor), Museu Paraense Emílio Goeldi, Belém, pp. 257–285

Rodrigues, Maira dos Santos; Conceição, Gonçalo Mendes da. 2014. Diversidade florística das diferentes fisionomias de cerrado do Parque Estadual do Mirador, Maranhão, Brasil. Brazilian Geographical Journal: Geosciences and Humanities research medium, Ituiutaba 5(1):139–156

Rodrigues, Ricardo Ribeiro; Martins, Sebastião Venâncio; Barros, Luiz Carlos de. 2004. Tropical rain forest regeneration in an area degraded by mining in Mato Grosso state, Brazil. Forest Ecology and Management 190: 323–333

Rojas, Natalia Alejandra Alvis. 2012. Patron de actividad, dieta, área de actividad y dispersion de semillas de *Alouatta seniculus* en un fragmento de Bosque en San Juan del Carare (Santander). Thesis, Universidad del Tolima, Facultad de Ciencias, Bogotá

Rondon Neto, Rubens Marques; Santos, Jeneffer Soares dos; Silva, Manoel Advíncula da; Koppe, Valdinei Cristi. 2010. Potencialidades de uso de espécies arbustivas e arbóreas em diferentes fisionomias de cerrado, em Lucas do Rio Verde/MT. Revista de Biologia e Ciências da Terra 10(2):113–126

Sampaio, F. A. 1969. História dos reinos vegetal, animal e mineral do Brasil, pertecente à medicina. Anais da Biblioteca Nacional 89:1–91

Santos, Josefa Silva dos; Guarim Neto, Germano. 2017. Plantas de quintais da área urbana de Comodoro, Mato Grosso, Brasil. Flovet 1(9):31–57

Santos, João Ubiratan M. dos; Rosário, Carlos da S. 1988. Levantamento da vegetação fixadora das dunas de Algodoal-PA. Boletim do Museu Paraense Emílio Goeldi, Série Botânica 4(1): 133–154

Santos, João Ubiratan M. dos; Amaral, Dário Dantas do; Gorayeb, Inocêncio de Sousa; Bastos, Maria de Nazaré do C.; Secco, Ricardo de S.; Neto, Salustiano Vilar Costa; Costa, Denise Cristina Torres. 2003. Vegetação da área de proteção ambiental Jabotitiua-Jatium, Município de Viseu, Pará, Brasil. Acta Amazonica 33(3):431–444

Santos, Gideão Costa; Tourinho, Manoel Malheiros; Mendes, Fernanda da Silva; Guimarães, Célia Maria Costa. 2014a. Fitossociologia e práticas de manejo tradicional em uma floresta de várzea em Santa Bárbara do Pará, Brasil. Revista de Ciências Agrárias 57(2):138–145

Santos, M. R. A.; Lima, M. R.; Oliveira, C. L. L. G. 2014b. Medicinal plants used in Rondônia, western Amazon, Brazil. Revista Brasileira de Plantas Medicinais 16(3):707–720

Santos, Leonardo Maracahipes; Lenza, Eddie; Santos, Josias Oliveira dos; Marimon, Beatriz Schwantes; Eisenlohr, Pedro V.; Marimon Junior, Ben Hur; Feldpausch, Ted R. 2015. Diversity, floristic composition, and structure of the woody vegetation of the cerrado in the cerrado-Amazon transition zone in Mato Grosso, Brazil. Brazilian Journal of Botany 38(4):877–887

Santos, Joielan Xipaia dos; Reis, Alisson Rodrigo Souza; Matos, Sterphane Araújo de; Leão, Fábio Miranda; Carvalho, Josiane Celerino de. 2016a. Caracterização etnobotânica de essências florestais com fins medicinais utilizadas pela etnia Xipaya, no município de Altamira-PA. Biota Amazônia 6(2):1–8

Santos, Otilene dos Anjos; Couceiro, Sheyla Regina Marques; Rezende, Alinne Costa Cavalcante; Silva, Márcia Daniela de Sousa. 2016b. Composition and richness of woody species in riparian forests in urban areas of Manaus, Amazonas, Brazil. Landscape and Urban Planning 150:70–78

Sanz-Biset, Jaume; Campos-de-la-Cruz, José; Epiquién, Mirbel A.; Cañigueral, Salvador. 2009. A first survey on the medicinal plants of the Chazuta valley (Peruvian Amazon). Journal of Ethnopharmacology 122:333–362

Schultes, R. E. and R. F. Raffauf. 1990. The Healing Forest: Medicinal and Toxic Plants of the Northwest Amazonia. Dioscorides Press, Portland, Oregon

Scudeller, Veridiana Vizoni; Souza, Adriana Mota Gomes de. 2009. Florística da mata de igapó na Amazônia Central. In: Biotupé: Meio Físico, Diversidade Biológica e Sociocultural do Baixo Rio Negro, Amazônia Central, Santos-Silva, Edinaldo Nelson; Scudeller, Veridiana Vizoni (eds), pp 97–108, UEA Edições, Manaus

Silva, Suzy Cristina Pedroza da. 2006. Os sistemas agroflorestais na Reserva de Desenvolvimento Sustentável Piagaçu-Purus: uma alternativa para a agricultura familiar. Master's thesis, Universidade Federal do Amazonas, Manaus

Silva, Maria Betânia Araujo da. 2008. Composição florística e estrutura fitossociológica de uma floresta ripária situada às margens do Rio Moju, Pará. Master's thesis, Universidade Federal Rural da Amazônia, Belém

Silva, Manoela F. F. da; Rosa, Nelson A. 1989. Análise do estrato arbóreo da vegetação sobre jazidas de cobre na Serra dos Carajás-PA. Boletim do Museu Paraense Emílio Goeldi, Série Botânica 5(2):175–206

Silva, M. F., P. L. B. Lisbôa, R. C. L. Lisbôa 1977. Nomes Vulgares de Plantas Amazônicas. INPA, Manaus

Souza, A. G. C., N. R. Sousa, S. E. L. Silva, C. D. M. Nunes, A. C. C. Canto, L. A. A. Cruz, 1996. *Fruteiras da Amazônia.* EMBRAPA (Empresa Brasileira de Agropecuária), Brasília

Souza, Alexandre Fadigas de; Andrade, Antonio Carlos Silva de; Ramos, Flavio Nunes; Loureiro, Marta Bruno. 1999. Ecophysiology and morphology of seed germination of the neotropical lowland tree *Genipa americana* (Rubiaceae). Journal of Tropical Ecology 15(5): 667–680

Stagegaard, Jesper; Sorensen, Marten; Kvist, Lars Peter. 2002. Estimations of the importance of plant resources extracted by inhabitants of the Peruvian Amazon flood plains. Perspectives in Plant Ecology, Evolution and Systematics 5(2):103–122

Stefanello, Daniel; Fernandes-Bulhão, Clarissa; Martins, Sebastião Venâncio. 2009. Síndromes de dispersão de sementes em três trechos de vegetação ciliar (nascente, meio e foz) ao longo do Rio Pindaíba, MT. Revista Árvore 33(6):1051–1061

Stevenson, Pablo R. 2000. Seed dispersal by woolly monkeys (*Lagothrix lagothricha*) at Tinigua National Park, Colombia: dispersal distance, germination rates, and dispersal quantity. American Journal of Primatology 50: 275–289

Sunquist, Mel E.; Sunquist, Fiona; Daneke, Dennis E. 1989. In: Advances In Neotropical Mammalogy, Redford, K.H.; Eisenbeg, J.F. (eds.), Sand Hill Crane Press, Gainesville, pp. 197–232

Takeuchi, Masayuki. 1960. A estrutura da vegetação na Amazônia II: as savanas do norte da Amazônia. Boletim do Museu Paraense Emílio Goeldi, Nova Série Botânica 7:1–14

Tobler, Mathias W.; Janovec, John P.; Cornejo, Fernando. 2010. Frugivory and seed dispersal by the lowland tapir *Tapirus terrestris* in the Peruvian Amazon. Biotropica 42(2):215–222

Van den Berg, M. E. 2010. Plantas Medicinais na Amazônia. Museu Paraense Emílio Goeldi, Belém

Van den Berg, M.E. and M.H.L. da Silva. 1986. Plantas medicinais do Amazonas. In: Anais do1°Simpósio do Trópico Úmido, Belém, PA, 12 a 17 November, vol. 2, Flora e Floresta, EMBRAPA, Brasília, pp. 127–133

Vásquez, Silvia Patricia Flores. 2014. Conhecimento, uso e conservação da diversidade vegetal em quatro comunidades ribeirinhas no Município Manacapuru, Amazonas. Doctoral dissertation, Programa de Pós-Graduação em Botânica, INPA, Manaus

Vásquez, Silvia Patricia Flores; Mendonça, Maria Silvia de; Noda, Sandra do Nascimento. 2014. Etnobotânica de plantas medicinais em comunidades ribeirinhas do Município de Manacapuru, Amazons, Brasil. Acta Amazonica 44(4):457–472

Veiga, J. B.; Scudeller, V. V. 2015. Etnobotânica e medicina popular no tratamento de malária e males asociados na comunidade ribeirinha Julião- baixo Rio Negro (Amazônia Central). Revista Brasileira de Plantas Medicinais 17(4):737–747

Velthem, L. H. and I. L. Linke 2010. *Livro da Arte Gráfica Wayana e Aparai*. Museu do Índio, FUNAI, Rio de Janeiro

Vidal, Lux. 1977. Morte e Vida de uma Sociedade Indígena Brasileira: Os Kayapó-Xikrin do Rio Cateté. Editora Hucitec/Editora da Universidade de São Paulo, São Paulo

Vieira, Diegodos Santos; Oliveira, Marcio Leles Romarco de; Gama, João Ricardo Vasconcelos; Machado, Evandro Luiz Mendonça; Görgens, Eric Bastos; Lafetá, Bruno Oliveira; Garcia, Josinei Silva. 2017. Phytosociology of a natural fragment of the floodplain forest in the lower Tapajós River, Brazil. Bosque 38(2):357–369

Villas Boas, Orlando; Villas Boas, Claudio. 1975. Xingu: The Indians, Their Myths. Souvenir Press, London

Viveiros de Castro, E. 1992. From the Enemy's Point of View: Humanity and Divinity in an Amazonian Society. University of Chicago Press, Chicago

Wang, Ellen; Donatti, Camila I.; Ferreira, Vanda L.; Raizer, Josué; Himmelstein, Jeffrey. 2011. Food habits and notes on the biology of *Chelonoidis carbonaria* (Spix 1824) (Testudinidae, Chelonia) in the southern Pantanal, Brazil. South American Journal of Herpetology 6(1):11–19

Williams, R. O.; Williams, Jr., R. O. 1951. The Useful and Ornamental Plants in Trinidad and Tobago. Port of Spain, Trinidad

Wittmann, Florian; Wittmann, Astrid de Oliveira. 2010. Use of Amazonian floodplain trees. In: Amazonian Floodplain Forests: Ecophysiology, Biodiversity and Sustainable Management, Junk, Wolfgang J.; Piedade, Maria T. F.; Wittmann, Florian; Schöngart, Jochen; Parolin, Pia (eds), Springer, Dordrecht, pp 389–418

Wittmann, Florian; Schöngart, Jochen; Queiroz, Helder L.; Wittmann, Astrid de Oliveira; Conserva, Auristela dos Santos; Piedade, Maria T. F.; Kesselmeier, Jürgen; Junk, Wolfgang J. 2009. The Amazon floodplain demonstration site: sustainable timber production and management of central Amazonian white-water floodplains. Ecohydrology & Hydrobiology 9(1):41–54

Zappi, Daniela C.; Milliken, William; Lopes, Celia Regina A. Soares; Lucas, Eve; Piva, José Hypólito; Frisby, Sue; Biggs, Nicola; Forzza, Rafaela Campostrini. 2016. Xingu State Park vascular plant survey: filling the gaps. Brazilian Journal of Botany 39(2):751–778

Chapter 45
Sapindaceae

45.1 *Paullinia alata*

Bolivia Barbasco colorado, barbasco negro; atsa dhere (Tacana), coyoj coyoj (Tsimane)

Brazil Guaraná bravo, mata fome, caferana, cipó de tres quinas, timbó, tingui, urarirana

Colombia Bejuco camándula, yoco

Ecuador Vejuco de tres esquinas; cuasisipacco'pifa (Cofán), taruga huasca (Kichwa), cae yoco (Siona-Secoya), teipahueme (Waorani)

Peru Seca boca, macota, macote, yurac macote, pava huasca; chanchak (Achuar), ikáncham wennu, pagki daék, sumpa daék (Aguaruna)

Status Wild

N. Smith, *Amazon Fruits: An Ethnobotanical Journey*, Ethnobiology,
https://doi.org/10.1007/978-3-031-12803-5_45

Ruben, 13 years, by a seca boca *Paullinia alata* liana in the forest. The fruits will turn red when they ripen. Near Miraflores, Rio Tigre, Loreto, Peru 7-1-06

One of at least two members of the Sapindaceae with edible fruits on floodplains, *Paullinia alata* is a woody vine bearing fruits that are edible when the skin is light tan colored but tastier when the pericarp turns red. The white aril surrounding the single black seed is typically eaten in the forest and the fruits are not marketed. Villagers from Miraflores at the mouth of the Tigre snack on the fruits when they appear during the rainy season. The Tacana in the Bolivian Amazon also snack on the fruits. The fruits of *Paullinia alata* are more accessible than those of *Paullinia pachycarpa*, which are also eaten, because they are found along the stem, as close as 30 cm from the ground, whereas those of *Paullinia pachycarpa* are born higher up, typically in the canopy. *Paullinia alata* can grow up to 15 m long and thrives in regrowth vegetation and forest on uplands and floodplains forests up to 1600 m in mountains. It is found from Costa Rica south to the Orinoco and Amazon Basins (Medeiros et al. 2016).

People use this plant for a variety of other purposes. The stem and roots are crushed and used as a piscicide in various areas (Beck 1990), hence the Spanish names for this plant in the Bolivian Amazon: barbasco colorado and barbasco negro. *Paullinia alata* is employed in various folk remedies in the Ecuadorian Amazon; for example, the Cofán apply ash from the burned bark to treat skin problems in peoples' nostrils and mouths (Cerón 1995:137). Also in the Ecuadorian Amazon, the Waorani obtain cordage from the slender stems to tie up their canoes (Cerón and Ayala 1998:156). And in the Peruvian Amazon, the black seeds are sometimes used as fish bait.

45.2 *Paullinia cupana*

Brazil Guaraná, guaranazeiro, cupana; waraná (Sateré-Maué)

Colombia Guaraná, cupana, yopo; guame kuiro (Desana)

Ecuador Chengrapi (Shuar)

Peru Guaraná, cupana

Venezuela Cupana

Status Wild, planted

Guaraná *Paullinia cupana* in fruit in a field. Near Maués, Amazonas, Brazil 11-17-02

Guaraná *Paullinia cupana* is a forest vine that was domesticated long ago in the Brazilian Amazon for its caffeine-rich fruits which are used traditionally to prepare a tonic. More recently guaraná has emerged as a key ingredient in some trendy energy drinks, pills, tea, as well as smoothies. Guaraná is the Amazon's answer to coffee, and more. Few people who quaff drinks containing guaraná in North America and Europe have any idea where the plant comes from.

Guaraná seeds contain more caffeine than any other plant in the world. Caffeine levels in guarana seeds range from 2% to 7.5% (Beck 2005b). Some 60 plant species contain caffeine, including coffee (*Coffea* spp.), distributed among 7 plant families. Guaraná seeds also contain traces of theobromine and theophylline, other alkaloids in the xanthine group. The evolutionary purpose of the xanthines is unclear, but they probably help protect the plants from attack by insect pests, other herbivores, and pathogens. In humans, xanthines stimulate the central nervous system, increase secretion of gastric acid, and serve as a bronchodilator and a diuretic.

Caffeine is also addictive, although it is not classified as a "drug" by government agencies charged with sorting out which plants and synthetic compounds are deemed a public menace. Guaraná contains about four times as much caffeine as coffee. Ironically, guaraná's high caffeine levels were once thought to be a detriment in promoting guaraná outside of Brazil because of health concerns about excessive caffeine consumption, but now this attribute has turned out to be a marketing asset (Smith and Atroch 2007).

Guaraná has long been used as a tonic and to treat various disorders in Brazil and abroad and became a national soda in Brazil about a century ago. In the last two decades or so, guaraná has emerged as a key ingredient in various "sports" and energy drinks. Instead of making it as a stand-alone soda abroad, guaraná has penetrated foreign markets as an ingredient in various "power" and aphrodisiac drinks and in pick-me-up pills. As preoccupation with how many cups of coffee, tea, or cans of cola one can safely drink has waned, prospects for guaraná have thus brightened.

Guaraná was domesticated in the interfluvial forests between the lower Tapajós and lower Madeira Rivers in the Brazilian Amazon. It appears that the Sateré-Maué, who live along the Maués River and its affluents, domesticated guaraná (Clement et al. 2010). The first European naturalists who explored the Amazon in the early nineteenth century noted that the Sateré-Maué were the original cultivators of the vine which even then was attracting the attention of settlers and entrepreneurs throughout the region (Monteiro 1965).

Trash bin modeled after a guaraná fruit. Maués, Amazonas, Brazil 7-9-02

Guaraná plays a prominent role in Sateré-Maué mythology, including a story about the very origin of their people. As recounted by the Sateré-Maué to the Brazilian ethnographer Nunes Pereira in 1939 (Pereira 1954), the genesis of guaraná involves rivalries between two brothers and their sister, Onhiamuaçabê. The brothers did not want anyone to marry their sister because she knew all the plants, and which ones were good for curing. She was also the owner of an enchanted place called Noçoquem where she had planted a Brazil nut tree. One day, a small snake took a fancy to her so he released a perfume along a trail used by Onhiamuaçabê. She liked the scent a lot. So the snake went ahead of her along the trail and touched her lightly on the leg as she passed by. She was immobilized so the snake took advantage of her and she was impregnated. The brothers were furious. Onhiamuaçabê gave birth to a beautiful boy, and when the boy was old enough, she took him to the enchanted place to eat Brazil nuts. An agouti noticed someone had made a fire at the base of the Brazil nut tree to roast nuts and he reported what he had seen to the brothers. They had guards posted at the enchanted place, and when the boy came the next day to eat some more nuts they decapitated him. His mother heard his cries of anguish, but it was too late. By the time she reached the enchanted place, he was dead. She resolved to pluck out the boy's left eye and planted it. But the plant that germinated was no good; it was called false guaraná. She then gouged out the right eye and planted it; this grew into true guaraná. She spoke out aloud, as if the child was still alive: "You, my son, will be the greatest force of Nature; you will do good for all men; you will be great; you will free men from some sicknesses and cure them of others." And out of the buried corpse of the boy itself arose the first Sateré-Maué.

The meaning of this myth recently became remarkably clearer as a result of genetic testing. The *sorbilis* variety of guaraná grown by the Sateré-Maué is a high level polyploid, with 210 chromosomes rather than the standard 24 for the genus (Freitas et al. 2007). In essence, this myth captures the guaraná domestication event, which occurred when the mythological woman recognized that a special type of guaraná had become available to her, as distinct from the more common and less useful "false guaraná", and that it should be planted for the benefit of future generations. The Sateré-Maué also believe in several spirit protectors of the forest including mãe do guaraná, the "mother" of guaraná (Yamã 2007:33).

Numerous other indigenous groups live in the vast area between the lower Tapajos and Madeira Rivers, including the important and once more numerous Munduruku just to the southeast of the Sateré-Maué, but none have any myths or legends regarding guaraná, nor do the early European naturalists mention other groups in the vicinity using guaraná.

Guaraná has long been esteemed for its energetic and curative properties. In the late seventeenth century, the Jesuit missionary João Felipe Bettendorf noted that the Sateré-Maué (then known as the Andirazes) valued guaraná as much as Europeans did gold because the drink gave them "so much energy, that when hunting, they could go from one day to the next without feeling hungry" (Bettendorf 1910:36). Father Bettendorf noted that guaraná was drunk to diminish fevers and cure headaches, and also served as a powerful diuretic. On the latter, he remarked that "With respect to its reputation for provoking the need to urinate, I can personally testify."

During the colonial period, guaraná was sold variously as a fortifier, stimulant, tonic, antidote to fever, a preventive against hardening of the arteries, and to treat migraines. Guaraná was deemed especially efficacious in the treatment of diarrhea and dysentery (Baena 2004:68; Castelnau 1850:232; Queirós da Silveira 1847; Spix and Martius 1976:105; Steinen 1886:19). But as with any product, overindulgence can trigger problems. Over two centuries ago, Francisco Xavier Ribeiro de Sampaio, a Brazilian judge, warned that excessive drinking of guaraná can lead to insomnia and even impotence (Sampaio 1825:6).

Two distinct types of guaraná are cultivated, one along the Middle Amazon in the vicinity of the Maués River and the other over a thousand kilometers away along the Upper Rio Negro in Brazil and Venezuela. The area devoted to guaraná is much smaller in the Upper Rio Negro; Alexander von Humboldt encountered the bitter brew at only one settlement along the Upper Rio Negro in Venezuela at the beginning of the nineteenth century (Humboldt and Bonpland 1821). However, the Yorkshire botanist Ricard Spruce noted its importance in the Upper Rio Negro portion of Venezuela in the mid-nineteenth century:

> I am now entering another Guaraná country. I have seen a few plants in the sitios, but it is across the frontier that it is cultivated and used in the greatest quantity. The Barré Indians of Venezuela drink it in immense quantities, especially the first thing in the morning, in place of coffee, and they use only the fresh berry, grated, without sugar. Their name for it is Cupana. (Spruce 1852)

By the early twentieth century, the crop had virtually disappeared along the Brazilian portion of the Rio Negro (Ducke 1937).

This disjunct distribution of guaraná cultivation has led to two opposing ideas about the crop's domestication and spread. Some suggest that guaraná was domesticated in the Upper Orinoco and was introduced later to the Amazon (Pereira 1967:121). Little guaraná is cultivated today along the Upper Negro and Orinoco. The more prevalent view is that guaraná was domesticated in the middle Amazon region and was then taken up by the Rio Negro, perhaps by the Baré (Lleras 1994).

The Maués River and surrounding area have long been the main center for guaraná production in the Amazon. Over a century ago, the indefatigable field geographer William Chandless remarked on how prevalent guaraná was along the Maués River and some of its affluents, such as the Abacaxis (Chandless 1870). The Maués River remains the epicenter of guaraná cultivation in the Amazon.

Farmer harvesting guaraná *Paullinia cupana*. Rio Limão, affluent of the Maués River, Amazonas, Brazil 11-17-02

Guaraná, like most crops, has been transformed under domestication. Farmers have selected for bushy growth to facilitate harvesting in contrast to wild forms of the gangly liana that drape the forest canopy. The fact that guaraná is planted in the open encourages shrubby growth because the plant does not have to climb to the light as it does in the forest. The plants are also pruned periodically. The Sateré-Maué gather spontaneous seedlings of guaraná in regrowth vegetation and forest to plant in their fields thereby maintaining a genetic diversity of the crop which helps resist pests and diseases (Henman 1982; Lorenz 1992). And whenever possible, the Sateré-Maué prefer anthropogenic dark earth (*terra preta de índio*) for planting guaraná.

Guaraná is grown on small and large farms, either as a mono-crop plantation or alongside other crops. Only small farmers practice agroforestry, the intercropping of guaraná with other perennials or annuals such as banana, buriti palm, pineapple, peach palm, mango, soursop, lime, sweet orange, cupuaçu, avocado, manioc, and the New World yam. As many as nine crops are interplanted with guaraná in the vicinity of Maués. In a 30-year-old guaraná field established by one Sateré-Maué family, several spontaneous wild fruit trees were protected including bacaba *Oenocarpus bacaba* and inajá *Attalea maripa* palms. In Rondônia, some pioneer farmers have intercropped their guaraná with either peach palm, Robusta coffee, or rubber.

Some of the trees in guaraná fields are spontaneous or survive from previous plantings, among them Brazil nut *Bertholletia excelsa*, caramuri *Neoxythece elegans*, and palms such as bacaba *Oenocarpus bacaba*, inajá *Attalea maripa*, and tucumã *Astrocaryum aculeatum*. Because those native trees produce valuable nuts or fruits, they are spared when weeding. On an upland farm on the outskirts of Itacoatiara in Amazonas State, for example, one grower has interplanted guaraná with sweet orange, mango, guava, pineapple, coconut, avocado, and jackfruit and left some Brazil nut and tucumã palm trees because of their nutritious nuts and fruits. A Sateré-Maué family spared a spontaneous sorva *Couma macrocarpa* tree in a guaraná field because it produces edible fruits and a milky latex that is taken to treat diarrhea.

When prices are low, growers may neglect guaraná for several years, and successionary growth envelops the plants. Guaraná fields thus become managed fallows, at least temporarily. Guaraná still yields well when competing with weeds, but it is more difficult to harvest because, as one farmer put it, guaraná reverts to its old ways and becomes a climbing vine. Indeed the liana-like growth habit of guaraná enables it to climb and obtain sufficient light, even when surrounded by quick-growing colonizing species such as *Cecropia*. When prices rebound, the volunteer plants are cut down and the guaraná reverts back to a more bush-like growth. During the clearing of the successionary vegetation, sections of guaraná vine are often cut and this accidental pruning actually helps increase yield by encouraging new growth.

Several birds apparently disperse guaraná seeds, some of which may germinate in disturbed areas (Prance 1997). Rural folk are pretty astute observers of natural history, but farmers near Maués have different opinions about the manner in which birds disperse the seeds. One retired farmer in Maués reported that silver-beaked tanagers *Ramphocelus carbo* swallow guaraná fruits and later defecate the seeds, whereas another farmer claims that the tanagers, known locally as pipira, clip off the white pulp (technically an aril) and leave the damaged fruit on the bush. Several farmers agree that toucans (*Ramphastos* spp.), and their smaller relatives the araçaris (*Pteroglossus* spp.), swoop down into guaraná fields from the surrounding forest and carry off the entire fruits. But one farmer suggested that the toucans then discard the seed before swallowing the pulp, whereas another grower asserted that toucans and araçaris ingest the entire fruit and later defecate the seeds. Guans (*Penelope* spp.) also raid guaraná fields and devour the whole fruits before returning to the cover of the forest.

In the Maués area, men, women, and children harvest the fruits by hand toward the end of the dry season and in the early part of the rainy period. The Sateré-Maué select seed for planting right after harvest, and if a new area is to be planted, a shaman is called to perform rituals to ensure future harvests. The ceremony for blessing a new guaraná planting also involves dancing to live music and generous libations of strong, locally-made manioc beer.

If the entire guaraná fruit bunch is ripe, it is either snipped off with scissors, clippers, or is broken off manually. If only a few fruits are ripe, they are picked individually. The fruits are then placed in a basket and carried back home. Before roasting the seeds, the red skin must be removed. The fruits are skinned by hand, left to soak in water, or simply stored for several days until the skin softens.

Sateré-Maué boy removing the red skins from guaraná *Paullinia cupana* fruits and placing the seeds with attached arils in a spathe of an inajá *Attalea maripa* palm. Igarapé Aruã, Rio Urupadi an affluent of the Maués River, Amazonas, Brazil 11-16-02

Workers on an estate with 280 hectares of guaraná near Maués also do not soak guaraná fruits. Rather, they store the fruits in buckets for 4 or 5 days until the skin deteriorates and the arils turn black. They then spread the fruits on the ground and tread on them to break off the fermenting arils and red pericarp prior to roasting. A farmer near Manaus holds harvested guaraná fruits in an old dugout canoe for 3 days until they are mushy (*mochado*) before spreading them on the ground and treading on them with bare feet. The cleaning of guaraná seeds is thus similar to the manner in which cacao beans are prepared for roasting.

On small farms in the Maués watershed, guaraná seeds are roasted on a griddle, preferably of clay to reduce the chances of burning. Clay griddles are about 2 m in diameter and are made at various households along the Maués River and some of its affluents, wherever suitable clay is found. The clay griddles are used to make manioc flour in the off-season. In most parts of the Brazilian Amazon, however, giant iron pans have long replaced clay ones to make manioc flour. Muruci (*Byrsonima* spp.), spontaneous fruit trees and colonizers of disturbed sites, are the preferred fuelwood for guaraná ovens. Itaubarana *Sweetia nitens*, another hardwood found along the sandy shores of black and clear waters in the Maués watershed, is also sought after to fuel guaraná ovens.

Guaraná seeds are roasted in an open-sided hut, but it is hot work and it takes several hours. It typically takes about 4 hours to roast the beans; if one tries to speed up the process by building a larger fire the seeds can be easily scorched, thus affecting the flavor. Women roast guaraná in Sateré-Maué society; they periodically stop stirring the seeds to pick some up and smell them to see if they are ready to come off the griddle. Many guaraná producers in the Amazon prefer to sell the raw seeds to dealers in towns, such as Maués and Manaus, where they are roasted in large, metal drums, similar to those used for roasting coffee.

Saterê-Mawé farmer roasting guaraná *Paullinia cupana* on an oven with a ceramic griddle. Igarapé Aruã, Rio Urupadi an affluent of the Maués River, Amazonas, Brazil 11-16-02

After the seeds are roasted, they are either sold or farmers along the Maués and its tributaries pound them in a wooden mortar until they are reduced to a meal-like consistency. The Sateré-Maué half bury the mortar in the ground to make it more stable. The wooden pestle used for pounding guaraná seeds is almost 2 m long, much bigger than those typically used to dehusk rice or remove the dried pulp and skin from coffee fruits.

Sateré-Maué farmer pounding roasted guaraná *Paullinia cupana* seeds in a wooden mortar (*pilão*) partially buried in the ground to make it more stable. Igarapé Aruã, Rio Urupadi an affluent of the Maués River, Amazonas, Brazil 7-10-02

Water is then added to the pulverized seeds to make a paste. The dough is then molded into cylinders (*bastão*), some 20–25 cm long, and dried in the sun until hard. In some cases, manioc flour or ground cacao beans are added to the paste before drying (Le Cointe 1922:476). Finally, the guaraná batons are put in a smokehouse for up to 40 days. The cylinders vary in diameter from as thin as a drumstick to the thickness of deli salami. The color also varies from black to dark brown, or reddish tan. The darker the color, the more soot the cylinder has acquired. Muruci is the preferred fuel for the smokehouse because the wood imparts an agreeable flavor to the guaraná. In Mato Grosso, the process of making *bastões* is semi-industrialized with machines grinding the toasted seeds and with molds employed to form the guaraná cylinders.

Docilene, 8 years, a Sateré-Maué girl holding a cylinder (*bastão*) of ground guaraná *Paullinia cupana* seeds. The bastão will be placed over a smoky fire for several weeks. Igarapé Aruã, Rio Urupadi an affluent of the Maués River, Amazonas, Brazil 7-10-02

Smokehouse for curing guaraná cylinders (*bastão*). The guarana sticks are placed on a rack above the slow-burning fire for several weeks. Some maize is hanging from a rafter to be used for planting. Docilene, 8 years, a Saterê-Mawé girl. Igarapé Aruã, Rio Urupadi an affluent of the Maués River, Amazonas, Brazil 11-16-02

Guaraná is traditionally consumed without sugar at room temperature. To prepare the drink, the hard cylinder is grated with the bony tongue of pirarucu *Arapaima gigas*, one of the largest fish in the Amazon. The powder falls directly into a calabash gourd or the cylinder is grated over a piece of paper and then the powder is stirred into a glass of water. Eventually, the piraructu tongues wear down; if the family does not have an adept fisher in their midst, they can buy a "new" tongue in urban markets for about US$3.50.

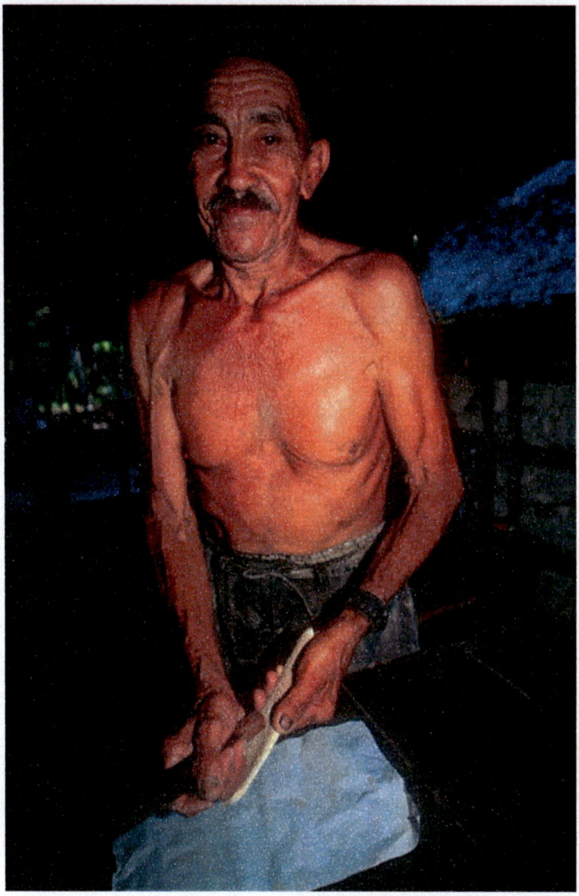

Luiz Neves, 74 years, using a dried pirarucu *Arapaima gigas* tongue to grate a guaraná *bastão*. Sitio Veracruz, Ilha Veracruz, Rio Maués, Amazonas, Brazil 7-9-02

Instead of a pirarucu tongue, the Sateré-Maué and some river dwellers near the indigenous reserve use a small piece of sandstone, obtained from the upper Maués watershed, to grate guaraná. A river dweller who lives along the Aruã stream, an affluent of the Urupadi River which in turn flows into the lower Maués, explained that the family prefers stone graters because it produces a finer powder. According to a Sateré-Maué family along the Aruã River, the sandstone is first heated and then dropped in water to "fix" it. A river dweller gave another explanation for boiling the sandstone grater before using it: to remove silt and dirt. The sandstone graters range in size from the palm of an adult hand to a fully extended hand.

Urban and rural folk in the Maués watershed typically take guaraná upon rising and wait an hour or two before a breakfast of black coffee and crackers. One 61-year-old guaraná worker near Maués reported that he rises before dawn each day to grate guaraná and drink it without food. The "kick" from guaraná is sufficiently strong for him to work through the morning and only break his fast at lunchtime. Francisco, a 78-year-old retired farmer now living in Maués, grates his guaraná at dawn and then eats a breakfast of boiled sweet manioc, crackers, or bread with black coffee about 2 hours later. Francisco explained that he prefers to grate his own guaraná rather than buy it in powdered form because it has a more robust flavor ("*tudo o sabor*") and has all its "vitamins" ("*toda vitamina*"). Because of his age, Francisco no longer takes guaraná during the day because he claims it weakens him and gives him the shakes. It is a custom among the Sateré-Maué to fast for an entire month when the chief or a member of his family passes away. Tribal members should only eat ants and drink water and guaraná; the latter evidently helps stave off hunger (Monteiro 1965:34).

Sateré-Maué woman grating a guaraná *bastão* with sandstone. The powder is being scraped into a calabash bowl containing water. Igarapé Aruã, Rio Urupadi an affluent of the Maués River, Amazonas, Brazil 7-10-02

The Sateré-Maué call the tan-colored guaraná drink *çapó*. In Sateré-Maué culture, only women prepare guaraná for drinking, whereas among non-indigenous folk, both men and women grate the cylinders. For more formal occasions, the Sateré-Maué use a small, hourglass-shaped stand (*patuai*), fashioned from forest vines, to support a calabash bowl when grating guaraná (Leacock 1964). The host and guests then pass the bowl around and periodically place it back on the stand.

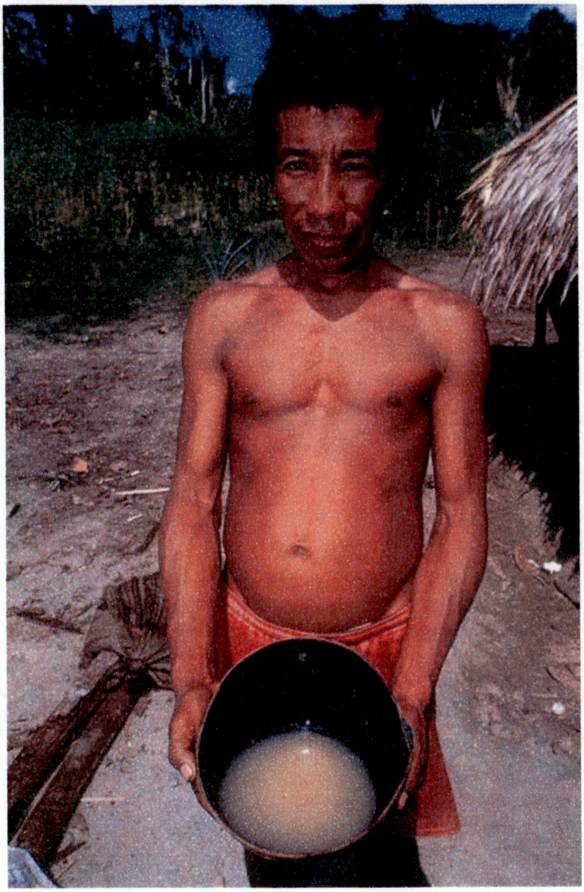

Sateré-Maué farmer holding a bowl of guaraná *Paullinia cupana* beverage that they call *çapó*. The bowl is fashioned from the cultivated calabash gourd *Crescentia cujete*. Igarapé Aruã, Rio Urupadi an affluent of the Maués River, Amazonas, Brazil 7-10-02

In the eighteenth century, guaraná's medicinal properties spurred its trade to other parts of Brazil and Bolivia, as well as Europe and the United States (Monteiro 1965). In the mid-nineteenth century, the Mura sold guaraná seeds along the lower Tapajós which were pounded and mixed with water and then drunk to treat diarrhea and intermittent fevers (Bates 1863:134). Also in the nineteenth century, a teaspoon of grated guaraná was mixed with water and drunk to alleviate mild attacks of ague especially along the Madeira River (Keller 1874:110). Belém, at the mouth of the Amazon, was exporting as much as 50 tons of guaraná to Europe each year in the early twentieth century, and most of that production came from the Maués area. Guaraná was popular in Belém itself until the 1870s when the juice of açaí *Euterpe oleracea* palm fruits supplanted it as the city's favorite beverage (Souza 1873:246).

One trade route pioneered in 1749 by gold merchants based in Cuiabá, the capital of Mato Grosso, entailed traveling by large canoes down the Tapajós (Maw 1829:330; Ferreira Penna 1869). Traders would proceed up the Amazon from the mouth of the Tapajós to Maués to pick up guaraná seed. Alternatively, they would purchase them from merchants in Santarém at the confluence of the Tapajós and Amazon rivers. Traders from Mato Grosso would bring hides, small diamonds, and gold to exchange for salt, iron, steel, lead, gunpowder, China, wine, and guaraná (Ferreira Penna 1869:89). It used to take 5 months for canoes to make the journey from the mouth of the Tapajós to Cuiabá because canoes had to be pulled overland around the numerous rapids along the upper Tapajós (Edwards 1847:102). By the early nineteenth century, guaraná had emerged as a major trade item along the Tapajós River and there was strong demand for it in Cuiabá, the capital of Mato Grosso (Florence 1941:202; Spix and Martius 1976:91). In the early twentieth century, some 30 metric tons of guaraná was being transported in such manner from Maués to the mouth of the Tapajós and then upriver to Mato Grosso. And the trade remained brisk until the middle of the twentieth century (Pires 1949).

Another colonial trade route followed the Madeira River from its headwaters in Bolivia to Itacoatiara (then Serpa) and Parintins (then Vila Bella) along the middle Amazon. On their return journey, Bolivian traders would also pass by Maués to stock up on guaraná for sale in the Beni, Santa Cruz de la Sierra, and Cochabamba (Agassiz and Agassiz 1896:304; Keller 1874:111; Souza 1873:257). Bolivian traders also faced hardships on their way back, especially at the treacherous Teotônio rapids upstream from Porto Velho, where the Madeira tumbles off the granitic shield. The long, heavily laden canoes had to be dragged around the rapids before the journey could resume. Those rapids have now been flooded by a hydroelectric dam.

Guaraná was transformed from an elixir and home-prepared beverage to a mass produced soda in the early twentieth century. The emergence of ice-making machines further propelled the soda industry in Brazil and elsewhere in the early 1900s. São Paulo-based Antarctica, which merged in 1999 with Brahma to form AmBev and was in turn purchased by InBev, a Belgian company, in 2003, introduced carbonated guaraná to Brazilians in 1909. At first guaraná soda was sold mainly in southern Brazil, but many independent bottlers began making their versions all over Brazil. By the 1940s, fizzy guaraná had become Brazil's national soda, about the same time that caffeine-based colas were dominating the US soda market.

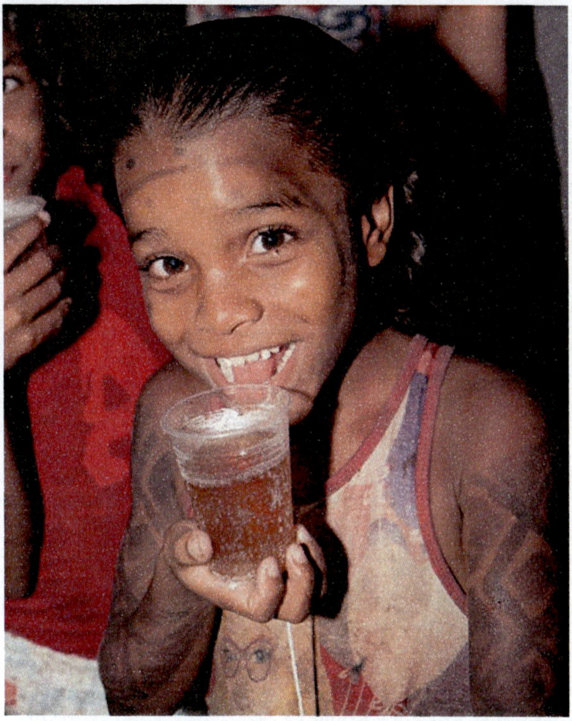

Arara girl drinking guaraná soda. She is wearing genipap *Genipa americana* **body paint. Terrawangã, middle Xingu River near its confluence with the Bacajá, Terra Indígena Arara de Volta Grande, Município Senador José Porfírio, Pará, Brazil 9-10-15**

The bottled guaraná industry has paralleled the story of many other enterprises in which many small operations eventually give way to a few big players. At one time, many towns in the Brazilian Amazon, especially in the state of Amazonas, had one or more bottling plants producing their own distinctive brands. Itacoatiara, some 250 km down the Amazon from Manaus, for example, once produced an especially appealing brand of bottled guaraná called *Guaraná Rio Negro*. This product, sold in small bottles, was particularly rich in guaraná: the strong flavor came through, unlike the rather insipid, cream soda flavor of popular national brands today. The bus company providing service between Itacoatiara and Manaus in the mid and late 1970s used to serve *Guaraná Rio Negro* on board to passengers who elected to pay a higher price for the seats because the bus did not make stops on the way. *Guaraná Antarctica*, the top-selling brand nationally, carries a "champagne" label, suggesting that it is rather light. The plant that produced the darker-colored, and more satisfying, *Guaraná Rio Negro* closed its doors in the early 1980s (Wesche and Bruneau

1990:62). One of the reasons that *Guaraná Rio Negro* went out of business is that local merchants began stocking more guaraná produced by Antarctica and Brahma because those companies also produce beer. Merchants make more profit selling beer than soda, but in order to qualify for the beer, they had to purchase crates of *Guaraná Antarctica*. And Maués, at the very center of the crop's origin, lost its unique bottled *Guaraná Maués* to competing brands from Manaus and southern Brazil in the mid-1970s.

Independent bottlers producing the more potent traditional guaraná soda survive only in Manaus, where people can still savor the rich, dark amber beverage produced by Baré, Magistral, Real, Regente, and Tuchauá. Even though Tuchauá and Baré were acquired by Coca Cola and Antarctica, respectively, they retain their original formulas. Two independent bottlers in Belém and Abaetetuba in Pará produce only the "light" version of guaraná soda. In 2003, a new traditional bottled guaraná appeared in the Manaus market, Tauá, an indication that demand is still strong for the more potent and savory flavor of the traditional soda. Indeed, the traditional guaraná soda is part of the cultural identity of many of the nearly two million inhabitants of Manaus.

The emergence of guaraná soda as a national drink prompted growers in other areas of Brazil to start planting guaraná, often with the support of Antarctica. Maués is a long way from the major population centers of Brazil. Bahia, on the other hand, is closer to the southern and central part of the country where most Brazilians live yet has a hot, humid climate similar to that of the central Amazon. Separated from the Amazon by a dry belt of savanna, Bahia has the added advantage of being relatively free of diseases and pests that trouble guaraná growers in the Amazon. Yields of guaraná in Bahia are about double those typically obtained in Amazonas, in part because of reduced disease and pest pressure (Tavares et al. 2005). Furthermore, Bahia has a better infrastructure, such as roads and power supplies, for the transportation and processing of guaraná. Accordingly, Bahia now accounts for about two-thirds of guaraná production in Brazil while Amazonas has slipped to a fifth of national production (Homma 1999).

Guaraná soda in Brazil has held its own against colas, produced under license for Coca Cola and Pepsi. The latter US firms even market their own brands of guaraná in Brazil. Coca Cola, for example, markets Kwat in Brazil, but not in the United States. Indeed, bubbly guaraná is virtually unknown in the United States. One reason is that for a long time the Brazilian government prohibited the bottling of guaraná abroad in an effort to promote value-added processing within the country. But in 1987, the Brazilian government relinquished this control after being pressured by Antarctica which realized that sales abroad would take off if the product could be bottled overseas (Griffin 1988).

In spite of this regulatory change, guaraná soda has been slow taking off outside of Brazil. In 1995, for example, Pepsi test marketed a guaraná soda called Josta in 24 US cities. Sporting a sleek, black panther on the label, Josta allegedly released "raw, primal power." In 1998, Pepsi discontinued Josta, to the dismay of some adherents who called themselves Jostafarians and resorted to websites to encourage protests (Moffet and Deogun 1999).

Only in some ethnic markets in cosmopolitan areas of the United States can one find guaraná soda, in canned form or now increasingly in 2-l plastic bottles, imported from Brazil. At about twice the cost of nationally produced sodas and with no advertising to promote the novel drink, it is not surprising that guaraná soda has not made many inroads into the crowded field of effervescent drinks in the United States.

Guaraná soda has met with more success in Europe, though still on a relatively modest scale. Sales of guaraná soda made by Santa Claudia, a Manaus-based company, started picking up in Italy in the late 1990s after it switched from plastic to glass bottles. Most guaraná soda sold in Brazil today comes in plastic bottles or aluminum cans.

In the Brazilian states of Amazonas and Bahia, guaraná popsicles are popular among poorer folk in urban areas. Known as *dindin* or *chopp*, the frozen pushups are a cottage industry and are sold by kids, usually boys, carrying Styrofoam containers slung across their shoulders. Women stir guaraná powder and sugar in water then pour the mixture into tubular plastic bags. The popsicles are kept in a freezer at home until ready for sale.

Ambulant vendor selling guaraná popsicles (*chopp*) that he keeps in a Styrofoam container slung from his shoulder. Maués, Amazonas, Brazil 11-16-02

A few decades ago, guaraná carts began appearing in some street markets in the Brazilian Amazon, such as Belém, Abaetetuba, Manaus, and Manacapuru. Such carts sell guaraná beverages prepared in a blender with crushed ice. The formula for the energy drink varies among vendors, but generally contains nuts, powdered milk, aphrodisiac roots or barks, and guaraná powder or syrup. The recipe of one vendor in Abaetetuba, for example, consisted of syrups of guaraná and catuaba, raw quail's eggs, peanuts, cashew, Brazil nuts, and powdered milk. Catuaba refers to numerous wild plants in various botanical families and is a renowned aphrodisiac in the Brazilian Amazon.

Cart selling smoothies featuring guaraná and a variety of other ingredients. Abaetetuba, Pará, Brazil 11-26-02

At a guaraná cart in Manacapuru in the Brazilian state of Amazonas, the vendor uses the following ingredients in his smoothies: guaraná powder and syrup, mirantã (unidentified forest plant), peanuts, granola, powdered milk, catuaba, banana cv Maçã, and avocado. Peanuts are thought to increase libido throughout Brazil. The calorie-laden and potent concoctions are served in 300–500 ml plastic cups for 30–50 US cents each, much cheaper than smoothies in the United States, which generally cost at least $3 and provide less kick.

Guaraná is an ingredient in several smoothies offered by a small restaurant just off the main plaza in Iquitos, Peru. One of the smoothies called Exterminador contains a variety of ingredients including ginseng, papaya, and plantain. The three other smoothies containing guaraná on the menu suggest aphrodisiac properties: Super Erectus a lo Ari's, Infiel, and Insaciable. Similarly, a snack bar in Maués, Amazonas, offers various smoothies containing guaraná including one called Turbinado.

A guaraná smoothie dubbed turbinado containing milk, peanuts, Brazil nuts, and mirantã powder, the latter reputedly an aphrodisiac. Maués, Amazonas, Brazil 7-10-02

As guaraná becomes a global commodity, more thought needs to be given to conserving and better utilizing the genetic resources of the crop. When breeders have a diverse gene pool to draw upon, they have a better chance of developing varieties that are genetically resistant to diseases and pests, thereby reducing or eliminating the need to apply pesticides. Wild populations of crops should be conserved in situ so that they can continue to evolve along with insects and pathogens.

Wild guaraná is of interest to crop breeders because guaraná plantations are attacked by a variety of pests and diseases, especially anthracnose which is triggered by *Colletotrichum guaranicola*, a fungal pathogen (Sefer 1961). Also, several bees and wasps pollinate guaraná (Erickson et al. 1984; Gondim 1984; Krug et al. 2015; Prance 1984), so insecticide spraying can hurt fruit set. Genetic resistance to pests is thus highly desirable, another trait that might be found in wild guaraná.

Guaraná occurs in wild form from the lower Tapajós west to Acre in the southwestern part of the Brazilian Amazon (Lleras 1994). In 1966, the late Paulo Cavalcante, a Brazilian botanist with intimate field knowledge of Amazonian fruits, discovered wild guaraná growing further east than previously recorded on the plateau behind Santarém at the confluence of the Tapajós and the Amazon (Cavalcante 1967). Cavalcante found the wild guaraná along a freshly opened road linking Santarém with the last falls along the Curuá-Una River, now the site of a hydroelectric dam. The wild guaraná vines soared 40 m into the forest canopy. It is possible that the wild guaraná found by Cavalcante is actually derived from a cultivated form once maintained by indigenous peoples. The Santarém area is rich in dark black soils formed by ancient indigenous settlements. And in the eighteenth century, some Sateré-Maué settled along the left bank of the lower Tapajós opposite Alter do Chão where they traded guaraná and likely cultivated the crop there. Guaraná is no longer cultivated along the lower Tapajós. Alternatively, as Cavalcante speculates, the guaraná found in the forest near Curuá-Una could be truly wild. As is the case for many Amazonian fruits, the boundary between wild, escaped, and cultivated is blurred.

Small-scale farmers in the Maués area resort occasionally to wild material when planting their guaraná fields, thereby enriching the genetic diversity of the crop. And the Sateré-Maué are aware that young leaflets have extrafloral nectaries that nurture ants, which in turn help protect the plants (Prance 1974). Such local strategies for sustaining crop productivity can provide valuable lessons for agricultural development but are typically ignored in industrial operations that favor purchased inputs to combat pests and raise yields.

Although the Brazilian agricultural research agency (EMBRAPA) established a field gene bank for guaraná in 1987 near Manaus, such living collections typically embrace a relatively small slice of the genetic diversity of a crop. The bulk of guaraná's genetic resources will have to be safeguarded in situ; in farmers' fields in the case of cultivated forms, and in the forest for wild guaraná. Guaraná's seeds perish if dried and frozen, so they cannot be stored in seed gene banks (Carvalho et al. 1982). The forest habitat for wild guaraná is under assault in many parts of its range, including the northward expansion of the soybean front along the Cuiabá-Santarém highway and the increased tempo of a forest clearing in Rondônia and Acre for cattle production.

Near relatives of guaraná could also be of interest to breeders if more is known about them or possibly help launch new crops in their own right. The growing international interest in guaraná may eventually turn the spotlight on a near relative, yoco *Paullinia yoco*, which has a caffeine-rich bark that has been gathered for millennia. Yoco is a jungle vine gathered in the southwestern part of the Colombian Amazon for its bark, which is used to make an infusion to boost energy and stave off hunger (Schultes and Raffauf 1992). Curiously, even though yoco is so much a part of daily life that indigenous people will even move their village when supplies run out, the plant has never been domesticated.

Sales of products containing guaraná are likely to continue increasing. The stimulating effects of guaraná are longer lasting than coffee because the caffeine in guaraná apparently binds with tannins (Rätsch 2005). In Belém, Pará, guaraná is sold from stalls and some stores to treat fatigue (Shanley and Lux 2003). Guaraná also has the luster of an exotic rainforest vine with a long history of use as a medicinal plant. Curiously, guaraná was consumed in the United States and Europe in the eighteenth and nineteenth centuries, and then dropped from the pharmacopeias in those regions until the late twentieth century when it re-emerged in carbonated energy drinks, aphrodisiac concoctions, and stay alert pills. The success of such guaraná products may provide momentum for soda manufacturers to launch guaraná soda in the industrial nations, perhaps with a stronger concentration of syrup typical of the robust sodas produced by local companies in Manaus, Brazil.

If attention is paid to such issues as fair trade, certification of origin, the Amazonia brand name, organic production and other value-added strategies, this expanded market could benefit Amazonian farmers, especially native farmers in Maués and adjacent municipalities. If these kinds of strategies are ignored, in a business-as-usual rush for profit, guaraná's genetic resources and associated traditional knowledge will continue to erode in its hearth, eventually succumbing to deforestation for pasture or soybean.

45.3 *Paullinia pachycarpa*

Brazil Timbó preto, mata fome, cipó cururú

Peru Seca boca, macota

Status Wild

Roy, 9 years, gathering seca boca *Paullinia pachycarpa* to snack on the white arils. Vine gathered from floodplain forests along the margin of a sediment-rich river. Rio Tigre, 2.3 km upstream from Miraflores, Rio Tigre, Loreto, Peru 7-1-06

Paullinia pachycarpa is a sprawling vine that drapes trees along river banks, such as the lower Tigre, and produces numerous fruits with a bright red skin (pericarp) that splits open to reveal a bright white pulp (aril). The aril that surrounds a single black seed is eaten, but the shiny seed is discarded. The fruits are thus easily spotted and locals can gather them from their canoes or by climbing up a tree supporting the vine and pulling off the fruit bunches. As with several other species in this genus, *Paullinia pachycarpa* stems are crushed and used as a piscicide (Beck 1990). *Paullinia pachycarpa* occurs in wetlands and on uplands throughout the Amazon and Orinoco Basins up to about 1400 m in mountains.

The genus *Paullinia* contains some 180 species, mostly confined to lowland forests of the Neotropics and several species are consumed in Amazonia. The edible fruits of *Paullinia pachycarpa* and *Paullinia alata*, both known as seca boca (dry mouth) along the lower Tigre, may represent the tip of an iceberg of potential new flavors that could eventually reach tables in urban areas. One of the common names for this fruit in Peru, seca boca, aptly describes the dry, farinaceous texture of the arils, and it leaves a somewhat astringent aftertaste. Nevertheless, the pulp is sufficiently sweet to attract locals. The best-known member of this genus is guaraná (*Paullinia cupana*), domesticated in the Brazilian Amazon thousands of years ago and cultivated on uplands for its seeds that are roasted and ground into powder to manufacture soda and a variety of dietary supplements.

45.4 *Talisia esculenta*

Bolivia Pitón, pitón dulce, mutuyubi

Brazil Pitomba, pitomba da mata, pitomba do cerrado, pitomba de cutia, pitomba de macaco, pitomba rana, pitombeira, pitombinha, kupai, angustura, olho de boi, cascudo, grão de galo; tsamyekawy (Rikbaktsa)

Peru Pitomba

Status Wild, spontaneous in cultural settings, planted

Pitomba *Talisia esculenta* fruits. Vila do Treme, Município de Bragança, Pará, Brazil 3-27-15

Pitomba *Talisia esculenta* is a small tree reaching up to 15 m tall that produces edible, tan-orange fruits during the rainy season (Guarim Neto et al. 2003). The slightly elongated fruits, roughly the size and shape of a grape, measure 2.5 cm by 2 cm. The tough skin is broken to access the slightly tart but delicious pulp. The thin, translucent mesocarp surrounds one to two light tan seeds. Pitomba fruits are encountered occasionally in street markets, such as in Santarém, Alenquer, and Belém in Pará (Smith 1996). They have evidently been consumed by humans in Amazonia for a long time as evidenced by carbonized seeds found in archaeological sites in sandstone caves near Monte Alegre that were occupied by hunters and gatherers some 9800–11,200 years ago (Roosevelt 1999; Roosevelt et al. 1996). Collared peccaries *Pecari tajacu* eat the fruits and seeds of pitomba so this fruit tree also contributes indirectly to the human diet because peccaries are popular game animals (Beck 2005a). Perfume from the flowers can be smelled some 20 m from the tree and the flowers are visited by some bees which undoubtedly pollinate them (Castro 2002; Kerr et al. 1986).

This plant is also used in some folk remedies. The fruit's pit (*caroço*) and leaves are used to make an infusion to treat diarrhea (Le Cointe 1934:377; Vásquez et al. 2014). In the Mercado do Porto in Cuiabá, the seeds of this fruit are sold to make a tea to treat dehydration and bleeding (Pinto et al. 2013). In Bom Jardim, a rural community 60 km south of Cuiabá in Mato Grosso, the leaves are employed to make a tea to treat kidney problems (Pasa 2011) while in parts of eastern Mato Grosso rural people macerate the fruits and eat them to treat kidney stones (Ribeiro et al. 2017). In the vicinity of Manacapuru along the Solimões near its confluence with the Rio Negro, rural folk use the leaves to make a treat to reduce high blood pressure (Vásquez 2014:20).

Irlene, 8 years, gathering pitomba *Talisia esculenta* fruits in a home garden in her village. Pontão, Rio Urubu near Silves, Amazonas, Brazil 4-4-15

Talisia esculenta occurs in regrowth vegetation and forest along rivers and streams and on uplands throughout the Amazon Basin, the Atlantic forest of Brazil, and south to Paraguay. It also occurs in some stunted woodlands (*campina*) growing on upland sandy soils, such as in the vicinity of Manaus (Takeuchi 1960). It is also found in some home gardens, such as in the São Raimundo district of Manaus, in the vicinity of Manaus, Manacapuru, and Parintins, in rural communities of the lower Purus River near its confluence with the Solimões, along the lower Madeira River, further down the Amazon near Óbidos and Santarém, in the municipality of Bragança in northeastern Pará, in the vicinity of Abaetetuba along the lower Tocantins in eastern Pará, and in São Luis, the capital of Maranhão (Akinnifesi et al. 2010a, b; Almeida and Gama 2014; Barbosa 2018:44; Garcia et al. 2017; Gonçalves and Lucas 2017; Guillaumet et al. 1990; Kawa et al. 2015; Klüppel 2006:24; Lourenço et al. 2009; Machado 2016:38; Major et al. 2005; Rosa et al. 2007; Santos 2006:54; Silva et al. 2017; Silva 2006:127; Souza 2010:32). River folk in the municipality of Barcelos along the middle Rio Negro in Amazonas plant this tree in their home gardens (Silva et al. 2007) as do some rural folk in the watershed of the Mazagão which flows into the lower Amazon in Amapá (Steward 2013). It is also planted in some home gardens in the southern fringe of the Amazon Basin, such as in the rural community of São Miguel in the municipality of Várzea Grande in Mato

Grosso (Mamede and Pasa 2018) and in the municipality of Cuiabá (Gonçalves and Pasa 2015). In home gardens of São Luis, Maranhão, on the eastern fringe of Amazonia, this tree is among the top ten fruit trees (Akinnifesi et al. 2010a, b). In home gardens, pitomba arises either by deliberate planting or after the seeds are spat out. A small farmer along the middle Xingu upstream from Altamira transplanted a seedling of pitomba he obtained along the Iriri River, an affluent of the Xingu, in his manioc field. The distribution and density of *Talisia esculenta* have thus likely been influenced by humans. On Marajó Island at the mouth of the Amazon, pitomba is one of several fruit trees found on indigenous mounds (Lisboa 2012:415).

Girl selling pitomba *Talisia esculenta* fruits in a street market. Santarém, Pará, Brazil 3-14-93

References

Agassiz, L. and E. C. Agassiz. 1896. A Journey in Brazil. Houghton, Mifflin and Company, Boston

Akinnifesi, F. K.; Sileshi, G.; Costa, Joselene da; Moura, E. G. da; Silva, Rosele F. da; Ajayi, O. C.; Linhares, J. F. P.; Akinnifesi, A. I.; Araujo, Maria de; Rodrigues, M. A. I. 2010a. Floristic composition and canopy structure of home-gardens in São Luís city, Maranhão state, Brazil. Journal of Horticulture and Forestry 2(4):72–86

Akinnifesi, Festus Kehinde.; Sileshi, Gudeta W.; Ajayi, Oluyede C.; Akinnifesi, Adetutu I.; Moura, Emanoel G. de; Linhares, Jairo F. P.; Rodrigues, Ivanilde 2010b. Biodiversity of the urban homegardens of São Luís city, northeastern Brazil. Urban Ecosystems 13(1):129–146

Almeida, Larissa Santos de; Gama, João Ricardo Vasconcellos. 2014. Quintais agroflorestais: estrutura, composição florística e aspectos socioambientais em área de assentamento rural na Amazônia brasileira. Ciência Florestal 24(4):1041–1053

Baena, Antônio Ladislau Monteiro 2004. Ensaio Corográfico sobre a Província do Pará. Edições do Senado Federal, Brasilia

Barbosa, Cristiano de Souza. 2018. Diversidade e uso de plantas úteis nas quintais do bairro de São Raimundo, zona oeste de Manaus, AM. Master's thesis, INPA, Manaus

Bates, H. W. 1863. *The Naturalist on the River Amazons*. John Murray, London, vol. 2

Beck, H. T. 1990. A survey of the useful species of *Paullinia* L. (Sapindaceae). Advances in Economic Botany 8: 41–56

Beck, H. 2005a. Seed predation and dispersal by peccaries throughout the Neotropics and its consequences: a review and synthesis. In: Seed Fate: Predation, Dispersal and Seedling Establishment, Forget, Pierre-Michel; Lambert, Joanna E.; Hulme, Philip E.; Wall, Stephen B. Vander (eds), CABI, Wallingford, pp 77–115

Beck, H. T. 2005b. Caffeine, alcohol, and sweeteners. In: *The Cultural History of Plants*. Prance, G.T. and Nesbitt, M (eds), Routledge, New York, 173–190

Bettendorf, J.F., 1910. Chronica da missão dos padres da Companhia de Jesus no Estado do Maranhão (Vol. 72, No. 1). J. Leite.

Carvalho, J. E. U., Frazão, D. A. C.; Figueiredo, F. J. C.; Oliveira, R. 1982. Conservação da viabilidade de sementes de guaraná *Paullinia cupana* var. *sorbilis* (Mart.) Ducke. EMBRAPA Centro de Pesquisa Agropecuária do Trópico Úmido, Circular Técnica, Belém

Castelnau, F. 1850. *Éxpedition dans les Parties Centrales de l'Amérique du Sud, de Rio de Janeiro à Lima, et de Lima au Pará, pendant les Années 1843–1847*. Bertrand, Paris, vol. 2

Castro, Marina Siqueira de. 2002. Bee fauna of some tropical and exotic fruits: potencial pollinators and their conservation. In: Pollinating Bees: The Conservation Link between Agriculture and Nature, Kevan, P.; Imperatriz Fonseca, V. L. (eds), Ministry of Environment, Brasilia, pp. 276–288

Cavalcante, P. B. 1967. O guaraná (*Paullinia cupana*, var. *Sorbilis*) em estado provávelmente espontâneo, no planalto de Santarém, Pará. *Boletim do Museu Paraense Emílio Goeldi, Nova Série, Botânica* 26: 1–5

Cerón, Carlos Eduardo. 1995. Etnobiologia de los Cofanes de Dureno, Provincia de Sucumbíos, Ecuador. Museo Ecuatoriano de Ciencias Naturales, Herbario Nacional, Serie Monografia 10(3), Quito

Cerón, Carlos Eduardo; Ayala, Consuelo Montalvo. 1998. Etnobotanica de los Huaorani de Quehueiri-ono Napo, Ecuador. Abya-Yala, Quito

Chandless, W. 1870. Notes on the Rivers Maué-assú, Abacaxis, and Canumá; Amazons. *Journal of the Royal Geographical Society* 40: 419–432

Clement, C. R., M. Cristo-Araújo, G. C. d'Echenbrugge, A. A. Pereira, and D. Picanço-Rodrigues. 2010. Origin and domestication of native Amazonian crops. *Diversity* 2:72–106.

Ducke, A. 1937. Diversidade dos guaranás. *Rodriguésia* 3(10): 155–156

Edwards, W. H. 1847. *A Voyage up the River Amazon, including a Residence at Pará*. John Murray, London

Erickson, H. T., Correa, M. P. F., and Escobar, J. R. 1984. Guaraná (*Paullinia cupana*) as a commercial crop in Brazilian Amazonia. *Economic Botany* 38(3): 273–286

Ferreira Penna, D. S. 1869. *A Região Occidental da Provincia do Pará*. Typographia do Diario, Belém

Florence, H. 1941. Viagem Fluvial do Tietê ao Amazonas de 1825 a 1829. Edições Melhoramentos, São Paulo

Freitas, D. B., Carvalho, C. R., Nascimento Filho, F. J., Astolfi Filho, S. 2007. Karyotype with 210 chromosomes in guaraná (*Paullinia cupana* 'Sorbilis'). *Journal of Plant Research* 120:399–404

Garcia, Bruna Naiara Rocha; Vieira, Thiago Almeida; Oliveira, Francisco de Assis. 2017. Tree and shrub diversity in agroforestry homegardens in rural community in eastern Amazon. Floresta 47(4):543–552

Gonçalves, Janaína Pinheiro; Lucas, Flávia Cristina Araújo. 2017. Agrobiodiversidade e etnoconhecimento em quintais de Abaetetuba, Pará, Brasil. Revista Brasileira de Biociências 15(3):119–134

Gonçalves, Karina Gondolo; Pasa, Maria Corette. 2015. A etnobotânica e as plantas medicinais na Comunidade Sucuri, Cuiabá, MT, Brasil. Interações 16(2):245–256

Gondim, Carlos José Esteves. 1984. Alguns aspectos da biologia reprodutiva do guaranazeiro *Paullina cupana* var. *sorbilis* (Mart.) Ducke, Sapindaceae. Acta Amazonica 14(1–2):9–38

Griffin, S. 1988. Sales in the area are fizzing for Brazilian soft drink firm. *Washington Post*, Washington, D.C., Business Section, 21 November

Guarim Neto, Germano; Santana, Santina Rodrigues; Silva, Josefa Valdete Bezerra da. 2003. Repertório botânico da "pitombeira" (*Talisia esculenta* (A. St.-Hil.) Radlk.- Sapindaceae). Acta Amazonica 33(2):237–242

Guillaumet, J-L.; Grenand, P.; Bahri, S.; Grenand, F.; Lourd, M.; Santos, A.A. dos; Gély, A. 1990. Les jardins-vergers familiaux d'Amazonie centrale: une exemple d'utilisation de l'espace. Turrialba 40(1):63–81

Henman, A. R. 1982. Guaraná (*Paullinia cupana* var. *sorbilis*): Ecological and social perspectives on an economic plant of the central Amazon basin. Journal of Ethnopharmacology 6:311–338

Homma, A.K.O. 1999. Biopirataria na Amazônia: como reduzir? Gazeta Mercantil, 13 July, p. 2

Humboldt, A. and A. Bonpland. 1821. Personal Narrative of Travels to the Equinoctial Regions of the New Continent during the Years 1799–1804. Longman, Hurst, Rees, Orme, and Brown, London, vol. 5

Kawa, Nicholas C.; Michelangeli, José A. Clavijo; Clement, Charles R. 2015. Household agrobiodiversity management on Amazonian dark earths, oxisols, and floodplain soils on the lower Madeira River, Brazil. Human Ecology 43:339–353

Keller, F. 1874. The Amazon and Madeira Rivers, Sketches and Descriptions from the Note-Book of an Explorer. Chapman and Hall, London

Kerr, Warwick Estevam; Absy, Maria Lúcia; Souza, Antonio Carlos Marques. 1986. Espécies nectaríferas e poliniferas utilizadas pela abelha *Melipona compressipes fasciculata* (Meliponinae, Apidae) no Maranhão. Acta Amazonica 16:145–156

Klüppel, Marina Pinheiro. 2006. Sistemas agrícolas e plantas medicinais em terra preta de índio da Amazônia Central. Master's thesis, INPA/UFAM, Manaus

Krug, Cristiane; Garcia, Marcos Vinicius; Gomes, Flávia Batista. 2015. A scientific note on new insights in the pollination of guarana (*Paullinia cupana* var. *sorbilis*). Apidologie 46:164–166

Le Cointe, P. 1922. *L'Amazonie Brésilienne: Le Pays-ses Habitants, ses Resources, Notes et Statistiques jusqu'en 1920*. Augustin Challamel, Paris, vol 1

Le Cointe, P. 1934. *Amazônia Brasileira III: Árvores e Plantas Úteis*. Livraria Classica, Belém

Leacock, S. 1964. Economic life of the Maué Indians. *Boletim do Museu Paraense Emilio Goeldi* (Nova Série, Antropologia) 19:1–30

Lisboa, P. L. B. 2012. *A Terra dos Aruã: Uma História Ecológica do Arquipélago do Marajó*. Museu Paraense Emílio Goeldi, Belém

Lleras, E. 1994. Species of *Paullinia* with economic potential. In: *Neglected Crops: 1492 From a Different Perspective*, H. Bermejo and J. León (eds). Food and Agriculture Organization, Rome, 223–228

Lorenz, S. S. 1992. Sateré-Maué: Os Filhos do Guaraná. Centro de Trabalho Indigenista, São Paulo

Lourenço, José Nestor de; Sousa, Silas Garcia de; Wandelli, Elisa Vieira; Lourenço, Francisneide de Sousa; Guimarães, Rosângela dos Reis; Campos, Lian da Silva; Silva, Roberto Lima da; Martins, Vivian Fernanda Carneiro. 2009. Agrobiodiversidade nos quintais agroflorestais em três assentamentos na Amazônia Central. Revista Brasileira de Agroecologia 4(2):965–969

Machado, Danilo de Oliveira. 2016. A agrobiodiversidade de quintais agroflorestais em propriedades agrícolas familiares na BR 174, Ramal do Pau-Rosa, Manaus, AM. Master's Thesis, INPA, Manaus

Major, Julie; Clement, Charles R.; DiTommaso, Antonio. 2005. Influence of market orientation on food plant diversity of farms located on Amazonian Dark Earth in the region of Manaus, Amazonas, Brazil. Economic Botany 59(1):77–86

Mamede, Jeneffer Soares dos Santos; Pasa, Maria Corette. 2018. Os quintais agroflorestais na comunidade rural em Várzea Grande, Mato Grosso, Brazil. Biodiversidade 17(2):54–70

Maw, H. L. 1829. *Journal of a Passage from the Pacific to the Atlantic, Crossing the Andes and Descending the River Marañon or Amazon*. John Murray, London

Medeiros, Herison; Forzza, Rafaela Campostrini; Acevedo-Rodríguez, Pedro. 2016. Wild relatives of guaraná (*Paullinia cupana*, Sapindaceae) in southwestern Brazilian Amazon. Systematic Botany 41(1):202–228

Moffet, M. and N. Deogun. 1999. Brazilians like coke, but what they love to drink is Guaraná: it, too, offers a caffeine boost, but does it pack a wallop of a rather different sort? *Wall Street Journal*, 8 July, pp. A1, A8

Monteiro, M. Y. 1965. *Antropogeografia do Guaraná*. Instituto Nacional de Pesquisas da Amazônia, Cadernos da Amazônia, No. 6, Manaus

Pasa, Maria Corette. 2011. Saber local e medicina popular: a etnobotânica em Cuiabá, Mato Grosso, Brasil. Boletim do Museu Paraense Emílio Goeldi, Ciencias Humanas 6(1):179–196

Pereira, N. 1954. Os Índios Maués. Organização Simões, Rio de Janeiro

Pereira, N. 1967. *MoronGuêta: Um Decameron Indígena*. Civilização Brasileira, Rio de Janeiro, vol. 1

Pinto, Andréia Zelenski de Lara; Assis, Ana Flávia Silva de; Pereira, Arildo Gonçalo; Pasa, Maria Corette. 2013. Levantamento etnobotânico de plantas medicinais comercializados no mercado do porto em Cuiabá, Mato Grosso, Brasil. Flovet 5:51–70

Pires, J. M. 1949. Guaraná e cupana. *Revista da Sociedade dos Agrônomos e Veterinários do Pará* 1 (3): 9–20

Prance, G. T. 1974. Introduction on knowledge and use of the Amazon forest. *Interciencia* 9(5): 275–279

Prance, G. T. 1984. Introduction on knowledge and use of the Amazon forest. *Interciencia* 9(5):275–279

Prance, G. T. 1997. The ethnobotany of the Amazon indians as a tool for the conservation of biological diversity. Monografías del Jardín Botánico de Córdoba 5:135–143

Queirós da Silveira, João de São José de 1847. Viagem e visita do sertão em o Bispado do Gram-Pará em 1762 e 1763. Revista Trimensal de História e Geografia 9: 43–107, 179–227, 328–375

Rätsch, C. 2005. *The Encyclopedia of Psychoactive Plants: Ethnopharmacology and its Applications*. Park Street Press, Rochester, Vermont

Ribeiro, Reginaldo Vicente; Bieski, Isanete Geraldini Costa; Balogun, Sikiru Olaitan; Martins, Domingos Tabajarade Oliveira. 2017. Ethnobotanical study of medicinal plants used by ribeirinhos in the north Araguaia microregion, Mato Grosso, Brazil. Journal of Ethnopharmacology 205: 69–102

Roosevelt, A. C. 1999. Twelve thousand years of human-environment interaction in the Amazon floodplain. Advances in Economic Botany 13:371–392

Roosevelt, A. C.; Costa, M. Lima da; Machado, C. Lopes; Michab, M.; Mercier, N.; Valladas, H.; Feathers, J.; Barnett, W.; Silveira, M. Imazio da; Henderson, A.; Silva, J.; Chernoff, B.; Reese, D. S.; Holman, J. A.; Toth, N.; Schick, K. 1996. Paleoindian cave dwellers in the Amazon: the peopling of the Americas. *Science* 272:373–384

Rosa, Leonilde dos Santos; Silveira, Erycélede Lima; Santos, Mônica Mota dos; Modesto, Rozi da Silva; Perote, James Richard Silva; Vieira, Thiago Almeira. 2007. Os quintais agroflorestais em áreas de agricultores familiares no município de Bragança-PA: composição florística, uso de espécies e divisão de trabalho familiar. Revista Brasileira de Agroecologia 2(2):337–341

Sampaio, F. X. R. 1825. Diário da Viagem que em Visita, e Correição, das Povações da Capitania de S. Joze do Rio Negro fez o Ouvidor, e Intendente Geral da Mesma, no Anno de 1774 e 1775. Typographia da Academia, Lisbon

Santos, Jerffereson Lobatodos. 2006. Uso e diversidade de espécies vegetais cultivadas na Reserva de Desenvolvimento Sustentável do Tupé, Manaus, Amazonas. Master's thesis, INPA/UFAM, Manaus

Schultes, R. E. and R. F. Raffauf. 1992. *Vine of the Soul: Medicine Men, their Plants and Rituals in the Colombian Amazonia*. Synergetic Press, Oracle, Arizona

Sefer, E. 1961. Catálago dos insetos que atacam as plantas cultivadas da Amazônia. *Boletim Técnico do Instituto Agrônomico do Norte* 43: 23–53

Shanley, Patricia; Lux, Leda. 2003. The impacts of forest degradation on medicinal plant use and implications for health care in eastern Amazonia. Bioscience 53(6):573–584

Silva, Suzy Cristina Pedroza da. 2006. Os sistemas agroflorestais na Reserva de Desenvolvimento Sustentável Piagaçu-Purus: uma alternativa para a agricultura familiar. Master's thesis, Universidade Federal do Amazonas, Manaus

Silva, Andréa Leme; Tamashiro, Jorge; Begossi, Alpina. 2007. Ethnobotany of riverine populations from the Rio Negro, Amazonia (Brazil). Journal of Ethnobiology 27(1): 46–72

Silva, Hanna Kassia Machado da; Gama, João Ricardo Vasconcellos; Sousa, Randerson José de Araujo; Lameira, Mahyanny Karoline da Silva; Costa, Daniele e Lima da; Oliveira, Douglas Valente de; Rocha, Jobert Silva da; Oliveira, Thiago Gomes de Sousa. 2017. Composição florística de quintais agroflorestais na vila Cuera, Bragança, Pará. Agroecossistemas 9(2):330–338

Smith, N. J. H. 1996. Relevance of Spruce's work to conservation and management of natural resources in Amazonia. In: *Richard Spruce (1817–1893): Botanist and Explorer*, M.R.D. Seaward and S.M.D. Fitzgerald (Editors), pp. 227–237. The Royal Botanic Gardens, Kew

Smith, N. J. H. and A. L. Atroch, 2007. Guaraná's journey from regional tonic to aphrodisiac and global energy drink. *Evidence-Based Complementary and Alternative Medicine*, Oxford University Press online journal https://doi.org/10.1093/ecam/nem162.

Souza, F. B. 1873. *Lembranças e Curiosidades do Valle do Amazonas*. Typografia do Futuro, Belém.

Souza, Célia Cristina Valero. 2010. Etnobotânica de quintais em três comunidades ribeirinhas na Amazônia central, Manaus, AM. Master's thesis, INPA, Manaus

Spix, J. B. and C. F. P. Martius, 1976. *Viagem pelo Brasil, 1817–1820*. Edições Melhoramentos, São Paulo, Vol.3

Spruce, R. 1852. Letter from Mr. Spruce to George Bentham, Esq. Hooker's Journal of Botany and Kew Garden Miscellany 4:305–312

Steinen, K. von den. 1886. *Durch Central-Brasilien: Expedition zur Erforschung des Shingú im Jahre 1884*. Brockhaus, Liepzig

Steward, A. 2013. Reconfiguring Agrobiodiversity in the Amazon Estuary: Market Integration, the Açaí Trade and Smallholders' Management Practices in Amapá, Brazil. Human Ecology 41:827–840

Takeuchi, Masayuki. 1960. A estrutura da vegetação na Amazônia III: a mata de campina na região do Rio Negro. Boletim do Museu Paraense Emílio Goeldi, Nova Série Botânica 8:1–13

Tavares, A. M., Atroch, A. L., Nascimento Filho, F.J., Pereira, J. C. R., Araújo, J. C. A.; Moraes, L. A. C., Santos, L. P., Garcia, M. V. B., Arruda, M. R., Sousa, N. R., and Angelo, P.C. 2005. *Cultura do Guaranazeiro no Amazonas*. EMBRAPA Amazonia Occidental, Manaus, Brazil

Vásquez, Silvia Patricia Flores. 2014. Conhecimento, uso e conservação da diversidade vegetal em quatro comunidades ribeirinhas no Município Manacapuru, Amazonas. Doctoral dissertation, Programa de Pós-Graduação em Botânica, INPA, Manaus

Vásquez, Silvia Patricia Flores; Mendonça, Maria Silvia de; Noda, Sandra do Nascimento. 2014. Etnobotânica de plantas medicinais em comunidades ribeirinhas do Município de Manacapuru, Amazons, Brasil. Acta Amazonica 44(4):457–472

Wesche, Rolf; Bruneau, Thomas. 1990. Integration and Change in Brazil's Middle Amazon. University of Ottawa Press, Ottawa

Yamã, Y. 2007. O Livro Sagrado do Povo Saterê-Mawê. Editora Peirópolis, São Paulo

Chapter 46
Sapotaceae

46.1 *Manilkara bidentata*

Bolivia Masaranduba

Brazil Balata, balateira, balateira quadradinha, maçaranduba balata, maçaranduba chicle, maçaranduba branca, maçaranduba vermelha, maçaranduba do igapó, maçaranduba do caranazal, maçaranduba da caatinga, maçaranduba da campina, pau garoto, maparajuba, maparajubinha da várzea; yrykywa-ju-'y (Ka'apor), nai hi (Yanomami)

Colombia Chicle, balata, balata roja, quinilla, nispero, nisperillo, caimitillo, caimito, caimito trapichero, caimo, caimo colorado, caucho rey, purgo, yugo de charapa, quinilla; namuru (Nukak), llidorana, meníñokiai jifaiño (Witoto)

Ecuador Quinilla

Peru Quinilla, quinilla colorada, quinilla roja, balata, pamasho, brea caspi, lechero, masaranduba, michino, árbol de la vida; pashiqui (Campa), barata (Kakataibo), kinilla (Kichwa), kose (Matses), shana (Shipibo)

Venezuela Balata, cuberu, purgo, purguo, chicle, acano, cohinillo; naɨ hi (Yanomami)

Status Wild

N. Smith, *Amazon Fruits: An Ethnobotanical Journey*, Ethnobiology,
https://doi.org/10.1007/978-3-031-12803-5_46

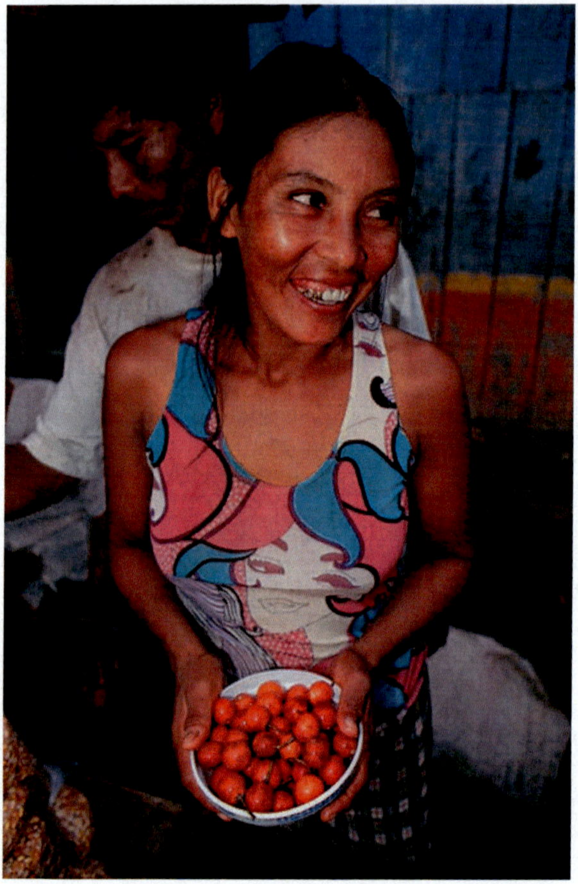

Quinilla ***Manilkara bidentata*** **fruit in a street market. Mercado Masusa, Iquitos, Loreto, Peru 3-5-04**

Quinilla *Manilkara bidentata* is a giant tree reaching up to 40 m in upland forests that produces edible fruits during the rainy season. It also occurs occasionally in floodplain forests, such as along the lower Purus River in Amazonas, Brazil (Assis et al. 2017), the Amazon estuary (Amaral et al. 2009), in southeastern Colombia (Cárdenas et al. 2009), and in the northern part of the Bolivian Amazon (Mostacedo et al. 2006). This tree is also found in some stunted woodlands (*campinarana*) growing on sandy soils that are subject to seasonal flooding, such as in the Viruá National Park in Roraima (Vale 2011:57) as well as sandy patches that are not subject to flooding (Demarchi et al. 2018, 2019; Ferreira 2009:44). This tree also occurs on sand dunes along the coast of Maranhão (Machado and Almedia Jr. 2019) and Pará (Valente et al. 2013). *Manilkara bidentata* is widespread in the Neotropics from Central America and the Caribbean south to the Guianas and the Amazon and Orinoco Basins (Pennington 2006). In the Andean foothills of the Peruvian Amazon, it can be found as high as 1000 m (Mendoza and Guerrero 2010). The fruits of a related species, *Manilkara achras*, are gathered in the wild in northeastern Mexico and are also cultivated in home gardens in many parts of Central and South America. Another related species, *Manilkara huberi* occurs in the forests of Amazonia, and the fruits are gathered for local consumption.

The round fruits typically 2 cm in diameter have a bright orange-red skin encasing a pale green pulp. The mesocarp surrounds 1 or 2 black flat seeds which people spit out after ingesting the juicy pulp. Several game animals also eat this fruit thus contributing indirectly to the diet of rural people. Woolly monkeys *Lagothrix lagotricha* and *Lagothrix cana*, spider monkeys *Ateles belzebuth*, *Ateles chamek* and *Ateles paniscus*, red howler monkeys *Alouatta seniculus*, and capuchin monkeys *Cebus apella* also eat the fruits and disperse some of the seeds, and these widespread monkeys in the Amazon Basin are hunted for their meat (Defler and Defler 1996; Levi and Peres 2013; Peres 1994; Ratiarison and Forget 2011; Regan et al. 2001; Van Roosmalen and Klein 1988; Voss and Fleck 2011). The bald uakari *Cacajao calvus*, monk saki *Pithecia monachus*, buffy saki *Pithecia albicans*, and the coppery titi *Callicebus cupreus*, all of which are confined to western Amazonia, also eat these fruits and disperse some of the seeds (Peres 1993; Voss and Fleck 2011), but they are not important game animals because of their small size and limited distributions. Squirrel monkeys *Saimiri sciureus* also eat these fruits (Voss and Fleck 2011), though they are not usually hunted because of their small size. In the Upper Tefé River in Amazonas, Brazil, gray woolly monkeys *Lagotricha cana* swallow the fruits of *Manilkara bidentat*a whole and the resulting gastrointestinal seed cleaning facilitates germination (Levi and Peres 2013). These monkeys are critical in dispersing the seeds from parent trees. Widespread capuchin monkeys *Cebus apella*, as well as the black-headed uakari *Cacajao melanocephalus melanocephalus* which is restricted to parts of northwestern Amazonia, also eat the fruits (Boubli 1999; Simmen and Sabatier 1996; Zhang 1995) and are likely involved in dispersing the seeds as well. The bearded saki *Chiropotes chiropotes*, which lives in upland forests in the northern part of the Brazilian Amazon and contiguous areas of Venezuela and the Guianas, eats this fruit and likely disperses some of the seeds (Barnett et al. 2012; Boyle et al. 2012; Shaffer 2013). The collared peccary *Pecari tajacu* and white-lipped peccary *Tayassu pecari* as well as tapirs *Tapirus terrestris* and tortoises, popular game animals throughout the Amazon and other Neotropical regions, also eat this fruit (Beck 2005; Cummings et al. 2018).

The fruits occasionally turn up in street markets, such as at the Masusa port in Iquitos, where 30 fruits were selling for the equivalent of 15 U.S. cents in March 2004. The Waimiri Atroari in Amazonas, Brazil, gather the fruits in the forest for consumption in their villages (Milliken et al. 1992:24). Because of the tree's great height, the fruits are usually gathered from the forest floor. *Manilkara bidentata* also grows in forest on some river banks, such as the black water Uaupés and Nanay, and in stunted woodlands on sandy soils called *campinarana* in Brazil and *varillal* in Peru such as in the vicinity of Iquitos (Tejada 2008:26).

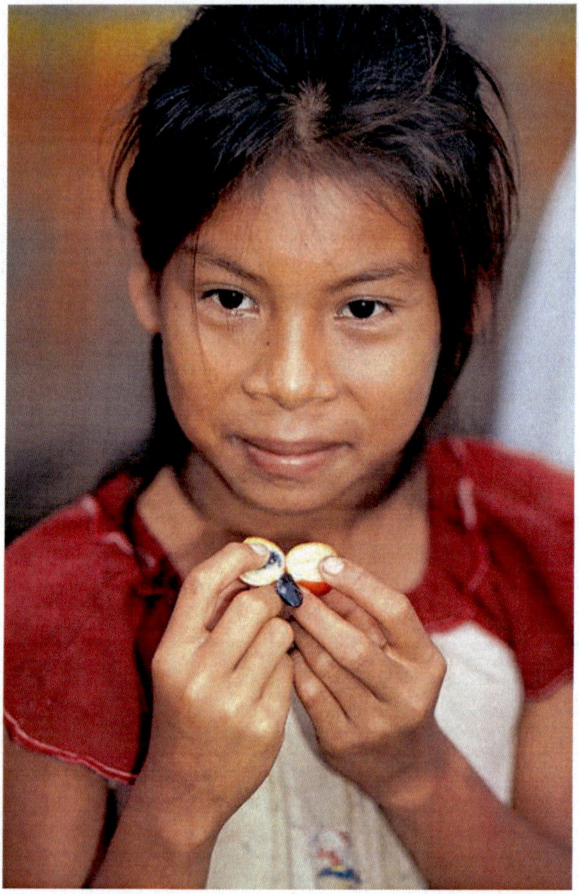

Girl eating quinilla *Manilkara bidentata* fruit. Mercado Masusa, Iquitos, Loreto, Peru 3-5-04

Manilkara bidentata is a much sought-after timber tree, so the availability of this fruit is diminishing (Guariguata et al. 2009). The hard, deep red wood has long been valued by indigenous people in the Amazon. Clubs and spears dated at 5000 years old have been found at an archaeological site in Itaituba along the Tapajós (Lisboa and Coirolo 1995). Yagua shamans in the Peruvian Amazon use the wood to make pipes they use for smoking tobacco (Chaumeil 1998:94). *Manilkara bidentata* is also tapped or sometimes cut down to obtain the latex to make various products including belts and golf balls. Trade in this latex started in Peruvian Amazon in the early twentieth century (Chibnik 1994:44). A related species, *Manilkara sapota*, was once tapped in Central America to obtain the latex for chewing gum (Coe 1980:87; Offen 1998).

46.2 *Micropholis acutangula*

Brazil Culhão de guariba, abiurana, abiu quadrado, abiu carambola, abiu mangabarana, guajará, curupixá; iwazu'yw-ran, iwazurã'yw (Tembé)

Status Wild, spontaneous in cultural settings

Arlinho, 10 years, picking culhão de guariba *Micropholis acutangula* fruit in upland successionary vegetation. A few kilometers inland from Carrazedo, confluence of the Xingu and Amazon Rivers, Município de Gurupá, Pará, Brazil 4-21-15

Culhão de guariba *Micropholis acutangula* is a liana-like tree in successionary vegetation and forests on both uplands and floodplains and is mostly confined to Amazonia but it also occurs in French Guiana (Pennington 2006). It produces edible fruits in both the rainy and dry seasons which can be picked by hand in early regrowth or boys toss sticks into the canopy of successionary forest to knock down the yellow-orange fruits. The fruits have a thick pericarp so they are not damaged when they fall to the ground. The fruits can float in water and so the seeds are probably dispersed by some species of fish as well as mammals and birds. This tree can be as tall as 25 m and capuchin monkeys *Cebus apella* eat this fruit in the vicinity of Manaus and disperse some of the seeds (Van Roosmalen and Garcia 2000).

Culhão de guariba ***Micropholis acutangula*** **fruit. Arlinho 11 years. Near Carrazedo, confluence of the Xingu and Amazon Rivers, Município de Gurupá, Pará, Brazil 11-2-16**

The ellipsoid fruits vary slightly in size from 2.4 cm × 1.4 cm to 2.5 cm × 2 cm. The skin encases 2–3 gray seeds, 1.5 cm × 1 cm, which are surrounded by a thin layer of translucent pulp. Although the mesocarp is meager, it has a sweet taste so rural people enjoy snacking on it. One of the common names for this fruit in the Brazilian Amazon is howler monkey balls, but this name does not discourage locals from eating them. The fruits do not appear to reach urban markets.

46.3 *Pouteria caimito*

Bolivia Aguaycillo, coloradillo de la bajura, coloradillo de bañao, gabetillo, lucumillo, lúcuma, lúcumo; quëo (Chácobo)

Brazil Abiu, abiu ferro, abiu cutite, abiu massa, abiu seco, abiu gigante, abiu vermelho, abiurana, abiurana amarela, abiurana vermelha, abiurana ferro, abiuzeira, abiuzeiro, abiorana seca, abieiro, abiaba, cabo de machado, carvãozinho, sapoti do mato; jaraí, paraju, guapeva, caimito; hemári, hemali (Baniwa), kanẽ (Desana), wa-yu-'ɨ (Guajá), ewa (Juruna), karamirinda'y, iratawa'y (Ka'apor), aju'ybárána (Piripkura), wɨra-mani'akaw-ran-'ɨ (Tembé), taü (Ticuna), kã're, kanẽ (Tukano), tysyhyna ipe, wairaky (Waimiri-Atroari), ari ãhi, pãxo hãtemohi, paxo wãtemo hi (Yanomami)

Colombia Caimo, caimo amarillo, caimo de charapa, caimotillo, caimito, zapote costeño; ta-su, tasúcoiché (Andoke), abil, kanea (Barasana), kãnea, guarijoague (Makuna), mútsúútsù'è (Miraña), mɨsese (Muinane), taú (Ticuna), jífikona, jipícona, jipikona (Witoto)

Ecuador Caimito, caimo, yaraso, yarasu, yarazo, abío, avío, abiyu, abiu, abiurana, cauje, cauco; yaasnumi (Achuar), cijica (Cofán), tarpu abio, tarpu aviju, avio, abiyu, abillu, apiu, japiyu, toro avio (Kichwa), yáas, yass, yarasu, yaraso (Shuar), toa (Siona-Secoya), ontogamowe, meñekawe, menika, meygue, menigue, meñigoe (Waorani)

Peru Caimito, caimito de monte, caimo, caimitillo, tocino caimito, cauje, maduraverde, quinilla; dúpi, yáas (Aguaruna), mutsɨtsɨhe (Bora), kaimíto (Campa), bata bi mia (Cashinahua), kehpa, kug'pa (Chayahuita), basaá (Ese Eja), yaás (Huambisa), segorikashi (Matsigenka), caimito (Shipibo), taxü (Ticuna), jifi-icona (Witoto)

Venezuela Temare, caimito; duwëna jyaï (Jodï)

Status Wild, spontaneous in cultural settings, planted

Abiu *Pouteria caimito* in a home garden. Near Anori, Rio Solimões, Amazonas, Brazil 6-22-07

Caimito *Pouteria caimito* was domesticated long ago for its sweet, succulent fruits in western Amazonia and has since spread out across the basin and into other lowland parts of South America, such as Bahia, Brazil, where it is known as abiu, as well as coastal Ecuador and Colombia. It was also taken into Venezuela, Trinidad, and Tobago, and Central America, most likely in pre-contact times. In the Brazilian portion of the Amazon Basin *Pouteria caimito* is found mostly under cultivation (Ducke 1946); this plant was likely domesticated in western Amazonia. "Wild" populations occasionally reported in the Brazilian Amazon, such as in Acre, Amazonas, eastern Pará, Rondônia, and Roraima, may be remnants of ancient plantings. The Witoto in the Caquetá valley in the Colombian Amazon, for example, plant *Pouteria caimito* in their agroforestry plots so that they can gather the fruits 3 or 4 years later (Veléz and Veléz 1992).

Caimito grows up to 35 m tall on the higher parts of floodplains as well as upland forest and successionary vegetation. It is found as high as 1500 m in the Andean foothills (Pennington 2006). Gathered in the wild in the Peruvian Amazon it is planted in both home gardens and cultivated fields in many parts of the Amazon Basin. The Andoke in the Colombian Amazon like to plant seeds of this tree close to charred tree stumps in their fields, presumably to provide organic matter for the seedlings (La Rotta 1983:74). Kids gather the fruits from younger trees by climbing the trunk and branches. *Pouteria caimito* starts producing fruit when only 2 m tall so some of the fruits can be reached from the ground. This species fruits just 3 years after planting (Ribeiro 1990), which is why youngsters can easily access the fruits. A long pole with a blade attached to one end is also used to cut down the fruits when the tree grows larger. A pole with a notch at one end is also used to yank off the fruits. Once the tree reaches some 20 m, however, the fruits are brought down by tossing a piece of wood into the branches. In the vicinity of Carrazedo at the mouth of the Xingu, abiu is planted around some manioc processing huts (*casa de forno*) inland from the village. Abiu is also cultivated around manioc processing huts near Jocojó, a village along a small river of the same name that flows into the Amazon a few km upstream from Gurupá.

Mailson, 8 years, using a notched pole to bring down an abiu *Pouteria caimito* fruit in his home garden. Urubu River near Caretas, Amazonas, Brazil 10-16-12

Pouteria caimito fruits sporadically throughout the year, although in the eastern Amazon abiu fruits during the rainy season. *Pouteria caimito* has long been cultivated by indigenous people. The Ticuna of the Solimões, for example, cultivate abiu in managed fallows, following a crop of manioc, their basic staple. The silver-beaked tanager *Ramphocelus carbo* feeds on the fruit in successionary vegetation and is probably involved in dispersing the seeds (Carvalho 1957). Woolly monkeys

Lagothrix lagotricha, spider monkeys *Ateles paniscus*, and white-lipped peccaries *Tayassu pecari* which are all widely hunted in the Amazon Basin for their meat, also consume the fruits of *Pouteria caimito* and disperse some of the seeds (Izawa 1975; Peres 1994; Salazar 2007; Stevenson 2000, 2007; Van Roosmalen and Klein 1988). Squirrel monkeys *Saimiri sciureus* have been observed feeding on this fruit in the vicinity of the Tucuruí reservoir along the Tocantins River in eastern Pará and although this monkey is not normally hunted it may disperse some of the seeds (Pinheiro et al. 2013). At least one species of stingless bee, *Melipona seminigra*, gathers pollen from this tree (Ferreira and Absy 2015, 2017) and the honey of this bee is widely appreciated in the Amazon Basin.

Nicivia, 9 years, with caimitillo *Pouteria caimito* fruit obtained from the tree behind her in a home garden. The tree was planted long ago from seed obtained from a wild *Pouteria caimito* in the nearby forest hence the fruit's relatively small size. Yarina, Rio Yanayacu, Reserva Nacional Pacaya-Samiria, Loreto, Peru 6-5-04

Fruit size varies between 3 and 10 cm long and 2.5 and 8 cm wide, and they usually weigh between 50 and 800 g (Falcão and Clement 1999). Wild forms, sometimes known as abiurana in Brazil, generally have smaller fruits. Considerable genetic diversity is maintained under cultivation because trees are propagated by planting seeds rather than grafting. Caimito starts bearing fruit 5 years after planting and a mature tree can easily produce 100 kg of fruit annually. Indigenous people along the Solimões between Benjamin Constant and Fonte Boa have selected especially large fruits, some weighing almost 2 kg (Kerr and Clement 1980; Kerr 1986).

Segundo, 11 years, eating caimito *Pouteria caimito* fruit gathered in his home garden. San Carlos, Rio Puinahua (a branch of the Ucayali), Reserva Nacional Pacaya-Samiria, Loreto, Peru, 8-5-03

A number of factors can influence fruit size, from soil type to the amount of light reaching the tree. As with many other fruits in Amazonia, it is difficult to draw the line between truly wild populations and feral or semi-domesticated forms. Wild fruit seeds are planted, or arise spontaneously, around homes and their resulting fruits may be intermediate in size between smaller wild forms and domesticated types. Some trees revert to the wild as homesites are abandoned, and if they manage to reproduce, their progeny may span the spectrum from small to large-fruited forms.

Shiny yellow when ripe, *Pouteria caimito* fruits are mostly eaten locally, but they also occasionally find their way to urban markets, such as in Macapá, the capital of Amapá (Adaime et al. 2014) and as I have observed in Yurimaguas along the Huallaga River. A pale yellow, somewhat sticky pulp surrounds from one to four large almond-shaped black seeds, the size of a date. The smooth, shiny fruits are not considered a premier fruit because of their white latex, which accounts for the somewhat gummy texture of the pulp.

Ironically, though, it is precisely this slippery quality that renders abiu a medicinal fruit among the Desana of the Upper Negro; women eat the viscous flesh when suffering difficult childbirth. The viscous pulp is thought to help the baby on its way (Buchillet 1988). Rural folk in the vicinity of Manacapuru along the lower Solimões near its confluence with the Rio Negro prepare a tea from the leaves to treat diarrhea and stomachache (Vásquez et al. 2014). In the Peruvian Amazon, the Chayahuita mince the leaves and stuff them in their nostrils at night to treat leishmaniasis (Odonne et al. 2013) while in the Loreto province of the Peruvian Amazon a decoction is prepared with a handful of leaves in 1 l of water which is drunk in the morning for 1 week to treat liver problems (Roumy et al. 2020). In the vicinity of Chazuta along the upper Huallaga River in the Peruvian Amazon, locals use the bark to make a decoction for expelling intestinal worms and to serve as a laxative (Sanz-Biset et al. 2009). The Witoto in the Colombian Amazon macerate and toast the leaves of this plant and then apply the paste to wounds as a disinfectant (Schultes and Raffauf 1986a, b). In the Bolivian Amazon, the Chácobo use the wood for fuel (Zambrana et al. 2017), as do some rural people along the Amazon River in Peru (Balbin and Samaniego 1996), while in Roraima in the northern part of the Brazilian Amazon the Yanomami use the wood to fashion posts for their communal houses (Milliken and Albert 1997).

Several other species of *Pouteria* are gathered in the wild or cultivated in home gardens and managed fallows. In the vicinity of Agrovila Nova Fronteira near Altamira on the Transamazon Highway, for example, hunters sometimes encounter the fragrant fruits of pariri *Pouteria pariry*, also known as frutão, on the forest floor. Lime-green pariri fruits are so heavy and fall from such a height, that they usually split open upon impact. The fruits are not usually brought home because they are often damaged and their skin easily breaks; rather they provide a welcome snack in the forest. The powerful, sweet odor of the pariri fruits doubtless evolved to entice seed dispersal agents, such as tapir and brocket deer. One family living on uplands overlooking the Amazon floodplain near Óbidos has planted a pariri tree in their backyard. For the most part, however, the family does not eat the fruits directly. Rather, they planted the tree to provide food for a dozen red-footed tortoises *Chelonoidis carbonaria* in their fenced yard. The tortoises are avid fruit eaters, and after sufficient fattening on pariri, other wild and cultivated fruits, and sweet manioc, they end up in the cooking pot. And the Waimiri-Atroari near Manaus gather the fruits of *Pouteria guianensis* and *Pouteria cuspidata* in the upland forest.

The fruits of another wild species of *Pouteria* are gathered in the Upper Rio Negro watershed. While traveling up the Uaupés River in 1852, the English naturalist Alfred Russel Wallace made the following comments on ocokí *Pouteria ucuqui*:

> The ocokí is a large pear-shaped fruit with a hard thick outer skin of almost a woody texture, then a small quantity of very sweet pulpy matter, and within a large black oval stone. The pulp is very luscious, but is so acrid as to make the mouth and throat sore, if more than two or three are eaten. When, however, the juice is boiled it loses this property; and when made into mingau with tapioca, is exceedingly palatable and very highly esteemed in the Upper Rio Negro, where it is abundant. (Wallace 1889:243)

Caimito is allelopathic, that is it releases an herbicide from the decaying leaves that prevents most plants from germinating underneath the tree. In this manner, the tree secures more nutrients for itself. This interesting chemical warfare conducted by caimito has even entered regional lore. In various parts of the Peruvian Amazon, such as along the Marañón River, rural folk believe that chullachaqui, a supernatural humanoid that lives in upland forests, cultivates small gardens of caimito and wild pineapple. Storytellers note that chullachaqui is fastidious in keeping his garden "clean" by weeding underneath the caimito. In another version of this story told in Yarina along the Yanayacu, caimitillo, a wild relative of caimito, is cultivated by chullachaqui. Not all rural folk, buy this story, however.

Caimito surfaces in a myth told by the Barasana along the Pirá-paraná River in the Colombian Amazon. In the beginning, there were no people, trees, or soil, just rock and Ayawa, three Thunder spirits. The Ayawa believed they were spirits from the sky. One day, they came upon a kanea *Pouteria caimito* tree with lots of fruit. They ejaculated into the fruit and then gave it to their grandmother to eat. Another woman warned her not to eat the fruit, but she ignored the warning and ate the fruit and became pregnant. The grandmother then gave birth to a boy who became a powerful shaman (Hugh-Jones 1979:267).

46.4 *Pouteria longifolia*

Ecuador Meego, domapue (Waorani)

Peru Lucma, quinilla

Status Wild

Children eating lucma *Pouteria longifolia* fruits in their village. The fruits were gathered in nearby floodplain forest. Veinte de Enero, Rio Yanayacu, Reserva Nacional Pacaya-Samiria, Loreto, Peru 4-23-06

A relative of the widely cultivated *Pouteria caimito*, *Pouteria longifolia* is a large tree of lowland forests, including floodplains, in the western Amazon from Ecuador south to Bolivia. The grapefruit-sized fruits, some 14 cm long and 12 cm wide, drop to the ground during the rainy season. Lucma, as the tree is known in the Reserva Nacional Pacaya-Samiria in Peru, is not common, and when encountered while clearing land for fields, it is typically spared, such as in the vicinity of Veinte de Enero along the lower Yanayacu.

People gather the fruits to eat the yellow pulp which has a slightly dry, mealy texture and an intriguing flavor, quite unlike any other fruit. Indeed, the tantalizing flavor of another relative, lúcuma *Pouteria lucuma*, which was domesticated in the Andes, has recently caught the attention of ice cream makers in Peru, including Danafrio, which is owned by Nestlé. Lúcuma desserts, such as mousses and tarts (*tartaleta de lucuma*) can now be found in restaurants throughout Lima and other cities in Peru. Furthermore, fruits of *Pouteria lucuma* are common in many markets in the montane forest zone of the Andes (*selva alta*) as well as coastal cities in Peru. The fruits of another near relative, *Pouteria multiflora*, which is native to lowland tropical forests from Central America to the Amazon Basin, are marketed in Iquitos. One day, *Pouteria longifolia* may too emerge from relative obscurity to become a popular flavor in ice creams, milkshakes, and desserts throughout the Amazon and beyond.

46.5 *Pouteria macrophylla*

Bolivia Lúcuma, lúcumo, lúcma, cucuma, coquino, lujma; yahë (Chácobo), bójo (Ese Eja), vicoi' (Tsimane)

Brazil Tuturubá, tuturubá de guariba, taturubá, taturubá amarelo, gotite, cutite, cutite pequeno, cutiti, cutitiribá, abiurana, abiurana preta, abiurana cutite, abiu cabeça de macaco, abiu folha grande, abiu cutite, abiu seco, acará, acará uba, gema de ovo, jarai; akuči-tɨr-ɨwa-'ɨwa (Asurini), akuči-ter-ewa-'ɨ (Guajá), akuŝitɨrɨwa'ɨ (Ka'apor), kamokô, kamokó (Kayapó), aju'ybá ipotýra (Piripkura), ɨwa-zu-ete (Tembé)

Peru Lucma, lucuma, quinilla blanca, quinilla caimitillo, sapote quinilla, sapote de la sachavaca, fruta huevo, yema de huevo, caimo, caimito

Status Wild, planted

Farmer touching the trunk of a young tuturubá *Pouteria macrophylla* tree he planted in his home garden. Archaeological site with anthropogenic dark earth on an upland bluff overlooking the Xingu floodplain. Sitio Senhor Heladio, left bank of the Rio Xingu about 50 km upstream from Altamira, Pará, Brazil 9-17-15

Tuturubá *Pouteria macrophylla* is a tree reaching 30 m tall in upland forests with edible fruit in many parts of the Amazon including open forests with lianas and palms in the eastern fringes of the basin in Maranhão (Gama et al. 2007) and seasonal forest in the Tocantins (Haidar et al. 2013). It occurs in both mature forests and successionary vegetation, including gallery forests in the Upper Tocantins and Araguaia watersheds in the southeastern part of the Brazilian Amazon (Balée 1993b; Coelho et al. 2012; Espírito-Santo et al. 2005; Marimon and Lima 2001; Pennington 2006; Pereira and Santos 2015; Scoles 2009; Silva 2015:62, 66; Silva et al. 2016; Souza 2019:12; Stefanello et al. 2009). It also occurs occasionally in other floodplain forests, such as along the sediment-rich Purus, Solimões, and Tefé rivers in Amazonas (Assis et al. 2017; Luize et al. 2015), the Upper Madeira in Rondônia (Casula 2012:68), the black water Uatumã in Amazonas (Lobo 2017:69), and in the Parque Nacional Madidi in Bolivia (Araujo-Murakami et al. 2012). The tree is widespread in the Amazon Basin from the Andean foothills up to 1500 m, such as in Bolivia (Arellano 2013:200) east to Maranhão in Brazil. It may well be native to the Andean foothills (Villachica et al. 1996:171). Although mostly found on uplands, it is also occasionally encountered in floodplain forests.

The size of a chicken egg, hence one of its common names in the Peruvian Amazon (*fruta huevo*), each creamy fruit contains from one to two round seeds. A soft, yellow pulp surrounds the seeds (Gordon et al. 2011). This fruit has significant antioxidant capacity and is a good alimentary source of polyphenols (Silva et al. 2012). The tree fruits during the rainy season and rural people, such as the Ka'apor in eastern Amazonia, enter the forest to gather the fruits which are yellow-green when ripe (Balée 1993a). The wood is also used locally for construction purposes, and the Tsimane in the Bolivian Amazon use the wood for fuel and to make charcoal (Reyes-García 2001:256).

The fruits are also procured by several game animals. Tapirs *Tapirus terrestris*, the largest terrestrial game animal in the Amazon, apparently eats the fruits of *Pouteria macrophylla* as suggested by one of its common names in Peru: sapote de la sachavaca. The fruit is evidently also popular with agoutis *Dasyprocta* spp., a diurnal rodent much procured for its meat, as suggested by the common name for this fruit tree among the Guajá: akuci-terewa-'ɨ which means agouti fruit tree (Balée 2013:185). Woolly monkeys *Lagothrix lagotricha* and spider monkeys *Ateles chamek*, which are hunted for their meat throughout the Amazon, are known to eat the fruits of this tree (Felton et al. 2008, 2010; Peres 1994). The red-nosed bearded saki *Chiropotes albinasus* and squirrel monkeys *Saimiri sciureus* also eat this fruit and although these monkeys are not usually hunted for their meat, they are likely involved in dispersing some of the seeds (Barnett et al. 2012; Pinto 2008:79; Santos et al. 2018). Tuturubá de guariba is one of the common names for this tree in Maranhão on the eastern fringe of the Amazon Basin suggesting that howler monkeys (*Alouatta*) eat this fruit and likely disperse some of the seeds.

Some people plant the seeds in their home gardens, such as rural folk along the middle Xingu River and in the town of Juriti along the middle Amazon. Some people on the floodplain of the lower Solimões near its confluence with the Rio Negro River also plant this species in their home gardens (Lima and Saragoussi 2000) as do upland farmers along the Ramal do Pau Rosa, a side road leading from the BR 175 Manaus-Boa Vista highway north of Manaus (Machado 2016:39), in the vicinity of São Domingos do Araguaia in southeastern Pará, and several rural communities in the municipality of Bonito in the Bragantina zone of northeastern Pará (Pereira et al. 2010; Vieira et al. 2012). Rural dwellers in the Mazagão watershed in Amapá plant *Pouteria macrophylla* in their yards (Steward 2013) as do some farmers in the vicinity of Macapá (Pereira et al. 2017). The Chácobo, who inhabit the Beni watershed in Bolivia, also plant *Pouteria macrophylla* in their home gardens (Boom 1989). Such trees become part of anthropogenic forests when the homesites are abandoned (Balée 2010). The Kayapó in the interfluvial area between the Tocantins and Xingu plant *Pouteria macrophylla* in some of their manioc fields so that they can return to those areas long after the tubers have been harvested to gather fruits (Posey 1985). This tree is also found in some cacao agroforestry plots in the municipality of São Félix do Xingu in southern Pará (Braga et al. 2019).

Apartment buildings that have replaced houses and their gardens. View from Edificio Maria Bernadete, Avenida Serzedelo Correa, 105, Distrito Batista Campos, Belém, Pará, Brazil 10-19-16

This tree was once common in home gardens in urban areas such as Belém, but they have diminished dramatically as more houses are cleared for apartment buildings and shopping centers (Cavalcante 2010:108). Many residents of Belém are shifting to apartments because house break-ins are increasing as crime rates rise. Even houses with metal grates over windows and doors are still subject to breaking and entering since thieves use jacks to pry them off. Neighborhood police patrols are virtually non-existent.

46.6 *Pouteria torta*

Bolivia Cojma, so'rocaj (Tsimane)

Brazil Abiurana, abiurana vermelha, abiurana do baixio, abiuarana paruru, abiurana de goma, abiurana curríola liso, abiu, abiu da várzea, abiu do cerrado, abiu cabeça de macaco, abiu piloso, acá ferro, amarelinho, maçaranduba, guapeba, guapeva, bacóide, curiola, pêssego do mato, grão de galo, pateiro; paadzawa (Baniwa), manakata (Paresí), owetíngapauhúa (Piripkura)

Colombia Caimitillo, caimo de guacamaya, caimo peludo, caimo de monte, mediacaro, zapotemonte

Ecuador Piast, iñaco, cacho, caimo de monte; chanange tetacho (Cofán), sacha avio (Kichwa), kamputin yaas (Shuar), meñegohue (Waorani)

Peru Caimitillo, caimito, caimito negro, quinilla, quinilla blanca, quinilla colorada, quinilla negra, quina quina, varilla de agua, lucuma, sachavaca papaya; dupí (Aguaruna)

Status Wild

Caimitillo *Pouteria torta* fruit gathered in regrowth forest on uplands. Near Miraflores, Rio Tigre, Loreto, Peru 6-11-04

Pouteria torta is a tree that soars to 40 m high in forests and produces edible fruits during the rainy season. The globular fruits, 3–7 cm in diameter, are typically picked up from the forest floor because the tree is too tall to climb safely. People savor the juicy, but slightly sticky, white pulp that surrounds the one to four shiny black seeds which are discarded after eating the mesocarp. The yellow fruits are consumed by rural people and do not enter urban markets. In the central Amazon, this tree bears fruit in April and May during the rainy season (Pennington 2006).

Red howler monkeys *Alouatta seniculus* also eat the fruits and disperse the seeds as do woolly monkeys *Lagothrix lagotricha* and capuchin monkeys (Julliot 1996a, b, 1997; Peres 1994; Regan et al. 2001; Simmen and Sabatier 1996). These monkeys are hunted for their meat throughout the Amazon, so the fruits also contribute indirectly to the diet of humans through game animals. Black-headed uakaris *Cacajao melanocephalus melanocephalus* also eat the fruits; these monkeys, which occur in bands of around 100 individuals, are restricted to the Upper Rio Negro in Amazonas, Brazil, and contiguous southern Venezuela (Boubli 1999). Although these monkeys are small, rarely weighing over 4 kg, they are likely hunted occasionally, especially by indigenous groups. Tapirs *Tapirus terrestris* and brocket deer (*Mazama* spp.) reportedly eat this fruit in the Ecuadorian Amazon (Cerón and Ayala 1998:160); these game animals are widespread in Amazonia.

The range of *Pouteria torta* extends from Belize in Central America to the tropical lowlands of South America as far south as Paraguay. In the Andean foothills, it occurs up to 2300 m (Arellano 2013:200). The versatile tree occurs in both wetlands and uplands, thrives on a wide range of soils from sand to clay, and grows in various environments from mature forest, gallery, and seasonal forest in the cerrado region bordering southern and southeastern Amazonia, to scrub woodland and grasslands (Araujo-Murakami et al. 2009; Batalha and Martins 2004; Costa et al. 2014; Haidar et al. 2013; Ivanauskas et al. 2004; Korning et al. 1994; Kunz et al. 2008, 2010a, b, 2014; Lorenzi et al. 2006:307; Marimon and Lima 2001; Marimon et al. 2002; Medeiros and Walter 2012; Medeiros et al. 2008; Miguel et al. 2011; Oliveira and Daly 1999; Oliveira et al. 2018; Ratter et al. 1996; Silva 2011:24; Stefanello 2008:19; Stefanello et al. 2010; Thomas 2009), hence one of its popular names in Brazil: abiu do cerrado.

In addition to its edible fruits, *Pouteria torta* has a variety of other uses. In the Bolivian Amazon, the Tsimane use the hard inner wood of this tree to fashion mortars for grinding (Reyes-García 2001:102). In the Ecuadorian Amazon, the wood is used for construction as well as for fuel, such as among the Cofán and Waorani (Cerón et al. 1994:67; Cerón and Ayala 1998:160) and for making furniture among other indigenous groups in southern Ecuador (Aguirre-Mendoza et al. 2015:80). Also in the Ecuadorian Amazon, the Waorani employ the seeds in a remedy for intestinal parasites, while the Kichwa use the leaves to make a topical application for treating leishmaniasis (Gachet et al. 2010).

46.7 *Sarcaulus brasiliensis*

Brazil Abiurana, abiu, culhão de guariba da várzea, guajará, guajará do igapó, jaraí, jaraí branco, jaraí vermelho, pau doce; kyxiyrakry (Waimiri-Atroari)

Ecuador Caimitillo, taracachihila, tuta pishsu hila, shimbi; toro avio (Kichwa), meñegoe, meñingohue, meyguhue (Waorani)

Peru Caimitillo, caimito, quinilla, tushmo amarillo, cachari; ujukan yáas, shákashak (Aguaruna)

Status Wild, cultivated

Culhão de guariba da várzea *Sarcaulus brasiliensis* fruit gathered by Wadrio, 11 years, in floodplain forest. Carrazedo, mouth of the Xingu River, Município de Gurupá, Pará, Brazil 7-24-16

Abiurana *Sarcaulus brasiliensis* is a tree growing up to 30 m high in floodplain and upland forests throughout the Amazon Basin including the tidal zone in the estuary (Braga and Jardim 2019; Braga et al. 2015; Carim et al. 2013; Ferreira et al. 2013; Fortini et al. 2006; Freitas et al. 2019; Gama et al. 2003; Kvist et al. 1995; Lemos et al. 2015; Maués et al. 2011; Myster 2018; Salomão et al. 2007a, b; Van Roosmalen and Garcia 2000). *Sarcaulus brasiliensis* ranges from Costa Rica south to the Amazon Basin and the Atlantic forest of Brazil. It can be found as high as 1500 m in the Andean foothills (Arellano 2013:200). It is also found in some successionary vegetation, such as in the tidal zone of the Amazon estuary (Bentes-Gama et al. 2002) and in the vicinity of Iquitos and Madre de Dios in the Peruvian Amazon (Macedo 2014:55; Robles 2017:49). This tree is also found in some agroforestry systems in the vicinity of Cametá and Tomé-Açu in northeastern Pará (Santos et al. 2016, 2019).

It is known as culhão de guariba da várzea (howler monkey testicles of the floodplain) at the mouth of the Xingu. It produces orange-yellow fruits which are gathered by locals during the rainy season. The shiny fruits, similar in shape and size to those of peach palm *Bactris gasipaes*, are eaten locally and do not reach markets. The thick skin (pericarp) is broken open with fingers to access the tan-colored mesocarp surrounding a single large seed. The pulp only accounts for about 10% of the fruit but is nevertheless appreciated by rural folk, such as in the vicinity of Afuá in northeastern Marajó Island (Gama et al. 2003). Along the lower Ucayali River in the Peruvian Amazon, rural people also eat the pulp of this fruit (Kvist et al. 2001). In the vicinity of Cametá along the lower Tocantins River, rural people use the wood for construction purposes and to serve as fuel (Santos et al. 2004), as do the Waorani in the Ecuadorian Amazon (Cerón and Ayala 1998:161). Extracts from this plant exhibit inhibitory activity toward human platelet aggregation (Oliveira et al. 2009).

In the Colombian Amazon, spider monkeys *Ateles belzebuth*, woolly monkeys *Lagothrix lagotricha*, and squirrel monkeys *Saimiri cassiquiarensis albigena* eat these fruits and may be involved in dispersing the single seed found in each fruit (Ahumada et al. 1998; Carretero-Pinzón et al. 2016; Stevenson and Link 2010; Stevenson et al. 1994) as could be the case with howler monkeys *Alouatta seniculus* (Neves and Rylands 1991). All three of those monkey species are hunted for their meat. The red-nosed bearded saki *Chiropotes albinasus* has also been recorded eating this fruit in the Tapajós National Forest south of Santarém in Pará (Pinto 2008:79); although this monkey is not regularly hunted for its meat, it is likely involved in dispersing some of the seeds. Peccaries, tapir *Tapirus terrestris*, and armadillos (several species and genera) also reportedly eat this fruit and these animals are hunted for their meat (Cerón and Ayala 1998:161).

Rodrigo, 10 years, eating an abiurana *Sarcaulus brasiliensis* fruit he gathered from a canoe in floodplain forest. Igarapé Divisa behind Carrazedo, confluence of the Xingu and Amazon Rivers, Município de Gurupá, Pará Brazil 4-17-15

References

Adaime, Ricardo; Jesus-Barros, Cristiane Ramos de; Souza-Filho, Miguel Francisco de. 2014. Moscas-das-frutas (Diptera: Tephritidae) obtidas de frutos comercializados em feiras públicas de Macapá, Amapá. Boletim de Pesquisa e Desenvolvimento, EMBRAPA 85:1–17

Aguirre-Mendoza, Zhofre; Loja, Ángel Fernando; Solano, Carmen; Aguirre, Nikolay. 2015. Especies Forestales más aprovechadas del Sur del Ecuador. Universidad Nacional de Loja, Ecuador

Ahumada, Jorge A.; Stevenson, Pablo R.; Quiñones, Marcella J. 1998. Ecological response of spider monkeys to temporal variation in fruit abundance: the importance of flooded forest as a keystone habitat. Primate Conservation 18:10–14

Amaral, Dário Dantas do; Vieira, Ima Célia Guimarães; Almeida, Samuel Soares de; Salomão, Rafael Paiva; Silva, Antonio Sérgio Lima da; Jardim, Mário Augusto Gonçalves. 2009. Checklist da flora arbórea de remanescentes florestais da região metropolitana de Belém e valor histórico dos fragmentos, Pará, Brasil. Boletim do Museu Paraense Emílio Goeldi, Ciencias Naturais 4(3):231–289

Araujo-Murakami, Alejandro; Paniagua, Narel; Cayola, L.; Jorgensen, Peter M.; Valdes, Ovidio; Macuapa, Eric; Calzadilla-Tomianovich, Miki. 2009. Diversidad y estructura florística del bosque de llanura y palmar de pantano Amazónico preandino en el sector de Chalalan, Parque Nacional Madidi, Bolivia. Kempffiana 5(1):3–27

Araujo-Murakami, Alejandro; Poma-Chura, Abraham; Palabral, Arely; Salvatierra, Rolando; Hurtado, Fidel. 2012. Composición florística de los bosques Amazónicos de tierra firme e inundable en las proximidades de las Pampas del Sonene (río Heath), Parque Nacional Madidi, Bolivia. Kempffiana 8(1):2–21

Arellano, Gabriel. 2013. Patrones de diversidad, distribución y dominancia de plantas leñosas en los bosques montanos del Parque Nacional Madidi, Bolivia. Doctoral dissertation, Departamento de Biologia, Facultad de Ciencias, Universidad Autónoma de Madrid

Assis, Rafael Leandro de; Wittmann, Florian; Luize, Bruno Garcia; Haugaasen, Torbjon. 2017. Patterns of floristic diversity and composition in floodplain forests across four southern Amazon river tributaries, Brazil. Flora 229:124–140

Balbin, Isabel Oré; Samaniego, Dedy Llapapasca. 1996. Huertas domesticas como sistema tradicional de cultivo en Moena Caño, Rio Amazonas, Iquitos, Peru. Folia Amazonica 8(1):91–110

Balée, W. 1993a. *Footprints of the Forest: Ka'apor Ethnobotany, the Historical Ecology of Plant Utilization by an Amazonian People*. Columbia University Press, New York

Balée, W. 1993b. Indigenous transformation of Amazonian forests: an example from Maranhão, Brazil. L'Homme 33(126–128):231–254

Balée, W. 2010. Contingent diversity on anthropic landscapes. *Diversity* 2:163–181

Balée, W. 2013. Cultural Forests of the Amazon: A Historical Ecology of People and Their Landscapes. University of Alabama Press, Tuscaloosa

Barnett, Adrian A.; Boyle, Sarah A.; Pinto, Liliam P.; Lourenço, Waldete C.; Almeida, Thais; Silva, Welma Sousa; Ronchi-Teles, Beatriz; Bezerra, Bruna M.; Ross, Caroline; MacLarnon, Ann; Spironello, Wilson R. 2012. Primary seed dispersal by three Neotropical seed-predating primates (*Cacajao melanocephalus ouakary*, *Chiropotes chiropotes* and *Chiropotes albinasus*). Journal of Tropical Ecology 28(6):543–555

Batalha, Marco Antônio; Martins, Fernando Roberto. 2004. Reproductive phenology of the cerrado plant community in Emas National Park (central Brazil). Australian Journal of Botany 52:149–161

Beck, H. 2005. Seed predation and dispersal by peccaries throughout the Neotropics and its consequences: a review and synthesis. In: Seed Fate: Predation, Dispersal and Seedling Establishment, Forget, Pierre-Michel; Lambert, Joanna E.; Hulme, Philip E.; Wall, Stephen B. Vander (eds), CABI, Wallingford, pp 77–115

Bentes-Gama, Michelliny de Matos; Scolforo, José Roberto Soares; Gama, João Ricardo Vasconcellos. 2002. Potencial produtivo de madeira e palmito de uma floresta secundária de várzea baixa no estuário Amazônico. Revista Árvore 26(3):311–319

Boom, B. M. 1989. Use of plant resources by the Chácobo. In: *Resource Management in Amazonia: Indigenous and Folk Strategies*, D. A. Posey and W. Balée (Editors), Advances in Economic Botany, vol. 7, New York Botanical Garden, Bronx, pp. 78–96

Boubli, J. P. 1999. Feeding ecology of black-headed uacaris (*Cacajao melanocephalus melanocephalus*) in Pico da Neblina National Park, Brazil. International Journal of Primatology 20(5):719–749

Boyle, Sarah A.; Zartman, Charles E.; Spironello, Wilson R.; Smith, Andrew T.; Beck, Harald. 2012. Implications of habitat fragmentation on the diet of bearked saki monkeys in central Amazonian forest. Journal of Mammalogy 93(4):959–976

Braga, Elayne Oliveira; Jardim, Mário Augusto Gonçalves. 2019. Florística, estrutura e formas de vida do estrato inferior de uma floresta ombrófila densa aluvial, Pará, Brasil. Ciencia Florestal, Santa Maria 29(3): 1048–1059

Braga, Elayne Oliveira; Silva, Julie Andrews de França e; Pantoja, Milenna Vasconcelos; Jardim, Mário Augusto Gonçalves. 2015. Florística, estrutura fitossociológica e formas de vida do estrato inferior em um floresta de várzea Amazônica. Biota Amazônia 5(3):59–65

Braga, Daniel P. P.; Domene, Frederico; Gandara, Flávio B. 2019. Shade trees composition and diversity in cacao agroforestry systems of southern Pará, Brazilian Amazon. Agroforestry Systems 93:1409–1421

Buchillet, D. 1988. Interpretação da doença e simbolismo ecológico entre os índios Desana. Boletim to Museu Paraense Emílio Goeldi, Série Antropologia 4 (1): 27–42

Cárdenas, Dairon; Castaño, Nicolás; Sua, Sonia. 2009. Flora de la estrella fluvial de Inírida (Guianía, Colombia). Biota Colombiana 10(1–2):1–30

Carim, Marcelo de Jesus Veiga; Guillaumet, Jean-Louis Bernard; Guimarães, José Renan da Silva; Tostes, Luciedi Cássia Leôncio. 2013. Composição e estrutura de floresta ombrófila densa do extremo norte do Estado do Amapá, Brasil. Biota Amazônia 3(2):1–10

Carretero-Pinzón, Xyomara; Defler, Thomas R.; Ruiz-García, Manuel. 2016. How does the Colombian squirrel monkey cope with habitat fragmentation? Strategies to survive in small fragments. In: Phylogeny, Molecular Population Genetics, Evolutionary Biology and Conservation of the Neotropical Primates, Ruiz-García, M.; Shostell, J. M. (eds), Nova Science, New York, chapter 15, 1–16

Carvalho, C. T. 1957. Notas sôbre a biologia do *Ramphocelus carbo* (Passeres, Thraupidae). Boletim do Museu Paraense Emilio Goeldi, Zoologia 5:1–20

Casula, Kátia Regina. 2012. Análise da similaridade florística e estrutural das formações florestais inundáveis em um trecho do alto Rio Madeira e de seus afluentes, estado de Rondônia, Brasil. Master's Thesis, Núcleo de Ciências Exatas da Terra, Universidade Federal de Rondônia, Porto Velho

Cavalcante, P. B. 2010. *Frutas Comestíveis na Amazônia*. Museu Paraense Emílio Goeldi, Belém

Cerón, Carlos Eduardo; Ayala, Consuelo Montalvo. 1998. Etnobotanica de los Huaorani de Quehueiri-ono Napo, Ecuador. Abya-Yala, Quito

Cerón, C. E.; Montalvo, Consuelo G.; Umenda, José; Umenda Enma Chica. 1994. Etnobotánica y Notas sobre la Diversidad Vegetal en la Comunidad Cofán de Sinangüé, Sucumbíos, Ecuador. EcoCiencia, Quito

Chaumeil, Jean-Pierre. 1998. Ver, saber, poder: el chamanismo de los Yagua de la Amazonía Peruana. CAAAP/IFEA/CAEA-CONICET, Lima

Chibnik, M. 1994. *Risky Rivers: The Economics and Politics of Floodplain Farming in Amazonia*. University of Arizona Press, Tucson

Coe, M. D. 1980. *In the Land of the Olmec*. University of Texas Press, Austin, vol. 2

Coelho, Roberta de Fátima Rodrigues; Miranda, Izildinha Souza; Mitja, Danielle. 2012. Caracterização do processo sucessional no Projeto de Assentamento Benfica, sudeste do estado do Pará, Amazônia oriental. Boletim do Museu Paraense Emílio Goeldi, Ciencias Naturais 7(3):251–282

Costa, Reginaldo Brito da; Silveira, Ediléia Patrícia da; Souza, Thaianny Rodrigues de. 2014. Estrutura da vegetação de cerrado sensu stricto em terra indígena no noroeste do estado de Mato Grosso. Multitemas 46: 95–110

Cummings, Anthony R.; Read, Jane M.; Fragoso, Jose M. V. 2018. Implications of forest type and land tenure diversity for the sustainability of ecosystem services provided by northern Amazonia's multiple-use tree species. Landscape Ecology 33:423–438

Defler, Thomas R.; Defler, Sara B. 1996. Diet of a group of *Lagothrix lagothricha lagothrica* in southeastern Colombia. International Journal of Primatology 17(2):161–190

Demarchi, Layon Oreste; Scudeller, Veridiana Vizoni; Moura, Livia Carvalho; Dias-Terceiro, Randolpho Gonçalves; Lopes, Aline; Wittmann, Florian Karl; Piedade, Maria Tereza

Fernandez. 2018. Floristic composition, structure and soil-vegetation relations in three white-sand soil patches in central Amazonia. Acta Amazonica 48(1):46–56

Demarchi, Layon Oreste; Scudeller, Veridiana Vizoni; Moura, Livia Carvalho; Lopes, Aline. 2019. Logging impact on Amazonian white-sand forests: perspectives from a sustainable development reserve. Acta Amazonica 49(4):316–323

Ducke, A. 1946. *Plantas de Cultura Precolombiana na Amazônia brasileira. Notas Sôbre as Espécies ou Formas Espontâneas que Supostamente lhes teriam dado Origem.* Instituto Agronômico do Norte, Belém, Boletim Técnico 8, pp. 1–24

Espírito-Santo, Fernando Del Bon; Shimabukuro, Yosio Edimir; Aragão, Luiz Eduardo Oliveira e Cruz de; Machado, Evandro Luiz Mendonça. 2005. Análise da composição florística e fitossociológica da floresta nacional do Tapajós com apoio geográfico de imagens satélites. Acta Amazonica 35(2):155–173

Falcão, Martha de Aguiar; Clement, Charles R. 1999. Fenologia e produtividade do abiu (*Pouteria caimito*) na Amazônia central. Acta Amazonica 29 (1): 3–11

Felton, Annika M.; Felton, Adam; Wood, Jeff T.; Lindenmayer, David B. 2008. Diet and feeding ecology of *Ateles chamek* in a Bolivian semihumid forest: the importance of Ficus as a staple food resource. International Journal of Primatology 29:379–403

Felton, A. M.; Felton, A.; Foley, W. J.; Lindenmayer, D. B. 2010.The role of timber tree species in the nutritional ecology of spider monkeys in a certified logging concession, Bolivia. Forest Ecology and Management 259:1642–1649

Ferreira, Carlos Alberto Cid. 2009. Análise comparativa de vegetação lenhosa do ecosistema campina na Amazonia Brasileira. Doctoral dissertation, INPA/UFAM, Manaus

Ferreira, Marcos Gonçalves; Absy, Maria Lucia. 2015. Pollen niche and trophic interactions between colonies of Melipona (Michmelia) seminigra merrillae and Melipona (Melikerria) interrupta (Apidae: Meliponini) reared in floodplain in the Central Amazon. Arthropod-Plant Interactions 9:263–279

Ferreira, Marcos Gonçalves; Absy, Maria Lucia. 2017. Pollen analysis of honeys of Melipona (Michmelia) seminigra merrillae and Melipona (Melikerria) interrupta (Hymenoptera: Apidae) bred in Central Amazon, Brazil. Grana 56(6):436–449

Ferreira, Leandro V.; Chaves, Priscilla P.; Cunha, Denise de A.; Matos, Darley C. L.; Parolin, Pia. 2013. Variação da riqueza e composição de espécies da comunidade de plantas entre as florestas de igapós e várzeas na Estação Científica Ferreira Penna-Caxiuanã na Amazônia oriental. Pesquisas Botânica 64:175–195

Fortini, L. B.; Rabelo, F. G.; Zarin, D. J. 2006. Mixed potential for sustainable forest use in the tidal floodplain of the Amazon River. Forest Ecology and Management 231:78–85

Freitas, João da Luz; Santos, Erick Silva dos; Cruz Junior, Francisco de Oliveira; Borges, César Henrique Alves; Santos, Adriano Castelo dos. 2019. Structural analysis and distribution patterns in lowland tropical forest, eastern Amazon. Journal of Agricultural Studies 7(4): 240–255

Gachet, María Salomé; Lecaro, Javier Salazar; Kaiser, Marcel; Brun, Reto; Navarrete, Hugo; Muñoz, Ricardo A.; Bauer, Rudolf; Schühly, Wolfgang. 2010. Assessment of anti-protozoal activity of plants traditionally used in Ecuador in the treatment of leishmaniasis. Journal of Ethnopharmacology 128:183–197

Gama, J. R. V., S. A. Botelho, M. Bentes-Gama, and J. R. S. Scolforo 2003. Estrutura e potencial futuro de utilização da regeneração natural de floresta de várzea alta no Município de Afuá, estado do Pará. *Ciencia Florestal, Santa Maria* 13(2):71–82

Gama, João Ricardo Vasconcellos; Souza, Agostino Lopesde Souza; Calegário, Natalino; Lana, Guilherme Campos. 2007. Fitossociologia de duas fitocenoses de floresta ombrófila aberta no município de Codó, estado do Maranhão. Revista Árvore 31(3):465–477

Gordon, André; Jungfer, Elvira; Silva, Bruno Alexandre da; Maia, José Guilherme S.; Marx, Friedhelm. 2011. Phenolic constituents and antioxidant capacity of four underutilized fruits from the Amazon region. Journal of Agricultural and Food Chemistry 59:7688–7699

Guariguata, M.R., Licona, J.C., Mostacedo, B, and Cronkleton, P. 2009. Damage to Brazil nut trees (Bertholletia excelsa) during selective timber harvesting in Northern Bolivia. Forest Ecology and Management, 258(5):788–793

Haidar, Ricardo Flores; Fagg, Jeanine Maria Felfili; Pinto, José Roberto Rodrigues; Dias, Ricardo Ribeiro; Damasco, Gabriel; Silva, Lucasde Carvalho Ramos; Fagg, Christopher William. 2013. Florestas estacionais e áreas de ecótono no estado do Tocantins, Brasil: parâmetros estruturais, classificação das fitofisionomias florestais e subsídios para conservação. Acta Amazonica 43(3):261–290

Hugh-Jones, S. 1979. The Palm and the Pleiades: Initiation and Cosmology in Northwest Amazonia. Cambridge University Press, Cambridge

Ivanauskas, Natália Macedo; Monteiro, Reinaldo; Rodrigues, Ricardo Ribeiro. 2004. Estrutura de um trecho de floresta Amazônica na bacia do alto rio Xingu. Acta Amazonica 34(3):275–299

Izawa, Kosei. 1975. Foods and feeding behavior of monkeys in the upper Amazon basin. Primates 16(3):295–316

Julliot, C. 1996a. Seed dispersal by red howling monkeys (*Alouatta seniculus*) in the tropical rain forest of French Guiana. International Journal of Primatology 17(2):239–258

Julliot, C. 1996b. Fruit choice by red howler monkeys (*Alouatta seniculus*) in a tropical rain forest. American Journal of Primatology 40:261–282

Julliot, C. 1997. The impact of seed dispersal by red howler monkeys *Alouatta seniculus* on the seedling population in the understorey of tropical rain forest. Journal of Ecology 85:431–440

Kerr, W. E. 1986. Agricultura e seleções genéticas de plantas. In: Suma Etnológica Brasileira, Vozes, Petrópolis, Part 1 E, pp. 159–171

Kerr, W. E. and C. R. Clement. 1980. Prácticas agrícolas de consequências genéticas que possibilitaram aos índios da Amazônia uma melhor adaptaçao às condiçoes ecológicas da região. Acta Amazonica 10(2):251–261

Korning, Jorgen; Thomsen, Karsten; Dalsgaard, Kristian; Nornberg, Per. 1994. Characters of three Udults and their relevance to the composition and structure of virgin rain forest of Amazonian Ecuador. Geoderma 63:145–164

Kunz, Sustanis Horn; Ivanauskas, Natália Macedo; Martins, Sebastião Venâncio; Silva, Elias; Stefanello, Daniel. 2008. Aspectos florísticos e fitossociológicos de um trecho de floresta estacional perenifólia na Fazenda Triarão, Bacia do rio das Pacas, Querência, MT. Acta Amazonica 38(2):245–254

Kunz, Sustanis Horn; Martins, Sebastião Venâncio; Ivanauskas, Natália Macedo; Stefanello, Daniel; Silva, Elias. 2010a. Fitossociologia de uma área de floresta estacional perenifólia na Fazenda Amoreiras, Querência, MT. Revista Árvore 34(4):713–721

Kunz, Sustanis Horn; Martins, Sebastião Venâncio; Ivanauskas, Natália Macedo; Silva, Elias; Stefanello, Daniel. 2010b. Estrutura fitossociológica de um trecho de floresta estacional perenifólia, bacia do Rio das Pacas, Querência, MT. Cerne 16(2):115–122

Kunz, Sustanis Horn; Ivanauskas, Natália Macedo; Martins, Sebastião Venâncio; Stefanello, Daniel; Silva, Elias. 2014. Fitossociologia do componente arbóreo de dois trechos de floresta estacional perenifólia, bacia do Rio das Pacas, Querência, MT. Ciência Florestal, Santa Maria 24(1):1–11

Kvist, Lars P.; Andersen, Martin K.; Hesselsoe, Martin; Vanclay, Jerome K. 1995. Estimating use-values and relative importance of Amazonian flood plain trees and forests to local inhabitants. Commonwealth Forestry Review 74(4):293–300

Kvist, Lars Peter; Andersen, Martin K.; Stagegaard, Jesper; Hesselsoe, Martin; Llapapasca, Consuelo. 2001. Extraction from woody forest plants in flood plain communities in Amazonian Peru: use, choice, evaluation and conservation status of resources. Forest Ecology and Management 150:147–174

La Rotta, C. 1983. Observaciones Etnobotanicas sobre algunas Especies Utilizadas por la Comunidad Indigena Andoque (Amazonas, Colombia). Universidad Nacional de Colombia, Bogotá

Lemos, D. A. N.; Ferreira, B. G. A.; Siqueira, J. D. P.; Oliveira, M. M.; Ferreira, A. M. 2015. Floristic and phytosociology in dense "terra firme" rainforest in the Belo Monte hydroelectric plant influence area, Pará, Brazil. Brazilian Journal of Biology 75(3):257–276

Levi, Taal; Peres, Carlos A. 2013. Dispersal vacuum in the seedling recruitment of a primate-dispersed Amazonian tree. Biological Conservation 163:93–106

Lima, Roberval M. B. de; Saragoussi, Muriel. 2000. Floodplain home gardens on the central Amazon in Brazil. In: The Central Amazon Floodplain: Actual Use and Options for a Sustainable Management, Junk, W. J.; Ohly, J. J.; Piedade, M. T. F.; Soares, M. G. M. (eds), Backhuys Publishers, Leiden pp 243–268

Lisboa, P. L. B. and A. D. Coirolo. 1995. Notas sobre implementos indígenas com madeira de 5000 da microregião do Tapajós, Pará. Boletim do Museu Paraense Emílio Goeldi, Série Botanica 11(1):7–17

Lobo, Guilherme de Sousa. 2017. A alteração do regime hidrológico afeta a composição florística e estrutura de florestas de igapó? Um estudo comparativo entre um rio regulado e outro prístino na Reserva de Desenvolvimento Sustentável do Uatumã, Amazônia Central. Master's thesis, INPA, Manaus

Lorenzi, H., L. Bacher, M. Lacerda, S. Sartori 2006. Brazilian Fruits and Cultivated Exotics (for Consuming in Natura). Instituto Plantarum de Estudos da Flora Ltda, Odessa, São Paulo

Luize, Bruno Garcia; Venticinque, Eduardo Martins; Silva, Thiago Sanna Freire; Novo, Evlyn Marcia Leão de Moraes. 2015. A floristic survey of angiosperm species occurring at three landscapes of the central Amazon várzea, Brazil. Checklist: The Journal of Biodiversity Data 11(6):1–15

Macedo, Erick Joe Dossantos. 2014. Almacenamiento de carbono en la biomasa aérea del bosque primario y bosque secundario de la parcela "Muro Huayra" en la Reserva Nacional Allpahuayo Mishana, Iquitos, Perú. Master's Thesis, UNAP, Facultad de Ciencias Forestales, Iquitos

Machado, Danilo de Oliveira. 2016. A agrobiodiversidade de quintais agroflorestais em propriedades agrícolas familiares na BR 174, Ramal do Pau-Rosa, Manaus, AM. Master's Thesis, INPA, Manaus

Machado, Monielle Alencar; Almedia Jr., Eduardo Bezerra de. 2019. Spatial structure, diversity, and edaphic factors of an area of Amazonian coast vegetation in Brazil. Journal of the Torrey Botanical Society 146(1):58–68

Marimon, Beatriz Schwantes; Lima, Edson de Souza. 2001. Caracterização fitofisionômica e levantamento florístico preliminar no Pantanal dos Rio Morte-Araguaia, Cocalinho, Mato GrossoActa Botanica Brasilica 15(2):213–229

Marimon, B. S.; Felfili, J. M.; Lima, E. S. 2002. Floristics and phytosociology of the gallery forest of the Bacaba stream, Nova Xavantina, Mato Grosso, Brazil. Edinburgh Journal of Botany 59(2):303–318

Maués, Bernardo Antônio Rodrigues; Jardim, Mário Augusto Gonçalves; Batista, Fábio de Jesus; Medeiros, Tonny David Santiago; Quaresma, Adriano da Costa. 2011. Composição florística e estrutura do estrato inferior da floresta de várzea na área de proteção ambiental Ilha do Combu, município de Belém, estado do Pará. Revista Árvore 35(3):669–677

Medeiros, Marcelo Brilhante de; Walter, Bruno Machado Teles. 2012. Composição e estrutura de comunidades arbóreas de cerrado stricto sensu no norte do Tocantins e sul do Maranhão. Revista Árvore 36(4): 673–683

Medeiros, Marcelo Brilhante de; Walter, Bruno Machado Teles; Silva, Glocimar Pereira. 2008. Fitossociologia do cerrado stricto sensu no município de Carolina, MA, Brasil. Cerne 14(4):285–294

Mendoza, Abel L. Monteagudo; Guerrero, Marleni Huamán. 2010. Catálogo de los arboles y afines de la selva central del Perú. Arnaldoa 17(2):202–242

Miguel, Aline; Marimon, Beatriz Schwantes; Oliveira, Edmar Almeida de; Marachaipes, Leandro; Marimon-Junior, Ben Hur. 2011. Mudanças na estrutura da vegetação lenhosa em três porções da mata de galeria do Córrego Bacaba (1999–2006), Nova Xavantina-MT. Revista Árvore 35(3):725–735

Milliken, W. and B. Albert. 1997. The construction of a new Yanomami round-house. *Journal of Ethnobiology* 17(2):215–233

Milliken, W., R. P. Miller, S. R. Pollard, and E. V. Wandelli. 1992. *The Ethnobotany of the Waimiri Atroari Indians of Brazil.* Royal Botanic Gardens, Kew

Mostacedo, Bonifacio; Balcazar, Julio; Montero, Juan Carlos. 2006. Tipos de bosque, diversidad y composición florística en la Amazonia sudoeste de Bolivia. Ecología en Bolivia 41(2):99–116

Myster, Randall W. 2018. What we know about the composition and structure of igapó forests in the Amazon Basin. Botanical Review 84:394–410

Neves, Ana M. S.; Rylands, Anthony B. 1991. Diet of a group of howling monkeys, *Alouatta seniculus*, in an isolated forest patch in Central Amazonia. In: *A Primatologia no Brasil – 3*, Rylands, Anthony B.; Bernardes, Aline T. (eds), Fundação Biodiversitas para a Conservação da Diversidade Biológica, Belo Horizonte, pp 263–274

Odonne, Guillaume; Valadeau, Celine; Alban-Castillo, Joaquina; Stien, Didier; Sauvain, Michel; Bourdy, Genevieve. 2013. Medical ethnobotany of the Chayahuita of the Paranapura basin (Peruvian Amazon). Journal of Ethnopharmacology 146:127–153

Offen, K. 1998. An historical geography of chicle and tunu gum production in northeastern Nicaragua. Year Book 1998 of the Conference of Latin Americanist Geographers 24:57–73

Oliveira, Alexandre A. de; Daly, Douglas C. 1999. Geographic distribution of tree species occurring in the region of Manaus, Brazil: implications for regional diversity and conservation. Biodiversity and Conservation 8:1245–1259

Oliveira, Viviana Maria Araújo; Carneiro, Ana Lucia Basilio; Cauper, Glaucia Socorro de Barros; Pohlit, Adrian Martin. 2009. In vitro screening of Amazonian plants for hemolytic activity and inhibition of platelet aggregation in human blood. Acta Amazonica 39(4):973–980

Oliveira, Lucicléia Mendes de; Sousa, Rhonan Martins de; Correa, Norma Elena Reynosa; Santos, André Ferreira dos; Giongo, Marcos. 2018. Florística e síndromes de dispersão de um fragmento de cerrado ao sul do estado do Tocantins. Scientia Agraria Paranaensis 17(1): 104–111

Pennington, T. D. 2006. Flora da Reserva Ducke, Amazonas, Brasil: Sapotaceae. Rodriguésia 57(2): 251–366

Pereira, Antonio Carlos; Santos, Eduardo Ribeiro dos. 2015. Frutas nativas do Tocantins com potencial de aproveitamento económico. Agri-Environmental Sciences 1(1):22–37

Pereira, Carmen Nazário; Maneschy, Rosana Quaresma; Oliveira, Pâmela Dias; Oliveira, Ilmaione Keiza de Souza. 2010. Caracterização de quintais agroflorestais no projeto de assentamento Belo Horizonte I, São Domingos do Araguaia, Pará. Agroecossistemas 2(1):73–81

Pereira, Luciano Araujo; Lima-Barbosa, Jackson Rodrigo de; Almeida, Maria Zélia de; Guimarães, Elsie Franklin. 2017. Diversidade de plantas em quintais quilombolas, conhecimento local sobre uso e cultivo de pimentas na Amazônia oriental, Brasil. Revista de Biologia Neotropical 14(1):56–72

Peres, C. A. 1993. Notes on the ecology of buffy saki monkeys (*Pithecia albicans*, Gray 1860): a canopy seed-predator. American Journal of Primatology 31:129–140

Peres, C.A. 1994. Diet and feeding ecology of gray woolly monkeys (*Lagothrix lagotricha cana*) in central Amazonia: comparisons with other Atelines. International Journal of Primatology 15(3):333–372

Pinheiro, Tatyana; Ferrari, Stephen F.; Lopes, Maria Aparecida. 2013. Activity budget, diet, and use of space by two groups of squirrel monkeys (*Saimiri sciureus*) in eastern Amazonia. Primates 54:301–308

Pinto, L. P. 2008. Ecologia alimentar do cuxiú-de-nariz vermelho *Chiropotes albinasus* (Primates: Pitheciidae) na Floresta Nacional do Tapajós, Pará. Doctoral dissertation, Universidade Estadual de Campinas, São Paulo

Posey, D. A. 1985. Native and indigenous guidelines for new Amazonian development strategies: understanding biological diversity through ethnoecology. In: Change in the Amazon Basin Volume 1: Man's Impact on Forests and Rivers, Hemming, John (ed), Manchester University Press, Manchester, pp 156–181

Ratiarison, Sandra; Forget, Pierre-Michel. 2011. Fruit availability, frugivore satiation and seed removal in 2 primate-dispersed tree species. Integrative Zoology 6:178–194

Ratter, J. A.; Bridgewater, S.; Atkinson, R.; Ribeiro, J. F. 1996. Analysis of the floristic composition of the Brazilian cerrado vegetation II: comparison of the woody vegetation of 98 areas. Edinburgh Journal of Botany 53(2):151–180

Regan, B. C.; Julliot, C.; Simmen, B.; Viénot, F.; Charles-Dominique, P.; Mollon, J. D. 2001. Fruits, foliage and the evolution of primate colour vision. Philosophical Transactions of the Royal Society London B 356: 229–283

Reyes-García, Victoria. 2001. Indigenous people, ethnobotanical knowledge, and market economy: a case study of the Tsimane' Amerindians in lowland Bolivia. Doctoral dissertation, University of Florida, Gainesville.

Ribeiro, Berta G. 1990. Classificação dos solos e horticultura Desâna. In: Ethnobiology: Implications and Applications, Proceedings of the First International Congress of Ethnobiology, Posey, Darrell A.; Overal, William Leslie; Clement, Charles R.; Plotkin, Mark; Elisabetsky, Elaine; Mota, Clarice Novaes da; Barros, José Flávio Pessôa (eds), Museu Paraense Emílio Goeldi, Belém, pp. 25–49

Robles, Rembrandt Ramiro Canahuire. 2017. Composición florística y estructura de la recuperación natural en un área degradado por minería en la comunidad nativa de Tres Islas, Tambopata, Madre de Dios. Master's thesis, Escuela Profesional de Ingeniería Forestal y Medio Ambiente, Puerto Maldonado, Perú

Roumy, Vincent; Ruiz, Lastenia; Macedo, Juan Celidonio Ruiz; Gutierrez-Choquevilca, Andrea-Luz; Samaillie, Jennifer; Encinas, Leonor Arévalo; Mesia, Wilfredo Ruiz; Cotrina, Hivelli Ericka Ricopa; Rivière, Céline; Sahpaz, Sevser; Bordage, Simon; Garçon, Guillaume; Dubuisson, Jean; Anthérieu, Sebastien; Seron, Karin; Hennebelle, Thierry. 2020. Viral hepatitis in the Peruvian Amazon: Ethnomedical context and phytomedical resource. Journal of Ethnopharmacology 255 (112735):1–11

Salazar, Romer Salvador Miserendino. 2007. Dieta de los chanchos de monte: taitetu (*Tayassu tajacu*) y tropero (*Tayassu pecari*) en el territorio indigena Siriono, Beni, Bolivia. Revista Boliviana de Ecología y Conservación Ambiental 21:43–56

Salomão, Rafael de Paiva; Vieira, Ima Célia Guimarães; Suemitsu, Chieno; Rosa, Nélson de Araújo; Almeira, Samuel Soares de; Amaral, Dário Dantas do; Menezes, Moirah Paulo Machado de. 2007a. As florestas de Belo Monte na grande curva do rio Xingu, Amazônia Oriental. Boletim do Museu Paraense Emílio Goeldi, Ciencias Naturais 2(3):57–153

Salomão, Rafael Paiva; Rosa, Nélson Araújo; Morais, Kácio Andrey Câmara. 2007b. Dinâmica da regeneração natural de árvores em áreas mineradas na Amazônia. Boletim do Museu Paraense Emílio Goeldi, Ciencias Naturais 2(2):85–139

Santos, Silvio Roberto Miranda dos; Miranda, Izildinha de Souza; Tourinho, Manoel Malheiros. 2004. Análise florística e estrutural de sistemas agroflorestais das várzeas do rio Juba, Cametá, Pará. Acta Amazonica 34(2):251–263

Santos, Silvio Roberto Miranda dos; Kato, Osvaldo Ryohei; Tourinho, Manoel Malheiros. 2016. Análise florística e finalidade de espécies cultivadas em sistemas agroflorestais no município de Cametá, Pará, Amazônia Oriental. Amazônia: Ciência e Desenvolvimento 13(23):9–22

Santos, Deborah Cruz dos; Corrêa Neto, José de Jesus; Alvarenga, Adriano Braga Brasileiro de. 2018. Frutos consumidos por macaco-de-cheiro *Saimiri sciureus collinsi* (Osgood, 1916) em fragmentos de florestas em uma unidade de conservação no arquipélago do Marajó, Amazônia Oriental, Brasil. Revista Brasileira de Zoociências 19(1):82–90

Santos, Silvio Roberto Miranda dos; Kato, Osvaldo Ryohei; Tourinho, Manoel Malheiros. 2019. Diversidade florística e estoque de carbono de sistemas agroflorestais em dois municípios do nordeste paraense, Brasil. Boletim do Museu Paraense Emílio Goeldi, Ciencias Naturais 14(1):31–42

Sanz-Biset, Jaume; Campos-de-la-Cruz, José; Epiquién, Mirbel A.; Cañigueral, Salvador. 2009. A first survey on the medicinal plants of the Chazuta valley (Peruvian Amazon). Journal of Ethnopharmacology 122:333–362

Schultes, Richard Evans; Raffauf, Robert F. 1986a. De plantis toxicariis e Mundo Novo tropicale commentationes XXXVII. Miscellaneous notes on medicinal and toxic plants of the northwest Amazon. Botanical Museum Leaflets, Harvard University 30(4):255–285

Schultes, Richard Evans; Raffauf, Robert F. 1986b. De plantis toxicariis e mundo novo tropicale commentationes XXXVIII: ethnopharmacological and alkaloidal notes on plants of the northwest Amazon. Caldasia 15:71–75

Scoles, Ricardo. 2009. El quintal y las frutas: recursos económicos y alimentares en la comunidad negra de Itacoã, Acará, Pará, Brasil. Acta Amazonica 39(1):1–12

Shaffer, Christopher A. 2013. Feeding ecology of northern bearded sakis (*Chiropotes sagulatus*) in Guyana. American Journal of Primatology 75:568–580

Silva, Silvestre. 2011. Fruits of the Brazilian Amazon. Metalivros, São Paulo

Silva, Wheriton Fernando Moreira da. 2015. Florística, dinâmica e estrutura da regeneração natural em floresta explorada na FLONA do Tapajós, PA. Master's Thesis, INPA, Manaus

Silva, Bruno Alexandre da; Gordon, André; Jungfer, Elvira; Marx, Friedhelm; Maia, José Guilherme S. 2012. Antioxidant capacity and phenolics of Pouteria macrophylla, an under-utilized fruit from the Brazilian Amazon. European Food Research and Technology 234:761–768

Silva, Wagner Matos da; Lolis, Solange e Fátima; Viana, Rodney Haulien Oliveira. 2016. Composition and structure of the gallery forest in the Taquaruçu Grande sub-basin, Municipality of Palmas, Tocantins State. Acta Scientiarum Biological Sciences 38(1):17–24

Simmen, Bruno; Sabatier, Daniel. 1996. Diets of some French Guianan primates: food composition and food choices. International Journal of Primatology 17(5):661–693

Souza, Victória Cristina Oliveira de. 2019. Avaliação do estágio sucessional utilizando a classificação em grupos ecológicos de uma floresta urbana na região metropolitana de Belém-Pará. Bachelor's thesis, Universidade Federal Rural da Amazônia (Ufra), Belém

Stefanello, Daniel. 2008. Florística, estrutura e dispersão de sementes de três trechos de floresta estacional perenifólia ribeirinha na bacia hidrográfica do Rio das Pacas, Querência, MT. Master's Thesis, Universidade Federal de Viçosa, Minas Gerais

Stefanello, Daniel; Fernandes-Bulhão, Clarissa; Martins, Sebastião Venâncio. 2009. Síndromes de dispersão de sementes em três trechos de vegetação ciliar (nascente, meio e foz) ao longo do Rio Pindaíba, MT. Revista Árvore 33(6):1051–1061

Stefanello, Daniel; Ivanauskas, Natália Macedo; Martins, Sebastião Venâncio; Silva, Elias; Kunz, Sustanis Horn. 2010. Síndromes de dispersão de diásporos das espécies de trechos de vegetação ciliar do rio das Pacas, Querência-MT. Acta Amazonica 40(1):141–150

Stevenson, Pablo R. 2000. Seed dispersal by woolly monkeys (*Lagothrix lagothricha*) at Tinigua National Park, Colombia: dispersal distance, germination rates, and dispersal quantity. American Journal of Primatology 50: 275–289

Stevenson, Pablo R. 2007. Estimates of the number of seeds dispersed by a population of primates in a lowland forest in western Amazonia. In: Seed Dispersal: Theory and its Application in a Changing World, Dennis, Andrew J.; Schupp, Eugene W.; Green, Ronda J.; Westcott, David A. (eds), CABI, Wallingford, pp 340–362

Stevenson, Pablo R.; Link, Andrés. 2010. Fruit preferences of *Ateles belzebuth* in Tinigua Park, northwestern Amazonia. International Journal of Primatology 31:393–407

Stevenson, Pablo R.; Quiñones, Marcela J.; Ahumada, Jorge A. 1994. Ecological strategies of woolly monkeys (*Lagothrix lagotricha*) at Tinigua National Park, Colombia. American Journal of Primatology 32:123–140

Steward, A. 2013. Reconfiguring Agrobiodiversity in the Amazon Estuary: Market Integration, the Açaí Trade and Smallholders' Management Practices in Amapá, Brazil. Human Ecology 41:827–840

Tejada, Ronald Manuel Panduro. 2008. Estructura, crecimiento, composición florística y potencial forestal del bosque varillal alto seco en la Reserva Allpahuayo-Mishana, Iquitos-Perú, con fines de manejo sostenible, 2007–2008. Doctoral dissertation, Universidad Nacional Trujillo, Peru

Thomas, Evert. 2009. New light on the floristic composition and diversity of indigenous territory and national park Isiboro-Sécure, Bolivia. Biodiversity Conservation 18:1847–1878

Vale, Gabriel Damasco do. 2011. O efeito do solo, inundação e topografia sobre as campinaranas inundáveis na Amazônia. Master's Thesis, INPA, Manaus

Valente, Deyvid Marques; Sousa, Julio dos Santos de; Bastos, Maria de Nazaré Carmo. 2013. Estudo taxonômico de Sapotaceae Juss. do litoral Paraense. Acta Amazonica 43(2):161–168

Van Roosmalen, M. G. M.; Garcia, Olímpia Mariada Cruz Gomes. 2000. Fruits of the Amazonian forest. Part II: Sapotaceae. Acta Amazonica 30(2):187–290

Van Roosmalen, Marc G. M.; Klein, Lewis L. 1988. The spider monkeys, genus *Ateles*. In: Ecology and Behavior of Neotropical Primates, Mittermeier, Russell A.; Rylands, Anthony B.; Coimbra-Filho, Adelmar F.; Fonseca, Gustavo A.B. da (eds), Ecology and Behavior of Neotropical Primates, World Wildlife Fund, Washington, DC, vol. 2, pp 455–537

Vásquez, Silvia Patricia Flores; Mendonça, Maria Silvia de; Noda, Sandra do Nascimento. 2014. Etnobotânica de plantas medicinais em comunidades ribeirinhas do Município de Manacapuru, Amazons, Brasil. Acta Amazonica 44(4):457–472

Veléz, Germán Alonso; Veléz, Antonio José. 1992. Sistema agroflorestal de "chagras" utilizado por las comunidades indígenas del medio Caquetá (Amazonia Colombiana). Colombia Amazonica 6(1): 101–134

Vieira, Thiago Almeida; Rosa, Leonilde dos Santos; Santos, Maria Marly de Lourdes Silva. 2012. Agrobiodiversidade de quintais agroflorestais no município de Bonito, Estado do Pará. Revista de Ciências Agrárias 55(3):159–166

Villachica, Hugo; Carvalho, José Edmar Urano de; Muller, Carlos Hans; Díaz, Camilo; Almanza, Martha; Mercado, Roberto Samanez; Izquierdo, Juan; Egg, Antônio Brack. 1996. Frutales y Hortalizas Promisorios de la Amazonia. Tratado de Cooperación Amazonica, Lima.

Voss, Robert S.; Fleck, David W. 2011. Mammalian diversity and Matses ethnomammalogy in Amazonian Peru Part 1: Primates. Bulletin of the American Museum of Natural History 351:1–81

Wallace, Alfred Russel. 1889. A Narrative of Travels on the Amazon and Rio Negro, with an Account of the Native Tribes, and Observations on the Climate, Geology, and Natural History of the Amazon Valley. Ward, Lock & Co., London

Zambrana, Narel Y. Paniagua; Bussmann, Rainer; Hart, Robbie E.; Huanca, Araceli L. Moya; Soria, Gere Ortiz; Vaca, Milton Ortiz; Álvarez, David Ortiz; Morán, Jorge Soria; Morán, María Soria; Chávez, Saúl; Moreno, Bertha Chávez; Moreno, Gualberto Chávez; Roca, Oscar; Siripi, Erlin. 2017. Traditional knowledge hiding in plain sight – twenty-first century ethnobotany of the Chácobo in Beni, Bolivia. Journal of Ethnobiology and Ethnomedicine 13:1–47

Zhang, S. 1995. Activity and ranging patterns in relation to fruit utilization by brown capuchins (*Cebus apella*) in French Guiana. International Journal of Primatology 16(3):489–507

Chapter 47
Simaroubaceae

47.1 *Simaba orinocensis*

Brazil Cajurana, marupá

Peru Crista huayo, sacha marupa, marupa negra

Venezuela Cerezo, cedrillo rebalsero

Status Wild

N. Smith, *Amazon Fruits: An Ethnobotanical Journey*, Ethnobiology,
https://doi.org/10.1007/978-3-031-12803-5_47

Simaba orinocensis **with unripe fruit along the margin of a floodplain lake. Cocha Grande, near San Carlos, Rio Puinahua (a branch of the Ucayali), Reserva Nacional Pacaya-Samiria, Loreto, Peru, 4-18-06**

A denizen of the floodplains of black, clear, and sediment-rich rivers throughout the Amazon and Orinoco basins, as well as the Guianas, *Simaba orinocensis* is a small tree that also occurs in gallery forests in savannas. Fishermen gather the red fruits during the high water season to make juice on the spot, such as on Cocha Pungal, a tannin-stained lake near the village of San Carlos on the Puinahua floodplain, a branch of the lower Ucayali. While waiting for fish to become entangled in gillnets, men crush the fruits into a bowl, usually fashioned from a calabash gourd, containing lake water and some sugar. The elliptical fruits, measuring some 3 cm by 1.8 cm, are green until they ripen toward the end of the rainy season. The flattened fruits contain a single seed. Fruits of crista huayo are rarely if ever taken home, and they never appear in markets.

In addition to their use in making a refreshing juice, fishermen in San Carlos along the Puinahua report that several fish important in commerce and for subsistence, such as sabalo (*Brycon*), gamitana *Colossoma macropomum*, palometa (*Mylossoma*), and several species of *Leporinus* also eat the fruit. According to a fisherman based in Itacoatiara along the middle Amazon, the fruits of a related species, *Simaba guianensis*, are also eaten by *Colossoma macropomum* which is known as tambaqui in Brazil. This observation has been confirmed by ichthyologists and ecologists (Honda 1974; Parolin et al. 2010; Silva et al. 2003) as well as a botanist (Ducke 1925). Another popular fish in the Amazon, pirapitinga *Piaractus brachypomus*, is also fond of the fruit (Lucas 2008). *Simaba orinocensis* is thus another example of an Amazon fruit that contributes indirectly as well as directly to the human diet.

47.2 *Simarouba versicolor*

Bolivia Chiriguaná

Brazil Pitomba, cajarana, mata cachorro, mata menino, mata piolho, morcegueira do cerrado, perdiz, pau de perdiz, pé de perdiz, paraíba, paparaúba, caixeta

Peru Paraiba, pitomba, marupa

Status Wild

Pitomba *Simarouba versicolor* tree in fruit. Near Ilha dos Bichos, a Marajoara mound, Cachoeira do Arari Municipality, Marajó Island, Pará, Brazil 5-17-99

The glistening purple-red fruits of *Simarouba versicolor*, known as cajarana along the Amazon and pitomba on Marajó Island at the river's mouth, resemble cherries of temperate climes in shape and texture. Fruits of *Simarouba versicolor* are 2–2.5 cm long (Cronquist 1944). Locals snack on the shiny, smooth fruits which they gather from canoes during the flood season. Rural folk have to take care when gathering the fruits, however. Some of the trees contain hornets' nests which can form stinging swarms if disturbed. Yellow-rumped caciques *Cacicus cela* sometimes build their hanging nests in pitomba trees with wasp nests for added protection. The wasps do not bother the noisy caciques, even when they build their woven nests close to the paper hive.

The fruits are a little tart, so rural folk in the vicinity of Lago Arari on Marajó sometimes prepare a thick treacle from the fruits called *doce de pitomba*. Homemade vinegar is also prepared from the juice-laden fruits by mashing them and adding salt. Pitomba vinegar is a tangy condiment on fish dishes. A family in Cachoeira do Arari on Marajó also enjoys pitomba mashed with manioc flour and sugar (Smith 2002:117).

Pitomba tree in fruit in a seasonally flooded savanna. Near Ilha dos Bichos, Marajó Island, Pará, Brazil 5-17-99

Simarouba versicolor is a medium-sized tree of seasonally flooded grasslands (campo), seasonal forests with a strong dry season, and scrub savannas (cerrado) in the Amazon Basin and central and southern Brazil (Ferreira and Pasa 2015; Furley et al. 1978; Haidar et al. 2013; Matos et al. 2020; Medeiros and Walter 2012; Mews et al. 2011; Miguel et al. 2016; Oliveira Filho and Martins 1991; Pereira et al. 2015; Pinto et al. 2014). In the cerrado region of Tocantins, leaf cutter ants *Atta sexdens* disperse some of the fruit seeds (Lopes et al. 2018). The tree forms dense stands in some places in the Amazon, such as along certain streams draining into Lago Arari. Arari lake was once densely settled by the Marajoara culture, which reached its apex about the same time as the classic Maya around 1000 AD, and it is possible that some of the pitomba groves can be attributed to humans.

The light, soft wood of *Simarouba versicolor* makes it ideal for a variety of purposes from boxes to matchsticks and plywood. It is highly resistant to termite attack and so is recommended for ceiling boards. Pitomba would thus make a good dual-purpose plantation tree on floodplains for timber and fruit production. The bark has medicinal properties; some rural folk prepare an infusion to treat lice, itching, and to promote healing (Ribeiro et al. 2017).

References

Cronquist, Arthur. 1944. Studies in the Simaroubaceae--II. The genus Simarouba. Bulletin of the Torrey Botanical Club 71(3): 226–234

Ducke, A. 1925. Plantes nouvelles ou peu connues de la région amazonienne (III partie). Archivos do Jardim Botanico do Rio de Janeiro 4:1–341.

Ferreira, André Luis de Souza; Pasa, Maria Corette. 2015. Aula de campo como metodologia de ensino em ecologia de florestas, Chapada das Guimarães-MT, Brasil. Biodiversidade 14(1): 49–62

Furley, P. A.; Ratter, J. A.; Gifford, D. R. 1978. Observations on the vegetation of eastern Mato Grosso, Brazil. III. The woody vegetation and soils of the Morro de Fumaça, Torixoreu. Proceedings of the Royal Society of London, B 203:191–208

Haidar, Ricardo Flores; Fagg, Jeanine Maria Felfili; Pinto, José Roberto Rodrigues; Dias, Ricardo Ribeiro; Damasco, Gabriel; Silva, Lucasde Carvalho Ramos; Fagg, Christopher William. 2013. Florestas estacionais e áreas de ecótono no estado do Tocantins, Brasil: parâmetros estruturais, classificação das fitofisionomias florestais e subsídios para conservação. Acta Amazonica 43(3):261–290

Honda, E. M. S. 1974. Contribuição ao conhecimento da biologia de peixes do Amazonas II. Alimentação de tambaqui, *Colossoma bidens* (Spix). *Acta Amazonica* 4(2):47–53

Lopes, D. R. S.; Tavares, R. C.; Batista, K. O. M.; Souza, P. B.; Nascimento, M. O.; Souza, D. J. 2018. *Simarouba versicolor* (Simaroubaceae) dispersal by the leaf-cutter ant *Atta sexdens*. Sociobiology 65(2):337–339

Lucas, C. M. 2008. Within flood season variation in fruit consumption and seed dispersal by two characin fishes of the Amazon. *Biotropica* 40(5): 581–589.

Matos, Vanda Fernandes de; Lolis, Solange de Fátima; Coimbra, Ronaldo Rodrigues; Viana, Rodney Haulien Oliveira. 2020. Phytosociology in areas of the Brazilian cerrado: cerrado stricto sensu, municipality of Porto Nacional, state of Tocantins. Acta Scientiarum Biological Sciences 42: e50353

Medeiros, Marcelo Brilhante de; Walter, Bruno Machado Teles. 2012. Composição e estrutura de comunidades arbóreas de cerrado stricto sensu no norte do Tocantins e sul do Maranhão. Revista Árvore 36(4): 673–683

Mews, Henrique Augusto; Marimon, Beatriz Schwantes; Maracahipes, Leandro; Franczak, Daniel David; Mamonon Junior, Ben Hur. 2011. Dinâmica da comunidade lenhosa de um cerrado típico na região Nordeste do Estado de Mato Grosso, Brasil. Biota Neotropica 11(1):1–10

Miguel, Eder Pereira; Rezende, Alba Valéria; Leal, Fabrício Assis; Pereira, Reginaldo Sérgio; Melo, Rafael Rodolfo de. 2016. Floristic-structural characterization and successional group of tree species in the cerrado biome of Tocantins state, Brazil. Revista Caatinga 29(2):393–404

Oliveira Filho, A. T.; Martins, F. R. 1991. A comparative study of five cerrado areas in southern Mato Grosso, Brazil. Edinburgh Journal of Botany 48(3):307–332

Parolin, P., D. Waldhoff, and M. T. Piedade. 2010. Fruit and seed chemistry, biomass and dispersal. In: *Amazonian Floodplain Forests: Ecophysiology, Biodiversity and Sustainable Management*. W. J. Junk, M. T. Piedade, F. Wittman, and J. Schöngart (eds), Springer, Dordrecht, pp. 243–258

Pereira, Firmino Cardos; Lenza, Eddie; Maracahipes-Santos, Leonardo; Mews, Henrique Augusto; Gomes, Letícia; Lima, Stefany Lorranyny; Santos, Kelyane Silva Melo. 2015. Comparação dos métodos de parcelas e pontos-quadrantes para descrever uma comunidade lenhosa de cerrado típico. Biotemas 28(2): 61–72

Pinto, José Roberto Rodrigues; Mews, Henrique Augusto; Jancoski, Halina Soares; Marimon, Beatriz Schwantes; Bomfim, Bárbara de Oliveira. 2014. Woody vegetation in a floodplain campo de murundus in central Brazil. Acta Botanica Brasilica 28(4): 519–526

Ribeiro, Reginaldo Vicente; Bieski, Isanete Geraldini Costa; Balogun, Sikiru Olaitan; Martins, Domingos Tabajarade Oliveira. 2017. Ethnobotanical study of medicinal plants used by ribeirinhos in the north Araguaia microregion, Mato Grosso, Brazil. Journal of Ethnopharmacology 205: 69–102

Silva, J. M., M. Pereira-Filho, and M. I. Oliveira-Pereira. 2003. Valor nutricional e energético de espécies vegetais importantes na alimentação do tambaqui. *Acta Amazonica* 33(4):687–700
Smith, N. J. H. 2002. *Amazon Sweet Sea: Land, Life and Water at the River's Mouth*. University of Texas Press, Austin.

Chapter 48
Solanaceae

Check for updates

48.1 *Brunfelsia guianensis*

Brazil Manacá, manacá de veado, jeratacá

Status Wild

N. Smith, *Amazon Fruits: An Ethnobotanical Journey*, Ethnobiology,
https://doi.org/10.1007/978-3-031-12803-5_48

Manacá de veado *Brunfelsia guianensis* in successionary vegetation. Joanderson 10 years. Jocojó, Rio Jocojó, Município de Gurupá, Pará, Brazil 10-22-16

Although people do not eat the fruits of manacá de veado *Brunfelsia guianensis* they are consumed by brocket deer *Mazama americana* and *M. gouazoubira*. The former has a reddish-brown coat whereas the latter isgray; both are widely hunted for their meat. Manacá de veado is a shrub generally under 4 m tall in successionary vegetation but it also occurs in forests so hunters can erect platforms to wait at night for the deer to come and eat. This fruit thus contributes indirectly to the diet of rural people. *Brunfelsia guianensis* is found in eastern Amazonia and the Guianas. The roots are used in preparing various folk remedies. The roots are used to make a laxative and to treat syphilis. Root scrapings are mixed with water and the solution is drunk twice a day to alleviate rheumatism (Le Cointe 1922:508).

48.2 *Physalis angulata*

Bolivia Toatillo, motojobobo embolsado; chichiruru (Aymara), camambú (Chiriguano), cama apu (Guarayo), cumamu (Paucerna), teini tumati (Tacana)

Brazil Camapu, canapu, camaru, camambu, juá, juá póca, juá de capote, tomate de capote, balãozinho, balão rajado, joá de balão, joá de capote, poca, bate testa, saco de bode, mata fome, bucho de rã, mullaca, timbó; kanapũ (Araweté), doopo iattite (Baniwa), kamamu (Ka'apor), comopó (Sateré-Maué), kamamuhu (Tembé), pikasoró (Tukano)

Colombia Girasol de restinga, ochuva, uchuva, vejigón, mullaca, tope torope, topotoropo, uva, uvilla, huilla, ajenje, pap gun

Ecuador Chuanmas, uvilla; saca cucuna (Kichwa), siri bia (Siona-Secoya)

Peru Mullaca, mullaca blanco, mullaca lisa, bolsa mullaca, yutuimas, capulí cimarrón; chuagmas, chuagmis, chuanmis, chuwagmas, cuagkatemc, yutuimas, yutagmis, kuanka teme (Aguaruna), wamashiki, jampashiki (Campa), awadan te ban (Cashinahua), mu'kara (Chayahuita), shinin (Kakataibo), mullaca (Kichwa), magoki (Matsigenka), muyáka mucháka, yuránkmis (Mayna), shimon (Shipibo)

Venezuela Tope-tope, topotopo

Status Wild, spontaneous in cultural settings, planted

Camapu *Physalis angulata* on an archaeological site with anthropogenic black earth. SCA-6, Palmeiras, left bank of the Xingu River upstream from Altamira, Pará, Brazil 9-25-15

The fruits of two species of *Physalis* are gathered in the Amazon: *Physalis angulata* and *P. pubescens*. The former is more common and can be distinguished by its larger leaves and the reddish-brown veins on the green "lantern" or husk (calyx) that envelops the fruit. Known as mullaca blanco in Peru and camapu in Brazil, the calyx covering the fruit is ten-angled (Smith et al. 2007:238). *Physalis pubescens* has smaller leaves, an entirely green husk, and tarter fruits. The calyx of *Physalis pubescens* has five angles and the plant is covered with numerous fine hairs, hence its species name. Both plants range widely from Argentina to the southern United States, but *Physalis angulata* is much more common in the Amazon. In upland locations *Physalis angulata* fruits in both the rainy and dry seasons whereas on floodplains it produces fruit during the low water period.

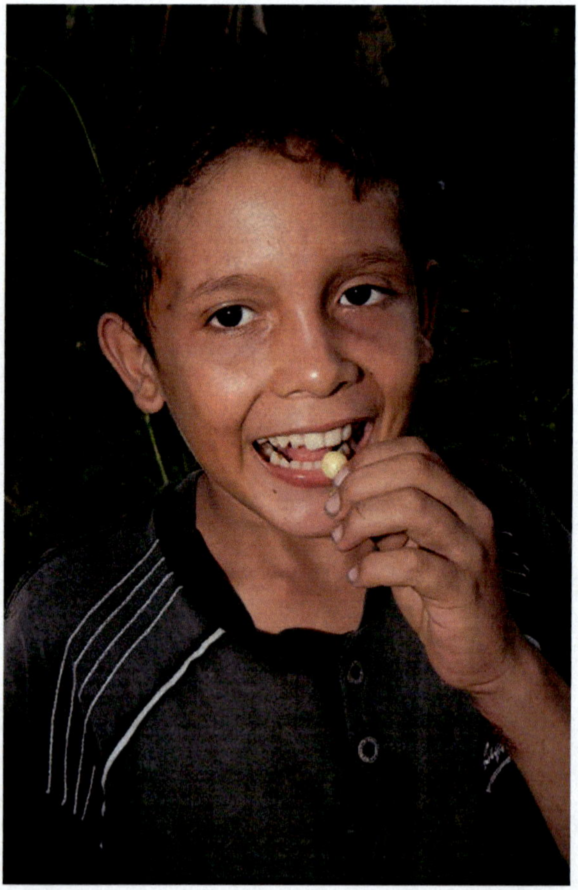

Edivan, 11 years, eating camapu *Physalis angulata* fruit with the husk (calyx) removed. He gathered the fruit along a path leading to his family's field on the Amazon floodplain. Ilha do Risco, Rio Amazonas near Itacoatiara, Amazonas, Brazil 12-8-06

Physalis angulata is an annual herb under 1.5 m high that grows in disturbed sites and on open alluvial areas exposed as water levels drop as well as in successionary vegetation in other areas, such as along sunlit paths on both uplands and floodplains. This species of *Physalis* is thought to have arisen in Amazonia (Silva 2011:116). It is found as a weed in cultivated fields, such as in some guaraná *Paullinia cupana* plantations in Amazonas, Brazil (Miléo et al. 2007), various crops in Roraima (Cruz et al. 2009), and some cattle pastures such as in the vicinity of Santana do Araguaia in southern Pará (Dantas and Rodrigues 1980). It also occurs in some home gardens, such as in the vicinity of Macapá, the capital of Amapá (Pereira et al. 2017). It is also a weed in some urban areas, such as Cuiabá in Mato Grosso (Pedrotti and Guarim Neto 1998). Seeds of this annual plant have been found in the soil of cattle pastures in the vicinity of Marabá along the Tocantins River in Pará (Miranda et al. 2009).

This fast-growing shrub often forms dense patches containing hundreds of individual plants, a real bonanza for fruit gatherers. The bulbous and slightly tapered calyx, measuring 2–3.5 cm long and 1.3–2.5 cm wide, is pulled off the fruits before eating. The round, shiny fruits measuring 1.3 cm by 1 cm contain numerous light brown seeds which are ingested along with the mesocarp. In the Peruvian Amazon, the moustached tamarin *Saguinus mystax* and the white-lipped tamarin *Saguinus labiatus* are known to disperse the seeds of this plant (Ique and Aquino 2015; Reátegui 2013:37).

The yellow-green fruits are popular because of their unusual flavor that combines sweetness with overtones of tartness. *Physalis angulata* is especially popular with kids who pop them in their mouths at every opportunity. Children can easily gather the fruits from the waist-high plants. Children of indigenous groups are also fond of the fruits, such as the Waorani in the Ecuadorian Amazon (Davis and Yost 1983). The fruits are also made into a dessert in some areas; in Macapá at the mouth of the Amazon, for example, some residents prepare a dish called *doce de camapu*. However, street markets for fruits and vegetables are becoming increasingly rare in the Brazilian Amazon as mayors prohibit the use of promenades for such purposes and more and more shoppers purchase fruits and vegetables, often from distant regions, in stores and supermarkets.

Camapu fruits mainly during the dry season, especially on floodplains when water levels have dropped thus allowing the plants to proliferate on the fertile alluvial soils. In some well-drained sites, *Physalis angulata* bear ripe fruits during the rainy months. The fruits are occasionally taken to market in both the Peruvian Amazon, such as Iquitos and Nauta, as well as in the Brazilian portion of the basin, including Macapá at the mouth of the Amazon (Smith 1999:63; Vásquez and Gentry 1989). In November 2016 at Gurupá along the lower Amazon camapu fruits with calyx were selling for the equivalent of US 30 cents per liter while fruits with the "lantern" removed were selling for twice as much.

Vendors peeling the calyx off mullaca *Physalis angulata* fruits in a street market. Mercado Belén, Iquitos, Peru 1-7-06

Mullaca is a bane or a blessing, depending on one's perspectives. For kids, it is a welcome snack, but for farmers, mullaca can become so prevalent in their fields that it must be weeded. In the vicinity of San Pedro along the lower Marañón, for example, maize growers sometimes hire workers to clear out mullaca in their chacras because they compete so vigorously for light and nutrients. On the other hand, the Cofán in the northern part of the Ecuadorian Amazon plant this annual because of its much-appreciated fruits (Cerón et al. 1994:68).

Physalis angulata is now pantropical (Nee 1991) and has even spread to the Mediterranean where it has recently turned up in fields of cotton, soybean, and maize in southern Turkey (Gönen et al. 2000). In areas where *Physalis angulata* has been introduced it is generally considered a weed rather than a resource.

Girls snacking on mullaca *Physalis angulata* fruits they gathered in the perimeter of their village. Mishana, Rio Nanay, Loreto, Peru 1-8-06

Physalis angulata is not cultivated, although it may have been in the past in some areas and then reverted to its wild form as many indigenous cultures were devastated by introduced diseases and slave raids during the colonial period. Andean peoples domesticated *Physalis peruviana* long before the Spanish arrived. Known as capulí and aguaymanto in the Peruvian highlands, *Physalis peruviana* has become a cash crop in several parts of the world where it is known as goldenberry or Cape gooseberry (Bernal 1991; Tindall 1983:359). Another example of *Physalis* domestication comes from Mexico where indigenous folk brought *Physalis philadelphica* into cultivation a long time ago judging by the large fruit size. Known as tomate, tomato verde, or tomatillo, this relative of the tomato is an important part of Mexico's cuisine, arguably more important even than the tomato itself (Montes and Aguirre 1994). Known as the husk tomato in English, the fruits are frequently found in supermarkets in the United States. People in North America eat tomatillo when they dip corn chips into some salsas. One day, *Physalis angulata* may emerge as a crop in its own right. Mullaca/camapu serves as a reminder that the next generation of crops to come out of the Amazon is just as likely to come from disturbed habitats as mature forests. The Araweté and Tembé in Pará use the fruit for fish bait as do rural folk along the Jauaperi River that separates the states of Roraima and Amazonas in the northern Brazilian Amazon to catch pacu (various species and genera; Pedrollo 2013:79).

Physalis angulata has other uses in the Amazon, especially in folk remedies. A tea is made from the roots to treat liver problems and malaria (Stasi et al. 1989;

Tomchinsky et al. 2017). Along the lower Urubu and its affluents, such as the Aneba, rural people use the roots to make a tea to treat headaches, stomachaches, and kidney stones. The roots are also used to treat malaria, such as in quilombola communities in the Trombetas watershed in Pará (Oliveira et al. 2015). Rural folk living in Quilombo Sangrador in the municipality of Presidente Juscelino, Maranhão, use the roots of this plant to make an infusion for treating breathing problems (Monteles and Pinheiro 2007). In the watershed of the Jaú River, an affluent of the middle Rio Negro, river dwellers prepare a decoction from the roots for treating inflammation (Rodrigues et al. 2010). An infusion prepared from the roots is also taken to treat bladder problems, as indicated when blood is passed in the urine. In Maranhão, on the eastern fringes of Amazonia, a tea is made from the roots to treat appendicitis (Van den Berg and da Silva 1986b). In Amapá, a decoction is made from the roots and leaves to treat cystitis, and sap from the plant is used as ear drops to alleviate otitis, an inflammation of the middle ear that triggers pain (Tenório et al. 1991). At Marudá, a village along the coast of Pará at the mouth of the Amazon, the roots are used to make a tea to treat hepatitis, anemia, bladder infections, stomachache, and kidney stones. The roots are also used to make an infusion to treat vaginal discharge (Coelho-Ferreira 2009). Along the Upper Rio Negro, rural folk use the whole plant to make a bathing solution which is washed over the body at the onset of malaria symptoms (Kffuri et al. 2016). Juice from the fruit is used to treat earaches while a tea prepared from the roots is drunk to alleviate rheumatism and liver problems (Le Cointe 1922:507).

This annual plant is also used in folk remedies in other parts of the Amazon Basin. The Mayna in the Peruvian Amazon crush the leaves and squeeze the juice out for use as a hair conditioner and to treat scalp fungal infections and the widely dispersed Kichwa in the Andean foothills make a decoction from the leaves for treating measles. In the Maynas province in the Peruvian Amazon, the sap and leaves are used to make a tea to treat post-partum infections while the fruit is macerated and applied to itchy skin (Jovel et al. 1996). In the Loreto province of the Peruvian Amazon the roots and some aerial parts of this plant as used to prepare a decoction to treat hepatitis (Roumy et al. 2020). In the Chazuta district along the Huallaga River in the foothills of the Andes in the Peruvian Amazon, the entire plant is sometimes used to make an infusion to treat malaria as well as coughs and bronchitis (Lossio et al. 1995:69; Sanz-Biset and Cañigueral 2011; Sanz-Biset et al. 2009). The Shipibo in the vicinity of Pucallpa along the Ucayali in the Peruvian Amazon prepare a poultice or fresh juice from the fruits of this plant to treat fungal infection and wounds (Polesna et al. 2011). An infusion is prepared from the leaves and a cup of this is drunk three times a day to serve as a diuretic in various parts of the Peruvian Amazon (Mejia and Rengifo 2000:46). The roots are mashed and placed in a bottle of sugarcane alcohol (*aguardiente*) for several days then honey is added and half a glass is drunk twice a day for 60 days to treat diabetes (Lossio et al. 1995:69). Leaves and fruits are cooked then rubbed on the skin to reduce inflammation and treat infections; the fresh leaves are also squeezed and the liquid applied directly to treat the same conditions. To treat wounds, unripe fruits are crushed and the pulp is applied to damaged areas for 7 days. In the Bolivian Amazon, the Tacana squeeze the whole plant into 1 l of water and then one cup of this solution is drunk per day to treat malaria and fever; two of the fruits are also squeezed in water to bathe infected skin (Arévalo-Lopéz et al. 2018). The Tacana wash1 kg of the roots of this plant then mash and boil them in 4 l of water for 20 minutes. This cooled liquid is then used to treat malaria (Ponz et al. 2005:92). This plant is also used in folk medicine in some areas where it has been introduced such as the tropics and subtropics of Asia where the plant is used to treat rheumatism, sore throat and gastrointestinal pain (Chiang et al. 1992).

48.3 *Physalis pubescens*

Bolivia Tomatillo; bojñota (Tsimane)

Brazil Camapu, balão, balãozinho, bucho de rã, joá de capote, tomate de capote, juá, juapoca, pimentaria, mata fome, bate testa, fisális, pipoco, saco de bode

Ecuador Uvilla, naranjilla de pájaro nocturno; tovija tton'tto (Cofán), yuranmis (Huambisa), tsitsin-cucuna (Kichwa), yurankmis (Shuar), huecamo (Waorani)

Peru Mullaca lanudo, bolsa mullaca, capulí; yutuimas, yuragmis, utuimas (Aguaruna), shemot-huare'shemot (Amuesha)

Venezuela Saposarapo, topotopo

Status Wild, spontaneous in cultural settings, planted

Modesto, 12 years, holding mullaca lanudo *Physalis pubescens* **(on right) and mullaca blanco** *Physalis angulata* **(on left with red veins on calyx). They were picked in a floodplain field and**

the fruits will be eaten. Opposite Veinte de Enero, Rio Yanayacu, Reserva Nacional Pacaya-Samiria, Loreto, Peru 3-12-04

Known as mullaca lanudo along the Yanayacu, an affluent of the lower Marañón in the Peruvian Amazon, *Physalis pubescens* is an herbaceous plant that grows on floodplains (Kalliola et al. 1991a, b). It also occurs as a weed in floodplain fields. The fruits are gathered and generally eaten on the spot by farmers and their children including indigenous groups such as the Wajãpi in northern Amapá and contiguous French Guiana (Grenand 1992) and the Kichwa in the Ecuadorian Amazon. The Cofán in the northern part of the Ecuadorian Amazon cultivate this annual in their manioc fields (Cerón et al. 1994:68). The fruits are nutritious because they have a high carotenoid content as well as being rich in phosphorus, potassium, zinc, and boron (El Sheikha et al. 2010).

Physalis pubescens has smaller leaves than its near relative *Physalis angulata*, and the husk is entirely green, unlike the red-veined calyx of *Physalis angulata*. Furthermore, the fruits are more acidic than those of *Physalis angulata*. The calyx of *Physalis pubescens* has five angles, and the plant is covered with numerous fine hairs, hence its species name. Both species often grow together, but *Physalis angulata* is much more common than *Physalis pubescens*. Both species originated in tropical America and are widespread there (Nee 1991). In the Andean foothills *Physalis pubescens* can be found as high as 2600 m. This species has become naturalized in disturbed habitats throughout much of Brazil (Lorenzi et al. 2006:312).

Physalis pubescens is also used in folk remedies. In the Bolivian Amazon, for example, the roots are used to make a decoction to treat back pain. In the Ecuadorian Amazon, the Shuar use the leaves to make a tea to treat diarrhea and they believe that eating the fruits helps one succeed at work and prevents illnesses (Bennett et al. 2002:265). In the foothills of the Andes in the central part of the Peruvian Amazon, the Amuesha mix the roots of this annual with those of one of the açaí palms, *Euterpe precatoria*, to prepare a concentrated decoction; women with ovary pain drink this infusion three times a day until they recover (Valadeau et al. 2010). In the Brazilian Amazon, rural folk use the sap to treat earache while the leaves are used to prepare a diuretic and to treat bladder inflammation and jaundice (Le Cointe 1934:83).

48.4 *Solanum coriaceum*

Brazil Rosa brava

Status Wild

Rosa brava *Solanum coriaceum* fruits gathered in floodplain forest along the Rio Inajá, a black water affluent of the lower Xingu. Iasmi, 5 years. Carrazedo, mouth of the Xingu River, Município de Gurupá, Pará, Brazil 10-26-16

One of the common names for this climbing shrub in the Brazilian Amazon is rosa brava (wild rose) on account of the cat claw-like spines on the stems. This vine-like bush, which can reach 5 m high, has sharp downward pointing spines which allow it to clamber on other vegetation. It grows along the edges of floodplain forests where it receives direct sunlight. It also occurs on uplands in successionary vegetation or forest edges bordering roads or trails and in stunted woodlands growing on iron-rich sedimentary rock (*canga*) on plateaus of the Serra dos Carajás in Pará (Giacomin and Gomes 2018; Mota et al. 2018). The fruits are orange when ripe, about the size of small cherry tomatoes. They are eaten by locals, such as along the lower Xingu and its affluents, but never appear in urban markets. This species ranges widely in the Brazilian Amazon into Venezuela and the Guianas.

48.5 *Solanum sessiliflorum*

Bolivia Cocona

Brazil Cubiu, cubiu manso, cubiu pequeno liso, maná cubiu, cubiuzeiro, cubiuzinho sem espinho, topiro, tomate de índio; kopiro (Baniwa), bere (Ticuna), extoá, ehtoa poá (Tukano)

Colombia Tupiro, lulo de árbol, lulo amazónico, cocona, naranjilla; falesú, fáresu (Andoke), mãrã (Makuna)

Ecuador Cocona, naranjilla de monte, naranjilla silvestre, sacha naranjilla, huevo de perro; ccujapacco tton'tton, tton'tto rande (Cofán), cocona, awas laran (Kichwa), kukúch (Shuar), kukuna, miucucuna (Siona-Secoya) dabuka, daboca (Waorani)

Peru Cocona, cocona grande, coconilla, lulo, lulo chocoano, kukush, topiro, tupiro, chope, tomate de índio; kishan kukus, kistian kukus, nantú kukúsh, siwankus, cucum, kunkun, kukush (Aguaruna), roolláhe (Bora), poetoki, popooki, tokona (Campa), uhkuna'pi (Chayahuita), popol (Ese Eja), pupú (Kakataibo), kokona (Matsigenka), popo (Shipibo), kukuna (Siona-Secoya), bere (Ticuna)

Venezuela Tupiro; hutukurimɨ mɨsɨkɨ (Yanomami)

Status Wild, spontaneous in cultural settings, planted

Spontaneous cubiu *Solanum sessiliflorum* in a manioc and banana field, evidently dispersed by birds. Santa Luzia, Lago Amanã, lower Japurá River, Amazonas, Brazil 9-3-12

Solanum sessiliflorum is planted widely in the Amazon and the Upper Orinoco below 1400 m for its fruits which are eaten fresh, made into juice and used in the preparation of sauces. Known as cocona in Peru and cubiu in Brazil this herbaceous shrub can attain 3 m in height when fully mature. It is particularly common in the Peruvian part of the basin. Equally, at home on uplands or floodplains, just about every settlement in the Peruvian Amazon, especially along the Ucayali and Marañón Rivers, has cocona bushes planted around the houses or in fields, often both. The Achuar along the Corrientes River, a tributary of the Tigre, also cultivate *Solanum sessiliflorum* in their home gardens. Cubiu also occurs in home gardens in various parts of the Brazilian Amazon, such as in the vicinity of Manaus and Parintins in Amazonas (Lourenço et al. 2009). One advantage of this relative of the tomato is that it thrives in nutrient-poor, acidic soils (Bueno et al. 2000). At least six species of bees pollinate the flowers (Storti 1988), and the *Mechanitis polymnia* butterfly in

Loreto province in the Peruvian Amazon also likely pollinates these flowers (Bardales et al. 2017).

Cubiu *Solanum sessiliflorum* in a home garden. São João do Mulato, Lago de Tefé, Amazonas, Brazil 8-29-12

Elsewhere in the Amazon, however, the cultivation of the peach tomato, as the plant is known in English, has declined following the introduction of tomato *Lycopersicon esculenta* and lime *Citrus aurantifolia* by Europeans because they have similar uses and are sometimes considered more prestigious (Silva Filho et al. 1993). Cocona has held its own in the Peruvian Amazon in large part because the penetration of European ethnic groups and culture has been less pronounced. *Solanum sessiliflorum* is the lowland equivalent of *Solanum quitoense* which is cultivated extensively to make juice in the Andes. In Peru, *Solanum quitoense* is known as Quito-Quito or naranjito Quito. In Ecuador, *Solanum quitoense* is called naranjilla whereas in Colombia it is referred to as lulo.

The precise origins of *Solanum sessiliflorum* are uncertain. One possibility is that the shrub was domesticated in the western part of the Amazon basin in the Andean foothills (Léon 1987:171; Villachica et al. 1996:99). Cocona grows wild in light gaps along the banks of the Yanayacu River an affluent of the lower Marañón; the wild plants can be easily distinguished from the cultivated forms because the former have spines along the stems and petioles. Also, wild *Solanum sessiliflorum* always produces orange fruits whereas some domesticated forms provide yellow fruits (Heiser 1992:65). Alternatively, *Solanum sessiliflorum* could have been domesticated in the upper Rio Negro watershed in Brazil, Colombia, and Venezuela where the plant is known as tupiro and then spread to the western part of the Amazon. Indigenous groups in the Upper Orinoco Basin cultivate *Solanum sessiliflorum* and the Guajibo assert that the spiney forms have better-tasting fruits (Volpato et al. 2004). Another possibility is that cubiu may have domesticated in several parts of its range. Given its popularity in the western part of the basin, however, Peru or Ecuador seems to be the more likely the origin of this crop. *Solanum sessiliflorum* sometimes arises spontaneously in manioc fields, evidently dispersed by birds, as around Lago Amanã along the lower Japurá; it is possible that after weeding it for some time, one or more indigenous groups decided to leave it alone because of its useful fruits and then eventually plant it as a crop.

Cocona arises spontaneously in disturbed sites such as along trails and in home gardens and fields where it is tolerated because of its edible fruits (Volpato et al. 2004). Cattle spread the seeds in pastures after digesting the fruits. Cocona is also dispersed in human feces; indeed, germination rates may be enhanced after passing through the digestive tract (Salick 1992).

Orlando, 10 years, picking cocona *Solanum sessiliflorum* planted in a field along with other crops including manioc. Yarina, Yanayacu River, Reserva Nacional Pacaya-Samiria, Loreto, Peru 7-24-10

Fruits vary in color from yellow to orange or red. Fruit size also varies from 3 to 6 cm long and 4 to 12 cm in diameter; some are round, others more flattened like a persimmon, whereas others are cylindrical (Silva et al. 1996; Silva Filho et al. 1989). Cocona fruits normally weigh between 25 and 250 g but can attain 400 g (Marx et al. 1998; Pahlen 1977). A great deal of variation in color and size of the fruit can sometimes be found in a single field, such as in the vicinity of Yarina along the Yanayacu River in the Reserva Nacional Pacaya-Samiria, indicating a great deal of genetic variation in the crop. No private or public seed companies have released "improved" varieties of *Solanum sessiliflorum* which has helped preserve the crop's rich gene pool in situ. Another reason that cocona's gene pool is relatively rich is that bees transfer pollen between cultivated plants and wild populations in some areas (Salick 1992).

Marianne, 6 years, with a cubiu *Solanum sessiliflorum* fruit she picked in her home garden. Comunidade Terra Preta, Rio Cariá, a left bank affluent of the lower Rio Pracupi, in turn an affluent of the Anapu River, Floresta Nacional Caxiuanã, Pará, Brazil 5-17-15

Cocona starts fruiting about 7 months after sprouting and produces for about 3 years. A bush can provide up to 14 kg of fruit a year depending on soil conditions and fruit size (Marx et al. 1998). Cocona is normally planted in patches, so yields on a per hectare basis have to be extrapolated. However, if planted on a larger scale yields are likely to be in the 30–60 tons/ha range, or even as high as 100 tons/ha (Silva Filho et al. 1999).

The fruits are eaten fresh after first rubbing off the short, fine bristles from the skin. While in the field or home garden, the large almost leathery leaves of the herb are used to remove the prickly hairs. Cocona fruits are full of juice and are eaten with their skin and a pinch of salt. Cocona flesh is pale yellow with lots of light-yellow seeds, the size of tomato seeds. Unlike tomato seeds, however, cocona seeds are not surrounded by a jelly-like substance. Cocona seeds are ingested along with the tart pulp. In the Ecuadorian Amazon, the Siona-Secoya cook the fruits to prepare a beverage (Vickers and Plowman 1984:31) while the Cofán use the fresh fruits to make juice (Cerón et al. 1994:69).

As in the case of tomatoes, which are also in the potato family and native to the New World, cocona fruits contain appreciable amounts of vitamin C (Salick 1992). One nutritional study of *Solanum sessiliflorum* fruits from the Amazon estuary suggests a relatively high phosphorus content of 383 mg/100 g (Marx et al. 1998). The fruits are also rich in iron and niacin (Silva Filho et al. 1993).

The fruits are also used to make juice in town and country. In Iquitos, street stalls sell the juice, called *cocona licuado*, in small plastic bags to take home or for drinking on the spot. Frutas del Bosque, a small juice factory in Iquitos, markets cocona juice in 250 ml bottles which are sold locally in stores. In San Ramón, a bustling town at 900 m in the Peruvian Andes, the Indalsa Company cans small quantities of cocona juice for sale in Lima and Italy. The Chayahuita in the Peruvian Amazon squeeze juice from fresh fruits onto the head to get rid of lice (Odonne et al. 2013).

Vendor drinking cocona *Solanum sessiliflorum* juice in a street market. Mercado Belen, Iquitos, Peru 5-9-03

In Peru, the fruits are also used to make *coconachado*, an inexpensive alcoholic beverage. Rural and urban folk prepare *coconachado* by mashing boiled cocona fruits by hand. The pulp is then sieved and sugar and *aguardiente* are added to the warm juice. The resulting alcoholic concoction is then stored in recycled glass bottles for sale or domestic consumption. In Moyobamba in the Mayo valley, the Recreo Licoreria Piscis liquor store sells *aguardiente* brews flavored with several native fruits, including cocona.

One small fruit processing factory in Iquitos, Persa, makes a cocona jam which is sold at its factory store as well as in supermarkets in town. Orquidea, a chocolate factory in Tarapoto started by a cooperative of cacao growers in 1997, also makes small quantities of cocona jam. Jams and marmalades are not particularly popular foods in the Peruvian Amazon, so only a tiny fraction of fruit production ends up as preserves.

Cocona *Solanum sessiliflorum* jam made by the PERSA company. Jack, 10 years, Iquitos, Loreto, Peru 8-2-03

One independent ice cream store by the Plaza de Armas, the main square of Iquitos, sells cocona ice cream, but as in the case of jam, ice cream accounts for a relatively insignificant portion of cocona consumption. Cocona ice cream is sold in other towns in the Peruvian Amazon, such as at Heladeria Alpamayo, one block from the Plaza de Armas in Yurimaguas.

Cocona fruits are also used in cooking. Along the Rio Negro in Brazil, for example, cocona fruits are squeezed on certain fish prior to cooking to temper their oily taste. Limes are also used for the same purpose throughout the Brazilian Amazon. In Iquitos and the vicinity, cocona juice is mixed with cilantro *Coriandrum sativum*, an exotic herb introduced from southern Europe and the Mediterranean to flavor a variety of dishes, including fish. A cocona salsa is served with boiled and peeled fruits of the pijuayo *Bactris gasipaes* palm at certain street stalls along Prospero, the main shopping avenue in Iquitos. Peach palm fruits are cut in half, the stones removed, and the salsa packed into the vacated cavities. A peach palm fruit smothered with cocona salsa sells for the equivalent of US $ 0.15 each, a healthy and affordable snack.

Solanum sessiliflorum fruits are also used to prepare desserts. In São João do Mulato, a small village along the lower Tefé River, women prepare a dessert by cooking the fruits and adding sugar (*doce de cubiu*). In Peru, cookies called Alfajor de Chocolate y Cocona contain layers of chocolate and cocona fruit paste (Gutsche 2008).

Cocona fruits find a ready market in urban centers throughout the Peruvian Amazon and more sporadically in Brazil, such as Portel in the Amazon estuary and Rio Branco in Acre (Lima 2014). Fruit production in the Amazon Basin south of the equator peaks in May at the end of the rainy season, but some plants with ripe fruits are encountered throughout the year. At the height of the season, the fruits are relatively cheap; in Jenaro Herrera along the Ucayali, for example, shoppers could buy 120 cocona fruits for the equivalent of US $ 1 in May 2003.

Cocona fruits also reach extra-regional markets, including supermarkets in Lima where they sell for about US $ 0.75/kg, several times more than in the Amazon, a reflection of the long-distance transportation costs. Cocona fruits are trucked to Lima from Tingo Maria or Pucallpa, a 3- to 4-day journey depending on road and weather conditions. From Lima, fruit wholesalers dispatch cocona fruits that have arrived from Pucallpa or Tingo Maria to various cities along the desert coast of Peru, including Chiclayo. The further the fruits have to travel, the higher the price. In Chiclayo in January 2005, for example, cocona fruits from Tingo Maria were selling for US $ 0.90/kg, about 20% more than in Lima because Chiclayo is about 700 km north of Lima. A major advantage of cocona for extra-regional markets is that the fruits remain fresh for as long as 40 days after harvest and the skin does not break easily.

Karina, 19 years, selling cocona *Solanum sessiliflorum* fruits. The fruits were trucked from Tingo Maria in the Peruvian Amazon. Mercado Modelo, Chiclayo, Peru 1-21-05

Solanum sessiliflorum is also considered a medicinal plant. Indeed, in May 2000 in Portel, Pará, cubiu fruits were found in a stall selling medicinal plants rather than in the area where fruits and vegetables are sold. In the Amazon estuary, the fruits are used in a body wash to heal lesions. In the vicinity of Tefé in Amazonas, Brazil, cubiu is reportedly useful to lower blood pressure and high levels of cholesterol. Some farmers near Manaus have increased plantings of cubiu in response to demand from Japanese buyers to treat high levels of cholesterol (Silva Filho et al. 1997). Members of Julião, a rural community in the Reserva de Desenvolvimento Sustentável (RDS) do Tupé some 25 km west of Manaus, eat the fruit to treat anemia (Veiga and Scudeller 2015). In some rural areas, liquid is squeezed from the leaves to treat diabetes (Van den Berg 2010:226; Van den Berg and da Silva 1986a). In Rio Branco, Acre, the leaves and fruit are employed to make tea and juice to treat amoeba infections and diarrhea (Siveiro et al. 2012). The Ticuna along the Solimões use the fruits, which they call bere, to treat itchy skin and kill lice (Lima 2006:151).

This plant is also used in folk remedies in other parts of Amazonia. The Waorani in the Ecuadorian Amazon use this plant to treat stings from stingrays and scorpions, as well as bites from other insects, including spiders (Davis and Yost 1983). The Waorani also treat dandruff by applying the juice from these fruits to their hair and massage the scalp which is then rinsed after about 10 minutes (Mondragón and Smith 1997:46). The Waorani also rub the fruit juice into their hair so that it becomes glossy. The juice is also drunk to prevent vomiting after scorpion stings and, after boiling, the juice is rubbed onto necrotic tissue resulting from spider bites (Schultes and Raffauf 1990:440).

The Shipibo who live in various villages in the vicinity of Pucallpa along the Ucayali River in the Peruvian Amazon use the fresh juice made from the fruits of this plant to treat burns and wounds and a decoction is drunk to ameliorate diabetes, diarrhea, neurosis, and high blood pressure (Polesna et al. 2011). In the Suni Miraño community in Maynas province in the Peruvian Amazon people use the fresh fruits and leaves which they apply to the skin to treat burns, tumors, and infections (Jovel et al. 1996). In the vicinity of Iquitos, both urban and rural folk wash their hair with a cocona solution made at home in order to invigorate the scalp. The juice of cocona is reputed to alleviate intestinal disorders. Throughout the Peruvian Amazon, rural people crush the leaves and apply the liquid to skin burns, while the juice expressed from the fruits is applied to infected skin (Mejia and Rengifo 2000:78; Tananta 2014:72). To treat diabetes, the juice of 12 fruits is squeezed and one cup of this liquid is drunk with meals or every 4 hours (Lossio et al. 1995:107). To treat burns, the juice is squeezed from two leaves and applied to the burns along with the squeezed leaves (Lossio et al. 1995:108). In the Colombian Amazon, the Tukano of the Apaporis River make a strong decoction of the leaves of *Solanum sessiliflorum* to purge intestinal worms (Schultes and Raffauf 1991) while rural folk in various parts of the Colombian Amazon use this fruit to treat snake bites, scorpion stings and skin infections (Lizcano et al. 2010).

48.6 *Solanum stramoniifolium*

Bolivia Manzanilla

Brazil Jurubeba, jurubeba roxa, juá, juúna, buieira, cubiu pequeno com espinho, cubiuzinho com espinho; madzawi (Baniwa), wɨra-rači (Guajá), juruwera'ir, yuruwe, yuruwe-piraĩ (Ka'apor), xu'yw, zuruwe-čĩ (Tembé), ehtoa poá (Tukano), yũ-piñũ (Wajãpi), baelium-aha-sik, pirima ahɨsi (Yanomami)

Colombia Lulillo, lulo, lulo pequeño, lulito, cubiu; kobuía (Cubeo), etó-paa (Miraña), e-tó-pa-a, ehtoa poá (Tukano)

Ecuador Cocona, tomate silvestre, naranjilla de monte, naranjilla roja, naranjilla silvestre, vagicho, baguichu, uchu lara; kukúch (Achuar), cu'a tton'tto (Cofán), uchu cucuna, uchu naranja, vagicho, baguichu, baguichu juchu cucuna (Kichwa), terench yaa kukúch (Shuar), toawi'ka (Siona-Secoya), dabokawe, dabuca, guiyad-aboca, meñibo, meñimo (Waorani)

Peru Coconilla, coconilla colorada, coconilla con espinas; yaakúch (Achuar, Mayna), shiwánkush, kukush, yeag kukush (Aguaruna), shemot-tsem po'shemoteñ (Amuesha), kokonaki, tokona (Campa)

Venezuela Boroboro, bú-boro, bobora, tupirito, tupinito, huevo de gato grande

Status Wild, spontaneous in cultural settings, planted

Spontaneous jurubeba *Solanum stramoniifolium* in a pasture, dispersed by cattle or birds. The fruits will turn red when ripe. Opposite Caretas, Rio Urubu, Amazonas, Brazil 10-16-12

Solanum stramoniifolium, known as coconilla (the little cocona) in Peru and jurubeba or juá in Brazil, is another member of the potato family that has been brought into domestication in western Amazonia for its fruits. Coconilla is a sprawling shrub reaching about 2 m when fully mature that bears red berries similar in size and shape to small cherry tomatoes. It is found in successional vegetation and people also cultivate coconilla in their fields and home gardens on both floodplains and uplands in the Peruvian Amazon, particularly in the vicinity of Iquitos. In the Ecuadorian Amazon, the Siona-Secoya and Shuar sow this plant in their home gardens for snack food (Bennett et al. 2002:55; Vickers and Plowman 1984) as do the Cofán in the northern part of the Ecuadorian Amazon (Cerón 1995:144). In the Colombian Amazon, the Miraña do the same (Schultes and Raffauf 1990:441). In the Brazilian Amazon, however, *Solanum stramoniifolium* is rarely cultivated although it occurs in some home gardens in the vicinity of Macapá, the capital of Amapá (Pereira et al. 2017).

Children enjoy snacking on the fruits picked from spontaneous plants in and around manioc fields. It also occurs as an invasive weed in cattle pastures created after clearing forests, such as in the vicinity of Belém and Tomé Açu in eastern Pará (Lacerda et al. 2013; Modesto Júnior and Mascarenhas 2001). This relative of the cultivated tomato also grows in grasslands with open woods along the coast of northeastern Pará (Amaral et al. 2008; Bastos et al. 1995). The wild forms of this species have spines, similar in appearance to rose thorns, as is the case with its close relative *Solanum sessiliflorum*.

Francisco, 7 years, son of a cowboy, picking fruits of juá *Solanum stramoniifolium* in his home garden. Ilha dos Ananás, Mexiana Island, Amazon estuary, Pará, Brazil 5-20-01

Coconilla ranges from the Amazon and Orinoco Basins into the Guianas and the Atlantic forest of Brazil. *Solanum stramoniifolium* has also penetrated the isthmus of Central America but only as far as Panama (Gentry and D'Arcy 1986). Its altitudinal limit is around 1000 m in the Andes. As with its close cousin cocona *Solanum sessiliflorum*, farmers have selected spineless forms as well as genotypes with larger fruits (Wahlen et al. 1981). *Solanum stramoniifolium* is pollinated by various stingless bees (Bezerra and Machado 2003; Ferreira and Absy 2018; Silva et al. 2004). Some stingless bees are raised in suspended logs under the eaves of rural homes to obtain honey, so this plant also contributes indirectly to the rural diet in some areas.

Gislaine, 10 years, a Tuyuka girl rubbing jurubeba *Solanum stramoniifolium* fruits between her palms to remove the bristles. The shrub arose spontaneously in her family's manioc field. Near São Pedro, Rio Tiquié, Amazonas, Brazil 11-8-12

Coconilla fruits toward the end of the rainy season and through the drier months. The berries, some 1.4–1.6 cm in diameter, are sweeter than those of cocona and the skin is less pubescent. For those fruits with bristly hairs, people roll them between their palms or rub them on their clothing to remove them. The fruits of domesticated forms of *Solanum stramoniifolium* in the Peruvian Amazon have fewer hairs. The numerous small seeds of *Solanum stramoniifolium* are ingested along with the juicy pulp. The light brown seeds are dispersed by birds, humans, and other mammals. Five species of birds eat these fruits and disperse the seeds in successionary vegetation in the vicinity of San Carlos de Rio Negro in the Venezuelan Amazon (Uhl et al. 1981).

Coconilla also has medicinal uses. On Centro America, a floodplain island along the Amazon a little upstream from Tamshiyacu in Loreto province, Peru, locals rub the fruits on skin rashes to clear them up. In the Andean foothills in the central Peruvian Amazon, the Amuesha boil the leaves and then apply them as a poultice to treat muscular pain (Valadeau et al. 2010). In the Colombian Amazon, the Miraña apply fruits of *Solanum stramoniifolium* to sore gums to stop the bleeding while the Tukano mix the sap obtained from stems of the plant to relieve painful nipples of nursing mothers (Schultes and Raffauf 1991). The Yanomamo prepare a chewing "tobacco" by rolling the green leaves and sticking the wads in their mouths. Several villages in the municipality of Barcelos along the middle Rio Negro in Amazonas, Brazil, prepare a decoction with the roots of this plant to treat malaria (Tomchinsky et al. 2017). In the Amazon estuary, rural folk make a tea with the roots to treat hepatitis and the fruits are eaten to alleviate anemia and expel intestinal worms (Palheta et al. 2017). In Surinam which borders the Brazilian state of Roraima in the central Amazon Saramaccan maroons, descendants of African slaves, brought to the coast of the Guianas in the colonial period to work on sugarcane plantations, give these fruits to children to eat to treat asthma (Ruysschaert et al. 2009). In Guyana, the leaves of this plant are used to make a tea to cleanse blood (Austin and Bourne 1992).

Marianne, 6 years, eating a jurubeba *Solanum stramoniifolium* fruit she picked at the edge of a manioc field. Comunidade Terra Preta, Rio Cariá, a left bank affluent of the lower Rio Pracupi, in turn an affluent of the Anapu River, Floresta Nacional Caxiuanã, Pará, Brazil 5-17-15

References

Amaral, Dário Dantas do; Prost, Maria Thereza; Bastos, Maria de Nazaré do Carmo; Costa Neto, Salustiano Vilar; Santos, João Ubiratan Moreira dos. 2008. Restingas do litoral amazônico, estados do Pará e Amapá. Boletim do Museu Paraense Emílio Goeldi, Ciencias Naturais 3(1):35–67

Arévalo-Lopéz, Diandra; Nina, Nélida; Ticona, Juan C.; Limachi, Ivan; Salamanca, Efrain; Udaeta, Enrique; Paredes, Crispin; Espinoza, Boris; Serato, Alcides; Garnica, David; Limachi, Abigail; Coaquira, Dayana; Salazar, Sarah; Flores, Ninoska; Sterner, Olov; Giménez, Alberto. 2018. Leishmanicidal and cytotoxic activity from plants used in Tacana traditional medicine (Bolivia). Journal of Ethnopharmacology 216:120–133

Austin, Daniel F.; Bourne, Godfrey R. 1992. Notes on Guyana's medical ethnobotany. Economic Botany 46(3):293–298

Bardales, Joel Vásquez; Gómez, Ricardo Zárate; Canaquiri, Percy Huiñapi; Jiménez, Julio Pinedo; Hernández, Juan José Ramírez; Lamas, Gerardo; García, Pedro Vela. 2017. Plantas alimentícias de 19 especies de mariposas diurnas (Lepidoptera) en Loreto, Perú. Revista Peruana de Biología 24(1): 35–42

Bastos, M. N. C.; C. S. Rosário; L. C. B. Lobato. 1995. Caracterização fitofisionômica da restinga de Algodoal, Maracanã, Pará, Brasil. Boletim do Museu Paraense Emílio Goeldi, Série Botanica 11(2): 173–197

Bennett, Bradley C.; Baker, Marc A.; Andrade, Patricia Gómez. 2002. Ethnobotany of the Shuar of Eastern Ecuador. New York Botanical Garden, Advances in Economic Botany, vol. 14.

Bernal, J. A. 1991. Agronomic aspects of the cultivation of the uchuva, *Physalis peruviana*, on the high plateau of the Colombian departments of Cundinamarca and Boyacá. In: *Solanaceae III: Taxonomy, Chemistry, Evolution*, Hawkes, J. G.; Lester, R. N.; Nee, M.; Estrada, N. (eds), Royal Botanic Gardens, Kew, pp. 459–460

Bezerra, Elisangela Lúcia de S.; Machado, Isabel Cristina. 2003. Biologia floral e sistema de polinização de *Solanum stramonifolium* Jacq. (Solanaceae) em remanescente de mata Atlântica, Pernambuco. Acta Bot Bras. 17 (2): 247–257

Bueno, Carlos Roberto; Yuyama, Kaoru; Noda, Hiroshi; Silva Filho; Danilo Fernandes;Machado, Francisco Manoares; Paiva, Waldelice Oliveira de. 2000. Non-conventional crops: a feastible alternative for Amazonian floodplains. In: The Central Amazon Floodplain: Actual Use and Options for a Sustainable Management, Junk, W. J.; Ohly, J. J.; Piedade, M. T. F.; Soares, M. G. M. (eds), Backhuys Publishers, Leiden, pp. 171–189

Cerón, Carlos Eduardo. 1995. Etnobiologia de los Cofanes de Dureno, Provincia de Sucumbíos, Ecuador. Museo Ecuatoriano de Ciencias Naturales, Herbario Nacional, Serie Monografia 10(3), Quito

Cerón, C. E.; Montalvo, Consuelo G.; Umenda, José; Umenda Enma Chica. 1994. Etnobotánica y Notas sobre la Diversidad Vegetal en la Comunidad Cofán de Sinangüé, Sucumbíos, Ecuador. EcoCiencia, Quito

Chiang, Hsuch-Ching; Jaw, Shiow-Mei; Chen, Chia-Fu; Kan, Wei-Song. 1992. Antitumor agent, physalin F from *Physalis angulata* L. Anticancer Research 12: 837–844

Coelho-Ferreira, Márlia. 2009. Medicinal knowledge and plant utilization in an Amazonian coastal community of Marudá, Pará state (Brazil). Journal of Ethnopharmacology 126:159–175

Cruz, Diego Lima de Souza; Rodrigues, Guilherme Silva; Dias, Flávio de Oliveira; Alves, José Maria Arcanjo; Albuquerque, José de Anchieta Alves de. 2009.

Dantas, Mário; Rodrigues, Irenice Alves. 1980. Plantas invasoras de pastagens cultivadas na Amazônia. EMBRAPA, Centro de Pesquisa Agropecuária do Trópico Úmido, Boletim de Pesquisa, vol. 1.

Davis, W. E. and J. J. Yost. 1983. The ethnobotany of the Waorani of Eastern Ecuador. *Harvard Botanical Museum Leaflets* 29(3):159–217

El Sheikha, Aly F.; Zaki, Mohammed S.; Bakr, Ali A.; El Habashy, Magida M.; Montet, Didier. 2010. Biochemical and sensory quality of Physalis (*Physalis pubescens* L.) juice. Journal of Food Processing and Preservation 34:541–555

Ferreira, Marcos Gonçalves; Absy, Maria Lucia. 2018. Pollen niche of Melipona (Melikerria) interrupta (Apidae: Meliponini) bred in a meliponary in a terra-firme forest in the central Amazon. Palynology 42(2):199–209

Gentry, J. L. and W.G. D'Arcy. 1986. Solanaceae of Mesoamerica. Solanaceae: Biology and Systematics, D'Arcy, William G. (ed), pp. 15–26. Columbia University Press, New York

Giacomin, Leandro Lacerda; Gomes, Emeli Susane Costa. 2018. Flora das cangas da Serra dos Carajás, Pará, Brasil: Solanaceae. Rodriguésia 69(3):1373–1396

Gönen, Oya; Yildirim, Aysegül; Uygur, F. Nezihi. 2000. A new record for the flora of Turkey: *Physalis angulata* L. (Solanaceae). Turk J Bot 24: 299–301

Grenand, P. 1992. The use and cultural significance of the secondary forest among the Wayapi Indians. In: *Sustainable Harvest and Marketing of Rain Forest Products*, M. Plotkin and L. Famolare (Editors), Island Press, Washington, DC, pp. 27–40

Gutsche, A. 2008. *Frutas Amazonicas: Postres Peruanas de Vanguardia*. Wust Ediciones, Lima

Heiser, C. B. 1992. Of Plants and People. University of Oklahoma Press, Norman

Ique, Carlos A.; Aquino, Rolando. 2015. Principales especies florísticas utilizadas y dispersión de semillas por *Saguinus labiatus* E. Geoffroy, 1812 en isla Muyuy, Loreto, Perú. Ciencia Amazónica 5(1):81–88

Jovel, E. M.; Cabanillas, J.; Towers, G. H. N. 1996. An ethnobotanical study of the traditional medicine of the mestizo people of Suni Miraño, Loreto, Peru. Journal of Ethnopharmacology 53: 149–156

Kalliola, Riso; Salo, Jukka; Puhakka, Maarit; Rajasilta, Marjut. 1991a. New site formation and colonizing vegetation in primary succession on the western Amazon floodplains. Journal of Ecology 79(4):877–901

Kalliola, R.; Puhakka, M.; Salo, J.; Tuomisto, H.; Ruokolainen, K. 1991b. The dynamics, distribution and classification of swamp vegetation in Peruvian Amazonia. Annales Botanici Fennici 28:225–239

Kffuri, Carolina Weber; Lopes, Moisés Ahkutó; Ming, Lin Chau; Odonne, Guillaume; Kinupp, Valderly Ferreira. 2016. Antimalarial plants used by indigenous people of the Upper Rio Negro in Amazonas, Brazil. Journal of Ethnopharmacology 178:188–198

Lacerda, Fernando; Miranda, Izildinha; Kato, Oswaldo Ryohei; Bispo, Carlos José Capela; Vale, Igor do. 2013. Weed dynamics during the change of a degraded pasture to agroforestry system. Agroforestry Systems 87:909–916

Le Cointe, P. 1922. *L'Amazonie Brésilienne: Le Pays-ses Habitants, ses Resources, Notes et Statistiques jusqu'en 1920*. Augustin Challamel, Paris, vol 1

Le Cointe, P. 1934. *Amazônia Brasileira III*: *Árvores e Plantas Úteis*. Livraria Classica, Belém

Léon, J. 1987. Botanica de los Cultivos Tropicales. Instituto Interamericano de Cooperación para la Agricultura, San José

Lima, D. 2006. Ngiã nüna tadaugü i torü naãne (vamos cuidar da nossa terra). Editora UFMG, Belo Horizonte

Lima, Ana Carolina Barbosa de. 2014. Flavors of the city: access to regional fruit and fruit consumption in the State of Acre, Brazil. Boletim do Museu Paraense Emílio Goeldi, Ciencias Humanas 9(1):79–92

Lizcano, Leandro J.; Bakkali, Fadil; Ruiz-Larrea, M. Begoña; Ruiz-Sanz, José Ignacio. 2010. Antioxidant activity and polyphenol content of aqueous extracts from Colombian Amazonian plants with medicinal use. Food Chemistry 119:1566–1570

Lorenzi, H., L. Bacher, M. Lacerda, S. Sartori 2006. Brazilian Fruits and Cultivated Exotics (for Consuming in Natura). Instituto Plantarum de Estudos da Flora Ltda, Odessa, São Paulo

Lossio, L.C., Delgado, H.S., Ruiz, A.G., Donayre, R.A., Ruiz, J.G., Panduro, M.P., Sifuentes, T.C. 1995. Plantas Medicinales de la Amazonía Peruana. Instituto Peruano de Seguridad Social/Instituto de Medicina Tradicional, Iquitos

Lourenço, José Nestor de; Sousa, Silas Garcia de; Wandelli, Elisa Vieira; Lourenço, Francisneide de Sousa; Guimarães, Rosângela dos Reis; Campos, Lian da Silva; Silva, Roberto Lima da; Martins, Vivian Fernanda Carneiro. 2009. Agrobiodiversidade nos quintais agroflorestais em três assentamentos na Amazônia Central. Revista Brasileira de Agroecologia 4(2):965–969

Marx, F.; Andrade, Eloisa Helena A.; Maia, José Guilherme. 1998. Chemical composition of the fruit of *Solanum sessiliflorum*. Zeitschrift für Lebensmitlel-Untersuchung 206: 364–366

Mejia, Kember; Rengifo, Elsa. 2000. Plantas Medicinales de Uso Popular en la Amazonía Peruana. Agencia Española de Cooperación Internacional, Lima

Miléo, L. J.; Silva, J. F.; Bentes, J. L. S.; Christoffoleti, P. J. 2007. Plantas daninhas hospedeiras alternativas de *Colletotrichum guaranicola* em cultivos de guaraná no estado do Amazonas. Planta Daninha 25(4):771–782

Miranda, I. S.; Mitja, D.; Silva, T. S. 2009. Mutual influence of forests and pastures on the seedbanks in the eastern Amazon. Weed Research 49:499–505

Modesto Júnior, M. S.; Mascarenhas, R. E. B. 2001. Levantamento da infestação de plantas daninhas asociada a uma pastagem cultivada de baixa produtividade no nordeste Paraense. Planta Daninha 19(1):11–21

Mondragón, Martha L.; Smith, Randall. 1997. Bete Quiwiguimamo Salvando el bosque para vivir sano: algunas plantas y árboles utilizadas por la Nacionalidad Huaorani de la Amazonía Ecuatoriana. Abya Yala, Quito

Monteles, Ricardo; Pinheiro, Claudio Urbano B. 2007. Plantas medicinais em um quilombo maranhense: uma perspectiva etnobotânica. Revista de Biologia e Ciências da Terra 7(2):1–12

Montes, S.; Aguirre, J. R. 1994. Tomatillo, husk-tomato (*Physalis philadelphica*). In: Neglected Crops: 1492 from a Different Perspective, Bermejo, J. E.; León, Jorge (eds.), Food and Agriculture Organization of the United Nations, Rome, pp. 117–122

Mota, Nara Furtado de Oliveira; Watanabe, Mauricio Takashi Coutinho; Zappi, Daniela Cristina; Hiura, Alice Lima; Pallos, Julieta; Viveros, Raquel Stauffer; Giulietti, Ana Maria; Viana, Pedro Lage. 2018. Cangas da Amazônia: a vegetação única de Carajás evidenciada pela lista de fanerógramas. Rodriguésia 69(3):1435–1488

Nee, Michael. 1991. The systematics of lesser known edible Solanaceae of the New World. In: Solanaceae III: Taxonomy, Chemistry, Evolution, Hawkes, J. G.; Lester, R. N.; Nee, M.; Estrada, N. (Editors), pp. 365–368. Royal Botanic Gardens, Kew

Odonne, Guillaume; Valadeau, Celine; Alban-Castillo, Joaquina; Stien, Didier; Sauvain, Michel; Bourdy, Genevieve. 2013. Medical ethnobotany of the Chayahuita of the Paranapura basin (Peruvian Amazon). Journal of Ethnopharmacology 146:127–153

Oliveira, Danilo R.; Krettli, Anoniana U.; Aguiar, Anna Caroline C.; Leitão, Gilda G.; Vieira, Mariana N.; Martins, Karine S.; Leitão, Suzana G. 2015. Ethnopharmacological evaluation of medicinal plants used against malaria by quilombola communities from Oriximiná, Brazil. Journal of Ethnopharmacology 173:424–434

Pahlen, A. 1977. Cubiu (*Solanum topiro* Humb. & Bonpl.), uma fruteira da Amazônia. Acta Amazonica 7(3):301–307

Palheta, Ivanete Cardoso; Tavares-Martins, Ana Cláudia Caldeira; Lucas, Flávia Cristina Araújo; Jardim, Mário Augusto Gonçalves. 2017. Ethnobotanical study of medicinal plants in urban home gardens in the city of Abaetetuba, Pará state, Brazil. Boletín Latinoamericano y del Caribe de Plantas Medicinales y Aromáticas 16(3):206–262

Pedrollo, Camilo Tomazini. 2013. Baixo Jauaperi: da farmacopeia ao sistema de saúde, um estudo etnobotânico em comunidades ribeirinhas. Master's thesis, INPA, Manaus

Pedrotti, Débora Eriléia; Guarim Neto, Germano. 1998. A flora ruderal da cidade de Cuiabá, Mato Grosso, Brasil. Acta Botanica Brasilica 12(2):135–143

Pereira, Luciano Araujo; Lima-Barbosa, Jackson Rodrigo de; Almeida, Maria Zélia de; Guimarães, Elsie Franklin. 2017. Diversidade de plantas em quintais quilombolas, conhecimento local sobre uso e cultivo de pimentas na Amazônia oriental, Brasil. Revista de Biologia Neotropical 14(1):56–72

Polesna, Lucie; Polesny, Zbynek; Clavo, Mirella Z.; Hansson, Anders; Kokoska, Ladislav. 2011. Ethnopharmacological inventory of plants used in Coronel Portillo Province of Ucayali Department, Peru. Pharmaceutical Biology 49(2):125–136

Ponz, Elizabeth; Carpio, Griceldo; Meo, Severo. 2005. La Medicina Tradicional de los Tacana y Machineri: Conocimientos Prácticos de las Plantas Medicinales. Investigaciones Regionales Pando, La Paz

Reátegui, Katia Wendy. 2013. Hábitos alimenticios de *Saguinus mystax* Spix "pichico barba blanca" en Padre Isla, Loreto, Peru. Master's thesis, UNAP, Iquitos

Rodrigues, Eliana; Duarte-Almeida, Joaquim M.; Pires, Júlia Movilla. 2010. Perfil farmacológico e fitoquímico de plantas indicadas pelos caboclos do Parque Nacional do Jaú (AM) como potenciais analgésicas. Part I. Revista Brasileira de Farmacognosia 20(6):981–991

Roumy, Vincent; Ruiz, Lastenia; Macedo, Juan Celidonio Ruiz; Gutierrez-Choquevilca, Andrea-Luz; Samaillie, Jennifer; Encinas, Leonor Arévalo; Mesia, Wilfredo Ruiz; Cotrina, Hivelli Ericka Ricopa; Rivière, Céline; Sahpaz, Sevser; Bordage, Simon; Garçon, Guillaume; Dubuisson, Jean; Anthérieu, Sebastien; Seron, Karin; Hennebelle, Thierry. 2020. Viral hepatitis in the Peruvian Amazon: Ethnomedical context and phytomedical resource. Journal of Ethnopharmacology 255 (112735):1–11

Ruysschaert, Sofie; Van Andel, Tinde; Van de Putte, Kobeke; Van Damme, Patrick. 2009. Bathe the baby to make it strong and healthy: Plant use and child care among Saramaccan Maroons in Suriname. Journal of Ethnopharmacology 121:148–170

Salick, J. 1992. Crop domestication and the evolutionary ecology of cocona (*Solanum sessiliflorum* Dunal). Evolutionary Biology 26: 247–284

Sanz-Biset, Jaume; Cañigueral, Salvador. 2011. Plant use in the medicinal practices known as "strict diets" in Chazuta valley (Peruvian Amazon). Journal of Ethnopharmacology 137:271–288

Sanz-Biset, Jaume; Campos-de-la-Cruz, José; Epiquién, Mirbel A.; Cañigueral, Salvador. 2009. A first survey on the medicinal plants of the Chazuta valley (Peruvian Amazon). Journal of Ethnopharmacology 122:333–362

Schultes, R. E. and R. F. Raffauf. 1990. The Healing Forest: Medicinal and Toxic Plants of the Northwest Amazonia. Dioscorides Press, Portland, Oregon

Schultes, R. E. and R. F. Raffauf. 1991. De plantis toxicariis e Mundo Novo Tropicale Commentationes XXXVI: phytochemical and Ethnopharmacological notes on the Solanaceae of the Northwest Amazon. In: Solanaceae III: Taxonomy, Chemistry, Evolution, Hawkes, J. G.; Lester, R. N.; Nee, M.; Estrada, N. (eds.), Royal Botanic Gardens, Kew, pp. 25–49

Silva, Silvestre. 2011. Fruits of the Brazilian Amazon. Metalivros, São Paulo

Silva, Danilo F. da; Anunciação, Clodoaldo J. da; Nodal, Hiroshi; Reis, Odemar V. dos. 1996. Variabilidade genética em populações naturais de cubiu da Amazônia. Horticultura Brasileira 14(1):9–14

Silva, Alexandre Coletto da; Kinupp, Valdely Ferreira; Absy, Maria Lucia; Kerr, Warwick Estevam Kerr. 2004. Pollen morphology and study of the visitors (Hymenoptera, Apidae) of *Solanum stramoniifolium* Jacq. (Solanaceae) in Central Amazon. Acta Botanica Brasilica 18(3):653–657

Silva Filho, Danilo F. da; Clement, Charles R.; Noda, Hiroshi. 1989. Variação fenotípica em frutos de doze introduções de cubiu (*Solanum sessiliflorum* Dunal) avaliadas em Manaus, AM, Brasil. Acta Amazonica 19: 9–18

Silva Filho, Danilo Fernandes da; Noda, Hiroshi; Clement, Charles R. 1993. Genetic variability of economic characters in 30 accessions of cubiu (*Solanum sessiliflorum*, Dunal, Solanaceae) evaluated in central Amazonia. Revista Brasileira de Genética 16(2):409–417

Silva Filho, Danilo Fernandes da; Noda, Hiroshi; Paiva, Waldelice Oliveira de; Yuyama, Kaoru; Bueno, Carlos Roberto; Machado, Francisco Manoares. 1997. Hortaliças não convencionais nativas e introduzidas na Amazônia. In: Duas Décadas de Contribuições do INPA à Pesquisa Agronômica no Trópico Úmido, Noda, Hiroshi; Souza, Luiz Augusto Gomes de (eds), Instituto Nacional de Pesquisa da Amazônia, Manaus

Silva Filho, Danilo Fernandesda; Andrade, Jerusa Souza de; Clement, Charles Roland; Machado, Francisco Manoares; Noda, Hiroshi. 1999. Correlações fenotípicas, genéticas e ambientais entre descritores morfológicos e químicos em frutos de cubiu (Solanum sessiliflorum Dunal) da Amazônia. Acta Amazonica 29 (4): 503–511

Siveiro, A.; Delunardo, T. A.; Haverroth, M.; Oliveira, L. C.; Mendonça, A. M. S. 2012. Plantas medicinais em quintais urbanos de Rio Branco, Acre. Revista Brasileira de Plantas Medicinais 14(4):598–610

Smith, N. J. H. 1999. *The Amazon River Forest*. Oxford University Press, New York.

Smith, N., R. Vásquez, and W. H. Wust. 2007. *Amazon River Fruits: Flavors for Conservation.* Amazon Conservation Association/Missouri Botanical Garden, Lima

Stasi, Luiz Claudio di; Santos, Elza Maria Guimarães; Santos, Claudenice Moreira dos; Hiruma, Clélia Akiko. 1989. Plantas Medicinais na Amazônia. Editora UNESP, São Paulo

Storti, Eliana Fernandez. 1988. Biologia floral de *Solanum sessiliflorum* Dun. Var. *sessiliflorum*, na região de Manaus, AM. Acta Amazonica 18(3–4):55–65

Tananta, Leslie Virginia. 2014. Análisis del conocimiento tradicional del uso de especies vegetales en tres comunidades de la cuenca baja del río Ucayali, Loreto, Peru. Master's thesis, UNAP, Iquitos

Tenório, Maria Alice Ramalho de Oliveira; Van den Berg, Maria Elisabeth; Menezes, Ordoenha Furtado de; Salles, Pedro. 1991. Fitoterapia: uma estratégica terapêuticas natural do Amapá. In: Medicinas Tradicionais e Medicina Occidental na Amazônia, Buchillet, Dominique (ed), Museu Paraense Emílio Goeldi, Belém

Tindall, H. D. 1983. Vegetables in the Tropics. Macmillan, London

Tomchinsky, Bernardo; Ming, Lin Chau; Kinupp, Valdely Ferreira; Hidalgo, Ari de Freitas; Chaves, Francisco Célio Maia. 2017. Ethnobotanical study of antimalarial plants in the middle region of the Negro River, Amazonas, Brazil. Acta Amazonica 47(3):203–212

Uhl, Christopher; Clark, Kathleen; Clark Howard; Murphy, Peter. 1981. Early plant succession after cutting and burning in the Upper Rio Negro region of the Amazon Basin. Journal of Ecology 69:631–649

Valadeau, Céline; Castillo, Joaquina Alban; Sauvain, Michel; Lores, Augusto Francis; Bourdy, Geneviève. 2010. The rainbow hurts my skin: medicinal concepts and plants uses among the Yanesha (Amuesha), an Amazonian Peruvian ethnic group. Journal of Ethnopharmacology 127:175–192

Van den Berg, M. E. 2010. Plantas Medicinais na Amazônia. Museu Paraense Emílio Goeldi, Belém

Van den Berg, M.E. and M.H.L. da Silva. 1986a. Plantas medicinais do Amazonas. In: Anais do1°Simpósio do Trópico Úmido, Belém, PA, 12 a 17 November, vol. 2, Flora e Floresta, EMBRAPA, Brasília, pp. 127–133

Van den Berg, M.E. and M.H.L. da Silva. 1986b. Contribuição ao conhecimento da flora medicinal do Maranhão. In: Anais do1°Simpósio do Trópico Úmido, Belém, PA, 12 a 17 November, vol. 2, Flora e Floresta, EMBRAPA, Brasília, pp 119–125

Vásquez, R. and A. H. Gentry. 1989. Use and misuse of forest-harvested fruits in the Iquitos area. *Conservation Biology* 3(4): 350–361.

Veiga, J. B.; Scudeller, V. V. 2015. Etnobotânica e medicina popular no tratamento de malária e males asociados na comunidade ribeirinha Julião- baixo Rio Negro (Amazônia Central). Revista Brasileira de Plantas Medicinais 17(4):737–747

Vickers, W. T. and T. Plowman. 1984. *Useful Plants of the Siona and Secoya Indians.* Field Museum of Natural History, Fieldiana Botany New Series, no. 15

Villachica, Hugo; Carvalho, José Edmar Urano de; Muller, Carlos Hans; Díaz, Camilo; Almanza, Martha; Mercado, Roberto Samanez; Izquierdo, Juan; Egg, Antônio Brack. 1996. Frutales y Hortalizas Promisorios de la Amazonia. Tratado de Cooperación Amazonica, Lima.

Volpato, Gabriele; Marcucci, Rossella; Tornadore, Noemi; Paoletti, Maurizio G. 2004. Domestication process of two Solanum section Lasiocarpa species among Amerindians in the Upper Orinoco, Venezuela, with special focus on Piaroa Indians. Economic Botany 58 (2): 184–194

Wahlen, M. D.; Costich, D. E.; Heiser, C. B. 1981. Taxonomy of Solanum section Lasiocarpa. Gentes Herbarum 12 (2): 41–129

Chapter 49
Urticaceae

49.1 *Pourouma cecropiifolia*

Bolivia Ambaibillo, ambaiba macho, ambaiba uva, imbaúba mansa, puruma, tahuariba, toren uva, uvilla, uvillo, mapati; quëxquixaquini, xaquitato (Chácobo), tahuapa (Tacana), movãij (Tsimane)

Brazil Cucura, cocura, mapatí, purumã, torém embaúba, embaúbarana, embaúba de cheiro, embaúba de vinho, embaúba de preguiça, embaúba mansa, imbaúba, uva da mata, uva da Amazônia, uvão, uvinha, uvilha, uvilla, sucuba, mapati; kahméro, kamhero (Baniwa), igui (Desana), buhúh (Hupd'äh), atwyrà krã krê (Kayapó), ö'sé, ëhsé, diajié, diagcê (Tukano), shõrõshõrõ ã hi (Yanomami)

Colombia Cucura, uva de monte, uva caimarona, uva amazónica, uva de sapo, uvilla, uvilla de monte, uvo, caimarón, caimarón silvestre, caimarona, sirpe macho, yarumo; ta-mi, tazu, yotemaní (Andoke), ö-hê' (Barasana), baako (Bora), igí (Desana), we-wit'-kat-ku (Maku), üye, icyé (Makuna), baacoba (Miraña), baku (Muinane), ühe (Tatuyo), tiêá (Ticuna), jirikorei, jiricona, jirikona, bubuna (Witoto), caamú (Yukuna)

Ecuador Uva, uva de monte, uva de montaña, uva de campo, uva caimonera, uva negra, uva de árbol, sacha uva, uvilla, uvilla de monte, shuiña, embaúba de vinho, chontilla, pumamaqui, hormiguero estrella; bocha tsaja, pochoa tsaja (Cofán), sacha uvilla, huilla (Kichwa), washi shuinia, yanat, pau shuinia (Shuar), a'ko cuiya'i, airo cuiya'i, yai quëopëa, uñe (Siona-Secoya), ikitoca, yohue, yohuehue, yehue (Waorani)

Peru Uvilla, uvilla colorada, uvilla del monte, uvilla Amazónica, sacha uvilla, ubilla, uva de monte, uva caimarona, uba uba, ubilla, caimaron, caima, baácohe, chiquima, cocura, shuvia, ambaibauva; shiwiá (Achuar), shuwiya, shuíya

N. Smith, *Amazon Fruits: An Ethnobotanical Journey*, Ethnobiology,
https://doi.org/10.1007/978-3-031-12803-5_49

(Aguaruna), suwiya (Awajún), baácohe (Bora), shewantoki (Campa), xankin (Kakataibo), maca ede ñi (Maijuna), shankon, o níi xankon (Shipibo), kwi ya'i (Siona-Secoya), chixã (Ticuna)

Venezuela Cocura, cocura montañera, cocura de monte, manasono; uli wejkao (Jodï)

Status Wild, planted

Delia, 12 years, eating uvilla *Pourouma cecropiifolia* she gathered in her home garden. Cabo Lopez, Rio Itaya, Loreto, Peru 1-7-06

Uvilla *Pourouma cecropiifolia* is a medium-sized tree reaching as high as 30 m with a broad, spreading crown. During the rainy season, the branches are loaded with generous bunches of round purple fruits some 2–4 cm in diameter (Castañeda

1961:37). Each bunch contains from 10 to 40 fruits. The skin is thick, similar to the Muscadine grape of Florida, hence its English name: Amazon tree grape. The skin is broken open with the fingers or teeth to access the slippery, light green pulp that surrounds the single seed. The tan-colored seed, the size of the tip of an adult thumb, only pear-shaped, is ejected onto the ground where it may later sprout. The pulp, which accounts for about half of the fruit by volume, is not particularly nutritious but is nevertheless sweet and juicy. Productivity averages some 24 kg of fruit per tree per year (Villachica et al. 1996:292). The male and female flowers are visited by several native bees (Falcão and Lleras 1980; Heard 1999) so this fruit tree also contributes indirectly to the diet of some rural people when they gather honey from the bee nests in the wild and in some cases raise the bees in sections of hollowed out logs suspended from the outside eaves of their homes.

Thought to have been domesticated along the Rio Negro or the Solimões (Cavalcante 2010:174), the tree is cultivated in the central and western parts of the Amazon Basin (Allen 1947; Cárdenas and Ramírez 2004; Coomes and Burt 1997; Ducke 1946; Falcão and Lleras 1980; La Rotta 1983:63; Machado 2016:39; Padoch et al. 1985; Pulido and Cavelier 2001; Schultes 1977). Tukanoan groups along the Uaupés River and its affluents in the northwestern part of the Brazilian Amazon and contiguous Colombia, as well as the Witoto along the Caquetá in the Colombian Amazon interplant *Pourouma cecropiifolia* in the middle of their manioc fields so that they can gather fruits after the root crop has been harvested (Dufour 1990; Postma and Verheij 1994; Triana-Moreno et al. 2006; Veléz and Veléz 1992). In addition to the Witoto, various other indigenous groups in the Putumayo and Caquetá river basins in the Colombian Amazon return to harvested manioc fields to gather uvilla fruits (Whiffen 1915:136). The Ticuna along the Solimões have selected several varieties of this fruit, indicating that they have cultivated it for a long time (Kerr and Clement 1980). Known as uvilla in Peru, it is particularly common in home gardens and is also planted in fields on both uplands and floodplains in the vicinity of Iquitos and further down the Amazon River in Yagua villages (Balbin and Samaniego 1996; Coomes and Ban 2004; De Jong 1996; Lamont et al. 1999). The Cofán in the northern part of the Ecuadorian Amazon plant this tree in some of their polycultural fields where manioc is the main crop (Cerón et al. 1994:11). The Kayapó in the interfluvial area between the Tocantins and Xingu also plant *Pourouma cecropiifolia* in some of their manioc fields so that they can return to those areas long after the tubers have been harvested to gather fruits (Posey 1985). This plant produces fruit within 3 years of planting the seeds. This tree is planted in some rural home gardens, such as along the lower Solimões near its confluence with the Rio Negro (Lima and Saragoussi 2000) as well as in several other rural communities in the Brazilian state of Amazonas (Guillaumet et al. 1990; Silva 2017:32). Fruit production is greater in home gardens compared to fields being taken over by successionary growth (Postma and Verheij 1994).

Boy picking uvilla *Pourouma cecropiifolia* fruits in a home garden. Yarina, Rio Yanayacu, Reserva Nacional Pacaya-Samiria, Loreto, Peru 3-6-04

Amazon tree grape bears fruit in both the rainy and dry seasons and the fruit bunches are dislodged with the use of a notched pole. Kids also scramble up the tree to harvest the fruits. On younger trees, the branches are low enough for people to reach the fruits from the ground. The fruits often make their way to markets, such as in Iquitos, Requena, and Pucallpa, where they are eaten fresh. In June 2014, uvilla fruits in Iquitos were selling for the equivalent of 270 per US$1.

Uvilla *Pourouma cecropiifolia* **fruits in a motorized canoe (*peque-peque*) destined for sale in Iquitos. Mishana, Rio Nanay, Loreto, Peru 1-8-06**

The fruits have also long been used to make juice in the Peruvian Amazon, as noted at Pebas along the Amazon River by a French naturalist in the nineteenth century (Castelnau 1850). In the Rio Negro watershed, some indigenous groups ferment the fruit juice to make an alcoholic beverage, such as the Tukano who live along the Uaupés and Tiquié rivers (Pereira 1967:514; Silva 1962:223). The fruits are also used to make a liqueur made with sugarcane alcohol (*aguardiente*), such as along the Monzón road near Tingo Maria in Peru. The seeds are sometimes toasted and ground as a substitute for coffee (Villachica et al. 1996:292).

Uvilla *Pourouma cecropiifolia* fruits in a street market. Masusa District, Iquitos, Loreto, Peru 6-17-04

Although most fruits are gathered from planted trees, *Pourouma cecropiifolia* is also encountered wild in woodlands and successionary vegetation on uplands and floodplains up to 1700 m in western Amazonia near the foothills of the Andes in Bolivia, Peru, Ecuador, and Colombia (Araujo-Murakami et al. 2012, 2015; Arellano 2013:86, 203; Gomez 2011:85; Irvine 1987; Killeen et al. 1993:555; Lemos et al. 2015; Mackensen et al. 2000; Moser 2013:87; Muñoz 2010:17, 30; Paz-Rivera and Putz 2009; Pennington et al. 2004:114; Phillips 1993; Pitman et al. 2001; Quisbert and Macía 2005; Robles 2017:49; Thomas 2009; Vester and Cleef 1998), as well as in transitional Amazonian forest bordering scrub savanna in the Upper Teles Pires watershed in Mato Grosso, Brazil (Araujo et al. 2009). It is more common, however, below 600 m. This tree also occurs in light gaps in some upland forests, such as in Acre (Maranho and Salimon 2015). The Huaorani in the

Ecuadorian Amazon gather fruits of wild *Pourouma cecropiifolia* in floodplain forests (Ima 2012). Wild forms of Amazon tree grape are found in regrowth forests on floodplains in the vicinity of Yarina along the Upper Yanayacu in the Reserva Nacional Pacaya-Samiria, but it is not clear whether they are individuals that have survived after fields have been abandoned. For example, the Bora who live along the Yaguasyacu River, an affluent of the Ampiyacu which flows into the Amazon River at Pebas, plant uvilla and other fruit trees in some of their manioc fields so that they become orchards once the root crop has been harvested (Denevan and Treacy 1987). Groves of fruit trees, including *Pourouma cecropiifolia*, are planted around many indigenous communal houses (*malocas*) in the Amazon portions of Colombia and Peru (Allen 1947; Prance 1985). When communities are abandoned, such orchards become part of the returning forest forming a blend of wild and planted trees.

In mature forest, *Pourouma cecropiifolia* is typically found in light gaps for it is a successional species similar to its look-a-like *Cecropia*, known as embaúba in Brazil, ambaiba in Bolivia, and Setico in Peru. Indeed in Bolivia, *Pourouma cecropiifolia* is called ambaiba uva or ambaibillo. Several monkey species are likely dispersal agents for wild and cultivated *Pourouma cecropiifolia*. The black-chested moustached tamarin *Saguinus mystax*, the mottle-faced tamarin *Saguinus inustus*, the saddle-backed tamarin *Saguinus fuscicollis*, the white-lipped tamarin *Saguinus labiatus*, the red-bellied tamarin *Saguinus labiatus*, Goeldi's marmoset *Callimico goeldii*, and titi monkey *Callicebus torquatus* feed on this fruit in successional vegetation and forests and raid uvilla trees with fruit in home gardens and fields in the Bolivian, Colombian, and Peruvian Amazon (Buchanan-Smith 1991a, b; Garber 1993; Kinzey 1981; Knogge and Heymann 2003; Knogge et al. 2003; Palacios et al. 2004; Porter 2001; Sánchez and Levy 2013:32, 93; Soini 1987). In the Peruvian Amazon, several other species of monkey are known to eat the fruits of *Pourouma cecropiifolia* including white-fronted capuchins *Cebus albifrons*, squirrel monkeys *Saimiri sciureus*, and the emperor tamarin *Saguinus imperator* (Terborgh 1983:69). The bearded saki monkey *Chiropotes chiropotes*, which lives in the northern part of the Brazilian Amazon and contiguous forests in Venezuela and the Guianas, also eats this fruit and likely disperses some of the seeds (Boyle et al. 2012). Woolly monkeys *Lagothrix lagotricha*, spider monkeys *Ateles belzebuth chamek*, and tufted capuchin monkeys *Cebus apella* also eat the fruits and disperse some of the seeds (Defler and Defler 1996; Palma and Stevenson 2010); these monkeys are widely hunted in Amazonia for their meat. Pygmy marmosets *Cebuella pygmaea* pierce the bark of this tree with their teeth to lick up the exudate (Soini 1988); however, these monkeys are too small to be hunted for their meat since they generally weigh under 140 g. In the Peruvian Amazon, the yellow-footed tortoise *Chelonoidis denticulata* is known to eat this fruit and disperse some of the seeds, and this tortoise is gathered for its meat (Guzmán and Stevenson 2008).

Pourouma cecropiifolia has a variety of other uses from medicine to construction materials. The Maku, who inhabit parts of the northwestern Amazon in Colombia and Brazil, employ the bark as a contraceptive. Women drink an infusion prepared by rubbing bark scrapings in water; excessive doses can reputedly cause permanent sterility (Schultes 1986). Other indigenous groups in the same region, such as the

Makuna and Barasana, make a snuff by burning the leaves of *Cecropia* or *Pourouma cecropiifolia* and mixing the resulting ash with dried and pulverized coca leaves (Carmichael et al. 1985:86; Plowman 1981; Hugh-Jones 1979:201; Schultes 1980). The Tacana of the Madidi watershed in the Bolivian Amazon use strips of bark from *Pourouma cecropiifolia* to lash poles together when making houses, while in the Ecuadorian Amazon, the trunks are sawn into planks for house flooring (DeWalt et al. 1999). The Tukano of the Upper Rio Negro use the wood to make charcoal (Pereira 1967:514) while the Tsimane in the Bolivian Amazon burn the wood in their house fires (Guèze et al. 2014). The Mayoruna use the rough leaves to smooth their bows made from *Bactris gasipaes* or *Oenocarpus mapora* palm wood (Romanoff et al. 2004). The Desana, a Tukanoan group in the Colombian Amazon, use the leaves to wrap manioc flour and bread (Jackson 1983:52). In the Ecuadorian amazon, the wood is used to make fires (Van den Eynden et al. 1999:148), as is the case among the Tsimane in the Bolivian Amazon (Reyes-García 2001:240). Also in the Bolivian Amazon, the Chácobo fashion planting sticks from the wood which are used to make a hole in the ground to plant seeds or cuttings (Zambrana et al. 2017). The Shuar in the southern part of the Ecuadorian Amazon use the trunk for building their homes and the branches for fuel (Bennett et al. 2002:146).

References

Allen, P. H. 1947. Indians of southeastern Colombia. Geographical Review 37(4):567–582

Araujo, Rosalia de Aguiare; Costa, Reginaldo Brito da; Felfili, Jeanine Maria; Kuntz, Ivani; Melo e Sousa, Roberto Antonio Ticle de; Dorval, Alberto. 2009. Florística e estrutura de fragmento florestal em área de transição na Amazônia Matogrossense no município de Sinop. Acta Amazonica 39(4):865–878

Araujo-Murakami, Alejandro; Poma-Chura, Abraham; Palabral, Arely; Salvatierra, Rolando; Hurtado, Fidel. 2012. Composición florística de los bosques Amazónicos de tierra firme e inundable en las proximidades de las Pampas del Sonene (río Heath), Parque Nacional Madidi, Bolivia. Kempffiana 8(1):2–21

Araujo-Murakami, Alejandro; Villarroel, Daniel; Pardo, Guido; Vos, Vincent A.; Parada, G. Alexander; Arroyo, Luzmila; Killeen, Timothy. 2015. Diversidad arbórea de los bosques de tierra firme de la Amazonía Boliviano. Kempffiana 11(1):1–28

Arellano, Gabriel. 2013. Patrones de diversidad, distribución y dominancia de plantas leñosas en los bosques montanos del Parque Nacional Madidi, Bolivia. Doctoral dissertation, Departamento de Biologia, Facultad de Ciencias, Universidad Autónoma de Madrid

Balbin, Isabel Oré; Samaniego, Dedy Llapapasca. 1996. Huertas domesticas como sistema tradicional de cultivo en Moena Caño, Rio Amazonas, Iquitos, Peru. Folia Amazonica 8(1):91–110

Bennett, Bradley C.; Baker, Marc A.; Andrade, Patricia Gómez. 2002. Ethnobotany of the Shuar of Eastern Ecuador. New York Botanical Garden, Advances in Economic Botany, vol. 14.

Boyle, Sarah A.; Zartman, Charles E.; Spironello, Wilson R.; Smith, Andrew T.; Beck, Harald. 2012. Implications of habitat fragmentation on the diet of bearked saki monkeys in central Amazonian forest. Journal of Mammalogy 93(4):959–976

Buchanan-Smith, H. 1991a. Field observations of Goeldi's monkey, Callimico goeldii, in northern Bolivia. Folia Primatologica 57:102–105

Buchanan-Smith, H. 1991b. A field study on the red-bellied tamarin, Saguinus l. labiatus, in Bolivia. International Journal of Primatology 12(3):259–276

Cárdenas, Dairon; Ramírez, Juan Guillermo. 2004. Plantas útiles y su incorporación a los sistemas productivos del Departamento del Guaviare (Amazonia Colombiana). Caldasia 26(1):95–110

Carmichael, Elizabeth; Hugh-Jones, Stephen; Moser, Brian; Tayler, Donald. 1985. The Hidden Peoples of the Amazon. British Museum, London

Castañeda, Rafael Romero. 1961. *Frutas Silvestres de Colombia*. Bogota, vol. 1.

Castelnau, F. 1850. *Éxpedition dans les Parties Centrales de l'Amérique du Sud, de Rio de Janeiro à Lima, et de Lima au Pará, pendant les Années 1843–1847*. Bertrand, Paris, vol. 2

Cavalcante, P. B. 2010. *Frutas Comestíveis na Amazônia*. Museu Paraense Emílio Goeldi, Belém

Cerón, C. E.; Montalvo, Consuelo G.; Umenda, José; Umenda Enma Chica. 1994. Etnobotánica y Notas sobre la Diversidad Vegetal en la Comunidad Cofán de Sinangüé, Sucumbíos, Ecuador. EcoCiencia, Quito

Coomes, Oliver T.; Ban, Natalie. 2004. Cultivated plant species diversity in home gardens of Amazonian peasant village in northeastern Peru. Economic Botany 58(3):420–434

Coomes, O. T.; Burt, G. J. 1997. Indigenous market-oriented agroforestry: dissecting local diversity in western Amazonia. Agroforestry Systems 37:27–44

De Jong, W. 1996. Swidden-fallow agroforestry in Amazonia: diversity at close distance. Agroforestry Systems 34: 277–290

Defler, Thomas R.; Defler, Sara B. 1996. Diet of a group of *Lagothrix lagothricha lagothrica* in southeastern Colombia. International Journal of Primatology 17(2):161–190

Denevan, W. M. and J. M. Treacy. 1987. Young managed fallows at Brillo Nuevo. In: *Swidden-Fallow Agroforestry In the Peruvian Amazon*, W. M. Denevan and C. Padoch (eds.), Advances in Economic Botany vol. 5, New York Botanical Garden, pp. 8–46

DeWalt, Saara J.; Bourdy, Genevieve; Michel, Lia R. Chávez de; Quenevo, Celin. 1999. Ethnobotany of the Tacana: quantitative inventories of two permanent plots of northwestern Bolivia. Economic Botany 53 (3): 237–260

Ducke, A. 1946. *Plantas de Cultura Precolombiana na Amazônia brasileira. Notas Sôbre as Espécies ou Formas Espontâneas que Supostamente lhes teriam dado Origem*. Instituto Agronômico do Norte, Belém, Boletim Técnico 8, pp. 1–24

Dufour, D. L. 1990. Use of tropical rainforests by native Amazonians: these sophisticated and complex agricultural systems can serve as models of sustainable agroecosystems. Bioscience 40(9):652–659

Falcão, M. and E. Lleras. 1980. Aspectos fenológicos, ecológicos e de produtividade do mapati (*Pourouma cecropiifolia* Mart.). Acta Amazonica 10(4):711–724

Garber, P. A. 1993. Seasonal patterns of diet and ranging in two species of tamarin monkeys: stability versus variability. International Journal of Primatology 14(1):145–166

Gomez, Tedi Pacheco. 2011. Valoración económica de purmas de diferentes edades en la comunidad nativa de Santa Mercedes, río Putumayo, Perú. Master's thesis, UNAP, Iquitos

Guèze, Maximilien; Luz, Ana Catarina; Paneque-Gálvez, Jaime; Macía, Manuel J.; Orta-Martínez, Martí; Pino, Joan; Reyes-García, Victoria. 2014. Are ecologically important tree species the most useful? a case study from indigenous people in the Bolivian Amazon. Economic Botany 68(1):1–15

Guillaumet, J-L.; Grenand, P.; Bahri, S.; Grenand, F.; Lourd, M.; Santos, A.A. dos; Gély, A. 1990. Les jardins-vergers familiaux d'Amazonie centrale: une exemple d'utilisation de l'espace. Turrialba 40(1):63–81

Guzmán, Adriana; Stevenson, Pablo R. 2008. Seed dispersal, habitat selection and movement patterns in the Amazonian tortoise, *Geochelone denticulata*. Amphibia-Reptilia 29:463–472

Heard, Tim A. 1999. The role of stingless bees in crop pollination. Annual Review of Entomology 44:183–206

Hugh-Jones, C. 1979. From the Milk River: Spatial and Temporal Processes in Northwest Amazonia. Cambridge University Press, Cambridge

Ima, M. O. 2012. *Saberes Waorani y Parque Nacional Yasuní: Plantas, Salud y Bienestar en la Amazonía del Ecuador*. Iniciativa Yasuní ITT, Ministerio Coordinador de Patrimonio,

Ministerio del Ambiente, Programa de las Naciones Unidas para el Desarrollo (PNUD) y Fondo para el Medio Ambiente Mundial (FMAM), Quito

Irvine, Dominique. 1987. El manejo del bosque secundario por los Quichua amazonicos. Hombre y Ambiente 2:121–142

Jackson, J. E. 1983. The Fish People: Linguistic Exogamy and Tukanoan Identity in Northwest Amazonia. Cambridge University Press, Cambridge

Kerr, W. E. and C. R. Clement. 1980. Prácticas agrícolas de consequências genéticas que possibilitaram aos índios da Amazônia uma melhor adaptaçao às condiçoes ecológicas da região. Acta Amazonica 10(2):251–261

Killeen, Timothy J.; García, Emilia; Beck, Stephan G. 1993. Guía de Árboles de Bolivia. Herbario Nacional de Bolivia/Missouri Botanical Garden, La Paz.

Kinzey, Warren G. 1981. The titi monkeys, genus *Callicebus*. In: Ecology and Behavior of Neotropical Primates, Coimbra-Filho, Adelmar F.; Mittermeir, Russell A. (eds), Academia Brasileira de Ciências, Rio de Janeiro, vol. 1, pp 241–276

Knogge, Christoph; Heymann, Eckhard W. 2003. Seed dispersal by sympatric tamarins, *Saguinus mystax* and *Saguinus fuscicollis*: diversity and characteristics of plant species. Folia Primatologica 74:33–47

Knogge, Christoph; Herrera, Emérita R. Tirado; Heymann, Eckhard W. 2003. Effects of passage through tamarin guts on the germination potential of dispersed seeds. International Journal of Primatology 24(5):1121–1128

La Rotta, C. 1983. Observaciones Etnobotanicas sobre algunas Especies Utilizadas por la Comunidad Indigena Andoque (Amazonas, Colombia). Universidad Nacional de Colombia, Bogotá

Lamont, Susan R.; Eshbaugh, W. Hardy; Greenberg, Adolph M. 1999. Species composition, diversity, and use of homegardens among three Amazonian villages. Economic Botany 53(3):312–326

Lemos, D. A. N.; Ferreira, B. G. A.; Siqueira, J. D. P.; Oliveira, M. M.; Ferreira, A. M. 2015. Floristic and phytosociology in dense "terra firme" rainforest in the Belo Monte hydroelectric plant influence area, Pará, Brazil. Brazilian Journal of Biology 75(3):257–276

Lima, Roberval M. B. de; Saragoussi, Muriel. 2000. Floodplain home gardens on the central Amazon in Brazil. In: The Central Amazon Floodplain: Actual Use and Options for a Sustainable Management, Junk, W. J.; Ohly, J. J.; Piedade, M. T. F.; Soares, M. G. M. (eds), Backhuys Publishers, Leiden pp 243–268

Machado, Danilo de Oliveira. 2016. A agrobiodiversidade de quintais agroflorestais em propriedades agrícolas familiares na BR 174, Ramal do Pau-Rosa, Manaus, AM. Master's Thesis, INPA, Manaus

Mackensen, Jens; Tillery-Stevens, Madeleine; Klinge, Rudolf; Fölster, Horst. 2000. Site parameters, species composition, phytomass structure and element stores of a terra-firme forest in East-Amazonia, Brazil. Plant Ecology 151:101–109

Maranho, Álisson Sobrinho; Salimon, Cleber Ibraim. 2015. Maranthaceae overabundance decreases richness and abundance of regenerating woody plants in natural gaps. Neotropical Biology and Conservation 10(2):53–62

Moser, Pamela. 2013. Vegetação Arbórea e sua Relação com Fatores Ambientais e Especiais em Florestas de Terra Firme no Noroeste de Rondônia, Brasil. Master's thesis, Faculdade de Tecnologia, Departamento de Engenharia Florestal, Programa de Pós-Graduação em Ciências Florestais, Universidade de Brasília

Muñoz, Marianella Guevara. 2010. Evaluación de purmas con fines de manejo en la comunidad nativa de Santa Mercedes, Rio Putumayo, Perú. Master's thesis, Universidad Nacional de la Amazonia Peruana, Iquitos

Padoch, C. J.; J. Chota Inuma; W. de Jong; and J. Unruh. 1985. Amazonian agroforestry: a market-oriented system in Peru. Agroforestry Systems 3:47–58

Palacios, Erwin; Rodríguez, Adriana; Castillo, Claudia. 2004. Preliminary observations on the mottled-face tamarin (*Saguinus inustus*) on the lower Rio Caquetá, Colombian Amazon. Neotropical Primates 12 (3): 123–126

Palma, Ana Cristina; Stevenson, Pablo R. 2010. Dispersión de semillas por monos araña en la Estación Biológica Cocha Cashu, Perú. In: Primatología en Colombia: Avances al Principio del Milenio, Pereira-Bengoa, Victoria; Stevenson, Pablo R.; Bueno, Marta Lucía; Nassar-Montoya, Fernando (eds), Fundación Universitaria San Martín, Bogotá, pp 19–35

Paz-Rivera, Clea; Putz, Francis E. 2009. Anthropogenic soils and tree distributions in a lowland forest in Bolivia. Biotropica 41(6):665–675

Pennington, T. D., C. Reynel, and A. Daza. 2004. *Illustrated Guide to the Trees of Peru*. David Hunt, Sherborne, England

Pereira, N. 1967. *MoronGuêta: Um Decameron Indígena*. Civilização Brasileira, Rio de Janeiro, vol. 2

Phillips, O. 1993. The potential for harvesting fruits in tropical rainforests: new data from Amazonian Peru. Biodiversity and Conservation 2:18–38

Pitman, Nigel C. A.; Terborgh, John W.; Silman, Miles R.; Núñez, Percy; Neill, David A.; Cerón, Carlos E.; Palacios, Walter A.; Aulestia, Milton. 2001. Dominance and distribution of tree species in upper Amazonian terra firme forests. Ecology 82(8):2101–2117

Plowman, Timothy. 1981. Amazonian coca. Journal of Ethnopharmacology 3: 195–225

Porter, L. M. 2001. Dietary differences among sympatric Callitrichinae in northern Bolivia: *Callimico goeldii, Saguinus fuscicollis* and *S. labiatus*. International Journal of Primatology 22(6):961–992

Posey, D. A. 1985. Native and indigenous guidelines for new Amazonian development strategies: understanding biological diversity through ethnoecology. In: Change in the Amazon Basin Volume 1: Man's Impact on Forests and Rivers, Hemming, John (ed), Manchester University Press, Manchester, pp 156–181

Postma, T. M.; Verheij, E. W. M. 1994. Growth and yield of *Bactris gasipaes* and *Pourouma cecropiifolia* in swidden fields of Amazon Indians, Colombia. Scientia Horticulturae 57:73–88

Prance, G. T. 1985 The increased importance of ethnobotany and underexploited plants in a changing Amazon. In: Change in the Amazon Basin, Volume I: Man's Impact on Forests and Rivers, John Hemming (ed), Manchester University Press, Manchester, pp. 129–136

Pulido, María T.; Cavelier, Jaime. 2001. Comercialización de productos vegetales en los mercados de Leticia y Florencia, Amazonía colombiana. In: Evaluación de Recursos Vegetables no Maderables en la Amazonía Noroccidental, Duivenvoorden, Joost F.; Balslev, Henrik; Cavelier, Jaime; Grandez, César; Tuomisto, Hanna; Valencia, Renato (eds), Institute for Biodiversity and Ecosystem Dynamics, Amsterdam, pp. 265–310

Quisbert, Javier; Macía, Manuel J. 2005. Estudio comparativo de la composición florística y estructura del bosque de tierra firme en dos sitios de tierras bajas de Madidi. Ecología en Bolivia 40(3):339–364

Reyes-García, Victoria. 2001. Indigenous people, ethnobotanical knowledge, and market economy: a case study of the Tsimane' Amerindians in lowland Bolivia. Doctoral dissertation, University of Florida, Gainesville.

Robles, Rembrandt Ramiro Canahuire. 2017. Composición florística y estructura de la recuperación natural en un área degradado por minería en la comunidad nativa de Tres Islas, Tambopata, Madre de Dios. Master's thesis, Escuela Profesional de Ingeniería Forestal y Medio Ambiente, Puerto Maldonado, Perú

Romanoff, S., D. M. Jiménez, F. S. Uaquí, and D. W. Fleck. 2004. *Matsesën Nampid Chuibanaid: La Vida Tradicional de los Matsés*. Centro Amazónico de Antropología y Aplicación Práctica (CAAAP), Lima

Sánchez, Judith Jacira Achong; Levy, Allison Licett Nuñez. 2013. Características de frutos consumidos por *Saguinus nigrifrons* (Primates, Callitrichidae), en la Estación Biológica Quebrado Blanco, Loreto, Perú. Master's thesis, Universidad Nacional de la Amazonía Peruana, Iquitos

Schultes, R. E. 1977. Diversas plantas comestíveis nativas do noroeste da Amazonia. Acta Amazonica 7(3):317–327

Schultes, R. E. 1980. Coca in the northwest Amazon. *Botanical Museum Leaflets* 28(1): 47–59.

Schultes, Richard Evans. 1986. The reason for ethnobotanical conservation. Bulletin of the Botanical Survey of India 28 (1–4): 203–224

Silva, A. B. A. 1962. A Civilização Indígena do Uaupés. Centro de Pesquisas de Iauareté, Missão Salesiano do Rio Negro, São Paulo

Silva, Edinei Santos da. 2017. Plantas alimentícias em comunidades agrícolas no município de Rio Preto da Eva-AM. Master's thesis, INPA, Manaus

Soini, P. 1987. Ecology of the saddle-back tamarin *Saguinus fuscicollis illigeri* on the río Pacaya, northeastern Peru. Folia Primatologica 49:11–32

Soini, P. 1988. The pygmy marmoset, genus *Cebuella*. In: Ecology and Behavior of Neotropical Primates, Mittermeier, Russell A.; Rylands, Anthony B.; Coimbra-Filho, Adelmar F.; Fonseca, Gustavo A.B. da (eds), Ecology and Behavior of Neotropical Primates, World Wildlife Fund, Washington, DC, vol. 2, pp 79–129

Terborgh, J. 1983. Five New World Primates: A Study in Comparative Ecology. Princeton University Press, Princeton

Thomas, Evert. 2009. New light on the floristic composition and diversity of indigenous territory and national park Isiboro-Sécure, Bolivia. Biodiversity Conservation 18:1847–1878

Triana-Moreno, Luz Amparo; Rodríguez, Nohra Cecilia; García, Jesús. 2006. Dinámica del sistema agroforestal de chagras como eje de la producción indígena en el Trapecio Amazónico (Colombia). Agronomía Colombiana 24(1):158–169

Van den Eynden, Veerle; Cueva, Eduardo; Cabrera, Omar. 1999. Plantas Silvestres Comestibles del Sur del Ecuador. Ediciones Abya-Yala, Quito

Veléz, Germán Alonso; Veléz, Antonio José. 1992. Sistema agroflorestal de "chagras" utilizado por las comunidades indígenas del medio Caquetá (Amazonia Colombiana). Colombia Amazonica 6(1): 101–134

Vester, Hans F. M.; Cleef, Antoine M. 1998. Tree architecture and secondary tropical rain forest development: a case study in Araracuara, Colombian Amazon. Flora 193:75–97

Villachica, Hugo; Carvalho, José Edmar Urano de; Muller, Carlos Hans; Díaz, Camilo; Almanza, Martha; Mercado, Roberto Samanez; Izquierdo, Juan; Egg, Antônio Brack. 1996. Frutales y Hortalizas Promisorios de la Amazonia. Tratado de Cooperación Amazonica, Lima.

Whiffen, T. 1915. The North-West Amazons: Notes of Some Months Spent Among Cannibal Tribes. Constable and Company, London

Zambrana, Narel Y. Paniagua; Bussmann, Rainer; Hart, Robbie E.; Huanca, Araceli L. Moya; Soria, Gere Ortiz; Vaca, Milton Ortiz; Álvarez, David Ortiz; Morán, Jorge Soria; Morán, María Soria; Chávez, Saúl; Moreno, Bertha Chávez; Moreno, Gualberto Chávez; Roca, Oscar; Siripi, Erlin. 2017. Traditional knowledge hiding in plain sight – twenty-first century ethnobotany of the Chácobo in Beni, Bolivia. Journal of Ethnobiology and Ethnomedicine 13:1–47

Chapter 50
Verbenaceae

50.1 *Lantana trifolia*

Bolivia Japutamu lila, uva; urn (Chácobo)

Brazil Cambará, cambará de três folhas, cambarazinho, camará, cangica, cidreira, uva de roçado, uvinha de campo, erva cidreira, erva de grileiro, falsa cidreira, milho de grilo, salvia do mato; maruipö, txãkarapu (Tiriyó), warasa (Yanomami)

Colombia Venturosa, mancheco, yerba de conejo

Ecuador Choclito; pakanmaya, piyushang (Achuar), muras (Shuar), ebokawe (Waorani)

Peru Ayamanchana, ubita, hierba mora, huayito morada, tunchi albaca, tunchi albahaca, aya manchana, matico, verbena; takun pui cama (Cashinahua), Yahua'tan huëron (Chayahuita), saputinaru, pishkuchini, cashaquihua (Kichwa), yona rao (Shipibo)

Venezuela Cariaquito, cariaquito morado; hekura henakɨ (Yanomami)

Status Wild, spontaneous in cultural settings

N. Smith, *Amazon Fruits: An Ethnobotanical Journey*, Ethnobiology,
https://doi.org/10.1007/978-3-031-12803-5_50

Ubita *Lantana trifolia* in successionary vegetation. Near Yarina, Rio Yanayacu, Reserva Nacional Pacaya-Samiria, Loreto, Peru 6-6-04

Lantana trifolia is an herbaceous weed that produces small, lavender fruits in a clump along a stalk (peduncle) at the top of the plant during the rainy season and into the early part of the dry period. Native to Tropical America the sun-loving plant generally grows in disturbed sites in rural and urban areas below 2800 m from Argentina to Mexico, including the Caribbean. This pioneer plant grows along trails, the edges of fields, in uplands, and on alluvial soils. At a little over 1.5 m when fully mature, *Lantana trifolia* is at a convenient height for kids to gather the fruits. Children snack on the sweet fruits and ingest the flesh-colored seeds about the size of tomato seeds; they are thus also involved in dispersing the plant in backyards and along trails since flushing toilets and latrines are rare in rural areas.

Christian, 8 years, gathering ubita *Lantana trifolia* fruits in successionary vegetation on the outskirts of his village. Yarina, Yanayacu River, Reserva Nacional Pacaya-Samiria, Peru 6-5-04

Lantana trifolia can be regarded as a resource or a nuisance, even in the same area. Along the lower Huallaga in Peru, for example, ranchers hire workers to rid pastures of weeds, including *Lantana trifolia*. Nearby, villagers snack on the fruits in successionary vegetation as they head to their fields or to the forest. Some indigenous groups also eat the fruits, such as the Chácobo in the Bolivian Amazon (Boom 1987:42). Lantanas are considered toxic, but once ripe, the fruits are usually safe to eat, at least in the case of *Lantana trifolia*. Offensive chemicals are likely relegated to leaves, stalks, and immature fruits, thus improving the chances that it will reproduce successfully.

The pungent quality of the plant is recognized by curers along the Yanayacu River, an affluent of the lower Marañón in the Peruvian Amazon. When treating cases of *manchari* in children, a supernatural condition in which zombie-like youngsters have lost their soul or spirit because of an evil spell, shamans mix the leaves of ayamanchana with tobacco to blow the smoke over their patients. Curing cigars are different from cigarettes; healers do not inhale the former. In Loreto Province in the Peruvian Amazon, the juice is squeezed from the leaves to treat conjunctivitis. Also in Loreto Province, the Chayahuita smoke the leaves over a fire until dry then they are crushed to form a powder that is sprinkled over cutaneous leishmaniasis (Odonne et al. 2009). The Achuar in the Ecuadorian Amazon chew the leaves to blacken their teeth (Schultes and Raffauf 1990:462), presumably for ceremonial purposes. The Tiriyó who inhabit forests and grasslands in the border region between northern Pará and Surinam crush the leaves and/or scrape the stems in water which is then boiled; once the solution cools, it is applied as a bathing solution to treat fever and headaches (Cavalcante 1973:93).

Ayamanchana, as with some other species of *Lantana*, such as *L. camara*, have found their way into the ornamental plant trade, particularly in the southern United States which has a mild climate. Ayamanchana is known by various names in the nursery industry including three-leaf shrub verbena and lavender popcorn. Another reason that lantanas are so popular with gardeners is that they attract butterflies.

Ayamanchana has been introduced to other parts of the world, such as Africa and tropical Asia, where it has become naturalized after escaping from gardens. Although generally considered a weed by agronomists, *Lantana trifolia* has found non-ornamental uses in foreign lands, such as in Rwanda where the leaves are used in folk remedies for a variety of ailments including gonorrhea, hepatitis, and angina (Cos et al. 2001). Its employment in treatments for hepatitis stems from its antiviral properties. Presumably, folk healers in Rwanda have learned to calibrate the dosages to avoid poisoning their patients.

References

Boom, B. M. 1987. Ethnobotany of the Chácobo Indians, Beni, Bolivia. The New York Botanical garden, Advances in Economic Botany, vol. 4

Cavalcante, P. B. 1973. A Farmacopéia Tiriyó: Estudo Étno-Botânico. Museu Paraense Emílio Goeldi, Publicações Avulsas No. 24

Cos, P.; Hermans, N.; De Bruyne, T.; Apers, S.; Sindambiwe, J.P.; Vanden Berghe, D.; Pieters, L.; Vlietinck, A.J. 2001. Further evaluation of Rwandan medicinal plant extracts for their antimicrobial and antiviral activities. Journal of Ethnopharmacology 79:155–163

Odonne, G.; Bourdy, G.; Castillo, D.; Estevez, Y.; Lancha-Tangoa, A.; Alban-Castillo, J.; Deharo, E.; Rojas, R.; Stien, D.; Sauvain, M. 2009. Ta'ta', Huayani: perception of leishmaniasis and evaluation of medicinal plants used by the Chayahuita in Peru. Part II. Journal of Ethnopharmacology 126:149–158

Schultes, R. E. and R. F. Raffauf. 1990. The Healing Forest: Medicinal and Toxic Plants of the Northwest Amazonia. Dioscorides Press, Portland, Oregon

Chapter 51
Vitaceae

51.1 *Cissus sicyoides*

Bolivia Tayuya; carabó coatí (Chácobo)

Brazil Cipó puçá, cipó puci, cortina de oxala, cortina dos pobres, anil trepador, anvil trepador, insulina trepadeira, insulina do mato, insulina vegetal, uva branca, uva brava, uva do mato, uvinha, tinta de índio, tinta dos gentios, bejuco caro, bejuco de porra, bejuco ubi; tukur-ɨwɨpo (Guajá)

Colombia Bejuca de agua, uvilla

Ecuador Epeme (Waorani)

Peru Palomenta huayo, huayo negro, sapo huasca

Status Wild

N. Smith, *Amazon Fruits: An Ethnobotanical Journey*, Ethnobiology,
https://doi.org/10.1007/978-3-031-12803-5_51

Cissus sicyoides **vine on the edge of a mangrove. Salinas do Roque, Bragança-Ajuruteua road, Pará, Brazil 3-24-15**

Cissus sicyoides is a vine in wetlands with dark purple fruits that are used for fish bait. One of its common names in the Peruvian Amazon is palometa huayo, which means fruit of the palometa *Mylossoma duriventre*. The fruits are also eaten by other fish, such as gamitana *Colossoma macropomum*. Moustached tamarins *Saguinus mys*tax and saddle-back tamarins *Saguinus fuscicollis* have been recorded eating this fruit in the northeastern part of the Peruvian Amazon and these small monkeys disperse the seeds (Knogge et al. 2003). The fruits, 0.5 cm × 0.8 cm, contain a purple pulp that surrounds a single seed, which is the size and shape of a grape seed (approximately 0.4 cm by 0.3 cm). The mesocarp leaves a purple stain on skin and clothes and is not eaten by people. Men and boys paddle up to the vine during the high-water season to retrieve the fruits to bait their hooks. In the Bolivian Amazon, the Chácobo eat the fruits (Zambrana et al. 2017) and the Waorani in the Ecuadorian Amazon do the same (Mondragón and Smith 1997:54). The Waorani also cut the stems of this vine and drink the liquid.

Palometa *Mylossoma* sp. and sardiña *Triportheus* sp. eating fruits of huayo negro *Cissus sicyoides* that drop into the water when ripe. The vine is climbing a Lagarto caspi tree. Drawing by Julio Zamora, 12 years, Veinte de Enero, Rio Yanayacu, Reserva Nacional Pacaya-Samiria, Loreto, Peru 2004

The vine has other uses. In the Bolivian Amazon, for example, the stems and leaves are added to the latex of rubber *Hevea brasiliensis* to thicken it (Boom 1987:42). In Colombia, some indigenous people use the fruits to dye cloth, although this practice is likely declining as manufactured clothing dyed artificially are increasingly common in the interior. Leaves are used to treat rheumatism and abscesses. The leaves appear in the Guamá market for herbal remedies in Belém, Pará, and are used to make a tea for treating headaches (Bitencourt et al. 2014). An infusion of the leaves of this plant is used to treat diabetes by some residents of Imperatriz, Maranhão, which lies close to the border of eastern Pará (Penido et al. 2016). Thin stems of this vine are used to make ligatures to alleviate joint pain (Pérez-Arbeláez 1956:747). This vine occurs from Argentina through the Amazon Basin to Central America as far north as Mexico. It is also found in parts of the Caribbean. It occurs as high as 1500 m in the Andean foothills.

Sapo huasca ***Cissus sicyoides*** **fruits used as fish bait to catch palometa** ***Mylossoma duriventre*****. Yanayacu River near Veinte de Enero, Reserva Nacional Pacaya-Samiria, Loreto, Peru 6-8-04**

References

Bitencourt, Bruna Letícia Gentil; Lima, Pedro Glécio Costa; Barros, Flávio Bezerra. 2014. Comércio e uso de plantas e animais de importância mágico-religiosa e medicinal no mercado público do Guamá, Belém do Pará. Revista Faculdade Santo Agostino 11(3):96–158

Boom, B. M. 1987. Ethnobotany of the Chácobo Indians, Beni, Bolivia. The New York Botanical garden, Advances in Economic Botany, vol. 4

Knogge, Christoph; Herrera, Emérita R. Tirado; Heymann, Eckhard W. 2003. Effects of passage through tamarin guts on the germination potential of dispersed seeds. International Journal of Primatology 24(5):1121–1128

Mondragón, Martha L.; Smith, Randall. 1997. Bete Quiwiguimamo Salvando el bosque para vivir sano: algunas plantas y árboles utilizadas por la Nacionalidad Huaorani de la Amazonía Ecuatoriana. Abya Yala, Quito

Penido, Alexandre Batista; Morais, Selene Maia de; Ribeiro, Alan Bezerra; Silva, Ana Zélia. 2016. Ethnobotanical study of medicinal plants in Imperatriz, state of Maranhão, northeastern Brazil. Acta Amazonica 46(4):345–354

Pérez-Arbeláez, Enrique. 1956. Plantas Útiles de Colombia. Liberia Colombiana, Bogotá

Zambrana, Narel Y. Paniagua; Bussmann, Rainer; Hart, Robbie E.; Huanca, Araceli L. Moya; Soria, Gere Ortiz; Vaca, Milton Ortiz; Álvarez, David Ortiz; Morán, Jorge Soria; Morán, María Soria; Chávez, Saúl; Moreno, Bertha Chávez; Moreno, Gualberto Chávez; Roca, Oscar; Siripi, Erlin. 2017. Traditional knowledge hiding in plain sight – twenty-first century ethnobotany of the Chácobo in Beni, Bolivia. Journal of Ethnobiology and Ethnomedicine 13:1–47

Appendix

Appendix 1 Voucher specimens of Amazonian fruits collected in Peru

Local name(s)	Scientific name	Family	Voucher
Aceitunilla	*Diclidanthera penduliflora*	Polygalaceae	IHC 7571
Aguajillo	*Mauritiella armata*	Arecaceae	IHC 7898
Ana huayo	*Plinia* sp.	Myrtaceae	IHC 7794, 7912
Anonilla	*Annona montana*	Annonaceae	IHC 7819
Anonilla	*Annona hypoglauca*	Annonaceae	NS 019, 023; IHC 7801
Ayamanchana, uvita	*Lantana trifolia*	Verbenaceae	NS 006
Azúcar huayo	*Maripa axilliflora*	Convolvulaceae	IHC 7621, 7903
Bombo huayo	*Salacia juruana*	Celastraceae	IHC 7889
Cacahuillo	*Theobroma obovatum*	Malvaceae	IHC 7832
Cacahuillo	*Herrania nitida*	Malvaceae	IHC 7582, 7837
Cacahuillo	*Herrania nycterodendron*	Malvaceae	IHC 7825
Caimito	*Pouteria caimito*	Sapotaceae	IHC 7786
Caimitillo	*Alibertia edulis*	Rubiaceae	NS 018, IHC 7780
Caimitillo	*Pouteria torta* subsp. *glabra*	Sapotaceae	IHC 7790, 7796
Caimitillo	*Pouteria* sp.	Sapotaceae	IHC 7745
Cashillo	*Cathedra acuminata*	Olacaceae	IHC 7789
Cashu	*Anacardium occidentale*	Anacardiaceae	IHC 7555
Charichuelo	*Garcinia brasiliensis*	Clusiaceae	IHC 7783, 7866
Charichuelo	*Garcinia macrophylla*	Clusiaceae	IHC 7570, 7902
Charichuelo	*Garcinia madruno*	Clusiaceae	NS 001
Charichuelo	*Garcinia gardneriana*	Clusiaceae	IHC 7626
Chontilla	*Bactris riparia*	Arecaceae	IHC 7567
Chope	*Gustavia longifolia*	Lecythidaceae	IHC 7908
Chope	*Gustavia augusta*	Lecythidaceae	IHC 7845
Ciruela	*Bunchosia armeniaca*	Malpighiaceae	IHC 7587
Cormiñón	*Vitex orinocensis*	Verbenaceae	IHC 7585

(continued)

N. Smith, *Amazon Fruits: An Ethnobotanical Journey*, Ethnobiology,
https://doi.org/10.1007/978-3-031-12803-5

Appendix 1 (continued)

Local name(s)	Scientific name	Family	Voucher
Coto huayo	*Anomospermum chloranthum*	Menispermaceae	IHC 7793
Cotoruntu	*Rudgea loretensis*	Rubiaceae	NS 017, IHC 782
Crista huayo	*Simaba orinocensis*	Simaroubaceae	IHC 7625
Granadilla	*Passiflora quadrangularis*	Passifloraceae	IHC 7821
Granadilla	*Passiflora nitida*	Passifloraceae	NS 013, IHC 7797
Granadilla	*Passiflora riparia*	Passifloraceae	IHC 7627
Guabilla	*Inga ingoides*	Fabaceae	NS 021, IHC 7594, 7593
Guabilla	*Inga laurina*	Fabaceae	IHC 7568, 7618
Guayaba del agua	*Psidium acutangulum*	Myrtaceae	IHC 7739, 7813
Guayaba de restinga	*Mouriri tessmannii*	Memecylaceae	NS 024, IHC 7860, 7574
Habilla	*Omphalea diandra*	Euphorbiaceae	IHC 7828
Huanchaco huayo	*Psychotria loretensis*	Rubiaceae	IHC 7865
Huevigato	*Cordia nodosa*	Boraginaceae	IHC 7788
Huitillo	*Duroia micrantha*	Rubiaceae	IHC 7843
Juanache	*Eugenia inundata*	Myrtaceae	IHC 7861
Lancetilla, lanza huayo	*Eugenia discreta*	Myrtaceae	IHC 7803, 7888
Lanza caspi	*Eugenia multirimosa*	Myrtaceae	IHC 7892
Lucma	*Pouteria longifolia*	Sapotaceae	IHC 7740, 7907
Manito de gato	*Perebea longepedunculata*	Moraceae	NS 014, IHC 7785
Mata pasto	*Clidemia septuplinervia*	Melastomataceae	NS 012, IHC 7763
Mata pasto	*Clidemia octona*	Melastomataceae	NS 003
Mata pasto	*Clidemia hirta*	Melastomataceae	NS 031, IHC 7761
Mata pasto	*Clidemia japurensis*	Melastomataceae	NS 030, IHC 7753
Mata pasto	*Miconia serrulata*	Melastomataceae	IHC 7765
Mata pasto	*Miconia nervosa*	Melastomataceae	IHC 7805
Meta huayo	*Moutabea aculeata*	Polygalaceae	IHC 7867
Mullaca	*Physalis angulata*	Solanaceae	IHC 7592, 7807
Mullaca lanudo	*Physalis pubescens*	Solanaceae	IHC 7811
Ñejilla	*Bactris bifida*	Arecaceae	NS 016, IHC 7890
Ñejilla	*Bactris brongniartii*	Arecaceae	IHC 7799, 7850, 7742
Ñejilla	*Bactris concinna* var. *Sigmoidea*	Arecaceae	NS 004, 026, IHC 7839
Ñejilla	*Bactris maraja*	Arecaceae	NS 002, IHC 7885
Ñejilla	*Bactris martiana*	Arecaceae	IHC 7630, 7822
Nispero	*Bellucia pentamera*	Melastomataceae	NS 009, IHC 7773, 7905
Palillo	*Campomanesia lineatifolia*	Myrtaceae	IHC 7629, 7827
Pampa sanango	*Tabernaemontana siphilitica*	Apocynaceae	IHC 7613, IHC 7820
Pampa tortuga	*Duguetia odorata*	Annonaceae	IHC 7738, 7791, 7840
Parinari	*Couepia chrysocalyx*	Chrysobalanaceae	IHC 7584

(continued)

Appendix 1 (continued)

Local name(s)	Scientific name	Family	Voucher
Parinari	*Couepia ulei*	Chrysobalanaceae	NS 022, IHC 7849
Paujillo	*Gnetum leyboldii*	Gnetaceae	IHC 7857
Pichico huayo	*Abuta grandifolia*	Menispermaceae	IHC 7873
Puricho huayo	*Salacia cordata*	Celastraceae	IHC 7577
Puropuro	*Passiflora serratodigitata*	Passifloraceae	IHC 7558, 7583, 7591
Quinilla	*Eugenia lambertiana*	Myrtaceae	NS 015, 025
Quinilla	*Eugenia patens*	Myrtaceae	IHC 7859
Rifari	*Miconia longifolia*	Melastomataceae	IHC 7851
Sachamangua	*Grias neuberthii*	Lecythidaceae	IHC 7844
Sacha pandicho	*Pachira aquatica*	Bombaceae	IHC 7890
Sacha uva	*Cordia collococca*	Boraginaceae	IHC 7806
Sanango	*Tabernaemontana sananho*	Apocynaceae	NS 005, 010, IHC 7769
Sapo shimbillo	*Inga auristellae*	Fabaceae	IHC 7766
Seca boca	*Paullina alata*	Sapindaceae	IHC 7751, 7887
Seca boca	*Paullinia pachycarpa*	Sapindaceae	IHC 7883
Shambo huayo	*Stylogyne cauliflora*	Myrsinaceae	IHC 7876
Shambo huayo	*Stylogyne longifolia*	Myrsinaceae	NS 028
Shimbillo	*Inga cinnamomea*	Fabaceae	NS 021, IHC 7594, 7593
Shimbillo	*Inga bourgonii*	Fabaceae	IHC 7814
Shimbillo	*Inga punctata*	Fabaceae	IHC 7798
Shimbillo de la cocha	*Inga leiocalycina*	Fabaceae	IHC 7624
Shimbillo	*Inga ruiziana*	Fabaceae	IHC 7562, 7595
Shimbillo	*Inga nobilis*	Fabaceae	IHC 7551
Shimbillo rabo de mono	*Inga acuminata*	Fabaceae	NS 029, IHC 7628, 7871
Tamamuri	*Brosimum lactescens*	Moraceae	IHC 7744
Tortuga huayo	*Duguetia spixiana*	Annonaceae	NS 020, IHC 7818, 7855
Tucunareñahui	*Clavija elliptica*	Theophrastaceae	NS 007, IHC 7793, 7830
Tumbo	*Passiflora quadrangularis*	Passifloraceae	IHC 7552, 7764
Ubos	*Spondias mombin*	Anacardiaceae	
Ungurahui	*Oenocarpus bataua*	Arecaceae	
Ushun	*Buchenavia grandis*	Combretaceae	IHC 7795, 7846
Uvilla	*Pourouma cecropiifolia*	Urticaceae	
Vaca paleta	*Inga macrophylla*	Fabaceae	
Vino huayo	*Coccoloba densifrons*	Polygonaceae	NS 008, IHC 7575
Yarina	*Phytelephas macrocarpa*	Arecaceae	
Yumanasa	*Muntingia calabura*	Muntingiaceae	IHC 7560
Zorro caspi	*Tapura amazonica*	Dichapetalaceae	IHC 7841

Voucher specimens were collected in 2005 and 2006 with the collaboration of Rodolfo Vasquez and deposited with the Herbario Nacional in Lima and herbaria of the Missouri Botanical Garden in Oxapampa, Peru, and in St. Louis, USA

Bibliography

Carvalho, José Edmar Urano de; Nazaré, Raimunda Fátima Ribeiro de; Nascimento, Walnice Maria Oliveira do. 2003. Características físicas e físico-químicas de um tipo de bacuri (Platonia insignis Mart.) com rendimento industrial superior. Revista Brasileira de Fruticultura 25(2):326–328

Jardim, Mário Augusto G.; Medeiros, Tonny David Santiago. 2006. Plantas oleaginosas do Estado do Pará: composição florística e usos medicinais. Revista Brasileira de Farmacognosia 87(4):124–127

Rahal, Sheila Canevese; Tourinho, Manuel Malheiros. 2012. Ethnoveterinary knowledge and practices at Colares island, Pará state, eastern Amazon, Brazil. Journal of Ethnopharmacology 144:346–352

N. Smith, *Amazon Fruits: An Ethnobotanical Journey*, Ethnobiology,
https://doi.org/10.1007/978-3-031-12803-5